SURVEYING

THEORY and PRACTICE

SURVEYING

THEORY and PRACTICE

RAYMOND E. DAVIS, M.S., C.E., D.Eng.

Professor of Civil Engineering, Emeritus
University of California

FRANCIS S. FOOTE, E.M.

Professor of Railroad Engineering, Emeritus
University of California

JOE W. KELLY, B.S.

Professor of Civil Engineering, Emeritus
University of California

FIFTH EDITION

McGRAW-HILL BOOK COMPANY

NEW YORK ST. LOUIS SAN FRANCISCO TORONTO LONDON SYDNEY

SURVEYING: THEORY AND PRACTICE

Library of Congress Catalog Card Number 64-66263

16-MAMM-7

PREFACE TO THE FIFTH EDITION

The fifth edition maintains the logical arrangement and thoroughness which make the book useful not only as a teaching text but also as a reference for practicing engineers and surveyors. One distinguishing characteristic continues to be emphasis throughout on precision and accuracy, without appreciation of which by the surveyor some measurements would not fulfill their purpose, whereas others would be unduly time-consuming and expensive.

The entire text is updated in critical detail. Modern instruments and methods are described, including levels, transits, electronic distance-measuring and location-finding methods, photogrammetric plotting instruments, and uses of photogrammetry and computers in route surveying. The chapters on route location, hydrographic surveying, and photogrammetric surveying have been rewritten by authorities in these fields.

Many sections have been rewritten, and others added. Numerous cross references relate various aspects of topics, and a selected list of references at the end of each chapter serves as a guide to more intensive study. For convenient reference, the listing of articles in the contents is by number and an ample index is provided. Noteworthy features of the new edition are as follows.

Probable error is used in evaluation of precision because it is simple to visualize, directly applicable, and ample for the purposes of this book. Its relationship to standard error, of which it is a special case, is explained.

Logarithms are described briefly and six-place tables are given, because in many situations computers are not available. Descriptions of arithmetical and slide-rule operations have been shortened, but principles and precision of computations are retained.

Field problems are included not only for direct use by classes but also because for others they contain many details on procedure not covered in the text. The original book of this series on surveying was in the form of a manual of field problems.

Even though new types of levels and transits are described, conventional types and their adjustment are retained because they will continue to be used in ordinary surveying for many years. The newer types are used in a manner similar to the older. Errors and precision of level and transit work are evaluated.

In the chapters on astronomy, reference to the "American Ephemeris" has been discontinued because it now employs ephemeris time, which differs from civil time. The examples are now based in some cases upon the "Nautical

Almanac" and civil time and in other cases upon the ephemeris of the U.S. Bureau of Land Management and apparent or "sun" time, which is convenient in land surveying.

New topics in the chapter on construction surveying are optical tooling and ship surveying.

The chapter on route location, written by L. L. Funk, makes particular reference to highway location and construction. It includes not only methods common to all route location but also modern practices and standards of the California Division of Highways.

The chapter on hydrographic surveying, written by Walter Hofmann and S. E. Rantz, includes modern equipment and methods used by the U.S. Geological Survey such as echo sounders, location finders, stage gages, current-velocity meters, and measurement and computation of discharge.

The chapter on photogrammetric surveying and mapping, revised by Sidney W. Smith, also incorporates current practices of the Geological Survey. Of particular interest are descriptions of the sequence of operations in topographic mapping, plotting instruments, automation of mapping instruments, and orthographic pictorial representation.

R. E. D.
F. S. F.
J. W. K.

ACKNOWLEDGMENTS

Account has been taken throughout of the many suggestions offered by users of the book, and the authors are grateful to them. Professors Sturla Einarsson and Milos Polivka, colleagues of the authors at the University of California, rendered valuable suggestions and counsel.

Many of the illustrations and tables have been taken or adapted from publications or directly furnished material of Federal agencies, especially the U.S. Coast and Geodetic Survey, U.S. Geological Survey, and U.S. Bureau of Land Management. Use has also been made of photographs and tabular information from the U.S. Air Force, U.S. Army, Corps of Engineers, Soil Conservation Service of the U.S. Department of Agriculture, U.S. Naval Observatory, and California Division of Highways.

Illustrative material was furnished by the following manufacturers of surveying equipment: Abrams Aerial Survey Corporation, Aero Service Corporation, Wm. Ainsworth & Sons, Benson-Lehner Corporation, C. L. Berger & Sons, Eugene Dietzgen Company, Fairchild Space and Defense Systems, W. & L. E. Gurley, Philip B. Kail Associates, Keuffel & Esser Company, A. Lietz Company, O. M. I. Corporation of America, Wild-Heerbrugg Instruments Inc., and Zeiss-Aerotopograph, Munich.

Credit is due to John Wiley & Sons, Inc., for permission to use Tables IX and X.

CONTENTS

GENERAL TABLES

FIELD AND OFFICE PROBLEMS

CHAPTER 1

CONCEPTS

1-1. Surveying. Surveying has to do with the determination of the relative location of points on or near the surface of the earth. It is the art of measuring horizontal and vertical distances between objects, of measuring angles between lines, of determining the direction of lines, and of establishing points by predetermined angular and linear measurements.

Concomitant with the actual measurements of surveying are mathematical calculations. Distances, angles, directions, locations, elevations, areas, and volumes are thus determined from data of the survey. Also, much of the information of the survey is portrayed graphically by the construction of maps, profiles, cross-sections, and diagrams.

Thus the process of surveying may be divided into the *field work* of taking measurements and the *office work* of computing and drawing necessary to the purpose of the survey.

1-2. Uses of Surveys. The earliest surveys known were for the purpose of establishing the boundaries of land, and such surveys are still the important work of many surveyors.

Every construction project of any magnitude is based to a greater or less degree upon measurements taken during the progress of a survey and is constructed about lines and points established by the surveyor. Aside from land surveys, practically all surveys of a private nature and most of those conducted by public agencies are of assistance in the conception, design, and execution of engineering works.

For many years the government, and in some instances the individual states, have conducted surveys over large areas for a variety of purposes. The principal work so far accomplished consists in the fixing of national and state boundaries, the charting of coast lines and navigable streams and lakes, the precise location of definite reference points throughout the country, the collection of valuable facts concerning the earth's magnetism at widely scattered stations, and the mapping of certain portions of the the interior particularly near the seacoasts, along the principal rivers and lakes, in the localities of valuable mineral deposits, and in the older and more thickly settled regions.

Thus surveys are divided into three classes: (1) those for the primary purpose of establishing the boundaries of land, (2) those providing information necessary for the construction of public or private works, and (3) those of large extent and high precision conducted by the government and to some extent by the states. There is no hard and fast line of demarcation between surveys of one class and those of another as regards the methods employed, results obtained, or use of the data of the survey.

1-3. The Earth a Spheroid. The earth is an oblate spheroid of revolution, the length of its polar axis being somewhat less than that of its equatorial axis. The lengths of these axes are variously computed, as follows:

	Polar axis, ft.	Equatorial axis, ft.
Clarke (1866)	41,710,242	41,852,124
Hayford (1909)	41,711,920	41,852,860
Adopted (1924) by International Geodetic and Geophysical Union	41,711,940[a]	41,852,860

[a] Computed from equatorial axis by assuming that the flattening of the earth is exactly 1 ÷ 297.

The lengths computed by Clarke have been generally accepted in the United States and have been used in government land surveys. Hayford's values are now regarded as being more nearly correct than those of Clarke. The values adopted by the International Geodetic and Geophysical Union are published by the U.S. Naval Observatory.

It is seen that the polar axis is shorter than the equatorial axis by about 27 miles. Relative to the diameter of the earth this is a very small quantity, less than 0.34 per cent. Imagine the earth as shrunk to the size of a billiard ball, still retaining the same shape. In this condition, it would appear to the eye as a smooth sphere, and only by precise measurements could its lack of true sphericity be detected.

Let us consider that the irregularities of the earth have been removed. The surface of this imaginary spheroid is a curved surface every element of which is normal to the plumb line. Such a surface is termed a *level surface*. The particular surface at the average sea level is termed *mean sea level*.

Imagine a plane as passing through the center of the earth, as in Fig. 1-1. Its intersection with the level surface forms a continuous line around the earth. Any portion of such a line is termed a *level line*, and the circle defined by the intersection of such a plane with the mean level of the earth is termed a *great circle* of the earth. The distance between two points on the earth, as A and B (Fig. 1-1), is the length of the arc of the great circle passing through the points, and is always somewhat more than the chord intercepted by this arc. The arc is a level line; the chord is a mathematically straight line.

If a plane is passed through the poles of the earth and any other point on the earth's surface, as A (Fig. 1-2), the line defined by the intersection of the level surface and plane is called a *meridian*. Imagine two such planes as passing through two points as A and B (Fig. 1-2) on the earth, and the section between the two planes removed like the slice of an orange, as in

Fig. 1-3. At the equator the two meridians are parallel; above and below the equator they converge, and the angle of convergency increases as the poles are approached. No two meridians are parallel except at the equator.

Imagine lines, normal to the meridians, drawn on the two cut surfaces of the slice. If the earth be regarded as a perfect sphere these lines converge at a point at the center of the earth. Considering the lines on either or both of

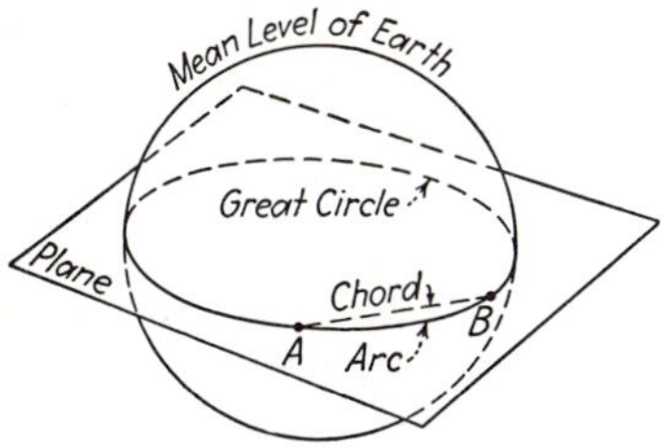

FIG. 1-1.

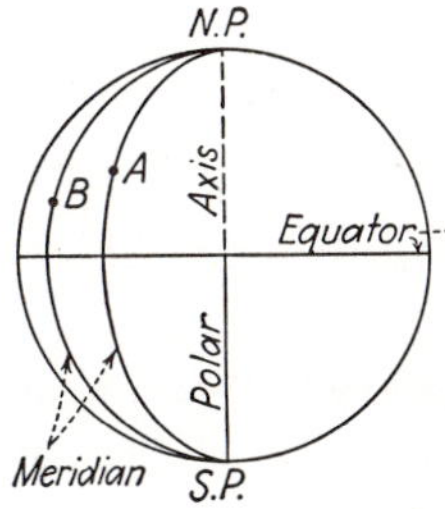

FIG. 1-2.

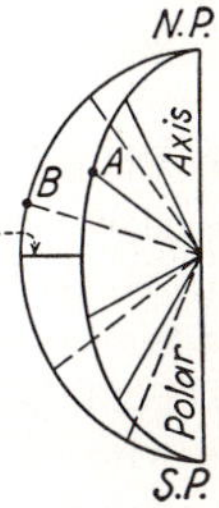

FIG. 1-3.

the cut surfaces, no two are parallel. The radial lines may be considered as vertical or plumb lines, and hence we arrive at the deductions that all plumb lines converge at the earth's center and that no two are parallel. (Strictly speaking, this is not quite true, owing to the unequal distribution of the earth mass and owing to the fact that normals to an oblate spheroid do not all meet at a common point.)

Consider three points on the mean surface of the earth. Let us make these three points the vertices of a triangle, as in Fig. 1-4. The surface within the triangle ABC is a curved surface, and the lines forming its sides are arcs of great circles. The figure is a spherical triangle. In the figure the dotted lines represent the plane triangle whose vertices are points A, B, and C. (Actually the "auxiliary plane triangle" of geodetic work has sides equal in length to the *arcs* of the corresponding spherical triangle.) Lines drawn tangent to the sides of the spherical triangle at its vertices are shown. The angles a, b, and c of the spherical triangle are seen to be greater than the corresponding angles a', b', and c' of the plane triangle. The amount of this excess would be small if the points were near together, and the surface forming the triangle would not depart far from a plane passing through the three points. If the points were far apart the difference would be considerable. Evidently the same conditions would obtain for a figure of any number of sides. Hence we see that angles on the surface of the earth are spherical angles.

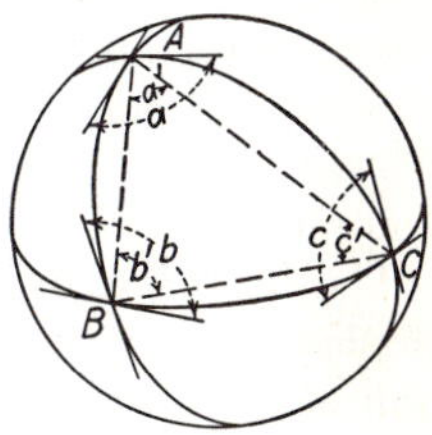

FIG. 1-4.

In everyday life we are not concerned with these facts, principally because we are dealing with only a small portion of the earth's surface. We think of a line passing along the surface of the earth directly between two points as being a straight line, we think of plumb lines as being parallel, we think of a level surface as a flat surface, and we think of angles between lines in such a surface as being plane angles.

As to whether the surveyor must regard the earth's surface as curved or may regard it as plane (a much simpler premise) depends upon the character and magnitude of the survey and upon the precision required.

1-4. Plane Surveying. That type of surveying in which the mean surface of the earth is considered as a plane, or in which its spheroidal shape is neglected, is generally defined as *plane surveying.* With regard to horizontal distances and directions, a level line is considered as mathematically straight, the direction of the plumb line is considered to be the same at all points within the limits of the survey, and all angles are considered to be plane angles.

By far the greater number of all surveys are of this type. When it is considered that the length of an arc 11½ miles long lying in the earth's surface is only 0.05 ft. greater than the subtended chord, and further that the difference between the sum of the angles in a plane triangle and the sum of those in a spherical triangle is only one second for a triangle at the earth's surface having an area of 75.5 sq. miles, it will be appreciated that the shape of the earth need be taken into consideration only in surveys of precision covering large areas.

Surveys for the location and construction of highways, railroads, canals, and, in general, the surveys necessary for the works of man are plane surveys, as are also the surveys made for the purpose of establishing boundaries, except state and national. The United States system of subdividing the public lands employs the methods of plane surveying but takes into account the shape of the earth in the location of certain of the primary lines of division.

The operation of determining *elevation* is usually considered as a division of plane surveying. Elevations are referred to a spheroidal surface, a tangent at any point in the surface being normal to the plumb line at that point. The curved surface of reference, usually mean sea level, is called a "datum" or, curiously and incorrectly, a "datum plane." The procedure ordinarily used in determining elevations automatically takes into account the curvature of the earth, and elevations referred to the curved surface of reference are secured without extra effort on the part of the surveyor. In fact it would be more difficult for him to refer elevations to a true plane than to the imaginary spheroidal surface which he has chosen. Imagine a true plane, tangent to the surface of mean sea level at a given point. At a horizontal distance of 10 miles from the point of tangency the vertical distance (or elevation) of the plane above the surface represented by mean sea level is 67 ft., and at a distance of 100 miles from the point of tangency the elevation of the plane is 6,670 ft. above mean sea level. Evidently the curvature of the earth's surface is a factor which cannot be neglected in obtaining even very rough values of elevations.

This book deals chiefly with the methods of plane surveying.

1-5. Geodetic Surveying. That type of surveying which takes into account the shape of the earth is defined as *geodetic surveying*. Surveys employing the principles of geodesy are of high precision and generally extend over large areas. Where the area involved is not great, as for a state, the required precision may be obtained by assuming that the earth is a perfect sphere. Where the area is large, as for a country, the true spheroidal shape of the earth is considered. Surveys of the latter character have been conducted only through the agencies of governments. In the United States such surveys have been conducted principally by the U.S. Coast and Geodetic Survey and the U.S. Geological Survey. Such surveys have also been conducted by the Great Lakes Survey, the Mississippi River Commission, several boundary commissions, and others. Surveys conducted under the assumption that the earth is a perfect sphere have been made by such large cities as Washington, Baltimore, Cincinnati, and Chicago.

Though only a few engineers and surveyors are employed in geodetic work, the data of the various geodetic surveys are of great importance in that they furnish precise points of reference to which the multitude of surveys of lower precision may be tied. For each state, a system of plane coordinates has been devised, to which all points in the state can be referred without significant error in distance or direction arising from the difference between the reference surface and the actual mean surface of the earth.

1-6. Operations of Surveying. The nature of the measurements made by the surveyor has been indicated in preceding articles.

In *land surveying* his work consists in:

1. Rerunning old land lines to determine their length and direction.
2. Reestablishing old or obliterated land lines from recorded lengths and directions and such other information as it is possible to secure.
3. Subdividing lands into parcels of predetermined shape and size.
4. Setting monuments to preserve the location of land lines.
5. Locating the position of such monuments with respect to permanent landmarks.
6. Calculating areas, distances, and angles or directions.
7. Portraying the data of the survey on a *land map*.
8. Writing descriptions for deeds.

A *topographic survey* is made to secure field data from which may be made a *topographic map* indicating the relief, or elevations and inequalities of the land surface, and the location of natural and artificial objects. The work consists in:

1. Establishing by angular and linear measurements the horizontal location of certain points for the skeleton of the survey, termed the *horizontal control.*

2. Determining the elevation of control points by the operation of leveling, termed the *vertical control.*

3. Determining the horizontal location and elevation of a sufficient number of ground points to provide data for the map.

4. Locating such other natural or artificial details as the requirements of the survey demand.

5. Calculating angles, distances, and elevations.

6. Plotting and finishing the topographic map (see also "Photogrammetric Surveying," later in this article).

Route surveying refers to those surveys necessary for the location and construction of lines of transportation or communication, such as highways, railroads, canals, transmission lines, and pipe lines. The preliminary work usually consists of a topographic survey. The location and construction surveys may further consist in:

1. Establishing the center line by setting stakes at intervals.
2. Running levels to determine the profile of the ground along the center line.
3. Plotting such profile, and fixing grades.
4. Taking cross sections.
5. Calculating volumes of earthwork.
6. Measuring drainage areas.
7. Laying out structures, such as culverts and bridges.
8. Locating right-of-way boundaries.

Hydrographic surveying refers to surveying bodies of water for purposes of navigation, water supply, or subaqueous construction. The operations of hydrographic surveying consist broadly in:

1. Making a topographic survey of shores and banks.

2. Taking soundings to determine the depth of water and the character of the bottom.

3. Locating such soundings by angular and linear measurements.

4. Plotting the hydrographic map showing the topography of the shores and banks, the depths of soundings, and other desirable details.

5. Observing the fluctuation of the ocean tide or of the change in level of lakes and rivers.

6. Measuring the discharge of streams.

In a sense, the surveys for drainage and for irrigation are hydrographic in character, but the principal work is essentially either topographic or route surveying.

Mine surveying makes use of the principles of land, topographic, and route surveying, with modifications in practice made necessary by altered conditions. Both surface and underground surveys are required. The work of the mine surveyor consists in:

1. Establishing (on the surface) the boundaries of claims for mineral patent and fixing reference monuments.

2. Locating (on the surface) shafts, adits, bore-holes, railroads, tramways, mills, and other details.
3. Making a topographic survey of the mine property.
4. Constructing the surface map.
5. Making underground surveys necessary to delineate fully the mine workings.
6. Constructing the underground plans showing the workings in plan, longitudinal section, and transverse section.
7. Constructing the geological plan.
8. Calculating volumes removed.

Cadastral surveying, a practically obsolete term, has particular reference to extensive urban or rural surveys made for the purpose of locating property lines and improvements in detail, primarily for use in connection with the extent, value, ownership, and transfer of land (see Art. 22-20). The term is sometimes applied to the primary control of the U.S. public-land surveys.

City surveying is the term frequently applied to the operation of laying out lots and to the municipal surveys made in connection with the construction of streets, water-supply systems, and sewers. There is no essential distinction between such surveys and those just described except that the degree of refinement observed in making measurements is made proportional to the value of the land with which the survey is concerned.

The term *city survey* has come to mean an extensive coordinated survey of the area in and near a city for the purposes of fixing reference monuments, locating property lines and improvements, and determining the configuration and physical features of the land (see Art. 22-19). Such a survey is of value for a wide variety of purposes, particularly for planning city improvements. The work consists in:

1. Establishing horizontal and vertical control, as for topographic surveying.
2. Making a topographic survey and topographic map.
3. Marking critical points such as street corners with suitable monuments referred to a common system of rectangular coordinates.
4. Making a property map, with layout and dimensions of properties.
5. Making a wall map.
6. Making a map, or maps, to show underground utilities.

Photogrammetric surveying is the application to surveying—usually topographic work—of the science of measurement by means of photographs. With specially designed cameras, photographs are taken either from airplanes or from ground stations. The principles of perspective are employed in the projection of details from the photographs onto maps drawn to scale. In connection with limited ground surveys made for the purpose of accurately establishing control points visible from the air, *aerial photogrammetry* is employed on many topographic surveys by making certain necessary adjustments and projections. Important advancements and simplifications in the technique of aerial photogrammetry have made this method by far the most rapid and accurate except perhaps where the ground is relatively flat, where forest growths are heavy, or where the area is small. The advantages of aerial photogrammetry are the speed with which the work is accomplished, the wealth of detail secured, and the use in locations otherwise difficult or impossible of access. The method is used not only for military purposes but also for general topographic surveys, route surveys, and surveys of agricultural areas. Considerable areas of the United States have already been photographed, and in many cases the photographs are available to surveyors and others for a nominal fee.

Terrestrial photogrammetry, or photographic surveying from ground stations, has been found a useful adjunct to other methods in the small-scale mapping of mountainous areas. The work consists in taking photographs from two or more control stations and in utilizing the photographs for the projection of details of the terrain in plan and elevation.

1-7. Kinds of Surveying and Mapping. In 1959 the American Society of Civil Engineers adopted the classification of surveying and mapping shown in Table 1-1 and adopted the policy that items 1 to 4 therein be classified as part of civil engineering. The Surveying and Mapping Division of ASCE has technical committees on cartographic surveying, engineering surveying, geodetic surveying, and land surveying. The American Congress on Sur-

TABLE 1-1. KINDS OF SURVEYING AND MAPPING

1. *Land or Property Surveying* (Cadastral)
   a. Property and boundary surveys
   b. Subdivision surveys and plats
   c. Public-lands surveys
   d. Surveys for plans and plats
2. *Engineering Surveys for Design and Construction*
   a. Design data surveys (including route surveys)
   b. Construction surveys
   c. Mine surveys
3. *Geodetic surveying, geodetic engineering, or geodesy*
   a. Control surveys, first- and second-order accuracy
   b. Geodetic astronomy
   c. Gravity surveys, magnetic declination surveys, figure-of-the-earth studies
4. *Cartographic surveying, cartographic engineering, or map and chart surveying*
   a. Control surveys, third- and fourth-order accuracy
   b. Topographic-planimetric surveys and maps
   c. Hydrographic surveys
5. *Aerial survey services*
   a. Aerial photography
   b. Electrical measurements for distances and position fixes
   c. Airborne magnetometer surveys
   d. Radar-altimeter profiles and elevations
6. *Cartography* (not requiring original surveys)
   a. Map design
   b. Compilation derived from existing source data
   c. Map editing
   d. Map reproduction

veying and Mapping has divisions concerned with property surveys, control surveys, topography, cartography, and instruments. Photogrammetry is within the province of the American Society of Photogrammetry.

1-8. Definitions. A *level surface* is a curved surface every element of which is normal to the plumb line. Disregarding local deviations of the plumb line from the vertical, it is parallel with the mean spheroidal surface of the earth. A body of still water provides the best example.

A *horizontal plane* is a plane tangent to a level surface.

A *horizontal line* is a line tangent to a level surface. In surveying, it is commonly understood that a horizontal line is straight.

A *horizontal angle* is an angle formed by the intersection of two lines in a horizontal plane.

A *vertical line* is a line perpendicular to the plane of the horizon. A plumb line is an example.

A *vertical plane* is a plane of which a vertical line is an element.

A *vertical angle* is an angle between two intersecting lines in a vertical plane. In surveying, it is commonly understood that one of these lines is horizontal, and a vertical angle to a point is understood to be the angle in a vertical plane between a line to that point and the horizontal plane.

In surveying, measured angles are either vertical or horizontal.

In plane surveying, distances measured along a level line are termed *horizontal distances*. The distance between two points is commonly understood to be the horizontal distance from the plumb line through one point to the plumb line through the other. Measured distances may be either horizontal or inclined, but in practically all cases the inclined distances are reduced to equivalent horizontal lengths.

The *elevation* of a point is its vertical distance above (or below) some arbitrarily assumed level surface, or datum.

A *contour* is an imaginary line of constant elevation on the ground surface. The corresponding line on the map is called a *contour line.*

The vertical distance between two points is termed the *difference in elevation.* It is the distance between an imaginary level surface containing the high point and a similar surface containing the low point. The operation of measuring difference in elevation is called *leveling.*

The *grade,* or *gradient,* of a line is its slope, or rate of ascent or descent.

Additional definitions pertaining to field work are given in Art. 3-9.

1-9. Units of Measurement. The operations of surveying entail both angular and linear measurements.

The units of angular measurement are the *degree, minute,* and *second.* A plane angle extending completely around a point equals 360 degrees; 1 degree = 60 minutes; 1 minute = 60 seconds. In France the *grad,* or *grade,* is the angular unit; 400 grads = 360 degrees. In military operations the *mil* is used; 6,400 mils = 360 degrees. On most surveys, measurement to the nearest minute is sufficiently exact. On precise surveys, frequently angles are determined to tenths of seconds.

In English-speaking countries the common units of linear measurement are the *yard*, *foot*, and *inch*. On most surveys in these countries, distances are measured in feet, tenths of feet, and hundredths of feet; and surveyor's tapes are usually graduated in these units. In laying out construction work for men of the building trades, the surveyor will often find it necessary to employ the foot, the inch, and the eighth of an inch. Most measurements in surveying need not be taken closer than hundredths of a foot, and often distances to the nearest foot or even to the nearest 10 ft. are sufficient for the purpose of the survey.

Formerly the *rod* and the *Gunter's chain* were units much used in land surveying, and the Gunter's chain as a unit of length is still employed in the subdivision of the United States public lands. The Gunter's chain is 66 ft. long and is divided into 100 links each 7.92 in. long. 1 mile = 80 chains = 320 rods, poles, or perches = 5,280 ft.

Many other countries employ the *meter* as the unit of length. 1 meter = 39.370 in. = 3.2808 ft. = 1.0936 yd. The meter is the unit of length employed by the U.S. Coast and Geodetic Survey.

The *vara* is a Spanish unit of linear measurement used in Mexico and several other countries falling under early Spanish influence. In portions of the United States formerly belonging to Spain or to Mexico, the surveyor will frequently have occasion to rerun property lines from old deeds in which lengths are given in terms of the vara. Commonly 1 vara = 32.993 in. (Mexico), 33 in. (California), or 33⅓ in. (Texas); but other somewhat different values of the vara have been used for many surveys.

In the United States the units of area commonly used are the *square foot* and the *acre*. Formerly the *square rod* and the *square Gunter's chain* were also used. 1 acre = 10 sq. Gunter's chains = 160 sq. rods = 43,560 sq. ft.

The units of volumetric measurement are the *cubic foot* and *cubic yard*.

1-10. Precision of Measurements. In dealing with abstract quantities, we have become accustomed to thinking largely in terms of exact values. At the start, the student of surveying ought to appreciate that he is dealing with physical measurements which are correct only within certain limits, owing to errors that cannot be completely eliminated. The degree of precision of a given measurement depends upon the methods and instruments employed and upon other conditions surrounding the survey. It is desirable that all measurements be made with high precision, but unfortunately a given increase in precision is usually accompanied by more than a directly proportionate increase in the time and effort of the surveyor. It therefore becomes his duty to maintain a degree of precision as high as justified by the purpose of the survey, but not higher. It is important, then, that he have a thorough knowledge of the sources and kinds of errors, of their effect upon field measurements, and of instruments and methods to be employed to keep the magnitude of the errors within allowable values. It follows that he must understand the intended use of the survey data.

Before beginning work, the surveyor ought to consider the following questions:

1. What is the purpose of the survey?
2. What degree of precision is required for that purpose?
3. With what precision must each kind of measurement be taken?
4. Can a higher degree of precision be obtained without appreciable additional cost?
5. What are the sources of error?
6. What methods must be employed to keep errors within allowable limits?
7. What instruments should be used?
8. How may the work be organized to reduce the labor to a minimum?
9. How is the correctness of the work to be verified?

1-11. Principles Involved. The underlying principles of plane surveying are not difficult. They involve a thorough knowledge of geometry and trigonometry, and to a less degree a knowledge of physics, of astronomy, and of the theory and methods of adjustment of errors. Such portions of the last three subjects as are necessary to the understanding of the text will be given in succeeding chapters as the need arises. Geodetic surveying requires an expert knowledge of all the above subjects.

1-12. Practice of Surveying. Like other arts based upon the sciences, the practice of surveying is complex; and no amount of theory will make a good surveyor unless he has the requisite skill in the art of observing and is versed in field and office procedures. The student should realize the importance of a knowledge of the practical phases of the subject and should seek to become as well-grounded in the practice as possible.

Often surveying is one of the first professional subjects studied by the engineering student. He may not expect to become a surveyor, but he ought to understand that the training he will receive in the art of observing and computing, in the study of errors and their causes and effects, and in the practice of mapping will directly contribute to success in other subjects, regardless of the branch of engineering in which he may be interested.

1-13. Requisites of a Surveyor. As the term "surveyor" is used here, it has reference not only to that individual who makes his chief livelihood from surveying and expects so to continue in the remote future but also to that individual of a large army of engineers to whom surveying is merely one of the arts of his profession, to whom the survey is incidental to the engineering problem.

A thorough knowledge of the theory of surveying and skill in its practice are principal requisites of the surveyor, but upon the evidence of employers themselves, it is also true that traits of character are far more potent factors in the success of the surveyor or engineer than is mere technical knowledge or skill. Therefore, it should be stated with all emphasis that, while mastering the theory and practice of surveying, the student will do himself a great benefit if he also develops appropriate traits of character and habits of mind by diligent application of the laws of habit formation, which are fairly well known. Some of the traits which should be possessed by the surveyor are that he should:

Be reliable.

Be of sound judgment.

Maintain the attitude of the scientist, that no measurement or result is trustworthy until every reasonable test of its accuracy has been applied.

Possess initiative and attack a problem with resourcefulness and energy.

Be thorough, not content with his work until it has been finished in a workmanlike fashion.

Be able to think without confusion and to reason logically without prejudice.

Be of good temper, thoughtful of those coming under his direction, commanding the respect of his associates, and watchful of the interests of his employer.

REFERENCES

1. AMERICAN SOCIETY OF CIVIL ENGINEERS, "Definitions of Surveying, Mapping, and Related Terms," *Manual of Engineering Practice* 34, 1954.
2. MITCHELL, HUGH C., "Definitions of Terms Used in Geodetic and Other Surveys," *U.S. Coast and Geodetic Survey, Spec. Pub.* 242, Government Printing Office, Washington, D.C., 1948.

CHAPTER 2

FEATURES OF SURVEYING INSTRUMENTS

2-1. Principal Instruments. The principal surveying instruments and accessories and their uses are listed below:

Tape. A graduated flexible ribbon used for measuring distance (see Figs. 7-2*a* and 7-2*b*).

Chaining Pins. Steel pins about 1 ft. long, for temporarily marking the location of the ends of the tape as distances are measured (see Fig. 7-3).

Plumb Bob. A pointed metal weight suspended from a string. Used to project the horizontal location of a point from one elevation to another.

Engineer's Level. A telescope to which is attached a spirit-level tube, all revolving about a vertical axis and mounted on a tripod (see Fig. 8-4). A level is employed for determining difference in elevation. Its use is termed leveling.

Level Rod. A graduated wooden rod which, in conjunction with the level, is used in determining difference in elevation. Graduations are usually in hundredths of feet. The rod may be either in a single piece or jointed. Common length when extended is 12 or 13 ft. (see Figs. 8-11 to 8-13).

Surveyor's Compass. A magnetic compass mounted on a tripod and equipped with sight vanes. Used for determining the direction of lines. Nearly obsolete except for rough surveys, as in forestry (see Fig. 12-13*a*).

Flag, Flagpole, or *Range Pole.* A pole, either of steel or of wood shod with a steel point, painted with bands of alternating red and white. Used as a sighting rod in connection with either angular or linear measurements (see Fig. 7-4).

Engineer's Transit. The universal instrument. Used principally for measuring horizontal and vertical angles, for measuring distances by stadia, and for prolonging straight lines. The transit has a telescope which may be revolved about either a horizontal or a vertical axis. It is usually equipped with a magnetic needle and is mounted on a tripod (see Fig. 13-1).

Plane Table. A drawing board mounted on a tripod, and an alidade, or straightedge equipped with a telescope, which can be moved about on the board. The plane table is used for mapping (see Fig. 17-1).

2-2. The Engineer's Level. Figure 2-1*a* is a diagram of the principal parts of the engineer's level. The level consists of the telescope *A* mounted upon the level bar *B* which is rigidly fastened to the spindle *C*. Attached to the telescope or the level bar and parallel to the telescope is the level tube *D*.

The spindle fits into a cone-shaped bearing of the leveling head E, so that the level is free to revolve about the spindle C as an axis. The leveling head is screwed to a wooden tripod F. In the tube of the telescope are cross-hairs at G, which appear on the image viewed through the telescope as illustrated by Fig. 2-1*b*. The bubble of the level is centered by means of the leveling screws H.

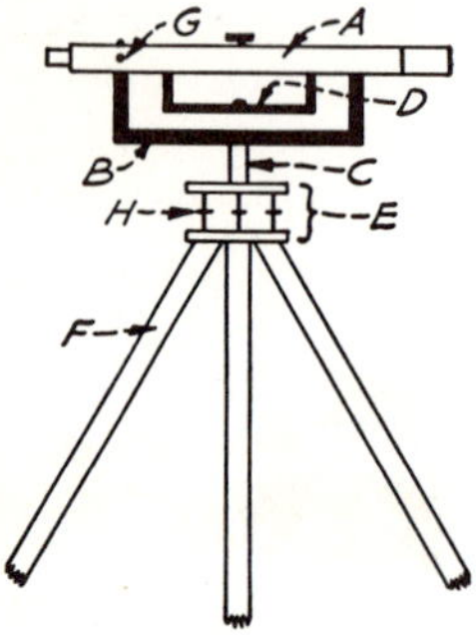

FIG. 2-1*a*.

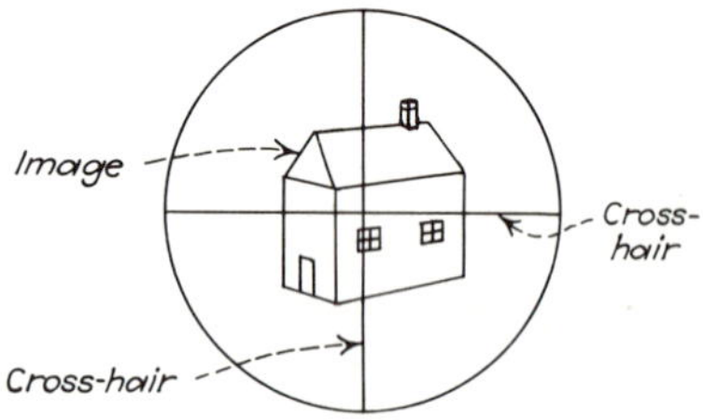

FIG. 2-1*b*.

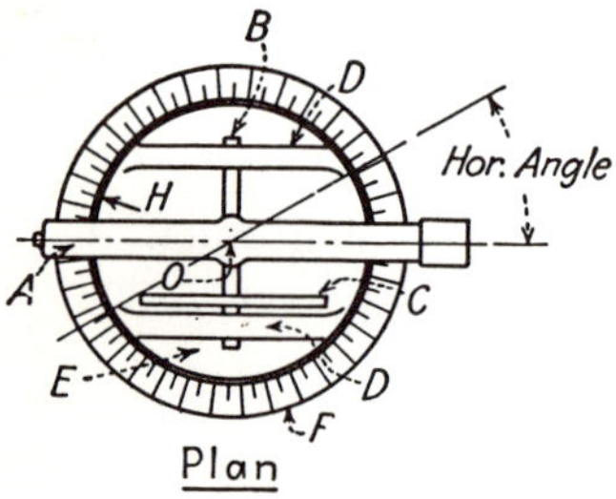

FIG. 2-2*a*.

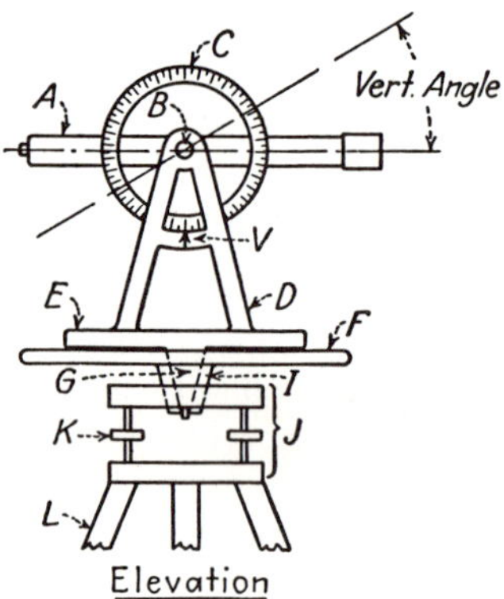

FIG. 2-2*b*.

2-3. The Engineer's Transit. Figures 2-2*a* and 2-2*b* illustrate, in plan and elevation, respectively, the principal parts of the engineer's transit. The transit consists of the telescope A, mounted on a horizontal axis B which is supported by standards D. Attached beneath the telescope is a spirit-level tube (not shown), similar to that for the level just described. Angles of rotation of the telescope in a vertical plane are indicated by the vertical circle C which is graduated in degrees and which is read by means of the index V attached to one of the standards. The standards rest on the upper plate E which is equipped with spirit levels (not shown) and which rotates about the vertical axis O on a spindle G called the inner spindle. The lower plate F revolves about the vertical axis on the outer spindle I; the rim of its upper face is a circle graduated in degrees and read by means of an index H on the upper circle. The spindles G and I are supported by the leveling head J, which is screwed to a wooden tripod L. The vertical axis is made vertical by means of three or four leveling screws K. A magnetic compass (not shown) is centered on the upper plate.

A detailed description of the transit is given in Chap. 13. For the present discussion it is sufficient to state that:

1. The instrument can be leveled by means of the plate levels and the leveling screws.
2. The telescope can be rotated about the horizontal axis to measure vertical angles, or about the vertical axis to measure horizontal angles.
3. Horizontal angles are measured by clamping the graduated lower plate and observing the rotation of the upper plate between pointings of the telescope.
4. The telescope can be leveled by means of the telescope level tube, and hence the transit can be employed for direct leveling.
5. Vertical angles are measured by reading the graduations on the vertical circle.
6. Small movements about the vertical axis and the horizontal axis are accomplished by means of clamps and tangent-screws, described in Art. 2-19.
7. Readings of each graduated circle are facilitated by the use of a vernier scale, described in Art. 2-18, at the index.
8. By means of the magnetic compass, directions can be observed and horizontal angles checked.

2-4. Essential Features. Essential parts of the engineer's level, the transit, and the plane-table alidade are the *level tube* and the *telescope*. *Verniers* are used on transits, alidades, leveling rods, sextants, and planimeters. The *magnetic compass* is used either as a separate instrument or as an adjunct to the transit or the alidade. In the following articles of this chapter will be discussed level tubes (Arts. 2-5 to 2-7), telescopes (Arts. 2-9 to 2-17), verniers (Art. 2-18), and other details of surveying apparatus.

2-5. Level Tube. The level tube used in surveying instruments (Fig. 2-3) is a glass vial with the inside ground barrel-shaped, so that a longitudinal line on its inner surface is the arc of a circle. The tube is nearly filled with sulphuric ether or with alcohol. The remaining space is occupied by a bubble of air which takes up a location at the high point in the tube. The tube is usually graduated in both directions from the middle; thus by observation of the ends of the bubble it may be "centered," or its center brought to the mid-point of the tube. The tube is set in a protective metal housing, usually with plaster of paris. The housing is attached to the instrument by means of screws which permit adjustment, as shown in the figure.

Some leveling instruments are equipped with a prismatic viewing device by means of which one end of the bubble appears reversed in direction and alongside the other end. The bubble is centered by matching its ends. Settings can be made more accurately by such coincidence than by matching the ends of the bubble with respect to the graduations on the level tube.

A longitudinal line tangent to the curved inside surface at its upper midpoint is called the *axis of the level tube* or *axis of the level.* When the bubble is centered, the axis of the level tube is horizontal.

A *reversion level* is one graduated on both top and bottom and so mounted that it can be used when the telescope is either normal or inverted.

A *striding level* consists of a level tube mounted in a metal frame having legs so that the level may be placed on a telescope which it is desired to level. A

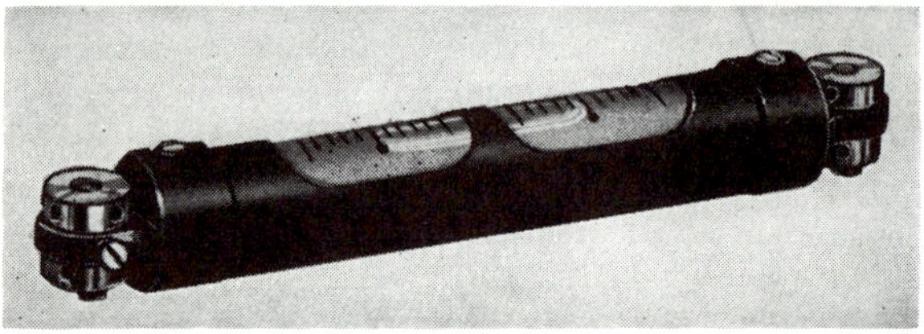

FIG. 2-3. Level tube.

striding level may also be used to level the horizontal axis of a transit telescope or the board of a plane table.

A *circular level,* also called a *universal level* or *bull's-eye level,* is circular in plan and has a spherical upper surface. When the bubble is concentric with respect to a small circle marked at the center of the upper surface, the instrument is level.

2-6. Sensitivity of Level Tube. If the radius of the circle to which the level tube is ground is large, a small vertical movement of one end of the tube will cause a large displacement of the bubble; if the radius is small, the displacement will be small. Thus the radius of the tube is a measure of its sensitivity. The sensitivity is generally expressed in seconds of the central angle whose arc is one division of the tube. The sensitivity expressed in this manner is inversely proportional to the number of seconds. For many instruments the length of a division is 2 mm., while for others it is 0.1 in. (2.5 mm.); the practice is not uniform among manufacturers of surveying instruments. For this reason, the sensitivity expressed in seconds of arc is not a definite measure unless the spacing of graduations is known. The values shown in Table 2-1 roughly represent common practice for various instruments.

A simple method of determining the radius of curvature in the field is explained in field problem 2, Art. 8-28.

The more sensitive the tube, the longer the time required to center the bubble. Hence, time is wasted if the tube is more sensitive than the device to which it is attached. For example, in a telescope level tube the first noticeable movement of the bubble should be accompanied by a barely apparent movement of the line of sight as indicated by the cross-hairs on an object at an average sighting distance from the instrument.

TABLE 2-1. SENSITIVITY OF LEVEL TUBE

Instrument	Radius of curvature, ft.	Seconds of arc for 2-mm. division of tube
Better grade of engineer's levels	68	20
Precise level (U.S. Coast and Geodetic Survey)	677	2
Engineer's transit:		
Telescope level	45	30
Plate levels	18	75
Plane table:		
Telescope level	30	45
Control level on vertical circle	23	60

2-7. Adjustment of Level Tube. The principle involved in bringing the axis of the level tube into the proper relation with the device to which it is attached is invariably that of *reversion*, or reversing the level tube end for end. There are two general cases: (1) when the tube is fixed to a telescope or plate which can be rotated about a vertical axis, as on a transit plate, and (2) when the tube can be lifted from the support and reversed end for end thereon, as on a plane-table alidade. However, the method of adjustment is the same in either case.

The steps involved in adjustment are shown in Fig. 2-4. In view (*a*) is shown the tube out of adjustment by the amount of the angle α, but with the bubble centered; the support is therefore not level, and the vertical axis is not vertical. In view (*b*) the level tube has been lifted and reversed end for end. (In the case of a support on a vertical axis the same relations would exist if the support were rotated 180° about the vertical axis.) The axis of the level tube now departs from the horizontal by 2α, or *double* the error of the setting. In view (*c*) the bubble has been brought back *halfway* to the middle of the tube by means of the adjusting screw C, without moving the support; the tube is now in adjustment. Finally, in view (*d*) the bubble is again centered by raising the low end (and/or lowering the high end) of the support; the support is now level and the adjustment may be checked by reversing the tube again.

If it were desired to level the support in the direction of the tube without taking time to adjust the tube, this could be accomplished by centering the bubble as in view (*a*), Fig. 2-4; reversing the tube as in view (*b*); and raising the low end (and/or lowering the high end) of the support until the bubble

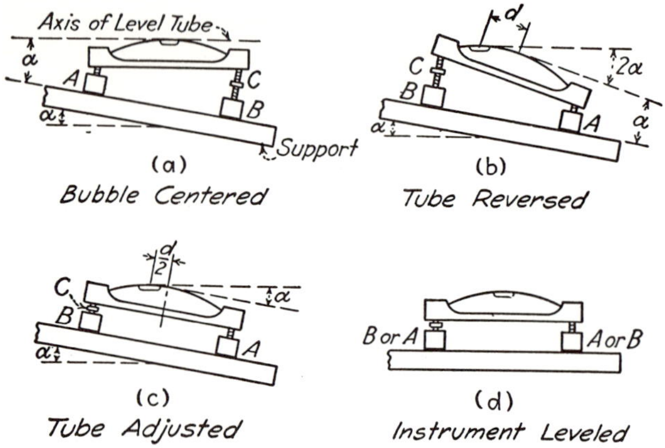

FIG. 2-4. Adjustment of level tube by reversion.

is brought halfway back to the center of the tube. This position of the bubble corresponds to the error of setting of the tube; and whenever the bubble is in this position the support will be level.

2-8. Leveling Head. On the level, transit, and one type of plane table, the head of the instrument is leveled by means of leveling screws, or foot screws. A simplified diagram is shown in Fig. 2-5, in which the spindle *A* revolves in the socket of the leveling head *B*. Near the bottom of the leveling head is a ball-and-socket joint *C*, which makes a flexible connection with the foot plate *D*. The leveling head has four radial arms, into each of which is threaded a leveling screw *E*. (Only two of the screws are shown in the figure.) The leveling screws bear on the foot plate, and by means of these screws the leveling head can be tilted.

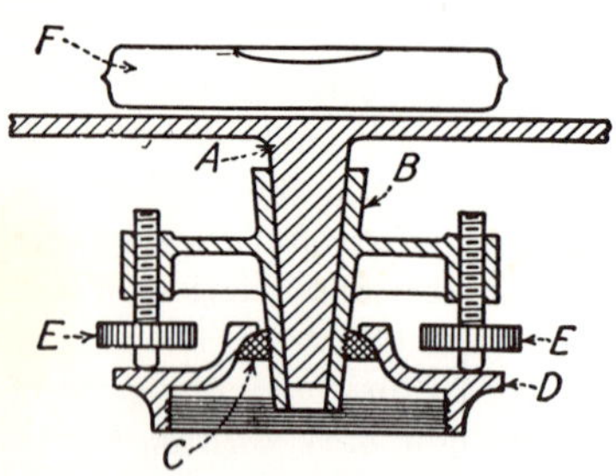

FIG. 2-5. Leveling head.

The engineer's level, which has only one level tube, is leveled as follows: The instrument is turned about the vertical axis until the level tube is approximately over one pair of opposite leveling screws, and the level bubble is brought approximately to the center by turning that pair of screws, keeping both screws lightly in contact with the foot plate and thus keeping the ball-and-socket joint lightly in bearing. It is convenient to remember that the bubble travels in the *same direction as the left thumb.* The instrument is then rotated 90°, and the level bubble is centered over the other pair of opposite leveling screws. This process is repeated alternately over the two pairs of screws until (if the level tube is in adjustment) the bubble will remain centered for any direction of pointing of the instrument.

If the instrument has two level tubes perpendicular to each other, the process of leveling is similar except that it is not necessary to rotate the instrument. Each level tube is alined with one pair of opposite leveling screws and is controlled by that pair. If the instrument has a circular type of level, the process of leveling is similar.

Some instruments are equipped with a three-screw leveling head by means of which it is relatively easy to center the bubble. However, this type of head changes in elevation when the screws are manipulated, whereas the four-screw leveling head does not. On instruments equipped with a three-screw leveling head, usually the circular type of level vial is used.

It is a waste of time to center the bubble exactly over one pair of leveling screws before bringing it approximately to center over the other pair. It is best to leave all four screws rather loose, or barely in bearing, until the instrument is almost level. If one pair of screws turns hard, the other pair should be loosened slightly. The final centering of the bubble will be facilitated by turning one screw rather than by attempting to manipulate two opposite screws at the same time, provided this movement neither loosens nor binds the instrument unduly.

Even if the level tube is out of adjustment, it is possible to use the instrument properly by use of the principle of reversion discussed in the preceding article. The bubble is centered over one pair of opposite leveling screws, the telescope is rotated end for end about the vertical axis, and the bubble is brought halfway back to center by means of the leveling screws. The process is repeated over the other pair of leveling screws. It will be found that the bubble will then remain in the same position regardless of the direction of pointing; and the vertical axis of the instrument will be truly vertical. This method of operation is sometimes used to avoid stopping the work to adjust the level tube.

2-9. Telescope. Figure 2-6*a* shows the principal parts of an *external-focusing* telescope. Rays of light emanating from an object within the field of view of the telescope are caught by the objective lens *A* and are brought to a focus and form an image in the plane of the cross-hairs *B*. The lenses of the eyepiece *C* form a microscope which is focused on the image at the cross-hairs. The objective lens is screwed in the outer end of the objective slide *D* which fits in the telescope tube *E*. The objective lens is focused by the screw *F* at the inner end of which is a pinion that engages the teeth of a rack fixed to the objective slide. The eyepiece slide G is held in position transversely by rings H and J, through which it may be moved in a longitudinal direction for focusing. By means of screws the ring J may be moved transversely so that the intersection of the cross-hairs will appear in the center of the field of view.

The line of sight is defined by the intersection of the cross-hairs and the optical center of the objective lens. The instrument is so constructed that the optical axis of the objective lens coincides (or practically coincides) with the axis of the objective slide; in other words, a given ray of light passing through the optical center of the objective always occupies the same position in the telescope tube regardless of the longitudinal position of the lens. The cross-hairs can be so adjusted that the line of sight and the optical axis coincide.

Another type, called the *internal-focusing* telescope, has increased greatly in use. It is shown in section in Fig. 2-6*b*. Its arrangement and operation are

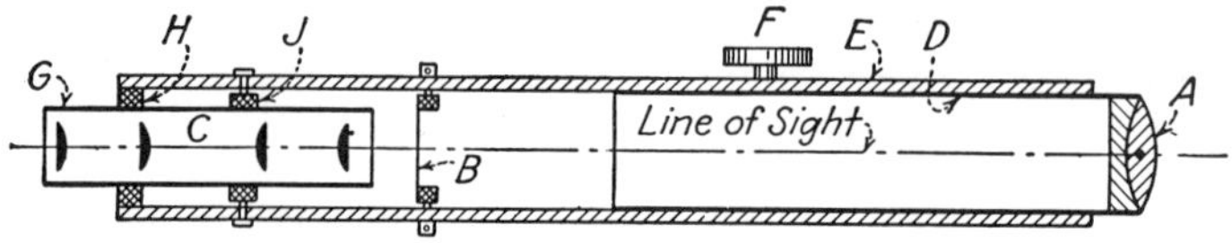

FIG. 2-6*a*. Longitudinal section of external-focusing telescope.

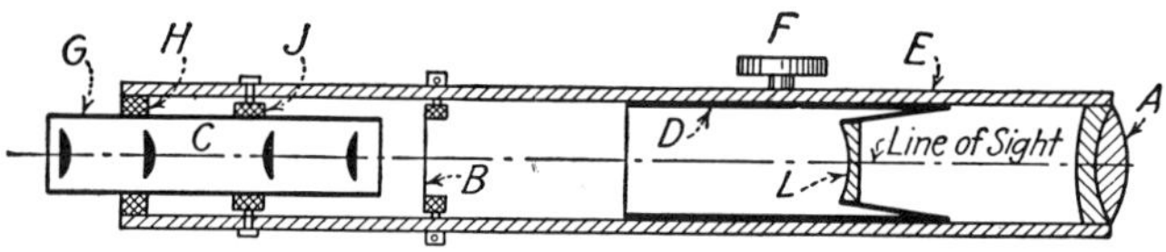

FIG. 2-6*b*. Longitudinal section of internal-focusing telescope.

similar to that just described, except that the objective lens A is fixed in the end of the telescope tube and that the slide carries a focusing lens L. The advantages of the internal-focusing type are as follows:

1. Because both ends of the telescope are closed, the focusing slide is practically free from grit which would cause wear; and the entire interior of the telescope is practically free from dust and moisture.

2. Since the focusing slide is light in weight and is located near the middle of the telescope, the telescope tends to balance well.

3. In making measurements by the stadia method (Chap. 15), an instrumental constant is eliminated and the computations thus simplified.

The disadvantages are that the extra lens required reduces the illumination and that the interior of the telescope is not so easily accessible for field cleaning or repairs.

The discussions hereinafter refer to the external-focusing type of telescope, except as specifically stated. With modifications in detail, they apply also to the internal-focusing type.

2-10. Focusing. When the telescope is to be used, the eyepiece is first moved in or out until the cross-hairs appear sharp and distinct. This adjustment of the eyepiece should be tested frequently, as the observer's eye becomes tired.

When an object is sighted, the objective slide is moved in or out until the image appears clear, when the image should be in the plane of the cross-hairs. If a slight movement of the eye from side to side produces an apparent movement of the cross-hairs over the image, the plane of the image and the plane of the cross-hairs do not coincide, and *parallax* is said to exist. Since parallax is a source of error in observations, it should be eliminated by refocusing the objective, the eyepiece, or both until further trial shows no apparent movement. The objective lens must be focused for each distance sighted. The nearer the object sighted, the greater must be the distance between the objective and the cross-hairs. Although for short sights there

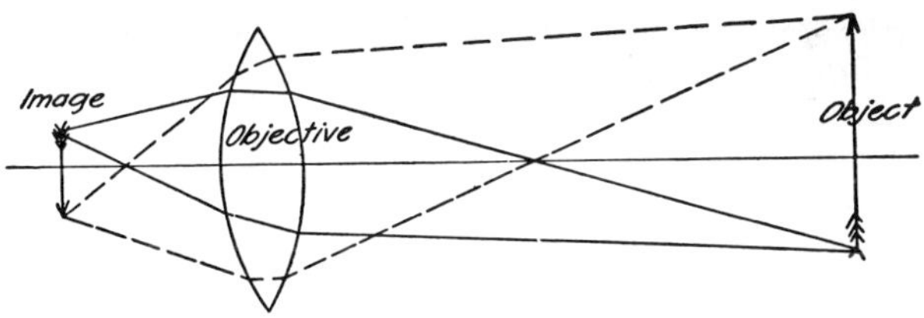

FIG. 2-7.

must be a considerable movement of the objective for a comparatively small change in distance, for the longer sights only a small movement of the objective is necessary regardless of the distance.

The telescope cannot be focused on objects closer than about 6 ft. from the center of the instrument, unless special short-focus lenses are employed.

Frequently the telescope will be so badly out of focus that the outline of the object cannot at first be detected. It often facilitates the work of focusing if the telescope is directed approximately in the proper direction by sighting along the outside of the tube. Some instruments are equipped with peep sights for this purpose.

The strain on the observer's eyes will be reduced if he learns to keep both eyes open while sighting.

Figure 2-7 illustrates the manner in which rays from an object are deviated by the objective and brought to a focus to form the image. It will be noted that the image is inverted.

2-11. Objective. The principal function of the telescope objective is to form an image for sighting purposes. For accuracy of measurements the objective should produce an image that is well lighted, accurate in form, distinct in outline, and free from discolorations. A single biconvex lens meets the first two of these requirements but is faulty in regard to the other two for the following reasons:

1. Rays entering the lens near its edge come to a focus nearer the objective than do those entering near its center. The image does not lie in a plane, but in the surface of a sphere. Hence, as viewed through the telescope, portions of the object are blurred. This defect is called *spherical aberration.*

2. Rays of the various colors of the spectrum are deviated by different amounts as they pass through the lens, hence the field of view appears discolored by lights of various hues. This is called *chromatic aberration.*

These two objectionable features of the single lens are nearly eliminated in most surveying instruments by providing an outer double-convex lens of crown glass and an inner concavo-convex lens of flint glass. The two lenses are usually cemented together with balsam but are sometimes separated by a thin spacer ring.

The *optical center* of the objective is that point in the lens through which a ray of light will pass without permanent deviation, regardless of the direction of the object from which the light emanates. In other words, the direction of the ray is the same after leaving the lens as before entering it. In a biconvex lens with faces of equal curvature the optical center and geometrical center coincide.

The *optical axis* is the line taken by a light ray that experiences no deviation either on entering or on leaving the objective. It passes through the optical center and the centers of curvature of the lens.

The *principal focus* is a point on the optical axis back of the objective where rays entering the telescope parallel with the optical axis are brought to a focus; or it is a point in front of the objective from which diverging light rays entering the lens emerge from it parallel with the optical axis. Stated in another form, the image of a point on the optical axis and an infinite distance away is at the principal focus back of the objective. If a point is at the principal focus in front of the objective, however, it will have no image.

The *focal length* of the objective is the distance from its optical center to the principal focus. When the telescope is focused on a distant point, the focal length is very nearly the distance from the optical center of the objective to the plane of the cross-hairs, for reasons which the preceding paragraph makes clear.

2-12. Objective Slides. Any lateral movement of the objective causes a deviation in the position of the optical axis and also in the line of sight, thereby introducing errors in measurements. The objective slide should therefore fit as neatly as possible and still admit readily of longitudinal movement for focusing. The workmanship on any good instrument is sufficiently precise to insure practical elimination of errors of this sort when the telescope is new, but in the course of long use, wear develops between the sliding parts and the slide becomes loose. This condition produces uncertainties in observation which no amount of adjustment can overcome.

For most instruments, the objective slide fits neatly into the telescope tube so that there is nearly perfect contact between these two parts for a considerable length near each end of the slide. Any wear that develops through use is therefore distributed over most of the length of the slide. Other instruments are provided with objective slides which are held in position by two metal rings as in Fig. 2-8. One ring is screwed in the end of the telescope tube. The other ring is placed in the rear of the rack and pinion and is held in position by four screws passing through the telescope tube. The inner ring is of somewhat smaller diameter than the telescope tube, so that by means of the screws just mentioned the objective slide may be adjusted laterally.

Particular care should be taken to protect the objective slide from dust, water, and other foreign matter. Many instruments are equipped with a guard which affords at least partial protection. If the slide is lubricated at

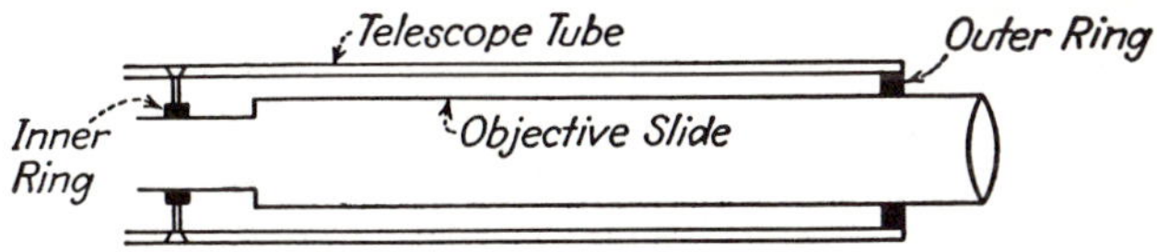

FIG. 2-8. Objective slide.

all, only a drop of the finest watch oil should be used and all excess oil should be removed with a soft cloth. If the objective lens is removed it should be replaced as nearly as possible in its original position, and after replacing it the adjustment should be checked.

2-13. Cross-hairs. The cross-hairs used to define the line of sight ordinarily consist of a vertical and a horizontal hair fastened to a metal ring called the *cross-hair ring* or *reticule*. The hairs are usually made of threads from the cocoon of the brown spider, but may be made of very fine platinum wire. In modern instruments usually the reticule consists of a glass plate on which are etched fine vertical and horizontal lines which serve as cross-hairs. Special patterns of additional lines are used on some instruments, for example, a circle for solar observations or double horizontal and vertical lines closely spaced for sighting between them precisely.

As shown in Fig. 2-9, the cross-hair ring is held in position by four capstan-headed screws which pass through the telescope tube and tap into the ring. The holes in the telescope tube are slotted so that when the screws are loosened, the ring may be rotated through a small angle about its own axis. To rotate the ring without disturbing its centering, two adjacent screws are loosened; and the same two screws are tightened after the ring has been rotated. The ring is smaller than the inside of the tube, and it may be moved either horizontally or vertically by means of the screws. Thus, to move it to the left, the right-hand screw is loosened and the left-hand screw is tightened. If the movement is to be large, first the top or bottom screw is loosened slightly; and after the movement to the left has been completed, the same (top or bottom) screw is tightened again.

Broken cross-hairs can be replaced in the field. Threads from ordinary spider webs are too rough, coarse, and dirty for use as cross-hairs. The best spider thread is freshly spun from a small spider, but commonly the thread from a cocoon—preferably a brown cocoon—is used. If the thread is stretched too tightly it will break easily; if too loose, it will sag in wet weather. The thread is handled by means of a pair of dividers or a forked stick. It is wetted, stretched moderately, and held securely in position on the marks of the cross-hair ring while a drop of shellac is placed on each end and left to dry.

The cross-hair ring is removed from the tube as follows: Two opposite capstan-headed screws are removed, and the ring is rotated 90° about the remaining two screws by means of a pointed stick inserted through the end of the telescope. The stick is then inserted in a screw hole, the remaining screws are taken out, and the ring is withdrawn without damage. The operations of replacing the cross-hair ring are in the reverse order of those employed in removing it.

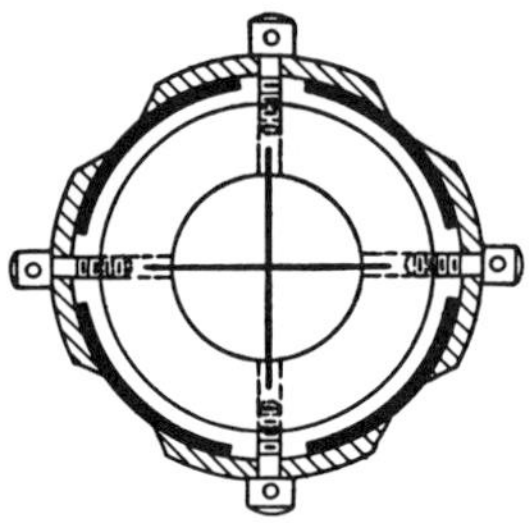

FIG. 2-9. Cross-hairs.

2-14. Stadia Hairs. Many telescopes are also equipped with two horizontal hairs called *stadia hairs*, one above and the other an equal distance below the horizontal cross-hair, for use in measuring distances by stadia (Chap. 15). Usually they are mounted in the same plane with the cross-hairs, and hence when the eyepiece is in focus all four hairs appear in the field of view. To prevent confusing the horizontal hairs with one another, in some cases two additional hairs in the form of an X are mounted on the cross-hair ring. Sometimes the stadia hairs are mounted in another plane so that when the cross-hairs are in focus the stadia hairs are invisible, or *vice versa;* the stadia hairs are then called *disappearing* hairs.

2-15. Eyepiece. Attention has previously been drawn to the fact that the image formed by the objective is inverted. Eyepieces are of two general types:

The *erecting* or *terrestrial* eyepiece, the more common of the two types, reinverts the image so that the object appears to the eye in its normal position. Usually it consists of four plano-convex lenses placed in a metal tube called the eyepiece slide (Fig. 2-10*a*). In the figure *A* represents the object, *B* the inverted image in the plane of the cross-hairs, *C* the image which is magnified by the lens nearest the eye, and *D* the magnified image as it appears to the eye.

The *inverting* or *astronomical* eyepiece simply magnifies the image without reinverting it. It is composed of two plano-convex lenses generally arranged as shown in Fig. 2-10*b*. The arrangement is seen to be identical with that of the two lenses farthest apart in the erecting eyepiece. The magnified image *D* is seen to be inverted, and the object as viewed through the telescope is upside down.

For either eyepiece the ratio of the angle at the eye subtended by the magnified image, to that subtended by the object itself, is the magnifying power of the telescope. If, in either Fig. 2-10*a* or Fig. 2-10*b*, *D* is the apparent length of the magnified image and *E* is the apparent length of the object as seen by the naked eye, the ratio of *D* to *E* is the magnifying power.

Each lens which is interposed between the object and the eye absorbs some of the light which strikes it. Hence, other things being equal, the

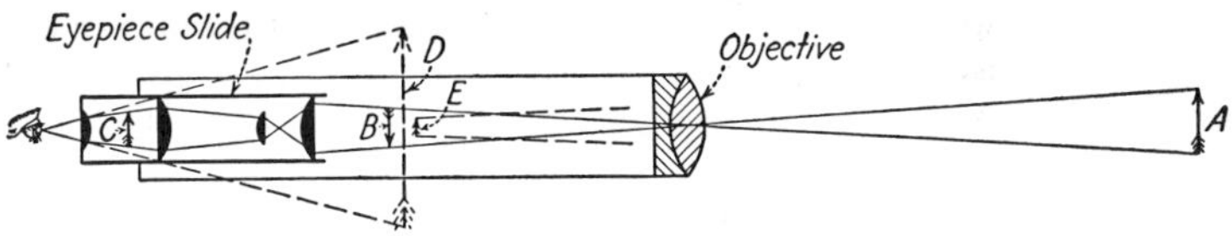

FIG. 2-10*a*. Erecting eyepiece.

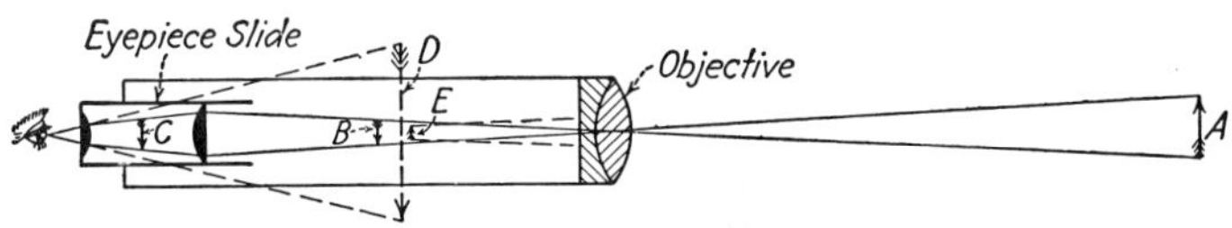

FIG. 2-10*b*. Inverting eyepiece.

object is more brilliantly illuminated when viewed through the inverting eyepiece; this is a great advantage, particularly when observations are made during cloudy days or near nightfall. Another important advantage of the inverting eyepiece is that the telescope is shorter and the instrument lighter in weight. The beginner experiences some inconvenience on viewing things apparently upside down, but this difficulty is overcome with a little practice.

The single advantage of the erecting eyepiece is that objects appear in their natural position, and this is the reason why its use is so common. Most American engineers and surveyors prefer the erecting eyepiece.

The slide of the erecting eyepiece is usually held in position by rings (Fig. 2-6*a*) similar to those described for the objective slide (Art. 2-12). In most instruments the slide is held tightly by spring friction, and it is focused by a screw-like motion. The slide of the inverting eyepiece is usually held by a single wide ring which is fixed in the end of the telescope tube and which admits of no lateral adjustment.

Various special eyepieces are available from instrument manufacturers. One type is equipped with a prism which permits the observer to sight transversely; it is useful for making sights that are steeply inclined.

2-16. Properties of the Telescope. The *illumination* of the image depends upon the effective size of the objective, the quality and number of lenses, and the magnifying power. Other conditions being the same, either the larger the objective or the smaller the magnifying power, the better the illumination; that is, the better lighted appears the object.

Distortion of the field of view so that it does not appear flat is mainly caused by what is termed the *spherical aberration* of the eyepiece (see also Art. 2-11). Although this introduces no appreciable error in ordinary measurements, it is not desirable when two points in the field are to be observed at the same time, as in stadia measurements.

The *definition* of a telescope is its power to produce a sharp image. It depends upon the quality of the glass, the accuracy with which the lenses are ground and polished, and the precision with which they are spaced and centered. Light rays passing through the lenses near their edges are particularly troublesome, and to improve the definition these rays are intercepted by diaphragms or screens placed between the lenses of the eyepiece and in the rear of the objective. The effect of these screens is to decrease the illumination somewhat.

The angular width of the field of view is the angle subtended by the arc whose center is nearly at the eye and whose length is the distance between opposite points of the field viewed through the telescope. For a particular instrument this angle may be readily determined by observation. It is independent of the size of the objective. In general the larger the telescope and the greater the magnifying power, the less the angle of the field of view. For most surveying of moderate precision, the work is greatly retarded if the instrument does not have a fairly large field of view, and this is one of the reasons why the telescopes are not usually made of high magnifying power. Usually the angular width of the field ranges from about 1°30′ for a magnifying power of 20, to 45′ for a magnifying power of 40.

The magnifying power of the telescope may be determined by observations as outlined in field problem 1, Art. 8-28. For the better grade of engineer's levels it is about 30 diameters. The U.S. Coast and Geodetic Survey type of precise level has a magnification of about 40 diameters. The magnification of transit telescopes is commonly 18 to 24 diameters.

2-17. Relation between Magnifying Power and Sensitivity. It is desirable that the sensitivity of the level tube be such that for the smallest noticeable movement of the bubble there is an apparent movement of the cross-hairs on a level rod held at an average sighting distance from the instrument; and likewise for the smallest noticeable movement of the cross-hairs there should be an observable movement of the bubble. The least noticeable movement of the cross-hairs depends to some extent upon the definition and illumination of the image, but principally upon the magnification of the telescope.

If the level tube is more sensitive than is necessary, time is wasted in centering the bubble. If the magnifying power is higher than it need be, unnecessary labor is expended by reason of the more limited field of view and by reason of the increased difficulties of focusing the objective properly. A satisfactory test may be conducted by one person sighting at a rod while a second person bears down slightly on one end of the telescope and at the same time observes the level tube. If the first noticeable movement of the bubble is accompanied by an apparent movement of the cross-hairs, there is a satisfactory balance between sensitivity and magnification. If the cross-hairs move first, a level tube of greater radius might properly be employed.

2-18. Verniers. A vernier, or vernier scale, is a short auxiliary scale placed alongside the graduated scale of an instrument, by means of which fractional parts of the least division of the main scale can be measured precisely; the length of one space on the vernier scale differs from that on the main scale by the amount of one fractional part. The precision of the vernier depends on the fact that the eye can determine more closely when two lines coincide than it can estimate the distance between two parallel lines. The scale may be either straight (as on a leveling rod) or curved (as on the horizontal and vertical circles of a transit). The zero of the vernier scale is the index for the main scale.

Verniers are of two types: (1) the *direct* vernier, which has spaces slightly shorter than those of the main scale, and (2) the *retrograde* vernier, which has spaces slightly longer than those of the main scale. The use of the two types is identical, and they are equally sensitive and equally easy to read. Since they extend in opposite directions, however, one or the other may be preferred because it permits a more advantageous location of the vernier on the instrument. Both types are in common use.

Direct Vernier. Figure 2-11*a* represents a scale graduated in hundredths of feet, and a direct vernier having each space 0.001 ft. shorter than a 0.01-ft. space on the main scale; thus each vernier space is equal to 0.009 ft., and 10 spaces on the vernier are equal to 9 spaces on the scale. The index, or zero of the vernier, is set at 0.400 ft. on the scale. If the vernier were moved upward 0.001 ft., its graduation numbered 1 would coincide with a graduation (0.41 ft.) on the scale, and the index would be at 0.401 ft.; and so on. It is thus seen that the position of the index is determined to thousandths of feet without estimation, simply by noting which graduation on the vernier coincides with one on the scale. Note that the coinciding graduation on the main scale does *not* indicate the main-scale reading.

The *least count*, or fineness of reading of the vernier, is equal to the difference between a scale space and a vernier space. For a direct vernier, if s

is the length of a space on the scale and if n is the number of vernier spaces of total length equal to that of $(n - 1)$ spaces on the scale, then the least count is s/n.

Retrograde Vernier. On the retrograde vernier shown in Fig. 2-11*b*, each space on the vernier is 0.001 ft. *longer* than a 0.01-ft. space on the main scale, and 10 spaces on the vernier are equal to 11 spaces on the scale. As before, the index is set at 0.400 ft. on the scale. If the vernier were moved upward 0.001 ft., its graduation numbered 1 would coincide with a graduation (0.39 ft.) on the scale; and so on. It is seen that, from the index, the retrograde vernier extends backward along the main scale, and that the vernier graduations are also numbered in reverse order; however, the retrograde vernier is read in the same manner as the direct vernier.

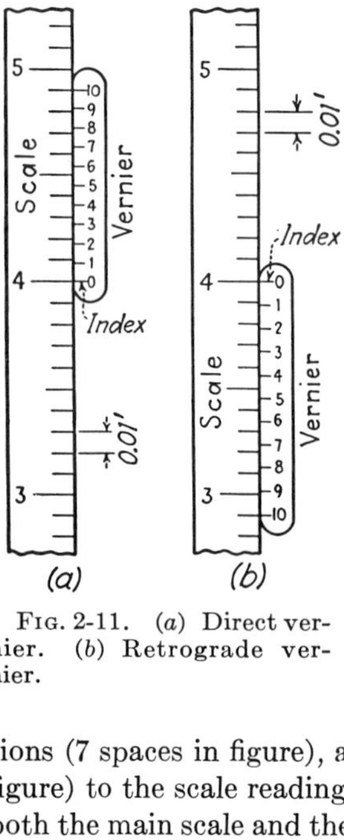

FIG. 2-11. (*a*) Direct vernier. (*b*) Retrograde vernier.

The least count of the retrograde vernier is equal to the difference between a scale space and a vernier space, as in the case of the direct vernier. For the retrograde vernier, if s is the length of a space on the scale and if n is the number of vernier spaces of total length equal to that of $(n + 1)$ spaces on the scale, then the least count is s/n.

Reading the Vernier. Figure 2-12 illustrates settings of direct verniers on the target (at left) and on the back (at right) of a Philadelphia leveling rod (Art. 8-15). The rod reading indicated by the target (4.347 ft. in figure) is determined by first observing the position of the vernier index on the scale to hundredths of feet (4.34 in figure), next by observing the number of spaces *on the vernier* from the index to the coinciding graduations (7 spaces in figure), and finally by adding the vernier reading (0.007 ft. in figure) to the scale reading (4.34 ft.). On the back of the rod (Fig. 2-12, right) both the main scale and the direct vernier read *down* the rod. The scale reading is 9.26 ft. and the vernier reading is 0.004 ft., hence the rod reading is 9.264 ft.

A helpful check in reading the vernier is to note that the lines on either side of the coinciding line should depart from coincidence by the same amount, in opposite directions. As a check against possible mistakes, it is advisable to estimate the fractional part of the main-scale division by reading the index directly.

Figure 2-13 illustrates a setting of a direct vernier on the horizontal circle of a transit. The vernier is of the double type, that is, the vernier on the left of the index is for reading clockwise angles while that on the right is for reading counter-clockwise angles. The circle is graduated in half degrees, or 30′. Each space on the vernier is 01′ less than a space on the circle, and 30 spaces on the vernier are equal to 29 spaces on the circle; the least count is 30′/30 = 01′. Considering counter-clockwise angles, the index is seen to lie between 17°00′ and 17°30′. The number of minutes greater than

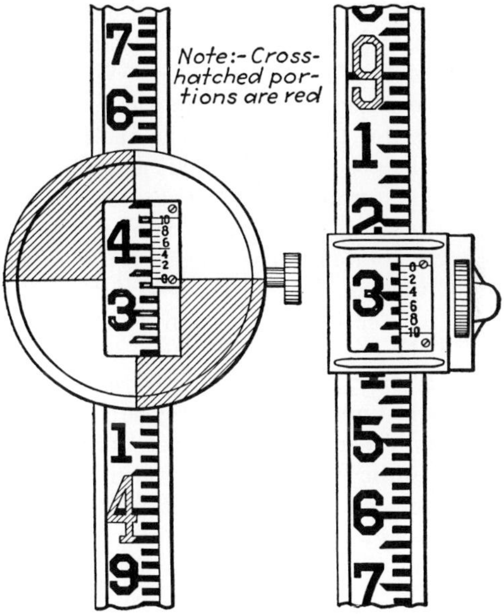

FIG. 2-12. Direct-vernier settings.

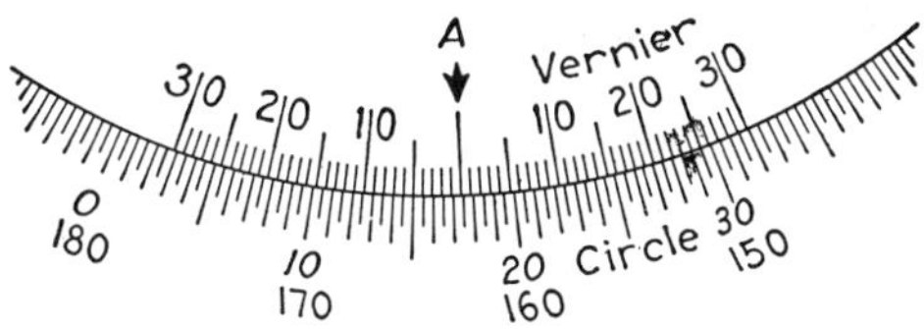

FIG. 2-13. Double direct-vernier reading to minutes.

17°00′, determined by observing the graduation on the right vernier that coincides with one on the circle, is seen to be 25′. Hence the angle is

$$17°00' + 25' = 17°25'$$

Similarly reading the left vernier, the clockwise angle is

$$162°30' + 05' = 162°35'$$

Settings of other types of vernier are shown in Art. 13-5.

2-19. Clamps and Tangent-screws. In order to make accurate settings of the telescope and to maintain a setting while an angle is being read, the horizontal axis and the vertical spindles of the transit are equipped with a device consisting essentially of a clamp with provision for controlled rotation through a small angle. A typical assembly is shown in Fig. 2-14, in which *A* is a spindle and *B* is a lug on the part of the instrument relative to which the movement of *A* is to be controlled. When the clamp-screw *C* is loosened, the split clamp *D* opens slightly, and the shaft is free to rotate. When the clamp-screw is tightened, the position of the shaft is fixed in relation to the lug *B*. By turning the tangent-screw *E* when the clamp-screw is tight, the clamp *D* is moved and therefore the shaft is rotated. The tangent-screw *E* is held firmly against the lug *B* at all times by means of the plunger *F* and coil spring *G* (shown in section) operating in a sleeve in the opposite leg of clamp *D*.

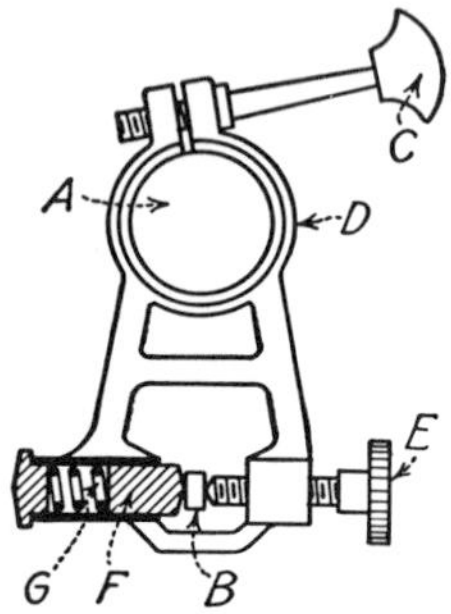

FIG. 2-14. Clamp and tangent assembly.

In making a setting by means of the tangent-screw, it is important that the last motion of the screw be clockwise—thus *compressing* the opposing spring—in order to eliminate lost motion. If the motion of the screw is reversed, the spring may not be powerful enough to move the parts at once, and later jarring may cause the setting to change slightly.

Clamps and tangent-screws are used not only on transits but also on plane-table alidades, sextants, and engineer's levels.

2-20. Gradienter. By means of a graduated drum attached to the tangent-screw of the vertical motion of a transit or a plane-table alidade, the number of revolutions of the screw can be observed. Such a device is called a *gradienter*. A gradienter attached to the vertical movement of a transit is shown in Fig. 2-15. The relation between graduations, pitch of the tangent-screw, and length of the arm is usually such that one division on the drum corresponds to a movement of the line of sight of 0.01 ft. at 100 ft. from the instrument; for other distances the movement of the line of sight is in proportion to the distance. This relation is useful in setting grades and sometimes in measuring distances and differences in elevation where sights are nearly level.

FIG. 2-15. Gradienter.

Example 1: It is desired to set off a 0.3 per cent grade with the transit. The telescope is leveled, and the gradienter drum is either read or set at zero. The tangent-screw is turned until 30 divisions of the drum are indicated by the index. The line of sight is then along the desired grade.

Example 2: It is desired to determine the distance from the instrument to a given point. A rod graduated in feet is held vertically on the point, and the line of sight is directed at some convenient mark (say, 3.00 ft.) on the rod. An initial reading of the gradienter is taken (say, 6 divisions). The tangent-screw is turned until the rod reading is some other convenient value (say, 5.00 ft.), and the gradienter is again read (say, 46 divisions). Then the number of divisions turned through is $46 - 6 = 40$, corresponding to a movement of the line of sight of 0.40 ft. at 100 ft. Since the movement of the line of sight on the rod is $5.00 - 3.00 = 2.00$ ft., the rod is $(2.00/0.40) \times 100 = 500$ ft. from the instrument. If desired, the tangent-screw could have been turned some convenient number of divisions on the drum, and the corresponding rod readings taken.

Some levels are equipped with a vertical micrometer screw at one end of the level bar, by means of which screw the telescope can be tilted through small vertical angles without moving the leveling screws. This screw is usually provided with a graduated drum and may be used as a gradienter in the manner just described.

Similarly, distances can be measured by sighting on a horizontal graduated rod with the transit and observing the horizontal angle subtended by a given distance on the rod.

2-21. Plumbing Devices. The type of plumb bob used in surveying is relatively slender, so that the point can be seen from almost directly above; the sides of the cone are inclined at an angle of about 10° with the axis. The bob is of precision manufacture. Usually it consists of a brass body weighing 10 to 16 oz., a replaceable tip of wear-resistant alloy steel, and a device to which the plumb-bob string can be attached centrally. For convenience in setting up a transit over a point, various devices for adjusting the length of the plumb-bob string are available.

The plumb bob is moved about by wind. In order that the transit can be centered over a point more rapidly and conveniently than with the plumb bob, some transits and practically all theodolites are equipped with an *optical plummet*, or *optical collimator*. This is a small telescope attached to the transit plate, with the line of sight through the center of the plate and perpendicular to the plane of the levels. It is so constructed that the observer looks into it horizontally. When the transit is level, the line of sight is vertical.

Another plumbing device free from the effects of wind is a telescoping metal rod with a pointed end which is placed on the station mark over which the transit is to be set; the rod is plumbed by means of an attached circular (bull's-eye) level.

2-22. Numerical Problems.

(For additional numerical problems and for field problems, see Chap. 8.)

1. What is the radius of curvature of a level tube graduated to 0.1 in. and having a sensitiveness of 30 seconds of arc per division?

2. What is the sensitiveness of a level tube graduated to 2 mm. and having a radius of curvature of 34 ft.?

3. What is the least count of a retrograde vernier divided into $\frac{9}{64}$-in. spaces when used on a scale graduated to $\frac{1}{8}$ in.? How many spaces are there in the vernier scale? Sketch the vernier applied to the scale.

4. What is the least count of a direct vernier on a circle which is graduated to $\frac{1}{4}°$ if 45 spaces on the vernier are equal to 44 spaces on the circle?

5. Design a direct vernier reading to 1 mm., applied to a scale graduated to 0.5 cm. For the same scale, design a retrograde vernier with least count of 0.2 mm. Sketch the verniers in relation to the scale.

6. A telescope is sighted on the 2.00-ft mark of a rod held on a distant point, and the corresponding reading of the gradienter is noted. The screw is turned until the 6.00-ft. mark on the rod is sighted, when it is observed that 84 divisions have been turned off. How far is the rod from the instrument?

REFERENCES

1. KIELY, R. E., "Surveying Instruments," Columbia University Press, New York, 1947.
2. Manuals issued by manufacturers of surveying instruments.
3. See also references at end of Chap. 3.

CHAPTER 3

FIELD WORK

3-1. General. The nature of surveying measurements has already been indicated. Field work consists in:

1. Adjusting instruments and caring for field equipment.
2. Determining the location of or establishing stakes or other markers for the control of the survey or for other purposes.
3. Fixing the horizontal location of objects or points by horizontal angles and distances.
4. Determining the elevation of objects or points by one of the methods of leveling.
5. Making a record of the field measurements, usually in the form of field notes in the field notebook, but sometimes directly in the form of a map drawn to scale.

On all surveys the field work is of primary importance. To become skilled in surveying operations requires a certain amount of experience in the field. The study of a text may serve to enlighten one as regards the underlying theory, the instruments and their uses, and the methods; but in surveying, as in other arts, mastery depends to a large degree upon the length, extent, and variety of actual experience.

3-2. Student Field Practice. In most courses in surveying a certain amount of field practice is given in connection with the study of the text. Field problems designed to give the student some practice in the elementary operations of surveying are outlined in later pages of this book.

It is not possible, in the ordinary field course in surveying, to develop the student into an expert instrumentman; it is expected, however, that the course will give the student a working knowledge of surveying instruments and their uses. In elementary field work no long surveys are attempted, but a number of short problems are taken up which in practice might become parts of extended surveys.

Members of the student field parties should from day to day alternately assume the various duties involved in the field work. The ability to hold the rod properly is as essential as the knowledge of how to manipulate the level, for a thorough understanding of details is necessary for intelligent direction.

3-3. Study the Problem. Before going into the field, the student should understand exactly what he is to do and why he is to do it. This understanding can be accomplished only by a thorough study of the problem, first noting its object and then conducting a critical examination of the course of procedure. In his mind he should go through the various steps involved so that in the field he may spend his time and attention in putting into practice that of which he has already learned the theory. He should prepare a list of the equipment necessary for its performance. All equipment should be examined as it is issued, and any defect or injury should be reported immediately.

3-4. Speed. Speed in field work depends to a large extent upon practice in handling instruments; but no amount of practice will secure rapid work and at the same time secure satisfactory results unless the work is carefully planned, systematized, and carried out with consistent accuracy and precision.

3-5. Correctness. No measurement should be regarded as correct until verified. So far as practicable, methods of verification should differ from the methods used in original measurements. All persons are liable to mistakes, but a mistake in field work becomes discreditable to the maker if he allows any other than himself to discover the discrepancy. Nothing, unless it is willful dishonesty, is so injurious to the reputation as habitual carelessness.

3-6. Consistent Precision. The precision of the measurements should be consistent with the purposes of the survey. Beginners often fail to comprehend the different degrees of precision necessary for the different kinds of work, or fail to maintain a consistent degree of precision throughout any one survey. There can be no fixed rules for the relative precision of different classes of surveys, for the objects and conditions are too many and too complicated; but one can always resort to common sense. Each survey is a problem in itself, for which the surveyor must establish the limits of error, using his own judgment and the experience of others to guide him. The best surveyor is not the one who is extremely precise, but the one who makes a survey with sufficient precision to serve its purpose without waste of time or money.

3-7. Relation between Angles and Distances. It is common practice before beginning a survey or any distinct portion of a survey to fix the permissible error of linear measurement; sometimes it is desired to locate a given point within a specified distance of its true location. If measurements are to be consistent, the precision of angles should correspond to the precision of related distances—in other words, the error in location of a point on account of error in angle should not be greatly different from its error in location on account of error in distance. Thus in Fig. 3-1, the location of the point B with respect to the line AC is represented by the point B, its true location, and B', its erroneous location resulting from the error e_d in the measured distance AB and the displacement e_a due to the error e_α in the measured angle.

Evidently a consistent relation between errors in angle and errors in distance would require that the distances e_a and e_d be equal or nearly so. The error in distance is expressed as a ratio, as, say, 1/2,000; thus if the distance AB were 2,000 ft., the distance e_d would be 1 ft. Similarly the distance e_a should equal 1 ft., and the tangent (or sine) of the error e_α in angle would be 1/2,000. Accordingly, it may be stated that a consistent relation between angles and distances will be maintained if the tangent or sine of the allowable error in angles equals the allowable error, expressed as a ratio, in the distances.

It is impossible to maintain an exact equality between these two ratios; but with some exceptions, which will be considered presently, surveys should be so conducted that the difference between precision of angle and precision of distance will not be large. Since the precision of a survey is often judged by the number of significant places in the recorded distance, it is always best

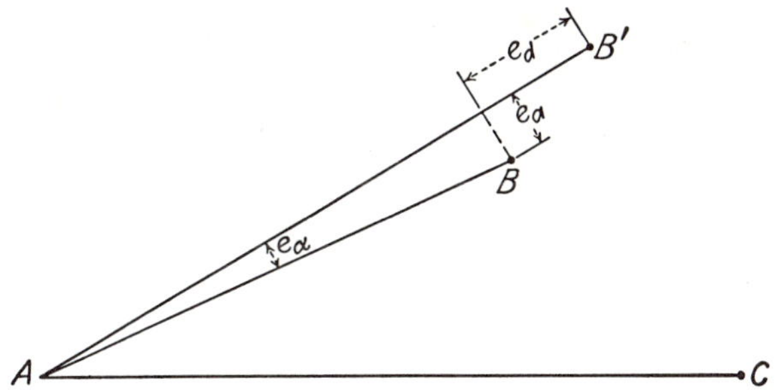

FIG. 3-1. Error in angle and distance.

to measure angles with a precision at least equal to the precision of the distances. Table 3-1 shows for various angular errors the corresponding ratios of precision and the linear errors for a length of 1,000 ft. For a length other than 1,000 ft. the linear error is in direct proportion. A convenient relation to remember is that an angular error of 01′ corresponds to a linear error of about 0.3 ft. per 1,000 ft.

To illustrate the use of the table, suppose distances are to be chained with a precision of 1/10,000; the corresponding permissible angular error is 20″. Again, suppose the distance from the instrument to a desired point is determined as 660 ft. with a probable error of 2 ft. For an angular error of 10′ the corresponding linear error is $0.66 \times 2.9 = 1.91$ ft. Therefore, the angle need be determined only to the nearest 10′.

The *exceptions* referred to earlier in this article are surveys in which the distances are roughly determined and the angles are measured with more than the required precision without increased effort or loss of time. For example, in rough chaining the ratio of precision might be 1/1,000, corresponding to an angular error of 03′. But with the ordinary transit, the angles could be determined to the nearest 01′ as readily as to the nearest 03′.

TABLE 3-1. CORRESPONDING ANGULAR AND LINEAR ERRORS

Angular error	Linear error in 1,000 ft.	Ratio of precision
10′	2.9089	$\frac{1}{344}$
5′	1.4544	$\frac{1}{688}$
1′	0.2909	$\frac{1}{3,440}$
30″	0.1454	$\frac{1}{6,880}$
20″	0.0970	$\frac{1}{10,300}$
10″	0.0485	$\frac{1}{20,600}$
5″	0.0242	$\frac{1}{41,200}$

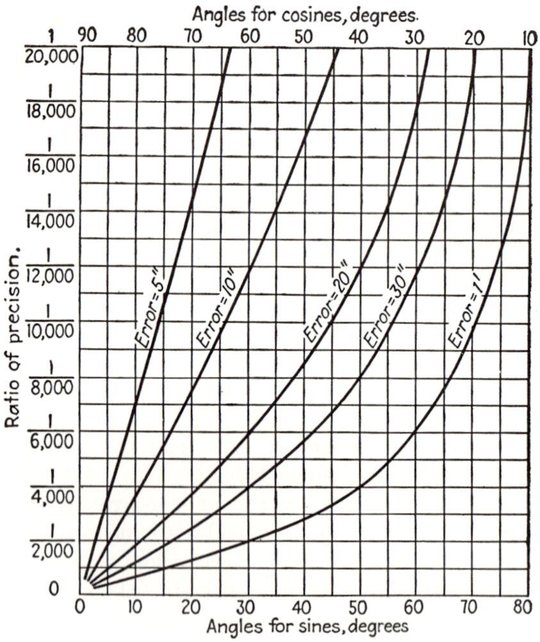

FIG. 3-2. Ratios of precision for sines and cosines.

3-8. Precision of Angular Measurement. Often field measurements are made the basis of computations involving the trigonometric functions, and it is necessary that the computed results be of a required precision. If the values of these functions were exactly proportional to the size of the angles—in other words, if any increase in the size of an angle were accompanied by a proportional increase or decrease in the value of a function—the problem of determining the precision of angular measurements would resolve itself into that explained in the preceding article. However, since the rates of change of the sines of small angles, the cosines of angles near 90°,

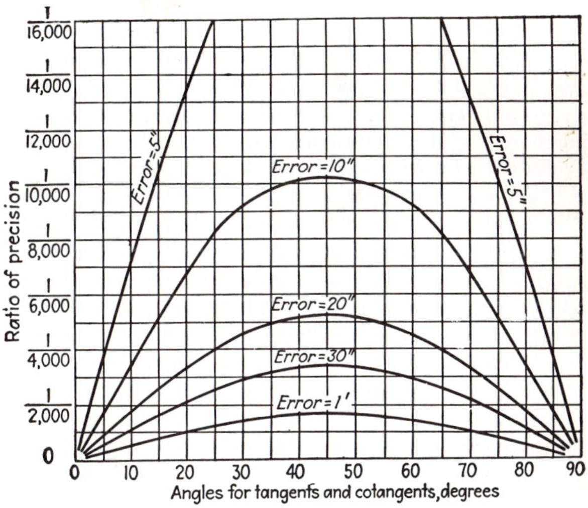

FIG. 3-3. Ratios of precision for tangents and cotangents.

and the tangents and cotangents of small and large angles are relatively large, it is evident that the degree of precision with which an angle is determined should be made to depend upon the size of the angle and upon the function to be used in the computations (see also Art. 4-6). Usually too little attention is paid to this important phase of the precision of measurements, even by experienced surveyors; as a consequence computed results are often assumed to be more precise than they really are. It is not practicable to measure each angle with exactly the precision necessary to insure sufficiently accurate computed values, but at least the surveyor should have a sufficiently comprehensive knowledge of the purpose of the survey and of the properties of the trigonometric functions to keep the angles within the required precision.

The curves of Figs. 3-2 and 3-3 show the ratios of precision corresponding to various angular errors from 05″ to 01′ for sines, cosines, tangents, and cotangents. For the function under consideration these curves may be used as follows:

1. To determine the ratio of precision corresponding to a given angular error and angle.

2. To determine the maximum or minimum angle that for a given angular error will furnish the required ratio of precision.

3. To determine the precision with which angles of a given size must be measured to maintain a required ratio of precision in computations.

The following examples illustrate the use of the curves:

1. An angle measured with a 1-min. transit is recorded as 32°00′. It is desired to know the ratio of precision of a computation involving the tangent of the angle, if the error of the angle is 30″. On the diagram of Fig. 3-3 it will be seen that the ratio of precision opposite the intersection of the curve $E = 30''$ and a line corresponding to 32° is 1/3,000.

2. In a triangulation system the angles can be measured with an error not exceeding 05″. Computations involving the use of sines must maintain a precision not lower than 1/20,000. It is desired to determine the minimum allowable angle. On the diagram of Fig. 3-2 (sines) the angle corresponding to a ratio of precision of 1/20,000 and an angular error of 05″ is about 26°.

3. In computations involving the use of cosines a ratio of precision of 1/10,000 is to be maintained. It is desired to know with what precision angles must be measured. On the diagram of Fig. 3-2 (cosines) opposite 1/10,000 it will be seen that for angles of about 76° the angular error cannot exceed 05″, for angles of about 64° the angular error cannot exceed 10″, and so on.

It should here be noted that linear as well as angular measurements must always be taken with the required precision of the computed results, for no result can be more nearly correct than the data from which it was obtained.

3-9. Definitions. For a better understanding of the following articles brief definitions of a few of the terms of surveying are appropriate.

Chaining. The operation of measuring horizontal or inclined distances with a tape. The persons who make such measurements are called "chainmen."

Flagman. A person whose duty it is to hold the flagpole, or range pole, at selected points, as directed by the transitman or other person in charge.

Rodman. A person whose duty it is to hold the rod and otherwise to assist the levelman or topographer.

Backsight. (1) A sight taken with the level to a point of known elevation. (2) A sight or observation taken with the transit along a line of known direction to a reference point, usually in the rear.

Foresight. (1) A sight taken with the level to a point the elevation of which is to be determined. (2) A sight taken with the transit to a point (usually in advance), along a line whose direction is to be determined.

Grade or *Gradient.* The slope, or rate of regular ascent or descent, of a line. It is usually expressed in per cent; for example, a 4 per cent grade is one which rises or falls 4 ft. in a horizontal distance of 100 ft. The term *grade* is also used to denote an established line on the profile of an existing or a proposed roadway. In such expressions as "at grade" or "to grade" it denotes the elevation of a point either on a grade line or at some established elevation as in construction work.

Hub. A transit station, or point over which the transit is set, in the form of a heavy stake set nearly flush with the ground, with a tack in the top marking the point.

Line. The path or route between points of control along which measurements are taken to determine distance or angle. To *give line* is to direct the placing of a flagpole, pin, or other object on line.

Turning Point. A fixed point or object, often temporary in character, used in leveling where the rod is held first for a foresight, then for a backsight.

Bench Mark. A fixed reference point or object, more or less permanent in character, the elevation of which is known. A bench mark may also be used as a turning point.

3-10. Signals. Except for short distances a good system of hand signals between different members of the party makes a more efficient means of communication than is possible by word of mouth. A few of the more common hand signals are as follows:

Right or *Left.* The corresponding arm is extended in the direction of the desired movement. A long, slow, sweeping motion of the hand indicates a long movement; a short, quick motion indicates a short movement. This signal may be given by the transitman in directing the chainman on line, by the levelman in directing the rodman for a turning point, by the chief of the party to any member, or by one chainman to another chainman.

Up or *Down.* The arm is extended upward or downward, with wrist straight. When the desired movement is nearly completed, the arm is moved toward the horizontal. The signal is given by the levelman.

All Right. Both arms are extended horizontally, and the forearms waved vertically. The signal may be given by any member of any party.

Plumb the Flagpole or *Plumb the Rod.* The arm is held vertically and moved in the direction in which the flagpole or rod is to be plumbed. The signal is given by the transitman or levelman.

Give a Foresight. The instrumentman holds one arm vertically above his head.

Establish a Turning Point or *Set a Hub.* The instrumentman holds one arm above his head and waves it in a circle.

Turning Point or *Bench Mark.* In profile leveling the rodman holds the rod horizontally above his head and then brings it down on the point.

Give Line. The flagman holds the flagpole horizontally in both hands above his head and then brings it down and turns it to a vertical position. If he desires to set a hub, he waves the flagpole (with one end of it on the ground) from side to side.

Wave the Rod. The levelman holds one arm above his head and moves it from side to side.

Pick Up the Instrument. Both arms are extended outward and downward, then inward and upward, as they would be in grasping the legs of the tripod and shouldering the instrument. The signal is given by the chief of the party or by the head chainman when the transit is to be moved to another point.

3-11. Care and Handling of Instruments. As the use of the various surveying instruments is discussed in the following chapters, suggestions for the care and manipulation of these instruments are given. In performing the field problems these should be heeded, not only because the student is responsible for the equipment he is using but also because he will establish the foundation of a very desirable qualification—that of carefulness.

Surveying Instruments. The following suggestions apply to such instruments as the transit, level, surveyor's compass, and plane table. More detailed information is given in the references at the end of this chapter.

1. Handle the instrument with care, especially when removing it from the box.

2. See that it is securely fastened to the tripod head.

3. Avoid carrying the instrument on the shoulder while passing through doorways or beneath low-hanging branches; carry it under the arm, with the head of the instrument in front.

4. Before climbing over a fence or similar obstacle, place the instrument on the other side, with the tripod legs well spread.

5. Whenever the instrument is being carried or handled, the clamp-screws should be clamped lightly so as to allow the parts to move if the instrument is struck.

6. Protect the instrument from impact and vibration.

7. If the instrument is to be shipped, pack paper or cloth around it in the case; and pack the case, well padded, in a larger box.

8. Never leave the instrument while it is set up in the street, on the sidewalk, near construction work, in fields where there are livestock, or in any other place where there is a possibility of accident.

9. Just before setting up the instrument, adjust the wing nuts controlling the friction between tripod legs and head, so that each leg when placed horizontally will barely fall from its own weight.

10. Do not set the tripod legs too close together, and see that they are firmly planted. Push *along* the leg, not vertically downward. So far as possible, select solid ground for instrument stations. On soft or yielding ground, do not step near the feet of the tripod.

11. While an observation is being made, do not touch the instrument except as necessary to make a setting; and do not move about.

12. In tightening the various clamp-screws, adjusting screws, and leveling screws, bring them only to a firm bearing. *The general tendency is to tighten these screws far more than necessary.* Such a practice may strip the threads, twist off the screw, bend the connecting parts, or place undue stresses in the instrument so that the setting may not be stable. Special care should be taken not to strain the small screws which hold the cross-hair ring.

13. For the plumb-bob string, learn to make a sliding bowknot that can be easily undone. Hard knots in the string indicate an inexperienced or slovenly instrumentman.

14. Before observations are begun, focus the eyepiece on the cross-hairs and (by moving the eye slightly from side to side) see that no parallax is present (Art. 2-10).

15. *When the magnetic needle is not in use, see that it is raised off the pivot.* While the needle is resting on the pivot, impact is apt to blunt the

point of the pivot or to chip the jewel, thus causing the needle to be sluggish.

16. Always use the sunshade. Attach or remove it by a clockwise motion, in order not to loosen the objective.

17. If the instrument is to be returned to its box, put on the dust caps for the objective and the eyepiece, and wipe the instrument clean and dry.

18. Have the rain hood available when the instrument is in use. If caught without it, place the dust cap on the objective and point the telescope up.

19. Remove grit from exposed movable parts such as the threads of tangent-screws by wiping them with an oiled rag. If the threads or slides work hard, clean them in gasoline or kerosene, and oil lightly. Use no abrasives.

20. Use only the best quality of clock or watch oil. Oil sparingly, and wipe off the excess oil. In cold weather, it may be necessary to use graphite for lubrication.

21. Never rub the objective or the eyepiece of a telescope with the fingers or with a rough cloth. Use a camel's-hair brush to remove dust, or use clean chamois or lint-free soft cloth if the dust is caked or damp. Occasionally the lenses may be cleaned with a mixture of equal parts of alcohol and water. Keep oil off the lenses.

22. Never touch the graduated circles and verniers with the fingers. Do not wipe them more than necessary, and particularly do not rub the edges.

23. Do not touch the level vials nor breathe on them, as unequal heating of the level tube will cause the bubble to move out of its correct position.

24. If the level vial becomes loose, it can be reset with plaster of paris, or wedged lightly with strips of paper or with toothpicks.

25. Minor repairs such as the replacement of a level vial (Art. 2-5) or broken cross-hairs (Art. 2-13) may be made in the field, but whenever possible, repairs should be made by an experienced instrument mechanic in the shop or factory.

26. It is advisable to carry a few spare parts such as a level vial, cross-hair ring, foot screw, and tangent-screw.

27. In cold weather, the instrument should not be exposed to sudden changes in temperature (as by bringing it indoors); mittens should be worn over the instrumentman's gloves; and the observer should be careful not to breathe on the eyepiece. Films of ice which may form on the lenses may be removed with a pointed piece of wood.

Suggestions regarding the adjustment of instruments are given in Art. 3-12.

Chaining Equipment. Keep the tape straight when in use; any tape will break when kinked and subjected to a strong pull. Steel tapes rust readily and for this reason should be wiped dry after being used. Some tapes are wound on a reel, but usually the tape is done up in 5-ft. lengths into a figure 8 and then "thrown" into the form of a circle with diameter about 10 in., as follows:

Stand beside the zero end of the tape, take the end of the tape in the left hand, and—allowing the tape to slide loosely through the right hand—extend the arms. As the 5-ft. mark is reached, grasp it with the right hand. Bring the hands together and lay the 5-ft. mark of the tape in the fingers of the left hand without permitting the tape to turn over. Then grasp this loop with the left hand, and again extend the arms for another 5-ft. length; and so on. When the last mark is reached, tie the loop tightly where the ends of the tape come together, by means of the rawhide thongs. Grasp the loop with the right hand, and at the opposite point with the left hand. Twist the loop in such a manner that it will be thrown into circular form, with diameter half that of the loop.

To undo the tape, reverse the operation of throwing; untie the thongs; remove the first loop in such a way as not to twist the tape; and walk in the direction of measurement, removing one loop at a time and watching for kinks.

Use special care when working near electric power lines. Fatal accidents have resulted from throwing a metallic tape over a power line.

Do not use the flagpole as a bar to loosen stakes or stones; such use bends the steel point and soon renders the point unfit for lining purposes.

To avoid losing pins, tie a piece of colored cloth (preferably bright red) through the ring of each.

Leveling Rod. Do not allow the metal shoe on the foot of the rod to strike against hard objects as this, if continued, will round off the foot of the rod and thus introduce a possible error in leveling. Keep the foot of the rod free from dirt. When not in use, long rods should be either placed upright or supported for their entire length; otherwise they are likely to warp. When not in use, jointed rods should have all clamps loosened to allow for possible expansion of the wood.

3-12. Adjustment of Instruments. By "adjustment" of a surveying instrument is meant the bringing of the various fixed parts into proper relation with one another, as distinguished from the ordinary operations of leveling the instrument, alining the telescope, etc.

The ability to perform the adjustments of the ordinary surveying instruments is an important qualification of the surveyor. Although it is a fact that the effect of instrumental errors may largely be eliminated by proper field methods, it is also true that instruments in good adjustment greatly expedite the field work. The operations of making the adjustments are not laborious, nor are the principles upon which the adjustments are based difficult to understand; yet a considerable number of surveyors regard the making of adjustments as something requiring the skill of an instrument maker. It is important that the surveyor:

1. Understand the principles upon which the adjustments are based.
2. Learn the method by which nonadjustment is discovered.
3. Know how to make the adjustments.
4. Appreciate the effect of one adjustment upon another.
5. Know the effect of each adjustment upon the use of the instrument.
6. Learn the order in which adjustments may most expeditiously be performed.

The frequency with which adjustments are required depends upon the particular adjustment, the instrument and its care, and the precision with which measurements are to be taken. Often in a good instrument, well cared for, the adjustments will be maintained with sufficient precision for ordinary surveys over a period of months or even years. On the other hand, blows that may pass unnoticed are likely to disarrange the adjustments at any time. On ordinary surveys, it is good practice to test the critical adjustments once each day, especially on long surveys where frequent checks on the accuracy of the field data are impossible. Failure to observe this simple practice sometimes results in the necessity of retracing lines which may represent the work of several days. Testing the adjustments with reasonable frequency lends confidence to the work and is a practice to be strongly commended. The instrumentman should, if possible, make the necessary tests at a time that will not interfere with the general progress of the survey party. Some adjustments may be made with little or no loss of time during the regular progress of the work.

The adjustments are made by tightening or loosening certain screws. Usually these screws have capstan heads which may be turned by a pin called an *adjusting pin*. Following are some general suggestions:

1. The adjusting pin should be carried in the pocket and not left in the instrument box. Disregard of this rule frequently leads to loss of valuable time.
2. The adjusting pin should fit the hole in the capstan head. If the pin is too small, the head of the screw is soon ruined.
3. Preferably make the adjustments with the instrument in the shade.
4. Before adjusting the instrument, see that no parts (including the objective) are loose. When an adjustment is completed, always check it before using the instrument.
5. When several interrelated adjustments are necessary, time will be saved by first making an approximate or rough series of adjustments and then by repeating the series to make finer adjustments. In this way, the several disarranged parts are gradually brought to their correct position. This practice does not refer to those adjustments which are in no way influenced by others.

3-13. Field Notes. No part of the operations of surveying is of greater importance, yet no part is more often neglected, than the field notes. In fact, the competency of the surveyor is reflected with much greater fidelity in the character of his field notes than in his use of the instrument. These notes should constitute a permanent record of the survey with data in such form as to be interpreted with ease by anyone having a knowledge of surveying. Unfortunately, this is often not the case. Many surveyors seem to think that their work is well done if the field record, reinforced by their own memories, is sufficiently comprehensive to make the field data of immediate use for whatever purpose the survey may have. On most surveys, however, it is impossible to predict to what extent the information gathered may become of value in the remote future. Often court proceedings involve surveys made long before. Often it is desirable to rerun, extend, or otherwise make use of surveys made years previously. In such cases it is quite likely that the old field notes will be the only visible evidence, and their value will depend largely upon the clearness and completeness with which they are recorded.

The notes consist of numerical data, explanatory notes, and sketches. Also, the record of every survey or student problem should include the date, the weather conditions, the names and duties of the members of the party, and a title indicating the location of the survey and its nature or purpose.

All field notes should be recorded *in the field book* at the time the work is being done. Notes made later, from memory or copied from temporary notes, may be useful, but they are not field notes. Notes should be neat. They are generally recorded in pencil, but they should be regarded as a permanent record and not as memoranda to be used only in the immediate future.

It is not easy to take good notes. The recorder should realize that the notes may be used by persons not familiar with the locality, who must rely entirely upon what he has recorded. Not only should the notebook contain all necessary information, but also the data should be recorded in a form which will admit of only the correct interpretation. A good sketch will help to convey a correct impression, and sketches should be used freely. The use to be made of the notes will guide the recorder in deciding what data are necessary and what are not. To make the notes clear, he should put himself in the place of one who is not on the ground at the time the survey is made. Before any survey is made, the necessary data to be collected should be considered carefully, and in the field all such data should be obtained, but no more.

Although convenient forms of notes such as those shown in this book are in common use, it will generally be necessary to supplement these, and in many cases it will be necessary for the surveyor to devise his own form of record. A code of symbols is desirable.

In some cases, as in locating details for mapping, the field notes may be supplemented by photographs taken with an ordinary camera.

3-14. Notebook. In practice the field notebook should be of good quality rag paper, with stiff board or leather cover, made to withstand hard usage, and of pocket size. Treated papers are available which will shed rain; some of these can be written on when wet.

Special field notebooks are sold by engineering supply companies for particular kinds of notes, such as cross-sections or earthwork. For general surveying or for students in field work where the problems to be done are general in character, an excellent form of notebook has the right-hand page divided into small rectangles with a red line running up the middle, and has the left-hand page divided into several columns; both pages have the same horizontal ruling. In general, tabulated numerical values are written on the left-hand page; sketches and explanatory notes on the right. This type is called a *field book* (see Fig. 14-5, etc.). Another common form, used in leveling, has both pages ruled in columns and has wider horizontal spacing than the field book; this is called a *level book* (see Fig. 9-3).

The field notebook may be bound in any of three ways: conventional, ring, or loose-leaf. The ring type, which consists of many metal rings passing through perforations in the pages, is not loose-leaf; it has the advantage over the conventional binding that the book opens quite flat and that the covers can be folded back against each other.

Loose-leaf notebooks are increasing in use because of the following advantages:

1. Only one book need be carried, as in it may be inserted blank pages of various rulings together with notes and data relating to the current field work.
2. Sheets can be withdrawn for use in the field office while the survey is being continued.
3. Carbon copies can be made in the field, for use in the field or headquarters office. Carbon copies are also a protection against loss of data. Duplicating books are available.
4. Notes of a particular survey can be filed together. Files can be made consecutive and are less bulky than for bound books.
5. The cost of binders is less than that for bound books.

The disadvantages are:

1. Sheets may be lost or misplaced.
2. Sheets may be substituted for other sheets—an undesirable practice.
3. There may be difficulty in establishing the identity of the data in court, as compared with a bound book. (This feature is important in land surveying.) When loose-leaf books are used, *each sheet* should be fully identified by date, serial number, and location.

Loose leaves are furnished in either single or double sheets. Single sheets are ruled on both sides and are used consecutively. Double sheets comprising a left-hand and a right-hand page joined together are ruled on one side only. Sheets for carbon copies need not be ruled.

3-15. Recording Data. A 4H pencil, well pointed, should be used. Lines made with a harder pencil are not so distinct; lines made with a softer pencil may become smeared. Reinhardt slope lettering (Fig. 6-3*a*) is commonly considered to be the best form of lettering for taking notes rapidly and neatly. Office entries of reduced or corrected values should be made in red ink, to avoid confusion with the original data.

The figures used should be plain; one figure should never be written over another. In general, *numerical data should not be erased;* if a number is in error, a line should be drawn through it, and the corrected value written above. Portions of sketches and explanatory notes may be erased if there is a good reason.

In tabulating numbers, the recorder should place all figures of the tens column, etc., in the same vertical line. Where decimals are used, the decimal point should never be omitted. The number should always show with what degree of precision the measurement was taken; thus a rod reading taken to the nearest 0.01 ft. should be recorded not as 7.4 ft. but as 7.40 ft. Notes should not be made to appear either more precise or less precise than they really are.

Sketches are rarely made to exact scale, but in most cases they are made approximately to scale. They are made freehand and of liberal size. The recorder should decide in advance just what the sketch is to show. A sketch crowded with unnecessary data is often confusing even though all necessary features are included. Large detailed sketches may be made of portions having much detail. Many features may be most readily shown by conventional symbols (Art. 6-11); special symbols may be adopted for the particular organization or job.

Explanatory notes are employed to make clear what the numerical data and sketches fail to do. Usually they are placed on the right-hand page in the same line with the numerical data that they explain. If sketches are used, the explanatory notes are placed where they will not interfere with other data and as near as possible to that which they explain.

If a page of notes is abandoned, either because it is illegible or because it contains erroneous or useless data, it should be retained, and the word "void" written in large letters diagonally across the page. The page number of the continuation of the notes should be indicated.

3-16. Student Field Notes. A neat title should be made either on the flyleaf or on the cover, showing the owner of the notebook, the number and name of the course, and the year in which the notes are taken. Two or three pages should be left in the front of the book for a table of contents, and the table of contents should always be kept up to date. The remaining pages of the book should be numbered, with one number assigned to each two facing pages, or "spread."

For each problem, at the top of the spread should be shown the number and name of the problem. Near the top of the right-hand page should be shown the date, weather, names of members of the party and the duty of each, and the equipment used.

In the field, the recorder should always record at once the uncorrected readings of the instrument and apply any necessary computed corrections afterward. All written field calculations should be made in the field notebook, usually on the right-hand page.

In order to develop an appreciation of the value of good field notes, the students may exchange field books for use in the office work of computing or plotting.

3-17. Safety. In addition to usual precautions for safety of the surveying party, such as avoiding electrical contacts with the metal tape, in the case of highway surveys additional precautions are needed because of the hazards of passing traffic. Members of the party should be specially instructed and should wear conspicuous clothing such as a red coat or shirt. Signs, traffic cones, and flagmen should be properly placed for the protection of the party, beginning several hundred feet in advance. As far as possible control points should be placed on sidewalks and shoulders.

3-18. Numerical Problems.

1. A point is to be established on the ground at a distance of about 600 ft. from a given point by means of one linear and one angular measurement. It is desired to establish the point within 0.1 ft. of its true location. With what precision need the angle be measured?
2. An angle of 40° is measured with a transit having a 30″ vernier. The maximum error is half the least count of the vernier. What is the ratio of precision if the sine of the angle is to be used in computations? The cosine? The tangent?
3. If tangents or cotangents are involved, what is the highest precision corresponding to single measurements (of angles of any size) with the transit of the preceding problem?
4. What is the minimum allowable angle that for computations involving sines will permit a ratio of precision of 1/10,000 to be maintained if the angular error is 10″?

REFERENCES

1. Bardsley, C. E., and E. W. Carlton, "Surveyor's Field-note Forms," 3d ed., International Textbook Company, Scranton, Pa., 1952.
2. Berger, L. H., "The Care of Surveying Instruments," *Civil Engrg.*, Vol. 9, pp. 422–425, 1939.
3. Berger, L. H., "Care of Surveying Instruments in Unfavorable Weather," *Military Engr.*, May, 1942; reprinted in *Surveying and Mapping*, pp. 20–23, July, 1943.
4. Pafford, F. William, "Handbook of Survey Notekeeping," John Wiley & Sons, Inc., New York, 1962.
5. Sargent, R. L., "Repair and Maintenance of Surveying Instruments," *Surveying and Mapping*, pp. 489–492, October-December, 1956.

CHAPTER 4

COMPUTATIONS

4-1. General. Calculations of one kind or another form a large part of the work of surveying, and the ability to compute with speed and accuracy is an important qualification of the surveyor. Computing is an art acquired only with practice, but no amount of experience will make an efficient computer unless (1) he possesses a knowledge of the precision of measurements and the effect of errors in given data upon the precision of values calculated therefrom, and (2) he is familiar with the algebraic and graphical processes and mechanical devices by means of which the labor of computing may be reduced to a minimum.

Herein *computations* are usually considered to be direct mathematical operations (such as multiplication) on given numerical data and yielding a definite result. *Calculations* are considered to be relatively broad general operations which may involve not only computation but also the exercise of judgment in such matters as the organization of related computations and the formulation of any necessary assumptions. The terms "computation" and "calculation" necessarily overlap somewhat in meaning.

Computations are made *algebraically* through the use of the simple arithmetical processes, logarithms, and the trigonometric functions; *graphically* by accurately scaled drawings; or *mechanically* by devices such as the slide rule and computing machines. The ability to make mental computations quickly is desirable.

The student should review the elementary relations of trigonometry; those which are most likely to be needed in surveying are included in Table VI.

Before making calculations of any great importance or extent, the computer should carefully plan a clear and orderly arrangement, using tabular forms so far as practicable. This is necessary to save time, to prevent mistakes, to make the calculations legible to others, to afford proper checks, and to facilitate the work of the checker. The work should not be crowded.

4-2. Office Computations. All computations should be preserved in a notebook for that purpose. This book may be the same as that used for field notes, but preferably it should be larger, say 8½ by 11 in. The pages should be cross-ruled to facilitate the making of tabulations and sketches.

In general, computations in the office are a continuation of some field work. They should be easily accessible for future reference; for this reason the pages of the computation book should be numbered and a table of contents made. Parts of problems separated by other computations should be cross-referenced.

Each problem should have a clear heading which should include the name of the survey, the kind of computations, the field book and number of page of the original notes, the name of the computer, the name of the checker, and the dates of computing and checking. Usually enough of the field notes should be transcribed to make computations possible without further reference to the field notebook. All transcripts should be checked.

4-3. Checking. In practice, *no confidence is placed in results that have not been checked,* and important results are preferably checked by more than one method. Rarely is a lengthy computation made without some mistake. The student should form the habit of checking his own work until he is certain that his results are correct. *Each student should depend upon himself.* Comparing each step with that of other students as the work proceeds is not true checking and would not be countenanced in practice.

Many problems can be solved by more than one exact method. Since the use of the same method in checking may cause the same error to occur, results should be checked by a different method when this is feasible. Approximate checks to discover large arithmetical mistakes may be obtained in many cases by the use of such mechanical devices as the slide rule, the planimeter, and the protractor. The slide rule is a valuable means of checking each step approximately. Of course, mistakes due to confusion of numbers or to wrong methods will not be shown. Often graphical methods may be used as an approximate check; they usually take less time than arithmetical or logarithmic solutions, and possible incorrect assumptions in the precise solution may be detected.

Each step in a long computation which cannot be verified otherwise should be checked by repeating the computation.

When work is being checked and a difference is found, the computation should be repeated before a correction is made, as the check itself may be incorrect.

In many cases, large mistakes such as faulty placing of the decimal point can be detected by inspection of the value to see whether it appears reasonable.

When work is being proofread by two persons, it is considered good practice to keep the listener alert by occasionally calling out an incorrect word or figure.

4-4. Significant Figures. The term *significant figures* is used to refer to those digits in a number which have meaning; that is, those digits the values of which are known. Confusion in the matter of computations involving measured quantities arises from the failure of the novice to distinguish between exact numbers and numbers which carry with them the inevitable errors of measured quantities. If we measure roughly a given distance with a steel tape, we may find it to be 732 ft., but if we measure it more carefully, we may find it to be 732.4 ft., or by still greater refinements we may determine it to be 732.38 ft. But we have not yet reached an exact number, nor can we, for whatever refinement may be used, there will always be an error of indeterminable amount. The number of significant figures in the three foregoing results is 3, 4, and 5, respectively. Obviously then, the number of digits that will have meaning and that may be used to indicate the length of this line is strictly limited by the precision with which the measurement has been made. The precision is not always apparent in measurements, as will now be explained.

Measurements are of two kinds, *direct* or *indirect*. A direct measurement is made when the observed quantity is compared with the scale directly, as, for example, when a carpenter measures the width of a board with his rule. An indirect measurement is made when the observed quantity is determined by several related and dependent observations.

When direct measurements are taken, the number of significant figures in the result is evident. Thus, suppose the surveyor measures a distance with the steel tape (graduated to hundredths of a foot) and finds it to be 37.42 ft. There is no question but that the number of significant figures in this result is four. Suppose the distance is much greater, however, so that the tape must be stretched several times over rough ground; and suppose that the total distance is determined as 623.58 ft. It is now very doubtful if the last digit is correct, even though it can be read with certainty on the tape, because the indirect measurement has introduced a number of sources of error (as marking the ends of the tape, keeping the tape level, etc.) which render the accuracy of the last digit, and possibly the last two digits, uncertain. Hence, it cannot be said offhand how many significant figures there are in any measured quantity until the character and magnitude of the errors have been examined.

It is not always easy to determine just the degree of uncertainty with which a measurement has been made, but in some cases it can be estimated or calculated with some precision and is expressed by a number called the *probable error* (Art. 5-8). We say, in such cases, that each digit is a significant figure until we reach that one for which the probable error equals or exceeds 5 units. Thus the number 623.58 $\pm$ 0.02 has five significant figures, but the number 623.58 $\pm$ 0.08 has only four significant figures and should properly be written as 623.6. Obviously if the last digit is uncertain by as many as 5 units, then the next to the last digit becomes uncertain and it would be absurd to assign values to any digits beyond one which itself is uncertain.

In a decimal the number of significant figures is not necessarily the number of decimal places, as the following examples illustrate:

0.0000065 contains two significant figures.
0.00000650 contains three significant figures.
10.00000650 contains ten significant figures.
0.08000650 contains seven significant figures.

In a number ending with one or more ciphers and having no decimal point, the number of significant figures is not definite but can be made so by using powers of 10 as a factor. Thus if 65,000 is written as 65.0×10^3, it is clear that there are three significant figures. If 0.0000065 is written as 6.5×10^{-6}, it is clear that there are two significant figures. For convenience in computation, usually powers of 10 are expressed in multiples of 3, corresponding to thousands, millions, etc.

If the maximum allowable error in a value is 1 per cent, the value must have at least three significant figures. To illustrate this relation, it is evident that if an error of 1 in the last place of a given whole number is not to exceed 1 per cent, the number must be at least 100. Similarly, an allowable error of 0.1 per cent requires at least four significant figures, and so on for other degrees of precision.

The demarcation between successive numbers of significant figures is not a sharp one. Thus, considering whole numbers, 999 has three significant figures and 1,001 has four, but an error of 1 in the last digit of 999 is practically the same in percentage as that for 1,001. For purposes of computation, numbers in the upper range of those having a given number of significant figures may appropriately be used with numbers in the lower range of those having one additional significant figure.

In computations it is advisable to carry out the intermediate results to one figure more than that desired in the final result.

Certain quantities are mathematically exact, and the corresponding numbers can be considered to have as many significant figures as necessary. For example, if a quantity containing several significant figures is to be taken twice, or multiplied by 2, the number 2 is mathematically exact, or *absolute*. Also mathematically exact are the number of degrees about a point and the sum of the angles of a closed figure. Numbers obtained by counting are absolute.

4-5. Precision of Computations. A proper regard for consistency between measured values and the computed results based upon them requires an understanding of the effects of the errors of measurement when combined in the operations of arithmetical computations.

Two important principles are as follows: In addition (or subtraction), the precision of the values is governed by the number of *places of figures,* whereas in multiplication (or division), the precision is governed by the number of *significant figures.* Application of these principles will be discussed in the following paragraphs.

Addition. Suppose that it is desired to add two (or more) quantities of earthwork which have been measured with precision appropriate to the values 37.2 and 468 cu. yd. The sum, 505.2, cannot properly be expressed to tenths because one (or more) of the quantities has not been measured to tenths; the sum should be written as 505 cu. yd.

To illustrate the fact that significant figures do not control the precision of addition, suppose that the taped distances 104.32 ft. and 0.64 ft. are to be added together, yielding the sum 104.96 ft. The result is properly expressed to hundredths of feet and contains five significant figures, even though one of the quantities used in the computation has only two significant figures.

Further, the amount of the total error in a sum may be shown by supposing that a considerable number of earthwork quantities are to be added, as shown, and that the probable error in each quantity is known to be ± 0.3 cu. yd. The sum is 501.7 $\pm$ cu. yd., but this number is affected by the probable error of each quantity of which it is composed. These separate probable errors, assumed here to be ± 0.3 cu. yd., are accidental in nature (Arts. 5-3 and 5-4). Hence they will probably combine in the sum as the square root of the number of times which they occur. In this example, then, since there are 10 numbers, the total probable error will be ± 0.3 cu. yd. $\times \sqrt{10}$ or about ± 1.0 cu. yd. But when the last digit is in doubt by more than 5 units, it has ceased to be a significant figure, and the result should more properly be written as 502 cu. yd. Whether it is so written or as 501.7 cu. yd., it has essentially three significant figures and no more.

37.2
45.6
53.1
63.2
43.7
45.2
63.8
72.1
36.4
41.4
501.7

Multiplication. The amount of the total error in a product resulting from errors in the factors may be shown by supposing the case of a rectangular field where the area is the product of two factors, the length and the width. Thus in Fig. 4-1, let

b = length of the field
a = width of the field
$A = ab$ = area
e_a = error in length of side a
e_b = error in length of side b
E_a = error in area due to e_a
E_b = error in area due to e_b
$E_A = E_{ab}$ = total error in area.

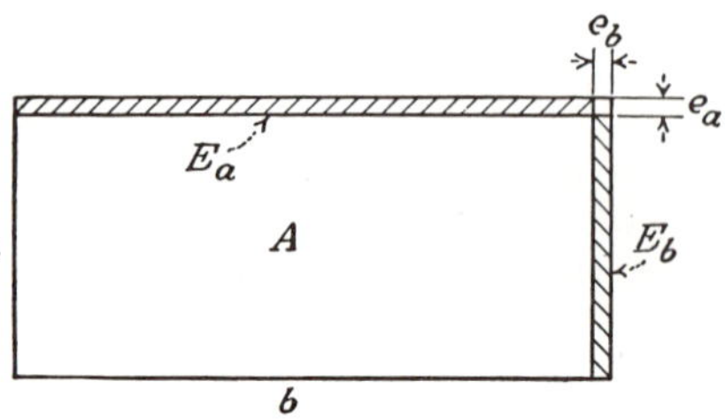

FIG. 4-1. Error in a product.

Evidently

$$\frac{e_b}{b} = \frac{E_b}{A} \qquad \text{and} \qquad \frac{e_a}{a} = \frac{E_a}{A}$$

but

$$\frac{e_b}{b} \text{ is the relative error in the side } b$$

and

$$\frac{E_b}{A} \text{ is the relative error in the area due to } e_b$$

also

$$E_A = E_b + E_a \text{ and the relative error in } A \text{ is given by}$$

$$\frac{E_A}{A} = \frac{E_b}{A} + \frac{E_a}{A} = \frac{e_b}{b} + \frac{e_a}{a}$$

or the relative error in the area is equal to the sum of the relative errors in the length and width. (To simplify this demonstration, the negligible error e_ae_b at the corner has been omitted.)

Hence, the relative error in a product is equal to the sum of the relative errors in the factors. And from this fact the important principle follows, that on a relative basis *the probable error of a product cannot be less than that of the least precise factor.*

Relation to Field Measurements. This principle must be kept in mind by the surveyor in the field if his computed results are to have the precision requisite to his purpose. For example, suppose he wishes to compute the area of a triangle (Fig. 4-2) whose sides a = 680.8 ft., b = 75.30 ft., and angle C = 132°02′ are obtained by field measurement. (Area = ½ab sin C.) If his purpose requires four significant figures in his result, say a permissible error of about 1/4,000,* then a, b, and C must be measured with such precision as to yield four significant figures; and side b must be measured to hundredths whereas side a needs to be measured to tenths only.

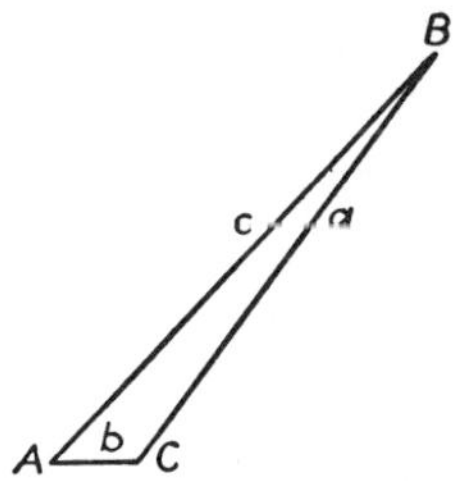

FIG. 4-2.

The precision required in the measurement of the angles is governed by the same considerations, and the ratio-of-precision curves (Figs. 3-2 and 3-3) will aid in determining the precision with which the angles should be measured. Thus, in the example cited above, the angle C must be measured in the field with such precision that the use of sin C will not introduce an error greater than 1/4,000. From Fig. 3-2 this allowable error is found to be slightly less than 01′.

Another consideration often misunderstood by computers is the fact that the precision of a result is entirely independent of the unit in which it is expressed. Thus, in the case of the triangle mentioned above, the area is 19,040 sq. ft., to four significant figures as justified by the given data. In acres, this value would properly be expressed as 0.4371 acre, again to four significant figures.

The following example illustrates the value of examining the given data and of knowing the purpose of the computation:

Notes of a land survey indicate that the distances were measured with a precision of 1/10,000, while angles were measured with a probable error of 01′. It is desired

* Four significant figures are used when the ratio of precision lies between 1/1,000 and 1/10,000.

that computations for area (involving sines and cosines) be as precise as the data warrant. Referring to the ratio-of-precision curves for sines and cosines (Fig. 3-2), it will be seen that such an angular error for angles of average size (near 45°) corresponds to a precision of about 1/3,000. It is this precision which will govern that of the computed area, and the computed area should contain four significant figures. Had it been presumed that the precision of angles was consistent with that of distances, doubtless five significant figures would have been shown in the result.

4-6. Computations for Angles and Distances. The relation between measurements of angles and distances is discussed in Art. 3-7, and the precision of angular measurements in Art. 3-8.

When computations for angles or distances may be made in one of several ways, each of which depends upon data of like precision, it is best to compute *angles* by using functions which change rapidly, that is, tangents or cotangents; and to compute *distances* by using functions which change slowly, that is, sines or cosines.

For example, the ratio-of-precision curves for tangents and cotangents (Fig. 3-3) show that if the precision of data is 1/10,000 the maximum error of angles computed by tangents or cotangents is about 10″; the ratio-of-precision curves for sines and cosines (Fig. 3-2) show that, with data of a similar precision, angles between 27° and 63° could not have been computed by either sines or cosines with a precision as great as 10″.

4-7. Trigonometric Tables. The number of places to be used in trigonometric tables will depend upon the ratio of precision of the particular value of the function involved rather than upon the angular error. The number of significant figures for a particular angle, angular error, and function may be readily determined either by inspecting Figs. 3-2 and 3-3 or from trigonometric tables. *Assuming that the precision of angle is the governing precision,* for tangents and cotangents three places will be sufficient for angles in error more than 02′; four places, for angles in error less than 02′ but more than 10″; and five places, for angles in error less than 10″ but more than 01″.

For sines and cosines of angles of *average size* (near 45°) it will be seen that four places are sufficient for angles in error not less than 20″. Correspondingly, five places are sufficient for angles in error less than 20″ but more than, say, 05″; and six places for angles in error less than 05″ but more than ½″. Since angles are likely to be other than average size, the preceding limits should, in general, be doubled, though for sines of very large angles and for cosines of very small angles the necessary number of places should be ascertained by first determining the ratio of precision corresponding to the given angular error. For example, the ratio of the tabular difference for 10″ to sin 80° (or cos 10°) is about 1/120,000, therefore six places will be necessary for angles as small as 10° when the error is 10″. As another example, from the ratio-of-precision curves for sines it will be seen that, when the angular error is 01′, five places will be required for angles greater than 70°.

For very small angles, the number of places required is affected by the number of significant figures in the function. For example, the sine of 02′ to five places is 0.00058, which contains but two significant figures. In such cases, the use of logarithms is desirable.

4-8. Graphical and Mechanical Methods. Graphical and mechanical methods are of particular value in approximately checking more precise computations. In general, results may be obtained graphically with less labor than by arithmetic, and mistakes are less likely to occur.

Frequently, combined graphical and mechanical methods may be employed in conjunction with algebraic processes. For example, earthwork cross-sections may be plotted to scale (graphical), the area of the cross-sections may be measured with the planimeter (mechanical), and the volume of earthwork may be determined by arithmetic.

Similarly, unknown lengths and angles which have been algebraically or mechanically computed may be approximately checked by plotting the known data, measuring the unknown lengths with a scale, and measuring the unknown angles with a protractor.

The most common mechanical aid available is the ordinary 10-in. *slide rule.* This rule greatly facilitates computations involving no more than three significant figures and is in every way the equivalent of a three-place table of logarithmic functions. The use of the slide rule is described in Art. 4-11.

For results of more than three significant figures the *computing machine* is used whenever it is available; the use of logarithms is now largely limited to computations in the field. As opportunity presents itself, the student should familiarize himself with the operation of a computing machine. The chances of mistakes in computations are greatly decreased. With most types the operations of multiplication, division, squaring, and taking the square root are instantly proved, so that further checking is not required. The details of operation vary with the type of machine, but they are simple and can be mastered in a few minutes of instruction. Essentially, multiplication is accomplished by automatic multiple addition, and division by multiple subtraction.

Electronic computers are coming into extensive use in surveying for repetitive computations such as those for traverses (Chaps. 18 and 19), cross-sections and volumes of earthwork (Chaps. 10 and 11), and applications of photogrammetry to highway location and construction (Chap. 26). It is not necessary for the surveyor to own or operate a computer, as standardized programs and commercial computing services are available. Most state highway departments and many other large engineering organizations have their own computing facilities.

Another device frequently utilized by the surveyor is the *polar planimeter* (Art. 4-12). It is of great value in finding the areas of figures plotted to scale. The precision with which results may be obtained depends upon a number of factors but principally depends upon the skill of the operator in traversing the lines of the drawing with the tracing point. In general, results may be determined to three significant figures, a precision in keeping with much of the field data upon which calculations of area are based. It is a simple instrument to operate and furnishes the most efficient means of determining the area of figures with irregular or curved boundaries.

4-9. Logarithms vs. Natural Functions. Whether logarithms should be used depends upon the computation under consideration if no computing machine is available. It takes less time to multiply two numbers of three digits each by arithmetic than by logarithms; for a single computation these numbers might be extended to four digits each. But for multiplying, dividing, squaring, cubing, or taking roots of numbers, arithmetic cannot be used advantageously beyond three or perhaps four significant figures. Where there are a number of similar computations, there is a decided advantage in using logarithms for even four significant figures, for not only will time be saved but also the liability of mistakes will be lessened and the mental strain on the computer will be decreased.

4-10. Use of Logarithms. The logarithm of a number is the power to which some base must be raised to produce the number. In computations made by the surveyor the *common* system of logarithms is employed, for which the base is 10.* Hence

$$\log 10 = \log 10^1 = 1; \log 100 = \log 10^2 = 2;$$
$$\log 1{,}000{,}000 = \log 10^6 = 6; \log 1 = \log 10^0 = 0;$$
$$\log 0.1 = \log \frac{1}{10} = \log \left(\frac{10^0}{10^1}\right) = 0 - 1 = -1 = 9 - 10$$

For any number (except 1) that is not a power of 10, the logarithm is a fractional quantity. For example

$$\log 1.5 = \log 10^{0.17609} = 0.17609; \log 15 = \log 10^{1.17609} = 1.17609;$$
$$\log 0.015 = \log \left(\frac{15}{1000}\right) = \log \left(\frac{10^{1.17609}}{10^3}\right)$$
$$= 1.17609 - 3 = 8.17609 - 10 = \bar{2}.17609$$

The whole number of the logarithm is called the *characteristic;* the decimal is called the *mantissa.* For a number greater than one the logarithm is a positive quantity and the value of the characteristic is one less than the number of places in the integer of the number.

For a number less than 1, the logarithm is a negative quantity, and to determine the characteristic the common practice is to deduct from 10 a number equal to one more than the number of ciphers to the right of the decimal point. When a logarithm is written in this manner, it is *understood*

* For *natural* logarithms the base is 2.71828+. The natural logarithm of any number is equal to the common logarithm multiplied by 2.30258+.

that 10 is to be deducted from it. Instead of writing the logarithm in this manner, the characteristic is often shown as a quantity one more than the number of ciphers to the right of the decimal point, and a negative sign is placed over it. The logarithm of 0.0435 may therefore appear either as 8.63849 or as $\bar{2}.63849$.

For the same sequence of figures, the mantissa remains unchanged regardless of the position of the decimal point. Thus the logarithm of 4,350 is 3.63849 and the logarithm of 0.00435 is 7.63849 or $\bar{3}.63849$.

The same considerations govern the number of places to be used in logarithmic computations as govern those of arithmetic. The number of significant figures in the final result should be consistent with its purpose or with the precision of the given data, as discussed in Arts. 4-5 to 4-7. The number of places in the mantissa of the logarithm should be equal to the number of significant figures in the corresponding number which is being multiplied or divided. The usual practice is to use a number of places of logarithms one more than the number of places desired in the final result.

In the ordinary work of the surveyor five places are usually sufficient, but sometimes six places are required. On the more precise surveys, seven places and occasionally eight places are necessary. Tables XIV and XV give logarithms of numbers and of the functions of angles to six places, but in using these tables the last figure should be dropped if only five places are required. In tables of logarithms of numbers (see Table XIV) only the mantissa is shown and the characteristic must be supplied by the computer. Tables of the logarithmic functions of angles (see Table XV) show both the characteristic and the mantissa.

4-10a. Finding Logarithm and Antilogarithm. The process of finding the logarithm of a number from tables is best illustrated by an example.

Example 1: Find the logarithm of 6,458.6 correct to the sixth place.

In Table XIV opposite the number 645 the mantissa in the column headed 8 is 810098, and in the column headed 9 is 810165. The difference between the two is 67. The last figure in the given number is 6 and hence the desired mantissa is $810098 + (0.6 \times 67)$. To facilitate the multiplication, the table of proportional parts at the bottom of the page is given. Opposite 67 the quantity in the 6 column is 40.2. Hence $0.6 \times 67 = 40.2$. The mantissa is, therefore, $810098 + 40 = 810138$. The number has four digits to the left of the decimal point, and hence the characteristic is 3 and the logarithm is 3.810138. Note that all logarithms between two adjacent horizontal broken lines have the same figures in their first two places, these figures being shown in the column headed 0.

The process of finding an *antilogarithm,* or number the logarithm of which is known, is the reverse of that just described.

Example 2: Find the number whose logarithm is 2.688544. The number is to have five significant figures.

By Table XIV it is seen that the mantissa of the logarithm of 4881 is 688509 and that of the logarithm of 4882 is 688598. The difference between these two mantissas is 688598 − 688509 = 89; the difference between the given mantissa and that for 4881 is 688544 − 688509 = 35. The required number is therefore $4881 + {}^{35}\!/_{89}$. By the table of proportional parts opposite 89 find the number nearest 35 (it is 35.6). At the head of the column the corresponding number is seen to be 4. The characteristic is 2, and the number is therefore 488.14.

If six places were required it would be necessary to interpolate between values given in the table of proportional parts.

Thus, for the preceding example, the difference between 35 and 26.7 is 8.3, and the difference between 35.6 and 26.7 is 8.9. But 8.3/8.9 = 0.9 (approximately). Therefore, the digit in the sixth place is 9, and the number is 488.139.

This interpolation may also be made by using the table of proportional parts directly, merely moving the decimal point one place. Thus the difference of 8.3 becomes 83, and the nearest value in the table of proportional parts opposite 89 is 80.1, in the column headed 9.

4-10b. Multiplication and Division. The product of two numbers is determined by adding their logarithms.

Example 1: $15 \times 12 = 10^{1.18} \times 10^{1.08} = 10^{2.26}$ = number whose log is 2.26 = 180.

One number is divided by another by subtracting the logarithm of the divisor from that of the dividend.

Example 2: $\dfrac{180}{12} = \dfrac{10^{2.26}}{10^{1.08}} = 10^{1.18}$ = number whose log is 1.18 = 15.

4-10c. Powers. A number is raised to a power by multiplying its logarithm by that power.

Example 1: $12^4 = 10^{1.08\times 4}$ = number whose log is 4.32 = 21,000 (to two significant figures).

Following is an example of a logarithmic process of raising a number less than one to a fractional power:

Example 2: $0.6324^{1.7180} = 10^{(9.8010-10)1.7180}$ = number whose log is (9.8010 − 10) 1.7180.

$$\begin{aligned} \log 9.8010 + \log 1.7180 &= 0.99127 + 0.23502 = 1.22629 \\ 9.8010 \times 1.7180 &= 16.838 \\ 0.6324^{1.7180} &= \text{number whose log is } 16.838 - (10 \times 1.7180) \\ &= \text{number whose log is } 9.658 \\ &= 0.455 \end{aligned}$$

In the preceding example the logarithm of a logarithm has been determined; for short this is called the *log log*. Similarly for brevity, the antilogarithm is often called the *antilog*.

4-10d. Trigonometric Functions. When the logarithm of a trigonometric function is to be found from Table XV, it should be noted that there are four angular values on each page, one at each corner of the table. Each angular value is used with the line of headings and the column of minutes which are nearest to it, as illustrated by the following example:

Example: From Table XV,

log sin 22°12′ is 9.577309–10
log sin 67°12′ is 9.964666–10
log sin 112°12′ is 9.966550–10
log sin 157°12′ is 9.588289–10

Functions of an angle α greater than 45° can be determined conveniently by use of the corresponding complement $(90° - \alpha)$ or supplement $(180° - \alpha)$ of the angle, with due regard to the sign of the corresponding function.

4-11. Use of the Slide Rule. Special books of instruction for each of the several varieties of slide rules are issued by the manufacturers. Herein only some of the more frequent computations which may be performed on all rules of the Mannheim type will be described briefly.

In Fig. 4-3 is shown one simple style of 10-in. slide rule, on the face of which are four graduated scales. The two scales on the body of the rule are lettered A and D, and those on the slide are lettered B and C. The rectangular glass runner may be moved to any position along the rule, its setting being indicated by a fine line etched on the glass at right angles to the axis of the rule.

The C and D scales are exactly alike and are graduated with numbers from 1 to 10. The A and B scales are similar to the C and D scales but the corresponding graduations are only one-half as great. All four of the scales are logarithmic, but the figures shown are for the numbers, not the logarithms. The distance from a graduation corresponding to a given number to the "1" at the left of the C or D scale (left index) represents to the scale of 1 = 25 cm. the mantissa of the logarithm of that number. If the distance from the "2" graduation to the left index were measured with a 25-cm. scale, it would be found to be 0.301 of the length of the scale, which is the value of the mantissa of the logarithm of 0.2, 2.0, 2,000, or any number having the same sequence of figures.

If for any computation the student is uncertain as to which scales should be employed, he may make a trial computation with simple numbers or with known values of the logarithms or trigonometric functions.

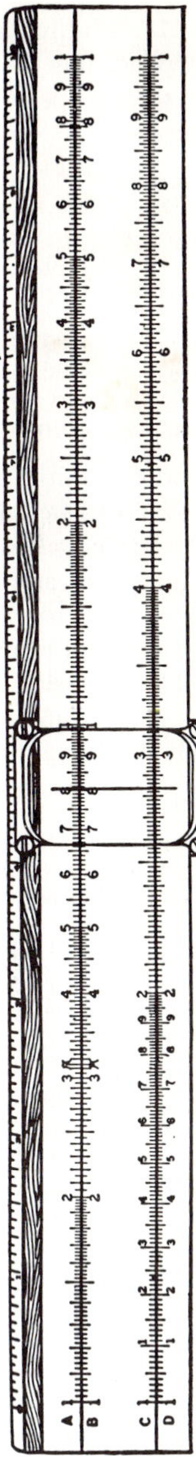

FIG. 4-3. Slide rule (Mannheim type).

4-11a. Multiplication. In logarithmic computations two numbers are multiplied together by adding their logarithms. This operation may be mechanically performed by using the slide rule as illustrated by the following examples:

Example 1: Multiply 4 by 2.
Set the runner at 2 on scale *D*; move the slide until its left index is at the runner. (Graphically the mantissa of 2 has now been laid off.) Move the runner to 4 on scale *C*. (Graphically the mantissa of 4 has been added to that of 2.) On scale *D* read 8.

Example 2: Multiply 8.2 by 7.3.
Set the runner to 8.2 on scale *D*; the right index to the runner; the runner to 7.3 on scale *C*; read the answer 59.9 on scale *D*. The position of the decimal point is determined by mental computation, using round numbers.

Had the initial setting been made with the left index at 8.2, the result would have been off the rule, at a *logarithmic* distance of 1 to the right of the final setting as determined in the example. For a logarithmic change of 1, the mantissa, and hence the sequence of numbers, is the same. Therefore settings may be made with either index, that one being chosen which will bring the final result within the length of the *D* scale.

4-11b. Division. Division of one number by another is accomplished by finding the difference between their logarithms. Recalling the manner in which the logarithmic scales are represented on the slide rule, the operation of division becomes self-evident.

Example 1: Divide 8 by 4.
Set the runner at 8 on scale *D*. (The distance from the left index on scale *D* to 8 on scale *D* represents the mantissa of the logarithm of 8.) Set 4 on scale *C* to the runner. (The distance from 4 to the left index of scale *C* represents the mantissa of the logarithm of 4.) Set the runner to the left index on scale *C*. Read the answer 2 on scale *D*. (The difference between the mantissas of the logarithms of the two numbers is represented by the distance from 2 on scale *D* to the left index on scale *D*.)

In the remaining examples the following abbreviations will be used:
A, *B*, *C*, and *D* refer to corresponding scales.
R is the runner.
LI is left index.
MI is middle index.
RI is right index.

Example 2: $\dfrac{48 \times 63}{97 \times 15}$.

R to 48 *D*; 97 *C* to *R*; *R* to 63 *C*; 15 *C* to *R*; *R* to *LI C*. Answer, 2.08 *D*.

4-11c. Squares and Square Roots. Scales *A* and *B* may be used for solving multiplications and divisions in exactly the same manner as are the

C and *D* scales, but are more generally employed in conjunction with the *C* and *D* scales for finding the squares and square roots of numbers.

The *logarithmic* scale of the *A* and *B* scales (1 = 12½ cm.) is one half of that of the *C* and *D* scales (1 = 25 cm.). Hence if the runner is set to a given number on the *D* scale, the square of the number is given by the runner reading on the *A* scale, for the effect has been graphically to multiply the mantissa by two.

Example 1: Square 6.23.
Set *R* to 6.23 *D*; read answer 38.8 *A*, or set *R* to 6.23 *C*; read answer 38.8 *B*.

Square root is obtained by setting the runner to the number on the *A* (or *B*) scale and reading the root on the *D* (or *C*) scale. If the *integer* of the number contains an odd number of places (as 4.83; 125; 17,536), the runner is set on the left scale; if it contains an even number of places (as 16; 42.8; 1,174), the runner is set on the right scale. If the number is a decimal without ciphers between the decimal point and the first finite figure (as 0.428; 0.87), or with an even number of ciphers between the decimal point and the first finite figure (as 0.0087; 0.000064), the runner is set to the number on the right scale; if the number is a decimal with an odd number of ciphers between the decimal point and the first finite figure (as 0.0426; 0.0000065), the runner is set to the number on the left scale. If in doubt, the correct scale to use may be found by rough trial, using a round number which is a perfect square and which is near to the number under consideration.

Example 2: Find the square root of 16.4.
Set *R* to 16.4 right *A*; read answer 4.05 *D*.

Example 3: Find the square root of 1.64.
Set *R* to 1.64 left *A*; read answer 1.28 *D*.

Example 4: $\dfrac{0.53(14.3)^2}{0.035\sqrt{1{,}178}}$.

Set *R* to 0.53 *D*; 0.035 *C* to *R*; *R* to 14.3 *C*; *LI* to *R*; *R* to 14.3 *C*; 1,178 right *B* to *R*; *R* to *RI*; read answer 90.4 *D*.

Example 5: $(12.2)^{5/2}$.
Set *R* to 12.2 *D*; *LI* to *R*; *R* to 12.2 *C*; *LI* to *R*; *R* to 12.2 on right *B*; read answer 520 *D*.

4-11d. Trigonometric Functions. On the back of the slide is a scale *S* for sines to the *logarithmic* scale of 1 = 12½ cm. and a scale *T* for tangents to the logarithmic scale of 1 = 25 cm. If the slide is removed, turned over, and replaced so that the indexes of the *S* and *T* scales coincide with those of the *A* and *D* scales, respectively, the values of the natural sines of angles between 0°34′ and 90° and of natural tangents between 5°43′ and 45° are read directly from the *A* and *D* scales, respectively. Thus by readings on the *A* scale the sine of 0°34′ is seen to be 0.0100, the sine of 5°44′ is seen to be 0.100, the sine of 30° is seen to be 0.500, etc., and by readings on the D scale the tangent of 5°43′ is seen to be 0.100, the tangent of 30° is seen to be 0.577, etc.

The tangent of any angle less than 5°43′ may be considered to equal the sine, within the precision of the slide rule. Sines of angles less than 0°34′ may be assumed to be proportional to the angle. On this assumption, since the sine of 0°34′ is 0.0100, the sine of 0°05′ would be $5/34 \times 0.0100 = 0.00147$. The correct value is 0.00145.

The values of other trigonometric functions may be obtained by the simple relationships of trigonometry.

4-11e. Logarithms. Almost every slide rule has a scale L divided into 100 equal major parts. On the rule shown in Fig. 4-3, this scale is on the back of the slide and its graduations are numbered from the right end. From this scale the logarithm of a number may be found as follows: Set the runner on the number on the D scale. Move the slide until the left index is at the number, as indicated by the runner. Read the mantissa from the scale of equal parts. This amounts to scaling the distance from the left-hand index to the number.

Example: Find the logarithm of 34.4. Set R to 34.4 D; LI to R; read mantissa 537 from scale of equal parts. The characteristic is 1; hence the complete logarithm is 1.537.

Some slide rules have a different arrangement of scales from that just illustrated. In any case, it is well to check the scales by finding the logarithm of some simple known number; for example, the logarithm of 5 is 0.699.

POLAR PLANIMETER

4-12. Polar Planimeter. Figure 4-4 shows a polar planimeter with adjustable tracing arm and adjustable pole arm. The planimeter is supported at the anchor point or pole P, the roller R, and the tracing point T. The arm carrying the anchor point is hinged to the frame of the planimeter. On the adjustable tracing arm A are graduations which give known relations between the readings of the planimeter and the area. The arm A is clamped in position by the screw at B, and fine settings are made by means of the tangent-screw C. The circumference of the drum D of the roller R is graduated into 100 parts. At E is a vernier for the roller. By means of a worm, the roller when revolving turns the graduated disk F in the ratio 10:1. The whole number of revolutions of the roller is read on the disk F by means of an index; hundredths of a revolution of the roller are indicated by the drum reading at the index of the vernier E; and thousandths are estimated by reading the vernier.

An alternative feature of the polar planimeter is a tracing lens having an index mark at the center, which serves as the tracing point. The lens magnifies the line to be followed. Other optional features are magnifying lenses over the vernier and the counting disk and a device to zero the roller and the counting disk quickly.

When a plotted area is to be determined, the sheet on which the figure is to be plotted is stretched flat and free from wrinkles. The needle tip of the anchor point is pressed into the paper in a convenient location, such that the entire area can be traversed, and is held down by the weight W. The tracing point is set at a definite point on the perimeter of the figure, and either the roller is set to read zero or preferably an initial reading is taken. The perimeter is then completely traversed until the tracing point is brought to its original location, and a final reading is taken. Particular care should be

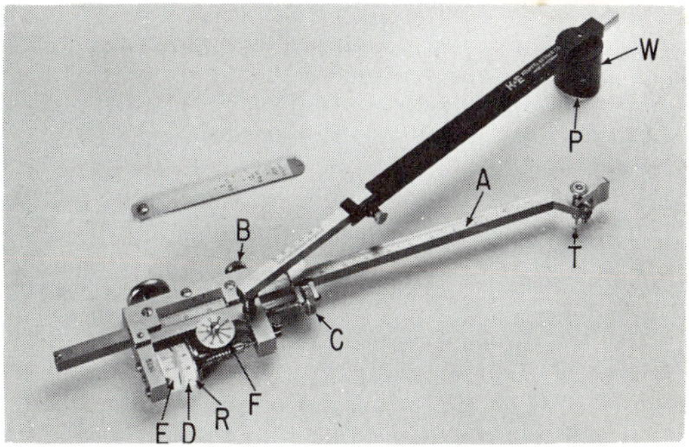

FIG. 4-4. Polar planimeter. (*Keuffel & Esser Co.*)

exercised in returning the tracing point exactly to the point of beginning before taking the final reading. The difference between the initial reading and the final reading is the net number n of revolutions of the roller:

$$n = \text{final reading} - \text{initial reading} \tag{1}$$

in which n is positive if the net rotation of the roller is forward and negative if the net rotation is backward. Account must be taken of the net number of times the zero graduation of the disk F may have passed the index. The area of the figure is computed as described in the following paragraphs.

For small areas, the anchor point is placed *outside* the figure. If the figure is larger than can be traced in one operation with the anchor point outside, it may be divided into smaller figures; however, if there are many large areas, faster progress can be made by placing the anchor point *inside* the figure.

In order to simplify the discussion, herein the direction of motion of the tracing point around the figure is taken as *always clockwise* unless specifically stated to the contrary. For counter-clockwise traversing, the rotation of the roller would be the opposite of that for clockwise traversing.

4-13. Area with Anchor Point outside Figure. The planimeter is so constructed that, when the anchor point is outside the figure and the perimeter is traversed clockwise, the final reading will be greater than the initial reading, and n will be positive. As shown in Art. 4-19, the area A of the figure is directly proportional to the number of revolutions, or

$$A = Cn \tag{2}$$

in which C, called the *planimeter constant,* indicates the area per revolution of roller. It is also shown in Art. 4-19 that the value of C is equal to the product of the length of the tracing arm and the circumference of the roller. If the length of the tracing arm is fixed, as on many planimeters, C is usually 10.00 sq. in., and the value is stated either on the top of the tracing arm or in the planimeter case. On some instruments, C is stated in metric units.

Example: The roller of a fixed-arm planimeter having a planimeter constant of 10.00 sq. in. is set at zero, and the perimeter of a figure is traversed clockwise with anchor point outside. The final reading is 2.367. Then the area of the figure is $10.00 \times (2.367 - 0.000) = 23.67$ sq. in.

4-14. Determination of Constant. The value of the planimeter constant can be determined by traversing the perimeter of a figure of known area, with anchor point outside. Preferably several trials should be made, and the average computed for use.

Example: The length of the tracing arm of a planimeter is so set that the roller registers 0.893 revolution when the perimeter of a 2 by 5-in. rectangle is traversed with anchor point outside. From Eq. (2),

$$C = \frac{A}{n} = \frac{2.00 \times 5.00}{0.893} = 11.20 \text{ sq. in.}$$

If the test figure has straight sides, the tracing point may be guided by a straightedge placed along each side. However, tracing the test figure freehand has the advantage that the calibration includes not only the setting of the instrument but also the tendency of the operator to keep the tracing point on one side or the other of the line. The same practice as that used in the calibration should be followed in tracing the perimeter of figures whose areas are to be determined.

An accessory furnished with some planimeters consists of a flat bar at one end of which is a needle point (Fig. 4-4); at the other end is a small hole into which the tracing point may be set. The distance between needle point and hole is equal to the radius of a circle whose area is 10.00 sq. in. The needle point is pressed into the paper, and the planimeter is quickly tested by tracing the circumference of the circle, the tracing point being held in the hole of the bar as it is revolved about the needle point.

If the tracing arm is adjustable, it is set so that some convenient relation exists between area and revolutions of roller. The accuracy of the setting is then tested by tracing the boundary of a figure of known area, as just described; and if necessary, the length of the arm is adjusted by trial until the desired relation is established.

It is not absolutely necessary to determine the instrumental constant. All that is necessary is to determine the difference in planimeter readings for a known area. Then by proportion, any required area is to the corresponding difference in readings as the figure of known area is to its difference in readings. However, the computation of the constant is so simple that it is usually made.

4-15. Area with Anchor Point inside Figure. When the tracing arm is held in such a position relative to the anchor arm that the plane of the roller passes through the anchor point, the tracing point can be made to describe completely the circumference of a circle without there being any revolution of the roller. This is called the *zero circle.* It can be shown that, when the perimeter of a figure is traversed with the anchor point within the figure, the indicated area ($A' = Cn'$) is equal to the difference between the area A of the figure and the area Z of the zero circle. The planimeter is so constructed that, for clockwise traversing with the anchor point inside the figure, the net rotation of the roller will always be $\left\{\begin{matrix}\text{forward}\\ \text{backward}\end{matrix}\right\}$ if the area of the figure is $\left\{\begin{matrix}\text{greater}\\ \text{less}\end{matrix}\right\}$ than that of the zero circle; hence the final reading will be $\left\{\begin{matrix}\text{greater}\\ \text{less}\end{matrix}\right\}$ than the initial reading, and n' will be $\left\{\begin{matrix}\text{positive}\\ \text{negative}\end{matrix}\right\}$. It follows that the area of the figure is

$$A = Cn' + Z \tag{3}$$

with due regard to the sign of n'.

Example 1: The perimeter of a cross-section is traversed clockwise with the anchor point inside the figure, with the length of the tracing arm so set that the planimeter constant is 10.00 and the area of the zero circle is 132.16 sq. in. The initial reading is 1.234, and the final reading is 8.703, the net rotation of the roller being forward. Then $n' = 8.703 - 1.234 = +7.469$; and from Eq. (3),

$$A = [10.00 \times (+7.469)] + 132.16 = 206.85 \text{ sq. in.}$$

Example 2: Conditions as in the preceding example, except that it was observed on the disk (F, Fig. 4-4) that the net rotation of the disk was *backward* and that the zero graduation of the disk had passed the index once. Then

$$n' = 8.703 - (10 + 1.234) = -2.531$$

and from Eq. (3),

$$A = [10.00 \times (-2.531)] + 132.16 = 106.85 \text{ sq. in.}$$

Example 3: If in the preceding example the operator has observed that the area of the figure was smaller than that of the zero circle and, starting with the same initial reading (1.234), had traversed the figure *counter-clockwise*, the net rotation of the roller would have been forward, and the final reading would have been 3.765. The negative area thus determined would be $10 \times (3.765 - 1.234) = 25.31$ sq. in.; and this area subtracted from that of the zero circle would be 106.85 sq. in. as before. Some surveyors prefer to use counter-clockwise rotation in order to avoid backward net rotation of the roller and thus to simplify the observations. Herein, however, the discussion and Eq. (3) are based on clockwise traversing.

4-16. Area of Zero Circle. The area of the zero circle can be determined by traversing the perimeter of a figure, once with the anchor point outside the figure and once with the anchor point inside. The first determination gives the area of the figure ($A = Cn$), and the second gives an indicated area Cn' representing the difference between the area of the figure and the area of the zero circle; it follows from Eq. (3) that

$$Z = Cn - Cn' = C(n - n') \tag{4}$$

where n' is $\begin{Bmatrix}\text{positive}\\\text{negative}\end{Bmatrix}$ if the area of the figure is $\begin{Bmatrix}\text{greater}\\\text{less}\end{Bmatrix}$ than that of the zero circle, as indicated by the direction of rotation of the roller.

Example: A given planimeter has a constant of 10.00 sq. in. A figure is traversed clockwise, first with anchor point outside and then with anchor point inside; the observed differences in planimeter readings are 2.124 and -9.537, respectively. Then by Eq. (4) the area of the zero circle is

$$Z = 10.00 \times [2.124 - (-9.537)] = 116.61 \text{ sq. in.}$$

If the length of the tracing arm is fixed, the area of the zero circle is usually stated on top of the tracing arm or in the planimeter case.

4-17. Figure Plotted at Other Than Full Scale. If a figure is plotted to scale other than full size, the required area is computed by multiplying the actual area by the product of the horizontal and vertical scaled relationships. For example, if a profile is plotted to the scale of 1 in. = 400 ft. (horizontal) and 1 in. = 20 ft. (vertical), each square inch on the paper represents $400 \times 20 = 8{,}000$ sq. ft.

4-18. Precision of Planimeter Measurements. If the relation between revolutions and area is established accurately, the errors of planimeter measurements are accidental (Chap. 5) and are due principally to the inability of the observer to follow exactly the boundary of the figure with the tracing point. For the same care and skill on the part of the observer, the smaller the area, the larger the relative error of measurement. Hence it is desirable that the areas be plotted to a scale consistent with the desired accuracy. Ordinarily, planimeter measurements of small areas may be expected to be correct within 1 per cent, and measurements of figures of considerable size may be correct within perhaps 0.1 or 0.2 per cent. In general, an area determined directly from a difference between initial and final planimeter readings may be determined to three (or the lower range of four) significant figures; summations of such areas (to a given number of decimal places) may have a greater number of significant figures.

In conformity with the precision of planimeter work, observations on the roller should be made to 0.001 revolution, and values of area (C, A, and Z) should be determined to 0.01 sq. in. The constants C and Z should be the mean of several observations.

4-19. Theory of Planimeter. Mathematical proofs of the theory of the polar planimeter by means of the calculus are given in various publications. A direct and simple geometric demonstration of the theory is as follows:

In Fig. 4-5*a*, the heavy solid lines represent a polar planimeter with tracing point T, wheel W, and anchor point or pole A. The tracing arm TW is hinged at H to the anchor arm AH. The length of the portion TH of the

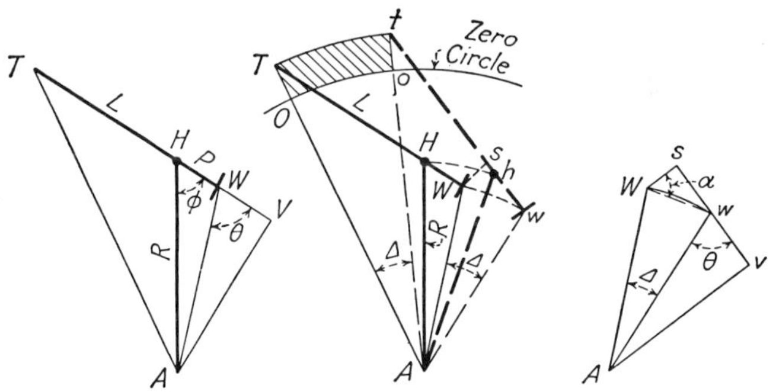

FIG. 4-5*a*. FIG. 4-5*b*. FIG. 4-5*c*.

tracing arm is designated as L, the wheel arm HW as P, and the anchor arm AH as R. (If, as in some designs, the wheel arm HW is folded back on the tracing arm, the relations herein demonstrated still apply.)

In Fig. 4-5*b*, consider the infinitesimal area $TtoO$, a portion of the sector TAt, which lies just outside the zero circle. If the tracing point is caused to traverse the perimeter clockwise, the only permanently recorded rotation of the wheel will be that due to the motion from T to t. This is because the movement from t to o is offset by the reverse movement from O to T, and because the wheel does not rotate while the tracing point is moved along the zero circle from o to O. The heavy dash lines represent the planimeter with tracing point at t.

As the tracing point moves from T to t, the wheel moves from W to w, partly by rolling and partly by sliding. The rolling component of this motion is represented by the line Ws, and the sliding component by the line sw. It will now be shown that the area $TtoO$ is in direct proportion to the roll of the wheel and is therefore equal to a constant C times the number of revolutions n. Reasons for the steps involved in Eqs. (5) to (14) are given immediately following the equations.

$$\text{Area } TtoO = \frac{\overline{AT} \cdot \overline{Tt}}{2} - \frac{\overline{AO} \cdot \overline{Oo}}{2} \tag{5}$$

$$= \frac{\overline{AT}^2 \cdot \Delta}{2} - \frac{\overline{AO}^2 \cdot \Delta}{2} \tag{6}$$

$$= \frac{\Delta}{2}(L^2 + 2LR \cos \phi + R^2 - 2PL - L^2 - R^2) \tag{7}$$

$$= L \cdot \Delta \cdot (R \cos \phi - P) \tag{8}$$

$$= L \cdot \Delta \cdot \overline{WV} \tag{9}$$

$$= L \cdot \Delta \cdot \overline{AW} \cdot \cos \theta \tag{10}$$

$$= L \cdot \overline{Ww} \cdot \cos \alpha \tag{11}$$

$$= L \cdot \overline{Ws} \tag{12}$$

$$= L \cdot c \cdot n \tag{13}$$

$$= C \cdot n \tag{14}$$

Eq. (5): The area of a sector is one half of the product of the radius and the subtended arc.

Eq. (6): An arc, in linear units, is equal to the product of the radius and the subtended angle in radians.

Eq. (7): The lines AT and AO can be expressed in terms of the dimensions of the planimeter, as follows:

In Fig. 4-5a,

$$\begin{aligned} \overline{AT}^2 &= \overline{TV}^2 + \overline{AV}^2 \\ &= (L + R \cos \phi)^2 + (R \sin \phi)^2 \\ &= L^2 + 2LR \cos \phi + R^2(\cos^2 \phi + \sin^2 \phi) \\ &= L^2 + 2LR \cos \phi + R^2 \end{aligned}$$

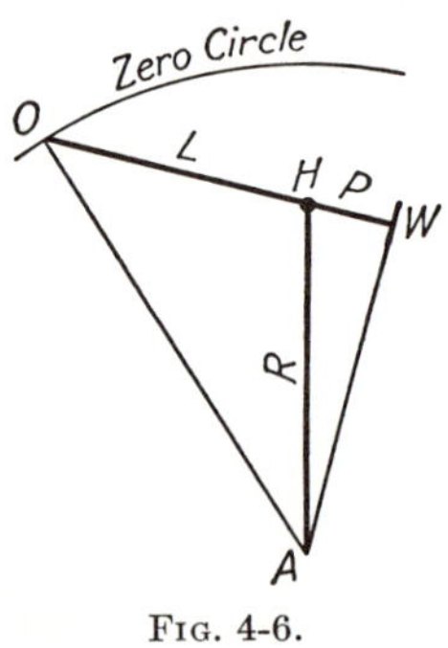

FIG. 4-6.

In Fig. 4-6,

$$\begin{aligned} \overline{AO}^2 &= \overline{OW}^2 + \overline{AW}^2 \\ &= (P + L)^2 + R^2 - P^2 \\ &= 2PL + L^2 + R^2 \end{aligned}$$

Eq. (8): Collect terms of Eq. (7).

Eq. (9): In Fig. 4-5*a*, $\overline{WV} = \overline{HV} - P = R \cos \phi - P$.

Eq. (10): From geometry of Fig. 4-5*a*.

Eq. (11): An arc, in linear units, is equal to the product of its radius and the subtended angle in radians, hence in Fig. 4-5*c* the arc $\overline{Ww} = \Delta \cdot \overline{AW} = \Delta \cdot \overline{Aw}$. Line Ws is perpendicular to line sv; in the limit, line Ww is arc Ww and is perpendicular to line Aw; hence $\cos \alpha = \cos \theta$.

Eq. (12): From geometry of Fig. 4-5*c*.

Eq. (13): The roll of a wheel is the product of the length of its circumference c and the number of revolutions n.

Eq. (14): The two characteristic dimensions of the instrument (L and c) are grouped into one instrumental constant C.

For a figure drawn at natural scale, the planimeter constant C is equal to the product of the length of the tracing arm and the circumference of the wheel. For figures drawn at other scales, the planimeter constant may include the factor for converting actual areas to areas at the given scale; in this case the constant is usually designated as C'.

4-20. Numerical Problems.

1. How many feet are there in $3\frac{1}{2}$ rods? In 6 chains and 8 links?

2. How many acres are there in a rectangular field 18.25 rods wide and 920.0 ft. long?

3. How many significant figures are there in each of the following numbers?

a. 0.208
b. 76248.0
c. 1.4800
d. 30.0×10^6
e. 0.006
f. 1.04 ± 0.02
g. 28.46 ± 0.08

4. Find the product of each of the following series of numbers, using logarithms:

a. $2.64 \times 79.18 \times 0.0767 \times 1.0028$
b. $(3.415)^2 \times \sqrt[3]{72.849}$
c. $(6.618)^{1/3} \times (68.627)^{1/2}$

5. Find the quotient for each of the following series of numbers, using logarithms:

a. $76.21 \div 49.20$ (5-place logs)
b. $0.4092 \div 0.006314$
c. $1.4823 \div 16.4917$ (6-place logs)
d. $8.472 \div \tan 42°21'$

6. Compute the log of 27, $\sqrt{15}$, and $\sqrt[3]{12}$, using only the logarithms of the following numbers: log 2 = 0.30103, log 3 = 0.47712, log 5 = 0.69897.

7. If the angles of an approximately equilateral triangle are measured with an angular error of 30″ and the distances with corresponding precision, how many significant figures will there be in the area computed by sines and cosines?

8. One side of a triangle in a small triangulation system (Chap. 16) is to be used as a base line from which the lengths of other sides are to be computed by means of sines. The measured length of the base line is 603.82 ft., and it is estimated that this length is within 0.05 ft. of the true value. With what precision should

the angles be measured to correspond, if the values of the angles lie between 50° and 70°? How many significant figures will there be in the computed lengths of the other sides?

9. The circumference of a circle 4 in. in diameter is traversed clockwise, with the anchor point of the planimeter outside the figure. The initial reading is 5.637 and the final reading is 6.939. What is the planimeter constant?

10. For a given planimeter set so that the instrumental constant is 10.22, the area of the zero circle is 116.24 sq. in. A figure is traversed clockwise with anchor point inside. The initial reading is 1.085 and the final reading is 9.632, the net rotation being backward and the zero graduation of the disk having passed the index once. What is the area of the figure?

11. The tracing arm of a planimeter is set so that the roller reads 0.583 revolution for 10.00 sq. in. The perimeter of an area is traversed clockwise first with anchor point outside and then with anchor point inside. The corresponding differences in readings are 2.095 and −7.786. What is the planimeter constant C? What is the area of the zero circle?

12. The perimeter of a figure is traversed clockwise, with the anchor point inside and with the tracing arm set as in the preceding problem. The difference in readings is −3.781. What is the area of the figure?

13. With the tracing arm of the planimeter set as in the preceding problem, a figure, for which the vertical scale is 1 in. = 8 ft. and the horizontal scale is 1 in. = 20 ft., is traversed clockwise with anchor point outside. The difference in planimeter readings is 1.932. What is the actual area in square inches? What area in square feet does it represent?

14. For a given polar planimeter, the circumference of the wheel is 2.094 in. For use on a map having a scale of 1 in. = 10 ft., it is desired to use a planimeter constant C' of 1,000. What is the required length of the tracing arm?

4-21. Office Problem.

Problem 1. Area with Planimeter

Object. With the polar planimeter, to determine the area of a figure plotted to scale.

Procedure. (1) Set the tracing arm so that one revolution of the roller will bear some simple relation to the given scale and unit of measurement. (2) Test the accuracy of the setting by traversing a figure of known area, say a 2 by 5-in. rectangle, three times. Record the readings to 0.001 revolution. If necessary, adjust the tracing arm until the desired relation is obtained. (3) With the anchor point outside, traverse clockwise the perimeter of the figure whose area is to be determined. Check the operation. (4) Convert the difference between readings into terms of area. (5) Determine the area of the zero circle as described in Art. 4-16. (6) Measure the given figure with anchor point *inside*, and compute the area. (7) Set the tracing arm so that the relation between revolutions of roller and area is unknown. Determine the difference in planimeter readings for the figure of known area and for the given figure. By proportion determine the area of the figure.

Hints and Precautions. Locate the anchor point so that the roller will stay on the paper as the tracing point is moved about the figure. Preferably have the tracing arm and the anchor arm nearly perpendicular to each other when the tracing point is near the center of the area, and never allow the angle between the arms to become small. See that the paper is free from wrinkles, and that the contact edge of the roller is free from dirt.

REFERENCES

1. Bruhns, C., "A New Manual of Logarithms to Seven Places of Decimals," 18th ed., D. Van Nostrand Company, Inc., Princeton, N.J., 1941.
2. Holman, S. W., "Computation Rules and Logarithms," The Macmillan Company, New York, 1906.
3. Ives, Howard C., "Natural Trigonometric Functions to Seven Decimal Places for Every Ten Seconds of Arc," 2d ed., John Wiley & Sons, Inc., New York, 1942.
4. McCormick, E. M., "Digital Computer Primer," McGraw-Hill Book Company, New York, 1959.
5. U.S. Bureau of Land Management, "Standard Field Tables," 8th ed., Government Printing Office, Washington, D.C., 1956.

CHAPTER 5

ERRORS

5-1. General. Reference has been made to the necessity of the surveyor's appreciating the errors involved in measurements (Art. 1-10) and computations (Arts. 4-5 and 4-6). Every observed or measured quantity contains errors of unknown magnitude due to a variety of causes, and hence a measurement is never exact. One of the important functions of the surveyor is to secure measurements which are correct within certain limits of error prescribed by the nature and purpose of the survey. This requires that he know the sources of errors, understand the effect of the various errors upon the observed quantities, and be familiar with the procedure necessary to maintain a required precision, with reasonably consistent precision as between angles and distances. Often surveyors of considerable experience have displayed an ignorance of this phase of their work which was both ludicrous and lamentable.

In dealing with measurements it is important to distinguish between *accuracy* and *precision*. As defined by the American Society of Civil Engineers, accuracy is "absolute nearness to the truth" whereas precision is "relative or apparent nearness to the truth." (Formerly ASCE defined precision as "degree of fineness of reading in a measurement, or, the number of places to which a computation is carried.") As defined by the U.S. Coast and Geodetic Survey, accuracy is "degree of conformity with a standard" whereas precision is "degree of refinement in the performance of an operation or in the statement of a result." From these mutually consistent definitions it follows that a measurement may be accurate without being precise, and *vice versa*. For example, a distance may be measured very carefully with a tape, to thousandths of a foot, and still be in error of several hundredths of a foot because of erroneous length of tape; the measurement is precise but not accurate.

5-2. Sources of Error. Errors arise from three sources:

1. From imperfections or faulty adjustment of the instruments or devices with which measurements are taken. For example, a tape may be too long or a level may be out of adjustment. Such errors are termed *instrumental errors*.

2. From the limitation of the human senses of sight and touch. For example, an error may be made in reading the angle on the graduated circle of a transit or in estimating the tension of a steel tape. Such errors are called *personal errors*.

3. From variations in the phenomena of nature such as temperature, humidity, wind, gravity, refraction, and magnetic declination. For example, the length of tape will become greater or smaller according as the temperature increases or decreases, and readings of the magnetic needle are affected by variations in the magnetic declination. Such errors are called *natural errors* or *external errors*.

5-3. Kinds of Error. A *mistake*, or *blunder*, is an unintentional fault of conduct arising from poor judgment or from confusion in the mind of the observer. It is quite distinct from the mathematical or physical meaning of error. Throughout this text this distinction will be observed. Mistakes have no place in a discussion of the theory of errors. They are detected and eliminated by checking all work.

The *resultant error* in a given quantity is the difference between the measurement and the true value. It is made up of individual errors from a variety of sources, some of the individual errors tending perhaps to make the measurement too large, and others to make it too small. If the measurement is too large, the error is said to be positive; if too small, the error is said to be negative. For a single quantity which has been determined by observation, neither the resultant error nor any of its individual parts can ever be determined exactly but can be fixed within certain probable limits.

A *discrepancy* is the difference between two measurements of the same quantity (see Art. 5-5).

A *systematic error* is one that, so long as conditions remain unchanged, always has the same magnitude and the same algebraic sign (which may be either positive or negative). If conditions do not change during a series of measurements, the error is termed a *constant* systematic error; for example, a line may be measured with a tape which is too short. If conditions change, resulting in corresponding changes in the magnitude of the error, it is termed a *variable* systematic error; for example, a line may be chained during a period in which the temperature varies. A systematic error always follows some definite mathematical or physical law, and a correction can be determined and applied. It is cumulative in effect. The error may be instrumental, personal, or natural.

An *accidental error* is an error due to a combination of causes beyond the ability of the observer to control and for which it is impossible to make correction; for each observation the magnitude and algebraic sign of the accidental error are matters of chance and hence cannot be computed as can the magnitude and algebraic sign of a systematic error. However, accidental errors taken collectively obey the law of probability (see Art. 5-6). Since each accidental error is as likely to be positive as negative, a certain compensative effect exists; and accidental errors are sometimes called "compensating" errors. Accidental errors are also termed "irregular errors" or "erratic errors." As an example of the occurrence of an accidental error, in chaining it is impossible to set the chaining pin exactly at the proper graduation on the tape. Accidental errors remain after mistakes have been eliminated by checking and systematic errors have been eliminated by correction.

5-4. Systematic and Accidental Errors Compared. The total systematic error in any given number of measurements is the algebraic sum of the individual errors of the individual measurements. Thus if a distance is measured with a tape which is too short, the systematic error due to the tape's not being of the standard length would be directly proportional to the length of the line.

Example 1: The length of a line as measured with a 100-ft. tape at 60°F. is 1,000.00 ft. Later the tape is compared with the standard length and is found to be 100.021 ft. long. The error in the recorded length of the line is $-0.021 \times 10 = -0.21$ ft., and the actual length of the line is 1,000.21 ft.

The example above illustrates the manner in which a *constant* systematic error increases with the number of observations. The example below illustrates the effect of a *variable* systematic error.

Example 2: A line measured with a 300-ft. tape is found to be 1,200.00 ft. long. Calculations based upon observations of temperature of the tape indicate that its probable length was 299.998 ft. for the first tape length, 300.001 ft. for the second tape length, 300.008 ft. for the third tape length, and 300.004 ft. for the fourth tape length. The total systematic error due to variation in temperature would therefore be the sum of the above errors, or, $+0.002 - 0.001 - 0.008 - 0.004 = -0.011$ ft., and the length of the line would be $1,200.00 + 0.01 = 1,200.01$ ft.

Often a systematic error from one source may be of opposite sign to that from another source, so that the resultant systematic error is perhaps smaller than any of the errors from individual sources. Thus under a given tension and at a given temperature a tape might be of the standard length. Suppose that for the conditions under which the measurement of a line was made a variation from standard in tension in the tape produced an error of -0.022 ft. per tape length and a variation from standard in temperature produced an error of $+0.018$ ft. per tape length; the resultant unit error due to variations in temperature and tension would be -0.004 ft.

For many observations the order of procedure is such that systematic errors are either eliminated or at least reduced to a negligible quantity. Thus in chaining, the error due to temperature change in a steel tape may be nearly eliminated by observing the temperature and making correction; and errors in leveling due to faulty adjustment of the level may be eliminated by balancing backsight and foresight distances.

Accidental errors, as the name signifies, are purely accidental in character, and there is no way of determining or eliminating them in the sense that we may determine or eliminate most systematic errors. Thus, while the effect of change of temperature upon the length of a tape can be approximately eliminated by calculations based upon physical measurements, there is no corresponding method of eliminating the accidental error due to marking the ends of the tape on the ground or due to reading the rod in leveling. Although accidental errors are as likely to be positive as negative, the error for one observation of a quantity is not likely to be the same as for the second observation.

According to the mathematical theory of probability, accidental errors tend to increase in proportion to the *square root* of the number of opportunities for error. Thus if the accidental error in measuring one tape length were ± 0.02 ft., the chances would be even that the total accidental error due to measuring 100 tape lengths would not exceed

$$\pm 0.02 \times \sqrt{100} = \pm 0.20 \text{ ft.}$$

A systematic error of the same magnitude would produce a total error of $0.02 \times 100 = 2.00$ ft. It is thus seen that for any connected series of observations of independent but related quantities, the accidental errors are of relatively small importance as compared with systematic errors *of the same magnitude.* Though accidental errors cannot be eliminated, they may be reduced to a small quantity through the use of proper instruments and methods. By taking a series of like observations of a single quantity, an estimate of the accidental error may be made, as will later be shown; but its true magnitude can never be determined.

The relative importance of systematic errors as compared with accidental errors depends upon the nature of the observations, the care exercised by the observer, and the instruments and methods of procedure employed. In general, the rougher the methods used, the larger the systematic errors as compared with the accidental errors.

5-5. Discrepancy. If a given quantity is measured twice, the difference between the two measurements is termed the *discrepancy.* Frequently, quantities measured by the operations of surveying are "checked" by a second measurement. If the discrepancy between two such measurements is small, it is an indication that no mistakes have been made and that the accidental errors are small, but it is not an indication that the systematic errors are small. For example, two tape measurements of a line a mile long might show a discrepancy of 0.3 ft., but the systematic errors due to such causes as temperature, sag, and slope of tape might be 3 ft.

5-6. Theory of Probability. It has been stated that, by employing proper methods, systematic errors may be largely eliminated. Although this is true, it is also true that for certain kinds of surveys, particularly those of low precision, it is unnecessary and impracticable even approximately to eliminate such errors. For the surveys of higher precision special effort is made to eliminate systematic errors, and the precision of a measured quantity is governed by the accidental error which it contains. To form a judgment of the probable value or the probable precision of a quantity, *from which systematic errors have been eliminated,* it is necessary to rely upon the theory of probability, which deals with accidental errors of a series of like or related measurements. It is assumed that

1. Small errors are more numerous than large errors.
2. Very large errors do not occur.
3. Errors are as likely to be positive as negative.
4. The true or ideal value of a quantity is taken as the mean of an infinite number of like observations.

In practice it is impossible either to eliminate systematic errors entirely or to take an infinite number of observations; hence the value of a quantity is never known exactly. However, in the discussions to follow, it is assumed that systematic errors are so far eliminated as to be a negligible factor.

A thorough understanding of the law of probability may be obtained only by the study of a text on least squares, but a few of the rules for simpler cases of the adjustment of observations and determination of probable values and probable errors will be stated here, with relation to the measurements of surveying. The theory of probability is useful in indicating the precision of results only in so far as they are affected by accidental errors and does not in any way determine the magnitude of systematic errors which may be present.

OBSERVATIONS OF EQUAL RELIABILITY

5-7. Probable Value. The *most probable value* of a quantity is a mathematical term used to designate that adjusted value which, according to the principles of least squares, has more chances of being correct than has any other. Determination of the most probable value from a series of measurements is the principal use which the surveyor makes of the theory of probability.

5-7a. Same Quantity. For a series of measurements of the same quantity made under identical conditions, the most probable value is the mean of the measurements.

Example: After all systematic errors have been eliminated, the several measured lengths of a line are 1,012.36, 1.012.35, 1,012.38, 1,012.32, 1,012.33, and 1,012.30 ft. The most probable value is the mean of the measurements, or 1,012.34 ft.

5-7b. Related Quantities. For related measurements taken under identical conditions, the sum of which should equal a mathematically exact quantity, the most probable values are the observed values corrected by an equal part of the total error. (This situation can arise only in the case of angles about a point or angles in a closed figure.) The correction is in proportion to the *number* of related measurements and not to the *magnitude* of the individual measurements.

Example 1: The angles about a point have the following observed values: $130^\circ15'20''$, $142^\circ37'30''$, and $87^\circ07'40''$. The sum of the measurements is $360^\circ00'30''$; therefore the total error is $30''$. Since there are three angles, the error is assumed to be $10''$ for each measurement. The most probable values are

$$\begin{array}{rcrcr}
130^\circ15'20'' & - & 10'' & = & 130^\circ15'10'' \\
142^\circ37'30'' & - & 10'' & = & 142^\circ37'20'' \\
87^\circ07'40'' & - & 10'' & = & 87^\circ07'30'' \\
\hline
360^\circ00'30'' & - & 30'' & = & 360^\circ00'00''
\end{array}$$

For related measurements taken under identical conditions, the sum of which should equal a single measurement taken under the same conditions, the most probable values are obtained by dividing the discrepancy equally among all the measurements, including the sum. If the correction is added to each of the related measurements, it is subtracted from the measurement representing their sum; and *vice versa*.

Example 2: Measurements of three angles about a point O are $AOB = 12^\circ31'50''$, $BOC = 37^\circ29'20''$, and $COD = 47^\circ36'30''$. The measurement of the single angle AOD is $97^\circ37'00''$. The discrepancy between the sum of the three measured angles and the measurement of the angle representing their sum is $40''$. Since the size of the errors is independent of the size of the angle, the discrepancy is divided into equal parts: $40/4 = 10''$. This correction is to be subtracted from measurements of each of the angles AOB, BOC, and COD and is to be added to the measurement of the angle AOD. The most probable values are:

$$\begin{array}{lcrcrcrl}
AOB & = & 12^\circ31'50'' & - & 10'' & = & 12^\circ31'40'' & \\
BOC & = & 37^\circ29'20'' & - & 10'' & = & 37^\circ29'10'' & \\
COD & = & 47^\circ36'30'' & - & 10'' & = & 47^\circ36'20'' & \\
\text{Sum} & & 97^\circ37'40'' & - & 30'' & = & 97^\circ37'10'' & \rbrace\ \textit{check} \\
AOD & = & 97^\circ37'00'' & + & 10'' & = & 97^\circ37'10'' &
\end{array}$$

It is desired to emphasize the fact that the methods illustrated by the preceding examples apply only to measurements each of which in a given example is made under the same conditions as are all the others, with the same instrument, observer, weather conditions, etc. In example 1, the probabilities are that the error of measuring any one of the angles is the same as that of any other. It is true that the chances of the error in one measurement being the same as in the others is very small; but it is more probable that the errors will each be the same magnitude than that they will be of any other assignable value.

In adjusting several measurements the sum of which should equal a single measurement, it should be noted that there is a distinction between observations like those cited in example 2 and those for which several operations are involved in the measurement of a single quantity. If the case illustrated by example 2 involved linear instead of angular measurements, application of the method would be limited to distances not greater than one tape length. If distances between adjacent points along a line of considerable length were measured with a tape, and then if the full length of the line were measured in the same manner, it is evident that the corrections should be made to depend upon some function of the number of applications of the tape, being greater for the full distance than for any of its parts and being greater for long segments of the line than for short ones.

5-8. Probable Error. If a series of like or related observations of a single quantity is made, a number of values of the quantity are obtained. The differences between these values furnish data from which the *probable error* can be determined. The probable error of a measurement is a mathematical quantity giving (inversely) an indication of precision and does not signify either the true error or the error most likely to occur. It is a valid measure of the precision of observations only with regard to accidental errors, that is, after systematic errors have been reduced to a negligible quantity.

Probable error is a plus or minus quantity within which limits the actual accidental error is as likely as not to fall. In other words, if the probable error of a measurement is both added to and subtracted from the observed value, the chances are even that the true value of the measured quantity lies inside (or outside) the limits thus set. Thus if 6.23 represents the mean of several measurements and 0.11 represents the probable error of the mean value, the chances are even that the true value lies between the limits $6.23 - 0.11 = 6.12$ and $6.23 + 0.11 = 6.34$. In this case, the quantity would be written 6.23 ± 0.11, which expresses the probable limits of precision. The *probable ratio of precision* of the measurement is $0.11 \div 6.23 = 1/57$ (approximately). Throughout this chapter the discussion of probable errors may, in the case of linear measurements, be applied to probable ratios of precision by converting one method of expressing the precision into terms of the other method as desired.

Another measure of the precision of a series of observations is *standard deviation*, or *standard error*. It is computed as indicated by Eq. (1) and the example of this article, by the principles of least squares. Probable error is simply a special case of standard deviation which corresponds to a 50 per cent probability that the accidental error will lie within (or without) the stated limits. The two measures are related by a numerical constant (probable error = 0.6745 standard deviation). For the purposes of this book the probable error, or 50 per cent error, is adequate and simple, and the discussions of this chapter are based on probable error rather than on standard deviation.

In the adjustment of observations, the probable error of the most probable value of each quantity can be estimated from a series of measurements of that quantity, and the probable errors can then be used in computing weights and/or the corrections to be applied to related quantities. Consideration of probable error is also useful in choosing methods of surveying to produce desired degrees of precision.

5-8a. Same Quantity. It has been stated that the mean, or average, of a series of like observations of a single quantity is the most probable value. For the purpose of determining the probable error, this mean value is mathematically regarded as being the most likely value (based on this series of observations), and the difference between each of the individual measurements and the mean value is determined. These differences are termed *residuals* or *deviations.* The theory of least squares demonstrates that the probable error is a function of the square root of the sum of the squares of the residuals.

Although no attempt to derive the following expressions will be made here, it is well to state that they are based upon the hypothesis that a large number of measurements of a single quantity has been taken. The results of experiments indicate, however, that they may be applied to a limited number of observations with good results. It seems doubtful if they can be consistently applied to a series of observations containing less than 10 measurements.

The probable error of a *single observation* is calculated by the equation

$$E = 0.6745\sqrt{\frac{\Sigma v^2}{n-1}} = 0.6745\sigma \tag{1}$$

in which Σv^2 is the sum of the squares of the residuals, n is the number of observations, and σ is the standard deviation. The probable error of a single observation is not used in the determination of the most probable value of related measurements, but it indicates the degree of precision which may be expected in any single observation made under the same conditions.

The probable error of a single observation may be calculated approximately by the expression

$$E = \frac{0.845\Sigma v}{\sqrt{n(n-1)}} \qquad \text{(approx.)} \tag{2}$$

in which Σv is the sum of the residuals without regard to signs.

The probable error of a single observation may be calculated with about the same degree of approximation by the equation

$$E = 0.845\bar{v} \qquad \text{(approx.)} \tag{3}$$

in which $\bar{v}$ is the mean or average value of the residuals without regard to signs. The term $\bar{v}$ is also called the *average deviation.*

Since the determination of probable error is at best an approximation, in many cases it is permissible (and usually conservative) to take the probable error of a single observation as being roughly equal to the average deviation, or

$$E = \bar{v} \qquad \text{(rough)} \tag{4}$$

The probable error of the *mean* of a number of observations of the same quantity is calculated by the equation

$$E_m = 0.6745\sqrt{\frac{\Sigma v^2}{n(n-1)}} = \frac{E}{\sqrt{n}} = 0.6745\frac{\sigma}{\sqrt{n}} \tag{5}$$

It is seen that the probable error of the mean is inversely proportional to the square root of the number of observations.

Equations (2) to (4) are more convenient to apply than is Eq. (1). Whether or not they may be properly used depend upon the number of observations, the distribution of the residuals, and the desired number of places in the probable error.

The following example illustrates the methods of applying the preceding equations and indicates the degree of approximation arising through the use of Eqs. (2), (3), and (4).

Example: Following is a series of 10 rod readings which were taken with a wye level under identical conditions. The day was calm and cloudy. The instrument was set up, and the target rod was held on a point 600 ft. away. Before each reading the target was moved and the instrument was leveled.

Rod reading, ft.	v, ft.	v^2
2.467	0.002	0.000004
2.460	0.005	0.000025
2.469	0.004	0.000016
2.465	0.000	0.000000
2.471	0.006	0.000036
2.461	0.004	0.000016
2.463	0.002	0.000004
2.466	0.001	0.000001
2.460	0.005	0.000025
2.468	0.003	0.000009
	$\Sigma v = 0.032$	$\Sigma v^2 = 0.000136$
Mean = 2.465	$\bar{v} = 0.0032$	$\sigma = \pm\sqrt{\frac{0.000136}{9}} = \pm 0.00389$

From the values in the tabulation, the probable error of a single observation is:

By Eq. (1), $E = \pm 0.6745\sqrt{\frac{0.000136}{9}} = \pm 0.00262$ ft.

By Eq. (2), $E = \pm\frac{0.845 \times 0.032}{\sqrt{10 \times 9}} = \pm 0.00285$ ft.

By Eq. (3), $E = \pm 0.845 \times 0.0032 = \pm 0.00270$ ft.

By Eq. (4), $E = \pm 0.00320$ ft.

By Eq. (5), using the value of E determined by Eq. (1), the probable error of the mean is

$$E_m = \pm \frac{0.00262}{\sqrt{10}} = \pm 0.00083 \text{ ft.}$$

The preceding example is illuminating not only as showing the steps made in calculating probable errors but also as indicating in a measure the degree of approximation introduced by using the approximate expressions for the probable error of a single observation. The results of the example have purposely been extended to more places than are consistent for the given data (see Art. 4-4), in order to make the comparison.

In work involving many calculations of probable errors, a decided saving in labor will be accomplished if one or another of the approximate formulas is used. Except for the observations taken on surveys of high precision, the approximate formulas (2) and (3) are sufficiently precise.

As soon as the residuals are computed, they should be examined in comparison with the average residual. Values corresponding to any unduly large residuals, say four or five times the average residual (excluding the doubtful value), should be rejected and the computation continued with the remaining values.

5-8b. Related Quantities. The probable error of the sum of observations, each having the same probable error, is equal to the probable error of a single observation multiplied by the square root of the number of observations (opportunities for error), or

$$E_s = E\sqrt{n} \tag{6}$$

Equation (6) corresponds to a special case of Eq. (11) in Art. 5-11, in which special case all quantities have the same probable error and therefore are of equal reliability. (See also "Addition" in Art. 4-5.)

Example: If the probable error in measuring one tape length were ± 0.01 ft., the probable error in the measured length of a line 1 mile long (assuming full 100-ft. tape lengths, without breaking chain) would be $\pm 0.01 \times \sqrt{53} = \pm 0.07$ ft.

OBSERVATIONS OF DIFFERENT RELIABILITY

5-9. Weight. In the foregoing discussion, it has been assumed that all observations are taken under the same conditions and consequently are equally reliable. Frequently in surveying, however, it is required to combine the results of measurements which are not made under similar conditions and which therefore have different degrees of reliability. In such cases it is necessary to consider the degree of reliability, or *weight* (as nearly as it can be determined), that applies to each of the separate measurements. For example, suppose that an angle has been measured perhaps at different times and by different observers but presumably with equal care; and suppose that the results are as follows:

$47°37'40''$ (one measurement)
$47°37'22''$ (four measurements)
$47°37'30''$ (nine measurements)

If it is assumed that each single reading was made with equal care, then it is a logical assumption that the second value ($47°37'22''$) has four times the reliability of the first value ($47°37'40''$) and that the third value ($47°37'30''$) has nine times the reliability of the first value. In general terms, *weights are proportional to the number of like observations.* For convenience, in the example a weight of unity is assigned to the least precise (in this case the first) value; then the second and third values have weights of 4 and 9, respectively. Weights are relative or comparative, not absolute; thus the numbers 2, 8, and 18 would represent the weights as well as the numbers 1, 4, and 9.

Often weights will be assigned to observations, not according to the number of observations, but arbitrarily according to the judgment of the observer. For example, he might judge the value of an elevation secured from a line of levels run on a calm, temperate day as being two or three times as reliable as that secured from another line of levels run over the same route but on a windy, cold day.

If the probable error is known instead of the number of observations, the weight can be computed as follows: For observations (of a given quantity) made with equal care, it has been stated that (1) weights vary directly with the number of observations and (2) probable errors (of the mean value) vary inversely with the square root of the number of observations. It follows that *weights are inversely proportional to the square of the corresponding probable errors,* or

$$\frac{W_1}{W_2} = \frac{E_2^2}{E_1^2} \tag{7}$$

in which W_1 and W_2 are the weights to be assigned given measurements and E_1 and E_2 are the corresponding probable errors. For any number of measurements, Eq. (7) may be expressed in the form

$$W_1E_1^2 = W_2E_2^2 = W_3E_3^2 \cdots \tag{7a}$$

5-10. Adjustment of Weighted Observations. With the weights known, as determined by any of the three methods just described, the most probable values can be determined. There are two cases: (1) various measurements of the *same quantity* and (2) measurements of *related quantities.*

5-10a. Same Quantity. The most probable value of a quantity for which measurements of different reliability have been made is the *weighted mean.* The weighted mean is computed by multiplying each value by its weight, adding the products, and dividing by the sum of the weights.

Example 1: It is desired to determine the most probable value of the angle discussed in the preceding article. For each value the number of observations is given, and hence the weight is known; the weights are 1, 4, and 9, respectively. In the following computation, the labor is reduced by employing only the seconds, which represent the *differences* between the observed values and the common value 47°37′.

$$\begin{aligned} 47°37'40'' \times 1 &= 47°37'40'' \\ 22'' \times 4 &= 88'' \\ 30'' \times 9 &= 270'' \\ \text{Sum}\quad 14 & \quad\quad 398'' \\ {}^{398}\!/\!_{14} &= 28'' \end{aligned}$$

The weighted mean is 47°37′28″, the most probable value.

Example 2: *Solution a.* Lines of levels to establish the elevation of a point are run over four different routes. The observed elevations of the point with probable errors are given below.

Line	Observed elevation, ft.
a.	721.05 ± 0.02
b.	721.37 ± 0.04
c	720.62 ± 0.06
d.	721.67 ± 0.08

Since the probable errors are given, the weights can be computed from Eq. (7*a*):

$$W_a 2^2 = W_b 4^2 = W_c 6^2 = W_d 8^2$$

or

$$W_a = 4W_b = 9W_c = 16W_d$$

Let $W_a = 1$; then $W_b = 1/4$, $W_c = 1/9$, and $W_d = 1/16$.

Each observed elevation is multiplied by its weight, as shown in the following tabulation:

Line	Observed elevation	Weight	Weighted observation
a.	721.05	1	721.05
b.	721.37	1/4	180.34
c.	720.62	1/9	80.07
d.	721.67	1/16	45.10
Sum.	. . . . . .	205/144	1,026.56

The most probable value of the elevation is the weighted mean, or the sum of the weighted observations divided by the sum of the weights:

$$\text{Weighted mean} = \frac{1{,}026.56 \times 144}{205} = 721.10 \text{ ft., most probable value}$$

Solution b. For problems in which the quantities are large, as in the solution above, it reduces the labor considerably if *differences* are weighted, rather than the observed values themselves. To illustrate, let us work example 2 by weighting the differences between 721.00 and the observed elevations. Also, instead of recording the weights as fractions, let us assign them in whole numbers in order to facilitate the work of computing.

Line	Observed elevation	Less 721.00	Weight	Weighted difference
a................	721.05	+0.05	144	+ 7.2
b................	721.37	+0.37	36	+13.3
c................	720.62	−0.38	16	− 6.1
d................	721.67	+0.67	9	+ 6.0
Sum............	......	......	205	+20.4

The most probable difference between 721.00 and the most probable value of the elevation is $(+20.4/205) = +0.10$ ft. Hence the most probable value of the elevation is $721.00 + 0.10 = 721.10$ ft., which is the same as that obtained by solution *a*.

In the foregoing example, solution *a* requires multiplications to five places, while solution *b* requires multiplications to only three places. If the slide rule is used, the problem may be solved by considering differences in about one-third the time that it takes for the solution in which the observed values are weighted directly.

Probable Error of Weighted Mean. By the principles of least squares, it is known that the probable error of the weighted mean is

$$E_{wm} = 0.6745 \sqrt{\frac{\Sigma(Wv^2)}{(\Sigma W)(n-1)}} \tag{8}$$

Example 3: It is desired to determine the probable error of the weighted mean computed in example 2. The use of Eq. (8) is considered sufficiently precise for most purposes of surveying, although in other fields a method of "propagation of error" would be used here.

The computation of ΣW and $\Sigma(Wv^2)$, based on solution *a*, is indicated by the successive columns of the following tabulation:

Line	Observed elevation, ft.	v	v^2	W	Wv^2
a........	721.05	0.05	0.0025	1	0.0025
b........	721.37	0.27	0.0729	¼	0.0182
c........	720.62	0.48	0.2304	1/9	0.0256
d........	721.67	0.57	0.3249	1/16	0.0203
	721.10, weighted mean			$\Sigma W = 205/144$	$\Sigma(Wv^2) = 0.0666$

Then, by Eq. (8),

$$E_{wm} = 0.6745 \sqrt{\frac{0.0666}{(205/144)(4-1)}} = \pm 0.08 \text{ ft.}$$

5-10b. Related Quantities. When the sum of measured values having different weights must equal a known value, either measured or exact, the most probable values are the observed values each corrected by an appropriate portion of the discrepancy or of the total error. *The corrections to be applied are inversely proportional to the weights,* or

$$\frac{C_1}{C_2} = \frac{W_2}{W_1} \tag{9}$$

where C is the correction to be applied to a measured value of a quantity to obtain its most probable value consistent with the related quantities. The measured value itself may have been obtained as the weighted mean of a number of observations of the same quantity. As before, the weights may be determined from the number of observations, from the probable errors, or arbitrarily.

For any number of related quantities, Eq. (9) may be expressed in the form

$$C_1W_1 = C_2W_2 = C_3W_3 \cdots \tag{9a}$$

Example 1: Two angles *AOB* and *BOC* and the single angle *AOC* are measured about a point *O* under identical conditions, with results as given in the following tabulation. It is desired to determine the most probable values.

Angle	Observed value	No. of measurements
AOB.........	23°46′00″	1
BOC.........	59°14′27″	4
AOC.........	83°01′07″	6

The discrepancy between the sum of angles *AOB* and *BOC* and the angle *AOC* is 40″. The weights are 1, 4, and 6, respectively; hence the comparative corrections are 1, 1/4, and 1/6, respectively. The sum of the comparative corrections is equal to $24/24 + 6/24 + 4/24 = 34/24$; in such cases it is said that there are 34 *parts* of the total correction. The total correction in seconds is divided among the angles in proportion to the individual comparative corrections (parts); thus the individual corrections are

$$\begin{aligned} C_{AOB} &= 24/34 \times 40'' = 28'' \\ C_{BOC} &= 6/34 \times 40'' = 07'' \\ C_{AOC} &= 4/34 \times 40'' = 05'' \end{aligned}$$

For angles *AOB* and *BOC* whose sum was smaller than *AOC*, the correction is to be added; for *AOC* the correction is to be subtracted. The most probable values are

$$\begin{aligned} AOB &= 23°46'00'' + 28'' = 23°46'28'' \\ BOC &= 59°14'27'' + 07'' = 59°14'34'' \\ \text{Sum} &\ldots\ldots\ldots\ldots = 83°01'02'' \\ AOC &= 83°01'07'' - 05'' = 83°01'02'' \end{aligned} \Bigg\} \textit{check}$$

Since corrections are inversely proportional to weights, and since weights are inversely proportional to the square of the corresponding probable errors, it follows that *corrections are directly proportional to the square of the corresponding probable errors*, or

$$\frac{C_1}{C_2} = \frac{E_1^2}{E_2^2} \quad \text{or} \quad \frac{C_1}{E_1^2} = \frac{C_2}{E_2^2} = \frac{C_3}{E_3^2} \cdots \tag{10}$$

Direct use of this relation obviates the determination of weights when probable errors are given, and thus simplifies the computations.

Example 2: Three angles about a point are each measured by a series of observations. The mean values with their probable errors are given in the following tabulation. Their sum should equal 360°. It is desired to determine the most probable value of the angles.

$$\begin{aligned} AOB &= 130°15'20'' \pm 02'' \\ BOC &= 142°37'30'' \pm 04'' \\ COA &= 87°07'40'' \pm 06'' \\ \text{Sum} &= 360°00'30'' \end{aligned}$$

The total error—therefore, the total correction to be made—is 30″; that is, $C_{AOB} + C_{BOC} + C_{COA} = 30''$. The successive columns of the following tabulation show the steps in computing the corrections.

Angle	Probable error E		E^2	Correction C	
	Absolute	Comparative		Comparative	Absolute
AOB.....	02″	1	1	1	$\frac{1}{14} \times 30'' = 02''$
BOC.....	04″	2	4	4	$\frac{4}{14} \times 30'' = 09''$
COA.....	06″	3	9	9	$\frac{9}{14} \times 30'' = 19''$
Sum...	...	...	...	14	$\frac{14}{14}$ 30″

The most probable values are, therefore,

$$\begin{aligned} AOB &= 130°15'20'' - 02'' = 130°15'18'' \\ BOC &= 142°37'30'' - 09'' = 142°37'21'' \\ COA &= 87°07'40'' - 19'' = 87°07'21'' \\ \text{Sum} &= 360°00'30'' - 30'' = 360°00'00''; \textit{ check} \end{aligned}$$

5-11. Errors in Computed Quantities. The probable error of the *sum* of independent measurements $Q_1, Q_2, \ldots, Q_n$ for which the probable errors are $E_1, E_2, \ldots, E_n$, respectively is

$$E_s = \sqrt{E_1^2 + E_2^2 + \cdots + E_n^2} \tag{11}$$

If $E_1 \cdots E_n$ are all equal, this equation reduces to $E_s = E\sqrt{n}$.

The probable error of the *difference* between two independent measurements Q_1 and Q_2 for which the probable errors are E_1 and E_2, respectively, is

$$E_d = \sqrt{E_1^2 + E_2^2} \tag{12}$$

The probable error of a *product* of a constant or known quantity K and a measured quantity Q, for which the probable error is E, is

$$E_p = KE \tag{13}$$

If E_1 and E_2 represent, respectively, the probable errors of lengths L_1 and L_2, the probable error of the area representing the *product* of these two lengths is

$$E_a = \sqrt{L_1^2E_2^2 + L_2^2E_1^2} \tag{14}$$

5-12. Summary of Principal Relations. In each of the four cases which arise in practice, the most probable value is as shown in the following tabulation.

Measurements	Most probable value	
	Same quantity	Related quantities
Of equal reliability	Mean	Each observed value corrected equally
Of different reliability	Weighted mean	Each observed value corrected by an amount inversely proportional to its weight

The corresponding probable errors are as follows:

Measurements	Probable error of most probable value	
	Same quantity	Related quantities
Of equal reliability	$E_m = 0.6745\sqrt{\dfrac{\Sigma v^2}{n(n-1)}} = \dfrac{E}{\sqrt{n}}$	$E_s = E\sqrt{n}$
Of different reliability	$E_{wm} = 0.6745\sqrt{\dfrac{\Sigma(Wv^2)}{(\Sigma W)(n-1)}}$	$E_s = \sqrt{E_1^2 + E_2^2 + \cdots + E_n^2}$

Weights to be used in the adjustment of weighted observations are determined (1) as being proportional to the number of like observations of a given quantity ($W \propto n$); (2) as being inversely proportional to the square of corresponding probable errors ($W \propto 1/E^2$); or (3) by arbitrary assignment.

For the adjustment of weighted observations of related quantities, the corrections are taken as being inversely proportional to the corresponding weights ($C \propto 1/W$).

5-13. Numerical Problems.

1. The following values were observed in a series of rod readings under identical conditions. What is the most probable value? Its probable error? What is the probable error of a single measurement (*a*) as nearly as can be determined and (*b*) as determined by the various approximate relations?

ROD READINGS, FT.

3.187	3.181	3.186	3.181
3.182	3.184	3.183	3.188
3.179	3.176	3.178	3.179

2. Adjust the following angles measured at station *O*:

Angle	Observed value
AOB	46°14′45″
BOC	74°32′29″
COD	85°54′38″
AOD	206°41′28″

3. The interior angles of a triangle are observed to be: $A = 28°53'58''$, $B = 61°05'50''$, and $C = 90°00'00''$. What is the most probable value of each of these angles?

4. The difference in elevation between two points is determined to be 117.843 ft., by leveling over a route in which 18 set-ups are required. It is estimated that the probable error of the difference in elevation determined at each set-up is 0.003 ft. What is the probable error of the total difference in elevation?

5. The difference in elevation between two points is observed by three independent measurements, with results as follows:

214.38 ± 0.09 ft.
214.19 ± 0.06 ft.
213.86 ± 0.15 ft.

What is the most probable value of the difference in elevation? Its probable error?

6. Adjust the angles of problem 2 if weights of 6, 1, 3, and 5, respectively, are assigned to the four angles.

7. Adjust the angles of problem 3 if weights of 1, 1½, and 3, respectively, are assigned to angles *A*, *B*, and *C*.

8. A base line is measured in three sections with probable errors of ±0.014,

±0.022, and ±0.016 ft., respectively. What is the probable error of the total length?

9. The sides of a rectangular field are 1193.6 ± 0.6 and 582.7 ± 0.4 ft., respectively. What is the probable error of the computed area?

REFERENCES

1. American Society for Testing and Materials, "ASTM Manual on Presentation of Data," Philadelphia, 1940.
2. Franklin, W. S., "An Elementary Treatise on Precision of Measurements," Franklin and Charles, Lancaster, Pa., 1925.
3. Goodwin, H. M., "Precision of Measurements and Graphical Methods," McGraw-Hill Book Company, New York, 1920.
4. Irish, Sumner B., "A New Philosophy for Making Measurements," *Surveying and Mapping,* Vol. 15, No. 3, pp. 341–351, July–September, 1955.
5. Kissam, Philip, Chap. 21, "Probability," and Chap. 22, "Least Squares," in "Surveying for Civil Engineers," McGraw-Hill Book Company, New York, 1956.
6. Leland, O. M., "Practical Least Squares," McGraw-Hill Book Company, New York, 1921.
7. Wright, T. W., and J. F. Hayford, "The Adjustment of Observations," D. Van Nostrand Company, Inc., Princeton, N.J., 1906.

CHAPTER 6

MAP DRAFTING

6-1. The Drawings of Surveying. It is assumed that the student is familiar with the use of the ordinary drafting instruments and with the elements of mechanical drawing. Much of the drafting with which the student is here concerned calls for a degree of skill and precision of execution quite unnecessary on dimensional plans. The beginner should realize from the start that a consistent relation between the field measurements and the map requires great care in plotting.

The drawings of surveying consist of maps, profiles, cross-sections, and (to some extent) graphical calculations; the usefulness of these drawings is largely dependent upon the accuracy with which points and lines are projected on paper. For the most part, few dimensions are shown, and the person who makes use of the drawings must rely either upon distances as measured with a scale or upon angles as measured with a protractor.

6-2. Map Projection. Since the surface of the earth is curved and the surface of the map is a plane, no map can be made to represent a given territory without some distortion. If the area is small, the earth's surface may be regarded as plane, and a map constructed by orthographic projection, as in mechanical drawing, will represent the relative location of objects without measurable distortion. The maps of plane surveying are constructed in this manner, points being plotted either by rectangular coordinates or by horizontal angles and distances.

As the size of the territory increases, this method becomes inadequate, and various forms of projection are employed to minimize the effect of map distortion. The points of control are plotted by spherical coordinates through the use of elaborate geographic tables. Since the spherical coordinates of a point are its latitude and longitude, it is customary to show meridians and parallels on the finished map. The maps of states and countries, as well as those of some smaller areas, are constructed in this manner. The various methods of map projection are discussed in Chap. 32.

Recently state plane-coordinate systems have been devised whereby, even over large areas, points can be mapped accurately without the direct use of spherical coordinates (Art. 16-30).

6-3. Information Shown on Maps. Maps are classified variously according to their specific use or type, but in general either they become a part of public records of land division or they form the basis of a study for the works of man. In general, the following information should appear on any map:

a. The direction of the meridian.

b. A graphical scale of the map with a corresponding note stating the scale at which the map was drawn.

c. A legend or key to symbols other than the common conventional signs (Fig. 6-5*a* to *d*).

d. An appropriate title.

e. On topographic maps, a statement of the contour interval (Chap. 24).

In addition, a map that is to become a part of a public record of land division should contain the following information:

1. The length of each line.
2. The bearing of each line or the angle between intersecting lines.
3. The location of the tract with reference to established coordinate axes.
4. The number of each formal subdivision such as a section, block, or lot.
5. The location and kind of each monument set, with distances to reference marks.
6. The location and name of each road, stream, landmark, etc.
7. The names of all property owners, including owners of property adjacent to the tract mapped.
8. A full and continuous description of the boundaries of the tract by bearing and length of sides; and the area of the tract.
9. The witnessed signatures of those possessing title to the tract mapped; and if the tract is to be an addition to a town or city, a dedication of all streets and alleys to the use of the public.
10. A certification by the surveyor that the plat is correct to the best of his knowledge.

Explanatory notes may be employed to give such information as the sources of data for the map, the precision of the survey, or the datum to which the elevations are referred.

Maps made the basis of studies may or may not show the relief of the ground. In general, such maps show very few dimensions (often not any), the value of the map depending upon the correct representation of the location of features of the land, rather than directly upon field measurements or computed values. In addition to items *a* to *e* above, generally such a map shows streams and shore lines, roads and railroads, political boundaries, significant property lines and structures, and perhaps the condition and culture of the land. If the relief of the ground surface is not to be indicated, the map is called a *plan,* a *plat,* or a *planimetric map.* If the relief is to be shown, the map is called a *topographic map.* Relief is usually indicated by *contour lines* drawn for stated equal intervals of elevation, each contour line joining points of equal elevation (Chap. 24).

In connection with lawsuits regarding automobile or train collisions, falls, injuries during construction, and other accidents, the surveyor is sometimes called on to prepare a large-scale map for exhibit in the courtroom. While the map for this purpose should be extremely simple, it should include all details that might have a bearing on the accident. Some of these details may be the grade and crown of the roadway; height of curb; depressions; location (at time of accident) of obstructions to traffic or to view such as trees, poles, signs, and parked automobiles; sources of light (if at night); and location of points (in plan and elevation) from which it is stated that the accident has been witnessed. Colors are sometimes employed to make the various features more intelligible to the layman.

6-4. Scales. The scale of a map is the fixed relation that every distance on the map bears to the corresponding distance on the ground. The scale must be shown on the map because dimensions are not given (except for boundary lines on land maps). It may be stated either by numerical relations or graphically, as follows:

1. One inch on the drawing represents some whole number of tens, hundreds, or thousands of feet on the ground, as, 1 in. = 200 ft. This type is called the *engineer's scale.* It is used for most maps for construction purposes. In another form of this scale often used on land maps, a whole number of inches on the drawing represents 1 mile on the ground, as, 6 in. = 1 mi.

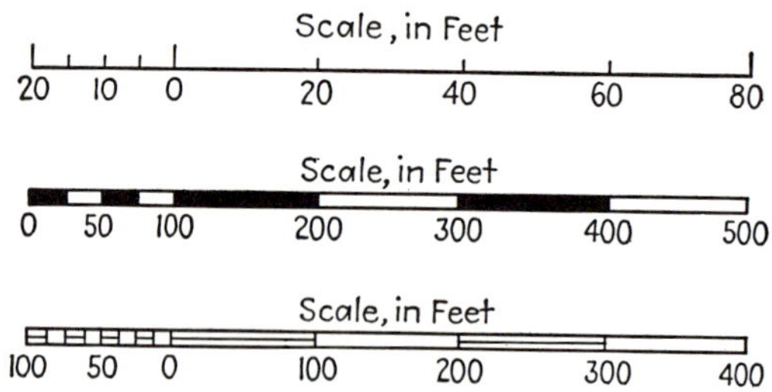

FIG. 6-1. Graphical scales.

2. One unit of length on the drawing represents a stated number of the *same* units of length on the ground, as 1/62,500. This ratio of map distance to corresponding ground distance is called the *representative fraction.* The scale is independent of the units of measurement. It is used extensively for geographic and military maps.

3. A *graphical scale* is a line subdivided into map distances corresponding to convenient units of length on the ground. Various forms of subdivision are shown in Fig. 6-1, in which the top line represents a scale of 1 in. = 40 ft. and the two lower scales represent 1 in. = 200 ft. In order to leave the main portion of the scale clear, sometimes a finely divided portion extends to the left from the zero of the main scale, as shown in the top and bottom scales of the figure. On a graphical scale the units of measurement should always be stated.

The numerical scales described above are subject to error if the drawing paper shrinks or swells, as often happens, but this error is not of consequence for many uses of the map. An important objection to the use of numerical scales alone, however, is that often maps are reproduced in other sizes by photographic means. If distances are to be determined accurately from the map, *a graphical scale should always be shown.* If, for convenience, a numerical scale is stated, it should be made clear that this is the scale at which the map was drawn or published; for example, "Original scale 1 in. = 200 ft." When a published map is a considerable enlargement, that fact should be stated on the enlargement.

The scale should be shown in or near the title of the map so that it will catch the eye readily.

As a general rule, the scale of a map should be no larger than is necessary to represent the location of details with the required precision. Maps for engineering projects have scales generally ranging from 1 in. = 20 ft. to 1 in. = 800 ft. Maps of land subdivisions have scales ranging from 6 in. = 1 mile to 1 in. = 1 mile. Geographic maps have scales of 1 in. = 1 mile to 1 in. = 20 miles or more. General topographic maps have scales ranging from 1/10,560 to 1/250,000.

For convenience in discussion, maps are herein divided arbitrarily into those of

Large scale: 1 in. = 100 ft. or less.
Intermediate scale: 1 in. = 100 ft. to 1 in. = 1,000 ft.
Small scale: 1 in. = 1,000 ft. or more.

6-5. Meridian Arrows. The direction of the meridian is indicated by a needle or feathered arrow pointing north, of sufficient length to be transferred with reasonable accuracy to any part of the map. The true meridian is usually represented by an arrow with *full* head; the magnetic meridian by an arrow with *half* head. When both are shown, the angle between them should be indicated. The general tendency is to make needles and arrows too large, blunt, and heavy. A simple design is shown in Fig. 6-2.

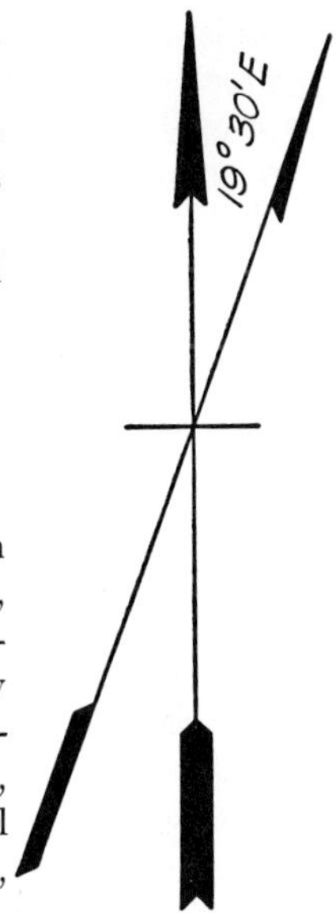

FIG. 6-2. Meridian arrows.

Preferably, the top of a map should represent north, although the shape of the area covered or the direction of some principal feature of a project may make another orientation preferable.

6-6. Profiles. Longitudinal sections made by projecting the ground line upon a vertical surface are known as *profiles* (Chap. 11). In conjunction with maps, they are of assistance to the engineer in fixing the grades and alinement of such works as sewers, railroads, highways, and canals. They are also of value in estimating volumes of earthwork. The data from which profiles are plotted consist of ground elevations at known distances apart along some line, as, for example, the center line of a highway. The ground profile is formed by a continuous line drawn through the plotted points. In addition to the ground profile there are also shown one or more grade profiles and other pertinent information. For example, the profile along the center line of a canal would show the ground line and the canal bed and probably also the water surface and top of bank.

Profiles are usually made on *profile paper*, obtainable in standard rulings. For each of these rulings, every fifth horizontal line and every tenth vertical line is accentuated by making it heavier than the intervening lines.

6-7. Cross-sections. The calculations of volumes of earthwork are frequently facilitated by plotting cross-sections of the earthwork to scale (Chap. 11). The area of a cross-section is then determined either by means of a planimeter or by dividing the figure into triangles and rectangles and computing the partial areas. The data for plotting consist of elevations and distances either measured in the field or computed.

Cross-sections may be shown on ordinary paper, but more often are plotted on cross-ruled paper called *cross-section paper*. The divisions are the same horizontally as vertically. The lines marking the half inches or inches are made heavier than the rest.

6-8. Lettering. Inasmuch as a drawing is likely to be judged by the quality of its lettering, it is important that the draftsman be able to form letters with at least a fair degree of skill and to assemble them in such form, size, and arrangement as to make the drawing clear and of pleasing appearance. In machine and structural drawing, simplicity and clearness are of primary importance; a considerable portion of the drawings made by the map draftsman require in addition a certain artistic quality. This requirement is particularly true of maps which are to be largely used by the public. The lettering should be of a style in keeping with the purpose of the drawing.

For *office drawings*, or drawings that are not to be used by the general public, the *Reinhardt* style of single-stroke freehand lettering is employed almost entirely in this country. The letters are constructed rapidly and are easy to read. Reinhardt letters are made either vertical (Fig. 6-3*b*) or inclined (Fig. 6-3*a*). Irregularities of lettering are not so apparent in the slope form as in the vertical form.

Variations of these forms in the matter of increasing or decreasing the horizontal dimensions of the letters (relative to the vertical dimensions) are frequently desirable. When the horizontal dimensions are reduced and the letters are close together, the lettering is said to be *compressed*. When the horizontal dimensions are elongated and the letters are some distance apart, the lettering is said to be *extended*. Frequently vertical and inclined lettering including the extended and compressed forms may be combined advantageously on a single drawing. Thus the names of streams might be shown in extended slope capitals, the names of streets in extended vertical capitals, the names of property owners in normal lower-case vertical letters, and notes in compressed lower-case inclined letters.

On drawings where their use is justified, *gothic, roman,* and *italic* letters may be employed. *Gothic* letters with a few exceptions are similar to vertical Reinhardt letters, but their lines are heavier. *Hair-line antique* lettering, a modification of the gothic style, is composed of very fine lines which make it more difficult to execute. *Roman* and *italic* letters are slow and difficult to construct because of the serifs and because many of the lines must be outlined and filled. Slight deviations in roman lettering catch the eye at once. Italics are inclined, and slight deviations in their form are not so noticeable.

In map drafting, where the details to be shown are many and are varied in character, it often renders the map clearer if a particular style of lettering

FIG. 6-3*a*. Reinhardt letters, slope form.

FIG. 6-3*b*. Reinhardt letters, vertical form.

is employed for each class of objects shown. Often the several styles just described might be employed on a single map. For example, the names of civil divisions might be shown in roman letters; the names of streams and lakes and other hydrographic features in italics; the names of mountains, valleys, and other land forms in vertical gothics; public works in inclined gothics; and marginal lettering in hair-line antique.

In general, letters should be drawn freehand but should be alined by means of guidelines and slope lines. Commercial devices are available for quickly and uniformly constructing these lines. Complete lettering guides are increasing in use because letters of regular form can be made quickly; if the letters are spaced properly, satisfactory lettering can be secured with these guides. Freehand and mechanical lettering should not be used on the same drawing.

Detailed information concerning lettering is to be found in textbooks on drawing, but it is perhaps appropriate here to offer a few suggestions on freehand lettering to the beginner.

1. Be sufficiently familiar with the construction of each letter so that its form will always appear the same.
2. It is very important that the slope of letters in a word, sentence, or paragraph be uniform. If this is accomplished, a good effect will be secured even though the separate letters may be faulty.
3. The inclination of slope letters should not be excessive. A common slope is 22°, or 2 in 5, from the vertical.
4. In order to avoid the appearance of "falling over on their faces," vertical letters should be sloped slightly backward, about 1 in 24.
5. Never seek to improve freehand letters by making straight portions by mechanical means.
6. Three common defects in the lettering of the beginner are: (*a*) letters of varying shape, (*b*) excessive spacing, and (*c*) the unequal or apparently unequal spacing of letters as they appear in words.
7. Avoid sharp angles in the rounded portions of letters. The curves should be smooth.
8. If spacing is important, do all lettering in pencil as neatly as possible before inking in.
9. Make all the elements of Reinhardt letters by single strokes. Use a pen that will produce a line of the required weight at the first stroke.
10. Make the letters of a size in keeping with their purpose. The names of the larger or more important objects should catch the eye quickly; notations concerning relatively unimportant details should be inconspicuous.
11. In lettering drawings which are to be reproduced to a reduced scale, make the size and weight of the letters conform to the requirements of the process of reproduction.
12. Leave a generous interval between the letters forming the names of elongated or large objects such as streams, streets, lakes, mountain ranges, counties, and railroads.

6-9. Titles. Titles should be so constructed that they will readily catch the eye. The best position for the title is the lower right-hand corner of the sheet, except where the shape of the map makes it advantageous to locate the title elsewhere. The space occupied by the title should be in proportion to the size of the map; the general tendency is to make the title too large. In general each line should be centered, and the distance between lines should be such that the title as a whole will appear well balanced. The different parts should be weighted in order of their importance, beginning with the principal object of the drawing or the name of the area. Only the common styles of letters should be used. A change in the style of lettering between different parts is permissible to accentuate the important parts of the title, but slope letters and vertical letters should not be included in the same title. For the two general types of maps, the items to be included in the title are stated in Art. 6-3. A simple form of title is shown in Fig. 6-4. Revisions of the map should be shown by dated notes at the left of the title.

6-10. Notes and Legends. Explanatory notes or legends are often of assistance in interpreting a drawing. They should be as brief as circumstances will allow, but at the same time should include sufficient information as to leave no doubt in the mind of the person using the drawing. A key to the symbols representing various details ought to be shown unless the symbols are conventional. The nature and source of data upon which the drawing is based ought sometimes to be made known. For example, the data for a map may be obtained from several sources, perhaps partly from old maps, partly from old survey notes, and partly from new surveys; the surveys have been made with a certain precision; the direction of the

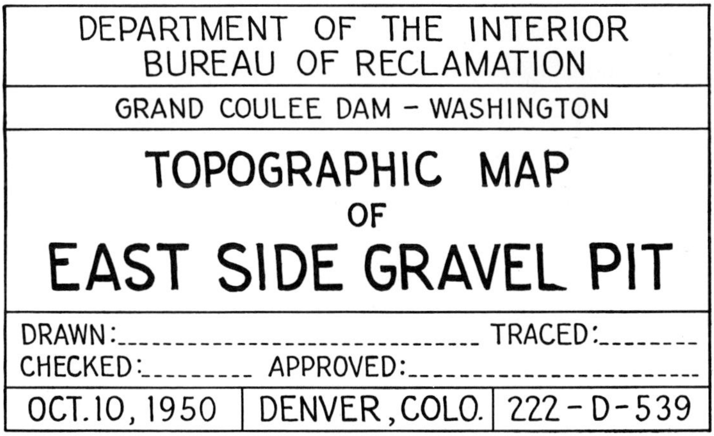

Fig. 6-4. Title for map.

meridian has been determined by astronomical observation; and elevations are referred to a certain datum as indicated by a certain bench mark of a previous survey.

Notes ought to be in such a position on the drawing as to catch the eye readily, conditions allowing. A favorable position is the lower portion to the left of the title.

6-11. Symbols. Objects are represented on a map by *symbols*, many of which are conventional. A symbol is a diagram, design, letter, or abbreviation which by convention or reference to a legend is understood to represent a specific characteristic or object. Standard symbols have been adopted for the United States, and these symbols are commonly used by other organizations. Some topographic map symbols published by the U.S. Geological Survey are shown in Fig. 6-5*a* to *d*. A chart showing these symbols is available from the Survey, without charge.

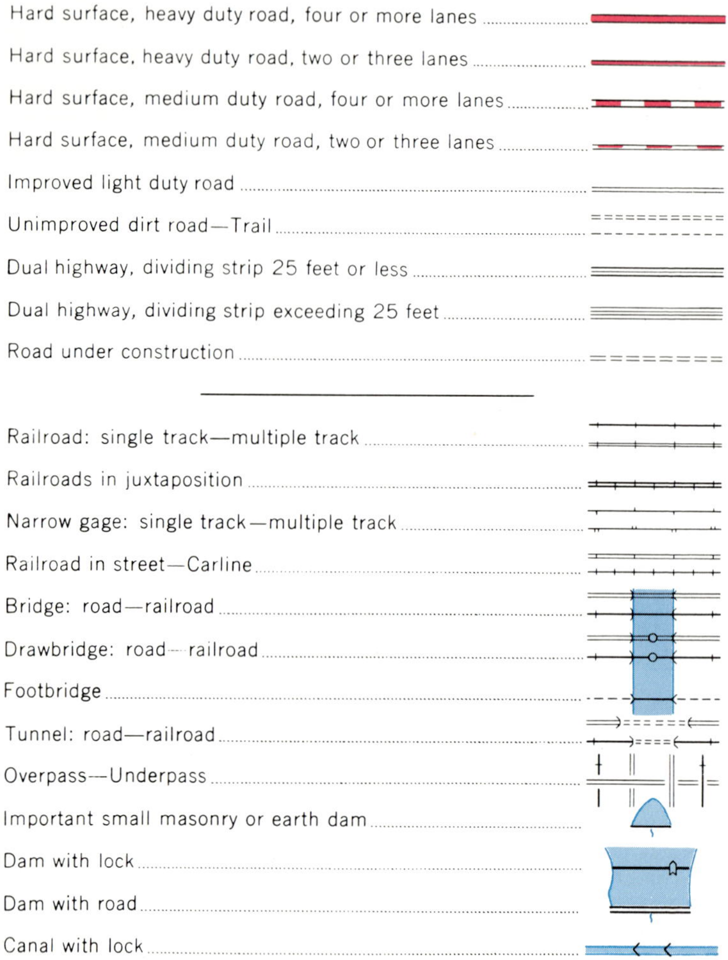

FIG. 6-5*a*. Symbols for roads, railroads, and dams.

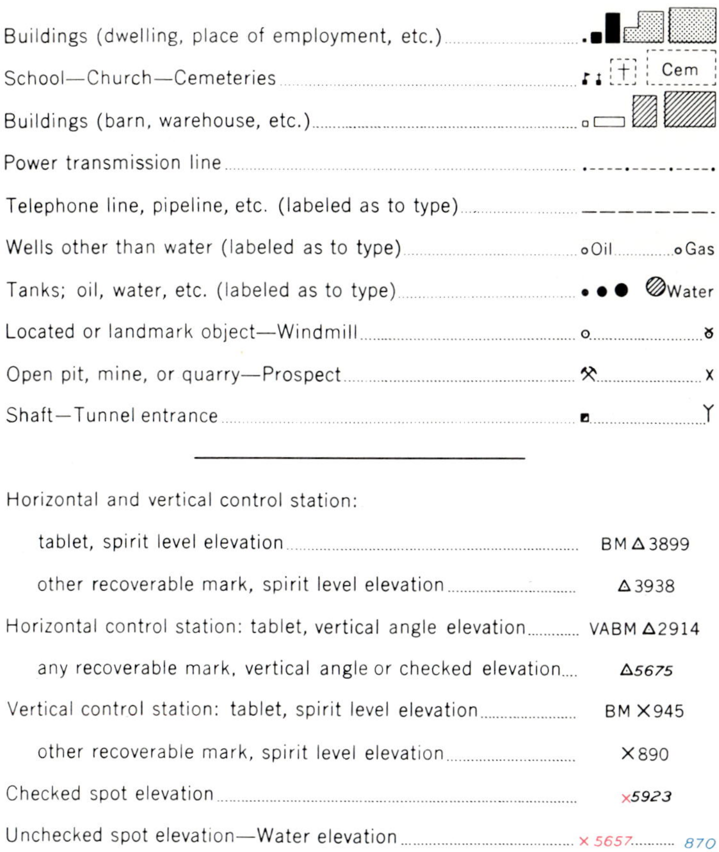

FIG. 6-5*b*. **Symbols for structures and stations.**

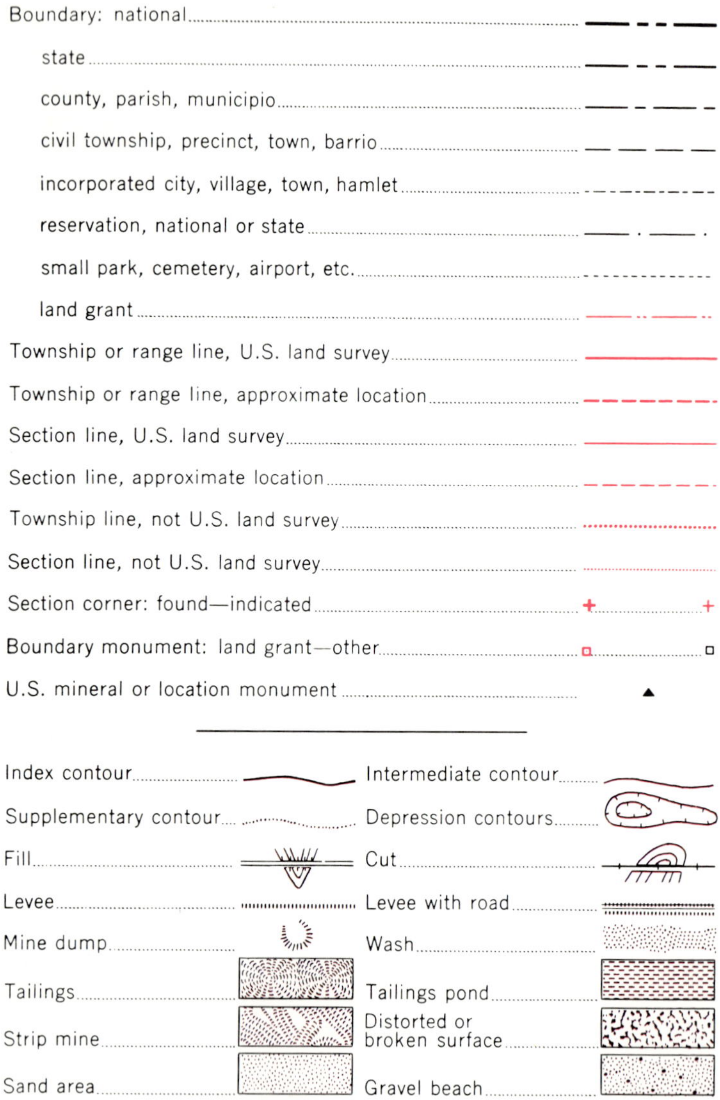

FIG. 6-5*c*. Symbols for boundaries, contours, and excavations.

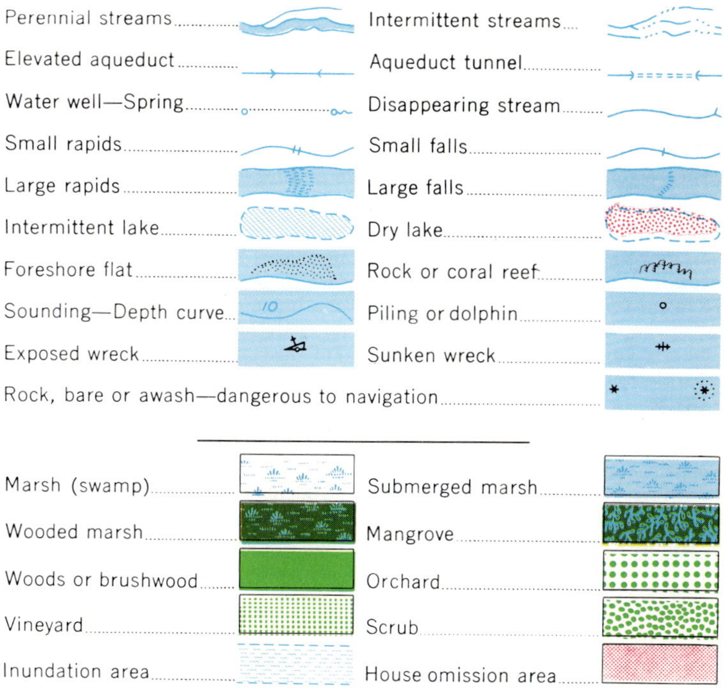

Fig. 6-5*d*. Symbols for hydrography and land classification.

Where feasible, the information shown on topographic maps is distinguished by colors in which the symbols are printed. Black is used for man-made or cultural features such as roads, buildings, names, and boundaries. Blue is used for water or hydrographic features such as lakes, rivers, canals, and glaciers. Brown is used to show the relief or configuration of the ground surface as portrayed by contours or hachures. Green is used for wooded or other vegetative cover, with typical patterns to show such features as scrub, vineyards, or orchards. Red emphasizes important roads and public-land subdivision lines and shows built-up urban areas. Where colors are not available or where the map is to be reproduced by contact or photographic printing and will therefore be in one color, black is used throughout; the conventional shapes of the symbols are the same as where color is used.

For many purposes, it would lessen the usefulness of the map if all the objects were shown for which symbols are available. The size of the symbols should be proportioned somewhat to the scale of the map.

6-12. Drawing Symbols. *Grass.* The methods described in this article are illustrated in Fig. 6-6. The symbol for grass consists of a series of lines drawn radially toward a point. The tops of the lines begin on an arc of a circle, and the bottoms of the lines terminate on a ground line parallel with the bottom border line of the map. The symbol may be composed of three, five, or seven lines or blades. The separate symbols are distributed evenly over the area, but they should be irregularly spaced so as not to give the appearance of rows.

Marsh. The symbol for marsh consists of the grass symbol beneath which the water surface is shown by either a single or a double line drawn slightly longer than the base of the grass tuft (Fig. 6-6).

Trees. Tree symbols may be drawn either in plan or in elevation. The latter practice is better adapted to reconnaissance sketches or elevation drawings of a terrain, whereas on most topographic drawings the symbol shown in plan is more suitable. It is common practice to differentiate between deciduous and evergreen trees.

The symbol in plan for *deciduous* trees is executed by first drawing an outline as a scalloped, broken line to represent the two or three main branches of a tree. The inside area is then sparingly filled in with small scalloped, broken lines. The symbol may be shaded as shown. In elevation the tree symbol is shown as a fairly even, symmetrical, scalloped outline, beneath which the trunk is represented by a heavy vertical line. Again, the symbol is shaded, and the shadow of the tree on the ground may also be sketched.

To represent *evergreen* trees in plan, the symbol is drawn as bold lines radiating from a central point. The separate symbols should each be composed of five or six lines and should be fairly symmetrical and uniform in shape. The representation in elevation is drawn as a series of closely spaced horizontal lines beginning with a dot at the top and gradually increasing in length toward the bottom. Beneath the last of these lines the trunk is shown as a heavy vertical line. The area in shadow may be sketched on the map.

The size of tree symbols is varied on maps of different scales. On large-scale maps, if many trees are to be drawn, the symbol obscures other features. Hence on such maps the outline only is drawn; or in some cases the trunk only may be indicated as a dot, and the diameter of the trunk and the kind of tree may be recorded beside it, as for example, 20-in. maple. On intermediate-scale maps the symbols can be made of size to show very nearly the horizontal projection of the tree represented, but on small-scale maps no attempt is made to draw the symbols to scale. If a large area is to be shown as covered with forest, or if many other details are to be shown on the map, a flat tint or uniform shading should be used.

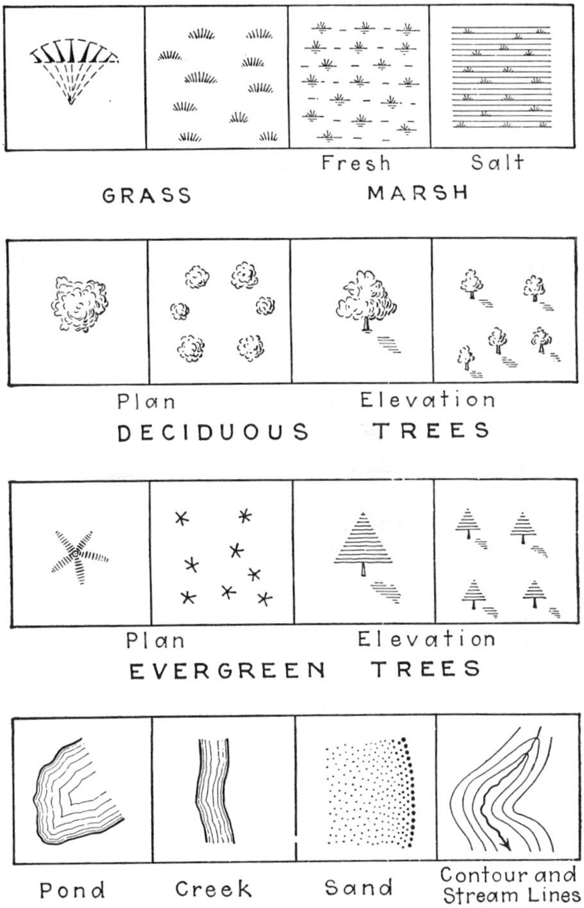

FIG. 6-6. Details of map symbols.

Water Lines. To draw the water-line symbol for lakes and ponds, the draftsman begins by sketching the shore line as a heavy line, then drawing a fine line as close as possible to the shore line. Fine lines are then drawn successively, each in close conformity with the preceding line, and the spacing between the lines is increased uniformly outward from the shore. The conformity between adjacent lines is effected by drawing each line so that it makes a series of smooth, intersecting arcs of curves, all of which arcs are drawn concave toward the shore. The excellence of the total effect depends upon the regularity and the rate with which the spacing increases from the shore outward, the uniformity of the space between two given lines, and the smoothness of the arcs. The lines may be drawn to fill in the water area completely, or if the space is large it may be left without lines near the center.

For rivers the method is the same as that for lakes, but for small streams the lines in the center may be broken occasionally to heighten the effect of an open water surface. The width of stream is drawn to scale, as nearly as may be.

Sand. A sand bar or flat is represented by dots evenly spaced over the area. The edge of the flat or the shore lines is indicated by a line of closely spaced, relatively large dots. Just behind this row, a line of dots, smaller in size, is drawn, each dot being placed in a space between two dots on the first line. A third row of dots still smaller in size is placed behind the second row. Beyond that line the appearance of rows should be avoided and the remaining area filled in with dots spaced more or less uniformly but being farther apart as the distance from the shore line increases.

Contour Lines. Contour lines are drawn as fine, smooth, freehand lines of uniform width (Chap. 24). Each fifth line is weighted slightly heavier than the others to facilitate the reading of elevations on the map. If the contours are spaced closely on the map only the fifth contour lines need be numbered; but on areas of low relief where the contour lines are spaced widely apart each line may be numbered with its elevation. The line is broken to leave a space for the number. So far as practical, the numbers are made to be read from the bottom and right-hand side of the map; some organizations require that the numbers be faced to read uphill. On large-scale maps where the contour lines extend over large areas, it is difficult to maintain an even weight of line unless a contour pen (Fig. 6-14) is used.

DRAFTING PAPERS AND REPRODUCTION MEDIA

6-13. Drafting Papers. Pencil drawings and temporary drawings are often made on a smooth manila *detail paper*, of which there are several grades and weights. For general map work a fairly smooth, tough *drawing paper* of uniform texture is desirable. The paper should take ink well and should stand erasures without its surface becoming fibrous. For permanent drawings a paper should be chosen that will not discolor or become brittle with age. Drawings to be subjected to hard usage may be constructed on paper that is mounted on muslin. Plane-table sheets may be mounted on aluminum in order to prevent shrinkage.

In consideration of the importance of the map and of the slight expense of the paper in relation to the survey as a whole, it may be considered false economy to use any but high-grade papers for mapping purposes.

6-14. Tracing Paper and Cloth. A tracing is a drawing in ink or pencil on a transparent sheet of paper or cloth for the purpose of reproduction.

Tracing paper comes in several grades, all suitable for pencil drawings. The better grades, usually processed, are also suitable for ink drawings. Ordinary tracing papers will not stand repeated erasures well and will become torn and cracked unless they are handled carefully, but improved tracing papers are available.

Original pencil drawings on tracing paper have recently come into common use for many purposes, including reproduction by photography. Clearer contact prints can be made by using oiled tracing paper, but oiling the paper reduces its toughness. Especially transparent tracing papers are available.

Tracing cloth is made from fine starch-filled cotton or linen cloth specially treated to render it firm, transparent, and smooth. It is used for permanent drawings or for drawings which will be subject to considerable handling. The *glazed* or *smooth* side is seldom used, although it will take ink and stands erasures well; the unglazed or *rough* side is preferred by most draftsmen. Pencil drawings on tracing cloth become smudged easily and are seldom made. Good grades of tracing cloth will not deteriorate with age, but the conventional tracing cloth turns white and wrinkles when touched by water. Waterproofed tracing cloth is available.

Preparatory to making a tracing in ink on cloth, the surface is dusted with powdered talc or chalk and is rubbed with a dry cloth; any excess powder is removed.

Erasures of ink on cloth are made with least damage to the surface by rubbing lightly with a soft pencil eraser, using an erasing shield. Ordinary ink erasers are too abrasive and produce a fibrous surface which does not take ink well. Some draftsmen employ a sharp knife to scrape or pick off the ink on the surface, then use the eraser to remove the ink impregnating the fibers of the cloth. A rough erased area can be smoothed and reconditioned by being rubbed with soapstone and then powdered as described above. Pencil lines are removed and the tracing is cleaned by rubbing either with artgum or with a cloth saturated in cleaning fluid.

6-15. Film. Recently polyester film has come into use for drawings and tracings because of its high transparency, dimensional stability, tearing strength, and resistance to heat and age. It is insoluble and waterproof. It takes either pencil or ink work, and either can be erased easily and clearly. It can be coated with semi-opaque material or in various colors for scribing, a process of forming lines with a sharp-pointed instrument which scratches through the coating. The points used in scribing are similar to phonograph needles, with steel or sapphire tips, and are available in various widths; they are fastened to handles similar to penholders.

6-16. Reproduction of Drawings. A drawing may be reproduced at the same scale by making a *contact print* on processed paper, resulting in a *blueprint*, a *brownprint* (vandyke), or a *direct blackline print*. By direct printing or by reprinting from a vandyke contact negative, any of these prints may be made with white lines on a dark background or with dark lines on a white background.

A drawing may be reproduced in black either at the same scale or at different scales (enlarged or reduced), from either drawings or tracings, by various photographic processes such as the *photostat* process, the *photo-offset* methods, *xerography*, or other methods. One advantage of reproducing drawings at a reduced scale is that they are easier to handle in the field, as on construction work. The California Division of Highways makes extensive use of photo-offset plans at one-half the scale at which they were drawn. Usually maps and other drawings for general distribution are either lithographed or printed from etchings.

Only the common methods of reproduction will be discussed herein.

6-17. Blueprints. The most common method of reproduction at the same scale is that of making from the tracing a blueprint, in which white lines appear on a blue background. To produce blue lines on a white background, the method herein described is followed, except that a vandyke negative (see Art. 6-18) is employed instead of the tracing. A blueprint is made by placing the inked side of a tracing next to a sheet of glass, placing the sensitized side of processed paper or cloth next to the tracing, exposing this side to light, and developing the exposed sheet in a bath of water.

Most blueprints are made on paper, but those likely to be subjected to rough handling are sometimes made on a sized cloth. What is said herein regarding paper applies also to cloth. The fresh, unexposed paper is a pale greenish yellow. In a moist climate, unless carefully protected the paper soon takes on a bluish tinge and becomes unfit for use even though unexposed to light. When the paper is handled, it should be protected from atmospheric moisture and from exposure to light except during the period of actual printing.

Although manufactured blueprint paper is so economical that ordinarily no one would consider making his own, emergencies may arise when a knowledge of the process of manufacture is of value. A satisfactory blueprint paper can be prepared by applying to paper a mixture of equal parts of the following solutions:

1 part (by weight) of prussiate of potash to 5 parts of water.

1 part (by weight) of citrate of iron and ammonia to 5 parts of water.

Either of the two solutions may be prepared in sunlight, but they should be combined and applied to the paper in a dark room or in a subdued light. The paper is placed on a flat surface and the sensitizing solution is spread with a sponge, cloth, or camel's-hair brush, employing long strokes first lengthwise and then crosswise of the sheet. Only enough of the solution is applied to wet the surface of the sheet. The paper is dried in a dark place and is ready for use.

Blueprints are made by exposure either to sunlight or to artificial light. The electric blueprinting machine is a part of the equipment of many large offices. In most cities are firms who specialize in the making of blueprints, and most engineers and surveyors having a limited amount of work find it more economical and more satisfactory to have their blueprints made by commercial firms.

The proper time of exposure depends upon the intensity of the light and upon the quality of the paper and is best determined by trial with small pieces of the paper. The rapid papers which are commonly used require an exposure of ½ to 1 min. in strong sunlight; others require an exposure of 2 to 3 min. On cloudy days, the time of exposure is longer. The sensitizing formula given herein will produce a slow paper, for which the necessary time of exposure to sunlight is 5 to 6 min. A larger proportion of citrate will make a faster paper. If underexposed, the body of the print when washed will be a pale blue; if overexposed or "burned," it will be a dark mottled blue, and the fine lines of the drawing will be indistinct or missing. During the period of exposure the sheet should be square to the rays of light.

An exposed print is developed by washing it in water, preferably in subdued light. To avoid streaks and indistinct lines, the print is quickly submerged and is agitated until the greenish tinge has disappeared. The print is then allowed to soak for 10 to 30 min. The blue of the print will be intensified by the addition to the bath of a small quantity of potassium dichromate; this solution also tends to overcome the results of overexposure. A commercial "no potash" paper is available which produces a deep blue color when washed in water only.

After being thoroughly washed and soaked, the print is removed from the bath and is hung to dry in a subdued light. Wrinkles may be removed by ironing.

Fig. 6-7. Blueprint frame.

Alterations on blueprints can be made with a weak solution of caustic soda. This is used as an ink to produce white lines. If a colored line is desired, the solution may be mixed with ink. Several solutions of this nature are on the market in bottled form and are known as *erasing fluids*. Alterations on blueprints may also be made by using "salts of sorrel" (potassium acid oxalate, a poison) as an ink. Special inks for alteration of blueprints are available in white, red, and yellow.

A blueprint frame of common design is shown in Fig. 6-7. The glass should be clear, preferably a fairly heavy plate. A thick felt or pneumatic pad is placed next to the paper. The spring clamps force the hinged back against the pad and insure close contact between tracing and paper.

6-18. Brownprints. A print made from a tracing on processed *vandyke* paper has white lines on a dark-brown background, which is nearly impervious to light. The color of the fresh paper is light yellow. The operation of exposing the paper is the same as in blueprinting, but the necessary time of exposure is considerably greater, usually about 5 min. in strong sunlight. The exposure is sufficient when the paper protruding beyond the tracing takes on a rich brown color. The exposed sheet is washed for about 5 min. in water (in subdued light), becoming lighter in color. It is then transferred to a fixing bath consisting of 2 oz. of hyposulphate of soda to 1 gal. of water. When the print takes on a deep brown, it is again washed thoroughly in clear water and is left to soak for 20 to 30 min.

The principal use of the vandyke process is in the making of a negative from which blueline positive prints or other contact prints may be made. To render the white lines more nearly transparent, the vandyke negative may be sponged with a banana-oil compound, but this compound should be thoroughly evaporated from the surface before blueprinting is attempted. Blueline positive prints are then made by placing the brown surface of the vandyke negative next to the sensitized sheet of blueprint paper, and by exposing and washing in the usual manner. Brownline vandyke prints and blackline prints are similarly made.

Prints with a white background are clearer than those with a dark background, and additional notations stand out well. As in the case of blueprints, alterations are evident because erasure of the lines damages the paper.

6-19. Blackline Prints. A blackline contact print on transparent paper, called a *blackline tracing,* is made from a vandyke negative by a process similar to that for blueline prints. Blackline tracings shrink during the process of manufacture, and the scale is thereby altered appreciably. However, they are economical and are useful for preliminary working drawings from which prints are to be obtained.

A *direct blackline print* is made from the tracing by contact printing in sunlight or artificial light, using a special sensitized paper. The print is developed by applying a chemical furnished by the manufacturer, after which it is thoroughly washed in water and is dried. This type of print is increasing in use, as it has the advantages of printing without a negative, a white background, freedom from excessive shrinkage, and difficulty of alteration without detection.

Blackline reproductions by other processes are described in Arts. 6-20 to 6-22.

6-20. Ozalid Prints. An *ozalid* or *diazo* print with red, blue, brown, or black lines is made on a special sensitized paper by direct contact printing in the usual manner. In the dry process, it is developed by being placed in a tight container, which is then filled with ammonia vapor. In the wet process, the exposed sensitized sheet is passed through a liquid ammonia solution.

6-21. Photostat Process. A drawing on any kind of paper or cloth may be reproduced to any scale by the photostat process, provided the lines are of a color which photographs well. The process is widely used, especially in the reproduction of pages from books.

The photostat machine is a modified form of camera. The drawing is strongly illuminated by artificial light, and a negative to the desired scale is made, in which black lines of the original appear white. By rephotographing, a positive is produced in which black lines of the original appear black on a white or gray background.

Large reproductions are considerably distorted near the edges, but in the usual sizes the distortion is not great. Reproductions up to 40 by 60 in. may be made by this process.

Photostat reproduction of a blueprint may be made in one operation, using the blueprint as the negative.

6-22. Photo-offset Process. The photo-offset process, known by various trade names, is useful when many prints of a drawing are required. This process consists in making a negative to the desired scale by photographing the original; from this negative a plate is prepared and mounted for use in an offset type of printing press. The prints may be made on any good grade of bond paper; they have distinct black lines. For large quantities the cost is low.

MAP DRAFTING INSTRUMENTS

6-23. Pencils. For drawings on hard, fine-grained papers the 6H pencil is widely used. For very thin lines and for permanent work the 8H is occasionally employed. Many drawings on smooth papers are made with the 4H pencil. On the soft profile and cross-section papers, lines made with a 2H pencil show up well. For coarse tracings made on tracing paper a pencil as soft as 2B is sometimes necessary. Drawings made directly on the rough side of tracing cloth will show up sufficiently well for inking if a 2H pencil is used. Lines that are to be traced should in any case be made sufficiently heavy to show clearly through the tracing cloth. Many draftsmen sharpen one end of the pencil to a wedge point for use in drawing straight lines and the other end to a conical point for sketching irregular lines and for lettering. For any but simple sketches of a few lines, care should be taken to choose a pencil which will not smudge readily.

6-24. Inks. The bottled inks commonly used by the map draftsman are black, brown, blue, green, and vermilion. For line drawings, the waterproof inks are satisfactory. Lines made with them will not smudge when rubbed with the moist hand and are not affected by the application of liquid tints or washes. Stick India inks are not much used but are preferred by some draftsmen for intricate drawings; they are prepared for use by being ground and mixed with water. Inks may be thinned by adding distilled water or a dilute solution of ammonia.

6-25. Scales. *Engineer's scales* are divided into 10, 20, 30, 40, 50, 60, 80, or 100 parts to the inch. Rules thus divided are either flat or triangular in shape and are obtainable in various lengths, commonly 6 and 12 in. The 12-in. triangular boxwood rule with 10, 20, 30, 40, 50, and 60-ft. scales on its three faces (Fig. 6-8) is most commonly used in mapping. It has the advantage of compactness.

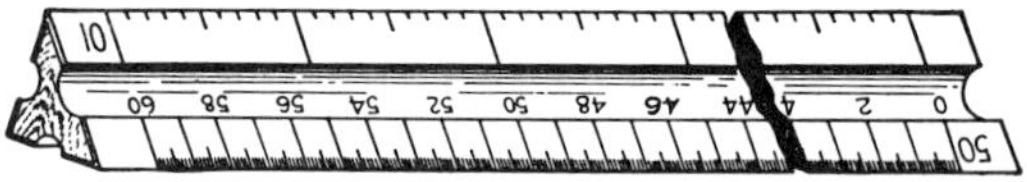

FIG. 6-8. Engineer's scale, triangular.

The flat rule with two scales on edges of opposite bevel (Fig. 6-9) is most satisfactory to use. Mistakes in plotting or in scaling distances through using the wrong scale are much less likely to occur than with the triangular rule.

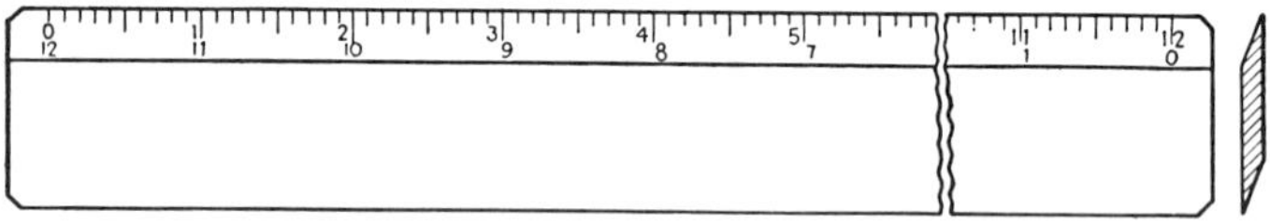

FIG. 6-9. Engineer's scale, flat with opposite levels.

The scale graduated to $\frac{1}{10}$ in. is intended for measuring to the scale of 10, 100, or 1,000 ft. to the inch, but the divisions are so large that it cannot be used for precise plotting. For such work it is better to use the scale divided into 50 parts to the inch. Probably $\frac{1}{100}$ in. is as close as distances can be plotted by the ordinary methods of drafting.

For precise drafting, points should be pricked with a fine needle, a reading glass should be employed, and distances should be plotted with the eye directly above the graduation to which the distance is measured. Under these conditions, points can be plotted to $\frac{1}{200}$ in. A special scale is available which has a sliding micrometer at one end, reading to 0.001 in.

6-26. Protractors. The protractor is a device for laying off and measuring angles on drawings. The usual form for mapping consists of a full circle or semicircular arc of metal, plastic, or paper graduated in degrees or fractions of a degree. Some have radial scales by means of which a distance and an angle may be plotted at one operation. Others have one or more radial arms with scales.

Figure 6-10 illustrates the most common form of semicircular metal or celluloid protractor. To lay off an angle, the center O of the protractor is placed at the vertex of the angle, with the edge of the bar coinciding with the line to which the angle is referred. A mark is then made on the drawing at the proper graduation of the protractor arc, the protractor is removed, and a line is drawn joining this mark with the vertex. An angle is measured in a similar manner.

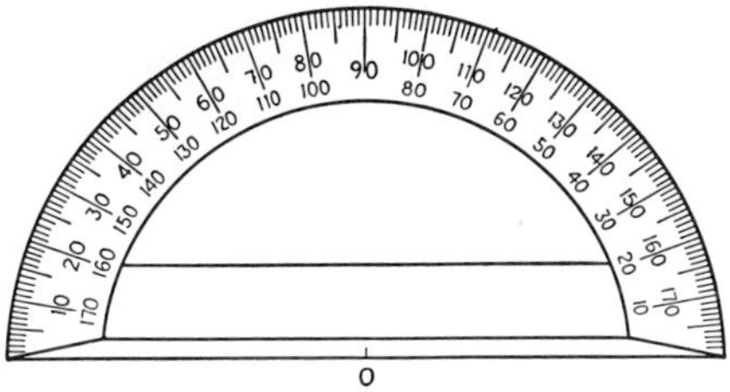

FIG. 6-10. Semicircular protractor.

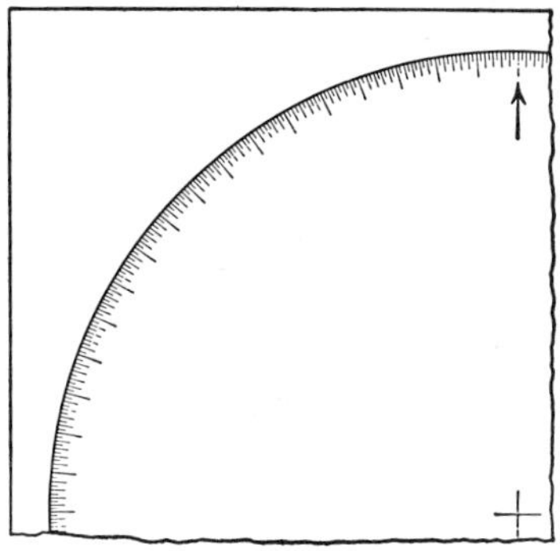

FIG. 6-11. Full-circle paper protractor.

Protractors with full circles are also in common use. Figure 6-11 shows part of a full-circle paper protractor. These are usually printed in 8 and 14-in. sizes on rectangular sheets of tough paper or bristol board, without the graduations being numbered. The 14-in. size is graduated to ¼ degrees and the 8-in. size to ½ degrees. To prepare such a protractor for use, its graduations are numbered as desired and it is cut either on a circle passing through the outer ends of the graduations or on a circle of somewhat smaller radius. In the former case the outer portion is discarded and angles are laid off by using the inner portion in the manner described for the semicircular metal protractor. In the latter case the inner portion is discarded and angles are laid off by means of a straightedge passing through the center of the protractor. The protractor is centered and oriented by means of two intersecting lines on the drawing, one line being at right angles to the other.

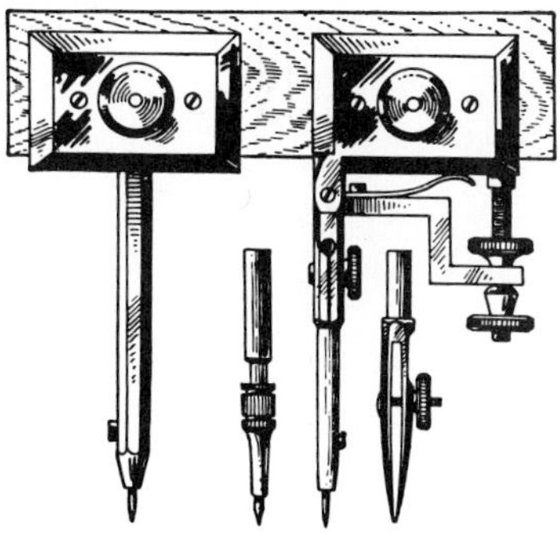

FIG. 6-12. Beam compass.

6-27. Beam Compass. This compass, illustrated in Fig. 6-12, is used for drawing the arcs of large circles, say of radius greater than 6 in. For the precise drawing of circles having radii greater than 6 in., the beam compass should be used.

A beam compass may easily be improvised from any thin strip of wood by driving a needle through the strip near one end and cutting a V-shaped notch in the edge of the strip at the required distance from the needle point. The point of a pencil or ruling pen is held in the notch, and the arc is drawn as with the regular beam compass. A stretched line or wire may be used in a similar manner.

6-28. Railroad Curves. These are thin strips of cardboard, wood, metal, hard rubber, or plastic, the edges of which are arcs of circles. A number on each curve indicates its radius in inches, and sometimes also an additional number indicates the degree of curvature for a given scale. With these curves, arcs of circles can be drawn without determining the center, and the arcs can be drawn with much larger radius than could be used with a beam compass.

FIG. 6-13. Road pen.

FIG. 6-14. Contour pen.

6-29. Road Pen. This pen, also called the *railroad pen,* is used principally for drawing two parallel lines either freehand or by means of a straightedge or curve. It consists of two ruling pens with spring shanks attached to a handle, the distance between the two pens being controlled by a screw passing through the shanks (Fig. 6-13). Its use greatly facilitates the drawing of parallel lines which are curved or irregular.

6-30. Contour Pen. This pen is useful for drawing contours or other freehand curves. The pen is connected rigidly to a shaft which turns freely in the handle (Fig. 6-14). The point of the pen is eccentric with the axis of the shaft so that the pen will turn in whatever direction it is being drawn on the paper. In use, the handle is held vertical, the fingernails of the third and fourth fingers of the right hand being in contact with the paper. The line is preferably drawn toward the draftsman.

6-31. Straightedge. For the drawings with which the surveyor is chiefly concerned, the T-square will not produce parallel lines with sufficient precision. Moreover, in map drafting the lines are seldom perpendicular or

FIG. 6-15. Drafting machine.

parallel to one another. The most satisfactory form of straightedge for general office use is of nickel-plated or stainless steel, with one edge beveled. Such a straightedge will lie flat on the drawing, its weight makes it less easily displaced than are those of wood, and it will not warp. For all-round use, the 42-in. length is satisfactory.

6-32. Proportional Dividers. For transferring distances from one map to another at a different scale, proportional dividers are useful. They consist of two legs, each pointed at both ends, which are held together by means of a central pivot. When the legs are opened in the form of an X, either end of the instrument forms a pair of ordinary dividers. The position of the pivot along the legs can be varied to produce any desired ratio of the distance between one pair of points to the distance between the other pair of points. Graduations along the legs facilitate the setting of the pivot to the desired ratio.

6-33. Drafting Machine. The drafting machine, one form of which is shown in Fig. 6-15, is increasing in use for map drafting, particularly for plotting details (Art. 18-19). It combines the functions of the T-square, straightedge, triangles, scales, and protractor.

Essentially, the drafting machine shown consists of a mechanical linkage bearing a pivoted *protractor head* to which are attached two mutually perpendicular graduated *arms*. The arrangement is such that each arm remains parallel to its original direction as the protractor head is moved about on the drafting table; thus, for example, a reference system of rectangular coordinates can be established on the drafting sheet by drawing lines along the edges of the arms. Further, the protractor head can be oriented quickly in any desired direction and clamped, thus enabling lines to be drawn or measurements to be made in oblique directions. By means of a vernier the protractor can be set to 05′ or, on some machines, to 01′. The arms are removable in order that engineer's scales of various graduations and lengths may be employed.

REFERENCES

1. French, Thomas E., and Charles J. Vierck, "A Manual of Engineering Drawing for Students and Draftsmen," 9th ed., McGraw-Hill Book Company, New York, 1960.
2. Sloane, R. C., and J. M. Montz, "Elements of Topographic Drawing," 2d ed., McGraw-Hill Book Company, New York, 1943.

CHAPTER 7

DISTANCE; TAPING

7-1. Distance. In surveying, the distance between two points is understood to mean the *horizontal* distance, regardless of the relative elevation of the points. Though frequently slope distances are measured, they are reduced to their equivalent horizontal projection for use. In geodetic surveying, horizontal distance is reduced to the equivalent at sea level (Art. 16-18). In plane surveying, horizontal distances are reduced to sea level only when it is desired to convert them into equivalent distances at another elevation, such as that of the state plane-coordinate system or the average elevation of a survey for which the variation in elevation over the area is large.

Various methods of determining distance are useful, depending upon the degree of precision required, the cost, and other conditions. On rough reconnaissance, for example, a precision of 1/100 or less may be sufficient for the purpose of the survey; on the other hand, certain base lines established by the U.S. Coast and Geodetic Survey have been measured with a probable error of about 1/2,000,000. On some surveys, a combination of methods may be used to advantage.

Table 7-1 classifies the principal methods of measuring distance according to the usual degree of precision obtained. The various methods are discussed further in the following articles.

METHODS OF MEASUREMENT

7-2. Pacing. Pacing furnishes a rapid means of approximately checking more precise measurements of distance. It is used on reconnaissance surveys and, in small-scale mapping, for locating details and traversing with the plane table. Pacing over rough country has furnished a precision of 1/200; under average conditions, a person of experience will have little difficulty in pacing with a precision of 1/100.

Either the natural pace (about 2¾ ft.), a 2½-ft. pace, or a 3-ft. pace may be employed, according to the preference of the surveyor. The authors favor the 2½-ft. pace. Since it is a little less than the natural step, allowance can be made for uneven ground by lengthening the pace without tiring, and a convenient relation exists between the pace and the foot; that is, 40 paces = 100 ft.

Each two paces or double step is called a *stride*. Thus, for the 2½-ft. pace the stride would be 5 ft., or there would be roughly 1,000 strides per mile.

TABLE 7-1. METHODS OF MEASURING DISTANCE

Method	Usual precision	Use	Instrument for measuring angles with corresponding precision
Pacing	1/100 to 1/200	Reconnaissance, small-scale mapping, checking tape measurements	Hand compass, peep-sight alidade
Stadia	1/300 to 1/1,000	Location of details for mapping, rough traverses, checking more precise measurements	Transit, telescopic alidade of plane table, surveyor's compass
Ordinary taping	1/1,000 to 1/5,000	Traverses for land surveys and for control of route and topographic surveys, ordinary construction work	Transit (angles doubled)
Precision taping	1/10,000 to 1/30,000	Traverses for city surveys, base lines for triangulation of intermediate precision, precise construction work	Transit (angles by repetition), theodolite
Base-line taping	1/100,000 to 1/1,000,000	Triangulation of high precision for large areas, city surveys, or long bridges and tunnels	Theodolite, direction instrument
Electronic measurements	±0.04 ft. ± 1/300,000	Traverses for control of precise surveys, base lines or triangle sides for triangulation	Theodolite

Paces or strides are usually counted by means of a tally register operated by hand or by means of a pedometer which is attached to the leg. In hilly country, rough corrections for slope can be applied.

The beginner should standardize his pace by walking over known distances on level, sloping, and uneven ground.

7-3. Stadia. The stadia method, described in detail in Chap. 15, offers a rapid indirect means of determining distances. It is used extensively and is particularly useful in topographic surveying. The telescope of the transit, level, or plane-table alidade is equipped with two horizontal hairs, one above and the other an equal distance below the horizontal cross-hair. The distance from the instrument to a given point is indicated by the intercept between these stadia hairs as shown on a graduated rod held vertically at the point. The precision of the stadia method depends upon the instrument, the observer, the atmospheric conditions, and the length of sights. Under average conditions the stadia method will yield a precision between 1/300 and 1/1,000.

7-4. Small Angles. Where sights are nearly level, the *gradienter* (Art. 2-20) can be used to measure distances in a manner similar to that for stadia and with about the same precision. Similarly, the transit may be used to measure the small horizontal angle subtended by a known distance on a rod mounted horizontally at right angles to the line of sight.

7-5. Chaining, or Taping. The common method of determining distance is by direct measurement with the tape. Formerly on surveys of ordinary precision it was the practice to measure the length of lines with the *engineer's chain* or the *Gunter's chain;* for measurements of the highest precision special bars were used.

The *engineer's chain* was 100 ft. long and was composed of 100 links each 1 ft. long. At every 10 links brass tags were fastened, notches on the tags indicating the number of 10-link segments between the tag and the end of the tape. Distances measured with the engineer's chain were recorded in feet and decimals.

The *surveyor's* or *Gunter's chain* was 66 ft. long and was divided into 100 links each 7.92 in. long; distances were recorded in chains and links. It is still the unit of measurement used in land surveying on account of the convenient relation between its length and the units of land measure.

1 (Gunter's) chain = 100 links = 4 rods
80 (Gunter's) chains = 1 mile
10 square (Gunter's) chains = 1 acre

Measuring with chains was originally called "chaining." The term has survived and is often applied to the operation of measuring lines with tapes, although the term "taping" is gradually succeeding it except in land surveying. Herein, as elsewhere, the two terms are used interchangeably.

The precision of distance measured with tapes depends upon the degree of refinement with which measurements are taken. On the one hand, rough taping through broken country may be less precise than the stadia. On the other hand, when extreme care is taken to eliminate all possible errors, measurements have been taken with a probable error of less than 1/1,000,000. In ordinary taping over flat, smooth ground, the precision is about 1/3,000 to 1/5,000.

7-6. Electronic Measurements. Recent scientific advances have led to the development of electronic instruments which are of great value to the surveyor for accurate measurements of distances. Of the three such instruments currently in use (1965), the *Geodimeter* utilizes light waves while the *Tellurometer* and the *Electrotape* employ radio waves. All three are tripod-mounted and are portable, each weighing well under 100 lb. The latest models require 10 to 30 min. for a distance measurement, and all have a maximum inherent instrumental error of approximately 0.04 ft. plus or minus 2 to 3 parts per million of the distance measured. A typical nighttime setup for use of the Geodimeter is shown in Fig. 7-1.

The basic principle of the Geodimeter is the indirect determination of the time required for a light beam to travel between two stations. The instrument, set up on one station, emits a modulated light beam to a passive reflector set up on the other end of the line to be measured. The reflector, acting as a mirror, returns the light pulse to the instrument, where a phase comparison is made between the projected and reflected pulses. The velocity of light is the basis for computation of the distance. A clear line of sight

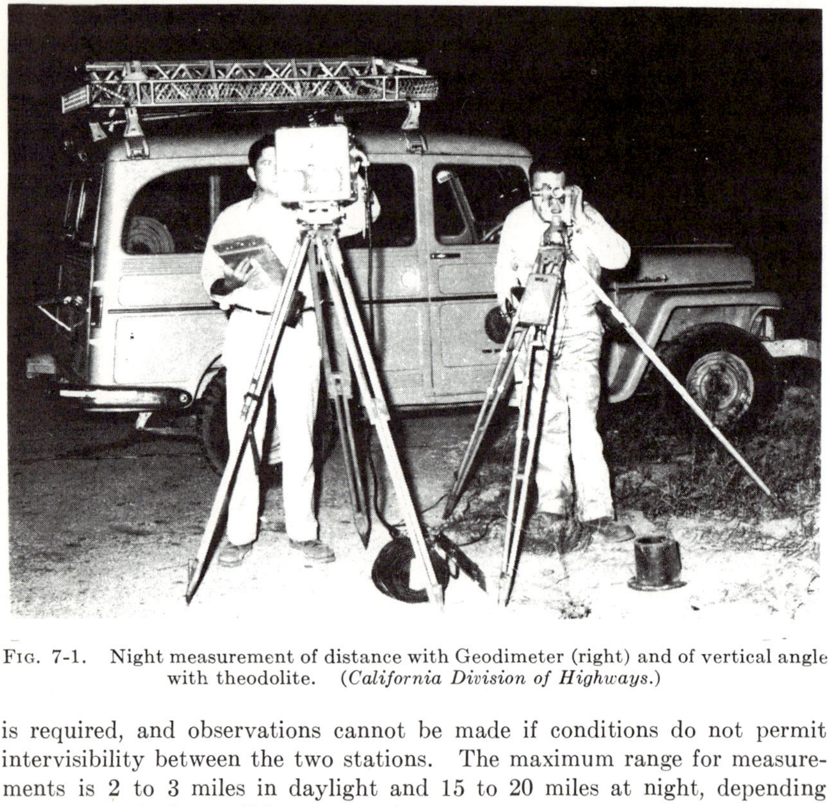

FIG. 7-1. Night measurement of distance with Geodimeter (right) and of vertical angle with theodolite. (*California Division of Highways.*)

is required, and observations cannot be made if conditions do not permit intervisibility between the two stations. The maximum range for measurements is 2 to 3 miles in daylight and 15 to 20 miles at night, depending upon atmospheric conditions.

The Tellurometer and the Electrotape both consist of two interchangeable instruments, one being set up on each end of the line to be measured. The sending instrument transmits a series of microwaves which are run through the circuitry of the receiving unit and are retransmitted back to the original sending unit, which measures the time required. Distances are computed on the basis of the velocity of the radio waves. A clear line of sight between the two instruments is necessary. However, intervisibility is not required, and therefore observations can be made in fog and other unfavorable weather conditions. Distances up to 40 or 50 miles can be measured under favorable conditions.

In addition to the inherent instrumental errors previously mentioned, electronic distance measurements are subject to errors due to atmospheric conditions and to reflections. As the velocities of both light and radio waves are affected by atmospheric conditions, it is necessary to make corrections for varying conditions where precision in measurements is necessary. Light waves are primarily affected by temperature and atmospheric pressure, the effects of humidity and light color being relatively minor; for most purposes, the mean of temperature and atmospheric pressure data taken at the instrument and reflector stations will suffice for the necessary corrections. In general the corrections to be applied to the velocity of radio waves are somewhat larger than those for light waves, with temperature, pressure, and humidity each having a large effect upon the refractive index. When radio waves are used, therefore, humidity data are necessary in addition to information regarding temperature and atmospheric pressure.

Reflections from buildings, paved streets, still water, etc., may result in reflected radio signals. For long distances the effects are usually negligible, but for short distances the reflections may affect the resulting accuracy appreciably. Where light waves are used, reflections are not a source of error because the light emitted by the instrument is polarized and the resulting measurements are not affected by unpolarized light.

In general the proportional accuracy of measurements by electronic instruments will tend to be greater for longer distances. Any of the instruments is capable of first-order results for measurements of several miles or more. Under favorable conditions they are usually capable of second-order accuracy for distances of somewhat less than ½ mile. However, for second-order surveys to qualify for and be incorporated into the national geodetic network, the U.S. Coast and Geodetic Survey includes the following conditions in its 1962 "Requirements for Second Order Surveying": "When electronic equipment employing radio waves is used in distance measurement on traverse surveys, the lines so measured shall never be less than one-half mile, seldom less than one mile, and preferably over two miles in length. When such equipment is used to measure base lines in triangulation, at least two lines in a single geometric figure, each at least five miles in length, shall be measured. The check between these measured lengths, as computed through the triangulation, shall be one part in 50,000 or better."

Measurements obtained by use of electronic instruments are *slope* distances, and they must be converted to horizontal distances for further use. The usual practice is to measure vertical angles with a theodolite, as exemplified in Fig. 7-1. Computation by exact methods, taking into account the curvature of the earth, is cumbersome and time-consuming. A method of approximation which provides a rapid and accurate solution is given in Ref. 5 at the end of this chapter. Further reductions to sea level and to the plane of the state coordinate systems are discussed in Art. 26-7.

Electronic distance-measuring instruments may be advantageously and economically employed in many kinds of surveys for traversing, trilateration, or measurement of base lines for triangulation. Their use for traversing is discussed in Art. 26-14. Trilateration is a procedure involving the computation of triangles from precise measurements of all three sides in lieu of measuring the angles and a base line as in the usual triangulation method (Chap. 16).

Where electronic instruments are used for traversing, the resulting measurements of distance will frequently be more precise than the angular measurements; in consequence, the methods of traverse adjustment given in Arts. 18-13 and 18-14 may not be applicable. In some cases the precision of distance measurements may even be greater than that of the network monuments on which the survey is based, particularly if the traverse involves a connection between two arcs of triangulation. In many cases, arbitrary adjustments based upon judgment and a knowledge of conditions will provide the best solution.

7-7. Other Methods. Distance may be measured by observing the number of revolutions of the wheel of a vehicle. The *mileage recorder* attached to the ordinary automobile speedometer registers distance to 0.1 mile and may be read by estimation to 0.01 mile. Special speedometers are available reading to 0.01 or 0.002 mile. The *odometer* is a simple device which can be attached to any vehicle and which registers directly the number of revolutions of the wheel. With the circumference of the wheel known, the relation between revolutions and distance is fixed. The distance indicated by either the mileage recorder or the odometer is somewhat greater than the true horizontal distance, but in hilly country a rough correction based on the estimated average slope may be applied.

Distances are sometimes roughly estimated by *time interval of travel* for person at walk, saddle animal at walk, or saddle animal at gallop.

By *mathematical* or *graphical* methods, unknown distances may be determined through their relation to one or more known distances. These methods are used in triangulation and plane-table work.

7-8. Choice of Methods. Most important lines, including land boundaries, traverses of horizontal control for extensive surveys, and the lines for the location and construction of the works of man, are measured with tapes because no other practicable method furnishes the required precision. However, much time has been wasted in chaining distances that could have been measured with all necessary precision by some less laborious method such as the stadia. Each of the methods mentioned in the preceding articles has a field of usefulness. On the survey for a single enterprise, the surveyor may find occasion to employ a combination of methods to advantage.

TAPING

7-9. Tapes. Tapes are made in a variety of materials, lengths, and weights. Those more commonly used by the surveyor are the heavy steel tape, sometimes called the engineer's tape, surveyor's tape, or chain tape, and the metallic tape. The link chain is obsolete.

The *metallic tape* (Fig. 7-2*a*) is a ribbon of waterproofed fabric into which are woven small brass or bronze wires to prevent its stretching. It is usually 50 or 100 ft. long and is graduated in feet, tenths, and half-tenths; it is usually ⅝ in. wide. It is used principally in earthwork cross-sectioning, in location of details, and in similar work where a light, flexible tape is desirable and where small errors in length are not of consequence. Recently a nonmetallic tape which is a nonconductor of electricity has been developed for use near power lines.

FIG. 7-2*a*. Metallic tape.

A tape of phosphor bronze is rustproof and is particularly useful for work in the vicinity of salt water.

For very precise measurements, such as those for base lines and in city work, the *invar tape* has come into general use. Invar is a composition of nickel and steel with a very low coefficient of thermal expansion, sometimes

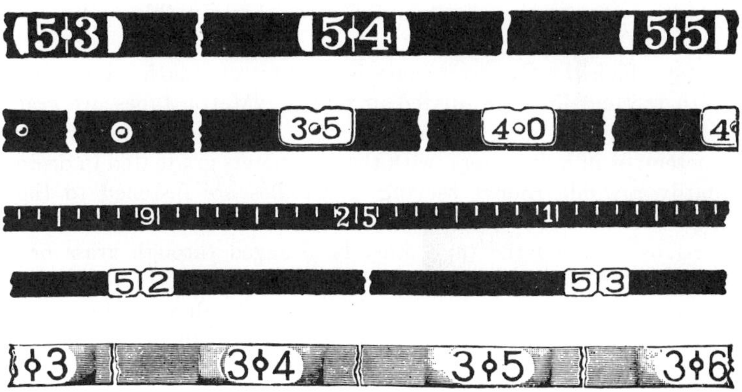

FIG. 7-2*b*. Steel tapes.

as small as one-thirtieth that of steel, and therefore affected but little by temperature changes. Since the compositions having the lowest coefficients of expansion may not remain constant in length over a period of time, it is customary to use a composition having a larger coefficient; the thermal expansion of commercial invar tapes is about one-tenth that of steel tapes. Invar is a soft metal, and the tape must be handled very carefully to avoid bends and kinks. This property and its high cost make it impracticable for ordinary use.

The *steel tape* is generally employed for the direct linear measurement of important survey lines. In the United States and Canada the length most commonly used is 100 ft., but tapes may also be obtained in lengths of 50, 200, and 300 ft.; 1, 2, 5, and 8 Gunter's chains; and 25, 30, 50, and 100 m. The common widths of 100-ft. tape are ¼ and 5⁄16 in.; longer tapes are usually narrower.

Tapes for which the foot is the unit of length are graduated in feet and decimals as follows: For small, lightweight box tapes and for some engineer's tapes, graduations are etched every 0.01 ft. throughout the length. Usually, however, the heavy engineer's tapes have graduations with numbers every foot, with only the end feet graduated in tenths or hundredths of feet. The graduations may be etched or may be stamped on babbitt metal or on brass sleeves. Some of the common graduations are shown in Fig. 7-2*b*. Some tapes have an extra graduated foot at one or both ends; some have ends graduated so that corrections for slope can be applied directly; and some have ends graduated so that temperature corrections can be applied directly. Usually the ribbon extends about 6 in. beyond the graduated portion of the tape, but for some tapes the ends of the rings mark the zero and last graduation. The latter type is not well adapted to precise measurements. Some tapes have shoulders at the zero and last graduation to assist in locating these points. However, the shoulders are objectionable when chaining through brush.

Tapes for which the Gunter's chain is the unit of length are graduated in links, with the end links graduated in tenths. Metric tapes are graduated in meters, decimeters, and centimeters, with the end decimeters graduated in millimeters; or in half-meters, with the end meters graduated in decimeters.

Ordinarily rawhide thongs serving as handles are fastened to the rings at each end of the chain tape. Wire handles are sometimes used but they are objectionable when the tape must be dragged through grass or brush. Detachable clamp handles are available for grasping the tape at any point.

The tape is usually wound on a reel but may be done up into a figure 8 and thrown into circular form with diameter about 10 in., as described in Art. 3-11.

The steel tape, being elastic, stretches when a pull is applied. It also expands or contracts as the temperature changes. Tapes when received from the manufacturer are usually quite close to the correct length when subjected to a given pull and a given temperature. For the 100-ft. tape some manufacturers attempt to furnish the standard length at 68°F. under a pull of 10 lb., the tape being horizontal and supported throughout its entire length. It is well to have a standard of length available to which the tape can be referred occasionally; many cities have such standards. For a small fee the National Bureau of Standards, Washington, D.C., will compare a tape with the standard for any specified pull and support and will issue a certificate of comparison stating its length under the conditions of the standardization test.

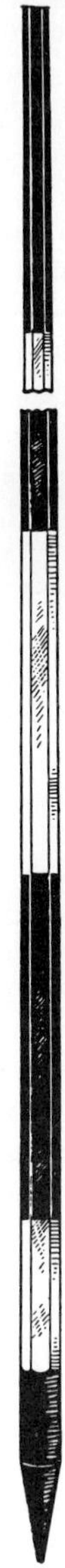

Fig. 7-4. Range pole.

Special spring scales are available for applying the proper tension to the tape in the field. Tape levels, plastic tape thermometers, and clamping handles are also available. Other accessories include equipment for repairing and splicing tapes.

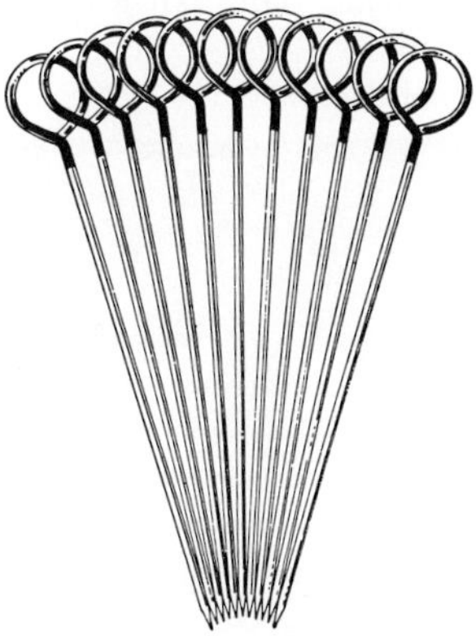

Fig. 7-3. Chaining pins.

Suggestions for the care and handling of tapes and chaining equiqment are given in Art. 3-11.

7-10. Markers. Steel chaining pins, also called taping arrows or surveyor's arrows, are commonly employed to mark the ends of the tape during the process of chaining between two points more than a tape length apart. They are usually 10 to 14 in. long. A set consists of 11 pins (see Fig. 7-3).

For more precise chaining or for future reference, nails may be driven into the earth. Marking tags, which can be stamped with identification as desired, may be threaded over the nails before they are driven. On paved surfaces, tape lengths and other points may be marked with keel, pencil, or spray paint; for precise work a short piece of opaque adhesive tape may be stuck on the pavement, and the point marked with a pencil or ballpoint pen.

Wooden stakes (Art. 10-1) or hubs (Art. 14-4) are driven into the ground to mark the significant points of profile-level and transit-traverse lines.

7-11. Range Poles. Metal, wooden, or fiber-glass range poles, also called flags, flagpoles, or lining rods, are used as temporary signals to indicate the location of points or the direction of lines. They are of octagonal or circular cross-section and are pointed at the lower end. Wooden and fiber-glass range poles are shod with a steel point. The common length is 8 ft. Usually the pole is painted with alternate bands of red and white 1 ft. long (see Fig. 7-4).

7-12. Taping on Smooth, Level Ground. The procedure followed in measuring distances with the tape depends to some extent upon the required precision and the purpose of the survey. The following description represents the usual practice when the measurements are of ordinary precision (say 1/5,000). Errors in taping are discussed in Art. 7-16, and mistakes in Art. 7-24.

The tape is supported throughout its length. If only the distance between two fixed points (as the corners of a parcel of land) is to be determined, the equipment will consist of one or more range poles, 11 chaining pins, and a 100-ft. heavy steel tape, say with the interval 0 to 1 ft. graduated in tenths of feet and the remainder of the tape graduated in feet. One range pole is placed behind the distant point to indicate its location.

FIG. 7-5. Taping on smooth, level ground.

The rear chainman *with one pin* stations himself at the point of beginning. The head chainman, with the zero (graduated) end of the tape and 10 pins, advances toward the distant point. When the head chainman has gone nearly 100 ft., the rear chainman calls "chain" or "tape," a signal for the head chainman to halt. The rear chainman holds the 100-ft. mark at the point of beginning and, by hand signals or by voice, lines in a chaining pin (held by the head chainman) with the range pole (or other signal) marking the distant point. During the lining-in process, the rear chainman is in a kneeling position on the line and facing the distant point; the head chainman is in a kneeling position to one side and facing the line so that he can hold the tape steady and so that the rear chainman will have a clear view of the signal marking the distant point. The head chainman with one hand sets the pin vertically on line and a short distance to the rear of the zero mark. With his other hand he then pulls the tape taut and, making sure that it is straight, brings it in contact with the pin. The rear chainman, when he observes that the 100-ft. mark is at the point of beginning, calls "stick" or "all right." The head chainman pulls the pin and sticks it at the zero mark of the tape, with the pin sloping away from the line (Fig. 7-5). As a check, he again pulls the tape taut and notes that the zero point coincides with the pin at its intersection with the ground. He then calls "stuck" or "all right," the rear chainman releases the tape, the head chainman moves forward as before, and so the process is repeated.

As the rear chainman leaves each intermediate point, he pulls the pin. Thus there is always one pin in the ground, and the number of pins held by the rear chainman at any time indicates the number of hundreds of feet, or *stations*, from the point of begining to the pin in the ground.

At the end of each 1,000 ft. (10 full stations), the head chainman has placed his last pin in the ground. He signals for pins, the rear chainman comes forward and gives the 10 pins to the head chainman, both check the tally, and the head chainman records it. The procedure is then repeated. The count of pins is important, as the number of tape lengths is easily forgotten owing to distractions.

When the end of the course is reached, the head chainman halts and the rear chainman comes forward to the last pin set. The head chainman holds the zero mark at the terminal point. The rear chainman pulls the tape taut and observes the number of *whole feet* between the last pin and the end of the line. He then holds the next *larger* foot mark at the pin, and the head chainman pulls the tape taut and reads the decimal by means of the finer graduations of the end foot. The decimal is counted from the 1-ft. mark; thus the distance between the last pin and the end of the line is 1 ft. *less* than that indicated at the pin, plus the decimal read by the head chainman. For example, with the rear chainman holding at 87 ft. and the head chainman reading 0.68 ft., the distance from the last pin is $87 - 1 + 0.68 = 86.68$ ft. The chainmen should agree on some system of checking to prevent mistakes.

For tapes having an extra graduated foot beyond the zero point, it is not necessary to subtract 1 ft. as just described. The rear chainman holds the next *smaller* foot mark at the pin, and the head chainman reads the decimal.

Where the transit is set up on the line to be measured, the transitman usually directs the head chainman in placing the pins on line.

On some surveys it is required that stakes be set on line at short intervals, usually 100 ft. Sometimes stakes are driven by the rear chainman after he has pulled the pin. On surveys of low precision no pins are used and the head chainman sets the stakes in the manner previously described for pins, measuring the distance between centers of stakes at their junction with the ground. On more precise surveys, measurements are carried forward by setting a tack in the head of each stake. When the tack has been driven, it is tested for line and distance.

7-13. Horizontal Taping over Sloping Ground. The process of taping over uneven or sloping ground or over grass and brush is much the same as that just described for smooth, level ground, except that the plumb bob is used. The tape is held horizontal, and a plumb bob is used by either, or at times by both, chainmen for projecting from tape to pin or *vice versa*. For rough work, plumbing can be accomplished with the range pole. One form of field notes is given in Fig. 7-15.

FIG. 7-6. Plumbing at downhill end of horizontal tape.

To secure anything like the same precision as in taping over level ground, considerable skill is required. Some experience is necessary to determine when the tape is nearly horizontal; the tendency is to hold the downhill end of the tape too low. The tape is unsupported between the chainmen, and either the pull must be increased to eliminate the effect of sag or a correction for sag must be applied. A firm stance is important; the chainman should place the plane of his body parallel to the tape, with his legs well apart. His forearm should be in line with the tape, and he should steady or snub it against his leg or body as required by the height at which the tape must be held.

Where the slope is less than 5 or 6 ft. in 100, the head chainman advances a full tape length at a time. If the course is downhill, he holds the plumb line at the zero point of the tape, with the tape horizontal and the plumb bob a few inches from the ground, then pulls the tape taut and is directed to line by the rear chainman. When the plumb bob comes to rest, he lowers it carefully to the ground (see Fig. 7-6) and then sets a pin in its place. As a check, the measurement is repeated. If the course is uphill, the rear chainman holds a plumb line suspended from the 100-ft. mark and signals the head chainman to give or take until the rear chainman's plumb bob comes to rest over the pin. The head chainman sets a pin, and the measurement is repeated.

Where the course is steeper and is downhill, the head chainman advances a full tape length and then returns to an intermediate point from which he can hold the tape horizontal. He suspends the plumb line at a foot mark, is lined in by the rear chainman, and sets a pin at the indicated point. The rear chainman comes forward, *gives the head chainman a pin,* and at the pin in the ground holds the tape at the foot mark from which the plumb line was previously suspended. The head chainman proceeds to another point from which he can hold the tape horizontal, and so the process is repeated until the head chainman reaches the zero mark on the tape. At each *intermediate* point of a tape length the rear chainman gives the head chainman a pin, but not at the point marking the full tape length. In this manner the tape is always advanced a full length at a time, the number of pins held by the rear chainman at each 100-ft. point indicates the number of hundreds of feet from the last tally, and the rear chainman's count of pins is not confused. The process is called "breaking chain" or "breaking tape."

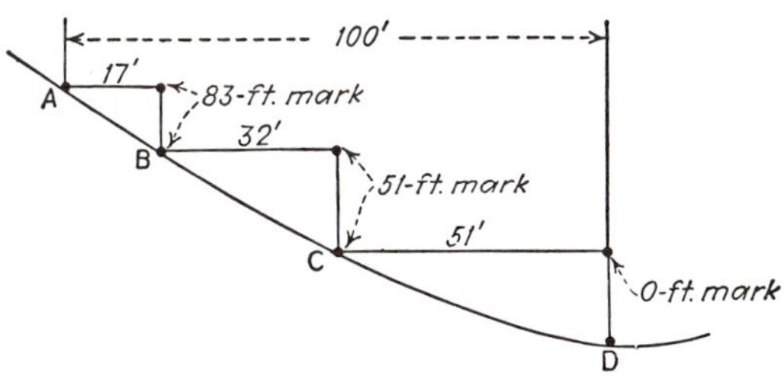

FIG. 7-7. Horizontal measurements on steep slope.

To illustrate, Fig. 7-7 represents the profile of a line to be measured in the direction of A to D, and A represents a pin marking the end of a 100-ft. interval from the point of beginning. The head chainman goes forward until the 100-ft. mark is at A, where the rear chainman is stationed. The head chainman then returns to B where he holds the tape horizontal and plumbs from the 83-ft. mark to set a pin at B. The rear chainman gives the head chainman a pin and holds the 83-ft. mark at B. The head chainman plumbs from the 51-ft. mark and sets a pin at C. The rear chainman gives the head chainman a pin and holds the 51-ft. mark at C. The head chainman plumbs from the zero mark to set a pin at D at the end of the full tape length. The rear chainman goes forward but keeps the pin which he pulled at C.

Some surveyors prefer to measure distances less than the tape length individually and add these measurements. However, this practice requires recording and may lead to mistakes in addition.

Usually the tape is estimated to be horizontal by eye. This practice commonly results in the downhill end's being too low, sometimes causing a significant error in horizontal measurement. The safe procedure in rough country is to use a hand level.

In horizontal measurements over uneven or sloping ground, the tape sags between supports and becomes effectively shorter. The effect of sag can be eliminated by standardizing the tape, by applying a computed correction, or by using the normal tension (Art. 7-21). In breaking chain or when the tape is supported for part of its length, the difference in effect of sag as between a full tape length and the unsupported length can be taken into account roughly by varying the pull.

In measurements over crops, vegetation, or other low obstructions, if the tape is partially supported between the chainmen, account should be taken of the effect of this support upon sag.

7-14. Slope Taping. Where the ground is fairly smooth, measurements on the slope may sometimes be made more accurately and quickly than horizontal measurements, and slope measurements are generally preferred. Some means of determining either the slope or the difference in elevation between successive 100-ft. points and/or breaks in slope is required. For surveys of ordinary precision, either the clinometer (for measuring slope) or the hand level (for measuring difference in elevation) may be used to advantage. If only the distance between the ends of the line is required, the procedure of taping is the same as on level ground, but a record is kept either of the slope or of the difference in elevation for each tape length (or less at breaks in slope). The horizontal distances are then computed from the distances measured on the slope.

Where stakes are to be placed every 100 ft., corrections to the slope distances may be applied as the chaining progresses, either by mental calculation or by use of the slide rule. Corrections are much more readily calculated than are the horizontal distances themselves, as will shortly be demonstrated. Unless the correction is greater than 1 ft., the 100-ft. tape with an extra foot on the zero end is particularly useful for measuring on slopes, in that the head chainman can first determine the correction for slope and in one operation can then lay off the true slope distance to give a horizontal distance of 100 ft. Tapes with slope-correction graduations at the end are also useful.

Measurements on the slope are preferred for the United States public-land surveys. The manual of the U.S. Bureau of Land Management states:

> The most approved method of measurement involves the use of steel ribbon tapes from 2 to 8 [Gunter's] chains in length; in its use in the public-land surveys the tape is properly alined and stretched, and the measurements are made on the slope at any convenient distance up to the length of the tape as determined by the topography. The vertical angles of the lesser slopes are determined by the use of clinometers in the hands of the chainmen, while the vertical angles of the particularly sharp slopes are determined with the transit . . . . It is not considered necessary to exhibit in the official field notes any but the true horizontal distances . . . .

7-15. Corrections for Slope. Where the slopes are measured with sufficient precision so to warrant, the corresponding horizontal distances may be computed by exact trigonometric relations. For measurements of ordinary precision where the slope is not greater than about 20 in 100, it is simpler and sufficiently precise to subtract an approximate correction from the slope distance to obtain the horizontal distance.

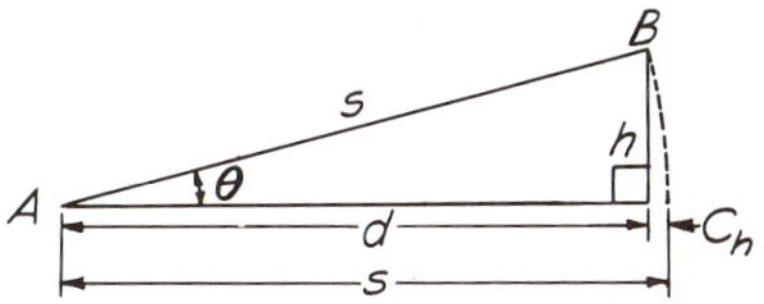

FIG. 7-8. Slope correction.

In Fig. 7-8 let s represent the slope distance between two points A and B, h the difference in elevation, and d the horizontal distance, all in feet. The correction is $C_h = s - d$. Then

$$h^2 = s^2 - d^2 = (s + d)(s - d)$$

Where the slope is not large, $s + d = 2s$ (approximate); making this substitution:

$$h^2 = 2s(s - d) \qquad \text{(approx.)}$$

and the correction is

$$C_h = s - d = \frac{h^2}{2s} \qquad \text{(approx.)} \tag{1}$$

For the usual case, where $s = 100$ ft., this formula can be solved mentally, and it is worth remembering. The error introduced through its use is negligible for ordinary slopes. The degree of approximation is shown in the following table:

Difference in elevation per 100 ft. of slope distance, ft.	Error due to using approximate formula in 100 ft. of slope distance, ft.
5	0.0001
10	0.001
15	0.007
20	0.02
30	0.1
40	0.3
60	2.0

For steep slopes, higher precision in making the slope correction can be obtained by using Eq. (1*a*) which includes the first two terms of a power series.

$$C_h = \frac{h^2}{2s} + \frac{h^4}{8s^3} \qquad \text{(approx.)} \tag{1a}$$

Where the *angle* of slope is determined, as when using the clinometer, Eq. (1) may still be used readily if it is remembered that for small angles the difference in elevation per 100 ft. is about 1.75 ft. times the slope angle in degrees (tan 1° = 0.0175). Hence, if $\theta°$ is the slope angle (Fig. 7-8) in degrees and S is the slope distance in *100-ft. stations,*

$$C_h = \frac{(1.75S\theta°)^2}{200S} = 0.015S(\theta°)^2 \qquad \text{(approx.)} \tag{2}$$

The distance S is expressed in 100-ft. stations rather than in feet in order to reduce the number of decimal places in the formula.

Where vertical angles are measured with sufficient precision to justify, the horizontal distances may be computed by exact trigonometric formula, most readily by determining the slope correction.

In Fig. 7-8, θ is the angle of slope and s, d, and C_h are as previously defined. Then

$$C_h = s - d = s - s \cos \theta = s(1 - \cos \theta) = s \text{ vers } \theta \qquad \text{(exact)} \tag{3}$$

If a table of versed sines is not available, they may be readily computed from a table of natural cosines (vers $\theta = 1 - \cos \theta$).

For most cases the correction can be calculated with sufficient precision by using the slide rule. Tables and charts giving slope corrections are published in various forms.

Having given the slope distance, to find its horizontal projection, the correction is *subtracted.*

Where it is desired to set points at a given horizontal distance (as 100 ft.)

apart or to compute a slope distance when the horizontal projection is known, the corresponding slope distance is *approximately* given by *adding* the same correction to the horizontal distance (see Fig. 7-8). If the correction is computed by Eq. (3), for a slope of 10 in 100 the error due to this approximation is about 0.002 ft. per 100 ft. and for a slope of 20 in 100 it is about 0.04 ft. per 100 ft.

7-16. Errors in Taping. Errors in taping may be attributed to the following causes:

1. *Tape Not Standard Length.* This produces a systematic error which may be practically eliminated by verifying the tape and applying the correction thus determined. The tape may be compared with a standardized tape or with some permanent standard of length which is established locally. Small differences from the correct length can be compensated by varying the pull, using a spring scale.

2. *Imperfect Alinement of Tape.* The head chainman is likely to set the pin sometimes on one side and sometimes on the other side of the correct line. This produces a variable systematic error, since the horizontal angle which the tape makes with the line is not the same for one tape length as for the next. The error cannot be eliminated, but it can be reduced to a negligible quantity by care in lining. Generally it is the least important of the errors of chaining, and extreme care in lining is not justified. The linear error when one end of the tape is off line a given amount can be computed by Eq. (1) in the same manner as the slope correction. For a 100-ft. tape, the error amounts to 0.005 ft. when one end with respect to the other is off line 1 ft. and to only 0.001 ft. when the error in alinement is 0.5 ft. Many surveyors use unnecessary care in securing good alinement without paying much attention to other more important sources of error. Errors in alinement tend to make the measured length greater than the true length and hence are positive.

3. *Tape Not Horizontal or Slope of Tape Not Correctly Determined.* The effect is to produce an error similar to that due to imperfect alinement. Often slopes are deceptive, even to experienced men; the tendency is to hold the downhill end of the tape too low. The authors have seen inexperienced chainmen keeping very careful alinement, yet taping what they thought were horizontal distances on a slope of perhaps 10 per cent. The corresponding error is 0.5 ft. per 100 ft. or 26 ft. per mile. In ordinary taping, this is one of the largest of contributing errors. It will not be eliminated by repeated measurements, but it can be reduced to a negligible amount by leveling the tape by means of either a hand level or a clinometer.

4. *Tape Not Straight.* In taping through grass and brush or when the wind is blowing, it is impossible to have all parts of the tape in perfect alinement with its ends. The error arising from this cause is systematic and variable and is of the same sign (positive) as that from measuring with a tape

that is too short. If the head chainman is careful to stretch the tape taut and to observe that it is straight by sighting over it, the error is not of consequence.

5. *Imperfections of Observing.* Errors in plumbing, reading the tape, and setting the pins are accidental errors; hence the probable error tends to vary as the square root of the number of tape lengths. Only the error due to plumbing is of real importance. In ordinary taping through rough country where it is necessary to break tape frequently, the probable error per tape length may perhaps amount to ± 0.05 or ± 0.1 ft. for each tape length. Using the maximum of ± 0.1 ft., the probable error would be about ± 0.7 ft. per mile. When the required precision is high, errors of plumbing can be avoided by chaining on the slope. The probable error of setting pins and of observing the tape graduations would perhaps be ± 0.01 ft. per tape length or ± 0.07 ft. per mile. Although these errors cannot be eliminated, their effect on the resultant error is usually not large. All the accidental errors can be kept reasonably low by care in taping.

6. *Variations in Temperature.* The tape expands as the temperature rises and contracts as the temperature falls. Therefore, if the tape is standardized at a given temperature and measurements are taken at a higher temperature, the tape is too long. For a change in temperature of 15°F., a 100-ft. steel tape will undergo a change in length of about 0.01 ft., introducing an error of about 0.5 ft. per mile. Under a change of 50°F. the error would be 1.5 ft. per mile. It is seen that, even for measurements of ordinary precision, the error due to thermal expansion becomes of consequence when the measurements are taken during cold or hot weather. A case is recalled where, at 30°F. below zero, measurements were very carefully established along the track of a railroad but no allowance was made for the extremely low temperature. Later, when the line was remeasured, the original error was found to be about 3 ft. per mile.

For temperature corrections, see Art. 7-18. Some tapes have a temperature scale at one end by means of which the correction for variation in temperature may be made without calculation.

7. *Variations in Tension.* The tape, being elastic, stretches when tension is applied. If the pull is greater than that for which the tape is standardized, the tape is too long; if the pull is less, the tape is too short. The error is systematic and of a magnitude depending upon the methods employed and the individuals who are taping. The error is negligible except in precise work. For the heavy 100-ft. tape a change in tension of 3 lb. changes the length of the tape about 0.001 ft. For tension corrections, see Art. 7-19.

8. *Sag.* This error occurs whenever the tape is supported at the ends or at intervals rather than throughout its full length. If the heavy 100-ft. tape weighing 3 lb. is standardized flat—that is, supported for its full length—and is used supported at the ends only, the systematic error per tape length

due to sag alone is as follows: for a tension of 10 lb., 0.37 ft.; 20 lb., 0.09 ft.; and 30 lb., 0.04 ft. For sag corrections, see Art. 7-20. Since the correction varies with the cube of the unsupported length, for short spans it is often negligible.

Normal Tension. The stretch of the tape partly offsets the effect of sag. For the heavy tape the resultant error for a 30-lb. pull is perhaps 0.03 ft. per 100 ft. or 1.5 ft. per mile. With the lighter tapes a pull which can be applied without undue exertion can be determined, either by computation or by experiment, at which the effect of sag will just offset the effect of increase in tension. This is usually called the *normal tension* (Art. 7-21).

7-17. Corrections. Table 7-2 summarizes the various errors which have been discussed, together with the procedures which can be adopted to eliminate or at least to reduce their effect. The various corrections are discussed in Arts. 7-18 to 7-22. It is remarkable that the apparently simple operation of linear measurement by taping is affected by so many factors.

It is seen that, in ordinary taping, the systematic errors are likely to be of much greater magnitude than the accidental errors. Hence the resultant error varies as the number of tape lengths or as the length measured.

When every possible device is employed to detect and to eliminate these systematic errors, as in precise base-line measurement, the accidental errors of observation become of relatively great importance; for this reason long tapes are usually employed. To make corrections for or to eliminate errors due to sag or to elongation of tape by tension, the pull is observed by spring by spring scales; to make corrections for thermal expansion, the temperature is observed by the use of thermometers attached to the tape.

When corrections are applied to the observed length of a line measured with a tape that is too long, the correction is *added.* When a required distance is laid out with a tape that is too long, the correction is *subtracted* from the required distance to determine the distance to be laid out. For a tape that is too short, of course, the corrections are opposite in direction to those just stated.

7-18. Correction for Temperature. The coefficient of thermal expansion of steel is approximately 0.00000645 per 1°F. If the tape is standard at a temperature of T_0 degrees and measurements are taken at a temperature of T degrees, the correction C_x for change in length is given by the formula

$$C_x = 0.00000645L(T - T_0) \tag{4}$$

in which L is the measured length.

Errors due to variations in temperature are greatly reduced by using an invar tape. If a steel tape is used, one or more plastic thermometers should be taped to it.

7-19. Correction for Tension. If the tension, or pull, is greater or less than that for which the tape is verified, the tape is elongated or shortened

TABLE 7-2. ERRORS AND CORRECTIONS IN CHAINING

NOTE: In *measuring* a distance with a tape that is *too long*, *add* the correction.

Error	Source		Amount	Error of 0.01 ft. per 100-ft. tape length caused by	Makes tape too	Importance in ordinary chaining	Procedure to eliminate or reduce	
Systematic	Erroneous length		. . .	. . .	Long or short	Usually small, but should be checked	Standardize tape and apply computed correction	
	Temperature		$C_x = 0.0000065L(T - T_0)$	15°F.	Long or short	Of consequence only in hot or cold weather	Measure temperature and apply computed correction. In precise work, chain at favorable times and/or use invar tape	
	Pull or tension (change in)		$C_p = \frac{(P - P_0)L}{AE}$	15 lb. (for 1½-lb. tape)	Long or short	Negligible	Apply computed correction. In precise work, use spring balance	May use normal tension, $P_n = \frac{0.204W\sqrt{AE}}{\sqrt{P_n - P_0}}$
	Sag		$C_s = \frac{w^2L^3}{24P^2} = \frac{W^2L}{24P^2}$	0.6 ft.	Short	Large, especially with heavy tape	Apply computed correction. May be avoided by chaining on slope	
	Slope		$C_h = h^2/2s$ (approx.) $= 0.015S(\theta°)^2$ (") $= s$ vers θ (exact)	1.4 ft.	Short	. . .	At breaks in slope, determine difference in elevation or slope angle; apply computed correction	
	Imperfect alinement	Horizontal	Same as slope	1.4 ft.	Short	Not serious	Use reasonable care in sighting. Keep tape taut and reasonably straight	
		Grass	Twice slope	0.7 ft.	Short	Not serious		
		Wind	Same as sag	0.6 ft.				
		Vertical	Same as slope	1.4 ft.	Short	Often large	Level tape	
Accidental	Manipulation	Plumbing	In rough country, breaking tape, 0.05 to 0.10 ft. per tape length		±	Large, but accidental	May avoid by slope chaining	
		Marking ends of tape; reading graduations	0.01 ft. per tape length		±	Not serious	. . .	Use reasonable care to reduce
		Error in pull	Same as pull	15 lb. (1½-lb. tape)	±	Not serious	. . .	
		Error in determination of slope	Amount varies with slope		±	Not serious	. . .	

accordingly. The correction for variation in tension in a steel tape is given by the formula

$$C_p = \frac{(P - P_0)L}{AE} \tag{5}$$

in which C_p = correction per distance L, in feet
P = applied tension, in pounds
P_0 = tension for which the tape is standardized, in pounds
L = length, in feet
A = cross-sectional area, in square inches
E = elastic modulus of the steel, in pounds per square inch

The National Bureau of Standards has found that the average modulus of elasticity of steel tapes submitted for verification is 28 million pounds per square inch. However, for many purposes the modulus is conveniently taken with sufficient precision as 30 million pounds per square inch. The cross-sectional area of the tape can be computed from the weight and dimensions, since steel weighs approximately 490 lb. per cubic foot. For the 100-ft. tape, it is usually sufficiently precise to take A, in square inches, as being equal to 0.003 times the weight of the tape in pounds.

Some idea of the effect of variation in tension can be obtained from the following example.

Example: It will be assumed that both a very heavy and a medium-weight tape are standard under a tension of 10 lb.; E = 30,000,000 lb per square inch. The cross-sectional area of the heavy tape is 0.010 sq. in. and of the light tape 0.005 sq. in. It is desired to determine the elongation due to an increase of tension from 10 to 30 lb. For the very heavy tape,

$$C_p = \frac{(30 - 10) \times 100}{30{,}000{,}000 \times 0.010} = 0.0067 \text{ ft.}$$

For the medium-weight tape,

$$C_p = \frac{(30 - 10) \times 100}{30{,}000{,}000 \times 0.005} = 0.0133 \text{ ft.}$$

The results show that in ordinary chaining the error due to variation in tension is of consequence only for light tapes.

7-20. Correction for Sag. When the tape sags between points of support, it takes the form of a catenary. The correction to be applied is the difference in length between the arc and the subtending chord. For the purpose of determining the correction, the arc may be assumed to be a parabola, and the correction is then given with sufficient precision for most purposes by the formula

$$C_s = \frac{w^2L^3}{24P^2} = \frac{W^2L}{24P^2} \tag{6}$$

in which C_s = correction between points of support, in feet
w = weight of tape, in pounds per foot
W = total weight of tape between supports, in pounds
L = distance *between supports*, in feet
P = applied tension, in pounds

The correction is seen to vary directly as the cube of the unsupported length and inversely as the square of the pull. Although the equation is intended for use with a *level* tape, it may be applied without error of consequence to a tape held on a slope up to approximately 10°.

Following is an example illustrating the variation in correction due to a variation in tension, weight of tape, and distance between supports.

Example: In Table 7-3 are results of calculations for 100-ft. tapes weighing 3 and 1½ lb., for pulls of 10 and 30 lb., and for distances between supports of 100 and 25 ft. The correction for sag of the 3-lb. tape for a pull of 10 lb. and distance

TABLE 7-3. COMPARISON OF TAPE CORRECTIONS

	Very heavy tape				Medium-weight tape			
Weight, lb.	3				1½			
Distance between supports, ft....	100		25		100		25	
Pull, lb..........	10	30	10	30	10	30	10	30
C_s (correction for sag) per 100 ft. of length, ft....	0.37	0.042	0.023	0.0026	0.094	0.0104	0.0059	0.00065
Change in C_s for 1-lb. variation in P, ft........	0.07	0.003	0.004	0.0002	0.018	0.0007	0.0011	0.00004
Elongation per 100 ft. due to change in P from 10 to 30 lb. (Art. 7-19), ft.............	0.0067		0.0067		0.0133		0.0133	
Normal tension (Art. 7-21), lb..	51.8		23.2		28.0		14.3	

between supports of 100 ft. is seen to be roughly 600 times as great as for the 1½-lb. tape for a pull of 30 lb. and distance between supports of 25 ft. Other conditions remaining the same, the effect of decreasing the pull from 30 to 10 lb. is to increase the sag correction ninefold, and the effect of increasing the distance between supports from 25 to 100 ft. is to increase the sag correction sixteen times. An error of 1 lb. in an assumed tension of 10 lb. introduces an error of 19 per cent in the calculated correction for sag. Although in measurements of moderate precision this error is not of much consequence for the lighter tape and shorter distance between supports, it becomes large for the heavy tape and longer distance between supports. Evidently in ordinary chaining, where the pull is applied entirely by estimation, it would rarely be the case that the chainmen could apply the tension without an error greater than 1 lb.

The example further illustrates the serious disadvantage of using a very heavy tape for horizontal measurements over sloping ground.

7-21. Normal Tension. By equating the right-hand members of Eqs. (5) and (6), the elongation due to increase in tension is made equal to the shortening due to sag; thus the effect of sag can be elimimated. The pull that will produce this condition, called *normal tension* P_n, is given by the formula

$$P_n = \frac{0.204W\sqrt{AE}}{\sqrt{P_n - P_0}} \tag{7}$$

This equation is solved by trial. The commonly used type of slide rule is convenient, as the following example shows.

Example: Given the heavy 100-ft. tape of the previous examples; distance between supports 100 ft.; $E = 30{,}000{,}000$ psi., $A = 0.01$ sq. in.; $P_0 = 10$ lb. Determine the tension at which the effect of sag will be eliminated by the elongation of the tape due to increased tension.

The numerator of the right-hand member is

$$0.204 \times 3\sqrt{0.010 \times 30{,}000{,}000} = 335; \text{ then } P_n = \frac{335}{\sqrt{P_n - 10}}$$

Set runner to 335 on D scale; move slide until by inspection $P_n - 10$ on right B scale is at runner, when the C index is at P_n on D scale. The runner reads 41.8 on B scale when the C index reads 51.8 on the D scale; hence $P_n = 51.8$ lb.

The four values of normal tension are shown in Table 7-3. It is seen that for a heavy tape over a long span the use of normal tension would not be practicable.

7-22. Combined Corrections. Whenever corrections for several effects such as tension, temperature, and sag are to be applied, for convenience they may be combined as a single net correction per tape length. Since the corrections are relatively small, the value of each is not appreciably affected by the others and each may be computed on the basis of the nominal tape length. For example, even if the verified length of a tape were 100.21 ft., the correction for temperature (within the required precision) would be found to be the same whether computed for the exact length or for a nominal length of 100 ft. The combined correction per tape length can be used as long as conditions remain constant; however, the temperature is likely to vary considerably during the day, and the ground may be such that it may be necessary to support the tape at various intervals of length.

Further, the method of *adding* (or subtracting) the small corrections encountered in taping is far more convenient than would be that of *multiplying* (or dividing) by correction factors, since fewer figures are required.

7-23. Precision of Taping. It would be valuable if a definite outline of procedure could be established to produce any desired degree of precision in taping. Unfortunately the conditions are so varied, and so much depends upon the skill of the individual, that the surveyor must be guided largely by his own experience and by his knowledge of the errors involved. Recommended specifications for taping to produce various degrees of precision in transit-tape surveys are given in Art. 14-16.

The usual practice in rough chaining through broken country is to take measurements with the tape horizontal, plumbing from the downhill end, breaking tape where necessary, applying tension by estimation, and making no corrections for sag, temperature, or tension. The tape is usually 100 ft. long and weighs about 2 lb. The discussion of the preceding articles makes it evident that the larger errors are likely to arise owing to (1) tape not level, (2) sag in tape, (3) variation in temperature, and (4) poor plumbing. Call these errors per tape length, respectively, e_h, e_s, e_t, and e_v. Of these, e_h and e_s are positive systematic errors, which increase directly with the number of opportunities for error, e_t is probably either positive or negative systematic, and e_v is accidental and would therefore increase in proportion to the square root of the number of opportunities for error. Let us assume values for these errors, neglecting entirely those arising from other sources, and estimate the limits of the resultant error e per 1,000 ft. From the preceding articles we might perhaps expect the following:

$$e_h = +0.04 \text{ ft.}, \qquad e_s = +0.03 \text{ ft.}, \qquad e_t = \pm 0.01 \text{ ft.}, \qquad e_v = \pm 0.05 \text{ ft.},$$
$$e = 10(+0.04 + 0.03 \pm 0.01) \pm 0.05\sqrt{10} = +0.44 \text{ to } +0.96 \text{ ft.}$$

The corresponding limits of precision are roughly 1/2,300 to 1/1,000. In a general way these limits correspond to those usually attained on work of the character specified above where no particular effort is made to secure precision. The measured lengths of such lines are nearly always considerably longer than their true lengths, since most of the systematic errors tend to make the tape effectively too short (Table 7-2). Hence it may be considered good practice either arbitrarily to deduct a reasonable quantity from such measurements or to offset the over-all effect by using a pull somewhat greater than that for which the tape is standardized. If the taping is done during extremely cold weather with inexperienced chainmen who fail to keep the tape taut and who do not use reasonable care in keeping the tape horizontal, the precision may be less than 1/500.

In ordinary taping over flat, smooth ground the principal errors are those due to variation in temperature and to inclination of tape. "Flat" and "smooth" are relative terms, as here used. Ordinarily the tape, when held to the ground, is neither perfectly straight nor horizontal. Frequently a difference in elevation of 2 or 3 ft. in 100 will go undetected if the slope is smooth. Assuming the error of setting pins as ± 0.007 ft. per tape length, that due to slope and uneven tape as $+0.02$ ft., and that due to temperature as ± 0.01 ft., the limits of precision are roughly 1/3,000 to 1/10,000. Usually 1/5,000 is considered good taping for the conditions stated, but to attain this precision a rough correction for variation in temperature must frequently be made. Under extreme weather conditions, change in temperature alone might introduce an error of 1/2,000 or even greater.

In taping along a smooth surface such as a paved highway, if slope measurements are taken with reasonable care, the principal error is that due to variation in temperature. For measurements of moderate precision when a rough correction for thermal expansion is made without actual observations of temperature, the systematic error per 100 ft. due to temperature variation might be between ± 0.006 and ± 0.01 ft. As compared with this, the accidental errors due to variations in tension and to marking the ends of the tape are of relatively small account. Under these conditions it might reasonably be expected that a precision of 1/10,000 to 1/15,000 could be maintained. In practice it is generally assumed that a precision of 1/10,000 is about the maximum that can be obtained without the aid of special apparatus.

For measurements of higher precision the temperature is determined by a thermometer attached to the tape, and the tension is regulated through the use of a spring scale. Generally the tapes used are light in weight so that, where they are unsupported, the uncertainty of the effect of sag will be small. When measurements are corrected for the accumulative effects of sag, slope, etc., the remaining errors are largely accidental, though for measurements in sunlight there is likely to be an appreciable difference between the observed temperature and the actual temperature even though the bulb of the thermometer is in contact with the tape. Assuming that the sum of all systematic errors in the same direction might be 0.004 ft. per 100 ft. and that all accidental errors might amount to ± 0.007 ft. per 100 ft., for a line 1 mile long we might expect a precision of 1/20,000 to 1/30,000. Experience indicates that lines of considerable length may measured under the conditions stated above with a precision as high as 1/30,000. Along a smooth course such as a railroad or a highway, where measurements are taken with the tape supported its full length, this precision may be maintained by using moderate care in setting the pins or in otherwise marking the location of the end of the tape on the ground. If the course is rough or overgrown, it is necessary to employ some device by means of which the tape may be suspended and yet held firmly in position while the tension is applied and the end marks of the tape are projected to the ground points by plumbing; also it is necessary to plumb with great care.

When steel tapes are used, measurements calling for a very high precision, say 1/100,000 to 1/500,000 or higher, are made at night or on cloudy days so that uncertainties regarding the temperature of the tape are greatly reduced. For work of this character, the tapes employed are usually 50 meters or more in length and are very carefully standardized. The successive positions of the forward end of the tape are marked by lines scratched on zinc or copper strips fastened to substantial posts. (See also "Base-line Measurement" in Chap. 16.)

Errors due to variations in temperature are greatly reduced by using an invar tape.

7-24. Mistakes in Taping. Some of the mistakes commonly made by inexperienced chainmen are:

1. Adding or dropping a full tape length. This is not likely to occur if both chainmen count the pins, or when numbered stakes are used, if the rear chainman calls out the station number of the rear stake in response to which the head chainman calls out the number of the forward stake as he marks it. A tape length may be added through failure of the rear chainman to give the head chainman a pin at breaks marking fractional tape lengths. A tape length may be dropped through failure of the rear chainman to take a pin at the point of beginning.

2. Adding a foot. This usually happens in measuring the fractional part of a tape length at the end of the line. This distance should be checked by the head chainman holding the zero mark on the tape at the terminal point and the rear chainman noting the number of feet and approximate fraction at the last pin set.

3. Other points incorrectly taken as 0 or 100-ft. marks on tape. The chainman should note whether these marks are at end of rings or on the tape itself, also whether there is an extra graduated foot at one end of the tape.

4. Reading numbers incorrectly. Frequently "68" is read "89" or "6" read as "9." As a check, good practice is to observe the number of the foot marks on each side of the one indicating the measurement, especially if the numbers are dirty or worn. Also the tape should be read with the numbers right side up.

5. Calling numbers incorrectly or not clearly. For example, 50.3 might be called "fifty, three" and recorded as 53.0. If called as "fifty, point, three" or preferably "five, zero, point, three," the mistake would not be likely to occur. When a decimal point or a zero occurs in a number, it should be called. When numbers are called to a recorder, he should repeat them as they are recorded.

Often large mistakes will be prevented or discovered if the chainmen form the habit of pacing distances or of estimating them by eye. If a transit is being used to give line, distances can be checked by reading the approximate stadia interval on a flagpole.

SURVEYS WITH TAPE

7-25. Surveys. The survey of a field with the tape is accomplished by dividing the field into triangles and obtaining sufficient measurements of the sides, altitudes, and/or angles of the triangles to permit the computations of remaining sides and angles required for plotting and for the calculation of areas. Typical field notes for the survey of a field with the tape are given in Figs. 7-16 and 7-17.

Where the measurement of angles is involved, surveying with the tape alone is too slow to be used to any great extent except on surveys covering small areas. However, often it is convenient to measure an angle or to erect a perpendicular with the tape.

7-26. Measurement of Angles. Angles are measured by the chord method, as follows: With the vertex A of the angle α as a center (Fig. 7-9), the tape is swung, and pins are set at points a and b where the arc intersects the sides AB and AC of the angle. The chord distance ab is measured. Then

$$\sin \frac{1}{2} \alpha = \frac{ab}{200} \tag{8}$$

An angle is laid off in a similar manner, making use of the chord relation of Eq. (8). For example, to lay off an angle α from line AB with A as the vertex (Fig. 7-9), arcs are swung from A and a to an intersection at b.

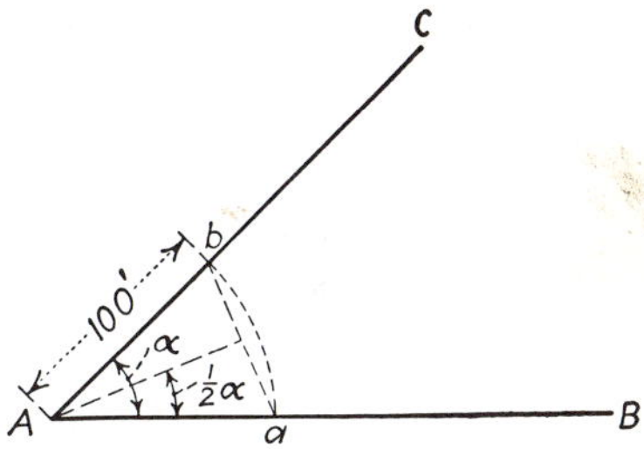

FIG. 7-9. Angle with tape.

Very small angles can be laid off more precisely with the tape than with the transit by making use of the tangent or sine of the angle as described in Art. 13-18. Other angles measured with the tape are usually considered as being less accurate than those measured with the transit.

7-27. Erecting Perpendicular to Line. For the purpose of locating the altitude of a triangle or of laying out a right angle as for a building corner, it is necessary to erect on the ground a perpendicular to an established line. This is usually done either by the 3:4:5 method or by the chord-bisection method. The 3:4:5 method requires less time, but the chord-bisection method is more precise. A prismatic sighting device is available which permits the quick and precise erection of perpendiculars without the use of a tape or transit.

3:4:5 *Method.* To erect a perpendicular to the line AB (Fig. 7-10a) that will include point C, a point a on line AB is assumed to be on the perpendicular, and a pin is set at a. With sides 3, 4, and 5 ft. or a multiple of these numbers, such as 24, 32, and 40 ft., a right triangle abc is constructed as follows: A pin is set on line AB at b, 32 ft. from a. The zero end of the tape is fixed with a pin at a, and the 100-ft. end at b. The head chainman moves to c and holds the 24-ft. and the 60-ft. marks of the tape in one hand, with the tape between these marks laid out so as to avoid kinking. He then sets a pin at c. The rear chainman moves from a to b as necessary to check the position of the tape at these points as c is established. He then sights along ac to C' beside C. If C does not lie on this prolongation, the perpendicular distance from C to the line aC' is measured, and the foot a of the perpendicular aC' is moved along the line AB by an equal amount, to the point

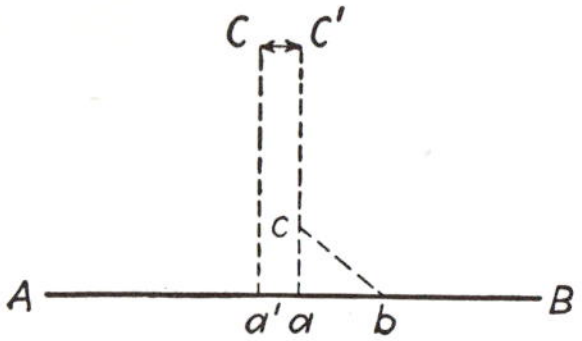

FIG. 7-10a. Perpendicular by 3:4:5 method.

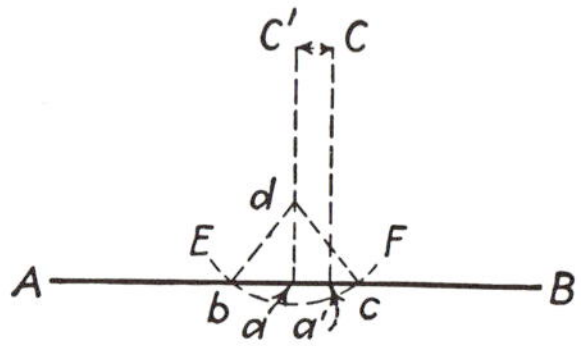

FIG. 7-10b. Perpendicular by chord-bisection method.

a'. If the trial perpendicular aC' fails to include the point C by several feet, the process is repeated for a', the new point; otherwise the location of a' may be assumed as correct.

If ground conditions are favorable, the point c may be established by striking arcs on the ground with $ac = 24$ ft. and $bc = 40$ ft. as a radius, using a chaining pin. The point c lies at the intersection of the arcs. This procedure avoids either fastening or bending the tape.

A good way of finding the approximate position of the perpendicular is to stand on the line AB with arms extended horizontally along the line. With eyes closed, bring the arms to the front, palms together; then sight along the line of the hands.

Chord-bisection Method. To erect a perpendicular to the line AB (Fig. 7-10b) that will include point C, the position of the perpendicular is estimated, and a pin is set at d on this estimated perpendicular, somewhat less than one tape length from the line AB. With d as center and length of tape as radius, the head chainman describes the arc EF of a circle, setting pins at the intersections b and c of the arc with the line AB. The rear chainman stationed at A or B determines the location of the intersections b and c on line. The point a is established midway between b and c. The line *ad* is prolonged to C' beside C, and the point a is moved if necessary, as described for the 3:4:5 method.

7-28. Irregular Boundary. Where a boundary line is irregular or curved, as along a shore line or a winding road, the usual procedure of locating the boundary is by means of perpendicular offsets from a straight line run as near the boundary as practicable. For straight portions of the boundary, offsets need be taken only at the ends. Where a curved boundary has many changes in direction, offsets should be taken at short intervals, which will usually be irregular. However, for convenience in calculating areas (Arts. 19-9 to 19-13), so far as possible the offsets are taken at regular intervals.

If the distance from the line to the boundary is not more than about 50 ft. and the boundary is fairly regular, usually it is sufficiently precise to erect the perpendiculars by estimation with the eye. For more precise work, the perpendicular may be erected with the tape (Art. 7-27), the transit, the compass, or a prismatic sighting device.

Typical field notes for determining offsets with the tape are given in Fig. 7-17.

7-29. Obstructed Distance. Often it becomes necessary to determine the distance between two points where direct chaining is impossible. If the points are intervisible, the distance may be determined by swing offsets, parallel lines, or similar triangles. If the points are not intervisible, the methods employing parallel lines are impossible.

Swing Offsets. To find the distance AB (Fig. 7-11) by the swing-offset method, the head chainman attaches the end of the tape to one end of the

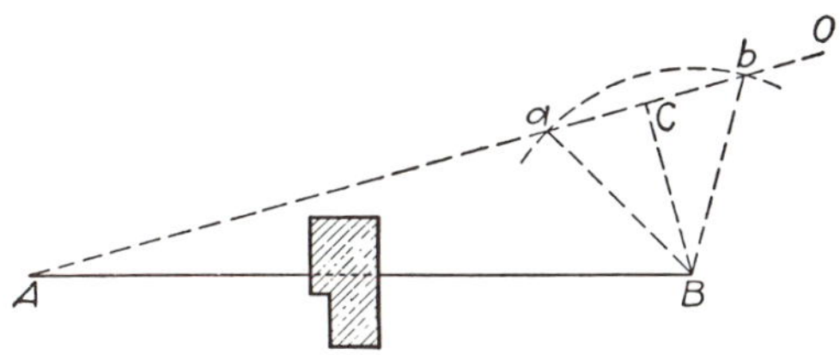

FIG. 7-11. Swing offset with tape.

line as at B and describes an arc with center B and radius 100 ft. The rear chainman stationed at A lines in the end of the tape with some distant object as O and directs the setting of pins at points a and b where the end of the tape crosses line AO. A point C midway between a and b lies on the perpendicular CB. The line ab is bisected, a pin is set at C, and the distances BC and CA are measured to obtain the necessary data for computing the length of AB.

Parallel Lines. If the necessary offset distance from the line AB is *short,* perpendiculars $AA' = BB'$ are erected by either method of Art. 7-27 to clear the obstacle (Fig. 7-12). The line $A'B'$ is then chained, and its length is taken as that of AB.

If a *long* offset is necessary, the method just described will be inaccurate because of the uncertainty of right angles measured with the tape. In such a case, a point C (Fig. 7-13) is established to clear the obstacle, such that the estimated value of α is less than 45°. The chord length of α for a radius of 100 ft. is determined. AC and CD are measured, CD being roughly perpendicular to AB. At C the angle α is laid off so that CE will be parallel to

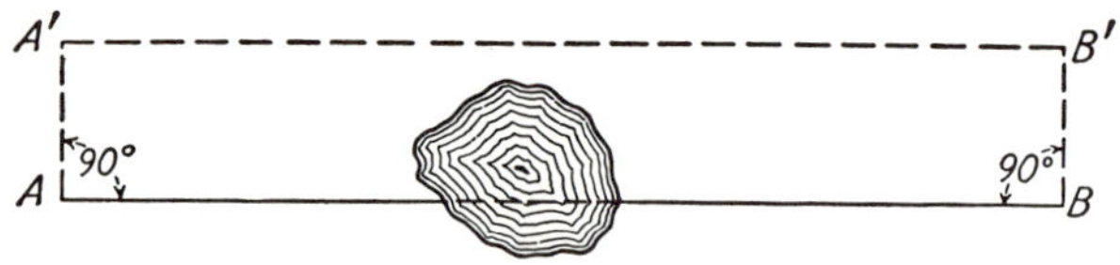

FIG. 7-12. Obstructed distance (short offset).

AB. CE is measured, E being any convenient point such that β will be less than 45°. EB and EF are measured, EF being roughly perpendicular to AB. The angle β is measured by determining its chord length for a radius of 100 ft. The right-angle triangles ADC and BFE are solved for AD and FB. Then $AB = AD + CE + FB$.

The precision of this method over that of the preceding method is due to the fact that the angles laid off are small and that the distances AD and BF are *computed* rather than measured. The reason for computing these distances will readily be seen when it is considered that any variation of the line DC from the true perpendicular will make little difference in its length as compared with the corresponding change of length of AD. To illustrate,

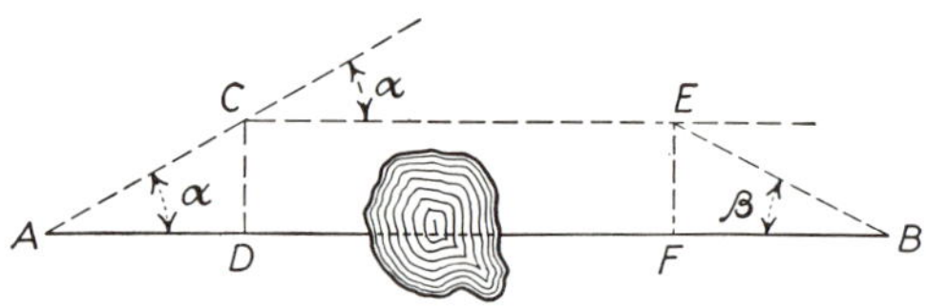

FIG. 7-13. Obstructed distance (long offset).

suppose α (Fig. 7-13) equals 45°, that the true length of AD and CD is 200 ft., and that D, supposedly on the perpendicular through C, is 10 ft. in error. As *computed* from AC and CD, AD is about 0.3 ft. in error, but its *measured* length would be 10 ft. in error.

Similar Triangles. Let C (Fig. 7-14) be a point from which A and B are visible. AC and BC are measured. CD and CE are laid off so that CD will bear the same relation to CA that CE bears to CB; that is, $CD/CA =$

CE/*CB*. It will generally be convenient to make this a simple ratio such as ½ or ⅓. The triangles *ACB* and *DCE* are similar. *DE* is measured, and *AB* is computed.

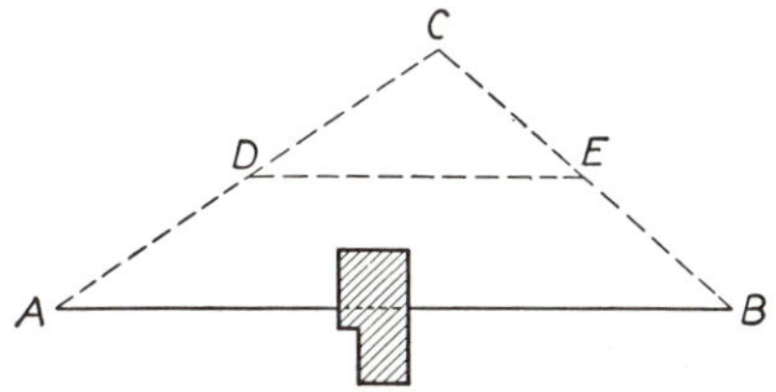

FIG. 7-14. Obstructed distance (similar triangles).

7-30. Numerical Problems.

1. The length of a line as measured with a 100-ft. steel tape is 1,012.3 ft. Afterward the tape is compared with the standard and is found to be 0.03 ft. too long. Compute the length of the line.
2. A building 80.00 by 160.00 ft. is to be laid out with a 50-ft. tape which is 0.016 ft. too long. What ground measurements should be made?
3. The actual distance between two marks at the City Hall is known to be 100.080 ft. When a field tape is held on this line, the observed distance is 100.03 ft. What is the actual length of the tape?
4. The slope measurement of a line is 800.0 ft. The differences in elevation between successive 100-ft. points, as measured with a hand level, are 1.0, 1.5, 2.5, 3.8, 4.6, 5.0, 7.5, and 6.2 ft. Determine the horizontal distance.
5. The slope measurement of a line is 1,246.5 ft. Slope angles measured with a clinometer are as shown below. Determine the horizontal distance, by exact and by approximate methods.

Chainage, ft.	0		300		800		1,000		1,246.5
Slope angle, degrees		½		1¼		2½		4	

6. Two points at a slope distance of about 100 ft. apart have a difference in elevation of 12.0 ft. What slope distance should be laid off to establish a horizontal distance of 100.00 ft.? Compute by exact and by approximate methods.
7. Compute the effect of sag per tape length for the two tapes of the example of Art. 7-20, using tensions of 20 and 40 lb. and distances between supports of 25 and 100 ft.
8. A 100-ft. tape weighing 2 lb. is of standard length under a tension of 12 lb., supported for full length. A line on smooth level ground is measured with the tape under a tension of 35 lb. and found to be 4,863.5 ft. long. E = 29,000,000 lb. per square inch; 3.53 cu. in. of steel weighs 1 lb. Make the correction for increase in tension.
9. A second line is measured with the tape of problem 8, the tape being supported at intervals of 50 ft. and the pull being 20 lb. The measured length is 1,823.6 ft. Compute the corrections for sag and variation in tension and determine the corrected length of the line.

10. Compute the normal tension for the tape of problem 8, the tape being supported at its ends.

11. Chainmen made two independent measurements of a line 10,000 ft. long. The ground was sloping and measurements were taken with the tape horizontal. The tape was 100 ft. long, and weighed 3 lb. One of the measurements of the line was made by two chainmen supporting the tape at the 0 and 100-ft. marks; the second measurement was made by three chainmen supporting the tape at the 0, 100, and 50-ft. points. The discrepancy between the two measurements was 11.7 ft. Several tests with a spring balance indicated that the average pull exerted by the chainmen was 20 lb. How much of the above discrepancy might be attributed to the different modes of supporting the tape?

12. For the purpose of establishing monuments in a city, a line along a paved street having a grade of 2.5 per cent is measured on the slope. The applied tension is 12 lb., and observations of temperature are made at each application of the tape. The measured length on the slope is 1,320.64 ft., and the mean of the observed temperatures is 87.4°F. The 100-ft. steel tape used for the measurements is standardized at 70°F., supported for full length, and is found to be 0.004 ft. too short under a tension of 12 lb. Determine the horizontal length of the line.

13. A line through rough country is taped by horizontal measurements and found to be 2,450 ft. long. On the average, it was necessary to use the plumb bob every 50 ft. If the probable error of plumbing from the end of the tape to the ground is ± 0.03 ft. in the direction of the line, compute the probable error due to inaccurate plumbing.

14. If, in problem 13, the average slope of the tape when measurements are taken is 2 in 100, what error from this source is introduced in the length of the line?

15. A line roughly 2 miles long along a railroad track is measured with a steel tape, and corrections are made for observed temperatures. What error will be introduced if the actual temperature of the tape is 2°F. higher than the observed temperature? State the error in fractional form with 1 as the numerator.

16. Assume that an invar tape having a coefficient of thermal expansion of 0.00000083 per 1°F. is used under the conditions of problem 15. Compute the error introduced.

17. A hedge along the line AB makes direct measurement impossible. A point C is established at an offset distance of 20 ft. from the line AB and roughly equidistant from A and B. The distances AC and CB are then chained; $AC = 1287.2$ ft., and $CB = 1353.0$ ft. By an approximate method, using the "slope correction," compute the length of the line AB.

18. It is desired to measure a distance of approximately 10 miles with a maximum permissible error of 1/10,000. The country is rolling (average slopes, 5 per cent) and wooded, so that for perhaps half the distance the tape must be held level, unsupported. Make any other necessary assumptions and write specifications (similar to those in Art. 14-16) for this work.

7-31. Field Problems.

Problem 1. Pacing

Object. To determine the length of normal pace, to test the reliability of 2½ and 3-ft. paces, and to determine an unknown distance by pacing.

Procedure. (1) Walk over an assigned course of known length ten times at an ordinary gait, counting the paces each time. Record each observed number in the field notebook. Compute the average length of the natural pace and average number of paces for 100 ft. (2) With a tape mark a course 30 ft. long, with every 3-ft. interval indicated. Walk over this course several times to obtain the proper stride; then pace the assigned course with this stride, recording the data and making computations as previously explained. (3) Follow a similar procedure for a 2½-ft. pace. (4) Walk over a course of unknown length several times at a natural pace, at a 2½-ft. pace, and at a 3-ft. pace. Estimate the distance by each method; and then find the true distance with a steel tape. Note the error.

Hints and Precautions. (1) In attempting to walk at a natural rate, avoid the general tendency to exceed that rate. (2) Count the paces carefully, estimating to the nearest one-tenth pace at the end of the course. (3) Reject observations that vary from the mean by more than 3 per cent. (4) Remember that field notes are a permanent record and should show clearly all the work done in the field. Plan an orderly form of notes and computations in advance. If an observation is rejected, draw a line through it, but do not erase.

Problem 2. Taping over Level Ground

Object. To measure with the 100-ft. steel tape over an approximately level course about 1,200 ft. long, and to check the distance by taping in the opposite direction.

Procedure. (1) Set a hub at each end of the line, and set a flagpole about 1 ft. beyond the far hub. Follow the procedure indicated in Art. 7-12. The distance should be read directly to tenths of feet and estimated to hundredths. Record the two lengths, and compute the ratio of discrepancy to length. Measurements should check within 1/5,000. As a rough check, measure the distance by pacing.

Hints and Precautions. (1) The rear chainmen should not hold the tape as he moves from station to station; otherwise if he moved too slowly, the head chainman would be retarded, and if he moved too fast the chain might become kinked. (2) Be careful not to disturb the "stuck" pin by allowing the tape to press against it. (3) Avoid injury to the tape; always keep it straight while in use. (4) Avoid inconsistent errors by checking every measurement.

Problem 3. Taping over Uneven Ground

Object. To standardize the 100-ft. steel tape; to find, by two methods, the horizontal length of an assigned course about 800 ft. long over uneven ground; and to correct for the error in length of tape as determined by the standardization tests.

Procedure. (1) Standardize the tape before and after the field work, by comparing it with the official standard of length. Maintain the required pull by means of a spring balance. Determine the error with a finely divided scale. (2) Chain the course by horizontal measurement after the manner described in Art. 7-13. Maintain a pull estimated to be equal to that used during standardization. For measurements in which the line slopes more than 5 or 6 ft. in 100, "break chain." (3) Correct for the error in length of tape by adding to or subtracting from the observed distance the observed error multiplied by the number of tape lengths in the course. Record the data in a form similar to that of Fig. 7-15. (4) Tape the course by measurement on the slope, recording the difference in elevation (to the nearest foot) of each 100-ft. tape length as determined by a hand level. If sharp breaks in slope occur between 100-ft. stations, treat each distance between breaks in the same manner as a full station. Correct the measurements for error in length of tape.

CHAINING OVER UNEVEN GROUND

5

Sta.	Length in feet Forward	B'kward	Mean	Correction	Corr. Length
351					
	311.73	311.65	311.69	+0.07	311.76
352					
	198.60 ~~198.40~~	198.62	198.61	+0.04	198.65
353					
	213.84	213.78	213.81	+0.05	213.86
354					

Standardization	
Actual length of standard	100.038'
Observed length of standard	
Before	100.017'
After	100.015'
Mean	100.016'
Actual length of tape	100.022'
Correction per tape length	+0.022'

Chaining Equipment	J. Pratt, H.C.
Locker No. 35	S. Fulton, R.C.
Chicago 100 Ft. Steel	Oct. 3, 1965 (2 hrs.)
Tape	Cold and Windy

Tk. in Stk. Ctr. Line M.C.R.R. at Haskell St.
East on Ctr. Line Haskell St. up Steep Hill
Nail in Pavement
North over Creek and R.R. Embankment
S.B.S.E. Corner Jones' Lot
North up Hill (Rough)
1" Iron Pipe N.E. Corner Jones' Lot

Tape compared with Standard at City Hall

FIG. 7-15. Notes for taping over uneven ground with tape horizontal.

Correct the measurements for slope, and determine the horizontal distance. (5) Compare the results obtained by the two methods of measurement.

Hints and Precautions. (1) In taping with the tape horizontal, avoid the general tendency to hold the downhill end of the tape too low, by (*a*) comparing the tape with some level line, (*b*) having the two ends in line with the horizon, or (*c*) estimating the angle between tape and plumb line. (2) It is usually more accurate to chain downhill, as the rear end of the chain can be held firmly on the ground.

PROBLEM 4. SURVEY OF FIELD WITH TAPE

Object. To collect sufficient data for calculating the area of a field having rectilinear boundaries, by two sides and included angle, by three sides, and by base and perpendicular methods.

Procedure. (1) Divide the field into triangles, avoiding as far as possible any construction that will result in forming a very acute angle; that is, make the triangles as nearly equilateral as the shape of the field will readily permit. (2) Measure the sides, the altitude, and one angle of each of the triangles, by the methods of Arts. 7-26 and 7-27. (3) Record the data in a form similar to that of Fig. 7-16.

Hints and Precautions. (1) If the perpendicular and the segments of the base on each side of its foot are measured, sufficient data will have been obtained for determining the angles by means of their tangents. However, the chord method of measuring angles is more precise, and the tangent method should be used only as a check. (2) Considerable care should be taken in lining-in points, and intersections should be determined as closely as the eye of the observer will allow.

SURVEY OF FIELD OF P. S. STROUT

With 100 ft. Steel Tape

Sta.	Dist.	Perpendicular	Chord Length	Sin ½ Angle	Angle
A					
	321.43				
D			168.00	0.8400	114°20′
	286.74				
B					
	511.08	164.32			
A					
B					
D			135.92	0.6796	85°38′
	184.01				
C					
	352.57	149.07			
B					
E					
	310.85				
D			139.64	0.6982	88°34′
A					
	441.50	226.80			
E					
D			116.94	0.5847	71°34′
C					
					360°06′

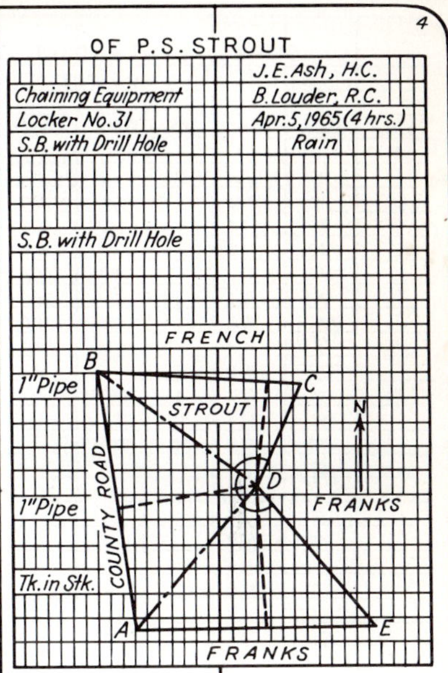

FIG. 7-16. Notes for survey with tape.

PROBLEM 5. IRREGULAR BOUNDARY BY OFFSETS WITH TAPE

Object. To collect sufficient data for calculating the area and for plotting an assigned field.

Procedure. (1) Divide as much of the field as its shape will permit into triangles as in problem 4, having in view simplicity of field operations. Lay out the triangles so that no long offsets will be necessary. (2) Collect data for finding the area of the triangles by one of the methods of problem 4. (3) From the triangle sides nearest the boundary, take offsets at such intervals as will insure sufficient precision for plotting and computation. (4) Record the data as shown in sample page of notes (Fig. 7-17).

Hints and Precautions. (1) To take the offsets correctly and quickly, set several pins at the desired intervals along the side of the triangle before measuring the offsets. (2) If the boundary changes abruptly at any point, be sure to take an offset at that point. (3) Do not measure the offsets with unnecessary precision. For example, if the sides of the triangles are measured to the nearest 0.01 ft., measure the offsets no closer than the nearest 0.1 ft. (4) Do not record offsets as being on the right of the line when they are on the left, or *vice versa*.

PROBLEM 6. DETERMINING OBSTRUCTED DISTANCE WITH TAPE

Object. To determine an obstructed distance between two points.

Procedure. (1) On an assigned course about 800 ft. long, assume that there is some obstacle which makes direct measurement and intervision impossible, and find

SURVEY OF LAND OF DAVID PARSONS, SMITHVILLE 3

With 100 ft. Steel Tape

Sta.	Dist.	Offset L	Sta.	Dist.	Offset L
A	0	0			
	25	21.4	D	0	0
	50	37.8		25	10.0
	75	44.6		50	26.2
	100	27.2		75	34.6
	111	0		100	34.5
	125	15.5		125	25.2
	140	16.7		150	14.0
B	159.06 0	0		175	4.1
	30	21.6	E	208.64 0	0
	50	42.0	A	275.98	
	90	51.4	AC	281.14	
	140	27.8	EC	273.61	
	165	13.0			
C	187.88 0	0			
	14	0			
	45	16.3			
	55	31.6			
	75	61.0			
	90	14.2			
	100	7.0			
D	109.89	0			

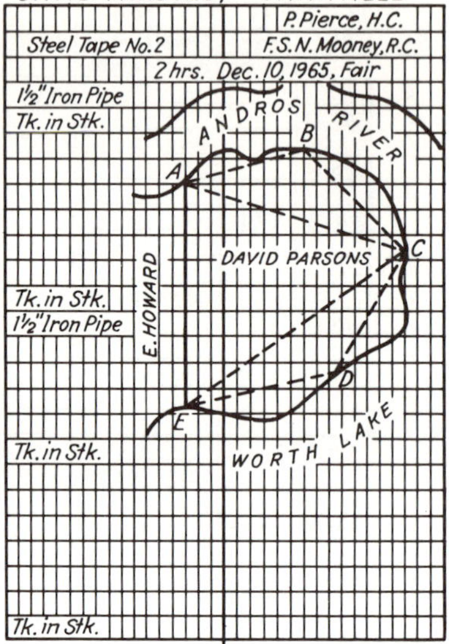

FIG. 7-17. Notes for offsets with tape.

the distance by the swing-offset method. (2) Find the distance by similar triangles. (3) Assume that the points are intervisible but that direct taping is impossible and determine the distance by the method of parallel lines, using offsets of 20 ft. and of 100 ft. (4) As a check, measure the distance directly. Compare the results.

REFERENCES

1. HARRISON, A. E., "Electronic Surveying: Electronic Distance Measurements," *J. Surveying Mapping Div., Am. Soc. Civil Engr.*, Vol. 89, No. SU3, pp. 97–116, October, 1963.
2. KONECNY, GOTTFRIED, "Control Surveys by Geodimeter and Tellurometer in Canada," paper presented at 43d annual meeting of Highway Research Board, January, 1964.
3. LAURILA, SIMO H., "Electronic Surveying: Accuracy of Tellurometer over Short Distances," *J. Surveying Mapping Div., Am. Soc. Civil Engr.*, Vol. 89, No. SU3, pp. 77–95, October, 1963.
4. NATIONAL BUREAU OF STANDARDS, "Calibration of Line Standards of Length and Measuring Tapes at the National Bureau of Standards," *Circ.* 572, Government Printing Office, Washington, D.C., 1956.
5. RADMANOVICH, D., "Evaluation of the Model 4 Geodimeter," paper presented at 43d annual meeting of Highway Research Board, January, 1964.

CHAPTER 8

LEVELING—METHODS AND INSTRUMENTS

METHODS

8-1. Definitions. The *elevation* of a point near the surface of the earth is its vertical distance above or below an arbitrarily assumed *level surface* or curved surface every element of which is normal to the plumb line. The level surface (real or imaginary) used for reference is called the *datum.* A *level line* is a line in a level surface.

The *difference in elevation* between two points is the vertical distance between the two level surfaces in which the points lie. *Leveling* is the operation of measuring vertical distances, either directly or indirectly, in order to determine differences in elevation.

A *horizontal line* is a line, in surveying taken as straight, tangent to a level surface.

A *vertical angle* is an angle between two intersecting lines in a vertical plane. In surveying, it is commonly understood that one of these lines lines is horizontal.

The datum most commonly used is mean sea level, particularly the 1929 Mean Sea Level Datum established by the United States Government. If merely the *relative* elevations of points for a particular survey are desired, the datum may bear no known relation to sea level. For example, the initial point in a survey may be assumed to have an elevation of 100 ft., and the elevations of all succeeding points computed accordingly. In order to avoid negative values of elevation, either the datum is taken below the lowest point of the survey or a positive value is assigned to it.

8-2. Curvature and Refraction. In leveling, it is necessary to consider the effects of (1) the curvature of the earth and (2) atmospheric refraction, which affects the line of sight. Usually these two effects are considered together.

In Fig. 8-1 is shown a horizontal line tangent at A to a level line near the surface of the earth. The vertical distance between the horizontal line and the level line is a measure of the earth's curvature. It varies approximately as the square of the distance from the point of tangency.

Owing to the phenomenon of atmospheric refraction, rays of light transmitted along the surface of the earth are refracted, or bent downward slightly. Thus, as viewed from point A (Fig. 8-1), an object actually at B would appear to be at C; the actual line of sight is along a curve AB which is concave downward. The amount of refraction is variable, but it is relatively small compared with the earth's curvature; therefore, for ordinary work it is taken as constant.

The combined effect of the earth's curvature and atmospheric refraction is given by the expression

$$h' = 0.57K^2 = 0.021M^2 \qquad \text{(approx.)} \tag{1}$$

in which K = distance from the point of tangency (station of the observer), in miles

M = same distance, in thousands of feet

h' = combined effect of the earth's curvature and atmospheric refraction, in feet

The effect of the earth's curvature alone is about $0.66K^2$ or $0.024M^2$. The effect of refraction alone is about $0.09K^2$ or $0.003M^2$ in the opposite direction.

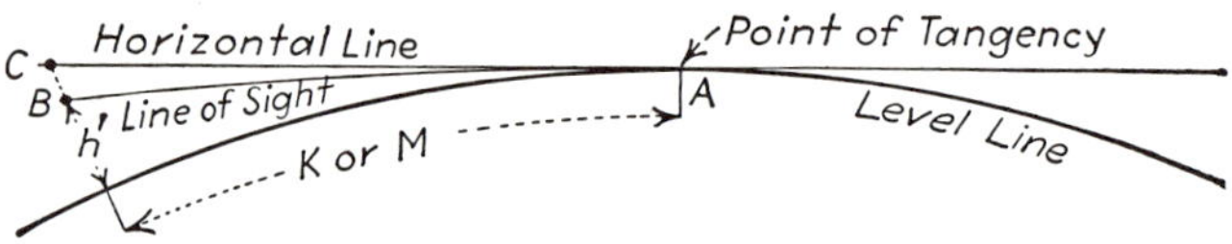

FIG. 8-1. Earth's curvature and atmospheric refraction.

8-3. Methods. Difference in elevation may be measured by the following methods:

1. *Direct* or *spirit leveling*, by measuring vertical distances directly. Direct leveling is the most precise method of determining elevations and is the one commonly used (Art. 8-4).

2. *Indirect* or *trigonometric leveling*, by measuring vertical angles and horizontal distances (Art. 8-5).

3. *Barometric leveling*, by measuring the difference in atmospheric pressure at various stations by means of a barometer (Art. 8-6).

Differential leveling (Chap. 9) is the operation of determining differences in elevation of points some distance apart or of establishing bench marks. Usually differential leveling is accomplished by direct leveling. *Precise leveling* is a precise form of differential leveling.

Profile leveling (Chap. 10) is the operation—usually by direct leveling—of determining elevations of points at short measured intervals along a definitely located line such as the center line for a highway or a sewer.

Direct leveling is also employed for determining elevations for cross-sections, grades, and contours.

A "ground elevation meter" developed by the U.S. Corps of Engineers consists of a vehicle on which is mounted a mechanical device that integrates the vertical component of any longitudinal movement. Its accuracy is within 1 ft. $\sqrt{\text{miles}}$, and it can travel up to 30 miles per hour.

8-4. Direct Leveling. In Fig. 8-2, A represents a point of known elevation and B represents a point the elevation of which is desired. In the method of direct or spirit leveling, the level is set up at some intermediate point as L, and the vertical distances AC and BD are observed by holding a leveling rod first at A and then at B, the line of sight of the instrument being horizontal. (Owing to refraction the line of sight is slightly curved as explained previously.)

If the difference in elevation between the points A and E is designated as H_a and between E and B as H_b,

$$H_a = h_a - h_a' \qquad \text{and} \qquad H_b = h_b - h_b'$$

in which h_a and h_b are the vertical distances read at A and B, respectively, and h_a' and h_b' are the effects of curvature and refraction for the horizontal distances LA and LB, respectively.

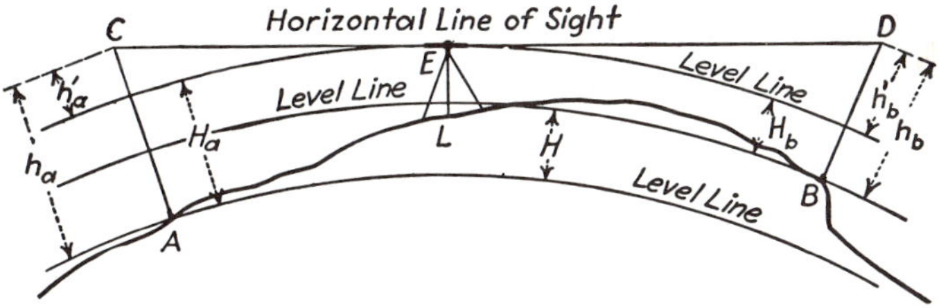

FIG. 8-2. Direct leveling. (Owing to refraction, the line of sight is slightly curved.)

The difference in elevation H between A and B is then

$$\begin{aligned} H &= H_a - H_b = (h_a - h_a') - (h_b - h_b') \\ &= h_a - h_b - h_a' + h_b' \end{aligned} \tag{2}$$

If the backsight distance LA is equal to the foresight distance LB, then $h_a' = h_b'$ and

$$H = h_a - h_b \tag{3}$$

Thus if backsight and foresight distances are balanced, the difference in elevation between two points is equal to the difference between the rod readings taken to the two points, and no correction for curvature and refraction is necessary. In direct leveling, usually the work is so conducted that the effect of curvature and refraction is reduced to a negligible amount (Art. 9-5).

On lines of direct levels the usual procedure is as follows. Let A and D be two established points some distance apart whose difference in elevation is desired. With the level in some convenient location, not necessarily on a line joining A and D, a backsight is taken to point A and a foresight is taken to some convenient point B. The level is then moved ahead, a backsight is taken to B, and a foresight is taken to some accessible point as C. And so the process is repeated until the terminal point is reached.

8-5. Indirect Leveling. In Fig. 8-3, A represents a point of known elevation and B a point the elevation of which is desired. In the method of indirect or trigonometric leveling, the vertical angle at A is measured, and

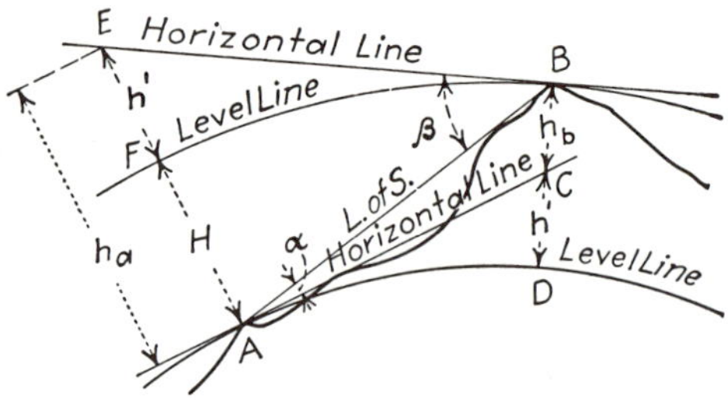

FIG. 8-3. Indirect, or trigonometric, leveling. (Owing to refraction, the line of sight is slightly curved.)

the distance AD is determined by some method of measurement. Within the limits of ordinary practice $AD = AC$ and $\angle BCA = 90°$. Therefore

$$h_b = AC \tan a$$

The correction h' for curvature and refraction is, by Eq. (1);

$$h' = 0.021M^2 = 0.021 \left(\frac{AC}{1{,}000}\right)^2$$

in which AC is in feet.

The difference in elevation H is, therefore,

$$H = h_b + h' = AC \tan \alpha + 0.021 \left(\frac{AC}{1{,}000}\right)^2 \tag{4}$$

If the vertical angle β is now taken from B to A, by a similar course of reasoning $h_a = EB \tan \beta$. For the horizontal distances employed on any ordinary survey, EB is the equivalent of AC. Therefore,

$$h_a = AC \tan \beta \tag{5}$$

and the difference in elevation is

$$H = h_a - h' = AC \tan \beta - 0.021 \left(\frac{AC}{1{,}000}\right)^2 \tag{6}$$

From Eqs. (4) and (6) it will be noted that when the vertical angle is upward or positive the curvature and refraction correction is added; and when downward or negative, the curvature and refraction correction is subtracted.

Adding Eqs. (4) and (6),

$$2H = (h_a - h') + (h_b + h') = h_a + h_b$$

or

$$H = \frac{h_a + h_b}{2} = \frac{AC}{2} (\tan \alpha + \tan \beta) \qquad (7)$$

From Eq. (7) may be deduced the general rule that *when vertical angles are measured from (and to) each of two points whose difference in elevation is desired, the difference in elevation is one half of the horizontal distance between them multiplied by the sum of the tangents of the angles,* and the effect of the earth's curvature and atmospheric refraction is thereby eliminated. In precise trigonometric leveling this is the procedure employed.

Uses. In ordinary surveying, indirect leveling furnishes a rapid means of determining the elevations of points in rolling or rough country. On reconnaissance surveys, angles may be measured with the clinometer and distances may be obtained by pacing. On more precise surveys, angles are measured with the transit and distances by the stadia. Indirect leveling is used extensively in plane-table work. Determining difference in elevation with the gradienter (Art. 2-20) is one form of indirect leveling.

Procedure. On lines of indirect levels for which angles are measured with the transit, the usual procedure is as follows. Let A and D be two points whose difference in elevation is desired. With the transit at some intermediate location T_1, not necessarily on line between A and D, the distance and vertical angle to A are determined by a *backsight,* and similar quantities are measured by taking a *foresight* to another intermediate point B at a convenient location not necessarily on line between A and D. The instrument is then moved ahead to T_2, and similar observations are taken to B and C. And so the process is repeated until the end of the line is reached. If the transit is *equidistant* from the points on either side of it to which sights are taken, the effect of curvature and refraction will be eliminated.

In practice, very little attention is paid to keeping the backsight and foresight distances balanced, since the effect of curvature and refraction is negligible (0.02 ft. for a distance of 1,000 ft.) as compared with the precision with which elevations can be determined by this method (generally not closer than tenths of feet). The error from such practice is accidental in that, while one backsight distance may be greater than the corresponding foresight distance, normally the next is just as likely to be smaller. Generally the transit is set at some convenient place where good sights can be obtained in both directions and which is about the same distance from adjacent points on which sights are to be taken.

In small-scale mapping, the method of indirect leveling is employed to determine the difference in elevation between the plane table and a point sometimes at a distance of several miles. In such cases the effect of curvature and refraction becomes large and the correction must be applied. For example, if the horizontal distance from the plane table to the point sighted were 10 miles, the correction would be 57 ft.

Errors. The errors of indirect leveling are chiefly accidental. The precision attainable depends upon the length of sight, the instrument used, and the magnitude of the vertical angles. With the transit, under average conditions the error may be expected to be not greater than 0.4 ft. times the square root of the distance in miles.

8-6. Barometric Leveling. Since the pressure of the earth's atmosphere varies inversely with the elevation, the barometer may be employed for making observations of difference in elevation. Barometric leveling is employed principally on exploratory or reconnaissance surveys where differences in elevation are large, as in hilly or mountainous country. Since atmospheric pressure may vary over a considerable range in the course of a day or even an hour, elevations determined by one ordinary barometer carried from one elevation to another may be several feet in error. However, by means of sensitive barometers and special techniques, elevations can be determined within a foot or so.

Usually barometric observations are taken at a fixed station during the same period that observations are made on a second barometer which is carried from point to point in the field. This procedure makes it possible to correct the readings of the portable barometer for atmospheric disturbances.

Computations. The difference in elevation between two points A and B is given by Eq. (8), assuming that the mean of the temperatures at A and B is 50°F. and neglecting the effects of humidity and of atmospheric disturbances.

$$z \text{ (uncorrected)} = 62{,}737 \log \frac{30}{h_a} - 62{,}737 \log \frac{30}{h_b} \tag{8}$$

where z is the difference in elevation in feet, and h_a and h_b are, respectively, the barometer reading (in inches of mercury) at points A and B. Each term of the second member represents the elevation of the corresponding point above a datum plane of barometric pressure 30 in., or approximately at mean sea level.

For a mean temperature at the two points other than 50°F. and *for average conditions of humidity*, a proportionate correction is added (algebraically). The amount of this correction is determined by multiplying the uncorrected difference in elevation, from Eq. (8), by the appropriate factor in the following tabulation:

CORRECTION FACTORS FOR ELEVATION BY BAROMETER

Mean temp., °F.	Factor	Mean temp., °F.	Factor	Mean temp., °F.	Factor
0	−0.1024	35	−0.0273	70	+0.0471
5	−0.0915	40	−0.0166	75	+0.0575
10	−0.0806	45	−0.0058	80	+0.0677
15	−0.0698	50	+0.0049	85	+0.0779
20	−0.0592	55	+0.0156	90	+0.0879
25	−0.0486	60	+0.0262		
30	−0.0380	65	+0.0368		

Example: Given barometer readings at A and B, respectively, of 26.850 and 28.315 in., and corresponding temperatures of 48 and 72°F. Determine the difference in elevation.

By Eq. (8)

$$z \text{ (uncorrected)} = 3{,}022.5 - 1{,}575.0 = 1{,}447.5 \text{ ft.}$$

From the table, the correction factor for a mean temperature of 60°F. is +0.0262.

$$1{,}447.5 \times +0.0262 = +37.9 \text{ ft.}$$
$$z \text{ (corrected)} = 1{,}447.5 + 37.9 = 1{,}485.4 \text{ ft.}$$

Some mercurial barometers have an auxiliary scale by means of which the correction is made mechanically.

Instruments and Methods. The mercurial barometer is accurate, but it is cumbersome and is suitable only for observations at a fixed station. For field use, an aneroid barometer is commonly used because it is light and is easily transported. The usual type has a dial about 3 in. in diameter, graduated both in inches of mercury and in feet of altitude (elevation); it is compensated for temperature. At a point of known altitude, the pointer can be set at the corresponding reading on the scale in order to place the instrument in adjustment. The aneroid barometer can be calibrated against a mercurial barometer by comparing values at a given station over a range of temperature.

In use, the barometer should be given time to reach the temperature of the air before an observation is made.

A single aneroid barometer is sometimes used by topographers on small-scale surveys where the contour interval is large. Stops are made at frequent intervals during the day, and the rate of change in atmospheric conditions is observed; suitable corrections are thus determined and are applied to the observed values. Where distances permit, it is preferable to return to the starting point and to correct the intermediate readings in proportion to the change in atmospheric pressure during the interval between observations.

Sensitive Barometers. Extremely sensitive barometers have been developed, with which elevations can be determined within a foot or so. One type used in topographic surveying employs two of the instruments at fixed bases and one or more instruments carried from point to point over the area being surveyed. One fixed instrument is located at a point of known elevation near the highest elevation of the area, and one near the lowest elevation; these instrument stations are called the *upper base* and *lower base*, respectively. A third instrument is carried to the point whose elevation is desired, and a reading is taken. Readings on the fixed instruments are taken either simultaneously (as determined by signaling) or at fixed intervals of time; in the latter case the readings at the desired instant are determined by proportion. The elevation of the portable instrument is then determined by interpolation. The horizontal location of each point at which a reading is taken is determined by conventional methods.

Example: Given elevation of upper base 275 ft., of lower base 56 ft.; the difference in elevation between the bases is, therefore, 275 − 56 = 219 ft. At a given instant, the three altimeter readings indicate that the difference in elevation of an intermediate point from the upper base is 209 ft. and from the lower base is 25 ft.; thus the indicated total difference in elevation between bases is 234 ft. The corrected differences in elevation are proportionately 219/234 × 209 = 195 ft. (from upper base) and 219/234 × 25 = 24 ft. (from lower base); as a check, the total computed difference in elevation between bases is now 195 + 24 = 219 ft. The elevation of the point is 80 ft., computed by difference from either base (275 − 195 = 80; or 56 + 24 = 80).

Another procedure is to employ one barometer at a fixed base and one barometer which is carried to points whose elevation is desired, simultaneous readings being taken. The carried barometer is finally brought either back to the starting point or to another point of known elevation; the computation of the elevation of each point then takes account of the corresponding change in atmospheric pressure at the fixed base.

INSTRUMENTS FOR DIRECT LEVELING

8-7. Kinds of Levels. Any instrument commonly used for direct leveling has as its essential features a *line of sight* and a *level tube* (Art. 2-5) or some other means of making the line of sight horizontal. The level tube is so mounted that its axis is parallel to the line of sight. The instrument used principally in the United States is the *engineer's level*, the essential features of which are described in Chap. 2. The two distinct types of the engineer's level are the *dumpy level* (Art. 8-8) for which the telescope tube is permanently fastened to the level bar and the *wye level* (Art. 8-9) for which the telescope is removable and rests in Y-shaped supports. Modern American and European levels, including *tilting* and *self-leveling* or *automatic* levels, are described in Art. 8-10. The *architect's level* or *construction level*, a modified form of the engineer's level but with a telescope of lower magnifying power and with a less sensitive level tube, is used in establishing grades for buildings. The *hand level* (Arts. 8-11 and 8-12) is a simple and useful device for determining differences in elevation roughly. Instruments which are frequently used for direct leveling but which are not primarily designed for this purpose are the *engineer's transit* (Chap. 13) and the telescopic alidade of the *plane table* (Chap. 17).

The measurements of difference in elevation are made by sighting on a graduated rod, called a *leveling rod* (Arts. 8-13 to 8-16). Other accessories sometimes used are a *rod level* (Art. 8-17) and a *turning point* (Art. 8-18).

Usually the leveling head of the engineer's level is equipped with four leveling screws, but the three-screw type is favored by some engineers and surveyors, particularly for instruments of high precision. Reasons for this preference are that the three-screw type can be leveled rapidly, requires the use of only one hand, and is relatively stable as compared with the four-screw type when the latter is not perfectly set. With the three-screw leveling head, on the other hand, when one of the screws is turned, the elevation of the telescope may change slightly and in some models the leveling head may rotate slightly.

8-8. Dumpy Level. Figure 8-4 shows the details of a conventional dumpy level with erecting eyepiece. The telescope A is rigidly attached to the

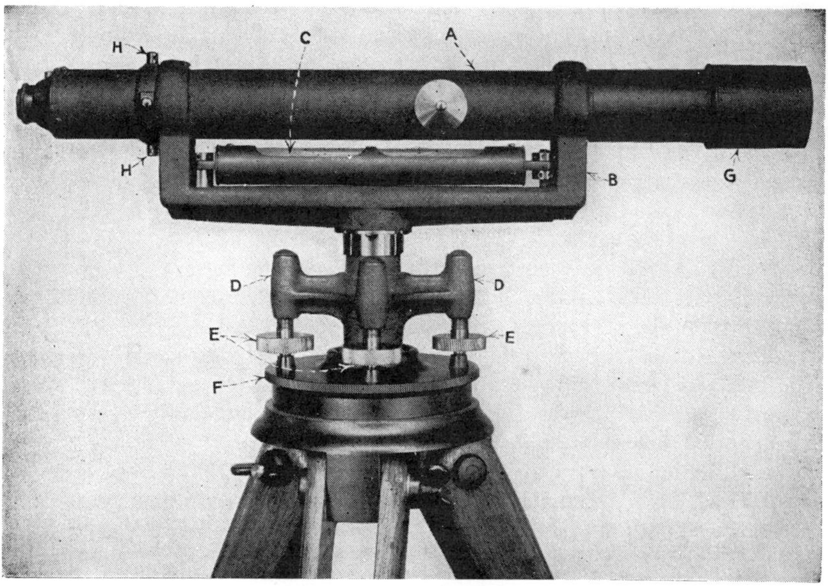

FIG. 8-4. Engineer's dumpy level.

level bar *B*, and the instrument is so constructed that the optical axis of the telescope is perpendicular to the axis of the center spindle. The level tube *C* is permanently placed so that its axis lies in the same vertical plane as the optical axis, but it is adjustable in altitude by means of a capstan-headed screw at one end. The spindle revolves in the socket of the leveling head *D*, which is controlled in position by the four leveling screws *E*. At the lower end of the spindle is a ball-and-socket joint which makes a flexible connection between the instrument proper and the foot plate *F*. When the leveling screws are turned, the level is moved about this joint as a center. The sunshade *G* protects the objective from the direct rays of the sun. The adjusting screws *H* for the cross-hair ring are near the eyepiece end of the telescope.

The telescope of the dumpy level usually has a magnifying power of about 30 diameters, and the level tube usually has a sensitiveness of 20 seconds of arc per graduation (2 mm.).

The name "dumpy level" originated from the fact that formerly this level was usually equipped with an inverting eyepiece and therefore was shorter than a wye level of the same magnifying power. Its advantages over the wye level are that it is simpler in construction, has fewer parts subject to wear, requires fewer adjustments, and stays in adjustment better. It is the type most commonly employed.

Precise Dumpy Levels. A modified form of the dumpy level, used for precise work, has the telescope hinged to one end of the level bar and resting on the point of a micrometer screw in the other end of the level bar. The level tube is either attached to the telescope in the usual manner or at the side or above. The *precise level* of the U.S. Geological Survey is a refined form of the dumpy level; its internal-focusing telescope has a magnifying power of about 40 diameters, and the level tube has a sensitiveness of 10 seconds of arc per graduation (2 mm.). The U.S. Coast and Geodetic Survey also has a precise level similar in its essentials to but differing in detail from the Geological Survey level.

8-9. Wye Level. Although the wye level is passing out of use, so many are continuing in service in the United States that the characteristics and adjustment of this type of level are described herein.

Figures 8-5 and 8-6 show the details of a wye level with erecting eyepiece. The telescope rests in Y-shaped bearings called the *wyes*. The leg of each wye passes through the level bar and is secured in position by capstan nuts. By means of the capstan nuts (30 and 31) on one of the wye legs, the wye can be raised or lowered. The telescope is secured in position by the wye clips or stirrups (32). When the clips are raised, the telescope may be revolved in the wyes or it may be lifted from the wyes and turned end for end. The enlarged cylindrical portions (39) of the telescope barrel which rest in the wyes are called the *rings* or *collars*. The line joining the centers of the

FIG. 8-5. Engineer's wye level.

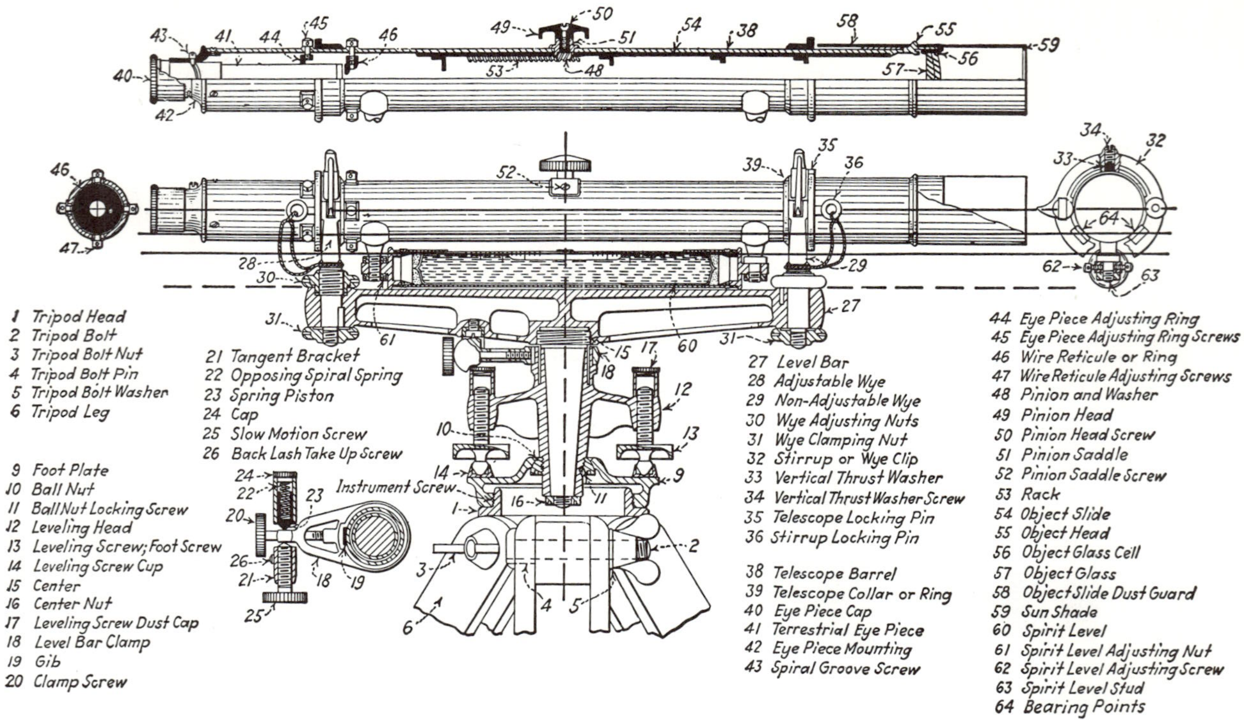

Fig. 8-6. Cross-section of wye level.

rings is defined as the *axis of the collars.* The *axis of the wyes* is a general term used to denote a reference line—sometimes actually the axis of the collars and sometimes actually the axis of the supports—which represents the alinement of the telescope tube in the wye supports. The telescope is held longitudinally by a flange on each ring, which bears against the side of the wye. When the clips are fastened, the telescope is held from turning about its axis by a lug on one of the clips.

The telescope of the wye level usually has a magnifying power of about 30 diameters, and the level tube usually has a sensitiveness of 20 seconds of arc per graduation (2 mm.).

The level tube is attached to the telescope and is adjustable in both a horizontal and a vertical plane. Other details are much the same as for the dumpy level previously described. In the instruments shown in Figs. 8-5 and 8-6, above the leveling head is a collar and a clamp-screw by means of which the spindle may be clamped to the leveling head. The tangent-screw controls small movements of the level about its vertical axis after the clamp-screw is tightened.

The distinguishing characteristics of the wye level are (1) that the telescope may be revolved about its own axis in the wyes and (2) that the telescope may be lifted from the wyes and turned end for end. These features are of no particular advantage in the work of leveling, but they facilitate the making of adjustments, provided the bearings are not worn. (The wye level can be adjusted by one man whereas the dumpy level requires two men.) Each collar of the telescope is in contact with the wye at two points (64, Fig. 8-6). At these points the collars become worn and flattened in use so that they are no longer cylindrical, nor are they likely to be of the same size or shape; furthermore, the bearing points of the wyes may become worn unevenly. Under these conditions it is impossible to adjust the instrument correctly by the usual methods, and the adjustments are the same as for the dumpy level.

8-10. Modern Types. Modern levels, although essentially of the "dumpy" type, incorporate several features which make them quite different from the conventional dumpy level described in Art. 8-8. They are small, light, and very accurate and permit of rapid setting up and observing. They are usually light in color to minimize temperature effects of sunlight. They need little or no adjustment in the field. They are equipped with a circular spirit level for approximate preliminary leveling. They may have three or four leveling screws; one type has a clamping ball-and-socket arrangement instead of leveling screws. The telescope is internal-focusing and has the crosslines, including stadia lines, etched on a glass diaphragm. The telescope level bubble is easily and accurately centered by bringing the images of the ends of the bubble into coincidence as viewed from the eyepiece end of the telescope through a system of prisms. Electric illumination can be provided for night observations. Some instruments are equipped with an optical micrometer, which raises or lowers the line of sight slightly without changing its direction and thus permits readings to be taken at even graduations on the rod; the micrometer reading is applied to the rod reading. Some levels are equipped

FIG. 8-7. K & E tilting level.

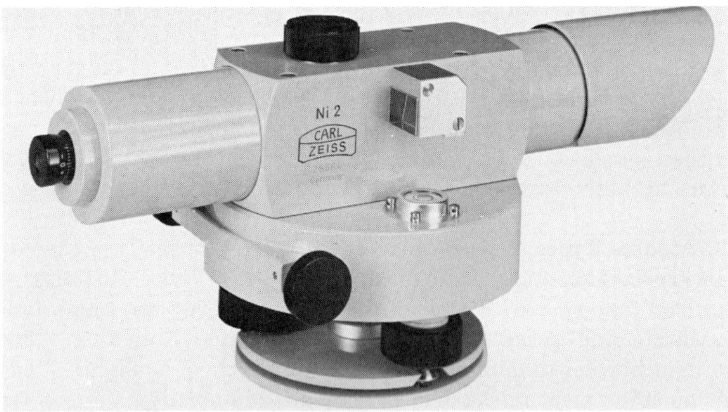

FIG. 8-8. Zeiss self-leveling level. (*Keuffel & Esser Co.*)

with a horizontal circle, with graduations on glass, for observing directions approximately.

Tilting Level. The distinctive feature of a *tilting level* (Fig. 8-7) is that the telescope is mounted on a transverse fulcrum at the vertical axis and on a micrometer screw at the eyepiece end of the telescope. After the instrument has been leveled approximately in the usual manner, perhaps by use of a circular spirit level, the telescope is pointed in the direction desired and is then "tilted," or rotated slightly in the vertical plane of its axis by turning the micrometer screw, until the sensitive telescope-level bubble is centered. The line of sight is then horizontal, even through the instrument as a whole is not exactly level. On the micrometer screw is mounted a graduated drum which may be used as a gradienter (Art. 2-20).

Self-leveling Level. A *self-leveling,* or *automatic,* level keeps its line of sight horizontal by means of some form of pendulum and a system of prisms and mirrors. The pendulum is damped magnetically, and its operation is quick and accurate. The level has no tilting arrangement and no spirit-level tube; at each set-up the instrument is leveled approximately by use of a circular spirit level, and the pendulum then maintains the line of sight horizontal. The telescope may be provided with an optical micrometer for measuring the difference in elevation between the line of sight and the nearest graduation on the leveling rod. One form of self-leveling level is shown in Fig. 8-8.

8-11. Locke Hand Level. The Locke hand level is widely used for rough leveling. It consists of a metal sighting tube about 6 in. long on which is mounted a level vial (Fig. 8-9). In the tube beneath the vial is a prism

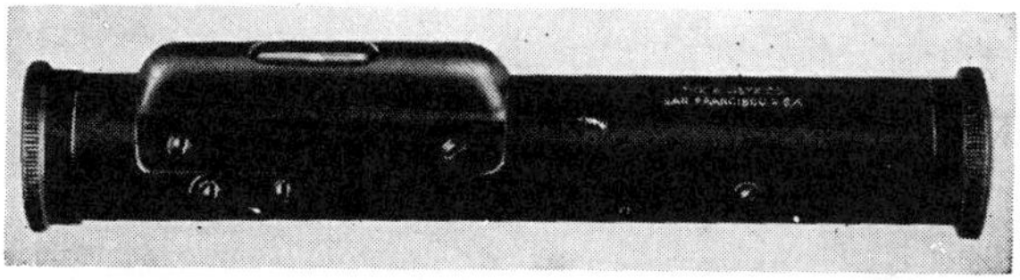

FIG. 8-9. Locke hand level. (*The A. Lietz Co.*)

which reflects the image of the bubble to the eye end of the level. Just beneath the level vial is a cross-wire which is adjustable by means of a pair of screws, the heads of which protrude from the case; one screw is loosened and the other is tightened. The eyepiece consists of a peephole mounted in the end of a slide which fits inside the tube and is held in a given position by friction. Mounted on the right half of the inner end of the slide is a semicircular convex lens which magnifies the image of the bubble and cross-wire as reflected by the prism. Both the object and the eye ends of the tube are closed by disks of plain glass so that dust will not collect on the prism and lens. The magnifying lens is focused by moving the eyepiece slide in or out.

In using the level the object is viewed directly through the left half of the sighting tube, without magnification, while with the same eye and at the same time the position of the bubble with respect to the cross-wire is observed in the right half of the field of view. The level is held with the level vial uppermost and is tipped up or down until the cross-wire bisects the bubble, when the line of sight is horizontal. After a little practice one may make observations with greater facility by keeping both eyes open. Some observers steady the hand level by holding it against, or fastening it to, a staff. Hand levels equipped with stadia hairs are available.

8-12. Abney Hand Level and Clinometer. As its name indicates, this level is suitable both for direct leveling and for measuring the angles of slopes. The instrument shown in Fig. 8-10 is graduated both in degrees and in percentage of slope, or grade. When it is used as a level, the index of the vernier is set at zero, and it is then used in the same way as the Locke hand level. When it is used as a clinometer, the object is sighted, and the level tube is caused to rotate about the axis of the vertical arc until the

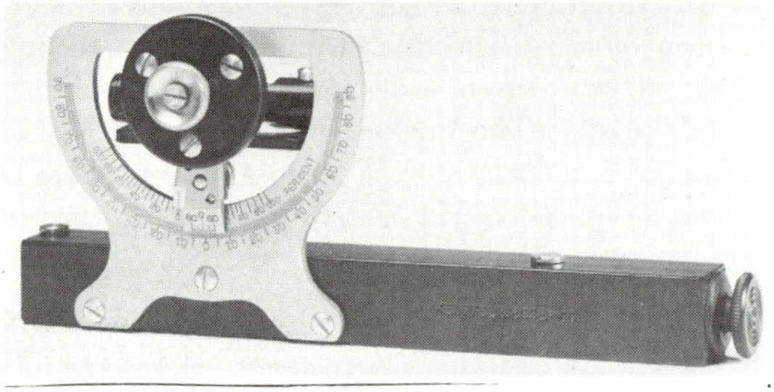

FIG. 8-10. Abney hand level and clinometer. (*Keuffel & Esser Co.*)

cross-wire bisects the bubble as viewed through the eyepiece. Either the slope angle or the slope percentage is then read on the vertical arc.

8-13. Leveling Rods. These are graduated wooden rods of rectangular cross-section by means of which difference in elevation is measured. The lower or ground end of the rod is shod with metal to protect it from wear and is usually the point of zero measurement from which the graduations are numbered.

The rod is held vertical, and hence the reading of the rod as indicated by the horizontal cross-hair of the level is a measure of the vertical distance between the point on which the rod is held and the line of sight.

Rods are obtainable in a variety of types, patterns, and graduations and are either in single pieces or in sections which are jointed together or slide past each other and are clamped together. Common lengths are 12 and 13 ft. It the United States, ordinarily rods are graduated in hundredths of a foot. On some government surveys the rods are graduated in decimals of the meter or the yard.

The two general classes of leveling rods are (1) *self-reading* rods, sometimes called *speaking* rods, which may be read directly by the leveler as he looks through the telescope of the level, and (2) *target* rods, for which a target sliding on the rod is set by the rodman as directed by the leveler. Under ordinary conditions, observations with the self-reading rod can be made with nearly the same precision and much more rapidly. The self-reading rod is the one commonly employed, even for precise leveling.

Suggestions for the care of leveling rods are given in Art. 3-11.

8-14. Self-reading Rods. With the self-reading rod, the rodman simply holds the rod vertical; the leveler observes the graduation at which the line of sight intersects the rod and records the reading. Observations closer than the smallest division on the rod are made by estimation.

The self-reading rod should be so marked that the graduations appear sharp and distinct for any normal distance between level and rod. Most commonly the background is white with graduations 0.01 ft. wide painted in black as shown in Figs. 8-11 and 8-12; the readings to 0.01 ft. are made on the edges of the graduations. Usually the rod is read to half-hundredths of a foot, by estimation. The numbers indicating feet are in red, and those indicating tenths of feet are in black. This style of graduation is satisfactory for self-reading when the length of sight is less than 400 or 500 ft. For sights of greater length such graduations become hazy, and rods with larger blocks of contrasting color are desirable. Many types of rods are designed for precise reading at short distances and at the same time for clear reading at long distances and are adapted for use either as leveling rods or as stadia rods; some of these are shown in Fig. 15-2. Figure 8-14 shows a graduation suitable for leveling with the hand level.

The *Philadelphia rod* (Fig. 8-11) is the most widely used of all rods. It is equipped with a target and is, therefore, also a target rod. It is usually in two sections or strips, the strips being held in contact by two brass sleeves. By means of a screw attached to the upper sleeve the two strips may be clamped together in any relative position desired. For readings of 7 ft. or less (on a 13-ft. rod) the back strip is clamped in its normal position. For greater readings, the rod is extended its full length; the graduations on the front face of the back strip are then a continuation of those on the front strip. When thus extended, the rod is called a "long" rod.

The *Chicago rod* (Fig. 8-12) is in three sections with slip joints.

The *California rod* has graduations similar to those of the Philadelphia rod, but on the front side only; it is made in either three or two sections.

The *Florida rod* is a one-piece rod 10 ft. long with a tapering rib fastened to its back. It is equally well adapted to short and to long sights.

Graduated flexible *ribbons* of enameled and waterproofed fabric are obtainable. Such a ribbon attached to a plain wooden strip makes a serviceable and accurate leveling rod.

Some rods are provided with a graduated strip of invar steel in order to minimize the effect of changes in temperature and of changes in the length of the wooden rod due to changes in humidity. The invar strip is fastened at the ends only and is kept taut by means of a spring. Invar scales may be mounted on conventional rods.

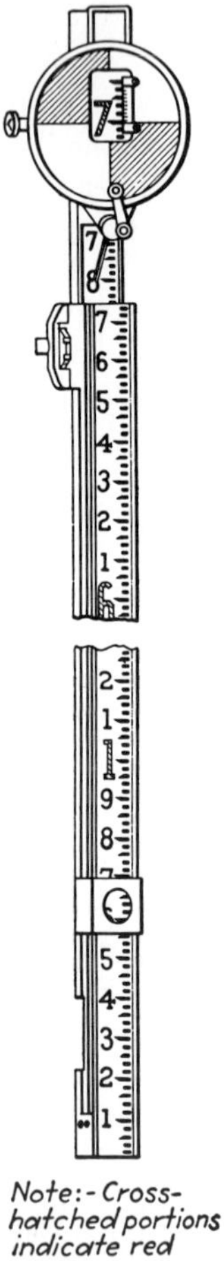

FIG. 8-11. Philadelphia rod.

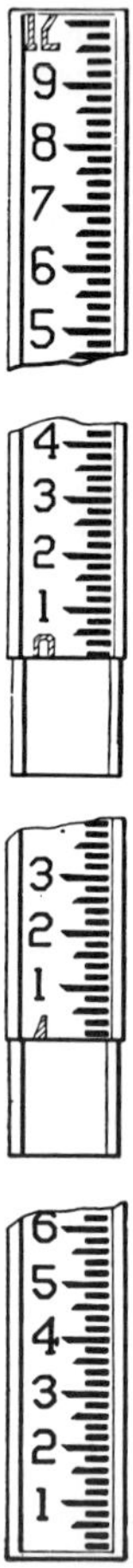

FIG. 8-12. Chicago rod.

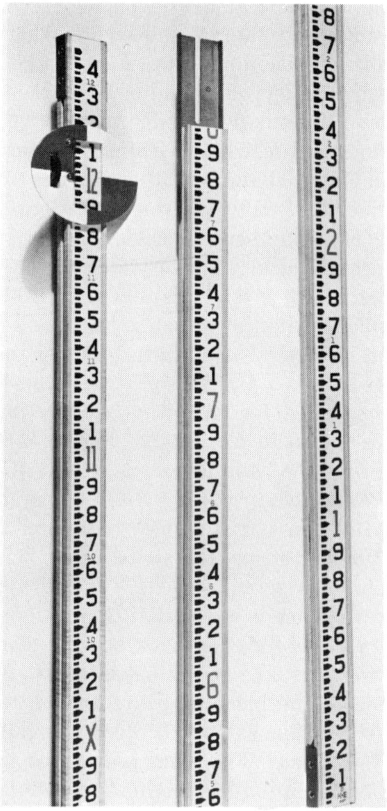

FIG. 8-13. Detroit rod. (*Keuffel & Esser Co.*)

One type of leveling rod carries an endless steel band which for a given elevation of the line of sight can be set so that elevations of points sighted from the same set-up can be read directly.

8-14a. Topographer's Rod. This self-reading rod (Fig. 8-14) is especially adapted for use on topographic surveys where the contours are located directly with the hand level. The distinguishing features of the rod are graduations numbered in both directions from a point near the middle of the rod and an adjustable base by means of which the zero point of the rod may be fixed at the height of the observer's eye above the ground. The rod is graduated in half-feet and numbered each foot. The numbers are large and the graduating marks are heavy so that they can be readily distinguished at considerable distances. Readings to tenths of feet are made by estimation. Usually the rod is a homemade device, and in some cases the feature of the adjustable base is not incorporated, the rod being so graduated that for one particular person the zero point is at the proper distance from the base.

The hand leveler, by standing with the base of the rod held at his toe, makes the proper adjustment to bring the zero point to his eye. The rod reading for any given location of the rod therefore indicates the distance above or below the point on which the hand leveler stands when the observation is made. Sometimes the hand level is fixed at the top of a stick about 5 ft. long.

Some surveyors prefer to use a rod similar to that just described except that it is graduated upward from the base.

8-15. Target Rods. With the target rod, the leveler signals the rodman to slide the target up or down until it is bisected by the line of sight. The target is then clamped, and the rodman, leveler, or both observe the indicated reading. Usually the target is equipped with a vernier (Art. 2-18) or other device by means of which fractional measurements of the rod graduations can be read without estimation. The principal advantage of the target rod is that mistakes are less likely to occur, particularly if both rodman and leveler read the rod. Under certain conditions its use materially facilitates the work, for example, where very long sights are taken, where the rod is partly obscured from view, or where it is necessary to establish a number of points all at the same elevation. However, under ordinary conditions its use retards progress without adding much, if anything, to the precision.

FIG. 8-14. Topographer's rod.

The *Philadelphia rod* (Fig. 8-11) previously described is designed as a self-reading rod but may also be used as a target rod. Lugs on the target engage in a groove on either side of the front strip. For readings on the lower half of the rod the reading is made to thousandths of feet by means of a vernier attached to the target. Graduations on the back of the rear strip are a continuation of those on the front strip and read downward. On the back of the top sleeve is a vernier employed for observations with the rod extended. For readings greater than can be taken with the "short" rod, the target is clamped at the same graduation on the face of the rod as the reading of the vernier on the back of the upper sleeve of the rod. The rod is then extended until the target is bisected by the line of sight. The vertical distance from the foot of the rod to the target is then indicated by the reading of the vernier on the back of the rod (Fig. 2-12, right).

The *New York rod*, used to some extent in building construction, is similar to the Philadelphia rod except in the manner in which it is graduated. The background of the scale is not painted, and distances from the foot of the rod are indicated by short, fine lines at intervals of 0.01 ft. and by longer, numbered lines at intervals of 0.1 ft. It is not a self-reading rod. The graduations for reading the rod in the extended position are on one side of the back strip and the accompanying vernier is cut in the wood of the front strip.

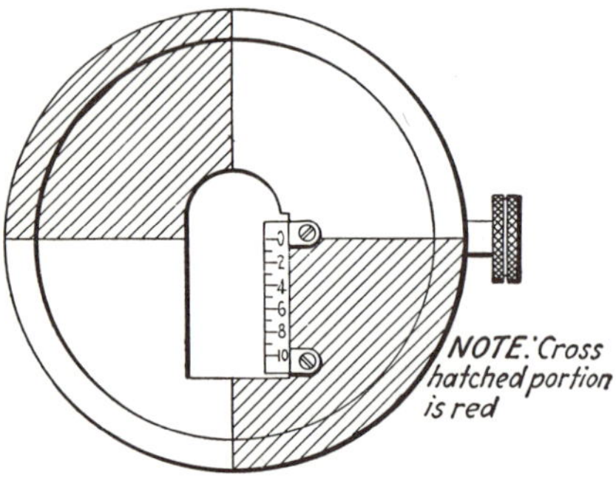

FIG. 8-15. Rod target.

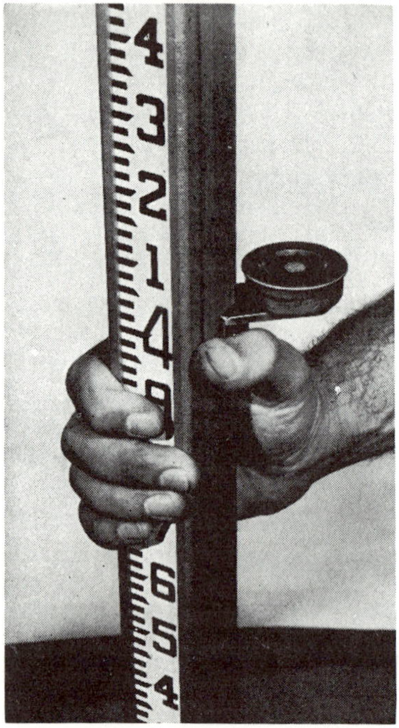

FIG. 8-16. Rod level.

The *Detroit rod* (Fig. 8-13) has pointed black graduations alternately 0.02 and 0.01 ft. wide. The points of the long, wide graduations indicate tenths of feet, and the points of the short wide graduations indicate half-tenths of feet. It is made in three sections for convenience of transporting.

The *architect's rod* or *builder's rod* is similar to the New York rod but is graduated in ⅛ in. and is equipped with verniers reading to $\frac{1}{64}$ in. Alternatively it may be graduated like the Philadelphia rod, in ⅛ in., with verniers reading to $\frac{1}{64}$ in. It use is confined to building construction.

8-16. Targets. The usual target (Fig. 8-15) is a circular or elliptical disk about 5 in. in diameter, with horizontal and vertical lines formed by the junction of alternate quadrants of white and red. A rectangular opening in the front of the target exposes a portion of the rod to view so that readings can be taken. The attached vernier (Art. 2-18) fits closely to the rod, its zero point or index being at the horizontal line of the target. In the figure a retrograde vernier is shown, but both retrograde and direct verniers are in common use. The *micrometer target* is equipped with a slow-motion device for fine settings (Fig. 8-11). The *angle target* faces in two directions, each half being perpendicular to the other.

8-17. Rod Levels. The rod level is an attachment for indicating the verticality of the leveling rod. One type (Fig. 8-16) consists of a circular or "bull's-eye" level vial mounted on a metal angle or bracket which either is attached by screws to the side of the rod or is held against the rod as desired.

Another type consists of a hinged casting on each wing of which is mounted a level tube which is held parallel to a face of the rod. When both of the bubbles are centered, the rod is plumb. The hinge makes it possible to fold the level compactly when it is not in use.

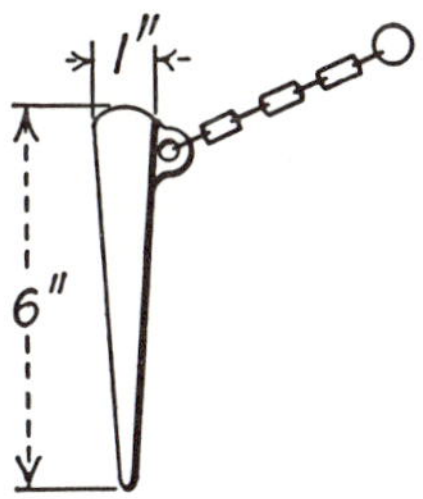

FIG. 8-17*a*. Turning point.

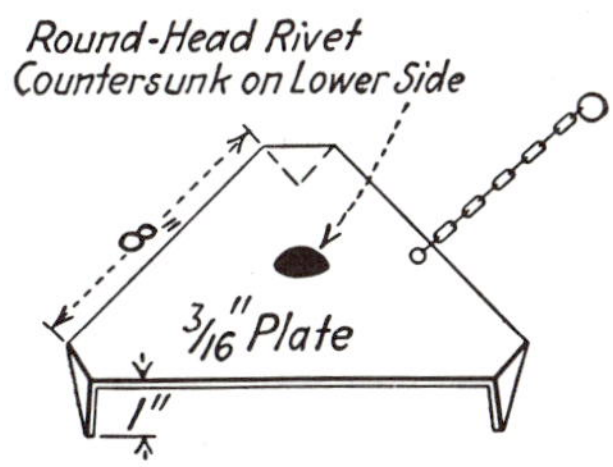

FIG. 8-17*b*. Turning plate.

8-18. Turning Points. A metal plate or pin which will serve temporarily as a stable object on which the leveling rod may be held at turning points is a useful part of the leveling equipment for lines of differential levels. The iron pin shown in Fig. 8-17*a* is adapted for use in firm ground. Often a railroad spike is used.

In soft ground the steel plate of Fig. 8-17*b* makes a satisfactory turning point. The plate is also adapted for use where the ground is so solid as to make driving the pin impossible or at least impracticable, as along highways. Under these conditions the plate with the dogs at its corners acts as a tripod, no special attempt being made to secure bearing between the lower surface of the plate and the ground.

USE OF LEVEL AND ROD

8-19. Setting up Engineer's Level. The engineer's level is placed in a desired location, with the tripod legs well spread and firmly pressed into the ground, with the tripod head nearly level, and with the telescope at a convenient height for sighting. If the set-up is on a slope, it is preferable to orient the tripod so that one of its legs extends up the slope. If the leveling head has four screws, the telescope is brought over one pair of opposite leveling screws, and the bubble is centered approximately; then the process is repeated with the telescope over the other pair. By repetition of this procedure the leveling screws are manipulated until the bubble remains centered, or nearly so, for any direction in which the telescope is pointed. If the instrument is in adjustment, the line of sight is then horizontal. If the leveling head has three screws, the process is similar, the telescope being brought successively over each of the screws.

Suggestions for the care and handling of the level are given in Art. 3-11, and suggestions for leveling the instrument, in Art. 2-8.

8-20. Reading Rod. For observations to hundredths or thousandths of feet, the rod is held on some well-defined point of a stable object. The rodman holds the rod vertical either by observing the rod level or by estimation. The leveler revolves the telescope about the vertical axis until the rod is about in the middle of the field of view, focuses the objective for

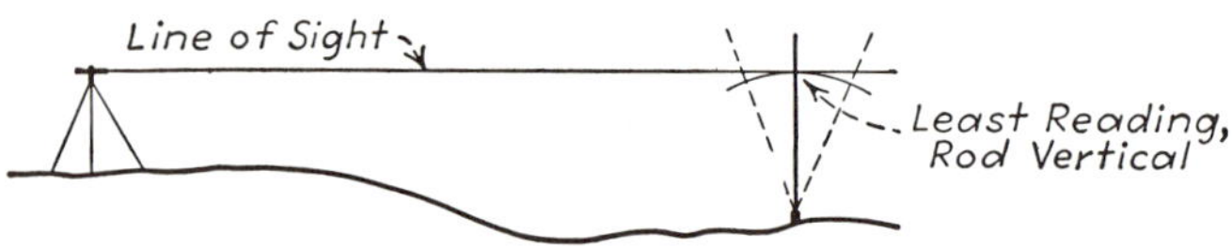

FIG. 8-18. Waving the rod.

distinct vision, and carefully centers the bubble. If the self-reading rod is used, the leveler observes and records the reading indicated by the line of sight, that is, the apparent position of the horizontal cross-hair on the rod. As a check he again observes the bubble and the rod. If the target rod is used, the procedure is identical except that the target is set by the rodman as directed by the leveler.

For leveling of lower precision, as when rod readings for points on the ground are determined to the nearest 0.1 ft., the observations usually are not checked, and proportionally less care is exercised in keeping the rod vertical and the bubble centered, always bearing in mind the errors involved and the precision with which measurements are desired.

If no rod level is used, in calm air the rodman can plumb the rod accurately by balancing it upon the point on which it is held. By means of the vertical cross-hair the leveler can determine when the rod is held in a vertical plane passing through the instrument, but he cannot tell whether it is tipped forward or backward in this plane. If it is in either of these positions, the rod reading will be greater than the true vertical distance, as illustrated by Fig. 8-18. To eliminate this error, the rodman *waves the rod,* or tilts it forward and backward as indicated by the figure, and the leveler takes the least reading, which occurs when the rod is vertical. The larger the rod reading, the larger the error due to the rod's being held at a given inclination; hence it is more important to wave the rod for large readings than for small readings. Further, whenever the rod is tipped backward about any support other than the front edge of its base, the graduated face rises and an error is introduced; for small readings this error is likely to be greater than that caused by not waving the rod.

ADJUSTMENT OF DUMPY AND WYE LEVELS

8-21. General. Regardless of the precision of manufacture, the dumpy and wye levels in process of use require certain field adjustments from time to time. It becomes an important duty of the surveyor to test his instrument at short intervals and to make with facility such adjustments as are found necessary. General features of the care and adjustment of instruments are given in Chaps. 2 and 3.

In some instances one adjustment is likely to be altered by, or depends upon, some other adjustment made subsequently. For example, lateral movement of the cross-hair ring may likewise produce a small rotation, and the lateral adjustment of the level tube depends upon the vertical adjustment. Hence if an instrument is badly out of adjustment, related adjustments must be repeated until they are gradually perfected.

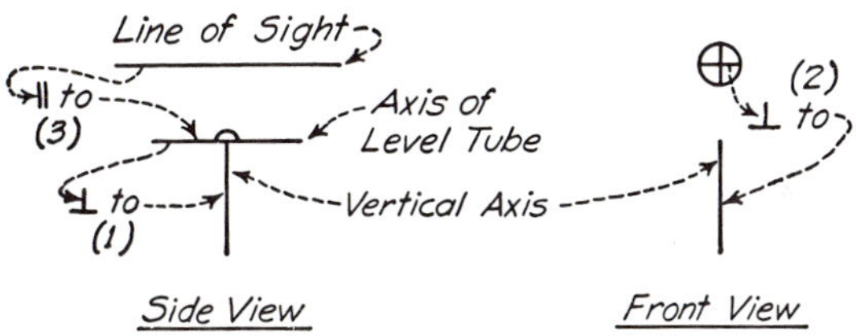

Fig. 8-19. Desired relations among principal lines of dumpy level.

8-22. Desired Relations in Dumpy Level. For a dumpy level in perfect adjustment the following relations should exist (see Fig. 8-19):

1. The axis of the level tube should be perpendicular to the vertical axis.
2. The horizontal cross-hair should lie in a plane perpendicular to the vertical axis, so that it will lie in a horizontal plane when the instrument is level.
3. The line of sight should be parallel to the axis of the level tube.

Also the optical axis, the axis of the objective slide, and the line of sight should coincide; but for the type of level commonly used in the United States the optical axis and the axis of the objective slide are fixed perpendicular to the vertical axis by the manufacturer, and no provision for further adjustment is made.

8-23. Adjustment of Dumpy Level. The parts capable of and requiring adjustment are the cross-hairs and the level tube. The basis for adjustments is the vertical axis. The adjustments are given in the following paragraphs. (For one form of student field notes, see Fig. 8-27.)

1. *To Make the Axis of the Level Tube Perpendicular to the Vertical Axis.* Approximately center the bubble over each pair of opposite leveling screws; then carefully center the bubble over one pair. Rotate the level end for end about its vertical axis. If the level tube is in adjustment, the bubble will retain its position. If the tube is not in adjustment, the displacement of the bubble indicates double the actual error, as shown by Fig. 8-20. If $(90° - \alpha)$ represents the angle between the vertical axis and axis of level tube, then when the bubble is centered the vertical axis makes an angle of α with the true vertical. When the level is reversed, the bubble is displaced

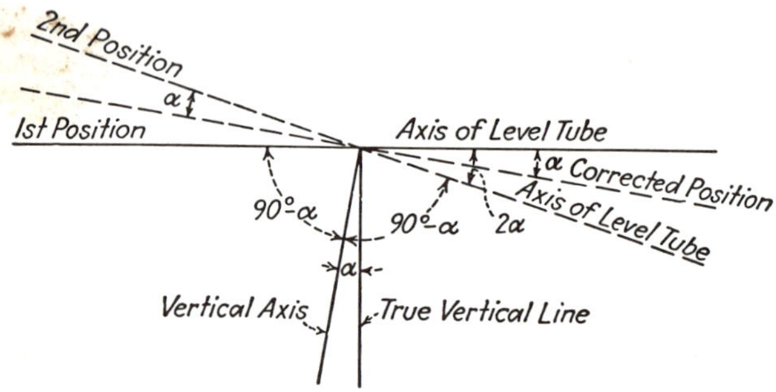

FIG. 8-20. Adjustment of axis of level tube of dumpy level.

through the arc whose angle is 2α. Hence the correction is the arc whose angle is α. Make the correction by bringing the bubble halfway back to the center by means of the capstan nuts at one end of the tube. Relevel the instrument with the leveling screws, and repeat the process until the adjustment is perfected. Usually three or four trials are necessary. As a final check, the bubble should remain centered over each pair of opposite leveling screws. This adjustment involves the principle of *reversion*.

2. *To Make the Horizontal Cross-hair Lie in a Plane Perpendicular to the Vertical Axis (and Thus Horizontal When the Instrument Is Level).* Sight the horizontal cross-hair on some clearly defined point (as A, Fig. 8-21) and rotate the instrument slowly about its vertical axis. If the point appears to travel along the cross-hair, no adjustment is needed.

If the point departs from the cross-hair and takes some position as A' on the opposite side of the field of view, loosen two adjacent capstan screws and rotate the cross-hair ring until by further trial the point appears to travel along the cross-hair. Tighten the same two screws. The instrument need not be level when the test is made.

3. *To Make the Line of Sight Parallel to the Axis of the Level Tube* (*Two-peg Test*). *Method A.* Set two pegs 200 to 300 ft. apart on approximately level ground. Set up and level the instrument in a location such that the eyepiece is ½ in. or less in front of the rod held on one of the pegs as at A, Fig. 8-22. With the rod held at A, take a rod reading a by sighting through the objective end of the telescope (with the eyepiece next to the rod.). The cross-hairs will not be visible, but the field of view will be so small (one or two hundredths of a foot) that its center may be determined within one or two thousandths of a foot by holding the point of a pencil on the rod.

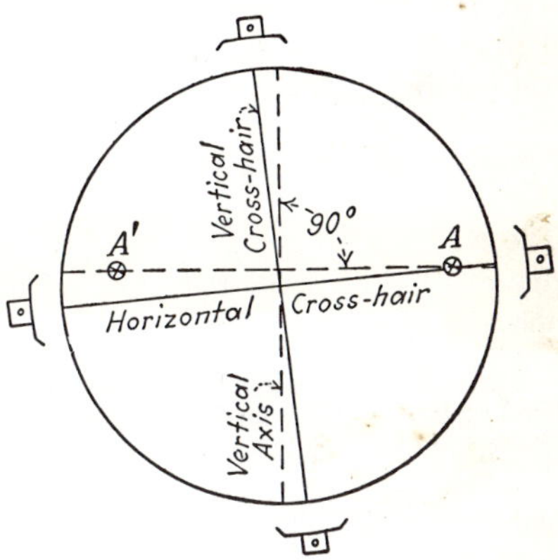

FIG. 8-21. Adjustment of horizontal cross-hair.

Move the rod to the other peg B, and take a rod reading b with level at A, in the usual manner.

Move the instrument to B, set up as before, and take rod readings c and d.

If $e = dd'$ represents the error in the line of sight for the distance A to B, then considering first the rod readings taken with the instrument at A, the true difference in elevation is

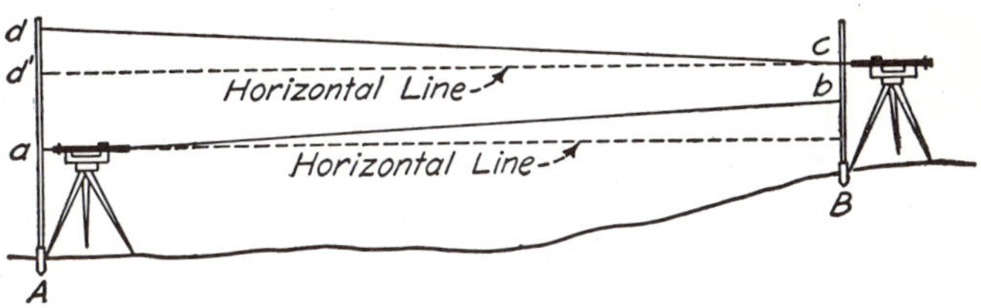

FIG. 8-22. Two-peg test, Method A.

$$\text{True diff. el.} = a - (b - e) \tag{9}$$

and considering the rod readings with the level at B is

$$\text{True diff. el.} = (d - e) - c \tag{10}$$

Adding Eqs. (9) and (10) there results

$$\text{True diff. el.} = \frac{(a - b) + (d - c)}{2} \tag{11}$$

Equation (11) shows that the true difference in elevation is the mean of the difference between rod readings taken with the instrument at A and the difference between those taken with the instrument at B.

If the two differences in elevation thus determined are equal, that is, if $(a - b) = (d - c)$, the line of sight is in adjustment. If not, the correct rod reading at A for instrument with position unchanged at B is

$$d' = c + \text{true diff. el.} = c + \frac{(a - b) + (d - c)}{2} \tag{12}$$

The adjustment is made by moving the cross-hair ring vertically until the line of sight cuts the rod at d'. The preceding steps are then repeated as a check.

Example 1: With level at A, observed readings are $a = 4.086$ and $b = 2.705$; with level at B, $c = 3.871$ and $d = 5.542$. Then by Eq. (11) the true difference in elevation is $\frac{(4.086 - 2.705) + (5.542 - 3.871)}{2} = 1.526$ ft., with B indicated as being higher than A. The correct rod reading for a horizontal line of sight with instrument still at B would be $3.871 + 1.526 = 5.397$ ft.

Equations (11) and (12) must be solved with due regard to signs; otherwise, if the error of adjustment should happen to be greater than the difference in elevation of the two pegs, the mean difference will not be the true difference in elevation. A sketch should always be drawn.

Strictly speaking, the effect of the earth's curvature and atmospheric refraction (see Art. 8-2) should be added to the correct rod reading d'. For a length of sight of 200 ft. this correction would amount to approximately 0.001 ft., and for 300 ft. approximately 0.002 ft. These quantities are so small as to be negligible in ordinary leveling.

Instead of viewing the near rod through the objective end of the telescope, the level may be set up a short distance (say, 6 or 8 ft.) beyond each near rod and the near-rod reading observed in the customary manner. The adjustment should be considered as a first approximation and should be repeated for precise results.

The advantages of Method A are: (1) the computations are relatively simple; and (2) the objective slide is in the same position for the two sights, thus a possible error in sighting is elimimated.

Method B. Set two pegs 200 to 300 ft. apart on approximately level ground, and designate as A the peg near which the second set-up will be made (Fig. 8-23); call the other peg B. Set up and level the instrument at any point M equally distant from A and B, that is, in a vertical plane bisecting the line AB. Take rod readings a on A and b on B; then $(a - b)$ will be the true difference in elevation, since any error would be the same for the two equal sight distances L_m. Due account must be taken of signs throughout the test.

Move the instrument to a point P near A, preferably but not necessarily on line with the pegs; set up as before, and measure the distances L_a to A and L_b to B. Take rod readings c on A and d on B. Then $(c - d)$, taken in the same order as before, is the indicated difference in elevation; if $(c - d) = (a - b)$, the line of sight is parallel to the axis of the level tube, and the instrument is in adjustment. If not, $(c - d)$ is called the "false" difference in elevation, and the inclination (error) of the line of

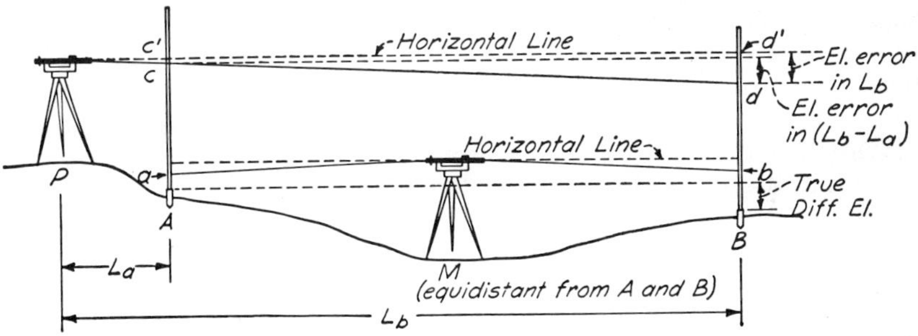

FIG. 8-23. Two-peg test, Method B.

sight in the net distance $(L_b - L_a)$ is equal to $(a - b) - (c - d)$. By proportion, the error in the reading on the far rod is

$$\frac{L_b}{L_b - L_a} [(a - b) - (c - d)]$$

Subtract algebraically the amount of this error from the reading d on the far rod to obtain the correct reading d' at B for a horizontal line of sight with the position of the instrument unchanged at P. Set the target at d' and bring the line of sight on the target by moving the cross-hair ring vertically.

Example 2: With level at M, the rod reading a is 0.970 and b is 2.986; the true difference in elevation $(a - b)$ is then $0.970 - 2.986 = -2.016$ ft., with B thus indicated as being lower than A. With level at P, the rod reading c is 5.126 and d is 7.018; the false difference in elevation $(c - d)$ is then $5.126 - 7.018 = -1.892$, with B again indicated as being lower than A. The distance L_a is observed to be 30 ft., and L_b to be 230 ft. The inclination of the line of sight in $(230 - 30 = 200)$ ft. is $(-2.016) - (-1.892) = -0.124$ ft. The error in elevation of the line of sight at the far rod is $(230/200) \times (-0.124) = -0.143$ ft. The correct rod reading d' for a horizontal line of sight is $7.018 - (-0.143) = 7.161$ ft.

As a partial check on the computations, the correct rod reading c' at A may be computed by proportion; the difference in elevation computed from the two corrected rod readings c' and d' should be equal to the true difference in elevation observed originally at M.

Example 3: In the preceding example, the error in elevation of the line of sight at the near rod is $(30/200) \times (-0.124) = -0.019$ ft. The correct rod reading c' is $5.126 - (-0.019) = 5.145$ ft. The "false" difference in elevation is $5.145 - 7.161 = -2.016$ ft., which is equal to the true difference in elevation; hence the computations are checked to this extent.

As in Method A, a sketch should always be drawn. Also, theoretically a correction for earth's curvature and atmospheric refraction should be added numerically to the final rod reading d', although in practice it is usually considered negligible.

Some surveyors prefer to set up at P within 6 to 8 ft. of A and to consider $[(a - b) - (c - d)]$ as being the *total* error in elevation, to be subtracted directly from d. This serves as a first approximation; the procedure for the set-up at P is then repeated. The amount of computation is thus reduced, but the amount of field work is increased.

The advantages of Method B are that in no case is it necessary to take a rod reading by projecting from the eyepiece, and that it is not necessary to set up on a line through AB.

8-24. Desired Relations in Wye Level. For a wye level in perfect adjustment, the following relations should exist (see Fig. 8-24):

1. The axis of the level tube should lie in the same plane with the axis of the wyes. (Herein the term "axis of the wyes" is used to denote a reference line—also called the "axis of the collars" or "axis of the supports"—which represents the alinement of the telescope tube in the wye supports.)

2. The axis of the level tube should be parallel to the axis of the wyes.

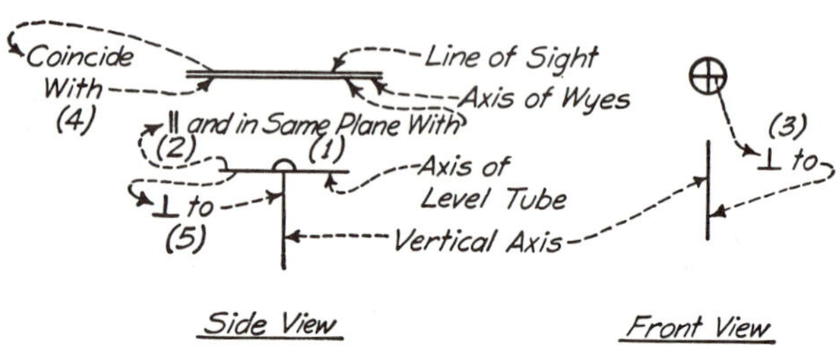

FIG. 8-24. Desired relations among principal lines of wye level.

3. The horizontal cross-hair should lie in a plane perpendicular to the vertical axis, so that it will be horizontal when the instrument is level.

4. The line of sight should coincide with the axis of the wyes, so that it will be parallel to the axis of the level tube.

5. For convenience in leveling, the axis of the wyes (and hence the axis of the level tube) should be perpendicular to the vertical axis.

Also, the optical axis and the axis of the objective slide should coincide with the line of sight. For the type of wye level commonly used in the United

States, the optical axis and the axis of the objective slide are fixed in the proper relation by the manufacturer, and no provision for further adjustment is made.

8-25. Adjustment of Wye Level. 1. *To Make the Axis of the Level Tube Lie in the Same Plane with the Axis of the Wyes.* Raise the wye clips, level the instrument, and rotate the telescope a few degrees in the wyes. If the desired relation exists, the bubble will remain centered.

If the bubble is displaced, bring it back to the center by means of the lateral adjusting screws at one end of the level tube.

2. *To Make the Axis of the Level Tube Parallel to the Axis of the Wyes.* Raise the wye clips, level the instrument carefully, lift the telescope from the wyes, and turn it end for end. If the desired relation exists, the bubble will remain centered.

If the bubble is displaced, the displacement is double the error (Fig. 8-20). Hence bring it back halfway to the center by means of the vertical adjusting nuts at one end of the level tube. Relevel the instrument by means of the leveling screws, and repeat the process until the adjustment is perfected.

3. *To Make the Horizontal Cross-hair Lie in a Plane Perpendicular to the Vertical Axis (and Thus Horizontal When the Instrument Is Level).* This adjustment is the same as adjustment 2 for the dumpy level. For some instruments the adjustment may be made by rotating the telescope in the wyes instead of rotating the cross-hair ring in the barrel of the telescope. The telescope is then fixed in the desired position by means of an adjustable stop attached to one of the wyes.

4. *To Make the Line of Sight Coincide with the Axis of the Wyes (and Thus Parallel to the Axis of the Level Tube).* Raise the wye clips, sight the intersection of the cross-hairs on some well-defined point, and clamp the vertical axis. Rotate the telescope 180° (about its own axis) in the wyes. If the line of sight still remains on the point, the desired relation exists.

If the line of sight is displaced from the point, adjust the cross-hair ring until the line of sight is midway between its two former positions (Fig. 8-25). Repeat the process until the proper relation is obtained. The adjustment is made by manipulating opposite screws, first bringing one cross-hair and then the other to what is estimated to be its correct position. If the required movement is large, however, it is best to loosen two adjacent screws slightly before attempting to aline the cross-hairs.

5. *To Make the Axis of the Level Tube (and Thus the Axis of the Wyes) Perpendicular to the Vertical Axis.* Inasmuch as the preceding adjustments have established parallelism or coincidence between the axis of the level tube, the line of sight, and the axis of the wyes, this adjustment does not add to the precision of observations. The adjustment makes it possible, however, to level the instrument so that the bubble will remain centered for any direction in which the telescope may be pointed.

Level the instrument, and rotate the telescope end for end about the vertical axis. If the bubble is displaced, its displacement is double the error (see Fig. 8-20). Bring it back halfway to the center by means of the capstan nuts controlling the vertical position of one of the wyes. The adjustment is seen to be identical with that of the level tube of the dumpy level, with the exception that, for the wye level, both the telescope and the level tube are moved in a vertical plane.

Worn Collars. When either the collars or the bearing points become worn, the adjustments just described are inadequate, and the two-peg test must be made as for the dumpy level. The common procedure is first to perform the adjustments for the cross-hairs in the usual manner, making the line of sight coincide as nearly as may be with the axis of the wyes. The true difference in elevation between two points having been determined by

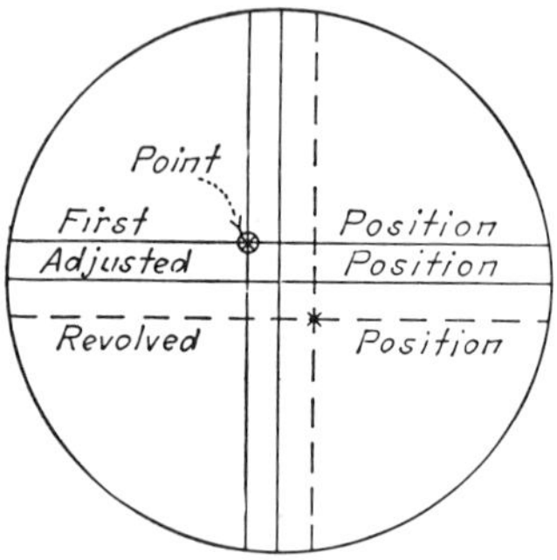

FIG. 8-25. Adjustment of line of sight of wye level.

the two-peg test, the line of sight is then set to the proper rod reading for a horizontal line by means of the leveling screws, and the bubble is centered by means of the capstan nuts at one end of the level tube.

8-26. Adjustment of Hand Level. The simplest procedure is to hold the hand level alongside an engineer's level that has been leveled and sighted at some well-defined point. The line of sight of the hand level should strike the same point when the bubble is centered.

The Locke hand level is adjusted by means of the screws at the ends of the level tube, which screws move the cross-wire defining the line of sight. The Abney hand level is adjusted by raising or lowering one end of the level tube until the bubble is centered, the index having first been set at zero on the graduated arc.

The hand level, even if not in adjustment, may be used to establish a horizontal line by employing the principle of the two-peg test (Art. 8-23). Let A and B (Fig. 8-26) be two posts, trees, or other convenient objects on nearly level ground. The level is held at C and with bubble centered is sighted to the point D. The level is then held at D, and the point E is established in a similar manner. The distance EC represents double the error. The point F established halfway between E and C is therefore on the horizontal line through D.

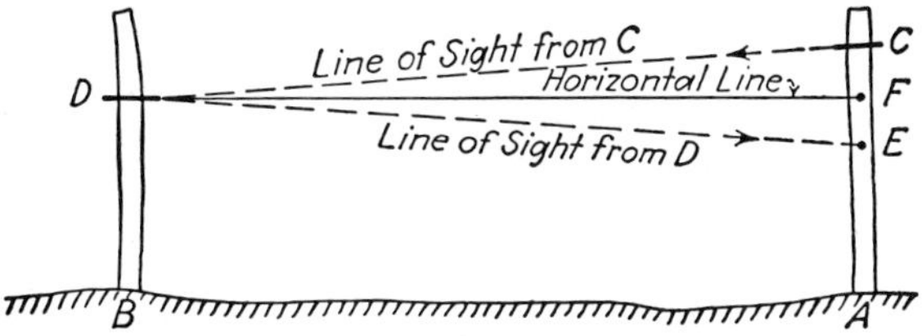

FIG. 8-26. Adjustment of hand level.

8-27. Numerical Problems.

1. What is the combined effect of the earth's curvature and mean atmospheric refraction in a distance of 300 ft.? In a distance of 3,000 ft.? In a distance of 6 miles? In a distance of 60 miles?

2. An observer standing on a beach can just see the top of a lighthouse 15 miles away. The eye height of the observer above sea level is 5.7 ft. What is the height of the lighthouse above sea level?

3. Two points, A and B, are each distant 2,000 ft. from a third point C, from which the measured vertical angle to A is $+3°21'$ and that to B is $+0°32'$. What is the difference in elevation between A and B?

4. Let A be a point of elevation 100.00 ft., and let B and C be points of unknown elevation. By means of an instrument set 4.00 ft. above B, vertical angles are observed, that to A being $-1°55'$ and that to C being $+3°36'$. If the horizontal distance AB is 1,500 ft. and the horizontal distance BC is 5,000 ft., what are the elevations of B and C, making due allowance for earth's curvature and atmospheric refraction?

5. Two points, A and B, are 1,000 ft. apart. The elevation of A is 615.03 ft. A level is set up on the line between A and B and at a distance of 250 ft. from A. The rod reading on A is 9.15 and that on B is 2.07. Making due allowance for curvature and refraction, what is the elevation of B? What would be the magnitude and sign of the error introduced if the correction for curvature and refraction were omitted?

6. Two points, A and B, are 400 ft. apart, and the elevation of A is 615.037 ft. A level is set up on line and distant 100 ft. from A in the direction of B. The rod reading on A is 5.812 and on B is 7.358. What is the elevation of B, neglecting the correction due to curvature and refraction? What are the magnitude and the sign of the error introduced by not considering the effect of curvature and refraction?

7. A sight is taken with an engineer's level at a rod held 300 ft. away, and an initial reading of 6.323 ft. is observed. The bubble is then moved through five spaces on the level tube, when the rod reading is 6.589 ft. What is the sensitiveness of the level tube in seconds of arc? What is the radius of curvature of the level tube if one space is 2 mm?

8. Through the telescope of a level the magnified image of the portion of the rod between 5.00 and 5.20 apparently covers the unmagnified image between 2.1 and 8.3. What is the magnifying power of the telescope?

9. Design (*a*) a direct vernier and (*b*) a retrograde vernier reading to thousandths of feet, each space on the rod being equal to 0.005 ft.

10. Design a direct vernier and a retrograde vernier, both reading to $\frac{1}{32}$ in., for an architect's rod graduated to $\frac{1}{8}$ in. Draw a neat sketch of each vernier and a portion of the rod for a reading of $2\frac{19}{32}$ in., with graduations shown and labeled.

11. On a rod graduated to 0.25 in., a retrograde vernier is to read to hundredths of inches. State the following: (*a*) length of one vernier space, (*b*) number of spaces on the vernier, (*c*) number of spaces on main scale corresponding to full length of vernier scale, (*d*) least count of vernier. For a rod reading of 10.37 in., what division on the vernier will be in line with a scale division?

12. The diameter of the field of view of a level is 5.25 ft. when the rod is 300 ft. from the objective. What is the angular width of the field of view?

13. If the rod were inclined forward 0.4 ft. in a length of 13 ft., what error would be introduced in a rod reading of 5.0 ft.? Compute by an approximate "slope" method and by an exact method, and compare results.

14. In the two-peg test of a dumpy level by Method A, the following observations are taken:

	Instrument at *A*	Instrument at *B*
Rod reading on *A*.........	4.937	3.077
Rod reading on *B*.........	6.736	4.752

What is the true difference in elevation between the two points? With the instrument in the same position at *B*, to what rod reading on *A* should the line of sight be adjusted? What is the error in the line of sight for the distance *A* to *B*?

15. In the two-peg test of a dumpy level by Method B, the following observations are taken:

	Instrument at *M*	Instrument at *P*
Rod reading on *A*.........	3.612	1.862
Rod reading on *B*.........	3.248	0.946

M is equidistant from *A* and *B*; *P* is 40 ft. from *A* and 240 ft. from *B*. What is the true difference in elevation between the two points? With the level in the same position at *P*, to what rod reading on *B* should the line of sight be adjusted? What is the corresponding rod reading on *A* for a horizontal line of sight? Check these two rod readings against the true difference in elevation, previously determined.

8-28. Field Problems.

Problem 1. Magnifying Power of Telescope

Object. To determine the number of diameters an object viewed through the telescope is magnified.

Procedure. (1) Sight at the rod held erect about 15 ft. in front of the instrument. (2) With both eyes open turn the instrument until the images, as seen by the naked eye and as seen through the telescope, appear to fall one upon the other. (3) Compare 0.1 ft. on the rod as seen through the telescope with a space as seen with the naked eye. The number of tenths apparently covered on the unmagnified image is the magnifying power of the telescope.

Hints and Precautions. (1) Some practice will probably be necessary before the student is able to see both images at the same time. First sight the image through the telescope; then, still keeping the image in sight, look at the rod with the other eye. After a little practice both images will appear distinct. Turning the level slightly, if necessary, will cause the unmagnified image to fall upon the magnified image. For observation select the tenth on the magnified image which is located wholly in the field of vision. Observe the reading of the upper line of this tenth on the unmagnified image, and then observe the lower. The difference of these readings in tenths of feet will be the magnifying power.

Problem 2. Radius of Curvature of Level Tube

Object. To determine, in the field, without the use of special apparatus, the radius of curvature of the level tube of transit or level.

Procedure. (1) Hold the rod on a solid point 300 ft. from the instrument. With one end of the bubble at a division near the end of the level tube, take a careful rod reading to the nearest 0.001 ft. Note the exact position of each end of the bubble. (2) Manipulate the leveling screws until the other end of the bubble falls near the other end of the tube. Take another rod reading, and measure the exact distance traversed by each end of the bubble. (3) Determine the bubble movement (this should be expressed to the nearest 0.001 ft.) and the difference between the two rod readings. (4) In this manner obtain a series of five bubble movements and their corresponding rod readings. (5) Compute the radius of curvature by the following formula: $R = (b/t)D$, in which R is the radius of curvature, D is the distance from the instrument to the rod, b is the mean of the five bubble movements, and t is the mean of the five differences in rod readings. (6) Compute the value of one division of the level tube in seconds of arc.

Problem 3. Adjustment of the Engineer's Level

Object. To make the field adjustments of the engineer's level.

Procedure. (1) Proceed as outlined in Art. 8-23 for the dumpy level or Art. 8-25 for the wye level. (2) Record all field observations in the field book. For each adjustment separately, state such items as the desired relation, the test, the condition of the instrument with regard to the adjustment, the method of adjusting, and the final test. Include sketches if necessary to make the method clear. One form

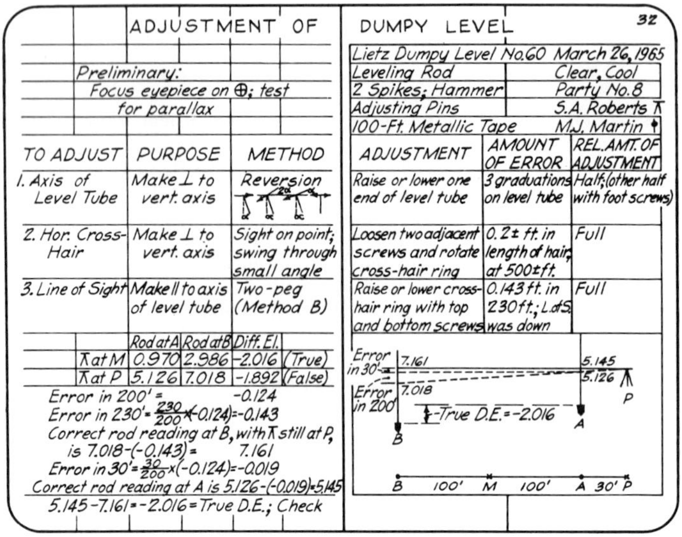

FIG. 8-27. Student field notes for adjustment of dumpy level.

of student field notes is shown in Fig. 8-27. In practice, knowledge of the amount by which an instrument is out of adjustment may be important as a basis for deciding whether prior work needs to be repeated.

REFERENCE

1. IRISH, SUMNER B., "Measuring the Precision of a Level," *J. Surveying and Mapping Div., Am. Soc. Civil Engr.*, pp. 35–44, 1963.

CHAPTER 9

DIFFERENTIAL LEVELING

9-1. General. Differential leveling is the operation of determining the elevations of points some distance apart. Usually this is accomplished by direct leveling. Differential leveling requires a series of set-ups of the instrument along the general route and, for each set-up, a rod reading back to a point of known elevation and forward to a point of unknown elevation.

9-2. Bench Marks. A *bench mark* (B.M.) is a definite point on an object, the elevation and location of which are known. It may be permanent (P.B.M.) or temporary (T.B.M.). Bench marks serve as points of reference

FIG. 9-1. U.S. Coast and Geodetic Survey bench mark.

for levels in a given locality. Their elevations are established by differential leveling, except that the elevation of the initial bench mark of a local project may be assumed.

Throughout the United States are permanent bench marks established by the U.S. Geological Survey and the U.S. Coast and Geodetic Survey. Similarly, bench marks have been established by various other Federal, state, and municipal agencies and by such private interests as railroads and water companies, so that the surveyor has not far to go before he can find some point of known elevation.

The Coast Survey bench marks (Fig. 9-1) consist of bronze plates set in stone or concrete and marked with the elevation above mean sea level. Those of the other agencies are similar. Other objects frequently used as bench marks are stones, pegs or pipes driven in the ground, nails or spikes in trees or pavements, and marks painted or chiseled on street curbs.

For any survey or construction enterprise, levels are run from some initial bench mark of known or assumed elevation to scattered points in desirable locations for future reference as bench marks.

In some areas, the elevation of bench marks may be altered by earth movements such as those caused by earthquakes, slides, lowering of water tables, pumping from oil fields, mining, or construction.

9-3. Definitions. A *turning point* (T.P.) is an intervening point between two bench marks upon which point foresight and backsight rod readings are taken. It may be a pin or plate (see Art. 8-18) which is carried forward by the rodman after observations have been made, or it may be any stable object such as a street curb, railroad rail, or stone. The nature of the turning point is usually indicated in the notes, but no record is made of its location unless it is to be reused. A bench mark may be used as a turning point.

A *backsight* (B.S.) is a rod reading taken on a point of known elevation, as a bench mark or a turning point. Usually it will be taken with the level sighting back along the line, hence the name. A backsight is sometimes called a *plus sight.*

A *foresight* (F.S.) is a rod reading taken on a point the elevation of which is to be determined, as on a turning point or on a bench mark that is to be established. A foresight is sometimes called a *minus sight.*

The *height of instrument* (H.I.) is the elevation of the line of sight of the telescope when the instrument is leveled.

In surveying with the transit, the terms backsight, foresight, and height of instrument have meanings different from those here defined.

9-4. Procedure. In Fig. 9-2, B.M.$_1$ represents a point of known elevation (bench mark), and B.M.$_2$ represents a bench mark to be established some distance away. It is desired to determine the elevation of B.M.$_2$. The rod is held at B.M.$_1$, and the level is set up in some convenient location, as L_1, along the general route B.M.$_1$ to B.M.$_2$. The level is placed in such a location that a clear rod reading is obtainable, but no attempt is made to keep on the direct line joining B.M.$_1$ and B.M.$_2$. A backsight is taken on B.M.$_1$. The rodman then goes forward and, as directed by the leveler, chooses a turning point T.P.$_1$ at some convenient spot within the range of the telescope along the general route B.M.$_1$ to B.M.$_2$. It is desirable, but not necessary, that each foresight distance, as L_1-T.P.$_1$, be approximately equal to its corresponding backsight distance, as B.M.$_1$-L_1. The chief requirement is that the turning point shall be a stable object at an elevation and in a location favorable to a rod reading of the required precision. The rod is held on the turning point, and a foresight is taken. The leveler then sets up the instrument at some favorable point, as L_2, and takes a backsight to the rod held on the turning point; the rodman goes forward to establish a second turning point T.P.$_2$; and so the process is repeated until finally a foresight is taken on the terminal point B.M.$_2$.

It is seen in Fig. 9-2 that a *backsight added to the elevation* of a point on which the backsight is taken gives the height of instrument, and that a *foresight subtracted from the height of instrument* determines the elevation of the point on which the foresight is taken. Thus if the elevation of B.M.$_1$ is 721.05 ft. and the B.S. is 7.11 ft., then the H.I. with the instrument set up at L_1 is $721.05 + 7.11 = 728.16$. And if the following F.S. is 1.24 ft., the elevation of T.P.$_1$ is $728.16 - 1.24 = 726.92$ ft. Also the difference

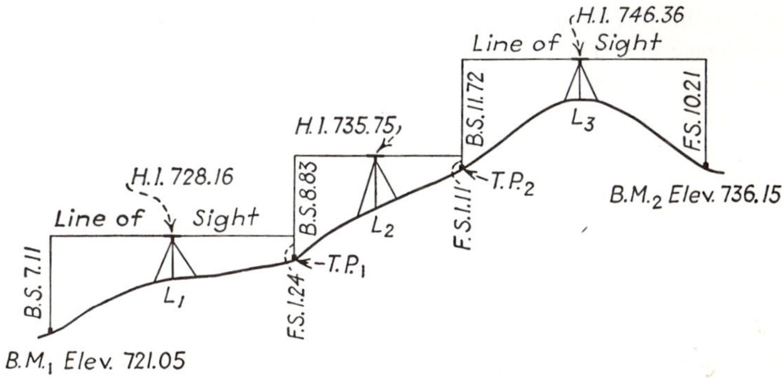

FIG. 9-2. Differential leveling.

between the backsight taken on a given point and the foresight taken on the following point is equal to the difference in elevation between the two points. It follows that the difference between the sum of all backsights and the sum of all foresights gives the difference in elevation between the bench marks.

Since normally the level is at a higher elevation than that of the points on which rod readings are taken, the backsights are often called "plus" sights and the foresights "minus" sights and are so recorded in the field notes. Sometimes, however, in leveling for a tunnel or a building it is necessary to take rod readings on points which are at a higher elevation than that of the H.I. In such cases the rod is held inverted, and in the field notes each such backsight is indicated with a minus sign and each foresight with a plus sign.

When several bench marks are to be established along a given route, each intermediate bench mark is made a turning point in the line of levels. Elevations of bench marks are checked sometimes by rerunning levels over the same route but more often by "tying on" to a previously established bench mark near the end of the line or by returning to the initial bench mark. A line of levels that ends at the point of beginning is called a *level circuit*. The final observation in a level circuit is therefore a foresight on the initial bench mark. If each bench mark in a level circuit is also a turning point and the circuit checks within the prescribed limits of error, it is regarded as conclusive evidence that the elevations of all turning points in the circuit, including all bench marks used as turning points, are correct within prescribed limits.

9-5. Balancing Backsight and Foresight Distances. In Art. 8-4 it has been shown that, if a foresight distance were equal to the corresponding backsight distance, any error in readings due to earth's curvature and to atmospheric refraction (under uniform conditions) would be eliminated. In ordinary leveling no special attempt is made to balance *each* foresight distance against the preceding backsight distance. Whether or not such distances are approximately balanced between bench marks will depend upon the desired precision. The effect of earth's curvature and atmospheric refraction is slight unless there is an abnormal difference between backsight and foresight distances. The effect of instrumental errors is likely to be of considerably greater consequence with regard to the balancing of these distances. The chances are that there is not absolute parallelism between the line of sight and the axis of the level tube, so that if the instrument were perfectly leveled, the line of sight would be inclined always slightly upward or always slightly downward. The error in a rod reading due to this imperfection of adjustment is proportional to the distance from the instrument to the rod and is of the same sign for a backsight as for a foresight. Since backsights are added and foresights are subtracted, this instrumental error is eliminated if, between bench marks, the *sum* of the foresight distances is made equal to the *sum* of the backsight distances.

In ordinary leveling no special attempt is made to equalize these distances *if there is assurance that the instrument is in good adjustment.* Normally, for levels run over flat or gently rolling ground, the line of sight will fall within the length of the rod regardless of the position of the instrument, and the distance between instrument and rod is governed by the optical qualities of the telescope. When the leveler moves forward, he generally paces or estimates by eye a distance about the proper maximum length of sight; and when the rodman moves forward, he similarly estimates the proper distance from the instrument.

For leveling of moderately high precision it is necessary to equalize backsight and foresight distances between bench marks; hence these distances are recorded. In less refined leveling, distances are usually determined by pacing; in precise leveling, they are usually measured with the stadia or the gradienter.

In leveling uphill or downhill the length of sight is usually governed by the slope of the ground. In order that maximum distances between turning points may be obtained and hence progress be most rapid, the leveler sets up the instrument in a position such that the line of sight will intersect the rod near its top if the route is uphill, or near its bottom if the route is downhill; and he directs the rodman to a similarly favorable location for the turning point. In leveling uphill or downhill, a balance between foresight and backsight distances can be obtained with a minimum number of set-ups by following a zigzag course.

Although between bench marks at the same (or nearly the same) elevation the backsight and foresight distances will tend to balance in the long run, regardless of the character of the terrain, for levels between two points having a large difference in elevation a very small inclination of the line of sight may normally be expected to produce a marked error unless some attempt is made to balance backsight and foresight distances.

Consider two points 40 miles apart, one being, say, at sea level and the other at an elevation of 5,000 ft.; the intervening ground is of a fairly uniform slope averaging say, $2\frac{1}{2}$ ft. in 100 ft., so that, starting from the lower point, the backsights would be near the top of the rod (say, at 11.5 ft.) and the foresights would be near the bottom (say, at 1 ft.). If the instrument were placed directly between turning points and the telescope were 4.5 ft. above the ground at each set-up, the backsight distance would on the average be about twice as great as the foresight distance, hence the effect of any error in the line of sight would be twice as great for backsights as for foresights. Suppose that, when the bubble was exactly centered, the line of sight was inclined upward 0.002 ft. in 100 ft. (fair adjustment). Then, due to this error alone, the sum of backsights would be about $2\frac{2}{3}$ ft. too large and the sum of foresights would be about $1\frac{1}{3}$ ft. too large; hence the calculated elevation of the terminal bench mark would be approximately 1.3 ft. too great. This is an error considerably larger than would ordinarily be permissible except in rough leveling.

It is also to be observed that the effect of earth's curvature and atmospheric refraction may be considerable under the conditions of the example just cited, even though the lengths of sight are within the range of those ordinarily used in leveling.

Thus, if the backsight distances were 300 ft. and the foresight distances were 150 ft., the constant error per set-up of the instrument due to curvature and refraction would be 0.0014 ft., or for the distance of 40 miles approximately 0.7 ft. This is about the maximum error that would be permissible in ordinary leveling (see Art. 9-12). For the example, both the error due to imperfect adjustment and that due to curvature and refraction combined happen to be of the same sign, hence the resultant error from these two sources is 2.0 ft.

The effect of earth's curvature and atmospheric refraction cannot be entirely eliminated by making the *sum* of the foresight distances equal to that of the backsight distances; rather it is necessary that *each* foresight distance be made equal to the corresponding backsight distance.

9-6. Differential-level Notes. For ordinary differential leveling when no special effort is made to equalize backsight and foresight distances between bench marks, usually the record of field work is kept in the form indicated by Fig. 9-3, in which the levels from B.M.$_1$ to B.M.$_2$ are the same as shown by Fig. 9-2. The left page is divided into columns for numerical data, and the right page is reserved for descriptive notes concerning bench marks and turning points. In the same horizontal line with each turning point or bench mark shown in the first column are all data concerning that point. The heights of instrument and the elevations are computed as the work progresses. Thus, when the backsight (7.11) has been taken on B.M.$_1$ it is added to the elevation (721.05) to determine the H.I. (728.16). The

LEVELS FOR BENCH MARKS ALONG RIDGE ROAD						Dec. 31, 1965 J.G. Sutter, ⊼ Fair W.R Knowles, Rod
Sta.	B.S.	H.I.	F.S.	Elev.		Remarks
B.M.$_1$	7.11	728.16		721.05		Top of Hydrant Cor. Oak St.
T.P.$_1$	8.83	735.75	1.24	726.92		Curb
T.P.$_2$	11.72	746.36	1.11	734.64		
B.M.$_2$	4.32	740.47	10.21	736.15		Spike in Pole North of Williams House Marked B.M 736.15
T.P.$_3$	3.06	733.57	9.96	730.51		
T.P.$_4$	2.74	727.40	8.91	724.66		Stone
T.P.$_5$	0.81	716.59	11.62	715.78		
B.M.$_3$			12.42	704.17		Concrete Monument No. of Road; County Line
ΣB.S.=	38.59	ΣF.S.=	55.47	721.05		
			38.59			
		Diff=	16.88 =	16.88	ck.	

FIG. 9-3. Differential-level notes.

height of instrument is recorded on the same line with the backsight by means of which it is determined. When the first foresight (1.24) is observed, it is recorded on the line below and is subtracted from the preceding H.I. (728.16) to determine the elevation of T.P.$_1$ (726.92). And so the notes are continued. Usually at the foot of each page of level notes the *computations* are checked by comparing the difference between the sum of the backsights and the sum of the foresights with the difference between the initial and the final elevation, as illustrated at the bottom of Fig. 9-3. Agreement between these two differences signifies that the additions and subtractions are correct but does not check against mistakes in observing or recording.

Bench marks should be briefly but definitely described and should be so marked in the field that they can be readily identified. They are usually marked with paint or with crayon that will withstand the effects of the weather. When the bench mark is on stone or concrete the position is often indicated by a cross cut with a chisel. A bench mark may or may not be marked with its elevation. Whenever there might arise any question as to the exact location of the point on which the rod was held, its nature should be clearly indicated in the notes. A description of turning points is of no particular importance unless the points are on objects that can be identified and might therefore become of some value in future leveling operations. Such points are usually marked with crayon and briefly described in the notes.

When backsight and foresight distances are to be balanced, the form of notes is the same, except that these distances are usually recorded in the last column of the left page, with the backsight distances in the left half of the column and the foresight distances in the right half. The cumulative excess or deficiency of foresight over backsight distance is noted for each turning point and bench mark.

9-7. Mistakes in Leveling. Some of the mistakes commonly made in leveling are:

1. Confusion of numbers in reading the rod, as for example, reading and recording 4.92 when it should be 3.92. The mistake is not likely to occur if the observer notices the numbers on both sides of the observed reading.
2. Recording backsights in foresight column and *vice versa.*
3. Faulty additions and subtractions; adding foresights and subtracting backsights. As a check, the difference between the sum of the backsights and the sum of the foresights should be computed for each page or between bench marks.
4. Rod not held on same point for both foresight and backsight. This is not likely to occur if the turning points are marked or otherwise clearly defined.
5. Not having the Philadelphia rod fully extended when reading the long rod. Before a reading on a turning point is taken, the clamp should be inspected to see that it has not slipped.
6. Wrong reading of vernier when the target rod is used.
7. When the long target rod is used, not having the vernier on the target set to read exactly the same as the vernier on the back of the rod when the rod is short.

9-8. Precise Differential Leveling. The subject of precise leveling as practiced on government surveys is not to be considered in detail here, as it is thoroughly covered in several U.S. Coast and Geodetic Survey manuals (see references at end of this chapter). However, it is appropriate to call attention here to certain refinements by means of which a relatively high degree of precision may be obtained with the ordinary wye or dumpy level and the self-reading rod.

For work of this nature the rod should be treated in some manner to prevent expansion or contraction through change in moisture content, and at intervals it should be compared with a standard length. Rods with a graduated strip of invar steel are preferable. The rod should have an attached rod level for plumbing. It is particularly important that turning points be on solid objects with rounded tops so that the base of the rod can be held in the same position for both backsight and foresight.

The level should be equipped with stadia hairs in addition to the regular cross-hairs or with the micrometer tilting device which can be used as a gradienter. Preferably it should be of the dumpy type with inverting eyepiece and reflecting mirror by means of which the bubble can be viewed at the instant the rod is read. To prevent unequal thermal expansion, the level should be protected from the sun's rays by an umbrella (Fig. 9-6). It should be set up very firmly so that no settlement will occur. To eliminate as far as possible the effects of any change in atmospheric refraction, settlement of the tripod, or warping of the level, it is desirable that the shortest possible

Sta.	Backsights			H.I.	Foresights			Elev.
	Hairs	Mean	Dist.		Hairs	Mean	Dist.	
	9.316							
	7.942							
B.M.$_1$	6.565	7.941	2.751	329.561				321.620
	11.742				4.112			
	10.635				2.911			
T.P.$_1$	9.528	10.635	2.214	337.283	1.716	2.913	2.396	326.648

FIG. 9-4. Precise-level notes.

time elapse between a backsight and the succeeding foresight. The backsight and foresight distances are determined preferably by stadia or gradienter but sometimes by pacing and are balanced very closely between bench marks. For stadia readings all three horizontal hairs should be read by estimation to thousandths of feet and the readings should be recorded. The mean of the readings for the three hairs is taken as the correct rod reading for each sight. The interval between the reading of the upper hair and that of the lower hair is a measure of the distance from instrument to rod.

Figure 9-4 shows a suitable form for numerical data, with distances observed by stadia. A portion of the right-hand page can be reserved for explanatory notes as in the notes of Fig. 9-3.

Two Rods. Excellent results have been obtained by employing two rods and two rodmen, each occupying alternate turning points (of the same set). In order further to eliminate possible systematic errors, the order of readings may be interchanged at alternate set-ups of the level; that is, at one set-up the backsight may be observed before the foresight, and at the next set-up the foresight may be determined before the backsight. This would be practicable only when two rodmen are employed.

In another method, which requires only one rodman and which permits checks to be made as the leveling progresses, readings are taken not only on the front face of the rod but also on the back face, on which is a scale with a different zero or different units. The scale on the back may be of the type which can be set so that elevations of points sighted from the same set-up can be read directly.

Double-rodding with Two Sets of Turning Points. This method was formerly used on some government surveys where two rods and two rodmen were generally employed. A single rod may be used with nearly as good results, although the speed will be considerably lessened. The advantage of the method does not lie so much in the increased precision over using one set of turning points as in checking the levels as the work progresses. Hence it is particularly useful in running levels that do not close on points of known elevation.

Two sets of turning points are established so that at each set-up of the level two independent backsights and two independent foresights are taken. The turning points on one line are usually a foot or more higher than corresponding points on the other line so as to eliminate the possibility of making the same mistake in reading the foot marks on both rods. When two rodmen are employed, one gives readings for points along the "high" line and the other for points along the "low" line.

An appropriate form of notes is illustrated by Fig. 9-5. The observations are seen to give two independent determinations for the height of instrument at each set-up. Were it not for errors of observation these H.I.'s should exactly agree. If at any set-up the discrepancy between the two H.I.'s shows a material variation from the discrepancy between H.I.'s for the preceding set-up, observations are repeated. In careful leveling the maximum allowable variation between the discrepancies at two successive set-ups is usually two or three thousandths of a foot. Normally the difference between H.I.'s may be expected to increase as the length of the line increases, and hence the two independent determinations of the elevation of a bench mark along the route may be expected to show a difference which in general will increase with the distance from the point of beginning. Thus in the notes, the discrepancies between the two H.I.'s are seen to be successively 0.002, 0.005, 0.007, 0.004, and 0.006. The variations between the discrepancies for succeeding set-ups are therefore 0.003, 0.002, 0.003, and 0.002. On the right-hand page of the notes are computations for checking the additions and subtractions. The difference between the total of all backsights and the total of all foresights is double the difference in elevation between the initial and final bench marks.

It is desirable that both rods be read on the terminal bench marks. Intermediate bench marks are employed as turning points on either of the two lines. When the discrepancy between H.I.'s becomes sufficiently large to be of consequence, the elevation of each bench mark is adjusted to conform to the mean of the two heights of instrument for the set-up from which a foresight to the bench mark was taken.

9-9. Geodetic Leveling. Geodetic leveling is direct or spirit leveling of a high order of accuracy and precision. It is based on the irregular surfaces associated with the geoid rather than the regular surface of a spheroid or an ellipsoid representing the surface of the earth. The geoid is the figure of the earth considered as a sea-level surface extended continuously through the continents. At every point, the geodetic surface is perpendicular to the plumb line; it is determined by observing deflections of the vertical.

LEVELS, DIXFIELD TO PERU — 26

Double-Rodded Line
Along P. & R.F.Ry.

Gurley Wye Level — F.K. Burton, ⊼
2-Phila. Rods — J.J. Hamel, Recorder
Lowe & Smith, Rods
July 9, 1965

Sta.	B.S.	H.I.	F.S.	Elev.	
B.M.$_1$	5.241	532.871		527.630	U.S.G.S. in Culvert — Fair & Warm
B.M.$_1$	5.239	532.869			800 ft. S. of Mile Post 13
T.P.$_1$ H	6.943	535.898	3.916	528.955	
T.P.$_1$ L	7.897	535.893	4.873	527.996	
T.P.$_2$ H	8.337	541.804	2.431	533.467	
T.P.$_2$ L	9.746	541.797	3.842	532.051	
T.P.$_3$ H	5.173	541.508	5.469	536.335	
T.P.$_3$ L	7.549	541.504	7.842	533.955	
T.P.$_4$ H	3.411	536.731	8.188	533.320	
B.M.$_2$ L	4.963	536.725	9.742	531.762	Spk. in Tel. Pole at Road to Abbotts Mills
T.P.$_5$ H	2.344	531.837	7.238	529.493	
T.P.$_5$ L	5.729	531.830	10.624	526.101	
B.M.$_3$ H	7.004	531.043	7.798	524.039	d.h. in Culvert at Alder Brook
T.P.$_6$ L	8.021	531.039	8.812	523.018	
	87.597		80.775		

		531.043	531.039
		527.630	527.630
87.597	3.413		3.409
80.775	3.409		
6.822 = 6.822 ck.			

FIG. 9-5. Notes for double-rodded line with two sets of turning points.

Geodetic leveling is of first-order precision (see Art. 25-4) and is usually conducted in connection with triangulation over large areas to furnish vertical control for topographic and other surveys.

The high order of precision required calls for instruments and methods which are ordinarily employed only by governmental agencies such as the U.S. Coast and Geodetic Survey and the U.S. Geological Survey. Following in this article are given briefly some features of first-order leveling, principally as followed by the Coast Survey, as an indication of the most refined practice. Detailed information is given in the references at the end of this chapter.

An extremely sensitive level of special design is used. The sensitivity of the level tube is 2 seconds of arc per 2-mm. division of tube. The tube and the middle portion of the telescope are protected from sudden and unequal changes of temperature by being encased in an outer tube, and the telescope and adjacent parts are of material having a low coefficient of expansion. The telescope is of high magnification and sharpness of image, and its large objective lens provides a relatively large field of view. The level bubble is sighted from the eyepiece end of the telescope.

The unit of vertical distance used by the Coast Survey is the meter. (The Geological Survey uses the yard.) The leveling rod is of the self-reading

FIG. 9-6. First-order leveling. (*U.S. Coast and Geodetic Survey.*)

type; it is about 10 ft. long and has a hardened-steel foot 1 in. in diameter with a slightly convex bottom face. As shown in Fig. 9-7, the front of the wooden rod is graduated in meters and decimeters; along a groove in the face of the rod extends a strip of invar metal which is graduated in centimeters and which permits readings to be taken (by estimation) to millimeters. The invar strip is fastened rigidly only at the bottom of the rod and is kept taut by means of a spring at the top; thus it is free to expand or contract independently of the wooden rod. For each sighting on the front of the rod three readings are taken—one of the horizontal cross-hair and one of each stadia hair. The back of the rod is graduated in feet and tenths; after each sighting on the front face, the horizontal cross-hair is read to 0.01 ft. on the back face as a check against mistakes. As the eyepiece of the telescope is inverting, the graduations of the rod are inverted. The rod is equipped with a rod level and a thermometer. Where other stable objects are not available as turning points, a steel pin driven into the ground is used.

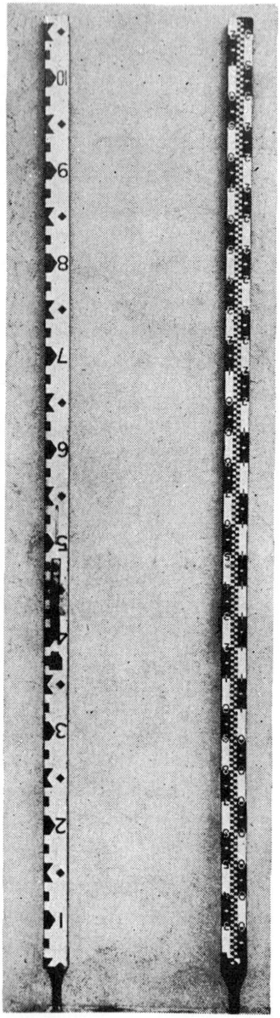

FIG. 9-7. U.S. Coast and Geodetic Survey leveling rod. (Front of rod shown at right.)

Two rods are employed in order that corresponding backsight and foresight readings can be taken as closely together as possible. At one set-up of the level, the backsight is observed before the foresight; at the next set-up, this order is reversed. Each foresight distance is kept within 10 m. of the corresponding backsight distance, and the cumulative difference between backsight and foresight distances is kept within 20 m. The maximum length of sight is 150 m. Between bench marks, the lines of levels are run both forward and backward; if these do not agree within specified limits (± 0.017 ft. $\sqrt{\text{distance in miles}}$), the line is rerun until two runnings in opposite directions do agree. The instrument is shaded from the direct rays of the sun (Fig. 9-6), even while it is being carried from set-up to set-up; and it is shielded from strong winds. Notes are kept in a form somewhat similar to that shown in Fig. 9-4.

Bench marks are established at intervals not exceeding 1 mile or, in certain cases, 2 miles. The level routes are almost invariably along railways or highways; therefore, the bench marks of the system of vertical control are usually not near triangulation stations of horizontal control, which stations are elevated for reasons of visibility. The usual form of bench mark is an inscribed bronze disk (Fig. 9-1) which is set solidly into a concrete post, a masonry structure, or rock. Engineers and surveyors are requested to see that the bench marks are preserved and to suggest needed changes in the published descriptions. If construction necessitates the moving of a bench mark, the director of the U.S. Coast and Geodetic Survey should be notified in advance, and when the change is authorized, the elevation should be transferred with great care to a new bench mark established nearby.

In the office, the field computations are checked and the values are transcribed to special computation forms. Corrections for calibrated length and temperature of rod are applied, and a correction depending on latitude and altitude is made for the effect of the geoidal shape of the earth. The circuit closures are tested for serious discrepancies. The final adjustment is then made to close the new circuits and fit them to the existing system. In the simpler cases, the circuits are closed by methods similar to those described in Arts. 9-13 to 9-15, but in the more complex cases the method of least squares is usually employed. The metric values of elevation are then converted to feet for the purpose of publication, and descriptions of bench marks are prepared.

9-10. Reciprocal Leveling. On occasion it becomes necessary to determine accurately the relative elevations of two intervisible points a considerable distance apart, between which points levels cannot be run in the ordinary manner. For example, it may be desired to transfer levels from one side to the other of a deep canyon, or from bank to bank of a wide stream. With certain modifications the method employed in getting the true difference in elevation between two points prior to the adjustment of the line of sight of the dumpy level (the two-peg test) may be utilized in situations of this kind.

If A and B are two such points, then the level is set up near A and one or more rod readings are taken on both A and B. Then the level is set up in a similar location near B, and rod readings to near and distant points are taken as before. The mean of the two differences in elevation thus determined is taken to be the true difference between the two points. Usually the distance between points is large (often a half mile or more) so that it is necessary to use a target on the distant rod. If precise results are desired, a series of foresights is taken on the distant rod and sometimes also a series of backsights on the near rod, the bubble being recentered and the target reset after each observation. The difference in elevation is then computed by using the mean of the backsights and the mean of the foresights. The three-wire method described in Art. 9-8 can be used here.

This method of reciprocal leveling assumes that the conditions under which observations are taken remain unaltered for the two positions of the level. Two factors which may appreciably alter the results, if the sights are long, are unequal expansion of the parts of the instrument and variations in atmospheric refraction. On this account it is best to make observations on cloudy days when atmospheric and temperature conditions do not vary greatly, or, if this is impossible, to protect the instrument from the sun's rays and to allow the minimum possible time to elapse between observations taken with the level in one location and those taken with it in the other. Corrections for curvature and refraction can be made as necessary.

When one point cannot be quickly reached from the other, the effect of variation in refraction may be eliminated by taking simultaneous observations with two instruments, one being set up near each point. The instruments are then interchanged, and simultaneous readings on near and far points are taken as before. All things being equal, the mean of the difference in elevation obtained with one level and that obtained with the other is assumed to be the true difference; if there is clear indication that one set of observations is inferior to the other, each set may be weighted (see Art. 5-9). Preferably the two instruments should have about the same magnifying power and same sensitiveness of level tube.

9-11. Errors in Leveling. In leveling, errors are due to some or all of the following causes (see also Table 9-1):

1. *Imperfect Adjustment of the Instrument.* In so far as results are concerned, the only essential relation is that the line of sight should be parallel to the axis of the level tube. Any inclination between these lines causes a systematic error, for then if the bubble were perfectly centered, the line of sight would be inclined always slightly upward or downward. Evidently the error in a rod reading due to this imperfection would be proportional to the distance from the instrument to the rod, and for a given distance would be of the same magnitude and sign for a backsight as for a foresight. Since backsights are added and foresights are subtracted, it is clear that the error in elevations will be eliminated to the extent that, between bench marks, the sum of the backsight distances is made equal to the sum of the foresight distances. Conversely, a systematic error will result to the extent that these distances are not equalized between any two bench marks. Often these distances will be sufficiently balanced in the long run, regardless of the terrain, to yield a satisfactory final result; but that fact does not insure a corresponding accuracy for the bench marks established along the line.

The amount of the error arising from this source for a particular case is computed in Art. 9-5.

The effect of imperfect adjustment of the instrument is minimized by adjusting the instrument and by balancing backsight and foresight distances. In precise leveling this error is also further reduced by computations.

2. *Parallax.* This condition produces an accidental error. It can be practically eliminated by careful focusing.

3. *Earth's Curvature.* This produces an error only when backsight and foresight distances are not balanced. Under ordinary conditions these distances do not tend to vary greatly, and whatever resultant error arises from this source is accidental in nature and in ordinary leveling is so small as to be of no consequence. When backsight distances are consistently made greater than foresight distances, or *vice versa*, a systematic error of

considerable magnitude is produced, particularly when the sights are long. The effect is the same as that due to the line of sight being inclined. The error varies as the square of the distance from instrument to rod and hence will be eliminated not merely by equalizing the sum total of backsight and foresight distances between bench marks but rather by balancing *each* length of foresight by a corresponding length of backsight.

4. *Atmospheric Refraction.* This varies as the square of the distance, but under normal conditions is only about one-seventh of that due to the earth's curvature, and its effect is opposite in sign. It is usually considered together with the earth's curvature, but though the effect of the latter will be entirely eliminated if each backsight distance is made equal to the following foresight distance, the atmospheric refraction often changes rapidly and greatly in a short distance. It is particularly uncertain when the line of sight passes close to the ground. Hence it is impossible to eliminate entirely the effect of refraction even though the backsight and foresight distances are balanced. In ordinary leveling its effect is negligible. In leveling of greater precision the change in refraction can be minimized by keeping the line of sight well above the ground (say, at least 2 ft.) and by taking the backsight and foresight readings in quick succession. In the long run the error is accidental, but over a short period, as a day, it may be systematic. So-called heat waves are evidences of rapidly fluctuating refraction. Errors from this source can be reduced by shortening the length of sight until the rod appears steady.

5. *Variations in Temperature.* The sun's rays falling on top of the telescope, or on one end and not on the other, will produce a warping or twisting of its parts and hence may influence rod readings through temporarily disturbing the adjustments. Although this effect is not of much consequence in leveling of ordinary precision, it may produce an appreciable error in more refined work. The error is usually accidental, but under certain conditions it may become systematic. It is practically eliminated by shielding the instrument from the rays of the sun.

6. *Rod Not Standard Length.* If the error is distributed over the length of the rod, a systematic error is produced which varies directly as the difference in elevation and bears no relation to the length of the line over which levels are run. The error can be eliminated by comparing the rod with a standard length and applying the necessary corrections. The case is analogous to measurement of distance with a tape that is too long or too short. If the rod is too long, the correction is added to a measured difference in elevation; if the rod is too short, the correction is subtracted.

Most manufactured rods are nearly of standard length, but where large differences in elevation are to be determined, few rods are near enough to the standard that corrections can be ignored in precise work.

If the rod is worn uniformly at the bottom, an erroneous height of instrument is shown at each set-up, but the error in backsight is balanced by that in the following foresight, and no error results in the elevation of the foresight point.

7. *Expansion or Contraction of the Rod.* Owing to change in moisture content or change in temperature the leveling rod may expand or contract. The resultant error is systematic. Wood when well-seasoned and painted will shrink or swell but little in the direction of the grain. Likewise its coefficient of thermal expansion is small. The error is of no particular consequence in ordinary leveling. For precise leveling, gage points may be established by inserting metal plugs in the rod, and corrections for shrinkage may be determined by observing any change in distance between the gage points. Corrections for thermal expansion may be based upon observed temperatures of the rod, as indicated by an attached thermometer, the temperature being recorded in the notes. An invar tape may be used on the rod.

8. *Rod Not Held Plumb.* This condition produces rod readings which are too large. In running a line of levels uphill or downhill it becomes a systematic error, inasmuch as the backsights are larger than the foresights, or *vice versa.* Over rolling or level ground the resultant error is accidental since the backsights are, on the average, about equal to the foresights. The error varies directly with the first power of the rod reading and directly as the square of the inclination. Thus, if a 10-ft. rod is 0.2 ft. out of plumb, the error amounts to 0.002 ft. for a 10-ft. reading and 0.0002 ft. for a 1-ft. reading; but if the rod is 0.4 ft. out of plumb, the corresponding errors are 0.008 and 0.0008 ft. It is therefore evident that appreciable inclinations of the rod must be avoided. The error can be eliminated by waving the rod, or by using a rod level.

9. *Faulty Turning Points.* An accidental error results when turning points are not well defined. Even a flat, rough stone, for example, does not make a good turning point for precise leveling for the reason that no definite point exists on which to hold the rod, which is not likely to be held in the same position for both backsight and foresight.

10. *Settlement of Tripod or Turning Points.* If the tripod settles in the interval that elapses between taking a backsight and the following foresight, the foresight will be too small and the observed elevation of the forward turning point will be too large. Similarly, if a turning point settles in the interval between foresight and backsight readings, the height of instrument as computed from the backsight reading will be too great. It is thus seen that by the normal leveling procedure, if either the level tripod or the turning point settles, as may occur to some extent when leveling over soft ground, the error will be systematic and the resulting elevations will always be too high.

Few occasions arise when turning points cannot be so selected or established as to eliminate the possibility of settlement, but care should be taken not to strike the bottom of the rod against the turning point between sights.

On the other hand, some settlement of the instrument is nearly certain to occur when leveling over muddy, swampy, or thawing ground or over melting snow. The errors due to such settlement can be greatly reduced by employing two rods and two rodmen, one rodman setting the turning point ahead while the other remains at the turning point in the rear. Backsight and foresight readings can then be made in quick succession. Small errors remaining from this source can be made accidental by reversing the order of sights at alternate set-ups, as described in Art. 9-8.

11. *Bubble Not Exactly Centered at Instant of Sighting.* This condition produces an accidental error which tends to vary as the distance from instrument to rod. Hence the longer the sight, the greater the care that should be observed in leveling the instrument.

12. *Inability of Observer to Read the Rod Exactly or to Set the Target Exactly on the Line of Sight.* This inability causes an accidental error of a magnitude depending upon the instrument, weather conditions, length of sight, and observer. It can be confined within reasonable limits through proper choice of length of sight.

Summary. From the errors just discussed it will appear that under normal conditions the important errors are accidental, provided the proper leveling procedure is observed. (See also the summary in Table 9-1.) Hence the resultant error may be expected to vary as the square root of the number of set-ups of the instrument or as the square root of the distance. Experience in general bears out this conclusion, and for this reason it is customary to express limiting errors of leveling in terms of the square root of the distance, usually in miles. It has been demonstrated, however, that on very long lines of precise levels, the errors are proportional to some power of the distance between one half and one, indicating that in spite of every precaution there are certain small systematic errors which cannot be eliminated by any known method of procedure.

9-12. Precision of Differential Leveling. The precision of differential leveling depends perhaps upon more factors than does any other operation of surveying. Although it is influenced by the instrument employed, it depends chiefly upon the care and skill of the leveler and upon the degree of refinement with which the work is executed. Other conditions remaining the same, the error for a given length of line will tend to vary as the number of set-ups above a certain minimum, hence the precision may be expected to be lower in hilly country where the sights are limited to short distances than in flat country where normal backsight and foresight distances are employed. Above a certain length of sight, however, the error of reading the rod increases very rapidly with the distance; therefore the precision will be lower for long sights than for those of normal length. Likewise, owing to erroneous length of rod, unequal refraction, and other causes, the precision of leveling between two points of large difference in elevation is likely to be lower than between two

TABLE 9-1. ERRORS IN LEVELING

Source	Type	Cause	Remarks	Procedure to eliminate or reduce
Instrumental	Systematic	Line of sight not parallel to axis of level tube	Error of each sight proportional to distance[a]	Adjust instrument; balance sum of backsight and foresight distances
		Rod not standard length (throughout length)[b]	May be due to manufacture, moisture, or temperature. Error usually small	Standardize rod and apply corrections, same as for tape
Personal	Accidental	Parallax	. . .	Focus carefully
		Bubble not centered at instant of sighting	Error varies as length of sight	Check bubble before making each sight
		Rod not held plumb	Readings are too large. Error of each sight proportional to square of inclination[a]	Wave the rod, or use rod level
		Faulty reading of rod or setting of target	. . .	Check each reading before recording. For self-reading rod, use fairly short sights
		Faulty turning points	. . .	Choose definite and stable points
Natural	Accidental	Temperature	May disturb adjustment of level	Shield level from sun
		Earth's curvature	Error of each sight proportional to square of distance[a]	Balance *each* backsight and foresight distance; or apply computed correction
		Atmospheric refraction	Error of each sight proportional to square of distance[a]	Same as for earth's curvature; also take short sights, well above ground, and take backsight and foresight readings in quick succession
	Systematic	Settlement of tripod or turning points	Observed elevations are too high	Choose stable locations; take backsight and foresight readings in quick succession, preferably alternating order of sights

[a] The error of *each sight* is systematic, but the resultant error is the difference between the systematic error for foresights and that for backsights; hence the resultant error tends to be accidental.

[b] Uniform wear of the bottom of the rod causes no error.

points the same distance apart, at or near the same elevation. Atmospheric disturbances also bear an important relation to the precision attainable.

Although conditions are so variable that no hard and fast rules can be laid down by means of which a desired precision can be maintained, experience indicates that under average conditions and with a level in good adjustment, the *maximum* error may be kept within the limits shown below. Usually the *average* error will be materially less.

1. *Rough leveling,* such as that practiced on rapid reconnaissance or preliminary surveys. Sights up to 1,000 ft. in length. Rod readings to tenths of feet. No particular attention paid to balancing backsight and foresight distances. Maximum error, in feet, $\pm 0.4\sqrt{\text{distance in miles}}$.

2. *Ordinary leveling,* such as that necessary in connection with the location and construction of highways, railroads, and most other engineering works. Sights up to 500 ft. in length. Rod readings to hundredths of feet. Backsight and foresight distances roughly balanced only when running for long distances uphill or downhill. Turning points on solid objects. Maximum error, in feet, $\pm 0.1\sqrt{\text{distance in miles}}$.

3. *Excellent leveling* for important city bench marks, or for the principal bench marks on extensive surveys. Sights up to 300 ft. in length. Rod readings to thousandths of feet with either the target rod or the self-reading rod. Backsight and foresight distances measured by pacing and approximately balanced between bench marks. Rod waved for large rod readings. Bubble carefully centered before each sight. Turning points on metal pin or plate, or on well-defined points of solid objects. Tripod set on firm ground. Maximum error, in feet, $\pm 0.05\sqrt{\text{distance in miles}}$.

4. *Precise leveling* for establishing bench marks with great accuracy at widely distributed points. High-grade level equipped with stadia hairs and with sensitive level tube. Adjustments carefully tested daily. Rod standardized frequently. Sights up to 300 ft. in length. Rod readings of three horizontal hairs to thousandths of feet. Level protected from the sun. Turning points on metal pin or plate. Two rodmen. Backsights and following foresights taken in quick succession. Bubble very carefully centered and under observation at instant of taking sight. Rod plumbed with rod level. Backsight and foresight distances balanced between bench marks by stadia readings. Level set up securely on firm ground. Levels not run when the air is boiling badly nor during high winds. Maximum error, in feet, $\pm 0.02\sqrt{\text{distance in miles}}$.

ADJUSTMENT OF ELEVATIONS

9-13. Intermediate Bench Marks. When a line of levels makes a complete circuit, almost invariably the final elevation of the initial bench mark as computed from the level notes will not agree with the initial elevation of this point. The difference is the true error of running the circuit and is called the *error of closure.* It is evident that elevations of intermediate bench marks established while running the circuit will also be in error, and there arises the problem of determining the probable errors for these intermediate points and of adjusting their elevations accordingly.

It has been shown that the principal errors of leveling are accidental, hence the probable error tends to vary as the square root of the number of opportunities for error, or as the square root of the number of set-ups. In the adjustment of elevations it will usually be sufficiently exact to assume that the number of set-ups per mile is the same for one portion of the circuit as for any other, and that therefore the probable error varies as the square root of the distance. Since corrections to such related quantities are proportional to the square of the probable errors (Art. 5-10*b*) it follows that the appropriate correction to the observed elevation of a given bench mark in the circuit is directly proportional to the distance to the bench mark from the point of beginning. Thus if E_c is the error of closure of a level circuit of length L, and if C_a, C_b, . . . , C_n are the respective corrections to be applied to observed elevations of bench marks A, B, . . . , N whose respective distances from the point of beginning are a, b, . . . , n, then

$$C_a = -\frac{a}{L}E_c; \quad C_b = -\frac{b}{L}E_c; \quad \cdots ; \quad \text{and} \quad C_n = -\frac{n}{L}E_c \tag{1}$$

Example: The accepted elevation of the initial bench mark B.M.$_i$ of a level circuit is 470.46 ft. The length of the circuit is 10 miles. The final elevation of the initial bench mark as calculated from the level notes is 470.76. The observed elevations of bench marks established along the route and the distances to the bench marks from B.M.$_i$ are as shown in the third and second columns of the accompanying tabulation. The most probable values of the elevations of these intermediate points are required.

$$E_c = 470.76 - 470.46 = +0.30 \text{ ft.}$$

By Eq. (1),

$$C_a = -2/10 \times 0.30 = -0.06 \text{ ft.}$$
$$C_b = -5/10 \times 0.30 = -0.15 \text{ ft.}$$
$$C_c = -7/10 \times 0.30 = -0.21 \text{ ft.}$$

Point	Distance from B.M.$_i$, miles	Observed el., ft.	Correction, ft.	Adjusted el., ft.
B.M.$_i$	0	470.46	0.0	
B.M.$_a$	2	780.09	−0.06	780.03
B.M.$_b$	5	667.41	−0.15	667.26
B.M.$_c$	7	544.32	−0.21	544.11
B.M.$_i$	10	470.76	−0.30	470.46

These corrections subtracted from the corresponding observed elevations give the adjusted elevations as tabulated. It is to be noted that, if the error of closure is positive, all corrections are to be subtracted, and *vice versa*.

If desired, corrections may be made directly proportional to the number of set-ups instead of the distances.

The principles described in this article apply also to the adjustment of elevations of bench marks on a line of levels run between two points whose difference in elevation has previously been determined by more accurate methods and is assumed to be correct.

9-14. Levels over Different Routes. A somewhat similar problem occurs in the adjustment of the elevation of a bench mark which is established by lines of levels run over several routes. For a point established in this manner there will be as many observed elevations as there are lines terminating at the point. Assuming that the probable error of each of the individual observed values varies as the square root of the length of the line of levels (or of the number of set-ups) by means of which the determination is secured, the weight to be applied to a given observed elevation will vary inversely as the length of the corresponding line (or of the number of set-ups). The most probable value of the elevation will then be the weighted mean of the observed values. The following example illustrates the procedure in securing the most probable value by weighting *differences* rather than by weighting the observations themselves (see Art. 5-10*a*, example 2, solution *b*).

Example: Lines of levels between B.M.$_1$ and B.M.$_2$ are run over four different routes. The length of the lines and the observed values of the elevation of B.M.$_2$ are shown in the accompanying tabulation. It is required to determine the most probable value of the elevation of B.M.$_2$.

Route	Length, miles	Observed el., B.M.$_2$	Diff. el., less 640.00	Weight	Weighted difference
a.........	2	640.72	0.72	1/2	0.36
b.........	4	640.56	0.56	1/4	0.14
c.........	10	641.08	1.08	1/10	0.11
d.........	20	640.26	0.26	1/20	0.01
				$\Sigma = 18/20$	$\Sigma = 0.62$

The weights may be represented by the reciprocals of the corresponding distances as shown. The products of the weights and the differences are shown in the last column of the table. The weighted mean is the sum of the weighted differences divided by the sum of the weights, or

$$\text{Weighted mean difference} = \frac{0.62}{18/20} = 0.69$$

and the most probable value of the elevation of B.M.$_2$ is

$$640.00 + 0.69 = 640.69 \text{ ft.}$$

If bench marks were established along any of the lines joining B.M.$_1$ and B.M.$_2$, the elevations of these bench marks in turn would require adjustment after the most probable value of the elevation of B.M.$_2$ had been determined. This would be done by assuming that the adjusted value of B.M.$_2$ represented the correct elevation and then by proceeding as explained in Art. 9-13 for a line closing on the point of beginning.

9-15. Level Net. Where elevations of bench marks in an interconnecting network of level circuits are to be adjusted, the method of least squares may be employed (see references at end of chapter). This method involves the solution of as many equations of condition as there are separate figures in the net.

A simpler and equally precise method is that of successive approximations. It consists in adjusting each separate figure in the net in turn, with the adjusted values for each circuit used in the adjustment of adjacent circuits; the process is repeated for as many cycles as necessary to balance the values for the whole net. Within each circuit the error of closure is normally distributed to the various sides in proportion to their lengths (or number of set-ups), as previously explained. The following example shows a method of solution suggested by Professor Bruce Jameyson.

Example: Figure 9-8 represents a level net made up of the circuits *BCDEB*, *AEDA*, and *EABE*. Along each side of the circuit is shown the length in miles and the observed difference in elevation in feet between terminal bench marks; the sign of the difference in elevation corresponds with the direction indicated by the arrows. Within each circuit are shown its length and the error of closure computed by summing up the differences in elevation in a clockwise direction.

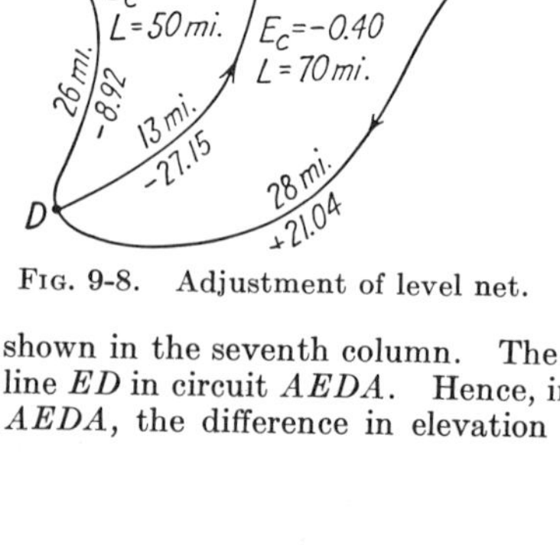

FIG. 9-8. Adjustment of level net.

Table 9-2 shows the computations required to balance the net. For each circuit are listed the sides, the distances (expressed in miles and in percentages of the total), and the differences in elevation. For circuit *BCDEB* the error of closure is -0.40 ft. This is distributed among the lines in proportion to their lengths; thus for the line *BC* the correction is $^{12}/_{70} \times 0.40$ or $0.17 \times 0.40 = 0.07$ ft., with sign opposite to that of the error of closure. The corrections are applied to the differences in elevation to obtain the values of "corrected difference in elevation" shown in the seventh column. The line *DE* in circuit *BCDEB* is the same as the line *ED* in circuit *AEDA*. Hence, in listing the differences in elevation for circuit *AEDA*, the difference in elevation for *ED* is taken, not as the observed value

(27.15), but as the adjusted value (27.08) from circuit *BCDEB*, with opposite sign. The error of closure for circuit *AEDA* is then +0.25 ft. The error is distributed as before. Similarly, in circuit *EABE* the differences in elevation listed for *EA* and *BE* are the adjusted values from the previous circuits. In Cycle II the process for Cycle I is repeated, always listing the latest values from previously adjusted circuits before computing the new error of closure. And so the cycles are continued until the corrections become zero or negligible.

The order in which the various circuits and lines are taken is immaterial, although the optimum order may reduce the number of cycles required. It is advisable to begin with the circuit having the largest error of closure. If desired, the computations may be based on the *corrections* rather than on the differences in elevation as shown. The sides of a given circuit, or a given circuit as a whole, may be weighted as desired. The elevations of intermediate bench marks are adjusted as described in Art. 9-13.

9-16. Numerical Problems.

1. A line of differential levels was run between two bench marks 20 miles apart, and the measured difference in elevation was found to be 2,163.4 ft. Later the rod whose nominal length was 13 ft. was found to be 0.003 ft. too short, the error being distributed over its full length. Correct the measured difference in elevation for erroneous length of rod.
2. Suppose that the levels of problem 1 had been run by using a rod which was 0.003 ft. too short owing to wear on the lower end. What would have been the error?
3. Suppose that the line of levels of problem 1 were continued to form a circuit closing on the initial bench mark. What error of closure due to erroneous length of rod would be expected?
4. Differential levels were run from B.M.$_1$ (el. 470.07 ft.) to B.M.$_2$, a distance of 100 miles. On the average, the backsight distances were 400 ft. in length and the foresight distances were 200 ft. in length. The elevation of B.M.$_2$ as computed from the level notes was 3,652.74 ft. Compute the error due to earth's curvature and atmospheric refraction, and correct the elevation of B.M.$_2$.
5. The levels of problem 4 were rerun using an average backsight distance of 200 ft. and an average foresight distance of 100 ft. The elevation of B.M.$_2$ as deduced from the level notes was 3,651.38. Compute the error due to curvature and refraction and correct the elevation of B.M.$_2$.
6. Suppose that the instrument used in running the levels of problems 4 and 5 was out of adjustment, so that when the bubble was centered the line of sight was inclined 0.001 ft. upward in a distance of 100 ft. Correct the observed results of problems 4 and 5 for inclination of line of sight.
7. A line of levels 10 miles long is run over thawing ground. Backsight and foresight distances average 300 ft. in length. What error would be introduced and what would be the sign of the correction to be applied to the elevation of the terminal bench mark if sights were taken in their normal order, and the average settlement of the instrument was 0.004 ft. between the backsight reading and the following foresight reading? Suppose that at alternate set-ups the order of reading were reversed, what would be the error of the line of levels?
8. If in running levels between two points the rod were inclined 0.3 ft. forward in

TABLE 9-2. ADJUSTMENT OF LEVEL NET BY SUCCESSIVE APPROXIMATIONS

Circuit	Side	Distance		Cycle I			Cycle II			Cycle III			Cycle IV		
		Miles	Per cent	DE[a]	Corr.[b]	Corr. DE[c]	DE[a]	Corr.[b]	Corr. DE[c]	DE[a]	Corr.[b]	Corr. DE[c]	DE[a]	Corr.[b]	Corr. DE[c]
BCDEB	*BC*	12	17	+10.94	+0.07	+11.01	+11.01	−0.02	+10.99	+10.99	−0.01	+10.98	+10.98	0	+10.98
	CD	28	40	+21.04	+0.16	+21.20	+21.20	−0.05	+21.15	+21.15	−0.01	+21.14	+21.14	−0.01	+21.13
	DE	13	19	−27.15	+0.07	−27.08	−27.02	−0.03	−27.05	−27.03	−0.01	−27.04	−27.03	0	−27.03
	EB	17	24	− 5.23	+0.10	− 5.13	− 5.06	−0.03	− 5.09	− 5.07	−0.01	− 5.08	− 5.08	0	− 5.08
	Total	70	100	− 0.40	+0.40	0	+ 0.13	−0.13	0	+ 0.04	−0.04	0	+ 0.01	−0.01	0
AEDA	*AE*	11	22	−17.91	−0.06	−17.97	−17.93	−0.01	−17.94	−17.93	0	−17.93	−17.93		
	ED	13	26	+27.08	−0.06	+27.02	+27.05	−0.02	+27.03	+27.04	−0.01	+27.03	+27.03		
	DA	26	52	− 8.92	−0.13	− 9.05	− 9.05	−0.04	− 9.09	− 9.09	−0.01	− 9.10	− 9.10		
	Total	50	100	+ 0.25	−0.25	0	+ 0.07	−0.07	0	+ 0.02	−0.02	0	0		
EABE	*EA*	11	26	+17.97	−0.04	+17.93	+17.94	−0.01	+17.93	+17.93	0	+17.93	+17.93		
	AB	15	35	−22.93	−0.06	−22.99	−22.99	−0.01	−23.00	−23.00	−0.01	−23.01	−23.01		
	BE	17	39	+ 5.13	−0.07	+ 5.06	+ 5.09	−0.02	+ 5.07	+ 5.08	0	+ 5.08	+ 5.08		
	Total	43	100	+ 0.17	−0.17	0	+ 0.04	−0.04	0	+ 0.01	−0.01	0	0		

[a] Difference in elevation.
[b] Correction.
[c] Corrected difference in elevation.

a height of 13 ft., what error would be introduced per set-up when backsight readings averaged 12 ft. and foresight readings averaged 1 ft.?

9. If levels are run from B.M.$_1$ (el. 2,000.00 ft.) to B.M.$_2$ (observed el. 3,000.00 ft.) and the rod is on the average 0.2 ft. out of plumb in a height of 12 ft., what error is introduced owing to the rod's not being plumb? What is the correct elevation of B.M.$_2$?

10. Suppose that in problem 9 both bench marks were at the same elevation. What would be the error?

11. The error of closure of a level circuit 100 miles long is 0.53 ft. The average length of sight is 250 ft. If all systematic errors have been eliminated, what is the probable error per set-up of the level? What is the probable error of a single observation of the rod?

12. If sights average 200 ft. in length and the probable error of a single observation is 0.004 ft., what is the probable error of running a line of levels 25 miles long? 100 miles long?

13. Complete the differential-level notes shown below. Perform the customary check:

Station	B.S.	H.I.	F.S.	El.
B.M.$_1$	6.11		....	416.23
T.P.$_1$	9.25		7.36	
T.P.$_2$	11.48		3.12	
T.P.$_3$	8.30		2.98	
B.M.$_2$	12.29		4.37	
T.P.$_4$	7.73		5.16	
T.P.$_5$	8.24		3.38	
T.P.$_6$	10.66		0.47	
B.M.$_3$	.....		4.33	

14. Complete the differential-level notes shown below. Determine the error of closure of the level circuit and adjust the elevations of B.M.$_2$ and B.M.$_3$, assuming that the error is a constant per set-up.

Station	B.S.	H.I.	F.S.	El.
B.M.$_1$	4.127		......	100.000
T.P.$_1$	3.831		9.346	
T.P.$_2$	4.104		10.725	
T.P.$_3$	2.654		12.008	
B.M.$_2$	4.368		7.208	
T.P.$_4$	6.089		6.534	
T.P.$_5$	8.863		4.736	
B.M.$_3$	12.356		2.100	
T.P.$_6$	10.781		3.662	
T.P.$_7$	12.365		4.111	
B.M.$_1$	......		9.059	

15. Lines of differential levels are run from B.M.$_1$ to B.M.$_2$ over three different routes. Following are the lengths of the routes and the observed elevations of B.M.$_2$. Determine the most probable value of the elevation of B.M.$_2$.

Route	Length, miles	El. of B.M.$_2$
a........	10	742.81
b........	16	742.58
c........	40	743.27

16. The following data are for a level net whose perimeter (reading clockwise) is *ABCDEFA*. Within the net, a line of levels extends from *B* to *F* and from *C* to *E*. The elevation of *A* is 100.00 ft. Adjust the elevations by the method of successive approximations.

Circuit	From	To	Distance, miles	Diff. el., ft.
	A	*B*	40	+17.47
ABFA	*B*	*F*	35	−10.87
	F	*A*	52	− 6.26
	B	*C*	33	+11.88
BCEFB	*C*	*E*	16	− 8.48
	E	*F*	26	−14.01
	F	*B*	35	+10.87
	C	*D*	27	−16.36
CDEC	*D*	*E*	34	+ 7.59
	E	*C*	16	+ 8.48

9-17. Field Problems.

Problem 1. Differential Leveling with Engineer's Level and Self-reading Rod

Object. Given the elevation of an initial bench mark, to determine the elevations of points in an assigned level circuit.

Procedure. Follow the procedure outlined in Art. 9-4. Keep notes in the form of the sample notes shown in Fig. 9-3, but estimate rod readings to thousandths of feet. Check each rod reading by taking a second observation. Close the circuit, and compute the error of closure.

Hints and Precautions. (1) Sight in the direction of the rod before centering the bubble exactly. (2) Center the bubble just before each reading, and check its position immediately afterward. (3) Test for parallax (see Art. 2-10). (4) Choose the turning points with an eye to simplicity of field operations, but roughly balance the backsight and foresight distances between bench marks; keep no record of these distances. (5) Use signals (Art. 3-10). (6) Keep the foot of the rod free from dirt. (7) Be sure that the rod is held vertical while a sight is being taken. When the "long rod" is used, wave the rod, and before each reading check the position of the back vernier. (8) Describe each bench mark in the field notes. (9) Check the computations.

Problem 2. Differential Leveling with Engineer's Level and Target Rod

Object. Given the elevation of an initial bench mark, to determine the elevations of points in an assigned level circuit.

Procedure. The procedure differs from that of the preceding problem only in that the rodman sets the target as directed by the leveler. Both men read the rod from the attached vernier to the nearest 0.001 ft. The field notes are kept by the leveler as explained in the preceding problem. In more precise leveling, the rodman in a separate book records the backsight and foresight distances (in paces) and the rod readings and, by observing the cumulative excess or deficiency of foresight distances over backsight distances as he goes along, roughly balances these distances between bench marks. The circuit is then closed, and the error of closure computed. The error of closure and the average time required per set-up are compared with those for the preceding problem.

Hints and Precautions. (1) The record of backsight and foresight distances is kept in the right-hand column of the left-hand page of the field notebook. The column is headed "Dist." and is subdivided into "B.S." and "F.S." A cumulative excess of 11 paces, for example, of foresight distances over backsight distances is noted as "+11" directly above each foresight distance. (2) When the long rod is used, the target must be clamped to read exactly the reading of the vernier on the back of the rod when the rod is short.

Problem 3. Reciprocal Leveling

Object. To determine precisely the difference in elevation between two points (B.M.$_a$ and B.M.$_b$) on opposite sides of a wide stream or ravine.

Procedure. (1) Set up the level in such a position that rod readings can be taken on each bench mark. (This usually necessitates the instrument's being much closer to one point than to the other.) Carefully take a series of five consecutive readings on B.M.$_a$. The mean of these is to be used as a backsight. (2) Take 10 careful readings on B.M.$_b$, the distant point. The mean of these readings is to be used as a foresight. Sight on all three horizontal hairs, if assigned. (3) Set up the level on the opposite side of the stream in such position that the distances from the instrument to a and b are respectively the same as the distances to b and a from the former position of the instrument. Take a series of readings on the near and distant points as before. (4) The difference between the mean of the backsight readings on b and the mean of the foresight readings on a from this series will also give a difference in elevation between the two points. (5) The mean of the two differences in elevation secured from the two settings of the instrument should be the correct difference. The precaution given in regard to the two-peg method (Art. 8-23), namely, that Eqs. (11) and (12) must be solved algebraically, should be given special attention in this case. (6) If the stream or ravine is imaginary, run a line of differential levels between the two points and note the discrepancy.

Hints and Precautions. (1) Be sure that the bubble is exactly centered at the time of each reading. The effect of bubble displacement will be particularly great on distant readings. For distant sights both the bubble and the target should be moved and reset after each observation. (2) If the instrument can be set up near the bench mark used as a backsight, only one observation need be taken to that point.

REFERENCES

1. BIRDSEYE, C. H., "Topographic Instructions of the U.S. Geological Survey," *U.S. Geol. Survey Bull.* 788, Government Printing Office, Washington, D.C., 1928.
2. DELL, G. H., "The Adjustment of a Level Net," *Trans. Am. Soc. Civil Engr.*, Vol. 101, pp. 834–856, 1936.
3. RAPPLEYE, HOWARD S., "Manual of Geodetic Leveling," *U.S. Coast and Geodetic Survey, Spec. Pub.* 239, Government Printing Office, Washington, D.C., 1948.
4. RAPPLEYE, HOWARD S., "Manual of Leveling Computation and Adjustment," *U.S. Coast and Geodetic Survey, Spec. Pub.* 240, Government Printing Office, Washington, D.C., 1948.
5. U.S. COAST AND GEODETIC SURVEY, "Control Leveling," *Spec. Pub.* 226, Government Printing Office, Washington, D.C., 1961.

CHAPTER 10

PROFILE LEVELING; CROSS-SECTIONS; GRADES

10-1. Profile Leveling. The process of determining the elevations of points at short measured intervals along a fixed line is called *profile leveling.* During the location and construction of highways, railroads, canals, and sewers, stakes or other marks are placed at regular intervals along an established line, usually the center line. Ordinarily the interval between stakes is 100 ft., 50 ft., or 25 ft. The 100-ft. points, reckoned from the beginning of the line, are called *full stations,* and all other points are called *plus stations.*

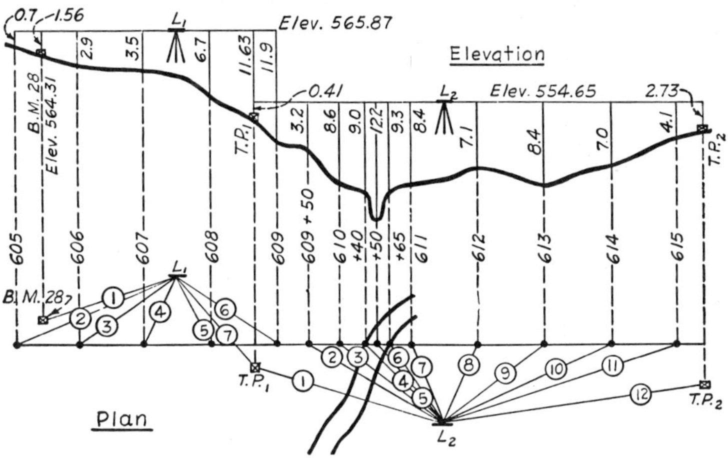

FIG. 10-1. Profile leveling.

Each stake is marked with its station and plus. Thus a stake set at 1,600 ft. from the point of beginning is numbered "16" or "16 + 00," and one set at 1,625 ft. from the point of beginning is numbered "16 + 25." Elevations by means of which the profile may be constructed are obtained by taking level-rod readings on the ground at each stake and at intermediate points where marked changes in slope occur.

Figure 10-1 illustrates in plan and elevation the steps in leveling for profile. In this case stakes are set every 100 ft. according to the common practice in highway and railway location. The instrument is set up in some convenient location not necessarily on the line (as at L_1), the rod is held on a bench mark (B.M. 28, El. 564.31), a backsight (1.56) is taken, and the height of instrument (565.87) is obtained as in differential leveling. Readings are then taken with the rod held on the ground at successive stations along the line. These rod readings, being for points of unknown elevation, are foresights regardless of whether they are back or ahead of the level. They are frequently designated as *intermediate foresights* to distinguish them from foresights taken on turning points or bench marks. The intermediate foresights (0.7, 2.9, . . . , 11.9) subtracted from the H.I. (565.87) give ground elevations of stations. When the rod has been advanced to a point beyond which further readings to ground points cannot be observed, a turning point (T.P.$_1$) is selected, and a foresight (11.63) is taken to establish its elevation. The level is set up in an advanced position (L_2), and a backsight (0.41) is taken on the turning point (T.P.$_1$) just established. Rod readings on ground points are then continued as before. The rodman observes where changes of slope occur (as 609 + 50, 610 + 40, . . . , 610 + 65), and readings are taken to these intermediate stations. The "plus," or distance from the preceding full station to the intermediate point, is measured by pacing or with a tape or the rod according to the precision required.

The care exercised in taking observations on turning points depends upon the distance between bench marks, the elevations of which have been determined previously, and upon the required precision of the profile. For a ground profile usually the backsights and foresights are read to hundredths, and no particular attention is paid to balancing backsight and foresight distances; the intermediate foresights to ground points are read to tenths of feet only. Occasions arise when it is desirable or necessary to determine intermediate foresights to hundredths of feet, for example, in securing the profile of railroad track or of the water grade in a canal; rod readings on turning points are then generally taken to thousandths of feet, and backsight and foresight distances are often balanced.

As the work of leveling for profile progresses, bench marks are generally established to facilitate later work; these are made turning points wherever possible. To check the elevation of turning points, it is necessary either to run levels back to the point of beginning or to run short lines of differential levels connecting with bench marks previously established by some other survey. The effect of an occasional error in the elevations of intermediate ground points on the profile is usually not of sufficient moment to justify the additional work which checking would make necessary, and if turning points are checked, it is regarded as sufficient.

10-2. Profile-level Notes. The notes for profile leveling may be recorded as shown in Fig. 10-2, in which foresights to turning points and bench marks are in a separate column from intermediate foresights to ground points. The values shown in the notes are the same as those illustrated in Fig. 10-1. The notes for turning points are kept in the same manner as for differential leveling.

The computations shown at the foot of the notes of Fig. 10-2 check all computations for H.I.'s and elevations of T.P.'s on the page and thus for the notes shown the difference between the sum of all backsights and the sum of all foresights is equal to the difference in elevation between B.M. 28 and the last H.I. Elevations of ground points are recorded only to the *number of decimal places contained in the intermediate foresights* regardless of the number of places in the H.I.

PROFILE LEVELS FOR I. N. RY. LOCATION

Cox Brook to Big Forks

Buff Dumpy Level — J.C. Brown, ⊼ — F. Graham, Rod — Sept. 16, 1965 — Fair

Sta.	B.S.	H.I.	I.F.S.	F.S.	Elev.	
B.M.28	1.56	565.87			564.31	On spruce root 50 ft. lt. Sta. 605.
605			0.7		565.2	
606			2.9		563.0	
607			3.5		562.4	
608			6.7		559.2	
609			11.9		554.0	
T.P.	0.41	554.65		11.63	554.24	On stone.
609+50			3.2		551.5	
610			8.6		546.1	
+40			9.0		545.7	Bank Cox Brook.
+50			12.2		542.5	Ctr. " " water 1.5 ft. deep.
+65			9.3		545.4	Bank " "
611			8.4		546.3	
612			7.1		547.6	
613			8.4		546.3	
614			7.0		547.7	
615			4.1		550.6	
T.P.	8.02	559.94		2.73	551.92	On plug.
616			9.7		550.2	
+40			6.3		553.6	Ctr. highway to St. Leonards.
	9.99		564.31	14.36		
			559.94	9.99		
			4.37 =	4.37	ck.	

FIG. 10-2. Profile-level notes.

The right-hand page is reserved for concise descriptions of bench marks and for other items of moment. For example, in the notes of Fig. 10-2 the stream and highway crossings are noted, either of which might materially influence the elevation of the roadbed at these points. Occasionally simple sketches are employed in conjunction with the explanatory notes.

CROSS-SECTIONS

10-3. Leveling for Earthwork. Four general situations arise in connection with field measurements to determine volumes of earthwork:

1. *Excavation to Predetermined Surface.* A given area is to be cut or filled to a predetermined surface, for example, in excavating the basement for a

building or in grading a piece of land. Cross-sections may be taken at short intervals in the manner described in Art. 10-4, though in some cases the sides of the squares may be as small as 10 ft. When the grade of the finished surface has been established, the cut or fill at each station will be known and the volume of earthwork can be calculated.

2. *Excavation for Trench.* A trench is to be excavated, as when a sewer or pipeline is to be laid. Profile levels are run along the proposed line. When the grade of the bottom of the trench has been fixed, the cut at each station can be computed. With the necessary width of trench at top and bottom and also its depth at each station known, the volume of excavation can be calculated.

3. *Borrow-pit Cross-sections.* An irregular mass of unknown volume is to be excavated at a given site, as, for example, to furnish material for a highway fill. Sufficient data for the calculation of volume can be obtained by taking cross-sections of the site before and after the material is removed. Usually a base line is staked out near one side, and crosslines are established at regular intervals, with a stake at each end of each crossline. Levels are run over the crosslines at distances measured from the base line. When the material has been removed, levels are rerun over the crosslines. The difference between the original cross-section and the final cross-section shows the area cut at each crossline, from which the volume can be determined.

4. *Road or Canal Cross-sections.* Earth must be cut or filled to a given grade line along some route, as a highway, railroad, or canal, and, furthermore, must have a prescribed shape of cross-section (see Arts. 10-6 to 10-10).

10-4. Cross-section Levels. Frequently in connection with problems of drainage, irrigation, grading of earthwork, location and construction of buildings, and similar enterprises, the shape of the surface of a piece of land is desired. This may be obtained by staking out the area into a system of squares and then determining the elevations of the corners and of other points where changes in slope occur. The side of the squares are usually 100, 50, or 25 ft. Directions of the lines may be obtained with either the tape or the transit, distances may be laid off either with the tape or by stadia, and elevations may be determined with either the engineer's level or the hand level, all depending upon the required precision. The elevations are carried forward, and the rod readings on ground points are determined as in profile leveling. The data may be employed in the construction of a contour map (see Art. 24-9).

Figure 10-3 illustrates a suitable form of notes. The sketch on the right-hand page shows the area divided into 100-ft. squares, the lines running in one direction being numbered and those running in the other direction being lettered. The coordinates of a given corner may then be stated as the letter and number of the two lines intersecting at the corner. Thus *A*-8 is a point at the intersection of the *A* line and the 8 line. The coordinates of a point not at the corner of a square are designated by its letter and plus in one direction and its number and plus in the other direction. Thus $(B + 50)$-7 is a point 50 ft. from line B toward line C and on line 7, and E-$(5 + 40)$ is a point on line E 40 ft. from line 5 toward line 6.

In rough, wooded country where long sights cannot be obtained and where approximate elevations (say to the nearest half-foot) will answer the purpose, more rapid progress may be secured by using the hand level and topographer's rod (see Arts. 25-13 and 25-14).

SURVEY OF PROPOSED RICE FARM, C.K. PERRY

(For Contour Map)

Scale 1" = 50' Contour Int. = 1 ft.

Sta.	B.S.	H.I.	F.S.	Elev.	
B.M.1	5.31	105.31		100.00	(Assumed)
A-8			7.3	98.0	
B-8			6.0	99.3	
C-8			4.1	101.2	
D-8			2.2	103.1	
E-8			1.4	103.9	
E-7			2.9	102.4	
D-7			4.2	101.1	
C-7			5.0	100.3	
(B+50)-7			5.6	99.7	
B-7			4.8	100.5	
A-7			3.2	102.1	
A-(7+60)			6.2	99.1	
A-6			2.5	102.8	
(A+50)-6			0.4	104.9	
B-6			1.9	103.4	
C-6			3.6	101.7	
D-6			4.2	101.1	
E-6			3.5	101.8	
E-(5+40)			6.7	98.6	
E-5			6.1	99.2	

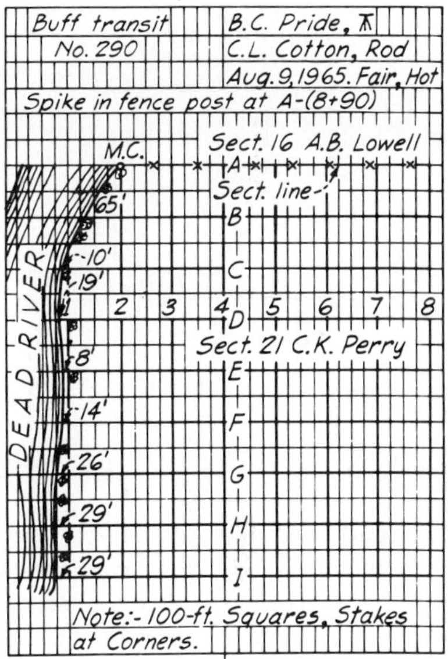

FIG. 10-3. Cross-section (checkerboard) notes.

10-5. Preliminary Route Cross-sections. Preliminary surveys for railroads, highways, and canals are often made by running a taped traverse line along the proposed route, stations being established by stakes set every 100 ft., as illustrated by the full line of Fig. 10-4 (see also Chap. 26). The elevations of the stations are then determined by profile leveling, as already described. To furnish data for location studies and for estimating volumes of earthwork, it is customary to determine the shape of the ground on both sides of the traverse line by running levels over crosslines at right angles to the traverse, usually at each station, as indicated by the dash lines of the figure. The direction of short crosslines is laid off by eye; that of long crosslines by means of a compass, transit, or right-angle mirror. Usually the elevations are determined with the hand level in rough country and with the engineer's level in flat country. For each crossline the height of instrument is established by a backsight on the ground at the center stake. The rod is then held on the crossline at breaks in the surface slope, and distances from the traverse line to these points are measured with the metallic tape.

Notes may be kept in the form shown in Fig. 10-5. The elevations of traverse stations are obtained from the profile levels. The center line of the

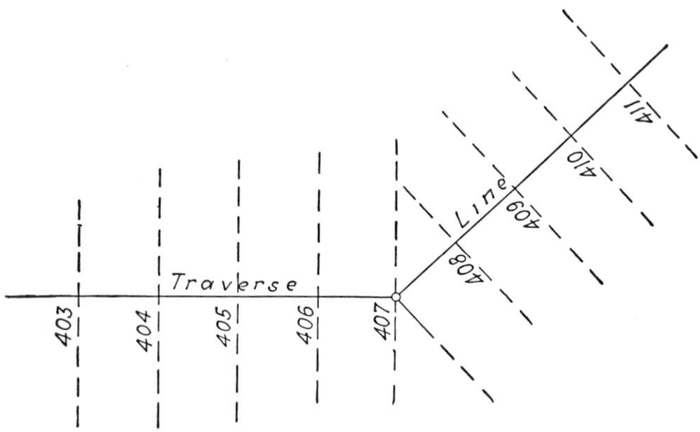

FIG. 10-4. Crosslines.

right-hand page of the notebook represents the traverse line, and to the right or left of this line are recorded the observed distances, rod readings, and the computed elevations. When a second H.I. is required to secure the necessary observations for a given crossline, a second line for that station is shown in the notes; thus station 405 occupies two lines in the notes, and station 406 occupies three. The location of fences, streams, etc., may be indicated by appropriate symbols or abbreviations.

An alternative form of preliminary cross-section notes suggested by Prof. Bruce Jameyson is shown in Fig. 10-6. This form is useful where many ground points are to be described, as in realinement of existing roads or streets. A separate line is used for each ground point.

10-6. Final Road Cross-sections. Figure 10-7 illustrates the elemental lines of typical highway or railroad cross-sections in cut and in fill. The rough subgrade is usually a plane surface, transversely level but on highway curves perhaps superelevated. On a given road the subgrade is of uniform but usually a smaller width in fill; still a third width may be used where the section is partly in cut and partly in fill. The finished cross-section may be

PRELIMINARY CROSS-SECTIONS C. & R. EXTENSION						O.H. Ellis, ⊼ C.O. Lord, Rod J.A. Crum, Tape		Jan. 20, 1965 Cold, Snow
Sta.	**B.S.**	**H.I.**	**F.S.**	**Elev.**		**Left**	**℄**	**Right**
405	12.4	633.0		620.6	(Dist.) (Elev.) (Rod)	300 210 123 80 632.1 630.8 627.0 626.7 0.9 2.2 6.0 6.3	620.6	
405	0.6	621.2		620.6				50 160 250 350 617.0 612.3 610.0 609.7 4.2 8.9 11.2 11.5
406	12.1	628.9		616.8		90 50 628.2 624.3 0.7 4.6	616.8	
406	11.5	639.7	0.7	628.2		280 200 120 638.3 635.6 632.2 1.4 4.1 7.5		
406	1.9	618.7		616.8				60 100 200 300 615.5 611.2 610.9 609.3 3.2 4.5 7.8 9.4
407	4.7	615.9		611.2		300 180 100 615.6 614.7 612.6 0.3 1.2 3.3	611.2	75 155 270 606.2 604.5 602.9 9.7 11.4 13.0
408	10.6	615.9		605.3		280 200 100 611.6 609.2 607.7 4.3 6.7 8.2	605.3	100 200 300 604.1 603.2 603.0 11.8 12.7 12.9

FIG. 10-5. Preliminary route cross-section notes.

PRELIMINARY CROSS-SECTIONS FOR REALINEMENT OF ROAD 18 — 7

April 14, 1965 — Fair, Warm — J.E. Ross ⊼ — B.L. Hart, Rod — H.O. Parker, Tape

Sta.	Dist. L	Dist. R	+ B.S.	H.I.	− Rod	Elev.	Remarks
23+00			5.4	409.4		404.0	℄ of Concrete Pavement at End
	10				5.6	403.8	Edge of " " " "
	18				6.2	403.2	
		10			5.6	403.8	" " " " " "
		18			6.1	403.3	
24+00			6.3	404.4		398.1	
	8				7.9	396.5	
		4			1.6	402.8	Edge of Existing Dirt Road
		16			1.3	403.1	" " " " "
25+00			4.7	396.4		391.7	
	72				10.8	385.6	
	T.P.		12.7	408.8	0.3	396.1	
		5			5.0	403.8	Edge of Existing Dirt Road
	T.P.	16	8.6	412.6	4.8	404.0	" " " " "
		32			2.9	409.7	
26+00			3.8	388.7		384.9	Edge of Existing Dirt Road
	23				12.5	376.2	
	95				10.7	378.0	
	T.P.	11	11.5	397.3	2.9	385.8	" " " " "
		32			0.9	396.4	

FIG. 10-6. Preliminary route cross-section notes, with ground points described.

sloped variously to provide shoulders, drainage, and rounded corners, as illustrated by Fig. 10-8.

The side slopes are plane surfaces of constant slope for a given material of excavation. The rate of the side slope is stated in terms of the number of units measured horizontally to one unit measured vertically; thus a 2 to 1 slope indicates a slope which in a horizontal distance of 2 ft. rises (or falls) 1 ft. The slope most commonly employed for cuts or fills through ordinary earth is 1½ to 1. For coarse gravel, the slope is often made 1 to 1; for loose rock, ½ to 1; for solid rock, ¼ to 1; for soft clay or sand, 2 or 3 to 1.

The preliminary survey is made, and the profile is plotted. With the profile and the data of the preliminary cross-sections as a guide, the grade line of the subgrade is established on the sheet with the plotted profile. The center cut or fill to be made at each station is then equal to the difference between the

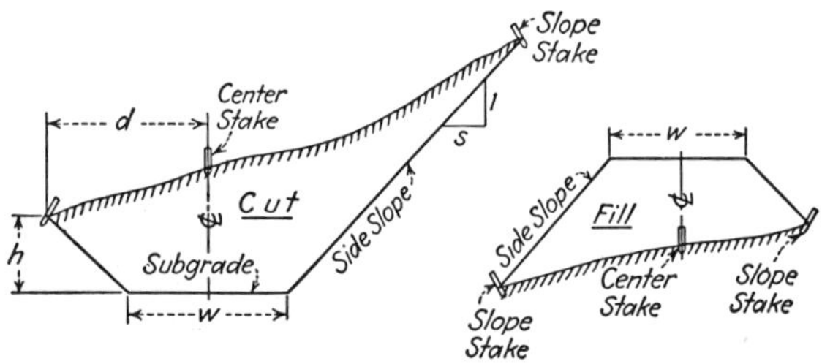

FIG. 10-7. Road cross-sections.

observed elevation of the ground line and the established elevation of the grade line.

Prior to actual construction, final cross-sections are taken, and *slope stakes* marking the intersection of the side slopes with the natural ground surface are set opposite each center stake (Art. 10-10).

Field Methods. Usually level readings for final cross-sections are taken with the engineer's level, and distances to the right or to the left of the center stake are measured with the metallic tape, all to tenths of feet. The hand level may be employed either for all observations in rough country or to extend the observations beyond the limits which can be observed from a given set-up of the engineer's level. In rough country a *slope board* consisting of a long board with a spirit level at each end may be used for leveling. Cross-sections may be taken by measuring slope distances and vertical angles (Art. 25-13).

Prior to going to the field the leveler secures a record of elevations of ground points as obtained from the profile levels, and also the elevation of the established grade at each station. In the field, the instrument is set up in any convenient location, and the H.I. is obtained by a backsight on a bench mark. At each station the rod is held on the ground (sometimes on a short wooden peg driven flush with the ground), a foresight is taken, the ground elevation is checked against that obtained by profile leveling, and the computed cut or fill is marked on the back of the center stake. A crossline through the station is established as described in Art. 10-5, and slope stakes are set as described in Art. 10-10.

If the ground is level in a direction transverse to the center line, the only rod reading necessary is that at the center stake, and the distance to the slope stake can be calculated once the center cut or fill has been determined; such a cross-section is called a *level section.* When rod readings are taken at each slope stake in addition to the reading taken at the center, as will normally be done where the ground is sloping, the cross-section is called a *three-level section.* When rod readings are taken at the center stake, the

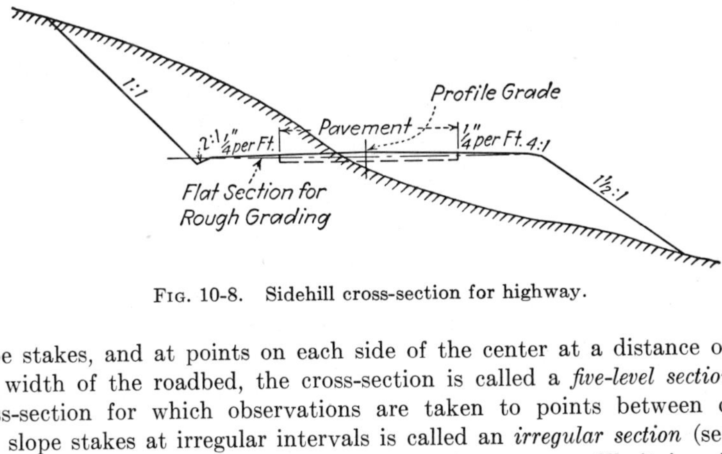

FIG. 10-8. Sidehill cross-section for highway.

slope stakes, and at points on each side of the center at a distance of half the width of the roadbed, the cross-section is called a *five-level section.* A cross-section for which observations are taken to points between center and slope stakes at irregular intervals is called an *irregular section* (see also Art. 11-8). Where the cross-section passes from cut to fill, it is called a *side-hill section* (Fig. 10-8), and an additional observation is made to determine the distance from center to the grade point; that is, the point where the subgrade will intersect the natural ground surface. A peg is usually driven to grade at this point, and its position is indicated by a guard stake marked "grade." In this case also, cross-sections are taken at additional plus stations as described in the following article.

10-7. Final Cross-section Notes. Figure 10-9 illustrates a suitable form of final cross-section notes. The left-hand page is essentially the same as for profile leveling except that a column is added for grade elevations. Some engineers prefer to have the notes read *up* the page, but in this case the computations are not so convenient as in the form shown. In some notebooks the columns for B.S., H.I., and F.S. are omitted and in the first three columns are shown the station, elevation, and grade. This arrangement provides additional space for classification of material or for computation of cross-section areas and earthwork volumes.

The notes are for a portion of the line for which profile-level notes are shown in Fig. 10-2. The cross-section levels are seen to check the elevations of stations as determined by the profile levels within 0.1 or 0.2 ft., which is

CROSS-SECTIONS FOR I.N.RY. FINAL LOCATION

Cox Brook to Big Forks — Dec. 4, 1965 — F.F. Smith, ⊼

Roadbed 20 ft. in Cut, 16 ft. in Fill — Cloudy — J. Richie, Rod

Slope 1½:1 — O. Byram, Tape

Sta.	B.S.	H.I.	F.S.	Elev.	Grade	Left	Ctr.	Right	Remarks
B.M.28	2.67	566.98		(564.31)					50 ft. Lt. Sta. 605
605			1.9	565.1	556.00	C8.6 / 22.9	C9.1	C11.2 / 26.8	Gravel in this hill.
606			4.0	563.0	555.60	C5.0 / 17.5	C7.4	C8.4 / 22.6	
607			4.5	562.5	555.20	C4.6 / 16.9	C7.3	C8.0 / 22.0	
608			8.0	559.0	554.80	C1.8 / 12.7	C4.2	C5.1 / 17.7	
+25			9.2	557.8	554.70	0.0 / 10.0	C3.1	C2.6 / 13.9	
T.P.	1.94	557.19	(11.73)	(555.25)					On plug.
+90			2.8	554.4	554.44	F1.8 / 10.7	0.0	C2.4 / 13.6	
609			3.4	553.8	554.40	F3.2 / 12.8	F0.6	0.0 / 4.0 C1.0 / 11.5	
+20			5.6	551.6	554.32	F3.6 / 13.4	F2.7	0.0 / 8.0	
610			11.1	546.1	554.00	F8.2 / 20.3	F7.9	F6.3 / 17.5	
+40			11.2	546.0	553.84	F8.0 / 20.0	F7.8	F7.8 / 19.7	Top of bank Cox Brook.
+45			14.6	542.6	553.82	F11.2 / 24.8	F11.2	F11.2 / 24.8	In brook.
+60			14.5	542.7	553.76	F11.1 / 24.7	F11.1	F11.1 / 24.7	" "
+65			11.6	545.6	553.74	F8.2 / 20.3	F8.1	F8.0 / 20.0	Top of bank.
611			10.9	546.3	553.60	F7.8 / 19.7	F7.3	F7.2 / 18.8	
612			9.7	547.5	553.20	F7.6 / 19.4	F5.7	F3.1 / 12.7	
613			10.8	546.4	552.80	F7.8 / 19.7	F6.4	F2.0 / 11.0	
T.P.	11.96	559.69	(9.46)	(547.73)					On stump.
	16.57	564.31	21.19						
		559.69	16.57						
		4.62=4.62 ck.							

FIG. 10-9. Cross-section notes for final location of roadway.

as close as can be expected. The cross-section notes on the right-hand page are in line with the station to which they refer. They illustrate a portion of a line passing from cut into fill where the slope is such that three-level sections are adequate. Cross-sections are taken at 608 + 25 where the left edge of the roadbed passes from cut to fill, at 608 + 90 where the center line passes from cut to fill, and at 609 + 20 where the right edge of the roadbed passes from cut to fill. The cross-sections at 608 + 90 and 609 are side-hill sections. Cuts or fills are shown, rather than elevations, and the cut or fill and the distance out for each point are expressed in a form resembling that of a fraction. The upper number is the cut or fill, and the lower number is the distance out from center. The values in columns marked "left" or "right" are for points at which the slope stakes are driven. If five-level or irregular sections were necessary, the values would be recorded between those for the center and the slope stakes.

10-8. Canal Cross-sections. Canal cross-sectioning is carried out in a manner similar to that for highways and railroads. Three cases commonly arise:

1. *Canal All in Cut.* In this case, no artificial banks are required. The field work is the same as for a road, slope stakes being set at the intersections of the side slopes with the actual ground surface.

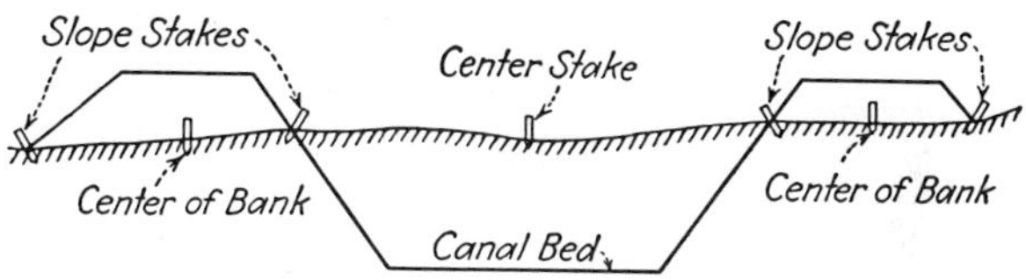

FIG. 10-10. Canal cross-section.

2. *Canal with Two Banks.* In this case, cross sections are partly in cut and partly in fill. Figure 10-10 illustrates this form of section. Stakes marking the center line of the canal are placed, and profile levels are run as in highway or railway work. The cross-section party takes cross levels and sets center stakes and slope stakes for each bank. The distance from canal center to bank center is a constant so long as the canal section remains unchanged. The distances from center to slope stakes depend upon the cut or fill and must be determined by trial.

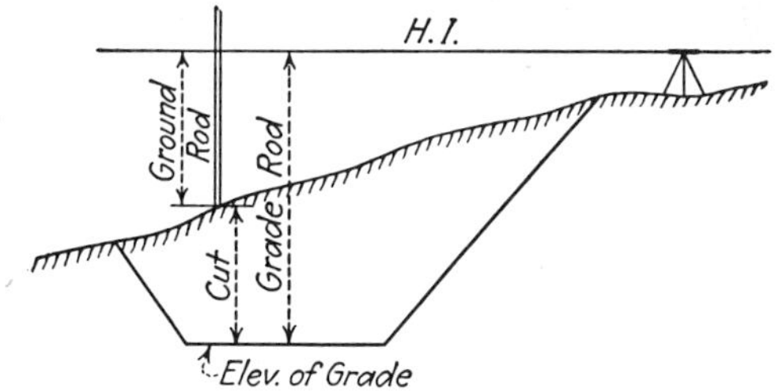

FIG. 10-11. Road cross-section in cut.

3. *Canal on Side Hill.* This case is similar to a highway or railway on a side hill, except that on the downhill side an artificial bank is required. The center and slope stakes for this bank are set as described above. The canal bed should always be in cut for its full width.

10-9. Cuts and Fills. Figure 10-11 shows the engineer's level in position for taking rod readings at a section *in cut.* The height of instrument (H.I.)

has been determined; the elevation of grade at the particular station is known. The leveler computes the difference between the H.I. and the grade elevation, a difference known as the *grade rod;* that is, H.I. − el. of grade = grade rod. The rod is held at any point for which the cut is desired, and a reading called the *ground rod* is taken. The difference between the grade rod and the ground rod is equal to the cut. Ordinarily, when the grade rod has been computed, the cut at any point is determined by mental computation.

Figure 10-12 is a similar illustration for a cross section *in fill.* If the H.I. is *above* grade (as at A) the fill is the *difference* between the grade rod and the

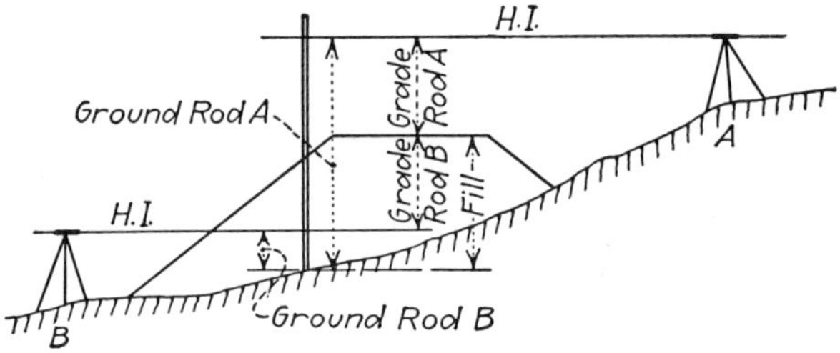

FIG. 10-12. Road cross-section in fill.

ground rod; if the H.I. is *below* grade (as at B) the fill is the *sum* of the grade rod and the ground rod.

10-10. Setting Slope Stakes. The process of setting slope stakes requires some additional explanation.

If w is the width of roadbed or canal bed, d the measured distance from center to slope stake, s the side-slope ratio (ratio of horizontal distance to drop or rise), and h the cut (or fill) at the slope stake, then, by Fig. 10-13, when the slope stake is in the correct position (at C),

$$d = \frac{w}{2} + hs \tag{1}$$

Example: The following numerical example for a cut illustrates the steps involved in establishing the correct location for a slope stake in the field; the same procedure is followed in fill.

Let w = 20 ft.; side slope = $1\frac{1}{2}$ to 1; grade rod = 15.2 ft. Suppose a slope stake is to be set on the left of the center stake (Fig. 10-13). As a first trial the rod is held at A; ground rod = 6.6 ft.

$$h_1 = \text{grade rod} - \text{ground rod} = 15.2 - 6.6 = 8.6 \text{ ft.}$$

The computed distance for this value of h_1 is $w/2 + h_1 s = 10.0 + 8.6 \times 3/2 =$ 22.9 ft. Measurement from the center stake shows d_1 to be 18.2 ft.; hence the rodman should go farther out.

For a second trial the rod is held at B; ground rod = 8.8 ft.; h_2 = grade rod − ground rod = 15.2 − 8.8 = 6.4 ft.; $w/2 + h_2 s = 10.0 + 9.6 = 19.6$ ft. The measured value of d_2 is 22.5 ft.; hence the rod is too far out.

Eventually, by trial, the rod will be held at C; ground rod $= 7.8$ ft.; $h = 15.2 - 7.8 = 7.4$ ft. The computed distance for this value of h is $w/2 + hs = 10.0 + 7.4 \times 3/2 = 21.1$ ft. The measured value of d is also 21.1 ft., hence this is the correct location for the slope stake. In the notes the coordinates of the slope stake are given by the expression $c7.4/21.1$ in the form of a fraction, but the trial observations are not recorded.

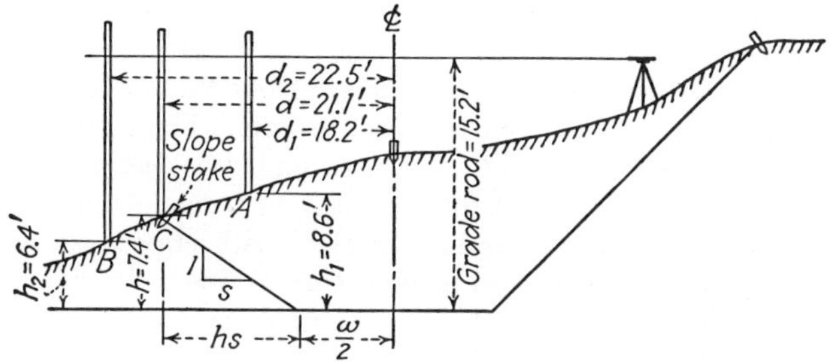

FIG. 10-13. Setting slope stakes.

Slope stakes are set side to the line, sloping outward in fill and inward in cut. On the back of the stake the station number is marked with crayon. On the front (side nearest the center line) are marked the cut or fill at the stake and sometimes the distance from center to slope stake. The numbers read down the stake.

In cut, some organizations set the slope stakes at a fixed distance, say 2 ft., back from the edge of the slope. The cut marked on the stake applies to the elevation of the ground at the stakes thus offset.

If cuts and fills are only a few feet deep, sometimes the slope stakes are omitted, and the stakes used for alinement are also employed as reference elevations for grade (see Art. 28-6).

A special "tape" rod has been devised, with a continuous metallic tape passing over rollers near the top and bottom of the rod, which renders unnecessary the subtraction of the ground rod from the grade rod. Furthermore, for a given width of roadbed and given slope, a specially graduated tape may be used which solves mechanically the expression $(w/2 + hs)$.

GRADES

10-11. General. In surveying, the term *grade* or *gradient* is used to denote the slope, or rate of regular ascent or descent, of a line. It is usually expressed in per cent; for example, a 4 per cent grade is one which rises or falls 4 ft. in a horizontal distance of 100 ft. The term *grade* is also used to denote an established line on the profile of an existing or a proposed roadway. In such expressions as "at grade" or "to grade" it denotes the elevation of a point either on a grade line or at some established elevation as in construction work.

The operation of leveling for grades is similar to profile leveling. After the grade line has been established on the profile, the grade elevation for each station is known. Leveling for grades is started from a bench mark and is carried forward by turning points. The grade rod to be employed in setting a given grade stake to grade is computed by subtracting the grade elevation from the H.I. The rodman starts the stake and holds the rod on its top. The leveler reads the rod and calls out the approximate distance which the stake must be driven to reach grade. The rodman drives the stake nearly the desired amount, and a second rod reading is taken; and so the process is continued until the rod reading is made equal to the grade rod. The top of the stake is usually marked with crayon to indicate that it is at grade. Sometimes the rod is moved up or down the side of the stake until the grade rod is read, when the position of grade is indicated by a crayon mark or a nail driven into the stake at the foot of the rod. When points are established a given distance above or below grade, the process is the same, except that the distance to grade is indicated either on the grade stake or on a guard stake nearby. Usually grade elevations are determined to hundredths of feet.

The notes are kept as in profile leveling except that the right-hand column of the left-hand page is for grade elevations. When the stakes are not driven to grade, a record is also kept of the cuts or fills (that is, the vertical distances from established points to the actual grade line).

The distance between points at which grade is established depends upon the character of the work and upon whether the grade is uniform (straight line in profile) or on a vertical curve. On railway track construction, grades are usually given at each 100-ft. station but on vertical curves are sometimes given every 50 ft. For street and highway pavements and for sewers, grade is usually established every 50 ft. if the grade is uniform and every 25 ft. or even every 10 ft. if the grade is on a vertical curve.

10-12. Shooting-in Grade. Unless the grade is level, the methods described in the preceding article necessitate computing the grade rod for every station which is established at grade. Likewise, if grade is established at any intermediate point the chainage of which has not been previously determined, the plus of the new station must be measured before the grade rod can be calculated.

When the line is tangent (straight in plan) and the grade is uniform for a considerable distance, the work of setting grade stakes may be facilitated by "shooting-in" the grade in the manner now to be described. Let A and B be two stations some distance apart (say, 800 to 1,000 ft.) on tangent, between which stations a uniform grade is to be established. A line of differential levels is run to include both A and B, and stakes are driven to grade, or to a fixed distance above or below grade, at these points. The level is then set up close to B with one pair of foot screws in line with A and, with the rod held on the stake at B, a reading is obtained by sighting through the objective end of the telescope. The rodman next holds the rod on *A*, and the leveler moves the telescope in a vertical plane (by means of the foot screws if a level is used, or by means of the altitude tangent-screw if the transit is used) until the horizontal cross-hair appears to cut the rod at the reading previously obtained with rod at *B*. Neglecting the effect of the earth's curvature and atmospheric refraction (which will be of no consequence for the distances suggested above), the line of sight is now a uniform distance above the uniform grade for all points between *A* and *B*. Hence the grade at any intermediate station is established by observing the same rod reading, the instrument, of course, remaining undisturbed for the interval during which the intermediate stakes are being set. As a check, it is well to sight again to *A* before the instrument is moved, in order to detect any displacement of the line of sight.

When the grade is nearly level, as is often the case for drainage works, the process described above may be simplified. The sensitiveness of the bubble may be determined as described in field problem 2 of Art. 8-28, and the number of spaces on the level tube corresponding to various rates of grade may be computed. By means of the bubble the line of sight may then be inclined at the same slope as the grade line.

Gradienter. When grades are to be established with a level equipped with a gradienter (Art. 2-20), the elevations are carried forward by direct leveling, as previously described, the instrument always being set up near a station on the line. After the H.I. for a given set-up has been determined, the gradienter drum is set to read zero when the instrument is level. The line of sight is then made parallel with the grade line by turning the gradienter until it reads the desired grade. Thus, if one turn of the screw inclines the line of sight 1 per cent and the gradienter drum is divided into 100 parts, a 1.2 per cent grade would be laid off by one full turn and 20 spaces as indicated by the drum graduations. Since the line of sight is parallel to the grade line, the grade rod for the station at which the level is set up is also the grade rod of any other station in the direction in which the instrument is pointed. For points in advance of the level, the gradienter must be set to a reading equal and opposite to that for points in the rear.

10-13. Contour Leveling. A contour is an imaginary line connecting points of equal elevation on the surface of the earth (see Art. 24-6). In topographic surveying the engineer's level, in conjunction with other instruments, is sometimes employed for the direct location of contours. Also in connection with the surveys for reservoir sites, levels are usually run to establish the proposed shore line. The process of establishing lines of this character consists in carrying a line of levels forward by turning points and in finding, by trial, a series of ground points at the required elevation.

Proposed shore lines are usually defined by stakes set at long intervals where the shore contour is straight and at short intervals where it is irregular or curved. A line of levels is run approximately along the contour. At each set-up the contour rod (the difference between the elevation of the H.I. and the elevation of the contour) is computed. The rodman proceeds along the line giving rod readings at critical points. At each point he is directed up or down the slope until the leveler reads the contour rod. Here a stake is set. In topographic surveying the process is essentially the same, except that no stakes are set (see Art. 25-15).

10-14. Establishing Grade Contours. In connection with the preliminary surveys for highways and canals in hilly or mountainous country where the general route lies along a side hill, levels may be run to establish points along some required grade. The irregular line (in plan) joining such points is termed a *grade contour*. If a level (or transit) with gradienter is available, the simplest method is to follow the same procedure as in contour leveling (Art. 10-13), except that for both turning points and ground points the gradienter is set so that the line of sight is at the required grade. For rough surveys, the clinometer may be used.

Generally rod readings are taken only to points where there is a noticeable change in the direction of the grade contour. The level should be in such a location that the ground points will not deviate greatly from the straight line joining the instrument and the adjacent turning points. Often in rough work the only intermediate ground point between turning points is at or near the level, and the rod readings for turning points are taken with rod held on the ground.

In more careful work, as when the grades are nearly level, the elevations may be carried forward as in differential leveling (that is, bubble centered when backsights and foresights are taken to turning points) and distances may be measured by stadia; at the same time the grade contour is located as described above. The stadia distances and the elevations determined by differential leveling offer a means of checking the grade contour at any point.

If the level is not equipped with a gradienter, unless the grade is nearly level so that the required slope can be laid off with the bubble, the grade contour must be established by the more laborious method of direct leveling, distances being measured by stadia or by pacing and a new grade rod being computed for each point at which a stake is driven, as described in Art. 10-11.

10-15. Vertical Curves. On highways and railways, in order that there may be no abrupt change in the vertical direction of moving vehicles, adjacent segments of differing gradient are connected by a curve in a vertical plane, called a *vertical curve*. Usually the vertical curve is the arc of a parabola, as this form is well adapted to gradual change in direction and as elevations along the curve are readily computed.

The length of a vertical curve depends upon several conditions, being in general greater for railways than for highways, and increasing with the difference in grade between adjacent segments. The maximum allowable change in grade per station is usually governed by specifications. The length of the vertical curve cannot be less than the algebraic difference in gradient between the two segments connected, divided by the maximum allowable change in grade per station. Usually it is an even number of stations in railway work or some convenient whole number of feet in highway work.

The station and plus of the vertex, or point of intersection of the two segments, and the elevations of stations along the uniform grade lines are determined from the grade profile. The length of the vertical curve is then computed or is taken at some convenient value which meets the specification requirements; and the stations and elevations of the beginning and end of curve are calculated. Then the offsets from the uniform gradients to the curve are computed, and the grade elevations at stations along the curve are thus determined. In the field the vertical curve is laid out by setting grade stakes at these stations, just as along a uniform grade.

One method of calculation for a vertical curve is as follows: The elevation of the mid-point of the "long chord" (Fig. 10-14) connecting the points of beginning and end of the vertical curve is computed. As the curve is a parabola, the elevation of the mid-point of the vertical curve is the mean of the elevation of the vertex and the elevation of the mid-point of the long chord. The tangent offsets to various points along the curve are then computed, employing the known property of a parabola that the tangent offset varies as the square of the distance from the tangent point.

Example: On a railroad a +0.8 per cent grade meets a −0.4 per cent grade at station 90 + 00 and at elevation 100.00 (Fig. 10-14). The maximum allowable change in grade per station is 0.2. It is desired to establish a vertical curve connecting the two grades.

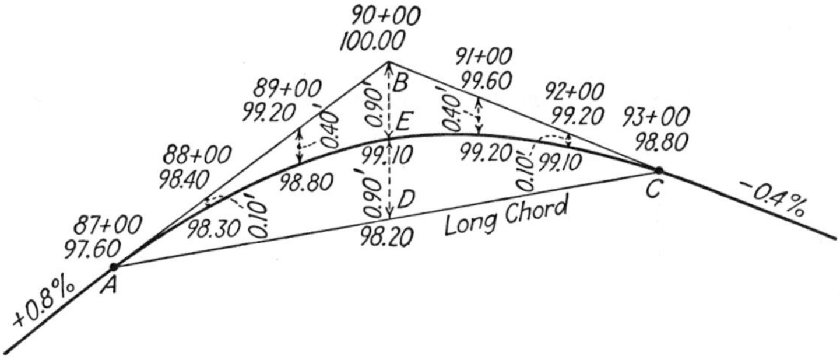

FIG. 10-14. Vertical curve.

The algebraic difference in gradient is $+0.8 - (-0.4) = 1.2$ per cent. The minimum length of curve is then $1.2/0.2 = 6$ stations or 600 ft. The length on either side of the vertex ($AB = BC$) is $^{600}\!/_{2} = 300$ ft. The station at A is therefore $90 - 3 = 87$, and the station at C is $90 + 3 = 93$. The elevation of A is $100.00 - 3 \times 0.80 = 97.60$, and the elevation of C is $100.00 - 3 \times 0.40 = 98.80$ ft.

The elevation of the mid-point D of the long chord AC is the mean of the elevations A and C:

$$\tfrac{1}{2}(97.60 + 98.80) = 98.20 \text{ ft.}$$

The mid-point E of the vertical curve is midway between D and the vertex B:

$$\tfrac{1}{2}(98.20 + 100.00) = 99.10 \text{ ft.}$$

The offset from vertex to curve is

$$100.00 - 99.10 = 0.90 \text{ ft.}$$

The tangent offsets at stations 89 and 91 are

$$\frac{2^2}{3^2} \times 0.90 = 0.40 \text{ ft.}$$

and the offsets at stations 88 and 92 are

$$\frac{1^2}{3^2} \times 0.90 = 0.10 \text{ ft.}$$

The elevations of points on curve are then determined as shown in the following tabulation:

Station	87 = *A*	88	89	90	91	92	93 = *C*
Elevation of tangent	97.60	98.40	99.20	100.00	99.60	99.20	98.80
Tangent offset	0.00	0.10	0.40	0.90	0.40	0.10	0.00
Elevation of curve	97.60	98.30	98.80	99.10	99.20	99.10	98.80

A convenient check is afforded by computing the "second differences" between the elevations of consecutive points on the curve, since for a parabola these should be a constant. For the example just given, the following tabulation illustrates the computations:

Stations	Elevation, ft.		Diff., ft.	Second diff., ft.
	Back station	Forward station		
87–88	97.60	98.30	+0.70	0.20
88–89	98.30	98.80	+0.50	0.20
89–90	98.80	99.10	+0.30	0.20
90–91	99.10	99.20	+0.10	0.20
91–92	99.20	99.10	−0.10	0.20
92–93	99.10	98.80	−0.30	

10-16. Location of Summit or Sag. The location and elevation of the summit (or of the lowest point of a sag) of a vertical curve can be computed as follows: The general equation of the parabola is

$$y = ax^2 \tag{2}$$

The rate of change in slope of the tangent to a parabola is the derivative of Eq. (2) with respect to x, or

$$\frac{dy}{dx} = 2ax \tag{3}$$

If x is in stations, the constant $2a$ is the change in grade per station.

The slope at the summit S is zero (Fig. 10-15), and the slope at the point of tangency C is equal to the grade G, so that

$$2ax = G \qquad \text{or} \qquad x = \frac{G}{2a} \tag{4}$$

The change in grade $2a$ either is known or can be readily computed from the data given for the curve.

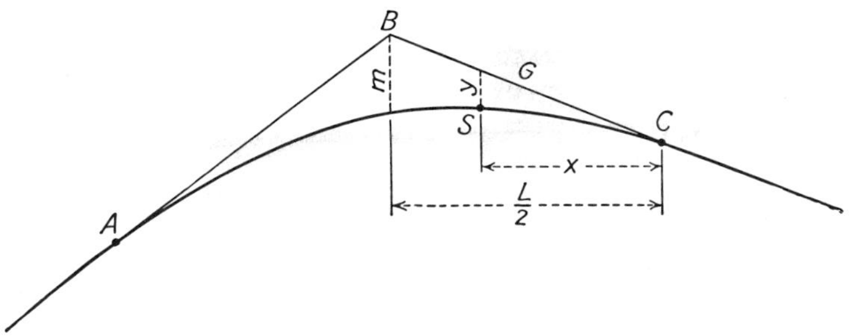

FIG. 10-15. Summit of vertical curve.

The tangent offset at the summit S is determined by solving the general equation for y at that point. The elevation of the summit is then computed by subtracting the tangent offset from the elevation of the corresponding point on the tangent.

Thus, in the example of Art. 10-15, the change in grade per station is $2a = 0.20$. By Eq. (4),

$$x = \frac{G}{2a} = \frac{0.40}{0.20} = 2.00 \text{ stations}$$

The station at the summit is $93 - 2 = 91$. By Eq. (2),

$$y = ax^2 = 0.10 \times 2.00^2 = 0.40 \text{ ft.}$$

The elevation at the summit is $99.60 - 0.40 = 99.20$ ft.

10-17. Numerical Problems.

1. Complete the profile-level notes shown below:

Sta.	B.S.	H.I.	I.F.S.	F.S.	El.
B.M. 10	6.32		....	.....	836.76
179	....	...	10.1	.....	
180	....	...	7.8	.....	
+35	....	...	12.6	.....	
181	....	...	4.7	.....	
182	....	...	3.4	.....	
T.P. 38	7.32		....	2.11	
183	....	...	8.5	.....	
+40	....	...	4.6	.....	
184	....	...	7.2	.....	
185	....	...	10.6	.....	
T.P. 39	5.93		....	11.49	
186	....	...	4.2	.....	

2. The width of roadbed of a proposed railroad is 24 ft. in cut, and the side slopes are 1½ to 1. At a given station the elevation of grade is 515.75. For obtaining the cross-section at the station the H.I. of the level is 528.32. The ground rod at the center stake is 6.5. Compute the grade rod and the center cut. The rod reading at the right slope stake is 1.2 and at the left slope stake is 10.9. Compute the cut and the distance out to each slope stake.

3. Make a page of cross-section notes (similar to Fig. 10-9) for a highway running from cut into fill. The grade of the highway is 4.0 per cent, width of roadbed 24 ft. in cut and 18 ft. in fill, and side slopes 1½ to 1. Show observations at center and slope stakes and at grade points.

4. Make a page of notes for establishing grades of top of rail from station 750 to station 762. Elevation of grade at station 750 is 381.60; grade from station 750 to station 758 is −0.6 per cent; grade from station 758 to station 762 is −0.4 per cent; elevation of bench mark near station 750 is 378.47. Do not consider the vertical curve.

5. The radius of curvature of the level tube of a level is 75 ft. and one space on the tube is equal to $\frac{1}{10}$ in. How many spaces will the bubble have to be displaced from the center to make the line of sight parallel with a grade rising 1.5 ft. per mile?

6. On a highway a −6.0 per cent grade meets a +4.0 per cent grade at station 67 + 50 and at elevation 516.32. The maximum allowable change in grade per station is 2.5. Compute the elevations of stations at 50-ft. intervals along a vertical curve connecting the two grades. Compute the station and elevation of the lowest point on the curve.

10-18. Field Problems.

Problem 1. Profile Leveling for a Roadway

Object. To determine the elevations necessary for plotting the profile of a line.

Procedure. (1) Lay out the assigned length of line with numbered stakes every 100 ft. (2) If no bench mark is given, select some permanent point as a bench mark, assuming an elevation such that no station will fall below the datum. (3) Adapt to the field conditions encountered the procedure indicated in Arts. 10-1 and 10-2. (4) Keep the notes in the form of the sample notes (Fig. 10-2).

Hints and Precautions. (1) Read the rod carefully to the nearest 0.01 ft. on bench marks and turning points, and quickly to the nearest 0.1 ft. on ground points. (2) Take rod readings on the ground at all full stations and at such other points on the line (plus stations) as are necessary to obtain a sufficiently accurate profile. In general these plus stations will be at points where the slope of the ground changes noticeably and at highways, railways, and streams. (3) If the line is long, establish bench marks every 1,500 or 2,000 ft. Bench marks should be placed at some distance to one side of the line, in such positions that they are not likely to be disturbed during construction; they should be well described in the notes. It is customary to mark the number and elevation on each bench mark at the time it is established. (4) Make the computations for turning points as the work progresses, and check each page of notes as soon as it is filled.

Problem 2. Profile Leveling for a Pipeline

Object. To prepare the line of a proposed sewer or water main for construction. It is assumed that the line has already been run and that center stakes marked with station and plus have been set every 25 or 50 ft. Ground pegs are to be set to give line and grade for ditchers and pipelayers.

Procedure. (1) Opposite each stake on the line and far enough from the line to insure its not being disturbed by the excavation, drive a short peg (or spike) flush with the ground, and beside this peg drive a stake marked (on the side away from the line) with the station number of the center stake and the offset of peg from center stake. (2) Start from a bench mark as in the preceding problem, and take profile readings on the ground pegs to the nearest 0.01 ft. Keep notes in the form of the sample notes of Fig. 10-2, except that additional columns are required for offsets of pegs from center line, grade elevations, and cuts. Complete the level work as in the preceding problem. (3) Roughly plot the profile; then fix the grade of the bottom of the trench, and determine the amount of cut at each station. (4) Mark the cut, expressed in feet and inches to the nearest ⅛ in., on the front of each side stake (facing the line).

Hints and Precautions. (1) Take rod readings with greater care on the turning points than on the ground pegs. (2) Mark all stakes to read down the stake. Center stakes should be driven with the marked side toward the beginning of the line. Side stakes should be driven side to the line in order that they will not be confused with center stakes. (3) In paved streets or hard roads it is impossible to drive stakes or ground pegs; and spikes, chisel marks, or paint marks are used instead. The spikes are driven flush with the road surface. In order that they may be found without difficulty, their position with respect to more prominent objects is carefully recorded.

Problem 3. Setting Slope Stakes; Cross-sections

Object. To prepare a proposed highway or railway for grading and to obtain data for calculating earthwork (see office problem 2, Art. 11-17).

Procedure. (1) From the level notes of problem 1 plot a profile and fix a grade such that the amount of cut will approximately balance the amount of fill. (2) Keep field notes in the form of Fig. 10-9. (3) Drive short pegs flush with the ground against the center stakes on the side farthest from the beginning of the line. Run profile levels over the line of pegs, checking the elevations obtained with those of problem 1, and mark on the back of each center stake the cut or fill at that point, as *C* 3.9 or *F* 4.7. (4) Assume a roadway 20 ft. wide with side slopes of 1½ to 1. Opposite each center stake, at right angles to and on both sides of the line, set slope stakes as described in Art. 10-10. The numbers should read down the stake. (5) Drive ground pegs at "grade points" where the center line and each edge of roadbed pass from cut to fill, and mark the location of these grade pegs by stakes marked "grade."

CHAPTER 11

PLOTTING PROFILES AND CROSS-SECTIONS; EARTHWORK

PROFILES AND CROSS-SECTIONS

11-1. Plotting Profiles. The profile is plotted from profile-level notes or from elevations taken from a contour map (Chap. 24). Usually it is plotted on regular *profile paper*, which is ruled with vertical lines at intervals of ¼ or ½ in. and with horizontal lines at intervals of $\frac{1}{20}$ or $\frac{1}{10}$ in. The horizontal and vertical scales to be employed depend upon the purpose of the profile. If the profile is to be used for fixing grades, as for a railroad or highway, a scale of 1 in. = 400 ft. (horizontal) and 1 in. = 20 ft. (vertical) is frequently used; if it is to be made the basis of earthwork calculations, as for a sewer or pipe line, a scale as large as 1 in. = 40 ft. (horizontal) and 1 in. = 4 ft. (vertical) may be required. The vertical scale is exaggerated because the vertical distances on the ground are relatively small as compared with the horizontal distances.

Figure 11-1 illustrates, to reduced scale, a portion of the profile for a proposed railroad. In this illustration only the horizontal lines are shown except for a small area at the top of the profile; the distance between these accentuated lines represents a difference in elevation of 5 ft. The space between vertical lines represents a horizontal distance of 100 ft. The numbered heaviest horizontal lines indicate multiples of 50 ft. in elevation, and the heaviest vertical lines indicate multiples of 10 stations counting from the beginning of the line. In the original drawing, the size of the small squares was ¼ in., the vertical scale was 1 in. = 20 ft., and the horizontal scale was 1 in. = 400 ft.

The general practice is to begin the profile at the left end of the sheet; the station numbers thus increase from left to right. The ground line is formed by drawing a line through the plotted points, usually freehand, as the elevations are plotted. The profile should not be succession of straight lines between adjacent points, because this does not represent the actual variation in the ground. On the other hand the profile should not be unduly rounded at summits and depressions, for on the drawing such points are exaggerated in sharpness on account of the exaggerated relation between horizontal and vertical scales. Figure 11-1 illustrates the irregular form of the profile.

Notes on the profile show the station and plus of important objects, as streams and roads, crossed by the line (Fig. 11-1). Such notes are placed

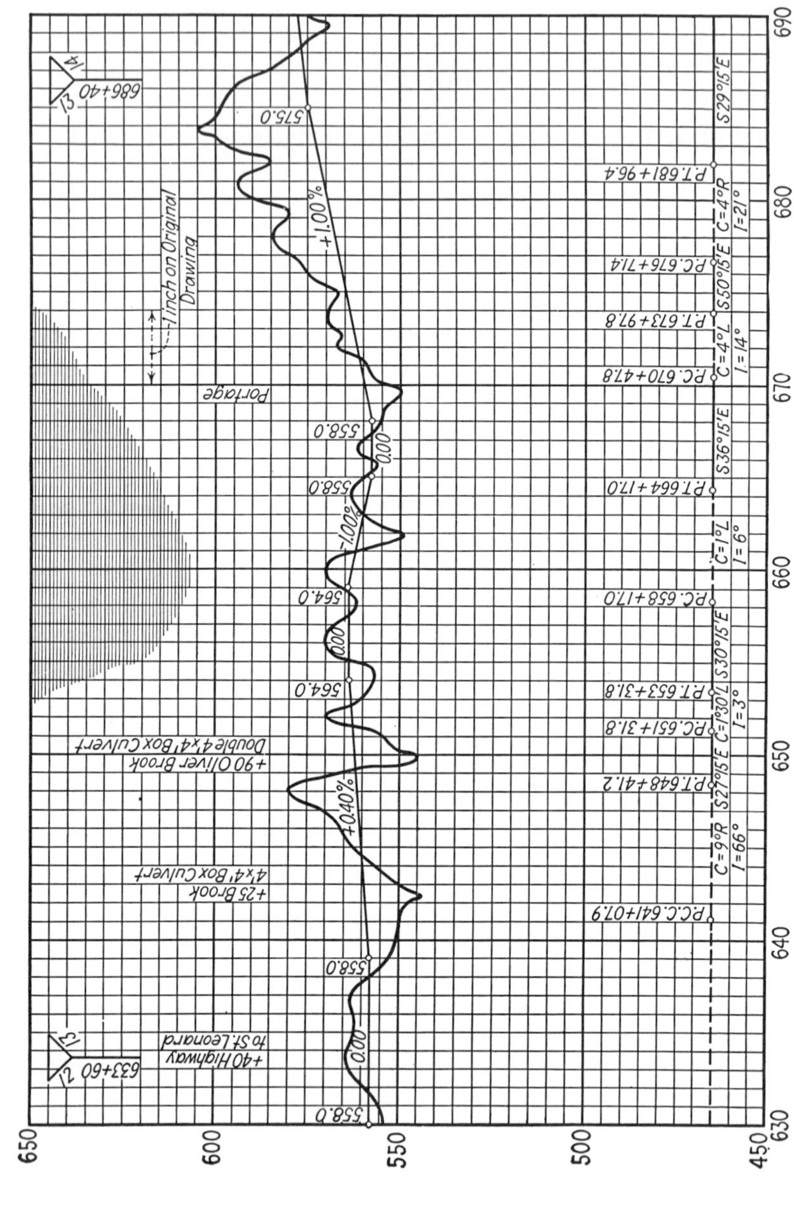

FIG. 11-1. Profile and alinement diagram of railroad.

directly above the points on the profile to which they refer. Usually an alinement diagram is drawn near the bottom of the sheet, with points on the diagram directly below corresponding points on the profile. The alinement diagram indicates the location of tangents, curves, and changes in direction of the line, but it is not a true plan view except when the line is straight. The diagram at the bottom of Fig. 11-1 is a suitable form for a railroad,

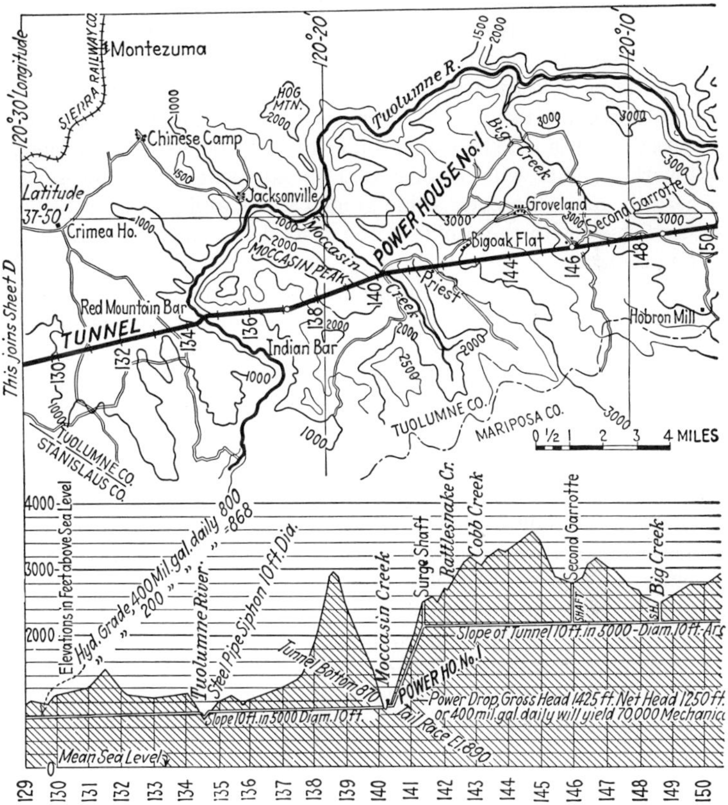

FIG. 11-2. General map and profile.

highway, or similar line. Sometimes the directions of land lines, streams, and other objects crossed by the line are indicated on the alinement diagram.

Sometimes general drawings show the profile on the same sheet with the map or plan of the line, as illustrated in Fig. 11-2. Such an arrangement is convenient for purposes of comparison in that the general relation between plan and profile may be seen at a glance.

In highway work it is common practice to plot the plan and profile on the same sheet, as illustrated in Fig. 26-1. The sheets, called "Federal aid sheets," are of a standard size (border line 22 by 33½ in.) with the profile portion ruled 4 by 20 or 2 by 10 to the inch.

11-2. Fixing Grades. The ground profile furnishes the basis for the study of economic grade elevation. Although the factors influencing the choice of grades will not be discussed here in detail, it is pertinent to mention some of them. In the location of highways or similar routes, maximum permissible rates of grade are usually established by considerations of traffic (Chap. 26). Likewise the elevation of grade is fixed within narrow limits at certain controlling points as at terminals and at stream and roadway crossings. Conforming to these limitations, between such controlling points the grade of the proposed road is fitted to the ground until as far as possible the volume of earthwork in cuts will balance that in adjacent fills. For sewers and drains certain minimum permissible grades are established by considerations of flow, and these with the profile of the ground determine the grade between controlling points.

Grade lines are established on the profile until by trial a satisfactory solution is obtained. Rates of grade and stations of points of change in grade are then fixed, and the corresponding elevations of points of change are computed. Field and office work are simplified if rates of grade are expressed in an exact decimal, as 2.5 per cent or 0.65 per cent, and not fractionally, as 2⅓ (2.333+) per cent or $^{16}/_{71}$ (0.225+) per cent. The grade line may be a succession of straight lines, or if the changes in grade are considerable, it may be a series of straight lines connected by vertical curves at the summits and depressions. The selected grade line is shown on the profile as illustrated by Fig. 11-1. The points of change are marked by small circles, and on vertical lines through these points are shown their elevations (and plusses if they do not fall at full stations). The rates of grade are shown just above the grade line or on dimension lines, as illustrated in the figure.

11-3. Finishing the Profile. The profile is finished in ink. If it is to be blueprinted, all lines are shown in black; in other cases the grade line and the numerical notes pertaining to it are usually shown in red and the remainder of the sheet is inked in black. Sometimes the alinement diagram with its accompanying notes is shown in a contrasting color, as blue or orange. The station numbers of the heavy vertical lines are placed at the bottom of the sheet. The elevations of the heavy horizontal lines are written at each end of the sheet, and at intermediate points if the profile is long. Numbers and explanatory notes which are written vertically on the profile are ordinarily placed so as to be read from the right end, though some engineers prefer them to read from the opposite end so that as the profile is unrolled with the zero end toward the body the notes will appear right side up.

11-4. Other Profiles. The profiles discussed in the preceding articles are of value in showing the relation between grade and the natural ground and are useful in fixing grades and estimating volumes of earthwork. In connection with construction, profiles are also frequently employed to portray graphically the progress of the work. For example, in highway construction an estimate is made of work done during each month; that is, the volume of earthwork moved, the amount of pavement laid, etc., are calculated from field measurements. When the monthly estimate has been completed, the progress profile is brought up to date by tinting the portions completed with a particular color assigned for the month, also by showing within or adjacent to each tinted area the volume of earthwork or other quantity involved. Pavement laid, right-of-way fences completed, culverts, bridges, etc., constructed may be designated by appropriate colored symbols on the alinement diagram.

For some kinds of work, as for subways and foundations, profiles showing not only the surface line but also the various subterranean strata are made. Points for the subsurface profiles are plotted from boring records, and full lines joining corresponding points indicate the upper and lower limits of the several subsoils. The various strata are then indicated in the profile either by tinting with contrasting colors or by using appropriate symbols.

11-5. Plotting Cross-sections. Irregular cross-sections for earthwork are commonly drawn to scale on regular cross-section paper which is ruled usually with 10 divisions to the inch in both directions. Governing points of the cross-section are plotted either from the cross-section notes or from the data of a topographic map (Chap. 24). These points are essentially coordinates, with the origin on the center line at the grade line. The surface may be indicated either by an irregular line or by a series of straight lines connecting the points; the fixed portion of the cross-section either may be plotted directly or may be drawn by means of a templet.

The scale to be used depends upon the precision with which it is desired to compute the cross-sectional area. For large cross-sections a horizontal and vertical scale of 1 in. = 10 ft. is common. If the cross-sections are shallow, sometimes the vertical scale is exaggerated, as for profiles.

The cross-section for the first station of the route is usually placed in the upper left corner of the sheet, and successive cross-sections are placed one below the other (Fig. 11-3). Below each cross-section is shown its station number. Within each cross-section is shown its computed area in square feet, and between successive cross-sections is shown the computed volume of earthwork in cubic yards. The coordinates may or may not be noted on the sheet. Some engineers place the first cross-section in the *lower* left corner of the sheet, show the elevation of the ground line at each station, and place the notations at locations other than those shown in the figure.

The cross-sections may be inked, but since they are generally used only in the office, they are commonly shown in pencil.

If volumes of earthwork are to be determined by means of computers, cross-sections need not be plotted.

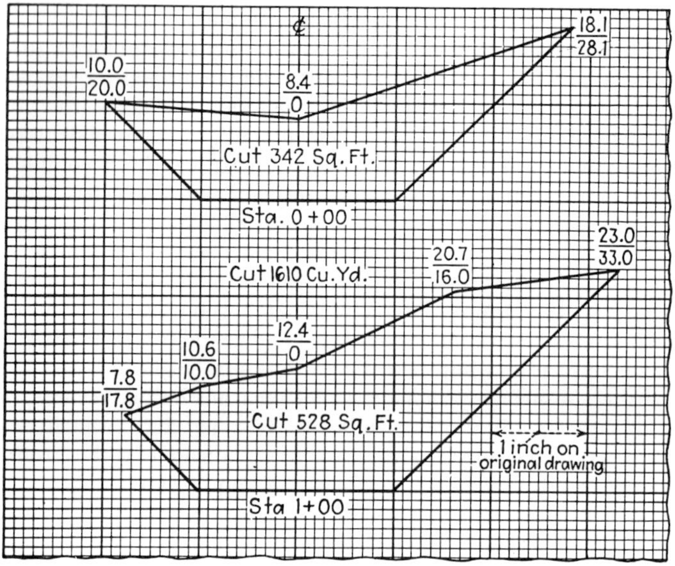

FIG. 11-3. Plotted cross-sections.

AREAS OF CROSS-SECTIONS

11-6. Regular Cross-sections; Level Section. Regular cross-sections are cross-sections for which levels are taken at one point on each side of the center line. Level and three-level sections are regular. Areas of regular cross-sections are readily determined by computation without plotting.

For a trench the cross-sectional area at any point is determined by multiplying the average of the top and bottom widths by the depth.

The same method may be applied to level cross-sections for highways and railroads; if d is the distance to either slope stake from the center, w is the width of the roadbed, and c is the center cut or fill, then the area A of the level section is

$$A = c\left(d + \frac{w}{2}\right) \tag{1}$$

11-7. Three-level Section. A three-level section may be divided into four triangles, as shown in Fig. 11-4. Then from the figure the area A is

$$A = \frac{1}{2} \cdot \frac{w}{2}(h_l + h_r) + \frac{1}{2}c(d_l + d_r)$$

or

$$A = \frac{w}{4}(h_l + h_r) + \frac{c}{2}(d_l + d_r) \tag{2}$$

Rule: Multiply the sum of the distances to the slope stakes by one-half the center cut or fill; to this add the product of one-fourth the width of the roadbed and the sum of the side heights. The result is the cross-sectional area.

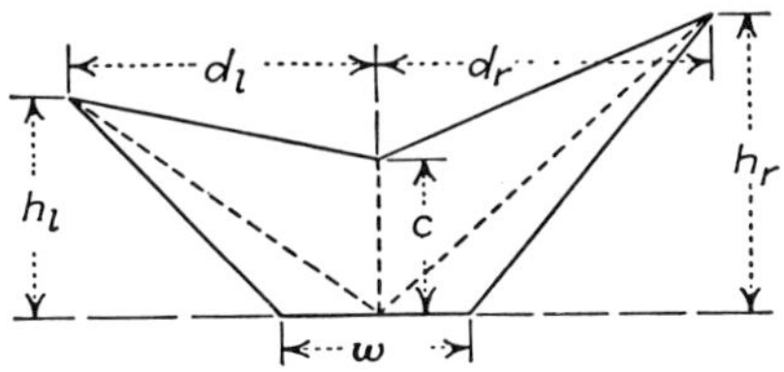

FIG. 11-4. Three-level section.

11-8. Irregular Road Cross-sections. Figure 11-5 represents an irregular road cross-section. The cross-section notes give C the cut (or fill) at the center stake A, and distances from center to, and cuts at, points E, F, G, and H. The method indicated here is an adaptation of the general method for computing areas by means of coordinates (Art. 19-4). In this case the cross-section notes provide the x and y coordinates for each vertex of the section (the origin being at 0) if the expression $0/(w/2)$ is supplied for the two vertices M and N, and if algebraic signs, plus and minus, are used to designate directions to the right and left of the origin, respectively.

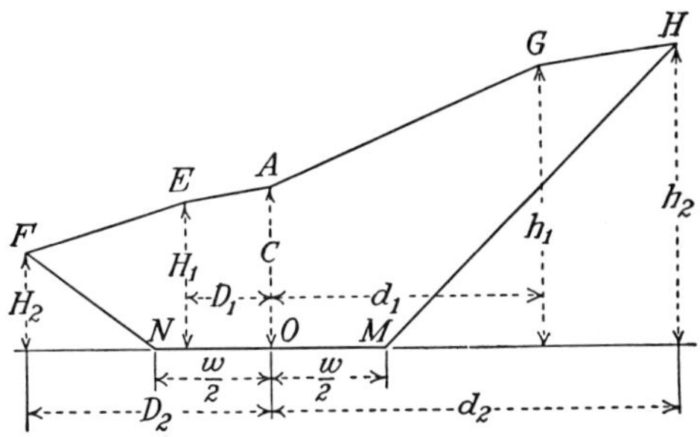

FIG. 11-5. Irregular section.

In the usual form the notes are recorded as follows:

$$\frac{H_2}{D_2} \quad \frac{H_1}{D_1} \quad \frac{C}{0} \quad \frac{h_1}{d_1} \quad \frac{h_2}{d_2}$$

Then, as just stated, if the algebraic signs and the coordinates of M and N are supplied, the coordinates of the section appear as follows:

$$\frac{0}{-\frac{w}{2}} \quad \frac{H_2}{-D_2} \quad \frac{H_1}{-D_1} \quad \frac{C}{0} \quad \frac{h_1}{+d_1} \quad \frac{h_2}{+d_2} \quad \frac{0}{+\frac{w}{2}}$$

The computation of the area will be made more convenient if now the opposite algebraic sign is placed on the opposite side of each lower term. The coordinates then appear thus:

$$\frac{0}{-\frac{w}{2}+} \qquad \frac{H_2}{-D_2+} \qquad \frac{H_1}{-D_1+} \qquad \frac{C}{0} \qquad \frac{h_1}{+d_1-} \qquad \frac{h_2}{+d_2-} \qquad \frac{0}{+\frac{w}{2}-}$$

The area may now be computed by the following rule:

Rule: Multiply each upper term by the algebraic sum of the two adjacent lower terms, using the signs facing the upper term. The algebraic sum of these products is double the area of the cross-section.

Example: Below are the notes for an irregular cross-section; the width of the roadbed is 20 ft.; the cross-sectional area is to be computed by the preceding rule. The coordinates $0/(w/2)$ are recorded, and the double opposite algebraic signs are supplied as shown:

$$\frac{0}{-10+} \qquad \frac{2.4}{-13.6+} \qquad \frac{4.2}{-10.0+} \qquad \frac{6.1}{0} \qquad \frac{8.3}{+15.0-} \qquad \frac{10.2}{+25.3-} \qquad \frac{0}{+10-}$$

$$\begin{aligned}
2.4 \times (+10.0 - 10.0) &= \ \ 0.0 \text{ sq. ft.}\\
4.2 \times (+13.6 - 0) &= +\ 57.1 \text{ sq. ft.}\\
6.1 \times (+10.0 + 15.0) &= +152.5 \text{ sq. ft.}\\
8.3 \times (0 + 25.3) &= +210.0 \text{ sq. ft.}\\
10.2 \times (-15.0 + 10.0) &= -\ 51.0 \text{ sq. ft.}\\
\text{Double area} &= +368.6 \text{ sq. ft.} \qquad \text{Area} = 184.3 \text{ sq. ft.}
\end{aligned}$$

For side-hill sections, where the road is partly in fill and partly in cut, the cross-sectional areas may be computed conveniently by dividing the section into partial areas consisting of trapezoids and triangles.

If cross-sections are bounded by curved lines or are very irregular, usually they are plotted as described in Art. 11-5, and the area is determined either by subdividing the cross-section into rectangles and triangles and computing the area of each or with greater facility by traversing the perimeter of each cross-section with a polar planimeter (Art. 4-12). Few, if any, cases arise where the areas of plotted cross-sections for earthwork cannot be determined by planimeter with a precision as high as that which is justified by the nature of the field measurements.

VOLUMES OF EARTHWORK

11-9. General. Volumes of earthwork are calculated by a variety of methods, depending upon the nature of the excavation and of the data. If cross-sections have been taken along a route, their areas are determined as described in preceding paragraphs, and the volumes of the prismoids between successive cross-sections are calculated either by the method of average end areas (Art. 11-11) or by the prismoidal formula (Arts. 11-12 and 11-13). The same procedure may be followed for borrow pits and similar excavations, or if elevations are observed at the same points before and after excavating, the volume may be computed by dividing it into vertical truncated prisms (Art. 11-10). Estimates for grading are frequently based upon a topographic map showing the contours for the undisturbed ground and contours for the ground as it will appear when grading has been completed; the volume is conveniently determined by dividing it into prismoids with horizontal bases and sloping sides. Volumes of earthwork may be computed by the use of contours as described in Art. 24-18. Total volumes are almost invariably expressed in cubic yards.

Because of the repetitive nature of computations of earthwork, including the areas of cross-sections, machine computation is highly desirable.

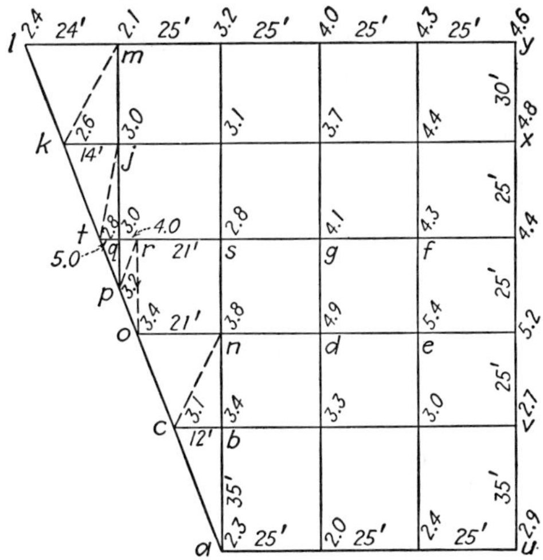

FIG. 11-6. Borrow pit.

11-10. Borrow Pit. The common method of determining the volume of a borrow pit is to cross-section the area before and after excavating (Arts. 10-3 and 10-4). The pit is then plotted in plan, and its area is divided into rectangles, triangles, and trapezoids, thus dividing the volumes into truncated prisms. Actually the upper and lower surfaces of the prisms are warped, but for earthwork computations they are assumed to be plane.

Figure 11-6 illustrates the plan view of a borrow pit, observations having been taken at the intersections of full lines. The numbers written diagonally are the cuts in feet.

The volume of a triangular truncated prism (as *abc*) is

$$V = \frac{A}{3}(h_1 + h_2 + h_3) \tag{3}$$

in which A is the horizontal sectional area and h_1, h_2, and h_3 are the corner heights.

Any rectangular prism (as *defg*) may be divided into two triangular prisms by either of two diagonal planes, but unless there are considerable variations in the corner heights, the error introduced by assuming the volume to be a rectangular truncated prism is inconsiderable as compared with errors due to undetected or neglected irregularities in the ground surface. Under these conditions the volume is determined with sufficient precision by multiplying the average of the corner heights by the horizontal sectional area. Where it is clear that the method just mentioned would introduce an error of consequence in the computed volume, the triangular prisms to be considered should be indicated in the field notes. Thus in the volume *jklm* the dotted line *km* is shown on the sketch to indicate that the shape of the ground warrants calculating volumes of the triangular prisms *jkm* and *klm*. For similar reasons, the volume *nopqrs* is divided into one rectangular and two triangular prisms.

If several adjacent rectangular prisms have the same horizontal section (that is, the same horizontal dimensions), they are computed as one solid, as follows: Multiply each corner height by the number of prisms of the same horizontal section in which it occurs; sum up the values thus determined, and multiply by the horizontal sectional area of a single prism. The product divided by four gives the volume of the solid.

Example: The accompanying tabulation shows the computations for the borrow pit of Fig. 11-6. The computations demonstrate the rule stated in the preceding paragraph. Thus *uvba* is seen to be made up of three 25 by 35-ft. rectangular prisms. Adding the corner heights by starting at *a* and proceeding clockwise around the figure, we have $2.3 + 3.4 + 2(3.3 + 3.0) + 2.7 + 2.9 + 2(2.4 + 2.0) = 32.7$ ft. as shown in the first line of the second column.

VOLUME OF BORROW PIT

Prism	Σ Cor. ht., ft.	Area, sq. ft.		Volume, cu. ft.
uvba	32.7	25 × 35	×¼	7,150
vxjqsb	161.9	25 × 25	×¼	25,300
xymj	59.9	25 × 30	×¼	11,230
nors	13.0	25 × 21	×¼	1,710
klm	7.1	12 × 30	×⅓	850
jkm	7.7	7 × 30	×⅓	540
kjt	8.5	7 × 25	×⅓	500
jqt	8.9	2.5 × 25	×⅓	190
tqp	9.1	2.5 × 14	×⅓	100
pqr	9.2	2 × 14	×⅓	90
opr	9.6	2 × 25	×⅓	160
con	10.3	10.5 × 25	×⅓	900
bcn	10.3	6 × 25	×⅓	520
abc	8.8	6 × 35	×⅓	620
			Total	49,860 cu. ft.
			or	1,847 cu. yd.

In the tabulation of the foregoing example the individual volumes are shown to the nearest 10 cu. ft., and the final volume is given to the nearest cubic yard. To compute volumes of earthwork to decimals of a cubic yard, as is sometimes done, is absurd when one considers that small irregularities in the ground surface between points at which elevations are taken would doubtless make a difference of several cubic yards between the actual and the computed volume. When volumes are large, calculations to the nearest 10 or even to the nearest 100 cu. yd. may be as exact as the nature of the field measurements will justify.

11-11. Average End Areas. The common method of determining volumes of excavation along the line of highways, railroads, canals, and similar works is that of *average end areas*. It is assumed that the volume between successive cross-sections is the average of their areas multiplied by the distance between them, or

$$V = \frac{l}{2}(A_1 + A_2) \tag{4}$$

in which V is the volume (cubic feet) of the prismoid of length l (feet) between cross-sections having areas (square feet) A_1 and A_2. If cross-sections are taken at full 100-ft. stations, the volume V_y in cubic yards between successive cross-sections A_1 and A_2 (square feet) is

$$V_y = 1.85(A_1 + A_2) \tag{5}$$

Formulas (4) and (5) are exact only when $A_1 = A_2$ but are approximate for $A_1 \gtrless A_2$. As one of the areas approaches zero, as on running from cut to fill on side-hill work, a maximum error of 50 per cent would occur if the formulas were followed literally. In this case, however, the volume is usually calculated as a pyramid; that is,

$$\text{Volume} = \tfrac{1}{3} \text{ area of base} \times \text{length}$$

Considering the fact that cross-sections are usually a considerable distance apart and that minor inequalities in the surface of the earth between sections are not considered, the method of average end areas is sufficiently precise for ordinary earthwork.

Where heavy cuts or fills occur on sharp curves, the computed volume of earthwork may be corrected for curvature, but ordinarily the correction is not large enough to be considered.

11-12. Prismoidal Formula. It can be shown that the volume of a prismoid is

$$V = \frac{l}{6}(A_1 + 4A_m + A_2) \tag{6}$$

in which l is the distance between end sections, A_1 and A_2 are the areas of the end sections, and A_m is the middle area or area halfway between the end sections, all in units of feet. A_m is determined by averaging the corresponding linear dimensions of the end sections and not by averaging the end areas A_1 and A_2. Equation (6) is one application of Simpson's Rule (Art. 19-11).

Example: In the accompanying tabulation are shown the three-level cross-section notes for two stations 100 ft. apart. The width of the roadbed is 20 ft., and the side-slope ratio is $1\frac{1}{2}$ to 1.

Station	Cross-section			Area, sq. ft.	Volume, cu. yd.
	L	*C*	*R*		
115	$\frac{c4.0}{16.0}$	$\frac{c6.0}{0}$	$\frac{c12.0}{28.0}$	212	
					575
116	$\frac{c2.0}{13.0}$	$\frac{c3.0}{0}$	$\frac{c8.0}{22.0}$	103	
Mid-section	$\frac{c3.0}{14.5}$	$\frac{c4.5}{0}$	$\frac{c10.0}{25.0}$	154	

The volume of earthwork between the two stations is to be computed by the prismoidal formula. Below the regular cross-section notes are shown those for the mid-section obtained by averaging the values given for sections at stations 115 and 116. In the column headed "Area, square feet," are areas of cross-sections computed by formula (2), Art. 11-7. Then by the prismoidal formula given above

$$V = \frac{100}{6}(212.0 + 4 \times 154.0 + 103.0) = 15{,}520 \text{ cu. ft. or } 575 \text{ cu. yd.}$$

Although the prismoidal formula gives the true volume of a prismoid, the difference between results obtained through its application and values obtained by the method of average end areas is not large except where the change in cross-section is abrupt.

For the foregoing example the volume computed by average end areas is 583 cu. yd., and the difference between the results obtained by the two methods is 8 cu. yd. or about 1.4 per cent. As an example of the magnitude of the error in volume introduced by apparently insignificant variations of the surface, suppose that between the two cross-sections given in the above example a sag takes place gradually until at station 115 + 50 it amounts to 0.5 ft. over a width of 20 ft., thus forming two wedges of error with a base of 10 sq. ft. and a length of 50 ft. The volume of these two wedges is $2 \times \frac{1}{2} \times 10 \times 50 = 500$ cu. ft. $= 18$ cu. yd., and the error is more than twice as great as the error due to computing the volume by average end areas instead of by the prismoidal formula. To one familiar with field conditions it is evident that much larger surface irregularities than those cited above are likely to go unnoticed unless more than the usual care is taken in field measurements.

As far as volumes of earthwork are concerned, the use of the prismoidal formula is justified only if cross-sections are taken at short intervals, if small surface deviations are observed, and if the areas of successive cross-sections differ widely. Usually it yields smaller values than those computed from average end areas. For excavation under contract, the basis of computation should be understood in advance; otherwise the contractor will usually claim (and obtain) the benefit of the common method of average end areas.

11-13. Prismoidal Correction. It can be shown that the difference between the volume computed by the prismoidal formula and that computed by average end areas for the prismoids defined by three-level sections is

$$C_v = 0.309(H_0 - H_1)(D_0 - D_1) \tag{7}$$

in which C_v = difference in volume, or correction, for a prismoid 100 ft. long in cubic yards

H_0 = center height at one end section, in feet

H_1 = center height at the other end section, in feet

D_0 = distance between slope stakes at the end section where the center height is H_0, in feet

D_1 = distance between slope stakes at the other end section, in feet

C_v is known as the *prismoidal correction;* it is *subtracted algebraically* from the volume as determined by the average-end-area method to give the more nearly correct volume as determined by the prismoidal formula.

Example: For the prismoid of Art. 11-12, the prismoidal correction is

$$C_v = 0.309(6.0 - 3.0)(44.0 - 35.0) = 8.3 \text{ cu. yd.}$$

which is consistent with the difference (583 − 575 = 8 cu. yd.) between the volumes obtained by the average-end-area method and the prismoidal formula, respectively.

11-14. Volumes from Road Profiles. Preliminary estimates of earthwork for highways, railroads, and canals are based upon the preliminary profile. If the side slopes were vertical, the volume of any cut or fill would be a direct function of the area between grade line and ground line on the profile. As the side slopes are inclined, the volume in cut or fill increases at a relatively greater rate than does the depth; hence (except for the purpose of rough estimates) the area representing cut or fill on the profile cannot be directly taken as a measure of volume, as would be the case for a trench.

For very rough estimates the profile area of any given cut or fill may be measured with the planimeter and divided by the length to obtain the average depth of cut (or fill) at the center line. The area of a level section having this average cut or fill is then computed, and the area is multiplied by the length of cut (or fill) to obtain the volume of earthwork.

Example: The length of a given cut is 1,650 ft., and the profile area between ground line and grade line is 18,500 sq. ft. The roadbed is 20 ft. wide and the side slopes are 1½ to 1. It is desired to determine roughly the volume of earthwork.

$$\text{Average depth of cut is } \frac{18{,}500}{1{,}650} = 11.2 \text{ ft.}$$

For a level section the distance to slope stake is

$$d = \frac{w}{2} + cs = 10.0 + 1\tfrac{1}{2} \times 11.2 = 26.8 \text{ ft.}$$

The cross-sectional area of the average section is

$$A = c\left(\frac{w}{2} + d\right) = 11.2(10.0 + 26.8) = 412 \text{ sq. ft.}$$

The total volume is, therefore

$$V = \frac{412 \times 1{,}650}{27} = 25{,}200 \text{ cu. yd.}$$

For less approximate computations the cut or fill at each full station is scaled from the profile, and the corresponding volume per station is computed for a level section whose depth is the scaled cut (or fill). The level section at each station is assumed to extend over a length of 100 ft. (50 ft. in advance and 50 ft. back of the station). The total volume for a given cut (or fill) is obtained by summing up the volumes per station. Tables of volumes (cubic yards) per 100 ft. for various widths of roadbed, side slopes, and depths of cut (or fill) are given in texts on highway and railroad surveying. If such a table is not available, volumes may be conveniently determined by constructing a diagram showing cuts or fills as ordinates and volumes in cubic yards per 100 ft. as abscissas. Volumes may also be determined by means of a scale graduated in cubic yards per 100 ft. of length for various depths of cut or fill, the scale being applied to the profile.

The foregoing method may be modified by constructing an earthwork diagram, either on the sheet with the profile or on a separate sheet, the ordinates of the diagram being in cubic yards per foot of length and the abscissas being distances along the line in feet. When the diagram is on the profile sheet, any convenient horizontal line is chosen as the base from which ordinates are measured, those above the base representing volumes in cut and those below representing volumes in fill. The horizontal scale is made the same as for the profile. The vertical scale is some convenient scale, as 1 in. = 20 cu. yd. per foot, depending upon the magnitude of the volumes involved. At each full station and at each plus where the direction of the profile changes abruptly, the distance between grade line and ground line is scaled, the volume per foot of length for a level section of corresponding depth is determined from tables or from a diagram, and this volume is plotted as an ordinate to the earthwork diagram. The diagram is completed by drawing an irregular line through the points thus plotted. The area under the diagram (at the scales used in plotting) gives the volume in cubic yards. Thus, if the horizontal scale is 1 in. = 400 ft. and the vertical scale is 1 in. = 20 cu. yd. per foot, then 1 sq. in. on the paper is the equivalent of 400 × 20 = 8,000 cu. yd.

11-15. Precision of Determining Volumes. It is instructive to consider the probable errors in the determination of earthwork quantities. These may be discussed in relation to each of the three general methods commonly used: (1) volumes computed from data obtained in setting slope stakes, (2) volumes computed from irregular cross-sections, and (3) volumes estimated from contour maps (Art. 24-18).

The measurements taken to compute earthwork quantities include horizontal measurements, usually taken with a metallic tape, and vertical measurements, taken with a level and rod. These measurements are subject to accidental errors due principally to holding and reading the tape, to reading the rod, and particularly to variations in the elevation of the ground surface where the rod is held. The size of these errors will vary greatly under various field conditions, but to illustrate the principles involved, a probable error of ±0.05 ft. will be assumed for each measurement, that is, for each horizontal (tape) and vertical (rod) reading. It is believed that this value is a reasonable assumption for average field conditions.

Since the horizontal distances are usually much greater than the vertical distances, it is evident that the *percentage* of error in horizontal dimensions is much less than in vertical dimensions. Hence, errors in computed volumes result for the most part from errors in cuts and fills. And since the magnitude of the errors is independent of the magnitude of the distances themselves, the percentage of error in the final result is greater for small volumes than for large volumes.

1. *Volumes from Slope-stake Data.* The principles stated above may be illustrated in the case of volumes computed from slope-stake data by the accompanying tabulation. The roadway is assumed to be 20 ft. wide, with side slopes of 1½ to 1. Volumes are computed for sections 100 ft. long.

Average cut or fill, ft.	Area, sq. ft.	Probable error in area		Volume, cu. yd.	Probable error in volume	
		Value, sq. ft.	Per cent		Value, cu. yd.	Per cent
2.0	46	±0.7	1.5	170	±1.8	1.1
4.0	104	±0.8	0.8	385	±2.1	0.5
5.5	155	±1.0	0.6	574	±2.6	0.5
12.5	485	±1.5	0.3	1,794	±3.9	0.2

An inspection of the tabulation shows: (1) that the percentage of error in the area and in the volume varies inversely with the depth of the cut or fill; (2) that the magnitude of the probable error is not important as compared with the probable errors due to variations over the ground surface; and (3) that the probable errors indicate an uncertainty of one or more in the last unit of the computed quantities. Hence it will be consistent to carry one decimal place in intermediate computations of areas and volumes; but it is absurd to record values beyond the last whole unit, either of areas or of volumes.

2. *Volumes from Irregular Cross-sections.* The relations in the preceding paragraph regarding roadway volumes apply equally well to borrow-pit volumes. Since the shapes of borrow pits are more irregular than those of roadways, however, and since two rod readings are required at each point, it may be expected that the computed volume for a shallow borrow pit will be affected by a larger percentage of error than a corresponding volume in a roadway. On the other hand, the readings are usually taken at small intervals (25 to 50 ft.), hence the errors due to irregularities in the ground surface are not so great as in the case of roadways; and since many readings are taken, the law of accidental errors tends to reduce the percentage of error in the total volume.

Assuming a probable error of ± 0.05 ft. for a single rod reading, the total probable error for the borrow pit shown in Fig. 11-6 is ± 2.4 cu. yd. The volume is 1,847 cu. yd. and the probable error is 0.1 per cent, which is about one half as large as the error in the roadway volume of 1,794 cu. yd. given above.

3. *Volumes from Contour Maps.* The errors of determining volumes from contours depend upon the scale of the map, the contour interval, and the precision with which contours are shown. The larger the scale and the smaller the contour interval, the more reliable are the computed volumes. Under the usual conditions, scales of 50 to 100 ft. to the inch and contour intervals of 1 or 2 ft. may render estimates correct within 5 or 10 per cent, depending upon the magnitude of the grading operations. Rough estimates are sometimes made from maps showing 5-ft. contours, but unless the cuts and fills are deep and the grading is on a large scale, the relative error involved is likely to be great. Where the ground is gently sloping and the cuts and fills are shallow, reliable estimates of volume cannot be made unless the contour interval is very small. A contour interval of ½ ft. is often employed for such work.

11-16. Numerical Problems.

1. Plot a profile for data given in numerical problem 1 of Chap. 10, between stations 179 and 186. Use a horizontal scale of 1 in. = 100 ft. and a vertical scale of 1 in. = 4 ft.

2. Following are the notes for cross-sections at stations 109 and 110. The width of the roadbed is 24 ft., and the side slopes are 2 to 1. Compute the areas of the two sections.

Station	Cross-section		
109	$\frac{c2.4}{16.8}$	$\frac{c1.2}{0.0}$	$\frac{c0.4}{12.8}$
110	$\frac{c12.2}{36.4}$	$\frac{c9.2}{0.0}$	$\frac{c4.8}{21.6}$

3. Following are the notes for an irregular road cross-section. The width of the roadbed is 24 ft. and the side slopes are 1½ to 1. Compute the cross-sectional area by the three methods of Art. 11-8. Compare the results.

$$\frac{c4.2}{18.3} \quad \frac{c6.8}{12.0} \quad \frac{c11.2}{0} \quad \frac{c14.4}{10.0} \quad \frac{c16.8}{25.0} \quad \frac{c18.4}{39.6}$$

4. Compute the volume in cubic yards between stations 109 and 110 of problem 2. Use both the average-end-area method and the prismoidal formula. Note the discrepancy in percentage between volumes as determined by the two methods; also compare the difference in volumes with that obtained by Eq. (7).

5. What error in volume between station 109 and station 110 of problem 4 would be introduced if the recorded cuts at centers and slope stakes were 0.1 ft. too great? What is the error in terms of percentage of the volume by average end areas?

6. Suppose that between the two stations in problem 4 a sag gradually takes over a width of 24 ft., becoming a maximum of 1 ft. at 109 + 50. What error, in cubic yards, is introduced in the computed volume?

7. Compute the volume in cut and in fill for the given cross-sections between stations 62 and 64. The roadbed is 24 ft. wide in cut and 20 ft. wide in fill, and the side slopes are 1½ to 1. Tabulate the data in the following form: "Station," "Cross-section," "Area," and "Volume." Use the prismoidal formula.

Station	Cross-section			
62	$\frac{c2.6}{15.9}$	$\frac{c4.8}{0.0}$	$\frac{c6.4}{21.6}$	
63	$\frac{0.0}{12.0}$	$\frac{c3.1}{0.0}$	$\frac{c4.4}{18.6}$	
63 + 25	$\frac{f4.6}{16.9}$	$\frac{0.0}{0.0}$	$\frac{c2.6}{15.9}$	
64	$\frac{f7.2}{20.8}$	$\frac{f4.8}{0.0}$	$\frac{0.0}{6.0}$	$\frac{c1.8}{14.7}$

8. Solve problem 7 by the method of average end areas, computing the volume of pyramids by the relation, volume = ⅓ (area of base times length).

9. In plan, a borrow pit is 75 by 135 ft. Before and after excavation, levels are run and offsets are measured from stations along one of the 135-ft. sides. The computed cuts are shown in the following table:

Offsets	Cut, ft.						
	Sta. 0	Sta. 0 + 30	Sta. 0 + 50	Sta. 0 + 75	Sta. 1 + 00	Sta. 1 + 15	Sta. 1 + 35
0	0.0	1.5	0.0	4.5	6.2	4.7	0.0
25	1.2	2.9	10.6	9.7	7.9	8.4	2.5
50	2.5	3.7	8.7	8.7	9.4	8.4	3.6
75	0.0	0.0	1.9	7.6	6.8	6.3	0.0

Compute the volume by the method described in Art. 11-10.

10. A given fill for a railroad is 1,350 ft. long. The profile is plotted to a horizontal scale of 1 in. = 400 ft and to a vertical scale of 1 in. = 20 ft. The perimeter of the area between ground line and grade line is traversed clockwise with anchor point outside, employing a planimeter set so that 1 revolution of the roller is equal to 10 sq. in. on the paper. The difference in readings is 0.269. What is the average depth of the fill in feet? Estimate roughly the volume of the fill in cubic yards, assuming a level section of the average depth, roadbed 18 ft. wide, and side slopes $1\frac{1}{2}$ to 1.

11-17. Office Problems.

Problem 1. Plotting Profile

Object. To plot a profile from level notes (field problem 1, Art. 10-18) and to fix the grade line for a road or pipe line.

Procedure. (1) Choose a horizontal and vertical scale in keeping with the purpose of the profile. (2) Examine the field notes to determine the range between points of maximum and minimum elevation. Number each of the heaviest horizontal lines of the profile paper with its elevation. Number each of the heaviest vertical lines with its station number. (3) From the profile-level notes plot the profile. Through the plotted points draw a freehand curve (see Fig. 11-1). Show the names of streams and roads crossed directly above their crossings on the profile. Check the profile, and ink it in *black*. (4) Fix the grade line, and ink it in *red*. At each change of grade draw a small circle and indicate the elevation (and plus if it does not fall at a full station) of each of these points on a vertical line directly under or over the small circle. On the grade line, record the rates of grade as shown in Fig. 11-1. (5) Near the bottom of the sheet indicate in *black* ink the horizontal alinement, using a scheme similar to that shown in the figure. (6) Make an appropriate title.

Hints and Precautions. (1) The profile should be checked by reading elevations and stations from the profile, not from the level notes. Two men can work together to good advantage—one reading the notes while the other plots the profile. When checking the profile, they should exchange places. (2) Avoid the common mistake of reading elevations of turning points and bench marks as ground elevations by enclosing the elevations of turning points and bench marks in the field notebook with a circle. (3) Ink the profile (freehand) with a ruling pen rather than with a lettering pen. Do not round off the summits and depressions unduly.

Problem 2. Plotting Cross-sections; Quantities of Earthwork

Object. To plot cross-sections of a roadway from field notes and to calculate quantities of earthwork. It is assumed that the cross-section notes give cut or fill rather than elevations (see field problem 3, Art. 10-18).

Procedure. (1) Beginning near the upper left corner of a sheet of cross-section paper, choose convenient heavy horizontal and vertical lines as grade and center lines. With these as coordinates plot the cross-section notes of the first station (see Fig. 11-3), counting the number of spaces out from the center and up from the grade line corresponding to these distances in the notes. Mark the plotted points with dimensions identical with those of corresponding points in the notes. Draw straight lines showing roadbed and side slopes of cut or fill and the original ground, thus enclosing the section. Below the cross-section, mark its station number. (2) At a convenient distance below and on the same center line, plot the next section in similar manner. (3) When the bottom of the sheet is reached, plot the next section a little farther to the right and at the top of the sheet, and in this way continue until all plotting is done. (4) Compute the area of each section, and show its value within the section (as 123 sq. ft.). Irregular sections may be planimetered. (5) Compute volumes by the prismoidal formula and by the average-end-area method, and show the volume of each prismoid (as 97 cu. yd.) between its end sections. (6) By each method find the total yardage in each cut and fill, and mark these totals conspicuously. (7) Make an appropriate title.

CHAPTER 12

ANGLES AND DIRECTIONS; THE COMPASS

12-1. Location of Points. As previously stated, the purpose of a survey is to determine the relative locations of points on or near the surface of the earth. The *location* of a point is fixed if measurements are made of (1) its direction and distance from a known point, (2) its direction from two known points, (3) its distance from two known points, or (4) its direction from one known point and its distance from another. If the relative locations of points as seen in horizontal projection are desired, the field operations involve the measurement of horizontal distances, as described in Chap. 7, and the determination of direction in the horizontal plane. In addition, if the relative elevations of points are required, they are determined by one of the methods of leveling described in Chaps. 8 to 10.

For horizontal projection or plan, the *direction* of any line (as fixed by two points) is defined by the horizontal angle between the line and some reference line. For vertical projection, the direction of one point with respect to another is defined by the vertical angle between the plane of the horizon and the line joining the two points. In general, therefore, the angular measurements of surveying are either horizontal or vertical, or approximately so.

When the *angle* between two points is mentioned, it is understood to mean the horizontal angle, or angle between the projections in the horizontal plane of two lines passing through the two points and converging at a third point. Thus at O (Fig. 12-1a) the angle between B and C is the horizontal angle BOC. The vertical angle to a point is its angle of elevation or depression from the horizontal; as measured from some point of reference, the angle is positive or negative according as the observed point is above or below the horizontal plane passing through the point of reference. Thus the vertical angle to a point B as measured from A is positive if B is higher than A. Measurement of vertical angles as applied to indirect leveling was briefly considered in Chap. 8 and will be described more in detail in Art. 13-19. This chapter treats only of angles and directions in the horizontal plane.

12-2. Meridians. The relative directions of lines connecting survey points may be obtained in a variety of ways. Figure 12-1a shows lines about a point. The direction of any line (as OB) with respect to an adjacent line (as OA) is given by the horizontal angle between the two lines (as a) and the direction of rotation (as clockwise). The direction of any line (as

OC) with respect to a line not adjacent (as OA) is not given by any of the measured angles but may be computed by adding the intervening angles (as $a + b$).

Figure 12-1*b* shows the same system of lines but with all angles measured from a line of reference OM. The direction of any line (as OA) with respect to the line of reference (as OM) is given by the angle between the lines (as 1) and its direction of rotation (as clockwise). The angle between any two lines (as AOC) is not given directly, but may be computed by taking the difference between the direction angles of the two lines (as $\angle 3 - \angle 1 = \angle AOC$).

The fixed line of reference may be any line in the survey, or it may be purely imaginary. It is termed a *meridian*. If it is arbitrarily chosen, it is called an *assumed meridian;* if it is a true north-and-south line passing through the geographical poles of the earth, it is called a *true meridian;* or if it

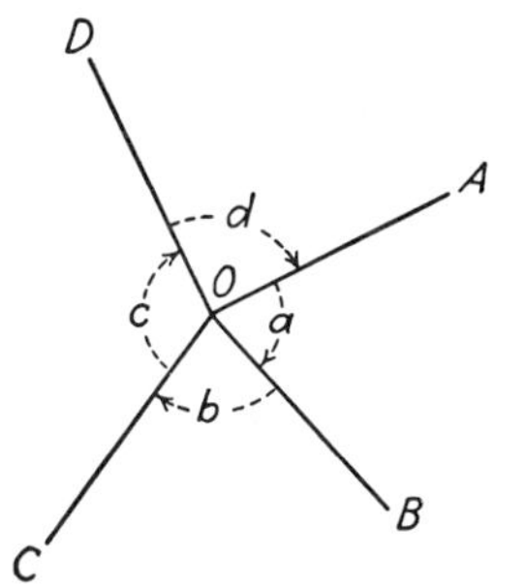

FIG. 12-1*a*. Directions by angles.

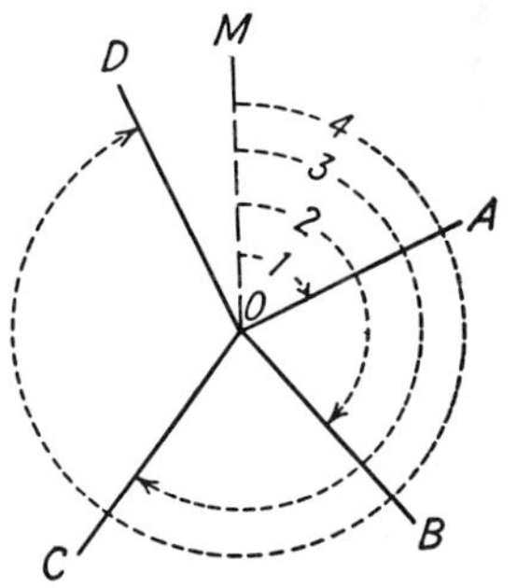

FIG. 12-1*b*. Directions referred to meridian.

lies parallel with the magnetic lines of force of the earth as indicated by the direction of the magnetic needle, it is called a *magnetic meridian.*

12-3. True Meridian. The *true meridian,* or *astronomic meridian,* is determined by astronomical observations as described in Chap. 21 (see also Art. 12-10). For any given point on the earth its direction is always the same, and hence directions referred to the true meridian remain unchanged regardless of time. The lines of most extensive surveys and usually the lines marking the boundaries of landed property are referred to the true meridian.

12-4. Magnetic Meridian. The direction of the *magnetic meridian* is that taken by a freely suspended magnetic needle. The magnetic poles are at some distance from the true geographic poles; hence in general the magnetic meridian is not parallel to the true meridian. The location of the magnetic poles is constantly changing; hence the direction of the magnetic meridian is not constant. However, the magnetic meridian is employed as a line of reference on rough surveys where a magnetic compass is used and often is employed in connection with more precise surveys in which angular measurements are checked approximately by means of the compass. It was formerly used extensively for land surveys. It is determined as described in Art. 12-10.

12-5. Magnetic Needle. Any slender symmetrical bar of magnetized iron when freely suspended at its center of gravity takes up a position parallel with the lines of magnetic force of the earth. In horizontal projection these lines define the magnetic meridian. In elevation, the lines are inclined downward toward the north in the Northern Hemisphere, and downward toward the south in the Southern Hemisphere. Since the bar takes a position parallel with the lines of force, it becomes inclined with the horizontal. This phenomenon is called the *magnetic dip*. The angle of dip varies from 0° at or near the equator to 90° at the magnetic poles. The needle of the magnetic compass rests on a pivot. To counteract the effect of dip, so that the needle will take a horizontal position when directions are observed, a counterweight is attached to one end (the south end in the Northern Hemisphere). The counterweight usually consists of a short piece of fine brass wire wound around the needle and held in place by spring action. So long as the needle is used in a given locality and so long as it loses none of its magnetism, it will remain balanced. When for any reason it becomes unbalanced, it is adjusted to the horizontal by sliding the counterweight along the needle. At the mid-point of the needle is a jewel which forms a nearly frictionless bearing for the pivot.

12-6. Magnetic Declination. The angle between the true meridian and the magnetic meridian is called the *magnetic declination,* or *variation.* If the north end the compass needle points to the east of the true meridian, the declination is said to be east (Fig. 12-13*b*); if it points to the west of the true meridian, the declination is said to be west.

If a true north-and-south line is established, the mean declination of the needle for a given locality can be determined by compass observations extending over a period of time. The declination may be estimated with sufficient precision for most purposes from an *isogonic chart* published by the U.S. Coast and Geodetic Survey; specific values for a particular locality can be obtained from the Survey.

12-7. Isogonic Chart. The isogonic chart of the continental United States shown in Fig. 12-2 applies to January 1, 1960. It is based upon observations made by the U.S. Coast and Geodetic Survey at stations widely scattered throughout the country. The solid lines are lines of equal magnetic declination, or *isogonic lines.* East of the heavy solid line of zero declination, or *agonic line,* the north end of the compass needle points west of north; west of that line it points east of north. The north end of the compass needle is moving eastward over the area of eastward annual change and westward elsewhere over the chart at an annual rate indicated by the lines of equal annual change.

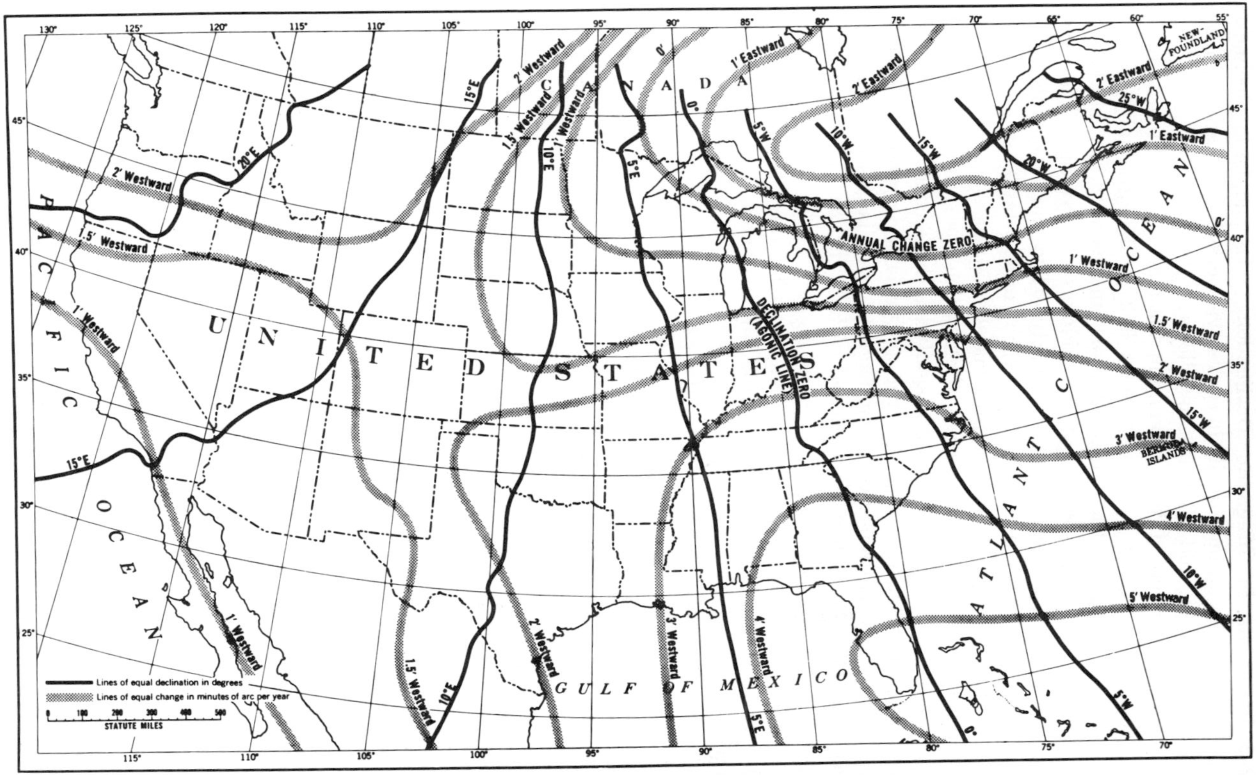

FIG. 12-2. Isogonic chart of the United States for 1960. (*U.S. Coast and Geodetic Survey.*)

12-8. Variations in Magnetic Declination. The magnetic declination changes more or less systematically in cycles over periods of (1) approximately 300 years, (2) 1 year, and (3) 1 day, as follows:

1. *Secular Variation.* Like a pendulum, the magnetic meridian swings in one direction for perhaps a century and a half until it gradually comes to rest and then swings in the other direction, and as with a pendulum the velocity of movement is greatest at the middle of the swing. The rate of change per year, however, varies irregularly. The causes of this secular variation are not well understood. In the United States, it amounts to several degrees in a half-cycle of approximately 150 years. In Fig. 12-2 are shown by dash lines the annual rates of change in the secular variation for the year 1960. On account of its magnitude, the secular variation is of considerable importance to the surveyor, particularly in retracing lines the directions of which are referred to the magnetic meridian as it existed years previously. When *variation* is mentioned without further qualification, it is taken to mean the secular variation.

2. *Annual Variation.* This is a small annual swing distinct from the secular variation. For most places in the United States, it amounts to less than 01′.

3. *Daily Variation.* This variation, also called *solar-diurnal* variation, is a periodic swing of the magnetic needle occurring each day. For points in the United States the north end of the needle reaches its extreme easterly swing at about 8 or 9 A.M. and its extreme westerly swing at about 1 or 2 P.M. The needle usually reaches its mean position between 10 and 11 A.M. and between 7 and 11 P.M. In Table VIII are given for each hour of the day the average values of variation for three widely separated locations at observatories in the United States. These values change slightly from year to year and are greater in summer than in winter. In general, the higher the latitude, the greater the range in the daily variation. The average range for points in the United States is less than 08′, a quantity so small as to need no consideration for most of the work for which the compass needle is employed. However, in the United States in summer, a line run 1,000 ft. by compass at 8 A.M. would end as much as 3 ft. to the right of the point where it would end if run at 1 P.M.

4. *Irregular Variations.* Irregular variations are due to magnetic disturbances usually associated with sunspots. They cannot be predicted but are most likely to occur during magnetic storms, when auroral displays occur and radio transmission is disturbed. They may amount to a degree or more, particularly at high latitudes.

12-9. Local Attraction. Objects of iron or steel, some kinds of iron ore, and currents of direct electricity alter the direction of the lines of magnetic force in their vicinity and hence are likely to cause the compass needle to deviate from the magnetic meridian. The deviation arising from such local sources is called *local attraction* or *local disturbance.* In certain localities, particularly in cities, its effect is so pronounced as to render the magnetic needle of no value for determining directions. It is not likely to be the same at one point as at another, even though the points be but a short distance apart. It is even affected by such objects as the steel tape, chaining pins, axe, and small objects of iron or steel that are on the person. Usually its magnitude can be determined, and directions observed with the compass can be corrected accordingly (Art. 12-23). Local attraction can usually be detected by observing the compass bearing of a line at two or more points on the line.

12-10. Establishing Meridian. The *true* meridian is established by astronomical observations as described in Chap. 21. Any of the celestial bodies whose astronomical position is known may be observed, but those commonly observed by surveyors are the sun and Polaris (the North Star). For surveys of ordinary precision the transit is employed.

On compass surveys, in order to determine the magnetic declination, sometimes the true meridian is established by ranging two plumb lines with Polaris, usually when the star is at elongation (farthest east or farthest west). If the time is accurately known, the observations are sometimes made when the star is at culmination (directly above or below the pole and hence on the meridian). One plumb line is suspended from some convenient high point, and a stake with tack representing the north point of the meridian is set beneath the bob. At a distance of 15 or 20 ft. south of the plumb line two stakes are set, one on each side of the estimated position of the meridian, and a piece of stout string is stretched between nails driven in their tops. A second plumb line is suspended from the stretched string. When the time of elongation or culmination approaches, the observer moves the second plumb line, keeping the two plumb lines in line with the star until the time of elongation or culmination has been reached. A stake with tack is set beneath the second plumb line. If the star is at culmination, the tacked stakes define the true meridian; if the star is at elongation, the true meridian is established by laying off an angle from the established line equal to the azimuth of the star as given in Table V. If the observation is made at western elongation, the angle is turned clockwise; if made at eastern elongation, the angle is turned counter-clockwise. The times of elongation and culmination of Polaris may be taken from published astronomical tables, the latitude and longitude having been approximately determined from a map.

The first plumb line will usually need to be illuminated with an artificial light. To dampen the swing, the plumb bob may be immersed in water. For some minutes preceding and succeeding the instant of elongation the star appears to move vertically, hence observations taken at elongation are not influenced by time errors and need not be hurried. If the star is at elongation and care is exercised in setting the ground points, the error in determining the meridian in this manner need not exceed 05′. At culmination, on the other hand, the star is moving east or west at the rate of about 06′ in 15 min. of time; hence the time must be known closely and the observation must be made quickly. The time of culmination of Polaris may be established by observing the meridian passage of other bright stars which are on line with Polaris and the pole.

A *magnetic* meridian can be established by setting up the compass over any convenient point and then sighting to set a point on a stake or other object that marks another point on the meridian. Several sights should be taken during the set-up. After each setting of the line of sight, the compass should be rotated a few degrees about the vertical axis and then moved back until the needle reads zero. The mean of the points thus established is assumed to be on the magnetic meridian, provided the observations are taken at a time of day when the declination is approximately at its mean value, otherwise corrections for daily variation may be made as indicated in Table VIII.

At many of the triangulation stations established throughout the United States by the U.S. Coast and Geodetic Survey, reference lines of known true direction have been established for use by local surveyors.

12-11. Angles and Directions. Angles and directions may be defined by means of *bearings, azimuths, deflection angles, angles to the right,* or *interior angles,* as described in the following articles. These quantities are said to be *observed* when obtained directly in the field, and *calculated* when obtained indirectly by computation. Conversion from one means of expressing angles and directions to another means is a simple matter if a sketch is drawn to show the existing relations.

12-12. Bearings. The direction of any line with respect to a given meridian may be defined by the *bearing.* Bearings are called *true (astronomic) bearings, magnetic bearings,* or *assumed bearings* according as the meridian is true, magnetic, or assumed. The bearing of a line is indicated by the quadrant in which the line falls and the acute angle which the line makes with the meridian in that quadrant. Thus in Fig. 12-3*a* the bearing of the line *OA* is read north 37° east and is written N37°E. The bearings of *OB*, *OC*, and *OD* are, respectively, S62°E, S50°W, and N20°W. In all cases, values of bearing angles lie between 0 and 90°. If the direction of the line is parallel with the meridian and north, it is written as NO° or *due north;* if perpendicular to the meridian and east, it is written as N90°E or *due east.*

In Fig. 12-3*b*, if the observed bearing of *OA* is N37°E and the angle *AOB* = 81°, then the calculated bearing of *OB* is S62°E.

12-13. Azimuths. The *azimuth* of a line is its direction as given by the angle between the meridian and the line measured in a clockwise direction usually from the south branch of the meridian. In astronomical observations azimuths are generally reckoned from the true south; in surveying, some surveyors reckon azimuths from the south and some from the north branch of whatever meridian is chosen as a reference, but on any given survey the direction of zero azimuth is either always south or always north. Herein, unless otherwise stated, azimuths are always reckoned from south. Azimuths are called *true* (*astronomic*) *azimuths*, *magnetic azimuths*, or *assumed azimuths* according as the meridian is true, magnetic, or assumed. Azimuths may have values between 0 and 360°.

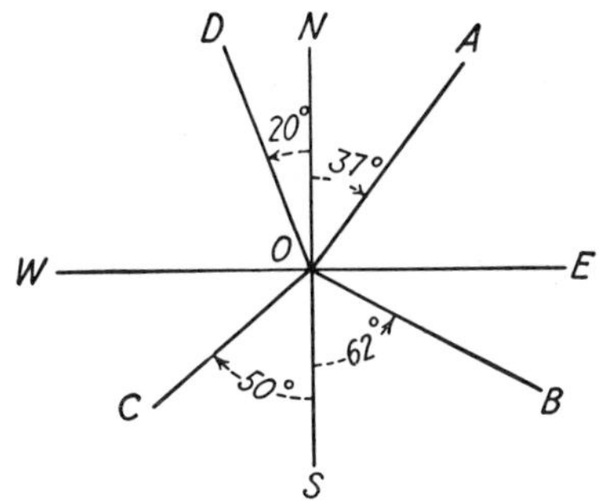

FIG. 12-3*a*. Bearings.

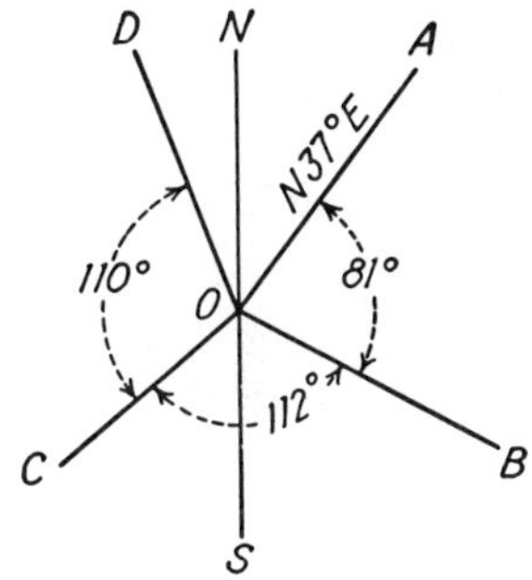

FIG. 12-3*b*. Angles and bearings.

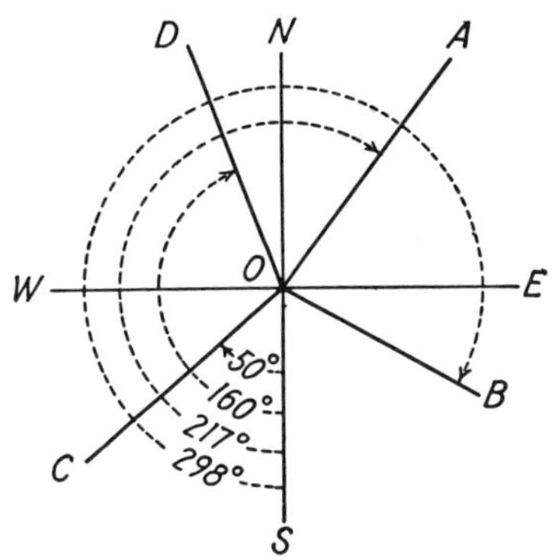

FIG. 12-4*a*. Azimuths from south.

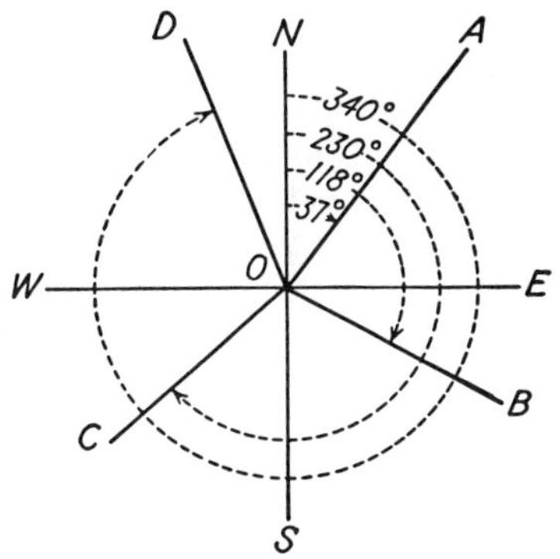

FIG. 12-4*b*. Azimuths from north.

In Fig. 12-4*a* azimuths measured from the south point are Az. $OA = 217°$, Az. $OB = 298°$, Az. $OC = 50°$, and Az. $OD = 160°$; or in Fig. 12-4*b*, in which are shown the same lines with azimuths measured from the north point, Az. $OA = 37°$, Az. $OB = 118°$, Az. $OC = 230°$, and Az. $OD = 340°$. In Fig. 12-4*a* if the observed azimuth of OA as reckoned from the south is 217° and the observed angle AOB is 81°, then the calculated azimuth of OB is 298°.

Azimuths may be calculated from bearings, or *vice versa*, preferably with the aid of a sketch. For example, if the bearing of a line is N16°E, its azimuth (from south) is 180 + 16 = 196°; and if the azimuth (from south) of a line is 285°, its bearing is 360 − 285 = S75°E.

In some special cases, the term "azimuth" is used in the sense of a bearing and, therefore, may be taken either clockwise or counter-clockwise, as in "azimuth of Polaris" (Art. 21-25) or "azimuth of the secant" (Fig. 23-7).

12-14. Deflection Angles. The angle between a line and the prolongation of the preceding line is called a *deflection angle*. Deflection angles are recorded as *right* or *left* according as the line to which measurement is taken lies to the right (clockwise) or left (counter-clockwise) of the prolongation of the preceding line. Thus in Fig. 12-5 the deflection angle at B is 22°R,

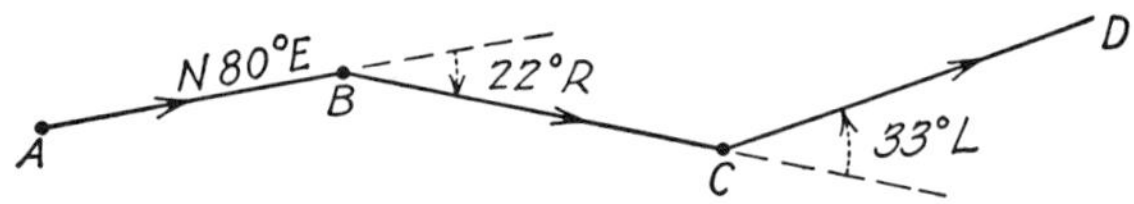

FIG. 12-5. Deflection angles.

and at C is 33°L. Deflection angles may have values between 0° and 180°, but usually they are not employed for angles greater than 90°. In any closed polygon the algebraic sum of the deflection angles (considering right deflections as of sign opposite to left deflections) is 360°.

If the bearing of any line is known and the deflection angles are observed, the bearings of other lines may be calculated. Thus in the figure the bearing of AB is given as N80°E, hence the bearing of BC is 180 − 80 − 22 = S78°E.

12-15. Angles to Right. Angles may be determined by clockwise measurements from the preceding to the following line, as illustrated by Fig. 12-6. Such angles are called *angles to right* or *azimuths from back line.*

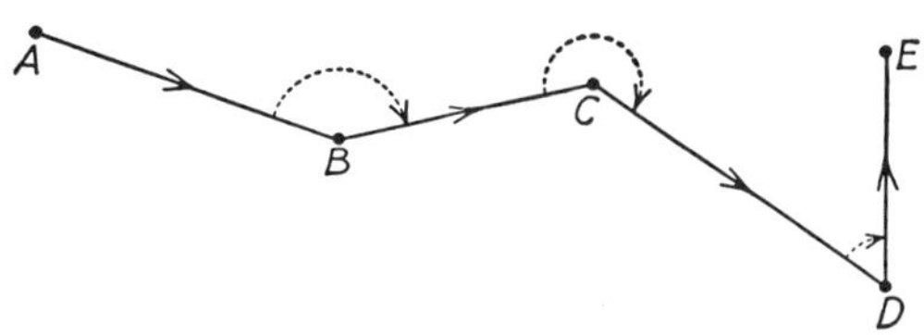

FIG. 12-6. Angles to right.

12-16. Interior Angles. In a closed polygon the angles inside the figure between adjacent lines are called *interior angles*. If n is the number of sides in a closed polygon, the sum of the interior angles is $(n - 2)$ 180°.

12-17. Traversing. The succession of straight lines connecting a succession of established points along the route of a survey is called a *traverse* or *traverse line*. The points defining the traverse line are called *traverse stations* or *traverse points.* Distances along the line between successive points are determined either by direct measurement or by stadia. At each point where the traverse changes direction an angular measurement is taken. If the traverse forms a closed figure, as, for example, the boundary of a parcel of land, it is called a *closed traverse;* if it does not form a closed figure, as, for example, the line for a highway, it is called an *open traverse* or *continuous traverse.* Traverses are also designated according to the purpose of the survey (as preliminary), the field instrument or method employed (as transit-tape), or the kind of angular measurements observed (as azimuth).

Usually deflection angles are employed on open traverses where the change in direction of the line is less than 90° at each traverse station. For the location of highways, railroads, canals, etc., angular measurements are nearly always taken by deflection angles. Azimuths are widely used on topographic surveys and similar surveys where a large number of details are located by angular measurements from the traverse stations. Ordinarily the interior angles of traverses run to establish the boundaries of land are observed. Traverses by bearings are rarely run except on rough surveys where the magnetic compass is employed, though magnetic bearings are generally observed as a rough check on deflection angles, azimuths, or interior angles determined by more precise methods.

For details of the methods of traversing, see Chap. 14. Traversing is the method of surveying in most common use where favorable routes are available.

12-18. Triangulation. Where the lines of a survey form triangular figures whose angles are measured and whose distances are determined by trigonometric computations, the operation of making the necessary field observations is called *triangulation.* The simplest case is that of a single triangle, one of whose sides is of known length; if any two angles of the triangle are measured, sufficient data are obtained for computing the length of the other two sides. Furthermore, if the third angle is measured, the angular measurements may be checked.

A *triangulation system* is made up of a series of triangles so connected that, having measured the angles of the triangles and the length of one line, the length of other lines may be computed. The line of known length, upon which all computed distances are based, is called a *base line.*

Triangulation is often necessary in connection with traversing where the direct measurement of one or more lines is impossible. Simple triangulation is also employed for the location of tunnel shafts and bridge piers. A simple chain of triangles or quadrilaterals affords a convenient means of locating points on opposite sides of a stream. Groups of polygons are suitable for the survey of an area. Generally an extensive triangulation system, such as that for a large city or a state, is composed of a combination of simple triangles, polygons, and quadrilaterals.

For details of the methods of triangulation, see Chap. 16. The advantage of triangulation over traversing lies in the small number of linear measurements that are necessary; the disadvantage lies in the greater amount of computing required. Triangulation is superior to the method of traversing where the terrain offers many obstacles (such as hills, vegetation, or marsh) to traverse work.

12-19. Methods of Determining Angles and Directions. Angles are commonly measured by means of the transit, but may also be measured by means of the tape, plane-table alidade, sextant, or magnetic compass (see also Table 7-1). Directions are observed with the magnetic compass.

1. *Transit.* The use and adjustment of the engineer's transit are described in Chap. 13. The essential features of the transit here to be considered are (1) a horizontal circle, graduated in degrees, which may be rotated and which may be clamped in any position, (2) a plate which may be rotated inside the graduated circle and which carries verniers for reading the graduated circle, and (3) a telescopic line of sight which is attached to the vernier plate,

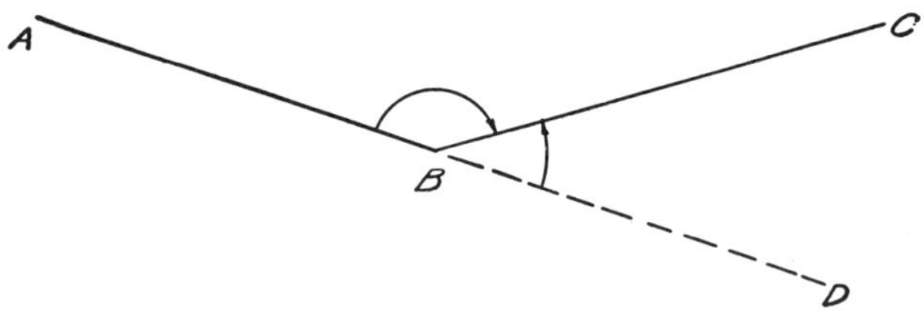

FIG. 12-7. Angle with transit.

rotates with it, and may be rotated in altitude. (See also Art. 2-3.) By means of the verniers the graduated circle on most instruments can be read to the nearest minute of arc, and on some precise transits to the nearest 10″.

If the angle ABC in Fig. 12-7 is to be measured, the transit is set at B. The index of the vernier is set at zero on the graduated circle, and a sight is taken to A. The graduated circle is clamped in position, and the line of sight is turned to C. Since the vernier plate is moved with the line of sight, it is rotated through the angle ABC and hence the vernier reads the angle. If the *azimuth* of BC is to be observed (and the circle is graduated from 0° to 360° in a clockwise direction), the backsight to A is taken with the vernier set to read the azimuth of the line BA. When the line of sight is rotated to C, the vernier reading will be the azimuth of the line BC. If the *deflection angle* at B is to be observed, a backsight is taken on A with the vernier set at 0°, the line of sight is rotated first in altitude (plunged) to point in the direction of BD, and then is rotated in azimuth until the point C is sighted; the vernier then reads the deflection angle DBC.

In running a traverse, as $ABCDE$ (Fig. 12-8), angular measurements are made at successive stations. If an error occurs in any angle, as ABC, then the observed or calculated directions of all succeeding lines in the traverse, as BC, CD, and DE, will be affected by the amount of the error. If the error at B is CBC', then BC', $C'D'$, and $D'E'$ indicate the observed or calculated directions of the following lines.

2. *Tape.* If the sides of a triangle are measured, sufficient data are obtained for computing the angles. Although for most surveying work the angles are measured directly, there are occasions where angles may be determined with sufficient precision by tape measurements; and for very small angles greater precision can be obtained by taping than by ordinary measurement with the transit. The simple method commonly employed is described in Art. 7-26. The error in the computed value of the angle depends on the care with which the points are established and on the precision with which the measurements are taken; for acute angles on level

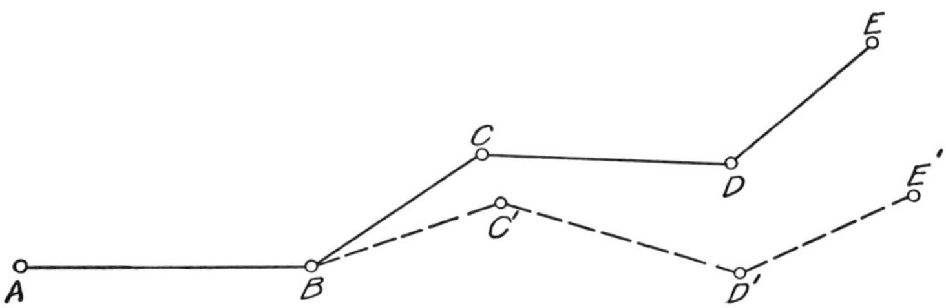

FIG. 12-8. Effect of angular error in transit traverse.

ground it need not exceed 05′ or 10′. For angles greater than 90° the corresponding acute angle is observed. The method is slow and is generally used only as a check or when other instruments are not available.

3. *Plane Table.* Angles may be graphically determined by means of the *plane table* and *alidade,* the use and adjustments of which are described in Chap. 17. The plane table consists essentially of a drawing board mounted on a tripod in such manner that it can be leveled and can be revolved in the horizontal plane. The essential feature of the alidade is a straightedge, parallel to which is a line of sight. Figure 12-9 illustrates the use of the plane table for the graphical determination of the angle AOB. The plane table, to which is fastened a sheet of drawing paper, is set over the ground point O. A point O on the paper is plotted over the point on the ground. With the straightedge through O, a sight is taken to A and a line is drawn. Point B is sighted and a similar line is drawn. The two lines on the paper are parallel to corresponding lines on the ground and hence define the angle AOB. The plane table is extensively employed in topographic mapping, for which work the graphical representation of angles is sufficient and the numerical values of the angles are not desired.

4. *Sextant.* The sextant, though used principally by the navigator, is sometimes employed by the surveyor, particularly on hydrographic surveys (see Chap. 30). The advantage of the sextant over other instruments lies in the fact that, through its use, an angle may be measured while the observer is on a moving object; hence, angles may be read from a boat from which soundings are taken. The angle measured is in the same plane as the two points sighted and the telescope, and hence, in general, is not a horizontal angle. In the situations where the sextant is used in surveying, however, the angle may generally be considered to be horizontal without appreciable error. The angle actually measured has its vertex not at the eye but at the intersection of the two sight rays; for small angles this intersection is at a considerable distance back of the observer. Hence the sextant is not an instrument of precision for small angles, say, less than 15°, nor for short distances, say, less than 1,000 ft.

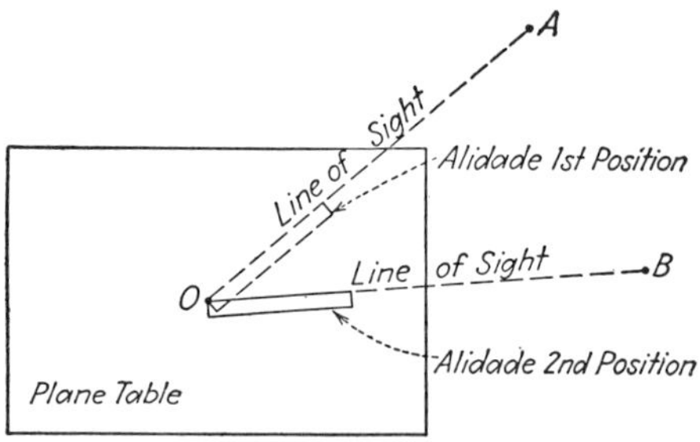

FIG. 12-9. Angle with plane table.

5. *Magnetic Compass.* The use of the magnetic compass is described in the following articles. The compass is useful alone in making rough surveys and retracing early land surveys and is useful on the transit as a means of approximately checking horizontal angles measured by more precise methods.

THE COMPASS

12-20. Magnetic Compass. The essential features of the magnetic compass used by the surveyor are (1) a compass box with circle graduated from 0° to 90° in both directions from the N and S points and usually having the E and W points interchanged as illustrated in Fig. 12-10, (2) a line of sight in the direction of the SN points of the compass box, and (3) a magnetic needle. When the line of sight is pointed in a given direction, the compass needle (when pivoted and brought to rest) gives the magnetic bearing. Thus in the figure the bearing of AB is N60°E. If the N point of the compass box is nearest the object sighted, the bearing is read by observing the north end of the needle.

The varieties exhibiting the three features just mentioned are:

1. Various *pocket compasses,* which are generally held in the hand when bearings are observed; used on reconnaissance or other rough surveys.

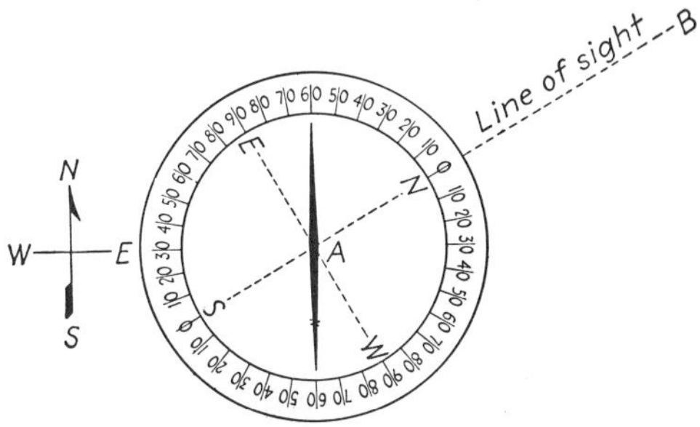

FIG. 12-10. Features of magnetic compass used in surveying.

2. The *surveyor's compass,* which is mounted usually on a light tripod or sometimes on a Jacob's staff (a pointed stick about 5 ft. long); formerly much used on all kinds of land surveying, but now little employed except for forest surveys or retracing early land surveys.

3. The *transit compass,* a compass box similar to that of the surveyor's compass, mounted on the upper or vernier plate of the engineer's transit (Chap. 13) and often used to check horizontal angles.

12-21. Pocket Compasses. Figure 12-11 illustrates one pattern of the pocket compass for which the line of sight is given by a line on the inside of the cover. An observation is taken by laying the cover back and holding the compass so as to sight along the line inside the cover. When this line of sight is in the proper direction, the needle is given time to come to rest. It is then raised and clamped in position by depressing a pin, the compass is lowered, and the bearing is read.

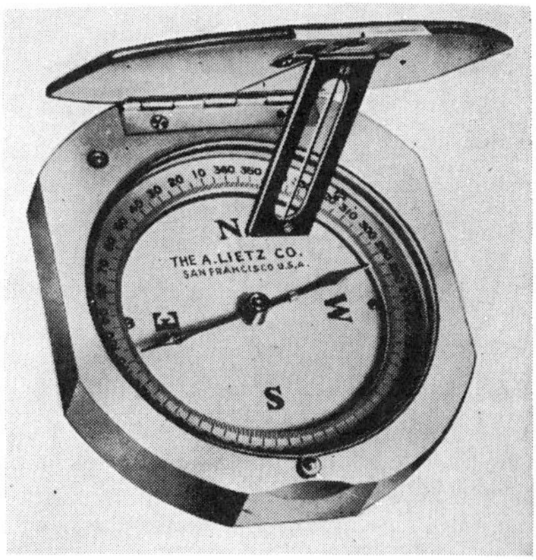

FIG. 12-11. Pocket compass. (*The A. Lietz Co.*)

The Brunton *pocket transit* shown in Fig. 12-12 is designed primarily as a hand instrument but may also be used on a tripod or a Jacob's staff. Inside the cover is a mirror *A*; at *B* is a folding peep sight. The cover is tilted until with the eye at *B* the reflected image of the compass circle is visible.

At C a portion of the mirror glass is without silver, and a line is etched on the glass in line with the N and S points of the compass and the peep sight B. The peep sight and the etched line define the line of sight of the instrument. An observation is taken by holding the peep sight to the eye and viewing the distant object through the plain glass at C. When the proper direction

FIG. 12-12. Brunton pocket transit. (*Wm. Ainsworth & Sons, Inc.*)

FIG. 12-13*a*. Surveyor's compass. (*The A. Lietz Co.*)

is obtained, the compass is leveled by centering the bubbles of the two level tubes as seen in the mirror. When the needle comes to rest, the bearing is observed by means of the mirror; or if desired the needle may be clamped by depressing the pin at D, and the compass may be lowered and read directly. Another method of observing a bearing is to hold the pocket transit in the hand a convenient distance below the eye, viewing it as in the figure and turning it about in a horizontal plane until the image of the object defining the far end of the line is bisected by the line etched on the mirror. The bearing is then read.

The Brunton pocket transit is also used as a hand level or a clinometer. When so employed, it is held on edge and one of the bubbles is centered by means of a thumb nut not visible in the figure. Vertical angles are read by means of the graduated arc F and the vernier G. Its use is nearly like that of the Abney level and clinometer (Art. 8-12).

The *prismatic compass* is similar in principle to the Brunton transit except that it has a floating card dial and that a prism is employed instead of a mirror.

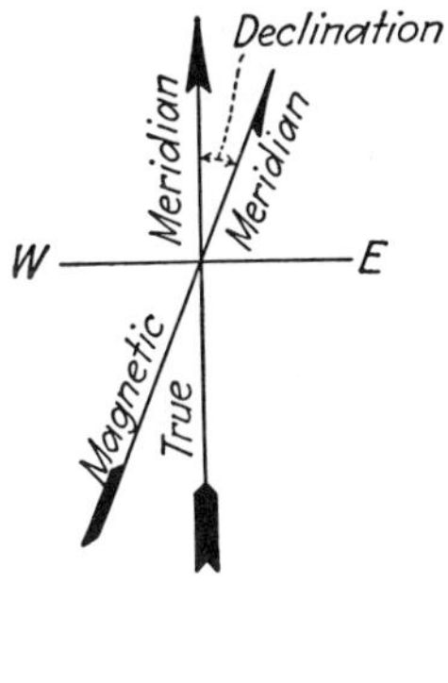

FIG. 12-13*b*. Declination east.

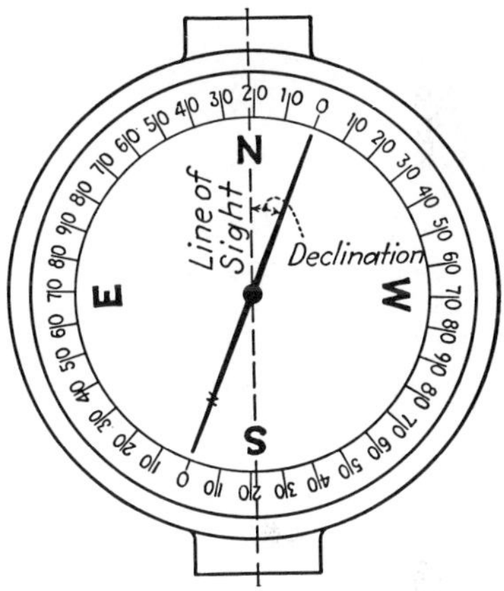

FIG. 12-13*c*. Declination set off on compass circle.

12-22. Surveyor's Compass. The *surveyor's compass* consists of a compass box to opposite sides of which are fastened vertical sight vanes (Fig. 12-13*a*). The box is rigidly connected to a vertical spindle which is free to revolve in a conical socket. Below the spindle is a leveling head consisting of a ball-and-socket joint by means of which the compass may be leveled; the upper portion of the socket is a thumb nut which is tightened until the ball is held securely by friction. The leveling head may be screwed onto a wooden tripod or a Jacob's staff. Within the compass box is a circular, or "bull's-eye," level. The compass is provided with a screw for lifting and clamping the needle and with a screw for clamping the vertical spindle. To counteract the effect of magnetic dip so that the needle will be horizontal while in use, the compass needle is counterweighted as described in Art. 12-5. The compass circle is graduated usually in half-degrees, and bearings may be read by estimation to 05′ or 10′.

In order that *true* bearings may be read directly, some compasses, as the one shown in the illustration, are so designed that the compass circle may be rotated with respect to the box in which it is mounted. When the circle is in its normal position, the line of sight as defined by the vertical slits in the sight vanes is in line with the N and S points of the compass circle, and the observed bearings are magnetic. If the magnetic declination is set off by means of the circle, the observed bearings will be true, as evident from Fig.

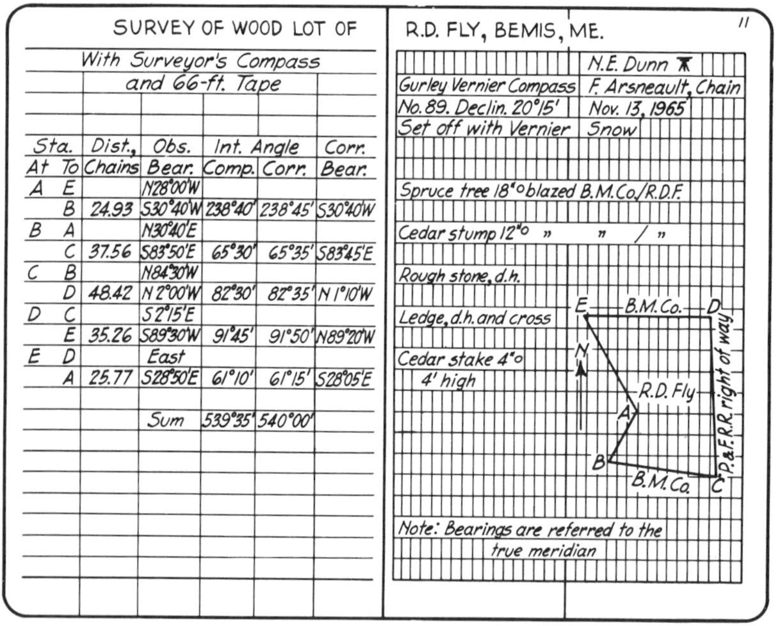

SURVEY OF WOOD LOT OF R.D. FLY, BEMIS, ME. 11

With Surveyor's Compass and 66-ft. Tape

Sta. At	To	Dist., Chains	Obs. Bear.	Int. Angle Comp.	Int. Angle Corr.	Corr. Bear.
A	E		N28°00'W			
	B	24.93	S30°40'W	238°40'	238°45'	S30°40'W
B	A		N30°40'E			
	C	37.56	S83°50'E	65°30'	65°35'	S83°45'E
C	B		N84°30'W			
	D	48.42	N2°00'W	82°30'	82°35'	N1°10'W
D	C		S2°15'E			
	E	35.26	S89°30'W	91°45'	91°50'	N89°20'W
E	D		East			
	A	25.77	S28°50'E	61°10'	61°15'	S28°05'E
			Sum	539°35'	540°00'	

FIG. 12-14. Notes for compass survey.

12-13*c*. If the declination is east, as in the figure, the circle is rotated clockwise with respect to the plate; if the declination is west, counter-clockwise.

When the direction of a line is to be determined, the compass is set up on line and is leveled. The needle is released, and the compass is rotated about its vertical axis until a range pole or other object on line is viewed through the slits in the two sight vanes. When the needle comes to rest, the bearing is read. Ordinarily the sight vane at the end of the compass box marked "S" is held next to the eye; in this case the bearing is given by the north end of the needle.

The following suggestions apply to compass observations: At each observation the compass box should be tapped lightly as the needle comes to rest, so that the needle may swing freely. In order not to confuse the north and south ends of the needle when taking bearings, the observer should always note the position of the counterbalancing wire (which is on the south end). Since the precision with which angles may be read depends on the delicacy of the needle, special care should be taken to avoid any jar between the jewel bearing of the needle and the pivot point. *Never move the instrument without making certain that the needle is lifted and clamped.*

Sources of magnetic disturbance such as chaining pins and axe should be kept away from the compass while a reading is being taken. Care should be taken not to produce static charges of electricity by rubbing the glass; a moistened finger pressed against the glass will remove such charges. Ordinarily the amount of metal about the person of the instrumentman is not large enough to deflect the needle appreciably, but a change of position between two readings should be avoided.

Surveying with the compass is usually by traversing (Art. 12-17). Only alternate stations need be occupied, but a check is secured and local attraction is detected if both a backsight and a foresight are taken from each station. Unlike a transit traverse, in which an error in any angle affects the observed or computed directions of all following lines, an error in the observed bearing of one line in a compass traverse has no effect upon the observed *directions* of any of the other lines. This is an important advantage, especially in the case of a traverse having many angles. Another advantage of the compass is that obstacles such as trees can be passed readily by offsetting the instrument a short measured distance from the line.

Field notes for a closed compass traverse are kept in a form similar to that of Fig. 12-14. The declination was set off on the compass so that bearings were referred to the true meridian. (See example of Art. 12-24 for related computations.)

12-23. Correction for Local Attraction. If local attraction from a fixed source exists at any station in a traverse, both the back and the forward bearings taken from that station will be affected by the same amount. Disregarding for the time being the accidental errors due to observing, it is probable that the terminal points of any line, as AB, are free from local attraction if the back bearing from B is the reverse of the forward bearing from A. Keeping in mind that the computed *angle* between the forward and back lines from any station can be determined correctly from the observed bearings taken from that station regardless of whether or not the needle is affected locally, the direction free from local attraction may be chosen as a basis, the traverse angles may be computed from observed bearings, and—starting from the unaffected line—the correct bearings of successive lines may be computed.

Example 1: From the observed bearings of an open traverse in the accompanying tabulation, it is assumed that points A and B are free from local attraction since the back and forward bearings of AB are opposite. Hence the correct forward bearing of BC is S60°E. The angle at C, computed from the observed bearings, is 180 − 62 − 31 = 87°; and this value of the angle is correct (excluding errors of observation) regardless of local attraction. The correct forward bearing of CD is therefore 180 − 60 − 87 = S33°W.

Station	Line	Observed bearing	Computed angle	Corrected bearing	Local attraction
A					0°
	AB	N45°E		N45°E	
	BA	S45°W		S45°W	
B			105°		0°
	BC	S60°E		S60°E	
	CB	N62°W		N60°W	
C			87°		2° clockwise
	CD	S31°W		S33°W	
	DC	N30°E		N33°E	
D					3° clockwise

Some surveyors find it more expedient to consider the magnitude and direction of the error due to local attraction and then to make corrections to observed bearings without computing the traverse angles.

Example 2: For the observed bearings of example 1 it is seen that the *correct* back bearing of BC is N60°W and that the *observed* back bearing is N62°W. The local attraction at C is, therefore, 2° clockwise (as may be seen from a sketch), and the correction to any observed bearing taken with the compass at C is 2° counterclockwise. The observed forward bearing of CD is S31°W, and the corrected forward bearing of CD is, therefore, 31 + 2 = S33°W. In similar manner the local attraction at D is found to be 3° clockwise.

Owing to errors of observation there are likely to be discrepancies between the observed forward and back bearings of lines, even though no local attraction exists. If the discrepancies are small and apparently not of a systematic character, it is reasonable to assume that the errors are due to causes other than local attraction.

12-24. Adjustment of Closed Compass Traverse. When the compass traverse forms a closed figure, the angle method of example 1, Art. 12-23, may be extended to include the effect of observational errors, as follows: The interior angle at each station is computed from the observed bearings; the computed value will be free from local attraction as previously described. The sum of the interior angles should equal $(n - 2)180°$ in which n is the number of sides in the traverse. Since the error of observing a bearing is accidental, the error of closure of the traverse (as indicated by the sum of the computed interior angles) is assumed to be distributed equally, and the interior angles are corrected accordingly. The bearings are then adjusted by starting from some line whose observed bearing is assumed to be correct and by computing the bearings of successive lines by means of the corrected interior angles.

Example: The observed bearings and computed interior angles for the compass traverse of Fig. 12-14 are shown in Fig. 12-15*a*, in which the short vertical lines represent the compass needle. The sum of the interior angles is 25′ less than the correct value of 540°00′; hence 05′ is to be added to each of the five interior angles to correct for observational errors. The corrected interior angles are shown in Fig. 12-15*b*, in which the short vertical lines represent the true meridian. Since line

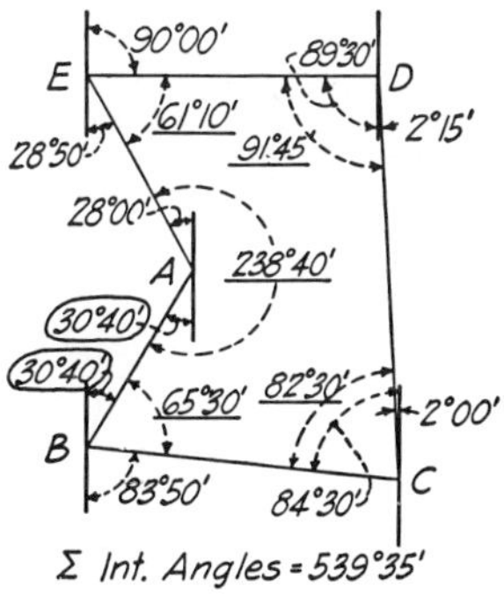

FIG. 12-15*a*. Observed bearings and computed interior angles.

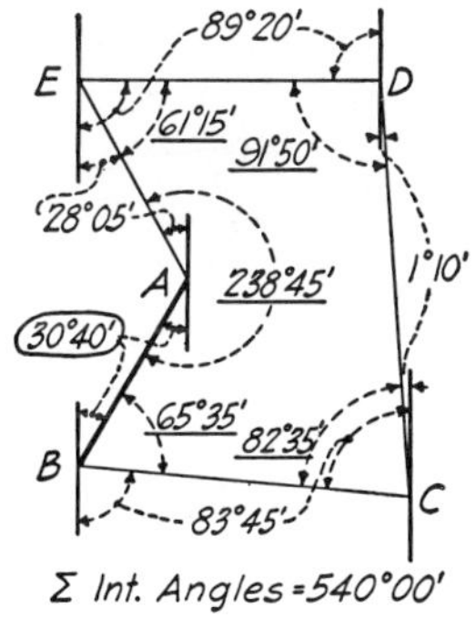

FIG. 12-15*b*. Corrected bearings computed from *AB* and corrected interior angles.

AB had the same observed back bearing as observed forward bearing, both ends of that line are assumed to be free from local attraction, and the bearing of *AB* is taken as being correct. Using the corrected interior angle at *B*, the corrected forward bearing of *BC* is then 180°00′ − 30°40′ − 65°35′ = S83°45′E. The corrected back bearing of *BC* must be the same as the forward bearing, just computed. At *C*, the corrected forward bearing of *CD* is 83°45′ − 82°35′ = N1°10′W. In this manner the computations are continued around the traverse. As a check, the forward bearing of the initial line *AB* is computed from the corrected back bearing *AE* of the preceding line and the corrected interior angle at *A*.

If the error in the sum of the interior angles is greater than 05′ or 10′ times the number of angles, it is probable that a mistake in reading the compass has occurred, and the field measurements should be repeated. If the error is within permissible limits but cannot be divided equally among the angles in amounts of 05′ or 10′, the greater corrections (in multiples of 05′) should be applied arbitrarily to those angles for which the conditions of observing were estimated to be the least favorable. The precision of compass measurements does not justify computations with a precision closer than multiples of 05′.

If two or more of the traverse lines appear to be free from local attraction, as indicated by the agreement between forward and back bearings, one of these lines is arbitrarily chosen as the "best line," and the computation of corrected bearings is referred to this line. If none of the lines is free from local attraction, that line is chosen which has the least discrepancy between forward and back bearings; and its forward bearing is assumed to be correct.

12-25. Sources of Error; Adjustment of Compass. 1. *Needle Bent.* If the needle is not perfectly straight, a constant error is introduced in all observed bearings. As shown by Fig. 12-16*a*, one end of the needle will read higher than the correct value whereas the other end will read lower; for each observation the error can be eliminated by reading both ends of the needle and averaging the two values. The needle can be straightened with pliers.

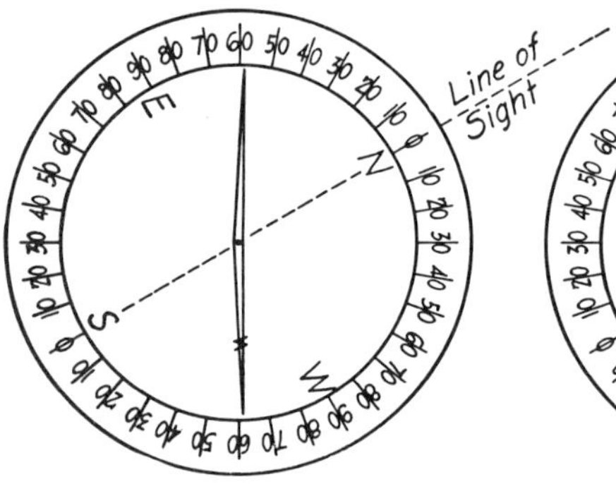

FIG. 12-16*a*. Bent needle.

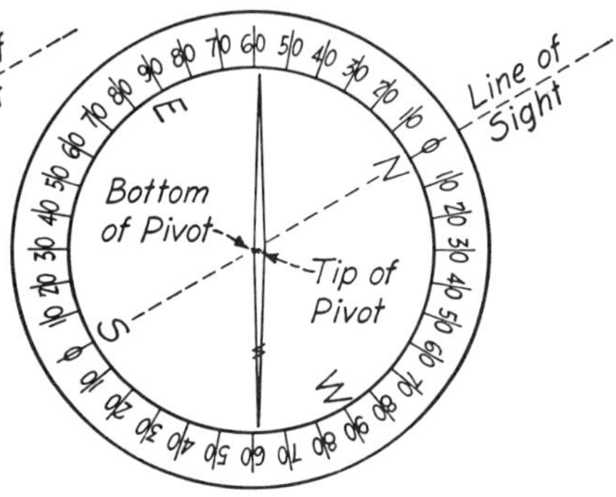

FIG. 12-16*b*. Bent pivot.

2. *Pivot Bent.* If the point of the pivot supporting the needle is not at the center of the graduated circle, there is introduced a variable systematic error, the magnitude of which depends upon the direction in which the compass is sighted. For one direction, the error is zero; for a direction 90° thereto, it is a maximum. In this case also, one end of the needle will read higher than the correct value whereas the other end will read lower (Fig. 12-16*b*); for each observation the error can be eliminated by reading both ends of the needle and averaging the two values. The instrument can be corrected by bending the pivot until the end readings of the needle are 180° apart for any direction of pointing.

3. *Plane of Sight Not Vertical, or Graduated Circle Not Horizontal.* This misalinement introduces a systematic error, but it is usually so small as to be of no consequence. However, the sight vanes may become bent so that, even though the instrument is leveled, an appreciable error is introduced particularly if the line of sight is steeply inclined when taking a bearing. The vanes may be tested by leveling the compass and sighting at a plumb line.

The adjustment of the level tubes may be tested by reversal, as described for the transit in Art. 13-30.

4. *Sluggish Needle.* The needle is not likely to come to rest exactly on the magnetic meridian. This lag produces an accidental error which is often large. As the needle comes nearly to rest, tapping the glass lightly will tend to prevent the needle from sticking to the pivot. If the needle is "weak," it may be remagnetized by drawing its ends over a bar magnet, from the center to the ends of the magnet. The south-seeking end of the compass needle is drawn over the north-seeking half of the bar magnet, and *vice versa.* On each return stroke the needle should be lifted well above the magnet. If the pivot point is blunt, it may be sharpened by rubbing it on a fine-grained oilstone.

5. *Reading Needle.* The inability of the observer to determine exactly the point on the graduated circle at which the needle comes to rest is generally the source of the most important and largest accidental error in compass work. The needle should be level, and the eye of the observer should be above the coinciding graduation and in line with the needle. If the needle dips perceptibly, its counterweight should be adjusted. Other conditions being equal, the longer the needle, the smaller the error of observing. With the 6-in. needle used on many surveyor's compasses, the probable error need not exceed ±05′; with the 3½- or 4-in. needle on the engineer's transit, the probable error is likely to be as much as ±10′.

6. *Magnetic Variations.* Undetected deviations of the magnetic needle from whatever cause are the source of the largest and most important systematic errors in compass work. Largely because of such variations, the compass, no matter how finely constructed, is not a suitable instrument for any except rough surveys. Deviations due to local attraction can be detected and corrections can be applied as described in the preceding articles. Particular care should be taken to keep iron or steel objects away from the compass while it is in use, and the observer should if possible remain on the same side of the instrument. Also static charges of electricity on the glass cover, which may attract the needle, may be removed by touching the glass with a moist finger.

12-26. Numerical Problems.

1. The magnetic bearing of a line is S47°30′W and the magnetic declination is 12°10′W. What is the true bearing of the line?

2. The true bearing of a line is N18°17′W and the magnetic declination is 7°12′E. What is the magnetic bearing of the line?

3. In an old survey made when the declination was 2°10′W, the magnetic bearing of a given line was N35°15′E. The declination in the same locality is now 3°15′E. What are the true bearing and the present magnetic bearing that would be used in retracing the line?

4. Following are the observed magnetic bearings of a compass traverse: *AB*, N37°45′E; *BC*, N84°30′E; *CD*, S66°40′E; *DE*, S79°00′E; *EF*, N55°15′E. Compute the deflection angles.

5. Following are deflection angles of traverse *A* to *F*: *B*, 37°21′L; *C*, 12°39′L; *D*, 63°31′R; *E*, 14°07′L. The true bearing of *AB* is S37°56′E. Compute the bearings of the remaining lines.

6. For the traverse of problem 4, the declination is 7°15′E. Compute the true azimuths reckoned from the north point.

7. For the traverse of problem 5 compute the true azimuths reckoned from the south point.

8. The interior angles of a five-sided closed traverse are as follows: *A*, 117°36′; *B*, 96°32′; *C*, 142°54′; *D*, 132°18′. The angle at *E* is not measured. Compute the angle at *E*, assuming the given values to be correct.

9. (*a*) What are the deflection angles of the traverse of problem 8? (*b*) What are the computed bearings if the bearing of *AB* is due north?

10. The following azimuths are reckoned from the north: *AB*, 187°12′; *BC*, 273°47′; *CD*, 318°48′; *DE*, 0°48′; *EF*, 73°00′. What are the corresponding bearings? What are the deflection angles?

11. Following are the deflection angles of a closed traverse: *A*, 85°20′L; *B*, 10°11′R; *C*, 83°32′L; *D*, 63°27′L; *E*, 34°18′L; *F*, 72°56′L; *G*, 30°45′L. Compute the error of closure. Adjust the angular values on the assumption that the error is the same for each angle.

12. In triangulating across a river a base line *AB* of the triangle *ABC* has a measured length of 536.27 ft., and the angles at *A* and *B* are respectively 87°32′ and 68°48′. Compute the distance *AC*.

13. The following are bearings taken for an open compass traverse. Correct for local attraction.

Line	Forward bearing	Back bearing
AB	N37°15′E	S36°30′W
BC	S65°30′E	N66°15′W
CD	S31°00′E	N31°00′W
DE	S89°15′W	N89°45′E
EF	N46°30′W	S46°45′E
FG	N15°00′W	S14°45′E

14. The following are bearings taken on a closed compass traverse. Compute the interior angles and correct them for observational errors. Assuming the observed bearing of the line *AB* to be correct, adjust the bearings of the remaining sides.

Line	Forward bearing	Back bearing
AB	S37°30′E	N37°30′W
BC	S43°15′W	N44°15′E
CD	N73°00′W	S72°15′E
DE	N12°45′E	S13°15′W
EA	N60°00′E	S59°00′W

12-27. Field Problems.

Problem 1. Determination of Magnetic Declination

Object. To determine the magnetic declination with the surveyor's compass.

Procedure. (1) See that the compass is in good adjustment. (2) Set the compass over one end of a true meridian that has been determined by astronomical observations, sight along the line, and clamp the compass in that position. (3) By means of the tangent-screw, move the compass circle until the needle reads zero. (4) From the declination arc read the declination to the nearest minute; record the declination and the time of observation. (5) Take observations as before, every 5 min. over a period of half an hour or more, resetting the line of sight and compass circle for each observation. (6) If possible take a series of observations at about the same time on each of several days. (7) Determine the most probable value of the declination for the hour of observation, and the probable error of a single observation and of the mean (see Art. 5-8*a*). (8) Determine the mean declination by adding to or subtracting from the observed declination the average daily variation (Table VIII) for the place nearest the place of observation.

Hints and Precautions. (1) Between 7 and 11 P.M. is usually the best time for taking magnetic observations, because at this time the magnetic declination reaches approximately its mean value for the day, as will be seen by examining Table VIII. Between 10 and 11 A.M. the declination also reaches its mean value, but the rate of change is more rapid at this time. (2) If the compass circle has been turned in a clockwise direction the declination is east. If the daily variation is positive for the time of the observation, the north end of the needle is deflected more to the west than when the declination is at a mean. Hence, the mean is determined by algebraically subtracting the daily variation from the observed declination, west being considered as positive and east as negative.

Problem 2. Survey of Field with Surveyor's Compass and Tape

Object. To find the true bearing and length of each side of an assigned field, using the surveyor's compass and 66-ft. chain tape. The data may be used in the office problems at the end of Chap. 18.

Procedure. (1) On the compass, set off the magnetic declination in order that true bearings may be read directly; if this cannot be done, observe the magnetic bearings and convert them to true bearings later. (2) Set up at one corner A of the field, lower the needle, and sight along the line AB, with the end of the compass box marked "S" next to the eye. As the needle comes nearly to rest, tap the compass lightly with a pencil. When the needle becomes stationary, read the north end, estimating the forward bearing of AB to the nearest 05′. (3) Take a back bearing from A. (4) Chain the line AB, and record the distance in chains to the nearest link. (5) Set up the compass at B. Observe the back bearing of the line AB and the forward bearing of the line BC. Chain the distance BC. (6) Continue around the field, taking both back and forward bearings from each point and chaining the lines. (7) Compute the interior angles of the field, and correct the observed bearings for local attraction and/or errors of observation (see sample notes, Fig. 12-14). (8) If magnetic bearings have been observed, apply the magnetic declination to convert them to true bearings; add a column in the field notebook for true bearings.

Hints and Precautions. (1) The upright sight vanes are usually unlike, the one to be attached nearer the north point of the compass box being marked for reading vertical angles and the one nearer the south point being fitted with peep sights. It is well to bear this in mind both when assembling the compass and when using it in the field. (2) Be sure to set off the declination in the correct direction.

PROBLEM 3. RETRACING SURVEY WITH COMPASS AND TAPE (TWO ADJACENT CORNERS KNOWN)

Object. To retrace property lines from the notes of an old compass survey for which two adjacent corners of the plot can still be identified. The bearings given in the original notes are magnetic, and the declination at the time of the original survey is unknown.

Procedure. (1) Measure the length of the known side, and compare it with the original. (2) By proportionate distances compute the lengths of the other sides in terms of the resurvey tape. (3) Set up the compass at one end of the known line, sight along the line, and clamp the compass. (4) Release the needle, and as it comes to rest, move the compass circle by means of the tangent-screw until the original bearing is read. (5) Proceed to lay out the field, chaining distances in terms of the resurvey tape as computed above and laying off the original bearings. Examine the ground for rotted stakes or other evidence that would have precedence over bearings and lengths of sides. (6) Reference the new corners by bearings and distances to nearby permanent objects.

REFERENCES

1. DEEL, SAMUEL A., "Magnetic Declination in the United States—1945," *U.S. Coast and Geodetic Survey, Serial* 664, Government Printing Office, Washington, D.C., 1946.
2. HOWE, H. HERBERT, and L. HURWITZ, "Magnetic Surveys," *U.S. Coast and Geodetic Survey, Serial* 718, 2d ed., Government Printing Office, Washington, D.C., 1956.

CHAPTER 13

ENGINEER'S TRANSIT

INSTRUMENTS

13-1. Conventional Transit. The engineer's transit is sometimes called the "universal surveying instrument" by reason of the wide variety of uses for which it is adapted. It may be employed for measuring and laying off horizontal angles, directions, vertical angles, differences in elevation, and distances and for prolonging lines.

Some of the modern types of transit differ considerably in design and construction from those long in use, but their essential features do not differ greatly, and their use not at all. Inasmuch as thousands of the conventional type are continuing in use, and as the conventional type requires field adjustment whereas some modern types require little or none, herein the conventional transit is described in detail (Arts. 13-1 to 13-6) and other types in less detail (Arts. 13-7 to 13-11). For a given type of transit, the details of design and construction differ somewhat as among the various instrument makers; however, their essential features are similar.

A Federal specification for a transit of ordinary precision, reading to 01′, is listed in the references at the end of this chapter.

Figure 13-1 is a photograph of a conventional engineer's transit of the type in common use; Fig. 13-2 is a vertical section of the same type of instrument. It is seen to consist of an *upper*, or *vernier, plate* to which are attached standards supporting the telescope and a *lower plate* to which is fixed a horizontal graduated circle. The upper and lower plates are fastened, respectively, to vertical inner and outer spindles, the two axes of rotation being coincident with and at the geometric center of the graduated circle. The outer spindle is seated in the tapered socket of the leveling head. Near the bottom of the leveling head is a ball-and-socket joint which secures the instrument to the foot plate yet permits rotation of the instrument about the joint as a center.

The outer spindle carrying the lower plate may be clamped in any position by means of the lower *clamp-screw*. Similarly, the inner spindle carrying the upper plate may be clamped to the outer spindle by means of the upper clamp-screw. After either clamp has been tightened, small movements of the spindle may be made by turning the corresponding *tangent-screw*. The axis about which the spindles revolve is called the *vertical axis* of the instrument.

Level tubes, called *plate levels,* are mounted at right angles to each other on the upper plate. Four *leveling screws,* or foot screws, are threaded into the leveling head and bear against the foot plate; when the screws are turned, the instrument is tilted about the ball-and-socket joint. When all four screws are loosened, pressure between the sliding plate and the foot plate is relieved, and the transit may then be shifted laterally with respect to the foot plate.

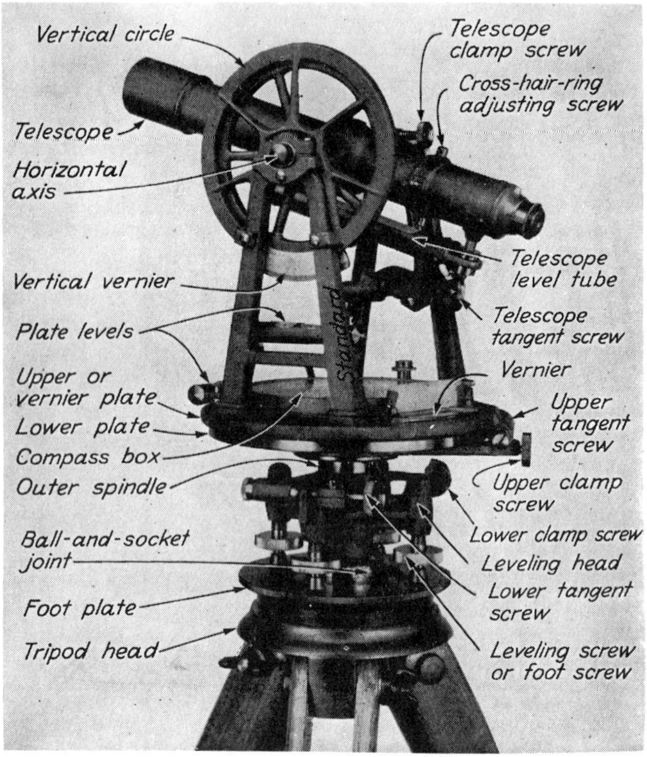

FIG. 13-1. Engineer's transit (conventional type).

From the end of the spindle and at the center of curvature of the ball-and-socket joint is suspended a chain with hook for the plumb line. The instrument is mounted on a tripod by screwing the foot plate onto the tripod head.

The *telescope* is fixed to a transverse *horizontal axis* which rests in bearings on the standards. The telescope may be rotated about this horizontal axis and may be fixed in any position in a vertical plane by means of the telescope clamp-screw; small movements about the horizontal axis may then be secured by turning the telescope tangent-screw. Fixed to the horizontal axis is the *vertical circle,* and attached to one of the standards is the vertical vernier. Beneath the telescope is the *telescope level tube.*

Attached to the upper plate is the *compass box,* the details of which are the same as for the surveyor's compass described in Art. 12-22. If the compass circle is fixed, its N and S points are in the same vertical plane as the

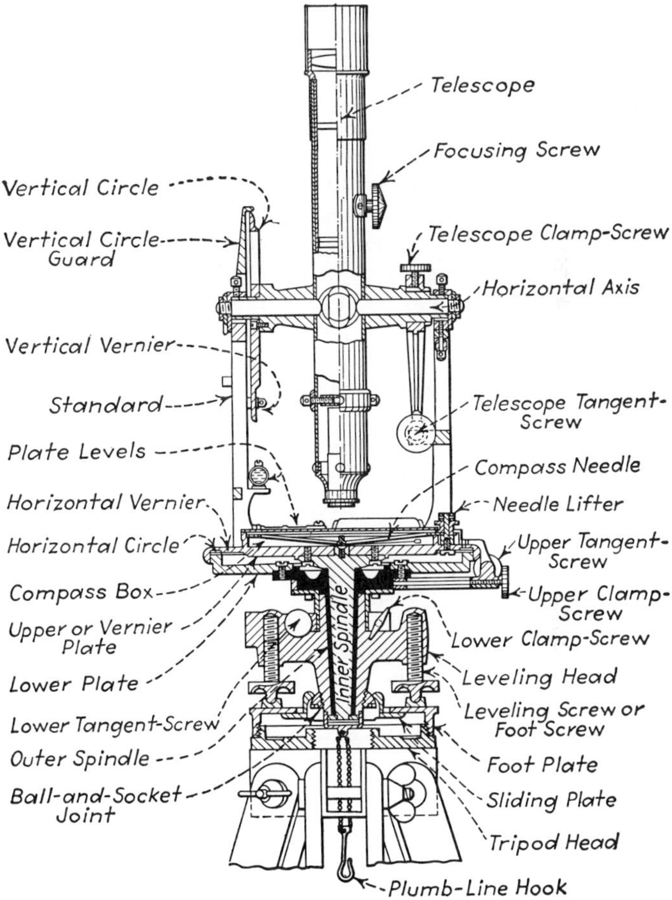

FIG. 13-2. Section of conventional transit.

line of sight of the telescope. The compass boxes of some transits are so designed that the compass circle may be rotated with respect to the upper plate, so that the magnetic declination may be laid off and true bearings may be read. At the side of the compass box is a screw, or needle lifter, by means of which the magnetic needle may be lifted from its pivot and clamped.

Summing up the several features: (1) the center of the transit can be brought over a given point by loosening the leveling screws and shifting the transit laterally; (2) the instrument can be leveled by means of the plate levels and the leveling screws; (3) the telescope can be rotated about either the horizontal or the vertical axis; (4) when the upper clamp-screw is tightened and the telescope is rotated about the vertical axis, there is no relative movement between the verniers and the horizontal circle; (5) when the lower clamp-screw is tightened and the upper one is loose, a rotation of the telescope about the vertical axis causes the vernier plate to revolve but leaves the horizontal circle fixed in position; (6) when both upper and lower clamps are tightened, the telescope cannot be rotated about the vertical axis; (7) the telescope can be rotated about the horizontal axis and can be fixed in any direction in a vertical plane by means of the telescope clamp-screws and tangent-screws; (8) the telescope can be leveled by means of the telescope level tube, and hence the transit can be employed for direct leveling; (9) by means of the vertical circle and vernier, vertical angles can be determined, and hence the transit is suitable for indirect leveling; (10) by means of the horizontal circle and vernier, horizontal angles can be measured; and (11) by means of the compass, magnetic bearings can be determined.

13-2. Level Tubes. The sensitiveness of the several spirit levels of the transit should be such as to produce a well-balanced instrument and hence should correspond to the fineness of graduation of the circles and the optical properties of the telescope. If the levels are more sensitive than necessary to maintain this balance, time is wasted in centering the bubbles; if less sensitive than necessary, the precision of measurements is less than it should be for the transit as otherwise designed.

The plate levels of the ordinary transit reading to 01′ usually are alike in sensitivity and have a value of about 60″ per 0.1-in. graduation, or about 75″ per 2-mm. graduation, of the level tube. When horizontal angles are measured between points nearly in the same horizontal plane, it can be shown that no appreciable error is introduced even if the bubbles are some distance off center. On the other hand, where there is a large difference in vertical angle between the points sighted, a small displacement of the bubble in the tube that is parallel to the horizontal axis causes a relatively large error in the horizontal angle. Some instruments are equipped with a 20″ or 30″ striding level which is employed for leveling the horizontal axis whenever sights are sharply inclined.

The telescope level has a sensitivity of about 30″ per 2-mm. graduation, depending upon the magnifying power of the telescope. The sensitivity of the vertical-vernier control bubble should depend upon the least reading of the vernier; for a vertical circle reading to 01′, a level tube having a sensitivity of 40″ per 2-mm. graduation is commonly employed. For further details concerning level tubes see Arts. 2-5 to 2-7.

13-3. Telescope. The telescope of the transit is similar to that of the engineer's level (Art. 2-9). When the transit is used as an instrument for either direct or trigonometric leveling, any point on the horizontal cross-hair may be used in sighting; when the transit is used for establishing lines, measuring angles, or taking bearings, any point on the vertical cross-hair may

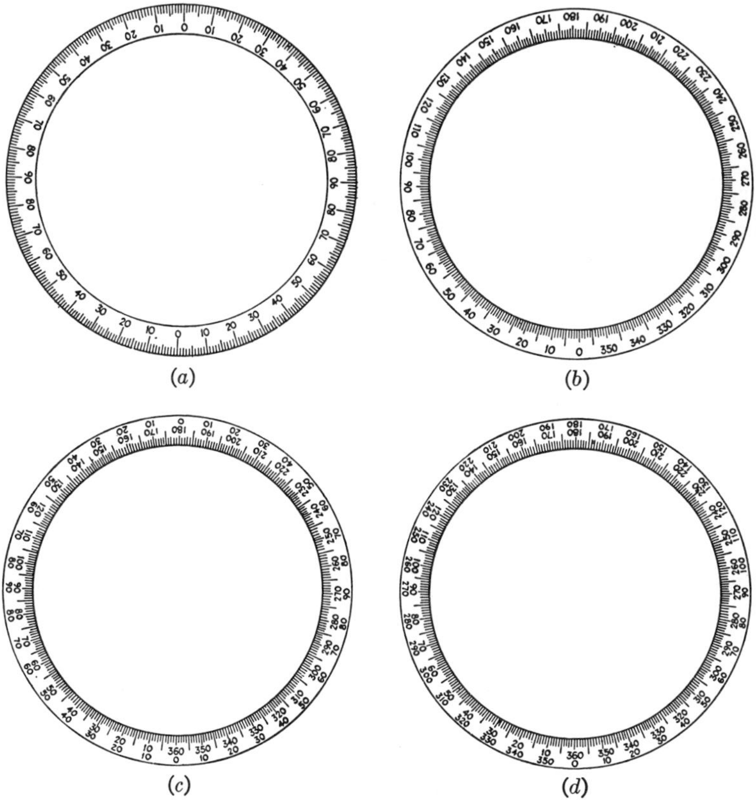

FIG. 13-3. Numbering of circles. (*a*) Vertical circle numbered in quadrants. (*b*) Horizontal circle numbered 0 to 360. (*c*) Horizontal circle numbered 0 to 360 and in quadrants. (*d*) Horizontal circle numbered 0 to 360 and 360 to 0.

be used. Most instruments are equipped with stadia hairs (Chap. 15), which are usually mounted in the same plane with the cross-hairs. The magnifying power is usually 18 to 24 diameters. For the transit, as for the level, usually the erecting eyepiece is employed; but the superior optical properties of the inverting eyepiece make it the favorite of some surveyors, and it is the type used in instruments of precision. For the relative merits of the inverting and erecting eyepieces, see Art. 2-15. Some telescopes are of the internal-focusing type (Art. 2-9).

13-4. Graduated Circles. The vertical circle has two opposite zero points and is graduated usually in half degrees, the numbers increasing to 90° in both directions from the zero points, as illustrated in Fig. 13-3*a*. When the telescope is level, the index of the vernier is at 0°.

The horizontal circle is likewise usually graduated in half-degrees, but may be graduated in third-degrees or quarter-degrees. It may be numbered from 0° to 360° clockwise (Fig. 13-3*b*), 0° to 360° clockwise and 0° to 90° in quadrants (Fig. 13-3*c*), or 0° to 360° in both directions (Fig. 13-3*d*). Most surveyors prefer the numbering system illustrated in Fig. 13-3*d*. Usually the numbers slope in the direction of reading.

The horizontal circles of transits designed for work of moderately high precision are graduated to 20′ or to 15′. Those for repeating theodolites are often graduated to 10′. In special cases, circles may be graduated in divisions based on the *mil* or the *grad* (Art. 1-9).

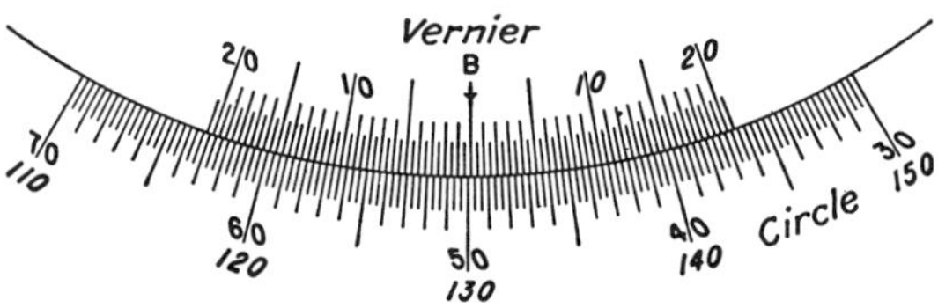

FIG. 13-4. Double direct vernier reading to 30 sec.

13-5. Verniers. The verniers employed for reading the horizontal and vertical circles of the transit are identical in principle with those for the target rod, described in Art. 2-18. Practically all transit verniers are of the direct type.

Figure 2-13 shows the usual type of double direct vernier reading to minutes. Figure 13-4 illustrates a double direct vernier on a circle graduated to 20′.

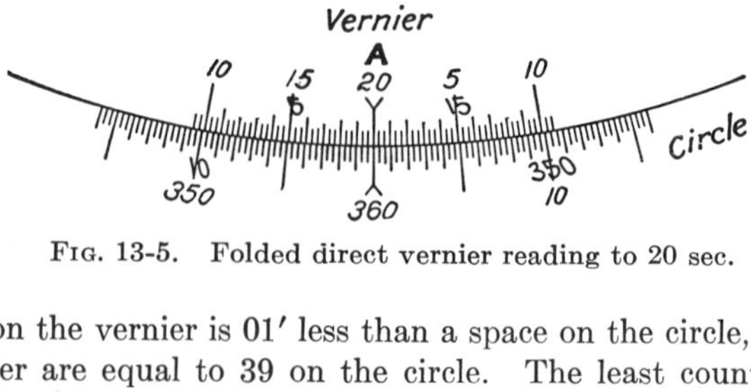

FIG. 13-5. Folded direct vernier reading to 20 sec.

Each space on the vernier is 01′ less than a space on the circle, and 40 spaces on the vernier are equal to 39 on the circle. The least count is, therefore, 20′/40 = 30″. Reading clockwise, the angle is 49°40′ + 10′30″ = 49°50′30″. Reading counter-clockwise the angle is 130°00′ + 09′30″ = 130°09′30″.

Figure 13-5 represents a folded direct vernier reading to 20″. The full length of the vernier is employed for reading angles in either direction. The circle is graduated to 20′, and 60 spaces on the vernier are equal to 59 on the circle. The vernier is read from the index toward either of the extreme divisions and then from the other extreme division in the same direction to the center. The index of the vernier and its 20′ mark are the same. In the illustration, the vernier reads 0°00′00″. The folded vernier is employed where the length of the corresponding double vernier would be so great as to make it impracticable.

13-6. Eccentricity of Verniers and Centers. Transits have two verniers for reading the horizontal circle, their indexes being 180° apart. The one nearest the upper clamp and tangent-screw is known as the *A* vernier, and the one opposite is known as the *B* vernier. The verniers are attached to the upper plate and are adjusted by the instrument maker so that they are much nearer to being truly 180° apart than their least count. Failure of the two verniers to register readings exactly 180° apart on the circle may be due to either or both of two following causes:

1. *Eccentricity of Verniers.* The verniers may have become displaced, so that a line joining their indexes does not pass through the center of rotation of the upper plate. The error will be the same for all parts of the graduated circle.

2. *Eccentricity of Centers.* The spindles may have become worn or otherwise damaged, so that the center of rotation of the upper plate does not coincide with the geometrical center of the graduated horizontal circle. The error varies according to the position of the verniers with respect to the horizontal plate. As indicated by Fig. 13-6, there will be one setting on the graduated circle for which the indexes are exactly 180° apart (first position), and 90° therefrom there will be another setting for which the verniers fail to register 180° apart by a maximum amount (second position).

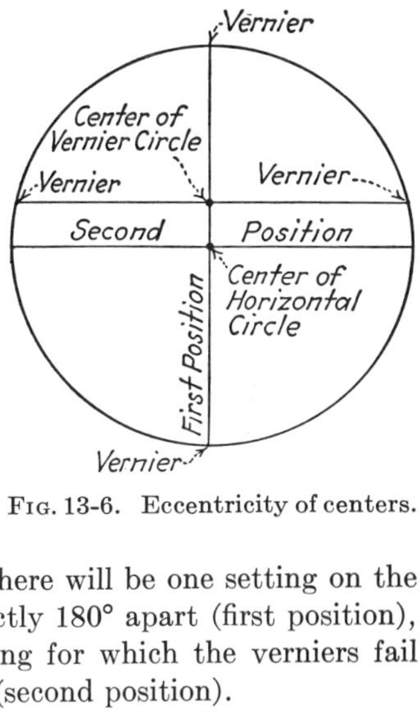

FIG. 13-6. Eccentricity of centers.

To correct either of these defects requires the services of an instrument maker, but neither defect limits the precision with which angles can be measured. Taking the mean of the two vernier readings (*A* and *B* verniers) eliminates errors due to either or both types of eccentricity. Furthermore, if the verniers only are eccentric, no error is introduced in an angle as long as the same vernier is used for making the final reading as for making the initial setting.

13-7. Other Transits and Features. A transit without vertical circle and telescope level tube is called a *plain transit.* One without compass and having U-shaped, one-piece standards instead of the A-shaped separate standards shown in Figs. 13-1 and 13-2 is often called a *city transit.* A *mountain transit* is relatively small and light. A *mining transit* (see Art. 29-6) has one or more of the following features: a small auxiliary telescope attached either to one end of the horizontal axis or to the top of the main telescope, the vertical circle graduated on the edge instead of the side, a vertical arc of 180° instead of a full circle, and a reversion telescope level which makes it possible to level the telescope with the level tube above as well as below the telescope.

Some transits have two opposite vertical verniers which may be rotated as a unit about the horizontal axis and which are equipped with a level tube for control. When the level bubble is centered, a line through the zeros of the two verniers is horizontal. This device enables vertical angles to be read correctly whether or not the plate-level bubbles are centered. Some transits are provided with a sensitive *striding level* which rests on the horizontal axis and insures that the line of sight will revolve in a vertical plane; this feature is particularly important in astronomical observations. On some transits the telescope tangent-screw is equipped with a gradienter (Art. 2-20).

13-8. Theodolites. A *theodolite* is essentially a transit of high precision. A *repeating theodolite* is so designed that successive measures of a horizontal angle may be accumulated on the graduated circle and the total arc passed through then divided by the number of repetitions, as described in Art. 13-17. Usually the precision of a repeating theodolite corresponds to that used in third-order traversing or triangulation (Art. 25-4).

A *direction theodolite* is so designed that the graduated horizontal circle remains fixed during a series of observations, the direction to each point being read on the circle only once, by means of micrometer microscopes, in terms of the angle from the zero of the circle. Several sets of observations are made, with the circle shifted between sets so that each direction is measured on different parts of the circle. The direction technique, the precision in reading the circle, the high precision of manufacture of the circle, and—for American instruments—the use of a relatively large circle result in greater precision of angular measurement with the direction instrument than with the repeating instrument. The direction instrument is used almost exclusively in first-order and second-order triangulation. It is mounted on a relatively stable base.

13-9. American Repeating Theodolite. A repeating theodolite of American manufacture is similar to the conventional engineer's transit but is larger and is of a higher grade of workmanship. Its horizontal circle is larger and more finely graduated, and its levels are more sensitive. Attached magnifying glasses are used for reading the verniers, which usually read to 10″. Ordinarily the instrument has no compass, and an optical plummet is used instead of a plumb line. A repeating vernier theodolite with 7-in. horizontal circle is shown in Fig. 13-7.

13-10. American Direction Theodolite. The Parkhurst direction theodolite, developed and long used by the U.S. Coast and Geodetic Survey, is shown in Fig. 13-8. The 9-in. horizontal plate has a single tangent motion, comparable to the upper motion of the ordinary transit, except that in the direction theodolite the graduated circle can be rotated while the telescope is clamped in a fixed direction. When angles are to be measured about a point, an initial circle reading is made when the instrument has been pointed on the first distant signal. This initial reading is a measure of the azimuth, or direction, of the first object sighted with respect to some (undefined) reference meridian the direction of which depends entirely on the chance position of the plate when it is fixed before sighting. The directions of all distant stations are then read successively without disturbing the horizontal circle, and from these readings the values of the angles (from one station to the next) are computed.

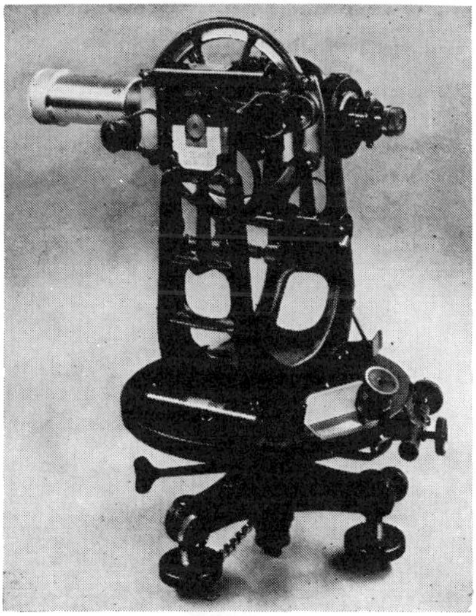

FIG. 13-7. Repeating vernier theodolite with 7-in. horizontal circle. (*U.S. Coast and Geodetic Survey.*)

FIG. 13-8. Nine-inch Parkhurst direction theodolite. (*U.S. Coast and Geodetic Survey.*)

Instead of a vernier, a micrometer microscope mounted on each side of the instrument is used to read the subdivisions of the graduated circle. The microscope has in the focal plane two closely spaced wires mounted on a movable slide. The slide is moved by means of a milled thumb-screw carrying a graduated drum called the micrometer head. When the telescope has been pointed toward an object and the horizontal motion has been clamped, the index or fiducial line of the micrometer lies ordinarily between two circle graduations. To determine the fractional part of a space, the line is moved until it coincides with a scale division, then the micrometer head is read. The direction is found by combining the micrometer reading and the scale reading. The micrometer may usually be read directly to 01″ and by estimation to 0.1″.

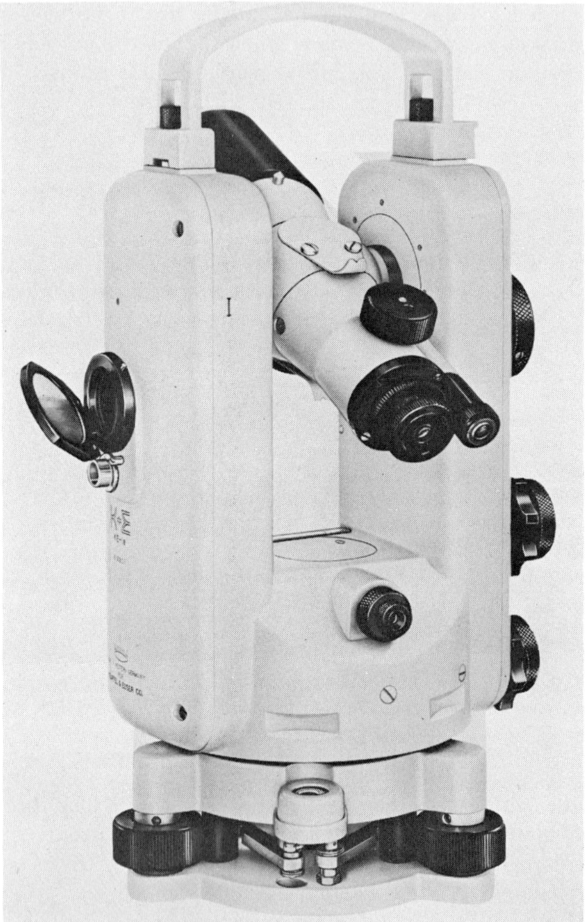

FIG. 13-9. Askania repeating theodolite. (*Keuffel & Esser Co.*)

13-11. European Theodolites. Theodolites of European manufacture combine a high degree of precision with facility of operation and lightness of weight. Their use is increasing in the United States not only for precise work but also for the ordinary operations of surveying.

In general construction and appearance, European direction theodolites are similar to repeating theodolites of the same manufacturer. The difference lies principally in the mounting of the horizontal circle; in the direction type the horizontal circle is fixed whereas in the repeating type it can be rotated (see Art. 13-8). Furthermore, usually the direction type is designed for

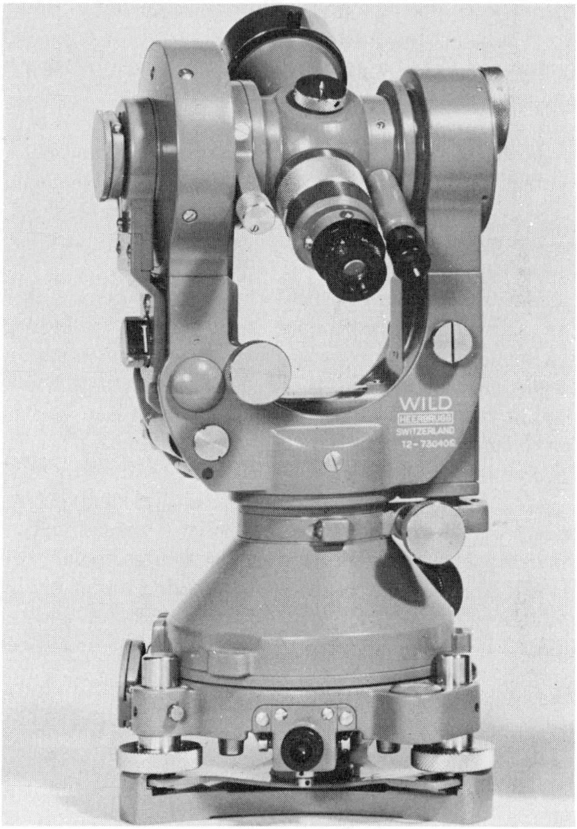

FIG. 13-10. Wild direction theodolite.

measurements of higher precision than those made with the repeating type. Two forms of European theodolites are shown in Figs. 13-9 and 13-10.

Although there are differences as among the various manufacturers, certain general features of European theodolites are fairly common. The weight (without tripod) is about 10 lb. as compared with about 15 lb. for the conventional American transit; the difference is due both to compact size and to the use of light materials in manufacture. The finish is light in color in order to minimize temperature effects of sunlight. There is no vernier plate, and the center is cylindrical and rotates on ball bearings. The plate level is circular and of moderate sensitivity, and the telescope level is tubular and of high sensitivity. There are three leveling screws. An optical plummet is used instead of a plumb line. All motions are enclosed, so that the instrument is dustproof and moistureproof. Interior lighting is accomplished by means of mirrors and prisms, and provision is made for night lighting. The only field adjustments are those of the levels, which adjustments are similar to those for the conventional American transit.

The telescope of the European theodolite is short and can be rotated completely about the horizontal axis. It is of the internal-focusing type, so that the stadia constant (Art. 15-7) is zero. The reticule is of glass with etched crosslines, and the stadia interval factor (Art. 15-8) is fixed at 100.0. For European use the eyepiece is of the inverting type; for American use it is usually of the erecting type.

The graduated horizontal and vertical circles are of glass and are relatively small; they are viewed simultaneously from the eye end of the telescope by means of a system of microscopes, prisms, and mirrors. Likewise the levels are viewed from the eye end of the instrument; it is not necessary for the observer to move around during an observation. The horizontal circle is read by means of an optical micrometer, in accordance with systems devised by the various manufacturers. Each reading is obtained by one observation which is the average of two readings at opposite sides of the circle and which is therefore free from errors due to eccentricity. Direction theodolites are usually read directly to 01″ and repeating theodolites to 20″ or 01′ and are read by estimation to one-tenth of the corresponding direct reading.

Some manufacturers produce instruments on which the theodolite, a sighting target, or a subtense bar (Art. 15-1) may be mounted interchangeably.

USE OF THE TRANSIT

13-12. General. The succeeding articles describe the elementary processes employed in running lines and in measuring horizontal and vertical angles with the transit. Transit surveys are considered in detail in the following chapter.

The process of taking magnetic bearings with the transit is the same as with the surveyor's compass. The transit may be employed for running direct levels in the same manner as with the engineer's level, the telescope level bubble being centered each time a rod reading is taken.

The operation of reversing the telescope by rotating it about the horizontal axis is called "plunging the telescope." When the telescope level tube is below, the telescope is said to be in the *normal*, or *direct*, position; when the level tube is above, the telescope is said to be in the *inverted*, or *reversed*, position.

Signals generally applicable to transit work are given in Art. 3-10. Suggestions for the care and handling of the transit are given in Art. 3-11.

13-13. Setting up Transit. Ordinarily the transit is set over a definite point, such as a tack in a stake. For centering the transit, a plumb line is suspended from the hook and chain beneath the instrument. First the transit is placed approximately over the point, with the telescope near the observer's eye level. Each tripod leg is then moved as required to bring the plumb bob within about ¼ in. of being over the tack, with the foot plate nearly level and with the shoe of each tripod leg pressed firmly into the ground. The instrument is leveled approximately by means of the leveling screws. Then two adjacent leveling screws are loosened, and the instrument is shifted laterally until the plumb bob is exactly over the tack. The length of the plumb line is changed as necessary to make the bob just clear the tack. The leveling screws are tightened to a firm, but not tight, bearing. The instrument is leveled by means of the leveling screws and the plate levels, each level tube being first brought parallel to a pair of opposite leveling screws (Art. 2-8). Both bubbles are brought approximately to center, and then each bubble is centered carefully. The telescope is tested for parallax (Art. 2-10) before observations are begun.

The operation of setting up and leveling the transit expeditiously requires on the part of the instrumentman a skill that is acquired only with practice.

Just before the transit is taken up, the instrument is centered on the foot plate, the leveling screws are roughly equalized, the upper motion is clamped, the lower motion is either unclamped or is clamped lightly, and the telescope is pointed vertically up and is clamped lightly.

13-14. Measuring Horizontal Angle. If a horizontal angle, as AOB, is to be measured, the transit is set up over O. The upper motion is clamped, with one of the horizontal verniers near zero, and by means of the upper tangent-screw one vernier is set at 0°. The telescope is sighted approximately to A, the lower motion is clamped, and by turning the lower tangent-screw the line of sight is set exactly on a range pole or other object marking the point. The upper clamp is loosened, and the telescope is turned until the line of sight cuts B. The upper clamp is tightened, and the line of sight is set exactly on B by turning the upper tangent-screw. The reading of the vernier which was initially set at 0° gives the value of the angle. It is convenient to consider the lower motion of the transit as a protractor, and the upper motion as a straightedge.

Following is a list of suggestions:

1. Make reasonably close settings by hand so that the tangent-screws will not need to be turned through more than one or two revolutions.
2. Make the last movement of the tangent-screw clockwise, thus compressing the opposing spring (see Art. 2-19).
3. When reading the vernier, have the eye directly over the coinciding graduation, to avoid parallax. It is also helpful to observe that the graduations on both sides of those coinciding fail to concur by the same amount.

4. As a check on the reading of one vernier, the other vernier may be read also. Or, check readings may be taken at each end of the vernier scale; these differ from the vernier reading by a value which is constant for the given type of vernier.

5. The plate bubbles should be centered before measuring an angle, but between initial and final settings of the line of sight the leveling screws should not be disturbed. When an angle is being measured by repetition (Art. 13-17) the plate may be releveled after each turning of the angle before again sighting on the initial point.

6. The flagman should stand directly behind the range pole, holding it lightly with the fingers of both hands, and balancing it on the tack or other mark indicating the point.

7. In sighting at a range pole the bottom of which is not visible, particular care should be taken to see that it is held vertical. When the view is obstructed for a considerable distance above the point to which the sight is taken, use a plumb line behind which a white card is held. For short sights a pencil or ruler held on the point makes a satisfactory target. Where the lighting is poor, the sight may be taken on a flashlight.

8. When a number of angles are to be observed from one point without moving the horizontal circle, the instrumentman should sight at some clearly defined object that will serve as a reference mark and should observe the angle. If occasionally the angle to the reference mark is read again, any accidental movement of the horizontal circle will be detected.

9. Whenever an angle is doubled, if the instrument is in adjustment, the two readings should not differ by more than the least count of the vernier. A greater discrepancy, if confirmed by repeating the measurement, will indicate that the instrument is out of adjustment.

13-15. Laying off Horizontal Angle. If an angle AOB is to be laid off from a line OA, the transit is set up at O, one vernier is set at 0°, and the line of sight is set on A. The upper clamp is loosened, and the vernier plate is turned until the index of the vernier is approximately at the required angle. The upper clamp is tightened, and the vernier is set exactly at the given angle by means of the upper tangent-screw. The point B is then established on the line of sight.

13-16. Common Mistakes. In measuring horizontal angles, mistakes often made are:

1. Turning the wrong tangent-screw.
2. Failing to tighten the clamp.
3. Reading numbers on the horizontal scale from the wrong row.
4. Reading angles in the wrong direction.
5. Dropping 30′ or 20′ by failure to take the full-scale reading before reading the vernier, for example, with a circle graduated to 30′ calling an angle 21°14′ when it is actually 21°44′, the vernier reading being 14′.
6. Reading the vernier in the wrong direction.
7. Reading the wrong vernier.

13-17. Measuring Angle by Repetition. By means of the transit, a horizontal angle may be mechanically multiplied and the product can be read with the same precision as the single value. Thus, with the ordinary transit having verniers reading to single minutes, an angle for which the true value is between the limits 30°00′30″ and 30°01′30″ will be read as 30°01′ and the limits of possible error will be ±30″. If the angle is multiplied six times on the horizontal circle, the product, likewise read to the nearest minute, might be 180°04′, its true value being within the limits 180°03′30″ and 180°04′30″; the limits of possible error, as far as reading the vernier is concerned, will likewise be ±30″. Dividing the observed product 180°04′ by 6, the single value becomes 30°00′40″ for which the limits of possible "reading" error are ±30″/6 = ±05″. This method of determining an angle is called *measurement by repetition.* The precision with which an angle can be measured by this method increases directly with the number of times the angle is multiplied or repeated up to six or eight; beyond this number the precision is not appreciably increased by further repetition on account of lost motion in the instrument and on account of accidental errors such as those due to setting the line of sight.

To repeat an angle, as AOB, the transit is set up at O and the single value of the angle is observed as previously described. The vernier setting is left unaltered, the instrument is turned on its lower motion, and a second sight is taken to the first point, as A. The upper clamp is loosened, and the telescope is again sighted to B. The angle has now been doubled. In this way the process is continued until the angle has been multiplied the desired number of times. The vernier is read, and the value of the angle is determined by dividing the difference between initial and final readings by the number of times the angle was turned. To obviate mistakes, this value is compared with the angle observed at the completion of the first turn.

The exact procedure to be employed in measuring an angle by repetition depends somewhat upon the desired precision. When the method is employed primarily as a check, the angle is doubled usually without reversing the telescope between repetitions.

When it is desired to increase the precision a moderate amount, usually the angle is multiplied four or six times, half of the observations being made with the telescope in the normal position and half with it in the inverted position, and both verniers are read. Certain instrumental errors, such as those due to eccentricity and to nonadjustment of the horizontal axis, are eliminated in this manner.

When a high degree of precision is necessary, several sets of perhaps five repetitions are taken with the telescope normal, and these sets are duplicated by others taken with the telescope inverted. To eliminate errors of graduation, settings are so made that readings are distributed over various parts of the circle and verniers; and to eliminate eccentricity, both verniers are read. Furthermore, special care is taken to manipulate the instrument in such a way that systematic errors due to lost motion in the clamps and to other causes will be eliminated.

If precise results are to be obtained, the instrument must be manipulated with care. The plate bubbles should be kept centered, but the leveling screws should not be disturbed except between repetitions. When turning on the lower motion, the hands should be in contact with the lower plate, and when turning on the upper motion, they should be in contact with the upper plate, and not the telescope. The last motion of the tangent-screws should be clockwise or against the opposing spring. To eliminate the effect of twist in the tripod, after each repetition the instrument should be rotated

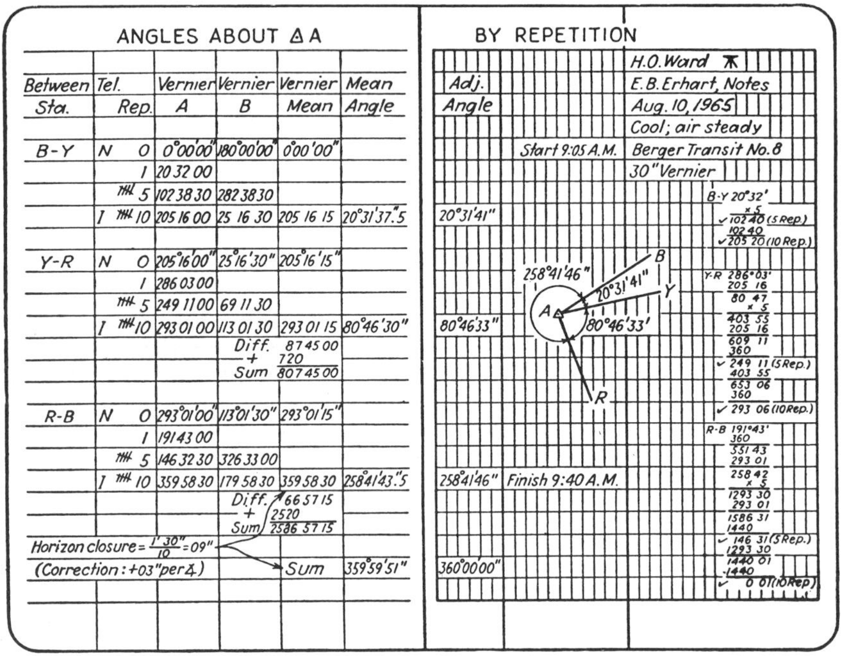

FIG. 13-11. Notes for measuring angles by repetition.

on its lower motion in the same direction that it was turned on its upper motion; that is, the direction of movement should be either always clockwise or always counter-clockwise. Owing to the possibilities of unequal settlement of the tripod and of unequal expansion of the parts of the telescope, it is desirable that the observations be made as rapidly as consistent with careful work. So far as possible, the instrument should be protected from sun and wind.

Sample notes for measuring the angles about a point by repetition are shown in Fig. 13-11. For each angle, five "repetitions" are taken with telescope normal and five with telescope inverted, always measuring clockwise.

The vernier is set at zero at the beginning but not thereafter; the error of closure (called the "horizon closure") is thus obtained directly as a check on the computations, and errors in setting the vernier are avoided. Rough computations on the right-hand page serve as a check on the number of repetitions and detect appreciable mistakes in turning the wrong tangent-screw. The recorded value for five repetitions is used only as a check; and the B vernier is used only as a check, except with regard to the number of seconds. The final adjusted values of the angles (to the nearest second) are recorded on the sketch, for ready reference in further computations.

13-18. Laying Off Angle by Repetition. If it is desired to establish an angle with a precision greater than that possible by a single observation, the methods of the preceding article may be employed in the following manner: In Fig. 13-12, OA represents a fixed line and AOB the angle which is to be laid off to establish the line OB. The transit is set up at O, the vernier is set at 0°, and a sight is taken to A. The vernier is set as closely as possible to the given angle and a trial point B' is established with the line of sight in its new position. The angle AOB' is then measured by repetition, and the line OB' is measured. The angle AOB' must be corrected by an angular amount $B'OB$ to establish the correct angle AOB. The correction, which is too small to be laid off accurately by angular measurement, is applied by offsetting the distance $B'B = OB'$ tan (or sin) $B'OB$, thus establishing the point B beside B'. It is convenient to remember that the tangent or sine of $1' = 0.0003$ (very nearly). As a check, the angle AOB is measured by repetition.

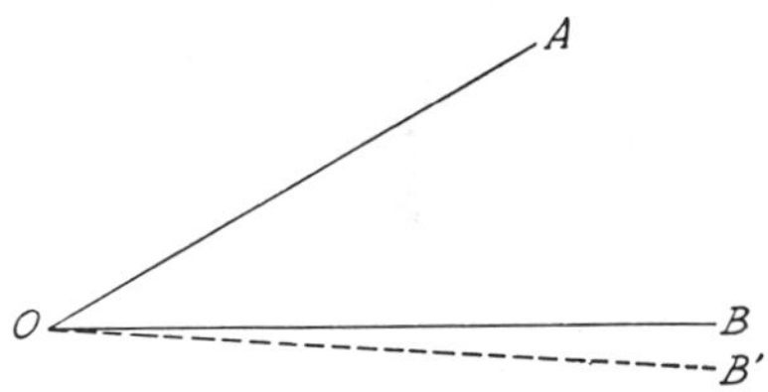

FIG. 13-12.

Example: Suppose that an angle of 30°00′ correct to the nearest 05″ is to be laid off and that the transit to be employed reads to the nearest 01′. Let the total value of AOB' after six repetitions be 180°02′, correct, say, to the nearest 30″. Then the measured value of AOB' is 180°02′ ÷ 6 = 30°00′20″ correct to the nearest 05″, and the correction to be applied to AOB' is 20″. Suppose that $OB' = 400$ ft. Then the length of the offset $B'B$ equals tan 20″ × 400 ft. = 0.0001 × 400 = 0.04 ft.

13-19. Measuring Vertical Angle. The vertical angle to a point is its angle of elevation (+) or depression (−) from the horizontal. The transit is set up and leveled as when measuring horizontal angles.

For a transit having a fixed vertical vernier, the plate bubbles should be centered carefully. The telescope is sighted approximately at the point, and the horizontal axis is clamped. The horizontal cross-hair is set exactly on the point by turning the telescope tangent-screw, and the angle is read by means of the vertical vernier.

For a transit having a movable vertical vernier with control level, the telescope is sighted on the point as described above, the vernier control bubble is centered, and the angle is read.

In ordinary trigonometric leveling, vertical angles are taken by sighting usually at a leveling rod, the line of sight being directed at a rod reading equal to the height of the horizontal axis of the transit above the station over which the transit is set up. In precise trigonometric leveling, the distance between stations is usually great, and vertical angles are measured with a theodolite by sighting at points defined by signals erected at the distant stations.

For leveling with the transit, for astronomical observations, or for measurement of horizontal angles requiring steeply inclined sights, usually it is desired to level the transit with greater precision than that which is possible through the use of the plate levels, and in such cases the vertical axis is made truly vertical by means of the telescope level as follows: First the transit is leveled by means of the plate levels in the usual manner. With the telescope over one pair of opposite leveling screws, the bubble of the telescope level is centered by means of the telescope tangent-screw. The telescope is rotated end for end about the vertical axis; then the bubble is brought halfway back to center by means of the leveling screws, the plate levels being disregarded. The process is repeated alternately for both pairs of opposite leveling screws until the bubble of the telescope level remains centered for any direction of pointing.

13-20. Double-sighting. For a transit having a full vertical circle, sights to determine vertical angles can be taken with the telescope either normal or inverted. The method of *double-sighting* consists in reading once with the telescope normal and once with it inverted, and taking the mean of the two values thus obtained. It eliminates the effect of certain instrumental errors (Art. 13-32) and reduces the personal error of observation.

The method of double-sighting is used, for example, in astronomical observations and in similar measurements of vertical angles to distant objects. In traversing, a similar result is obtained by measuring the vertical angle of each traverse line from each end, with the telescope the *same side up* for the two observations, and taking the mean of the two values thus obtained.

13-21. Index Error. *Index error* is the error in an observed angle due to (1) lack of parallelism between the line of sight and the axis of the telescope level, (2) displacement (lack of adjustment) of the vertical vernier, and/or (3) for a transit having a fixed vertical vernier, inclination of the vertical axis. If the instrument were in perfect adjustment and were leveled perfectly for each observation, there would be no index error; however, in practice these conditions seldom exist.

The effect of index errors due to lack of adjustment of the instrument can be eliminated either by double-sighting for each observation or by applying to each observation a correction determined (by double-sighting) for the instrument in its given condition of adjustment. For the common type of transit having a fixed vertical vernier, the effect of imperfect leveling cannot be eliminated by double-sighting, but—provided the line of sight is in adjustment—for each direction of pointing a correction can be determined (as described later) and applied. Often it is more convenient to apply the correction than to ensure that the instrument is perfectly adjusted and leveled.

The index correction is equal in amount but opposite in sign to the index error. Thus, if the observed vertical angle is $+12°14'$ and if the index error is determined to be $+02'$, the correct value of the angle is

$$+12°14' - 02' = +12°12'$$

Methods of determining the index error (and, therefore, the correction) are given in the following paragraphs:

1. *Lack of Parallelism between Line of Sight and Axis of Telescope Level.* If the axis of the telescope level is not parallel to the line of sight and if the vertical vernier reads zero when the bubble is centered (Fig. 13-13*a*), an error in vertical angle results. This error can be rendered negligible for ordinary work by careful adjustment of the instrument (Art. 13-30, adjust-

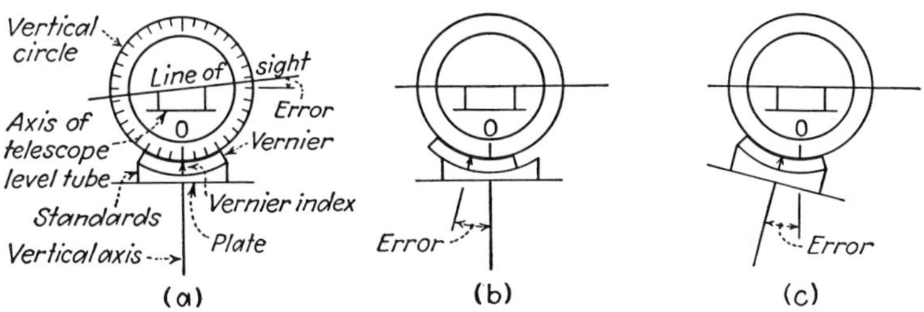

FIG. 13-13. Sources of error in measurement of vertical angles.

ment 5). The combined error due to this cause and to displacement of the vertical vernier (see following paragraph) can be eliminated by double-sighting. The index error due to the two causes can be determined by comparing a single reading on any given point with the mean of the two readings obtained by double-sighting to the same point. Thus, if the observed vertical angle to a point is $+2°58'30''$ with telescope normal and is $+2°55'30''$ with telescope reversed, the index error for readings with telescope normal is

$$(+2°58'30'' - 2°55'30'')/2 = +1'30''$$

2. *Displacement of Vertical Vernier.* Displacement of the vertical vernier (Fig. 13-13*b*) introduces a constant index error. The error can be rendered negligible by careful adjustment (Art. 13-30, adjustments 6 and 6*a*). The

combined index error due to this cause and to lack of parallelism between the line of sight and the axis of the telescope level can be eliminated by double-sighting; or the combined error can be determined as described in the preceding paragraph. For a transit having a fixed vertical vernier, the error due to displacement of the vertical vernier alone can be determined —provided the line of sight is in adjustment—by leveling the transit carefully, leveling the telescope, and reading the vertical vernier. For a transit having a movable vertical vernier with control level, the error due to displacement of the vertical vernier alone can be determined by leveling both the telescope level and the vernier level, and reading the vertical vernier.

3. *Inclination of Vertical Axis.* For a transit having a fixed vertical vernier, any inclination of the vertical axis (Fig. 13-13*c*) due to erroneous leveling of the instrument introduces an index error which varies with the direction in which the telescope is pointed and which is equal in amount to the angle through which the fixed vertical vernier is displaced about the horizontal axis while the instrument is directed toward the point. This index error can be rendered negligible by careful leveling of the transit before each observation, making sure that the plate-level bubbles remain in position for any direction of pointing. It is not eliminated by double-sighting, since the condition causing the error is not changed by reversal (and plunging) of the instrument (see Fig. 13-13*c*). If the line of sight and the vertical vernier are in adjustment, the index error due to inclination of the vertical axis alone can be determined for each direction of pointing by leveling the telescope and reading the vertical vernier.

When a series of horizontal and vertical angles is to be measured from a given station, recentering the plate bubbles necessitates taking a new backsight (and thus additional work) before additional horizontal angles can be measured correctly, yet the plate bubbles may be considerably displaced without appreciably affecting the correctness of *horizontal* angles. Often it involves less work to make the index correction for vertical angles than to relevel the instrument each time the plate-level bubbles are seen to be displaced.

For a transit having a movable vertical vernier with control level, any moderate inclination of the vertical axis does not introduce an appreciable error in vertical angles, provided the instrument is in adjustment and provided the vernier control-level bubble is centered each time an observation is made. On topographic surveys or similar work where many horizontal and vertical angles are to be observed, the use of the movable vertical vernier with control level results in a considerable saving of time.

13-22. Prolonging Straight Line. If a straight line as *AB* (Fig. 13-14) is to be prolonged to *P* (not already defined on the ground), which is beyond the limit of sighting distance or is invisible from *A* and *B*, the line is extended by establishing a succession of stations *C*, *D*, etc., each of which is occupied by the transit. Any of the following three methods may be employed, but the second method is most common and usually most convenient. Lines may also be prolonged by means of a prismatic sighting device, with lower precision than that obtainable with the transit.

Method 1. The transit is set up at A, a sight is taken to B, and a point C is established on a line beyond B. The transit is moved to B, a sight is taken to C, and point D is set on line beyond C. Thus the process is continued until the desired distance is traversed and point P is set.

Method 2. The transit is set up at B, and a backsight is taken to A. With both upper and lower motions clamped, the telescope is plunged, and a point C is set on line. If the line of sight is perpendicular to the horizontal axis, as it will be if the instrument is in perfect adjustment, it will generate a vertical plane as the telescope is revolved, and the point C will lie on the

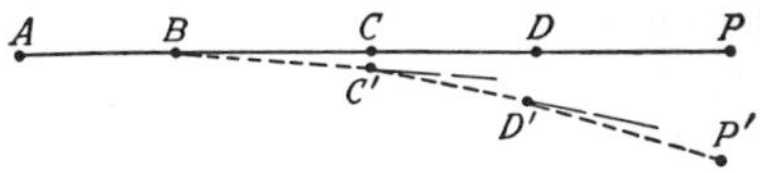

FIG. 13-14. Prolonging a straight line.

prolongation of AB. The transit is moved to C, a backsight is taken to B, a point D is set in similar manner, and thus the process is repeated until point P is set.

If the line of sight is not perpendicular to the horizontal axis of the transit, as the telescope is plunged (say from the inverted to the normal position), the line of sight will generate a portion of a cone whose vertex is at the center of the instrument and two of whose elements are AB and BC', and C' will not lie on the true prolongation of AB. If the instrument is set up at C', a backsight taken to B with the telescope inverted as before, and the telescope plunged to the normal position, a second and similar cone is generated and D' will not lie on the prolongation of BC'. Thus, if the line is extended by the method outlined and all backsights are taken with the telescope in one position (either normal or inverted), the points established will lie along a curve instead of a straight line, and each segment of the line will be deflected in the same direction (to the right or to the left) by double the error of adjustment of the line of sight. On the other hand if, say, at the even-numbered stations B, D, F, etc., backsights were taken with the telescope inverted and at odd-numbered stations, C, E, etc., backsights were taken with the telescope normal, a zigzag line would be established with some points on one side of the line joining the terminals and some perhaps on the other. By the first procedure, the angular error becomes systematic in character, and by the second procedure it becomes accidental. Generally where only a few setups are required and the instrument is known to be in reasonably good adjustment, no particular attention is paid to the procedure to be followed; but where there is a large number of set-ups and the line is long, the latter procedure is employed.

Method 3. This method, known as "double-sighting," or "double-centering," is employed if the instrument is in poor adjustment or if it is desired to establish the line with high precision. If the line AB (Fig. 13-15) is to be prolonged to some point P, the transit is set up at B and a backsight is taken to A with the telescope in, say, its *normal* position. The telescope is plunged,

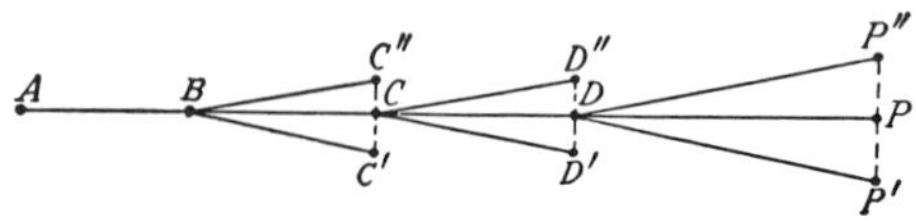

FIG. 13-15. Double-sighting to prolong line.

and a point C' is set on line. The transit is then revolved about its vertical axis, and a second backsight is taken to A with telescope *inverted*. The telescope is plunged, and a point C'' is established on line beside C'. It is evident that C' will be as far on one side of the true prolongation of AB as C'' is on the other. Midway between C' and C'', a point C is set defining a point on the correct prolongation of AB. In a similar manner the next point D is established by setting up at C, double-sighting to B, and setting points at D', D'', and D. Thus the process is repeated until the desired distance is traversed.

13-23. Prolonging a Line past an Obstacle. Figure 13-16 illustrates one method of prolonging a line AB past an obstacle where the offset space is limited. The transit is set up at A, a right angle is turned, and a point C is established at a convenient distance from A. Similarly the point D is established, the distance BD being made equal to AC. The line CD, which

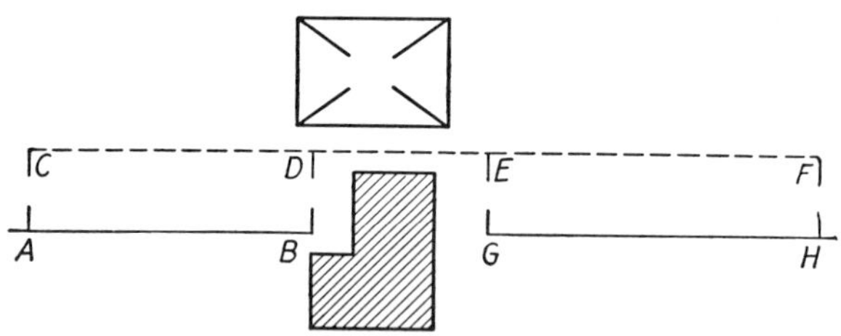

FIG. 13-16. Prolonging line past obstacle by perpendicular offsets.

is parallel to AB, is prolonged; and points E and F are established in convenient locations beyond the obstacle. From E and F right-angle offsets are made, and G and H are set as were C and D; GH then defines the prolongation of AB. The distance AH is determined by measuring the length of the lines AB, DE, and GH. If the chainage is to be carried forward with precision, it is necessary to erect the perpendiculars AC, BD, etc. with greater than ordinary care; and if the line is to be prolonged precisely, it is essential not only that the offset distances be measured carefully but also that *AB* and *EF*, the distances between offsets, be as long as practicable.

Another method of prolonging a line *AB* past an obstacle is illustrated by Fig. 13-17. A small angle α is turned off at *B*, and the line is prolonged to some convenient point *C* which will enable the obstacle to be cleared. At *C*, the angle 2α is turned off in the reverse direction, and the line is prolonged to *D*, with *CD* made equal to *BC*. The point *D* is then on the prolongation of *AB*; and *DE*, the further prolongation of *AB*, is established by turning off the angle α at *D*. If there were another obstacle between *D* and *E*, as there might often be in wooded country, the line *CD* is prolonged to some point, as *F*, from which the obstacle can be cleared; and so a zigzag course is followed until it is possible to resume traversing on the direct prolongation of the main line *AB*. As compared with the method of perpendicular offsets, this method is more convenient in the field, but it requires computation to determine the length *BD*. However, if the angle α is small, say, not

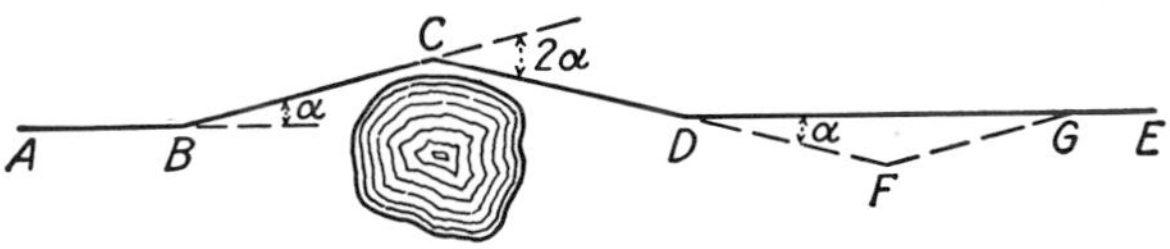

FIG. 13-17. Prolonging line past obstacle by angles.

greater than a degree or so, often it will be sufficiently precise to take the distance along the main line as equal to that along the auxiliary lines.

13-24. Running a Straight Line between Two Points. If the terminal points *A* and *B* of a line are fixed and it is desired to establish intervening points on the straight line joining the terminals, the method to be employed depends upon the length of the line and the character of the terrain. Three common cases are considered below:

Case 1. *Terminals Intervisible.* The transit is set up at *A*, a sight is taken to *B*, and intervening points are established on line.

If the intervening points thus established were to lie in the same plane with the center of the instrument and the terminal point *B*, they would define a truly straight line regardless of whether or not the horizontal axis of the transit were truly horizontal. If the horizontal axis is inclined with the horizontal, the line of sight will not generate a vertical plane as the telescope is revolved; and thus if it is necessary to rotate the telescope about the horizontal axis in order to set the intervening points, the points thus established will not lie on a truly straight line (as seen in plan view) joining the terminals.

Ordinarily the vertical angles through which it is necessary to rotate the telescope will be small, and if the horizontal axis is in fair adjustment and the plate bubbles are centered, the error arising from this source is negligible.

Occasionally, however, when the intervening points are to be set with high precision or when the adjustment of the instrument is uncertain and the vertical angles are large, the intervening stations are set by double-sighting.

Case 2. Terminals Not Intervisible, but Visible from an Intervening Point on Line. The location of the line at the intervening point C is determined by trial, as follows: In Fig. 13-18, A and B represent the terminals both of which can be seen from the vicinity of C. The transit is set up on the estimated location of the line near C, a backsight is taken to A, the telescope is plunged, and the location of the line of sight at B is noted. The amount that the transit must be shifted laterally is estimated; and the process is repeated until, when the telescope is plunged, the line of sight falls on the point at B. This process is known as "balancing in." The location of the transit should then be tested by double-sighting; the test will also disclose whether or not the line of sight and the horizontal axis are in adjustment.

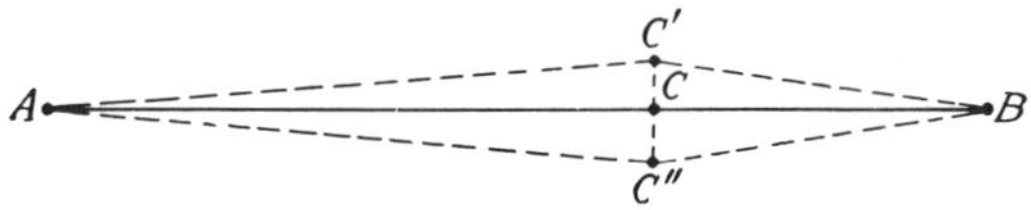

FIG. 13-18. Balancing in.

If the instrument were in perfect adjustment, its center would be on the true line joining AB. On the other hand, if the line of sight were not perpendicular to the horizontal axis, a cone would be generated by the line of sight when the telescope was plunged, as explained in Art. 13-22; also if the horizontal axis were not truly horizontal, the line of sight in its rotation would not generate a vertical plane, as explained under case 1. Hence, if the transit were not in adjustment, its center might be at C' and still the line of sight would bisect B when the telescope is plunged.

To locate the intermediate point truly on line, trials are made first with the telescope in, say, its normal position for backsights to A, until an intermediate point, as C', is determined. Then a second series of trials is made with the telescope inverted for backsights to A, until the corresponding point C'' is located. Then, for reasons previously explained, the true line is at C, halfway between C' and C''. Other intermediate points may then be established by setting up the transit at C and proceeding as in case 1.

Case 3. Terminals Not Visible from Any Intermediate Point. AB (Fig. 13-19) represents a straight line along which it is desired to establish intermediate points, the character of the ground being such that it is not possible to find any one intermediate location from which both terminals are visible. By one of the methods previously explained, a straight line, called a *random line,* is run from A in the estimated direction of B. In the figure, AX is such a random line, and C and D are stations established during the process of extending it. The transit is set up at D and sighted toward X. The tape is swung about B as a center, and the offset distance BE from the point B to the line DX is determined by sighting at the tape through the transit and taking the least reading (that is, by a swing offset).

The exact location of point E on the line AX is then established (see next paragraph), the length AE is measured, and the angle α which the random line makes with AB is computed by the equation $\tan \alpha = BE/AE$. The transit is again set up at A, a sight is taken to C, the computed value of α is laid off, and the line is run toward B, intermediate stations as F and G being established at desired points. With the transit at G, a backsight is taken to F, the telescope is plunged, and the linear offset error at B is noted. If the error is sufficiently large to be of importance, the points at F and G are corrected by linear measurements so as to place them on the true line, the correction being made proportional to the distance from the terminal A

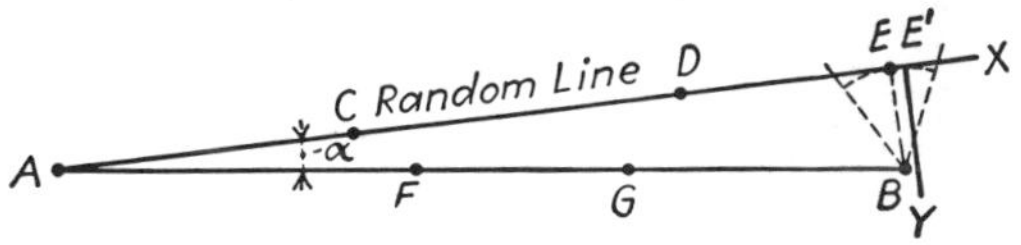

FIG. 13-19. Random line.

to the point. Thus the offset correction at F is to AF as the error at B is to AB. Points on the random line, as C and D, can be transferred to the true line by the same method.

If the offset distance BE is short compared with the length of the line AB, the location of E on the line may be estimated with sufficient precision. If the offset distance BE is long, E needs to be established by a less approximate method. The transit is set up at E', the estimated location of E; a perpendicular $E'Y$ is erected from the line AX; and the distance from this perpendicular $E'Y$ to B is measured by a swing offset. This offset gives the distance from E' to E along the line AX so that a perpendicular at E will pass through B.

The angle α which the random line makes with AB may be computed by the equation $\tan \alpha = BE/AE$. The transit is again set up at A, a sight is taken to C, the computed value of α is laid off, and the line is run toward B, intermediate stations as F and G being established at desired points. With the transit at G, a backsight is taken to F, the telescope is plunged, and the linear offset error at B is noted. If the error is sufficiently large to be of importance, the points at F and G are corrected by linear measurements so as to place them on the true line, the correction being made proportional to the distance from A to the point. Thus the offset correction at F is to AF as the error at B is to AB.

It is not necessary to compute the angle α, as the direction of AB can be established by an offset from the random line at any convenient distance from A. As α is usually very small, either of the approximate slope formulas (Art. 7-15) will usually yield sufficiently precise values of the length of AB.

13-25. Determining Inaccessible Distance. This involves triangulation (see Art. 12-18). Three simple methods are described below:

Method 1. AB (Fig. 13-20*a*) represents a line whose length cannot be measured directly, B being visible from A and vicinity. The transit is set up at A, a sight is taken to B, an angle of 90° is laid off, and C is set at any convenient point. The line AC is measured. The transit is set up at C, and the angle ACB is observed. Then $AB = AC \tan ACB$.

Method 2. An approximate method sometimes used in reconnaissance where the distant point A is accessible is as follows: With the transit at B (Fig. 13-20*a*) any convenient distance AC (as 100 ft.) is laid off from A, with the angle BAC made a right angle either by estimation or by one of the methods of Art. 7-27. The angle ABC is measured. Then $AB = AC \cot ABC$.

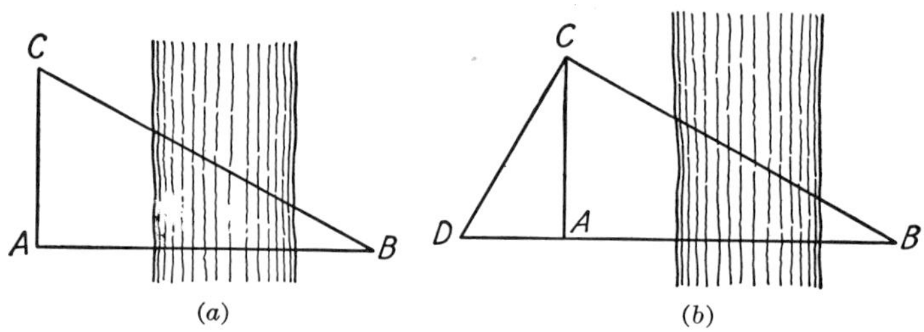

FIG. 13-20. Inaccessible distance.

Method 3. This method is applicable when trigonometric tables are not available. In Fig. 13-20*b*, AB represents the line whose length is to be determined. AC is established as in Method 1. The transit is set up at C, a sight is taken to B, and the direction of CD is fixed by laying off an angle of 90°. The point D is established at the intersection of this line and the prolongation of the line BA, as described in Art. 13-26. The lengths AC and AD are measured. By geometry ΔABC is similar to ΔACD. Hence $AB = (AC)^2/AD$.

For Methods 1 and 3 it is desirable that the distance AC be not less than one-half the distance AB; otherwise the errors of measurement will produce a relatively large error in the computed distance.

13-26. Intersection of Lines. The point of intersection of two lines as AB and CD (Fig. 13-21) is established as follows: One of the lines AB is prolonged (Art. 13-22), and points P' and P'' are established a short distance on opposite sides of the estimated location of the prolongation of CD. A string is stretched between P' and P''. The line CD is prolonged until it intersects the string at P.

13-27. Setting Monument. Often it is desired to set a subsurface monument to mark permanently a point on a transit survey. In such a case the location of the surface point established by the survey is well referenced (Art. 14-17), preferably by the intersection of two lines. When the monument is set and the subsurface mark is to be established, a string is stretched along each of the two reference lines. The location of the mark on the monument is projected below the surface by plumbing from the intersection of the two strings. If desired, a batter board or other frame may be set over the surface mark for the purpose of plumbing from a surface mark to a subsurface mark, or *vice versa*. Detailed information regarding the construction and setting of monuments is given in Ref. 2 of Chap. 25.

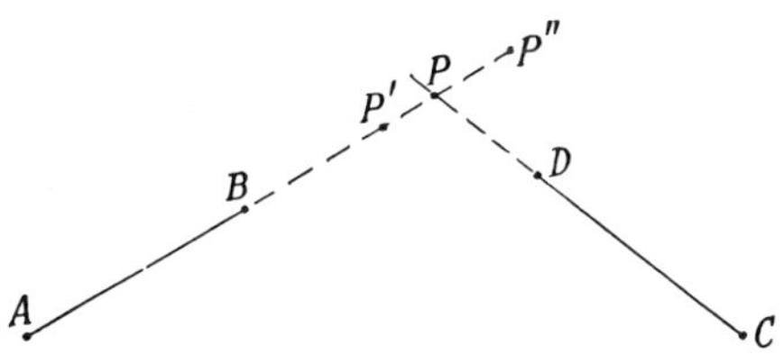

FIG. 13-21. Intersection of lines.

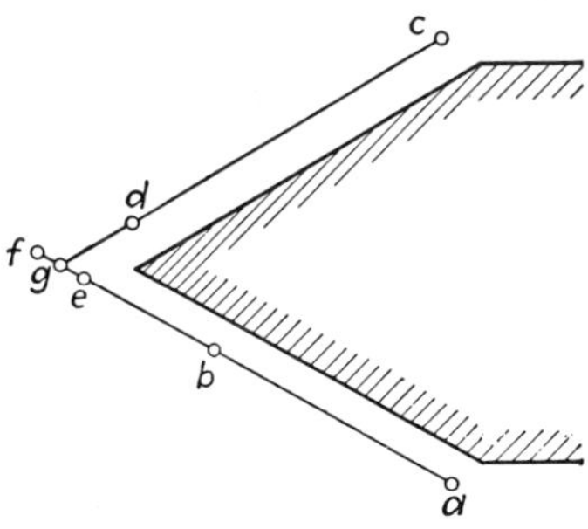

FIG. 13-22. Offset angle.

13-28. Measuring Angle When Transit Cannot Be Set at Vertex. Figure 13-22 illustrates a typical case where it is required to determine the angle between walls of a building or between fence lines.

The point a is established at any convenient distance from the wall. The perpendicular distance from the wall is determined by holding the tape on point a and swinging the end of the tape through an arc, varying the radius until the arc becomes tangent to the wall (that is, by a swing offset). Similarly a second point b is established at the same distance from the wall as a; then ab is parallel to the wall. In similar manner points c and d are established. The point of intersection g of lines ab and cd is determined as described in Art. 13-26. The transit is set up at g, and the angle agc, which is equal to the angle between the walls, is measured in the usual manner.

ADJUSTMENT OF CONVENTIONAL TRANSIT

13-29. Desired Relations. For a transit in perfect adjustment the relations stated in the following paragraphs should exist. The number of each paragraph is the same as that of the corresponding adjustment described in the following article. For adjustments 1 to 5, Fig. 13-23 shows the desired relations among the principal lines of the transit.

1. The vertical cross-hair should lie in a plane perpendicular to the horizontal axis so that any point on the hair may be employed when measuring horizontal angles or when running lines.

2. The axis of each plate level should lie in a plane perpendicular to the vertical axis so that when the instrument is leveled the vertical axis will be truly vertical; thus horizontal angles will be measured in a horizontal plane, and vertical angles will be measured without index error due to inclination of the vertical axis.

3. The line of sight should be perpendicular to the horizontal axis at its intersection with the vertical axis. Also, the optical axis, the axis of the objective slide, and the line of sight should coincide. If these conditions exist, when the telescope

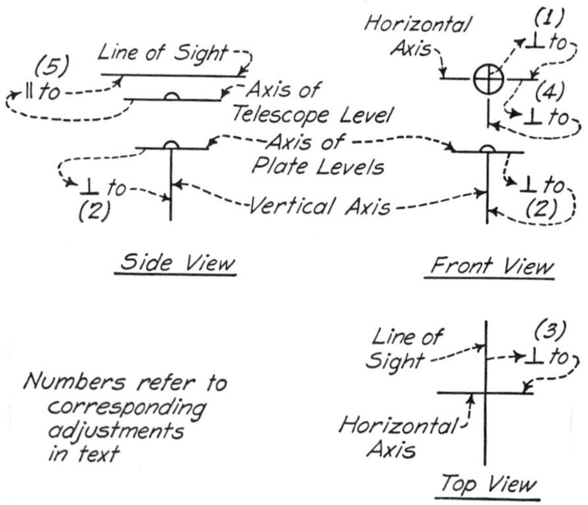

FIG. 13-23. Desired relations among principal lines of transit.

is rotated about the horizontal axis the line of sight will generate a plane when the objective is focused for either a near sight or a far sight, and that plane will pass through the vertical axis.

4. The horizontal axis should be perpendicular to the vertical axis so that when the telescope is plunged the line of sight will generate a vertical plane.

5. The axis of the telescope level should be parallel to the line of sight so that the transit may be employed in direct leveling and so that vertical angles may be measured without index error due to lack of parallelism.

6. If the transit has a fixed vernier for the vertical circle, the vernier should read zero when the plate bubbles and telescope bubble are centered, in order that vertical angles may be measured without index error due to displacement of the vernier.

6*a*. If the vertical vernier is movable and has a control level, the axis of the control level should be parallel to that of the telescope level when the vernier reads zero.

7. The optical axis and the line of sight should coincide (see 3, above).

8. The axis of the objective slide should be perpendicular to the horizontal axis (see 3, above).

9. The intersection of the cross-hairs should appear in the center of the field of view of the eyepiece.

10. If the transit is equipped with a striding level for the horizontal axis, the axis of the striding level should be parallel to the horizontal axis. Thus when the bubble of the striding level is centered and the instrument is plunged, the line of sight (if in adjustment) will generate a vertical plane.

13-30. Adjustments. In the description of the following adjustments (except 7 and 9) it is assumed that the objective slide does not admit of adjustment, but that it is permanently fixed in the telescope tube so far as lateral motion is concerned; and that the maker has so constructed the instrument that the optical axis and the axis of the objective slide coincide and are perpendicular to the horizontal axis. This ideal construction is never exactly attained; but in most instruments the departure is so slight that it need not be considered in ordinary transit work, and in precise surveying the resulting errors are eliminated by methods of procedure.

For those adjustments which involve sighting through the telescope, particular attention should be given to proper focusing of both the eyepiece and the objective prior to testing the adjustments.

The adjustments of the transit are more or less dependent upon one another. For this reason, if the instrument is badly out of adjustment, time will be saved by first making corrections roughly for related adjustments until all the tests have been tried and then repeating the tests and corrections in the same order. The plate levels will not be disturbed by other adjustments and should be corrected exactly before other adjustments are attempted.

Any movement of the screws controlling the cross-hair ring is likely to produce both lateral displacement and rotation of the ring; hence any considerable adjustment of the line of sight is likely to disturb the vertical hair so that it will no longer remain on a point when the telescope is rotated about the horizontal axis. The adjustment of the telescope level depends upon the unaltered position of the horizontal cross-hair and hence should not be tested until the line of sight and horizontal axis have been corrected.

The transit adjustments commonly made are 1 to 6*a* following. Adjustments 7 to 10 may be required occasionally for some instruments. Some general suggestions regarding adjustments are given in Art. 3-12.

1. *To Make the Vertical Cross-hair Lie in a Plane Perpendicular to the Horizontal Axis.*

Test. Sight the vertical cross-hair on a well-defined point not less than 200 ft. away. With both horizontal motions of the instrument clamped, swing the telescope through a small vertical angle, so that the point traverses the length of the vertical cross-hair. If the point appears to move continuously on the hair, the cross-hair lies in a plane perpendicular to the horizontal axis (see Fig. 13-24).

Correction. If the point appears to depart from the cross-hair, loosen two adjacent capstan screws and rotate the cross-hair ring in the telescope tube until the point traverses the entire length of the hair. Tighten the same two screws. This adjustment is similar to adjustment 2 of the dumpy level (Art. 8-23), with the terms *vertical* and *horizontal* interchanged.

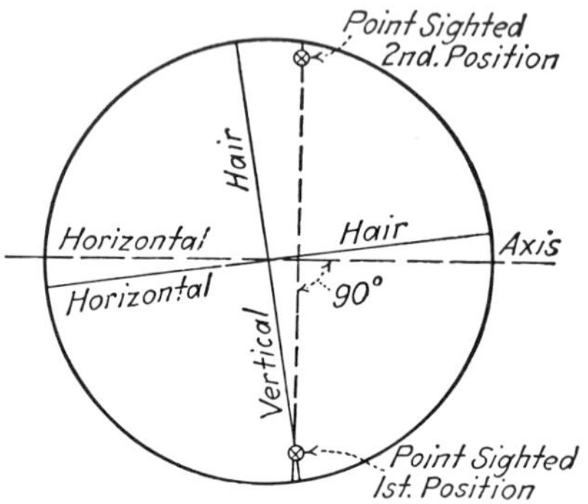

FIG. 13-24. Adjustment of vertical cross-hair of transit.

2. *To Make the Axis of Each Plate Level Lie in a Plane Perpendicular to the Vertical Axis.*

Test. Rotate the instrument about the vertical axis until each level tube is parallel to a pair of opposite leveling screws. Center the bubbles by means of the leveling screws. Rotate the instrument end for end about the vertical axis. If the bubbles remain centered, the axis of each level tube is in a plane perpendicular to the vertical axis (see Fig. 8-20).

Correction. If the bubbles become displaced, bring them *halfway* back by means of the adjusting screws. Level the instrument again and repeat the test to verify the results. This is the method of reversion (Art. 2-7).

3. *To Make the Line of Sight Perpendicular to the Horizontal Axis.*

Test. Level the instrument. Sight on a point *A* (see Fig. 13-25) about 500 ft. away, with telescope normal. With both horizontal motions of the

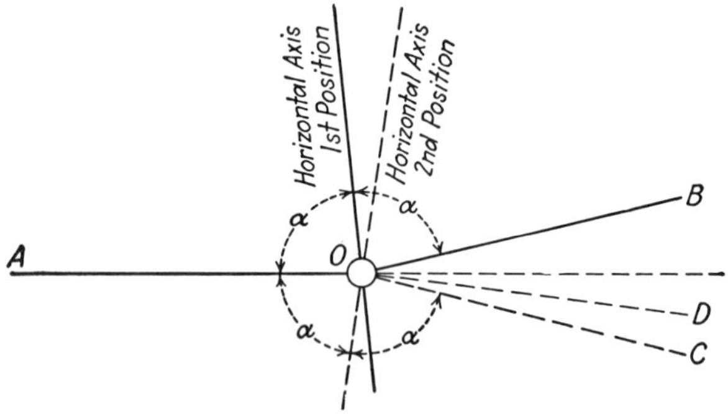

FIG. 13-25. Adjustment of line of sight of transit.

instrument clamped, plunge the telescope and set another point *B* on the line of sight and about the same distance away on the opposite side of the transit. Unclamp the upper motion, rotate the instrument end for end about the vertical axis, and again sight at A (with telescope inverted). Clamp the upper motion. Plunge the telescope as before; if B is on the line of sight, the desired relation exists.

Correction. If the line of sight does not fall on B, set a point C on the line of sight beside B. Mark a point D, *one quarter* of the distance from C to B, and adjust the cross-hair ring (by means of the two opposite horizontal screws) until the line of sight passes through D. The points sighted should be at about the same elevation as the transit.

4. *To Make the Horizontal Axis Perpendicular to the Vertical Axis.*

Test. Set up the transit near a building or other object on which is some well-defined point A at a considerable vertical angle. Level the instrument very carefully, thus making the vertical axis truly vertical. Sight at the high point A (see Fig. 13-26) and, with the horizontal motions clamped, depress the telescope and set a point B on or near the ground. If the horizontal axis is perpendicular to the vertical axis, A and B will be in the same vertical plane. Plunge the telescope, rotate the instrument end for end about the vertical axis, and again sight on A. Depress the telescope as before; if the line of sight falls on B, the horizontal axis is perpendicular to the vertical axis.

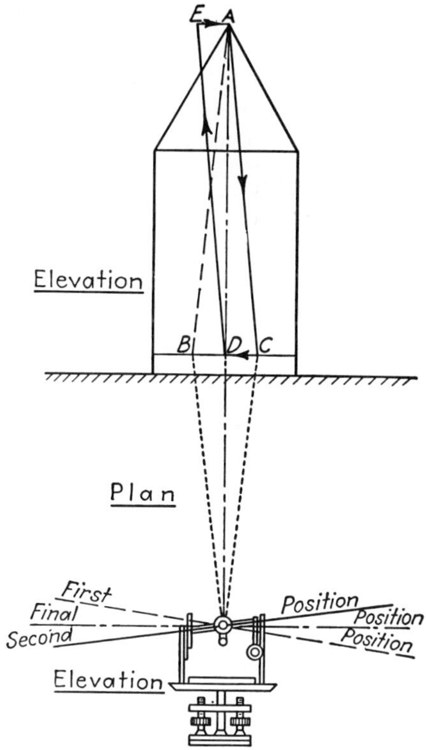

FIG. 13-26. Adjustment of horizontal axis of transit.

Correction. If the line of sight does not fall on B, set a point C on the line of sight beside B. A point D, halfway between B and C, will lie in the same vertical plane with A. Sight on D; elevate the telescope until the line of sight is beside A; loosen the screws of the bearing cap, and raise or lower the adjustable end of the horizontal axis until the line of sight is in the same vertical plane with A.

The high end of the horizontal axis is always on the same side of the vertical plane through the high point as the point last set.

In readjusting the bearing cap, care should be taken not to bind the horizontal axis, but it should not be left so loose as to allow the objective end of the telescope to drop of its own weight when not clamped.

5. *To Make the Axis of the Telescope Level Parallel to the Line of Sight.*

Test and Correction. Proceed the same as for the two-peg adjustment of the dumpy level (Art. 8-23, adjustment 3), except as follows: With the line of sight set on the rod reading established for a horizontal line, the correction is made by raising or lowering one end of the telescope level tube until the bubble is centered.

6. (For Transit Having a Fixed Vertical Vernier) *To Make the Vertical Circle Read Zero When the Telescope Bubble Is Centered.*

Test. With the plate bubbles centered, center the telescope bubble and read the vertical vernier.

Correction. If the vernier does not read zero, loosen it and move it until it reads zero. Care should be taken that the vernier will not bind on the vertical circle as the telescope is rotated about the horizontal axis.

6*a*. (For Transit Having a Movable Vertical Vernier with Control Level) *To Make the Axis of the Auxiliary Level Parallel to the Axis of the Telescope Level When the Vertical Vernier Reads Zero.*

Test. Center the telescope bubble, and by means of the vernier tangent-screw move the vertical vernier until it reads zero.

Correction. If the bubble of the control level which is attached to the vertical vernier is not at the center of the tube, bring it to the center by means of the capstan screws at one end of the tube.

13-30a. Special Adjustments. In addition to the usual adjustments just described, the following adjustments may be made as required by the type or the condition of the transit:

7. *To Make the Line of Sight, In So Far as Defined by the Horizontal Cross-hair, Coincide with the Optical Axis.*

Test. Set two pegs, one about 25 ft. and the other 300 or 400 ft. from the transit. With the vertical motion clamped, take a rod reading on the distant point, and without disturbing the vertical motion read the rod on the near point. Plunge the telescope, rotate the instrument about the vertical axis, and set the horizontal cross-hair at the last rod reading with the rod held on the near point. Sight to the distant point. If the desired relation exists, the first and last readings on the distant rod will be the same.

Correction. If there is a considerable difference between the rod readings, move the horizontal cross-hair by means of the upper and lower adjusting screws until it has apparently traversed over several times the apparent error. Repeat the process, gradually reducing the movement of the cross-hair as the rod readings to the distant point approach each other, until by successive approximations the error is reduced to zero. The rod when held on the near point should be read with great care, for a small difference in the position of the cross-hair on the near rod will be sufficient to indicate a considerable error on the distant rod.

If the transit has an erecting eyepiece which is permanently centered, adjustment 7 may usually be made with sufficient precision for ordinary direct or trigonometric leveling by simply moving the horizontal cross-hair until it appears in the center of the field of view. If the transit has an inverting eyepiece, the cross-hair ring limits the field of view, and the horizontal cross-hair will appear in the center whether or not it intersects the axis of the objective slide.

8. (For Transit Having an Adjustable Objective Slide) *To Make the Axis of the Objective Slide Perpendicular to the Horizontal Axis.* As stated in Art. 2-12, some telescopes have objective slides which move in adjustable rings. Ordinarily objective slides of this type require no further adjustment after leaving the factory, but it is well to test the adjustment occasionally, and the instrumentman should be able to make corrections if necessary.

The horizontal adjustment of the objective slide is made as follows: Having performed the adjustment of the vertical cross-hair (adjustment 3) for an average length of sight, focus the vertical cross-hair on a distant point. Move the objective out, and bring it to a focus on some well-defined point near the instrument. Plunge the telescope, rotate it about the vertical axis, and again set the vertical cross-hair on the near point. Sight toward the distant point. If the objective slide is in adjustment, the line of sight will strike the first point sighted. If the line of sight does not strike the distant point, move the ring controlling the objective slide by means of the screws on the sides of the telescope until by estimation the line of sight has moved one half of the apparent error at the distant point. The relation between the adjustments of the vertical cross-hair and optical axis is such that the adjustments must be repeated alternately until both are found to be correct.

The vertical adjustment of the objective slide may be performed in a similar manner with reference to the horizontal cross-hair, but it is usually best to make corrections with the horizontal cross-hair as described in adjustment 7, unless one is reasonably sure that the factory adjustment, through accident or otherwise, has been altered.

9. (For Transit Having an Adjustable Eyepiece Slide) *To Center the Eyepiece Slide.* A telescope which shows objects erect usually has an eyepiece slide, one end of which moves through an adjustable ring. When the transit has otherwise been adjusted, the cross-hairs may not appear in the center of the field of view owing to lack of coincidence of the axis of the eyepiece slide with the optical axis. This coincidence is a convenient relation but is unnecessary so far as the proper working of the transit is concerned. To center the slide, the adjustable ring is moved by means of four screws between the eye end of the telescope and the cross-hair ring.

10. (For Transit Equipped with Striding Level) *To Make the Axis of the Striding Level Parallel to the Horizontal Axis.*

Test. By means of the leveling screws center the striding-level bubble. Lift the level from its supports and turn it end for end. If it is in adjustment, the bubble will again be centered.

Correction. If the bubble is displaced, bring it halfway back to the center by means of the capstan screw at one end of the level tube (Fig. 8-20). Relevel the instrument by means of the leveling screws and repeat the test until the adjustment is perfected.

ERRORS IN TRANSIT WORK

13-31. General. Except in field astronomy, a measured angle is always closely related to a measured distance; and as previously stated (Arts. 3-6 and 3-7) in general there should be a consistent relation between the precision of measured angles and that of measured distances. From the standpoint of both precision and expediting the work, it is important (1) that the surveyor be able to visualize the effect of errors in terms of both angle and distance, (2) that he appreciate what degree of care must be exercised to keep certain errors within specified limits, and (3) that he know under what conditions various instrumental errors can be eliminated.

On surveys of ordinary precision it usually requires much more care to keep *linear* errors within prescribed limits than to maintain a corresponding degree of *angular* precision. Often undue attention is paid to securing precision in angular measurements, and at the same time large and important errors in the measurement of distances are overlooked.

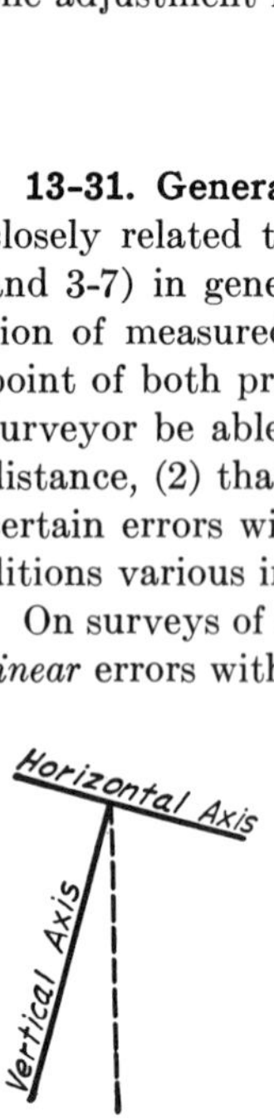

FIG. 13-27.

Errors in transit work may be instrumental (Art. 13-32), personal (Art. 13-33), or natural (Art. 13-34).

13-32. Instrumental Errors. These errors are caused by imperfections in the instrument itself. The adjustments, even though carefully made, are never exact. Likewise the graduations are not perfect, and the centers are not absolutely true.

As stated in the last paragraph of this article, errors due to instrumental imperfections and/or nonadjustment are all systematic, and they can either be eliminated or reduced to a negligible amount by proper methods of procedure.

1. *Errors in Horizontal Angles Caused by Nonadjustment of Plate Levels.* When the bubbles of plate levels in nonadjustment are centered, the vertical axis is inclined, and hence measured angles are not truly horizontal angles. Also the horizontal axis is inclined to a varying degree depending upon the direction in which the telescope is sighted. There will be one vertical plane which will include the vertical axis in its inclined position; this is illustrated by Fig. 13-27, in which the horizontal axis and the vertical axis are in the plane of the paper. When the line of sight is in the plane of the paper, how-

ever, the horizontal axis is truly horizontal and the line of sight will generate a vertical plane when the telescope is plunged; hence no error in direction is introduced regardless of the angle of elevation to the point sighted. As the transit is rotated about the vertical axis, the horizontal axis becomes inclined, making a maximum angle with the horizontal when it reaches the plane of the paper. With the horizontal axis in this position, the line of sight generates a plane making an angle with the vertical equal to the error in the position of the vertical axis; and with the line of sight inclined at a given angle, the maximum error in determining the direction of a line is introduced. The larger the vertical angle, the greater the error in direction. The error cannot be eliminated by double-sighting.

The diagram of Fig. 13-28 shows for various vertical angles (values of a) the errors introduced in horizontal angles due to an inclination of 01′ in the vertical axis

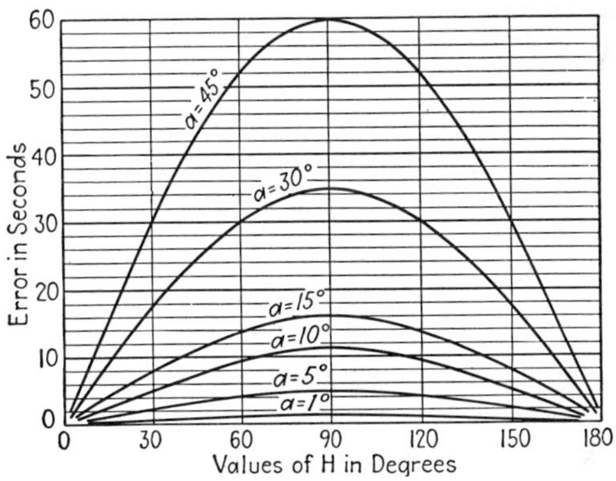

FIG. 13-28. Errors in horizontal angles for a 1-min. inclination of the vertical axis (a = vertical angle).

or one space on the plate levels of the ordinary transit. The values of H are the horizontal angles which the line of sight makes with the vertical plane in which lies the vertical axis in its inclined position (that is, with the plane of the paper, Fig. 13-27). The curve for $a = 0°$ is not shown by reason of the small scale, but the maximum error occurs when $H = 45°$ or $135°$ and is about $\frac{1}{20}''$. Within reasonable limits the error in horizontal angle varies directly as the inclination of the vertical axis, hence a similar diagram for an inclination of 02′ would show ordinates twice as great as those of Fig. 13-28.

Although the diagram may not be of much practical value, it serves to illustrate some noteworthy facts.

(a) For observations of ordinary precision taken in flat country where the vertical angles are rarely greater than 3° and usually much less, the plate bubble may be out several spaces without appreciably affecting the precision of horizontal angular measurements. For example, if an angle were measured between $H = 0°$ and $H = 90^{c}$, with bubble out two spaces and $a = 3°$, the error would be about 06″; or if in prolonging a straight line the telescope were plunged from the position

$H = 90°$ to $H = 270°$ with both backsight and foresight taken at a vertical angle of $+3°$, the error would be doubled and for the bubbles out two spaces (vertical axis in error 02′), the angular error introduced in the direction of the line would be 12″.

(*b*) For angular measurements of higher precision, such as when measuring an angle by repetition, the plate levels must be in good adjustment and the bubbles must be centered with reasonable care even though the survey is conducted over fairly smooth ground. For example, if a horizontal angle were measured between the positions $H = 0°$ and $H = 90°$, $a = 5°$, and the vertical axis were inclined 30″, the error in horizontal angle would amount to 02.2″.

(*c*) In rough country where the vertical angles are large, even for surveys of ordinary precision the plate levels must be in good adjustment and the bubbles must be carefully centered if errors in horizontal angles or in the prolongation of lines are to be kept within negligible limits. For example, if a line were prolonged by plunging the telescope from the position $H = 90°$ to $H = 270°$, a for both backsight and foresight being $+30°$ and the vertical axis being inclined 01′, the diagram shows that the error introduced is $2 \times 34.6'' = 01'09.2''$. In other words, the angle at the station at which the instrument was set instead of being a true 180° would be 180°01′09.2″, and beyond the station the established line would depart from the true prolongation about 0.1 ft. in each 300 ft.

2. *Errors in Vertical Angles Due to Nonadjustment of Plate Levels.* These errors obviously vary with the direction in which the instrument is pointed. With the fixed vertical vernier they are eliminated by observing (for each sighting) the index error of the corresponding observed vertical angle (Art. 13-21). With the movable vernier having a control level, the errors are eliminated by keeping the control-level bubble centered as described in Art. 13-19.

It may be noted further that nonadjustment of the plate levels causes an inclination of the plane of the vertical arc. This source of error may be considered negligible.

3. *Line of Sight Not Perpendicular to Horizontal Axis.* If the telescope is not reversed between backsight and foresight, if the sights are of the same length so that it is not required to refocus the objective, and if both points sighted are at the same angle of inclination of the line of sight, no error is introduced in the measurement of horizontal angles even though this adjustment be badly out. If the instrument is plunged between backsight and foresight, the resultant error in the observed angle is double the error of adjustment. If there is a considerable movement of the objective between sights, an appreciable error may be introduced, owing to the fact that the line of sight does not make a constant angle with the horizontal axis for both sights.

With the line of sight out of adjustment by a given amount, the effect of the error depends on the vertical angle to the point sighted. In Fig. 13-29, OA and DB are horizontal and are perpendicular to the horizontal axis OH of the instrument; e is the angle between the nonadjusted line of sight and a vertical plane normal to OH (that is, e is the error in direction for a horizontal sight OB); E is the error in direction for an inclined sight OC; h is the actual vertical angle to C, the point sighted; and OB is made equal to OC. Then

$$\begin{aligned}\sin e &= \frac{AB}{OB}\\ &= \frac{AB}{OB}\cdot\frac{OD}{OD} = \frac{AB}{OD}\cdot\frac{OD}{OB} = \frac{AB}{OD}\cdot\frac{OD}{OC}\\ &= \sin E \cos h\end{aligned}$$

or,

$$\sin E = \sin e \sec h \tag{1}$$

If a is the observed vertical angle, it can similarly be shown that

$$\tan E = \tan e \sec a \tag{2}$$

For all ordinary cases Eq. (1) or Eq. (2) may be taken as

$$E = e \sec h = e \sec a \text{ (approx.)} \tag{3}$$

For two direct pointings (the backsight and the foresight) there will be a value of E for each, and the error in the angle is the difference between them. In the measurement of a deflection angle by the method in which the telescope is inverted between backsight and foresight, the error in angle is the *sum* of the two values of E.

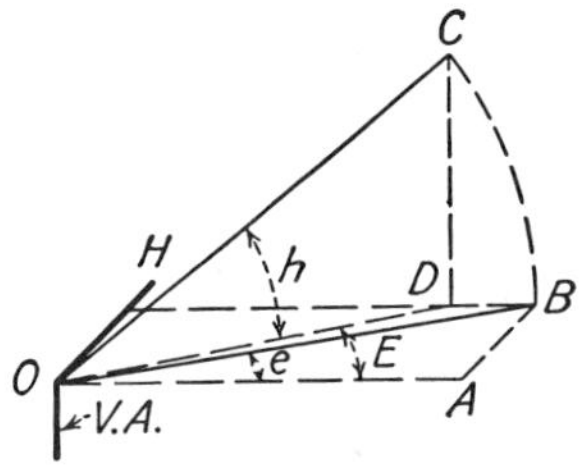

FIG. 13-29.

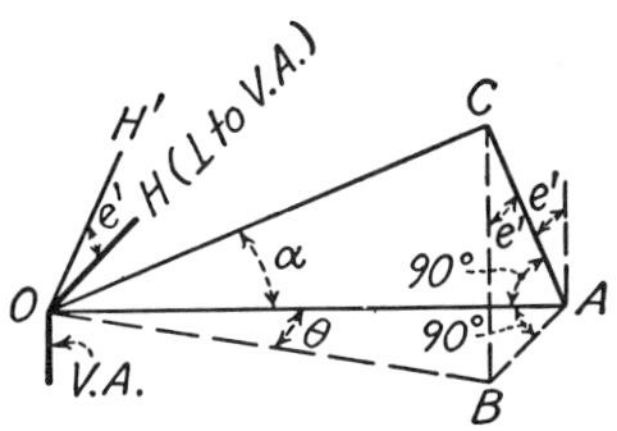

FIG. 13-30.

The error may be eliminated by taking the mean of two angular observations, one with the telescope in the normal position and the other with the telescope inverted. In prolonging a line, errors are avoided by the method of double-sighting described in Art. 13-22.

4. *Horizontal Axis Not Perpendicular to Vertical Axis.* No error is introduced in horizontal angles so long as the points sighted are at the same angle of inclination of the line of sight. The angular error in the observed direction of any line depends both on the angle by which the horizontal axis departs from the perpendicular to the vertical axis and on the vertical angle to the point sighted. In Fig. 13-30, OH is perpendicular to the vertical axis;

OH' is the horizontal axis in nonadjustment with the vertical axis by the angular amount e'; OA is horizontal and is perpendicular to OH and OH'; a is the observed vertical angle to C, the point sighted; B is directly beneath C and in the same horizontal plane with OA; angles OAB, OAC, and ABC are right angles; and θ is the angular error in direction. From the figure,

$$\tan\theta = \frac{BA}{OA}$$

$$= \frac{BA}{OA}\cdot\frac{CA}{CA} = \frac{BA}{CA}\cdot\frac{CA}{OA}$$

$$\tan\theta = \sin e' \tan a \tag{4}$$

or with sufficient precision,

$$\theta = e' \tan a \text{ (approx.)} \tag{5}$$

Thus, for example, if a horizontal angle were measured between D, to which the vertical angle is $-30°$, and E, to which the vertical angle is $+15°$, the horizontal axis being inclined $02'$, then the error in horizontal angle is

$$\theta = 02'(\tan 15° - (-\tan 30°)) = 32'' + 01'09'' = 01'41''$$

The preceding example is sufficient to show that the error in an observed horizontal angle may become large. Obviously the sign of the error in a horizontal angle depends upon the direction of displacement of the horizontal axis from its correct position. Hence if any angle is measured with the telescope first in the normal and then in the inverted position, one value will be too great by the amount of the error and the other will be correspondingly too small; thus the error is eliminated by taking the mean of the two values.

5. *Effect of Lack of Coincidence between Line of Sight and Optical Axis.* Under these conditions if the line of sight is perpendicular to the horizontal axis for one position of the objective, it will not be perpendicular for other positions, but will swing through an angle as the objective is moved in or out. If an angle is measured without disturbing the position of the objective, no error is introduced. For most instruments the error from this source is not sufficiently large to be of consequence in ordinary transit work. It is eliminated by taking the mean of two angles, one observed with the telescope normal and the other with it inverted.

6. *Errors Due to Eccentricity.* With the transit in good condition, errors due to eccentricity of verniers and/or eccentricity of centers are of no consequence in the ordinary measurement of angles. In any case, such errors are eliminated by taking the mean of readings indicated by opposite verniers.

7. *Imperfect Graduations.* Errors from this source are of consequence only in work of high precision. They can be reduced to a negligible amount by taking the mean of several observations for which the readings are distributed over the circle and over the vernier.

8. *Lack of Parallelism between Axis of Telescope Level and Line of Sight.* This introduces an error in leveling (see Art. 9-11) and in measuring vertical angles (see Art. 13-19).

9. *Nonadjustment of the Vertical Vernier.* This produces a constant error in the measurement of vertical angles. If the transit is equipped with a full vertical circle, the error can be eliminated by taking the mean of two values, one observed with the telescope normal and the other with the telescope inverted.

Summary. Summing up, it is seen that with regard to instrumental errors:

1. Errors in horizontal angles due to nonadjustment of plate levels or of horizontal axis become large as the inclination of the sights increases.

2. The maximum error due to nonadjustment of the line of sight is introduced when the telescope is plunged between backsight and foresight readings. When the telescope is not plunged between backsight and foresight readings, no error is introduced if these distances are equal and if the inclination of the line of sight is the same for both readings.

3. Errors due to instrumental imperfections and/or nonadjustment are all systematic, and without exception they can be either eliminated or reduced to a negligible amount by proper procedure. In general, this procedure consists in obtaining the mean of two values—one observed before and one after a reversal of the horizontal plate by plunging the telescope and rotating it about the vertical axis. One of these values is as much too large as the other is too small. An exception is the error in either horizontal or vertical angle due to inclination of the vertical axis, which cannot be so eliminated but which can be eliminated, so far as its *systematic* character is concerned, by releveling the plate bubbles in addition to the reversal of the plate. However, for precise work the usual practice would be to make the vertical axis truly vertical by means of the telescope level, and then to proceed in the ordinary manner.

13-33. Personal Errors. Personal errors arise from the limitations of the human eye in setting up and leveling the transit and in making observations.

1. *Effect of Not Setting Up Exactly over the Station.* This produces an error in all angles measured at a given station, the magnitude of the error varying with the direction of pointing and inversely with the length of sight. It is convenient to remember that 1 in. is the arc whose angle is 01′ when the radius is approximately 300 ft. Thus, if the transit were offset ½ in. from the end of a line 50 ft. long, the error in the observed direction of the line would be 03′, but if the line were 600 ft. long, the error would be only 15″. In general the error may be kept within negligible limits by reasonable care, but many instrumentmen waste time by exercising needless care in setting up when the sights to be taken are long.

2. *Effect of Not Centering the Plate Bubbles Exactly.* This produces an error in horizontal angles after the manner described in the preceding article

for plate levels out of adjustment. The error from this source is small when the sights are nearly level, but may be large for steeply inclined sights (see Fig. 13-28). The average transitman does not appreciate the importance of careful leveling for steeply inclined sights; on the other hand, he often uses unnecessary care in leveling when sights are nearly horizontal. Since the error in horizontal angle is caused largely by inclination of the horizontal axis, the striding level is a necessity on precise work.

3. *Errors in Setting and Reading the Vernier.* These are functions of the least count of the vernier and of the legibility of scale and vernier lines. For the usual 01′ transit the probable error is less than 30″; for the 30″ transit the probable error is about 15″. The use of a reading glass enables closer reading, particularly for finely graduated circles. Also in reading the vernier it is helpful to observe the position of the graduations on both sides of the ones that appear to coincide, and to note that the unmatched graduations appear to lack coincidence by the same amount.

4. *Not Sighting Exactly on the Point.* This is likely to be a source of rather large error on ordinary surveys where sights are taken on the range pole of which often only the upper portion is visible from the transit. The effect upon a direction is, of course, the same as the effect of not setting up exactly over the station. For short sights greater care should be taken than for long sights, and the plumb line should be employed instead of the range pole.

5. *Imperfect Focusing (Parallax).* The error due to imperfect focusing is always present to a greater or less degree, but with reasonable care it can be reduced to a negligible quantity. The manner of detecting parallax is described in Art. 2-10.

Summary. All the personal errors are accidental and hence cannot be eliminated. They form a large part of the resultant error in transit work. Of the personal errors, those due to inaccuracies in reading and setting the vernier and to not sighting exactly on the point are likely to be of greater magnitude.

13-34. Natural Errors. Sources of natural errors are (1) settlement of the tripod, (2) unequal atmospheric refraction, (3) unequal expansion of parts of the telescope due to temperature changes, and (4) wind, producing vibration of the transit or making it difficult to plumb correctly.

In general, the errors resulting from natural causes are not large enough to affect appreciably the measurements of ordinary precision. However, large errors are likely to arise from settlement of the tripod when the transit is set up on boggy or thawing ground; in such cases the instrument may be kept relatively stable through the use of extra-long tripods or of small platforms for the tripod legs. Settlement is usually accompanied by an angular movement about the vertical axis as well as linear movements both vertically and horizontally. When horizontal angles are being measured, usually a larger error is produced by the angular displacement of the circle between backsight and corresponding foresight than by the movement of the transit laterally from the point over which it is set. Errors due to adverse atmospheric conditions can usually be rendered negligible by choosing appropriate times for observing.

For measurements of high precision the methods of observing are such that instrumental and personal errors are kept within very small limits, and natural errors become of relatively great importance. Natural errors are generally accidental, but under certain conditions systematic errors may arise from natural causes. On surveys of very high precision, special attempt is made to establish a procedure which will as nearly as possible eliminate natural systematic errors. Thus the instrument may be set up on a masonry pier and protected from sun and wind; also certain readings may be made at night when temperature and atmospheric conditions are nearly constant.

13-35. Precision of Angular Measurements. Many factors influence the angular precision of transit work, and no rigid rules can be formulated to insure a required precision. It is clear from the preceding articles that, with proper methods, the important systematic errors can be practically eliminated and that the resultant error is largely accidental. No matter how precisely the transit may be adjusted nor how carefully it may be set up, there yet remain the errors of sighting and of reading the angle, and these are of major importance in nearly all surveying. The angular precision with which the line of sight may be directed obviously depends upon the length of sight and the character of the target or other object used to mark the point sighted, as well as upon the quality of the instrument and the skill of the observer. The precision with which an angle may be read depends upon the character of the graduations of the circle.

The following values represent, in a general way, the *maximum* error likely to occur in measuring a horizontal angle under average conditions of practice, instruments being in fair condition and in fair adjustment except as otherwise stated. The *average* error will of course be materially less. Also, as the errors are largely accidental, the resultant error in the sum of a series of measured angles may be expected to vary as the square root of the number of angles involved.

Case 1. Short sights, point indicated by range pole obscured near ground. Range pole plumbed by eye. Single observation of angle. Maximum error 02′ to 04′.

Case 2. Long sights, but otherwise as stated for case 1. Maximum error 01′ to 02′.

Case 3. Unobscured but steeply inclined sights; no special attention given to making horizontal axis truly horizontal; single measurement of angle. Maximum error 01′ to 02′.

Case 4. Unobscured sights on well-defined points; sights not steeply inclined. Single observation of angle, vernier reading to minutes. Maximum error 30″ to 01′.

Case 5. As for case 4, but transit in excellent condition and in good adjustment. Angles estimated to ½ min. Maximum error 20″ to 30″.

Case 6. As for case 4 but angle doubled, the telescope being plunged between sights. Maximum error 15″ to 30″.

Case 7. Unobscured sights on well-defined points. Sights not steeply inclined. Verniers reading to 30″. Single observation of angle represented by mean of readings of both verniers. Transit in excellent condition and in good adjustment. Maximum error 15″ to 30″.

Case 8. As for case 7, but verniers reading to 10″. Also instrument set up with great care. Maximum error 10″ to 15″.

Case 9. Unobscured sights on well-defined points. Instrument set up with great care. Sights not steeply inclined. Transit in excellent condition and adjustment. Verniers reading to 30″. Angles repeated six times with telescope normal and six times with it inverted. Maximum error 02″ to 04″.

Case 10. As above, but transit reading to 10″. Observations taken at favorable times. Maximum error 01″ to 02″.

13-36. Numerical Problems.

1. Thirty spaces on a transit vernier are equal to 29 spaces on the graduated circle, and 1 space on the circle is 15′. What is the least count of the vernier?
2. Sixty spaces on a transit vernier are equal to 59 on the graduated circle, and 1 space on the circle is 15′. What is the least count of the vernier?
3. A transit for which the circle is graduated 0° to 360° clockwise is used to measure an angle by 10 clockwise "repetitions," 5 with telescope normal and 5 with telescope inverted. Compute the most probable value of the angle from the following data:

Telescope	Reading	Vernier *A*	Vernier *B*
Normal.........	Initial	48°46′	228°46′
Normal.........	After first turning	161°09′	...
Inverted........	After tenth turning	92°41′	272°42′

4. In laying out the lines for a building, a 90° angle was laid off as precisely as possible with a 01′ transit. The angle was then measured by repetition and found to be 89°59′40″. What offset should be made at a distance of 250 ft. from the transit to establish the true line?
5. The following observations were made to determine an index correction: Vertical angle to point $A = +7°16'$ with telescope direct and $+7°14'$ with telescope inverted. Compute the index correction for observations with telescope direct.
6. A vertical angle measured by a single observation is $-12°02'$, and the index error is determined to be $+06'$. What is the correct value of the angle?
7. A line *AB* is prolonged to *F* by setting up the transit at succeeding points *B*, *C*, *D*, and *E*; backsighting to *A*, *B*, *C*, and *D*, respectively; and plunging the telescope. If the line of sight made an angle of 10″ with the normal to the horizontal axis and the procedure were such that each backsight was taken with the telescope normal, what would be the angular error in the segment *EF*? What would be the offset error (approximate) in the position of *F* if the segments *AB*, *BC*, etc., were each 400 ft. long?

8. Two points A and B, 5,280 ft. apart, are to be connected by a straight line. A random line run from A in the general direction of B is found by computation to deviate 03′18″ from the true line. On the random line at a distance 1,250.6 ft. from A an intermediate point C is established. What must be the offset from C to locate a corresponding point D on the true line?

9. In Fig. 13-17, a straight line AX is run at random from A in the general direction of B, point B not being visible from A. A swing offset is measured from B to line AX and found to be 63.40 ft. The transit is set up at E', and $E'Y$ (perpendicular to DX) is erected. The swing offset from B to $E'Y$ is 1.1 ft. Also, the distance AE' is 2,633.9 ft. Compute the angle α which must be laid off from the random line in order to establish points on the straight line AB and determine the length AB. What must be the precision of α in order that the line established shall fall within 0.1 ft. of the point B?

10. What error would be introduced in the computed value of the angle α of problem 9 if the swing offset distance from B to $E'Y$ had been neglected and AE' had been assumed to be the base of a triangle of which AB is the hypotenuse?

11. Given the data of problem 9, it is proposed to establish points on the line AB by perpendicular offsets from C and D. What must these offsets be if $AC = 937.6$ ft. and $AD = 1{,}932.0$ ft.?

12. In Fig. 13-20*a*, suppose that the distance AC is 317.2 ft. and the angle ACB is 67°13′. What is the distance AB?

13. In Fig. 13-20*b*, if the length of the line AC is measured and found to be 517.2 ft. and the length of AD is found to be 315.5 ft., what is the distance AB?

14. In prolonging a straight line the transit is set at B, a backsight is taken to A, and the telescope is plunged to set C 1,000 ft. in advance of B. If the vertical axis were inclined 01′ with the true vertical in a vertical plane making 90° with the direction of the line, what would be the linear offset error in the located position of C: (*a*) If A and B are at the same elevation, but the vertical angle from B to C is +15°00′? (*b*) If A, B, and C are all at the same elevation? (*c*) If the vertical angle from B to A and from B to C is +15°00′?

15. What error would be introduced in the measurement of a horizontal angle, with sights taken to points at the same elevation as the transit if, through nonadjustment, the horizontal axis was inclined (*a*) 03′? (*b*) 3°? (*c*) If the horizontal axis was inclined 03′, what error would be introduced if both sights were inclined at angles of +30°? (*d*) If one sight was inclined at +30° and the other at −30°?

16. In measuring a horizontal angle the error of setting up the transit is 0.03 ft., the direction of displacement being such as to produce a maximum angular error. What error is introduced in a 60° angle if the length of sights is (*a*) 50 ft.? (*b*) 1,000 ft.?

17. If the ratios of linear precision to be maintained on the various parts of a survey are 1/1,000, 1/5,000, 1/20,000, and 1/40,000, about how closely should the corresponding horizontal angles be observed in order that a consistent relation may exist between precision of angles and precision of distances?

18. It is desired to determine by computation the length of a side of a right triangle, the angle opposite and the hypotenuse being measured. If the angle is 20°, with what precision should it be measured in order that the ratio of precision in the computed length be 1/10,000?

19. It is desired to determine by computation the length of a side of a right triangle, the angle opposite and the side adjacent thereto being measured. If the angle is 20°, with what precision should it be measured in order that the ratio of precision in the computed length be 1/10,000?

13-37. Field Problems.

PROBLEM 1. MEASUREMENT OF HORIZONTAL ANGLES WITH TRANSIT

Object. To measure several angles about a point with the transit, and to check the values of the angles by the use of magnetic bearings.

Procedure. (1) Set up and level the transit at any point O. (2) Set six chaining pins at about 150 ft. from the transit, forming six angles about the station O. (3) According to the procedure of Art. 13-14, measure each of the angles, using the A vernier only and resetting the vernier on each backsight. (4) The sum of the measured angles should not differ from 360° by more than ±03′. (5) If this difference is exceeded, the angles should be remeasured until the sum falls within the stated limits. (6) Release the compass needle, sight on each point, and, according to the method of Art. 12-22, read and record the magnetic bearing to each pin. (7) Compute the angles by bearings, and compare with the transit angles. The discrepancy between any transit angle and the same angle by bearings should not exceed 30′.

Hints and Precautions. The pins should be set as nearly vertical as possible with reasonable care. They may be plumbed by the vertical cross-hair of the transit. If each pin is run through a piece of paper, piercing it in several places, the paper will form an excellent background for sighting the pin.

PROBLEM 2. MEASUREMENT OF ANGLES BY REPETITION

Object. To obtain a more precise determination of the horizontal angles between various stations about a point than would be possible by a single measurement (see Art. 13-17).

Procedure. (1) Set up the transit very carefully over the point. (2) Set the A vernier at zero, read the B vernier, and record the readings. (3) Keep notes in a form similar to that of Fig. 13-11. (4) With the telescope *normal*, measure one of the angles clockwise, and record both vernier readings to the least reading of the vernier. (5) Leaving the upper motion clamped, again set on the first point and again measure the angle clockwise (thus doubling the angle). (6) Continue until five "repetitions" (observations) have been secured. Record both vernier readings and the total angle turned. (7) In like manner, without resetting the vernier, measure the angle (five repetitions) with the telescope *inverted*, always measuring clockwise. (8) Go through the same process for all other angles about the point. (9) Compute the value of each of the angles for the 10 repetitions, and compare with the single measurement. (10) For a transit reading to single minutes, the error of horizon closure should not exceed $10'' \sqrt{\text{number of angles}}$. (11) Adjust the angles so that their sum will equal 360° by distributing the error equally among the mean values.

Hints and Precautions. (1) Level the transit very carefully before each repetition but do not disturb the leveling screws while a measurement is being made. (2) Be careful not to loosen the wrong clamp-screw. (3) Do not become confused when computing the total angle turned. Observe how the horizontal limb is graduated and do not omit a full turn. (4) The instrument should be handled very carefully. When the lower motion is being turned, the hands should be in contact with the *lower* plate, not the upper motion. When making an exact setting on a point, the last movement of the tangent-screw should be *clockwise* or against the opposing spring. (5) After each repetition the instrument should be turned on its lower motion in the same direction as that of the measurement. (6) Do not walk around the transit to read the second vernier; rotate it to you, always turning the instrument clockwise in order to avoid possible errors due to slackness and twist. (7) The single measurement is taken as a check on the number of repetitions; it should agree closely but not necessarily exactly with the mean value.

Problem 3. Laying Off an Angle by Repetition

Object. To lay off a given horizontal angle more precisely than is possible with a single setting of the vernier (see Art. 13-18).

Procedure. (1) Drive and tack two stakes about 500 ft. apart. (2) Carefully set up the transit over one end of the line. Sight at the point at the other end, and lay off the given angle. (3) Set a stake on the line of sight about 500 ft. from the instrument (distance by pacing), and carefully set a tack. (4) By repetition measure the angle laid off, as in the previous problem, making five "repetitions" with telescope normal and five with it inverted. (5) Find the difference between the angle laid off and the required angle, and by trigonometry compute the linear distance that the tack must be moved perpendicular to the line of sight. (6) Set the tack accordingly.

Problem 4. Measurement of Vertical Angles with Transit

Object. To determine the height of a building above the water table, by measurement of vertical angles with the transit.

Procedure. (1) Set up the transit at A, at a distance from the building approximately twice the height of the building. Level the telescope, and note the location of the horizontal hair on the building; mark the point sighted, and measure the distance h_a above or below the water table of the building. (2) By double-sighting, determine the index error of the vertical circle. Sight on the high point T of the building, and record the vertical angle. (3) Depress the telescope, and set a point B in the same vertical plane with A and T, about halfway between A and the building. (4) Set up the transit at B, and measure h_b and the vertical angle to point T, as at A. (5) Measure the horizontal distance AB. (6) Draw a sketch, and compute the difference in elevation (a) between T and the horizontal line of sight from either transit station, and (b) between T and the water table.

Hints and Precautions. The index error may be read either before or after observing a vertical angle; while this reading is being taken, the line of sight should be in the same vertical plane as the point sighted.

Problem 5. Prolongation of Line by Double-sighting with Transit

Object. To prolong a straight line with precision, setting stakes at intervals of about 300 ft. (see Art. 13-22).

Procedure. (1) Set two points about 300 ft. apart in such location as to afford an open view for 1,000 ft. or more in advance. (2) Set up the instrument on the forward point. Backsight with the telescope inverted. (3) Plunge the telescope, and set a stake on the line 100 paces in advance. Mark a point on the stake exactly on line. (4) Take a second backsight on the rear stake in the same manner but with the telescope normal. Plunge the telescope again, and mark a point on the advance stake. (5) If this point does not coincide with the first point set, a point midway between them is on the line. (6) Set up the transit over this point, and advance by the same process, backsighting on the nearest point in the rear. Continue in this way for the desired distance. (7) Check the work by setting the instrument over the first point, sighting carefully on the next point, and then noting the linear error of the points set by double-sighting, without moving either horizontal motion of the instrument.

Problem 6. Prolongation of Line past Obstacle

Object. To prolong a line AB past an obstacle where the conditions are such as to limit the lengths of the offsets.

Procedure. (1) As outlined in the first paragraph of Art. 13-23. (2) The lengths of offsets should be measured very carefully. If the instrument is in good adjustment, the points E and F may be set by a single reversal; or if a clear sight can be obtained, the transit may be set up at C and the points E and F established without reversal. (3) If the obstacle is imaginary, check the accuracy of the work by setting the transit at A and locating G and H by the direct prolongation of AB.

Problem 7. Running a Straight Line between Two Points Not Intervisible

Object. To establish points along a straight line joining two given points not intervisible.

Procedure. (1) As outlined in Art. 13-24, cases 2 and 3. (2) Under case 2, take two points on opposite sides of a hill. To check the located position of C (Fig. 13-18) set up the transit at that station and by the method of double-sighting prolong the line AC to B. Note the error at B. (3) Under case 3, determine offsets from the intermediate points on the line AX (Fig. 13-19) to the line AB, and establish the corresponding points on AB by tape measurements. Then lay off the angle α from AX, and establish a second set of points on the line AB by the method of double-sighting. Note the discrepancies.

Problem 8. Determination of Inaccessible Distance

Object. To obtain the distance between two points on opposite sides of a river.

Procedure. As outlined in Art. 13-25. The transit points should be tacked stakes. If the river is imaginary, after the distance has been computed by each method check it by direct measurement.

Problem 9. Intersection of Lines with Transit

Object. To bisect the three angles of a triangle and to mark the point of concurrence of the bisectors.

Procedure. (1) Drive three stakes, A, B, C, at the vertices of a roughly equilateral triangle having sides about 300 ft. in length. Tack each stake. (2) Set up the transit, and measure the angle at A. Lay off one half of the measured amount, thus establishing the bisector of angle A. On the bisecting line of sight and on an estimated bisector of angle B drive a stake o and drive a tack halfway. Set two more tacked stakes m and n on the bisecting line of sight about 10 ft. from and on opposite sides of *o*. (3) Set up the transit at *B*, and locate the position of the bisector as at *A*. Drive a stake on this line and under a cord stretched from *m* to *o* or *n* to *o*, as the conditions require. Tack the exact point of intersection *p*. (4) Set up the transit at *C*, measure the angle, and bisect as at *A* and *B*. (5) Measure the discrepancy between this bisector and the point of intersection of the first two bisectors at *p*, to hundredths of feet. Also measure the angular discrepancy; it should not exceed 02′.

Problem 10. Measurement of Angle When Transit Cannot Be Set at Vertex

Object. To measure the angle between two walls of a building.

Procedure. As outlined in Art. 13-28. The points *a*, *b*, etc., should be tacks in stakes driven at appropriate locations.

Problem 11. Adjustment of the Transit

Object. To make the field adjustments of the engineer's transit.

Procedure. As outlined in Art. 13-30.

REFERENCES

1. Foster, N. B., and L. C. Higbee, "Collimators Permit Rapid and Accurate Adjustment of Surveying Instruments," *Engrg. News-Record*, Vol. 135, No. 16, pp. 502–503, Oct. 18, 1945.
2. Keuffel, Carl W., "American Surveying and Mapping Equipment," *Surveying and Mapping*, pp. 358–362, October–December, 1951.
3. Larsen, Harold R., "Gurley Four-inch Army Theodolite," *Trans. Am. Geophys. Union*, Vol. 26, Part 1, pp. 19–27, 1945.
4. Rayner, W. H., "Two Investigations of Transit Instruments," *Univ. Ill., Engrg. Exp. Sta. Circ.* 32, 1938.
5. "Transits, One-minute; and Transit Tripods," *Federal Specification* GG-T-621, Government Printing Office, Washington, D.C., 1948.
6. Catalogs of manufacturers of surveying instruments.

CHAPTER 14

TRANSIT–TAPE SURVEYING

14-1. General. This chapter treats of the general methods utilized on the large variety of surveys of ordinary precision for which the transit is employed for the measurement of horizontal angles and the tape is used for the measurement of distances. With minor modifications, these practices are common to land, city, topographic, and hydrographic surveys and to location surveys such as those for highways and railroads.

14-2. Transit Party. The transit party is usually composed of a *transitman,* a *head chainman,* and a *rear chainman.* The transitman directs the party, operates and cares for the transit, and keeps the notes. The head chainman performs the duties of that position as described in Arts. 7-12 and 7-13, gives line as directed by the transitman, and is responsible for the accuracy and speed of the taping operations. Where stakes are set, he attends to their proper marking. The rear chainman carries out the duties of that position as described in Arts. 7-12 and 7-13, gives backsights as directed by the transitman, often carries and drives stakes, and assists in removing obstructions to the vision of the transitman. In wooded country, axemen are employed to clear the transit line of trees and brush; the head chainman keeps in close communication with the axemen and assists in their direction. Where sights are long, a rear flagman may be employed to give backsights to the transitman. Where many observations are taken, the notes may be kept by a recorder, who may also act as the chief of party.

14-3. Equipment of Transit Party. The equipment of the transit party usually consists of a transit, 100-ft. steel tape, two range poles, stake bag, stakes, tacks, axe or hammer, two or three plumb bobs, field notebook, chaining pins, and marking crayon. Also included are devices for marking stations such as ordinary or special nails, a cold chisel, spray paint, and colored plastic flagging or tape. A light tripod with plumb bob is useful for sighting on transit stations where vision is obscured near the ground. Portable sending and receiving radios, called "walkie-talkies," may be used to expedite communication.

14-4. Transit Stations. Any temporary or permanent point of reference over which the transit is set up is called a *transit station.* On most surveys the transit station is a peg, called a *hub,* driven flush with the ground and having a tack driven in its top to mark the exact point of reference for measurements. On pavements the transit station may be a driven nail or a cross cut in the pavement or curb. In land surveying, the stations are often iron pipes, stones, or other more or less permanent monuments set at the corners. In mountainous country, often the station marks are cut in the natural rock.

The location of a hub is usually indicated by a flat *guard stake* extending above the ground and driven sloping so that its top is over the hub. This guard stake carries the number or letter of the transit station over which it stands. Usually the number is marked with keel or lumber crayon on the underside of the guard stake and reads down the stake. The hubs are usually square, say 2 by 2 in., and the guard stakes are usually flat, perhaps ¾ by 3 in.

In order to avoid the necessity for using a rear flagman to give backsights, as far as practical each transit station is marked by some temporary signal such as a lath or a stick set on line, with a piece of paper, cloth, or tape attached.

14-5. Transit Lines. Lines connecting transit stations are called *transit lines.* If a system of lines run with the transit forms a traverse (as described in Art. 12-17), it is called a *transit traverse* to distinguish it from traverses run with other instruments. The transit lines forming a triangulation system are called *lines of triangulation,* and the points at which the transit is set up are called *triangulation stations.*

Both in the field notes and as a part of the identification mark left in the field, a traverse-station number is preceded by the symbol ⊙ and a triangulation-station number or name by the symbol △. In most cases the transit stations are identified by consecutive numbers as the survey progresses. Sometimes triangulation stations are given names suggested by the locality where each is established.

For many open traverses where lengths are measured with the tape, distances are referred to the point of beginning of the survey and stakes marked with the distance from the initial point are set every 100 ft. These 100-ft. points are called *full* stations, and intermediate points are called *plus* stations. The distance to any plus station is indicated as the number of *hundreds of feet* from the initial point to the preceding full station plus the distance in feet from the preceding full station to the point in question. Thus, a full station at, say, 1,200 ft. from the initial point would be numbered 12 + 00, or simply 12, and a plus station at, say, 1.927.2 ft. from the initial point would be numbered 19 + 27.2. The stations intermediate between transit stations are marked usually by flat stakes driven vertically, with the number on the side toward the initial point and with the number reading down the stake. Where stakes cannot be driven, nails or marks may be employed.

Transit stations are guarded and marked as described above; thus the guard stake for a transit station 1,216.3 ft. from the initial station would be marked as ⊙12 + 16.3. Often the transit stations are referenced (Art. 14-17).

14-6. Transit Surveys. Surveys have for their object either (1) the location of certain features of the terrain or (2) the establishment of points and lines of predetermined length and direction, which are to be employed as a guide to the future enterprises of man. Often a single survey may accomplish both objects.

The field work of surveying with the transit may be divided into:

1. Establishing transit stations and lines by angular and linear measurements. The transit lines may be said to form the skeleton of the survey and are called the *control* or *horizontal control.*

2. Locating objects and points with respect to the transit lines, thus furnishing the *details* with which the transit lines are clothed (see Art. 14-18).

For some surveys the amount of detail secured from the transit lines is little. For example, for surveys to establish the boundaries of land the transit stations are usually at corners of the property and, if the boundaries are straight, few if any measurements to details are required. For some other surveys the location of features not on the transit lines forms the greater portion of the work. For example, in certain topographic surveys, for every transit station there may be as many as 50 points to which measurements must be taken to secure adequate information for the construction of the map.

On some surveys the collection of details may take place as the work of laying out the transit lines proceeds. On others the system of transit lines is first established, and after it has been checked, the details are obtained; this procedure is most likely to be employed where the survey covers a considerable territory and where the methods and instruments used in running the transit lines are not those used in the collection of details.

The simplest survey to be made with the transit and tape employs a single transit station over which the instrument is set. Angles and distances to surrounding points and objects are observed. This is called the method of *radiation* (Art. 14-7). By itself it is applicable only to surveys covering small areas. It is generally employed in the location of details on extensive surveys.

A survey nearly as simple employs two transit stations connected by a single transit line called the base line. From each station, angles with respect to the transit line are observed to objects which it is desired to locate. Thus any point is defined by the two angles taken from the transit stations and by the length of the base line. This is called the method of *intersection* (Art. 14-8) and is a form of triangulation. It is employed on surveys of small areas and in the location of details from transit traverses.

On surveys of any considerable extent the method of *traversing* (Art. 14-9) is usually employed to establish most of the transit lines, and the two preceding methods together with others later to be described are used in locating details. On surveys of ordinary precision, transit traverses in one form or another probably make up more than nine-tenths of the systems of control.

The *triangulation* system (Chap. 16) as a means of providing control on surveys of ordinary precision is not often used except for hydrographic surveying and for topographic surveying in rough open country. Where conditions are favorable, it is very economical of time in the field, and it could be employed to advantage more generally than at present. Precise triangulation is employed extensively on surveys covering wide areas, particularly in connection with topographic and hydrographic work.

14-7. Radiation. The transit is set up at any convenient station from which can be seen all points that it is desired to locate. The distance from the transit station to each of the points is measured, and the horizontal angle is observed. The angles between successive points may be measured, or

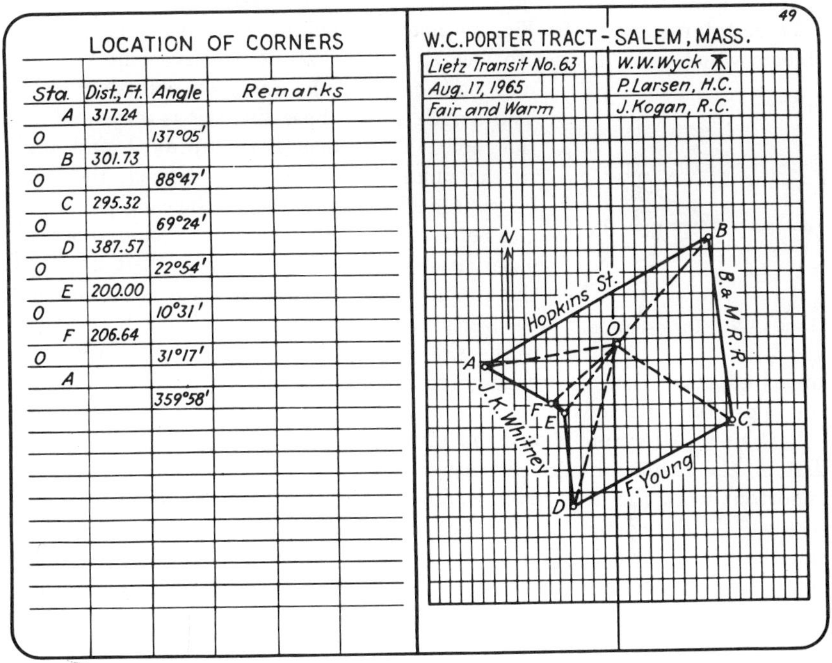

FIG. 14-1. Notes for survey by radiation.

the true, magnetic, or assumed bearing or azimuth of each of the lines joining the points with the transit station may be observed.

Where points only need to be located, as where the survey is made for a map, the method is excellent. But if the lengths and directions of lines other than the radial lines are to be determined, trigonometric computations are necessary.

For example, in Fig. 14-1, the field notes give for each of the triangles into which the figure is divided two sides and the included angle at O. To determine the length of any unknown side (as EF) and the value of any unknown angle (as OFE), it is necessary to use (either directly by right-angle triangles or indirectly by oblique triangles) the sine of the angle at *O* (as sin *EOF*). A disadvantage of the method lies in the weakness of the computed values when the measured angle is small (see Fig. 3-2).

Thus, if the angles in Fig. 14-1 were measured with an error of 30″, the error in the computed length *EF* would correspond to the low ratio of precision of 1/1,260 while the ratio of precision of the computed length *CD* would be practically 1/20,000.

Inasmuch as there is likely to be at least one small measured angle in any figure, the method is not commonly used for property-line surveys.

14-8. Intersection. In Fig. 14-2 let the points *A*, *B*, *C*, etc., represent objects which it is desired to locate, and let *OP* be a convenient line from both ends of which the unknown points are visible. The length of the base line *OP* is measured with the tape. The transit is set up at *O*, and angles to the unknown points are observed; these may be expressed as azimuths, as bearings, or as angles between successive points. A similar series of observations is made with the transit at *P*. In this manner each of the unknown points becomes the vertex of a triangle of which the base line *OP* is the side of measured length and in which the angles adjacent thereto are observed values; the locations of the unknown points are thus defined.

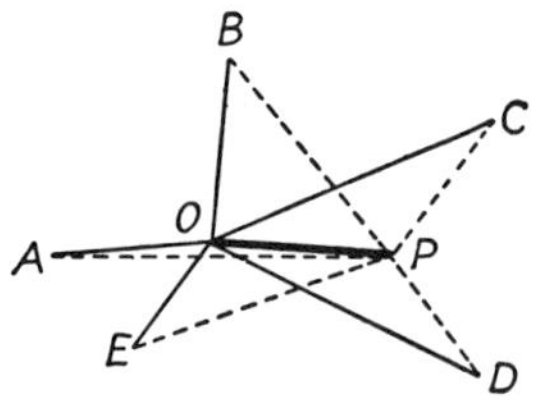

FIG. 14-2. Intersection.

Particular attention should be given to securing well-shaped triangles; that is, the angles should be neither too large nor too small. Usually on important work the attempt is made to secure angles between 30° and 150°. In Fig. 14-2 the triangle *AOP* is weak, the angles at *A* and *P* being too small and the one at *O* being too large; hence the uncertainty of the position of point *A* would be large. The triangle *BOP* is well shaped, and the computed position of *B* would be much more certain for a given precision of angular measurement than would that of *A*. Thus, if the angles were measured with an error of 30″, and the values were $OAP = 5°$ and $OBP = 45°$, the ratios of precision as governed by angular errors would be about 1/700 for the computed distance *AO* and about 1/7,000 for the computed value of *BO*.

In general, the method of intersection is not well suited to surveys made for the purpose of determining the lengths and directions of boundaries because of the large amount of computing necessary and because of the uncertainties attached to the computed values when triangles are weak. For example, in order to determine the length and direction of *EA*, it would be necessary to solve triangles *AOP*, *EOP*, and *AOE*, and the weakness of the triangle *AOP* would, of course, be reflected in the computed direction and length of *EA*.

The method is not usually employed alone but is used in conjunction with other methods, particularly traversing. Where well-defined details at a considerable distance are visible from two or more transit stations, they may usually be located by this method much more expeditiously than by angle and distance. Angles taken to some landmark from several successive stations provide a means of checking the traverse.

Sights may be taken simultaneously by means of two transits set up at opposite ends of a measured base line. In hydrographic surveying, this method of intersection is useful in locating soundings. For the construction of bridges and dams, the method is used to establish points for piers and other structural parts which are difficult of access.

A variation of the method of intersection is the method of *resection*, by means of which the transit may be set up at a station of unknown location and its location determined by sighting on points of known location. The principle of resection employing the transit is as described for the plane table in Art. 17-11 and is illustrated in Art. 16-29.

14-9. Traversing. A traverse may be run for the purpose either of locating features already existing in the field or of establishing points and lines in accordance with predetermined measurements.

Closed Traverse. Following is a general description of the work of running a *closed* traverse, the transit stations being established in advantageous locations as the survey progresses and distances being measured between successive transit stations. In Fig. 14-3 let A and B be selected locations for transit stations marking the first line of a traverse. Hubs are set defining the points. The transit is set up at B, the horizontal vernier is set to a given angular value, a backsight is taken on a range pole at A, and the lower motion is clamped. The line AB is then taped (Chap. 7), the head chainman usually being lined in by the transitman rather than by the rear chainman. The distance AB is recorded. The location of C is selected, and the transit point is established. The transit is turned on the upper motion until a foresight is secured on C. The upper motion is clamped, and the angular value is read and recorded. The distance BC is chained and recorded in a manner similar to that for AB. The transit is moved forward to C; a backsight is taken to B; the point D is chosen; a foresight is taken to D; and the angle is read. The line CD is chained. The process is repeated for point E, etc., until the traverse is finally brought to a closure on the initial point A.

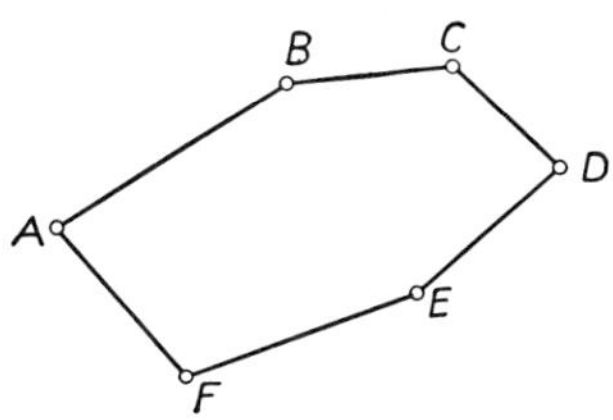

FIG. 14-3. Closed traverse.

As a means of checking the angular measurements against mistakes or large errors, usually magnetic bearings are observed on both backsight and foresight from each station and are compared with bearings computed from the measured angles as described in Chap. 12. Also as a check, after the traverse is brought to a closure the initial station is occupied by the transit and the angle between the first and last lines of the traverse is measured; in this way the angular error of closure is determined. Where the traverse is not long, a sketch of the traverse, together with any other desired features of the survey, is usually shown on the right-hand page of the notebook, and the measurements are tabulated on the left-hand page, the numerical notes being kept down the page in order of the observations (Figs. 14-4 and 14-6).

	DEFLECTION-ANGLE				
		Sigas, N. B.			
Sta. At-To	Dist., Ft. –	Defl. Ang.	Mag. Bear.	Defl. Ang. by Bear.	Calc. Bear.
A E		57°54'R	N28°W	58°R	
B	507.65		S30°W		S29°37'W
B A		113°38'L	N29½°E	113½°L	
C	784.68		S84°E		S84°00'E
C B		98°15'L	N84°W	98¼°L	
D	994.60		N2¼°W		N2°15'W
D C		88°19'L	S2°E	88½°L	
E	739.72		S89½°W		S89°26'W
E D		117°43'L	N89¼°E	117½°L	
A	526.00		S28¼°E		S28°17'E
		360°01'			

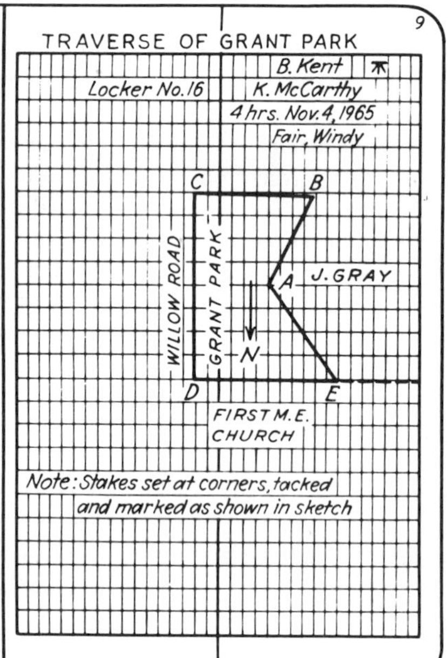

FIG. 14-4. Notes for short closed deflection-angle traverse.

Even though measurements to details are not included, it is usually customary to show on the sketch the approximate location of important objects.

Open Traverse. An *open* or continuous traverse may be run in exactly the same manner, except that of course there is no closure. However, the open traverse may begin and close on previously established lines, the directions and locations of which are known; or on long open traverses the true meridian may be determined at intervals (by astronomical observations) and the direction of the traverse line thus checked, taking due account of convergency of meridians (Art. 23-11).

Where stakes are set every 100 ft., the chainage is referred to the point of beginning; and if the traverse is of considerable length, the notes are usually taken *up* the page, and for sketches the traverse line is considered as being on the center line of the right-hand page (Fig. 14-5). Also in the notebook a line is given to each station and plus. This manner of keeping notes facilitates sketching, since objects on the sketch will appear in the same relative position as they appear to the recorder proceeding along the line.

Where stakes are set every 100 ft., those marking intermediate points between transit stations are not usually tacked except where they mark the

PRELIMINARY SURVEY OF I.N. RY., C. TO ST. LEONARDS

(Defl. Ang. Traverse)

Sta.	Defl. Ang.	Mag. B.	Cal. B.	Remarks
12				
+82				Water Line
+17				" "
11				
10				
9				
⊙8+36.3	23°14'L	N 35°W	N34°47'W	
8				
+99.8				
+68				℄ Due E to St.L.
+33.1				
7				
6+00				Ctr. of Brook
5				
4				
3				
2				
+23.6				
1				
⊙0+00		N11°W.	N11°33'W	on C.L. C.P.Ry. 523'N. M'l. Post No. 26

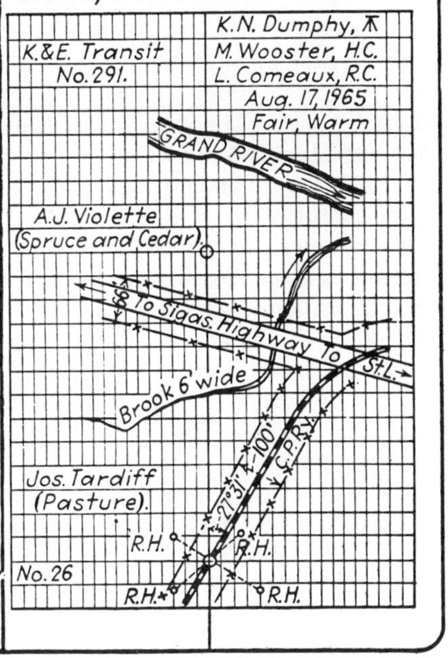

FIG. 14-5. Notes for open deflection-angle traverse.

final location of some very definite line, as for example, the center line of a railroad track. On rough surveys, distance is measured by the rear chainman holding his end of the tape to the center of base of the last stake driven and the head chainman thrusting the end of the range pole in the ground at the other end. The rear chainman before leaving the point calls out the number of the station as marked on the stake, and the head chainman replies with the number of the station at the range pole, at the same time marking the stake which is to be set at the new station. He then removes the range pole from the ground and puts the stake in its place; the rear chainman, when he comes forward, drives the stake and checks its number. Often the head chainman carries a notebook in which he records the number of each station as soon as the stake is marked and set. On more precise surveys, linear measurements are carried forward by means of chaining pins as described in Chap. 7, each stake being driven by the rear chainman as he pulls the pin. If tacks are set in intermediate stakes, the chainage is carried forward from tack to tack.

General. Usually when the transit is to occupy a station for any considerable length of time, there is more or less likelihood of its being disturbed. To detect any movement of the lower motion, the transitman often observes the angle to some prominent feature of the landscape just after having taken a backsight to the preceding transit point, and occasionally thereafter he sights again at the reference mark and notes the angle. If no change is observed, it is proof that there has been no accidental rotation of the horizontal circle. The transitman should invariably apply this check as the last operation before leaving any station.

Where the traverse is through wooded country, it is impossible to establish transit stations in advance of the instrument until the line has been cleared. Consequently, a foresight is taken in the general direction of the proposed station and the clearing proceeds until a favorable location for the advance transit station is encountered. If 100-ft. stations are to be established, the stakes are driven on a fixed line as fast as clearing will allow.

One or more transit stations may be necessary between points at which angles are turned. Usually the line is prolonged beyond such stations by plunging the telescope rather than by turning 180° on the upper motion. Where a number of stations lie between two adjacent angle points, it is good practice to backsight at one station with the telescope in the normal position and at the next station with the telescope inverted. When there is any doubt as to the correctness of adjustments or when the precision demands it, the line should be prolonged by the method of double-sighting (see Art. 13-22). In the notes the stations thus established are indicated in the same manner as are the transit points at which angles are turned.

The angles of the traverse may be measured by observing *deflection angles* (Art. 14-10), azimuths (Art. 14-11), *angles to the right* (Art. 14-12), or *interior angles* (Art. 14-13), as desired. For rough traverses, the angles may be measured by means of the transit compass alone, the bearings of the lines being observed.

14-10. Deflection-angle Traverse. This method of running traverses is probably more commonly employed than any other, especially on open traverses where only a few details are located as the traverse is run. It is used almost entirely for the location surveys for roads, railroads, canals, and pipelines. It is employed to a less extent in land surveying and in establishing control traverses for topographic and hydrographic surveys.

Successive transit stations are occupied, and at each station a backsight is taken with the *A* vernier set at zero and the telescope inverted. The telescope is then plunged, the foresight is taken by turning the instrument about the vertical axis on its upper motion, and the deflection angle is observed. The angle is recorded as right *R* or left *L*, according to whether the upper motion is turned clockwise or counter-clockwise.

Figure 14-4 illustrates the field notes for a short *closed* deflection-angle traverse. It will be seen that they are kept *down* the page. Magnetic bearings have been observed forward and backward from each station. For any closed traverse the summation of the deflection angles, considering those turned to the right as being of opposite sign to those turned to the left, should equal 360°. The actual sum indicated in these notes shows that there existed an angular error of closure of 01′. Deflection angles computed from the observed magnetic bearings are shown in the fifth column. These values will be free from local attraction and should agree closely with the corresponding observed deflection angles. In the last column are the bearings as computed from the observed deflection angles, assuming the computed and observed bearings of *BC* to be the same. Usually one or the other of the last two columns is omitted. The values given in both are used as a rough check as the work of running the traverse progresses. The computed bearings are often shown on land plots and are useful in computing areas and coordinates.

Figure 14-5 shows the notes for a portion of an *open* deflection-angle traverse where stakes are set every 100 ft. The center line of the right-hand page represents the traverse. The notes read *up* the page, and observations are checked against mistakes and large errors by the close agreement between computed bearings and observed magnetic bearings. The notes are typical of the form used on highway, railroad, and other route surveys. Some surveyors prefer a form of notes having separate columns for left and right deflection angles.

The method just described is open to the objection that plunging the telescope from the inverted to the normal position introduces a constant error in each observed angle if the line of sight is not in perfect adjustment. If there are a large number of lines in the traverse, the total angular error introduced in this manner may become considerable, even through the error at each individual station is less than the least reading of the vernier. For this reason, it is better practice to set the *A* vernier at 180°, take the backsight, and then turn the instrument on its upper motion (instead of plunging) to the foresight. If the practice of plunging the telescope is followed, one backsight should be taken with the telescope normal and the next backsight with the telescope inverted.

The precision of measurements is increased somewhat and each observation is checked by doubling the angle. To eliminate instrumental errors, the first backsight from the station is taken with the telescope normal and the second backsight is taken with the telescope inverted. If angles are doubled, usually magnetic bearings are not observed.

14-11. Azimuth Traverse. An advantage of the azimuth method is that the simple statement of one angular value, the azimuth, fixes the direction of the line to which it refers. The method is used extensively on topographic and other surveys where a large number of details are located by angular and linear measurements from the transit stations. Any angular error of closure of a traverse becomes evident by the difference between initial and final observations taken along the first line. The reference meridian may be either true or assumed.

First Method. Successive stations are occupied, beginning with the line of known or assumed azimuth. At each station the transit is "oriented" by setting the *A* vernier to read the back azimuth (forward azimuth ±180°) of the preceding line and then backsighting to the preceding transit station. The instrument is then turned on the upper motion, and a foresight on the following transit station is secured. The reading indicated by the *A* vernier is the azimuth of the forward line.

Figure 14-6 illustrates the notes for a short closed traverse for which the azimuths were observed as just described. As indicated in the notes, each line has both an observed forward azimuth and a set-off back azimuth whose values differ by 180°. The traverse is started from the line 1-5 the azimuth of which is 270°28′ reckoned from true north. The forward azimuth of line 1-2 is found to be 350°30′. When the transit is set up at station 2, the back azimuth is computed by subtracting 180° from the forward azimuth (350°30′ − 180° = 170°30′), and this value is set on the vernier before the backsight to station 1 is taken. Magnetic bearings are observed, and a check against mistakes or large errors is secured by noting that the computed bearings vary from corresponding magnetic bearings by about 9°10′, the amount of the magnetic declination.

Second Method. Another method, faster but more liable to mistake, consists in leaving the vernier setting unchanged between a foresight and the following backsight and plunging the telescope between each backsight and the corresponding foresight. For example, if *AB* is some transit line in the traverse and a foresight reading has been taken from *A* to *B* with telescope normal, the transit is brought forward to *B* without disturbing the vernier setting, and a backsight on *A* is taken with the telescope inverted. It is then plunged to the direct position (which orients the transit), and the foresight to *C* is obtained by turning on the upper motion. The reading of the *A* vernier gives the azimuth of the line *BC*. The vernier should always be read before a backsight is taken to make sure that no slip between plates has occurred while the transit was being brought forward.

The advantage of this method lies in increased speed and in probably smaller accidental errors of reading the vernier. No error is introduced beyond the initial setting of the vernier. The disadvantages are that a mistake may be made in reading an angle without its becoming evident at closure of the traverse and that, unless the line of sight is in perfect adjustment, a constant angular error is introduced at each set-up.

Third Method. In a third method of running an azimuth traverse, the upper motion is left clamped after each foresight until the transit has been carried forward and the backsight taken at the next station. The vernier opposite to that registering the forward azimuth is employed for the backsight

AZIMUTH TRAVERSE AT HIGH-WATER LINE

Proposed Mill Pond, El. 741.36
Silver Creek, Penn.
(For Land Damage Est.)

Sta.	Obj.	Dist.	Azimuth	Mag. B.	Cal. Bear.
1	5		270°28'	N 80½°W	N 89°32'W
	2	689.32	350°30'	N	N 9°30'W
2	1		170°30'	S	S 9°30'E
	3	509.66	303°05'	N 48°W	N 56°55'W
3	2		123°05'	S 48¼°E	S 56°55'E
	4	678.68	236°13'	S 65½°W	S 56°13'W
4	3		56°13'	N 65¼°E	N 56°13'E
	5	572.50	177°58'	S 7°W	S 2°02'E
5	4		357°58'	N 7¼°E	N 2°02'W
	1	1082.71	90°29'	S 80°E	S 89°31'E
			Error =	01'	

Gurley transit No. 191

J. Stanbois ⊼
F. Lowe
June 15, 1965
Cloudy, Warm

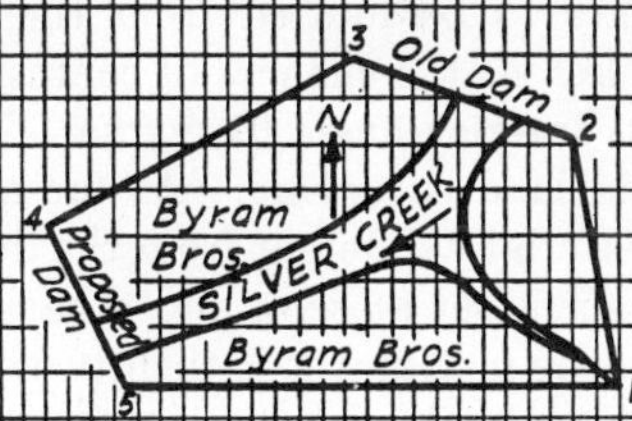

True azimuth of line 1-5 found by solar observation.
Mag. declination = 9° 10' W

Fig. 14-6. Notes for short closed azimuth traverse.

and for all other readings at a given station, and the telescope is not plunged between backsight and following foresight. The method depends upon the fact that, if there is no slip between plates as the transit is moved from one station to the next, the opposite vernier will indicate the back azimuth and can be used for the backsight. No resetting of a vernier is required; thus the method eliminates the accidental error which is inherent in the first method. Furthermore, since the telescope is not plunged while the transit is set up at a given station, there is no error due to the line of sight's not being perpendicular to the horizontal axis, which error is inherent in the second method. Eccentricity of verniers would introduce an error, but this error is infrequent and is easily detected.

For convenience in reading the opposite vernier at each alternate station, usually the telescope is plunged between foresight and following backsight. When this is done, the vernier to be read is that nearest the eyepiece end of the telescope. To guard against reading the wrong vernier or sighting with the telescope in the wrong position, the station numbers in the notes may be alternately marked A and B.

14-12. Traverse by Angles to the Right. This method is similar to the azimuth method, except that at each station the backsight to the preceding transit station is taken with the A vernier set at zero. The instrument is turned on the upper motion, a foresight is taken to the following station, and the clockwise angle is read on the A vernier. Angles to the right are often called *azimuths from back line.* Notes are kept in much the same form as for deflection angles (Figs. 14-4 and 14-5). Traverse angles are checked either by magnetic bearings or by doubling.

The method is used mostly on open traverses, particularly where many details are to be located from the traverse stations. For such work the method lessens the chances of confusion.

14-13. Interior-angle Traverse. This method of traversing is used principally in land surveying. As far as the field operations are concerned, it is not materially different from the deflection-angle method. At each station the vernier is set at zero, and a backsight to the preceding transit station is taken. The instrument is then turned on its upper motion until the advance station is sighted, and the interior angle is read. Either the notes may be kept in the form of a sketch, or the numerical values may be tabulated in form similar to that of Fig. 14-4. Angles may be checked by the geometrical relation that in any polygon having n sides the sum of the interior angles is $(n - 2)180°$.

14-14. Checking Traverses. The errors involved in traversing are of two kinds, angular and linear. Magnetic bearings offer an excellent means of checking observed angles against mistakes or large errors. On traverses of ordinary precision this check is usually applied by reading the compass needle on the transit and then comparing the observed bearing with the bearing computed from observed transit angles.

On important traverses, often the angular values are checked by doubling the angles and the linear measurements are checked by taping forward and back over each line.

Closed Traverses. For a closed traverse it is possible by known geometrical relations to determine the angular error of closure. Thus, for a deflection-angle traverse the algebraic sum of the deflections should be 360°, for an interior-angle traverse the sum of the interior angles should equal $(n - 2)180°$, and for an azimuth traverse the azimuth of the first line as observed at closure should equal the known or assumed azimuth of the same line at the beginning of the survey.

The actual total error in measured distances cannot be determined. But the coordinates of the first point as computed by successive angles and distances around the traverse should equal the coordinates of the same point as used at the beginning of computations, and this condition renders it possible to calculate the *linear error of closure* due to errors in both angles and distances (Chap. 18). (However, a closed traverse may close perfectly even though the tape is not of the correct length.)

It is customary to compute the angular error of closure in the field. When the linear error is determined, as on all important traverses, the calculations are made in the office. Where graphical methods are of sufficient precision, a short traverse may be checked expeditiously by plotting with protractor and scale or by other methods later to be described, but for long many-sided traverses the cumulative errors of plotting are likely to be large.

Open Traverses. For an open traverse no means are available for checking the measurements as a whole. Although linear errors cannot be detected, angular errors can be determined closely by astronomical observations taken at intervals as the traverse progresses, taking the convergency of meridians (Art. 23-11) into account if the distances are considerable. Where conditions allow, *cut-off lines* run between certain intermediate points are established to check parts of the traverse. Thus in Fig. 14-7 the line $ABC \cdots N$

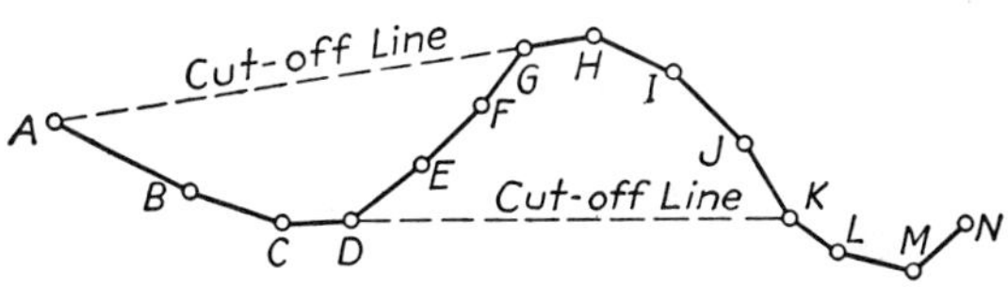

FIG. 14-7. Cut-off lines.

represents an open traverse. If the direction of the cut-off line AG is observed at both A and G, the angular measurements of the traverse from A to G may be checked, and further if the distance AG is measured the linear error of closure may be computed. Similar measurements for the cut-off line DK would enable the checking of the traverse up to K. The method really amounts to running a succession of closed traverses. Conditions favorable to establishing a cut-off line are met only occasionally, and often then it is not convenient to measure the length of the line.

By observing the angle to some distant landmark from each of several stations, several determinations of the rectangular coordinates of the mark may be obtained. If these values agree closely, it is good evidence that all angular and linear measurements involved are free from serious errors. If there is disagreement between computed coordinates, there is no way of telling with certainty just where the error lies, but usually its location may be fixed approximately.

In some cases the open traverse begins and ends at points the locations of which have been accurately determined by previous field operations (for example, the triangulation stations of the U.S. Coast and Geodetic Survey or the monuments set in connection with state plane-coordinate systems). In such cases the error of closure of traverse on the known point may be determined by comparing its established or accepted coordinates with those computed from the traverse observations. The error of closure obtained by this method contains the combined effect of angular and linear errors.

A series of traverse lines between stations some distance apart may be projected (by dropping perpendiculars) onto a single straight line which connects the stations, and this line may then be treated as if it were a measured line of the traverse.

14-15. Precision of Transit-tape Traverses. The precision of transit-tape traverses is affected by both linear and angular errors of measurement. The precision of linear measurements with the tape is discussed in Art. 7-23; the precision of angular measurements with the transit is discussed in Art. 13-35. Under ordinary conditions the *angular* errors are largely accidental, hence the probable error in the direction of any line in a traverse with respect to any other may be expected to vary as the square root of the number of set-ups between the two lines. On the other hand the important *linear* errors on traverses of ordinary precision are likely to be systematic. The precision of the location of any transit point, therefore, is generally influenced much more by the systematic linear errors than by the accidental angular errors; and, except on surveys of high precision, the precision is usually found to vary approximately as the length of the traverse lines. In stating limits of error in transit work the ratio of linear precision (as 1/5,000) is used.

The relation between the precision of angles and that of related distances is discussed in Arts. 3-7, 3-8, and 4-5. In ordinary surveying, transit angles are usually read to 01′; the corresponding linear error per 1,000 ft. is 0.291, and the ratio of precision is 1/3,440. Hence in ordinary work the consistent precision of chaining should be about 1/3,000. Similarly, if angles are read to 30″ or if angles read to 01′ are doubled, the corresponding precision of chaining should be about 1/6,000.

Unless a traverse either forms a closed figure or begins and ends on points previously established by measurements known to be practically correct, its precision is indeterminate; but if the proper procedure is employed (see Arts. 7-23 and 13-35), and if the several angles and distances making up the traverse are checked by a second measurement or by other methods previously discussed, there is reasonable assurance that the precision will not be below a fixed standard. The required precision of a traverse, whether open or closed, depends upon the character, purpose, and extent of the survey, and for any given case is a value to be fixed after due consideration of the factors involved.

On important or extensive surveys it is common practice for those in charge to issue definite instructions covering in detail the procedure to be followed by the several members of the surveying staff and specifying the maximum allowable discrepancies between check measurements or otherwise signifying the precision which it is desired to maintain. Where a traverse forms a closed figure, the angular error of closure is ordinarily determined through fixed geometrical relations already mentioned, and the linear error of closure is determined by trigonometric computations which will be described in Chap. 18. Often limits are placed upon both the angular and the linear error of closure. Typical instructions and specifications for various degrees of precision are given in the following article.

14-16. Specifications for Traversing. Limitations as regards the skill of surveying personnel, quality of instruments, and field conditions are so numerous as to make any statement of the precision to be attained in traversing with the transit of only very general value. The following specifications have been prepared with this in mind, and the values therein given are not by any means considered fixed. The specifications give *maximum* values; and, if the surveys are executed by well-trained men, with instruments in good adjustment and under average field conditions, in general the error of closure should not exceed *half* the specified amount. Even in rough chaining, the effects of the systematic errors can be greatly reduced by easily calculated corrections applied to the measured values (Arts. 7-15 to 7-23). The specifications apply to traverses of considerable length. It is assumed that a standardized tape is used.

Class 1. Precision sufficient for many preliminary surveys, for horizontal control of surveys plotted to intermediate scale, and for land surveys where the value of the land is low.

Transit angles read to the nearest minute. Sights taken on a range pole plumbed by eye. Distances measured with a 100-ft. steel tape. Pins or stakes set within 0.1 ft. of end of tape. Slopes under 3 per cent disregarded. On slopes over 3 per cent, distances either measured on the slope and corrections roughly applied, or measured with the tape held level and with an estimated standard pull.

Angular error of closure not to exceed $1'30'' \sqrt{n}$, in which n is the number of observations. Total linear error of closure not to exceed 1/1,000.

Class 2. Precision sufficient for most land surveys and for location of highways, railroads, etc. By far the greater number of transit traverses fall in this class.

Transit angles read carefully to the nearest minute. Sights taken on a range pole carefully plumbed. Pins or stakes set within 0.05 ft. of end of tape. Temperature corrections applied to the linear measurements if the temperature of air differs more than 15°F. from standard. Slopes under 2 per cent disregarded. On slopes over 2 per cent, distances either measured on the slope and corrections roughly applied, or measured with the tape held level and with a carefully estimated standard pull.

Angular error of closure not to exceed $1' \sqrt{n}$. Total linear error of closure not to exceed 1/3,000.

Class 3. Precision sufficient for much of the work of city surveying, for surveys of important boundaries, and for the control of extensive topographic surveys.

Transit angles read twice with the instrument plunged between observations. Sights taken on a plumb line or on a range pole carefully plumbed. Pins set within 0.05 ft. of end of tape. Temperature of air determined within 10°F. and corrections applied to the linear measurements. Slopes determined within 2 per cent and corrections applied. If tape is held level, the pull kept within 5 lb. of standard and corrections for sag applied.

Angular error of closure not to exceed $30'' \sqrt{n}$. Total linear error of closure not to exceed 1/5,000.

Class 4. Precision sufficient for precise surveying in cities and for other especially important surveys.

Transit angles read twice with the instrument plunged between readings, each reading being taken as the mean of both A and B vernier readings. Verniers reading to 30″. Instrument in excellent adjustment. Sights taken with special care. Pins set within 0.02 ft. of end of tape. Temperature of tape determined within 5°F. and corrections applied. Slopes determined within 1 per cent, and corrections applied. If tape is held level, the pull kept within 3 lb. of standard, and corrections for sag applied.

Angular error of closure not to exceed $15'' \sqrt{n}$. Total linear error of closure not to exceed 1/10,000.

14-17. Referencing Transit Stations. Many of the hubs marking the location of highways, railroads, and other works of man are bound to be uprooted or covered during the progress of construction; and they must be replaced, often more than once, before construction is completed. Such hubs marking the transit stations are usually tied by angular and/or linear measurements to temporary wooden hubs called *reference hubs,* or to other objects that are not likely to be disturbed. A transit station is said to be *referenced* when it is so tied to nearby objects that it can be replaced readily. The manner of referencing a transit station is indicated in the notes by an appropriate sketch. The precision of referencing should be comparable to that of the measurements between transit stations. Particularly on land surveys, the corners should be tied to nearby objects which can be found readily, which are not likely to be moved or obliterated, and which are of a more or less permanent character.

Often in land surveying, a corner is incorrectly said to be "witnessed" when angular and linear measurements are taken to nearby objects of the character just mentioned. This unfortunate designation leads to the confusion of reference marks for a corner which has been established with *witness corners* (Art. 23-24), which are markers set on one or more of the land lines leading to a corner when the corner falls in a place where it would be either impossible or impracticable to establish or to maintain a monument.

Figures 14-8 to 14-11 illustrate several methods of referencing a transit station. The station shown in Fig. 14-8 is tied by linear measurements to two reference hubs. While the original measurements can be made very expeditiously, the field work connected with replacement of the station is not so simple, and if either of the reference hubs is disturbed, the station cannot be relocated. The tie lines should intersect at a favorable angle (preferably about 90°); otherwise the station cannot be relocated with certainty.

Figure 14-9 shows a station referenced by setting the two reference hubs on a line passing through the station and by taking linear measurements from these hubs to the station. Replacement is accomplished by setting

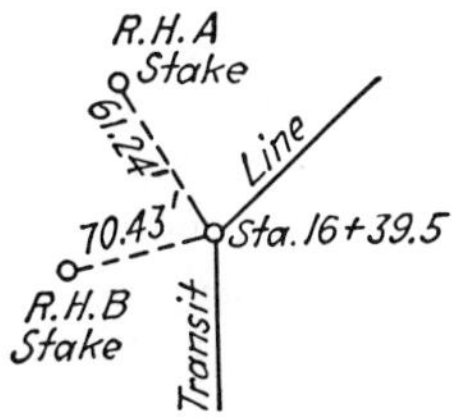

FIG. 14-8.

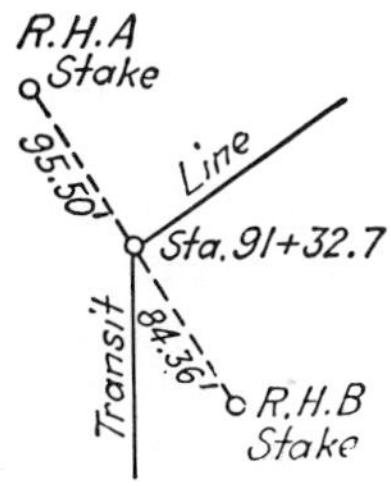

FIG. 14-9.

up the transit at R.H. *A* and sighting to R.H. *B* (or *vice versa*) and then laying off along the line thus established the given distance from reference hub to station; only one of the linear measurements is necessary, but the other is desirable as a check. If either of the reference hubs is disturbed, the station cannot be relocated.

Figure 14-10 represents a third method, often employed when there is any likelihood of the reference hubs being accidentally lost. The station

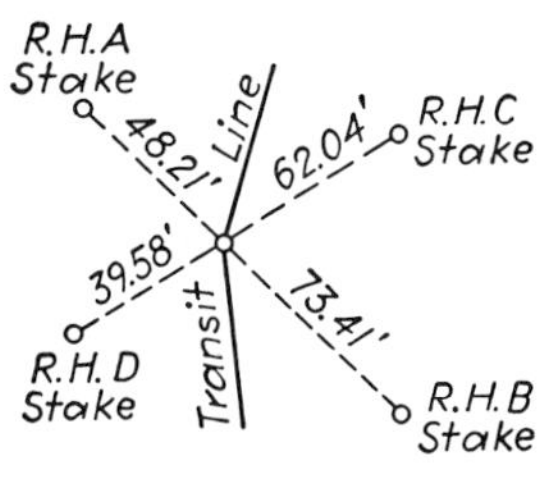

FIG. 14-10.

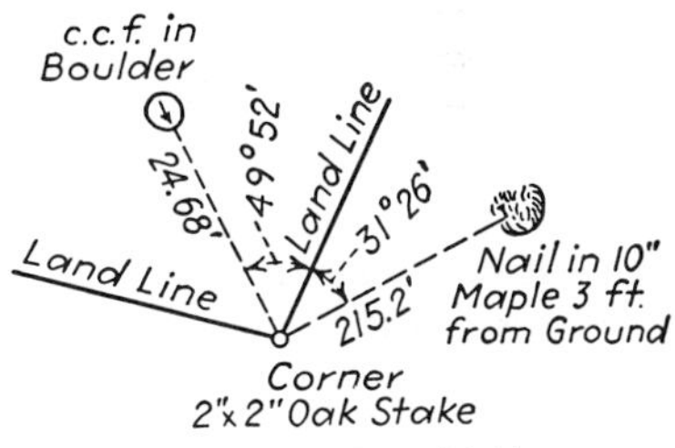

FIG. 14-11.

is at the intersection of the line *AB* and the line *CD*. With the measurements as shown, any two reference hubs may be destroyed and still the station may be relocated. Sometimes it is more convenient to place all the reference points to one side of the transit line and to locate the station by angular measurement only.

Figure 14-11 is typical of the manner in which corners are referenced in land surveying. Both angular and linear ties are employed.

14-18. Details from Transit Lines. On nearly all transit surveys certain *details,* or natural and artificial features of the terrain, are located with respect to the transit lines. The nature and number of details to which measurements are taken depend both upon the purpose of the survey and upon the character of the country through which the survey runs. On property-line surveys, only a few important objects close to the transit lines would need to be located. On the other hand a topographic survey would include the location of many features of the terrain.

The precision with which details are located likewise depends upon the purpose of the survey. When property lines are retraced, the actual lines may be obstructed by hedges or buildings, so that the actual corners must be located by measurements from other transit lines; such measurements should be taken with a precision as great as that for the transit line. If details are located solely for map-making purposes, usually the required precision of measurements to details is less than that for the transit lines.

All well-defined objects should be correctly shown within the scale of the map, bearing in mind that points cannot be plotted within less than perhaps 0.01 in. Thus if the scale of the map were 1 in. = 1,000 ft. there would be no particular advantage in taking measurements closer than the nearest 10 ft. from a transit line to details, but if the scale were 1 in. = 10 ft., measurements should be taken to 0.1 ft.

Angular measurements to details are usually made with the transit, with angles read to minutes. Often the angles used in mapping details are estimated to the nearest 05′ without the aid of the vernier in order to save time. Where details are located with respect to stations intermediate between transit stations, often some hand instrument is employed to measure the angles or directions.

Linear measurements to details are made with the 100-ft. steel tape, with the 50-ft. metallic tape, by stadia (Chap. 15), or sometimes by pacing. Where distances are long and where a high precision is required, the steel tape is employed. Where a considerable number of short distances are to be measured, measurements with the metallic tape are sufficiently precise for most purposes. In securing data for a map, the stadia method is sufficiently precise except for very large scales. Distances to details of indefinite outline (for example the bank of a stream or the edge of a wood) are sometimes determined by pacing.

On most surveys the details are located as the traverse progresses; thus at any transit station the foresight to the following transit station is observed and then the transit is left in position until angles to details have been read. These observations with the transit are called *side shots* to distinguish them from the traverse angles.

14-19. Locating Details. Following are descriptions of the common methods of locating details with the transit and tape. The method or combination of methods is chosen which requires the least time in the particular case. In general, the *dimensions* of an object (as a building) should be determined by direct measurement. For rectilinear objects such as lots which may not be square, the diagonals may be measured.

Usually the notes show by sketches the character and general position of the objects located. Where details are numerous, each observed point is given a number on the sketch, and angles and distances are tabulated opposite corresponding numbers on the left-hand page of the notebook. If angular values are not placed on the sketch, care should be taken that the notes make evident the manner in which the angles were observed and the transit line to which the angles are referred. Where details are numerous, azimuths rather than deflection angles are observed for the reason that the chances of later confusion are much less.

1. *By Angle and Distance from Transit Station.* As illustrated by Fig. 14-12, any given point on the object is located by an angle and a distance. The

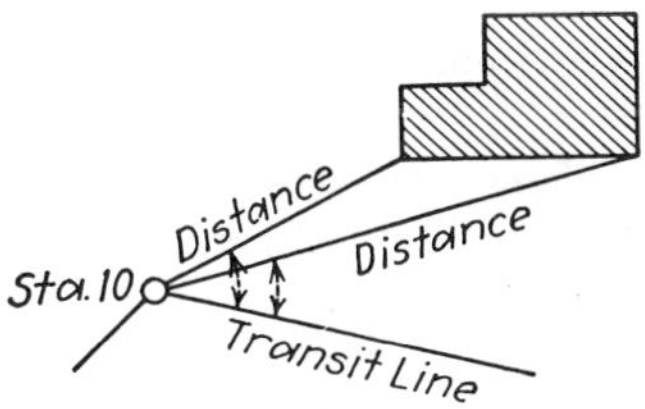

FIG. 14-12.

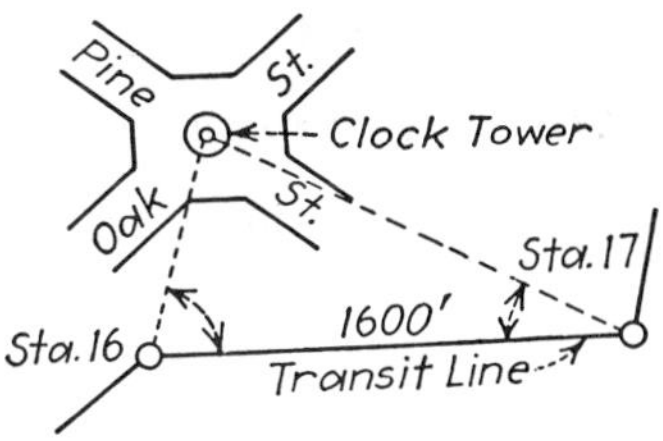

FIG. 14-13.

method is widely used, particularly where the details are close to transit stations.

2. *By Angles from Two Transit Stations.* As shown by Fig. 14-13, the point is located with respect to the transit line by angles taken from at least two stations, that is, by the method of intersection. An advantage of the method is that no linear measurements to details are required; hence the field work is reduced. A disadvantage is that the distance from transit station to point sighted can be determined only by computation (except that a rough value can be scaled from a map). Also the location of the point becomes indefinite as the angle at the point approaches 0° or 180°. The method is useful for distant or inaccessible points.

3. *By Distances from Two Stations.* On transit lines for which stakes are set every 100 ft. and where the details are close to the traverse line yet distant from the nearest transit station, they are conveniently located by linear ties from two traverse stations. Thus, in Fig. 14-14 the stone bound at the fence corner is located by distances from stations 4 + 00 and 5 + 00, neither of which is a transit station.

4. *By Angle from One Station and Distance from Another.* Occasionally angles are measured from a transit station and distances are measured from intermediate stations. Figure 14-15 illustrates a situation where the method would prove advantageous, the lake making impossible the direct measurement of the distance from the transit station at 16 + 40.5 to the corner of the building, but there being no obstacle to the measurement of the distance

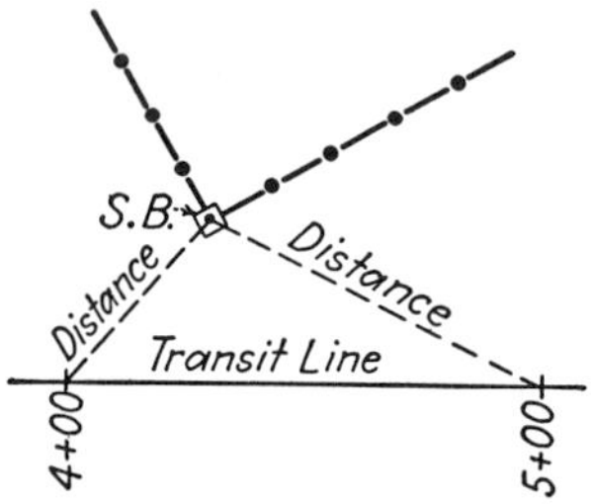

FIG. 14-14.

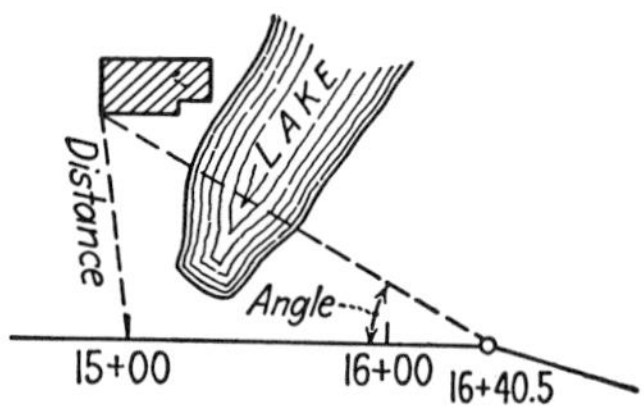

FIG. 14-15.

from station 15 + 00 to the corner. Care must be taken to secure good intersections.

5. *By Ranges, Range Ties, and Swing Offsets.* If the features to be located are buildings, the work of location may be facilitated by *ranging*, or sighting along one or more sides of the building and finding the points of intersection of lines thus defined with other lines such as the transit line, a fence, or the side of another building. The station and plus of such a point of intersection

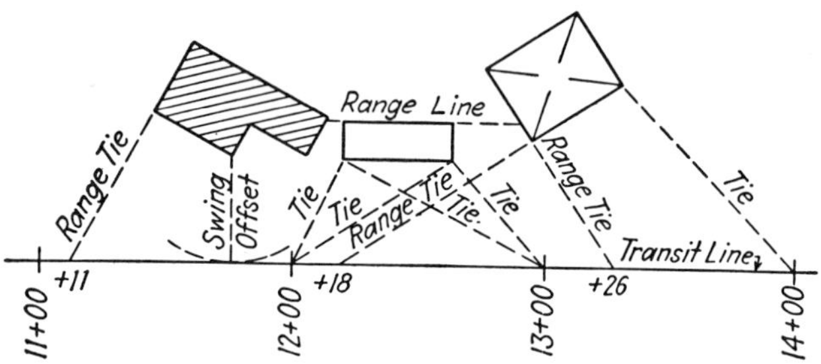

FIG. 14-16. Details by ranges, ties, and offsets.

with a transit line is called a *range,* and a range together with the distance along the range line to a corner of the building is called a *range tie.*

A *swing offset,* or perpendicular distance from a transit line to a point, is determined by swinging the tape about the point as a center and taping the least distance from the point to the transit line.

Figure 14-16 shows how a group of buildings near a traverse line may be located by tape measurements only. The building at the right is completely located by two range ties, one intersecting the traverse line at 12 + 18 and the other at 13 + 26. The location of the building is checked (assuming that the lengths of sides are measured) by the check tie to station 14 + 00. The location of the middle building of the group is established by ties to station 12 + 00 and station 13 + 00, three ties being sufficient to locate the building and the fourth tie being taken to check the location. The location of the building on the left is fixed by the range tie intersecting the traverse line at 11 + 11 and by the swing offset shown. The locations are checked by the range line tying the three buildings together.

6. *By Perpendicular Offsets from the Transit Line.* This method is adapted to the location of irregular or curved boundaries, streams, and roads that closely parallel the transit lines. As indicated by Fig. 14-17, a point is

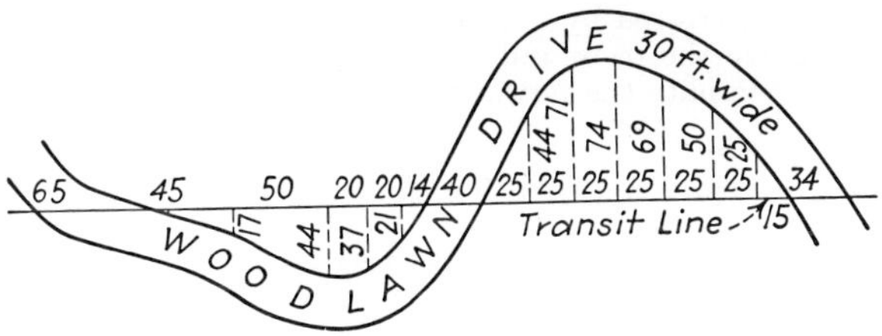

Fig. 14-17. Details by perpendicular offsets.

located by measuring the distance along the transit line to the foot of a perpendicular offset through the point and then measuring the length of the perpendicular offset. Features such as those mentioned are located sometimes by offsets at regular intervals, but more often at critical points which will make the offsets come at irregular intervals. Where stakes are placed every 100 ft. along the transit line, the station and plus of the foot of each perpendicular is secured, rather than the distance between offsets as shown.

Usually the direction of the perpendiculars is estimated by eye except where offsets are long; in this case the tape, magnetic compass, sextant, right-angle mirror, or pentagonal sighting prism is employed. Where extreme precision is required, the offsets are laid off with the transit.

14-20. Field Problems.

Problem 1. Open Deflection-angle Traverse with Transit and Tape

Object. To locate a section of an assigned route, setting stakes at full stations. This problem may be employed in connection with field problems 1 (Art. 10-18) and 1 (Art. 26-30). For plotting, see office problems 1 and 2 (Art. 18-26).

Procedure. (1) Stake out a route perhaps 1,500 ft. long with three or four changes in direction. (2) Set up the transit over the first stake marking a change in direction. With the *A* vernier set at zero and with the telescope inverted, sight on the stake at the beginning of the line. Read and record the magnetic bearing of this backsight. (3) Tape the line from the beginning (station 0 + 00) to the transit station, and record the length to the nearest 0.01 ft. The transitman lines in the head chainman. As the rear chainman pulls each pin, he replaces it with a stake on which he has written the station number. (4) Plunge the telescope, unclamp the upper motion, and sight on the next forward transit station. Record the reading of the *A* vernier and *R* or *L* to indicate whether the deflection is right or left from the prolongation of the preceding line. Also record the magnetic bearing of the forward line. (5) Before the transit is moved, compute the deflection angle from the magnetic bearings taken at the station and compare with the deflection angle indicated on the horizontal circle, as a rough check. (6) Tape the forward line. (7) Set up the transit at succeeding stations and observe the deflection angle at each. Give line for the chainmen on each forward line. (8) Keep notes in the form of the sample notes (Fig. 14-5). Include sketches of streams, roadways, fences, etc., crossed by the line. (9) Assume as correct the magnetic bearing of some course the back and forward bearings of which have the same angular value, and compute the forward bearing of each of the other courses.

Problem 2. Closed Azimuth Traverse with Transit and Tape

Object. To collect data sufficient for plotting the boundaries and determining the area of a field, employing the azimuth method of traversing. The data may be used later, as in the following field problem (problem 3) and in office problem 1 or 2 (Art. 18-26).

Procedure. (1) Stake out an irregular field having perhaps five sides and containing an acre or more. (2) Measure the sides with the steel tape, to the nearest 0.01 ft. (3) Set up the transit at one corner. Set the *A* vernier at 0°, and turn the instrument on the lower motion to sight along the magnetic meridian either north or south according as azimuths are to be reckoned from north or south. Clamp the lower motion in this position. (4) Unclamp the upper motion, and turn the instrument to sight the next corner forward. The angle turned off in a clockwise direction and read from the *A* vernier is the azimuth of the forward line. Record the azimuth. (5) Record the magnetic bearing. Compare this with the bearing computed from the azimuth of the line. (6) Compute the back azimuth of the line by adding 180° to the forward azimuth. (7) Set up the instrument on the next corner forward, and with the *A* vernier set on the back azimuth of the line, backsight on the corner previously occupied. The instrument is now oriented. (8) Turn the instrument on the upper motion, and sight on the next corner forward; the azimuth of the forward line is then indicated by the *A* vernier. (9) Proceed in this manner until each corner has been occupied. Also set up again at the first corner, and take readings as for the other corners. (10) Keep notes in the form of the sample notes (Fig. 14-6). (11) Note the angular error of closure, which should not exceed $30'' \sqrt{\text{number of sides}}$. Distribute the error equally among the angles.

Problem 3. Details with Transit and Tape

Object. To obtain sufficient data for plotting a detailed map of a portion of the campus. The data may be used in office problem 1 or 2 (Art. 18-26).

Procedure. (1) Run a closed azimuth traverse, as described in the preceding problem, through the area to be mapped. Locate the lines and corners of the traverse so that linear and angular measurements necessary to fix the position of details with respect to the traverse may be taken with the least labor. Reference the corner hubs. (2) Obtain the location of buildings, streets, walks, hydrants, fences, etc., by the several methods of Arts. 14-18 and 14-19. (3) Sketch indefinite details such as trees, streams, and shrubbery without measurements other than by pacing. (4) When the traverse is plotted (Art. 18-26), plot the details according to the manner in which they were secured.

Hints and Precautions. (1) Details may be taken as the traverse is run. (2) Determine the angular error of closure of the traverse. If this exceeds

$$30'' \sqrt{\text{number of sides}}$$

rerun the traverse until the mistake is discovered. (3) The location of important details should be checked, preferably by a different method. (4) Sketches should not be overcrowded with measurements. Many measurements can be tabulated on the left-hand page and referred to on the sketch by a single letter or number.

REFERENCES

1. Birdseye, C. H., "Topographic Instructions of the United States Geological Survey," *U.S. Geol. Survey, Bull.* 788, Government Printing Office, Washington, D.C., 1928.
2. See also references for Chap. 25.

CHAPTER 15

STADIA SURVEYING

15-1. Tacheometry. Distances can be measured indirectly by sighting through a small angle at a distant scale transverse to the line of sight and in the plane of the angle. The angle may be fixed and the length it subtends on the scale measured, or a length on the scale may be fixed and the angle measured. The general process is called *tacheometry* or *tachymetry;* various forms are described in the reference at the end of this chapter.

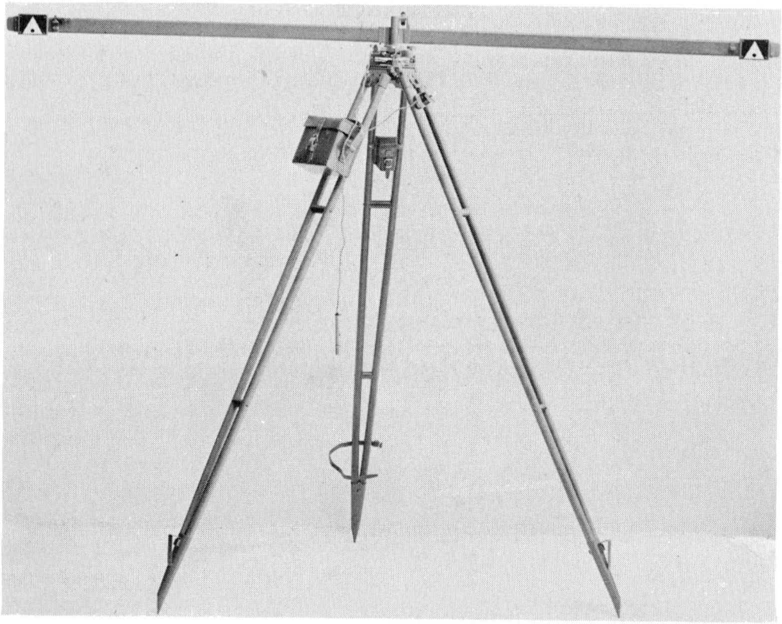

FIG. 15-1. Wild invar subtense bar.

In European practice, usually a horizontal rod of fixed length, called a *subtense bar*, is used. Sighting marks usually 2 m. apart are supported on a bar which is mounted on a tripod (Fig. 15-1). The tripod is centered over the station to which it is desired to measure, the bar is leveled, and the bar is oriented perpendicular to the line by sighting through an attached telescope toward the angle-measuring station. The horizontal angle subtended by the 2-m. length is then read precisely by means of a theodolite or a transit, and the distance is computed by trigonometry. The horizontal angle is independent of the inclination of the line of sight; however, for most purposes the measured inclined distance must be reduced to the equivalent horizontal distance.

Although the subtense bar is used to some extent in the United States, common practice is to employ the *stadia* method of measuring distances indirectly. In this method the sighting telescope of the transit or plane-table alidade has a fixed small angle defined in the vertical plane, and the length which this angle subtends is read on a graduated rod held vertical (whether or not the line of sight is inclined) on the distant station.

15-2. Stadia Method. The stadia method is employed extensively on topographic, hydrographic, and other surveys conducted for the purpose of securing data for the plotting of maps (see Art. 15-13) It is far more rapid than taping, and under certain conditions is as precise. It is a useful means of checking more precise measurements.

The equipment for stadia measurements consists of a telescope with two horizontal hairs called *stadia hairs* and a graduated rod called a *stadia rod.*

The process of taking a stadia measurement consists in observing through the telescope the apparent locations of the two stadia hairs on the rod, which is held vertical. The interval thus determined, called the *stadia interval* or *stadia reading,* is a direct function of the distance from the instrument to the rod, as demonstrated in Art. 15-6. For most instruments the ratio of distance to stadia interval is 100.

15-3. Stadia Hairs. The telescopes of transits, theodolites, plane-table alidades, and many levels are furnished with stadia hairs in addition to the regular cross-hairs, one stadia hair above and the other an equal distance below the horizontal cross-hair. Stadia hairs are usually mounted on the same ring and in the same plane as the horizontal and vertical cross-hairs. Under these conditions the stadia hairs are not adjustable with respect to each other, and hence the distance between hairs remains unchanged. Both stadia hairs and cross-hairs are simultaneously visible and in focus. The advantage of fixed stadia hairs is that the interval between them cannot be accidentally altered. The disadvantage is that the hairs may be so placed as to produce an inconvenient stadia interval factor (ratio of distance to interval), but considering the precision with which the hairs are usually placed by the manufacturer, the interval factor is so nearly 100 that frequently it may be so considered without appreciable error.

To eliminate the possibility of confusing the stadia hairs with the horizontal cross-hair in transit or level work not involving the stadia method, the stadia hairs are sometimes mounted in a plane a short distance in the rear of the cross-hairs. Under these conditions the stadia hairs and cross-hairs are not simultaneously visible, and it is necessary to change the focus of the eyepiece to render visible the stadia hairs when the ordinary cross-hairs have been in use. Stadia hairs mounted as just described are called *disappearing* stadia hairs.

Formerly instruments were manufactured with adjustable stadia hairs so that the interval between hairs could be regulated to make the interval factor any desired quantity. In general, the adjustable feature is not regarded with favor, owing to the fact that the adjustment is likely to be accidentally disturbed.

15-4. Stadia Rods. The stadia rod is usually graduated in decimals of a foot but may be graduated in decimals of a meter or a yard. Any self-reading

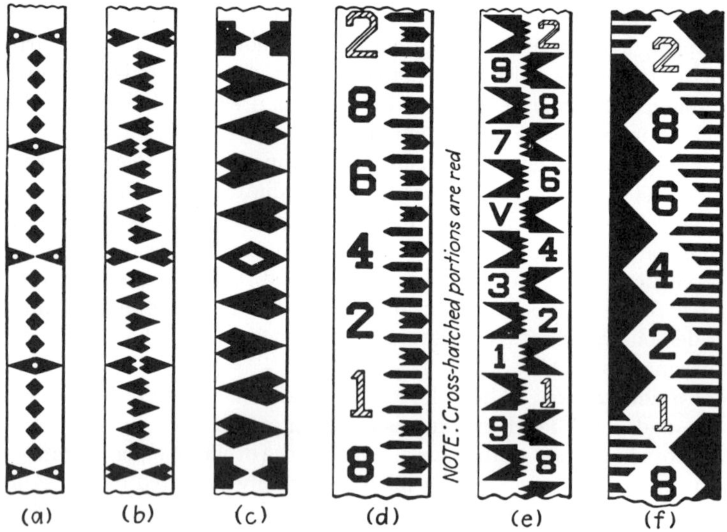

FIG. 15-2. Stadia rods.

leveling rod may be used as a stadia rod, but the common leveling rod graduated in hundredths of feet (as illustrated by Figs. 8-11 and 8-12) is suitable only for short sights, say less than 400 ft. For longer sights a rod with larger and heavier graduations is necessary. The graduations shown by Fig. 15-2*a* are suitable for distances up to 700 ft. Figure 15-2*e* and *f* illustrates patterns which combine fine graduations for accurate readings at short distances and heavy graduations for observations at long distances. These rods are equally adapted for use in stadia surveying and in leveling. For stadia work alone the finer graduations are usually omitted and numbers indicating feet and tenths are often not shown.

The U.S. Coast and Geodetic Survey leveling rod shown in Fig. 9-7 is suitable for stadia readings. It is graduated in foot units on one side and metric units on the other. This arrangement may be used in stadia work where it is desired to take readings in one unit of measurement and to check them by means of readings taken in the other unit.

For general stadia work where the length of sight may be 1,500 ft. or greater, the rods are usually 3 or 4 in. wide and 10 to 15 ft. long and the finest division is 0.05 ft. Stadia rods are usually made in one piece with graduations painted on the rod. For ease of transportation they are sometimes hinged or made in sections with a sleeve on one end of each section. Although rods of the patterns shown in Fig. 15-2*a* to *f* are procurable from manufacturers, many stadia rods are "home made" to the specifications of the individual surveyor. It is therefore not surprising that the number of patterns is large.

The wood for a rod should be well seasoned and should be from a light, tough, straight-grained species such as white spruce. The paint should be one which will withstand weather yet one which does not have a high gloss. The lacquer paints which can be applied with a brush or spray yet which dry quickly are suitable. If the rod is varnished, it may be rubbed to a dull finish with powdered pumice or a similar abrasive.

So-called *flexible* stadia rods, consisting of graduated ribbons of coated woven fabric, are on sale by instrument manufacturers. When such a ribbon is tacked to a board, it makes a satisfactory stadia rod. When removed from the board and rolled up, it occupies little space and is easily transported.

15-5. Observation of Stadia Interval. On transit or plane-table surveys the stadia interval is usually determined by setting the lower stadia hair on a foot mark and then reading the location of the upper stadia hair. Thus the stadia interval is mentally computed more easily and with less chance of mistake than if the lower hair were allowed to take a random position on the rod. When the vertical angle is taken to a given mark on the rod, the corresponding stadia interval is observed with the lower hair on the foot mark that renders a minimum displacement of the horizontal cross-hair from the mark to which the vertical angle is referred.

Thus, if a vertical angle were taken with the line of sight cutting the rod at 4.9 ft. and for this position of the horizontal cross-hair the lower stadia hair fell at 2.3 ft., the telescope would be rotated about the horizontal axis until the lower hair was at 2.0 ft., when the horizontal cross-hair would fall at 4.6 ft.

Whenever the stadia interval is in excess of the length of the rod, the separate half intervals are observed and their sum is taken.

For precise stadia work, the readings may be made by means of two targets on the rod, one target on, say, the 2-ft. mark and the other set by the rodman as directed by the instrumentman. To avoid excessive effects of atmospheric refraction, the intercept of the lower cross-hair with the rod should not fall nearer the ground than necessary.

15-6. Principle of the Stadia. Figure 15-3 illustrates the principle upon which the stadia method is based. In the figure the line of sight of the telescope is horizontal and the stadia rod is vertical. The stadia hairs are indicated by the points a and b; the distance between the stadia hairs is i. The apparent locations of the hairs on the rod are A and B, and the stadia interval is s.

In optics it is shown that a ray of light passing through the optical center of a lens remains undeviated in direction and, further, that rays which are parallel on one side of the lens are all brought to a focus at a fixed point on the optical axis. This point is called the *principal focus,* and its distance from the optical center is called the *focal length* of the lens.

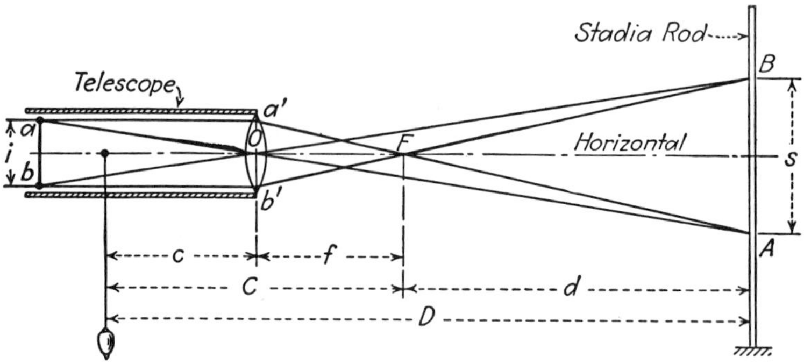

FIG. 15-3. Horizontal stadia sight.

Imagine that aa' and bb' in the figure are parallel rays emanating from the stadia hairs a and b. Then F is the principal focus through which these rays pass after emerging from the objective, f is the focal length of the objective, and the emerging rays take the positions $a'FA$ and $b'FB$. Also imagine that aOA and bOB are rays emanating, respectively, from a and b that pass undeviated through the optical center O.

As all rays from a given point are brought to a focus on the opposite side on the objective, it follows that if rays from a passing through O and F are brought to a focus at A, the reverse is true, namely, that rays from A which pass through F and O are brought to a focus at a. These two points a and A, or any other corresponding points as b and B, are called *conjugate foci,* and their distances from the optical center of the objective measured along the optical axis are called the *conjugate focal distances.*

As $ab = a'b'$, by similar triangles

$$\frac{f}{i} = \frac{d}{s}$$

Hence the horizontal distance from the principal focus to the rod is

$$d = (f/i)s = Ks$$

in which $K = f/i$ is a coefficient called the *stadia interval factor* which for a particular instrument is a constant so long as conditions remain unchanged. Thus for a horizontal sight the distance from principal focus to rod is obtained by multiplying the stadia interval factor by the stadia interval. The horizontal distance from center of instrument to rod is then

$$D = Ks + (f + c) = Ks + C \qquad (1)$$

in which C is the distance from center of instrument to principal focus. This formula is employed in computing horizontal distances from stadia intervals when sights are horizontal.

15-7. Stadia Constants. The focal distance f is a constant for a given instrument. It can be determined with all necessary accuracy by focusing the objective on a distant point and then measuring the distance from the cross-hair ring to the objective. The distance c, though a variable depending upon the position of the objective, may for all practical purposes be considered a constant. Its mean value can be determined by measuring the distance from the vertical axis to the objective when the objective is focused for an average length of sight.

Usually the value of $C = f + c$ is determined by the manufacturer and is stated on the inside of the instrument box. For external-focusing telescopes, under ordinary conditions C may be considered as 1 ft. without error of consequence. Internal-focusing telescopes (Art. 2-9) are so constructed that C is zero or nearly so; this is an important advantage of internal-focusing telescopes for stadia work.

15-8. Stadia Interval Factor. The nominal value of the stadia factor $K = f/i$ is usually 100. The interval factor can be determined by observation. The usual procedure is to set up the instrument in a location where a horizontal sight can be obtained and with a tape to lay off, from a point distant $C = f + c$ in front of the center of the instrument, distances of 100 ft., 200 ft., etc., up to perhaps 1,000 ft., stakes being set at the points established. The stadia rod is then held on each of the stakes, and the stadia interval is read. The stadia interval factor is computed for each sight by dividing the distance from the principal focus to the stake by the corresponding stadia interval, and the mean is taken as the most probable value. Owing to errors in observation and perhaps to errors from natural sources, the values of K for the several distances are not likely to agree exactly.

To overcome any prejudicial tendencies on the part of the instrumentman, observations may be made on the rod held on stakes at random distances from the instrument, these distances being measured later with the tape.

For use on long sights, where the full stadia interval would exceed the length of the rod, the stadia interval factor may be determined separately for the upper stadia hair and horizontal cross-hair and for the lower stadia hair and horizontal cross-hair.

With adjustable stadia hairs the interval factor is made 100 by moving the hairs until their rod intercept is $\frac{1}{100}$ of the distance from the principal focus to the rod, this distance being measured with a tape. The stadia hairs are so adjusted that the horizontal cross-hair bisects the space between them, each in its turn being moved vertically until the distance between it and the horizontal hair is the proper half-interval, as indicated by the rod intercept.

15-9. Inclined Sights. In stadia surveying, most sights are inclined, and usually it is desired to find both the horizontal and the vertical distances from

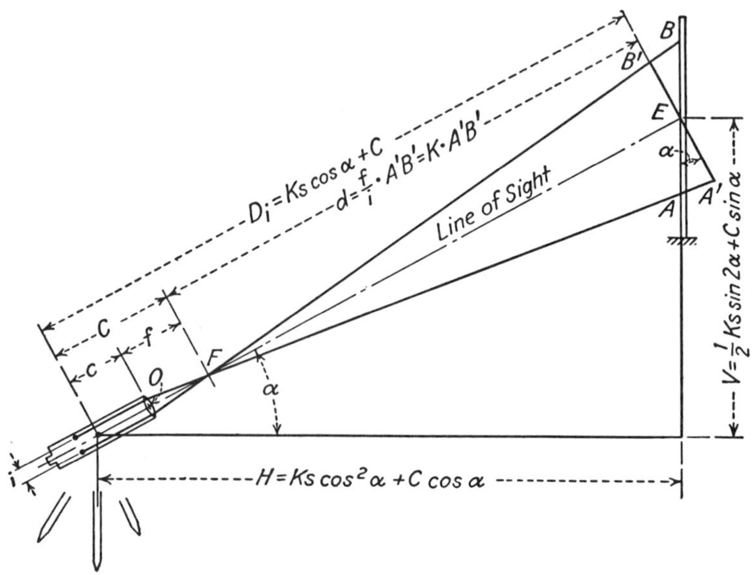

FIG. 15-4. Inclined stadia sight.

instrument to rod. The problem therefore resolves itself into finding the horizontal and vertical projections of an inclined line of sight. For convenience in field operations the rod is always held vertical.

Figure 15-4 illustrates an inclined line of sight, AB being the stadia interval on the vertical rod and $A'B'$ being the corresponding projection normal to the line of sight. The length of the inclined line of sight from center of instrument is

$$D_i = \frac{f}{i} \cdot A'B' + C \tag{2}$$

For all practical purposes the angles at A' and B' may be assumed to be 90°. Let $AB = s$; then $A'B' = s \cos \alpha$. Making this substitution in Eq. (2), and letting $K = f/i$, the inclined distance is

$$D_i = Ks \cos \alpha + C \tag{3}$$

The horizontal component of this inclined distance is

$$H = Ks \cos^2 \alpha + C \cos \alpha \tag{4}$$

which is the general equation for determining the horizontal distance from center of instrument to rod, when the line of sight is inclined.

The vertical component of the inclined distance is

$$V = Ks \cos \alpha \sin \alpha + C \sin \alpha$$

The equivalent of $\cos \alpha \sin \alpha$ is conveniently expressed in terms of double the angle α, or

$$V = \tfrac{1}{2}Ks \sin 2\alpha + C \sin \alpha \tag{5}$$

which is the general equation for determining the difference in elevation between the center of the instrument and the point where the line of sight cuts the rod. To determine the difference in ground elevations, the height of instrument and the rod reading of the line of sight must be considered.

Equations (4) and (5) are known as the *stadia formulas for inclined sights.*

15-10. Permissible Approximations. More approximate forms of the stadia formulas are sufficiently precise for most stadia work. Usually distances are computed only to feet and elevations to tenths of feet. Under these conditions, for side shots where vertical angles are less than 3°, Eq. (4) for horizontal distances may properly be reduced to the form

$$H = Ks + C \tag{6}$$

which is the same as for horizontal sights (Art. 15-6). But for traverses of considerable length, owing to the systematic error introduced, this approximation should not be made for vertical angles greater than perhaps 2°.

Owing to unequal refraction and to accidental inclination of the rod, observed stadia intervals are in general slightly too large (see Art. 15-19). To offset the systematic errors from these sources, frequently on surveys of ordinary precision the constant C is neglected. Hence in any ordinary case Eq. (4) may with sufficient precision be expressed in the form

$$H = Ks \cos^2 \alpha \text{ (approx.)} \tag{7}$$

Also Eq. (5) may often be expressed with sufficient precision for ordinary work in the form

$$V = \tfrac{1}{2}Ks \sin 2\alpha \text{ (approx.)} \tag{8}$$

However, the error in elevation introduced through using Eq. (8) may not be negligible, as for large vertical angles it amounts to several tenths of a foot.

Equations (7) and (8) are simple in form and are most generally employed. When K is 100, the common practice is to multiply mentally the stadia interval by 100 at the time of observation, and to record this value in the field notebook. This distance Ks is often called the *stadia distance*. Thus, if the stadia interval were 7.37 ft., the stadia distance recorded would be 737 ft.

For external-focusing telescopes, the degree of approximation in using Eqs. (7) and (8) may be greatly reduced either by adding 0.01 ft. to the observed stadia interval s or—when K is 100—by adding 1 ft. to the observed stadia distance Ks. It is convenient to add the correction mentally and to record the corrected value in the field notebook. The notes should state that corrected values are recorded.

15-11. Stadia Reductions. Ordinarily in practice the horizontal distances and the differences in elevation are not computed by actually solving the stadia formulas but are obtained by the use of a table, diagram, stadia slide rule, or stadia arc on the vertical circle of the transit or alidade (Art. 15-12), all these devices being based upon the formulas.

The precision of stadia surveying is such that ordinarily horizontal distances are determined to the nearest foot, and vertical distances are determined to the nearest 0.1 ft. Readings and computations are usually taken to three significant figures and in the lower range of four significant figures; slide-rule computations are sufficiently precise.

Table IX gives, for each 02′ of vertical angle up to 30°, the horizontal distances (from principal focus to rod) and differences in elevation for $Ks = 100$ ft., computed from equations $H = Ks \cos^2 \alpha$ and $V = \frac{1}{2}Ks \sin 2\alpha$ [see Eqs. (4), (5), (7), and (8)]. For any other value of Ks, the tabular quantities are to be multiplied by the value of Ks in hundreds of feet. The table also gives the horizontal distances and differences in elevation for three values of $C = (f + c)$, indicated as C in the table. Tables arranged differently from Table IX will be found in numerous publications.

Example: The following data were obtained by stadia observation: vertical angle = $+8°10'$, $s = 2.50$ ft. The stadia interval factor is known to be 95.0 and $C = 0.75$ ft.

In Table IX, the value given under "Hor. dist." is 97.98; hence the horizontal distance from principal focus to rod is $97.98 \times (95.0/100) \times 2.50 = 232.7$ ft. To this must be added the horizontal distance from principal focus to center of instrument, which is given at the bottom of Table IX as 0.74 ft. $232.7 + 0.7 = 233.4$, say, 233 ft.

Similarly, in Table IX the value given under "Diff. elev." is 14.06 ft., and at the bottom of the table 0.11 ft.

$$14.06 \times \frac{95.0}{100} \times 2.50 + 0.11 = 33.5 \text{ ft.}$$

Diagrams showing graphically the quantities $Ks \cos^2 \alpha$ and $\frac{1}{2}Ks \sin 2\alpha$ for all ordinary distances are published in a variety of forms. It is a simple matter to prepare such a diagram, and some surveyors prepare diagrams of their own design. The use of a stadia reduction diagram is considerably faster than the use of tables. The relative precision of the two methods depends upon the scale of the diagram, but values taken from the usual stadia diagrams are sufficiently precise for the ordinary purposes of stadia surveying.

The commonly used stadia slide rule is constructed like the ordinary slide rule, except that on the slide are given values of $\cos^2 \alpha$ and $\frac{1}{2} \sin 2\alpha$, these quantities being plotted to a logarithmic scale. This type is illustrated by Fig. 15-5. The upper and lower scales on the body of the rule represent values of distance (horizontal, vertical, or stadia), with values between 10 and 100 ft. common to both scales. The "$\cos^2 \alpha$" scale is at the right end of the upper scale of the slide; and the "$\frac{1}{2} \sin 2\alpha$" scale occupies the remainder of the upper edge of the slide and all the lower edge. To use

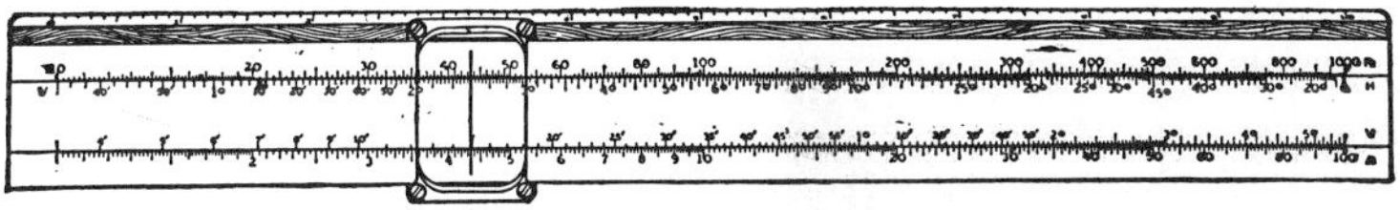

FIG. 15-5. Stadia slide rule.

the rule, first the index of the slide (at right end of upper scale) is set at the observed value of the stadia distance Ks on the main scale. Then the horizontal distance from principal focus to rod ($Ks \cos^2 \alpha$) is found by setting the runner at the observed vertical angle on the "$\cos^2 \alpha$" scale, and the corresponding difference in elevation is found by setting the runner to this same angle on the "$\frac{1}{2} \sin 2\alpha$" scale.

15-12. Beaman Stadia Arc. The Beaman stadia arc, in modified form known also as the *stadia circle,* is a specially graduated arc on the vertical circle of the transit or the plane-table alidade. It is used to determine distances and differences in elevation by stadia without reading vertical angles and without the use of tables, diagrams, or stadia slide rule. The stadia arc has no vernier, but settings are read by an index mark.

Horizontal Distance. In the type of stadia arc shown in Fig. 15-6*a*, the graduations for determining distances are at the left, inside the vertical circle. When the telescope is level (vertical vernier reading zero as shown), the reading of the arc is 100, indicating that the horizontal distance is 100 per cent of the observed stadia distance. When an inclined sight is taken, the observed stadia distance is multiplied by the reading of the "Hor." stadia arc, expressed as a percentage, to obtain the horizontal distance from principal focus to rod. For example, if the stadia distance is

411 ft. and the reading of the stadia arc is 99, the horizontal distance is 411 × 0.99 = 407 ft. The ordinary slide rule is sufficiently precise for the multiplication.

Another type of stadia arc (Fig. 15-6*b*) is graduated to give the *correction,* in per cent, to be subtracted from the observed stadia distance. Thus, for

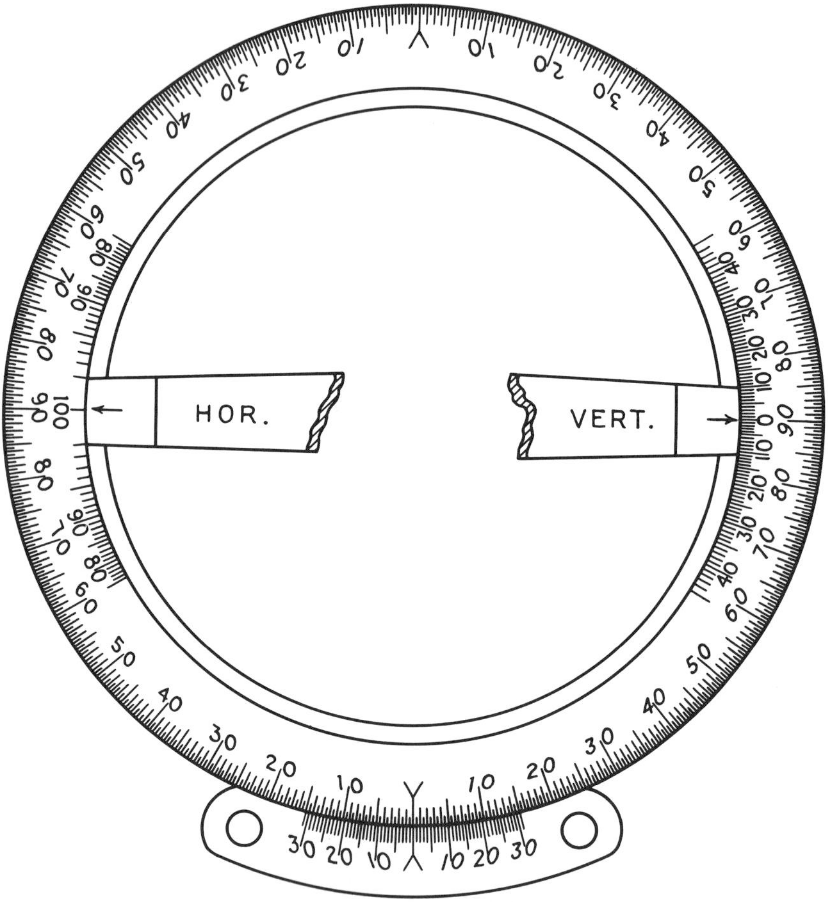

FIG. 15-6*a*. Stadia arc (multiplier type).

the foregoing example the reading of the stadia arc would be 1, and the horizontal distance would be 411 − (0.01 × 411) = 407 ft. Since the value of the correction is small, the multiplication can be performed mentally; but the computation involves both a multiplication and a subtraction.

Difference in Elevation. In Fig. 15-6*a*, the graduations for determining differences in elevation are at the right, inside the vertical circle. When the telescope is level (vertical vernier reading zero as shown), the reading of the arc is zero. When an inclined sight is taken, first the stadia distance is observed in the usual manner, that is, with the lower stadia hair on a foot mark of the rod. The telescope is then either elevated or depressed slightly until the nearest graduation on the "Vert." scale of the stadia arc coincides

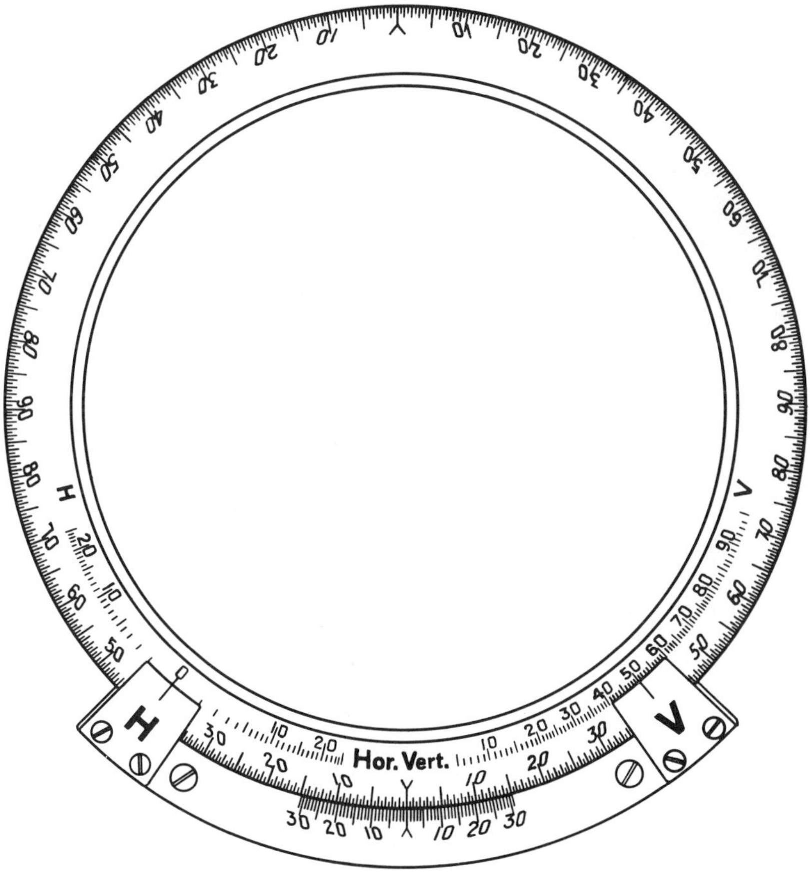

FIG. 15-6*b*. Stadia arc (horizontal scale subtraction type; vertical scale index at 50).

with the index of the arc (in order to avoid interpolation), and a rod reading is taken at the point where the line of sight strikes the rod. The observed stadia distance is multiplied by the reading of the "Vert." stadia arc, expressed as a percentage, to obtain the vertical distance from center of the instrument to point last sighted on the rod. This difference in elevation, combined with the height of instrument and the rod reading, gives the difference in elevation between the instrument station and the point on which the rod is held.

Example: The observed stadia interval with the rod held on a given point is 4.11 ft.; with the index of the stadia arc at −9 (the minus sign indicating that the line of sight is depressed) the line of sight falls at 3.3 ft. on the rod; the height of instrument (vertical distance from the instrument station to the center of the telescope) is 4.5 ft. The difference in elevation between the instrument station and the point on which the rod is held is to be found.

The difference in elevation between the center of the telescope and the point sighted on the rod is

$$-9 \times 4.11 = -36.99, \text{ say } -37.0 \text{ ft.}$$

The difference in elevation between the instrument station and the point on which the rod is held is

$$4.5 - 37.0 - 3.3 = -35.8 \text{ ft.}$$

The other type of stadia arc (Fig. 15-6*b*) is so graduated that, when the telescope is level, the reading of the "Vert." stadia arc is 50 instead of 0; when the telescope is elevated, the reading is greater than 50, and when the telescope is depressed, the reading is less than 50. In all cases, 50 is subtracted from the reading of the stadia arc, and the remainder (positive or negative as determined by the subtraction) is multiplied by the observed stadia distance to obtain the vertical distance from center of the telescope to the point sighted on the rod. This arrangement of the scale on the stadia arc avoids mistakes of reading a positive value for a negative value, or *vice versa*, but it introduces an additional step (subtraction) in the computations.

General. Observations with the Beaman stadia arc do not include the effect of the instrumental constant $C = f + c$. If, for an external-focusing telescope, more precise results are desired than are yielded by the approximate formulas [Eqs. (7) and (8)], the observed values should be corrected, particularly if the vertical angles are large. The simplest method is to add 0.01 ft. to the observed stadia interval.

The stadia arc is merely a mechanical device for quickly laying off angles for which the function $\frac{1}{2} \sin 2\alpha$ bears a simple relation to the difference in elevation. In Table IX it will be seen that the differences in elevation are, for example, respectively, 1, 2, 10, and 20 ft. per 100 ft. for vertical angles of 0°34′, 1°09′, 5°46′, and 11°47′. The arc facilitates the setting of these and other angles for which $\frac{1}{2} \sin 2\alpha$ is a simple multiple of 0.01. For the "Hor." scale a similar relation exists.

The use of the stadia arc makes unnecessary the reading of vertical angles and reduces the determination of difference in elevation to the very simple process of multiplying the stadia interval by a whole number, which, if the number is small, can be done mentally. On the other hand, it requires more time to set the arc than it does to read the vertical angle; hence from the standpoint of rapidity of operation in the field, the vertical-angle method has the advantage.

15-13. Uses of the Stadia. Uses of the stadia are as follows:

1. In differential leveling, the backsight and foresight distances are balanced conveniently if the level is equipped with stadia hairs.

2. In profile leveling and cross-sectioning, the stadia is a convenient means of finding distances from level to points on which rod readings are taken.

3. In rough trigonometric, or indirect, leveling with the transit, the stadia method is more rapid than any other. The line of trigonometric levels is run as described in Art. 8-5, except that stadia intervals are observed and that differences in elevation are computed by the stadia formula. The method is described in Art. 15-14.

4. On transit surveys of low precision where only horizontal angles and distances are required, the stadia method is more rapid than taping. It may be used either in running traverse lines or in locating details from traverse lines. Stadia intervals are observed as each point is sighted. Horizontal angles are measured, but vertical angles are observed only when of sufficient magnitude to make the horizontal distance appreciably different from the stadia distance (say, when greater than 3°) and then are estimated without reading the vernier. The transit-stadia method of running such surveys is described in Art. 15-15.

5. On transit surveys of low precision—particularly topographic surveys—where both the relative location of points in a horizontal plane and the elevation of these points are desired, the stadia is useful. Both horizontal and vertical angles are measured, and the stadia interval is observed, as each point is sighted; these three observations define the location of the point sighted. The transit-stadia method of making observations when both the horizontal location and the elevation are desired is described in Art. 15-16.

6. Where the plane table is used (see Chap. 17), stadia observations are made with the telescopic alidade in the same manner as with the transit, but horizontal distances and differences in elevation are computed in the field and are plotted immediately instead of being recorded in the form of notes.

15-14. Indirect Leveling by Stadia. Where the required precision is low and the country is rolling or rough, the stadia method of indirect leveling is rapid. The transit should preferably be provided with a sensitive control level for the vertical vernier in order that index error may be readily eliminated. With the ordinary transit having a vertical circle reading to single minutes, differences in elevation are usually computed only to the nearest 0.1 ft. The average length of sight in stadia surveying is usually considerably greater than in differential leveling.

In running a line of levels by this method, the transit is set up in a convenient location. A backsight is taken on the rod held at the initial bench mark, first by observing the stadia interval and then by measuring the vertical angle to some arbitrarily chosen mark on the rod. A turning point is then established in advance of the transit, and similar observations are taken, the vertical angle being measured with the horizontal cross-hair set on the same mark as before. The transit is moved to a new location in advance of the turning point, and the process is repeated. The stadia distances and vertical angles are recorded, also the rod reading which is used as an index when vertical angles are measured. If it is impracticable to sight at this chosen index reading, the vertical angle is measured to some

STADIA LEVELS FOR INDIAN						TRAIL RECONNAISSANCE	48
Sta.	Backsight Obs.	Backsight Diff. Elev.	Foresight Obs.	Foresight Diff. Elev.	Elev.	Rod Index 6.0'	J. L. Black, ⊼ V. A. Alden, Rod
B.M.42	-7°31' 518	+67.1			5972.4	Stone at John's Camp	July 24, 1965 Fair and Hot
T.P.146	-6°04' 349	+36.7	+9°12' 641	+101.1	6140.6		
T.P.147	-17°28' 197	+56.5	+15°07' 215	+54.1	6231.4		
T.P.148	-5°31' 502	+48.0	+4°36'(8.0) 797	+61.6	6349.5		
T.P.149	-12°36' 460	+98.2	+18°41' 744	+225.4	6622.9		
B.M.43	+2°05' 820	-29.8	+3°12' 275	+15.3	6736.4	Stone at Summit at 15" Fir	
T.P.150	+5°30' 644	-61.5	-10°42' 723	-131.7	6574.9		
T.P.151	+23°24' 925	-337.2	-4°46' 842	-69.8	6443.6		
T.P.152	+4°16' 475	-35.2	0°(11.4) 215	-5.4	6101.0		
B.M.44			-2°15' 612	-24.2	6041.6	Stone at Bottom of Canyon near Adit	

FIG. 15-7. Stadia level notes.

other graduation, usually a whole number of feet above or below the index, and this rod reading is given in the notes.

Figure 15-7 shows a suitable form of notes, the arrangement being somewhat the same as for differential-level notes. Opposite a particular bench mark or turning point in the notes are given the observed values for both backsight and foresight and the computed differences in elevation as determined by the approximate stadia formula, Eq. (8). In the notes the rod index or the rod graduation on which sights are generally taken is shown as 6.0 ft. In the columns headed "Observations" are recorded both the vertical angle and the stadia distance, one above the other, in the line opposite each station. Thus for T.P.147 the foresight stadia distance is 215 ft., and the foresight vertical angle is +15°07′.

For any set-up, the difference in elevation determined from either the backsight or the foresight observation is the difference in elevation between the index mark on the rod and the center of the instrument, and the algebraic sum of the backsight and foresight differences is the total difference in elevation between the two positions of the index mark. So long as the index mark to which vertical angles are taken is unchanged, the difference in elevation determined as just described gives also the difference in elevation between the two points on which the rod is held, the actual value of the index reading being of no consequence. Thus the difference in elevation between B.M.42 and T.P.146 is given by the sum of the backsight difference +67.1 and the following foresight difference +101.1, and the elevation of T.P.146 is, therefore, 5,972.4 + 67.1 + 101.1 = 6,140.6 ft., since for both sights the vertical angle was taken with the line of sight directed to 6.0 ft. on the rod.

For the line of levels for which notes are shown, when a foresight was taken to T.P.148, it was found expedient to sight at the 8.0-ft. mark instead of the 6.0-ft. mark. This is shown in the notes by the value of 8.0 in parentheses following the vertical angle +4°36′. Since the vertical angle is measured to a point 2.0 ft. above the adopted index mark and the sight is a foresight, the difference in elevation is taken as 2.0 ft. less than that given by the vertical angle and stadia distance (63.6 − 2.0 = 61.6).

When conditions are favorable it is preferable to read the rod with line of sight horizontal, as in differential leveling, because this practice eliminates the necessity of stadia reduction. Thus the foresight to T.P.152 is taken with the telescope level, the rod reading being 11.4. Since this is 11.4 − 6.0 = 5.4 ft. above the adopted index mark, the foresight difference in elevation is −5.4 ft.

15-15. Transit-stadia Surveying: Elevations Not Required. Where only the horizontal location of objects and lines is desired—as for certain reconnaissance surveys, preliminary surveys, rough surveys for the location of boundaries, and detailed surveys for maps—the transit-stadia method is sufficiently precise and considerably more rapid and economical than corresponding surveys made with transit and tape. The field party consists of a transitman, one to three rodmen, and usually a recorder. In general, the surveying procedure parallels that when the tape is used. Stadia intervals and horizontal angles (or directions) are observed as each point is sighted. Vertical angles, however, are observed only if large enough to make the horizontal distance appreciably different from the stadia distance (say, when greater than 3°), and then are estimated without reading the vernier. Horizontal distances are expressed to the nearest foot.

When the survey consists of a traverse, it is customary to observe the stadia interval both forward and backward from each set-up of the transit. In this way two independent stadia observations are made for each line or distance in the traverse. The closeness of agreement between the two values for each line is a check against mistakes, and the mean is taken as the most probable value.

Figure 15-8 is a sample page of notes for a short closed traverse. The recorded value of the interval factor is 100.2. The directions of the lines are

STADIA TRAVERSE OF GREEN ESTATE

For Horizontal Control

Sta.	Obj.	Az.	Mag. B.	Rod Int.	Vert. Ang.	Hor. Dist.
B	A	279°00′	N81°W	9.09	+2°40′	908
	C	356°14′	N4°W	8.94		
C	B	176°14′	S3°30′E	8.98		899
	D	296°56′	N63°W	13.45		
D	C	116°56′	S63°E	13.50		1351
	E	221°49′	S42°W	8.49	+4°40′	
E	D	41°49′	N41°E	8.47	-4°40′	845
	A	127°57′	S52°E	11.90		
A	E	307°57′	N52°W	12.00		1199
	B	98°58′	S81°E	9.05	-2°40′	

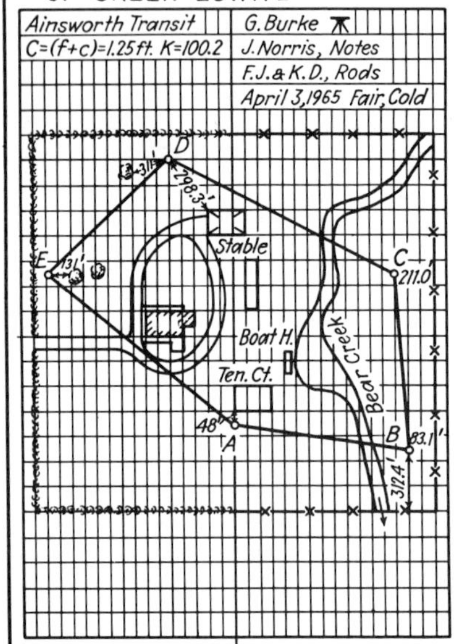

FIG. 15-8. Stadia traverse notes.

determined by azimuths and are checked by magnetic bearings. The stadia ("rod") intervals are given rather than stadia distances, on account of the fact that K is not 100. Vertical angles are observed to the nearest 10′. The vertical angle is recorded only for courses AB and DE, for which the angles are large enough to make a horizontal correction necessary.

Where details are to be located by stadia measurements from transit stations, the observations may be made either at the same time that the transit station is established or later. Figure 15-10 is a suitable form of notes for the former case, except that when elevations are not required, only those vertical angles of sufficient magnitude to make horizontal corrections necessary are measured. Figure 15-9 shows similar notes where numerous details are observed from a single transit station after it has been established, as when the horizontal control is a triangulation system or a taped distance.

The precision with which side shots are taken, of course, depends upon the intended use of the data. Thus, if they were to be used in plotting a map to the scale of 1 in. = 1,000 ft., distances to the nearest 10 ft. would be sufficiently precise, and horizontal corrections for angles below 6° could be neglected without affecting the precision of the map. Where details are varied in character, carefully prepared sketches are necessary for the correct interpretation of the notes.

15-16. Transit-stadia Surveying: Elevations Required. On topographic and similar surveys, both the elevation of each point and its horizontal location are desired. The field procedure consists in observing directions usually by azimuths and distances by stadia, as described in the preceding article. In addition, differences in elevation are determined either by direct leveling where practicable or more usually from observed vertical angles or stadia-arc readings and stadia distances. The field party consists of an observer, one or more rodmen, and usually a recorder.

In topographic surveying, this method may be employed merely for the collection of details, the horizontal and vertical control being established by other means, or it may be employed for establishing control as well as for details.

Details Only. The instrument is set up at a traverse or triangulation station, the elevation and location of which are known. The height of the instrument (H.I.) above the station over which it is set is measured with a rod or a tape or by swinging the plumb bob alongside a scale which has been laid off on a leg of the tripod. (As used here, the term "height of instrument" has a meaning different from that for direct leveling.) The transit is oriented by backsighting along a line the azimuth of which is known, this azimuth having been set off on the horizontal circle. The upper motion is unclamped, and sights to desired points are taken.

Figure 15-9 shows notes of observations taken from station *C* of a traverse, the elevation of the station having been previously determined as 423.9. The H.I. is 4.4 ft. The transit is oriented by sighting to *B*, the azimuth of the line *CB* being set off on the horizontal circle prior to taking the sight. In the first column are given the numbers of the side shots, their locations being shown on the sketch. In the second column are the azimuths of the several points sighted; in the third column are the rod intervals. In the case illustrated by the notes, the interval factor was not 100. Had it been, the intervals might have been replaced by the corresponding stadia distances. In the fourth column are the observed vertical angles, and the following columns show, respectively, the computed horizontal distances, differences in elevation, and elevations.

In measuring vertical angles it is customary, when practicable, to sight at a rod reading equal to the height of instrument above the station over which the transit is set. In this way, the difference in elevation between the center of instrument and the H.I. on the rod is the same as the difference in elevation between the station over which the transit is set and the point on which the rod is held. When the line of vision is obstructed so that the H.I. cannot be sighted, the sight is taken on some other graduation of the rod, usually a whole number of feet above or below the H.I., and this difference in rod readings is recorded in the notes.

TOPOGRAPHIC DETAILS, BLACK ESTATE 32

Inst. at C; El. 423.9; H.I.=4.4

Wisconsin Transit; (f+c)=1.25; K=100.2
G. Burke, ⊼; M.D. Rand, Notes; F.J. & K.D., Rods; Apr. 4, 1965; Cloudy, Cold

Obj.	Az.	Rod Int.	Vert. Ang.	Hor. Dist.	Diff. El.	Elev.	
B	176°14'						
1	10°21'	7.23	-3°11'	723	-40.2	383.7	Water's Edge-Corner
2	3°14'	7.02	-3°17'	702	-40.2	383.7	" " -On Line
3	352°45'	5.64	-4°11'	563	-40.9	383.0	" "
4	7°18'	5.76	-4°04'	575	-40.9	383.0	" "
5	349°10'	-(7.14×4) on 7.7		717	-31.9	392.0	Line (Intervals)
6	16°55'	5.50	-2°50'	551	-27.3	396.6	"
7	315°20'	-(7.86×5) on 1.9		789	-36.8	387.1	" (")
8	349°15'	4.13	-5°46'	410	-41.4	382.5	Water's Edge
9	339°30'	5.40	-4°22'	539	-41.1	382.8	Bank Brook 6' wide
10	0°05'	3.71	-4°12'	371	-27.2	396.7	
11	344°40'	4.85	-4°54'	484	-41.4	382.5	" " 15' "
12	25°00'	2.86	0° on 3.2	288	+1.2	425.1	Direct Levels
13	307°45'	4.88	-4°56'	487	-42.0	381.9	Water's Edge
14	319°10'	4.02	-5°56'	400	-41.6	382.3	" "
15	309°45'	5.80	-3°00'	581	-30.7	393.2	
16	318°25'	3.27	-4°36'	327	-26.3	397.6	
B	176°15'	ck.					
17	340°00'	6.34	-3°08'	635	-34.7	389.2	
18	278°35'	2.51	-5°43'	250	-25.0	398.9	
19	276°20'	3.07	-7°56'	303	-42.3	381.6	Water's Edge
20	277°40'	4.24	-5°40'	422	-41.9	382.0	" "

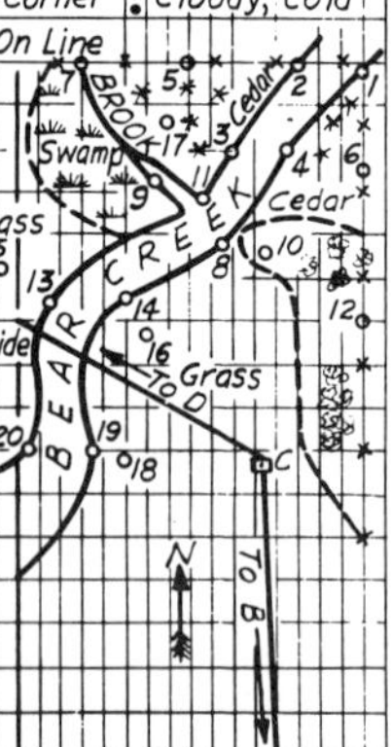

FIG. 15-9. Stadia notes for location of details, with elevations.

When the detail to be observed is at nearly the same elevation as the point over which the transit is set, there is a marked advantage in determining difference in elevation by direct leveling. The notes of Fig. 15-9 show that object 12 was observed in this manner.

Shots 5 and 7 of the stadia notes of Fig. 15-9 illustrate the stepping method (Art. 15-17).

Where measurements are made solely for plotting a map, usually horizontal angles or directions are estimated to 05′ without reading the vernier. Vertical angles are usually measured to minutes, and differences in elevation are computed to tenths of feet. Where elevations to the nearest foot are sufficiently precise, the vertical angles (except for long shots) may properly be read by estimation without use of the vernier.

The usual procedure of observing is to sight on the rod, the lower stadia hair being set on a foot mark such that the horizontal cross-hair falls somewhere near the H.I., and to read the stadia interval. The horizontal cross-hair is then set on the H.I., the horizontal angle is read without clamping the upper motion, and the vertical angle is observed.

PRELIMINARY (STADIA) SURVEY OF I.N.RY., BRIGHTON TO CAMBY

Obj.	Az.	Mag.B.	Stad.Dist.	Vert.Ang.	Hor.Dist.	Diff.Elev.	Elev.	Nov. 27, 1965 J.C.Clark, ⊼
	Inst. at Sta. P49; H.I.= 4.7						785.1	Cold T.N.Tillman, Notes
P48	169°34'	S10°30'E	637	-2°27'	636	-27.2	757.9	W.W.&H.H., Rods
P50	38°21'	N38°15'E	681	+1°14'	681	+14.6	799.7	On slope.
491	151°10'		366	-7°21'	360	-45.6	739.5	West bank Green River.
492	126°35'		418	-5°59'	413	-43.3	741.8	East " " "
493	78°05'		385	-5°36'	381	-37.4	747.7	West " " "
494	81°20'		387	-5°40'	383	-38.0	747.1	East " " "
495	298°55'		214	+6°34'	211	+24.3	809.4	Top Slope.
	Inst. at Sta. P50; H.I.= 4.9						799.7	
P49	218°21'	S38°30'W	683	-1°13'	683	-14.5	785.2	
501	294°40'		415	+4°38'	412	+33.4	833.1	Top slope.
502	16°00'		308	0°on 2.1	308	+ 2.8	802.5	On slope.
503	137°35'		374	-6°36'	369	-42.8	756.9	West bank Green River.
504	136°10'		486	-5°52'	481	-49.4	750.3	East " " "
505	5°45'		322	+7°36'	316	+41.8	841.5	Top slope.
P51	59°38'	N59°30'E	529	0°on 10.1	529	- 5.2	794.5	On slope.
506	94°25'		487	-3°36'	485	-30.5	769.2	West bank Green River.

FIG. 15-10. Stadia notes for preliminary route survey.

When numerous observations are to be taken from a single station, sights to some object the azimuth of which is known are taken at intervals in order to make sure that there is no undetected movement of the lower motion of the transit. The notes of Fig. 15-9 show a check measurement of this kind taken to station *B* after observations to object 16 had been completed.

Control and Details. Where the required precision is not high, the stadia traverse with elevations of traverse stations determined by vertical angle and stadia distance is a rapid means of establishing both horizontal and vertical control. The procedure is the same as that already described in

Art. 15-15; in addition, vertical angles are observed for both the backsight and the foresight from each station, the telescope being sighted at a rod reading equal to the height of instrument above the transit station over which the transit is set up. The details may be located at the same time.

Figure 15-10 shows a page of notes for a stadia traverse for which side shots are taken as the work of running the traverse progresses. The elevation of station *P*49 has previously been determined as 785.1. Directions of the traverse lines are determined by azimuths and roughly checked by observed magnetic bearings, and stadia distances are recorded rather than the rod intervals, the interval factor being practically 100. The backsight from station *P*50 to *P*49 checks reasonably close with the foresight from *P*49 to *P*50. The sights to points 502 and *P*51 are horizontal; the rod reading is shown in the notes, and the difference in elevation is determined by direct leveling.

In computing the length of a traverse line and the difference in elevation between its terminal points, the mean of the vertical angles and the mean of the stadia distances observed from each end are employed.

15-17. Stepping Method. Where the slope of the ground is so small as to make the horizontal distance practically equal to the inclined distance,

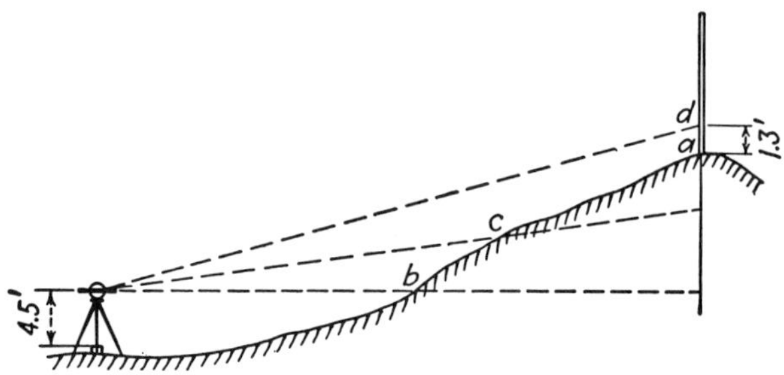

FIG. 15-11. Stepping method.

instead of reading the vertical angle and computing the difference in elevation as described in the preceding articles, the practice of determining the difference in elevation directly by the so-called *interval* or *stepping* method is sometimes followed. To illustrate the method, suppose that in Fig. 15-11 the difference in elevation between the point over which the transit is set up and the point *a* on which the rod is held is desired. The transit is sighted at the rod, and the stadia interval is observed as, say, 4.55 ft. The line of sight is then rotated in a vertical plane until the telescope is level, and the position of the horizontal line of sight on some clearly defined object of the landscape is noted at *b*. The telescope is raised until the lower stadia hair cuts *b*, and the position of the upper stadia hair on the landscape is noted at *c*.

The telescope is again rotated about the horizontal axis until the lower hair cuts *c*, when the upper hair is seen to intersect the rod at *d*, which is at a rod reading of, say, 1.3 ft. Then if the height of instrument (H.I.) is, say, 4.5 ft., the difference in elevation between the instrument station and *a*, assuming that no error is introduced by reason of the line of sight's being inclined when the stadia interval was observed, is $4.5 - 1.3 + (2 \times 4.55) = 12.3$ ft. A similar procedure is followed for negative vertical angles.

For vertical angles of sufficient magnitude to cause an appreciable difference between the observed stadia interval and that which would be observed with sight horizontal, this method would not be applicable. In practice it is generally limited to sights for which the inclination is less than 2°, or where the number of steps does not exceed three. The method is generally used for side shots in fairly flat country where direct leveling is impracticable. The precision obtainable is higher than might be expected, considering the fact that points marking the successive steps are not established by the surveyors but are chosen from among the objects of the landscape upon which the stadia hairs happen to fall. If the transitman keeps the reference mark constantly in view through the telescope while revolving the telescope about the horizontal axis between steps, the error need be little, if any, more than that of reading the rod at the corresponding distance, even though the object sighted may not be readily identified when once the eye has left it. Thus a point of reference might be the stem of a leaf, a pebble in a clod of earth, the tip of a weed, a seam in a rock, a flower, or even the edge of a cloud.

Shots 5 and 7 of the stadia notes of Fig. 15-9 further illustrate the stepping method. For object 5 the horizontal distance is $(7.14 \times 100.2) + 1.25 = 717$.

15-18. Surveying with Plane Table and Stadia. The alidade of the plane table is usually equipped with stadia hairs. The stadia is used extensively in plane-table work (Chap. 17), especially for the location of details in topographic surveying (Chap. 25).

15-19. Errors in Stadia Surveying. Many of the errors of stadia surveying are those common to all similar operations of measuring horizontal angles (Arts. 13-31 to 13-34) and differences in elevation (Art. 9-11). Sources of error in horizontal and vertical distances computed from observed stadia intervals are as follows:

1. *Stadia Interval Factor Not That Assumed.* This condition produces a systematic error in distances, the error being proportional to that in the stadia interval factor. The case is parallel to that of the tape which is too long or too short. When the value of the interval factor is closely determined by observations as described in Art. 15-8 and the stadia measurements are taken under parallel conditions, the error from this source may be made negligible.

2. *Rod Not Standard Length.* If the spaces on the rod are uniformly too long or too short, a systematic error proportional to the stadia interval is produced in each distance. Errors from this source may be kept within

narrow limits if the rod is standardized and corrections for erroneous length are applied to observed stadia intervals. Except for stadia surveys of more than ordinary precision, errors from this source are usually of no consequence.

3. *Incorrect Stadia Interval.* An accidental error occurs owing to the inability of the instrumentman to observe the stadia interval exactly. Following the theory of probability, in a series of connected observations (as a traverse) the error may be expected to vary as the square root of the number of sights. This is the principal error affecting the precision of distances. It can be kept to a minimum by proper focusing to eliminate parallax, by taking observations at favorable times, and by care in observing. Where high precision is required, stadia measurements may be taken by sighting on a rod with two targets, one fixed and the other movable.

4. *Rod Not Plumb.* This condition produces a small error in the vertical angle. It also produces an appreciable error in the observed stadia interval and hence in computed distances, this error being greater for large vertical angles than for small angles. It can be eliminated by using a rod level.

5. *Unequal Refraction.* Unequal refraction of light rays in layers of air close to the earth's surface affects the sight on the lower stadia hair more than the sight on the upper stadia hair and thus introduces systematic positive errors in stadia measurements. Although errors from this source are of no consequence in ordinary stadia surveying, they may be important on the more precise surveys. The periods most favorable for equal refraction are at times when it is cloudy or, if the sun is shining, during the early morning or late afternoon. On precise stadia surveys where it is necessary to work under a variety of atmospheric conditions, it is appropriate to determine the stadia interval factor for each condition and to apply the factor to all observations taken under that condition. Whenever atmospheric conditions are unfavorable, the sights should not be taken near the bottom of the rod.

15-19a. Effect of Error in Vertical Angles. Although the sources just listed produce errors in both horizontal distances and differences in elevation, errors in vertical angles also have their effect upon these computed values. From a study of the stadia formulas it is evident that errors in vertical angles, within the usual range of values, are relatively unimportant in their effect upon computed *horizontal distances.* Thus the ratio of precision corresponding to a 01′ error in angle is about 1/20,000 when the angle is 5° and about 1/6,000 when the angle is 15°. This makes it clear that, so far as precision of horizontal distances is concerned, the governing quantity is likely to be the observed stadia interval rather than the observed vertical angle. On the other hand, the errors in vertical angles are relatively important in their effect upon the precision with which *differences in elevation* are determined. For example, an error of 01′ in vertical angle within the usual range produces an error in elevation of nearly 0.1 ft. in a distance of 300 ft.

With the ordinary transit, vertical angles are read to single minutes, with an error of perhaps ½′, by means of the vernier, or to 05′, with an error of perhaps 03′, by estimation without the vernier. For an average length of sight of, say, 500 ft., the error in stadia distance need not exceed 2 ft. if care is taken in observing, though it might under some conditions amount to as much as 5 ft. For most stadia work vertical angles are less than 5°, but in rough country vertical angles may be 15° or more. For the 500-ft. distance the error in difference in elevation corresponding to the angular error of ½′ is less than 0.1 ft., and that corresponding to the angular error of 03′ is about 0.4 ft., regardless of the magnitude of the vertical angle within the ordinary range of values.

Under normal conditions, where vertical angles are small, the effect of observational errors in vertical angles may be expected to be somewhere near the same as that due to observational errors in stadia intervals; but where vertical angles are large, the errors in observed stadia intervals are likely to have the greater effect upon the precision of computed differences in elevation. To maintain a given precision in computed values of difference in elevation, stadia intervals must be observed with much greater refinement where vertical angles are large than where they are small.

15-20. Precision of Stadia Surveying. For surveys of ordinary precision made with the transit and tape, where only horizontal angles and distances are measured, it has been shown that the principal errors are the systematic errors of taping; for this reason the precision of such surveys is likely to vary directly with the distance.

In transit-stadia surveying, say, for a traverse, where horizontal and vertical angles and stadia intervals are observed, the precision with which the relative locations of points in plan and elevation are determined is dependent upon errors from each of these three sources. It has been shown that the principal errors of both horizontal and vertical angular measurements are likely to be accidental. If the stadia rod is standardized and proper corrections are applied for erroneous length, and if the interval factor is accurately determined and the rod is carefully plumbed, the principal error is that of observing the stadia interval. Under such circumstances, the errors are largely accidental, and hence it is to be expected that these errors will in the long run tend to vary as the square root of the distance. This is one of the important advantages of transit-stadia surveying over transit-tape surveying, and it explains why the precision obtained on extensive transit-stadia surveys compares favorably with the precision obtained on similar transit-tape surveys.

Unless care is taken to eliminate the systematic errors mentioned, the resultant error in plan is likely to vary more nearly as the distance, as is the case with tape measurements. Under these circumstances the advantage just mentioned is lost. If observed vertical angles are small, the resultant error may be expected to vary nearly as the square root of the distance, but if vertical angles are large, the error may be expected to vary directly with the distance.

Many factors influence the precision of stadia surveying. The quality of the instrument, the accuracy of graduation of the rod, the character of the country, the skill of the observer, the care with which the rod is held, the length of sight, and the condition of the weather all affect the results. Following are estimates believed to be fairly representative of several classes of stadia work, these estimates being based upon experience on surveys run under a variety of conditions. It should be borne in mind that the conditions surrounding no two surveys are alike and that no definite statement of the precision for a given procedure can be made.

1. For side shots where a single observation is taken with sights steeply inclined and with no particular care taken to insure the rod's being plumb, horizontal distances may have a precision lower than $\frac{1}{100}$, and individual differences in elevation may be in error 2 ft. or more per 1,000 ft. of horizontal distance.

2. Under the same conditions as in (1) but with small vertical angles and reasonable care used in approximately plumbing the rod and with lengths of sight between 200 and 1,500 ft., the precision of horizontal distances should be not lower than $\frac{1}{200}$; differences in elevation per 1,000 ft. of horizontal distance need not be in error more than 0.3 ft. if vertical angles are observed to 01′, or more than 1 ft. if vertical angles are estimated to 05′.

3. For a rapid stadia traverse of considerable length run through rough country with numerous long sights, angles being measured to minutes but without special precaution to eliminate systematic errors, the error of closure may be as low as 25 ft. per mile in plan and 3 ft. per mile in elevation.

4. For conditions as in (3) but for country fairly level so that all vertical angles are small, the error of closure ought not to exceed 15 ft. per mile in plan and 0.5 ft. $\sqrt{\text{distance in miles}}$ in elevation.

5. For rough country with vertical angles up to 15°, angles to minutes, rod standardized, rod plumbed with level, sights limited to 1,500 ft. and taken forward and backward from each transit station, and interval factor carefully determined, the error of closure may be less than 15 ft. $\sqrt{\text{distance in miles}}$ in plan and 1 ft. $\sqrt{\text{distance in miles}}$ in elevation.

6. For conditions as in (5) but for level country so that all vertical angles are small, the error of closure may be as small as 6 ft. $\sqrt{\text{distance in miles}}$ in plan and 0.3 ft. $\sqrt{\text{distance in miles}}$ in elevation.

7. For conditions as in (5) but stadia intervals determined by use of a target rod with two targets and observations made during cloudy days,

the error of closure will probably not exceed 4 ft. $\sqrt{\text{distance in miles}}$ in plan and 0.5 ft. $\sqrt{\text{distance in miles}}$ in elevation.

15-21. Numerical Problems.

1. With line of sight horizontal, a stadia reading is taken on a rod held at a chained distance of 600.0 + C ft. from the transit station. The rod reading of the lower stadia hair is 0.82 ft. and of the upper stadia hair is 6.77 ft. What stadia interval factor is indicated by this observation?

2. To determine the stadia interval factor, a transit is set up at a distance C back of the zero end of a level base line 800.0 ft. long, the base line being marked by stakes set every 100.0 ft. A rod is then held at successive stations along the base line. The stadia interval and each half-interval observed at each location of the rod are tabulated below. Compute the lower, upper, and full interval factor for each distance, and find the average value for the lower interval, the upper interval, and the full interval.

Distance $-C$, ft.	Lower interval		Upper interval		Full interval	
	Feet	Factor	Feet	Factor	Feet	Factor
100	0.49		0.50		0.99	
200	0.98		0.99		1.97	
300	1.47		1.48		2.95	
400	1.97		1.98		3.96	
500	2.46		2.47		4.94	
600	2.95		2.97		5.92	
700	3.45		3.47		6.91	
800	3.94		3.96		7.89	
Average	....		....		....	

3. A stadia interval of 6.31 ft. is observed with a transit for which the stadia interval factor is 98.5 and C is 1.00 ft. The vertical angle is $+7°42'$. Determine the horizontal distance and difference in elevation by means of (*a*) the exact stadia formulas for inclined sights, (*b*) the approximate formulas, and (*c*) Table IX.

4. The following observations are taken with a transit for which the interval factor is 100.0 and C is 1.00 ft.

Observation	Stadia interval, ft.	Vertical angle
a	10.00	$+\ 0°30'$
b	10.00	$+10°00'$
c	10.00	$+25°00'$

By means of Eqs. (4) and (5), Art. 15-9, compute the horizontal distances and differences in elevation. By means of the approximate Eqs. (7) and (8), Art. 15-10, determine the same quantities and note the errors introduced by the approximations.

5. What would be the amount and sign of error introduced in each computed horizontal distance and difference in elevation if the observations of the preceding problem were taken (*a*) with a 12-ft. rod which was unknowingly 0.5 ft. out of plumb with top leaning toward the transit? (*b*) With the top of the rod leaning 0.5 ft. away from the transit? (*c*) What conclusions may be drawn from these results?

6. What error will be introduced in each computed horizontal distance and difference in elevation if in the observations of problem 4 (*a*) each vertical angle contains an error of 01′? (*b*) Each stadia interval is in error 0.10 ft.? (*c*) What conclusions may be drawn from these results?

7. In determining the difference in elevation and the distance between two points A and B, a transit equipped with a stadia circle is set up at A and the following data are obtained: $V = +10$, $H = 98.0$, stadia interval = 3.50 ft., H.I. = 4.5 ft., line of sight at 4.5 ft. on rod. The instrumental constants are $K = 100.0$ and $C = 0$ (internal focusing telescope). The stadia circle has index marks of $H = 100$ and $V = 0$ for a horizontal line of sight. Compute the distance and difference in elevation between the two points A and B.

8. In determining the elevation of point B and the distance between two points A and B a transit equipped with a stadia arc is set up at A and the following data are obtained: $V = 38$, $H = 3.0$, stadia interval = 4.30 ft., H.I. = 4.2 ft., line of sight at 8.6 ft. on rod. The instrument constants are $K = 100.0$ and $C = 1.00$ ft. The stadia arc has index marks of $H = 0$ and $V = 50$ for a horizontal line of sight. The elevation of point A is 125.6 ft. Compute the distance AB and the elevation of point B.

9. Following are the notes for a line of stadia levels. The elevation of B.M.$_1$ is 637.05 ft. The stadia interval factor is 100.0 and $C = 1.25$ ft. Rod readings are taken at height of instrument. By use of Table IX determine the elevations of remaining points.

Station	Backsight		Foresight	
	Stadia interval, ft.	Vertical angle	Stadia interval, ft.	Vertical angle
B.M.$_1$	4.26	−3°38′	....	......
T.P.$_1$	2.85	−1°41′	3.18	+2°26′
T.P.$_2$	3.30	+0°56′	2.71	−4°04′
T.P.$_3$	2.26	+2°09′	4.45	−0°38′
B.M.$_2$	....	......	3.09	+7°27′

10. Following are stadia intervals and vertical angles for a transit-stadia traverse. The elevation of station A is 418.6 ft. The stadia interval factor is 100.0, and $C = 1.00$ ft. Rod readings are taken at height of instrument. Compute the horizontal lengths of the courses and the elevations of the transit stations, using Table IX.

Station	Object	Stadia interval, ft.	Vertical angle
B	*A* *C*	8.50 4.37	+0°48′ +8°13′
C	*B* *D*	4.34 12.45	−8°14′ −2°22′
D	*C* *E*	12.41 7.18	+2°21′ −1°30′

11. Following are stadia intervals and vertical angles taken to locate points from a transit station the elevation of which is 415.7 ft. The height of instrument above the transit station is 4.6 ft., and rod readings are taken at 4.6 ft. except as noted. The stadia interval factor is 100.0, and $C = 1.00$ ft. Compute the horizontal distances and the elevations.

Object	Stadia interval, ft.	Vertical angle
43	7.04	−0°58′
44	−(8.25 × 3) on 2.1	Intervals
45	7.56	−0°44′ on 9.2
46	−(7.25 × 2) on 6.0	Intervals
47	3.72	−5°36′

12. A transit equipped with a stadia arc is used in locating points from a transit station the elevation of which is 765.7 ft. The stadia arc has index marks of $H = 100$ and $V = 50$ for a horizontal line of sight. The instrument constants are $K = 100.0$ and $C = 0$ (internal focusing telescope). The height of instrument above the transit station is 4.5 ft. Compute the horizontal distances and the elevations.

Object	Stadia interval, ft.	Rod reading, ft.	Stadia arc readings	
			V	*H*
114	3.26	3.6	18	88.3
115	7.84	5.8	35	97.7
116	2.18	4.7	39	98.8
117	1.66	4.3	76	92.6
118	8.14	6.4	69	96.2

15-22. Field Problems.

PROBLEM 1. DETERMINATION OF STADIA INTERVAL FACTOR

Object. To determine the stadia interval factor $K = f/i$ of transit or level.

Procedure. (1) As described in Art. 15-8, employing a line about 800 ft. long. (2) Determine f and c by measurement (Art. 15-7). (3) For each observation, read the rod for lower, middle, and upper hairs. (4) Compute K for each distance for lower half interval, upper half interval, and full interval; take the mean of all computed values as the factor for the instrument. Discard any readings that differ widely from the others.

Hints and Precautions. (1) On fair days the line of sight defined by the lower hair should be at least 2 ft. above the ground. (2) It is convenient to set the lower hair on the nearest foot mark, and this may be done without appreciable error.

PROBLEM 2. PRELIMINARY TRAVERSE OF ROUTE WITH TRANSIT AND STADIA

Object. To obtain data for plotting a topographic map of a proposed highway route between two governing points. The data may be used in office problem 1 (Art. 24-23).

Procedure. (1) Run a stadia azimuth traverse between the two points, establishing stadia stations at advantageous points near where it appears that the line will eventually be placed. (2) Determine the distance between adjacent stations by observing the stadia interval on both backsights and foresights. (3) Observe the vertical angle between instrument stations on both backsights and foresights. Record the H.I. at each set-up. (4) Make the available checks before moving the transit. (5) While running the traverse, take side shots 200 to 600 ft. on each side of the traverse line, as necessary to define the configuration of the land and the location of objects that might affect the proposed line. (6) Note the type of soil, any indications of rock near the surface, and the type of cover. (7) Keep notes in the form of the sample notes (Fig. 15-10).

Hints and Precautions. (1) In determining the differences in elevation and the horizontal distances between traverse points, use the mean of the two vertical angles and the mean of the two stadia readings taken along the line joining these points. (2) Before taking side shots about a station occupied, set the next stadia station in advance. (3) In running the traverse, the magnetic bearing of each line should be recorded and immediately compared with the bearing computed from the azimuth of the line. (4) Inclined distances with vertical angles of less than 3° may be considered as horizontal without appreciable error. (5) Observe vertical angles to the nearest minute. Observe azimuths of traverse lines to the nearest minute, and azimuths of sights to details to the nearest 05′ without the use of the vernier. Read the rod intercept to the nearest 0.01 ft. (6) Many shots can be taken with the telescope leveled as in direct leveling. (7) The observer should form the habit of judging distances by eye, in order to avoid large mistakes. The middle cross-hair should not be mistaken for one of the stadia hairs.

PROBLEM 3. TRAVERSE AND LOCATION OF DETAILS WITH TRANSIT AND STADIA

Object. To collect sufficient data for making a topographic map of an assigned tract. The data may be used in office problem 1 (Art. 24-23).

Procedure. (1) Make a rapid reconnaissance of the tract, selecting the most advantageous points for instrument stations from which areas comprising the entire area can be observed. (2) Run a closed azimuth traverse through the selected points, observing the stadia intervals and vertical angles. The allowable angular error of closure should not exceed $01' \sqrt{\text{number of sides}}$. The error of closure in elevation should not exceed $0.3 \text{ ft.} \sqrt{\text{distance in thousands of feet}}$. (3) Occupy each of the traverse stations and with the instrument correctly oriented observe the azimuth, stadia distance, and vertical angle to all changes in ground slope and to other natural and artificial features which are within range of the instrument. (4) Include in the notes a sketch drawn approximately to scale. (5) By means of Table IX, a stadia slide rule, or a stadia reduction diagram, determine the horizontal distance and the elevation of each side shot.

Hints and Precautions. (1) See Hints and Precautions of the preceding field problem. (2) If the elevation of a point is not required for mapping, often the point can be located advantageously by the method of intersection, the azimuth being observed from two or more traverse stations. (3) It is sometimes advantageous, particularly if there are a large number of details, to plot the map in the field as the work progresses.

REFERENCE

1. Mussetter, William, "Tacheometric Surveying—Methods and Instruments," *Surveying and Mapping*, Vol. 16, Nos. 2 and 4, pp. 137–156 and 473–487, April–June and October–December, 1956.

CHAPTER 16

TRIANGULATION

16-1. General. Triangulation is employed extensively as a means of control for topographic and similar surveys. A *triangulation system* consists of a series of triangles in which one or more sides of each triangle are also sides of adjacent triangles, as illustrated in Figs. 16-1 to 16-3. The lines of a triangulation system form a network tying together the *triangulation stations* at the vertices of the triangles.

If it were possible to measure one side and all the angles in a triangulation system with absolute precision, no further linear measurements would be necessary. Unavoidable errors in the field measurements, however, make it desirable that the lengths of two or more lines in each system be measured as a means of checking the computed distances. The lines whose lengths are measured are called *base lines.*

The arrangement of the triangles in most systems affords many different geometrical figures for each of which the theoretical value of the sum of the included angles is known. Also, the sum of the angles about any station should equal 360°, and in any triangle the lengths of the sides should be proportional to the sines of the angles opposite. These known conditions serve as a measure of the precision of the angle measurements and as a means of adjusting the errors so as to secure the most probable values of the measured quantities.

If two angles in any triangle are measured the value of the third can be readily computed. This procedure, however, does not permit the application of the known conditions as a measure of the precision of the measurements or as a means of adjusting the errors; therefore, it is customary to measure all angles.

The recent development of electronic distance-measuring devices (Chap. 7) has made feasible the method of *trilateration,* or the measurement of all sides of the triangles. Use of the method has been limited, principally because of the complexity of the computations for adjustment of a system of triangles, but it is increasing. One method of trilateration adjustment is given in Ref. 13 at the end of this chapter. Except for the measurement of lines instead of angles, the field procedures of trilateration are similar to those discussed herein for triangulation.

There is a quality of triangulation corresponding to every degree of precision used in traversing. Thus, triangulation may be used for a simple topographic survey covering but a few acres, or it may be used to extend control of the highest order across the continent. The relative merits of the triangulation method and the traverse method are based upon the character of the terrain and not upon the degree of precision to be attained. If favorable routes are available, the method of traversing is superior to the method of triangulation, but if the terrain offers many obstacles to traverse work (such as hills, vegetation, or marsh), triangulation is superior.

The most notable example of triangulation is the precise transcontinental system established by the U.S. Coast and Geodetic Survey. The system is being developed to form a network to establish a control for the entire domain of the United States. A permanent reference point for the datum, called the "North American Datum of 1927," has been established at Meade's Ranch in Osborne County, Kansas, and to this point the precise surveys of the United States, Canada, and Mexico are referred.

Because of the character of the terrain near shore lines, the method of triangulation is extensively used in surveys for hydrographic charts and for maps of coast lines and navigable rivers.

16-2. Precision of Triangulation Systems. Triangulation systems are classified with respect to precision according to (1) the average angular error of closure in the triangles of the system and (2) the discrepancy between the measured length of a base line and its length as computed through the system from an adjacent base line.

The Federal Board of Surveys and Maps (composed of representatives of the Federal bodies engaged in surveying and mapping) has classified triangulation for the extensive surveys of the United States Government as shown in Table 16-1.

First-order triangulation furnishes the primary horizontal control for small-scale mapping operations, the triangle sides often being many miles in length. The system which extends across the continent and from Canada to Mexico is of this order. First- and second-order triangulation call for methods of high precision not often necessary except on very extensive surveys.

Third-order triangulation establishes points of horizontal control at short intervals in advantageous locations for detail mapping. This order is often employed in intermediate- and large-scale surveys of limited extent. It calls for methods of intermediate precision, although the requirements may sometimes be met by methods of ordinary precision. Fourth-order triangulation, not included in the Federal classification, calls for methods of ordinary precision.

The classification by orders relates more particularly to surveys for small-scale maps which cover relatively large areas. For the surveys with which most surveyors and engineers deal, many prefer to retain the designations of *primary*, *secondary*, etc. to indicate the relative degrees of precision in the work. As in the case of traverse work (Chap. 25), both primary and secondary

TABLE 16-1. CLASSIFICATION AND STANDARDS OF PRECISION FOR TRIANGULATION

	First order			Second order		Third order
	Class I (special)	Class II (optimum)	Class III (standard)	Class I	Class II	
Principal uses	Urban surveys, scientific studies	Basic network of U.S.	All other (state, private)	Area networks and supplemental cross arcs in national net	Coastal areas, inland waterways, and engineering surveys	Topographic mapping
Base measurement:						
Probable error not to exceed 1 part in	1,000,000	1,000,000	1,000,000	1,000,000	500,000	250,000
Triangle closure:						
Average error not to exceed	1″	1″	1″	1.5″	3″	5″
Length closure:						
Discrepancy between measured and computed length of base line not to exceed 1 part in	100,000	50,000	25,000	20,000	10,000	5,000

(sometimes tertiary and quaternary) triangulation may be used on the same survey; also, triangulation and traverse work may be combined to meet best the field conditions (see also Art. 25-4 and Table 25-1).

16-3. Triangulation Figures. In a narrow triangulation system a chain of figures is employed, consisting of *single triangles, polygons, quadrilaterals,* or combinations of these figures. A triangulation system extending over a wide area is likewise divided into figures irregularly overlapping and intermingling, as illustrated by the system of Fig. 25-3. The computations for such a system can be arranged to afford checks on the computed values of most of the sides. As many sides as possible are included in the routes through which the computations are carried from one base line to the next.

1. *Chain of Triangles.* In the chain of single triangles (Fig. 16-1) there is but one route by which distances can be computed through the chain. If AB is the base line whose length is measured, and if all the angles of the

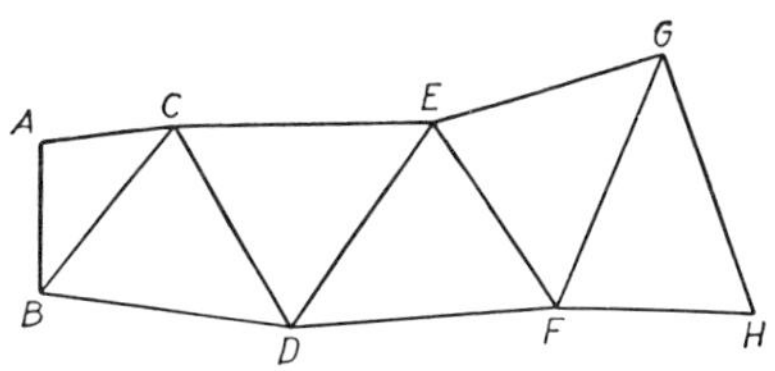

FIG. 16-1. Chain of single triangles.

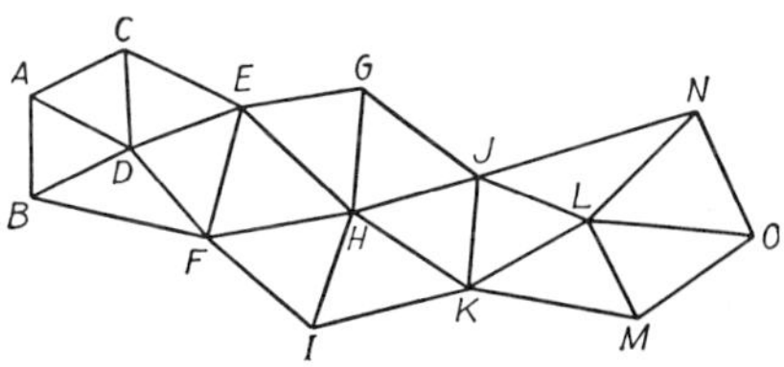

FIG. 16-2. Chain of polygons.

triangles are observed, the length of the triangle sides in the chain (as AC, BC, BD, etc.) may be calculated progressively along the chain from the measured base line to the triangle side farthest removed from the base line. If two lines are measured as base lines, one at each end of the system, the calculations may be carried from each toward the other to a triangle side somewhere between them.

As the sum of the measured angles in each triangle normally will not equal exactly 180°, the angles are adjusted to satisfy this requirement before the distances are computed. The method of making the adjustment is described in Art. 16-24.

2. *Chain of Polygons.* In triangulation, a polygon, or "central-point figure," is composed of a group of triangles, the figure being bounded by three or more sides and having within it a station which is at a vertex common to all the triangles. A chain of such composite figures is illustrated in Fig. 16-2, in which $BACEF$ is a five-sided polygon with D as the central point, and $FEGJKI$ is a six-sided polygon with H as the central point.

The sum of the measured angles in each triangle of the polygon should equal 180°; also, the sum of the angles about the central point should equal 360°. Further, the length of any side may be computed by two routes, and these two computed lengths should agree. Assume, for example, that *AB* is the base line. With the length of that line known, the length of *EF* can be found either by way of the triangles *ABD*, *ACD*, *CDE*, and *DEF*, or by way of the triangles *ABD*, *BDF*, and *DEF*. If all the angles were known exactly, the computed value of the distance *EF* would be the same by one route as by the other. The observed angles are so adjusted by computation that this condition exists and, further, that the sum of the three angles of each triangle equals 180° and that the sum of the angles about the central point equals 360°. A similar adjustment is made for the other polygons in the triangulation chain.

3. *Chain of Quadrilaterals.* Figure 16-3 illustrates another type of triangulation figure, in which the individual triangles more of less overlap one another. This type usually occurs in the form of the quadrilateral, of which the figures *ABDC*, *DCEF*, etc., are examples. In the individual quadrilateral there is no triangulation station at the intersection of the diagonals. Consider one of the quadrilaterals, as *ACDB*. The measured angles give values for four triangles *ACD*, *CBA*, *DBA*, and *BCD*, in each of which the sum of the angles should equal 180°. In addition, the length of any line should be the same when computed by one route as when computed by another.

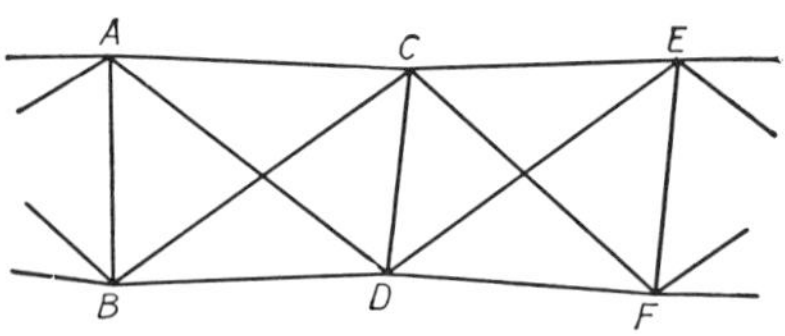

FIG. 16-3. Chain of quadrilaterals.

For example, consider *AB* as the side of known length and *CD* as the side whose length is required. There are four ways in which the required distance *CD* may be found: (1) by use of triangle *DBA* for the length of *AD* and triangle *ACD* for the length of *CD*; (2) by the use of triangle *DBA* for the length of *BD*, then of *BCD* for the length of *CD*; (3) using triangles *CBA* and *BCD*; and (4) using triangles *CBA* and *ACD*. The four values of *CD* should agree and will agree if the angles are precisely known. The adjustment of angles must be so carried out as to make their adjusted values satisfy this requirement as well as to make the sum of the three angles of each triangle equal 180°.

The adjustment of a quadrilateral is described in Art. 16-25.

16-4. Choice of Figure. Of the three forms of chains of triangulation figures, the chain of single triangles is the simplest, requiring the measurement of fewer angles than does either of the other two. This type of system, however, has the obvious weakness that the only check (except between bases) is in the sum of the angles of each triangle considered by itself. To reach the same precision in the determination of lengths, base lines would need to be placed closer together. This type of chain is not employed in work of high precision, but it is satisfactory where less precise results are required.

For more precise work, quadrilaterals or polygons are used; quadrilaterals are best adapted to long, narrow systems and polygons to wide systems.

16-5. Strength of Figure. It has been shown in Fig. 3-2 that values computed from the sine of angles near 0° or 180° are subject to large ratios of error. Since in triangulation computations nearly always the sine is used, it follows that angles near 0° and 180° are undesirable. It has been found in practice that satisfactory results can be secured for most purposes if the angles *used in the computations* fall between 30° and 150°. However, many angles measured in the field are not used in computing the length of the sides in the system. Such angles may be near 0° or 180° without impairing the excellence of the system as a whole.

This and other principles may be made clear by reference to Fig. 16-4. In the figure let AB represent a side whose length is known. This side and all others in the system whose lengths are desired are shown by heavy lines. The sine law, used in computing the lengths, states that in any triangle the sides are proportional to the sines of the angles opposite; accordingly, the angles affecting the computed lengths of sides in each triangle are those opposite the known and computed sides.

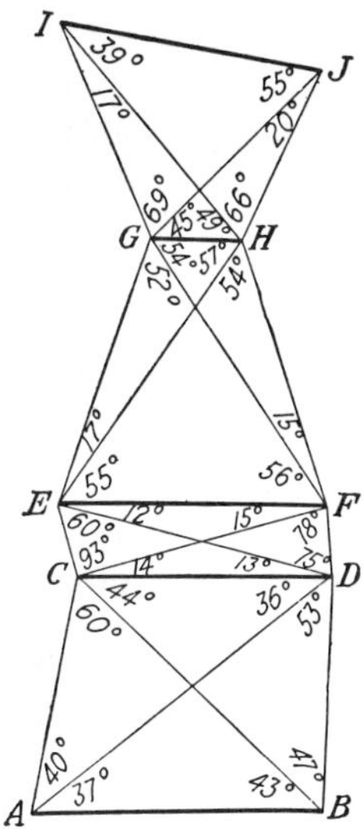

FIG. 16-4. Strength of figure.

Consider the quadrilateral $CEFD$. The length of the side CD results from calculations carried through the quadrilateral $ACDB$. Then the length of CF, in triangle CDF, is computed by use of the known side CD and the angles 78° and 88° (13° + 75°); and EF, in triangle CEF, may be computed by use of the known side CF and the angles 93° and 72° (60° + 12°). In these two computations involving the small angles (12° and 13°) it is seen that neither one is used separately and so neither, by itself, has any effect upon the length of the side EF. Similarly, the side ED, in triangle CED, is computed by the use of the side CD, and again, neither of the small angles (14° and 15°) is used separately in computing the length of EF. Thus it is seen that the side EF can be computed by two independent series of computations neither of which is affected detrimentally by the small angles involved. As a matter of fact, the quadrilateral $CEFD$ is a stronger figure than is $ACDB$ in which no angle less than 36° occurs.

By a similar analysis of the quadrilateral $EGHF$, it will be found, however, that it is impossible to compute the length of the side GH without making use in one series of computations of the angle 15° separately and in the other of the angle 17° separately. Therefore, by any possible means, the computed length of the side GH must be affected by the large ratios of error resulting from the use of small angles separately. The large degree of uncertainty thus introduced into the computed length of the side GH will be effective in all dependent values computed therefrom, as, for example, the length of the side IJ in the system shown.

As an aid in checking which of several alternative figures (or chains of figures) is to be used in triangulation, the relative strengths of the figures can be determined by computations based upon the size of the angles, the number of directions to be observed, and the number of geometric conditions that must be satisfied. Considerations of economy may render one figure more desirable than another even though it may be the weaker of the two. For example, in Fig. 16-4 the quadrilateral $ACDB$ may be more desirable than $CEFD$ because the progress of the work is advanced more rapidly by the former than by the latter, the ratio of progress being about that of the lengths BD/DF.

Computation of R. The relative strength of figure can be evaluated quantitatively in terms of a factor R based on the theory of probability; *the lower the value of R, the stronger the figure.* Strength of figure is a factor to be considered in establishing a triangulation system for which the computations can be maintained within a desired degree of precision. For example, for third-order triangulation it is desirable that R for a single figure not exceed 25 and that R between two base lines not exceed 125. In some cases it may not be necessary to occupy all the stations of the system or to observe all the lines in both directions. Furthermore, by means of computed strengths of figure, alternative routes of computation (chains of elemental triangles) can be compared and the best route chosen. The methods are described in detail in Ref. 7 at the end of this chapter. The following brief treatment gives the essential relations for computing R.

Let

C = number of conditions to be satisfied in figure.

n = total number of lines in figure, including known line.

n' = number of lines observed in both directions, including known line if observed.

s = total number of stations.

s' = number of occupied stations.

D = number of directions observed (forward and/or back), excluding those along known line.

δ_A, δ_B = respective logarithmic differences of the sines, expressed in units of the sixth decimal place, corresponding to a change of 1 second in the "distance angles" A and B of a triangle. The distance angles are those opposite the known side and the side required.

$\Sigma(\delta_A^2 + \delta_A\delta_B + \delta_B^2)$ = summation of values for the particular chain of triangles through which the computation is carried from the known line to the line required. Values of $(\delta_A^2 + \delta_A\delta_B + \delta_B^2)$ for a triangle are given in Table VII.

Then

$$C = (n' - s' + 1) + (n - 2s + 3) \tag{1}$$

$$R = \frac{D - C}{D} \sum (\delta_A^2 + \delta_A\delta_B + \delta_B^2) \tag{2}$$

Example: It is desired to compute the strength of the quadrilateral *ACDB* in Fig. 16-4 for computation of the side *CD* from the known side *AB* when all lines are observed in both directions. From Eq. (1)

$$C = (6 - 4 + 1) + (6 - 8 + 3) = 4$$

$$\frac{D - C}{D} = \frac{10 - 4}{10} = 0.60$$

The computation may be carried through any of four chains of triangles, as indicated in the accompanying tabulation.

It is seen that the strongest chain consists of triangles *BAC* and *BCD* and that the relative strength of the quadrilateral is 3.

By similar computations, for the remaining quadrilaterals, the least values of *R* are found to be *CEFD*, 0; *EGHF*, 29; and *GIJH*, 20. Therefore, the strongest quadrilateral is *CEFD* and the weakest is *EGHF*, as previously discussed. The strength of the figure as a whole (for *IJ* computed from *AB*) is represented by a value of *R* of 52, which is the sum of the lowest values for the four consecutive quadrilaterals in the chain.

Common side	Chain of triangles	Distance angles, deg.	$(\delta_A^2 + \delta_A\delta_B + \delta_B^2)$ Each	$(\delta_A^2 + \delta_A\delta_B + \delta_B^2)$ Σ	R
AC	*ACB* *ACD*	60; 43 40; 36	9.8 22.2	32.0	19
AD	*ADB* *ACD*	90; 53 104; 40	2.4 5.2	7.6	5
BC	*BAC* *BCD*	77; 60 89; 47	2.0 3.7	5.7	3
BD	*BAD* *BCD*	53; 37 47; 44	15.2 12.8	28.0	17

16-6. Base Nets. In a system of triangles, long sides (within limits) are more economical than short sides. It is difficult and expensive to measure long base lines by taping; hence, in practice, the base lines are usually much shorter than the average length of the sides of the triangles in the main triangulation system. In order to obtain the required precision in the computed length of the sides of the main triangles, it is necessary to expand the effect of the base line through a group of smaller triangles called the *base net.*

At the left in Fig. 16-5 is an example of an excellent base net affording quick and accurate expansion of the base line to the longer sides of the system. The form of base net suggested by the quadrilaterals *GHFE* and *GHJI* (Fig. 16-4) is satisfactory if it can be so laid out as to avoid the small angles there shown. This form is also shown at the right in Fig. 16-5.

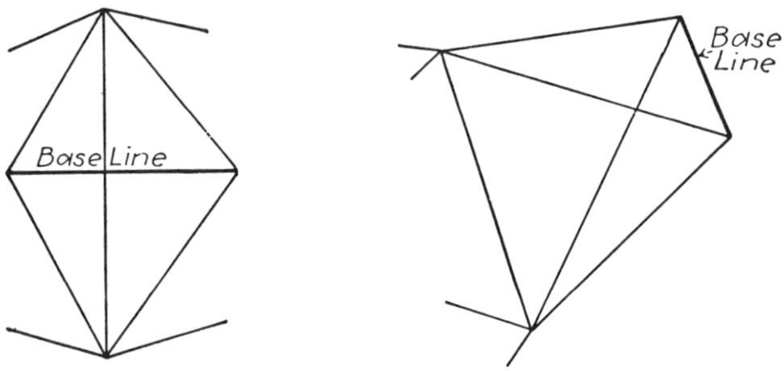

FIG. 16-5. Base nets.

The number of base lines required will depend on the excellence of the shapes of the triangles in the system. In practice, work of ordinary precision can be carried through a chain of 20 to 60 triangles, depending upon the strength of the figures.

METHODS

16-7. General. The work of triangulation consists of the following steps:

1. Reconnaissance, to select the location of stations.

2. Erection of signals and, in some cases, tripods or towers for elevating the signals and/or the instrument.

3. Measurement of angles between the sides of triangles (for trilateration, measurement of lengths of the triangle sides).

4. In most cases, astronomical observations at one or more triangulation stations, in order to determine the true meridian to which azimuths are referred; also in extensive systems to determine the geographical coordinates (latitude and longitude) of all points in the system.

5. Measurement of the base lines.

6. Computations, including the adjustment of the observations, the computation of the length of each triangle side, and the computation of the coordinates of the stations.

Herein the description of methods will be concerned principally with triangulation of ordinary precision (corresponding roughly to third- or fourth-order triangulation). Triangulation of high or low precision differs from that of ordinary precision as follows:

1. *Triangulation of High Precision.* The reconnaissance may amount to a preliminary survey. Extensive use is made of tall towers and signals and of signaling devices for reflecting sunlight or for night work. Angles are measured with either the repeating theodolite or the direction instrument (Art. 16-12). The angles of a system are adjusted by the method of least squares, and account is taken of spherical excess (Art. 16-23). The computations for latitude and longitude of the various stations take into account the curvature of the earth.

2. *Triangulation of Low Precision.* There is practically no reconnaissance, and often the stations are selected as the work progresses. The stations are marked with a stake, pole, or portable tripod. The base line is measured by the ordinary methods of taping or sometimes even by stadia. The angles of the triangles are not necessarily adjusted to meet the known geometric and trigonometric conditions. No correction is made when the instrument is not set up exactly over the station. No astronomical observations are made. Often the method of graphical triangulation with the plane table is employed (Art. 17-10).

16-8. Reconnaissance. Because of its influence on the accuracy and economy of the work, the reconnaissance is of the greatest importance. Reconnaissance consists in the selection of stations, and it determines the size and shape of the resulting triangles, the number of stations to be occupied, and the number of angles to be measured. In this connection are considered the intervisibility and accessibility of stations, the usefulness of stations in later work, the strength of figures, the cost of necessary signals, and the convenience of base-line measurements.

After a preliminary study of available maps and information, the chief of the party examines the terrain, choosing the most favorable sites for stations. If the information is not available from maps, angles and distances to other stations are estimated or measured roughly en route, so that the suitability of the system as a whole can be examined before the detailed work is begun. Angles are determined either directly by use of the prismatic compass or similar hand instrument or graphically by use of the plane table. Distances are determined either directly by pacing or odometer or graphically by use of the plane table. Where forest growth is present, the observer must make use of standing trees or guyed ladders or poles to establish visibility with adjacent stations.

In open, hilly regions, stations can often be located on summits such that the instrument for measuring angles can be set up on the ground. Under adverse conditions, however, the instrument must be elevated to a height sufficient to enable all adjacent stations to be observed. Above each station is placed a signal, such as one or more square flags attached to a center pole. Stations and signals are described in Art. 16-9.

Reconnaissance for triangulation of low precision is either very limited in extent or is omitted entirely, the stations being selected as the work progresses.

16-9. Signals and Instrument Supports. Each triangulation station is marked by a signal visible from stations from which it is to be sighted. The form of the signal depends on the locality and the available materials. If the station is not to be used as an instrument station, but is merely to be sighted, a relatively simple structure is used. This signal may be one constructed for the purpose or it may be an object already in place such as a chimney or powerline pole. A pole set vertically in the ground or held firmly vertical by a pile of stones, or by guys or bracing, makes an excellent signal

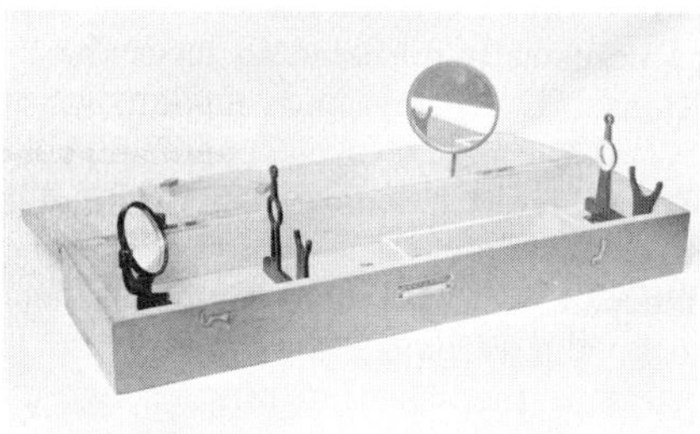

FIG. 16-6. Heliotrope, box type. (*U.S. Coast and Geodetic Survey.*)

on a bare summit or in open country. A white paint mark on a rock cliff is sometimes all that is required. To increase the visibility of a pole or tree signal, two rectangular targets are sometimes attached, being placed at right angles to each other.

Signals should be of such shape and orientation that they will be as nearly free as possible from phase, or asymmetry of sighting. If the signal is illuminated from the side, the observer tends to sight to one side of its center.

When the sun is shining, heliotropes of various designs are used as signals for long distances at which flags or poles are invisible. Essentially a heliotrope consists of a mirror so arranged as to flash sunlight to the distant station. Figure 16-6 shows the box type in which the line of sight of the telescope is fixed parallel with the axis of the open rings. Therefore, if the heliotrope telescope (not shown; rests on forked supports shown) is sighted at a distant station and by mirror adjustments the beam of light is reflected along the axis of the two rings, the ray is directed to the same station. An automatic heliotrope has been developed.

The best time for observing is in the late afternoon or at night. For night observations, an electric lamp is used as a signal.

When manual heliotropes are used, there must be an attendant to operate them. Communication between the observer and the attendants is established by the use of code signals flashed back and forth or by means of a portable sending-receiving radio set. Lamps may be turned on and off at the desired times by means of clockwork.

At an instrument station it is desirable to have a signal of a type that will permit placing the instrument directly over the station when angles are to be measured. In a small triangulation system with triangle sides only a few hundred feet in length, and with but few angles to be measured, a temporary signal which is readily moved may be all that is necessary. A light tripod to which is attached a plumb line may be centered over the station mark. If the stations are to be used over a longer period, the station may be marked by an iron pipe set vertically in the ground, in which pipe is placed a range pole or similar rod. When the station is occupied, the pole is temporarily removed. Where a tall mast is necessary for visibility, it may be supported in position by three guy wires attached to it near the top. Provision is made for swinging the bottom of the mast to one side when it is desired to place an instrument at the station. It is necessary that the guyed top be accurately centered over the station.

FIG. 16-7. Tripod signal. (*U.S. Coast and Geodetic Survey.*)

Where a more permanent signal is required that does not need to be moved to provide for setting up the instrument, a large tripod or tower like that shown in Fig. 16-7 may be used. Such a signal may be constructed either of round poles or of sawed lumber, with the vertical mast projecting upward from the junction with the legs. The signal should be solidly built and firmly anchored, with the vertical mast centered accurately over the station and made as nearly vertical as possible. For high visibility, the cloth may be of the fluorescent type.

The field instructions of the U.S. Geological Survey and the U.S. Coast and Geodetic Survey (Refs. 4 and 7 at the end of this chapter) give excellent instructions for the construction of signals.

Where the instrument must be elevated to secure visibility, a combined observing tower and signal like that shown in Fig. 16-8 is built of wood or steel. A central tripod supports the instrument. Around this, but entirely separate from it, is a three- or four-sided structure supporting the platform upon which the observer stands. Thus the instrument tower is free from the vibrations caused by movements of the observing party. The Bilby steel tower shown in Fig. 16-8 is sectional and can be quickly and easily erected to any height up to 126 ft.

FIG. 16-8. Bilby steel triangulation tower. (*U.S. Coast and Geodetic Survey.*)

For graphical triangulation (Art. 17-10) and for ordinary triangulation of low precision, it is not necessary that the instrument be placed exactly beneath the signal; and some stations are not designed to be occupied by the instrument. For such conditions a single staunch mast may be used.

16-10. Station Marks. For the extensive triangulation systems of the U.S. Coast and Geodetic Survey and the U.S. Geological Survey, every triangulation station is permanently marked with a metal tablet (similar to that in Fig. 9-1) which is fastened securely in rock or in a concrete monument. These stations are of great value as reference points for local surveys.

16-11. Angle Measurements. Thus far the term *triangulation station* has generally been used to designate *major stations*, that is, points where the instrument is set up to measure angles of the main system. In most triangulation systems, secondary control is established by observations to stations in the vicinity of the major stations, but these secondary stations, called *minor stations*, are not used in the extension of the main system of triangles. Obviously, the angle measurements of such stations may be made with a lower degree of precision than is required in the main system.

Major Stations. In work of ordinary precision, the average error of closure of the triangles should not exceed 5″ (Art. 16-2). For surveys where lower precision is permissible, corresponding modifications should be made in the measurements of the angles.

The instrument is set up at each major station, and angles with vertex at the station are measured. The method of procedure with the direction instrument is described in the following article. The method of repetition, commonly used in triangulation of ordinary precision, is described in Art. 13-17; the procedure is stated and the form of notes is shown under field problem 2 with the transit, Art. 13-37. The following suggestions are added to those there given: The instrument should be protected from wind and sun; good visibility is necessary, that is, the air should be free from smoke, mist, or heat waves; after the instrument has been set up, centered, and leveled, the tripod wing nuts should be loosened to free the tripod from any torsion developed while planting it in the ground, and the nuts should then be tightened to a firm bearing; if the stations observed are of some difference in elevation, the horizontal axis of the transit should be leveled with a striding level.

Minor Stations. These may be definite objects of prominence suitably located for control purposes, such as lone trees, church spires, flagstaffs, and chimneys; or they may be signals erected at desirable locations. The observations should be made with much the same care as those for major stations, but ordinarily the method of repetition need not be employed. Each angle should be measured, however, once with the telescope normal and once with it inverted, both verniers being read. Minor stations should be observed from at least three stations, if possible, to provide a check on the computed or plotted locations of these secondary points.

16-12. Instruments for Measuring Angles. For triangulation of somewhat higher than ordinary precision, the angles in the system are measured by means of a *repeating instrument* or *repeating theodolite* (Arts. 13-9 and 13-11). For triangulation of ordinary precision, the ordinary transit may be used.

For triangulation of high precision, either a repeating instrument or a *direction instrument* (Arts. 13-10 and 13-11) is used. In designing both types of instruments, it was once thought that greater precision could be secured by increasing the size of the circles, but it has been found that because of lost motion, mechanical errors in graduating the circles, etc., nothing is gained by increasing the diameter of the circle of the American direction instrument beyond about 10 in.

With the American direction instrument, the angles about a station are measured as follows: The circle is clamped in position, and the telescope is pointed toward the initial station and clamped in that position. By means of the tangent-screw, the signal is observed precisely, and the initial reading is taken by reading the micrometers. The instrument is then turned clockwise, the cross-hairs are set on the next station, and the micrometers are read. Each station is thus observed in turn until the horizon has been closed and the initial station is again sighted. The telescope is then lifted from its Y supports and plunged (the pivots, after plunging, resting in the same supports as before), and when the telescope is revolved about the vertical axis, the initial station is again sighted. The direction of each station is now observed as before, but the stations are sighted in reverse order; that is, the alidade is turned in a counter-clockwise direction from one station to the next. These two series of readings constitute one set. Before a second set is begun, the circle is shifted a number of degrees so that the readings for the next set will be observed on different parts of the circle. In work of the highest precision, 16 sets are observed, the circle being shifted approximately 11° between sets.

16-13. Azimuth Determinations. In computing the coordinates of triangulation stations a meridian of reference, either true or assumed, is used. The azimuth of a triangle side is determined at any convenient station, and the azimuths of all other sides are computed from corrected angles. If the system is extensive, an azimuth determination is made at intervals of 20 to 30 figures as a check on the angle measurements; account must be taken of the convergency of meridians and the spheroidal shape of the earth. Solar or stellar observations may be used, depending upon the field conditions; stellar observations are by far the more accurate. Methods of determining azimuth are described in Chap. 21.

BASE-LINE MEASUREMENT

16-14. Tape. For base-line measurements of ordinary precision either the steel tape or the invar tape may be employed, but for measurements of higher precision the invar tape is always used. Invar is a nickel-steel alloy for which the coefficient of thermal expansion may be as low as 0.0000002 per 1°F. (about one-thirtieth that of steel). The invar tapes commonly used have a coefficient of expansion one-eighth to one-tenth that of steel. Often a "long tape" (length 50 m. to 500 ft.) is used.

The length of the tape should be precisely determined by comparison with a standard of known length. The National Bureau of Standards at Washington, D.C., for a nominal fee, will compare the tape with a length which has been precisely determined and will issue a certificate showing the actual length of the tape under stated conditions as regards tension, temperature, and supports (see Art. 16-17). It is desirable to have the tape compared with the standard under the conditions of tension and support that will be employed in the field work so that no corrections for sag or stretch will be necessary.

A tape that has been compared with the standard at Washington may itself serve as a standard for other tapes in the field, but for work of the greatest precision all the tapes used should be compared with the standard at Washington.

If a tape is kinked in handling, its length will be appreciably changed. Invar is relatively soft and bends easily. Hence, tapes of this metal should be handled with great care and when not in use should be kept on a reel not less than about 15 in. in diameter. In the best practice, two or three tapes are provided and, when in use, are compared daily, thus to detect sudden changes in length due to whatever cause. In any case, a tape should have its length again compared with some standard upon the completion of the work.

Where electronic distance-measuring equipment is available, it may be used for base-line measurement, except for high orders of precision. Instead of measuring merely the base line, it would probably be advantageous to measure all the lines of the system electronically.

16-15. Measuring the Base Line. Where the base line is along a paved highway or a railroad, usually measurements are made with the tape supported over its entire length and at a time when the temperature of the supporting surface (highway or rail) is not appreciably different from that of the surrounding air.

Where the base line is over uneven ground, end supports for the tape are provided, usually by the of substantial posts, perhaps 2 by 4 in. or 4 by 4 in., driven firmly into the ground. These are placed on a transit line at intervals of one tape length as nearly as can be determined by careful preliminary measurements. A strip of copper or zinc is tacked to the top of the post to receive the markings. Portable tripods are also used to some extent as tape supports. Profile levels are run over the tops of the end supports to determine the gradient from support to support.

The tape is usually supported at one, two, or three points between the end supports. These intermediate points are placed accurately on the grade line between the tops of the two adjacent end posts, usually by driving nails at grade in 1 by 2-in. stakes placed on line at the proper intervals. These supports should, preferably, be provided at the same intervals as those used in the standard comparison, and the nails should be so driven that the tape will not become pinched between nail and stake.

The equipment for base-line measurement of ordinary precision where rather precise results are desired includes at least one verified tape (on important work, two tapes are essential); two stretcher devices for applying tension, one of which is equipped with a spring scale or a weight and pulley; two or three thermometers; a finely divided pocket scale; dividers; and a needle or a marking awl.

The party consists of four to six men whose duties are indicated by the following description of the procedure: The proper tension is applied to the tape by means of the stretchers, with the spring scale (or weight and pulley) attached to the forward end of the tape (Fig. 16-9). When the rear end of the tape is observed to coincide with the previously established mark and when the proper tension is applied, the position of the forward end of the tape is marked by a fine line engraved by means of a needle or marking awl on the metal strip on top of the post (Fig. 16-10). Thermometers fastened to

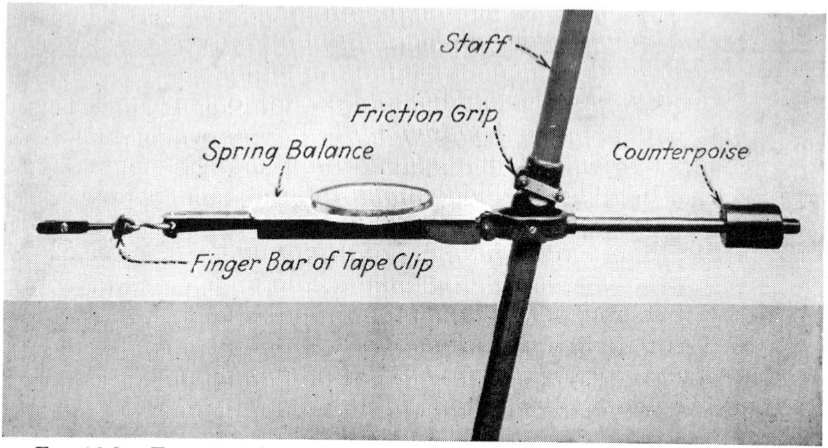

FIG. 16-9. Tape stretcher and spring scale. (*U.S. Coast and Geodetic Survey.*)

FIG. 16-10. Base-line measurement: making forward contact. (*U.S. Coast and Geodetic Survey.*)

the tape, one near each end and sometimes one near the middle, are read at the time that the tape length is marked on the forward post.

The tape is then carried forward without allowing it to drag on the ground, and the process is repeated. After a few measurements, the end of the tape will probably fall either beyond or short of the limits of the metal strip of the next forward post because of variations in temperature or because of inaccurate placement of the posts. Accordingly, it will be necessary occasionally to use either *set backs* or *set forwards* (also called *set-ups*) as may be necessary to keep the tape ends on top of the posts. These are measurements of small distances made by means of a finely divided pocket scale and a pair of dividers. A record is kept of all conditions and observations, as shown in Fig. 16-11.

MEASUREMENT OF EAST BASE

From Stake No.	To Stake No.	Thermometers No. 212 Deg. F.	Thermometers No. 213 Deg. F.	Set Forward Ft.	Set Back Ft.
E. End	32	55.0	54.5		
32	31	55.0	55.0		
31	30	54.5	54.5	0.103	

Nov. 16, 1965 — T. Wiley, Recorder
2:30 P.M. — P. Smith, Marker
Sky overcast — M. Moore, Observer
Measuring backward
Tension 12 lbs.
Tape, B.S. No. 3862
Spring Balance No. 4 (Tested)
Light East Breeze
Tape supported at 0, 50, and 100 ft.

FIG. 16-11. Notes for base-line measurement.

16-16. Errors in Base-line Measurement. The nature of the various sources of error in tape measurements as they affect work of ordinary precision is treated in Art. 7-16, but the special procedure used in base-line measurements causes differences in sources of error which should be noted.

1. *Tape Not Standard Length.* If the tape is carefully handled, the comparison made by the Bureau of Standards will be sufficiently precise to render the error from this source negligible for the duration of a field season. Special care should be taken not to kink the tape and to keep it from dragging on the ground, which causes wear. Experience has shown that a tape changes in length over a period of years even though it is not used. Hence, a tape should be verified in the same season in which it is used.

2. *Variations in Temperature.* The effect of temperature is the most serious source of error in precise linear measurements. For example, an error of 0.5°F. in the mean temperature of a steel tape introduces an error of 1/300,000 in the measured length, which alone is greater than the total probable error permitted in precise work (Tables 25-1 and 25-2). The magnitude of the error due to variations in temperature is greatly reduced by the use of an invar tape.

Field conditions favorable to a small variation in the thermal expansion are those in which the air and ground are at nearly the same temperature on cloudy days or at night. Measurements taken at such times have indicated probable errors as low as 1/3,000,000.

A pavement or railroad track may have a temperature considerably different from that of the air; hence, before a base-line measurement is made along such a surface, observations should be taken to make sure that air and surface temperatures agree closely.

3. *Tape Not Horizontal.* This source of error is rendered negligible as follows: The difference in elevation of adjacent posts is determined by a line of profile levels run with a transit or an engineer's level. The gradient for each tape length is then readily computed. The corrections for grades up to 5 per cent may be computed by the approximate formula (1) of Art. 7-15, or by the less approximate formula

$$C_h = \frac{h^2}{2s} + \frac{h^4}{8s^3}$$

Steeper grades may be used if necessary, but the corrections should then be computed by the exact formula (3) of Art. 7-15. The same correction is applied to measurements of base lines along a highway or railroad, the grade of the supporting surface being obtained by leveling.

4. *Variations in Tension; Sag in Tape.* Variations in tension are not important if only stretch in the tape is considered, as where the tape is supported throughout its length. The effect on the amount of sag, if the tape is supported at intervals, is much more serious. The less the number of supports, the more important this effect. Other things being equal, the shortening due to sag varies inversely as the square of the tension. The amount of the resulting error for varying conditions is given in Table 7-3, Art. 7-20. If the tension is determined correctly within 1 oz., the resulting error is negligible for the class of work under consideration.

A normal tension (see Art. 7-21) is used where conditions are favorable.

5. *Wind.* If a strong wind is blowing normal to the base line and if the tape is not supported throughout its length, there will be a lateral displacement of the unsupported portions of the tape, thus producing an effect similar to that of sag. Wind also sets up vibrations which render the tape unsteady. It is impossible to compute corrections for wind effects; accordingly, precise measurements should not be attempted if the tape is unsupported throughout its length and if a strong cross wind is blowing.

6. *Marking the Tape Lengths.* The magnitude of the errors resulting from imperfect marking of tape lengths depends upon the fineness of the tape graduation and of the line cut on the metal strip, and on the precision with which these lines are made to coincide when the tape lengths are being marked.

The lines marking the ends of ordinary steel tapes are relatively coarse, but makers of invar tapes use lines not exceeding 0.002 in. in width. The Bureau of Standards is careful to state which edge of the tape is used in making the comparison with the standard length. For a given tape, careful manipulation is the only means of reducing errors from this source.

16-17. Corrections to Measured Length. The methods of computing and applying corrections to tape measurements are given in Arts. 7-15 to 7-23.

Following is an example showing the corrections applied to the measured length of a base line:

Example: The length of a base line is recorded as 3,243.063 ft., and the average observed temperature is 59.7°F. In the field, the tape is supported and the tension is maintained the same as when the tape was compared with the standard. The standardization data are as follows: Length at 68°F. = 100.0214 ft. (tape supported at 0, 50, and 100-ft. marks; tension = 10 lb.); coefficient of thermal expansion = 0.00000645 per degree Fahrenheit. Corrections (including corrections for slope) are as follows:

	Feet
Recorded length	3,243.063
Length correction	+0.694
Total set forwards	+0.364
Total set backs	−0.158
Temperature correction	−0.174
Total slope correction	−0.364
Length of base	3,243.425

16-18. Reduction to Sea Level. It is sometimes necessary to reduce the length of the base line to the equivalent length at mean sea level. The correction C_l to be subtracted from the actual length is given by the equation

$$C_l = \frac{LA}{R} \tag{3}$$

in which L is the length of the base line, A is the mean altitude of the base line above sea level, and R is the radius of the earth (mean R = 20,889,000 ft., log R = 7.31992). For convenience, the correction may be determined from published tables. For a distance of 1,000 ft. and a difference in elevation of 1,000 ft., it is 0.0478 ft., and as shown by Eq. (3) it is directly proportional both to distance and to difference in elevation.

Conversely, to compute the length at the elevation of a survey corresponding to a given length at sea level, Eq. (3) may be solved with the sea-level length as L, and the correction added.

16-19. Discrepancy between Bases. Experience indicates that for first-, second-, and third-order triangulation, the precision attained in a base line computed from a measured base through a chain of approximately 20 figures will be reduced to about one fifth of that of the measured base line, provided the angles of the system are measured with a precision corresponding to that of the accidental errors in the base measurement. Thus, if the probable error of a measured base is, say, 1/25,000, the probable error of a base computed from the measured base through a chain of 20 figures is about 1/5,000. This relation makes it possible to estimate in advance the required precision of base-line measurements to produce a check on base which will meet the requirements of a given specification.

16-20. Specifications for Base-line Measurement. Following are the essential requirements of the U.S. Coast and Geodetic Survey for measurement of base lines (Refs. 6 and 7 at the end of this chapter).

	First order	Second order	Third order
Actual error of base not to exceed......	1/300,000	1/150,000	1/75,000
Probable error of base not to exceed......	1/1,000,000	1/500,000	1/250,000
Discrepancy between two measurements of a section not to exceed............	10 mm. $\sqrt{\text{kilometers}}$	20 mm. $\sqrt{\text{kilometers}}$	25 mm. $\sqrt{\text{kilometers}}$

The length of each base line is determined by at least two complete measurements with each of two standardized tapes (three tapes for first-order triangulation). Invar tapes 50 m. long are used. The standard tension is 15 kg., and the error in tension is not permitted to exceed 100 g. for first- and second-order triangulation or 150 g. for third-order work. The temperature at each end of the tape is observed for each tape length. Precautions are taken against, or corrections are made for, errors due to grade, alinement, temperature, sag, stretching, erroneous tension, method of support, change in weight of tape (due to adherent moisture), friction, wind, marking, and elevation above sea level.

Low Precision. For measurements of low precision the systematic errors are likely to become more important than in refined measurements, and for this reason a somewhat higher degree of precision in the measurements must be maintained than would otherwise be necessary. A detailed analysis of the interrelation of the errors in triangulation work is beyond the scope of this text, but the following general specifications for three degrees of ordinary and low precision are given as applicable to average conditions. It should be realized that field conditions vary widely and that they appreciably influence the precision of results.

Discrepancy between base lines not to exceed		1/3,000	1/1,000	1/500
Specifications	Minimum length of each base line, ft	2,500	1,500	1,000
	Probable error of base not to exceed	1/20,000	1/10,000	1/5,000
	Length of triangle sides	½ to 3 mi.	¼ to 2 mi.	1/10 to 1 mi.
	Average closing error of triangles not to exceed	8″	15″	30″

COMPUTATIONS

16-21. General. In triangulation of low precision, the measured angles and base line may be used, without correction or adjustment, for computation of the lengths of the remaining sides. In triangulation of ordinary and higher precision, however, the observed angles are corrected before the lengths of the sides are computed. If sights have been taken from, or to, any point which is not exactly at a triangulation station, the angles at that point are corrected for such eccentricity (Art. 16-22). The angles about each station are adjusted to total 360°. In precise work involving large triangles, the angles of each triangle are corrected for spherical excess (Art. 16-23). The system—which may consist of triangles (Art. 16-24) or quadrilaterals (Art. 16-25)—is adjusted to make the angles meet the known geometric and trigonometric conditions. The lengths of the triangle sides are then computed from the corrected angles and the base line, and the coordinates (plane or geographic) of the stations are computed.

16-22. Reduction to Center. At certain triangulation stations it is difficult, if not impossible, to place the instrument vertically beneath the object which has been observed from adjacent stations. At such a place, the instrument is set over any convenient point near the principal station, and angles to the adjacent stations are measured with the same precision as other angles in the system. These angles will not be the same as those which would be observed if the instrument were occupying the exact location of the station. To obtain the corresponding values for the main station, corrections are computed and applied to the measured angles. This procedure of correcting the observed angles is termed *reduction to center*.

In addition to the measurement of the angles to adjacent stations, measurements are made of (1) the distance from the main station to the occupied station, and (2) the (clockwise) angle at the occupied station between the main station and an adjacent station in the system. The situation is illustrated by Fig. 16-12, in which O represents a main station in the system $OABCD$, and T represents the point occupied by the instrument; the distance d and the angle ATO are measured. The lengths of all sides in the main system, as for example $t_1 = 8{,}659$ ft., are known approximately from the angles which have been measured at the stations A, B, C, D and from the known sides AB, BC, etc.

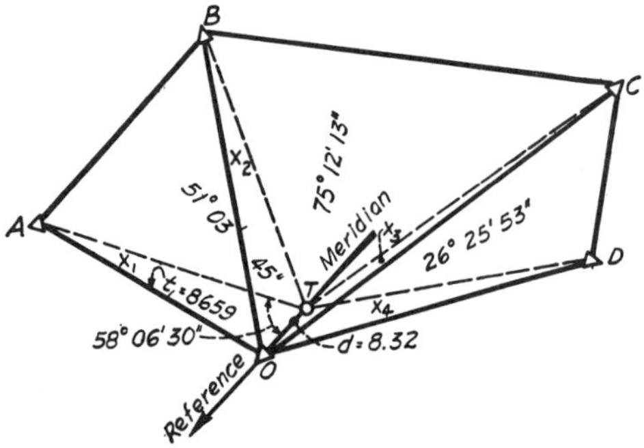

FIG. 16-12. Reduction to center.

In the triangle AOT, the angle ATO (herein called T_1) and the two sides t_1 and d are known, hence the angle x_1 can be computed. This angle is seen to be the difference in direction at station A between the lines AT and AO. Therefore, if the direction (azimuth) of AT is known with respect to any reference meridian, the direction of AO with respect to the same meridian can be computed. Likewise, the directions of the lines BO, CO, and DO can be found, and since these directions are referred to the same meridian, the correct angles at O between these stations can be determined.

The value of x_1 in triangle AOT is given by the equation

$$\sin x_1 = \frac{d \sin T_1}{t_1} \tag{4}$$

and since the angle x_1 is a small angle for which the sine is nearly equal to the arc, the value of x_1 will be given in seconds of arc if both members of the equation are divided by the sine of 1″, ($\sin 1'' = 0.000004848$); thus

$$x_1'' = \frac{d \sin T_1}{\sin 1'' t_1} \text{ (approximate)} \tag{5}$$

It will be noted that ($d/\sin 1''$) is a constant for a given station, so that once its value has been determined the successive correction angles x_1, x_2, etc., can be computed by a single multiplication. Since the correction angles are usually small, the slide rule will ordinarily render values correct to seconds.

Example: Using the slide rule, by Eq. (4) $\sin x_1 = (8.32)(0.849)/8659 = 0.000816$, and from Table XVI, $x_1 = 2.8'$. By Eq. (5),

$$x_1'' = (0.000816)(0.00000485) = 168'' = 2.8'.$$

The method of correcting angles for which one of the sights has been taken to an eccentric signal is the same as that just described for an eccentric instrument station.

16-23. Spherical Excess. Since the measured angles are spherical angles, each triangle will contain more than 180°. The amount greater than 180° is termed the *spherical excess* and is about 1″ for each 75 sq. miles of area of triangle. More precisely,

$$E = \frac{a}{C}(1 - e^2 \sin^2 \phi)^2 \tag{6}$$

in which E = spherical excess, in seconds
a = area, in square miles
ϕ = latitude at center of triangle
$\log e^2 = 7.8305026 - 10$
$\log C = 1.8787228$

It is clear that no correction for spherical excess will be necessary unless the triangles are very large and then only in the most precise work. One-third of the correction is subtracted from each of the angles.

16-24. Adjustment of a Chain of Triangles. A single chain of triangles is adjusted in two steps: (1) the *station adjustment*, or *local adjustment*, to make the sum of the angles about each point equal 360°, and (2) the *figure adjustment*, to make the sum of the three angles in each triangle equal 180° (plus the spherical excess).

In precise triangulation the station adjustment and the figure adjustment are made in one operation by the method of least squares, but the following approximate solution yields results that are sufficiently precise for most cases of triangulation of ordinary precision.

To make the sum of the angles about each point equal 360°, the observed angles are added together and the sum is subtracted from 360°. The resulting difference is divided by the number of angles, and the quantity so found is added algebraically to each angle. To make the sum of the angles in each triangle equal 180°, a similar plan is followed, using the values obtained by the station adjustment; that is, the three angles of each triangle are added together, and their sum is subtracted from 180°. One-third of the difference is added algebraically to each of the three angles. If certain angles are measured with a higher precision than others, the method may be modified readily by weighting the observations of the several angles within the system, either arbitrarily or by the method of least squares, as described in Chap. 5.

Following is an example of the adjustment of the angles in a simple chain of three triangles, all the observed angles being assumed to be of equal precision.

Example: In the first accompanying tabulation are listed the measured and adjusted angles at each station in the chain of triangles shown in Fig. 16-13. The station adjustment is made aproximately by dividing the errors equally among the

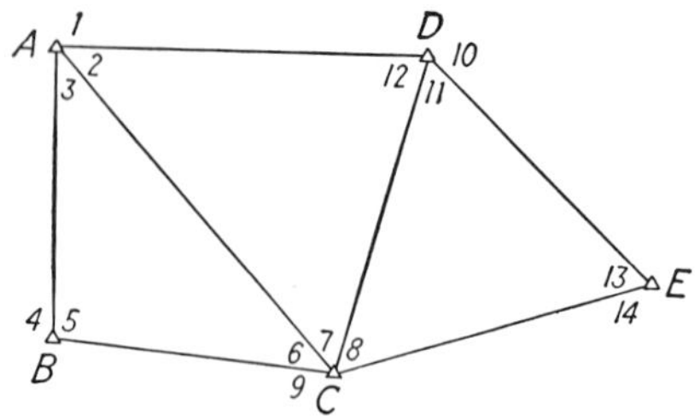

FIG. 16-13. Adjustment of a chain of triangles.

angles. In the second tabulation are listed for each triangle the values from the station adjustment and finally the values from the figure adjustment computed by dividing the errors approximately equally among the angles.

STATION ADJUSTMENT (CHAIN OF TRIANGLES)

Station	Angle	Measured value	Adjusted value
A	1	240°19′00″	240°18′50″
	2	73 31 10	73 31 00
	3	46 10 20	46 10 10
	Sum	360°00′30″	360°00′00″
B	4	267°12′20″	267°12′25″
	5	92 47 30	92 47 35
	Sum	359°59′50″	360°00′00″
C	6	41°02′00″	41°01′55″
	7	63 10 40	63 10 35
	8	74 43 10	74 43 05
	9	181 04 30	181 04 25
	Sum	360°00′20″	360°00′00″
D	10	260°33′00″	260°32′50″
	11	56 09 00	56 08 50
	12	43 18 30	43 18 20
	Sum	360°00′30″	360°00′00″
E	13	49°07′50″	49°07′50″
	14	310 52 10	310 52 10
	Sum	360°00′00″	360°00′00″

FIGURE ADJUSTMENT (CHAIN OF TRIANGLES)

Triangle	Angle	Value from station adjustment	Value from figure adjustment
	3	46°10′10″	46°10′16″
ABC	5	92 47 35	92 47 42
	6	41°01 55	41 02 02
	Sum	179°59′40″	180°00′00″
	2	73°31′00″	73°31′02″
	7	63 10 35	63 10 37
ACD	12	43 18 20	43 18 21
	Sum	179°59′55″	180°00′00″
	8	74°43′05″	74°43′10″
	11	56 08 50	56 08 55
CDE	13	49 07 50	49 07 55
	Sum	179°59′45″	180°00′00″

16-25. Adjustment of a Quadrilateral. As in the case of the chain of triangles, the angles around each station of a quadrilateral are adjusted to total 360° before the figure adjustment is made. In the figure adjustment, two conditions are considered: (1) the *geometric condition* that the sum of the interior angles of a plane rectilinear figure is equal to $(n - 2)180°$, in which n is the number of sides of the figure, and (2) the *trigonometric condition* that in any triangle the sines of the angles are proportional to the lengths of the sides opposite.

First, the station adjustment is made; then the geometric condition is satisfied by adjustment of the angles of the four overlapping triangles forming the quadrilateral. Then the trigonometric condition is satisfied by means of computations involving the sines of the angles, the angles being adjusted so that the computed length of an unknown side opposite a known side will be the same regardless of which of the four possible routes is used.

1. *Geometric Condition.* When all angles in a quadrilateral are measured, there are four overlapping triangles. These are shown as triangles *ABC*, *ACD*, *ABD*, and *BCD* in Fig. 16-14. In each of these triangles the sum of the three angles must be 180°. Hence from the figure,

$$b + c + d + e = 180° \tag{7a}$$
$$a + f + g + h = 180° \tag{7b}$$
$$a + b + c + h = 180° \tag{7c}$$
$$d + e + f + g = 180° \tag{7d}$$

Also the sum of the eight lettered angles in the figure must equal 360°, since they form the interior angles of a closed figure of four sides. This may be derived also by the addition of Eqs. (7*a*) and (7*b*) or (7*c*) and (7*d*).

$$a + b + c + d + e + f + g + h = 360° \tag{8}$$

Further, since the opposite angles at the intersection of the diagonals must be equal, it follows that

$$b + c = f + g \tag{9}$$

$$d + e = h + a \tag{10}$$

Equation (9) is the equivalent of Eq. (7*b*) minus Eq. (7*c*), and Eq. (10) is the equivalent of Eq. (7*b*) minus Eq. (7*d*).

If any three of these seven equations, called "angle equations," are satisfied, the other four must of necessity be satisfied also. Equations (8), (9), and (10) are the most convenient to use.

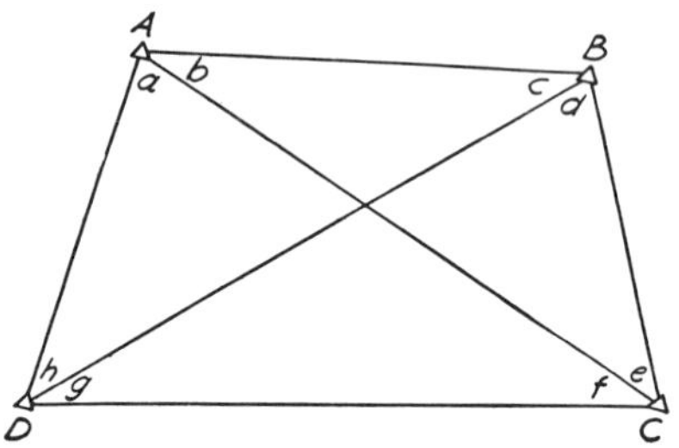

FIG. 16-14. Adjustment of a quadrilateral.

The following procedure is suggested for triangulation of ordinary precision:

A. Make the station adjustment as follows: Adjust the angles around each point to make their sum equal 360° by distributing the error equally (or approximately so) among the several angles.

B. Using the values resulting from the station adjustment *A*, add the eight angles *a*, *b*, *c*, *d*, *e*, *f*, *g*, and *h*, and subtract their sum from 360°. Divide the difference by 8, and algebraically add the result to each of the eight angles, thus satisfying the conditions of Eq. (8).

C. Using the adjusted values from *B*, find the difference between the sums $(b + c)$ and $(f + g)$ and divide that difference by four. Apply the result as a correction to each of the four angles, increasing each of the two whose sum is the smaller and decreasing each of the two whose sum is the larger, thus making these angles satisfy Eq. (9) without disturbing the adjustment for Eq. (8). Proceed in the same way with each of the four angles involved in Eq. (10).

2. *Trigonometric Condition.* If the length of one line, as *AB*, is known and the length of the opposite side *CD* is to be computed, the computer may

select one or another series of triangles for use in accomplishing this result. For example, a solution of triangle ABC gives the length of AC, then from triangle ACD the required length of CD is found; or in the triangle ABC the length BC is found, then in BCD the length CD is computed. There are four possible choices of route through the figure. It now remains to be seen whether the angles, as so far adjusted, are so related as to make the value of the length of a computed side independent of the route used. The equations used are called "side equations." Assume that the length of AB is known and the length of CD is to be found.

$$AD = AB \frac{\sin c}{\sin h} \tag{11}$$

$$CD = AD \frac{\sin a}{\sin f} = AB \frac{\sin a \sin c}{\sin f \sin h} \tag{12}$$

Similarly,

$$CD = AB \frac{\sin b \sin d}{\sin e \sin g} \tag{13}$$

Equating these two values of CD,

$$\frac{\sin a \sin c}{\sin f \sin h} = \frac{\sin b \sin d}{\sin e \sin g} \tag{14}$$

or

$$\frac{\sin a \sin c \sin e \sin g}{\sin b \sin d \sin f \sin h} = 1 \tag{15}$$

Expressed in logarithmic form, this is

$$(\log \sin a + \log \sin c + \log \sin e + \log \sin g) - (\log \sin b + \log \sin d + \log \sin f + \log \sin h) = 0 \tag{16}$$

The angles are tested for satisfaction of this equation by adding the logarithmic sines in the two groups as indicated and by finding the difference between the two sums.

Various adjustments by which this difference may be reduced to zero are possible. The least-squares adjustment gives the most probable values to the adjusted angles (see Ref. 11 at the end of this chapter); but it is somewhat more elaborate than is necessary for most surveys. A simple approximate method which gives an equal correction to each angle and does not disturb the geometric condition is as follows (see adjustment D of the following example):

(*a*) Record the log sines as shown in the example.

(*b*) For each angle, record the tabular logarithmic sine difference for 1″ opposite each logarithm.

(*c*) Find the average required change (α) in log sine by dividing the difference between the sums by 8.

(*d*) Find the average difference (β) for 1″.

(*e*) The ratio α/β gives the number of seconds of arc to be applied as a correction to each angle.

(*f*) Add this correction to each of the four angles the sum of whose log sines is the smaller, and subtract it from each of the angles the sum of whose log sines is the larger, and thus the corrected values are obtained.

If one or more of the angles is greater than 90°, adjustment *D* is made as just described, without disturbing the geometric relations. However, since the sine of an obtuse angle decreases as the angle increases, the corresponding log sines will be changed in the direction opposite to that desired. Usually the error introduced by this condition will be negligible for this approximate adjustment; if not, adjustment *D* should be repeated.

Example: Given the angles as measured in the quadrilateral of Fig. 16-14, for which the station adjustment (adjustment *A*) has been made (see second column of accompanying tabulation). Find the adjusted angles for the geometric condition and then the trigonometric condition.

ADJUSTMENT OF QUADRILATERAL

Angle	Station adjustment	Figure adjustment		
		Geometric condition		Trigonometric condition
	Adjustment *A*	Adjustment *B*	Adjustment *C*	Adjustment *D*
a	38°44′06″	38°44′05″	38°44′06″	38°44′08″
b	23 44 38	37	35	33
c	42 19 09	08	06	08
d	44 52 01	00	51 59	51 57
e	69 04 21	20	20	22
f	39 37 48	47	49	47
g	26 25 51	50	52	54
h	75 12 14	13	13	11
Sum	360°00′08″	360°00′00″	360°00′00″	360°00′00″

Adjustment B. The sum of the angles *a*, *b*, *c*, etc., resulting from the statio[n] adjustment *A*, is found to differ from 360° by the amount of 08″. This amoun[t] divided by the number of angles gives the amount of the correction (01″) to b[e] subtracted from each angle as shown in the third column in the table.

Adjustment C. $b + c = 66°03'45''$
$f + g = 66°03'37''$

Dividing this difference by 4, the correction to each angle is found to be 02″, to b[e] subtracted from the angles *b* and *c*, and added to the angles *f* and *g*. In like manne[r] the correction to each of the angles d, e, h, and a is found to be 00.5″, to be added to h and a, and subtracted from d and e. (Since these computations are carried out only to seconds, 01″ is added to a and 01″ is subtracted from d.) The resultant angles are shown in the fourth column of the table.

Adjustment D. Trigonometric Condition. The logarithmic sines of the angles as given by adjustment C and as indicated in Eq. (16) are recorded as shown in the following tabulation, and the tabular difference for 1″ is recorded for each angle.

		Difference for 1″
log sin a	9.796380	2.6
log sin c	9.828176	2.3
log sin e	9.970361	0.8
log sin g	9.648478	4.2
Sum	9.243395	
log sin b	9.604912	4.8
log sin d	9.848470	2.1
log sin f	9.804706	2.6
log sin h	9.985354	0.6
Sum	9.243442	8)20.0
		2.5 = β
	9.243395	
Difference	8)47	
	5.9 = α	

The difference between the two sums is 47 units of the six places of logarithms used. This value, divided by 8, gives the average required change in log sine, $5.9 = \alpha$. The average tabular difference for 1″ is $2.5 = \beta$. Hence $\alpha/\beta = 5.9/2.5 = 2''$ (nearly), which is the average correction to be applied to each angle. Obviously, it will be added to angles a, c, e, and g and subtracted from angles b, d, f, and h.

Since this adjustment is applied with opposite sign to alternate angles, it does not disturb the geometric condition. The final adjustment is given in the fifth column of the tabulation.

If the triangulation system consists of a chain of quadrilaterals, each quadrilateral is adjusted in the manner just described. The computations for length of the various lines are then carried through the chain from the base line.

16-26. Adjustment of Chain between Two Base Lines. If two base lines are measured, an additional condition is introduced, namely, that the length of each side in the connecting chain of triangles or quadrilaterals must be the same when computed from one base line as when computed from the other. An exact solution is possible by the method of least squares, but the following approximate methods may be used in triangulation of ordinary precision in the case of a single chain of figures.

The figures (triangles or quadrilaterals) are adjusted individually as previously described. The lengths of the sides are then computed from each base line to a common side about midway between them. This common side may then be corrected to reconcile the two computed values of its length, with equal or different weights being assigned to the two computed values as desired, based on the known conditions. The effect of this correction may then be carried back through each half of the chain, as follows:

If the precision of the *angular* measurements is relatively high as compared with that of the linear measurements, the lines of each half of the chain are corrected in proportion to their lengths as compared with the length of the common line, leaving the angles unchanged. In effect, this procedure shrinks one entire half of the chain (including its base line) by a fixed proportion and swells the other half (including its base line) by a fixed proportion.

If, however, the precision of the *linear* measurements is relatively high, the lengths of the base lines may be assumed to be correct. In this case, the correction is tapered off from the full amount at the common line to zero at each base line, the correction to each line being not only proportional to the length of the line but also roughly proportional to the relative distance of the given line from the base line. This procedure changes the values of the angles, and the new values of the angles are used in further computations.

Between these two extremes, the procedure depends upon the relative precision of the angular and the linear measurements, and weights may be assigned accordingly.

16-27. Computation of Triangles and Coordinates. In computing the lengths of the sides and the coordinates of the stations in a triangulation system, it is desirable to follow an orderly procedure to expedite the work and to avoid mistakes. Convenient arrangements for these computations for plane triangulation are given in Figs. 16-15 and 16-16.

Triangles. A sketch of the figure is drawn (Fig. 16-15), and the vertices are lettered as A, B, and C in a clockwise direction, beginning with the side the length of which is known. The sides opposite the vertices are indicated by the corresponding lower-case letters, as a, b, and c. The sine relation states that

$$b = c\frac{\sin B}{\sin C}, \qquad \text{or} \qquad \log b = \log c - \log \sin C + \log \sin B \qquad (17$$

Accordingly, if the logarithms are recorded in the column of logarithms i the order log c, colog sin C, log sin A, and log sin B, then log a is found b covering log sin B with a narrow strip of paper and adding the other thre values. Also log b is found by covering log sin A and adding the other thre values. Finally the distances a and b are found as the numbers correspondin to their respective logarithms. An example of the computations is shown i Fig. 16-15, for which the known data are given on the sketch of the triangl

Coordinates. Figure 16-16 shows a form and example for computing th coordinates of a station C from each of the stations B and A. The know plane coordinates (total latitude and total departure) of B and A and th

Station or line	Angle or distance	Logarithm	Figure
c	1,432.58 ft	3.156119	Given: 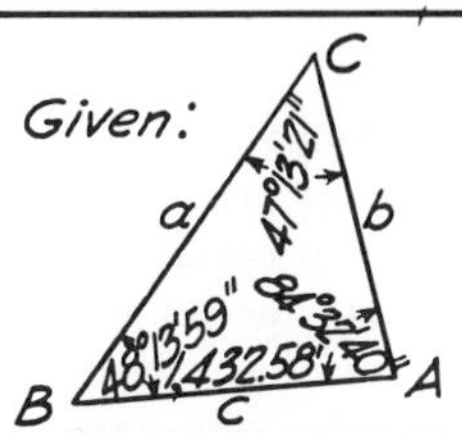
C	47°13'21"	0.134306 (colog)	
A	84°32'40"	9.998028	
B	48°13'59"	9.872657	
a	1,942.94 ft	3.288453	
b	1,455.74 ft	3.163082	

FIG. 16-15. Computation of triangle.

From Station B	From Station A
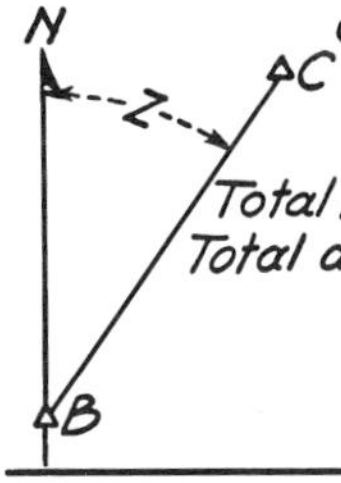 Given: Z = 34°32'54"; BC = 1,942.94; Total lat. B = +661.36; Total dep. B = −1,590.94	Given: Z = 12°40'27"; AC = 1,455.74; Total lat. A = +841.37; Total dep. A = −169.71 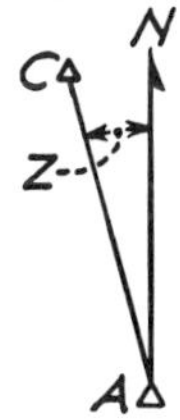

Mean Total Latitude C = +2,261.64

(Check)

Total lat. C	+2,261.64	Total lat. C	+2,261.63
Total lat. B	+661.36	Total lat. A	+841.37
L cos Z	+1,600.28	L cos Z	+1,420.26
log L cos Z	3.204195	log L cos Z	3.152369
log cos Z	9.915742	log cos Z	9.989287
log L	3.288453	log L	3.163082
log sin Z	9.753661	log sin Z	9.341249
log L sin Z	3.042114	log L sin Z	2.504331
L sin Z	+1,101.83	L sin Z	−319.40
Total dep. B	−1,590.94	Total dep. A	−169.71
Total dep. C	−489.11	Total dep. C	−489.11

(Check)

Mean Total Departure C = −489.11

FIG. 16-16. Computation of coordinates.

known bearings and lengths of BC and AC are shown at the top of the figure. The computation is carried out as indicated, with due regard to signs. Beginning with log L in the tabulation, computations for total latitude are made reading upward, and computations for total departure are made reading downward.

16-28. Computation of Geodetic Position. Geodetic position is computed only for triangulation of high precision or over large areas, where it is necessary to consider the curvature of the earth. The adjustment of the observations is accomplished by the method of least squares and is too elaborate for treatment here. The angles of the system are adjusted, and the lengths of the sides are computed. From these data are computed the geodetic coordinates, that is, the latitude and longitude of the stations included in the system.

The geodetic position of a station is computed from that of a station of known latitude and longitude, having given the length and the azimuth (at the known station) of the connecting line. Owing to the convergency of meridians, the azimuth of the connecting line at the unknown station will not differ by exactly 180° from that at the known station. In the following formulas, azimuths are measured clockwise from south.

Let ϕ = latitude of the known station
λ = longitude of the known station
α = azimuth of the connecting line at the known station
ϕ', λ', α' = corresponding quantities for the unknown station
s = length of line, in meters
N' = length of the normal at the unknown station, produced to the earth's polar axis, in meters
a = semidiameter of equatorial axis = 6,378,206.4 m.
b = semidiameter of polar axis = 6,356,583.8 m., both for the Clarke spheroid of 1866, on which the published tables are based
e^2 = square of eccentricity = $\dfrac{a^2 - b^2}{a^2}$ = 0.006,768,658

From these quantities are computed the factors A', B, C, and D, using the following formulas:

$$A' = \frac{1}{N' \text{ arc } 1''} = \frac{(1 - e^2 \sin^2 \phi')^{1/2}}{a \sin 1''} \quad (1\ldots$$

$$B = \frac{(1 - e^2 \sin^2 \phi)^{3/2}}{a(1 - e^2) \sin 1''} \quad (1\ldots$$

$$C = \frac{(1 - e^2 \sin^2 \phi)^2 \tan \phi}{2a^2(1 - e^2) \sin 1''} \quad (2\ldots$$

$$D = \frac{3/2 e^2 \sin \phi \cos \phi \sin 1''}{1 - e^2 \sin^2 \phi} \quad (2\ldots$$

Then

$$\phi - \phi' \text{ (in seconds)} = s \cos \alpha \cdot B + s^2 \sin^2 \alpha \cdot C + (s \cos \alpha \cdot B)^2 \cdot D \tag{22}$$

$$\lambda' - \lambda \text{ (in seconds)} = \frac{s \sin \alpha \cdot A'}{\cos \phi'} \tag{23}$$

$$(\alpha - \alpha' + 180°) \text{ (in seconds)} = (\lambda' - \lambda) \sin \tfrac{1}{2}(\phi' + \phi) \tag{24}$$

Special care must be taken with regard to the signs of the azimuth functions.

These formulas give sufficiently precise results for a distance s not greater than 25 km., or about 15 miles. For the more precise formulas used on longer lines and for examples of the use of these formulas, see Ref. 14 at the end of this chapter.

Inverse Solution. Sometimes the latitudes and longitudes of the two stations are given, and the problem is to find the length and direction of the connecting line.

Knowing $(\phi - \phi')$ and $(\lambda' - \lambda)$, find the value of the term $s \cos \alpha \cdot B$ from Eq. (22) by subtracting the values of the small C and D terms from the known value of $(\phi - \phi')$. Divide $s \cos \alpha \cdot B$ by B to find the value of $s \cos \alpha$. Then

$$s \sin \alpha = \frac{(\lambda' - \lambda) \cos \phi'}{A'} \tag{25}$$

$$\tan \alpha = \frac{(\lambda' - \lambda) \cos \phi'}{A'(s \cos \alpha)} = \frac{s \sin \alpha}{s \cos \phi} \tag{26}$$

Knowing α, s is found from $s \cos \alpha$ or from $s \sin \alpha$.

Finally, $(\alpha - \alpha')$ is found as before, from Eq. (24).

16-29. Three-point Problem. When the main triangulation has been completed, frequently it is desired to determine the location of additional points which are to be used as instrument stations of a topographic survey or for other purposes. In triangulation work the location of an instrument station as O (Fig. 16-17) is determined by measuring each of the two angles subtended by three visible stations, as A, B, and C, and by solving the triangles involved. Thus, in Fig. 16-17, all parts of the triangle formed by the stations A, B, and C are known. The angles α and β are measured at the station O. The problem is solved when the values of the angles x and y have been determined, for the remaining parts in each of the triangles ABO and ACO can then be computed. A check is afforded if the same value for the side AO results from each of these triangles.

The problem is indeterminate if the station O lies on or near the great circle passing through the stations A, B, and C. This condition will be evidenced by the condition that $\alpha + \beta + A = 180°$.

There are many solutions for this problem. The one presented here follows that given by the U.S. Coast and Geodetic Survey (Ref. 11 at the end of this chapter). For solution of the three-point problem in plane-table work by resection, see Art. 17-14.

Solution: Given the sides b and c and the angle A, also the observed angles α and β (Fig. 16-17). Required to find the angles x and y. If station

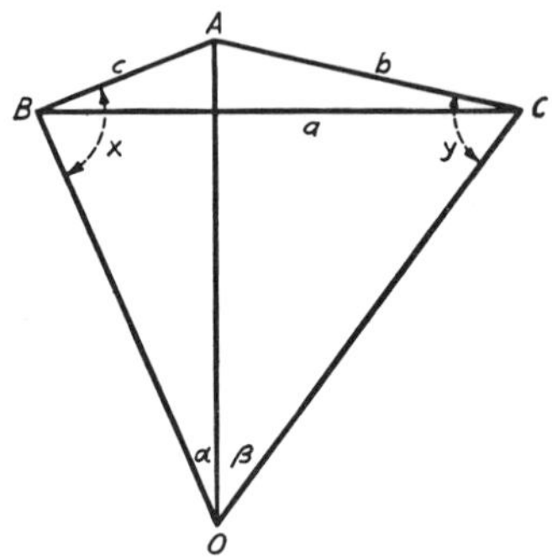

FIG. 16-17. Three-point problem.

O is either outside the triangle ABC and opposite station A (Case 1) or within the triangle (Case 2), let

$$S = 180° - \tfrac{1}{2}(A + \alpha + \beta) = \tfrac{1}{2}(x + y) \tag{27}$$

If station O is outside the triangle ABC and on the same side of side a as station A (Case 3), let

$$S = \tfrac{1}{2}(A - \alpha - \beta) = \tfrac{1}{2}(x + y) \tag{27a}$$

for which case the solution by this method is impossible when $\alpha + \beta = A$. Also let

$$\tan \phi = \frac{c \sin \beta}{b \sin \alpha} \tag{28}$$

and let

$$\Delta = \tfrac{1}{2}(x - y) \tag{29}$$

then

$$\tan \Delta = \cot (\phi + 45°) \tan S \tag{30}$$

If $\tan \Delta$ is positive, $x = S + \Delta$ and $y = S - \Delta$ (31)
If $\tan \Delta$ is negative, $x = S - \Delta$ and $y = S + \Delta$ (31a)

Example: Given

$c = 6{,}672.5$ ft. $\qquad \alpha = 20°05'53''$
$b = 12{,}481.7$ ft. $\qquad \beta = 35°06'08''$
$A = 152°23'22''$

This falls under Case 1. The following computations solve for S by Eq. (27), then for ϕ by Eq. (28), then for Δ by Eq. (30), and finally for x and y by Eq. (31).

$$\begin{aligned} A &= 152°23'22'' \\ \alpha &= 20°05'53'' \\ \beta &= 35°06'08'' \\ &\quad 2)\overline{207°35'23''} \\ &\quad 103°47'42'' \\ S &= 76°12'18'' \end{aligned}$$

$$\begin{aligned} \log c &= 3.824288 & \log b &= 4.096274 \\ \log \sin \beta &= 9.759696 & \log \sin \alpha &= 9.536088 \\ \log (c \sin \beta) = \text{sum} &= 3.583984 & \log (b \sin \alpha) = \text{sum} &= 3.632362 \end{aligned}$$

$$\log \tan \phi = \log (c \sin \beta) - \log (b \sin \alpha) = 9.951622$$

$$\phi = 41°48'55''$$

$$\begin{aligned} \phi + 45° &= 86°48'55'' \\ \log \cot (\phi + 45°) &= 8.745396 \\ \log \tan S &= 0.609894 \\ \log \tan \Delta &= 9.355290 \\ \Delta &= 12°46'06'' \end{aligned}$$

$$\begin{aligned} S &= 76°12'18'' & S &= 76°12'18'' \\ \Delta &= +12°46'06'' & \Delta &= -12°46'06'' \\ x &= \mathbf{88°58'24''} & y &= \mathbf{63°26'12''} \end{aligned}$$

STATE PLANE COORDINATES

16-30. State Systems. One activity of the Federal agencies engaged in triangulation is the establishment throughout the country of monuments whose geographic coordinates, or latitude and longitude, are known. It is desirable to refer local surveys, which employ plane coordinates, to such monuments in order as far as possible to avoid discrepancies at the edges of adjacent surveys and in order to coordinate the work of surveying as a whole.

For many years, information has been available whereby the geographic coordinates of available triangulation stations may be converted into plane coordinates of a local system, and *vice versa* (Ref. 10 at the end of this chapter). The projection is on a plane tangent to the spheroid representing the earth (Chap. 32), and its use is limited to areas not farther than about 20 miles from the origin of the local system. The tangent-plane projection has been employed on surveys of several large cities but not to a great extent elsewhere.

A far greater opportunity for use of the national triangulation system has come about through the adoption of state systems of plane coordinates, whereby one set of plane rectangular coordinates is made to serve the whole area of a small state or a portion (usually half) of the area of a large state. Many additional triangulation stations have been established and monumented by the U.S. Coast and Geodetic Survey. A map projection has been chosen for the state or portion thereof such that the errors of projection will not exceed 1/10,000 in a width of 158 miles and, therefore, will be negligible for most local surveys. For states of greater extent east and west a Lambert conformal conic projection (Art. 32-12) is used, while for states of greater extent north and south a transverse Mercator projection (Art. 32-13) is employed. Both projections are conformal; that is, at any point on the projection the angles are the correct angles. Along the two standard parallels of the Lambert projection and along the two standard meridional lines of the Mercator projection the scale of the projection is exact. Variations from exact scale with distance from the standard lines are given in published tables, and full allowance may be made for these scale errors. Also, distances at the elevation of the survey can be converted to or from equivalent sea-level distances (Art. 16-18) on which the coordinate systems are based, or to grid distances. Reference axes for each zone are such that the x and y coordinates of all points within the area will be positive; the location of any point can be designated by simply stating these two coordinates, as illustrated in Fig. 22-6.

At a distance of ¼ to 2 miles from each triangulation station is established an *azimuth mark*, or monument, which can be sighted from the station. The plane coordinates of the station and the plane azimuth to the mark have been computed by methods described in Ref. 3 at the end of this chapter and are published for the information of engineers and surveyors. (Care must be taken not to confuse true or geodetic azimuths, which take account of the convergency of meridians, with plane or grid azimuths, which are referred to a single meridian for the zone.) In addition to the monuments for horizontal control, a system of bench marks for vertical control has been established throughout the country.

To make use of the state system for a local survey, the surveyor sets up the transit at a nearby station of the triangulation system, orients it on the line of known plane azimuth, and runs a survey (by traversing or triangulation) to the area under consideration. The coordinates of any point in the local survey can then be conveniently computed in terms of the state system by the ordinary methods of plane surveying. Preferably the survey should be checked by traversing either back to the original station or to another triangulation station. If elevations are determined, they are referred to the established system of bench marks.

Use of state plane coordinates in route surveying is described in Chap. 26, and detailed computations in Ref. 1 at the end of this chapter.

An important advantage of the state-wide systems of coordinates is that the location of obliterated monuments can be reestablished with certainty and checked from various control points. Already most of the states have legalized the use of the state coordinate system for establishing and describing the monuments which mark the boundaries of land. For general mapping purposes, the coordinate system ensures reasonable agreement between maps of adjacent or overlapping areas. For extensive surveys such as those for routes, waterways, or municipal areas, the coordinate system facilitates checking, unifies the surveys of various portions of the project, and permits economies to be made in the conduct of the work. The use of state coordinates is simple and should be more widely adopted by surveyors.

16-31. Numerical Problems.

1. For the measurement of a base line the following data are given: The Bureau of Standards certificate states that the tape has a length of 99.942 ft. at 68°F., when supported at the 0 and 100-ft. points and under a tension of 10 lb.; the coefficient of thermal expansion of the tape is 0.00000645 per degree Fahrenheit; the tape weighs 1½ lb. The field records give the measured length as 1,418.314 ft.; the average temperature was 63.6°F.; the stakes were set on a 2 per cent grade; the sum of the set forwards was 0.234 ft.; the sum of the set backs was 0.114 ft. The interval and tension were the same as those used for the standard comparison. Compute the length of the line.

2. For the conditions given in problem 1, compute the normal tension.

3. For the conditions given in problem 1, assume that the interval between supports in the field is 50 instead of 100 ft. Compute the corresponding change in the distance between end marks of the tape.

4. For a given triangle ABC, the observed angles are as follows:

Station	Interior angle	Other angles about station
A	78°30′28″	281°29′36″,
B	54°17′30″	78°45′03″, 95°06′11″, 131°51′12″
C	47°12′16″	110°27′15″, 202°20′32″

Determine the most probable value of the interior angles at A, B, and C.

5. The angles in a quadrilateral $ABCD$, resulting from the station adjustments, are as follows: CAD = 45°30′55″, CAB = 42°11′39″, ABD = 41°54′40″, DBC = 62°40′53″, ACB = 33°12′51″, ACD = 28°05′30″, CDB = 56°00′50″, BDA = 50°22′51″.

(*a*) Compute the adjustment for the geometric condition.

(*b*) Compute the adjustment for the trigonometric condition.

6. In measuring the angles at a triangulation station O, it was necessary to set the transit over another point T, at a distance of 13.25 ft. from O. The angle measured at T from O to the first distant station W, was 95°10′30″. The angles between the distant stations W, X, Y, and Z were as follows: WTX = 39°37′48″; XTY = 69°04′20″; YTZ = 83°16′08″. The distances to the stations are found to be: OW = 8,949 ft.; OX = 14,334 ft.; OY = 5,647 ft.; and OZ = 7,326 ft.

Correct the angles measured at station T, to those which would have been measured if the transit had been set at station O.

7. In the quadrilateral $ABCD$, of Art. 16-25, assume that the side AB has a length of 13,100.3 ft. Use the finally adjusted angles and compute the length of the side CD by two independent series of computations.

8. For the quadrilateral of problem 7, assume that the azimuth of AB from north is 102°35′18″ and that the coordinates of station A are: total latitude = +50,000 ft., total departure = +40,000 ft. Compute the coordinates of stations B, C, and D.

9. Given the data listed below. Assume the instrument at station P, within the triangle ABC (Fig. 16-18), and solve the three-point problem (Case 2) for the angles x and y.

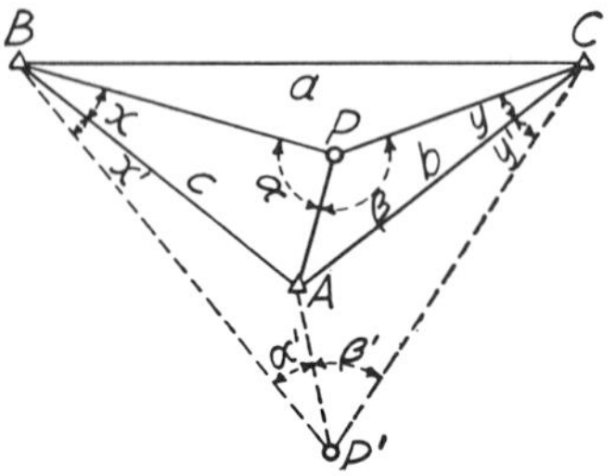

FIG. 16-18.

$A = 102°45'20''$ $\alpha = 89°15'30''$
$b = 6{,}883.4$ ft. $\beta = 128°20'10''$
$c = 6{,}605.3$ ft.

10. Given the data of problem 9 and the additional data shown below. Assume the instrument at station P' (Fig. 16-18), outside the triangle ABC, and solve the three-point problem (Case 3) for the angles x' and y'.

$\alpha' = 26°34'50''$
$\beta' = 44°15'15''$

11. Compute the relative strength of figure, as indicated (inversely) by R, for each of the quadrilaterals $CEFD$, $EGHF$, and $GIJH$ of Fig. 16-4.

16-32. Field Problem.

PROBLEM 1. MEASUREMENT OF BASE LINE

Object. To measure a short base line with the steel or invar tape. It is assumed that the base-line site has already been chosen and that permanent marks have been established at the end points.

Procedure. (1) Install strips of zinc or copper, perhaps ½ by 5 in., along the base line at intervals of the length of tape with which measurements are to be taken. If the base line is not to be measured along a smooth surface, as a paved highway or a railroad track, install substantial posts and intermediate stakes at grade as described in Art. 16-15. Line in the strips with the transit. (2) If the base line is to be measured on posts, build a substantial table over each end of the base line, and tack a metal strip directly above the end point. Carefully project each end point of the base line to the strip as follows: Set up the transit at about 25 ft. from the table, and sight to the end point on the ground. Elevate the telescope, and mark two points on the line of sight a few inches apart on the metal strip above. Scratch a straight line between these two points. Repeat this procedure with the transit set up so that the line of sight is approximately at a right angle with the first line of sight. (3) Run levels over the line, determining the elevation of all marking strips and intermediate supports. (4) Measure the line, following the procedure of Art. 16-15. At the end of the line, generally there will be a fractional part of a tape length; mark on the strip at the tape division that falls nearest the end point, and measure the remaining distance with a finely divided scale. (5) In the same manner measure the base line at least four times. (6) Make the necessary reductions to determine the correct length.

Hints and Precautions. (1) Measurements should not be taken with the steel tape in sunlight, or with a suspended steel or invar tape when a cross wind is blowing. (2) If measurements are to be taken with the invar tape in sunlight, at no place should the grade line come closer to the ground than 1 ft. (3) When the required tension has been applied, the tape should be set in vibration long enough to allow the amount of tension to become uniform throughout its length. The tapeman should take particular care to see that the device for applying tension is free from friction; and the device should be held at such a height that the tape will barely rest on the adjacent support. (4) Unless the spring scale is already adjusted for weighing in a horizontal position, it should be so calibrated.

REFERENCES

1. AMERICAN SOCIETY OF CIVIL ENGINEERS, "State Plane Coordinates," Chap. 1 in "Report on Highway and Bridge Surveys," *Manual of Engineering Practice* 44, 1962.
2. AMERICAN SOCIETY OF CIVIL ENGINEERS, "Technical Procedure for City Surveys," *Manual of Engineering Practice* 10, 1963.
3. ADAMS, O. S., and C. N. CLAIRE, "Manual of Plane-coordinate Computation," *U.S. Coast and Geodetic Survey, Spec. Pub.* 193, Government Printing Office, Washington, D.C., 1935.
4. BIRDSEYE, C. H., "Topographic Instructions of the U.S. Geological Survey," *U.S. Geological Survey, Bull.* 788, Government Printing Office, Washington, D.C., 1928.
5. BRITTAIN, J. H., "Computation of Traverse by Plane Coordinates," *U.S. Coast and Geodetic Survey, Serial* 624, Government Printing Office, Washington, D.C., 1940.
6. HODGSON, C. V., "Manual of First-order Triangulation," *U.S. Coast and Geodetic Survey, Spec. Pub.* 120, Government Printing Office, Washington, D.C., 1935.
7. HODGSON, C. V., "Manual of Second and Third Order Triangulation and Traverse," *U.S. Coast and Geodetic Survey, Spec. Pub.* 145, Government Printing Office, Washington, D.C., 1935.
8. MITCHELL, HUGH C., and LANSING G. SIMMONS, "The State Coordinate Systems: A Manual for Surveyors," *U.S. Coast and Geodetic Survey, Spec. Pub.* 235, Government Printing Office, Washington, D.C., 1945.
9. NATIONAL BUREAU OF STANDARDS, "Calibration of Line Standards of Length and Measuring Tapes at the National Bureau of Standards," *Circ.* 572, Government Printing Office, Washington, D.C., 1956.
10. REYNOLDS, W. F., "Relation between Plane Rectangular Coordinates and Geographic Positions," *U.S. Coast and Geodetic Survey, Spec. Pub.* 71, Government Printing Office, Washington, D.C., 1938.
11. REYNOLDS, W. F., "Manual of Triangulation Computation and Adjustment," *U.S. Coast and Geodetic Survey, Spec. Pub.* 138, Government Printing Office, Washington, D.C., 1955.
12. SHARP, H. OAKLEY, "Geodetic Control Surveys," John Wiley & Sons, Inc., New York, 1943.
13. STIPP, DONALD W., "Trilateration Adjustment," *Surveying and Mapping,* Vol. 22, No. 4, pp. 575–580, December, 1962.
14. U.S. COAST AND GEODETIC SURVEY, "Formulas and Tables for Computation of Geodetic Positions," *Spec. Pub.* 8, 7th ed., Government Printing Office, Washington, D.C., 1933.
15. U.S. GOVERNMENT (Federal agencies), "Classification and Standards of Accuracy of Geodetic Control Surveys," *Surveying and Mapping,* Vol. 19, No. 2, pp. 219–224, June, 1959.
16. WHITMORE, G. D., "Advanced Surveying and Mapping," International Textbook Company, Scranton, Pa., 1949.
17. WRIGHT, T. W., and J. F. HAYFORD, "The Adjustment of Observations," D. Van Nostrand Company, Inc., Princeton, N.J., 1906.

CHAPTER 17

THE PLANE TABLE

17-1. General. A plane table consists essentially of (1) a *drawing board* mounted on a tripod and (2) an *alidade* having the vertical plane of the line of sight fixed parallel to a straightedge which rests upon but is not attached to the board (Figs. 17-1 and 17-2). A sheet of drawing paper, called a *plane-table sheet,* is fastened to the board.

The location of any object is determined as follows (Fig. 17-6): With the straightedge through the plotted point o representing the station O occupied by the instrument, the line of sight is directed to the object A, and a line oa of indefinite length is drawn along the straightedge on the plane-table sheet; this line represents the direction from station to object. The measured distance OA between station and object is then plotted to scale from o, thus locating A on the map at a.

The term "plane table" is somewhat ambiguous, being used sometimes to designate only the board with its supporting tripod and sometimes (more generally) both the table proper and its accompanying alidade.

By means of the plane table, points on the ground to which observations are made are immediately plotted in their correct relative positions on the drawing, all angles being plotted graphically. The plane-table method is especially adapted to securing the details of the map; on extensive surveys the primary points of horizontal and vertical control are generally established by other more precise methods. It is a valuable and commonly used means of completing the compilation of maps from aerial photographs (Chap. 31), especially where the ground is hidden by overgrowth.

17-2. Comparison with Transit. It is helpful to consider the similarity between angular measurements made with the transit, together with the office procedure of plotting the notes, and the corresponding operations with the plane table.

The plane-table board may be said to correspond to the graduated horizontal circle of the transit. After the board is oriented, the direction of any line passing through the instrument station is observed by turning the alidade about the plotted location of the instrument station until the line of sight coincides with the line (just as with the transit, the upper motion is rotated); the direction of the line is given graphically by the position of the straightedge on the paper (just as with the transit, the numerical value of the azimuth is given by the vernier reading). The corresponding line on the paper is then established by drawing a line along the straightedge and by laying off to scale the measured length of the line on the ground. Thus with the plane table there is employed a combination of transit and drafting-room methods, but no record of numerical values is secured. The plane table is therefore useful for mapping only.

17-3. Tables. Three distinct types of table (board and tripod) are in common use: the *Coast Survey*, the *Johnson*, and the *traverse* table.

1. *Coast Survey Table.* This is the most stable of the three types. It is suitable for triangulation work (Art. 17-10) with sights possibly several miles in length, for the relatively high degree of precision required by the U.S. Coast and Geodetic Survey on its shore-line charts, and for large-scale city work. The board, 24 by 31 in., (Fig. 17-1) is securely attached to a

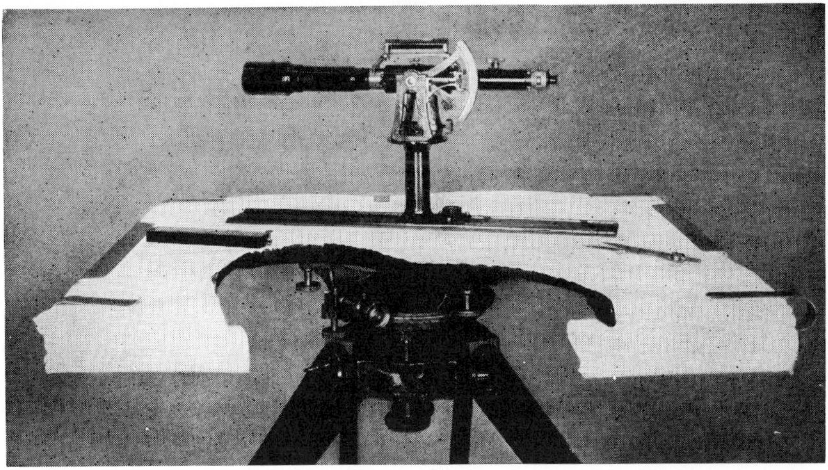

FIG. 17-1. U.S. Coast and Geodetic Survey plane table and alidade.

metal casting below, so arranged that the board can be leveled accurately by means of three leveling screws. By means of a clamp and tangent-screw the board can be fixed in any position in azimuth. The plane-table sheet is held in position by metal spring clamps, thus permitting the use of a sheet larger than the board. The tripod is of heavy and rigid construction.

2. *Johnson Table.* This table, shown in Fig. 17-2, was devised by Willard D. Johnson of the U.S. Geological Survey. The drawing board is either 18 by 24 in. or 24 by 31 in. Into the underside of the board is gained a circular brass plate, shown as *D* in Fig. 17-3. By means of the threaded opening in the plate, the board can be screwed to the upper casting *E* of the tripod head. The head comprises a ball-and-socket joint and a vertical spindle; the cup *F* is supported by the tripod (not shown). When clamp *A*

is loosened, the grip of parts G and H on the cup is released, and thus the board can be leveled. The clamp is then tightened to fix the board in a horizontal plane. When clamp B is loosened, the board can be rotated about the vertical axis and can thus be oriented. The plane-table sheet is held in position by countersunk screws in the top of the board.

FIG. 17-2. Johnson plane table with U.S. Geological Survey alidade.

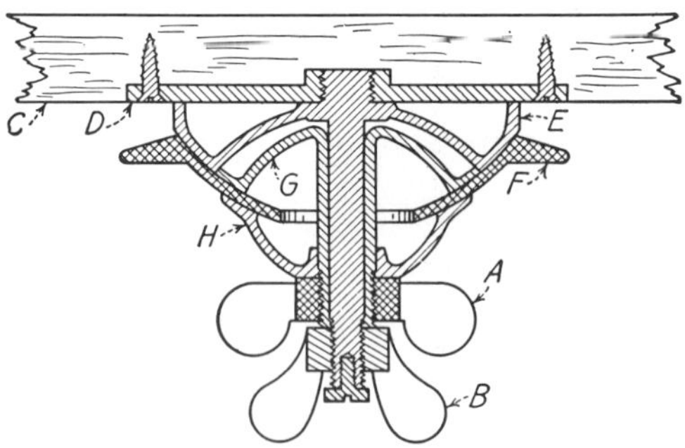

FIG. 17-3. Johnson tripod head for plane table.

3. *Traverse Table.* The traverse table (Fig. 17-4) consists of a small drawing board, usually 15 by 15 in., mounted on a light tripod in such manner that the board can be rotated about the vertical axis and can be clamped in any position. The table is leveled by adjusting the tripod legs, usually by estimation with the eye. A compass is fixed into a recess in the board. Ordinarily a peep-sight alidade is used with the traverse table. The traverse table is suitable for military reconnaissance sketches, traverses for small-scale

maps, and the mapping of relatively inaccessible areas to fill in a topographic map being drawn on a larger sheet.

17-4. Alidades. A plane-table alidade is a combined sight and straightedge ruler.

Peep-sight Alidade. One type of alidade consists of a peep sight mounted on a ruler (Fig. 17-4). The peep sight is formed by two vertical sight vanes, either fixed or folding, similar to those employed on the surveyor's compass. The ruler usually consists of a brass plate, 6 to 10 in. long, one edge of which is beveled and graduated to a suitable scale. The peep-sight alidade is convenient for sighting details while sketches are being made and is often employed by topographers as an auxiliary to the telescopic alidade.

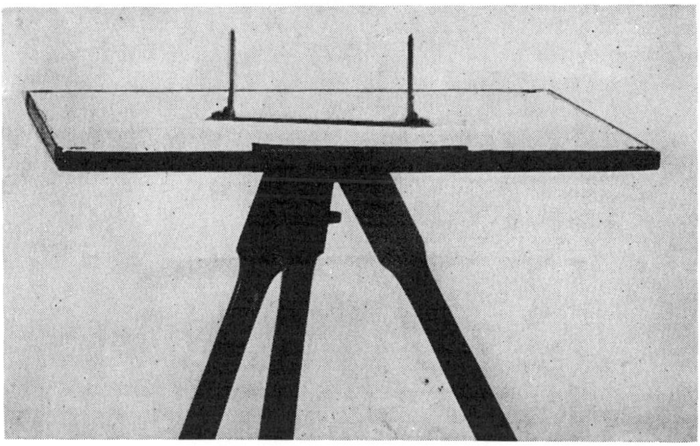

Fig. 17-4. Traverse table with peep-sight alidade.

Conventional Telescopic Alidade. The telescopic alidade is designed to afford greater precision in the control of the table and especially to make possible the stadia method of measuring distances. The base, or plate, of the alidade consists of a brass ruler or straightedge usually 3 by 18 in., beveled on one edge and chrome-plated on the bottom. Upon one end of the plate is mounted either a circular level or a pair of level tubes at right angles to each other. Upon the other end of the plate is mounted a trough compass consisting of a magnetic needle mounted in a narrow box with a short graduated arc at the end. In the center of the plate is mounted a column which supports a telescope similar to that of the transit. The telescope is equipped with stadia hairs, a vertical arc, and either a striding level or an attached level. In addition, many instruments are provided with a Beaman stadia arc, a vernier-control level, and a gradienter as previously described for the transit.

The telescope tube is fitted into a cylindrical sleeve which is rigidly attached to the horizontal axis. The telescope can be turned about its axis in the sleeve much as the telescope of the wye level can be turned in its wyes. On the telescope are turned two shoulders perhaps 5 in. apart, upon which rests a striding level.

The vernier of the vertical arc may be fixed or movable, or there may be an auxiliary level tube attached to the vernier arm as with the transit. Because the plane table is relatively unstable as compared with the transit, a control level mounted on the movable vernier arm greatly facilitates the measurement of vertical angles, rendering unnecessary an initial reading and index correction.

Self-indexing Alidade. Recently the U.S. Geological Survey has been active in the development of a self-indexing alidade, in which a damped pendulum

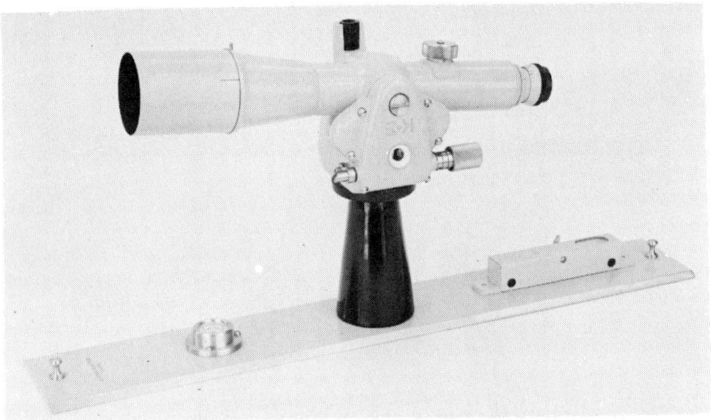

FIG. 17-5. Self-indexing alidade. (*Keuffel & Esser Co.*)

automatically brings the index of the vertical arc to the correct scale reading even if the plane-table board is not quite level. The form shown in Fig. 17-5 incorporates other modern features, as follows: All three scales (degree, H, and V) of the vertical arc are visible simultaneously through a microscope; the degree scale can be read to ½′ by estimation and requires no vernier. The telescope is internal-focusing; therefore the stadia constant $(f + c)$ is zero. The reticule is fixed in the telescope, and the telescope is fixed to its axis; thus neither the line of sight nor the axis of the telescope needs adjustment. No striding level or collars are required. The design is simple, and most parts are enclosed. A telltale circle on the bull's-eye level indicates when the blade is tilted beyond the range of the pendulum. A detachable elbow eyepiece is provided. The only field adjustment is to zero the index of the vertical arc when the line of sight is horizontal, as follows: With the line of sight on the rod reading established for a horizontal line by means of the two-peg test (Art. 8-23, adjustment 3), the index is brought to the zero mark (or equivalent scale graduation) by means of a capstan-head screw which controls its setting.

17-5. Plane-table Sheet. As the plane-table sheet is exposed to outdoor conditions, specially prepared papers are required to avoid undue expansion or shrinkage. Only the best drawing papers should be used. The paper can be seasoned, that is, rendered more resistant to changes in humidity of the air, by exposing it alternately to very moist and very dry atmospheres for a number of cycles.

The drawing paper can be mounted on muslin or on each side of a sheet of muslin with the grain of the paper of one sheet laid transverse to the grain of the other. These forms of plane-table sheet are excellent but are not sufficiently flexible to be rolled under the edge of the plane-table board if a sheet larger than the board is desired. For accurate work, such as graphical triangulation, the drawing paper may be mounted on a thin aluminum sheet. A sheet of celluloid with roughened surface is sometimes used for work in light rains; the details thus plotted are later transferred to the regular sheet. Cellulose-acetate or drafting film sheets are used in compilation of data from and in connection with aerial photographs.

If the plotting is to be continued for several days, the map sheet is protected by a cover sheet of tough paper which is torn away piece by piece to expose the map sheet as the work progresses.

Sharply pointed, hard (6H to 9H) pencils are used for drawing lines and plotting details, and a fine needle is used for plotting control stations. Special care should be taken not to smear the drawing; the alidade should be lifted instead of slid into position.

17-6. Setting Up and Orienting the Table. The plane table is set up approximately waist-high, so that the topographer can bend over the board without resting against it. The tripod legs are spread well apart and planted firmly in the ground. The board is leveled by whatever device is provided; but since few tables are sufficiently rigid to remain level as the alidade is shifted about, no special attempt is made to see that the board is perfectly level each time an observation is made.

For plotted angles to be theoretically correct, the plotted location of the station at which the plane table is set should be exactly over the corresponding point on the ground. Practically, the degree of care exercised in bringing the plotted point over the ground point depends upon the scale of the map. For map scales smaller than perhaps 1 in. = 50 ft., the plane table is set up over the station without any attempt to place the plotted point vertically above the station point. For maps of larger scale, the table is set up roughly and oriented approximately, and then it is shifted bodily until the point on the paper is practically over the station point, as indicated by plumbing. Either a hook-shaped plumbing arm may be used to support the plumb line under the plotted point, or the plumb line and point may be sighted from two directions approximately at right angles to each other. In any case, the aim is to set up with sufficient care so that the plotted position of lines drawn from the station will be shown correctly within the scale of the map.

The table may be oriented (1) by use of the *magnetic compass,* (2) by use of the *Baldwin solar chart,* (3) by *backsighting,* (4) by solving the *three-point problem* (Art. 17-14), or (5) by solving the *two-point problem* (Art. 17-15). As soon as the table is oriented, it is clamped in position and all mapping at the station is carried on without disturbing the board.

1. *Orientation by Compass.* For rough mapping at small scale, often orientation by the magnetic compass is sufficiently precise. This method is susceptible to the same errors as those encountered when using the surveyor's compass, but an error in the plotted direction of one line introduces no systematic errors in the lines plotted from succeeding stations.

If the compass is fixed to the drawing board, the board is oriented by rotating it about the vertical axis until the fixed bearing (usually magnetic north) is observed. If the compass is attached to the alidade or to a movable plate, the edge of the ruler or the plate is alined with a meridian previously established on the plane-table sheet, and the board is turned until the needle reads north.

2. *Orientation by Solar Chart.* In regions where the local conditions will not permit orientation by the magnetic compass, the table may be oriented by means of the Baldwin solar chart. This chart consists of a series of elliptical lines indicating the sun's path for various latitudes intersected by a series of straight lines indicating various sun times. When the chart is laid on the leveled plane table and is turned until the shadow cast by a plumb line passes through two designated points on the chart, an orientation arrow on the chart will point true north. The chart and its use are described in Ref. 1 at the end of this chapter.

3. *Orientation by Backsighting.* For mapping at intermediate or large scale, usually the board is oriented by backsighting along an established line, the direction of which has been plotted previously but the length of which need not be known. The method is equivalent to that employed in azimuth traversing with the transit. Greater precision is obtainable than with the compass, but an error in direction of one line is transferred to succeeding lines.

The plane table is set up as at B (Fig. 17-7) on the line AB which has previously been plotted as ab, the straightedge of the alidade is placed along the line ba, and the board is oriented by rotating it until the line of sight falls at A. The length of AB or ab need not be known. Preferably the longest line available should be used, for precision in orienting.

17-6a. Sighting. When a station or object is sighted and a ray is drawn through the plotted location of the station occupied toward the station or object, the sight is called a *foresight.* When the straightedge of the alidade is placed along a previously plotted line passing through the plotted location of the station occupied and that of another station or object, and then the board is turned until the line of sight cuts the station or object, the sight is called a *backsight.* When a known station is sighted and a line is drawn through the plotted location of that station toward the station occupied, the sight is called a *resection.* Resection is also a general term applied to the process of determining the location of the station occupied (Art. 17-11).

17-7. Radiation. When the table has been oriented, the direction to any object in the landscape may be drawn on the map by pivoting the alidade about the plotted location of the plane-table station, pointing the alidade toward the distant object, and drawing a line along the straightedge.

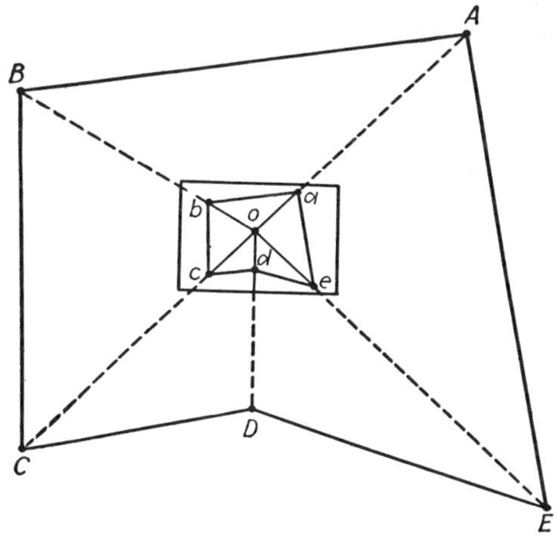

Fig. 17-6. Radiation with plane table.

Thus in Fig. 17-6 the plane table is shown in position over station O in the center of the field. The plotted location of the plane-table station is indicated at o on the plane-table sheet. The alidade is pivoted about this point; and as sights are taken to points as A, B, and C, rays are drawn along the edge of the ruler. The distances are measured and are then plotted to scale along the corresponding rays, thus locating the points A, B, and C on the map at a, b, and c. This procedure is called *radiation.*

17-8. Traversing. Traversing with the plane table involves the same principles as traversing with the transit. As each successive station is occupied, the table is oriented, sometimes with the compass but usually by backsighting on the preceding station. A foresight is then taken to the following station, and its location is plotted as in the radiation method just described. The distances between successive stations must be measured.

Plane-table traversing is used for the control of small surveys and for extension of control of larger surveys in order to locate points from which map details can be plotted advantageously.

In Fig. 17-7 a series of traverse stations *A*, *B*, *C*, etc., is represented. The plane table is set up as at *A*, and *a* representing *A* is plotted in such location that the other stations will fall within the limits of the sheet. With the straightedge passing through *a* a foresight is taken to station *B*, and *b*, its location on the map, is plotted by radiation as just described. The instrument is then set up at station *B* and is oriented by backsighting on station *A* as described in Art. 17-6. A foresight is taken to station *C*, and its location is plotted at *c*. By a similar procedure the locations of the remaining stations

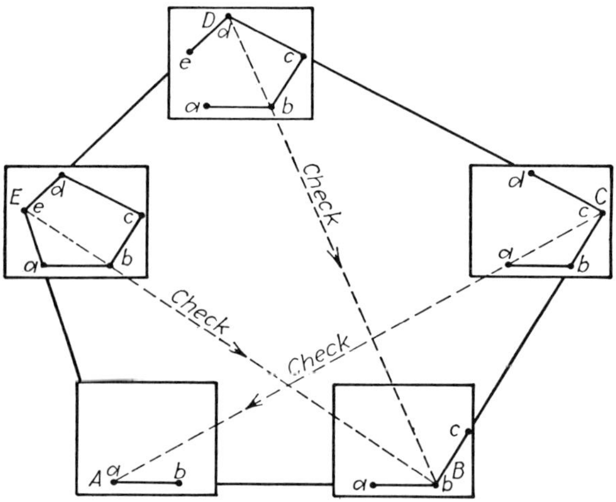

FIG. 17-7. **Traverse with plane table.**

are plotted. If the traverse forms a closed figure, any error of closure will become apparent on the plane-table sheet when the initial station is again plotted at the end of the traverse. With reasonable care, the linear error of closure can be kept within $\frac{1}{500}$ of the length of the traverse.

At any station a portion of the traverse may be checked if two or more of the preceding stations are visible and are not in the same straight line with the station occupied. Thus, if the plane table at station *C* is oriented by sighting at *B* and, in addition, station *A* can be seen, a ray drawn from *c* toward station *A* should pass through *a*, provided the traverse between the two stations is correctly drawn. In order that the check be reliable, the angle between the traverse line and the check line should not be small.

17-9. Intersection. This method is similar to that employed with the transit in locating an object by angles taken from each end of a line of known length (Art. 14-8), and makes use of the principle that if the angles and one side of a triangle are known, the remaining sides can be determined. It is useful for locating objects when distances to them are not otherwise conveniently obtainable. The location of an object is determined by sighting at the object from each of two plane-table stations (previously plotted) and by drawing rays as in the method of radiation; the intersection of the two rays thus drawn marks the plotted location of the object. No linear measurements are required except to determine the length of the line joining the two plane-table stations.

Thus the locations of the objects *A*, *B*, *C*, etc. (Fig. 17-8) may be plotted as follows: The plane table is set up at station *M*, a foresight to station *N* is

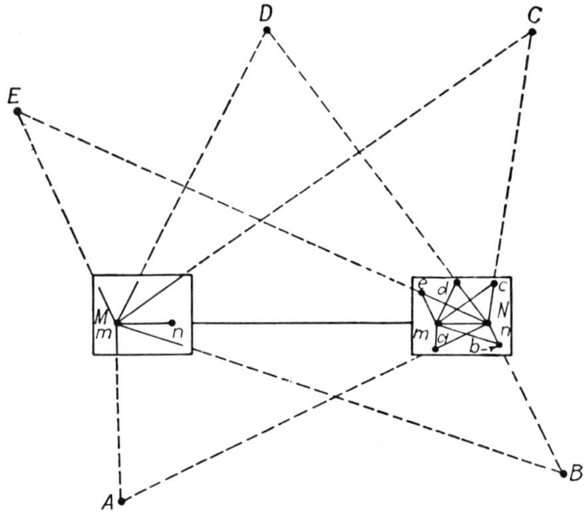

Fig. 17-8. **Intersection with plane table.**

taken as in the method of traversing, and the line *mn* representing the line *MN* is drawn to scale. Rays of indefinite length are drawn from *m* toward the objects *A*, *B*, *C*, etc. The plane table is then set up at station *N* and is oriented by backsighting to station *M*. Rays are drawn from *n* toward the same objects. The intersections of these rays with the corresponding rays drawn from *m* mark the plotted locations of the objects at *a*, *b*, *c*, etc. Distances to the objects are not measured but may be scaled from the map. If the angle between the intersecting rays is small, the location will be indefinite.

17-10. Graphical Triangulation. Graphical triangulation achieves the same results as triangulation with the transit (Chap. 16), but the procedure differs in that the plotted locations of the distant signals are determined graphically on the plane-table sheet instead of by the use of transit angles, office computations, and plotting methods. It is used for control of surveys over a small area and for extension of control of larger surveys. It is especially employed in intermediate-scale or small-scale mapping. As compared with plane-table traversing, graphical triangulation is most advantageous where the terrain offers unobstructed sights, considerable relief, and many well-defined objects.

The particular advantage of the use of the principles of intersection (Art. 17-9) and of resection (Art. 17-11) in plane-table mapping makes it desirable to locate many definite landmarks which are widely visible and suitably

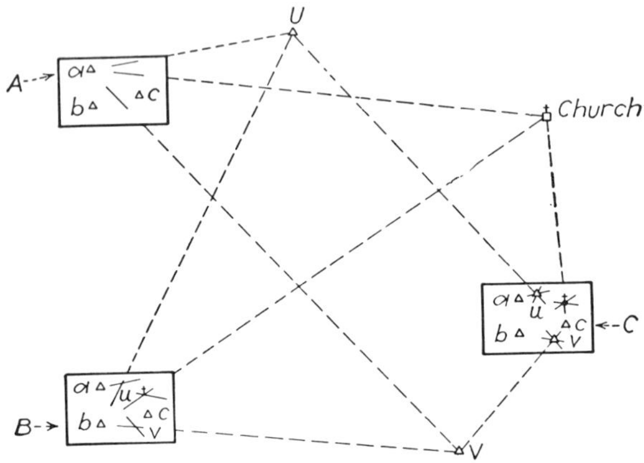

FIG. 17-9. Graphical triangulation.

situated, such as flagstaffs, church spires, and lone trees. Accordingly, while the topographer is locating the signals of distant stations, he also locates landmarks which are not be be used as instrument stations but which will be useful in subsequent work.

Two (preferably three) plane-table stations as *A*, *B*, and *C* (Fig. 17-9) the locations of which are known must be capable of being occupied by the instrument and must be marked by signals. Prior to the beginning of the work, the locations of these stations are plotted on the plane-table sheet at *a*, *b*, and *c*. The field procedure is then as follows: The plane table is set up as at *A* and is oriented by sighting at *B* and *C*; and rays are then drawn toward other stations such as *U* and *V* and, say, the church spire. The table is then set up and oriented at stations *B* and *C* in succession, and the same objects are sighted again. The correct plotted location of each station is determined by the intersection of two rays, as by those drawn from stations *A* and *B*, but the location is proved if the three rays drawn toward a given object are found to pass through a point, as shown by the third rays drawn from station *C*. The series of control stations may be extended as needed.

RESECTION AND ORIENTATION

17-11. Resection. Resection is the process of determining the plotted location of a station occupied by the instrument by means of sights taken toward known points the locations of which have been plotted. Resection enables the topographer to select advantageous plane-table stations which have not been plotted previously.

With the plane table oriented over the desired station of unknown map location, two or more objects of known location are sighted; as each object is sighted, a line of indefinite length is drawn through the plotted location of that object on the map. The intersection of these lines marks the plotted location of the station occupied by the plane table. For precision, it is desirable to resect from nearby stations rather than distant stations.

The table may be oriented by any of the methods stated in Art. 17-6. It is emphasized that, for the methods of orientation by magnetic compass and solar chart and by backsighting, resection can be accomplished only *after the board has been oriented.* If resection is by the three-point problem or the two-point problem, orientation and resection are accomplished in the same operation.

17-12. Resection after Orientation by Compass. If the plane table has been oriented by means of the compass, the method of resection is as follows: Let P be the station of unknown location occupied by the plane table, and let A and B be two visible stations which have been plotted on the sheet at a and b. Then the plotted location of P is determined by drawing a line, or resecting, through a in the direction of A and resecting through b in the direction of B. The point p where the two (or more) lines cross, or *resect,* marks the plotted location of the instrument station P. The method is utilized only for small-scale or rough mapping for which the relatively large errors due to orienting with the compass needle would not impair the usefulness of the map.

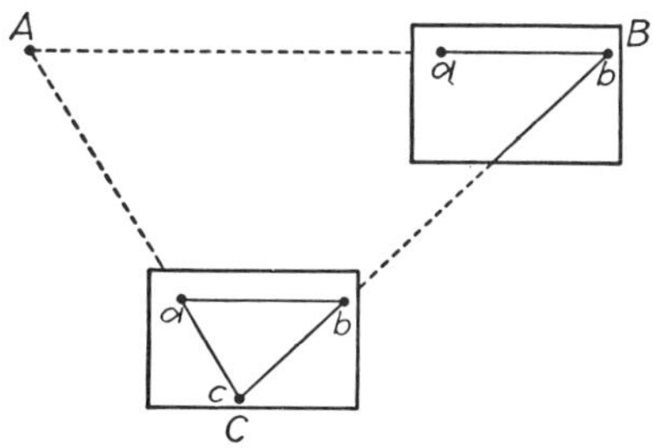

FIG. 17-10. Resection after orientation by backsighting.

17-13. Resection after Orientation by Backsighting. If the table can be oriented by backsighting along a previously plotted foresight line, the location of the plane-table station can be determined by drawing a line through the plotted location of another known station to an intersection with the backsight line.

Thus, suppose that the topographer wishes to occupy station C (Fig. 17-10) the location of which has not been plotted but from which can be seen two points as A and B the locations of which have been plotted at a and b. He orients the board at one of the known stations as B, takes a foresight to C, and draws through b a ray of indefinite length. He then sets up the plane table at station C, orients it by backsighting to B, and resects from A through a. The intersection c of the ray from b and the resection line from a is the plotted location of the plane-table station C, since the triangles abc and ABC are similar.

If the angle between the ray and the resection line is small, the location will be indefinite; for strong location, the acute angle between these lines should be greater than 30°.

17-14. Resection and Orientation: Three-point Problem. Frequently the topographer wishes to occupy an advantageous station which has not been located on the map and toward which no ray from located stations has been drawn, and at the same time orientation by use of the compass is not sufficiently accurate. If three located stations are visible, the three-point problem offers a convenient method of orienting and resecting in the same operation. There are several solutions of the three-point problem. In the United States, experienced topographers commonly employ a method of direct trial, guided by rules (Art. 17-14*a*). The mechanical or tracing-cloth solution (Art. 17-14*b*) is simpler to understand but is not so satisfactory nor so expeditious under the usual field conditions. For solution of the three-point problem by computation, employed in transit work, see Art. 16-29.

17-14a. Trial Method. The plane table is set up over the station of unknown location and is oriented approximately either by compass or by estimation. Resection lines from the three stations of known location are drawn through the corresponding plotted points. These lines will not intersect at a common point unless the trial orientation happens to be correct. (An exception to this statement occurs when the station of unknown location happens to fall on the circumference of a circle passing through the three stations of known location, as discussed later in this article.) Usually a small triangle called the *triangle of error* is formed by the three lines.

Thus in Fig. 17-11, suppose that the plane table has been set up over a ground point P and oriented approximately. Resection lines are drawn from A, B, and C through the corresponding plotted points a, b, and c, respectively, forming a triangle of error. The correct plotted location p of the plane-table station, called the *point sought,* is then determined more closely.

One method is to draw arcs of circles through the points as shown (through a, b, and point ab; b, c, and bc; and a, c, and ac); the circles will intersect at p, the point sought.

Usually the correct location of the point sought is estimated more conveniently by means of Rules 1 and 2, given below.

The board is then reoriented by backsighting through the estimated location of p toward one of the known stations (preferably the most distant);

and the orientation is checked by resecting from the other two known stations. If the three lines still do not meet at a point, the process is repeated until they do; the orientation is then correct, and the common intersection of the three lines is the correct plotted location of the plane-table station.

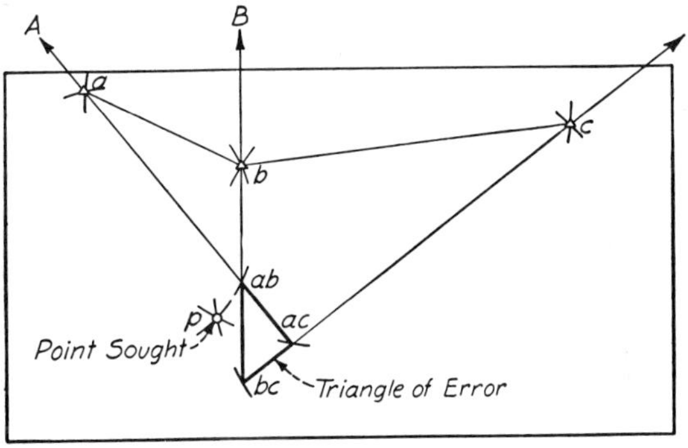

FIG. 17-11. Three-point problem with plane table.

Rule 1. *The point sought is on the same side of all resection lines.* That is, it lies either to the right of each line (as the observer faces the corresponding station) or to the left of each line.

Rule 2. *The distance from each resection line to the point sought is proportional to the length of that line.* By "length" is meant either the actual distance from plane-table station to known station or the corresponding plotted distance.

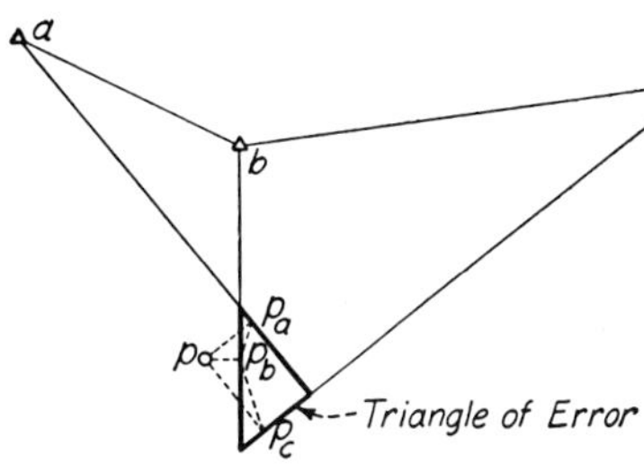

FIG. 17-12. Triangle of error.

Rule 2 can be proved by reference to Fig. 17-12, which shows the triangle of error as in Fig. 17-11. The triangle of error is actually small, and distances from a, b, and c to any point of it may be taken as the distance to p without error of consequence. Lines pp_a, pp_b, and pp_c are respectively perpendicular to the resection rays from a, b, and c. Then, by similar triangles,

$$\frac{pp_a}{pa} = \frac{pp_b}{pb} = \frac{pp_c}{pc} \tag{1}$$

Rules 1 and 2 are general and apply to any location of the plane-table station except on the *great circle* passing through the three known stations (Fig. 17-13). In this case, regardless of the orientation of the table, the lines will meet in a common point (see points 2 in the figure) which will not necessarily be the point sought. If it is suspected but not known that the plane-table station is on the great circle, either the great circle should be plotted on the plane-table sheet or the orientation of the board should be changed slightly and a second trial made. If it is found that the station is on the great circle, one (or more) of the three known stations must be replaced by a known station (or stations) suitably located.

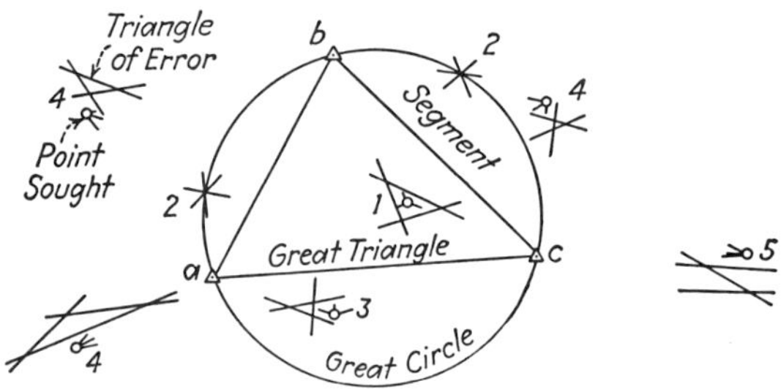

FIG. 17-13. Three-point problem, solution by trial.

Auxiliary Rules. Rules 1 and 2 are supplemented by the following auxiliary rules which apply to particular locations of the plane-table station:

(*a*) If the new station is inside the great triangle, the point sought is within the triangle of error and is in the same position relative to the triangle of error that the triangle of error occupies in the great triangle (Fig. 17-13, point 1).

(*b*) If the new station is in one of the three segments of the great circle formed by the sides of the great triangle, the point sought is on the *opposite* side of the resection line through the middle known point from the intersection of the other two lines (Fig. 17-13, point 3).

(*c*) If the new station is outside the great circle, the point sought is always on the *same* side of the resection line from the most distant point as the point of intersection of the other two lines (Fig. 17-13, points 4).

(*d*) If the new station is outside the great triangle, of the six sectors formed by the resection lines there are only two in which the point sought can be on the same side (right or left) of all lines.

(*e*) If the new station is so located that the triangle of error is not formed within the limits of the plane-table sheet, that is, if two of the resection lines are almost parallel (Fig. 17-13, point 5), the foregoing rules still apply.

(*f*) If the new station is on line between two of the known stations, the resection lines drawn from those two stations will be parallel. The foregoing rules still apply.

Strength of Determination. The strength of the determination varies with the location of the plane-table station, as described below. The strength of determination should be considered not only in selecting the most favorable of the available known stations but also in deciding whether the three-point problem can satisfactorily be used with the plane table at a given location.

(*a*) When the new station is inside the great circle, the nearer the new station to the center of gravity of the great triangle, the stronger the determination.

(*b*) When the new station is on the great circle, its location is indeterminate.

(*c*) When the new station is near the great circle, the determination is weak.

(*d*) When the new station is outside the great circle, for given angles the nearer the new station to the middle known station, the stronger the determination.

(*e*) Either when one angle is small or when the new station is on line with two known stations, the larger the angle to the third known station (up to 90°), the stronger the determination. The two known stations near or on line should not be near each other.

17-14b. Tracing-cloth Method. A simple solution of the three-point problem, known as the *tracing-cloth method,* is as follows: A piece of tracing cloth or tracing paper is fastened on the plane table over the map. Any convenient point on the tracing cloth is chosen to represent the unknown station over which the plane table is set, and from it rays are drawn toward the three known stations or objects. Then the cloth is loosened and is shifted over the map until the three rays pass through the corresponding plotted points. The intersection of the rays marks the plotted location of the plane-table station. It is pricked through onto the map, and the table is oriented by backsighting on one of the known stations (preferably the most distant).

The three-armed protractor described in Art. 30-20 can be used in a similar manner, the arms of the protractor being set to form the three rays.

17-15. Resection and Orientation: Two-point Problem. The purpose of the two-point problem is to orient the table and to locate the station occupied when only two stations are visible and when it is impossible or undesirable to occupy either of them, as when the signals are inaccessible or at a considerable distance. To accomplish this, it is necessary that two set-ups be made, the first at a convenient distance from the station to be occupied, and the second at that station. Owing to the length of the procedure, this method is not practical except where other methods cannot be used.

One graphical solution of the two-point problem is as follows: The locations of the two known stations *A* and *B* are plotted on the plane-table sheet at *a* and *b* (Fig. 17-14). The plane table is set up over some ground point C from which can be seen A, B, and the point P whose plotted location is desired. The board is oriented as nearly as possible either by compass or by estimation. A point c' corresponding to C on the ground is plotted by estimation on the plane-table sheet (Fig. 17-14, left). Foresights are taken from c' on A, B, and P. The distance from C to P is estimated, and the corresponding estimated location of P is plotted at p'. The table is taken to station P and is oriented (tentatively) by a backsight on C. Foresights are taken from p' on A and B; the intersections of these rays with corresponding rays from c' are a' and b' (Fig. 17-14, right). The line $a'b'$ is parallel

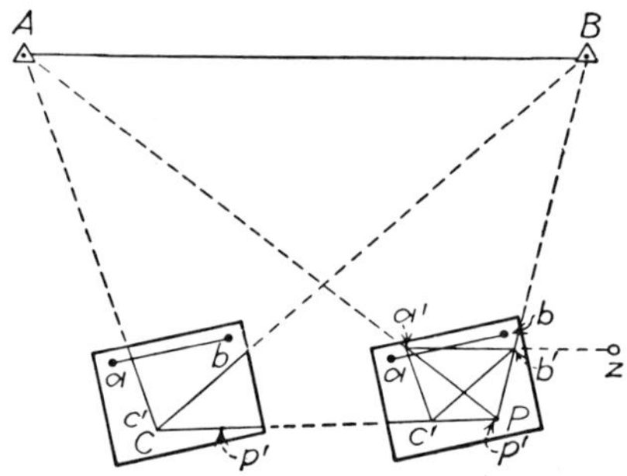

FIG. 17-14. Two-point problem, graphical solution.

to a line between the corresponding actual points A and B. With the straightedge of the alidade along the line $a'b'$, a point Z at some distance from the table is set (or selected) on the line of sight. The alidade is moved to the line ab, and the board is turned until the same point Z is sighted. The plane table is now oriented correctly, and by resection through a and b the correct location of the plane table is plotted at p.

USE AND ADJUSTMENT OF PLANE TABLE

17-16. Measurement of Difference in Elevation. The methods of plotting the *horizontal projection* of ground points have been described. As the plane table is used principally in the work of topographic mapping, many *elevations* of ground points are determined by methods closely similar to those for leveling by means of the transit, namely, by direct leveling and trigonometric leveling including stadia leveling. However, two conditions peculiar to the plane table should be mentioned.

First, the table is not so precise or so stable in its controls as the horizontal plate of the transit. Accordingly, unless either a control level is attached to the vernier arm or a self-indexing alidade is used, in measuring the vertical angles in stadia or trigonometric leveling, the index correction is likely to be much larger with the plane table, and it is necessary to determine the index error of each sight by leveling the telescope.

Second, the elevation of a distant point the location of which has been plotted by the method of intersection may be secured readily by the method of trigonometric leveling (Art. 8-5), in which the horizontal distance may be scaled directly from the map. The conditions as to distance and as to the accuracy required will determine whether or not corrections for curvature of the earth and atmospheric refraction should be applied.

The precision of plane-table leveling, whether direct or trigonometric, is not high; however, with reasonably careful work the maximum error in elevation, in feet, may be kept within 0.2 or 0.3 $\sqrt{\text{miles}}$.

17-17. Details with Plane Table. The operation of plotting details by radiation with the plane table is commonly known as taking *side shots*. The general procedure is described in Art. 25-12, for an alidade not equipped with a stadia arc.

Table 17-1 gives the sequence of operations for taking a side shot when the alidade is equipped with a stadia arc but not with a vertical control level. It is assumed that the party consists of a plane-table man (called a *topographer*), one or more rodmen, and a computer. In actual practice, usually no record is kept of the side shots. In this article, however, it is assumed that a recorder is employed for purposes of training, checking, or record, and his operations are also listed. Each line of the table reads from left to right, and the operations in each line follow those of the preceding line. It is evident that the progress of the party depends largely upon systematic operation and upon the dispatch with which the topographer performs his duties.

If the alidade is equipped with a vertical vernier control level, it is not necessary for the topographer to level the telescope; instead he centers the control-level bubble. If the alidade is not equipped with a stadia arc or vertical control level, he levels the telescope, reads the index error, and then reads the vertical angle; the details of the procedure are modified accordingly.

If field notes are taken, the form is similar to that for transit-stadia surveying (Chap. 15). Column headings will differ according to the type of vertical arc and type of telescope. With an external-focusing telescope not equipped with a stadia arc, pertinent column headings would include object, stadia interval, rod reading, vertical angle, stadia distance, horizontal distance, difference in elevation, elevation, and remarks. Such notes are often kept for the sights to control stations, whether or not the side shots are recorded.

Aerial photographs may be used in connection with the plotting of details on the plane table (Chap. 31). By examination of both photographs and terrain, the topographer can fill in details not visible from the air, can check against mistakes, and can expedite the work. If the scale of the photograph is at or near the scale of the map, an overlay sheet of transparent cellulose

TABLE 17-1. SEQUENCE OF OPERATIONS FOR SIDE SHOT WITH PLANE TABLE
Alidade equipped with stadia arc and fixed vertical vernier

Rodman	Topographer	Computer	Recorder
Sets rod on ground point	Sights on rod, with lower cross hair on a foot mark		
	Reads and calls stadia interval	Sets index of slide rule on stadia distance	Records stadia interval (or distance)
	Draws ray		
	Levels telescope, and sets stadia arc to zero (or 50)[a]		
	Sights on rod, with stadia index set on some mark of *V* arc		
	Calls *V* reading, *H* reading, and reading of middle crosshair of rod	. . . .	Records *V*, *H*, and rod reading
	Waves rodman ahead	Computes and calls horizontal distance	Records horizontal distance
Starts to new ground point	Plots point	Computes and calls vertical distance	Records vertical distance
		Computes and calls elevation	Records elevation
	Records elevation on sheet		
	Interpolates and sketches contours and features		

[a] If the alidade is equipped with a vertical vernier control level, the topographer centers the control-level bubble instead.

acetate may be used on which to plot certain details, keeping in mind that the photograph is not an orthographic projection. Another technique is to print the compilation sheet obtained by photogrammetry onto a plane-table sheet sensitized with blueprint emulsion, then in the field to plot the contours on this sheet.

17-18. Adjustment of Conventional Alidade. The adjustments of the conventional tube-in-sleeve telescopic alidade are similar to those previously described for the wye level and the transit. Observations made with the plane table are not required to be so precise as those made with the engineer's transit; accordingly, in general the adjustments need not be so refined and in one or two cases they are omitted entirely. The telescope of the alidade is never inverted as is that of the transit; hence no large error is introduced through any lack of perpendicularity between the line of sight and the horizontal axis, nor any error through lack of parallelism between the line of sight and the straightedge. It may be assumed that the telescope collars are circular and of the same diameter, so that the axis of the striding-level support on the collars is parallel to the axis of the telescope sleeve. Also, it may be assumed without appreciable resultant errors that the edges of the ruler are straight and parallel, and that the horizontal axis is parallel to the plane of the ruler. The edge of the ruler is not in a vertical plane through the line of sight, but the error from this source is negligible because, in plotting, the offset is multiplied by the scale fraction.

1. *To Make the Axis of Each Plate Level Parallel to the Plate.* Center the bubble (or bubbles) of the plate level (or levels) by manipulating the board. On the plane-table sheet mark a guide line along one edge of the straightedge. Turn the alidade end for end, and again place the straightedge along the guide line. If the bubble is off center, bring it back *halfway* by means of the adjusting screws. Again center the bubble by manipulating the board, and repeat the test. For a circular level, the procedure is similar.

2. *To Make the Vertical Cross-hair Lie in a Plane Perpendicular to the Horizontal Axis.* Sight the vertical cross-hair on a well-defined point not less than 200 ft. away, and swing the telescope through a small vertical angle. If the point appears to depart from the vertical cross-hair, loosen two adjacent screws of the cross-hair ring, and rotate the ring in the telescope tube until by further trial the point sighted traverses the entire length of the hair. Tighten the same two screws.

3. *To Make the Line of Sight Coincide with the Axis of the Telescope Sleeve.* Sight the intersection of the cross-hairs on some well-defined point. Rotate the telescope in the sleeve through 180° (usually the limits of rotation are fixed by a shoulder and a lug). If the cross-hairs have apparently moved away from the point, bring each hair halfway back to its original position by means of the capstan screws holding the cross-hair ring. The adjustment is made by manipulating opposite screws, bringing first one cross-hair and then

the other to its estimated correct position. Again sight on the point, and repeat the test.

4. *To Make the Axis of the Striding Level Parallel to the Axis of the Telescope Sleeve (and Hence Parallel to the Line of Sight).* Place the striding level on the telescope, and center the bubble. Remove the level, turn it end for end, and replace it on the telescope tube. If the bubble is off center, bring it back halfway by means of the adjusting screw at one end of the level tube. Again center the bubble (by means of the tangent-screw), and repeat the test.

5. (For Alidade Having a Fixed Vertical Vernier) *To Make the Vertical Vernier Read Zero When the Line of Sight is Horizontal.* With the board level, center the bubble of the telescope level. If the vertical vernier does not read zero, loosen it and move it until it reads zero.

5*a*. (For Alidade Having a Movable Vertical Vernier with Control Level) *To Make the Axis of the Vernier Control Level Parallel to the Axis of the Telescope Level When the Vertical Vernier Reads Zero.* Center the bubble of the telescope level, and move the vernier by means of its tangent screw until it reads zero. If the control-level bubble is off center, bring it to center by means of the capstan screws at the end of the control-level tube.

17-19. Sources of Error. In the main, the sources of error in plane-table work are the same as those which affect transit work and plotting, and the discussion relating to those subjects need not be repeated here. However, the following three sources of error should be considered:

1. *Setting Over a Point.* Because plotted results only are required, it is not necessary to set the plotted location of the plane table over the corresponding ground point with any greater precision than is required by the scale of the map. (See Art. 17-6.)

2. *Drawing Rays.* The accuracy of plane-table mapping depends largely upon the precision with which the rays are drawn; consequently the rays should be of considerable length. To avoid confusion, however, only enough of each ray is drawn to insure that the plotted point will fall upon it, with one or two additional dashes drawn near the end of the alidade straightedge to mark its direction. Fine lines are desirable both for precision and for legibility.

3. *Instability of the Table.* If it is manipulated with care, the plane table can be oriented with considerable precision; however, a principal source of error in its use arises from the fact that its position is subject to continual disturbance by the topographer while he is working. Errors from this source can be kept within reasonable limits (*a*) by planting the tripod firmly in the ground, (*b*) by setting the table approximately waist-high so that the topographer can bend over it without leaning against it, (*c*) by avoiding undue pressure upon or against the table, and (*d*) by testing the orientation of the board occasionally and correcting its position if necessary. This test is always applied before a new instrument station is plotted.

17-20. Field Checks. An important advantage of the plane-table method is that it provides many convenient opportunities for verifying the plotted locations of points on the map. The previously plotted location of any visible object is verified during a subsequent set-up if a ray drawn toward the object passes through the plotted point. The plotted location of the plane table itself may be verified by resecting from distant visible objects the plotted locations of which are known to be correct. Checks of this sort should be applied at each station in order to guard against faulty orientation and mistakes in observations.

17-21. Advantages and Disadvantages. As compared with other methods of mapping, the plane-table method has these advantages: (1) Relatively few points need be located because the map is drawn as the survey proceeds; (2) contours and irregular objects can be represented accurately because the terrain is in view as the outlines are plotted; (3) as numerical values of angles are not observed, the consequent errors and mistakes due to reading, recording, and plotting are avoided; (4) as all plotting is done in the field, omissions in the field data are avoided; (5) the useful principles of intersection and resection are made convenient; (6) checks on the location of plotted points are obtained readily; and (7) the amount of office work is relatively small.

The disadvantages are: (1) The plane table and accessories are cumbersome; (2) considerable time is required for the topographer to gain proficiency; (3) the time required in the field is relatively large; and (4) the usefulness of the method is limited to relatively open country (that is, where visibility is fair) and to weather conditions relatively free from rain, cold, and high wind.

With a long-legged tripod the table may be raised above low brush or cornstalks. In wet weather a sheet of celluloid may be used instead of drawing paper. The plane table may be both raised and protected from rain by being mounted in a covered automobile truck having sides which can be opened as desired; during a set-up the truck is braced to keep the table stationary.

Comparing the plane table with the transit: If a large number of *points* (such as buildings, poles, and isolated trees) are to be plotted on a map, the transit method is superior; but if *irregular lines and areas* (such as drives, streams, contours, and woods) are to be represented, the plane-table method is superior in accuracy, speed, and economy.

17-22. Office Problems.

1. Verify the solutions of the two-point and three-point problems in the drafting room, using a large (say, 24 by 30-in.) sheet of paper to represent the field and a small (say, 3 by 5-in.) sheet to represent the plane table. A long straightedge will serve as a line of sight to connect stations.

2. For each of the plane-table set-ups sketched in Fig. 17-15, indicate by estimation (in accordance with Rules 1 and 2) the location of the point sought, and state the direction in which the table should be rotated in order to orient it properly. In the sketches, the triangle of error is exaggerated for clearness.

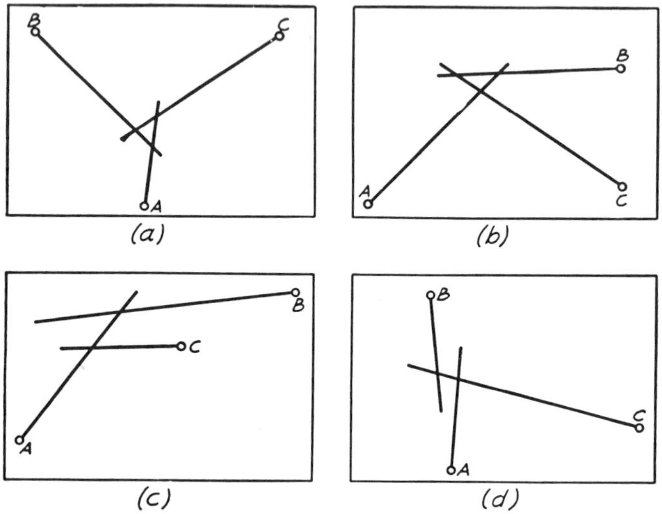

FIG. 17-15. Problems in orientation of plane table.

17-23. Field Problems.

PROBLEM 1. ADJUSTMENT OF THE PLANE-TABLE ALIDADE

Object. To make the field adjustments of the telescopic alidade.
Procedure. As outlined in Art. 17-18.

PROBLEM 2. TRAVERSE WITH PLANE TABLE

Object. To run a closed traverse and to determine the accuracy of the plane-table method.

Procedure. (1) Set up the plane table over the first station of the assigned traverse, with the board oriented by magnetic compass and with the point which is taken as the plotted location of the station directly over the mark on the ground. (2) In sighting, keep the center of the alidade opposite the plotted point. (3) Take a foresight to the second traverse station, determine (by stadia) the distance to that station, and plot the distance to the assigned scale of the map. (4) For purposes of checking, draw rays toward several distant points that can be seen from three or more traverse stations. (5) Set up the plane table at the second traverse station and orient the board by backsighting, with the plotted point over the ground point. As a check, determine by stadia the distance from the first traverse station. (6) Take a foresight to the third station, and determine and plot the distance to that station. (7) Draw rays toward the points selected for checking. (8) Continue in the same manner around the traverse. (9) Note the error of closure and the amounts by which the rays drawn from three or more stations toward a given ground point fail to meet at a common point. These indicate (inversely) the accuracy of the work.

Problem 3. Plane-table Survey of Field

Object. To make a plane-table survey of an assigned portion of the campus by a combination of the methods of radiation, intersection, and traversing. Either a plain map or a topographic map may be made.

Procedure. (1) Stake out an irregular field having perhaps five sides. (2) Set up the plane table at one corner of the field, and so locate the plotted position of the station that the area to be covered will fall within the limits of the plane-table sheet. (3) Draw a magnetic meridian on the plane-table sheet. (4) Locate and sketch all conveniently accessible objects by the method of radiation (Art. 17-7), employing the stadia. Also, draw and identify rays toward relatively inaccessible objects which can also be seen from another point on the traverse. (5) Move to the next corner of the field, orient the board by backsighting, check the orientation by the magnetic compass, and similarly locate objects by radiation and intersection. (6) Continue in this manner around the field to form a closed traverse, and plot sufficient data to construct a complete map. At each station, check the location of the plane table by resection (Art. 17-20). (7) If a marked error becomes evident, the orientation of the board should be checked, and, if necessary, one or more of the previous stations should be reoccupied. (8) If it is convenient to occupy some station other than a traverse station, locate the station occupied by resection, employing the method of backsighting (Art. 17-13). A ray must have been drawn previously toward the station to be occupied. (9) If a topographic map is to be made, locate a sufficient number of controlling ground points to enable contours to be drawn in the field. (10) Note the error of closure of the traverse; if the error is considerable, the work should be repeated. (11) Finish the map.

Problem 4. Orientation and Resection by Three-point Problem

Object. With three stations (or objects) of known location assigned, to plot the location of a fourth station occupied by the plane table and simultaneously to orient the board.

Procedure. (1) Plot the three known points on the plane-table sheet. (2) Set up the plane table at another assigned station and solve the three-point problem by the trial method as described in Art. 17-14*a*. (3) If other known points are visible, plot their locations and check the location and orientation of the plane table by sighting on these points. (4) Change the orientation of the board slightly and again orient the board by the tracing-cloth method (Art. 17-14*b*). (5) Compare the results obtained by the two methods.

Hints and Precautions. If the station occupied is at or near a great circle passing through the three known points, the solution is indeterminate.

Problem 5. Orientation and Resection by Two-point Problem

Object. With two stations (or objects) of known location assigned, to plot the location of a third station occupied by the plane table and simultaneously to orient the board.

Procedure. (1) Plot the two known points on the plane-table sheet. (2) Solve the two-point problem as described in Art. 17-15.

Hints and Precautions. The location of C should be selected so as to give intersection angles from P and C to A and B between 30° and 150°.

REFERENCES

1. BIRDSEYE, C. H., "Topographic Instructions of the United States Geological Survey," *U.S. Geol. Survey Bull.* 788, Government Printing Office, Washington, D.C., 1928.
2. DENAULT, M. F., "Plane Table Versus Transit for Topographic Mapping," *Engr. News-Record,* May 4, 1944; reprinted in *Surveying and Mapping,* Vol. 4, No. 3, pp. 7–12, July 1944.
3. LOW, JULIAN W., "Plane Table Mapping," Harper & Row, Publishers, Incorporated, New York, 1952.
4. SWAINSON, O. W., "Topographic Manual," *U.S. Coast and Geodetic Survey, Spec. Pub.* 144, Government Printing Office, Washington, D.C., 1928.

CHAPTER 18

MAP PLOTTING

18-1. General. The methods of plotting described in this chapter are those employed in mapping areas of limited extent where the earth's surface is assumed to be plane. These methods are applicable to surveys for highways, railroads, and irrigation and drainage systems; to many topographic and hydrographic surveys; and to rural and urban land surveys.

18-2. Process of Making a Map. The mechanical processes of the preparation of maps are described in some detail in the chapter on map drafting (Chap. 6). Maps are usually so plotted that features are shown in the same relative location that they occupy on the ground, at a given scale. The operations of plotting are in a sense the reverse of the operations of surveying.

Some maps, notably those made for the purpose of delineating the boundaries of land, show numerically the dimensions of the main features, and their worth is unimpaired if they are not drawn to true scale. Most maps, however, depend upon the accuracy with which details are shown, and the general aim is to plot the more definite features within proper limits of error.

The process of mapping involves (1) the plotting, by more precise methods, of horizontal *control,* which are generally transit stations and which may be traverse points, triangulation points, or both, and (2) the plotting, by less precise methods, of the map *details,* employing angular and linear measurements from the lines and points in the horizontal control system.

Most maps are plotted wholly in the office, but where the objects to be shown are numerous, the details may be plotted more expeditiously in the field. In such cases, usually the primary horizontal control is plotted in the office and points of secondary horizontal control are fixed on the ground as necessary to expedite the location of details.

Methods of plotting details are given in Art. 18-19. Methods of plotting profiles and cross-sections related to mapping are described in Chap. 11.

18-3. Methods of Plotting Control. Horizontal control may be plotted (1) by use of the protractor (Art. 18-4), (2) by the tangent method (Art. 18-5), (3) by the chord method (Art. 18-6), or (4) by the coordinate method (Arts. 18-8 to 18-18). In any case, distances are measured with the engineer's scale; for precise work the points are pricked with a needle and a reading glass is employed. Traverse and triangulation stations are indicated by appropriate symbols (see Fig. 6-5*b*), and control lines are drawn carefully with a hard pencil having a fine point. Points are numbered or lettered to conform to the system employed in the field work. Usually the control is not shown on the finished map.

The precision of plotting the control depends upon the instrument used and upon the care with which the work is done. With a well-sharpened hard pencil and with the use of reasonable care, points can be plotted within perhaps $\frac{1}{100}$ in. With a fine needle and a reading glass and with the use of great care, with the eye directly above the needle, points can be plotted within perhaps $\frac{1}{200}$ in.

18-4. Protractor Method. Where the control system is not extensive and the map is small, a fairly large protractor provides a sufficiently precise means of plotting angular values (see Art. 6-26). However, for the 6-in. protractor (a size in common use), the accidental error of plotting may amount to 15′, and for the largest protractor commonly manufactured (14 in.), the probable error is not less than 05′ with even the most careful plotting. The protractor method is sufficiently precise for rough surveys over small areas, and it is often useful in checking angles plotted by other methods.

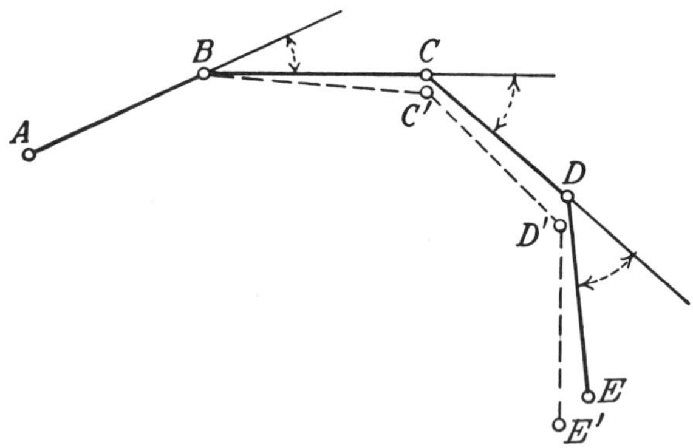

FIG. 18-1. Plotting traverse by protractor.

When a deflection-angle traverse is to be plotted by protractor, the position of the first line is fixed by estimation and its length (as *AB*, Fig. 18-1) is laid off by measurement. The protractor is oriented at the forward point (as *B*), the deflection angle to the succeeding line is laid off, and a light line of indefinite length is drawn. Along this line is laid off the given distance (as *BC*) to the succeeding traverse point (as *C*), and so on.

An objection to the foregoing procedure is that any error in the direction of one line affects to a like degree the directions of all succeeding lines, and thus the linear error in the location of succeeding points increases with the distance. Thus in Fig. 18-1 if *CBC*′ represents the error made in laying off the angle at *B*, then the corresponding linear errors at the succeeding points are *CC*′, *DD*′, and *EE*′, and the magnitude of these errors varies directly with the distance from the point at which the angular error occurs.

Use of Meridians. When the directions of lines of a traverse are given either as azimuths or as bearings, a meridian line is drawn through each station and the direction of the succeeding line is laid off with respect to the meridian. A better method, however, is to erect a meridian near the middle of the sheet and, with any desired point on the meridian as an origin, to lay off the directions of all the traverse lines. As illustrated by Fig. 18-2*a*, the line designated as a/b is laid off at an angle with the meridian equal to the bearing or azimuth of AB on the ground; and so on, for other lines in the traverse. The traverse is then plotted by transferring the directions thus determined to the appropriate positions on the sheet and laying off the lengths of the traverse lines. Thus the direction a/b (Fig. 18-2*a*) is transferred to the position AB (Fig. 18-2*b*) and the length of AB is laid off; the

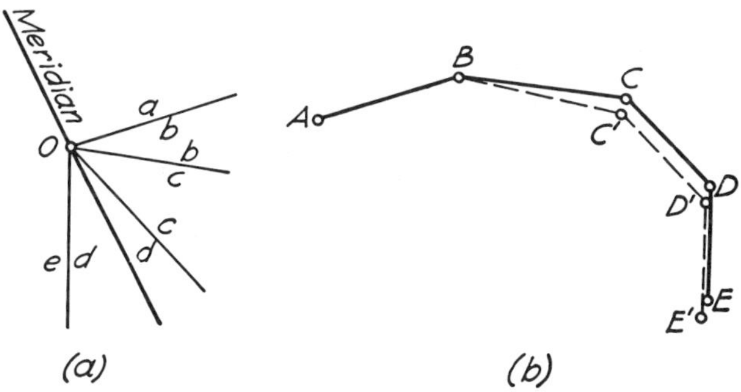

FIG. 18-2. Plotting traverse from meridian line.

direction b/c is transferred to a line of indefinite length passing through B, and the distance BC is scaled along this line from B to establish C; and so on.

Any angular error made in plotting a given line does not affect the plotted *directions* of succeeding lines, although it results in a constant linear error in the *locations* of succeeding traverse points. Thus, if CBC' (Fig. 18-2*b*) is the error introduced in the direction of BC, then the linear error in the location of C is CC'. Since $C'D'$ is given by the direction c/d, it will, if it is assumed to be without error, be parallel to its true position; hence $DD' = CC'$ and by similar reasoning $EE' = CC'$. The method of plotting from a meridian is likely to produce better results than that first described in spite of the fact that the accidental errors of plotting the individual lines are likely to be larger when the directions are transferred. Where traverses are established by deflection angles, the bearings or azimuths may be computed as described in Chap. 12, and the computed values may then be plotted from the central meridian.

18-5. Tangent Method. This method is employed principally for plotting transit traverses for the control of maps of moderate size where the number of stations is small. In principle, it is similar to the protractor method described in Art. 18-4, with the difference that an angle is laid off by a linear measurement which is a constant times the natural tangent of the angle. In Fig. 18-3, AB represents the plotted position of the initial line of a traverse, and α is the deflection angle at B which it is desired to lay off in order to determine the direction of BC. By the tangent method, the line last established (in this case AB) is prolonged some convenient distance, usually 10 in., to form a base line Bb, at the end of which a perpendicular bb' is erected; and the distance bb' is laid off equal to the length of the base line Bb multiplied by the natural tangent of α. A line drawn from B through b' defines the direction of BC. Then the point C is plotted by laying off to scale the given distance BC. Thus the process is repeated for succeeding lines. If the traverse approaches the edge of the sheet, the base line may be measured *back* along the preceding line, as at Dd in Fig. 18-3.

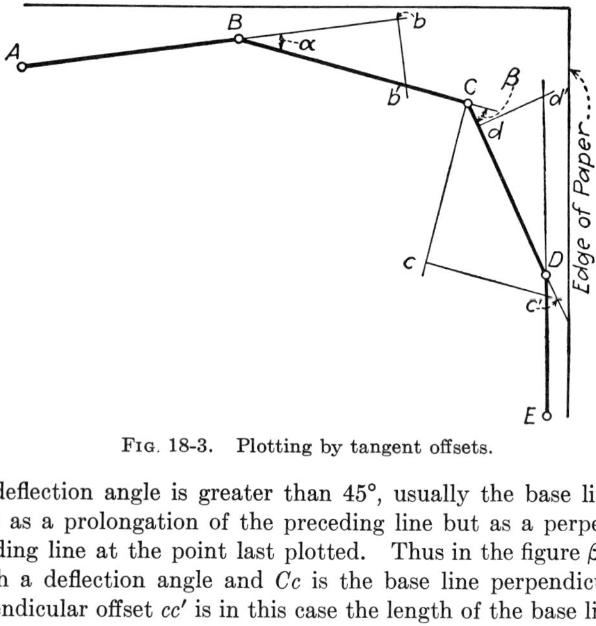

FIG. 18-3. Plotting by tangent offsets.

If the deflection angle is greater than 45°, usually the base line is established not as a prolongation of the preceding line but as a perpendicular to the preceding line at the point last plotted. Thus in the figure β is assumed to be such a deflection angle and Cc is the base line perpendicular to BC. The perpendicular offset cc' is in this case the length of the base line Cc times the natural cotangent of the angle β. The line Cc' fixes the direction of CD, and the point D is then plotted by linear measurement from C.

When a traverse is to be plotted by the method of tangent offsets, the deflection angles and distances are tabulated, and the tangents of angles less than 45° and cotangents of angles greater than 45° are recorded. The traverse is plotted roughly to small scale, the protractor being used for laying off the angles; from the small-scale sketch a suitable position for the first line of the traverse is estimated; and the line is plotted.

The precision with which angles may be laid off by this method depends upon the care used in plotting and upon the length of base line employed. For precise results all points should be pricked with a needle, all lines should be drawn with a hard, sharp pencil, and perpendiculars should be erected with great care. When this careful procedure is followed, the error in laying off an angle from a 10-in. base line need not exceed 03′. This is about the error that might be expected in using a protractor having a diameter of 20 in., if such a size were available.

Plotting deflection angles by tangents is open to the same objections as laying off deflection angles with the protractor, namely, that any error in the direction of one line affects to a like degree the directions of all succeeding

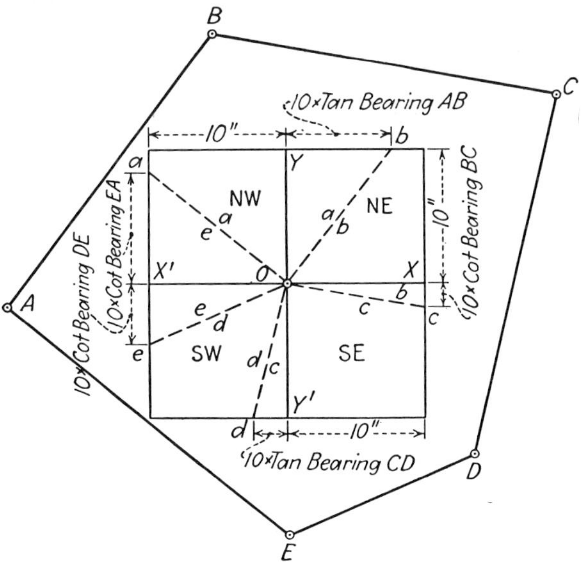

FIG. 18-4. Plotting by tangents from meridian and base line.

lines. Hence for long traverses some persons prefer to compute the bearings or azimuths of the several lines and to lay off the tangent distances from a meridian and base line, as described in the following paragraphs. The process of checking is simplified, however, if deflection angles are used.

Use of Meridians. When the directions of lines are given by bearings or azimuths, a meridian and base line may be established at each transit point and the succeeding line may be plotted by methods similar to those described for deflection angles. In this way any error in the direction of one line does not affect the direction of succeeding lines.

When the traverse is many-sided, the method illustrated by Fig. 18-4 is preferable. A 20-in. square composed of four 10-in. squares with sides perpendicular to or parallel with the meridian is constructed near the middle of the sheet, as shown. The tangent offsets of bearing angles less than 45°, or of azimuths corresponding to such bearing angles, are scaled east or west of the line YY' along the outer sides of squares forming corresponding quadrants; in a similar manner the cotangent offsets of bearing angles greater than 45°, or of azimuths corresponding to such bearing angles, are scaled north or south of the line XX'. Lines joining these points with the intersection of the X and Y axes have the same bearings as corresponding lines of the traverse.

For example, the line AB has a bearing of N37°E. Its tangent offset, 7.54 in., is, therefore, scaled from Y along the north side of the NE quadrant, thus locating b. The plotted position of AB must be parallel to Ob. The line BC has a bearing of S80°E, which is greater than 45°. The point c is located by scaling the cotangent distance, 1.76 in., south from the point X along the east side of the SE quadrant, and BC must be made parallel to Oc. In this way the directions of all lines in the traverse are laid off, the direction a/b is then transferred to its proper position on the sheet, and the distance AB is plotted to scale; the direction b/c is plotted; and so on, around the traverse.

18-5a. Tangent Protractor. A useful laborsaving device known as the *tangent protractor* consists of a scale subdivided into tangent distances (for a 10-in. base) corresponding to angles of multiples of 10′ between 0° and 45°. The values of angles, not the tangent distances, are numbered on the protractor, 45° being 10 in. (10 tan 45°) from the zero end, 30° being 5.77 in. (10 tan 30°) from the zero end, etc. The use of the protractor is identical with that of the ordinary scale, except that in plotting the tangent offsets the draftsman is guided by the angular values marked on the protractor rather than by the actual tangent offsets. To facilitate the plotting of angles between 45° and 90° by cotangent distances, a second set of angular values is placed under the first, the cotangent scale decreasing from 90° to 45° as the tangent scale increases from 0° to 45°. The use of the protractor eliminates the necessity of determining the numerical values of tangents, but the protractor is made only for a 10-in. base, and it cannot be conveniently used for any other distance.

18-6. Chord Method. This method is much like that of plotting by tangents as just described, except that instead of erecting a perpendicular at the end of a 10-in. base line, an *arc* of 10-in. radius is struck. The chord distance for the given angle is then scaled from the point of intersection between arc and base line to a point on the arc.

The chord method is generally regarded as being somewhat less precise than the tangent method though it is not clear that there should be any appreciable difference between the two methods. The chord method is not in as general use as the tangent method, probably for the reason that it requires more time to compute the chord distances than it does to compute the tangent offsets. If a table of chords is available, however, the method is considerably quicker than the tangent method.

18-7. Checking. Checking the correctness of the plotted horizontal control is quite as essential as the verifying of field measurements. The methods of checking are essentially the same for the protractor, tangent, and chord methods of plotting just described. The process of checking the plotted location of points of horizontal control may proceed as the work of plotting progresses, but in any case it is completed before the work of plotting the details is begun. Frequently the process of checking serves not only to verify the work of plotting but also to verify the field measurements.

A traverse that has been brought to closure in the field should likewise close on paper, and if it does, a check is secured on the correctness of both the field work and the drafting. If the field work is assumed as correct, any error of closure on paper will be due to inaccuracies in laying off angles and distances. A traverse of considerable length will rarely close on paper when plotted by the methods just described, no matter how careful the work of plotting.

If the error of closure is small, it is usually assumed to have accumulated gradually, and the traverse is made to close by a progressive change in the position of the plotted lines. Thus in Fig. 18-5 the full line $ABC'D'E'A'$ represents a traverse as first plotted, $A'A$ being a small error of closure, and the dash line $BCDEA$ shows the adjusted portion of the traverse.

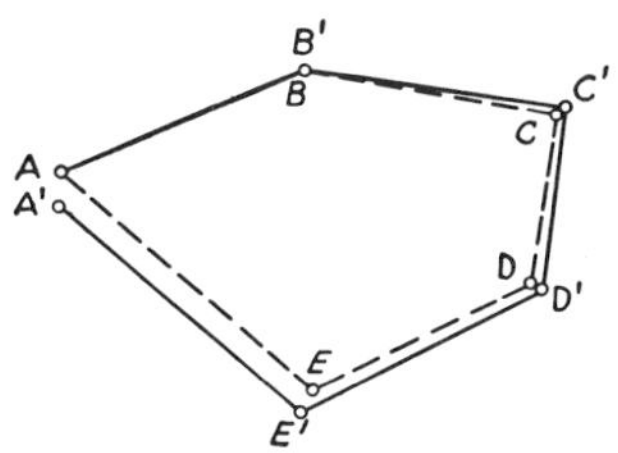

FIG. 18-5.

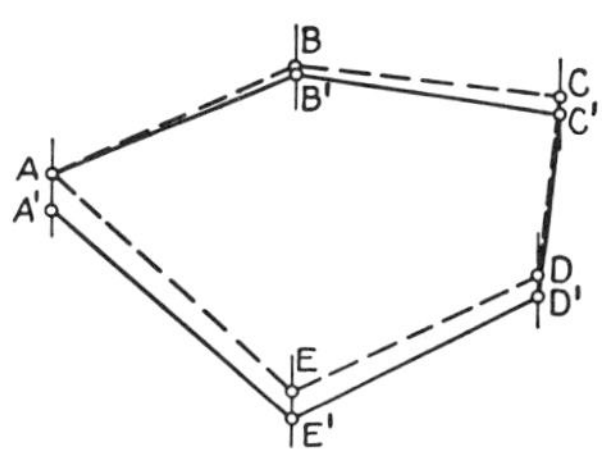

FIG. 18-6.

Another method of closure is illustrated in Fig. 18-6. Each traverse point is moved in a direction parallel to the side of closure, by an amount proportional to the distance along the traverse from the initial point to the given point. For example,

$$C'C = AB'C'/AB'C'D'E'A' \times A'A$$

Essentially this method is an application of the Compass Rule (Art. 18-13).

If the error of closure is large, a mistake in laying off an angle or a distance has occurred; sometimes the mistake can be located quickly by observing the simplest way of bringing the traverse to a closure. Thus, in Fig. 18-6, with the error of closure $A'A$ as shown, one might expect that a mistake had been made in the length of the side $C'D'$ which is nearly parallel to the side of closure; and the first step in locating the error would be to scale this distance. Another possibility is a mistake of 1° or 10° in one of the azimuths or angles, in which case a perpendicular bisector of the line joining the starting and closing points of the traverse will pass nearest the station where the mistake was made. If the cause of the error of closure cannot be detected readily by measurement of the plotted angles and distances, the traverse may be plotted in the reverse direction, starting from the same initial point, until a point coincides with the corresponding point for the traverse as originally plotted. Thus, Fig. 18-7, $ABCD'E'A'$ represents a traverse as originally plotted and $AEDC$ represents the traverse plotted in the opposite direction from the initial point A. The two traverses come together at C, and the corrected line is, therefore, $ABCDEA$.

Open Traverses. When the horizontal control is an *open* or continuous traverse, there is no absolute check on the precision of plotting as in the case of the closed figure. A satisfactory way of checking distances is as

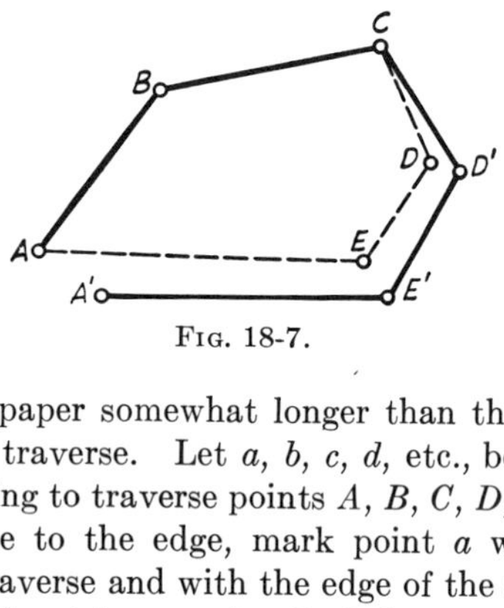

FIG. 18-7.

follows: Cut a strip of paper somewhat longer than the combined length of the several lines in the traverse. Let a, b, c, d, etc., be points to be marked on this strip corresponding to traverse points A, B, C, D, etc., on the drawing. Near one end and close to the edge, mark point a with a needle. Place this mark at A of the traverse and with the edge of the strip along AB, mark point b on the edge of the strip opposite B of the traverse. Move the strip to BC with b opposite the corresponding point of the traverse, and mark point c. Continue in this way around the entire traverse. Scale the total length marked off on the strip. This scaled distance should agree closely with the sum of the lengths in the notes used for plotting.

If deflection angles are used in plotting, the direction of any line can be checked by computing its azimuth or bearing and observing whether the line makes this azimuth or bearing with an established meridian. This check does not need to be applied to every line of the traverse but in any case should be performed for the two end lines. If the relative directions of these two lines are found to be correct, it may be assumed that the directions of all other lines are without appreciable error, by reason of the fact that the direction of each line depends upon that of the preceding line. For traverses with numerous sides, however, this check is usually applied every five or six courses as the work of plotting progresses, and adjustment is made if appreciable error is found.

If bearings or azimuths are used in plotting, the deflection angles at the several points can be computed, and the directions of the courses can be checked by measuring the deflection angles either with the protractor or by scaling the tangent offsets or chord distances. For this case *all* courses should be checked, since the direction of any course may be in error without its affecting the direction of any other course.

18-7a. Cut-off Lines. When, during the process of running an open traverse, cut-off lines (Art. 14-14) are established, both the field work and the plotting may be checked. Each cut-off line may be considered as forming the closing side of an independent traverse. Where the length of the cut-off line is measured, both angles and distances within the length intercepted by the cut-off line are checked if it closes on paper. If it fails to close, the procedure of checking is the same as for any other closed traverse; and when large errors appear to have been eliminated, the portion intercepted by the cut-off line may be adjusted until the cut-off line actually closes the plotted circuit. Usually the length of the cut-off line is not measured in the field, in which case an absolute check on the plotted length of the intercepted portion of the traverse cannot be obtained.

Figure 18-8 represents a portion of a main traverse with cut-off line of unmeasured length passing through A and E. The full line $ABC'D'E'$ shows the portion as originally plotted for which the cut-off line does not pass through A, thus indicating an error in angle or distance or both. The line $ABCDE$ shows the traverse after an adjustment in angle has been made at B so that the cut-off line passes through

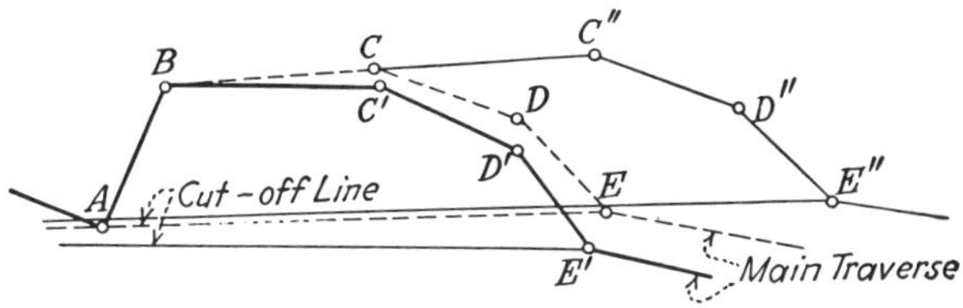

FIG. 18-8. Checking by cut-off lines.

A. The line $ABC''D''E''$ represents the same portion of the main traverse but with a gross error in the length of the second line, BC being the correct length and BC'' the erroneously plotted length. The cut-off line is seen to pass nearly through A as before, showing that a large error may be made in plotting a distance within the intercepted portion of the traverse and still have the cut-off line pass through the points of intercept or nearly so, provided the error is in a line that is nearly parallel to the cut-off line.

Thus the cut-off line of unmeasured length but of observed direction provides a means of checking the angles of the intercepted portion of the traverse and provides a means of checking against large errors in distance, except for those lines which have the same general direction as the cut-off line.

18-7b. Intersecting Lines. When angles to a given station not on the traverse are observed from a number of stations of the traverse, both the field work and the plotting may be regarded as correct if the plotted lines to the given station intersect at a common point.

For example, in Fig. 18-9, the full line $ABC'D'E'$ represents a portion of an open traverse, and the full lines intersecting at O and O' represent lines to the station not on the traverse, the angles OAB, OBC', $O'C'D'$, etc., being laid off in the same manner as are the angles of the traverse. The fact that the lines do not intersect at a common point indicates that an error exists in angle or in distance. The

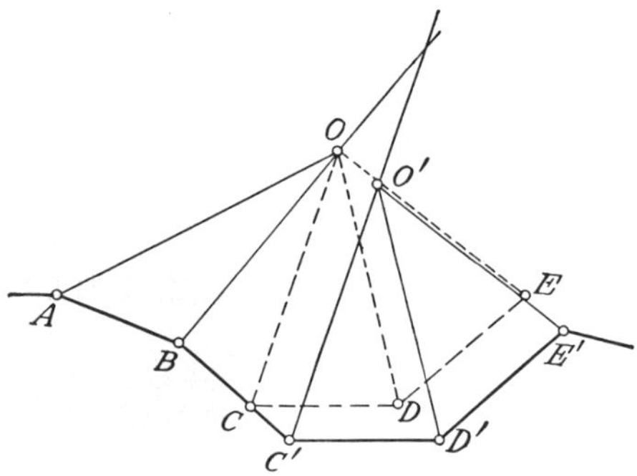

FIG. 18-9.

location of the error, and sometimes also its nature, may be determined by inspection. Thus for the case shown in the figure, OO' is nearly parallel to BC, hence it might be expected that the length of BC was in error by an amount approximately equal to OO'. If this were found to be so, then the traverse would be adjusted as shown by the line $ABCDE$ and the intersecting lines from C, D, and E (shown dotted) would pass through or approximately through the point O.

18-8. Method of Rectangular Coordinates. This method, also known as the *method of total latitudes and departures*, is recognized as the most reliable method of plotting. It is the only practical method for plotting extensive systems of horizontal control. It is always employed for plotting triangulation figures except those of the simplest character, and it is considerably more accurate than any of the methods so far described for plotting traverses (see Art. 18-18). Rectangular coordinates are employed not only for plotting maps but also frequently for calculating areas, as described in Art. 19-4.

The coordinate axes are a *reference meridian* (true, magnetic, or assumed) and a line at right angles thereto called a *reference parallel*. The intersection of these lines, marking the origin, may be any point in the survey or may be outside the survey. The azimuths or bearings of all survey lines are either given directly or computed from observed angles. With the direction of each line determined and its length known, the lengths of its orthographic projection upon the meridian and upon the parallel are computed. The projection upon the meridian is termed the *latitude* of the line; the projection upon the parallel is called the *departure* of the line.

The origin having been chosen, the coordinates for the several control points are computed by using the latitudes and departures. The coordinate of a point measured normal to the $\left\{\begin{matrix}\text{parallel}\\ \text{meridian}\end{matrix}\right\}$ is called the $\left\{\begin{matrix}\textit{total latitude}\\ \textit{total departure}\end{matrix}\right\}$ or the $\left\{\begin{matrix}\textit{parallel distance}\\ \textit{meridian distance}\end{matrix}\right\}$ of the point. $\left\{\begin{matrix}\text{Total latitudes}\\ \text{Total departures}\end{matrix}\right\}$ are positive or negative according to whether the corresponding points lie $\left\{\begin{matrix}\text{north or south}\\ \text{east or west of}\end{matrix}\right.$ $\left.\begin{matrix}\text{of the reference parallel}\\ \text{the reference meridian}\end{matrix}\right\}$. With the coordinate axes established on paper, a point is plotted by laying off its total latitude and its total departure to the required scale. Articles 18-9 to 18-17 are devoted to a more detailed discussion of the processes involved.

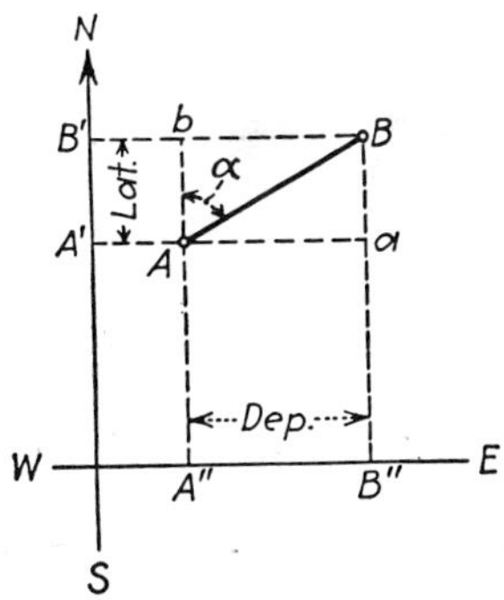

FIG. 18-10. Latitude and departure of line AB.

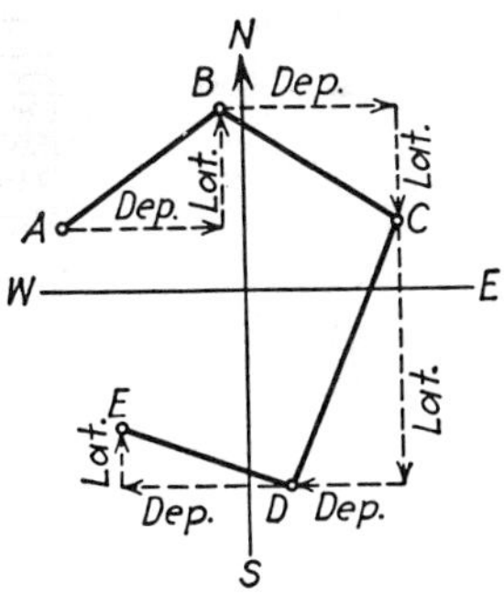

FIG. 18-11. Directions of latitudes and departures.

Before the latitudes and departures of the traverse lines are computed, the angular error of closure of the traverse is determined by the known geometrical conditions, and the angles or bearings are so adjusted or corrected that the known geometrical conditions will be fulfilled (Art. 18-12).

18-9. Latitudes and Departures. In Fig. 18-10, AB represents any line the latitude and departure of which it is desired to determine, and the lines NS and EW represent any meridian and any parallel. The line AB makes the angle α with the meridian.

As the latitude of a line is the orthographic projection of the line upon a meridian, the latitude of AB is $A'B' = Ab = AB \cos \alpha$.

And as the departure of a line is its orthographic projection upon a parallel, the departure of AB is $A''B'' = Aa = AB \sin \alpha$.

Stated in the form of a general rule applicable to any line:

$$\text{Latitude} = \text{length} \times \text{cosine bearing angle} \quad (1)$$
$$\text{Departure} = \text{length} \times \text{sine bearing angle} \quad (2)$$

Latitudes are designated as *North* or *positive* for all lines having a northerly bearing, and *South* or *negative* for all lines having a southerly bearing. Departures are designated as *East* or *positive* for lines having an easterly bearing, and *West* or *negative* for lines having a westerly bearing. Thus, in Fig. 18-11, for the line AB the latitude is North or + and the departure is East or +. For BC the latitude is South or − and the departure is East or +. For CD the latitude is South or − and the departure is West or −. For DE the latitude is North or + and the departure is West or −.

If the latitudes and departures are determined solely for purposes of map construction, the number of places to be used in computations should be such that points may be plotted correctly within the scale of the map. Thus for a scale of 1 in. = 800 ft., distances can hardly be plotted closer than to the nearest 10 ft., but to insure all coordinates being correct to the nearest 10 ft. the latitudes and departures would probably be determined to the nearest foot, and to insure the latitudes and departures being correct to the nearest foot, intermediate computations might be carried out to tenths of feet.

Computations are made either with a computing machine or by logarithms. "Traverse tables" found in various publications, such as Ref. 3 at the end of this chapter, give for various bearing angles the latitudes and departures for various lengths of line.

If the computing machine is employed, the data may be kept in the following form:

Line	Bearing	Length, ft.	Cos bearing	Sin bearing	Latitude, ft.		Departure, ft.	
					N	S	E	W

First the bearings and lengths of the lines are tabulated. Then the natural cosines and sines of the bearing angles are recorded. And finally for each side the latitude and departure are computed, the length being set on the machine as the multiplicand.

If logarithms are used, the form of the following example is convenient:

Example: For a given course in a traverse the computed bearing is N34°21′W and the observed length is 1,215.3 ft. The traverse is to be plotted to the scale of 1 in. = 100 ft. It is desired to compute the latitude and departure.

At the given scale, distances may be plotted within 1 or 2 ft., and therefore latitudes and departures should be correct to perhaps the nearest quarter foot if the traverse has many sides. Five-place logarithms will be used. The computations are given in the accompanying tabulation. Beginning with "log distance" in the tabulation, computations for latitude are made reading upward, and computations for departure are made reading downward.

Latitude	1,003.3 ft.	
log latitude		3.00145
log cos bearing		9.91677
log distance		**3.08468**
log sin bearing		9.75147
log departure		2.83615
Departure	685.7 ft.	

18-10. Error of Closure. In any closed traverse it is obvious that the sum of the north latitudes should equal the sum of the south latitudes, and that the sum of the east departures should equal the sum of the west departures. In other words, for any closed traverse the algebraic sum of the latitudes (ΣL) should be equal to zero, and the algebraic sum of the departures (ΣD) should be equal to zero. Similarly, in some open traverses, the location of the end stations relative to each other is known from other sources; such traverses can be treated as closed traverses with the line connecting the two end stations forming the closing line.

However, owing to errors in field measurements of both angles and distances, in general an unadjusted traverse will not close on paper, even though the plotting be without error. The conditions stated in the previous paragraph, however, make it possible to determine the error of closure by means of the computed latitudes and departures.

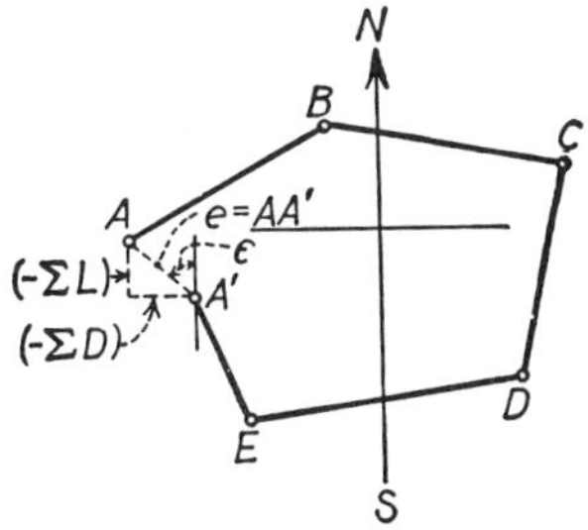

FIG. 18-12. Error of closure of traverse.

Figure 18-12 shows a traverse that does not close, the line $A'A = e$ being the side of error. In order that the algebraic sum of the latitudes and the algebraic sum of the departures shall each be equal to zero, the latitude of the side of error must be $-\Sigma L$ and its departure must be $-\Sigma D$, considered algebraically. These two quantities form the base and altitude of the right-angle triangle of which the side of error is the hypotenuse; hence the linear error of closure is

$$e = \sqrt{(\Sigma L)^2 + (\Sigma D)^2} \tag{3}$$

The direction of the side of error is given by the relation

$$\tan \epsilon = \frac{-\Sigma D}{-\Sigma L} \tag{4}$$

with due regard to sign, an equation which is satisfied by bearings differing by 180°. The data will make it apparent in which quadrant the bearing lies.

Example: In a given closed traverse the sum of the south latitudes exceeds the sum of the north latitudes by 9 ft. and the sum of the east departures exceeds the sum of the west departures by 12 ft. What is the linear error of closure and the bearing of the side of error?

The linear error of closure is $e = \sqrt{(9)^2 + (12)^2} = 15$ ft.

Since for the given sides the south latitudes exceed the north latitudes, the latitude of the side of error must be north, and by similar reasoning the departure of the side of error must be west; hence the bearing angle is in the northwest quadrant.

$$\tan \epsilon = \frac{-\Sigma D}{-\Sigma L} = \frac{-12}{+9} = -1.33$$

$$\epsilon = -53°08'$$

$$\text{Bearing} = \text{N}53°08'\text{W}$$

18-11. Balancing the Survey. This discussion deals with the practice of adjusting field observations, the principles of which are discussed in Chap. 5.

When the error of closure has been determined as described in the preceding article, usually corrections are made so that the traverse will form a mathematically closed figure, and the corrections are applied to the latitudes and departures in such manner as to make their algebraic sum equal zero. This operation of distributing corrections to eliminate the error of closure is called *balancing the survey* or *balancing the traverse.* There is, of course, no possible means of determining the true magnitude of the errors in angle and distance which occur throughout the traverse; but if conditions surrounding the field measurements have been uniform, it is fair to assume that errors have gradually accumulated, and corrections should be made accordingly.

Electronic computers are used increasingly for computation of latitudes and departures from the lengths and directions observed in the field and for balancing the traverses and computing the coordinates of the stations. Standard computer programs are available in many localities.

If the error of closure is excessive, it indicates that a mistake in field work or in plotting has been made, and the work should be checked (Art. 18-7).

There are several rules for distribution of errors, each of which will produce a mathematically closed figure and each of which is assumed to be adapted to certain conditions as regards measurements. However, many surveyors rely on their own judgment with little or no regard for any established rule, and distribute the error arbitrarily in accordance with their estimation of the field conditions.

If certain courses are over rough ground, the error of chaining these courses would be expected to be relatively large, and the correction to the observed distances should be correspondingly large. Where sights are steep or short and where visibility is poor, larger angular errors would be expected than where conditions of observing are more nearly ideal; and hence in balancing the survey it is fair to assume that the larger changes in directions should be in the courses where conditions surrounding the observations were relatively unfavorable.

18-12. Adjustment of Angular Error. The values of the measured angles of a traverse are checked by the known geometrical conditions. The total angular error thus determined is distributed among the angles or computed bearings of the traverse *before computations of the latitudes and departures are made.* Often this adjustment is arbitrary, based upon a knowledge of the field conditions; but, if all angles have been measured under like conditions, the error is distributed equally to each angle in the traverse.

The total angular error is the result of both accidental and systematic errors which have affected the work. The angular adjustment eliminates the effects of all systematic errors to the extent that they have been constant and equal in their effect upon each angle measured. It does not yield true values, for each measured angle is subject to accidental errors whose sign and magnitude are unknown. However, it meets the known geometrical conditions; and the adjusted values are the most probable values that can be assigned.

18-13. Compass and Transit Rules for Balancing a Survey. The rules generally used to balance a traverse are the *Compass Rule* and the *Transit Rule.* (See also the Crandall Method, Art. 18-14.) The Compass Rule is the rule most commonly employed.

The Compass Rule states that the correction to be applied to the $\left\{\begin{matrix}\text{latitude}\\ \text{departure}\end{matrix}\right\}$ of any course is to the total correction in $\left\{\begin{matrix}\text{latitude}\\ \text{departure}\end{matrix}\right\}$ as the *length* of the course is to the length of the traverse. It is based upon the assumptions (1) that the errors in traversing are accidental and, therefore, vary with the square root of the lengths of the sides, thus making the correction to each side proportional to its length, and (2) that the effects of the errors in angular measurements are equal to the effects of errors in chaining.

The Transit Rule states that the correction to be applied to the $\left\{\begin{matrix}\text{latitude}\\ \text{departure}\end{matrix}\right\}$ of any course is to the total correction in $\left\{\begin{matrix}\text{latitude}\\ \text{departure}\end{matrix}\right\}$ as the $\left\{\begin{matrix}\textit{latitude}\\ \textit{departure}\end{matrix}\right\}$ of that course is to the arithmetical sum of all the $\left\{\begin{matrix}\text{latitudes}\\ \text{departures}\end{matrix}\right\}$ in the traverse.

It is based upon the assumptions (1) that the errors in traversing are accidental and (2) that the angular measurements are more precise than are those of chaining.

The adjustment of observations subject to accidental errors lies properly within the province of the method of least squares, and in the light of these principles the following comments may be made:

1. It can be shown that the Compass Rule is logical for the assumptions made.

2. The Transit Rule is merely a rule of thumb which, it is found, does not apply successfully to many cases. In fact it meets the assumptions upon which it is based only to the extent that each side is parallel to one or the other of the coordinate axes.

As an example of the effects of the two rules, the following example is given for a particular case.

Example: In the accompanying tabulation are given the bearings, lengths, latitudes, and departures for a closed traverse of six sides. The survey is to be balanced by both of the rules given in this article.

Line	Bearing	Length, ft.	Latitude, ft.		Departure, ft.	
			N	S	E	W
AB	N	500.0	500.0	....	0.0	....
BC	N45°00′E	848.6	600.0	....	600.0	....
CD	S69°27′E	854.4	....	300.0	800.0	....
DE	S11°19′E	1,019.8	....	1,000.0	200.0	....
EF	S79°42′W	1,118.0	....	200.0	....	1,100.0
FA	N54°06′W	656.8	385.0	....	....	532.0
Sum	....	4,997.6	1,485.0	1,500.0	1,600.0	1,632.0

The error of closure in latitude is 15.0 ft. and in departure is 32.0 ft. The sum of the lengths of the sides is 4,997.6 ft. or practically 5,000 ft. The length of the side of closure (linear error of closure) is $\sqrt{15^2 + 32^2} = 35.3$, and the relative linear error of closure is $35.3/5{,}000 = 1/142$. (This error is too large to be permitted in practice, but was made large for the purpose of this example.)

By the Compass Rule the amount of the correction in latitude of the course *CD* is $(15/5{,}000) \times 854.4$ ft. $= 2.6$ ft. As the south latitudes are too large in this example, the correction is to be subtracted, that is, the correction is -2.6 ft. The correction to the departure of *CD* is $(32/5{,}000) \times 854.4 = 5.4$ ft.; as the east departures are too small, the correction is to be added. The corrections to the other courses are computed in a similar manner, with due regard to sign.

For use with the Transit Rule, the arithmetical sum of the latitudes is 2,985.0 ft. and the arithmetical sum of the departures is 3,232.0 ft. The correction to the latitude of *CD* is $(-15/2{,}985) \times 300 = -1.5$ ft., and the correction in departure is $(+32/3{,}232) \times 800 = +8.0$ ft.

In the accompanying tabulation are given the corrections in feet as determined by the two rules, together with the corresponding corrections by the Crandall Method which is discussed in the following article. As a check, the arithmetical sum of the corrections in latitude or departure (given in the last line of each column) should equal the total error in latitude or departure. This is a check on the correctness of the computations. When the corrections have been applied, the algebraic sum of the latitudes and of the departures must be zero, and hence the survey is balanced.

Line	Correction in latitude, ft.			Correction in departure, ft.		
	Compass Rule	Transit Rule	Crandall Method	Compass Rule	Transit Rule	Crandall Method
AB	1.5	2.5	3.7	3.2	0	0
BC	2.6	3.0	8.3	5.4	6.0	8.3
CD	2.6	1.5	−2.6	5.4	8.0	7.0
DE	3.1	5.0	4.9	6.6	2.0	−1.0
EF	3.3	1.0	2.7	7.2	10.8	14.6
FA	1.9	2.0	−2.1	4.2	5.2	2.9
Sum	15.0	15.0	14.9	32.0	32.0	31.8

18-14. Crandall Method of Balancing a Survey. There are conditions where the accidental errors of linear measurements are likely to be greater than those in the measurement of angles, as, for example, in stadia traversing, or even in careful tape measurements where some of the systematic errors are rendered accidental in nature by reason of corrections and of special methods applied to the field measurements. In such cases we may be warranted in assuming, after any small angular error of closure has been distributed among the bearings of the traverse, that the bearings are without appreciable error; and the adjustments should properly be made to the linear measurements only. For these cases Professor C. L. Crandall has applied the method of least squares and has arranged the solution in such form that it can be carried through with the use of a slide rule only (see Ref. 1 at the end of this chapter). As applied to the example given above, the computations are as follows:

In the formulas of this example the following notation is used:

L = latitude of any course
D = departure of any course
l = length of any course
v = correction to be applied to any quantity
q_L = total error in latitude
q_D = total error in departure

A and B are factors connecting various quantities in the computations, and Σ is the sign of summation.

As a matter of convenience, the distances are expressed in tape lengths by dividing the lengths l by 100 throughout.

$$A = \frac{q_D\left(\Sigma \dfrac{LD}{100l}\right) - q_L\left(\Sigma \dfrac{D^2}{100l}\right)}{\left(\Sigma \dfrac{D^2}{100l}\right)\left(\Sigma \dfrac{L^2}{100l}\right) - \left(\Sigma \dfrac{LD}{100l}\right)^2} \tag{5a}$$

$$B = \frac{q_L\left(\Sigma \dfrac{LD}{100l}\right) - q_D\left(\Sigma \dfrac{L^2}{100l}\right)}{\left(\Sigma \dfrac{D^2}{100l}\right)\left(\Sigma \dfrac{L^2}{100l}\right) - \left(\Sigma \dfrac{LD}{100l}\right)^2} \tag{5b}$$

Then

$$\left.\begin{aligned} v_{l1} &= L_1A + D_1B \\ v_{l2} &= L_2A + D_2B \end{aligned}\right\}, \text{ etc.} \tag{6}$$

$$v_{L1} = v_{l1}\frac{L_1}{100l_1} = A\frac{L_1^2}{100l_1} + B\frac{L_1D_1}{100l_1}, \text{ etc.} \tag{7}$$

$$v_{D1} = v_{l1}\frac{D_1}{100l_1} = A\frac{D_1L_1}{100l_1} + B\frac{D_1^2}{100l_1}, \text{ etc.} \tag{8}$$

Line	Bearing	Length	Latitude N	Latitude S	Departure E	Departure W	$\frac{L^2}{100l}$	$\frac{D^2}{100l}$	$\frac{LD}{100l}$
			+	−	+	−			
AB	N	500.0	500.0	....	0.0	....	5.00	0.00	+0.00
BC	N45°00′E	848.6	600.0	....	600.0	....	4.25	4.25	+4.25
CD	S69°27′E	854.4	....	300.0	800.0	....	1.05	7.50	−2.81
DE	S11°19′E	1,019.8	....	1,000.0	200.0	....	9.80	0.39	−1.96
EF	S79°42′W	1,118.0	....	200.0	....	1,100.0	0.36	10.81	+1.97
FA	N54°06′W	656.8	385.0	....	....	532.0	2.25	4.31	−3.11
		Sum	1,485.0	1,500.0	1,600.0	1,632.0	22.71	27.26	−1.66
			$q_L = -15.0$		$q_D = -32.0$				

$$A = \frac{(-32.0) \times (-1.66) - (-15.0 \times 27.26)}{(27.26 \times 22.71) - (-1.66)^2} = \frac{+53.1 + 409.0}{621.7 - 2.8} = +0.747$$

$$B = \frac{(-15.0) \times (-1.66) - (-32.0 \times 22.71)}{(27.26 \times 22.71) - (-1.66)^2} = \frac{+24.9 + 726.7}{621.7 - 2.8} = +1.214$$

$$\begin{aligned}
v_{L1} &= (5.00 \times 0.747) + (0.00 \times 1.214) = +\ 3.74 + 0.00 = +\ 3.74 \\
v_{L2} &= (4.25 \times 0.747) + (4.25 \times 1.214) = +\ 3.17 + 5.16 = +\ 8.33 \\
v_{L3} &= (1.05 \times 0.747) - (2.81 \times 1.214) = +\ 0.78 - 3.41 = -\ 2.63 \\
v_{L4} &= (9.80 \times 0.747) - (1.96 \times 1.214) = +\ 7.32 - 2.38 = +\ 4.94 \\
v_{L5} &= (0.36 \times 0.747) + (1.97 \times 1.214) = +\ 0.27 + 2.39 = +\ 2.66 \\
v_{L6} &= (2.25 \times 0.747) - (3.11 \times 1.214) = +\ 1.68 - 3.78 = -\ 2.10 \\
&\text{Check: } -q_L = +14.94
\end{aligned}$$

$$
\begin{aligned}
v_{D1} &= (0.00 \times 1.214) + (0.00 \times 0.747) = +\ 0.00 + 0.00 = +\ 0.00 \\
v_{D2} &= (4.25 \times 1.214) + (4.25 \times 0.747) = +\ 5.16 + 3.17 = +\ 8.33 \\
v_{D3} &= (7.50 \times 1.214) - (2.81 \times 0.747) = +\ 9.11 - 2.10 = +\ 7.01 \\
v_{D4} &= (0.39 \times 1.214) - (1.96 \times 0.747) = +\ 0.47 - 1.46 = -\ 0.99 \\
v_{D5} &= (10.81 \times 1.214) + (1.97 \times 0.747) = +13.13 + 1.47 = +14.60 \\
v_{D6} &= (4.31 \times 1.214) - (3.11 \times 0.747) = +\ 5.23 - 2.32 = +\ 2.91 \\
&\qquad\qquad\qquad\qquad\qquad\qquad \text{Check: } -q_D = +31.86
\end{aligned}
$$

These computations are more laborious than those required by the Compass Rule or the Transit Rule, but the labor is not excessive. This solution meets the desired assumptions and distributes the error of closure in the lengths of the lines only. A convenient check is supplied at the close, if the sum of the separate corrections equals the total error with opposite sign. Another check is afforded if the corrected lengths of the lines are computed from the corrected latitudes and departures. The Crandall Method is well suited for electronic computation.

18-15. Summary of Methods. The discussion relating to methods of balancing a survey may be summarized by the following statements:

1. In many cases a careful arbitrary distribution of errors on the basis of a knowledge of the field conditions is the best adjustment that can be made.
2. If systematic errors are believed to be present in the linear measurements they can be eliminated only by applying proper computed corrections to the field measurements before the survey is balanced, since systematic errors are not subject to distribution by any general rule.
3. If the error of closure is subject to accidental errors affecting angular and linear measurements equally, the Compass Rule is valid.
4. In most cases, the Transit Rule cannot properly be used.
5. If the accidental errors in linear measurements are assumed to be much larger than those in angles, the Crandall Method is valid.

18-16. Computation of Coordinates. When it is desired to plot points of horizontal control by the method of coordinates, the latitudes and departures of the control lines are computed, and in the case of a closed traverse the survey is balanced. The origin is chosen, and the coordinates, or total latitudes and departures, of the several points in the survey are computed by summing algebraically the latitudes and departures, respectively, of lines between that point and the origin.

The total $\left\{\begin{matrix}\text{latitude}\\\text{departure}\end{matrix}\right\}$ of any point equals the algebraic sum of $\left\{\begin{matrix}\text{latitudes}\\\text{departures}\end{matrix}\right\}$ of lines lying between that point and the $\left\{\begin{matrix}\text{reference parallel}\\\text{reference meridian}\end{matrix}\right\}$ passing through the origin.

In the accompanying tabulation are given (1) the latitudes and departures of the lines in the traverse of the preceding example, adjusted by the Compass Rule; (2) the computed total latitudes and departures, or coordinates, referred to station A as the origin; and (3) the adjusted bearings and lengths of the traverse lines.

Sta.	Adjusted lat., ft.		Adjusted dep., ft.		Total lat., ft.		Total dep., ft.		Adjusted bearing	Adjusted length, ft.
	N	S	E	W	N	S	E	W		
A					0	0	0	0		
	501.5	....	3.2	....					N 0°22′E	501.5
B					501.5	.....	3.2	...		
	602.6	....	605.4	....					N45°08′E	854.2
C					1,104.1	.....	608.6	...		
	....	297.4	805.4	....					S69°44′E	858.6
D					806.7	.....	1,414.0	...		
	....	996.9	206.6	....					S11°42′E	1,018.1
E					....	190.2	1,620.6	...		
	....	196.7	....	1,092.8					S79°48′W	1,110.4
F					....	386.9	527.8	...		
	386.9	....	....	527.8					N53°45′W	654.4
A					0	0	0	0		
Sum	1,491.0	1,491.0	1,620.6	1,620.6	....	.....	....	...	....	4,997.2

In a closed traverse the additions are verified if on making the complete circuit the total latitude and total departure of the initial point check these quantities as assumed at the beginning. In the tabulation this check has been performed.

18-17. Plotting Control by Coordinates. When a system of horizontal control is to be plotted by rectangular coordinates, the size of the enclosing rectangle is determined from an examination of the total latitudes and departures. This rectangle is one whose east and west sides are meridians passing through the most easterly and westerly points of the survey, respectively, and whose north and south sides are parallels passing through the most northerly and southerly points of the survey, respectively. In order to insure the proper location of points on the map sheet, usually the enclosing rectangle and the principal points of control are first plotted roughly to small scale. On the drawing paper the enclosing rectangle is drawn to the required scale, its position being fixed by estimation from the small-scale sketch. Preferably the top of the map should represent north, although the shape of the area may make another orientation preferable.

Let *HJKL* (Fig. 18-13) be the enclosing rectangle for the traverse of the preceding article. The rectangle should be drawn with great care and checked by scaling the lengths of the diagonals. Perpendiculars constructed by use of the ordinary triangle are not necessarily accurate; they should be checked by reversing the triangle, or the perpendicular should be erected by a 3:4:5

method or by the use of a drafting compass. The engineer's drafting machine (Art. 6-37) is useful for this work, or a recently developed coordinate-plotting machine may be available. The locations of the reference meridian and reference parallel are determined by scaling along the sides of the enclosing rectangle. For example, in Fig. 18-13 the reference parallel is located by scaling JA and KM equal to 387 ft., the total latitude of the most southerly point of the survey. The location is checked by scaling AH and ML equal to 1,104 ft., the total latitude of the most northerly point of the survey.

If the drawing is large and many control points are to be plotted, other meridians and parallels are constructed to divide the area into squares, the

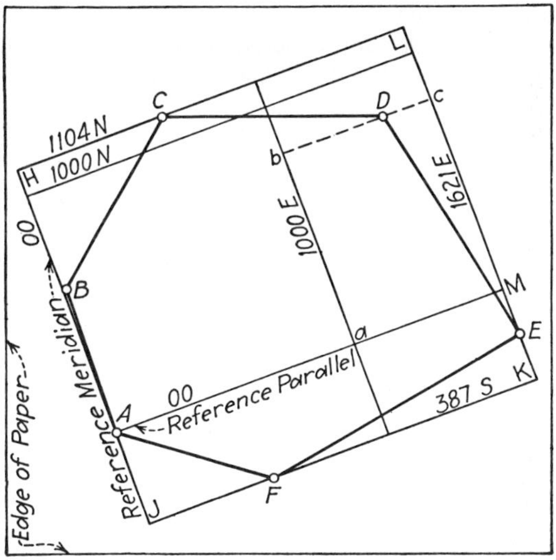

FIG. 18-13. Plotting control by coordinates.

sides of which are less than the length of the scale used and which represent some whole number of hundreds or thousands of feet at the given scale. Thus, in the figure, a meridian is drawn 1,000 ft. east of the reference meridian, and a parallel is drawn 1,000 ft. north of the reference parallel. If the scale of the map is 1 in. = 100 ft., the actual length of the sides of the resulting squares is 10 in.

Each point of horizontal control is located on the sheet by plotting its total latitude and departure, and its location is verified by scaling the length of the preceding course. For example, point D (Fig. 18-13) has a north total latitude of 807 ft. and an east total departure of 1,414 ft. The point D is plotted by laying off to scale above the reference parallel the distances ab and Mc each equal to 807 ft., then drawing the line bc, and finally laying off to scale the distance $bD = 1{,}414 - 1{,}000 = 414$ ft. As a check, the distance CD as scaled from the drawing should agree with the length of this line as computed from the adjusted latitude and departure ($CD = 858.6$ ft. by the foregoing tabulation).

When the angle between a given course and the meridian is small, a considerable error may be made in plotting the total departure of either end without appreciably altering the plotted length of the line. The same is true of any course making a small angle with the reference parallel when an error is made in plotting the total latitude of either end. For this reason, where two or more such courses are contiguous, the scaled latitude or departure (whichever is the smaller) of each course should be compared with the value from which the coordinates are computed.

Long traverses that do not form closed figures may be plotted without constructing the enclosing rectangle. The reference meridian and reference parallel are drawn as accurately as possible, and with these as a basis other meridians and parallels forming squares of convenient size are drawn where such lines are necessary for plotting the traverse points.

18-18. Advantages and Disadvantages of Coordinate Method. The coordinate method is recognized as the most reliable method of plotting control. Its principal advantages are as follows: (1) the size and shape of the drawing can be determined accurately beforehand; (2) the accuracy of the location of any point does not depend upon the accuracy with which previous lines in the control system are plotted; (3) the method of checking is simple and is unlike the method of plotting; (4) for closed traverses the field measurements are checked and the survey is balanced before plotting is begun; and (5) shrinkage or expansion of the paper is easily detected (by measuring the side of a control square) and proportionate corrections made.

The single disadvantage of the coordinate method as compared with the method of tangents or the method of chords is the greater amount of computation required preliminary to plotting. However, often in the case of a closed traverse the latitudes and departures are necessarily computed for the purpose of determining the area (Art. 19-5); and with these values determined, the labor of computing coordinates is not great.

18-19. Plotting Details. The processes of plotting details on the map are similar to those employed in making the field measurements, but are in the reverse order. The work of plotting details is less refined than that of plotting the horizontal control, but the aim is still to plot objects of definite size and shape within the allowable limit of error. The angles are laid off with a protractor, or the engineer's drafting machine (Art. 6-33) may be used.

The protractor shown in Fig. 18-14 may be used rapidly. The linear scale is so designed that distances are measured outward from the center along the diametrical edge. The angle numbers are so arranged that azimuths, bearings, or deflection angles may be laid off by rotating the protractor about the station as a center until the graduation on the protractor representing the given angle coincides with the reference line. The point is then plotted at the given distance on the diametrical scale. Thus in the figure, the point *a* is plotted at an azimuth of 50° and *b* at 230°.

If the directions to details are given in azimuths or bearings measured from a given station, a reference meridian is drawn through the station, and angles are laid off from this meridian.

If an object has been located in the field by the method of radiation, that is, by distance from a control station and angle from a reference line, on the map a line is drawn from the plotted location of the instrument station in a direction corresponding to that on the ground; the distance from station to object is then scaled off along this line.

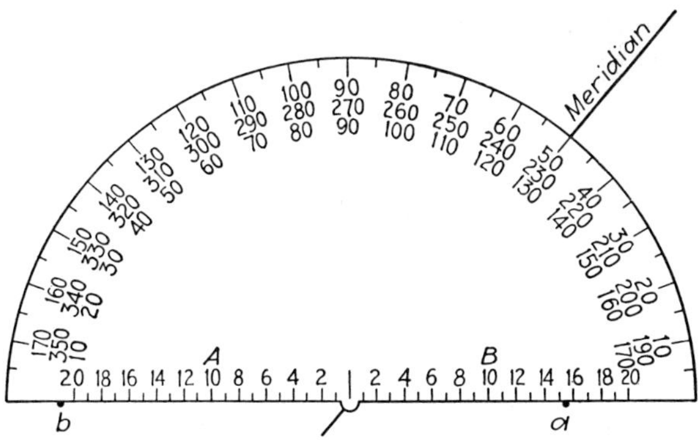

FIG. 18-14. Protractor for plotting details.

If an object has been located in the field by the method of intersection, that is, by angles from two instrument stations, on the map the corresponding angles are laid off from the two stations and the point is plotted at the intersection of the two lines thus defined. Where many points are located by angles from a pair of stations, time will be saved by using simultaneously two protractors of the type shown in Fig. 18-14; the point of intersection between the diametrical edges of the two protractors defines the plotted location of the object. (See also Art. 30-20.)

If an object has been located in the field by linear measurements from two stations, it is plotted at the intersection of arcs of the given radii drawn with the drafting compass.

If objects of somewhat indefinite form have been located in the field by perpendicular offsets from a traverse line, on the map the perpendiculars may usually be estimated with the eye.

The location of all important details should be checked by actual map measurements, but the location of less important details may be checked simply by inspection of the map. If it happens that mistakes in the field work give an object more than one location or that a part of the field notes is confusing, it is advisable to proceed with the plotting of adjacent portions, for these when mapped may help in clearing up the doubtful points.

OMITTED MEASUREMENTS

18-20. General. When for any reason it is impossible or impractical to determine by field observations the length and bearing of every side of a closed traverse, the missing data may generally be calculated, provided not more than two quantities (lengths and/or bearings) are omitted. (If only one measurement is omitted, a partial check is obtained on the work.) It must be assumed that the observed values are without error, and hence all errors of measurement are thrown into the computed lengths or bearings. Measurements which may be supplied in this manner are:

1. Length and bearing of one side (Art. 18-21).
2. Length of one side and bearing of another (Art. 18-22).
3. Lengths of two sides for which the bearings have been observed (Art. 18-23).
4. Bearings of two sides for which the lengths have been observed (Art. 18-24).

There are three general cases: (1) length and bearing of one side unknown, (2) omitted measurements in *adjoining* sides of the traverse, and (3) omitted measurements in *nonadjoining* sides. In case 3, the solution involves changing the order of sides in the traverse in such a way as to make the two partly unknown sides adjoin.

When one of the sides is known in direction but unknown in length, the solution can be facilitated by assuming that side to lie on the reference meridian.

Methods of parting land, which involve the calculation of lengths and bearings of unknown sides of a traverse, are described in Arts. 19-15 to 19-19.

18-21. Length and Bearing of One Side Unknown. The problem of determining the length and bearing of one side of a closed traverse is exactly the same as that of computing the length and bearing of the side of error in any closed traverse for which field measurements are complete. The latitudes and departures of the known sides are computed. For reasons explained in Art. 18-10, if the algebraic sum of the latitudes and the algebraic sum of the departures of the known sides are designated by ΣL and ΣD, respectively, then the length S of the unknown side is

$$S = \sqrt{(\Sigma L)^2 + (\Sigma D)^2} \tag{9}$$

and the tangent of the bearing angle α is

$$\tan \alpha = \frac{-\Sigma D}{-\Sigma L} \tag{10}$$

with due regard to sign.

18-22. Length of One Side and Bearing of Another Side Unknown. Figure 18-15 represents a closed traverse for which the direction of the line $DE = d$ and the length of the line $EA = e$ are not determined by field measurements. Let an imaginary line extend from D to A, cutting off the unknown sides from the remainder of the traverse. Then $ABCDA$ forms a closed traverse for which the side $DA = f$ is unknown in both direction and length. By the method of the preceding article,

$$\text{tan brg. angle of } f = \frac{\text{dep. } f}{\text{lat. } f} = \frac{-(\text{dep. } a + \text{dep. } b + \text{dep. } c)}{-(\text{lat. } a + \text{lat. } b + \text{lat. } c)} \tag{11}$$

and

$$\text{Length of } f = \frac{\text{lat. } f}{\text{cos brg. angle of } f} = \frac{\text{dep. } f}{\text{sin brg. angle of } f} \tag{12}$$

In computing the length of f by Eq. (12), it is desirable to use the larger of the two quantities, latitude or departure.

The angle between the lines e and f in triangle ADE is

$$\angle DAE = \text{azimuth of } AE - \text{azimuth of } AD \tag{13}$$

In the triangle ADE the length of the two sides d and f and one angle DAE are known. By the relation that sines of angles are proportional to sides opposite,

$$\sin DEA = \sin DAE \frac{f}{d} \tag{14}$$

With angle DEA known, angle ADE can be computed, and the remaining unknown length is given by the expression

$$e = f \frac{\sin ADE}{\sin DEA} = d \frac{\sin ADE}{\sin DAE} \tag{15}$$

Also,

$$\text{Azimuth of } DE = \text{azimuth of } DA - \angle ADE \tag{16}$$

Unknown Courses Not Adjoining. The preceding method of solution is generally applicable even though two partly unknown courses are not adjoining. Obviously the latitude and the departure of any line of fixed direction and length are the same for one location of the line as for any other. In other words, a line may be moved from one location to a second location parallel with the first, and its latitude and departure will remain unchanged. Since this is the case, then it must also be true that the algebraic sum of the latitudes and the algebraic sum of the departures of any system of lines forming a closed figure must be zero, regardless of the order in which the lines are placed. Thus the courses which are shown in the order *a*, *b*, *c*, *d*, *e*, in Fig. 18-15 are given in the order *a*, *e*, *b*, *c*, *d* in Fig. 18-16. If now it is assumed that the direction of *d* and the length of *a* (Fig. 18-16) are unknown, the problem of determining these unknown quantities is seen to be identical with that explained in the preceding article for the case where the partly unknown sides were adjoining.

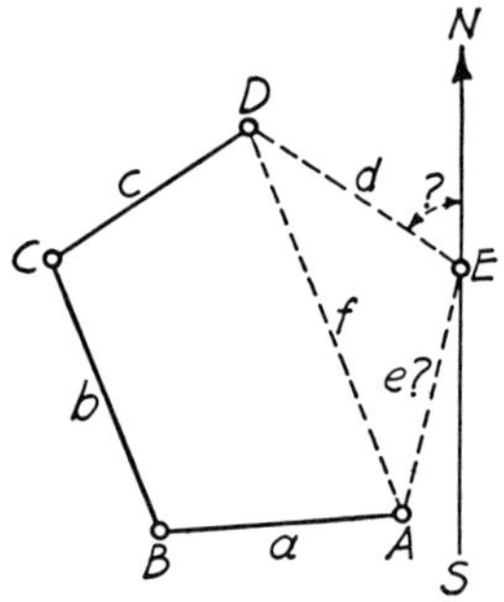

FIG. 18-15.

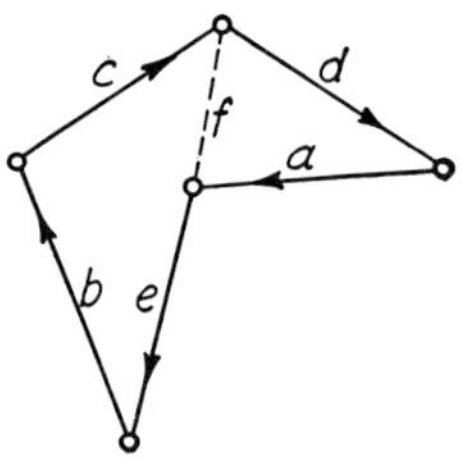

FIG. 18-16.

Hence, when two partly unknown sides of a closed traverse are not adjoining, one of the sides is considered as moved from its location to a second location parallel with the first, such that the two partly unknown sides adjoin. The solution then becomes identical with that just described. To simplify the problem, the data are usually plotted roughly to small scale. Following is the solution of a typical problem.

Example: In the accompanying tabulation are given the measured lengths and bearings for the courses of a closed traverse *a* to *f*, together with the latitudes and departures of the known sides. The length of *b* and the bearing of *e* are not

Line	Length, ft.	Bearing	Latitude		Departure	
			N	S	E	W
a	500.0	N	500.0	....	0.0	....
b	unknown	N45°00′E				
	(889.9)	....	(629.3)	....	(629.3)	....
c	854.4	S69°27′E	....	300.0	800.0	....
d	1,019.8	S11°19′E	....	1,000.0	200.0	....
e	1,118.0	unknown				
		(S78°57′W)	....	(214.3)	....	(1,097.3)
f	656.8	N54°06′W	385.0	....	....	532.0
g	(625.5)	(N48°26′W)	(415.0)	....	....	(468.0)

observed. The general direction of e is southwest. It is desired to compute the unknown length and direction. Quantities in parentheses are derived from following calculations.

In Fig. 18-17 the lines a, c, d, and f are the courses for which the length and bearing are known. The line g is the closing side of the figure formed by the known courses. From the tabulated quantities, the sum of the known latitudes is

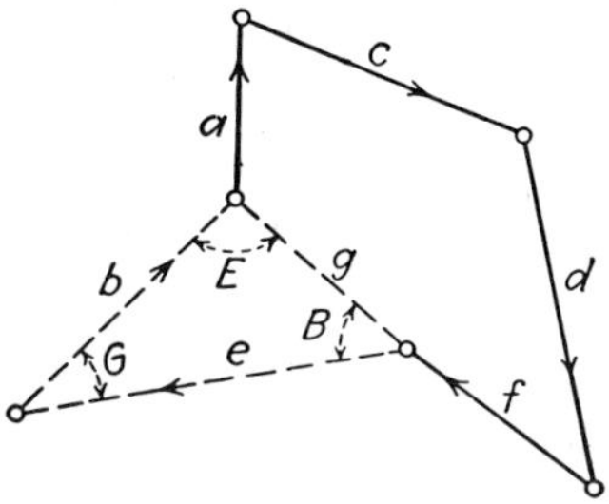

FIG. 18-17.

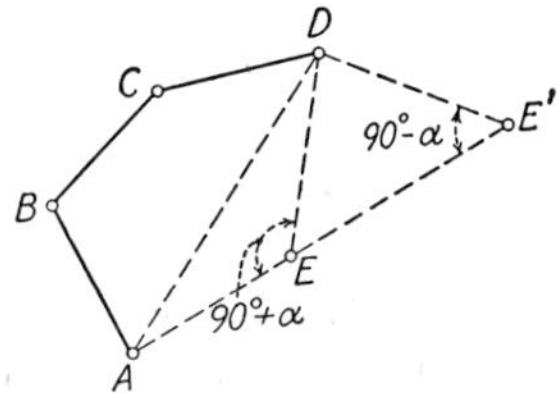

FIG. 18-18.

1,300S + 885N = 415S, hence the latitude of g is 415N. Similarly the departure of g is 1,000 − 532 = 468W.

$$\tan \text{bearing } g = \frac{468.0}{415.0} = 1.127$$

and the bearing of g = N48°26′W.

$$\text{Length of } g = \frac{\text{dep. } g}{\sin \text{brg. } g} = \frac{468.0}{0.7482} = 625.5 \text{ ft.}$$

Since the bearing of b is N45°00′E,

$$\angle E = 45°00' + 48°26' = 93°26'$$

$$\sin G = \frac{g}{e} \sin E = \frac{625.5}{1{,}118} \times 0.9982 = 0.5585$$

$$\angle G = 33°57'$$

$$\angle B = 180°00' - 93°26' - 33°57' = 52°37'$$

$$\text{Length } b = \text{length } e \cdot \frac{\sin B}{\sin E} = 1{,}118 \times \frac{0.7946}{0.9982} = 889.9 \text{ ft.}$$

$$\text{Bearing of } e = 180° - 48°26' - 52°37' = \text{S}78°57'\text{W}$$

As a check on the calculations, the latitudes and departures of the lines b and e are computed and the values are shown in parentheses. The sum of the latitudes and the sum of the departures for courses a, b, c, d, e, and f are found to be zero, and hence the computations for determining the unknown length and bearing are correct.

Choice of Values. When the length of one side and the bearing of another are unknown, the solution will generally render two values of each of the unknowns; often it is impossible to tell which are the correct values unless the general direction of the side of unknown bearing is observed. In Fig. 18-18, $ABCD$ is the known portion of a traverse and DA is the closing side for this portion. A line of unknown length but known direction extends from A in the direction of E and E'. A line of unknown direction but known length extends from D to intersect the line of known direction at E or E', DE being equal to DE'. Evidently $\sin \angle DEA = \sin \angle DE'A$, since $\sin (90° + \alpha) = \sin (90° - \alpha)$, and further, solving the triangle for the length of the unknown side gives the value AE or AE' depending on which of the preceding angular values is used.

As the angle between the line of unknown length and the line of unknown bearing approaches 90°, the solution becomes weak. This fact is shown in Fig. 18-19, for which DA is the closing side of the traverse formed by the known lines, DEE' is the direction of the side of unknown length, and $AE = AE'$ is the length of the side of unknown direction. It will be recalled (see Art. 3-8) that the sines of angles near 90° change slowly. Since the angle at E or E' is nearly 90°, and its value is computed through the use of its sine, a small error in computation may make a relatively large change

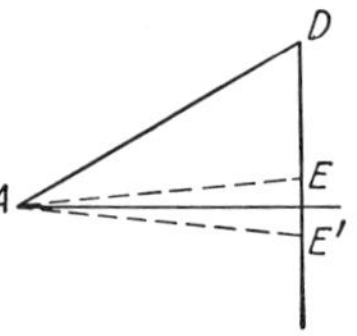

FIG. 18-19.

in the angle (a difference of 0.00001 in the sine effects a change of 0°11′, on either side of 90°). Also the angle at E is used in computing the angle at A, and hence any error in the computed value of $\angle E$ is transmitted to the computed value of $\angle A$. Now if $\angle A$ is of average size, the change in the sine will be much more rapid (on either side of 45°, a change of 0.00001 in the sine produces a change of only $\frac{1}{20}'$ in the angle). Since the side DE is computed by using $\sin \angle A$, the relative error in the computed value of DE is the same as that in $\sin \angle A$ which may be many times that in $\sin \angle E$.

Also when the angle between the partly unknown lines is nearly 90°, it is impossible to distinguish which of the two determinations is the correct one, even though the general direction of the line of unknown bearing has been observed.

Thus the solution becomes weak when the partly unknown lines approach perpendicularity, and the solution becomes strong as the angle between these lines becomes small.

18-23. Length of Two Sides Unknown. This problem commonly occurs where angular observations are taken from two or more points in the main traverse to some landmark, the measurements being introduced as a check. It occasionally occurs on main traverse lines where there are obstacles to the direct measurement of length but where angles are observed. The solution is nearly identical with that for the case where the direction of one side and the length of another are unknown (Art. 18-22).

In Fig. 18-20, $ABCD$ represents the portion of a closed traverse for which the courses are known in both direction and length, and the lines DE and EA represent courses for which the direction is known but the length is unknown. From the latitudes and departures of the known sides, the length and bearing of the closing line DA are computed; and in the triangle ADE the angles A, D, and E are computed from the known directions of the sides. The lengths DE and EA are determined through the relation

$$\frac{DE}{\sin A} = \frac{EA}{\sin D} = \frac{DA}{\sin E} \tag{17}$$

If the two lines are not adjoining, the problem may be solved as though they were, as explained in Art. 18-22. As the angle between the partly unknown lines approaches 90°, the solution becomes strong, and as the angle

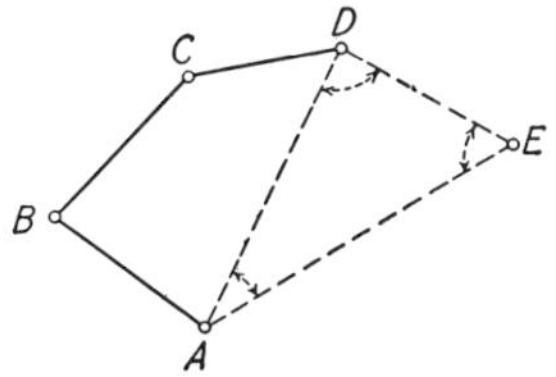

FIG. 18-20.

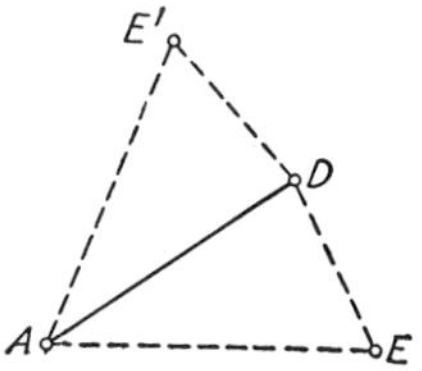

FIG. 18-21.

approaches 0° or 180°, the solution becomes weak, the problem being indeterminate when the lines are parallel.

18-24. Direction of Two Sides Unknown. The solution is similar to that described in the preceding article. In Fig. 18-20, if DA is the closing side of the known portion of the traverse, its direction and length are computed; then the lengths of the three sides of the triangle ADE are known, and the angles A, D, and E can be computed.

The general direction of at least one of the partly unknown lines must be observed, as the values of the trigonometric functions merely determine the shape of the triangle but do not fix its position. Thus in Fig. 18-21, if DA is the closing line of the known portion of the traverse forming the base of the triangle of which the courses of unknown direction but of known length are the legs, then it is evident that the vertex may fall at either E or E'.

When the partly unknown lines are parallel but not of the same length, their direction is that of the closing side of the figure formed by the known courses. When the partly unknown sides are parallel and of the same length, the problem is indeterminate, since the length of the closing side of the figure formed by the known courses becomes a point.

18-25. Numerical Problems.

1. With the protractor, plot to the scale of 1 in. = 200 ft. the closed deflection-angle traverse for which the following notes are given. Measure the linear error of closure and record it on the sheet. Distribute the error as suggested in Art. 18-7.

Station	Deflection angle	Length, ft.
A	....	
		338
B	9°00′L	
		307
C	76°45′R	
		792
D	74°45′R	
		822
E	102°00′R	
		624
F	23°30′R	
		620
A	92°00′R	

2. Assuming the course *AB* in the traverse of problem 1 to be of zero azimuth, compute the azimuths of the remaining courses. Using a protractor, lay off the azimuths from a central meridian and plot the traverse to a scale of 1 in. = 200 ft., as described in Art. 18-4. Measure the error of closure and record it on the sheet. Distribute the error of closure as suggested in Art. 18-7.

3. Plot the following open deflection-angle traverse to a scale of 1 in. = 400 ft. by the method of tangents, using a 10-in. base. Lay off successive lines by deflection angles as described in Art. 18-5. Assume the direction of the first course to be north, and compute the bearings of the other courses. Check the accuracy of the plotting by methods described for open traverses in Art. 18-7.

Station	Deflection angle
118 + 75.0	....
98 + 95.6	39°47′L
73 + 01.4	17°28′L
70 + 13.5	14°08′L
49 + 41.3	3°11′L
40 + 00	49°59′L
37 + 18.8	32°18′R
18 + 26.0	18°44′R
5 + 03.2	7°31′L
0 + 00	....

4. Plot the following open azimuth traverse to a scale of 1 in. = 400 ft. by the method of tangents, establishing the directions of the several lines from a centrally located meridian, as described in Art. 18-5. From the azimuths compute the deflection angles at the several points, and check the accuracy of the plotting as described in Art. 18-7.

Course	Azimuth	Distance, ft.
AB	142°08′	815.3
BC	181°37′	1146.0
CD	296°13′	520.8
DE	323°46′	816.5
EF	249°51′	726.4
FG	214°03′	1862.0
GH	195°45′	2795.5
HJ	191°28′	2463.7
JK	138°42′	586.4

5. Given the notes of problem 1. Assume the original direction of *AB* to be north and compute the bearings of the several courses. Compute the latitudes and departures to tenths of feet, assuming that the given lengths are correct to tenths of feet, and using four-place logarithms. Compute the error of closure, and balance the survey by the Compass Rule (Art. 18-13).

6. Solve the preceding problem approximately, using the slide rule.

7. Given the following notes of a transit survey. Compute the latitudes and departures, using five-place logarithms. Compute the error of closure, and balance the survey by the Crandall Method (Art. 18-14).

Course	Bearing	Distance, ft.
AB	N48°20′E	529.6
BC	N87°43′E	592.0
CD	S 7°59′E	563.6
DE	S82°12′W	753.4
EA	N48°12′W	428.2

8. Given the following notes for a closed traverse. (1) Compute the latitudes and departures, using five-place logarithms, and compute the error of closure. Balance the survey by each of the rules given in this text, and for each course find the change in length and direction caused by the application of each of the rules. (2) Add 40°00′ to each of the given azimuths, and make computations called for in (1). (3) Compare the results of (2) with those of (1), and explain reasons for variations.

Course	Azimuth	Distance, ft.
AB	0°42′	1,221.2
BC	94°03′	541.3
CD	183°04′	795.4
DA	232°51′	646.8

9. Given the data of problem 4. Calculate the coordinates of the several points in the survey, assuming that the origin is at *A*. Plot the traverse by the coordinate method, using a scale of 1 in. = 400 ft. (See also office problem 1.)

10. Given the following data for a closed traverse. Compute the length and bearing of the unknown side, using the slide rule.

Course	Bearing	Distance, ft.
AB	N82°W	461
BC	unknown	unknown
CD	N68°15′E	829
DA	N80°45′E	441

11. Given the following data for a closed traverse, for which the length *DE* and the azimuth of *EA* have not been observed in the field. Determine the unknown quantities, using five-place logarithms. The general direction of *EA* is easterly.

Course	Azimuth (from north)	Distance, ft.
AB	106°13′	1,081.3
BC	195°14′	1,589.5
CD	247°07′	1,293.7
DE	332°22′	unknown
EA	unknown	1,737.9

12. Solve the preceding problem by use of the slide rule, with corresponding precision (to 1 ft. and to 05′).

13. Given the following data for a closed traverse, for which the lengths of *BC* and *DE* have not been measured in the field. Compute the unknown lengths, using five-place logarithms.

Course	Bearing	Distance, ft.
AB	N9°30′W	689.32
BC	N56°55′W	unknown
CD	S56°13′W	678.68
DE	S2°02′E	unknown
EA	S89°31′E	1,082.71

14. Solve the preceding problem by use of the slide rule, with corresponding precision (to 1 ft. and to 05′).

18-26. Office Problems.

PROBLEM 1. PLOTTING BY TANGENTS; MAP CONSTRUCTION

Object. To plot a given traverse by the method of tangents. The data of field problems of Chaps. 12 and 14 may be used.

Procedure. (1) Tabulate angles and distances of the given traverse in the computation book. If angles are given as azimuths or bearings, change them to deflection angles. (2) Tabulate and check the natural tangent of each deflection angle less than 45° and the cotangent of each angle greater than 45°. (3) Plot the traverse roughly to small scale, using the protractor for angles; note its general form. (4) Carefully plot the first line of the traverse on drawing paper to the required scale, estimating the position of the line by means of the small-scale sketch. The line should lie so that the drawing, when finished, will be symmetrical with the sheet. (5) Plot the remaining lines of the traverse by the method described in Art. 18-5. Verify all measurements as soon as they have been plotted. (6) Check the traverse by the methods of Art. 18-7. If it is a closed traverse, distribute the error of closure as indicated by the conditions of the problem and perhaps by the direction of the side of closure. (7) Plot the details by methods corresponding to those used in the field. Use symbols wherever applicable. Show the meridian. Make a title.

Hints and Precautions. (1) Particular care should be taken in scaling the lengths of the lines of the traverse and the tangent distances. Points should be pricked with a needle, a reading glass should be used, and the eye should be above each point as it is plotted. (2) Lengths of lines and tangent distances plotted by estimation using the 10 scale are not sufficiently accurate; the 50 scale should be used. (3) A perpendicular should not be erected with the corner of the triangle at the point of intersection of base line and perpendicular, as the triangle corner may be rounded. The hypotenuse of the right triangle should be fitted to the base line, and a straightedge (or another triangle) placed in contact with the base of the triangle. Then the triangle should be turned through 90°, and its third edge placed in contact with the straightedge and moved along it until the hypotenuse passes through the point of intersection. (4) It is well to test each perpendicular by measuring the hypotenuse of a 45° right triangle having sides perhaps 8 in. long; in this case the length of the hypotenuse is 11.31 in. (5) If the construction lines would otherwise fall off the edge of the paper, measure back along the line as shown for the plotting of *DE* from *CD* in Fig. 18-3.

Problem 2. Plotting by Coordinates; Map Construction

Object. To plot a given traverse by the method of coordinates. The data of field problems of Chaps. 12 and 14 may be used.

Procedure. (1) Transcribe the given data in a computation book in the form shown by the first three columns of the tabulation in Art. 18-13. If directions of lines are given as azimuths or deflection angles, compute either the true or the assumed bearings. (2) Compute the latitudes and departures of the traverse lines; and if it is a closed traverse, balance the survey by the Compass Rule as described in Art. 18-13. (3) Assume one of the traverse points as the origin of coordinates, and compute the total latitudes and departures as described in Art. 18-16. (4) Check all computations. (5) To small scale, plot roughly the traverse and enclosing rectangle (Art. 18-17) and note their relative positions. (6) On drawing paper plot the enclosing rectangle to the required scale, estimating its position by means of the small-scale sketch. The traverse, not the rectangle, should be symmetrical with the sheet. (7) Test the accuracy of the plotting by scaling the length of the diagonals. (8) Plot the reference meridian, reference parallel, and any supplementary meridians and parallels as described in Art. 18-17. Check the plotting. (9) Locate each traverse point by plotting its total latitude and total departure. Check by scaling the length of the preceding traverse line. (10) Plot the details by methods corresponding to those used in the field. Use conventional signs wherever applicable. Finish the map without erasing construction lines. Label each traverse line with its corrected length *inside* the traverse and its corrected bearing *outside*. Show the meridian. Make a title.

Hints and Precautions. For details of plotting, see "Hints and Precautions" 1 and 2 of the preceding office problem.

REFERENCES

1. CRANDALL, C. L., "The Adjustment of a Transit Survey as Compared with That of a Compass Survey," *Trans. Am. Soc. Civil Engr.*, Vol. 45, pp. 453–464, 1901.
2. RAPPLEYE, H. S., "Adjustment of Transit and Stadia Traverses," *Trans. Am. Soc. Civil Engr.*, Vol. 95, p. 232, 1931.
3. U.S. BUREAU OF LAND MANAGEMENT, "Standard Field Tables," Government Printing Office, Washington, D.C.

CHAPTER 19

CALCULATION OF AREAS OF LAND

19-1. General. One of the primary objects of most land surveys is to determine the area of the tract. A closed traverse is run, the lines of the traverse being made to coincide with property lines where possible. Where the boundaries are irregular or curved or where they are occupied by objects which make direct measurement impossible, they are located with respect to the traverse line by appropriate angular and linear measurements. The lengths and bearings of all straight boundary lines are determined either directly or by computation, the irregular boundaries are located with respect to traverse lines by perpendicular offsets taken at appropriate intervals, and the radii and central angles of circular boundaries are obtained. The following articles explain the several common methods by means of which these data are employed in calculating areas.

In ordinary land surveying, as discussed herein, the area of a tract of land is taken as its projection upon a horizontal plane, and it is not the actual area of the surface of the land. For precise determinations of the area of a large tract, such as state or nation, the area is taken as the projection of the tract upon the earth's spheroidal surface at mean sea level.

19-2. Methods of Determining Area. The area of a tract may be determined by any of the following methods:

1. By plotting the boundaries to scale as described in Chap. 18; the area of the tract may then be found by use of the planimeter as described in Arts. 4-13 to 4-17, or it may be calculated by dividing the tract into triangles and rectangles, scaling the dimensions of these figures, and computing their areas mathematically. This method is useful in roughly determining areas or in checking those that have been calculated by more exact methods. Its advantage lies in the rapidity with which calculations can be made.

2. By mathematically computing the areas of individual triangles into which the tract may be divided (Art. 19-3). This method is employed when it is not expedient to compute the latitudes and departures of the sides.

3. By calculating the area from the coordinates, or meridian distances and parallel distances, of the *corners* of the tract (Art. 19-4).

4. By calculating the area from the double meridian distances and the latitudes of the *sides* of the tract (Arts. 19-5 to 19-7), or similarly from the double parallel distances and the departures of the sides (Art. 19-8).

5. For tracts having irregular or curved boundaries, the methods of Arts. 19-9 to 19-14 are employed.

For computation of the areas of cross-sections, see Arts. 11-6 to 11-8.

19-3. Area by Triangles. Table VI gives the relations between the area of a triangle and its angles and lengths of sides. When the lengths of two sides and the included angle of any triangle are known, its area is given by the expression

$$\text{Area} = \tfrac{1}{2}ab \sin C \tag{1}$$

When the lengths of the three sides of any triangle are given, its area is determined by the equation

$$\text{Area} = \sqrt{s(s-a)(s-b)(s-c)} \tag{2}$$

in which $s = \frac{1}{2}(a + b + c)$.

In surveying small lots as for a city subdivision, it is common practice to omit the determination of the error of closure of each lot, and hence the computation of latitudes and departures is unnecessary. Under such circumstances the area may be calculated by dividing the lot, usually quadrangular in shape, into triangles, as illustrated by Fig. 19-1, for each of which two

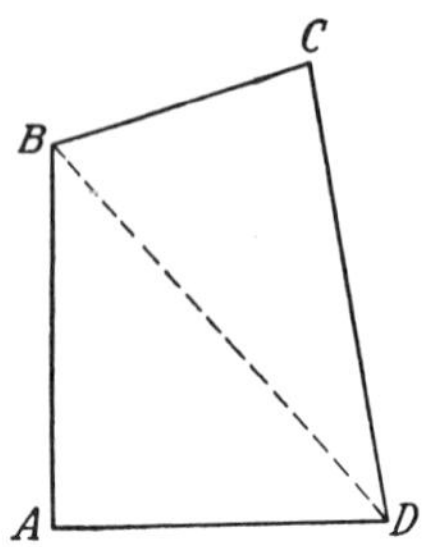

FIG. 19-1.

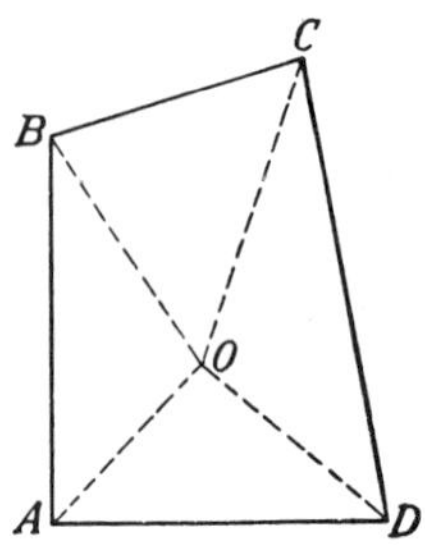

FIG. 19-2.

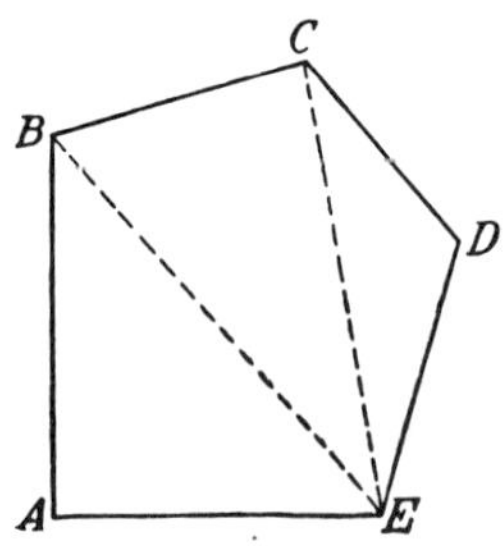

FIG. 19-3.

sides and the included angle have been measured. By Eq. (1), the areas of *ABD* and *BCD* are computed; the sum of these two areas is the area of the lot. The area thus found can be checked independently by computing the areas of the two triangles *ABC* and *CDA* formed by dividing the quadrilateral by a line from *A* to *C*.

The accuracy of the field work may be investigated by computing the lengths of the diagonals. Thus *BD* can be determined by solving either triangle *ABD* or triangle *BCD*. The field measurements are without error if the length of *BD* computed by solving one of the triangles is the same as that computed by solving the other.

Figure 19-2 illustrates a survey made by a single set-up of the transit at *O*, such as might be the case for a small lot where the property lines *ABCD* are obstructed or where the transit can not be set up at the corners. Under these circumstances the angles about *O* and the distances *OA*, *OB*, *OC*, and *OD* are measured in the field. Since in each triangle two sides and the included angle are known, the area can be determined by use of Eq. (1). If in addition to the above measurements, the lengths of the sides *AB*, *BC*, etc., are measured, the area of the lot can be checked independently by solving each triangle by Eq. (2).

In Fig. 19-3 the figure *ABCDE* represents the boundary of a tract surveyed by simple triangulation, each of the angles of the three triangles into which the figure is divided being measured, but only the distance *AB* being measured in the field. In order to determine the length of the unknown boundaries, it is necessary to solve in succession the triangles *ABE*, *BEC*, and *ECD*, the lengths of all sides being determined by the relation that the length of the side of a triangle is proportional to the sine of the opposite angle. When the lengths of the sides have thus been found, the areas of the individual triangles may be computed by either Eq. (1) or Eq. (2).

19-4. Area by Coordinates. Often the rectangular coordinates of the points defining the corners of a tract of land with reference to some arbitrarily chosen coordinate axes are computed before the directions and lengths of the connecting lines are known. For example, the corners may be located by a system of triangulation, neither the direction nor the length of any of the boundaries being measured directly. Or a traverse may be inside a given tract, and the corners of the property may be located by direction and distance from traverse points. In either case, the coordinates of the corners are useful not only in finding the lengths and bearings of the boundary lines but also in calculating the area of the tract. Essentially the calculation is that of finding the areas of trapezoids formed by projecting the lines upon one of a pair of coordinate axes, usually a true meridian and a parallel at right angles thereto.

In Fig. 19-4, *ABCDF* represents a tract the area of which is to be determined, *SN* a reference meridian, and *WE* a reference parallel. The coordinates of *A*, *B*, . . . , *F* are known; for any point the abscissa is the perpendicular distance from the reference meridian, defined as the *total departure* or the *meridian distance;* and the ordinate is the perpendicular distance from the reference parallel, defined as the *total latitude* or the *parallel distance.* Thus, for *A* the meridian distance is $aA = m_1$ and the parallel distance is $a'A = p_1$. Meridian distances are regarded as positive or negative according to whether they lie east or west of the reference meridian; parallel distances are regarded as positive or negative according to whether they lie north or south of the reference parallel. In the figure all meridian and parallel distances are positive.

The area of the tract can be computed by summing algebraically the areas of the trapezoids formed by projecting the lines upon the reference meridian; thus

$$\text{Area } ABCDF = \text{area } BCcb + \text{area } CDdc - \text{area } DFfd - \text{area } FAaf - \text{area } ABba \quad (3)$$

or

$$\text{Area} = \tfrac{1}{2}(m_2 + m_3)(p_2 - p_3) + \tfrac{1}{2}(m_3 + m_4)(p_3 - p_4) - \tfrac{1}{2}(m_4 + m_5)(p_5 - p_4) - \tfrac{1}{2}(m_5 + m_1)(p_1 - p_5) - \tfrac{1}{2}(m_1 + m_2)(p_2 - p_1) \quad (4)$$

By multiplication and a rearrangement of terms in Eq. (4), there is obtained

$$2 \cdot \text{area} = -[p_1(m_5 - m_2) + p_2(m_1 - m_3) + p_3(m_2 - m_4) + p_4(m_3 - m_5) + p_5(m_4 - m_1)] \quad (5a)$$

Rule. *To determine the area of a tract of land when the coordinates of its corners are known, multiply the parallel distance, or ordinate, of each corner by*

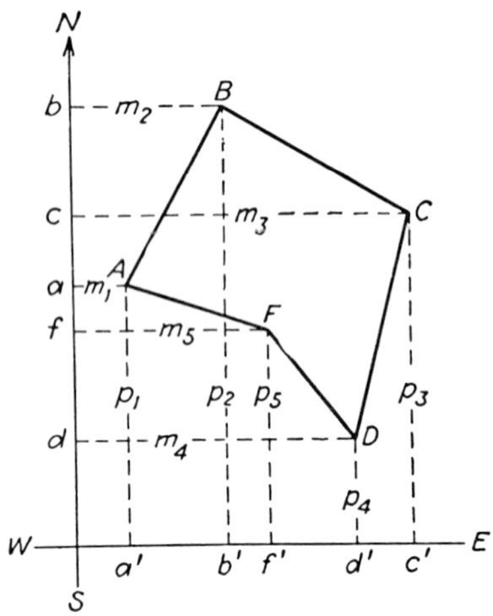

FIG. 19-4. Area by coordinates.

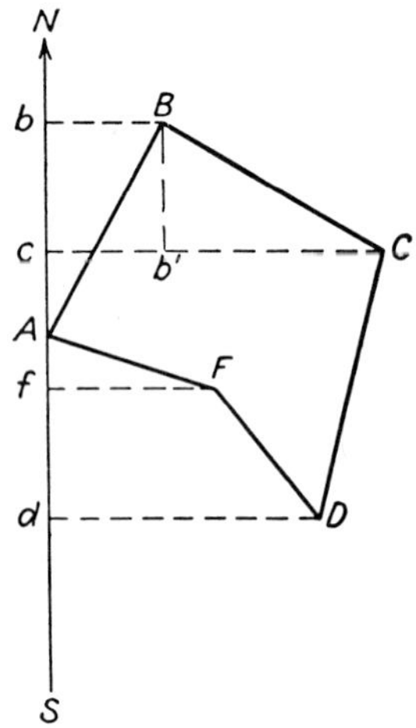

FIG. 19-5. Area by double meridian distances.

the difference between the meridian distances, or abscissas, of the following and the preceding corners, always algebraically subtracting the following from the preceding. One half of the algebraic sum of the resulting products is the required area.

A result identical except for sign would be obtained by always subtracting the preceding from the following. The sign of the area is not significant.

Example: Given the following data. Find the required area by applying the foregoing rule:

Corner	1	2	3	4	5
Meridian distance, ft.	300	400	600	1,000	1,200
Parallel distance, ft.	300	800	1,200	1,000	400

$$\begin{aligned} 2 \cdot \text{area} &= -[300(800) + 800(-300) + 1{,}200(-600) + 1{,}000(-600) + 400(700)] \\ &= -240{,}000 + 240{,}000 + 720{,}000 + 600{,}000 - 280{,}000 \\ &= 1{,}040{,}000 \text{ sq. ft.} \end{aligned}$$

$$\text{Area} = \frac{1{,}040{,}000}{2} = 520{,}000 \text{ sq. ft.}$$

Equation (4) can also be expressed in the form

$$\begin{aligned} 2 \cdot \text{area} = m_2p_1 + m_3p_2 + m_4p_3 + m_5p_4 + m_1p_5 \\ - m_1p_2 - m_2p_3 - m_3p_4 - m_4p_5 - m_5p_1 \end{aligned} \qquad (5b)$$

When this form is employed, computations can be made conveniently by tabulating each parallel distance below the corresponding meridian distance as follows:

$$\frac{m_1}{p_1} \times \frac{m_2}{p_2} \times \frac{m_3}{p_3} \times \frac{m_4}{p_4} \times \frac{m_5}{p_5} \times \frac{m_1}{p_1} \qquad (5c)$$

Then in Eq. (5*c*) the difference between the sum of the products of the coordinates joined by full lines and the sum of the products of the coordinates joined by dotted lines is equal to twice the area of the tract.

The foregoing example may be quickly checked by this method.

19-5. Principles of Double-meridian-distance Method. The D.M.D. method of calculating area is a convenient form of the method of coordinates just described, but the computations do not involve the direct use of coordinates. The latitudes and departures of all the courses are determined as described in Art. 18-9, and the survey is balanced. A reference meridian is then assumed to pass through some corner of the tract, usually for convenience the most westerly point of the survey; the double meridian distances of the lines are computed as described herein; and double the areas of the trapezoids or triangles formed by orthographically projecting the several traverse lines upon the meridian are computed. The algebraic sum of these double areas is double the area within the traverse.

The meridian distance of a point has been defined; thus in Fig. 19-5 the meridian distance of B is Bb and is positive. The meridian distance of a straight line is the meridian distance of its mid-point. The *double meridian distance* of a straight line is the sum of the meridian distances of the two extremities; thus the double meridian distance of BC is $Bb + Cc$. It is clear that if the meridian passes through the most westerly corner of the traverse, the double meridian distance of all lines will be positive, which is a convenience (although not a necessity) in computing.

As explained in Art. 18-9, the length of the orthographic projection of a line upon the meridian is the latitude of the line; thus in Fig. 19-5 the latitude of BC is bc and is negative, and that of DF is df and is positive.

From the figure it is seen that each projection trapezoid or triangle, for which a course in the traverse is one side, is bounded on the north and south by meridian distances and on the west by the latitude of that course. Thus the projection trapezoid for BC is $BCcb$. Therefore, the double area of any triangle or trapezoid formed by projecting a given course upon the meridian is the product of the double meridian distance of the course and the latitude of the course, or

$$\text{Double area} = \text{D.M.D.} \times \text{latitude} \tag{6}$$

In computing double areas, account is taken of signs. If the meridian extends through the most westerly point, all double meridian distances are positive; hence the sign of a double area is the same as that of the corresponding latitude. Thus in the figure the double areas of AbB, $DdfF$, and FfA are positive, the latitudes Ab, df, and fA being positive; while the double areas of $CcbB$ and $DdcC$ are negative, the latitudes bc and cd being negative. Since the projected areas *outside* the traverse are considered once as positive and once as negative, the algebraic sum of their double areas is zero. Therefore, the algebraic sum of all double areas is equal to twice the area of the tract *within* the traverse.

Whether this algebraic sum of the double areas is a positive or negative quantity is determined solely by the order in which the lines of the traverse are considered. If the reference meridian passes through the most westerly corner, then a clockwise order of lines, as in the figure, results in a negative double area, and a counter-clockwise order results in a positive double area. The sign of the area is not significant.

19-6. Computation of D.M.D. When the reference meridian passes through a traverse point, there is an intimate relation between the departures and double meridian distances. Thus, again referring to Fig. 19-5, it is seen that the D.M.D. of AB is bB, which is equal to the departure of the course in both magnitude and sign. And the D.M.D. of BC is equal to $bB + cb' + b'C$, which is equal to the D.M.D. of AB, plus the departure of AB, plus the departure of BC. Similar quantities make up the D.M.D.'s of CD and DF. For the last line FA the D.M.D. is fF, which is equal in magnitude but opposite in sign to the departure of FA.

Following are three convenient rules for determining D.M.D.'s, which are deduced from the relations just illustrated:

1. The D.M.D. of the first course (reckoned from the point through which the reference meridian passes) is equal to the departure of that course.

2. The D.M.D. of any other course is equal to the D.M.D. of the preceding course, plus the departure of the preceding course, plus the departure of the course itself.

3. The D.M.D. of the last course is numerically equal to the departure of the course but with opposite sign.

The first two rules are employed in computing values. The third rule is useful as a check on the correctness of the computations. Assuming the departures as balanced, the D.M.D. of the last line, as found by computing the D.M.D.'s in succession around the traverse from the first line, should be numerically equal to the departure of the last line if no mistake in addition or subtraction has been made.

19-7. Area within Closed Traverse by D.M.D. Method. Following is a summary of the steps employed in calculating by the D.M.D. method the area within a closed traverse when the lengths and bearings of the sides are known:

1. Compute the latitudes and departures of all courses as described in Art. 18-9.

2. Find the error of closure in latitude and in departure as described in Art. 18-10.

3. Balance the latitudes and departures, usually in accordance with one of the rules of Art. 18-13.

4. Assume that the reference meridian passes through the most westerly point in the survey and compute the D.M.D.'s by the rules of the preceding article, using the corrected departures.

5. Compute the double areas by multiplying each D.M.D. by the corresponding corrected latitude.

6. Find the algebraic sum of the double areas, and determine the area by dividing this sum by two.

Example: The foregoing steps are illustrated in the computations shown by Fig. 19-6 in which the area of a tract within a transit traverse is determined. It is seen that distances are in 66-ft. chains, and that computations are made by the use of logarithms. The survey is balanced by the Transit Rule, and the corrected latitudes and departures are recorded.

The point *C* is the most westerly point in the survey, and the double meridian distances are computed by beginning with the line *CD* for which the D.M.D. and corrected departure are of the same magnitude and sign. The D.M.D.'s for lines *DE*, *EA*, *AB*, and *BC* are computed by the second of the rules given in Art. 19-6. The D.M.D. of *BC*, the last line in the traverse, is seen to be numerically equal to the corrected departure of this line but with opposite sign, hence the D.M.D. computations are correct.

Below the logarithmic computations for latitudes and departures are given the logarithmic computations for double areas, each D.M.D. being multiplied by the corresponding corrected latitude.

In the last two columns of the upper tabulation are recorded the positive and negative double areas. These are algebraically added, and the result is divided by two as shown; the result is the area in square chains.

Although computations by logarithms have been shown, a computing machine would ordinarily be used in preference to logarithms.

7

AREA OF BALSAM PARK, ISLAND POND, VERMONT
by D.M.D. Method

Field Notes
Book No. 3
Page 47

Computations
Aug. 17, 1965
Computed and
Checked by J. D. M.

Line	Calc. Bearing	Dist., 66-ft. Ch.	Latitudes N	Latitudes S	Departures E	Departures W	Corrected Lats.	Corrected Deps.	D.M.D.'s	Double Areas +	Double Areas −
AB	S 80°29½′ W	34.464		5.694		33.991	−5.693	−33.990	61.812		351.89
BC	S 33 04 W	25.493		21.364		13.911	−21.361	−13.911	13.911		297.15
CD	S 33 46¾ E	33.934		28.205	18.867		−28.201	+18.867	18.867		532.06
DE	N 87 58¼ E	28.625	1.013		28.607		+1.013	+28.608	66.342	67.21	
EA	N 0 27 E	54.235	54.234		0.426		+54.242	+0.426	95.376	5173.51	
		176.751	55.247	55.263	47.900	47.902	ΣL = 0	ΣD = 0		5240.72	1181.10
				55.247		47.900				1181.10	
				0.016		0.002				2 | 4059.62	
										2029.81 Sq. Ch.	
										or 202.98 Ac.	

$E = \sqrt{0.016^2 + 0.002^2} = \underline{0.016}$ Chains

$E. of C. = \frac{0.016}{176.751} = \frac{1}{\underline{11{,}000}}$

Line	AB	BC	CD	DE	EA
Lat.	5.694	21.364	28.205	1.013	54.234
Log Lat.	0.75542	1.32968	1.45032	0.00584	1.73427
Log cos	9.21805	9.92326	9.91969	8.54899	9.99999
Log Dist.	1.53737	1.40642	1.53063	1.45674	1.73428
Log sin	9.99399	9.73689	9.74509	9.99973	7.89535
Log Dep.	1.53136	1.14331	2.27572	1.45647	9.62964
Dep.	33.991	13.911	18.867	28.607	0.426
Log Cor. Lat.	0.75534	1.32962	1.45026	0.00584	1.73434
Log D.M.D.	1.79107	1.14336	1.27570	1.82179	1.97944
Log D.A.	2.54641	2.47298	2.72596	1.82763	3.71378
Double Area	351.89	297.15	532.06	67.21	5173.51

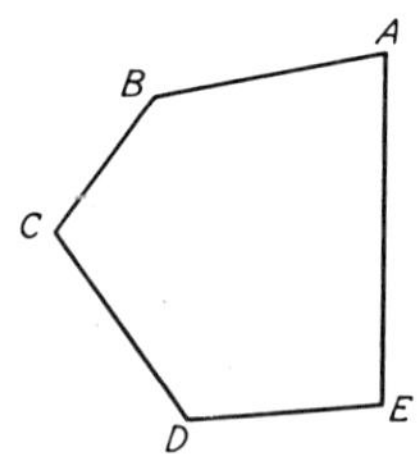

Note:
Survey Balanced by Transit Rule.

FIG. 19-6. Computations for area by D.M.D. method.

The correctness of the computations for latitudes, departures, and double areas cannot be readily checked except by recomputations of similar character. The corrections applied to latitudes and departures are checked if the algebraic sums of the corrected latitudes and departures, respectively, are zero. The application of the third of the rules for D.M.D.'s, as already explained, serves to verify the computations for D.M.D.'s.

19-8. Double Parallel Distances. Determining area within a closed traverse by the method of double parallel distances (D.P.D.'s) is essentially the same as the D.M.D. method described in the preceding articles. The only difference is that the courses are projected upon a parallel instead of upon the meridian.

Although the D.P.D. method possesses all the advantages of the D.M.D. method, it is used very little in practice. It is occasionally employed as an independent method of checking areas which have been computed by the D.M.D. method.

19-9. Area of Tract with Irregular or Curved Boundaries. If the boundary of a tract of land follows some irregular or curved line, such as a stream or road, it is customary to run a traverse in some convenient location near the boundary and to locate the boundary by offsets from the traverse line. Figure 19-7 represents a typical case, AB being one of the traverse lines. The determination of area of the entire tract involves computing the area within the closed traverse, by methods which have already been described, and adding to this the area of the irregular figure between the traverse line AB and the curved boundary. The offset distances are aa', bb', etc., and the corresponding distances along the traverse line are Aa, Ab, etc. Where the boundary is irregular, as from a' to f', it is necessary to take offsets at points of change and hence generally at irregular intervals. Where a segment of the boundary is straight, as from f' to g', offsets are taken only at the ends.

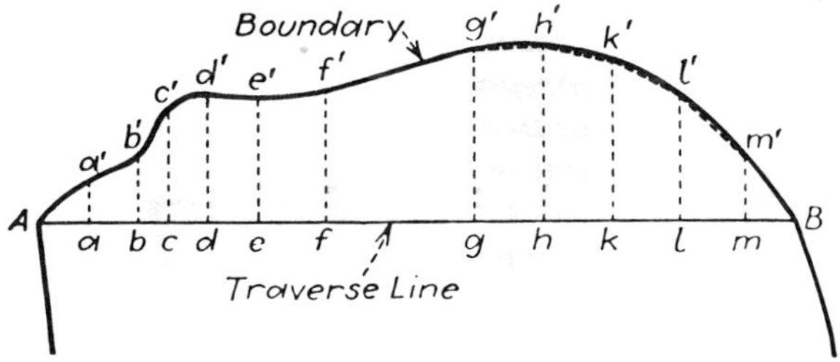

FIG. 19-7. Irregular boundary.

Where the boundary is a gradual curve, as from g' to m', ordinarily the offsets are taken at regular intervals.

If the offsets are taken sufficiently close together, the error involved in considering the boundary as straight between offsets is small as compared with the inaccuracies of the measured offsets. When this assumption is made, the assumed boundary takes some such form as that illustrated by the dotted lines $g'h'$, $h'k'$, etc., in Fig. 19-7, and the areas between offsets are of trapezoidal shape. Under such an assumption, irregular areas are said to be calculated by the *Trapezoidal Rule* (Art. 19-10).

Where the curved boundaries are of such definite character as to make it justifiable, the area may be calculated somewhat more accurately by assuming that the boundary is made up of segments of parabolas as first suggested by Simpson. Under this assumption, irregular areas are said to be computed by *Simpson's One-third Rule* (Art. 19-11).

19-10. Offsets at Regular Intervals: Trapezoidal Rule. Let Fig. 19-8 represent a portion of a tract lying between a traverse line AB and an irregular boundary CD, offsets h_1, h_2, . . . , h_n having been taken at the regular intervals d. The summation of the areas of the trapezoids comprising the total area is

$$\text{Area} = \frac{h_1 + h_2}{2} \cdot d + \frac{h_2 + h_3}{2} \cdot d + \cdots + \frac{h_{(n-1)} + h_n}{2} \cdot d$$

$$\text{Area} = d \left(\frac{h_1 + h_n}{2} + h_2 + h_3 + \cdots + h_{(n-1)} \right) \quad (7)$$

Equation (7) may be expressed conveniently in the form of the following rule:

Trapezoidal Rule. *Add the average of the end offsets to the sum of the intermediate offsets. The product of the quantity thus determined and the common interval between offsets is the required area.*

Example: By the Trapezoidal Rule find the area between a traverse line and a curved boundary, rectangular offsets being taken at intervals of 20 ft., and the

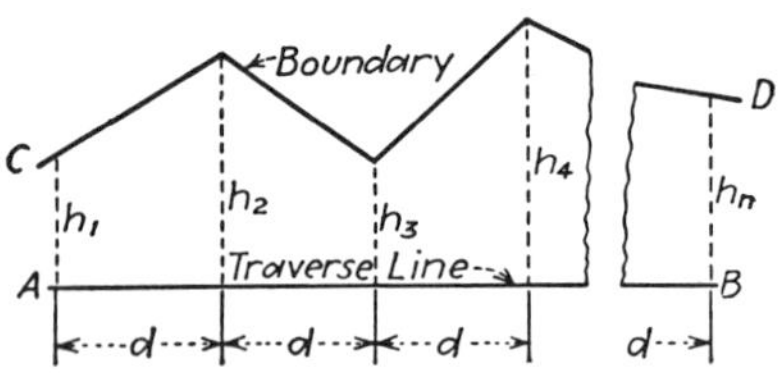

FIG. 19-8. Area by Trapezoidal Rule.

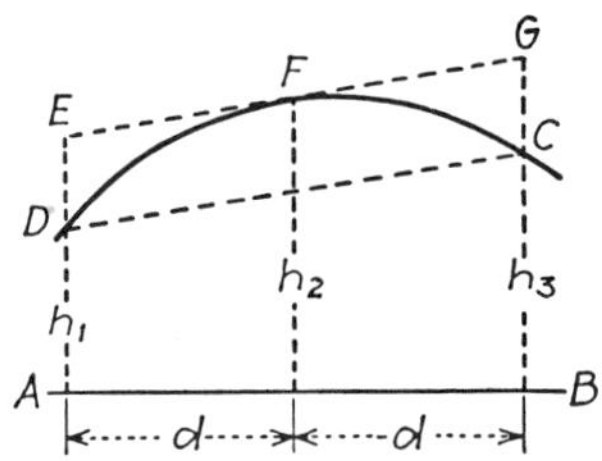

FIG. 19-9. Area by Simpson's Rule.

values of the offsets in feet being $h_1 = 3.2$, $h_2 = 10.4$, $h_3 = 12.8$, $h_4 = 11.2$, and $h_5 = 4.4$. By the foregoing rule,

$$\text{Area} = 20 \left(\frac{3.2 + 4.4}{2} + 10.4 + 12.8 + 11.2 \right) = 764 \text{ sq. ft.}$$

19-11. Offsets at Regular Intervals: Simpson's One-third Rule. In Fig. 19-9 let AB be a portion of a traverse line, DFC a portion of the curved boundary assumed to be the arc of a parabola, and h_1, h_2, and h_3 any three consecutive rectangular offsets from traverse line to boundary taken at the regular interval d.

The area between traverse line and curve may be considered as composed of the trapezoid $ABCD$ plus the area of the segment between the parabolic arc DFC and the corresponding chord DC. One property of a parabola is that the area of a segment (as DFC) is equal to two-thirds the area of the enclosing parallelogram (as $CDEFG$). Then the area between the traverse line and curved boundary within the length of $2d$ is

$$A_{1,2} = \frac{h_1 + h_3}{2} 2d + \left(h_2 - \frac{h_1 + h_3}{2} \right) 2d \cdot \frac{2}{3}$$

$$= \frac{d}{3} (h_1 + 4h_2 + h_3)$$

Similarly for the next two intervals

$$A_{3,4} = \frac{d}{3}(h_3 + 4h_4 + h_5)$$

The summation of these partial areas for $(n - 1)$ intervals, n being an odd number and representing the number of offsets, is

$$\text{Area} = \frac{d}{3}[h_1 + h_n + 2(h_3 + h_5 + \cdots + h_{(n-2)}) + 4(h_2 + h_4 + \cdots + h_{(n-1)})] \quad (8)$$

Equation (8) may be expressed conveniently in the form of the following rule, which is applicable if the number of offsets is odd.

Simpson's One-third Rule. *Find the sum of the end offsets, plus twice the sum of the odd intermediate offsets, plus four times the sum of the even intermediate offsets. Multiply the quantity thus determined by one third of the common interval between offsets, and the result is the required area.*

Example: By Simpson's One-third Rule find the area between the traverse line and the curved boundary of the example of Art. 19-10. By Simpson's Rule,

$$\text{Area} = {}^{20}\!/_3[3.2 + 4.4 + 2(12.8) + 4(10.4 + 11.2)] = 797 \text{ sq. ft.}$$

If the total number of offsets is *even*, the partial area at either end of the series of offsets is computed separately, in order to make n for the remaining area an odd number and thus to make Simpson's Rule applicable.

Simpson's Rule is also useful in other applications, such as finding centers of areas. The prismoidal formula for computing volumes of earthwork (Art. 11-12) embodies Simpson's Rule for area and a factor for the third dimension.

19-12. Trapezoidal and Simpson's Rules Compared. Results obtained by using Simpson's Rule are greater or smaller than those obtained by using the Trapezoidal Rule, according as the boundary curve is concave or convex toward the traverse line. Some appreciation of the variations between the two methods will be gained by studying the foregoing examples. It will be seen that the two results differ by more than 4 per cent. Under average conditions the difference will be much less than this, but in an extreme case it may be much larger.

Simpson's Rule is always more accurate, but it is not so easily applied as the Trapezoidal Rule. The latter approaches the former in accuracy to the extent that the irregular boundary has curves of contrary flexure thereby producing the compensative effects mentioned above.

19-13. Offsets at Irregular Intervals. The method of coordinates described in Art. 19-4 may be applied to this problem by assuming the origin as being in the traverse line and at the point where the first offset is taken. The coordinate axes are then the traverse line and a line at right angles thereto. The rule of Art. 19-4 may then be modified to the following:

Rule. *Multiply the distance (along the traverse) of each intermediate offset from the first by the difference between the two adjacent offsets, always subtracting the following from the preceding. Also multiply the distance of the last offset from the first by the sum of the last two offsets. The algebraic sum of these products, divided by two, is the required area.*

Area Between Meander Line and Bog Brook
on
Brigham Farm

Field Notes
Drawer 11,
Book T6
Pages 37-39

Computations
Aug. 16, 1965
Comp'td by J.E.S.
Checked by P.Q.D.

Line E-F				
Dist. from E~ft.	Length of Offset-ft	Difference	Products −	Products +
0=E	0.0			
20	14.3	− 23.1	460	
65	23.1	+ 4.7		310
87	9.6	+ 18.5		1610
100	4.6	− 8.1	810	
131	17.7	− 13.9	1820	
148	18.5	− 19.4	2870	
160	37.1	− 6.5	1040	
200	25.0	+ 12.9		2580
225	24.2	+ 2.0		450
250	23.0	+ 5.8		1450
300	18.4	+ 5.0		1500
350	18.0	− 20.4	7140	
385	38.8	+ 9.3		3580
400	8.7	+ 38.8		15520
413.7=F	0.0	+ 8.7		3600
			14,140	30,600
				14,140
			2	16,460
				8,230 sq. ft.
			or	0.189 Ac.

Note: For area computations within traverse see page 17 of this book

FIG. 19-10. Computations for area by offsets at irregular intervals.

The application of the rule to a specific problem is illustrated by the computations of Fig. 19-10, where the area between a meander line and a stream is determined.

19-14. Area of Segments of Circles. A problem of frequent occurrence in the surveying of city lots and of rural lands adjacent to the curves of highways and railways is that of finding the area where one or more of the lines of the boundary is the arc of a circle.

In Fig. 19-11, *ABCDEQF* may be taken as a boundary of this character for which it is convenient to run a traverse along the straight portions o the boundary and to make the chord EF the closing side of the traverse, the length of the chord $EF = L$ and the middle ordinate $PQ = M$ being measured in the field.

In calculating the area, it is convenient to divide the tract into two parts: (1) that within the polygon formed by the traverse $ABCDEF$, for which the area is found by the coordinate method or the double-meridian-distance method, and (2) that between the chord EPF and the arc EQF, which is the segment of a circle. The area of this segment is found exactly by subtracting the area of the triangle $OEPF$ from the area of the circular sector $OEQF$.

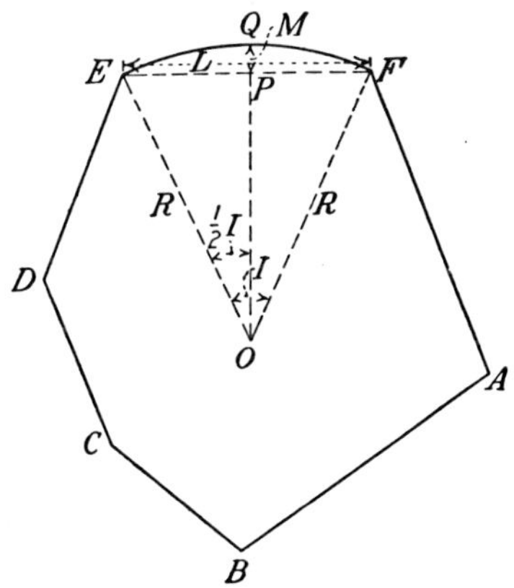

FIG. 19-11.

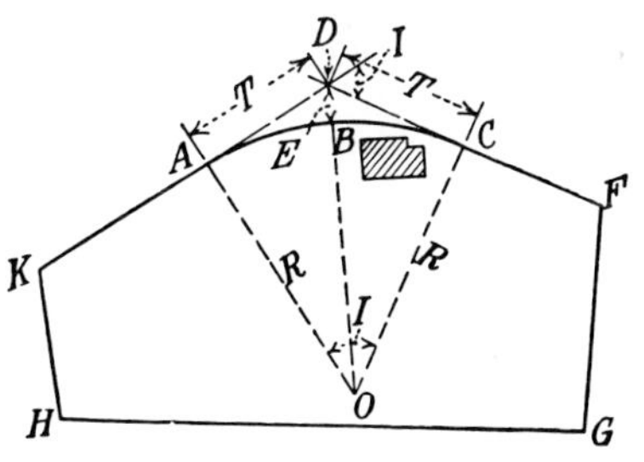

FIG. 19-12.

If I is the angle and R is the radius whose arc is EQF, then by Art. 27-4,

$$\tan \frac{1}{4} I = \frac{2M}{L} \tag{9}$$

and

$$R = \frac{L}{2 \sin \frac{1}{2} I} = \frac{M}{\text{vers } \frac{1}{2} I} \tag{10}$$

The area of the circular sector $OEQF$ is $A_s = \pi R^2 I°/360$, in which $I°$ is expressed in degrees.

The area of the triangle OEF is

$$A_t = \frac{R^2}{2} \sin I$$

The area of the segment is, then, exactly

$$A = A_s - A_t = R^2 \left(\frac{\pi I°}{360} - \frac{\sin I}{2} \right) \tag{11}$$

Example 1: Find the area of a circular segment when the chord length is 275.0 ft. and the middle ordinate is 31.35 ft.

By Eq. (9)

$$\tan \frac{1}{4} I = \frac{2 \times 31.35}{275.0} = 0.2280$$

$$\frac{1}{4} I = 12°51'; \; I = 51°24' = 51°.40$$

By Eq. (10)

$$R = \frac{L}{2 \sin \frac{1}{2} I} = \frac{275.0}{2 \times 0.4337} = 317.0 \text{ ft.}$$

By Eq. (11)

$$A = (317.0)^2 \left(\frac{3.142 \times 51.40}{360} - \frac{0.7815}{2} \right) = 5{,}810 \text{ sq. ft.}$$

An alternative method of finding the area of the tract *ABCDEQF* is to divide the area into a rectilinear polygon *ABCDEOF* and the circular sector *OEQF*, and to add the two areas. The polygon has one more side than the one used above, but there is no need to compute the area of a circular segment.

Approximation by Parabolic Segment. The area of a parabolic segment is

$$A_p = \tfrac{2}{3} LM \qquad (12)$$

in which the letters have the same significance as before. This expression may be employed for finding the approximate areas of circular segments, the precision decreasing as the size of the central angle I increases. The following example illustrates the error involved in applying this expression to the conditions of example 1.

Example 2: By Eq. (12) find the approximate area of the circular arc of example 1, and determine the percentage of error introduced through using this approximate expression.

$$A_P = \tfrac{2}{3} \times 275.0 \times 31.35 = 5{,}750 \text{ sq. ft.}$$

This value is

$$\frac{5{,}810 - 5{,}750}{5{,}810} 100 = 1.0 \text{ per cent too low}$$

When the central angle is small, the error involved in using Eq. (12) for circular arcs is often negligible; thus, when $I = 30°$, the error is less than 0.2 per cent. But for large values of I, the error introduced is so great as to render the approximate expression of little use; thus, when $I = 90°$, the error is about 3 per cent, and when $I = 180°$, the error is about 15 per cent.

Alternative Method. When tangents to the curve are property lines, it is sometimes more convenient to establish the traverse as illustrated by Fig. 19-12. Here *KA* and *FC*, which are tangent to the curve *ABC*, are run to an intersection at *D*, and the distances *AD* and *CD* and the angle I are measured. Also E is usually measured as a check.

The work of finding the area is conveniently divided into two parts: (1) that of calculating the area within the polygon *ADCFGHK* by the coordinate method or the double-meridian-distance method, and (2) that of calculating the external area between the arc *ABC* and the tangents *AD* and *CD*. The latter area subtracted from the former is the required area.

The external area may be found by subtracting the area of the circular sector $OABC = A_s = \pi R^2 I°/360$ from the area $OADC = TR$, in which T is the tangent distance $AD = CD$, and R is the radius of the curve. If R is unknown, it may be found by the relation

$$R = \frac{T}{\tan \frac{1}{2}I} = \frac{E}{\text{exsec } \frac{1}{2}I} \quad \text{(see Art. 27-4)} \tag{13}$$

PARTITION OF LAND

19-15. General. Four of the simpler cases frequently encountered in the subdivision of irregular tracts of land will be described in the succeeding articles. Methods of subdividing the U.S. public lands are given in Chap. 23.

Where a given tract is to be divided into two or more parts, a resurvey is run, the latitudes and departures are computed, the survey is balanced, and the area of the entire tract is determined. The corrected latitudes and departures are further employed in the computations of subdivision.

19-16. Area Cut Off by a Line between Two Points. In Fig. 19-13 let *ABCDEFG* represent a tract of land to be divided into two parts by a line extending from *A* to *D*. A survey of the tract has been made, the latitudes and departures have been balanced, and the area has been computed.

It is desired to determine the length and direction of the cut-off line *AD* without additional field measurements, and to calculate the area of each of the two parts into which the tract is divided.

Either of the two parts may be considered as a closed traverse with the length and bearing of one side *DA* unknown. Considering the part *ABCDA*, the latitudes and departures of *AB*, *BC*, and *CD* are given; hence the latitude, departure, length, and bearing of *DA* can be determined as described in Art. 18-21. The area of either part can then be found by the D.M.D. method (Art. 19-7).

A check on the field work and computations is obtained by actually measuring the length and direction of the line *DA* and noting the agreement between observed and calculated values. A further check is secured by noting that the sum of the areas of the two parts, each calculated independently, is equal to the calculated area of the entire tract.

19-17. Area Cut Off by a Line Running in a Given Direction. In Fig. 19-14, *ABCDEFG* represents a tract of known dimensions, for which the corrected latitudes and departures are given; and *DH* represents a line running in a given direction which passes through the point D and divides the tract into two parts.

It is desired to calculate from the given data the lengths DH and HA and the area of each of the two parts into which the tract is divided.

Either of the two parts may be considered as a closed traverse for which the lengths of two sides are unknown; these lengths can be computed as described in Art. 18-23. Considering the part $ABCDHA$, the latitudes and departures of AB, BC, and CD are known; from these the length and bearing of DA are computed. In the triangle ADH the lengths of the sides DH and HA are found, and their latitudes and departures are computed. The area of $ABCDHA$ is then calculated by the D.M.D. method.

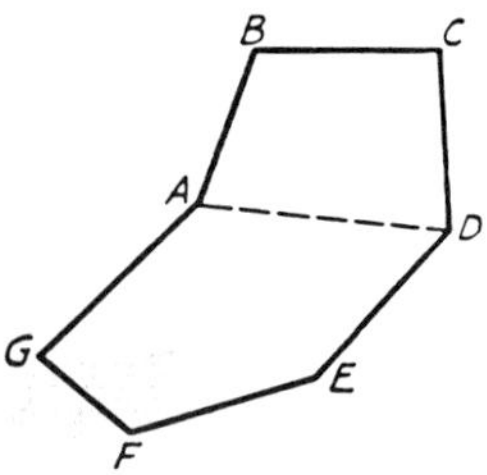

FIG. 19-13.

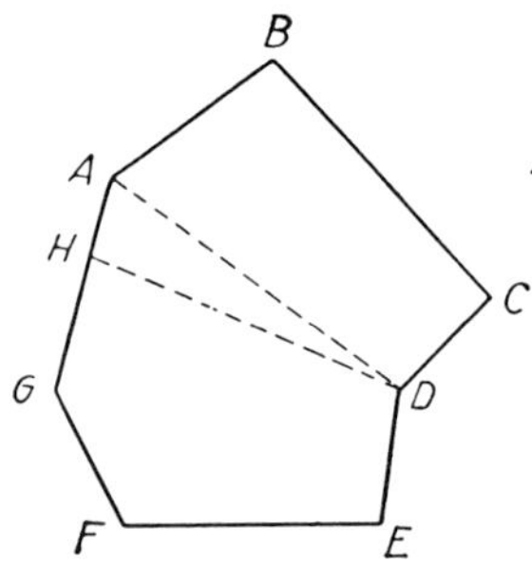

FIG. 19-14.

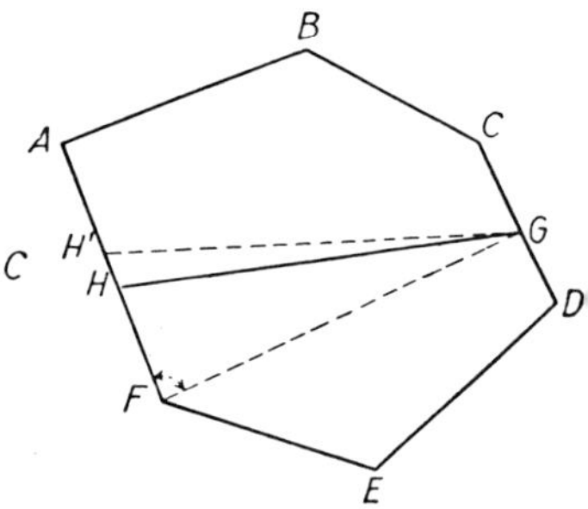

FIG. 19-15.

In the field the length and direction of the side DH are laid off from D, and a check on field work and computations is obtained if the point H thus established lies on the line GA and if the computed distance HA agrees with the observed distance. The computations are further verified by seeing that the algebraic sums of the latitudes and of the departures of AB, BC, CD, DH, and HA are equal to zero. (This is on the assumption that the latitudes and departures of both DH and HA are based upon the lengths of these lines as computed from the triangle ADH.) The area computations may be checked by observing that the sum of the areas of the two parts, each computed independently, is equal to the area of the entire tract.

19-18. To Cut Off a Required Area by a Line through a Given Point. In Fig. 19-15, $ABCDEF$ represents a tract of land of known dimensions, for which the corrected latitudes and departures are given; and G represents a point on the boundary through which a line is to pass cutting off a required area from the tract. It is assumed that the area within the tract has been calculated by the D.M.D. method and that a sketch of the tract has been prepared.

To find the length and direction of the dividing line the procedure is as follows:

A line *GF* is drawn to that corner of the traverse which, from inspection of the sketch, will come nearest being on the required line of division. The latitude and departure of *CG* are computed. Then in the traverse *ABCGFA* all sides are known except *GF*. By the methods of Art. 18-21, the latitude, departure, length, and bearing of *GF* are determined. By the D.M.D. method the area of *FABCG*, the amount cut off by the line *FG*, is calculated. The difference between this area and that required is found.

In the figure it is assumed that *FABCG* has an area greater than the desired amount, *GH* being the correct position of the dividing line. Then the triangle *GFH* represents this excess area; and as the angle *F* may be computed from known bearings, there are given in this triangle one side *FG*, one angle *F*, and the area. The length *HF* is computed from the relation, area = $\frac{1}{2}ab \sin C$, given in Table VI; that is,

$$HF = \frac{2 \times \text{area } GFH}{FG \sin F} \tag{14}$$

The triangle is then solved for angle *G* and length *GH*. From the known direction of *GF* and the angle *G*, the bearing of *GH* is computed. The latitudes and departures of the lines *FH*, *GH*, and *HA* are computed.

In the field the length *GH* is laid off in the required direction, and a check on field work and computations is obtained if the point *H* thus established falls on the line *FA* and if the computed distance *HF* or *HA* agrees with the measured distance.

Sometimes the tract in question will be of such shape that a line drawn from the given point in the boundary to any corner will cut off an area nowhere near that required. Under these circumstances or when the traverse has a large number of sides, it is advisable to plot the traverse with protractor and scale and to establish a trial line of subdivision such as *GH′* in Fig. 19-15. The planimeter (Art. 4-12) may be used to advantage for finding the area cut off by this trial line, and the line may be shifted until the area cut off agrees closely with that required. The scaled distance *AH′* may be used in the computations. It will be seen that the method of solution is now identical with that just described for the case where the trial line is drawn to a corner.

Figure 19-16 shows the computations for the division of a tract into equal parts by a line passing through the corner *A*. The tract is the same as that for which area computations are shown in Fig. 19-6.

Taking the computations of Fig. 19-16 in order down the page, there are: (1) tabulations for finding the area of the entire tract by the D.M.D. method, (2) tabulations for finding the area cut off by the trial line *AD*, (3) tabulations for checking the computations by determining the area cut off by the true line of division *AF*, and (4) logarithmic computations for (*a*) calculating the length and bearing of the trial line *DA*, (*b*) solving the triangle *ADF*, and (*c*) computing the latitudes, departures, D.M.D.'s, and double areas

COMPUTATIONS FOR BALSAM PARK DIVISION
To Divide Into Two Equal Parts
by a Line Through "A"

55

Field Notes
Book No.3, Page 47.
Area Computations
Page 7 this Book

Comp't'd by D.P.F. Aug.17, 1965
Checked by N.D.C. Aug.17, 1965

Total Area (see P.7)

Line	Calculated Bearing	Dist., 66-ft.Ch.	Latitudes N	Latitudes S	Departures E	Departures W	D.M.D.'s	Double Areas +	Double Areas −
AB	S 80°29½' W	34.46		5.69		33.99	61.81		351.89
BC	S 33 04 W	25.49		21.36		13.91	13.91		297.15
CD	S 33 46¾ W	33.93		28.20	18.87		18.87		532.06
DE	N 87 58¼ E	28.63	1.01		28.61		66.34	67.21	
EA	N 0 27 E	54.24	54.24		0.42		95.38	5173.51	
								5240.72	1181.10
							Total Area	202.98 Ac.	
					Area	ABCD			
AB				5.69		33.99	61.81		351.89
BC				21.36		13.91	13.91		297.15
CD				28.20	18.87		18.87		532.06
DA	N 27°43' E	62.41	55.25		29.03		66.77	3688.78	
						Area	ABCD	125.38 Ac.	1181.10
					Area	ABCF			
AB				5.69		33.99	61.81		351.89
BC				21.36		13.91	13.91		297.15
CF	S 33°46¾' E	25.22		20.96	14.03		14.03		294.08
FA	N 35 12 E	58.75	48.01		33.87		61.93	2973.10	
									943.12
						Area	ABCF	101.50 Ac;	Check

Formulas

$$FD = \frac{2\,ADF}{AD \sin\theta}$$

$$\tan\beta = \frac{FD \sin\theta}{DA - DF\cos\theta}$$

$$FA = \frac{FD \sin\theta}{\sin\beta}$$

Area ABCD = 125.38 Ac.
Half Park = 101.49
Area ADF = 23.89 Ac.

CD = 33.93
FD = 8.712
CF = 25.22

Line	DA	Log 2ADF	2.67928	Log DA-DF cos θ	1.76534	Line	CF	FA
Dist.	62.412	Log DA	1.79527	Log DF sin θ	0.88401	Departure	14.028	33.873
Log Dist.	1.79527	Log sin θ	9.94391	Log tan β	9.11867	Log Dep.	1.14687	1.52978
Log cos Br.	9.94704	Log DF	0.94010	β	7°29'	Log sin	9.74509	9.76075
Log Lat.	1.74231	DF	8.712	Bear. DA	N27°43'E	Log Dist.	1.40148	1.76903
Log D.M.D.	1.82457	Log DF sin θ	0.88401	Bear. FA	N35°12'E	Log cos	9.91969	9.91230
Log D.Ar.	3.56688	Log cos θ	9.67860	Log FD	0.94010	Log Lat.	1.32147	1.68133
D. Area	3688.78	Log DF cos θ	0.61870	Log sin θ	9.94391	Latitude	20.961	48.015
Log Dep.	1.46291	DF cos θ	4.156	Log sin β	9.11498	Log D.M.D.	1.14700	1.79189
Log tan Br.	9.72060	DA	62.412	Log AF	1.76903	Log D.Ar.	2.46847	3.47322
Bearing	N27°43'E	DA−DF cos θ	58.256	AF	58.753	D. Area	294.08	2973.10

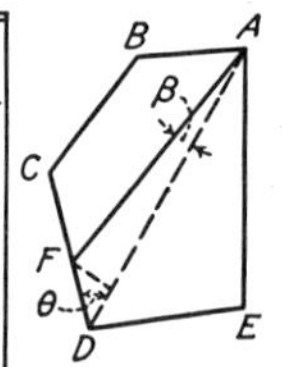

FIG. 19-16. Computations for partition of land.

for the lines CF and FA, which values are employed in determining the area $ABCF$ given in (3).

19-19. To Cut Off a Required Area by a Line Running in a Given Direction. In Fig. 19-17, $ABCDEF$ represents a tract of land of known dimensions and area, which is to be divided into two parts, each of a required area, by a line running in a given direction. The figure is assumed to be drawn at least roughly to scale, and the corrected latitudes and departures are known.

Through the corner that seems likely to be nearest the line cutting off the required area, a trial line DG is drawn in the given direction. Then in the closed traverse $GBCDG$ the latitudes and departures of BC and CD and the bearings of DG and GB are known, and the lengths of two sides DG and GB are unknown. By the methods of Art. 18-23, these unknown quantities are found, and the latitudes and departures of the courses are determined. The area cut off by the trial line is calculated. The difference between this area and that required is represented in the figure by the trapezoid $DGJH$ in which the side DG is known. The angles at D and G can be computed from the known bearings of adjacent sides, and in this way θ and ϕ are determined. Then

$$\text{Area of trapezoid} = DG \cdot x + \frac{x^2}{2} (\tan \theta + \tan \phi) \qquad (15)$$

in which $\tan \theta$ or $\tan \phi$ is positive or negative according as θ or ϕ lies within or without the trapezoid, and x is the altitude of the trapezoid. (In the figure both angles lie without the trapezoid and hence both tangents are negative.) The value of x is found by solving this equation. Then $GJ = x \sec \theta$; $DH = x \sec \phi$; and $JH = DG + x(\tan \theta + \tan \phi)$ in which the signs of $\tan \theta$ and $\tan \phi$ are as given above.

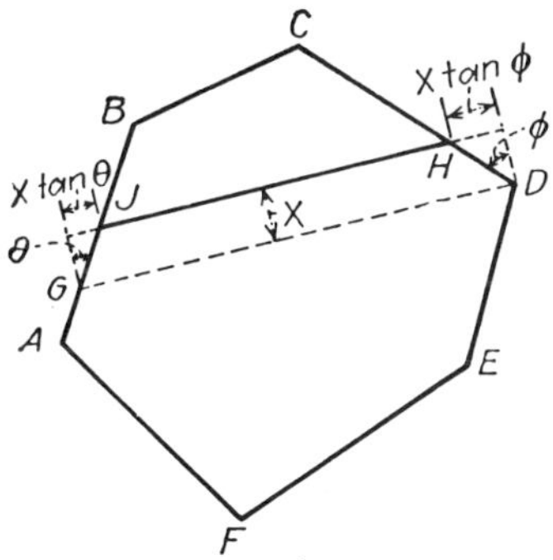

FIG. 19-17.

In the field the points H and J are established on the lines CD and AB, at the calculated distances from the adjacent corners. The side JH is then measured. If this measured value agrees with the computed value, the field work and portions of the computations are verified. A further check on the computations is introduced by calculating the area of $BCHJ$ and comparing it with the required area of this figure.

19-20. Numerical Problems.

1. A square field contains 40 acres. What are its dimensions in chains, in rods, and in feet?
2. How many acres are there in a rectangular tract 50 × 100 ft.? In a tract 400 × 400 ft.? In a tract 2,640 × 2,640 ft.?
3. What is the area of a triangle having sides of length 219.0, 317.2, and 301.6 ft.? Of a triangle having two sides of length 1,167.1 and 392.7 ft. and an included angle of 39°46′?
4. Given the notes shown in Fig. 7-16. Calculate the area of the field by using the two sides and included angle of each triangle. Check by using the three sides of each of the oblique triangles.
5. The mutually bisecting diagonals of a four-sided field are 480 and 360 ft. The angle of intersection between the diagonals is 100°. Find the interior angles and the lengths of the sides.

6. In the following tabulation are given total latitudes and total departures of a closed traverse. Calculate the area by the coordinate method.

Corner	*A*	*B*	*C*	*D*
Total latitude, ft.	+50.5	+203.4	−49.5	−75.0
Total departure, ft.	−102.5	0	+100.3	0

7. Given the notes tabulated below, for a closed traverse. Compute the latitudes and departures, and balance the survey by the Compass Rule. Assume that the coordinates of *C* are 267.3N and 580.8E, and compute the coordinates of all other corners. Calculate the area by the coordinate method.

Course	Bearing	Length, ft.
AB	N48°20′E	529.6
BC	N87°43′E	592.0
CD	S 7°59′E	563.6
DE	S82°12′W	753.4
EA	N48°12′W	428.2

8. In the following tabulation are given the latitudes and departures of a balanced closed traverse. Calculate the area (*a*) by the D.M.D. method and (*b*) by the coordinate method, using five-place logarithms.

Course	Latitude, ft.	Departure, ft.
AB	S198.7	W213.6
BC	N181.1	W174.4
CD	N334.1	E 89.2
DE	N224.9	E110.7
EA	S541.4	E188.1

9. Find the error of closure of the following traverse, balance the survey by the Compass Rule, and calculate the area in acres by the D.M.D. method using four-place logarithms:

Course	Bearing	Length, ft.
AB	S45°45′E	294.4
BC	N65°30′E	263.4
CD	N35°15′E	313.6
DE	N64°15′W	392.0
EF	S59°00′W	197.2
FA	S26°15′W	240.0

10. In the following table are the notes for a transit traverse, distances being in Gunter's chains. Compute the latitudes and departures, balance the survey by the Transit Rule, and calculate the area in acres by the D.M.D. method. Use four-place logarithms.

Course	Bearing	Length, chains
AB	S58°08′E	10.24
BC	S67°07′E	9.32
CD	S9°39′W	24.00
DE	S84°22′W	24.92
EF	N6°21′E	18.92
FA	N29°52′E	18.80

11. A traverse *ABCD* is established inside a four-sided field, and the corners of the field are located by angular and linear measurements from the traverse stations, all as indicated by the following data:

Course	Bearing	Length, ft.
AB	S89°58′E	296.4
AE	N20°00′W	34.2
BC	S43°20′W	333.9
BF	N35°20′E	16.9
CD	S80°21′W	215.6
CG	S73°00′E	27.6
DA	N27°24′E	314.2
DH	S36°30′W	15.7

Compute the latitudes and departures, and balance the traverse by the Compass Rule. Compute the coordinates of each transit point and of each property corner, using *D* as an origin of coordinates. Compute the length and bearing of each side of the field *EFGH*, and tabulate results. Calculate the area of the field by the coordinate method.

12. Given the following offsets from traverse line to irregular boundary, measured at points 25 ft. apart.

Distance, ft.	Offset, ft.	Distance, ft.	Offset, ft.
0	0.0	125	28.2
25	16.6	150	11.9
50	35.1	175	30.7
75	39.3	200	43.4
100	42.0	225	22.5

By the Trapezoidal Rule (Art. 19-10) calculate the area between traverse line and boundary.

13. Given the data of the preceding problem. Calculate the required area by Simpson's One-third Rule. Note that the number of offsets is even.

14. Following are offsets taken at intervals of 50 ft., to the right and to the left of a traverse line:

Offset left, ft.	Distance, ft.	Offset right, ft.
34.8	0	32.9
44.2	50	26.1
61.5	100	18.6
51.1	150	32.7
31.3	200	49.8
12.7	250	56.9
8.5	300	47.2

By the Trapezoidal Rule calculate the area between boundaries thus defined.

15. Given the data of the preceding problem. Calculate the required area by Simpson's One-third Rule.

16. Following are offsets from a traverse line to an irregular boundary, taken at irregular intervals:

Distance, ft.	Offset, ft.	Distance, ft.	Offset, ft.
0	18.5	100	44.1
25	37.7	170	53.9
60	58.2	200	46.0
70	40.5	220	34.2

Calculate the area between traverse line and boundary by means of the rule of Art. 19-13.

17. Given the notes shown in Fig. 7-17. Calculate the area of the tract surveyed, including the irregular areas between traverse *ABCDE* and river or lake.

18. In Fig. 19-11, what is the area of the circular segment *EQF* if the length of the chord L is 817.2 ft. and the middle ordinate M is 89.17 ft.?

19. In Fig. 19-11, what is the area of the circular segment *EQF* if the chord length L is 600 ft. and the middle ordinate M is 7.85 ft.?

20. Solve problems 18 and 19 using the approximate expression [Eq. (12)] of Art. 19-14. Compare the results with those of problems 18 and 19, and for each case compute the percentage of error introduced through use of the approximate expression.

21. A curved corner lot is similar in shape to that shown in Fig. 19-12. The tangent distances T are each 50.0 ft. and the intersection angle I is 40°. What is the area between the circular curve *ABC* and the tangents *AD* and *CD*? What is the external distance E?

22. Given the data of problem 10. Find the area north of a line running from F to a point G on the line *CD* and distant 10.00 chains from C. Calculate the length and bearing of *FG*.

23. Given the data of problem 10. Find the area of each of the two parts into which the tract is divided by a meridian line through the point B.

24. Given the data of problem 10. Find the length and direction of a line that runs through F and divides the tract into two equal parts.

25. Given the data of problem 10. The tract is to be divided in two equal parts by an east-west line. Compute the length of the dividing line, and compute the distances from the ends of the line to adjacent traverse stations.

19-21. Office Problem.

Problem 1. Area of Field Surveyed with Tape

Object. To determine the area of a field surveyed with the tape. The data of field problem 4, Art. 7-31, may be used. For other methods of calculating areas, see the numerical problems of Art. 19-20.

Procedure. (1) Decide upon a convenient and systematic form of computation for each of the following methods, using four-place logarithms where possible; and transcribe the necessary data from field book to computation book. (2) By the protractor method, plot the boundaries of the field to a scale commensurate with the precision of the field measurements. (3) Determine the area of each part and of the entire field by use of the planimeter (Art. 4-12). (4) Calculate the area of the triangles and the total area in square feet and acres, following each method through before beginning another. Check the results with a slide rule. (5) Make the computations (*a*) by using the two sides and included angle of each triangle, (*b*) by using the three sides of the oblique triangles, and (*c*) by using the measured altitude and base of each triangle. (6) Calculate by the method of offsets the area of any portions of the field having an irregular boundary. (7) Compare the results obtained through the use of the various methods.

REFERENCE

Dowling, Paul J., "Circular Area Segment Formulas," *Surveying and Mapping*, Vol. 20, No. 1, pp. 32–33, March, 1960.

CHAPTER 20

PRINCIPLES OF FIELD ASTRONOMY

20-1. General. The surveyor should be familiar with the astronomical and trigonometric principles upon which the observations and computations of field astronomy are based. In this chapter are given certain fundamentals which are applicable to all astronomical observations. However, the discussions in this and the following chapter are intended to be applied only to surveys of moderate precision.

The science of astronomy offers the surveyor a means of determining the absolute location of any point or the absolute location and direction of any line on the surface of the earth. The absolute location of a point is given by its latitude and longitude, and the absolute direction of a line is defined by the angle which the line makes with the true meridian.

The azimuth of a line is established by angular observations on some celestial body, most commonly on the sun or on Polaris, the North Star or polestar. For the purpose of computing the azimuth from an astronomical observation, it is necessary that the latitude of the place be known. Also for certain observations it is essential that the longitude be roughly determined. If the survey is through a territory for which there is a reliable map, the latitude and longitude may ordinarily be determined with sufficient precision by scaling from the map.

In *geodetic* surveying it is necessary to determine the latitude and longitude of certain points with great precision, the work involving observations on numerous stars and requiring instruments of high precision. The requirements of *plane* surveying, however, are met if the true azimuth or bearing of the survey lines is established with a degree of precision at least equal to that with which the angles between survey lines are measured. For plane surveying of ordinary precision, the use of the engineer's transit and the methods described herein will yield sufficiently precise results.

20-2. The Celestial Sphere. In making observations on the sun and stars, the surveyor is not interested in the distance of these celestial bodies from the earth but merely in their angular position. It is convenient to imagine their being attached to the inner surface of a hollow sphere of infinite radius of which the earth is the center. This imaginary globe is called the *celestial sphere.* It is also helpful to imagine the earth as being fixed, and to consider the celestial sphere as rotating from east to west, its axis being the prolongation of the axis of the earth. Thus to the naked eye the polestar appears to remain stationary, but the sun (and similarly the stars near the equator) appears above the horizon in the general direction of east, follows a curved path (convex southward) across the heavens, and disappears below the horizon in the general direction of west.

The portion of the celestial sphere seen by the observer is the hemisphere above the plane of his own horizon. The reference plane passes through the center of the earth parallel with the observer's horizon plane, but the radius of the earth is so small with relation to the distances to the stars that the error in vertical angle to a star is negligible. In the case of the sun the error produced by this assumption is much larger than for any of the stars, amounting under certain conditions to about 9 seconds of arc and requiring an appropriate correction to the observed vertical angle (see Art. 21-6). In any case, a refraction correction to observed vertical angles is necessary (Art. 21-7).

A vertical line at the location of the observer coincides with the plumb line and is normal to the observer's horizon plane. The point where this vertical line pierces the celestial sphere above the head of the observer is called the *zenith,* and the corresponding point in the opposite hemisphere, directly below the observer, is called the *nadir.*

The *celestial poles* are the points where the earth's axis prolonged pierces the celestial sphere.

The *celestial equator* is the great circle formed by the intersection of the earth's equatorial plane with the surface of the celestial sphere.

Figure 20-1 represents the celestial sphere, the point O being the earth and $NES'W$ being the horizon of an observer, with letters standing for the points of the compass. Figure 20-2 may be taken as an enlarged view of the earth in the same position as that assumed in Fig. 20-1. A is an observer in the Northern Hemisphere, the line N_aS_a being in his horizon plane. Evidently he views everything above the horizon plane or that portion of the celestial sphere (Fig. 20-1) which is shown by full lines. B is an observer in the Southern Hemisphere, at a point on the earth diametrically opposite A; the portion of the celestial sphere which he views above his horizon plane N_bS_b will be the opposite hemisphere to that seen by A, or that portion of Fig. 20-1 which is shown by dash lines. Since the size of the earth is negligible as compared with that of the celestial sphere, it may be considered that either N_aS_a or N_bS_b in Fig. 20-2 coincides with NS' in Fig. 20-1.

Assuming the observer to be in the Northern Hemisphere (Fig. 20-1), Z is the zenith; P and P' are the celestial poles, P being the visible or *elevated* pole; and $EQWQ'$ is the celestial equator, of which the portion EQW is visible to the observer.

Since we are, for the sake of simplicity, assuming that the celestial sphere is rotating and the earth remains stationary, N, E, S', W, and Z are regarded as fixed points with respect to any given station on the surface of the earth. If $S'N$ is a meridian line in the plane of the horizon passing through the station of the observer, then a vertical plane of which this line is an element cuts the celestial sphere in the great circle $S'ZPNZ'P'$, which is called the meridian circle or, more often, simply the meridian. At a given instant the meridian for one station does not occupy the same position in the celestial sphere as

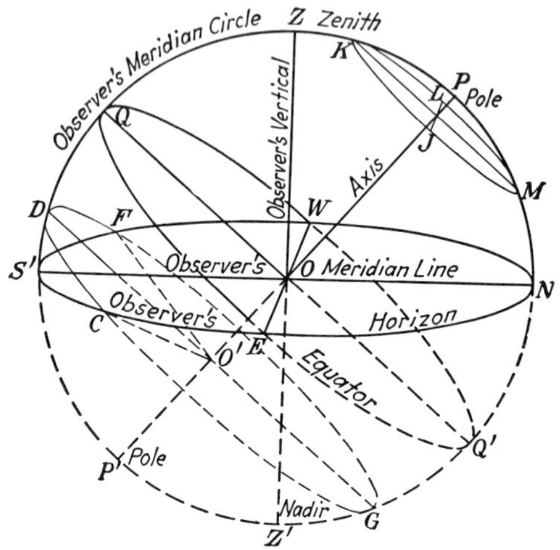

FIG. 20-1. Celestial sphere.

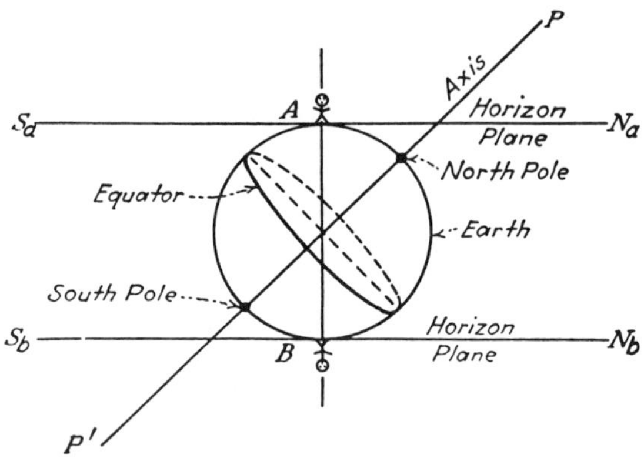

FIG. 20-2. Observer's horizon.

does the meridian for another station, unless the two stations are at the same longitude.

Any star which is below or south of the equator will follow some path as $CDFG$. It will become visible at C, will pass over the meridian at D, and will disappear from view at F. It will be above the horizon for a less length of time than it will be below, or the angle whose arc is *CDF* (angle *CO'F*) is less than 180°. From the figure it is evident that, if any star is sufficiently far below the equator, it will never appear above the observer's horizon.

Similarly, any star which is above or north of the equator will be above the horizon for a greater length of time than it is below. If it is far enough above the equator, it will be continuously visible to an observer in a northern latitude and will, during the course of a single revolution of the celestial sphere, follow some path as *JKLM*. When it is at the highest point of its apparent path, at *K*, it is said to be at *upper culmination;* when it is at the lowest point, at *M*, it is said to be at *lower culmination.*

20-3. Observer's Location on the Earth. The location, or *position,* of any point on the surface of a sphere may be fixed by angular measurement from two planes of reference at right angles to each other passing through the center of the sphere; these measurements are called the *spherical coordinates* of the point. The spherical coordinates of any station on the surface of the earth are designated as the *latitude* and *longitude* of the station. Figure 20-3 represents the earth, *PP'* being the axis and *QUVQ'* being the equator. Let *S* be the station of an observer. Then *PSUP'* is a *meridian circle* through the station. Also *RSR'* is a *parallel* passing through the station, the plane of *RSR'* being parallel to that of the equator.

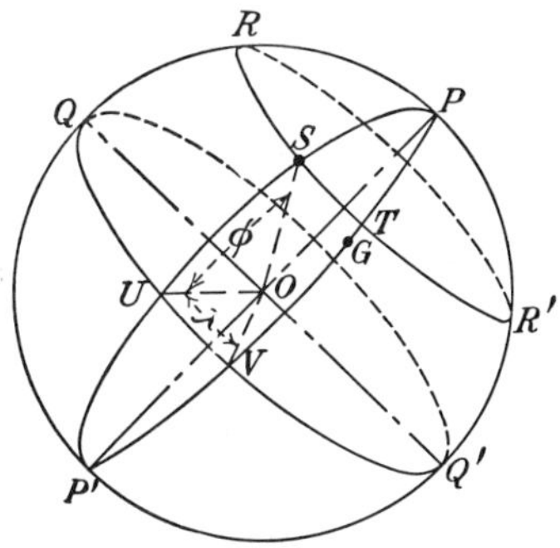

FIG. 20-3. Observer's location on the earth.

The latitude of a place may, for all practical purposes, be defined as the angular distance of the place above or below the equator. When the station is above the equator, the latitude is north and its sign is positive; when below the equator, the latitude is south and its sign is negative. Hence in the figure the latitude of *S* is given by the angle ϕ or by the angular distance, measured along any meridian circle, between the equator and the parallel passing through *S*, such as *US*, *VT*, *QR*, etc.; and the latitude is north or positive. The latitude of a place is stated in degrees. Thus the latitude of the equator is 0° and that of the North Pole is +90°, or 90°N.

The longitude of a place is defined as the angular distance measured along the arc of the equator between a reference meridian and the meridian circle passing through the station. The reference meridian is called the *primary meridian.* The primary meridian most generally used is that of Greenwich, England. (Prior to 1912, a Washington, D.C., meridian was extensively used in the United States.) Hence in the figure if the point *G* represents Greenwich, *PGP'* is the primary meridian, and the longitude of *S* is given by the angle λ or by the angular distance *VU*. Longitudes are expressed either in degrees of arc or in hours of time (15° = 1 hr.) and are measured either east or west of the Greenwich meridian.

In general, the discussions herein are intended to apply in the Northern Hemisphere and for longitudes west of Greenwich.

20-4. Right-ascension Equator System. In Fig. 20-4 is shown the celestial sphere in a position similar to that of the earth in Fig. 20-3, S being a celestial body whose position is to be fixed by spherical coordinates. Comparable with the meridian circles or meridians of longitude of the earth are the *hour circles* of the celestial sphere, all of which converge at the celestial poles. The arc PSU is a portion of the hour circle passing through S. Comparable with the parallels of latitude of the earth are the *parallels of declination* of the celestial

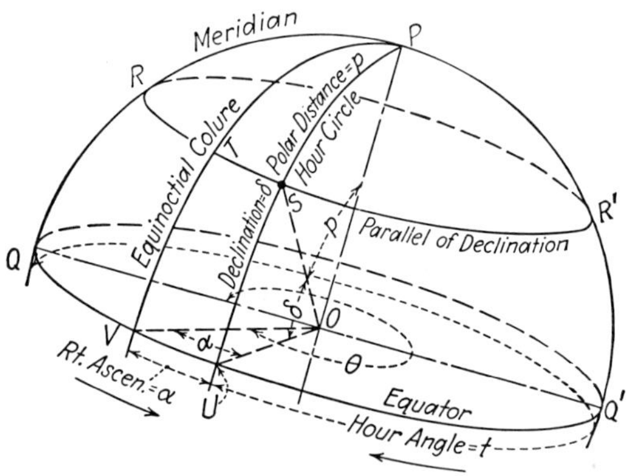

FIG. 20-4. Equator systems of spherical coordinates.

sphere. RSR' is the parallel of declination passing through S. And comparable with the prime meridian through Greenwich is the *equinoctial colure* of the celestial sphere which passes through the *vernal equinox*, an imaginary point among the stars where the sun apparently crosses the equator on March 21 of each year. In the figure, V represents the vernal equinox and PTV is the equinoctial colure.

The *right ascension* of the sun or any star is the angular distance measured along the celestial equator between the vernal equinox and the hour circle through the body. It is comparable with the longitude of a station on the earth. Right ascensions are measured *eastward* from the vernal equinox and may be expressed either in degrees of arc (0° to 360°) or in hours of time (0^h to 24^h). Thus, in the figure, the right ascension of S is given by the angle α in the plane of the equator or by the arc VU.

The *declination* of any celestial body is the angular distance of the body above or below the celestial equator. It is comparable with the latitude of a station on the earth. If the body is above the equator, its declination is said to be north and is considered as positive; if it is below the equator, its declination is said to be south and is considered as negative. Declinations are expressed in degrees and cannot exceed 90° in magnitude. Thus in the figure, the declination of S is given by the angle δ or by the arc of any hour circle between the equator and the parallel of declination RSR', such as US, VT, QR, etc.

The *polar distance* or *codeclination* of any celestial body is $p = 90° - \delta$ with due regard to the sign of the declination. In the figure, it is given by the angle p or by the arc PS. Polar distances are always positive. For computations referred to the North Pole, when the declination is north, the polar distance is the complement of the declination; but when the declination is south, as in the case of the sun during the winter months, the polar distance is greater than 90°. In defining the position of a star near either pole, often the polar distance is given instead of the declination.

For present purposes it may be considered that the vernal equinox is a fixed point on the celestial equator, just as Greenwich is a fixed point on the earth. But while stations on the earth maintain practically an unvarying location with respect to the equator and the meridian of Greenwich, the coordinates of celestial bodies with respect to the celestial equator and the equinoctial colure change more or less with the passage of time. The fixed stars, or those outside the solar system, alter their positions in the celestial sphere but slightly from month to month and from year to year, the annual change being less than a minute of arc in either right ascension or declination. These variations are due to (*a*) *precession* or the slow change in the direction of the earth's axis due to attraction of the sun, moon, and planets, and (*b*) *nutation* or small inequalities in the motion of precession, similar to the oscillation of a spinning top.

As the earth actually travels around the sun but not around the stars, the sun appears to move more slowly than do the stars, making in one year 365 apparent revolutions (approximately) while the stars make 366 apparent revolutions (approximately); thus the sun apparently makes a complete circuit of the heavens once each year, its right ascension changing from 0^h (or 0°) on March 21 to 12^h (or 180°) on September 22 and continuing to 24^h (or 360°) on the following March 21, when a new cycle begins. Further, as the axis of rotation of the earth is not normal to the plane of the earth's orbit, the path apparently traced by the sun among the stars on the celestial sphere, called the *ecliptic*, is a continuous curved line; each year the sun crosses the equator northward on March 21, reaches a maximum positive declination (about N23½°) on June 21, crosses the equator southward on September 22, and reaches a maximum negative declination (about S23½°) on December 21.

20-5. Hour-angle Equator System. In many of the problems of field astronomy it is necessary not only that a star's position in the celestial sphere be known but also that its position with respect to the meridian through a given station on the surface of the earth be determined. In Fig. 20-4 let $QRPR'Q'$ represent the meridian of some station on the earth, say that of the observer, and let S be some heavenly body, say a star, whose position with respect to the meridian $QRPR'Q'$ and the equator $QQ'UV$ it is desired to establish. The spherical coordinates of the star are given by (1) the angular distance of the star above or below the equator, which in the figure is given by the arc US, defined in the preceding article as the declination, and (2) the angular distance measured along the equator between the meridian and the hour circle through the star. When this angular measurement is from east to west, it is called an *hour angle.*

The hour angle of any celestial body may then be defined as the angular distance measured westward along the equator from the meridian of reference to the hour circle through the body. Thus, in the figure, the hour angle of S is given by the angle t or by the angular distance $QQ'U$. Hour angles are expressed either in hours of time or in degrees of arc. In the figure the hour angle is more than 12^h or more than 180°. When no qualification is stated, it is understood that an hour angle is measured from the upper branch of the meridian, that is, the branch above the station or above the observer's head.

In connection with the definition of civil time, hour angles are reckoned from the lower branch of the meridian. In the figure, if the hour angle were reckoned from the lower branch, it would be defined by the angular distance $Q'U$, and would be 12^h more or less than that given by the arc $QQ'U$, which is the hour angle reckoned from the upper branch.

Sometimes the hour angles of stars east of the meridian are reckoned eastward from the upper branch of the meridian, rather than westward. When an hour angle is expressed in this way, it is preceded by a minus sign. Thus if the hour angle of S (Fig. 20-4) were reckoned eastward, it would be given by the angular distance QU.

20-6. Equator Systems Compared. The system of coordinates described in Art. 20-5 is seen to be similar to that described in Art. 20-4 with this difference, that in the hour-angle system the angular distance along the equator is measured (westward) from a *fixed meridian,* while in the right-ascension system the angular distance along the equator is measured (eastward) from the *vernal equinox,* which is a point on the celestial equator that rotates with the celestial sphere. Thus, while right ascensions of fixed stars have annual variations of but a few seconds, hour angles of the stars change as rapidly as the celestial sphere apparently rotates (24^h or 360° for each 23^h 56^m of our civil time), and hour angles of the sun change approximately 24^h or 360° for each 24^h of our civil time.

The two systems are called *equator systems of coordinates,* since in each case the primary plane of reference is the celestial equator. The position of a celestial body above or below the equator is given by the declination δ which is the same in one system as in the other.

Let θ be the hour angle of the vernal equinox represented in Fig. 20-4 by the angular distance $QQ'V$ measured along the equator. At any instant of time, if the hour angle of the vernal equinox with respect to a given meridian is known, and if the right ascension α of a heavenly body S is known, then the hour angle t of the body can be computed, since by the figure

$$t = \theta - \alpha \qquad \text{or} \qquad \theta = t + \alpha$$

This equation is, therefore, an expression by means of which the coordinates of one system may be transposed to those of the other.

20-7. Astronomical Tables Used by the Surveyor. By means of astronomical observations and calculations, the positions of many of the celestial bodies are predicted, and values of their right ascensions and declinations for various dates are available in various publications. The position of a celestial body at any time can be obtained by interpolation.

The publication most widely used by *astronomers* in the United States is "The American Ephemeris and Nautical Almanac" (about 500 pages); herein it is called the "American Ephemeris." It is published one or two years in advance for each year by the Nautical Almanac Office, U.S. Naval Observatory. Formerly its tables were based on "universal" (civil) time (Art. 20-18), but beginning in 1960 the tables are now based on "ephemeris" time; it is not convenient for use in surveying.

Most of the astronomical data used by surveyors are presented in "The Nautical Almanac" (about 300 pages), which is also published annually in advance by the Nautical Almanac Office, U.S. Naval Observatory, primarily for use in navigation.

In condensed form is the "Ephemeris of the Sun, Polaris, and Other Selected Stars" (about 30 pages), herein called the "Ephemeris of the Sun and Polaris." It is published annually in advance by the U.S. Bureau of Land Management. This ephemeris lists for each day of the current year the position of the sun and of Polaris, by means of which the surveyor can compute from his field observations the latitude and longitude of the point of observation, the time of observation, or the azimuth of a reference line. It is in a form useful for land surveying. The major points of difference between the arrangement of these tables and that of the tables published by the Nautical Almanac Office are explained in Arts. 20-20 and 21-9.

All the foregoing publications are sold by the Superintendent of Documents, Government Printing Office, Washington 20402, D.C.

Useful condensed tables of data regarding the sun and Polaris are furnished to surveyors free of charge by various manufacturers of surveying instruments.

20-8. Horizon System of Spherical Coordinates. In the ordinary operations of surveying, the angles are measured in horizontal and vertical planes; to use other planes would be inconvenient. Likewise, in astronomical field work the angular location of a celestial body at a given instant is determined by measuring its vertical angle (referred to the horizon plane) and its horizontal angle (referred to a given line on the ground).

Figure 20-5 represents a portion of the celestial sphere in which O represents both the earth and the location of the observer, $NES'W$ the observer's horizon, and $S'ZN$ the meridian plane passing through his location. The point Z on the celestial sphere directly above the observer is called the *zenith.* The point S represents a celestial body, and BSZ is part of a great circle, called a *vertical circle,* through the body and the zenith. In this *horizon system* of spherical coordinates, the angular location of a celestial body is defined by its azimuth and altitude.

The *azimuth* of a celestial body is the angular distance measured along the horizon in a clockwise direction from the meridian to the vertical circle through the body. Azimuths may be reckoned from either the south point or the

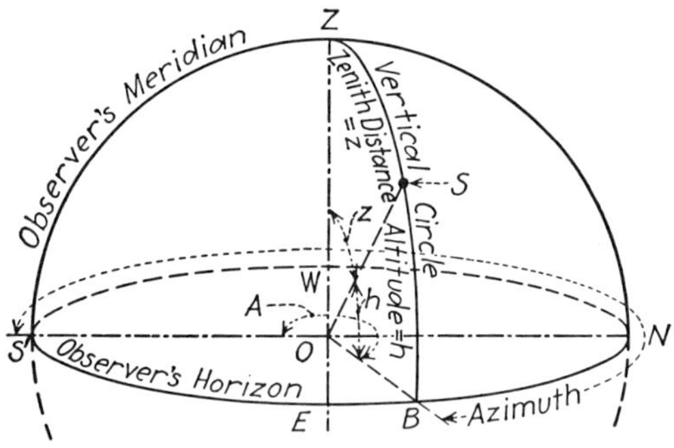

FIG. 20-5. Horizon system of spherical coordinates.

north point of the meridian; in astronomical work azimuths are customarily reckoned from south through 360°, except that for circumpolar stars they are often reckoned from north. Also in trigonometric computations the azimuths of stars west of north or east of south are often expressed as counter-clockwise angles from the meridian and are considered as negative values. In Fig. 20-5 the azimuth of S reckoned in the customary manner is given by the angle A or by the angular distance $S'NB$, an arc of the horizon. If the azimuth of S were reckoned from north, it would be given by the angle $(A - 180°)$ or by the angular distance NB. The negative azimuth reckoned from south is given by the arc $S'EB$.

The *altitude* of a celestial body is the angular distance measured along a vertical circle, from the horizon to the body; it corresponds to the vertical angle of ordinary surveying. It is expressed in degrees of arc. The altitude of S (Fig. 20-5) is given by the vertical angle h or by the angular distance BS, the arc of a vertical circle passing through the zenith. Except in rare instances, celestial objects are observed when above the true horizon, when the sign of the altitude is positive. It is seen that positive altitudes may vary between 0° and 90°.

The complement of the altitude is called the *zenith distance* or *coaltitude.* It is the angular distance from the zenith to the celestial body measured along a vertical circle. In the figure the zenith distance is given by the angle z or by the angular distance ZS. Thus $z = 90° - h$. Zenith distances are always positive.

Since the celestial sphere is apparently rotating about its axis, while the meridian, horizon, and zenith are imagined as remaining fixed in position, it is clear that in general both the azimuth and the altitude of a star are changing continuously.

20-9. Relation between Latitude, Altitude, and Declination. Figure 20-6 represents a section of the earth through the poles and the station of

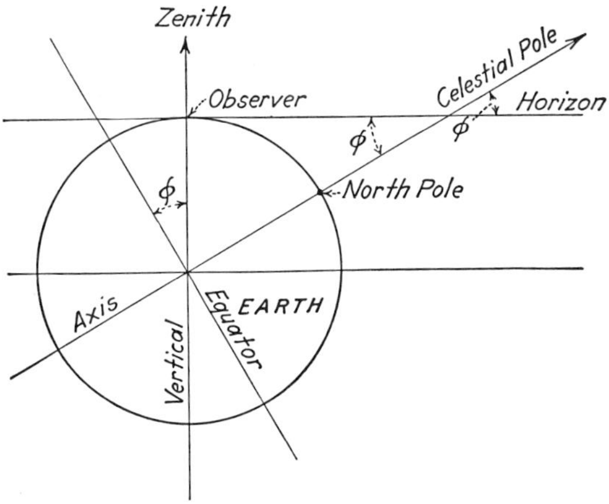

FIG. 20-6. Latitude of observer.

an observer. Since the latitude of a place is its angular distance from the equator measured along a meridian of longitude, the latitude of the observer is given by the angle ϕ between the equator and a vertical line through the observer's station, the angle being measured in the plane of the meridian. Also, from similar triangles, the latitude is given by the angle ϕ between the axis and the horizon, likewise measured in the plane of the meridian; this angle is the altitude of the elevated pole.

Similarly Fig. 20-7 represents a section of the celestial sphere through the celestial poles and observer's zenith. For reasons just explained,

$$\angle QOZ = \angle NOP = \phi = \text{latitude of place}$$

Hence the latitude of a place is given by the angular distance NP which is the altitude of the pole, or by the angular distance QZ which is the zenith distance or coaltitude of the equator. The angular distance from the pole to the zenith is $90° - \phi = c$ which is the colatitude of the place.

In a northern latitude if any heavenly body S whose declination is δ is on the meridian and south of the zenith, then from Fig. 20-7

$$\phi = (90° - h) + \delta$$

Similarly, for any star north of the zenith

$$\phi = h \pm (90° - \delta) = h \pm p$$

in which the sign preceding p, the polar distance, is positive or negative according as the star is below or above the pole.

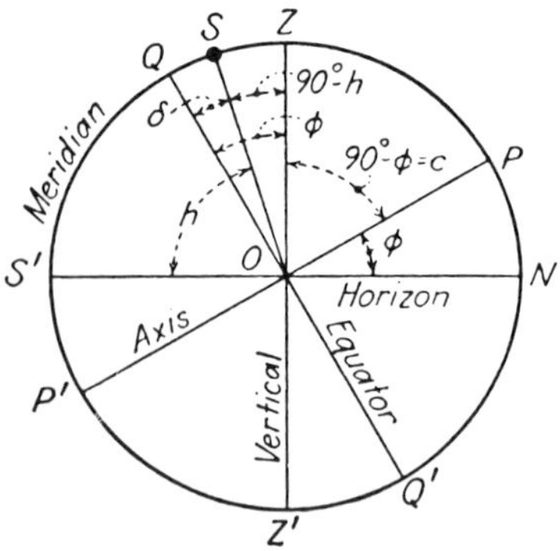

FIG. 20-7. Relation between latitude, altitude, and declination.

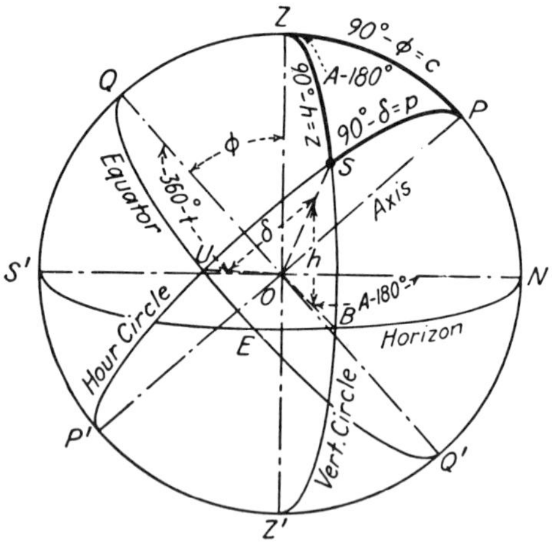

FIG. 20-8. Horizon and hour-angle equator systems combined.

20-10. Horizon and Hour-angle Equator Systems Combined. The relation between the coordinates of the horizon system and those of the hour-angle equator system described in Art. 20-5 is shown, for a star S not on the meridian, by Fig. 20-8. The meridians of the two systems coincide. The place of observation is assumed to be north of the equator at a latitude ϕ, as given either by the angle between the equator and the zenith or by the angular distance NP between the horizon plane and the celestial axis. The star is a position east of the meridian and above the celestial equator.

In the horizon system the coordinates of S are A, the azimuth measured from the *south* point of the horizon ($A - 180°$, the azimuth from north, is shown in the figure), and h, the altitude. In the equator system the coordinates are t, the hour angle measured westward from the upper branch of the meridian ($360° - t$ is shown in the figure), and δ, the declination. The colatitude ($90° - \phi = c$), the zenith distance ($90° - h = z$), and the polar distance ($90° - \delta = p$) define a spherical triangle the vertices of which are the pole P, the zenith Z, and the celestial body S. This triangle is called the *PZS triangle* or the *astronomical triangle*. Most of the problems of field astronomy involve transposing from one system of spherical coordinates to the other and solving the PZS triangle for unknown coordinates, having certain coordinates in one or both systems known or observed.

In the figure, the celestial body is shown as above the horizon and above the equator. If the body is below the horizon or below the equator, the sides of the PZS triangle are defined in a manner similar to that just described but account is taken of the algebraic sign of the altitude or the declination.

In the figure, the celestial body is shown as east of the observer's meridian; the angle Z of the spherical PZS triangle is, therefore, its azimuth from south minus 180°, or $Z = A - 180°$. Also, the angle P of the PZS triangle is equal to $360° - t$. By means of a sketch it can be shown readily that, when the body is west of the meridian, $Z = 180° - A$ and that $P = t$.

20-11. Spherical Trigonometry. The solution of a spherical triangle depends upon the principles of spherical trigonometry, of which the surveyor should have some knowledge. A derivation of the fundamental equations of spherical trigonometry follows:

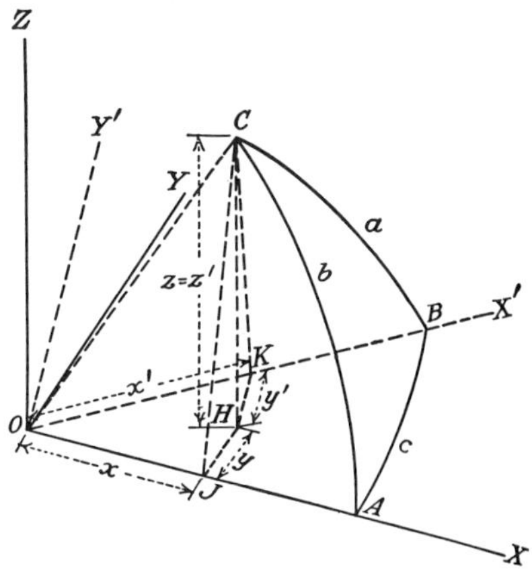

FIG. 20-9. Relations in spherical trigonometry.

In Fig. 20-9, let OX, OY, and OZ be the X, Y, and Z axes of rectangular coordinates, and let ABC be a spherical triangle on the surface of a sphere of unit radius of which O is the center, the side c being in the XY plane.

Since the radius of the sphere is unity, each of the distances OA, OB, and OC is unity, and the arcs a, b, and c are measures, respectively, of the central angles BOC, COA, and AOB.

Let H mark the projection of C on the XY plane and let JH be constructed parallel to OY. Then, since the plane of the plane triangle CJH is parallel to the YZ plane, $\angle CJH$ is equal to $\angle A$ of the spherical triangle ABC. The coordinates of C are $x = OJ$, $y = JH$, and $z = HC$. Then, since the radius of the sphere is unity

$$x = \cos b \tag{1}$$
$$y = \sin b \cos A \tag{2}$$
$$z = \sin b \sin A \tag{3}$$

Let the ZX and ZY planes be rotated about the Z axis through

$$\angle AOB = \angle c$$

the new position of the Y axis being OY' and the new position of the X axis being OX', passing through B. As before, the projection of C on the

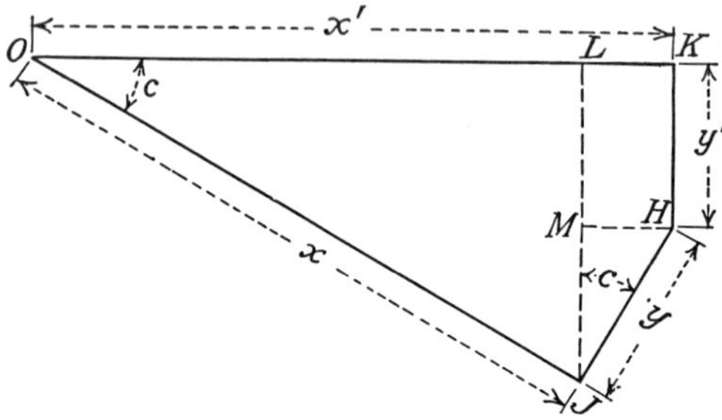

FIG. 20-10. Plan view of XY plane.

XY plane is H. Construct HK parallel to OY'. Then $\angle CKH$ of the plane triangle equals $\angle B$ of the spherical triangle. The coordinates of C with respect to the new position of the axes are $x' = OK$, $-y' = KH$, and $z' = HC$. Then

$$x' = \cos a \tag{4}$$
$$y' = -\sin a \cos B \tag{5}$$
$$z' = \sin a \sin B \tag{6}$$

From Fig. 20-9, it is clear that $z = z'$. In Fig. 20-10 is shown a plan view of the XY plane of Fig. 20-9. The angle between the two positions of the X axis is c, and the projection of the point C upon the XY plane is H. From the figure,

$$x' = OL + LK = x \cos c + y \sin c$$

Also

$$-y' = JL - JM = x \sin c - y \cos c$$

The three equations for transforming the coordinates given by Eqs. (4) to (6) to those given by Eqs. (1) to (3) are, therefore,

$$x' = x \cos c + y \sin c \tag{7}$$
$$y' = y \cos c - x \sin c \tag{8}$$
$$z' = z \tag{9}$$

Substituting in Eqs. (7) to (9) the values of the coordinates given by

Eqs. (1) to (6), there results

$$\cos a = \cos b \cos c + \sin b \sin c \cos A \tag{10}$$
$$\sin a \cos B = \cos b \sin c - \sin b \cos c \cos A \tag{11}$$
$$\sin a \sin B = \sin b \sin A \tag{12}$$

Since these equations connect the values of all angles and sides in any spherical triangle, they may be regarded as the fundamental expressions which give the relations that exist in any spherical triangle regardless of its size, shape, or position.

20-12. Solution of the *PZS* Triangle. In surveying, the astronomical triangle is solved in connection with determinations of azimuth. Observations are made on the sun or on some star that can be readily identified. The altitude of the celestial body is measured, its declination at the instant of observation is determined from published tables, and the latitude of the place of observation is known or is determined by separate observation. Hence, the three sides of the astronomical triangle are known (Fig. 20-11). The determination of azimuth of the celestial body involves the computation of the angle at Z; and determinations of longitude or time involve the computation of the angle at P as a measure of the hour angle.

In Fig. 20-11, let PZS be the astronomical triangle of Fig. 20-8, for which the sides are $90° - \phi$ (the colatitude), $90° - h$ (the coaltitude or zenith distance), and $90° - \delta$ (the codeclination or polar distance). Imagine that the spherical triangle of Fig. 20-9 is rotated in position so that its vertices A, B, and C coincide, respectively, with Z, S, and P of the astronomical triangle; then $a = 90° - \delta$, $b = 90° - \phi$, $c = 90° - h$, and $A = Z$. Substituting these values in Eq. (10), there results

$$\cos Z = \frac{\sin \delta}{\cos h \cos \phi} - \tan h \tan \phi \tag{13}$$

which is a general expression for determining azimuth from north when the three sides of the astronomical triangle are known, Z being considered positive if the star is east of the meridian and negative if the star is west of the meridian, and being less or greater than 90° according as the sign of $\cos Z$ is positive or negative.

When azimuths are reckoned from *south*, Eq. (13) takes the following form:

$$\cos A = \tan h \tan \phi - \frac{\sin \delta}{\cos h \cos \phi} \tag{13a}$$

in which A is the azimuth measured from south, clockwise if the celestial body is leaving the upper branch of the meridian, and counter-clockwise if approaching. The azimuth A is less than 90° if the sign of $\cos A$ is found to be positive, and greater than 90° if $\cos A$ is negative.

Equation (13) may also be expressed in terms of the versed sine (1 minus the cosine), as follows:

$$\text{vers } Z = [\text{vers } p - \text{vers } (\phi - h)] \sec \phi \sec h \tag{13b}$$

in which $p = 90° - \delta$ = polar distance. This is a convenient form when tables of versed sines are available.

An equation similar to Eq. (13) may be developed for the unknown angle at P. By assuming the vertices A, B, and C of the spherical triangle of Fig. 20-9 to coincide with P, S, and Z, respectively, of the astronomical

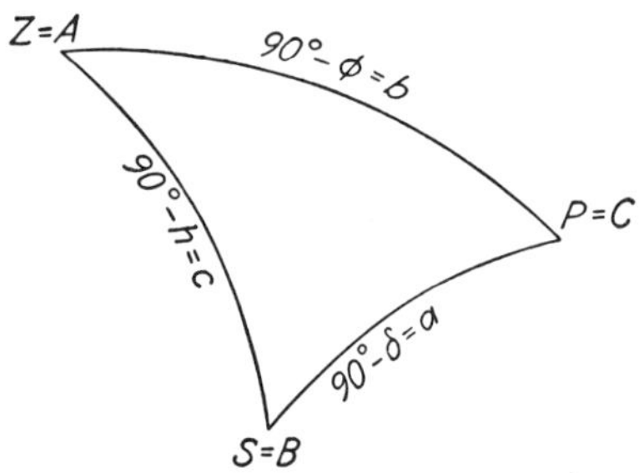

FIG. 20-11. *PZS* triangle (solution for *Z*).

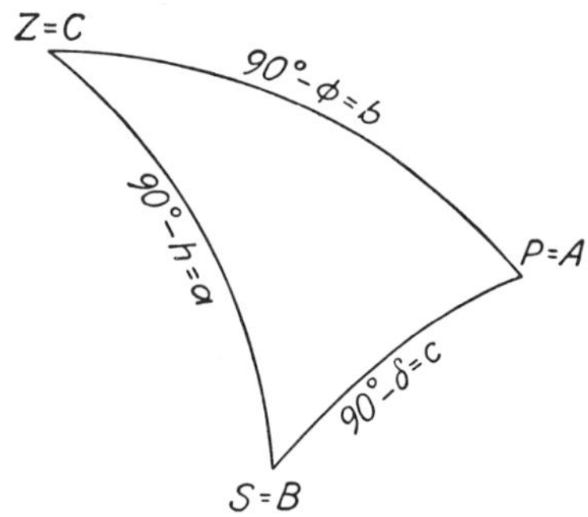

FIG. 20-12. *PZS* triangle (solution for *P*).

triangle of Fig. 20-8, then as shown by Fig. 20-12, $90° - h = a$, $90° - \phi = b$, and $90° - \delta = c$. Making these substitutions in Eq. (10) and letting $P = t$ = hour angle in either direction from the meridian,

$$\cos t = \frac{\sin h}{\cos \delta \cos \phi} - \tan \delta \tan \phi \tag{14}$$

which is a general expression for determining the hour angle of any celestial body when the three sides of the astronomical triangle are known.

The preceding equation may be expressed in the form

$$\text{vers } t = \sec \phi \sec \delta [\text{vers } z - \text{vers } (\phi - \delta)] \tag{14a}$$

in which $z = 90° - h$ = zenith distance.

20-13. Alternative Forms of Solution. *Cosines.* Equations (13) and (14) in the form given are not always as suitable nor as convenient as some other forms. When logarithmic computations are employed, the solution of either of these expressions involves the use of both logarithmic and natural trigonometric functions. Also when the unknown angle either is small or is near 180°, a relatively small error in the computed value of the cosine will produce a relatively large error in the angle itself, since the magnitude of the cosine is changing slowly. For this reason, *in so far as errors of computation are involved,* the above equations are not suitable for precisely computing azimuth and hour angle when the observed celestial body is near the meridian. On the other hand, when the unknown azimuth or hour angle is near 90° or 270°, its cosine is changing rapidly and hence Eqs. (13) and (14) are most suitable for precise computation. The following examples illustrate the point under discussion. It is seen that the error in the computed angle of example 1 is nearly eight times that of the angle of example 2.

Example 1: In determining the azimuth of a star at a given instant by Eq. (13) the errors of the computations are such that the ratio of precision of the computed cosine is 1/5,000. It is desired to know the error in the corresponding angle, the azimuth being approximately 20°. By Fig. 3-2, the angular error is approximately 01′50″, found by extrapolation at the intersection of the line representing 1/5,000 and that for 20° for cosines.

Example 2: Same conditions as example 1, but azimuth approximately 70°. The angular error in the azimuth is 15″, found by interpolation at the intersection of the 1/5,000 line with the line for 70° for cosines.

Tangents. By a series of substitutions which will not be given here but which may be found in any treatise on spherical trigonometry, Eq. (13) may be changed to the form

$$\tan^2 \frac{1}{2} Z = \frac{\sin (s - h) \sin (s - \phi)}{\cos s \cos (s - p)} \tag{15}$$

and Eq. (14) may be changed to the form

$$\tan^2 \frac{1}{2} t = \frac{\cos s \sin (s - h)}{\cos (s - p) \sin (s - \phi)} \tag{16}$$

In these two equations $p = 90° - \delta =$ polar distance, $s = \frac{1}{2}(h + \phi + p)$, and the remaining letters have the same significance as in Eqs. (13) and (14). In some cases $(s - p)$ will be negative, but the result will not be affected since the cosine of a negative angle has the same value and the same sign as the cosine of a positive angle of equal size.

When logarithmic computations are employed, Eqs. (15) and (16) are in more convenient form than Eqs. (13) and (14).

For a given angular value the tangent changes more rapidly than the cosine. Thus for a given error of computation of the trigonometric function, Eqs. (15) and (16) will generally render a closer determination of azimuth and hour angle than will Eqs. (13) and (14). For angles near 90° and 270° the difference between the rate of change of the tangent and of the cosine is not large. But when the object is near the meridian, that is, when the azimuth is near 0° or 180°, Eqs. (15) and (16) will for given errors of computation render possible closer determinations of angles than will Eqs. (13) and (14). This fact is illustrated by the following examples, a continuation of examples 1 and 2 of the preceding article.

Example 3: In determining the azimuth of a star by Eq. (15) it is desired to know what angular error will be introduced if the ratio of precision of the computed quantity $\tan^2 \frac{1}{2}Z$ is 1/5,000. If the error in $\tan^2 \frac{1}{2}Z$ is 1/5,000, then the error in $\tan \frac{1}{2}Z$ is approximately 1/10,000.

If Z is 20°, then $\frac{1}{2}Z$ is 10°. From Fig. 3-3, the corresponding angular error in $\frac{1}{2}Z$ is approximately 03½″; hence the error in the calculated value of Z is about 07″.

If Z is 70°, then $\frac{1}{2}Z$ is 35°. From Fig. 3-3, the corresponding angular error in $\frac{1}{2}Z$ is 09½″; therefore the error in the calculated value of Z is 19″.

Examples 1 to 3 are summarized in the following tabulation:

Form of equation	Error in computed azimuth	
	$Z = 20°$	$Z = 70°$
Cosine.....	110″	15″
Tangent...	7″	19″

It should be understood that the preceding discussion regarding the relative advantages of the tangent and cosine forms of expressions for determining azimuth and hour angle refers to the *errors of computation*. In effect, it means that the required precision of angle might be obtained with the tangent form with, say, a five-place table, when to obtain the same precision of computation it might be necessary to use, say, a six-place or seven-place table if the cosine form were used, all depending upon the magnitude of the calculated angle. It does not mean that a given error in an observed quantity, say the altitude, will have any less effect upon the computed value if the tangent form of equation is employed. Obviously, since the fundamental trigonometric relations are the same for one form of equation as for the other, the error introduced in a computed value on account of a given error in an observed quantity will be the same, regardless of the form of the equation used in determining the angle.

Azimuths from South. When azimuths are reckoned from south, Eq. (15) takes the following form, A being the azimuth measured either clockwise or counter-clockwise from south:

$$\cot^2 \frac{1}{2} A = \frac{\sin (s - h) \sin (s - \phi)}{\cos s \cos (s - p)} \tag{15a}$$

When $\cot \frac{1}{2}A$ has been determined, the computations for hour angle are somewhat reduced if Eq. (16) is modified as follows:

$$\tan \frac{1}{2} t = \frac{\sin (s - h)}{\cot \frac{1}{2}A \cos (s - p)} \tag{16a}$$

Haversines. Azimuths can be computed conveniently from Eq. (13c) which involves the use of *haversines*. The haversine of an angle is equal to half the versed sine, which in turn is equal to 1 minus the cosine. Tables giving values of haversines are published in Ref. 1 at the end of Chap. 21.

$$\text{hav } Z = [\text{hav } p - \text{hav } (\phi - h)] \sec \phi \sec h \tag{13c}$$

In the equation, Z is the azimuth from north (in northern latitudes), measured to the east if the sun is east of the meridian and to the west if the sun is west; the remaining symbols are as used previously

The haversine relation may be used as an independent check on computations made by means of Eq. (13).

20-14. Azimuth at Elongation. The most favorable position for determining azimuth by observation on any star which crosses the upper branch of the meridian north of the zenith occurs when it is farthest east or farthest west of the pole, when the star appears to be traveling vertically for some time. In this position it is said to be at eastern or western *elongation* according as it is east or west of the meridian. At the instant of elongation, since

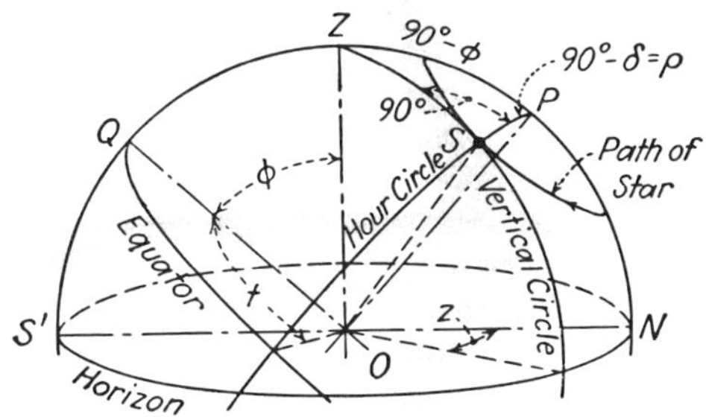

FIG. 20-13. Star at elongation.

the star appears to be traveling vertically, its apparent path in the celestial sphere is tangent to the vertical circle through the observer's zenith, as illustrated at S in Fig. 20-13. Therefore, the angle S between the plane of the hour circle and the plane of the vertical circle is a right angle. For azimuth determinations of this sort, the latitude of the place of observation is known, and either the declination or the polar distance of the star for the given date is obtained from published tables. At the instant of elongation, there are then known in the astronomical triangle the side $ZP = 90° - \phi$, the side $PS = 90° - \delta$, and the angle $S = 90°$.

If the spherical triangle ABC of Fig. 20-9 is made to coincide with the astronomical triangle of Fig. 20-13, so that $a = 90° - \delta = p$, $b = 90° - \phi$, and the vertices A, B, and C coincide, respectively, with Z, S, and P; then (remembering that the sine of S is unity when S is 90°) by substituting in Eq. (12) there is obtained

$$\sin Z = \frac{\sin (90° - \delta)}{\sin (90° - \phi)}$$

or

$$\sin Z = \frac{\sin p}{\cos \phi} \tag{17}$$

which is the general expression employed for determining the azimuth of a circumpolar star when at elongation, Z being the azimuth reckoned east or west of north according as the star is at eastern or western elongation.

By considering the spherical triangle ABC of Fig. 20-9 as taking the position PSZ in Fig. 20-13, and substituting the proper values in Eq. (11), there is derived

$$\cos t = \tan \phi \tan p \tag{18}$$

which is an expression for finding the hour angle of a star at the instant of elongation, the hour angle t being reckoned east or west of the upper branch of the meridian, depending upon the position of the star. The equation is useful in determining the time at which elongation will occur on any given date.

20-15. Azimuth of Circumpolar Star at Any Position. If the spherical triangle ABC of Fig. 20-9 is made to coincide with the astronomical triangle of Fig. 20-8, so that $a = 90° - h$, $b = 90° - \delta$, $c = 90° - \phi$, and the vertices A, B, and C coincide, respectively, with P, Z, and S, then by substituting in Eqs. (12) and (11), dividing Eq. (12) by Eq. (11), and dividing the result thereof by $\sin \delta$, there results

$$\tan Z = \frac{\sin t}{\cos \phi \tan \delta - \sin \phi \cos t} \tag{19}$$

This expression is commonly used in finding the azimuth of Polaris or any other circumpolar star when the star is not at elongation and when the hour angle of the star is precisely known.

20-16. Altitude of a Star. When a star cannot be readily identified through the transit telescope, the process of bringing it into the field of view is considerably expedited if its approximate altitude is computed prior to the observation and laid off on the vertical circle of the transit. Also a check on the correctness of observations and computations for azimuth and hour angle is obtained if the computed value of the altitude agrees with the observed value.

Again referring to the derivation of the fundamental equations of Art. 20-11, if the PZS triangle is substituted for Fig. 20-9 in such manner that $a = 90° - h$, $b = 90° - \phi$, $c = 90° - \delta$, and the vertices P, S, and Z lie, respectively, at A, B, and C, then by substituting in Eq. (10) there is obtained

$$\sin h = \sin \phi \sin \delta + \cos \phi \cos \delta \cos t \tag{20}$$

in which t is the hour angle at a given time, and h is the altitude at the same instant.

By trigonometric substitutions there may be derived from Eq. (20) the expression

$$\sin h = \cos (\phi - \delta) - \cos \phi \cos \delta \text{ vers } t \tag{21}$$

a form more suitable for precise determinations of the altitude.

TIME

20-17. Time; Solar and Sidereal Day. As the earth rotates about its axis in its travel through space, all celestial bodies apparently rotate about the earth (or about its axis) from east to west. Since the earth in its orbit travels about the sun but does not travel about the fixed stars, which are far outside its orbit, once each year the sun apparently encircles the celestial sphere along a path called the *ecliptic*, which twice cuts the celestial equator during this interval. The point among the stars where the sun in its apparent travel northward cuts the celestial equator on March 21 of each year is called the *vernal equinox*, which is a point of reference whose position on the celestial sphere is unchanging. There is no star at that point, but it is helpful to imagine that the vernal equinox is an invisible celestial body rigidly fastened in its position on the celestial sphere, while each of the so-called "fixed" stars slowly moves along a path of extremely small compass on the surface of the sphere, and the sun travels rapidly along the ecliptic in a direction opposite to that of the rotation of the celestial sphere.

The vernal equinox is referred to in ephemerides as "The First Point of Aries" or simply "Aries"; it is often represented by the zodiacal symbol ♈.

Because the sun is apparently traveling from west to east among the stars, while the rotation of the celestial sphere about the earth is apparently from east to west, the angular velocity of the sun about the axis of the celestial sphere is less than that of the fixed stars or of the vernal equinox, just as the angular velocity of a passenger walking toward the rear of a train on a circular track is less than that of the train. At a given meridian the hour angle of the sun and that of the vernal equinox will agree at some instant on March 21, but thereafter it will be less for the sun than for the vernal equinox. Six months later, on September 22, when the sun has covered one half of its annual journey, the hour angle of the sun will be 180° or 12^h less than that of the vernal equinox; and 1 year later the hour angle of the sun will be 360° or 24^h less than that of the vernal equinox, and hence the hour angles will again agree.

In the course of a tropical year as measured by the time taken by the sun apparently to make a complete circuit of the ecliptic, there actually occur 366.2422 revolutions of the earth, or apparently a like number of revolutions of the vernal equinox about the earth. For reasons just explained, the sun during this interval will have traveled through a total hour angle 360° or 24^h

less than that traversed by the vernal equinox, hence during a tropical year the sun apparently revolves about the earth 365.2422 times.

The interval of time occupied by one apparent revolution of the sun about the earth is called a *solar day*, the unit with which we are all familiar. The interval of time occupied by one apparent revolution of the vernal equinox is called a *sidereal day*, a unit much used by astronomers. Since 366.2422 sidereal days occupy the same period of time as 365.2422 solar days, the sidereal day is a shorter interval of time than the solar day.

When any celestial body, real or imaginary, apparently crosses the upper branch of a meridian, it is said to be at *upper transit* or *upper culmination;* when any celestial body crosses the lower branch of the meridian it is said to be at *lower transit* or *lower culmination.*

The beginning of a sidereal day at a given place occurs at the instant the vernal equinox is at upper transit.

The solar day is considered as beginning at the instant of lower transit of the sun (midnight), as does the civil day. (Prior to 1925, for astronomical purposes the solar day was considered as beginning at noon.)

Both sidereal and solar days are divided into 24 hr. each of 60 min. duration. For surveying purposes, the hours are reckoned consecutively from 0 to 24.

20-18. Civil (Mean Solar) Time. On account of the elliptical shape of the earth's orbit, the apparent angular velocity of the sun that we see, called the *apparent sun* or the *true sun*, is not constant; during four periods of each year it is greater, and during four intervening periods less, than the average velocity. Hence the days as indicated by the apparent travel of the true sun about the earth are not of uniform length. To make our solar days of uniform length, astronomers have invented the *mean sun*, a fictitious body which is imagined to move at a uniform rate along the celestial equator, making a complete circuit from west to east in one year. The time interval as measured by one daily revolution of the mean sun is called a *mean solar day*, which is the same as the civil day. The mean solar day begins at midnight, as does the civil day, and the *mean solar time* at any place is given by the hour angle of the mean sun plus 12^h. Thus, if the hour angle of the mean sun is $-15° = -1^h$, the mean solar time is $-1^h + 12^h = 11^h$. With regard to time, the terms "mean" and "civil" are interchangeable.

Civil time has the same meaning as *mean solar time* or *mean time* or *universal time* and, in the form of *standard time* (Art. 20-23), is the time in general use by the public. *Local civil time* is that for the meridian of the observer. Civil time for any other meridian is designated by name; for example, *Greenwich civil time.* Civil time for any meridian can be converted into terms of civil time for any other meridian by computations involving the longitude of the two meridians, as described in Art. 20-22; 1^h civil time corresponds to 1^h or 15° of longitude.

20-19. Apparent (True Solar) Time. The time interval as measured by one apparent revolution of the true sun about the earth is called an *apparent solar day*. The apparent solar day begins at midnight and the *apparent solar time* at any place is given by the hour angle of the true sun plus 12^h. Thus, if the hour angle of the true sun is $45° = 3^h$, the apparent solar time is $3^h + 12^h = 15^h$. With regard to time, the terms "true" and "apparent" are interchangeable.

Apparent time has the same meaning as *apparent solar time*. *Local apparent time* is that for the meridian of the observer. Apparent time for any other meridian is designated by name; for example, *Greenwich apparent time*. Apparent time for any meridian can be converted into terms of apparent time for any other meridian by computations identical with those for civil time, as described in Art. 20-22; 1^h apparent time corresponds to 1^h or 15° of longitude.

20-20. Equation of Time. When the apparent (true) sun is ahead of the mean sun, apparent time is faster than mean (civil) time; when behind, slower. The difference between apparent time and civil time at any instant is called the *equation of time*. It is used to convert civil time at any instant into apparent time and *vice versa*.

The maximum value of the equation of time is only about 16^m; hence for work in which its only use is for the determination of change in declination, it is sometimes neglected.

The equation of time may be obtained either from an ephemeris which gives values at given instants of civil time or from one which gives values for apparent time, as follows:

1. In the "Nautical Almanac," the equation of time, to 01^s, is given for each day at 0^h and 12^h Greenwich civil time. The civil time of meridian passage of the apparent (true) sun is also given. If the civil time of meridian passage is before noon, the equation of time should be added to civil time to obtain apparent time; if after noon, subtracted.

2. In the "Ephemeris of the Sun and Polaris" of the U.S. Bureau of Land Management, the equation of time, to 0.01^s, is given for each day at the instant of Greenwich apparent noon. The column headings state directly whether the equation of time is to be added to or subtracted from the apparent time when the civil time is desired.

To find the equation of time at any instant other than that for which a value is tabulated, it is necessary to interpolate, adding to or subtracting from the tabulated value of the equation of time the change in the equation of time since the instant to which the tabulated value applies.

Example 1: It is desired to determine by use of the "Nautical Almanac" the equation of time at the instant of $3^h30^m45^s$ P.M. Greenwich civil time on December 15, 1964. Greenwich civil time $= 12^h + 3^h30^m45^s = 15.51^h$.

From the "Nautical Almanac" the equation of time at 12^h G.C.T. is 04^m44^s. The change in the equation of time in 12 hr (to 0^h December 16) is

$$04^m29^s - 04^m44^s = -15^s$$

The change in the equation of time up to the given instant is

$$\frac{15.51 - 12}{12} \times (-15) = -4.4^s$$

The equation of time at the given instant is

$$04^m44^s - 4.4^s = 04^m39.6^s$$

The Greenwich civil (mean) time of meridian passage of the apparent (true) sun is given as 11^h55^m; therefore the equation of time is to be added to mean time to obtain apparent time or subtracted from apparent time to obtain mean time.

Example 2: It is desired to determine by use of the "Ephemeris of the Sun and Polaris" the Greenwich civil time (G.C.T.) at the instant of $9^h00^m15^s$ Greenwich apparent time (G.A.T.) on October 10, 1964. The time that will elapse before G.A. noon is 3.00^h.

The equation of time at G.A. noon is $13^m00.8^s$, to be subtracted from apparent time. The change in one day is $13^m00.8^s - 12^m44.8^s = 16.0^s$.

The change before G.A. noon is $\dfrac{3.00}{24} \times 16.0 = 2.0^s$.

$$\text{Eq. time for } 9^h00^m15^s \text{ G.A.T.} = 13^m00.8^s - 2.0^s = 12^m58.8^s$$

$$\begin{aligned} \text{G.A.T.} &= 9^h00^m15^s \\ \text{Eq. time} &= 12^m58.8^s\text{, to be subtracted from apparent time} \\ \text{G.C.T.} &= 8^h47^m16.2^s \end{aligned}$$

By inspecting the tabulated values of the equation of time as given in the ephemerides it will be seen that in February the true sun is as much as 14^m behind the mean sun and that in November the true sun is more than 16^m ahead of the mean sun, while on about the dates April 15, June 15, September 1, and December 25, the equation of time is zero and hence the hour angle of the true sun is for an instant the same as that of the mean sun.

20-21. Sidereal Time. The *sidereal time* at any place is the hour angle of the vernal equinox at that place; and the beginning of the sidereal day, occurring when the vernal equinox crosses the upper branch of the meridian, is called *sidereal noon.* Twenty-four-hour clocks regulated to keep sidereal time are called *sidereal clocks.* The vernal equinox, also called Aries, is an imaginary point and cannot be observed like the sun; but the right ascensions of stars are referred to the vernal equinox, and therefore the sidereal time can be obtained by determining the hour angle of any star the right ascension of which is known. Then if θ is the sidereal time, $\theta = t + \alpha$, as explained in Art. 20-6.

The sidereal day is shorter than the mean solar day by $3^m55.9^s$ mean solar time, or $3^m56.6^s$ sidereal time. The sidereal hour is shorter than the mean solar hour by 9.830^s mean solar time, or 9.856^s sidereal time.

The following example illustrates the use of the "Nautical Almanac" in the computation of the sidereal time at a given place at a given instant G.C.T., for which instant the hour angle of the apparent sun has been determined.

Example 1: At a given place the hour angle of the apparent sun at 10 A.M. Central standard time (4 P.M. or 16^h Greenwich civil time) on July 3, 1964, is $-32°15'45''$. It is desired to know the sidereal time at the given instant.

By the "Nautical Almanac" the Greenwich hour angle (G.H.A.) of Aries (the vernal equinox) for 16^h G.C.T. is $161°42.1'$. Also, the Greenwich hour angle of the apparent sun at that instant is given as $58°57.6'$.

The Greenwich hour angle of the place is $58°57.6' - (-32°15.75') = 91°13.35'$, which is the longitude of the place (see Art. 20-22).

The local hour angle of the vernal equinox, which is the sidereal time, is

$$161°42.1' - 91°13.35' = 70°28.75'$$

The sidereal time expressed in hour units is

$$\frac{70°28.75'}{15} = 4^h41^m55^s$$

The "Nautical Almanac" is useful for converting sidereal time into civil (mean solar) time, and *vice versa*. The following example illustrates one method of determining the Greenwich sidereal time (G.S.T.) corresponding to a given instant for which the G.C.T. is known.

Example 2: It is desired to know the Greenwich sidereal time corresponding to $15^h30^m15^s$ G.C.T. on August 1, 1964. Mean solar time interval since

$$0^h \text{ G.C.T.} = 15.504^h$$

By the "Nautical Almanac" the Greenwich hour angle of the vernal equinox (Aries) is

G.H.A. at 15^h G.C.T.	$175°14.7'$
G.H.A. at 16^h G.C.T.	$190°17.2'$
Difference in 1^h	$15°\ 2.5' = 902.5'$
Difference in 0.504^h	$454.9' = 7°34.9'$

$$\text{G.H.A. at } 15^h30^m15^s \text{ G.C.T.} = 175°14.7' + 7°34.9' = 182°49.6'$$

The sidereal time expressed in hour units is

$$\frac{182°49.6'}{15} = 12^h11^m18^s$$

20-22. Relation between Longitude and Time. As the sun apparently makes a complete revolution (360°) about the earth in one solar day (24 hr.), and as the longitudes of the earth range from 0° to 360°, it follows that in 1 hr. the sun apparently traverses $360/24 = 15°$ of longitude. The same statement applies equally well to the sidereal day and the vernal equinox. It follows that at any instant, the *difference in local time* between two places, whether the time under consideration be sidereal, mean solar, or apparent solar, is equal to the *difference in longitude* between the two places, expressed in hours. This relation is used to determine the difference in time when the difference in longitude between two places is known, or *vice versa.*

Some solar ephemerides are for the meridian of Greenwich, and a problem of frequent occurrence is to find the local time corresponding to a given instant Greenwich time, or *vice versa.* The local time (L.T.) of a place at a given instant is obtained by adding to or subtracting from the Greenwich time (G.T.) the difference in longitude ($\Delta\lambda$), expressed in hours, between the two places. If the place is east of Greenwich, the difference in longitude is added; if the place is west, the difference in longitude is subtracted.

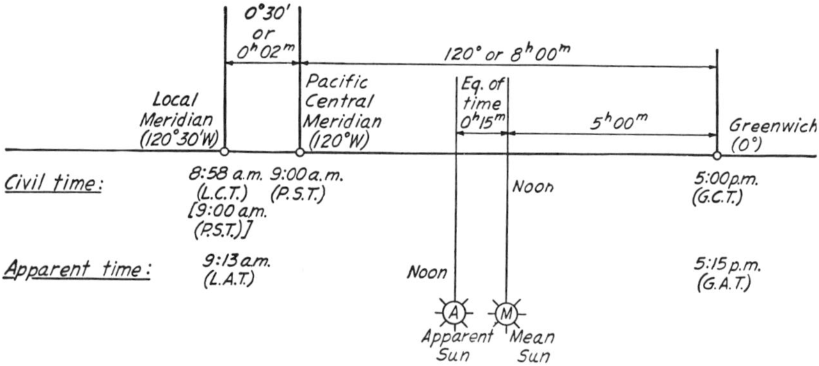

FIG. 20-14. Relations between longitude and time at a given instant.

Example 1: An observation on the sun is taken at $9^h52^m56^s$ local apparent time (L.A.T.). The longitude of the place is $7^h12^m36^s$ west of Greenwich. What is the Greenwich apparent time (G.A.T.)?

$$\begin{aligned} \text{G.A.T.} = \text{L.A.T.} + \Delta\lambda &= 9^h52^m56^s + 7^h12^m36^s \\ &= 17^h05^m32^s \end{aligned}$$

Example 2: On a given date the mean sun crosses the lower branch of the Greenwich meridian at $3^h52^m48.6^s$, Greenwich sidereal time. At that instant it is desired to find the local sidereal time at a place whose longitude is $5^h12^m24.2^s$ west of Greenwich.

$$\begin{aligned} \text{L.S.T.} = \text{G.S.T.} - \Delta\lambda &= 3^h52^m48.6^s - 5^h12^m24.2^s + 24^h \\ &= 22^h40^m24.4^s \text{ of the previous day} \end{aligned}$$

Example 3: At the instant of $18^h48^m15^s$ Greenwich civil time, the local civil time of a place is $10^h37^m42^s$. It is desired to determine the longitude of the place with respect to Greenwich.

$$\begin{aligned} \Delta\lambda &= 18^h48^m15^s - 10^h37^m42^s = 8^h10^m33^s \\ &= 122°38'15'' \text{ west of Greenwich} \end{aligned}$$

Example 4: It is desired to find the local civil time at longitude 122°38′15″ W, at the instant of $18^h48^m15^s$ Greenwich civil time.

The difference in longitude, in hours, is equal to the difference in longitude, in degrees, divided by 15.

$$\text{Local civil time} = 18^h48^m15^s - \frac{122°38'15''}{15} = 10^h37^m42^s$$

Sketches are a valuable aid in the solution of problems involving longitude and time, as they enable the surveyor to visualize the relations. A simple "straight-line" type of sketch is shown in Fig. 20-14, for the instant of 9:00 A.M. Pacific standard time. For clarity, values are given only to 01^m; the actual computations of a surveying problem would be more precise.

20-23. Standard Time. In order to eliminate the industrial confusion attendant upon the use of local time by the public, the United States has been divided into belts, each of which occupies a width of approximately 15° or 1^h of longitude. In each belt the watches and clocks that control civil affairs all keep the same time, called *standard time*, which is the local civil time for a meridian near the center of the belt. The time in any belt is a whole number of hours slower than Greenwich civil time, as follows:

Standard time	Abbreviation	Hours slower than Greenwich civil time	Central meridian	Where used
Atlantic...	A.S.T.	4	60°W	Maritime provinces of Canada
Eastern...	E.S.T.	5	75°W	Maine to central Ohio
Central...	C.S.T.	6	90°W	Central Ohio to central Nebraska
Mountain.	M.S.T.	7	105°W	Central Nebraska to western Utah
Pacific....	P.S.T.	8	120°W	West of Utah
Yukon....	Y.S.T.	9	135°W	Eastern Alaska
Alaska....	A.S.T.	10	150°W	Central Alaska, Hawaii
Bering Sea.	B.S.T.	11	165°W	Western Alaska

The exact boundaries of the time belts are irregular and can be determined only from a map.

Correct standard time can be obtained either from a clock known to be closely regulated or from radio signals, preferably those broadcast by the U.S. Naval Observatory.

In certain localities, "daylight saving time" is employed during the summer months. Daylight saving time is 1^h faster than standard time.

Computations. The Greenwich civil (mean) solar time is found by adding to the standard time the longitude (expressed in hours) of the meridian for which standard time is also local mean time.

Example 1: At a given instant the Central standard time is 9^h00^m A.M. It is desired to find the Greenwich civil time. The longitude of the meridian to which Central standard time is referred is 90° or 6^h west of Greenwich. The Greenwich civil time is $9^h00^m + 6^h00^m = 15^h00^m$, or 3^h00^m P.M.

If the longitude of a place is known, the standard time of the belt in which the place is situated can be determined by adding algebraically to the local mean time the difference in longitude (expressed in hours) between the given place and the meridian for which standard time is also local mean time.

Example 2: By observation on a star, the local mean time at a given instant is found to be $18^h37^m46^s$. The longitude of the place is $\lambda_l = 89°49'30'' = 5^h59^m18^s$. The standard time at the given instant is to be found. The place is evidently in the Central time belt for which the standard time (C.S.T.) is local time for the 90th meridian. The longitude of this meridian expressed in hours is $\lambda_s = 90°/15 = 6^h$.

$$\begin{aligned} \lambda_l &= 5^h59^m18^s \\ \lambda_s &= 6^h00^m00^s \\ \hline \Delta\lambda &= -0^m42^s \end{aligned}$$

$$\text{C.S.T.} = \text{L.M.T.} + \Delta\lambda = 18^h37^m46^s + (-0^m42^s) = 18^h37^m04^s = 6^h37^m04^s \text{ P.M.}$$

20-24. Numerical Problems.

1. When the local apparent time is $8^h17^m12^s$ at a place whose longitude is 96°15′10″W, what is the Greenwich apparent time?
2. On a given data 0^h Greenwich civil time occurs at $4^h17^m32^s$ Greenwich sidereal time. At that instant what is the local sidereal time at a place whose longitude is $7^h17^m43^s$W?
3. When it is $15^h31^m12^s$ Greenwich civil time, it is $10^h16^m37^s$ local civil time at a given place. What is the longitude of the place?
4. What is the Greenwich civil time when it is 3^h15^m P.M. Central standard time?
5. If the local civil time at a place is $16^h23^m22^s$ and the longitude of the place is 78°36′20″W, what is the Eastern standard time?
6. From an ephemeris find the equation of time for the instant of $4^h15^m00^s$ P.M. Pacific standard time on April 21 of the current year. If the longitude of the place is $7^h46^m03^s$W, calculate the local civil and local apparent times.
7. From an ephemeris, find the equation of time for the instant of $3^h19^m30^s$ P.M. apparent time July 4 of the current year, at a place whose longitude is $6^h15^m30^s$W. Compute the corresponding local civil time.
8. At a given place the hour angle of the true sun at 11^h30^m P.M. Greenwich civil time on January 12 of the current year is 42°36′30″. What is the local sidereal time?

REFERENCES

See references for Chap. 21.

CHAPTER 21

AZIMUTH, LATITUDE, LONGITUDE, AND TIME

21-1. General. In this chapter are described the rough methods commonly used in the United States on surveys of ordinary precision where the engineer's transit or the repeating theodolite is employed for angular measurements. For more precise methods, such as those necessary on precise geodetic surveys, the reader is referred to texts on geodesy and engineering astronomy (see references at end of chapter). The methods discussed herein are based upon the relations given in Chap. 20.

Since most observations are taken on the sun and Polaris, the discussion is concerned chiefly with these two bodies; but the principles involved are the same for any star.

Measurements to the sun cannot be made with the same degree of precision as to a star, hence the probable error in computed values is larger than when a fixed star is chosen. However, the sun may be viewed at convenient times, and solar observations are suitable for determinations of azimuth, latitude, and longitude with sufficient precision for most ordinary surveys.

Polaris, being near the pole, changes its position slowly. It is the most favorably located of all bright stars for precise determinations of latitude and azimuth, but owing to its slow change in azimuth it is not suitable for longitude or time observations.

21-2. Measurement of Angles. Whenever observations are made to determine azimuth, a part of the field work consists in measuring the horizontal angle between the celestial body and a reference mark on the earth's surface. As the sights to the celestial body are in general steeply inclined, it is highly important that the horizontal axis of the transit be in adjustment with respect to the vertical axis and that the transit be very carefully leveled (Art. 13-32). Even though the horizontal axis is in perfect adjustment, it will be inclined unless the vertical axis is truly vertical; and the error due to such inclination will in general not be eliminated by a reversal of the telescope between sights. For precise observations the transit should be equipped with a sensitive striding level by means of which the horizontal axis may be leveled prior to each sight. With the ordinary transit not so equipped, the plate should be leveled by means of the telescope level when other than rough observations are being made. Also sights should be taken with the telescope in both the normal and the inverted positions in order that the mean of horizontal angles may be free from other instrumental errors.

Whenever altitudes are observed, the index error of the vertical circle should be determined at the time of the observation (see Art. 13-21). The errors due to the line of sight's not being parallel to the axis of the level tube and due to the vertical vernier's being displaced can be eliminated by double-sighting (Arts. 13-20 and 13-21); however, any error due to inclination of the vertical axis will not be eliminated by double-sighting and, therefore, the transit should be leveled with great care, preferably by means of the telescope level.

The transit should be supported firmly; if the set-up is on soft ground, pegs should be driven to support the tripod legs.

SOLAR OBSERVATIONS

21-3. Observations on the Sun. To observe the sun directly through the telescopic eyepiece may result in serious injury to the eye. A piece of colored or smoked glass may be held between the eye and the eyepiece. Some transits are equipped with a colored *sun glass* that may be attached to the eyepiece.

Good observations can be made by bringing the sun's image to a focus on a white card held several inches in the rear of the telescopic eyepiece. A rough pointing on the sun is made by sighting over the telescope. The eyepiece is then drawn back, and the objective is focused until the sun's image and the cross-hairs are clearly seen on the card. If the eyepiece of the telescope is erecting, the image on the card will be inverted; if the eyepiece is inverting, the image will be erect. The cross-hairs are visible only on the image of the sun. As the angle between lines of sight defined by the stadia hairs is 34′, while the diameter of the sun is only 32′, not all of the three horizontal cross-hairs are visible on the sun's image at the same time. A common blunder is to mistake one of the stadia hairs for the middle cross-hair. This mistake can be avoided by rotating the telescope slightly about the horizontal axis until all three hairs have been seen.

When the vertical angle of sighting is large, it is impossible to look directly through the transit telescope. The *prismatic eyepiece* is a device which, when attached to the telescopic eyepiece, reflects the rays through an angle of 90° with the axis of the telescope. The image appears upside down, but it is not reversed horizontally. The prismatic eyepiece is equipped with a sun glass. The sun may be sighted at high altitudes by means of a white card held in the rear of the eyepiece, as described in the preceding paragraph.

The *solar screen* is a device utilizing the principle of the card. It consists of a piece of ground white glass fixed to a metal arm which is screwed or clamped to the eyepiece end of the telescope. The sun and the cross-hairs are brought to a focus on the ground glass as previously described for the white card.

21-4. Semidiameter Correction. As the sun is large (apparent angular diameter about 32′), its center cannot be sighted precisely with the ordinary transit, and it is customary to bring the cross-hairs tangent to the sun's image. When the horizontal cross-hair is brought tangent to the lower edge of the sun, the sight is said to be taken to the sun's *lower limb*, and this fact is indicated in the notes by the symbol $\underline{\odot}$. Similarly the symbol $\overline{\odot}$ indicates a sight to the sun's *upper limb*, $\odot|$ a sight with the vertical cross-hair to the sun's *right limb*, and $|\odot$ a sight to the sun's *left limb.*

When a vertical angle is measured to the sun's upper limb, it is necessary to subtract the sun's semidiameter in order to obtain the observed altitude of the sun's center; when to the lower limb, add the semidiameter. The

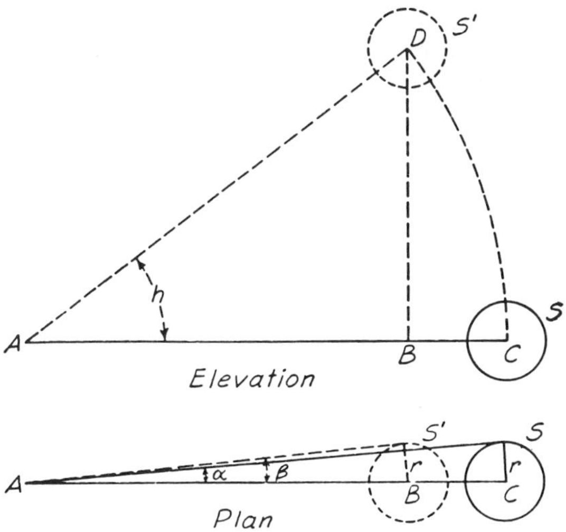

FIG. 21-1. Semidiameter correction to horizontal angle.

"Ephemeris of the Sun and Polaris" and other solar ephemerides give values of the semidiameter of the sun for each day of the year. The semidiameter varies from about 15′46″ in July to about 16′18″ in January; for rough calculations it may be taken as 16′.

When a horizontal angle is measured to the sun's right or left limb, a correction equal to the sun's semidiameter times the secant of the altitude is applied. Thus if the altitude h is 60° and the semidiameter is 16′, the correction to a horizontal angle is $16' \sec h = 32'$. As the sun approaches the zenith, the correction becomes very large (approaching 90°) and readings should not be taken to one limb only.

The semidiameter correction to a horizontal angle for the sun at any altitude is illustrated in Fig. 21-1 in which A is the station of an observer on the earth,

S is the sun at the horizon, S' is the sun at some altitude h above the horizon, r is the radius of the sun, and α and β are the small horizontal angles (semidiameter corrections) subtended by the radius of the sun at S and S', respectively. In the plan view, $\alpha \cdot AC = r$ and $\beta \cdot AB = r$; hence $\alpha \cdot AC = \beta \cdot AB$, or $\beta = \alpha \cdot AC/AB$. But in the elevation view $AC = AD$ and $AD/AB = \sec h$. Therefore,

$$\beta = \alpha \sec h \tag{1}$$

For most solar observations, an equal number of sights are taken to opposite limbs of the sun; the mean of the horizontal angles and the mean of the vertical angles at the mean of the times are taken, and no corrections for semidiameter are necessary.

21-5. Procedure of Sighting. The process of bringing both cross-hairs tangent to the sun is illustrated by Fig. 21-2, which shows the position of the

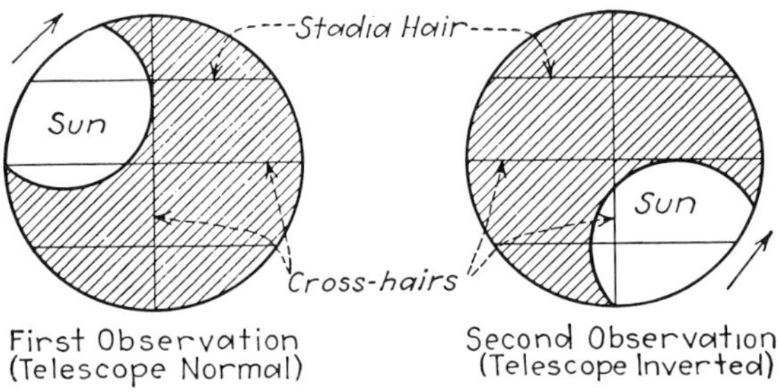

FIG. 21-2. Position of sun in field of view just prior to tangency in morning.

sun in the field of view of an erecting telescope just prior to tangency, in the morning. The portion of the field of view covered by the sun in the position shown depends on the angle of the field of view. For telescopes of low magnifying power, the full disk may be visible. If the position of the sun is determined by a card held in the rear of the eyepiece, the image on the card will include that portion of the sun's disk shown in the figure, and the cross-hairs on the shaded portion of the field of view will not be visible.

For the first of a pair of observations, the horizontal cross-hair is sighted a short distance above the sun's lower limb as illustrated. As the altitude of the sun is increasing, the horizontal cross-hair approaches tangency owing to the sun's apparent movement. At the same time, the vertical cross-hair is kept continuously on the sun's western limb by means of the upper-motion tangent-screw. At the instant when the vertical and horizontal cross-hairs are simultaneously tangent to the sun's disk, the motion of the telescope is stopped, the time is observed, and the horizontal and vertical circles are read. The telescope is then plunged, and the second observation is taken with the sun in the lower right-hand quadrant as shown in Fig. 21-2, as follows: The vertical cross-hair is set a short distance to the right of the sun's eastern limb. As the sun is traveling westward, the vertical cross-hair approaches tangency owing to the sun's apparent movement. At the same time, the horizontal cross-hair is kept continuously on the sun's upper limb by means of the telescope tangent-screw. As before, observations are taken for the instant when the horizontal and vertical cross-hairs are simultaneously tangent to the sun's disk. The procedure is such that the final setting for either observation requires the manipulation of only one tangent-screw and that the cross-hair which is approaching tangency is visible on the sun's disk. Also the procedure of double-sighting eliminates certain instrumental errors.

For afternoon observations in northern latitudes the procedure is the same as that just described, except that the sun is sighted first in the upper right-hand quadrant and then in the lower left-hand quadrant.

If the transit is not equipped with a full vertical circle, it cannot be plunged between sights, but otherwise the procedure may be as just described, and the means for the two observations taken as the observed angles.

Simplex Solar Shield. A special device for observing the sun without correction for semidiameter is the *Simplex solar shield* developed by Professor C. H. Wall of Ohio State University. It consists of a symmetrically perforated shield (Fig. 21-3) which is mounted between the eyepiece of the transit and a plate upon which the sun's image is focused. The perforations and other sighting points are so arranged that when a selected pair is brought tangent to the sun's image, the center of the sun's image is on the horizontal (or vertical) cross-hair. The illustration shows the sun's image in five successive positions as it travels (from left to right) for a distance of one full diameter in azimuth while the horizontal motion is clamped and the vertical tangent-screw is operated by the observer. At each position the time and the vertical angle are observed. The telescope is then plunged, and a second series of five readings is taken. The device can be used for observations of either azimuth or latitude.

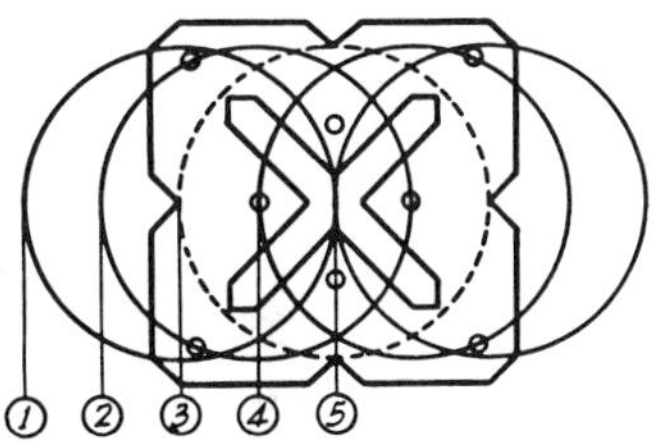

FIG. 21-3. Sun's image in successive azimuth positions with relation to Simplex solar shield.

Solar Prism. Another sighting device which requires no correction for semidiameter consists of a prism attachment whereby four overlapping images of the sun are simultaneously formed into a symmetrical square pattern the center of which is at the intersection of the cross-hairs. The device and its use are described in Ref. 5 at the end of this chapter.

21-6. Parallax Correction. In the previous discussions, it has been assumed that the celestial sphere is of infinite radius and that a vertical angle measured from a station on the surface of the earth is the same as it would be if measured from a station at the center of the earth. For the fixed stars this assumption yields results that are sufficiently accurate for the work described herein; but the distance between the sun and the earth is relatively small, and for solar observations a *parallax correction* is added to the observed altitude to obtain the altitude of the sun from the center of the earth.

In Fig. 21-4*a*, h' is the altitude of the sun above the horizon of an observer at A, and h is the altitude of the sun above the celestial horizon. The parallax correction is equal to the difference between these two angles. As h is always larger than h', the correction must be added to the observed altitude.

The parallax correction can be computed, since the distance to the sun is known. The magnitude of the correction depends upon the altitude,

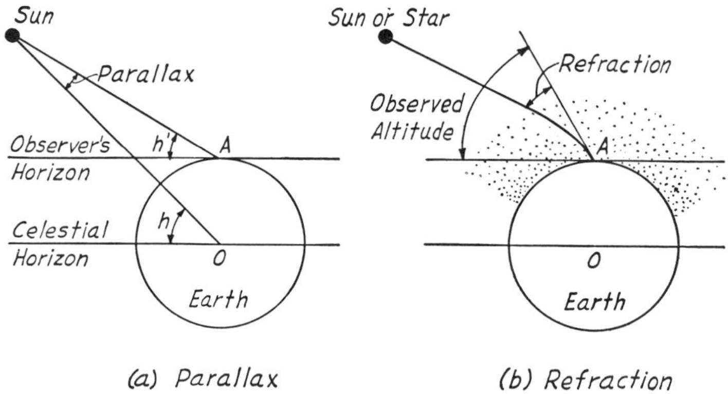

FIG. 21-4. Parallax and refraction.

being zero when the sun is directly overhead, and being a maximum when the sun is on the observer's horizon. When the sun is on the observer's horizon, the correction C_h is called the *horizontal parallax*. It can be readily demonstrated that the parallax correction C_p for any observed altitude h' is

$$C_p = C_h \cos h' \tag{2}$$

The horizontal parallax is always slightly less than 09″; hence the parallax correction at any altitude cannot exceed 09″. Values of the sun's parallax correction at various altitudes are given in the "Ephemeris of the Sun and Polaris."

Corrections for parallax and refraction are usually made together (see Art. 21-8).

21-7. Refraction Correction. When a ray of light emanating from a celestial body passes through the atmosphere of the earth, the ray is bent downward, as illustrated in Fig. 21-4*b*. Hence the sun and stars appear to be higher above the observer's horizon than they actually are. The angle of deviation of the ray from its direction on entering the earth's atmosphere to its direction at the surface of the earth is called the *refraction* of the ray. A *refraction correction* C_r of amount equal to the refraction is subtracted from the observed altitude to determine the actual altitude h' above the observer's horizon.

The magnitude of the refraction correction depends upon the temperature and barometric pressure of the atmosphere and upon the altitude of the ray, varying as the cotangent of the altitude. It does not depend upon the distance to the body from which emanates the ray. Under normal conditions the refraction correction is about 34′ when the sun or star is on the observer's horizon, about 05′ when the altitude is 10°, about 01′ when the altitude is 45°, and zero when the altitude is 90°. Table II herein gives values of refraction corrections for a barometric pressure of 29.5 in. (which may be assumed with sufficient precision for practical purposes), for various temperatures, and for altitudes between 10° and 90°.

Owing to the uncertainties of the refraction correction for low altitudes, observations for precise determinations are never taken on a celestial body which is near the horizon.

21-8. Combined Correction. For solar observations, refraction and parallax corrections are usually made together. The refraction correction, which is subtractive, is many times larger than the parallax correction, which is additive; hence the combined correction is of the same sign as the refraction correction. Table I herein gives corrections for the combined effect of refraction and parallax, to be subtracted from observed altitudes of the sun to determine the true altitudes above the celestial horizon.

21-9. Declination of the Sun. For the determination of azimuth, latitude, or longitude by solar observations, it is necessary that the declination of the sun at the instant of sighting be known. The declination at a given instant is obtained by interpolating between values given in a solar ephemeris for the current year. Either of the two common types of ephemeris may be used.

One type gives the apparent declination for each day of the year and is especially adapted for use when the standard time or the Greenwich civil time is known. The "Nautical Almanac" is an example of an ephemeris of this kind. It gives declinations for each hour of Greenwich civil time.

The other type gives the apparent declination for each day of the year at the instant of Greenwich apparent noon and is especially adapted for use when the longitude of the place and the local apparent time of the observation are known. The "Ephemeris of the Sun and Polaris" is of this sort. It is widely used in land surveying.

Abbreviated solar ephemerides, either for civil time or for apparent time, are published annually in the form of pamphlets by various manufacturers of surveying instruments and are furnished to surveyors free of charge.

The following examples illustrate the use of each of these types of ephemeris to determine declination:

Example 1: An observation is taken on the sun at 10^h00^m A.M. Eastern standard time, on December 15, 1964. It is desired to determine the declination at the given instant.

The Greenwich civil time at the instant of observation is $10^h00^m + 5^h = 15^h00^m = 15.00^h$. By ephemeris for 0^h Greenwich civil time the declination for 0^h on December 15 is $-23°08.4'$ and the change per hour is $-0.16'$. The change in declination since 0^h G.C.T. is $-0.16 \times 15.00 = -2.4'$. The declination at the instant of observation is $-23°08.4' - 2.4' = -23°10.8'$. The "Nautical Almanac" gives $-23°10.9'$ directly for 15.00^h G.C.T.

Example 2: An observation is taken on the sun as it crosses the meridian on November 16, 1964, at a place whose longitude is 87°49′30″ west of Greenwich. It is desired to determine the apparent declination at the given instant.

$$\text{G.A.T.} = \frac{87°49'30''}{15} = 5^h51^m18^s \text{ (after apparent noon)} = 5.86^h$$

From the "Ephemeris of the Sun and Polaris," the declination at Greenwich apparent noon is S18°48′56″. The average difference for $1^h = -37.2''$, the minus sign indicating that south declinations are increasing. The change in declination since Greenwich apparent noon is $37.2'' \times 5.86 = 03'38''$. The declination at local apparent noon at the place $= 18°48'56'' + 03'38'' =$ S18°52′34″.

Example 3: It is desired to determine the apparent declination of the sun at the instant of 1^h00^m P.M. Eastern standard time on November 18, 1964, from a solar ephemeris giving values for Greenwich apparent noon.

The difference between Eastern standard time and Greenwich mean time is 5^h; hence the instant of observation is 6^h00^m after Greenwich mean noon. At Greenwich apparent noon the equation of time as given in the ephemeris is $-14^m45.2^s$, to be subtracted from Greenwich apparent time to give Greenwich mean time. It follows that apparent time is faster than mean time, and the Greenwich apparent time is roughly $6^h00^m + 15^m = 6.25^h$. The daily rate of change in the equation of time is given by the difference between the equation of time for November 18 and that for November 19, or $14^m45.2^s - 14^m32.0^s = 13.2^s$. The change in the equation of time since Greenwich apparent noon is $(6.25 \times 13.2)/24 = 3.4^s$. The equation of time is decreasing, and hence the equation of time for the given instant is $14^m45.2^s - 3.4^s = 14^m41.8^s$. The interval since Greenwich apparent noon is $6^h00^m + 14^m41.8^s = 6^h14^m41.8^s = 6.245^h$. At Greenwich apparent noon the apparent declination is S19°18′02″; the average difference for 1^h is 35.5″. The change in apparent declination since Greenwich apparent noon is $35.5'' \times 6.245 = 3'42''$. South declinations are increasing; hence the apparent declination at the given instant is $19°18'02'' + 3'42'' =$ S19°21′44″.

In example 3 the equation of time has been determined for the given instant. For all practical purposes the equation of time for Greenwich apparent noon might have been employed, since the small error of 3.4^s in time would have no effect upon the computed change in the declination unless declinations were carried out to tenths of seconds.

If the equation of time were neglected entirely, the error introduced in the computed value of the apparent declination would be but 09″, not sufficiently large to be of consequence in rough calculations. Similarly, an observation of time for the sole purpose of determining declination need not be exact.

21-10. Latitude by Observation on Sun at Noon. The latitude of a given station can be determined with a fair degree of precision by observing with the engineer's transit the altitude of the sun at local apparent noon, when the sun crosses the meridian. If the longitude of the place is roughly known it is unnecessary to observe the time, but if the longitude is unknown the standard time of the observation must be taken. It is not necessary that the direction of the meridian be known. The problem consists in determining the true altitude h of the center of the sun above the celestial horizon and computing the apparent declination δ of the sun at the instant of sighting. Then as explained in Art. 20-9 the latitude ϕ is

$$\phi = 90° - h + \delta \tag{3}$$

The accuracy obtainable ordinarily depends upon the precision of the instrument. As the maximum rate of change of declination is only about 01′ per hour, a considerable error in time will affect the declination but slightly. With the ordinary transit having a vertical circle reading to 01′, the latitude may be determined in this manner with an error not greater than 01′. The mean of a series of observations on different days will, of course, render a closer result.

The usual procedure is as follows: The transit is set up and very carefully leveled. The horizontal cross-hair is sighted continuously on either the lower limb or the upper limb of the sun until the sun reaches its maximum altitude and begins its apparent descent. At that instant the watch time is observed. The maximum altitude and the watch time are recorded. With the telescope still approximately in the plane of the meridian, the index error is determined, preferably by the method of double-sighting on a mark as described in Art. 21-2. The watch is compared with a timepiece keeping correct standard time, and the error is noted. The Greenwich civil time of the observation is calculated, and the declination is found in a solar ephemeris as illustrated by example 1, Art. 21-9. The true altitude of the sun's center is determined by applying to the observed altitude the corrections for index error, semidiameter, and refraction and parallax. The latitude is then determined by Eq. (3). It should be noted that the sign of the refraction and parallax correction as given in Table I is always negative. The sign of the declination is negative from September to March and is positive for the remainder of the year.

When desired, a second sight may be taken on the opposite limb of the sun with the telescope inverted. The mean of the two vertical angles is taken as the altitude of the sun's center at the mean of the two times of observation, no correction for index error being necessary. If the time between sights does not exceed 3 or 4 min., the mean altitude may be considered as the altitude at apparent noon. The latitude is then calculated as described in the preceding paragraph.

If an ephemeris giving values at Greenwich apparent noon is to be used, the longitude of the place being unknown and the standard time being known, the

LATITUDE OF TOWN HALL

(Observation on sun at apparent noon)

Field work

Circle	Obj.	Time	V.Circle	Index E.	
L	A		+2°58'30"		h_1
R	A		+2°55'30"		h_2
				+0°01'30"	
L	☉	$11^h34^m01^s$	47°51'00"		h'

Computations

Watch time	$11^h34^m01^s$
" slow	29^s
G.C.T. (E.S.T.+5^h)	$16^h34^m30^s$
Eq. of time (from Ephemeris)	+ 4^m51^s
G.A.T.	$16^h39^m21^s$
δ_0 (decl. at G.A.noon)	+ 2° 54' 48"
$\Delta\delta$ (57.8" × 4.66^h)	− 4' 29"
δ	+ 2° 50' 19"
Obs. h' on ☉	47° 51.0'
Index correction	− 01.5'
Refr. and parallax (table I)	− 00.7'
Sun's semidiameter	+ 15.9'
h (corrected altitude)	48° 04.7'
δ	2° 50.3'
$\phi = 90° - h + \delta$	44° 45.6'

14

Wm. Bolton
H.L. Brown } Observers
Sept. 15, 1964
Remarks — Fair, Warm, Calm
"A" is mark on barn 400 ft south
See note — B&B Transit No. 142
By watch — Waltham watch
29^s slow E.S. Time

Note: Index error found by
16.66^h reversal on point "A"

$$\text{Index } E. = \frac{+(2°58'30") - (2°55'30")}{2}$$

" " = +0°01'30" with circle left

Ephemeris gave values for G.A. noon

Latitude of Town Hall

FIG. 21-5. Latitude by observation on sun at noon.

Greenwich apparent time of the observation is determined and the declination at the given instant is found as illustrated by example 3, Art. 21-9.

The field notes and computations are made in a form similar to that shown in Fig. 21-5. For these observations the longitude of the place was unknown, and the standard time was recorded. Only the sun's lower limb was sighted. The index error was determined by double-sighting at a mark; the letters "L" and "R" in the column headed "Circle" indicate whether the vertical circle was left or right and, therefore, whether the telescope was normal or inverted. As the available ephemeris gave values of declination at Greenwich apparent noon, the watch time was converted into Greenwich apparent time. In the line beginning "$\Delta\delta$," the change in declination during the 4.66 hr. that elapsed since Greenwich apparent noon was computed by multiplying the elapsed time by the variation per hour (57.8″) taken from the ephemeris.

If the longitude of the place is known, it is assumed that the instant of observation is local apparent noon. Hence the Greenwich apparent time is taken as the longitude of the place. The procedure in the field is identical with that described earlier in this article, except that time is not observed; however, the approximate standard time is calculated in advance. The declination is most conveniently found from an ephemeris giving values for Greenwich apparent noon, as illustrated by example 2, Art. 21-9.

21-11. Azimuth by Direct Solar Observation. The azimuth of a line can be determined by a single observation of the sun at any time when it is visible, provided the latitude of the place is known. (See Art. 21-14 for simultaneous determination of azimuth and longitude.)

At a known instant of time the sun is observed, and the altitude of the sun and the horizontal angle from the sun to a given reference line are measured. The declination of the sun at the given instant is found from a solar ephemeris. With the declination δ, latitude ϕ, and altitude h known, the *PZS* triangle can be solved as described in Art. 20-12, the azimuth of the sun being given by any of several expressions, one of which is

$$\cos Z = \frac{\sin \delta}{\cos h \cos \phi} - \tan h \tan \phi \qquad (4)$$

in which Z is the azimuth from north. The azimuth A from south is given by

$$\cos A = \tan h \tan \phi - \frac{\sin \delta}{\cos h \cos \phi} \qquad (5)$$

Equations (4) and (5) are the same as Eqs. (13) and (13*a*), respectively, of Chap. 20. The azimuth of the line is readily computed from the azimuth of the sun and the observed horizontal angle.

The usual procedure is as follows: The transit is set up and very carefully leveled over one end of the line. The *A* vernier is set at zero on the horizontal circle, and the reading of the *B* vernier is noted. A sight is taken along the given line with the telescope in, say, the normal position, and the lower motion is clamped. The upper motion is loosened, and sights are taken on the sun as described in Art. 21-5. The field work is completed by again sighting along the line and reading the horizontal circle, this time with the telescope still inverted. The watch is compared with a timepiece keeping correct standard time, and the error is noted. The mean of each pair of observed angles is taken as the angle to the sun's center at the mean of the observed times. The observed values are corrected, and the azimuth of the sun is computed as previously described, five-place logarithms being used. The azimuth of the line is computed by subtracting algebraically the mean of the horizontal angles from the azimuth of the sun, angles taken in a clockwise direction from line to sun being considered positive.

To the mean of the observed altitudes are applied corrections for refraction and parallax (Table I). The sun's apparent declination is found from a solar ephemeris as described in Art. 21-9. The azimuth of the sun at the

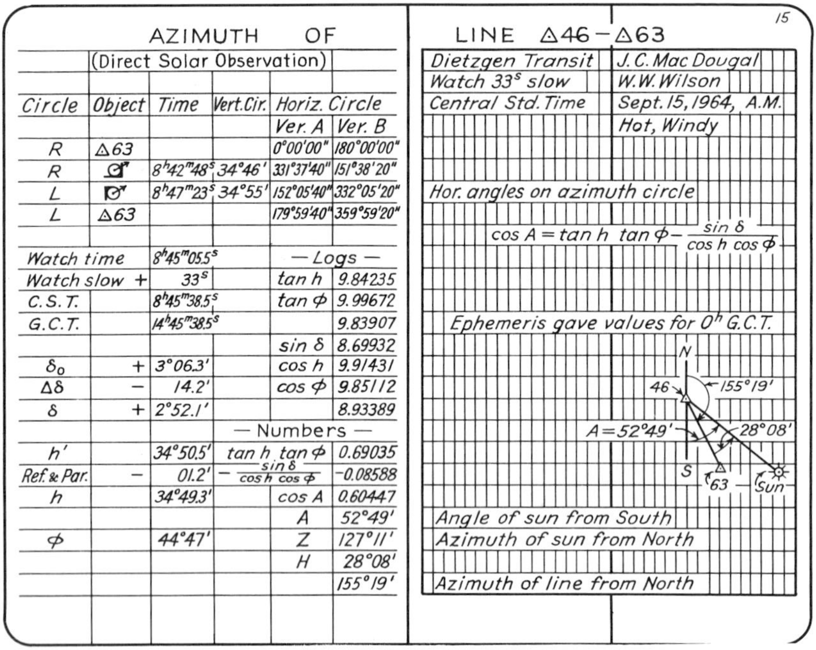

AZIMUTH OF LINE △46−△63 (15)

(Direct Solar Observation)

Circle	Object	Time	Vert.Cir.	Horiz. Circle Ver. A	Ver. B
R	△63			0°00'00"	180°00'00"
R	☉	$8^h42^m48^s$	34°46'	331°37'40"	151°38'20"
L	☉	$8^h47^m23^s$	34°55'	152°05'40"	332°05'20"
L	△63			179°59'40"	359°59'20"
Watch time		$8^h45^m05.5^s$		— Logs —	
Watch slow	+	33^s		tan h	9.84235
C.S.T.		$8^h45^m38.5^s$		tan ϕ	9.99672
G.C.T.		$14^h45^m38.5^s$			9.83907
				sin δ	8.69932
δ_0	+	3°06.3'		cos h	9.91431
$\Delta\delta$	−	14.2'		cos ϕ	9.85112
δ	+	2°52.1'			8.93389
			— Numbers —		
h'		34°50.5'	tan h tan ϕ		0.69035
Ref. & Par.	−	01.2'	$-\dfrac{\sin\delta}{\cos h\cos\phi}$		−0.08588
h		34°49.3'		cos A	0.60447
				A	52°49'
ϕ		44°47'		Z	127°11'
				H	28°08'
					155°19'

Dietzgen Transit	J. C. Mac Dougal
Watch 33ˢ slow	W. W. Wilson
Central Std. Time	Sept. 15, 1964, A.M.
	Hot, Windy

Hor. angles on azimuth circle

$$\cos A = \tan h \tan\phi - \frac{\sin\delta}{\cos h\cos\phi}$$

Ephemeris gave values for 0^h G.C.T.

Angle of sun from South

Azimuth of sun from North

Azimuth of line from North

FIG. 21-6. Azimuth by observation on sun.

given instant can be computed by solving Eq. (4) or (5), either by using a computing machine or by logarithms. When observations are of ordinary precision, as those taken with a transit reading to minutes, five places are sufficient for computations. In solving the equations, it should be noted that, when declinations are south, the sign of sin δ is negative. A field sketch showing relative positions of the transit, sun, and line is helpful. The angle Z, Eq. (4), is measured clockwise or counter-clockwise from the north to the sun; the angle A, Eq. (5), is measured clockwise or counter-clockwise from the south to the sun.

Figure 21-6 illustrates a suitable form for field notes, the observations being taken with a transit having a vertical circle reading to 01′ and a horizontal circle reading to 20″. It is seen that horizontal angles are read from the azimuth circle. An ephemeris giving declinations for 0^h Greenwich civil time was used.

Great care should be taken in leveling the instrument, for reasons explained in Art. 21-2. Unless it is equipped with a striding level, the final test prior to observing the sun should be to see that the telescope bubble remains centered as the instrument is revolved about the vertical axis. Plunging the telescope between observations will not eliminate the effect of inclination of the vertical axis.

If the transit is not equipped with a full vertical circle, it cannot, of course, be plunged between sights, but otherwise the procedure may be as just described. If the transit does not have a full vertical circle, care should be taken to see that it is in good adjustment.

If for any reason it is desired to obtain the azimuth of a line by a single pointing, the sun may be brought tangent in any of the quadrants. The altitude and azimuth corrections for the sun's center will then be made as described in Art. 21-4.

When it is assumed that at the mean of the two times the sun is at a position given by the mean of the vertical angles and the mean of the horizontal angles, it is equivalent to saying that the sun is apparently traveling in a straight line, which of course is not true. Within a period of 10 min., however, the error introduced is so small as to be of little or no consequence.

The watch time need not be observed closely, as it is used only for the purpose of determining declination.

Precision. Under ordinary conditions the azimuth of a line can be determined within about 01′. The precision depends not only upon the precision of field observations and the exactitude of the corrections in altitude, but also upon the shape of the astronomical triangle. The *PZS* triangle becomes weak as the sun approaches the meridian, and the solution becomes indeterminate at the instant of apparent noon. On the other hand, the refraction correction becomes large and very uncertain for low altitudes, particularly for those less than 10°. For these reasons, when possible, observations within the latitudes of the United States are usually taken between the hours of 8 and 10 A.M. or 2 and 4 P.M.

The effect of errors in the sides of the astronomical triangle upon the precision of the computed azimuth of the sun at the instant of observation is given by the accompanying table, in which are shown the changes in azimuth of the sun due to changes of 01′ in the latitude, declination, and altitude at the latitude of 40°, which is about the mean for the United States. The changes in azimuth have been computed by means of Eq. (5). The values are approximate and are given for comparative purposes only. The months named are those during which the declination is not greatly different from the value given in the second column. Thus during the period of November, December, and January, the declination varies from $-15°$ to $-23°$. The hour angles give the approximate time interval before or after local apparent noon. It will be seen that, when the hour angle is 1^h30^m, a 01′ error in latitude, declination, or altitude produces an error of about 03′ in the azimuth. When the hour angle has increased to 3^h, an error of 01′ in latitude or altitude produces an error of about 01′ in the azimuth. Under the given conditions, the effect of an error in declination is greater than the effect of an error of the same magnitude in either latitude or altitude.

APPROXIMATE CHANGE IN CALCULATED AZIMUTH OF SUN FOR 01′ CHANGE IN LATITUDE, DECLINATION, OR ALTITUDE, FOR LATITUDE 40°

Months	Declination	Hour angle	Altitude	Azimuth	Change in azimuth for 01′ change in		
					Latitude	Declination	Altitude
November, December, January	−20°	3^h15^m	15°	46½°	1′10″	1′45″	1′25″
March, September	0°	4^h40^m	15°	77°	35″	1′25″	55″
	0°	3^h20^m	30°	61°	1′10″	1′45″	1′20″
	0°	1^h30^m	45°	33°	3′00″	3′20″	3′00″
May, June, July	+20°	5^h45^m	15°	104°	04″	1′20″	45″
	+20°	4^h30^m	30°	88°	35″	1′25″	50″
	+20°	3^h10^m	45°	77½°	1′10″	1′45″	1′00″
	+20°	1^h45^m	60°	56°	2′40″	2′55″	2′10″

21-12. Time by Observation on Sun at Noon. If the longitude of a station is known and the direction of the meridian has been established, the standard time may be determined accurately by observing the sun as it crosses the meridian at local apparent noon. In determining time in this manner, the transit is set up and carefully leveled over the north end of the meridian line, and a sight is taken along the meridian. The line of sight is elevated to intercept the path of the sun, and, at the instant of tangency between the west limb and the vertical cross-hair, the time is noted. The telescope is quickly plunged, and a second sight is taken along the meridian. The line of sight is again elevated to intercept the path of the sun, and the time of tangency between the vertical cross-hair and the east limb of the sun is observed. The mean of the two times thus observed is the watch time of upper transit of the sun's center, which is local apparent noon.

TIME BY OBSERVATION ON SUN AT NOON — 16

Meridian at Student Observatory, Berkeley, Calif.

Circle	Object	Time
L	△20	
L	☉|	$11^h52^m42^s$
R	△20	
R	|☉	$11^h54^m58^s$

Time of passing of sun's center		$11^h53^m50^s$
L.A.T. of observation		$12^h00^m00^s$
Eq. of time at G.A. Noon	−	$15^m38.8^s$
Longit. x hourly ch. (8.15 x 0.38)	+	3.1^s
L.C.T. of L.A. Noon		$11^h44^m24.3^s$
Change to standard time	+	$09^m02.7^s$
Standard time of L.A. Noon		$11^h53^m27.0^s$
Watch time of L.A. Noon		$11^h53^m50^s$
Watch fast		23^s
Corr. to watch time to get std. time		-23^s

Lietz Transit	J.C. Williams
Watch	E. Moore
	Nov. 13, 1964
So. end of meridian	Fair, Warm
" " " "	

Note:- Longitude of Observatory $8^h09^m02.7^s$ Pacific standard time

From Ephemeris (Mean sun behind true sun)

FIG. 21-7. Time by observation on sun at noon.

The longitude of the place expressed in hours, for reasons explained in Art. 20-22, is the Greenwich apparent time reckoned from noon. From the "Ephemeris of the Sun and Polaris," or other ephemeris giving values for Greenwich apparent noon, the equation of time at the instant of observation is determined as illustrated by example 2, Art. 20-20. Since the equation of time is the difference between mean and apparent time at the instant of observation, it is clear that 12^h plus or minus the equation of time is the local mean time of local apparent noon, the sign being plus or minus according as mean time is faster or slower than apparent time, as indicated by the ephemeris. The local mean time of local apparent noon is changed to standard time by subtracting the difference in longitude (in hours) between the place of observation and the standard-time meridian if the place is east of that meridian and by adding if west. The computed standard time of local apparent noon is the correct watch time of the observation, and a comparison with the observed watch time indicates the correction to be applied to the watch time. Figure 21-7 is an example in which the correction for Pacific standard time is found.

If the ephemeris used gives values for 0^h Greenwich civil time, an exact determination of the equation of time necessitates changing the Greenwich apparent time of observation (reckoned from noon) to Greenwich civil time (reckoned from 0^h) by means of an approximate equation of time. The true equation of time is then found by interpolation between values given in the ephemeris. Since the equation of time is always less than 17^m and since its rate of change is small, for rough determinations of time the error will be negligible if, for the purpose of calculating the equation of time, mean and apparent time are assumed to agree, as the following example illustrates.

Example: It is desired to determine the equation of time at 18^h G.A.T. on October 6, 1964.

By the "Ephemeris of the Sun and Polaris," the equation of time at Greenwich apparent noon is $11^m54.44^s$ and the daily change is $+17.21^s$. The increase since noon is $17.21 \times 6/24 = 4.30^s$, and the exact equation of time at the given instant is $11^m54.44^s + 4.30^s = 11^m58.74^s$.

By the "Nautical Almanac," the equation of time at 12^h G.C.T. October 6 is 11^m55^s and at 0^h G.C.T. October 7 is 12^m03^s; by interpolation, at 18^h G.C.T. it is 11^m59^s. Thus, neglecting the difference between mean time and apparent time would result in a difference of only a fraction of a second.

If the equation of time in this type of ephemeris were given with sufficient precision so to warrant, the approximate equation of time just found could be subtracted from 18^h G.C.T. to give a value closer to 18^h G.A.T. and a second interpolation made. However, the "Nautical Almanac" gives values only to whole seconds.

The length of time taken by the sun in crossing the meridian depends somewhat upon the sun's declination and semidiameter, but it is approximately 2½ min. It is therefore clear that no time can be lost in plunging the telescope for a second sight along the meridian preparatory to observing the east limb of the sun.

If for any reason it is impracticable to observe more than one limb, the time in seconds earlier or later than the sun's center passes the meridian may be taken as approximately $4S/\cos\delta$, in which S is the sun's semidiameter in minutes of arc (approximately 16′) and δ is the sun's declination.

21-13. Longitude by Observation on Sun at Noon. If the standard time is known precisely and the meridian has been established, the longitude of a place may be determined by an observation on the sun at local apparent noon, the field procedure being in all respects identical with that just described for finding time.

With the standard time of passage of the center of the true sun (local apparent noon) known, the Greenwich civil time of local apparent noon can be computed, and the equation of time at the instant of local apparent noon can be found readily from an ephemeris giving values for 0^h Greenwich civil time. The standard time of local mean noon (meridian passage of the mean sun) differs from the standard time of local apparent noon by an amount equal to the equation of time, being greater if the mean sun is behind the true sun and less if ahead of it. The difference between the standard time of local mean noon and 12^h standard time is the difference in longitude $\Delta\lambda$ (in time units) between the meridian of the place and the standard-time meridian; if local mean noon occurs before 12^h standard time, the place is east of the standard meridian; if after, west.

Example: The sun's center is observed to pass the meridian at a given place at $11^h30^m12.2^s$ A.M. Pacific standard time on December 2, 1964. The longitude of the place is desired.

The difference between Pacific standard time and Greenwich civil time is 8^h; hence the G.C.T. is $19^h30^m12.2^s = 19.503^h$. From the "Nautical Almanac" the equation of time at 12^h G.C.T. December 2 is 10^m29^s and at 0^h December 3 is 10^m17^s, a change in 12 hr. of -12^s. The change since 12^h G.C.T. is $12 \times (7.503/24) = -04^s$. The equation of time at local apparent noon is $10^m29^s - 04^s = 10^m25^s$. The ephemeris indicates that apparent time is faster than mean time; hence local mean noon occurs at a standard time later than that for local apparent noon by an amount equal to the equation of time, and the Pacific standard time of local mean noon is $11^h30^m12^s + 10^m25^s = 11^h40^m37^s$. Then $\Delta\lambda = 11^h40^m37^s - 12^h = -19^m23^s = -4°50'45''$. Pacific standard time is local mean time for the 120° meridian; hence the longitude of the place is $120° - 4°50'45'' = 115°09'15''$.

21-14. Azimuth and Longitude by Solar Observation. By modification of the method described in Art. 21-11, the azimuth and longitude at a station can be determined closer than single minutes. If the horizontal circle can be read to 20″ and the vertical circle to 30″, the probable error in azimuth or longitude should not exceed 00.2′. The method is well adapted for use with a repeating theodolite. It consists in taking two series of observations, each series consisting of two sets of four observations each. One series is observed in the morning and the other in the afternoon; for most precise results the sun should be approximately the same distance from the meridian in the afternoon series as it was in the morning series. Half of the observations are taken with the telescope normal, and half with it inverted. The azimuth of the reference line is computed independently for each set of observations, and the mean of the four values thus obtained is taken as the most probable value. Similarly, the longitude of the station is computed independently for each set, the hour angle being determined by one of the forms of Eq. (14) of Chap. 20; and the mean of the four values of longitude is taken as the most probable value.

21-15. Solar Attachments. A solar attachment is a device which, when attached to the transit, furnishes a means of determining the direction of the meridian by mechanically solving the *PZS* triangle. There are several varieties of solar attachment differing widely in appearance but being alike in principle. Each type has a polar axis and a line of collimation. The solar attachment can be rotated about the polar axis just as the transit can be rotated about its vertical axis. Means are provided for laying off the latitude (or the colatitude) of the place and the declination of the sun. When these quantities have been laid off, the instrument is oriented by bringing the sun's image into position in the solar reticule; the line of sight of the transit telescope is then in the plane of the meridian.

The use of a solar attachment makes it possible to determine the azimuth of a line more quickly than by direct observation and numerical computation, but the precision of the azimuth determination is likely to be lower. The solar attachment is little used except on certain mining and public-land surveys of moderate precision, where several azimuth determinations must be made daily.

All that has been said regarding favorable hours for taking direct observations applies equally well to observations with a solar attachment. Also, errors in latitude and in declination settings will have the same effect as when the azimuth is computed mathematically.

The three types of solar attachment which have been used most are the *Smith*, the *Saegmuller*, and the *Burt*, each of which will be briefly described. The Smith is in most general use.

21-16. Smith Solar Attachment. This attachment consists of a telescope of low magnifying power, called the *solar telescope*, mounted in collar bearings which are attached to a graduated vertical arc called the *latitude arc.* This assembly is mounted on a horizontal axis, called the axis of the latitude arc, which is attached to one of the standards of the transit as shown in Fig. 21-8. The solar telescope can be rotated about its own axis in the collar bearings and can also be revolved about the axis of the latitude arc. The latitude arc is read by means of a vernier which is fixed to the standard of the transit. In front of the objective is a plane mirror, called the *reflector*, which reflects the sun's rays into the solar telescope. The mirror is mounted (in bearings) on an axis which is normal to the line of sight of the solar telescope. From the axis of the mirror an arm carrying a vernier leads to a graduated arc, called the *declination arc,* attached to the barrel of the telescope. Movements of both latitude and declination arcs are controlled by clamps and tangent-screws. Near the eyepiece end of the solar telescope is an *hour circle,* the index of which is on one of the collars. The solar telescope is equipped with cross-hairs and with a set of *equatorial hairs* parallel to the axis of the reflector and at a distance apart approximately equal to the apparent diameter of the sun.

When the latitude is laid off on the latitude arc and the main telescope (in the normal position) is pointed in the direction of north, the axis of the solar telescope points toward the pole. Further, for the position just specified, if the declination of the sun, corrected for refraction, is laid off on the declination arc, the mirror is so tilted that the course of the sun may be followed by rotating the solar telescope about its own axis, the image of the sun being reflected from the mirror to the objective and thence to a focus at the cross-hairs of the solar telescope. When the sun is viewed in this manner, the index of the hour circle reads the local apparent time.

The advantage of the Smith solar attachment lies in a construction which permits the use of the main telescope without disturbing the latitude and declination settings. Since the latitude is constant for a given locality and

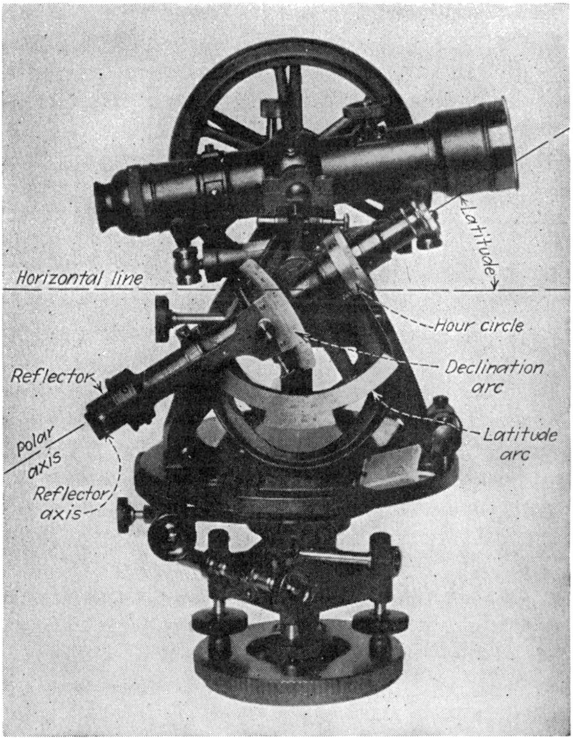

FIG. 21-8. Smith solar attachment.

the declination changes but slowly, this is a decided advantage when frequent observations are necessary; and for this reason the Smith solar attachment may be regarded as preferable to either of the other attachments described herein. It has been adopted by the U.S. Bureau of Land Management as the standard instrument for use on public-land surveys.

21-17. Azimuth with Smith Solar Attachment. In using the Smith solar attachment, the latitude and declination settings are made on the appropriate arcs, with the declination setting corrected for refraction. The transit is set up and carefully leveled, and the horizontal circle is set at zero. The local apparent time is then set off on the hour circle by rotating the solar telescope in its collar bearings. The transit is revolved on the lower motion until the image of the sun appears between the equatorial hairs of the solar telescope, when the line of sight of the transit telescope lies in the plane of the meridian, and the axis of the solar telescope points toward the celestial pole. Since the declination changes slowly, if no mistake has been made it will be possible to keep the image of the sun between the equatorial hairs for some time by simply rotating the telescope in its collar bearings, keeping the hour circle set at the apparent time. When a sight is taken along a line with the transit telescope, the reading of the clockwise scale gives the azimuth.

This type of solar attachment is particularly convenient for checking azimuth traverses, since the latitude and declination settings may be maintained without interfering with the normal functions of the transit telescope. At any station the azimuth can be checked simply by setting the horizontal circle back to zero, rotating the solar telescope until the hour circle reads the apparent time, and making any slight change in the declination setting that may have occurred in the declination since the preceding observation.

21-18. Adjustment of Smith Solar Attachment. In adjusting the Smith solar attachment, the latitude of the place should be known precisely, and the instrument should be set up in a location where objects a mile or more away may be viewed when the telescope is level. The following relations should exist:

1. *The equatorial hairs should be parallel to the axis of the reflector.* With all settings made as for a solar observation for azimuth, the transit is turned about the vertical axis until the sun's image is precisely spaced between the equatorial hairs, the vertical axis is clamped in this position, and the solar telescope is rotated about its own axis, causing the sun's image to travel across the field of view. If the limbs of the sun follow the equatorial hairs, the hairs are parallel to the axis of the reflector. If the sun's limbs depart sensibly from the hairs, adjustment is made by loosening the cross-hair ring and rotating it through a small angle, as for the corresponding adjustment of the dumpy level.

2. *The line of sight of the solar telescope should coincide with the axis of the collar bearings.* The test and adjustment is performed in a manner identical with the corresponding adjustment of the wye level. The mirror is swung to give an unobstructed view through the solar telescope, and the intersection of the cross-hairs is focused on some distant point. The telescope is then rotated about its axis through an hour angle of 12^h (180°). If the intersection has moved from the point, it is brought halfway back to its original position by means of the adjusting screws controlling the cross-hair ring.

3. *The line of sight of the solar telescope, and hence the polar axis, should be perpendicular to the axis of the latitude arc.* Further, when the solar telescope is rotated about the axis of the latitude arc, its line of sight should generate a plane parallel to that generated by the line of sight of the main telescope when it is revolved about the horizontal axis. Some patterns have no provision for this adjustment

the desired relation being established by the manufacturer and being assumed to remain permanently fixed. In the Bureau of Land Management pattern, provision is made for this adjustment, the axis of the latitude arc being of considerable length, and a striding level for this axis being furnished with the instrument. For the Bureau of Land Management pattern, the transit is very carefully leveled, the main telescope is sighted at a distant point, and the solar telescope is sighted toward the same point. The striding level is placed on the latitude axis and the latter is made level by means of the lower pair of capstan nuts on the base frame of the attachment. If the line of sight of the solar telescope falls to one side of the distant point, it is brought to the point by means of the left-hand upper pair of capstan nuts on the base frame of the attachment. When these two relations have been established, both telescopes are plunged, the transit is revolved 180° in azimuth, and a sight is taken to the distant point as before. If the line of sight falls to one side of the point, one half of the apparent error is due to the line of sight's not being perpendicular to the latitude axis, and the correction is made by bringing the line of sight halfway to the point by means of the capstan nuts at one end of the telescope and the remaining distance by means of the left-hand upper pair of capstan nuts on the base frame of the attachment. This procedure should be repeated until the adjustment is verified.

4. *The latitude arc should read zero when the line of sight of the solar telescope, or the polar axis, is horizontal.* The transit is leveled very carefully, and the main telescope is leveled and sighted at some distant point of the landscape. The solar telescope is revolved about the latitude axis until its line of sight strikes the same point (the polar axis is now horizontal), when the vernier of the latitude arc should read zero. If it does not read zero, either the index error may be observed and applied to future latitude settings or the vernier may be loosened and moved laterally until it does read zero. If the solar telescope is equipped with a striding level, the index error may be observed by leveling the solar telescope, and a sight to the distant point is unnecessary.

5. *The declination arc should read the true declination of the sun, corrected for refraction.* A short time before apparent noon the transit is set up and carefully leveled over one end of an established meridian. The latitude of the place is laid off on the latitude arc, the hour circle is set at 12^h, and the main telescope is sighted along the meridian. At the instant of apparent noon, the image of the sun is brought between the equatorial hairs of the solar telescope by rotating the reflector about its axis, and the reading on the declination arc is observed. The difference between this reading and the declination of the sun at the given instant (as determined from the solar ephemeris), corrected for atmospheric refraction, is the index error of the declination arc. This may be applied to future declination settings, or the error may be corrected by loosening the vernier and moving it along the arc until it reads the calculated declination setting.

On public-land surveys where the Smith solar attachment is in constant use, it is the practice to make this test daily, since it serves to check not only the adjustment of the declination vernier but also the adjustment of the latitude vernier and the collimation of the solar telescope.

6. *The reading of the hour circle at any instant should give the local apparent time.* The watch time of local apparent noon is determined by observing with the main telescope the instant when the sun's center crosses the meridian, as described in Art. 21-13. At any convenient time thereafter, the main telescope is pointed along the meridian, the latitude and declination are set off on their respective arcs, the sun's image is brought to the center of the field of view by rotating the solar telescope about its own axis, and the watch time is observed. If the time interval since the

observed passage of the sun over the meridian, which is the correct local apparent time, does not agree with the reading of the hour circle, the set-screw clamping the hour circle to the barrel of the telescope is loosened, and the hour circle is rotated until the index reads the correct value.

7. *The collar bearings should be free from inequality or roughness,* which will become apparent when the solar telescope is rotated in its collar bearings. Further, *the spacing of the equatorial wires should be uniform throughout and should fit the outside line of the inner circle.* Defects with regard to these relations can be corrected only by the manufacturer.

21-19. Saegmuller Solar Attachment. This attachment consists of a telescope of low power, called the *solar telescope,* mounted between standards which revolve about a polar axis. The assembly is mounted on top of the transit telescope, the polar axis being normal to the plane defined by the line of sight of the main telescope and the horizontal axis of the transit. The solar telescope is equipped with cross-hairs defining the line of sight as do those of the main telescope, and in addition is provided with four hairs forming a square the sides of which are approximately equal to the apparent diameter of the sun. Attached to the solar telescope is a level tube the axis of which is parallel to the line of sight of the solar telescope. The solar telescope is equipped with a prismatic eyepiece. The movement of the solar telescope about the polar axis and about the axis of the standards, or the equatorial axis, is controlled by clamps and tangent-screws.

When the image of the sun appears inside the square formed by the four hairs, the line of sight of the solar telescope is directed toward the sun's center. When the main telescope is elevated through an angle equal to the colatitude of the place and is pointed along the meridian in a southerly direction, it is clear from the discussion of Art. 20-10 that the line of sight of the main telescope is in the plane of the equator, and that the polar axis of the solar attachment points to the celestial pole. Further, with the main telescope in this position, if the line of sight of the solar telescope makes an angle with the polar axis equal to the sun's codeclination or polar distance, then at any time when the sun is above the horizon and the solar telescope is pointed in the direction of the sun, it should be possible to bring the line of sight to the sun's center simply by rotating the solar telescope about the polar axis. Also, when the lines of sight of both telescopes are in the same vertical plane, it is clear that the angle between them is equal to the sun's declination.

The methods of adjusting the Saegmuller solar attachment are similar to those governing the adjustment of the transit.

21-20. Burt Solar Attachment. The Burt solar attachment is attached to the main telescope in the same manner as is the Saegmuller, but the solar telescope is replaced by biconvex lenses and metallic screens in duplicate, these being rigidly mounted at opposite ends of a bar which forms the vernier arm for a graduated arc on which declinations may be set off. The line of collimation of the attachment is defined by the optical center of one of the biconvex lenses and the intersection of lines etched upon the opposite metallic screen, these lines corresponding to cross-hairs in the transit telescope.

When the line of collimation is pointed at the sun's center, the sun's unmagnified image appears at the intersection of the cross-hairs. With a magnifying glass the line of collimation can be directed more exactly at the sun's center by bringing the image inside a square etched on the metallic screen, this square corresponding to that formed by the four hairs in the solar telescope of the Saegmuller attachment. Regardless of whether the sun is above or below the equator, declinations are laid off in the same direction on the arc. When the attachment is in use, if the declination is north or positive the line of collimation is pointed at the sun with the declination arc nearer the sun; if the declination is south or negative the line of collimation is pointed with the declination arc nearer the observer. Surrounding the base of the polar axis is an *hour circle* graduated in hours and reading to 5 min. of time. If the colatitude is laid off on the vertical circle of the transit and the main telescope is pointed south, the index of the hour circle reads the local apparent time when the line of collimation of the attachment points at the sun. The use of the Burt attachment is identical with that of the Saegmuller attachment, except as indicated by the preceding description.

21-21. Declination Settings for Use with Solar Attachment. If any of the varieties of solar attachment is to be used frequently, a table is prepared giving the declinations to be laid off when using the solar attachment at various hours of the day. The time may be either local apparent or standard time, but is usually the former. The apparent declination for a given time is found from a solar ephemeris, as described in Art. 21-9. Since the sun appears to be higher than it really is, due to atmospheric refraction, it is evident from a study of the *PZS* triangle (Fig. 20-8) that the declination setting for a sight to the apparent position of the true sun at a given instant must be algebraically greater than the true declination of the true sun, which is the value obtained from the ephemeris. If the refraction correction in altitude is known, the refraction correction in declination for a given latitude, hour angle, and declination may be computed by solving the spherical triangle. Table III herein gives such values throughout the year for latitude 40°. Any number in the second column gives the hour angle of the true sun on either side of the meridian—or in other words, gives the local apparent time before or after apparent noon—to which the refraction correction given in the third column applies. For a latitude other than 40°, the correction of Table III is multiplied by the appropriate latitude coefficient of Table IV.

Example: The accompanying table of declination settings is computed for November 3, 1964, at a place where the latitude is 34°30′ and the longitude is 7^h48^m west of Greenwich. The hours are local apparent time. The setting for 8^h A.M. is computed as follows.

From the "Ephemeris of the Sun and Polaris," the apparent declination at Greenwich apparent noon is $-15°09'29''$ and the change for 1 hr. is $-47''$. At 8 A.M. local apparent time, it is $8^h + 7^h48^m - 12^h = 3^h48^m = 3.8^h$ after Greenwich apparent noon. Hence the declination at 8 A.M. local apparent time is $-15°09'29'' - (3.8 \times 47'') = -15°12'27'' = -15°12.5'$.

Local apparent time	Declination	Refraction correction	Declination setting
8^h A.M.	−15°12.5′	+2.7′	−15°09.8′
9	−15°13.3′	+1.7′	−15°11.6′
10	−15°14.1′	+1.3′	−15°12.8′
11	−15°14.9′	+1.1′	−15°13.8′
12 M.	−15°15.7′	+1.0′	−15°14.7′
1 P.M.	−15°16.5′	+1.1′	−15°15.4′
2	−15°17.3′	+1.3′	−15°16.0′
3	−15°18.1′	+1.7′	−15°16.4′
4	−15°18.9′	+2.7′	−15°16.2′

From Table III, at latitude 40° the refraction correction for November 3 and an hour angle of 4^h (equivalent to 8 A.M. or 4 P.M. local apparent time) is $+3'21''$. By Table IV the latitude coefficient for latitude 34°30′ is 0.80. The declination correction for 8 A.M. at the given latitude is, therefore, $3'21'' \times 0.80 = +2'41''$.

The declination setting at 8 A.M. L.A.T. is, therefore, $-15°12.5' + 2.7' = -15°09.8'$. Settings for the other hours of the day are computed similarly.

When an observation is to be made, the watch time is noted and the local apparent time is roughly calculated. The declination setting is then taken from the table, interpolating if necessary. Even by careful estimation the declination can be set only to half minutes of arc; hence values of the settings do not need to be interpolated with great accuracy. In fact, under certain conditions the declination setting may not change appreciably for several hours, as illustrated by the afternoon values in the example.

OBSERVATIONS ON STARS

21-22. General. In general, the methods of determining azimuth, latitude, longitude, and time by direct solar observations are, with slight modifications, applicable to observations on the stars. If a high degree of precision is not required, the same procedure may be followed for stellar observations as for solar observations. Usually, however, it is expected that a higher degree of precision will be obtained; consequently a corresponding degree of refinement is necessary, and special care is taken to eliminate systematic errors.

The measurement of horizontal and vertical angles to celestial bodies is discussed in Art. 21-2. The refraction correction is described in Art. 21-7. For observations on stars, no correction is required for parallax or semidiameter.

As there are fixed stars in all parts of the heavens, it is an easy matter to select a star or stars in a celestial region favorable to a precise determination of the quantity sought. Thus, conditions favorable to a precise determination of latitude by measuring the altitude of a star at culmination are (1) a fairly high altitude, in order that the uncertainty of the refraction correction be small, and (2) a rate of apparent movement that is small, in order that a series of observations may be taken without an appreciable change in the altitude. Within the latitudes of the United States, stars near the pole satisfy these conditions.

Likewise for precise determination, an observation for azimuth by measured altitude and known declination and latitude should be taken on a star in the east or west far enough above the horizon to eliminate the uncertain refraction but not so near the meridian as to produce a weak astronomical triangle. An observation for azimuth with the hour angle, declination, and latitude known should be taken on a circumpolar star—the nearer the pole the better—since the azimuth of such a star changes more slowly in a given length of time than does the azimuth of a star near the equator, and hence any error in time will have less effect.

For determinations of longitude or time, stars should be chosen near the equator because they are apparently traveling more rapidly than those near the pole.

The right ascension (in terms of sidereal hour angle) and declination for many stars are given in the "Nautical Almanac." In the "Ephemeris of the Sun and Polaris" are given the G.C.T. and the declination for many stars at the instant of their upper transit at the meridian of Greenwich, at intervals of one-half month. Since for the fixed stars these coordinates change very slowly, it is not necessary to determine values for the hour of observation, as with the sun. The sidereal time corresponding to any given solar time can be found, and the hour angle can be computed by the expression $t = \theta - \alpha$ as explained in Art. 20-21.

Stars can be identified by means of charts which show the various constellations. For many stellar observations, however, the published direction and altitude of the star can be set off on the transit with sufficient precision that the star will be brought into the field of view at a given time, and it is not necessary to distinguish the star from among its neighbors. In fact, observations are often taken on stars during daylight hours near the hour of darkness, even when the stars are invisible to the naked eye.

In sighting on a star, the objective should be focused until the star appears as a fine, brilliant point of light. Prior to looking for a star just before sunset or after sunrise, the objective may be focused approximately by sighting at a distant object in the landscape. The proper position of the objective slide for focus on a star may be permanently marked on the barrel of the transit telescope.

During the hours of darkness, artificial illumination is required to make visible the cross-hairs of the transit. Some instruments are equipped with a reflector sleeve which slips over the objective as does the sun shade. When a flashlight is held to one side of the reflector, the field of view is faintly illuminated and both the cross-hairs and the star can be seen. With a transit not so equipped, the cross-hairs can be illuminated sufficiently by holding the light several inches in front of the objective and a little to one side of the telescope barrel, thus causing the rays to enter the telescope diagonally. A position of the light will be found where both the cross-hairs and the star are visible. A better diffusion of light will be given by a drop of paraffin wax at the center of the objective lens, the wax being shaved to a thin layer. A piece of thin paper with a hole in the middle, the paper being secured over the objective by means of a rubber band, answers the same purpose.

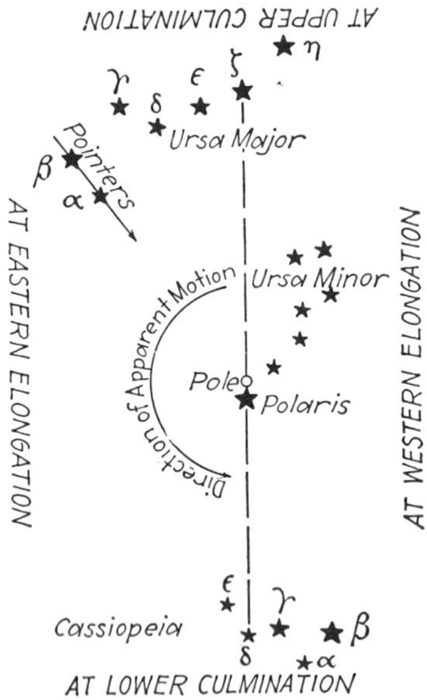

FIG. 21-9. Positions of constellations near the North Pole when Polaris is at culmination and elongation.

The location of any terrestrial mark that may be used in observing is indicated by a light. Usually the mark is a slit in the side of a box in which there is a lamp. The mark may be a strongly illuminated target, the source of illumination being shielded from the observer.

21-23. Polaris. The polestar, Polaris (α Ursa Minor), is the star more than all others on which observations for latitude and azimuth are taken in the latitudes of the United States. Its distance from the pole is approximately 1°. Its annual change in polar distance (or in declination) is less than 01′, and its maximum daily change in polar distance is less than ½″. It is a second-magnitude star the position of which is readily identified by the neighboring constellations of Ursa Major and Cassiopeia. Figure 21-9 shows the position of Polaris with respect to the pole and to these constellations. The seven most brilliant stars in the constellation of Ursa Major are known as the Great Dipper; and the two stars forming the part of the bowl farthest from the handle are called the *pointers* because a line through these stars points very nearly to the celestial north pole. It will be noted that the constellation of Cassiopeia is on the same side of the pole as Polaris, so that when Cassiopeia is above the pole, Polaris is near upper culmination; when Cassiopeia is west of the pole, Polaris is near western elongation; and so on. The position of Polaris relative to the pole may be quite closely estimated by noting the positions of δ Cassiopeia and ζ Ursa Major. A line joining these two stars passes nearly through the pole and Polaris. The line is nearly vertical when the star is at either culmination, and nearly horizontal when the star is at either elongation.

Observations to determine latitude are usually made when Polaris is at upper or lower *culmination* (Art. 20-17), when the star appears to be moving almost horizontally for some time. Observations to determine azimuth are usually made when Polaris is at eastern or western *elongation* (Art. 20-14), when it appears to be traveling vertically for some time.

Data concerning Polaris can be found from a variety of sources. The "Ephemeris of the Sun and Polaris" gives the declination of Polaris and the Greenwich mean time of its culmination and elongation at the meridian of Greenwich for each day of the current year. The data in the "American Ephemeris" and the "Nautical Almanac" are not in convenient form for use by the surveyor. Convenient tables are to be found in the ephemerides published by the manufacturers of surveying instruments.

Since Polaris, in common with other fixed stars, travels at an angular rate more rapid than that of the sun, it follows that at a given meridian it arrives at culmination a little earlier by mean solar (civil) time each day than it did the day before, the amount earlier being approximately equal to the gain of sidereal time on mean solar time for a 24^h interval, or approximately 3^m56^s per day or 9.86^s per hour (Art. 20-21). Furthermore, the time of culmination depends upon the longitude of the place. To determine the local mean time of upper culmination of Polaris at any given meridian on any given date, the value for the meridian of Greenwich is taken from the table and is reduced to the longitude of the place by means of the variation per hour.

Example: It is desired to find from the "Ephemeris of the Sun and Polaris" the Eastern standard time of upper culmination of Polaris on December 9, 1964, at a place the longitude of which is $5^h15^m45^s$ west of Greenwich.

On December 9, 1964, U.C. at Greenwich occurs at	$20^h45.5^m$
Change in time for $\Delta\lambda$ is $-51.26 \times 9.86^s = -52^s =$	-0.9^m
Local civil time of U.C. at place	$20^h44.6^m$
$\Delta\lambda = 5^h15^m45^s - 5^h = +15^m45^s$	$= +15.8^m$
E.S.T. of upper culmination at place	$= 21^h00.4^m$
	$= 9^h00^m24^s$ P.M.

21-24. Latitude by Observation on Polaris at Culmination. As shown in Art. 20-9, the latitude of a place is equal to the altitude of the elevated pole or, if h is the true altitude of any circumpolar star as it crosses the meridian, then the latitude ϕ is

$$\phi = h \pm p \tag{6}$$

in which the sign preceding the polar distance p is positive or negative according as the star is at lower or upper culmination. By this method the latitude of a station is determined by measuring the altitude of Polaris when it is at either upper or lower culmination, and by applying to this altitude, corrected for refraction, the star's polar distance as given in Polaris tables. Inasmuch as the star is apparently traveling in a horizontal line when at either of these two positions, it is not essential to the precision of the latitude determination that the time of culmination be found, but in any case it facilitates the work of observing if the approximate time is known.

Further, it is not essential that the altitude of the star be observed at the instant it crosses the meridian. For some minutes before and after culmination, the star travels in so nearly a horizontal line that, with the ordinary transit, vertical movement cannot be detected. Within the period 6^m before to 6^m after culmination, the maximum change in altitude is only 01″, and within 12^m before to 12^m after culmination the maximum change in altitude is only 0.1′.

The procedure to be employed in making an observation depends somewhat upon the precision with which the latitude is to be determined and upon the precision with which the time is known. For an observation with the ordinary transit having a vertical circle reading to minutes, when the watch time or the longitude of the place may be in doubt by a few minutes, the standard time of culmination at the given station is determined from an ephemeris as illustrated in the preceding article. A few minutes before the estimated time of culmination, the transit is set up and is leveled very carefully; as a final test the telescope bubble should remain centered as the transit is revolved about the vertical axis. The star is found with the naked eye by noting its position with respect to the neighboring constellations shown in Fig. 21-9. The telescope is focused for a star. If the latitude is known approximately, its estimated value, plus or minus the star's polar distance, is set off on the vertical circle to facilitate finding Polaris. The telescope is sighted at Polaris. When the star has been brought within the field of view, the cross-hairs are illuminated if necessary, and the star is continuously bisected with the horizontal cross-hair. When during a period of 3 or 4 min. Polaris no longer appears to move away from the hair but moves horizontally along it, the star is practically at culmination. The vertical angle is read with dispatch, the transit is carefully releveled, the telescope is plunged, and a second observation on the star is taken with the telescope inverted. Usually the instrument is releveled and a second pair of observations is made. The mean of the observed altitudes, corrected for refraction (Table II) and index error, is taken as the true altitude of the star. The polar distance can be found from any ephemeris giving either declinations or polar distances of Polaris for the days of the current year. Finally the latitude is computed by applying to the true altitude the polar distance with proper sign. Under ordinary conditions, by this method the latitude can be determined within about 01′, or less if the mean of several observations is taken.

Precise Determination. When it is desired to determine the latitude within a few seconds and the standard time and longitude of the place are known within a minute or so, the watch time of culmination may be precisely computed as illustrated in the example of Art. 21-23, and a series of observations may be taken when the star is near culmination. To find the time of lower culmination, 12^h minus one half of the variation per day (practically one half of 3^m56^s) is added to or subtracted from the time of upper culmination. The number of observations will depend upon the precision with which the latitude is to be determined. The observing program is usually arranged so that an equal number of observations will be taken before and after culmination. The observations are begun at a given time interval (usually not more than 10^m) before the calculated time of culmination, and at each sighting of the star, the watch time and the altitude are observed. Half of the observations are taken with the telescope normal and half with it inverted, and between pairs of observations the telescope is carefully releveled. The observed altitudes of the star for positions other than culmination are reduced to the altitude at culmination by applying a correction which, for altitudes within the United States, is given approximately in the accompanying table.

CORRECTIONS TO BE APPLIED TO ALTITUDES OF POLARIS NEAR CULMINATION TO GIVE ALTITUDE AT CULMINATION

Interval from culmination, minutes of time	Change in altitude from culmination, seconds of arc
3	00
6	01
9	03
12	06
15	09
18	12
21	17
24	22
30	34

When the star is at lower culmination, the correction is subtracted; when at upper culmination, the correction is added. The mean of the altitudes reduced to culmination is corrected for refraction (Table II), and the polar distance is found and the latitude is computed as for the case described in the preceding article.

21-25. Azimuth by Observation on Polaris at Elongation.[1] The azimuth of a line can be determined conveniently by an observation on Polaris at eastern or western elongation, provided the latitude of the place is known. As shown in Art. 20-14, the azimuth of any star at elongation is given by

[1] For rough determination of meridian by ranging plumb lines on Polaris, see Art. 12-10.

the expression

$$\sin Z = \frac{\sin p}{\cos \phi} \tag{7}$$

in which Z is the azimuth east or west of north according as the star is at eastern or western elongation, p is the star's polar distance, and ϕ is the latitude of the place. The azimuth of Polaris at elongation is given in ephemerides.

The direction of Polaris from the observer's station at the time of elongation is established by projecting a vertical plane from the star to the earth. The terrestrial line thus established has the same azimuth as the star at elongation, hence the azimuth of any connecting line can be found if the horizontal angle between the two lines is measured.

The star's polar distance is found approximately by ephemeris giving values of either the declination or the polar distance of Polaris for the days of the year in which the observation is made. The latitude is determined by observation, as explained in preceding articles. Inasmuch as the star is apparently traveling in a vertical line when at either elongation, it is not essential to the precision of azimuth determination that the time of elongation be found, but in any case it facilitates the work of observing if the approximate time is known. The time of elongation may be determined from an ephemeris for the current year.

Further, it is not essential that the direction to Polaris be observed at the exact instant of elongation. For some minutes before and after elongation the star travels in so nearly a vertical line that, with the ordinary transit, horizontal movement cannot be detected. For latitudes of the United States, within the period 4^m before elongation to 4^m after elongation, the maximum change in the azimuth of Polaris is less than 01″, and within 10^m of elongation the maximum change in azimuth is only 0.1′.

The procedure to be followed depends somewhat upon the precision with which azimuth is to be determined and upon the precision with which the time of elongation is known. When the watch time or the longitude of the place may be in doubt by a few minutes, the standard time of elongation is determined by use of an ephemeris. A few minutes before the estimated time of elongation, the transit is set up over a given station and is leveled very carefully. The telescope is focused for a star, the latitude of the place is laid off on the vertical circle to facilitate finding the star, and the transit is revolved about the vertical axis until Polaris comes within the field of view. The horizontal and vertical motions are then clamped, the cross-hairs are illuminated if necessary, and the star is continuously bisected with the vertical cross-hair. When during a period of 2 or 3 min. Polaris no longer appears to move away from the hair but moves vertically along it, the star is practically at elongation. The telescope is depressed, and a point on the line of sight is marked on a stake or other reference monument 300 ft. or more away. The telescope is then plunged, and another sight is taken on Polaris. The line of sight is again depressed, and a second point is set on the stake beside the first. Usually the transit is releveled and a second pair of observations is made.

Later the mean of the points is found and marked on the stake. The line joining the occupied station with the established mean point defines the direction of Polaris at elongation. Its azimuth either is computed by Eq. (7) or is found directly from tables. It is given to seconds by the tables in the annual "Ephemeris of the Sun and Polaris." The azimuth of any other line through the station can be determined by measuring the horizontal angle between the two lines, by the method of repetition (Art. 13-17) if necessary to secure the required precision. A true meridian can be established by a perpendicular offset from the established point on the stake, as illustrated in the following example:

Example 1: In taking an observation on Polaris at western elongation, the reference point marking the azimuth of the star is 400 ft. from the transit. The azimuth from north of the star at elongation is $-1°40'45''$. A point of the true meridian through the transit station is to be established by a perpendicular offset from the reference mark, using the relation that the perpendicular offset distance is equal to the distance from the instrument to the mark multiplied by the tangent of the offset angle.

$$\begin{aligned} \log 400 &= 2.60206 \\ \log \tan 1°40'45'' &= \underline{8.46710} \\ \log \text{offset} &= 1.06916 \\ \text{Offset} &= 11.725 \text{ ft.} \end{aligned}$$

The precision of azimuth determination by this method necessarily depends upon the quality of the instrument, the care and skill of the observer, and the number of observations; but, for the procedure described, under ordinary conditions the error should not exceed 10″.

It should be noted that a given error in latitude produces a relatively small error in azimuth. For latitudes of the northern part of the United States, an error of 01′ in latitude produces an error of about 02″ in azimuth, and for lower latitudes the effect is less. Since the latitude can easily be determined within 20″ with the ordinary transit, it is evident that the principal error in azimuth is likely to be due, not to errors in the computed value of the azimuth of Polaris, but to the field operations of projecting the direction of the star to the earth. If the transit is equipped with a full vertical circle, the procedure is such that practically all instrumental errors of projecting the direction of the star to the ground, except that due to the vertical axis not being truly vertical, are eliminated. For reasons explained in Art. 13-32, if precise observations are to be obtained, it is important that the transit be leveled with great care; and in order that the error may be of an accidental rather than of a systematic nature, the instrument should be releveled at least for each set of two observations. When the transit is equipped with a striding level for the horizontal axis, the bubble should be centered prior to each sight on the star.

Precise Determination. When the azimuth of a line is to be established within 02″ or 03″, either a theodolite or a high-grade transit with a telescope of large magnifying power and with a sensitive striding level for the horizontal axis should be employed, and the standard time and longitude should be known within a minute or so in order that the time of elongation may be computed with precision. A series of several observations can then be taken, at known times, during the interval just before and just after the instant of elongation; and the observations for times other than that of elongation can be reduced to elongation by applying a small correction.

The hour angle of the star when at elongation can be precisely determined as explained in Art. 20-14 by the equation $\cos t = \tan \phi \tan p$. The hour angle t expressed in time is practically the sidereal time interval between upper culmination and eastern or western elongation. The corresponding mean solar time interval is found by deducting from the computed hour angle a correction of 9.83^s per hour which, as explained in Art. 20-21, is the difference in solar time between the sidereal hour and the mean solar hour. The standard time of upper culmination of the star is found as shown in the example of Art. 21-23. The standard time of eastern or western elongation is then determined by adding to or subtracting from the time of culmination the mean time interval between upper culmination and elongation. Following is a numerical example:

Example 2: It is desired to find precisely the Eastern standard time of western elongation of Polaris occurring in the early morning hours of December 10, 1964, at a place where the longitude is $5^h15^m45^s$ west of Greenwich and the latitude is 50°00′00″ north.

$$\begin{aligned}
\text{From an ephemeris, } \delta &= 89°06'17'' \\
p &= 90° - \delta = 53'43'' \\
\log \tan p &= 8.193855 \\
\log \tan \phi &= \underline{10.076186} \\
\log \cos t &= 8.270041 \\
t &= 88°55'59'' \\
\text{Hour angle at elongation} = t &= 5^h55^m44^s = 5.93^h \\
-5.93 \times 9.83^s &= \underline{-58^s} \\
\text{Mean time interval from U.C.} &= 5^h54^m46^s
\end{aligned}$$

From example, Art. 21-23,

$$\begin{aligned}
\text{E.S.T. of U.C. on December 9, 1964} &= 9^h00^m24^s \text{ P.M.} \\
\text{E.S.T. of western elongation on December 10, 1964} &= 2^h55^m10^s \text{ A.M.}
\end{aligned}$$

In the "Ephemeris of the Sun and Polaris," the mean time of elongation for the meridian of Greenwich and latitude 40° is given for each day. To find the standard time of elongation for any other meridian and latitude 40°N, the procedure is the same as explained in Art. 21-23 for finding the mean or standard time of culmination. For latitudes other than 40°, to the time of western elongation is added 0.10^m for every degree south of 40° and from the time of western elongation is subtracted 0.16^m for every degree north of 40°. These operations are reversed to determine time of eastern elongation.

When it is desired to determine azimuths with precision and the watch time of elongation of Polaris is precisely known, a series of observations is taken on the star as previously described, the program being so timed that approximately one half of the observations will occur in an interval of a few minutes before the instant of elongation, and the remainder will occur in a like period after elongation. If the transit is equipped with a striding level, the horizontal axis is leveled each time the star is sighted. If not so equipped, the transit is carefully leveled with the telescope bubble prior to each set, which consists of two observations, one with the telescope normal and the other with the telescope inverted.

For each set of points marked on the distant reference monument, a mean is taken, and the azimuth of the star at the mean of the times of the two observations comprising each set is found by applying a slight correction to the azimuth at elongation, this correction being found in Table V. It will be noted that for the average latitude of the United States this correction amounts to less than 01″ when the star is 4^m from elongation, and about 05″ when the star is 10^m from elongation.

The distance from reference monument to transit station is measured. The mean mark for each set of two observations is corrected to give the equivalent mark at the instant of elongation, by calculating the linear offset (for the distance from transit to mark) corresponding to the angular correction found in Table V. If it were not for the accidental errors connected with the observations, the points thus determined would coincide. The mean of the group is taken as the point which gives the most probable direction of the star when at elongation. By measuring the linear variations from the mean and transforming these into angular variations, the probable angular error in the direction of the line defined by the transit station and the mean mark at the reference monument can be computed, and thus the reliability of the observations can be ascertained.

For observations of this character the stability of the transit during the course of the measurements is of the greatest importance. Preferably the transit should be removed from the tripod and placed on a concrete pier. Changes in temperature may also seriously affect the relations between the fundamental lines of the instrument, and hence the transit should be allowed to come to the temperature of the air before observations are begun. Also, the transit should be protected from wind.

Example 3: The azimuth of Polaris is to be computed for the time of western elongation at latitude 50°00′00″N and longitude $5^h15^m45^s$W on December 10, 1964. From an ephemeris, the declination δ for the given date is 89°06′17″.

$$\begin{aligned} p &= 90° - \delta = 53'43'' \\ \log \sin p &= 8.193802 \\ \log \cos \phi &= 9.808067 \\ \hline \log \sin Z &= 8.385735 \\ \text{Azimuth from north} &= 1°23'34'' \end{aligned}$$

This value is also given directly in the "Ephemeris of the Sun and Polaris."

Example 4: With conditions as given in example 3, an observation is taken 10^m after western elongation. What is the azimuth of Polaris at the given instant?

By example 3, the azimuth at western elongation	=	−1°23′34″
By Table V herein, the correction	=	05″
Azimuth from north of the star at given instant	=	−1°23′29″

Example 5: For the observation of example 4 the telescope is depressed, and a point in the same vertical plane as the star is marked on a monument which is 400 ft. from the transit. What are the amount and direction of a linear offset to be measured from this point to establish an equivalent point for Polaris when at western elongation?

The angular correction is 5″ as stated in example 4. The offset is 400 tan 5″ = 400 × 0.0000242 = 0.010 ft.

As the star has reached its most westerly position and is traveling east, the offset is made to the west.

21-26. Azimuth by Observation on Polaris at Any Time. Although elongation is the most favorable time for precise determination of the azimuth of Polaris, it is often inconvenient or impossible to view the star when it is in this position. Under these circumstances, if the standard time and the longitude of the place are precisely known, the hour angle of the star at any instant can be found and the azimuth of the star at any instant can be determined, as described in Art. 20-15, by the expression

$$\tan Z = \frac{\sin t}{\cos \phi \tan \delta - \sin \phi \cos t} \tag{8}$$

in which Z is the azimuth east or west of north, and t is the hour angle reckoned from 0° to 180° before or after upper culmination. It should be noted that for angles between 90° and 180° the sign of cos t is negative. Values of azimuth of Polaris for various hour angles, declinations, and latitudes are given in ephemerides. By interpolation the azimuth for any hour angle and declination can be found.

The field observation consists in measuring the horizontal angle between a terrestrial mark and the star. For a rough determination correct, say, within ½′, a single set of two observations, one with the telescope normal and the other with it inverted, should be taken, the time of the passing of the star across the vertical cross-hair being noted at each setting. The hour angle of the star is then found for the mean of the two times of observation, and the azimuth of the star (for this hour angle and the proper declination) is computed by means of the preceding equation or is found in an ephemeris. This azimuth combined with the mean of the two observed horizontal angles gives the azimuth of the reference line.

To find the hour angle of Polaris at any observed watch time, the watch is compared with a timepiece keeping correct standard time, and the observed time is corrected accordingly. The correct standard time is changed to local mean time by adding or subtracting the difference in longitude expressed in hours between the place of observation and the standard meridian (meridian where standard time is also local mean time), adding if the place of observation is east of the meridian, and subtracting if west. The local mean time of the culmination nearest the time of observation is determined from the ephemeris, as illustrated by the example of Art. 21-23. The difference between this value and the local mean time of observation gives the mean solar time interval before or after upper culmination. For reasons previously explained (Art. 20-21) Polaris is gaining on the mean sun at practically the same rate that sidereal time is gaining on mean solar time. Since in a mean solar hour there are $60^m + 9.856^s$ of sidereal time, it follows that the hour angle of the star expressed in time is greater than the mean solar time interval by 9.856^s or nearly 10^s per solar hour.

Example: What is the hour angle of Polaris December 9, 1964, at $10^h30^m15^s$ P.M. Eastern standard time at a place the longitude of which is $5^h15^m45^s$? The standard meridian is 5^hW. The place is west of the standard meridian and hence local time is slower than E.S.T. by the difference in longitude. The local mean time of observation is, therefore,

$10^h30^m15^s - 15^m45^s$	=	$10^h14^m30^s$ P.M.
		12^h
The mean time from 0^h	=	$22^h14^m30^s$
By the example of Art. 21-23, the local mean time of U.C.	=	$20^h44^m36^s$
Mean time interval since U.C.	=	$1^h29^m54^s = 1.50^h$
Gain of sidereal on mean = 1.50×9.86	=	15^s
Hour angle of Polaris	=	$1^h30^m09^s$ west of meridian

In the "Ephemeris of the Sun and Polaris," azimuths of Polaris at all hour angles are given, the argument being the mean time interval before or after upper culmination. When this table is used, it is unnecessary to determine the actual hour angle of the star.

When the azimuth of a line is to be precisely determined by observation on Polaris not at elongation, a series of observations is taken. Each angle from the star to the line is added to the sum of the preceding angles as when measuring an angle by repetition, and sights are taken first with the telescope normal and then with it inverted. The total angle turned divided by the number of observations gives the horizontal angle from the mean position of the star to the line. The azimuths of Polaris for the mean of the times for the two observations of each set are determined from tables in an ephemeris. The average of the azimuths thus determined for the several sets forming the series is considered to be the azimuth of the mean position of the star; and this value, combined with the mean horizontal angle from the star to the line, gives the azimuth of the line.

Precision. The precision to be obtained depends upon the position of the star, the precision of observations of time, the number of observations, the quality of the instrument, and the care and skill of the observer. When the star is near upper or lower culmination, the azimuth changes at a relatively rapid rate, this change amounting to about 01′ of arc in 3^m of time for latitude 40°N. For this reason the method should not be expected to give precise results when Polaris is near culmination unless the time is observed precisely. It is also important that systematic errors due to inclination of the vertical axis be eliminated by carefully releveling the instrument. If the transit is equipped with a striding level, the horizontal axis is leveled just before each observation when the telescope is pointed in the direction of the star. The ordinary transit not so equipped should be carefully leveled before each set, making the final test with the telescope level as described elsewhere. The stability of the instrument likewise plays an important part in determining the precision, and for precise measurements the transit should be set on a pier or other substantial object.

21-27. Observations on Other Stars. Latitude and azimuth by observation on any circumpolar star can be determined by methods identical with those for Polaris. The other stars near the pole are of less magnitude than Polaris and are therefore not so readily identified; but if the approximate direction of the meridian is known and the hour angle and declination of the star are known, the approximate altitude can be laid off and the telescope can be pointed so that the star will come within the field of view.

Latitude, azimuth, time, and longitude can be determined by observation on stars distant from the pole by methods similar to those described for the sun.

In the ephemerides for each year are tables giving the right ascensions and declinations of selected stars for the upper transit at Greenwich. As the right ascension of the fixed stars changes but a small fraction of a second per day, through the relation $\theta = t + \alpha$ (Art. 20-6) it is seen that the hour angle t of a star at any meridian other than Greenwich is readily found if the sidereal time is known, and that the sidereal time of upper culmination or upper transit of the star is equal to its right ascension α. Further, knowing the longitude of the place of observation and having given the sidereal time of 0^h Greenwich civil time, the relation at a given instant between sidereal time and local civil or standard time can be determined as explained in the preceding chapter. It is therefore possible to find (by aid of an ephemeris) not only the declination of a star but also its hour angle at any instant of mean solar or standard time. Solutions are expedited by using tables for the conversion of mean solar into sidereal time interval or *vice versa,* which tables are given in the "Nautical Almanac."

Example 1: What is the Pacific standard time of the upper transit of Betelgeuse (α Orionis) on February 27, 1964, at a place the longitude of which is 122°15′30″W?

From the "Ephemeris of the Sun and Polaris," by interpolation the Greenwich mean (civil) time of upper transit of Betelgeuse at Greenwich is $7^h26.6^m$ P.M.

The sidereal time interval between transit at Greenwich and transit at the place is $122°15'30''/15 = 8^h09.0^m$.

To change to mean solar time interval, subtract $(8.15 \times 9.83^s) = 1.3^m$.

Mean time interval between transit at Greenwich and transit at place is $8^h09.0^m - 1.3^m = 8^h07.7^m$.

Greenwich mean time of upper transit at place is $7^h26.6^m + 8^h07.7^m = 15^h34.3^m$.

To change to Pacific standard time, subtract 8^h.

Pacific standard time of upper transit is $15^h34.3^m - 8^h = 7^h34.3^m$ P.M.

Example 2: What is the hour angle of Betelgeuse at $11^h00.0^m$ P.M. Pacific standard time on February 27, 1964, at the same place as in example 1?

From example 1, the Pacific standard time of upper transit at the place is $7^h34.3^m$ P.M. The mean time interval since upper transit is therefore $11^h00.0^m - 7^h34.3^m = 3^h25.7^m$. Sidereal time gains on mean time at the rate of 9.86^s per hour. The hour angle of the star is

$$3^h25.7^m + (3.43 \times 9.86) = 3^h26.3^m$$

21-28. Determination of Latitude. To determine latitude by an observation on a star at upper transit, the star's declination and right ascension are found in an ephemeris, and the approximate standard time of upper transit at the given place is determined. Before this time, the transit is set up and the estimated altitude of the star is laid off on the vertical circle. (The latitude will be roughly known, the declination is known, and hence the altitude can be estimated with sufficient precision to bring the star within the field of view.) The telescope is pointed approximately along the meridian, and the instrument is revolved about the vertical axis back and forth through a small angle until the star is sighted. The star is followed with the horizontal cross-hair until the maximum altitude is reached; then the vertical angle is read. The latitude is determined as described in Art. 21-10.

In this way several stars whose times of upper transit differ by short intervals can be observed, and the latitude can be computed by taking the mean of the values thus found.

21-29. Determination of Time. To determine time by observing the upper transit of any star, the direction of the meridian and the longitude of the place being known, the standard time of upper transit is calculated. The star's declination is found from the ephemeris, and its altitude is roughly calculated, the latitude of the place being at least approximately known. Before the estimated time of upper transit, the instrument is set up, a sight is taken along the meridian, and the horizontal motion is clamped. The estimated altitude of the star is laid off on the vertical circle, and the course of the star is followed until it crosses the vertical hair. At this instant, time is observed. The difference between this time and the calculated time is the error of the timepiece. Time determinations should be made on stars near the celestial equator.

For more precise determinations a succession of observations such as that just described may be made on stars whose calculated times of upper transit differ from each other by only a few minutes. Most instrumental errors will be eliminated if the instrument is plunged between two successive observations. The average clock error thus determined is considered to be the error of the timepiece.

It is evident that time and latitude observations may be made simultaneously if, in addition to the observations just described, the vertical angle to each star as it crosses the meridian is measured.

21-30. Determination of Longitude. To determine longitude by observing the upper transit of any star, the direction of the meridian and the standard time of the place being known, the standard time of upper transit is calculated for a longitude estimated to be that of the place. The star is found as described in Art. 21-28, and the standard time of its upper transit is observed. The interval between the calculated time and the observed time of upper transit, changed to sidereal time, is the difference between the estimated longitude and the true longitude of the place.

21-31. Determination of Azimuth. In some cases it is impossible or impractical to observe circumpolar stars for azimuth, either because of clouds or because the star is too near the horizon or too near the zenith. To determine azimuth by observation on any star other than a circumpolar star, the same general procedure is followed as for solar observations of this character described in Art. 21-11. To determine the approximate position of a given star at a given time so that the star may be brought within the field of view, the right ascension and declination are found from an ephemeris, and the hour angle of the star is calculated. The approximate altitude is then computed by Eq. (21), Art. 20-16, and the approximate azimuth is computed by Eq. (19), Art. 20-15. With these two approximate values, the star is readily brought within the field of view if the direction of the meridian is roughly known. The intersection of the cross-hairs is then sighted at the star, the time is observed, and the horizontal and vertical circles are read. If a star chart is available, the position of a given star may be readily determined with the eye, and it may be sighted through the telescope at once without first laying off the approximate azimuth and altitude.

Where the azimuth is to be determined with a higher degree of precision, usually a procedure similar to that described for the sun in Art. 21-14 is followed, and a star is chosen which is in a favorable position.

Preferably several stars are observed in pairs, one star being nearly east and the other being nearly west of the observer, at about the same altitude in order to equalize any error in the correction for refraction. The altitude should be not less than about 20° nor more than about 40°.

Still another method of determining azimuth, which requires little computation but which requires pairs of observations several hours apart, is as follows: A star, preferably southward from the observer, is sighted in the eastern sky at the instant it rises to a fixed altitude which should be not less than about 20° nor more than about 50°. Either a mark is set under the star or its azimuth is observed with respect to some reference line on the ground. Later the same star is sighted similarly in the western sky at the instant it descends to the same altitude. The bisector of the horizontal angle between the two directions of pointing lies on the meridian of the observer. Several stars may be observed in this manner, and the mean of the observations used.

21-32. Numerical Problems.

1. In connection with a solar observation, sights to determine index error are taken on a mark in the general direction of the sun. The vertical angles to the mark are −3°17′00″ with telescope normal and −3°18′30″ with telescope inverted. With the telescope still inverted, a sight is taken to the sun and the observed vertical angle is +36°02′30″. Correct the observed angle for index error of the vertical circle.

2. The mean radius of the earth is 3,956 miles, and the mean distance to the sun is 92,900,000 miles. What is the sun's mean horizontal parallax? What is the parallax correction when the altitude is 30°?

3. The observed altitude of a star is 23°15′20″. The temperature is 90°F. By Table II herein find the refraction correction, and compute the true altitude of the star.

4. The observed altitude of the sun's center is 15°07′30″. The temperature is 15°F. By Table I herein find the correction for parallax and refraction and compute the true altitude of the sun.

5. Find the apparent declination of the sun for the instant of 9^h0^m A.M. Central standard time on February 15 of the current year, using an ephemeris giving values for 0^h Greenwich civil time.

6. Find the apparent declination of the sun for the instant of local apparent noon at a place whose longitude is $5^h52^m54^s$W for the date of July 21 of the current year, using an ephemeris giving values for Greenwich apparent noon.

7. Find the apparent declination of the sun at 3^h18^m P.M. Pacific standard time on November 4 of the current year, using an ephemeris giving values for Greenwich apparent noon and taking into consideration the equation of time. Determine the effect of neglecting the equation of time.

8. The observed altitude of the lower limb of the sun as it crosses the meridian at a given place is 55°31′30″. The observation is made at $11^h34^m20^s$ A.M. Eastern standard time on May 16 of the current year. The temperature is 55°F. Calculate the latitude of the place.

9. The observed altitude of the upper limb of the sun as it crosses the meridian at a given place is 52°13′. The longitude of the place is 7^h32^mW. The date is March 4 of the current year. The temperature is 47°F., and the index error of the transit is +1′30″. What is the latitude of the place?

10. At a place the latitude of which is 41°58′10″N, the observed altitude of the sun's center (taken as the mean obtained by double-sighting) at 3^h12^m P.M. Central standard time on October 21 of the current year is 20°04′30″. The horizontal angle measured clockwise from a reference line to the sun is 81°32′20″. The temperature is 40°F. What is the azimuth (measured from south) of the sun at the given instant and what is the azimuth of the reference line? Compute the azimuth of the sun by Eq. (5).

11. On August 1 of the current year the observed altitude of the sun at a given place is 30°51′45″ at $7^h42^m20^s$ A.M. local apparent time. The latitude of the place is 37°18′20″N, and the longitude is 102°17′30″W. The temperature is 75°F. The horizontal angle (measured clockwise) from reference line to sun is 89°39′15″. What is the azimuth of the sun measured from north, and what is the azimuth of the reference line? Compute the azimuth of the sun by Eq. (4).

12. Compute the changes in azimuth of sun due to a 01′ change in latitude, in declination, and in altitude for latitude 50°, declination 0°, and altitudes of 15° and 30°. Use Eq. (4) as a basis for computations. Compare results with corresponding quantities for latitude 40° given in table of Art. 21-11.

13. Same as problem 12 but for latitude 30°.

14. When an azimuth observation is taken on the sun, the vertical axis of the transit is inclined 01′ to the true vertical, the inclination being in a plane normal to the sight plane when the line of sight is directed toward the sun. The altitude of the sun is 60°. Owing to this inclination, what error will be introduced in the horizontal angle from reference line to sun?

15. At Orono, Maine, on December 5 of the current year the sun's center is observed to cross the meridian at a watch time of $11^h24^m21^s$ A.M. The longitude of the place is $4^h34^m40.3^s$W. What is the watch correction to give local mean time? What is the watch correction to give Eastern standard time?

16. At a given place the center of the true sun crosses the meridian at $11^h41^m37^s$ A.M. Pacific standard time on September 15 of the current year. What is the longitude of the place?

17. By a series of observations the true altitude of the sun's center at a given station is 24°28′44″ at the instant of $4^h13^m12^s$ P.M. Mountain standard time on March 14 of the current year. The clockwise horizontal angle from reference line to sun is 312°16′37″. The latitude of the place is 39°01′42″N. By Eqs. (15*a*) and (16*a*), Art. 20-13, compute the azimuth and hour angle of the sun at the given instant. Determine the longitude of the place and the azimuth of the reference line reckoned from south.

18. Compute the declination settings for a solar attachment at local apparent times of 1^h, 2^h, 3^h, and 4^h after noon on October 13 of the current year for a place whose latitude is 59°06′N and whose longitude is 118°36′45″W.

19. On September 7 of a given year, upper culmination of Polaris at the meridian of Greenwich occurs at $2^h35^m29^s$ Greenwich civil time. What is the Eastern standard time of upper culmination on September 10 of the same year at a place whose longitude is 78°30′15″W?

20. At a given place, on January 12 of the current year the observed altitude of Polaris at upper culmination is 44°36′25″. The temperature is 15°F. What is the latitude of the place?

21. Find the Central standard time of upper culmination of Polaris on December 7 of the current year at Des Moines, Iowa (longitude $6^h14^m30.6^s$W).

22. The altitude of Polaris is observed 20^m after the time of upper culmination and found to be 48°32′20″. The polar distance is 1°01′35″. What is the latitude of the place?

23. Compute the azimuth and hour angle of Polaris when at elongation, the polar distance being 1°01′35″ and the latitude of the place being 43°00′49″N.

24. What is the time of western elongation of Polaris at a given place when upper culmination occurs at $2^h15^m20^s$ P.M. Eastern standard time and the latitude is 42°22′47″N?

25. Find the Pacific standard time of eastern elongation of Polaris on August 18 of the current year for latitude 37°52′24″N and longitude $8^h9^m3^s$W.

26. Find the azimuth of Polaris when at elongation on August 18 of the current year for a place whose latitude is 37°52′24″N.

27. From Table V, determine the azimuth correction to be applied to an observation on Polaris 15^m after elongation to reduce to elongation, the azimuth at elongation being 1°34′12″. Compute the corresponding perpendicular offset to the meridian, at the reference monument beneath the star, when the monument is 600 ft. from the station occupied.

28. At a given place upper culmination of Polaris occurs at $3^h15^m20^s$ P.M. Central standard time on a given date. On the same date an azimuth observation is made at $7^h0^m20^s$ P.M. The latitude of the place is 41°15′30″N, and the polar distance of the star is 54′12″. Compute the hour angle and azimuth of the star.

29. Determine the Central standard time of upper transit of α Canis Minoris on January 8 of the current year at a place whose longitude is $6^h15^m12^s$W.

30. What is the hour angle of α Leonis at $2^h0^m0^s$ A.M. Eastern standard time on April 30 of the current year at a place whose longitude is $4^h49^m8^s$W?

21-33. Field Problems.

Problem 1. Latitude by Observation on Sun at Noon

Object. To determine the latitude of the place by an observation on the sun at local apparent noon, using the engineer's transit.

Procedure. Follow the procedure outlined in Art. 21-10, assuming that the longitude of the place is unknown. If the transit has a full vertical circle, use the method of double-sighting to determine the mean vertical angle to the sun's upper and lower limbs. If the transit is not equipped with a full vertical circle, the altitude correction for semidiameter (which may be taken as 16′) must be applied.

Hints and Precautions. (1) See Art. 21-5. (2) Pay particular attention to the algebraic sign of each quantity and of each correction. (3) If the longitude of the place is known approximately, the approximate standard time of upper transit of the sun may be calculated in advance as a guide in observing.

Problem 2. Azimuth by Direct Solar Observation

Object. To determine the true azimuth of a line by an observation on the sun with the engineer's transit.

Procedure. (1) Follow the procedure outlined in Art. 21-11. If the transit is not equipped with a full vertical circle, the correction for semidiameter (taken as 16′) must be applied to the altitude; further, either sights must be taken to both right and left limbs of the sun, or the correction for semidiameter (taken as 16′ sec h) must be applied to the observed horizontal angle. (2) As a check, observe the magnetic bearing of the line.

Hints and Precautions. (1) See Art. 21-5. (2) Pay particular attention to algebraic signs.

PROBLEM 3. LATITUDE BY OBSERVATION ON POLARIS AT CULMINATION

Object. To determine the latitude of the place by observing Polaris at upper or lower culmination.

Procedure. Follow the procedure outlined in Art. 21-24. If the transit is not equipped with a full vertical circle, make two observations with the telescope normal, releveling the instrument between observations.

Hints and Precautions. (1) See Art. 21-22. (2) Pay particular attention to algebraic signs. (3) As a check, the mean of the times of observation should agree (within a few minutes) with the computed time of culmination.

PROBLEM 4. AZIMUTH BY OBSERVATION ON POLARIS AT ELONGATION

Object. To determine the azimuth of a line by observation on Polaris at eastern or western elongation.

Procedure. (1) Follow the procedure outlined in Art. 21-25. If the transit is not equipped with a full vertical circle, make two observations with the telescope normal, releveling the instrument between observations. (2) As a check, observe the magnetic bearing of the established line.

Hints and Precautions. (1) See Art. 21-22. (2) Pay particular attention to algebraic signs. (3) As a check, the mean of the times of observation should agree (within a few minutes) with the computed time of elongation.

REFERENCES

1. BOWDITCH, NATHANIEL, "Useful Tables from the American Practical Navigator," *U.S. Hydrographic Office,* No. 9, 1934.
2. HOSKINSON, A. J., and J. A. DUERKSON, "Manual of Geodetic Astronomy," *U.S. Coast and Geodetic Survey,* Spec. Pub. 237, Government Printing Office, Washington, D.C., 1948.
3. HOSMER, G. L., and J. M. ROBBINS, "Practical Astronomy," 4th ed., John Wiley & Sons, Inc., New York, 1948.
4. NASSAU, J. J., "Textbook of Practical Astronomy," 2d ed., McGraw-Hill Book Company, New York, 1948.
5. ROELOFS, R., "A New Device for Solar Observations," *Surveying and Mapping,* Vol. 8, No. 4, pp. 187–193, October–December, 1948.
6. ROELOFS, R., "Astronomy Applied to Land Surveying," N. V. Wed. J. Ahrend & Zoon, Amsterdam, 1950.
7. U.S. BUREAU OF LAND MANAGEMENT, "Manual of Instructions for the Survey of the Public Lands of the United States," Government Printing Office, Washington, D.C., 1947.
8. U.S. BUREAU OF LAND MANAGEMENT, "Ephemeris of the Sun, Polaris, and Other Selected Stars for the Year ____," Government Printing Office, Washington, D.C., annual (in advance).
9. U.S. NAVAL OBSERVATORY, "The American Ephemeris and Nautical Almanac for the Year ____," Nautical Almanac Office, Government Printing Office, Washington, D.C., annual (in advance).
10. U.S. NAVAL OBSERVATORY, "The Nautical Almanac for the Year ____," Nautical Almanac Office, Government Printing Office, Washington, D.C., annual (in advance).
11. Condensed tables of data regarding the sun and Polaris, various manufacturers of surveying instruments, annual (in advance).

CHAPTER 22

LAND SURVEYING—RURAL AND URBAN

22-1. General. Land surveying deals with the laying off or the measurement of the lengths and directions of lines forming the boundaries of real or landed property. Land surveys are made for one or more of the following purposes:

1. To secure the necessary data for writing the legal description and for finding the area of a designated tract of land, the boundaries of the property being defined by visible objects.
2. To reestablish the boundaries of a tract for which a survey has previously been made and for which the description as defined by the previous survey is known.
3. To subdivide a tract into two or more smaller units in accordance with a definite plan which predetermines the size, shape, and location of the units.

Whenever real estate is conveyed from one owner to another, it is important to know and state the location of the boundaries, particularly if there is a possibility of encroachment by structures or roadways.

The functions of the land surveyor are to carry out field surveys, to calculate dimensions and areas, to prepare maps showing the lengths and directions of boundary lines and areas of lands, and to write descriptions by means of which lands may be legally conveyed, by deed, from one party to another.

The land surveyor must be familiar not only with technical procedures but also with the legal aspects of real property and boundaries. Usually he is required to be licensed by the state, either directly or as a civil engineer. Technical standards and equitable fees for property surveys are discussed in Refs. 1 and 2 at the end of this chapter. Proposals for changes in land-surveying practices are given in Refs. 12, 13, and 15.

In this chapter, practices as applied to both rural and urban properties are described, and some of the legal aspects of land surveying are discussed. In Chap. 23 the United States system of subdividing the public lands is outlined. Methods of calculating and subdividing areas are discussed in Chap. 19.

22-2. Kinds of Land Surveys. In accordance with the purposes listed in the preceding article, land surveys may be classified as follows:

1. *Original surveys*, to measure the unknown lengths and directions of boundaries already established and in evidence. Surveys of this character are usually of rural lands. For example, Adams may purchase from Brown

a certain parcel of land bounded or defined by features or objects such as fences, roads, or trees. In order that the deed may contain a definite description of the tract, a survey is necessary.

2. *Resurveys,* to reestablish the boundaries of a tract for which a survey has previously been made. The surveyor is guided by a description of the property based upon the original survey, and by evidence on the ground. The description may be in the form of the original survey notes, an old deed, or a map or plat on which are recorded the measured lengths and bearings of sides and other pertinent data. When, land is transferred by deed from one party to another, often a resurvey is made.

3. *Subdivision surveys,* to subdivide land into more or less regular tracts in accordance with a prearranged plan. The division of the public lands of the United States into townships, sections, and quarter sections is an example of the subdivision of rural lands into large units. The laying out of blocks and lots in a city addition or subdivision is an example of the subdivision of urban lands.

22-3. Instruments and Methods. Nearly all land surveys are run with the transit and tape, by methods described in Chaps. 7, 13, and 14. The directions of lines are usually referred to the true meridian, and angular measurements are transformed to bearings. Ordinarily, distances are measured to feet and decimals, and angles are measured to minutes or fraction thereof. On the United States public-land surveys all distances are in Gunter's (66-ft.) chains, as prescribed by law, measurements usually being taken with a tape graduated to read chains and links (1 chain = 100 links).

Formerly the surveyor's compass and 66-ft. link chain were used extensively, particularly in rural surveying; and the directions and lengths of lines contained in many old deeds are given in terms of magnetic bearings and Gunter's chains. In retracing old surveys of this character, allowance must be made for change in magnetic bearing since the time of the original survey. Also, it must be kept in mind that the compass and link chain used on old surveys were relatively inaccurate instruments and that great precision was not regarded as necessary since usually the land values were low. Further, for many years the United States public lands were surveyed under contract, at the low price of a few dollars per mile. Many of the lines and corners established by old surveys are not where they theoretically should be; nevertheless these boundaries legally remain fixed as they were originally established.

Wherever possible, the field procedure is such that the lengths of boundary lines and the angles between boundaries are determined by direct measurement. Therefore the land survey is in general a traverse, the transit stations being at corners of the property, and the traverse lines coinciding with property lines. Where obstacles render direct measurement of boundaries impossible, a traverse is run as near the property lines as practicable, and measurements are made from the traverse to property corners; the lengths and directions of the property lines are then calculated. Where the boundary is irregular or curved, the traverse is established in a convenient location, and offsets are taken from the traverse line to points on the boundary; the length of the boundary is then calculated.

In general, the required precision of land surveys depends upon the value of the land, being higher in urban than in rural areas. The possibility of increase in land values should also be considered. Distances are usually measured with the tape horizontal. On urban surveys, frequently the distances are measured on the slope and are then reduced to the horizontal. On rural surveys if slopes are steep, often measurements are made on the slope, vertical angles being observed with a clinometer.

22-4. Corners, Monuments, and Reference Marks. It is customary to mark the corners of landed property by visible monuments. The term *corner* is applied to a point established by a survey or by an agreement; the term *monument* or *corner mark* is applied to an object placed to mark the corner point upon the surface of the earth. For early original surveys, many of the corners were marked by natural objects such as trees and large stones already in place before the survey was made. In general, however, the corner monuments are established by the surveyor either to mark the intersections of boundaries already in existence or to define new boundaries. Unfortunately, many monuments (such as wooden stakes) are temporary, and many resurveys are necessitated by the obliteration of such temporary markers. So far as possible, the surveyor should establish permanent monuments.

Examples of markers of a more permanent character are an iron pipe or bar driven in the ground; a concrete or stone monument with drill hole, cross, or metal plug marking the exact corner; a stone with identifying mark, placed below the ground surface; charcoal placed below the surface; a mound of stones; a mound of earth above a buried stone; and a metal marker set in concrete below the surface, reached through a covered shaft. Monuments for city lots are usually set nearly flush with the ground. Subsurface stones were commonly used for corner monuments in localities where roads followed section lines. On many old governmental surveys, through wooded country where stones were not available, corners were established by building up a mound of earth over a quart of charcoal or a charred stake, or by building a mound around a tree at which the corner fell. The U.S. Bureau of Land Management has adopted as the standard for the monumenting of the public-land surveys a post made of iron pipe filled with concrete, the lower end of the pipe being split and spread to form a base, and the upper end being fitted with a brass cap with identifying marks (Art. 23-25). Metal caps for use as property markers are commercially available.

Damage to public and private survey monuments, or interference with the proper use of such monuments, is usually prohibited by law.

If there is a possibility that a corner monument will become displaced, the corner should be *referenced,* or connected to nearby objects of more or less permanent character in such manner that it may be readily replaced in case of loss (Art. 14-17). Usually the recorded measurement is called a *connection,* and the object is called a *reference mark* or a *corner accessory;* essentially it is a part of the monument. Examples of corner accessories are trees, mounds, pits, large stones, and buildings. In many large cities, systems of permanent monuments are established and all surveys are referred to them. On public-land surveys, the bearing and distance from a corner to a tree are taken where possible, the tree being blazed and so marked as to identify the section on which the tree stands, the mark terminating with the letters "B.T." signifying bearing tree. The Bureau of Land Management specifies that every corner established in the public-land surveys shall be referenced by one or more objects of any of the following classes: (*a*) "bearing trees, or other natural objects . . . ; permanent improvements; and memorials; (*b*) mound of stone; and (*c*) pits."

If the location of a corner within reasonable accuracy can be determined beyond reasonable doubt, the corner is said to *exist;* otherwise it is said to be *lost.* If the monument marking an existing corner cannot be found, the corner is said to be *obliterated,* but it is not necessarily lost.

Where a corner falls in such location as to make it impossible or impracticable to establish a monument in its true location, it is customary to set a point on one or more of the boundary lines leading to the corner, as near to the true corner as practicable. A point thus established is called a *witness corner.* Everything that has been said concerning monuments at the true corners also applies to witness corners. Witness corners are necessary where the true corner falls in a roadway or body of water, within a building, or upon a precipitous slope. Under certain circumstances, as when boundaries are in roads, it is impossible to place the witness corner on any of the property lines approaching the true corner, in which case the witness corner is established in any convenient location.

The field notes should give detailed information concerning the character, size, and location of all monuments and reference marks; and the data should be recorded in such manner that there will be no possibility of misinterpretation. So far as possible, all points established in the field should be clearly marked to indicate the object which they represent.

22-5. Meander Lines. In United States public-land surveys where regular corners fall in water, traverses called *meander lines* are run roughly following the bank of stream or shore of lake (see also Art. 23-20). The process of establishing such a line is called *meandering.* Meander lines are for surveying and mapping purposes only and are not property lines except in the rare cases where they are specifically stated as property lines in a deed.

22-6. Boundary Records. Descriptions of the boundaries of real property may be found from deeds, official plats or maps, or notes of original surveys. Typical descriptions are given in Arts. 22-13 and 22-17.

Unfortunately, many descriptions are inadequate or incorrect, and these faulty descriptions are a frequent source of confusion and expense. An adequate description should include, or be accompanied by, (1) the name of the author, (2) the date, (3) the source of the survey data, (4) the identity of the property, (5) ties to at least two durable monuments, (6) all dimensions and bearings of property lines, and (7) a plat or a reference to a recorded plat or map (Ref. 13 at the end of this chapter).

Records of the transfer of and from one owner to another are usually kept in the office of the county registry of deeds, exact copies of all deeds of transfer being filed in deed books. These files are open to the public and are a frequent source of information for the land surveyor in search of boundary descriptions when it is inconvenient or impossible to secure permission of the owner to examine the original deed.

In connection with the register an alphabetic index is kept, usually by years, giving in one part the names of *grantors* or persons selling property, and in the other the names of *grantees* or persons buying property. It is, therefore, a simple matter to find a given deed if either or both parties to the transfer and the approximate date of transfer are known. Usually the preceding transfer of the same property is noted on the margin of the deed.

Many states already have special "land courts" where title to land can be confirmed by simple procedure and at nominal cost, and the number is increasing (Refs. 12 and 16 at the end of this chapter). Land courts, together with the state systems of plane coordinates (which in some states may be used as the legal basis for description of land), are gradually simplifying and rendering more certain the registration and transfer of land titles.

In most cases the deeds of transfer of city lots give only the lot or block number and the name of the addition or the subdivision. The official plat or map showing the dimensions of all lots and the character and location of permanent monuments is on file either in the office of the city clerk or in the county registry of deeds; copies are also on file in the offices of city and county assessors.

Some organizations, usually called *title companies*, for a fee will search the records for boundary descriptions and will guarantee the title against possible defects in description, legal transfer, and certain types of claims such as those for right of way. Title insurance does not necessarily mean that the property corners are correctly marked on the ground; and if assurance is desired, a survey should be made.

As the United States public lands are subdivided, official plats are prepared showing the dimensions of subdivisions and the character of monuments marking the corners. When the surveys within a state have been completed, records are given to the state. An exception is Oklahoma, for which the United States survey records are filed with the Director of the Bureau of Land Management at Washington, D.C. States in possession of records have them on file at the state capitol. Usually information concerning these records can be secured from the state secretary of state. Photographic copies of the official plats are obtainable at nominal cost.

22-7. Legal Terms. Following are definitions, quoted from an early edition of Bouvier's "Law Dictionary" (Ref. 6 at the end of this chapter), of a few of the more common legal terms having to do with the conveyance of landed property.

Adverse Possession. The enjoyment of land, under such circumstances as indicate that such enjoyment has been commenced and continued under an assertion of right on the part of the possessor, is adverse possession.

When such possession has been actual and has been adverse for 20 years (or less under certain conditions), the law raises the presumption of a grant.

Where one enters into possession of real property by permission of the owner, without any tendency whatever being created, possession being given as a mere matter of favor, he can never acquire title by adverse possession, no matter how long continued against the true owner thereof.

The adverse possession must be "actual, continued, visible, notorious, distinct, and hostile."

The title by adverse possession for such a period as is required by statute to bar an action, is a fee-simple title, and is as effective as any otherwise acquired.

Alluvium. That increase of earth on a bank of a river, or on the shore of the sea, by the force of the water, as by a current or by waves, or from the recession of water in a navigable lake, which is so gradual that no one can judge how much is added at each moment of time, is known as alluvium. The proprietor of the bank which is increased by alluvium is entitled to the addition, this being regarded as the equivalent for the loss he might sustain from the encroachment of the waters upon his land.

Avulsion. The removal of a considerable quantity of soil from the land of one man and its deposit upon or annexation to the land of another, suddenly and by the perceptible action of water, is avulsion. In such case the property belongs to the first owner. Avulsion by the Missouri River, the middle of whose channel forms the boundary line between the states of Missouri and Nebraska, works no change in such boundary, but leaves it in the center line of the old channel.

Color of Title. Color of title, for the purposes of adverse possession under the statute of limitations, is that which has the semblance or appearance of title, legal or equitable, but which in fact is no title.

A writing which upon its face professes to pass title but which does not in fact do so, either from a want of title in the person making it or from the defective conveyance used, is also known as color of title. The term is also applied to a title that is imperfect, but not so obviously that it would be apparent to one not skilled in the law.

Fee. The word fee signifies that the land or other subject of property belongs to its owner and is transmissible, in the case of an individual, to those whom the law appoints to succeed him under the appellation of heirs.

Fee Simple. An estate of inheritance is a fee simple. The word "simple" adds no meaning to the word fee standing by itself, but it excludes all qualifications or restrictions as to the persons who may inherit it as heirs.

High-water Mark. The high-water mark is wherever the presence of the water is so common as to mark on the soil a character, in respect to vegetation, distinct from that of the banks; it does not include low lands which, though subject to periodic overflow, are valuable for agricultural purposes.

That part of the shore of the sea to which the waves ordinarily reach when the tide is at its highest is also known as the high-water mark.

Low-water Mark. Low-water mark is that part of the shore of the sea to which the waters recede when the tide is lowest, that is, the line to which the ebb tide usually recedes; or it is the ordinary low-water mark unaffected by drought. It has been said to be the point to which a river recedes at its lowest stage.

Parol. Parol is a term used to distinguish contracts which are made verbally, or in writing not under seal, which are called parol contracts, as distinguished from those which are under seal, which bear the name of deeds or specialties.

Patent. A patent is the title deed by which a government, either state or Federal, conveys its lands.

Reliction. An increase of the land by the retreat or recession of the sea or a river is known as reliction.

22-8. Legal Interpretation of Deed Description. The descriptions of the boundaries of a tract include the objects that fix the corners, the lengths and directions of lines between the corners, and the area of the tract. A deed description may contain errors or mistakes of measurement, or mistakes of calculation or record, thus introducing inconsistencies which cannot be reconciled completely when retracement becomes necessary. In such cases, where uncertainty has arisen as to the location of property lines, it is a universal principle of law that the endeavor is to make the deed effectual rather than void, and to execute the intentions of the contracting parties. The following general rules are pursuant to this principle:

1. *Monuments.* It is presumed that the visible objects which marked the corners when a conveyance of ownership was made indicated best the intentions of the parties concerned; hence it is agreed that a corner is established by an existing material object or by conclusive evidence as to the previous location of the object. A corner thus established will prevail against all other conflicting evidence, provided there is reason to believe that the monument was set in accordance with the original intention and that its location has not been disturbed. The kinds of evidence which are valid in relocating obliterated corners are stated in Art. 23-28.

2. *Distance and Direction vs. Area.* In the case of discord between the described courses and the calculated area of a tract, the deed-description requirements, or "calls," for distances or directions of courses will prevail against the call for area, again on the assumption that the boundary lines are more visible and actual evidence of the intentions of the parties than is the calculated area of the tract.

3. *Mistakes.* It is a well-established principle that a deed description which taken as a whole plainly indicates the intentions of the parties concerned will not be invalidated by evident mistakes or omissions. For example, such obvious mistakes as the omission of a full tape length in a dimension or the transposition of the words "northeast" for "northwest" will have no effect on the validity of a description, provided it is otherwise complete and consistent or provided its intention is manifest.

4. *Purchaser Favored.* In the case of a description that is capable of two or more interpretations, that one will prevail which favors the purchaser.

5. *Ownership of Highways.* Land described as being bounded by a highway or street conveys ownership to the center of the highway or street. Any variation from this interpretation must be explicitly stated in the description.

6. *Original Government Surveys Presumed Correct.* Errors found in original government surveys do not affect the location of the boundaries established under those surveys, and the boundaries remain fixed as originally established.

22-9. Riparian Rights. An owner of property that borders on a body of water is a riparian proprietor and has riparian rights (pertaining to the use of the shore or of the water) which may be valuable. Because of the difficulties arising from the irregularity of such boundaries, it is important that the surveyor be familiar with the general principles relating to riparian rights and with the statutes and precedents established in his particular state. For example, as regards the ownership of the bed of a navigable river, the two states of Iowa and Illinois bordering on the same river have very different laws. Clark (Ref. 8 at the end of this chapter) states: "It is a rule of property in Illinois, that the fee of the riparian owner of land in that state bordering on the Mississippi River extends to the middle line of the main channel of the river," whereas the Iowa courts hold "that the bed of the Mississippi River and the banks to the high-water mark belong to the state, and that the title of the riparian proprietor extends only to that line."

In establishing the property lines of riparian owners many dissimilar and complex situations are encountered, but the principles which usually apply are stated below under six general cases.

1. *Meander Lines.* It is a well-established principle that government patents of land bordering on meandered streams or lakes convey ownership, not to the meander line, but to the thread of a nonnavigable stream, or to the bank of a navigable stream, or to the shore of a lake.

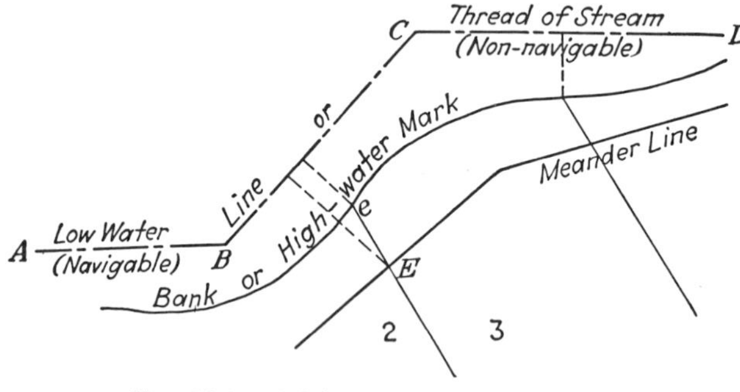

FIG. 22-1. Origin of riparian boundaries.

2. *Origin of Dividing Lines.* There are two opposing lines of decisions. Under one it is held that a dividing line has its origin at the high-water line of a river, or at the shore line of a lake, and not at the meander line. Thus in Fig. 22-1, the dividing line between lots 2 and 3 would be made perpendicular to the thread of

the stream, beginning at e (on the high-water line) and not at E (on the meander line). Under the other line of decisions, the reverse is held.

3. *Alluvium and Reliction.* The direction of the property lines dividing areas created by alluvium or by reliction is determined by the proportional lengths of the old and of the new shore lines. The extremities of these lines are fixed either by definite bends, as A, F, A', and F' (Fig. 22-2) or by the intersections of the old and

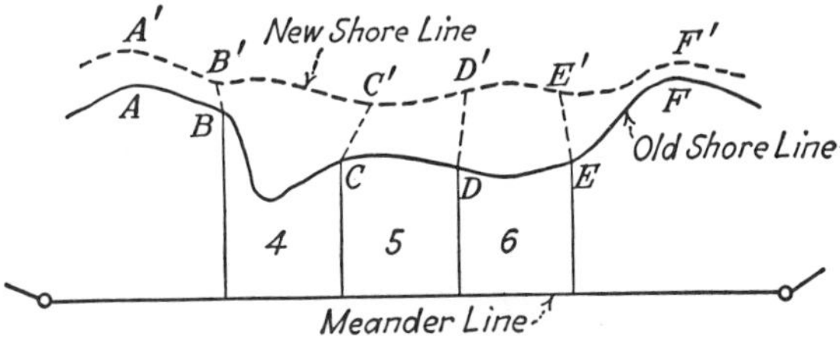

FIG. 22-2. Riparian boundaries for areas added by alluvium between definite bends in shore line.

new lines, as A and F (Fig. 22-3). The general rule is to measure along the old shore line between the old extremities, as A and F; measure along the new shore line between the new extremities, as A' and F'; and divide the new line ($A'F'$) into parts proportional in length to those of the old line (AF). Thus for lots 4 and 12 by proportion $B'C'/BC = A'F'/AF$. The area $BB'C'C$ represents the area added by alluvium to lots 4 and 12.

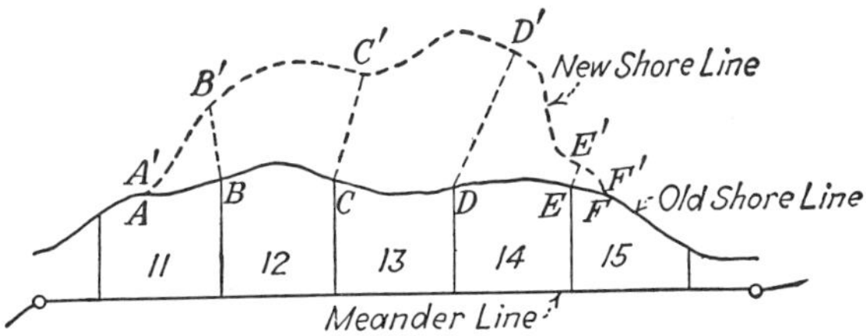

FIG. 22-3. Riparian boundaries for areas added by alluvium between intersection of old and new shore lines.

4. *Bays or Coves.* Property lines fixing riparian rights in bays or coves sometimes are established by lines beginning at the extremities of the property lines on shore and having a direction perpendicular to a line connecting the adjacent headlands of the bay or cove. Thus the lines BB', CC', etc., for lots 1, 2, and 3 (Fig. 22-4) are established perpendicular to the line AF, which connects the two headlands A and F.

Other court decisions have fixed the lines according to the following rule: Divide the straight line joining the headlands (AF in Fig. 22-4) into parts proportional to the lengths of the shore line held by each owner; the property line of inundated land is determined by joining the extremities of the property lines on shore and the corresponding points of subdivision on the line between headlands. These are shown in the figure by lines BB'', CC'', etc.

5. *Streams and Rivers.* The lines fixing the riparian rights of owners of property bordering on streams and rivers are established by lines perpendicular to the thread of the stream if nonnavigable, or to the low-water line (sometimes to the middle of the channel) if navigable. Thus the lines for lots 2 and 3 of Fig. 22-1 are established perpendicular to the line *ABCD*.

6. *Lakes.* The riparian property lines are established perpendicular to the center line of the lake; or, in the case of a circular shore line, by lines to the center of the lake. Thus in Fig. 22-5, the lines for lot 7 are established by the boundary *ABCDE*, and for lot 4 by the boundary *FGCHI*. Where the shore line is circular at the end of the lake, the land lines terminating at *J* and *K* are drawn to *O*, the center of the circular shore line.

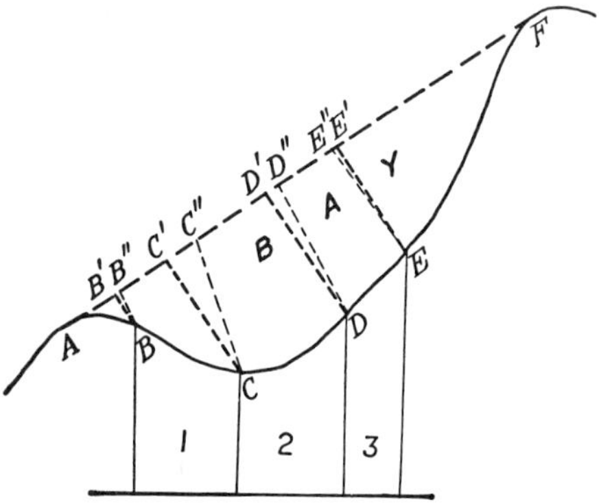

FIG. 22-4. Riparian boundaries in bay.

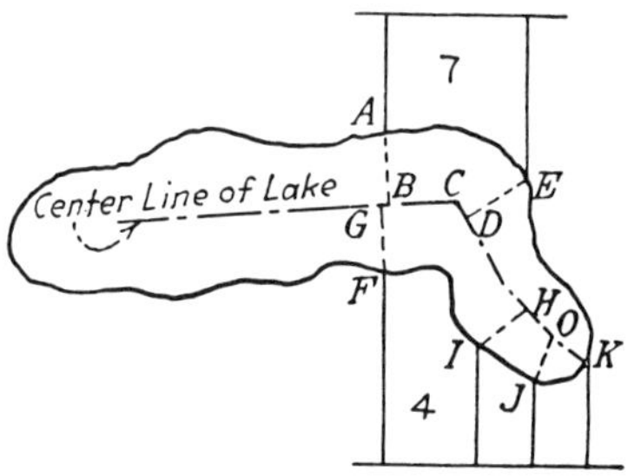

FIG. 22-5. Riparian boundaries in lake.

22-10. Adverse Possession. The many legal aspects of adverse possession cannot be treated here, but it is desirable to direct attention to the important fact that property lines may be fixed by continued possession and use of the land (usually for 20 years) as against original survey boundaries. The conditions and the period of time necessary to gain title are fixed by statute in the various states.

According to the definition given in Art. 22-7, adverse possession, to become effective, must be plainly evident to the owner, without his permission, to his exclusion, and hostile to his interests. Such possession may be evidenced by fencing, cultivation, erection of buildings, etc.

Right to title by adverse possession may be acquired by individuals, corporations, and even by the state. But the statute does not run *against* the state; that is, property in a street or highway cannot be acquired by adverse possession.

Under this principle, if a person should use the land up to a fence and should recognize it as a boundary line, to the exclusion of the owner, for the statutory period, the fence then becomes the legal property line even though it may be shown later that it is not on the true and original line. However, if the possession of the land has not been held adversely, that is, to the exclusion of the owner, and if the fence has merely served the convenience of the persons concerned, both parties recognizing that it was probably not on the true line, title cannot be claimed.

It is therefore clear that the application of the principle of adverse possession is entirely a matter of intention and belief. If land is held openly and notoriously with the intent to acquire title, or with the belief that the occupation is proper and right, then title will be granted if and when the statutory requirements are fulfilled. But if by parol agreement or by actions it is manifest that the parties concerned had no intention to occupy beyond the true line, at the same time knowing that the location of the true line was uncertain, then title cannot be gained adversely.

Adverse possession under "color of title" will "ripen into title" under the statute of limitations in some jurisdictions in half the time required without color of title; for example, if title may be gained without color of title in 20 years, it may be gained in 10 years with it.

22-11. Legal Authority of the Surveyor. A resurvey may be run to settle a controversy between owners of adjoining property. The surveyor should understand that, although he may act as an arbiter in such cases, it is not within his power legally to fix boundaries without the mutual consent and authority of all interested parties. In the event of a dispute involving court action, he may present evidence and argument as to the proper location of a boundary, but he has no authority to establish such a boundary against the wishes of either party concerned. A competent surveyor by wise counsel can usually prevent litigation; but if he cannot bring his clients to an agreement, the boundaries in dispute become valid and defined only by a decision of the court. In boundary disputes the surveyor is an expert witness, not a judge.

The right to enter upon property for the purpose of making public surveys is usually provided by law, but there is no similar provision regarding private surveys. The surveyor (or his employer, whether public or private) is liable for damage caused by cutting trees, destroying crops or fences, etc.

Permission should be obtained before entering private property, in order to avoid a possible charge of trespassing. If permission is not granted, it may be necessary to traverse around the property or to resort to aerial photographs. In certain cases, as, for example, where a boundary is in doubt or dispute, a court order may be obtained.

22-12. Liability of the Surveyor. It has been held in court decisions that county surveyors and surveyors in private practice are members of a learned profession and may be held liable for incompetent services rendered. Thus Clark (Ref. 8 at the end of this chapter), quoting from court decisions, states: "If a surveyor is notified of the nature of a building to be erected on a lot, he may be held liable for all damages resulting from an erroneous survey;

and he may not plead in his defense that the survey was not guaranteed." Similarly, it has been held that in any case where the surveyor knows the purpose for which the survey is made, he is liable for damages resulting from incompetent work.

The general principle invoked in such cases is that the surveyor is bound to exhibit that degree of prudence, judgment, and skill which may reasonably be expected of a member of his profession. Thus in the following quotations from Clark a Connecticut court says "the gist of the plaintiff's cause of action was the negligence of the defendant in his employment as a civil engineer. Having accepted that service from the plaintiff, the defendant . . . was bound to exercise that degree of care which a skilled civil engineer would have exercised under similar circumstances." Also, a Kansas court declares, "reasonable care and skill is the measure of the obligation created by the implied contract of a surgeon, lawyer, or any other professional practitioner." But Ruling Case Law says, ". . . yet a person undertaking to make a survey does not insure the correctness of his work, nor is absolute correctness the test of the amount of skill the law requires. Reasonable care, honesty, and a reasonable amount of skill are all he is bound to bring to the discharge of his duties."

RURAL–LAND SURVEYS

22-13. Description of Rural Land. In the older portions of the United States, nearly all of the original land grants were of irregular shape, many of the boundaries following stream and ridge lines. Also, in the process of subdivision the units were taken without much regard for regularity, and it was thought sufficient if lands were specified as bounded by natural or artificial features of the terrain and if the names of adjacent property owners were given. Thus a description of a tract as recorded in a deed reads:

> Bounded on the north by Bog Brook, bounded on the northeast by the irregular line formed by the southwesterly border of Cedar Swamp of land now or formerly belonging to Benjamin Clark, bounded on the east by a stone wall and land now or formerly belonging to Ezra Pennell, bounded on the south and southeast by the turnpike road from Brunswick to Bath, and bounded on the west by the irregular line formed by the easterly fringe of trees of the wood lot now or formerly belonging to Moses Purington.

1. *By Metes and Bounds.* As the country developed and land became more valuable, and as many boundaries such as those listed in the preceding description ceased to exist, land litigations became numerous. It then became the general practice to determine the lengths and directions of the boundaries of land by measurements with the link chain and surveyor's compass, and to fix the locations of corners permanently by monuments. The lengths were ordinarily given in rods or chains, and the directions were expressed

as bearings usually referred to the magnetic meridian. Surveys of this character are now usually made with the transit and tape, distances being recorded in feet or chains, and directions being given in true bearings computed from angular measurements. In describing a tract surveyed in this manner the lengths and bearings of the several courses are given in order, and the objects marking the corners are described; if any boundary follows some prominent feature of the terrain, the fact is stated; and the calculated area of the tract is given. When the bearings and lengths of the sides are thus given, the tract is said to be described by *metes and bounds*. Within the limits of the precision of the original survey, it is possible to relocate the boundaries of a tract if its description by metes and bounds is available, provided at least one of the original corners can be identified and the true direction of one of the boundaries can be determined.

Later in this article is an illustrative description by metes and bounds typical of rural lands in the eastern states and of isolated grants in the western states where the subdivision of lands has been outside the rectangular system of the public-land surveys. (See also Art. 22-17.)

2. *By Subdivisions of Public Land.* The type of description employed for lands which have been divided in accordance with the rectangular system of the Bureau of Land Management is described in detail in Chap. 23. The records and plats of the United States surveys are a part of the permanent public records and are accessible to anyone desiring to consult them. In conveying by deed a United States subdivision or fraction thereof, no doubt can at any time exist as to the tract involved if it is described by stating its sectional subdivision, section number, township, range, and name of the principal meridian on which the initial point is located (Figs. 23-2 and 23-3). Following is an example of the legal description of a 40-acre tract comprising a full quarter-quarter section:

> The north-east quarter of the south-west quarter of section ten (10), Township four (4) South, Range six (6) East, of the Initial Point of the Mount Diablo Meridian, containing forty (40) acres, more or less, according to the United States Survey.

3. *By Coordinates.* In some states, the locations of land corners are legally described by their coordinates with respect to the state-wide plane-coordinate system. Although practice is not as yet uniform in this regard, the following description by the Tennessee Valley Authority illustrates the description of land both by metes and bounds and by coordinates, with further reference to corners and lines of the United States public-land survey. The public-land survey is referred to the Huntsville principal meridian. A map of the tract is shown in Fig. 22-6.

> A tract of land lying in Jackson County, State of Alabama, on the left side of the Tennessee River, in the South Half (S ½) of the Northwest Quarter (NW ¼) of

Section Three (3), Township Six (6) South, Range Five (5) East, and more particularly described as follows:

Beginning at a fence corner at the southwest corner of the Northwest Quarter (NW ¼) of Section Three (3) (coordinates N 1,470,588; E 416,239), said corner being North six degrees twenty-four minutes West (N6°24′W) twenty-six hundred (2600) feet from the southwest corner of Section Three (3) (N 1,468,004; E 416,529),

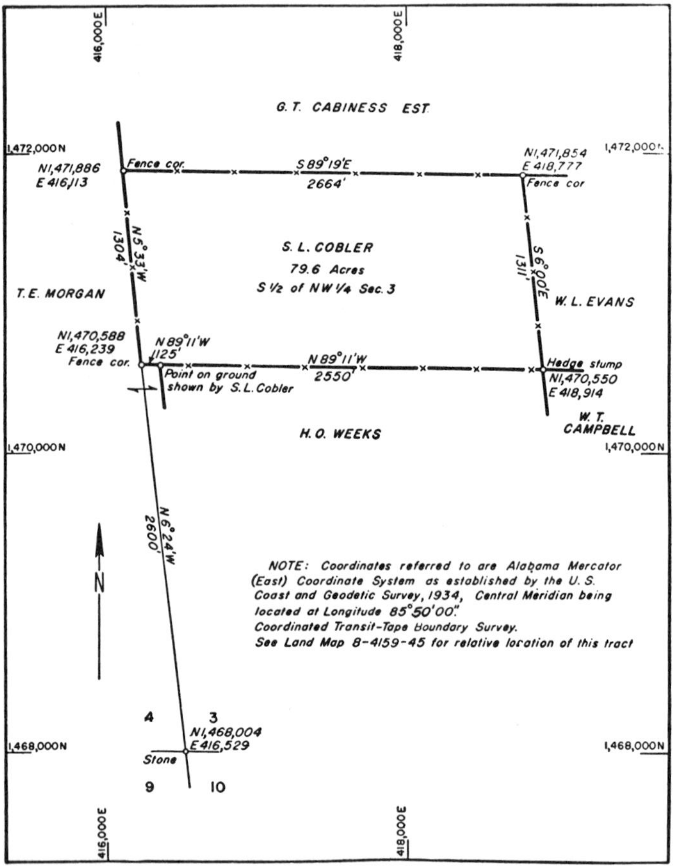

FIG. 22-6. Land map.

and a corner to the land of T. E. Morgan; thence with Morgan's line, the west line of Section Three (3), and a fence line, North five degrees thirty-three minutes West (N5°33′W) thirteen hundred four (1304) feet to a fence corner (N 1,471,886; E 416,113), a corner of the lands of T. E. Morgan, and the G. T. Cabiness Estate . . . thence with Weeks' line, the south line of the Northwest Quarter (NW ¼) of Section Three (3), and a fence line, North eighty-nine degrees eleven minutes West (N89°11′W) two thousand five hundred fifty (2550) feet to a point on the ground shown by S. L. Cobler, a corner of the lands of H. O. Weeks and T. E. Morgan; thence with Morgan's line, the south line of the Northwest Quarter (NW ¼) of Section Three (3), and the fence line North eighty-nine degrees eleven minutes West (N89°11′W), one hundred twenty-five (125) feet to the point of beginning.

The above described land contains seventy-nine and six-tenths (79.6) acres more or less, subject to the rights of a county road which affects approximately five-tenths (0.5) acres, and is known as Tract No. GR 275, as shown on Map No. 8–4159–45, prepared by the Engineers of the Tennessee Valley Authority.

The coordinates referred to in the above description are for the Alabama Mercator (East) Coordinate System as established by the U. S. Coast and Geodetic Survey, 1934. The Central Meridian for this coordinate system is Longitude eighty-five degrees (85°) fifty minutes (50′) no seconds (00″).

22-14. Original Survey. The need for an original survey usually arises when one person desires to transfer to another a tract of land which has not been previously surveyed but which is defined by certain natural or artificial features of the terrain.

With the desired boundaries of the land given, the surveyor establishes monuments at the corners and runs a closed transit traverse around the property, measuring the lengths of lines and the angles between intersecting lines. Where boundaries are not straight, offsets from transit line to curved boundary are measured at known intervals; and where obstructions make direct measurement along boundaries impossible, the traverse is run as close to the boundary as convenient and measurements are taken from transit stations to corners of the tract. Angular measurements may be taken by any of the methods described in Chap. 14, but usually the interior angles are observed. Preferably the corners should be referenced to permanent objects. Also the direction of the true meridian should be determined, usually by a solar observation (see Art. 21-11).

The information thus obtained is recorded in the surveyor's notebook, the angles and distances of the main traverse being tabulated, and the remaining data being recorded in the form of a sketch. The bearings of the sides are then computed, properly with respect to the true rather than the magnetic meridian.

A description of the tract, by metes and bounds and/or by coordinates, is prepared. Usually a plat is drawn, the boundaries being plotted by one or another of the methods described in Chap. 18, and details being shown as suggested by Fig. 22-6 and Art. 6-3. The area is calculated as described in Chap. 19. In the process of calculation, the error of closure of the traverse is determined and thus a check on the reliability of the survey is obtained. A copy of the description and a tracing and/or prints of the plat are submitted to the person for whom the survey is made.

22-15. Resurvey. The resurvey of lands is attended with greater difficulty than is usually appreciated by those inexperienced in work of this character. Particularly is this true in the older sections of the United States where the early surveys were not of the rectangular system and were not under the control of the U.S. Bureau of Land Management. The proper relocation of old lines calls for greater ingenuity and broader experience on the part of the surveyor than does any other kind of surveying.

The purpose of the resurvey is to reestablish boundaries in their original locations. To guide him the surveyor has available the description contained in the deed or obtained from old records, and descriptions of adjoining property.

If the description of the property were without error and one or more of the original corners were in evidence, and, further, if the resurvey could be run without error, the problem would be as simple as running the original survey. When the lengths and directions given in the description had been laid off, the surveyor could say with assurance that the reestablished corners were in their original location. The facts are, however, that the original survey *did* contain errors and probably rather large ones if it was made during the era of the compass and link chain. Further complications may be added by directions in the description being given by magnetic bearings and the declination at the time of the original survey being unknown, or by no statement having been made as to whether the bearings of the original survey were referred to the magnetic or to the true meridian. Often large mistakes are made in transposing from one record to another or are present in the measurements of the original survey. Loss of corners, lack of reference measurements, removal or alteration of physical boundaries, conflicting testimony of persons having knowledge concerning the position of boundaries, conflicts with adjoining property, and numerous other factors may add to the uncertainties of the problem.

As a first step the surveyor critically examines the descriptions for gross errors; he then computes the latitudes and departures of the several courses as given in the description, determines the error of closure, and plots the boundaries of the tract to scale.

If original bearings are magnetic, the magnetic declination at the time of the original survey is found, and true bearings are computed. If true bearings cannot be found in this manner (as when the date of the original survey is unknown) and if one or more boundaries can be positively identified, observations are made to determine the true bearings of these known lines; and by a comparison of true and original magnetic bearings the declination at the time of the original survey is estimated and the true bearings of the other lines are computed.

1. *One or More Boundaries Evident.* If one or more boundary lines can be identified from the monuments or from reliable reference marks, a comparison is obtained between the length of the chain or tape used on the original survey and that to be used on the resurvey; the proportionate lengths of other sides of the tract are then computed.

With the computed directions and proportionate lengths, the surveyor starts from a known corner and reruns the courses; at each estimated location of a corner he seeks physical evidence of the location of the original corner. If such evidence is found and if the old monument is not in good condition, he sets a new monument. Thus if a stake had originally been set at the corner, careful slicing of the top soil with a shovel might reveal rotted wood, a hole in the ground, or even discolored earth, which might be considered rather positive evidence of the old location. Ties to bearing trees or other objects to which reference measurements were taken would also prove useful in finding the probable location of an obliterated monument.

At any point where the surveyor finds what he regards as positive evidence as to the original location of the corner and this location does not agree with the relocation measurements derived from the description of the property, a monument is set at the original location and new measurements of angles and distances are made to refer to the mark thus established.

At any point where physical evidence as to the original location of the corner is entirely lacking, the corner is located temporarily by measurements derived from the description of the property. The survey is then continued until either positive evidence of the location of a succeeding corner is found or the traverse is brought to a closure at the initial point. In the former case, the temporary monuments established between two corners which are located with certainty are regarded as correctly located and are replaced by more permanent markers if the points established by the angles and distances of the resurvey fall at the true location of corresponding corners as indicated by visible evidence. If the points do not so fall, the error is determined and the locations of intermediate temporary corners are adjusted by proportionate measurements, the discrepancy between the original survey and the resurvey being assumed to have accumulated gradually.

If no physical evidence except one boundary is found, the survey is run to the point of beginning, and the linear error of closure is measured. The survey is then balanced as described in Art. 18-11, and the computed corrections are applied by moving the preceding temporary monuments and establishing them as permanent. Finally the lengths and bearings of the adjusted courses are measured in the field.

2. *One Corner Evident.* Where only a single corner can be found, the process of reestablishing boundaries is not so simple, particularly when the bearings of the original survey were observed with a compass and were referred to the magnetic meridian without the declination being given. Usually an estimate of the amount of the magnetic declination at the time of the original survey can be obtained by consulting old records, but often the date of the survey from which the description is derived is unknown and cannot be closely determined.

By means of the estimated declination, the magnetic bearings are changed to true bearings, the latitudes and departures of the boundaries are computed, and the linear error of closure of the original survey is determined. If this error is reasonably small (say, not greater than $\frac{1}{300}$ if the old survey was run with a compass), it is indicated that there are no mistakes in the lengths and bearings given in the description. About the only course then open to the surveyor is to establish the true meridian and to rerun the survey in accordance with the old description, and to consider the corners as being relocated to the best of his ability if the error of closure of the resurvey is no larger than that of the original survey. This error of closure is distributed proportionally among the several courses as described in the preceding article.

It is evident that there might be a large error in the length of the chain or tape used in making the original survey and still the traverse would close. Inasmuch as there is no way of making a comparison between the original and resurvey lengths, distances laid off during the resurvey may be considerably different from corresponding ground distances measured during the original survey. Hence although the resurveyed tract may have the same shape as the original tract, and its boundaries may maintain the same direction as the corresponding boundaries of the original, yet the actual area of the resurveyed tract may be considerably different from that of the original. Also, by a similar course of reasoning, it is evident that any error in the estimated declination will result in a resurvey figure which, although it may close, will be composed of lines each of which will make a constant angle with the corresponding boundary of the original tract. The surveyor should realize that, for a case such as this, a small error of closure of the resurvey is not conclusive evidence of the closeness with which corners are reestablished with respect to their original locations.

3. *No Corner Evident.* If a description of the tract is available but all evidence of the location of original corners is lost, the surveyor will find it expedient to search the records for descriptions of adjoining property and by means of these descriptions to reestablish by measurement as many corners of the tract in question as seems feasible. It is possible that these locations may be considerably in error. A corner may be reestablished by measurements from several sources, each resulting in a different location; in such cases the surveyor is called upon to exercise his judgment as to the most probable location.

Sometimes it is possible to determine the location of an obliterated corner through evidences of previously existing lines such as fences and roads. Thus, if the surveyor has reason to believe that a fence once stood on the line, he may be able to find evidences of rotted posts in the ground. Differences in the ground surface, or even differences in vegetation along a definite line, are valuable clues. Occasionally the surveyor may find it desirable to consult old settlers who were familiar with the original boundaries; but although such persons are usually very positive in their opinions, the information is seldom of much value and is frequently misleading.

Having thus tentatively fixed the location of one or more corners, the surveyor attempts to reconcile these locations with the description of the given tract, the resurvey being conducted somewhat as just described. Readjustments are made to conform to the judgment of the surveyor in light of the information that he obtains as the survey progresses.

Report. When a resurvey has been completed, it is the duty of the surveyor to render a report to his client stating exactly what he found and what course of procedure he employed in attempting to reestablish missing corners. The report should be accompanied by a plat similar to that of an original survey (see Art. 22-14). In addition, it should indicate which are original monuments and which are monuments established at the time of the resurvey. Mistakes in the original description should be pointed out, but the surveyor should clearly understand that it is his function to reestablish boundaries of a given tract in as nearly as possible their original location.

22-16. Subdivision Survey of Rural Land. A subdivision survey of rural land implies a survey which is conducted for the purpose of subdividing into two or more tracts, in accordance with some prearranged plan, an area whose boundaries are already established. In such cases, a resurvey of the tract is run, new monuments are established on the new boundary lines, and a new plat and description are prepared as in the case of an original survey.

Public Lands. The public lands are divided into townships, sections, and quarter sections by United States land surveyors, in a manner prescribed by law. In general the United States surveys establish the boundaries of sections and establish quarter-section corners on section lines; and any further subdivision is made after the lands have passed into the hands of private individuals, the work being carried out by surveyors in private practice. Subdivisions of this kind are described in detail in Arts. 23-17 to 23-19.

Irregular Subdivisions. Surveys of this class are conducted for a variety of purposes. The following examples serve to illustrate the procedure for certain cases:

Example 1: A railroad is to traverse the land belonging to Black, and the railroad company desires to secure title to a right of way of a definite width on either side of the center line which has already been surveyed and marked with stakes. A description of Black's tract has been secured.

The right-of-way surveyor reruns the boundaries of Black's tract that are intersected by the railroad line, establishes the directions of the right-of-way boundaries parallel with the center line, and sets monuments at the intersections of these boundaries with those of Black's tract. He then makes a survey of the tract thus defined, securing sufficient data so that the lengths and directions of the boundaries of the right-of-way tract now within Black's tract are obtained. He also ties right-of-way corners which he has established to the nearest old corners of Black's property.

With these data the area of the right-of-way tract is calculated, and a description of the tract is prepared as for an original survey (see Art. 22-13). The point of beginning is referred to one of the old monuments marking the original tract, and not to the center line of the railroad.

Example 2: It is stipulated in the will of Green that his New England farm is to be divided equally among his three sons, each to have an equal frontage on the highway which forms one of the boundaries of the tract. The farm is of irregular shape and has not been surveyed for many years.

In a case of this kind, the surveyor first makes a resurvey of the entire tract, angles being measured probably to minutes, and distances being measured probably to tenths of feet. From the data thus obtained the area is calculated. In connection with the resurvey, subdivision corners are established on the highway.

The simplest division is one for which the subdividing lines are straight, each cutting off the required area from the given tract. With the area of the entire tract known, the area each son is to receive is calculated, and the length and bearing of each of the lines rendering the subdivision are computed as described in Art. 19-18.

Finally, from each of the two subdivision corners already established on the highway frontage, the surveyor lays off the computed direction of the subdividing line through that point and establishes the remaining unknown corner at the point where this subdividing line intersects the opposite boundary. The distances from this latter corner to adjacent corners are measured, and the survey is considered as checked if these measured distances agree closely with the computed lengths of the same courses. A plat is drawn as for an original survey, the lengths and bearings of all lines and the area of each subdivision being shown; and a description of each of the three subdivisions is prepared.

URBAN–LAND SURVEYS

22-17. Description of Urban Land. The manner of legally describing the boundaries of a tract of land within the corporate limits of a city depends upon conditions attached to the survey by which the boundaries of the tract were first established, as indicated by the following classification:

1. *By Lot and Block.* If the boundaries of the tract coincide exactly with a lot which is a part of a subdivision or addition for which there is recorded an official map, the tract may be legally described by a statement giving the lot and block numbers and the name and date of filing of the official map. Most city property is described in this way. Following is a description of this character occurring in a deed:

Lot 15 in Block 5 as said lots and blocks are delineated and so designated upon that certain map entitled *Map of Thousand Oaks, Alameda County, California,* filed August 23, 1909, in Liber 25 of Maps, page 2, in the office of the County Recorder of the said County of Alameda.

2. *By Metes, Bounds, and Lots.* If the boundaries of a given tract within a subdivision for which there is a recorded map do not conform exactly to boundaries shown on the official map, the tract is described by metes and bounds (Art. 22-13), with the point of beginning referred to a corner shown on the official map. Also, the numbers of lots of which the tract is composed are given. Following is an example of a description of this kind:

Beginning at the intersection of the Northern line of Escondido Avenue, with the Eastern boundary line of Lot 16, hereinafter referred to; running thence Northerly along said Eastern boundary line of Lot 16, and the Eastern boundary line of Lot 17, eighty-nine (89) feet; thence at right angles Westerly, fifty-one (51) feet; thence South 12°6′ East, seventy-five (75) feet to the Northern line of Escondido Avenue; thence Easterly along said line of Escondido Avenue, fifty-three and 13/100 (53.13) feet, more or less, to the point of beginning.

Being a portion of Lots 16 and 17, in Block 5, as said lots and blocks are delineated and so designated upon that certain map entitled *Map of Thousand Oaks, Alameda County, California,* filed August 23, 1909, in Liber 25 of Maps, page 2, in the office of the County Recorder of the said County of Alameda.

3. *By Metes and Bounds to City Monuments.* Some of the larger and older cities of the United States have, by precise surveys, established a system of reference monuments and have determined the coordinates of these monuments with respect to an arbitrarily selected initial point. If the tract cannot be defined by descriptions such as the preceding, the point of beginning may be definitely fixed by stating its direction and distance from an official reference monument and by describing the monument that marks the corner. The boundaries of the tract may then be described by metes and bounds.

The location of corners may also be defined by rectangular coordinates referred to the origin or initial point of the city system and/or the state system, as described in Art. 22-13.

If the tract is within an area not so monumented, the point of beginning of the boundary description may be referred by direction and distance to the intersection of the center lines of streets. It is not good practice to refer to the intersection of sidewalk or curb lines, for these are apt to be changed from time to time. In sections of the country within the rectangular system of United States surveys, the point of beginning of a boundary description may properly be referred to section lines and corners.

22-18. Subdivision Survey of Urban Land. As a city or town develops, unimproved lands are subdivided into lots which are placed on sale as residential or business property. In most instances such extensions are the result of the activities of real-estate operators who acquire a tract of rural land of considerable area, develop a plan of subdivision which is approved by the authorities of the municipality to which the tract is to be attached, and cause surveys to be made for the purpose of establishing the boundaries of individual lots. A tract thus divided according to an acceptable plan is known as an *addition* or *subdivision.*

For large and important developments the work of originating the general plan is often carried out by persons specializing in city planning and landscape architecture, under whose direction the surveyor works. Such developments require a high degree of skill, and usually extensive surveys (particularly in hilly sections) are carried out before the actual plan of subdivision can be decided upon. Problems of this character can be adequately discussed only in treatises on city planning. However, it is appropriate to state here that the preliminary studies should consider the probable future character of the district; the probable location of business sections; the probable magnitude, direction, and character of future traffic; the topography of the land; the location, width, grade, and character of paving of streets; the size and shape of lots and blocks; the location and size of storm and sanitary sewers; and the disposition of electric and telephone wires and cables.

For the ordinary real-estate development the owner usually calls for the services of an engineer or surveyor who has had experience in such work. The surveyor confers with the owner, and they discuss a general plan. The surveyor makes a resurvey of the entire property; and if the character of the topography is irregular, he usually makes certain preliminary surveys for the purpose of finding the location and elevation of the governing features of the terrain. In some cases a complete topographic survey may be made. With the general plan fixed, and having studied the results of the field investigation and having considered the items listed in the previous paragraph, the surveyor works out a detailed plan on paper, showing on the drawing the names of all streets and the numbers of all blocks and lots, the dimensions of all lots, the width of streets, the length and bearing of all street tangents, and the radius and length of all street curves. He also prepares a report which, in addition to a discussion of the plan of subdivision, may consider the cost of subdividing, including not only the establishing of boundaries but also the work of grading, paving, constructing sewers, and landscaping.

This detailed plan, when approved by the owner, is submitted to the governing body in the municipality. If it meets with the requirements of this body, it is approved.

Upon the authority of the owner, the surveyor then proceeds to execute the necessary subdivision surveys, including the laying out of roads, walks, blocks, and lots. Often the lot and block corners are marked with permanent monuments; but in many cases, contrary to what may be considered good practice, the lot corners are marked by wooden stakes. When the surveys are completed, the map of the subdivision is revised to show minor changes made during the survey, together with the location and character of permanent monuments. A tracing is submitted to the municipality, and this, when duly signed by those in authority, becomes the official map of the subdivision. It then becomes a part of the public records and is usually filed in the registry of deeds of the county in which the municipality lies. Upon this approval, if the subdivision is outside the corporate limits of the municipality, they are extended to include it.

22-19. City Surveying. The term *city surveying* is frequently applied to the surveying operations within a municipality with regard to mapping its area, laying out new streets and lots, and constructing streets, sewers and other public utilities, and buildings. (See also Art. 22-19*a*.) Although the principles of city surveying are not different from those of ordinary surveying, there are some differences in the details of the methods employed. Some features pertinent to city surveying are as follows:

1. Measurements are made with a greater degree of refinement than for land of less value.

2. Some cities maintain a standard of length with which tapes may be compared.

3. Usually the horizontal control of the survey for the map of a city is by triangulation rather than by traversing, which would be employed for an equal area outside the city.

4. A system of reference points and bench marks is established, usually by traversing, at points a few blocks apart, usually at street intersections. Preferably this system is tied in with the United States precise surveys and with the state plane-coordinate system. Points are located in the street, at the curb, or on the sidewalk, one such point being sufficient for each chosen intersection. On subsequent surveys, it is good practice to tie in to more than one of these established points, as monuments may have been moved. (For a description of monuments and reference marks, see Art. 22-4.)

5. The established points are monumented and are well referenced (see Art. 14-17) to more or less permanent objects such as building corners, curb or walk lines, centers of street intersections, and manhole covers. In undeveloped districts, these points are referenced to stakes.

6. Maps showing the location of proposed sewers, street extensions, and other improvements usually show, to scale and in figures, the exact dimensions of adjacent lots and of all other lots that will be benefited by, or assessed for, the proposed improvement.

7. Sometimes separate maps are made of surface and underground utilities such as car lines, sewers, water lines, gas lines, electric power and telephone lines and conduits, tunnels, etc., both for convenient reference and in order to avoid interference in the location of new projects.

The subdivision of urban lands is discussed in Art. 22-18, and typical descriptions of urban lands are given in Art. 22-17. The usual methods of keeping records are described in Art. 22-6.

The operations of surveying in connection with the construction of buildings, bridges, sewers, pipe lines, pavements, and railroads are described in Chap. 28, and the building-site survey in Art. 28-15. Some details of running lines and locating details, pertinent to urban surveys, are given in Arts. 13-23, 13-28, and 14-19.

Details regarding the width of streets, size of blocks and lots, location of utilities, etc., are to be found in texts and manuals on city planning, highway engineering, and sanitary engineering.

22-19a. City Survey. The term *city survey* has come to mean an extensive coordinated survey of the area in and near a city for the purposes of fixing reference monuments, locating property lines and improvements, and determining the configuration and physical features of the land. In Ref. 5, *Manual* 10, at the end of this chapter, it is characterized as "an inventory of the physical facts relating to the land and its occupation." Such a survey is of value for a wide variety of purposes, particularly for planning city improvements. The technical procedure for city surveys is described in detail in Ref. 5. Briefly, the work consists in:

1. Establishing *horizontal and vertical control*, as described for topographic surveying. The primary horizontal control is usually by triangulation, supplemented as desired by precise traversing. Secondary horizontal control is by traversing of appropriate precision. Primary vertical control is by precise leveling.

2. Making a *topographic survey* and *topographic map*. Usually the scale of the topographic map is 1 in. = 200 ft. The map is divided into sheets which cover usually 5,000 ft. of longitude and 4,000 ft. of latitude. Points are plotted by rectangular plane coordinates.

3. *Monumenting* a system of selected points at suitable locations such as street corners, for reference in subsequent surveys. These monuments are referred to the plane-coordinate system and to the city datum.

4. Making a *property map*. The survey for the map consists in (*a*) collecting recorded information regarding property, (*b*) determining the location on the ground of street intersections, angle points, and curve points, (*c*) monumenting the points so located, and (*d*) traversing to determine the coordinates of the monuments. Usually the scale of the property map is 1 in. = 50 ft. The map is divided into sheets which cover usually 1,250 ft. of longitude and 1,000 ft. of latitude, thus bearing a convenient relation to the sheets of the topographic map. The property map shows the length and bearing of all street lines and boundaries of public property, coordinates of governing points, control, monuments, important structures, natural features of the terrain, etc., all with appropriate legends and notes.

5. Making a *wall map* which shows essentially the same information as the topographic map but which is drawn to a smaller scale; the scale should be not less than 1 in. = 2,000 ft. The wall map is reproduced in the usual colors—culture in black, drainage in blue, wooded areas in green, and contours in brown.

6. Making an *underground map*. Usually the scale and the size of the map sheets are the same as those for the property map. The underground map shows street and easement lines, monuments, surface structures and natural features affecting underground construction, and underground structures and utilities (with dimensions), all with appropriate legends and notes.

22-20. Cadastral Surveying. Cadastral surveying is a general term referring to extensive surveys relating to land boundaries and subdivisions, whether they are city surveys as described in the preceding article or surveys of rural land. The term is applied to the United States public-land surveys (Chap. 23) by the U.S. Bureau of Land Management.

A cadastral map shows individual tracts of land with corners, length and bearing of boundaries, acreage, ownership, and sometimes the cultural and drainage features. The surveying methods are the same as those described for topographic surveying for maps of intermediate and large scale (Chap. 25).

In Ref. 5, *Manual* 34, of the American Society of Civil Engineers, a cadastral survey is defined as follows:

> A survey relating to land boundaries and subdivisions, made to create units suitable for transfer or to define the limitations of title. Derived from "cadastre" meaning register of the real property of a political subdivision with details of area, ownership, and value. The term cadastral survey is now (1954) used to designate the surveys of the public lands of the United States, including retracement surveys for the identification and resurveys for the restoration of property lines; the term can also be applied properly to corresponding surveys outside the public lands, although such surveys are usually termed *land surveys* through preference.

22-21. Field and Office Problem.

PROBLEM 1. SURVEY OF TRACT FOR DEED DESCRIPTION

Object. To obtain sufficient data for a proper legal description of a tract and to prepare such a description (see also field problem 3 of Art. 12-27).

Procedure. (1) Around the assigned field run an azimuth traverse with the transit, measuring the sides with a steel tape and setting hubs at the corners. The angular error of closure in minutes should not exceed $\frac{1}{2} \times \sqrt{\text{number of sides}}$. Distribute the error of closure among the angles of the traverse. Refer azimuths to the true meridian. (2) Compute the latitudes and departures and the linear error of closure; the linear error of closure should not exceed 1/5,000. (3) Balance the survey, and calculate the area of the tract by the coordinate method. (4) Determine the location of one corner of the traverse from an established reference point (Art. 22-17), and beginning at this corner, write a description of the tract by metes and bounds.

REFERENCES

1. AMERICAN CONGRESS ON SURVEYING AND MAPPING, "Technical Standards for Property Surveys," *Surveying and Mapping*, Vol. 6, No. 3, pp. 210–213, July–September, 1946.
2. AMERICAN CONGRESS ON SURVEYING AND MAPPING, "Equitable Fees for Property Surveys," *Surveying and Mapping*, Vol. 9, No. 3, pp. 165–167, July–September, 1949.
3. AMERICAN CONGRESS ON SURVEYING AND MAPPING, "Minimum Detail Requirements for Land Title Surveys," *Surveying and Mapping*, Vol. 22, No. 2, pp. 339–341, June, 1962.
4. AMERICAN SOCIETY OF CIVIL ENGINEERS, "Land Surveys and Titles," Report of Joint Committee, *Proc. Am. Soc. Civil Engr.*, Vol. 67, No. 6, pp. 1065–1079, June, 1941.
5. AMERICAN SOCIETY OF CIVIL ENGINEERS, "Technical Procedure for City Surveys," "Location of Underground Utilities," "Land Subdivision," and "Definitions of Surveying, Mapping, and Related Terms," *Manuals of Engineering Practice*, 10 (1963), 14 (1937), 16 (1939), and 34 (1954), respectively.

6. BOUVIER, JOHN, "Bouvier's Law Dictionary," edited by William Edward Baldwin, Banks-Baldwin Publishing Company, Cleveland, 1948.
7. BROWN, CURTIS M., "Boundary Control and Legal Principles," John Wiley & Sons, Inc., New York, 1957.
8. CLARK, F. E., "Law of Surveying and Boundaries," 3d ed., Bobbs-Merrill Company, Inc., Indianapolis, 1959.
9. COOLEY, THOMAS, M., "The Judicial Functions of Surveyors," reprinted in *Surveying and Mapping*, Vol. 14, No. 2, pp. 161–168, April–June, 1954.
10. HODGMAN, F., "Land Surveying," The F. Hodgman Co., Climax, Mich., 1907.
11. HOLT, A. H., "The Surveyor and the Law," *Trans. Am. Soc. Civil Engr.*, Vol. 107, pp. 1379–1400, 1942.
12. KISSAM, PHILIP, "Proposed Improvements for Land Surveys and Title Transfers," *Trans. Am. Soc. Civil Engr.*, Vol. 105, pp. 731–743, 1940.
13. KISSAM, PHILIP, "Proposed Standards for Land Descriptions," *Civil Engrg.*, Vol. 10, No. 11, pp. 696–697, November, 1940.
14. LARRABEE, BRUCE M., "The Role of Monuments in Surveying," *Surveying and Mapping*, Vol. 14, No. 2, pp. 196–200, April–June, 1954.
15. MCENTYRE, JOHN G., and ARTHUR J. MCNAIR, "Land Surveying and Land Registration," *J. Surveying Mapping Div., Am. Soc. Civil Engr.*, Vol. 89, No. SU 1, pp. 59–75, February, 1963.
16. SCHOFIELD, L. T., "Massachusetts Land Court," *Surveying and Mapping*, Vol. 8, No. 4, pp. 180–186, October–December, 1948.
17. SKELTON, R. H., "Legal Elements of Boundaries and Adjacent Properties," Bobbs-Merrill Company, Inc., Indianapolis, 1931.
18. WATTLES, W. C., "Property Surveys Must Fit Their Titles," *Trans. Am. Soc. Civil Engr.*, Vol. 115, pp. 260–274, 1950.

CHAPTER 23

U.S. PUBLIC-LAND SURVEYS

23-1. General. This chapter deals with the methods of subdividing the public lands of the United States in accordance with regulations imposed by law. The public lands are subdivided into townships, sections, and quarter sections—in early years by private surveyors under contract, later by the Field Surveying Service of the General Land Office, and currently by civil-service employees of the Bureau of Land Management which succeeded the General Land Office in 1946. Further subdivision of such lands is made after the lands have passed into the hands of private owners, the work being carried out by surveyors in private practice.

The methods described herein are those now in force, but with minor differences they have been followed in principle since 1785, when the rectangular system of subdivision was inaugurated. Under this system, the public lands of 30 states have been or are in progress of being surveyed (Art. 23-3). In general, these methods of subdividing land do not apply in the 13 original states and in Hawaii, Kentucky, Tennessee, Texas, and West Virginia. As the progress of the public-land surveys has been from east to west, the details in states east of the Mississippi River differ somewhat from those of present practice.

The laws regulating the subdivision of public lands and the surveying methods employed are fully described in the "Manual of Instructions for the Survey of the Public Lands of the United States," published by the Bureau of Land Management (Ref. 6 at the end of this chapter). From the manual is drawn much of the material for this chapter.

Field notes and plats of the public-land surveys may be examined in the regional offices of the Bureau, and copies may be procured for a nominal fee.

23-2. Laws Relating to Public-land Surveys. Beginning with an ordinance passed by the Continental Congress in May, 1785, which provided for townships 6 miles square, each containing 36 sections 1 mile square, laws regulating the surveying, marking, and disposal of the public lands of the United States have from time to time been enacted by Congress. Following are the provisions of the public-land laws in which the surveyor is principally interested:

1. All responsibility for the surveying and sale of the public lands of the United States is placed in the hands of the Director of the Bureau of Land Management, who under the direction of the Secretary of the Interior is authorized to carry into execution every part of the public-land laws not otherwise specially provided for.

2. When the surveys and records of a state are completed, all the field notes, maps, and records pertaining to land titles are delivered to the secretary of state of that state.

3. Any agent of the United States, acting upon the authority of the Director of the Bureau of Land Management, has free access to public records delivered to any state, but no transfer of such records is made to any state until the state has enacted legislation providing for the safekeeping of such records and for the allowance of free access thereto by authorities of the United States.

4. It is required that all surveys and resurveys of public lands under the supervision of the Director of the Bureau of Land Management are to be made by surveyors selected by the Bureau of Land Management. (Prior to 1910, surveys were made by contract.) The field work is now performed by a permanent corps of engineers under civil-service regulations.

5. It is provided that resurveys may be made by the Government under certain conditions.

6. Boundaries of public lands, when established by duly authorized surveyors and when approved by the Director, are unchangeable.

7. The original corners established by the surveyors stand as the true corners they were intended to represent, whether in the place shown by the field notes or not. The primary purpose of the public-land surveys is to *mark the boundaries on the ground;* the field notes and plats are subordinate.

8. The unit of length is the Gunter's (66-ft.) chain divided into 100 links, each 7.92 in. long.

9. Quarter-quarter-section corners not established by the original surveys are to be on the line joining the section and quarter-section corners and midway between them, except in the northern and western half miles of the township.

10. The center lines of sections are to be straight between opposite quarter-section corners.

11. In a fractional section where no opposite quarter-section corner has been or can be established, the center line of such section is to be run from the proper quarter-section corner as nearly in a cardinal direction as due parallelism with section lines will permit to the meander line, reservation, or other boundary of such fractional section.

12. Lost or obliterated corners of the approved surveys are to be restored to their original location, if possible.

23-3. Historical Notes. The first surveys of the public lands of the United States, made under the ordinance of May, 1785, divided lands north of the Ohio River. Only the exterior lines of the townships were run, but section corners were established at intervals of 1 mile on the township lines, and the plats were marked into subdivisions 1 mile square. These surveys were made under the direction of the Geographer of the United States.

The act of Congress of May, 1796, provided for a surveyor general and directed the survey of lands northwest of the Ohio River and above the mouth of the Kentucky River. Under this law it was provided that the sections be numbered according to the plan in operation at the present time.

In 1800 an act of Congress provided for the subdivision of lands into half sections and required that excesses or deficiencies in measurement should be placed in the sections or half sections in the most northerly or westerly half miles of each township.

In 1805 an act of Congress directed that the public lands should be divided into quarter sections, and provided that all corners marked in the public surveys should be established as the proper corners which they were intended to designate, and that corners of half and quarter sections should be placed as nearly as possible equidistant from the two adjacent section corners on the same line.

In 1812 the General Land Office was established as a branch of the Treasury Department, and the office of Commissioner of the General Land Office was created.

In 1820 an act of Congress provided for the sale of public lands in half-quarter sections and required that the line of division of the quarter section should in every case run north and south.

In 1832 an act of Congress directed the subdivision of the public lands into quarter-quarter sections and required that the line of division of the half-quarter section should in every case run east and west. This act also provided that fractional sections be subdivided in accordance with regulations prescribed by the Secretary of the Treasury.

In 1849 the Department of the Interior was created, and the control of the General Land Office was transferred from the Department of the Treasury to the Department of the Interior.

In 1909 it was provided by act of Congress that resurveys may be made at the discretion of the Secretary of the Interior, if such resurveys are essential to mark properly the boundaries of the public lands previously surveyed but remaining undisposed of, provided such resurvey shall not be so executed as to impair the rights of entrymen or owners of lands affected.

In 1910 the contract system of surveying the public lands was abolished.

By act of Congress in 1918, resurveys may be made of public lands which are in private ownership upon application of the owners of three fourths of the privately owned lands in any township covered by public-land surveys, when more than 50 per cent of the area of such township is privately owned, provided there be deposited a sum equal to the estimated cost of the resurveys. Any portion of the deposit which may remain after the work is completed is repaid pro rata to the persons making the deposit.

In 1925 the office of surveyor general of the several districts was abolished, and all activities were transferred to the Field Surveying Service, under the jurisdiction of the U.S. Supervisor of Surveys.

In 1946 the Bureau of Land Management was created in the Department of the Interior, succeeding the General Land Office and the U.S. Supervisor of Surveys.

Under the regulations imposed by Congress, surveys of the public lands have either been completed or are in progress, in the following states:

Alaska	Iowa	New Mexico
Alabama	Kansas	North Dakota
Arizona	Louisiana	Ohio
Arkansas	Michigan	Oklahoma
California	Minnesota	Oregon
Colorado	Mississippi	South Dakota
Florida	Missouri	Utah
Idaho	Montana	Washington
Illinois	Nebraska	Wisconsin
Indiana	Nevada	Wyoming

The original records and plats of completed surveys have been transferred to the respective states except those for Oklahoma, which are on file in the Bureau of Land Management, Washington, D.C. Copies of most of these state records are also on file in Washington. The original records of surveys still in progress are on file in regional field offices of the Bureau of Land Management.

It must be kept in mind that the early surveys made under contract were made with relatively crude instruments and often under unfavorable field conditions; some were incompletely or even fraudulently executed. Hence, often the lines and corners will be found in other than their theoretical locations. However, the original corners as established legally stand as the true corners, and the surveyor must be guided by them in making resurveys or subdivisions, regardless of irregularities in the original survey. It is, therefore, important that he be familiar with the methods used in the original survey.

23-4. General Scheme of Subdivision. The regulations for the subdivision of public lands have been altered from time to time; hence, the methods employed in surveying various regions of the United States show marked differences, depending upon the dates when the surveys were made. In general principle, however, the system has remained unchanged, the primary unit being the *township*, bounded by meridional and latitudinal lines and as nearly as may be 6 miles square. The township is divided into 36 secondary units called *sections*, each as nearly as may be 1 mile square. Because the meridians converge (Art. 23-11), it is impossible to lay out a square township by such lines; and because the township is not square, not all the 36 sections can be 1 mile square even though all measurements are without error.

23-5. Standard Lines. Since the time of the earliest surveys, the townships and sections have been located with respect to principal axes passing through an origin called an *initial point;* the north-south axis is a true meridian called the *principal meridian*, and the east-west axis is a true parallel of latitude called the *base line*.

The principal meridian is given a name to which all subdivisions are referred. Thus the principal meridian which governs the rectangular surveys (wholly or in part) of the states of Ohio and Indiana is called the First Principal Meridian; its longitude is 84°48′50″W, and the latitude of the base line is 41°00′00″N. The extent of the surveys which are referred to a given initial point may be found by consulting a map, published by the Bureau of Land Management, entitled "United States, Showing Principal Meridians, Base Lines, and Areas Governed Thereby," or from Ref. 6 at the end of this chapter.

Secondary axes are established at intervals of 24 miles east or west of the principal meridian and at intervals of 24 miles north or south of the base line, thus dividing the tract being surveyed into quadrangles bounded by true meridians 24 miles long and by true parallels, the south boundary of each quadrangle being 24 miles long, and the north boundary being 24 miles long less the convergency of the meridians in that distance. (In some early surveys, these distances were 30 or 36 miles.) The secondary parallels are called *standard parallels* or *correction lines*, and each is continuous throughout its length. The secondary meridians are called *guide meridians*, and each is broken at the base line and at each standard parallel.

The principal meridian, base line, standard parallels, and guide meridians are called *standard lines*.

A typical system of principal and secondary axes is shown in Fig. 23-1. The base line and standard parallels, being everywhere perpendicular to the direction of the meridian, are laid out on the ground as curved lines, the rate of curvature depending upon the latitude. The principal meridian and guide meridians, being true north-and-south lines, are laid out as straight lines but converge toward the north, the rate of convergency depending upon the latitude.

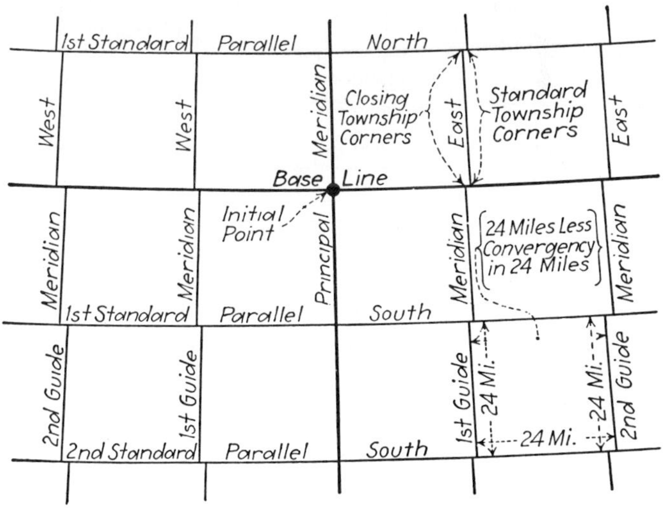

FIG. 23-1. Standard lines.

Standard parallels are counted north or south of the base line; thus the *second standard parallel south* indicates a parallel 48 miles south of the base line. Guide meridians are counted east or west of the principal meridian; thus the *third guide meridian west* is 72 miles west of the principal meridian.

23-6. Townships. The division of the 24-mile quadrangles into townships is accomplished by laying off true meridional lines called *range lines* at intervals of 6 miles along each standard parallel, the range line extending north 24 miles to the next standard parallel; and by joining the township corners established at intervals of 6 miles on the range lines, guide meridians, and principal meridian with latitudinal lines called *township lines*.

The plan of subdivision is illustrated by Fig. 23-2. A row of townships extending north and south is called a *range;* and a row extending east and west is called a *tier*. Ranges are counted east or west of the principal meridian, and tiers are counted north or south of the base line. Usually for purposes of description the word "township" is substituted for "tier." A township is designated by the number of its tier and range and the name of the principal meridian. For example, T7S, R7W (read *township seven south, range seven west*) designates a township in the seventh tier south of the base line and the seventh range west of the principal meridian.

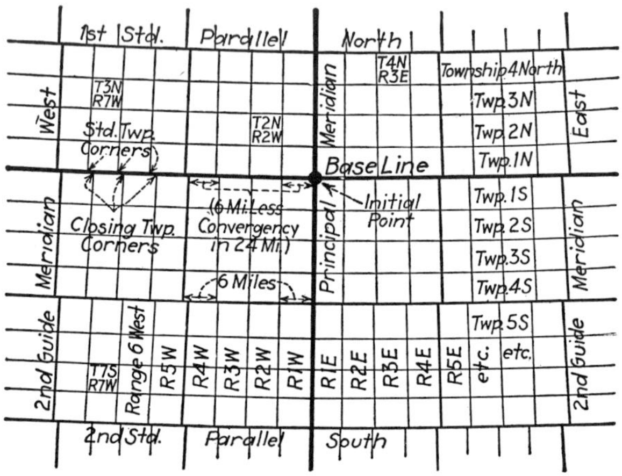

FIG. 23-2. Township and range lines.

23-7. Sections. The division of townships into sections is performed by establishing, at intervals of 1 mile, "meridional" lines parallel to the east boundary of the township and by joining the section corners established at intervals of 1 mile with straight latitudinal lines. (These lines are not exactly meridional, but are parallel to the east boundary of the township, which is a meridional line.) These lines, called *section lines*, divide each township into 36 sections, as shown in Fig. 23-3. The sections are numbered consecutively from east to west and from west to east, beginning with No. 1 in the northeast corner of the township and ending with No. 36 in the southeast corner. Thus Section 16 is a section whose center is $3\frac{1}{2}$ miles north and $3\frac{1}{2}$ miles west of the southeast corner of a township.

A section is legally described by giving its number, the tier and range of the township, and the name of the principal meridian; for example, Section 16, T7S, R7W, of the Third Principal Meridian.

On account of the convergency of the range lines (true meridians) forming the east and west boundaries of townships, the latitudinal lines forming the north and south boundaries of townships are less than 6 miles in length, except for the south boundary of townships that lie just north of a standard parallel.

As the north-south section lines are run parallel to the *east* boundary of the township, it follows that all sections except those adjacent to the west boundary will be 1 mile square, but that those adjacent to the west boundary will have a latitudinal dimension less than 1 mile by an amount equal to the convergency of the range lines within the distance from the section to the nearest standard parallel to the south.

The subdivision of sections is described in Arts. 23-17 to 23-19.

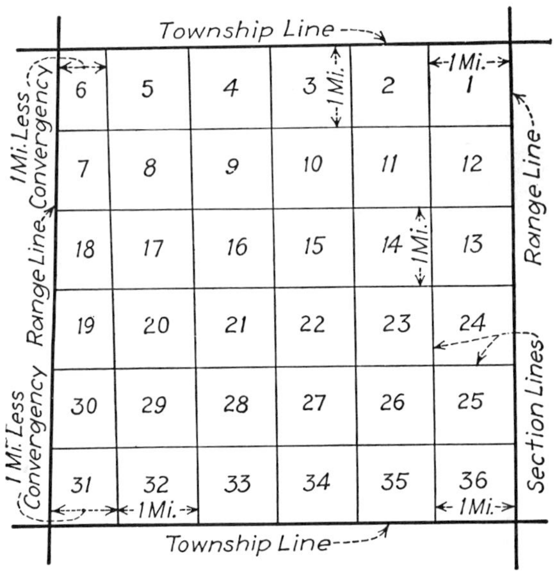

FIG. 23-3. Numbering of sections.

23-8. Standard Corners. Corners called *standard corners* are established on the base line and standard parallels at intervals of 40 chains; these standard corners govern the meridional subdivision of the land lying between each standard parallel and the next standard parallel to the north. Other corners called *correction corners* or *closing corners* are later established on the base line and standard parallels during the process of subdivision; these corners fall at the intersection of the base line or standard parallel either with the meridional lines projected from the standard township corners of the next standard parallel to the south (see Fig. 23-2) or with the intermediate section and quarter-section lines. Standard parallels are also called *correction lines.*

23-9. Irregularities in Subdivision. It should be understood that the plan of subdivision just described is the one which is carried out when conditions allow. There are, of course, always present the errors of measurement, so that the actual lengths and directions established in the field do not entirely agree with the theoretical values. But in addition, conditions met in the field often make it inexpedient or impossible to establish the lines of the survey in exact accordance with the specified plan. Thus there are numerous instances of standard parallels and guide meridians having been originally established at intervals of 30 and 36 miles, under old regulations; and of regions having been only partly surveyed. Later, under present regulations, meridians have been established between the old guide meridians; and recent subdivisions are, therefore, referred to standard lines many of which are less than 24 miles apart. Also the presence of large bodies of water, mountain ranges, Indian reservations, etc., may greatly modify the method of division, many townships and sections being made fractional.

23-10. Establishing the Standard Lines. *Principal Meridian.* The principal meridian is established as a true meridian through the initial point, either north or south, or in both directions, as conditions require. Permanent quarter-section and section corners are established alternately at intervals of 40 chains (½ mile), and regular township corners are placed at intervals of 480 chains (6 miles).

Independent linear measurements are taken either by two sets of chainmen or, when this is not possible, by the duplication of each measurement by one set of chainmen. When the discrepancy between two sets of measurements taken in the prescribed manner exceeds 20 links per mile, it is required that the line be remeasured to reduce the difference. The corners are set at the mean distances. When successive independent tests of the alinement, as determined by astronomical observations, indicate that the line has departed from the true meridian by more than 03′, it is required that the necessary correction be made to reduce the deviation in azimuth.

Base Line. From the initial point the base line is extended east and west on a true parallel of latitude, standard quarter-section and section corners being established alternately at intervals of 40 chains (½ mile) and standard township corners being placed at intervals of 480 chains (6 miles). The manner of taking the linear measurements of the base line and the required precision of both linear measurements and alinement are the same as for the survey of the principal meridian. Any of the three methods described in Art. 23-12, for laying out the true latitude curve, may be used.

Standard Parallels. At intervals of 24 miles north and south of the base line, true parallels of latitude called *standard parallels* or *correction lines* are run east and west from the principal meridian, these lines being established in a manner identical with that prescribed for the survey of the base line.

Guide Meridians. The guide meridians are extended north from the base line and standard parallels at intervals of 24 miles east and west of the principal meridian. Each guide meridian is established as a true meridian in a manner identical with that employed in laying off the principal meridian. The guide meridians terminate at the points of their intersection with the standard parallels, and hence are broken lines, each segment being theoretically 24 miles long. Errors of measurement are placed in the most northerly half mile of each 24-mile segment. At the point of intersection of the guide meridian and standard parallel, a township corner, called a *correction corner* or *closing corner*, is established by retracing the standard parallel between the first standard corners to the east and to the west of the point for the closing corner; and the distance from the closing corner to the nearest standard corner on the standard parallel is measured.

23-11. Convergency of Meridians. In Fig. 23-4, let ACP and BDP represent two meridians, P being the north pole of the earth, O the center of the earth, and AB an arc of the equator intercepted by the two meridians; and let

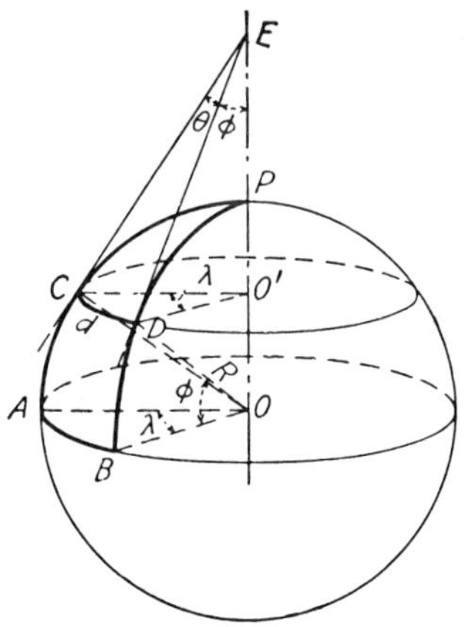

FIG. 23-4. Convergency of meridians.

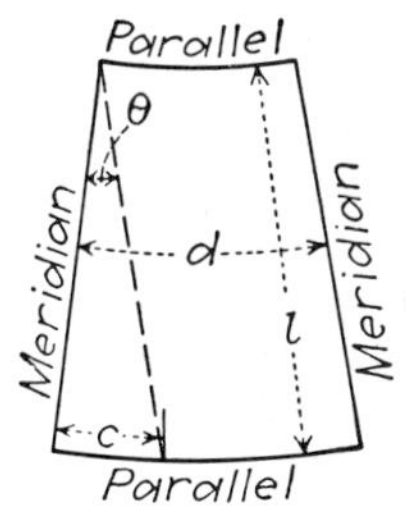

FIG. 23-5. Linear convergency between two meridians.

CD be the arc of a parallel of latitude at a latitude $\phi = COA = DOB$ at which it is desired to determine the angular and linear convergency of the meridians. Consider the earth as a perfect sphere.

The difference in longitude between the two meridians is

$$\lambda = \frac{CD}{CO'} \qquad \text{or} \qquad CD = CO' \cdot \lambda$$

The latitude of the arc CD is

$$\phi = DOB = DEO'$$

Then,

$$\sin \phi = \frac{DO'}{DE} \qquad \text{or} \qquad DE = \frac{DO'}{\sin \phi}$$

With negligible error the angle of convergency is

$$\theta = \frac{CD}{DE}$$

By substitution of the values for CD and DE obtained above

$$\theta = \lambda \sin \phi \tag{1}$$

Let the distance between two meridians measured along a parallel be $d = CD$, and let the radius of the earth at the parallel be R. Then from the figure

$$\lambda = \frac{CD}{CO'} = \frac{d}{R \cos \phi}$$

By substitution of this value in Eq. (1), there results

$$\theta = \frac{d \sin \phi}{R \cos \phi} = \frac{d \tan \phi}{R} \tag{2}$$

in which θ is in radians.

If d is in miles and R = 20,890,000 ft., the approximate mean radius of the earth, then θ in seconds is, from Eq. (2),

$$\theta'' = 52.13d \tan \phi \tag{3}$$

In Fig. 23-5 let l be the length of the meridian between two parallels and let θ be the mean angle of convergency of two meridians whose mean latitude is ϕ and whose mean distance apart measured on a parallel is d. Also let the linear convergency of the two meridians, measured along a parallel, be c. Then, with small approximation, $\theta = c/l$.

By substitution of this value in Eq. (2), there results

$$c = \frac{dl \tan \phi}{R} \tag{4}$$

which gives with sufficient precision for land surveying the linear convergency between two meridians. If d and l are in miles and R is the mean radius of the earth, then c in feet is given approximately by the expression

$$c_f = \tfrac{4}{3}\, dl \tan \phi \tag{5}$$

which is derived from Eq. (4) and which gives the linear convergency with sufficient precision for land surveying. The convergency in 66-ft. chains, where d and l are in miles, is then

$$c_c = 0.0202dl \tan \phi \tag{6}$$

Example 1: Find the angular convergency of two guide meridians 24 miles apart at latitude 43°20′. By Eq. (3),

$$\theta'' = 52.13 \times 24 \tan 43°20' = 1{,}182''$$
$$\theta = 19'42''$$

Example 2: Find the convergency in chains of two guide meridians 24 miles apart and 24 miles long at a mean latitude of 43°20′. By Eq. (6),

$$c_c = 0.0202 \times 24 \times 24 \tan 43°20' = 10.95 \text{ chains}$$

It will be noted in example 1 that the convergency of the two guide meridians is nearly a third of a degree. In example 2 the linear convergency is nearly 11 chains in the 24 miles; this would represent the jog at the correction line in the first guide meridian east or west, or one half of the jog in the second guide meridian. The north boundary of a township 24 miles north of a correction line at the given latitude is approximately 2¾ chains less than 6 miles.

In Table XI are given, for each degree of latitude, the linear and angular convergency of meridians 6 miles long and 6 miles apart. The linear convergency represents the correction to be applied to the north boundary of a regular township in computing the error of closure about the township. This value likewise represents double the amount of the offset from the tangent to the parallel at a distance of 6 miles from the point of tangency (see Art. 23-12).

Table XI also gives for the various latitudes the difference in longitude for 6 miles in both angle and time, and the difference in latitude for both 1 and 6 miles in angular measure.

Meridional Section Lines. In the subdivision of townships into sections, the establishment of section lines parallel to the east boundary of the township necessitates a correction in azimuth of these section lines on account of the angular convergency of the meridians. While meridional section lines are being run north, they are made to deflect to the left or west of the true meridian by an angle equal to the convergency in the distance to the section line from the east boundary. Hence, ⅙, ⅓, ½, ⅔, and ⅚ of the angles of convergency given in Table XI represent, respectively, the deflections from the true meridian for section lines respectively 1, 2, 3, 4, and 5 miles west of the east boundary of the township.

23-12. Laying Off a Parallel of Latitude. As the base line, standard parallels, and latitudinal township lines are true parallels of latitude, they are curved lines when established on the surface of the earth. This is evident from the fact that meridians converge and that a parallel of latitude is a line whose direction at any point is perpendicular to the direction of the meridian at that point. Its projection on the surface of the earth is the base element of a cone whose vertex is at the earth's center, and the radius of whose base is $R \cos \phi$, in which R is the earth's radius and ϕ is the latitude (see Fig. 23-4). It is defined by a plane at right angles to the earth's polar axis cutting the earth's surface on a circle whose radius is less for higher latitudes. The rate of curvature within the latitudes of the United States is so small that two points a quarter of a mile apart on the same parallel of latitude will, for all practical purposes, define the direction of the curve at either point; but the continuation of a line so defined in either direction would describe a great circle of the earth, gradually departing southerly from the true parallel. The great circle tangent to the parallel at any point along the parallel is called the *tangent to the parallel*, and it coincides with the true latitude curve only at the point of origin.

Though the tangent to the parallel is a straight line in plan, its bearing is not constant but varies with the distance from the point of tangency, the deflection from true east or true west being equal to the angle of convergency of the meridians within the distance from the point of tangency to the given point. Hence the angles of convergency given in Table XI also represent the deviation in azimuth of the tangent from the parallel in a distance of 6 miles, and 1/6, 1/3, 1/2, 2/3, and 5/6 of the tabulated angles represent the changes in azimuth of the tangent in distances respectively 1, 2, 3, 4, and 5 miles from the point of tangency.

Within the limits of precision necessary in land surveying, the offset from tangent to parallel at any distance from the point of tangency is one half of the linear convergency of the meridians within the same distance. This can be seen from Fig. 23-6, in which the angle of convergency is exaggerated; actually the angle θ and the distances c and a are relatively small. Hence values one half as great as the values of the linear convergency in 6 miles given in Table XI represent the offset from tangent to parallel, measured along the meridian, at a distance of 6 miles from the point of tangency. With small error a parallel of latitude may, within the limits of distance here considered, be assumed to behave as a parabola. Hence, the offset from tangent to curve at any point, for all practical purposes, may be said to vary as the square of the distance from the point of tangency; and the offsets at 1/2, 1, 1 1/2, 2, etc., miles from the point of tangency would bear to the offset at 6 miles the ratios 1/144, 1/36, 1/16, 1/9, etc., respectively.

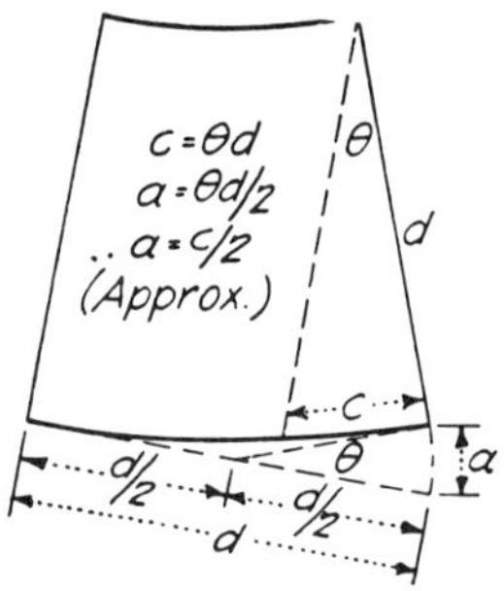

FIG. 23-6. Relation between tangent offset and linear convergency.

There are three general methods of establishing a true parallel of latitude, which may be employed independently to arrive at the same result: (1) The solar method, (2) the tangent method, and (3) the secant method. The secant method is most commonly employed.

Solar Method. By this method a solar attachment to the engineer's transit is employed as described in Art. 21-17. If the instrument is in good adjustment, the true meridian may be established with sufficient precision at each transit station, and the true parallel may be established by turning an angle of 90° in either direction from the meridian. If sights taken with the telescope pointing in the latter direction are not longer than 20 to 40 chains, the line thus defined will not depart appreciably from the true parallel.

Tangent Method. This method consists in determining the true meridian at the point of tangency, from which the tangent to the parallel is established by laying off an angle of 90°. The tangent is extended in a straight line for a distance of 6 miles, and as each 40 chains is laid off along the tangent, the corresponding section or quarter-section corner is established on the parallel by laying off along the meridian the appropriate offset from tangent to parallel.

At the end of 6 miles a new tangent is laid off, and the process just described is repeated. The values of the offsets may be found from Table XI, as previously suggested.

Secant Method. This is a modification of the tangent method, in which the secant is a straight line 6 miles in length forming the arc of a great circle, which intersects the true parallel at the end of the first and fifth miles from the point of beginning, as illustrated by Fig. 23-7. For the latitude of the given parallel, the offsets (in links) from secant to parallel are given in the figure, at intervals of ½ mile. From the figure it is clear that the secant is parallel with a tangent to the parallel at the end of the third mile (240 chains); hence, the offset south

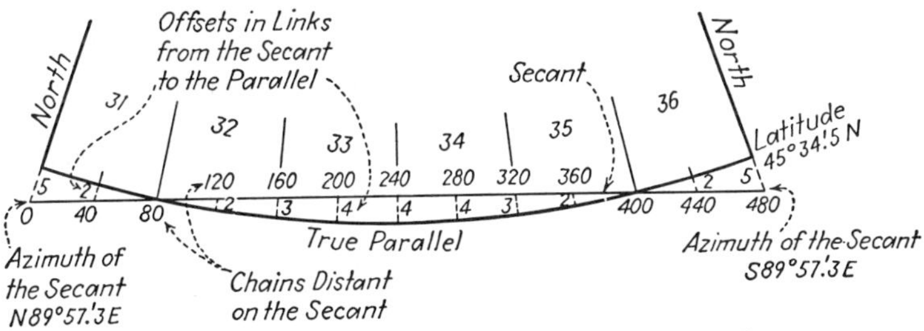

FIG. 23-7. Parallel of latitude by secant method.

from the third-mile point on the secant line to the corner on the true parallel is the same as the offset from the tangent to the parallel in a distance of 2 miles. Also, it is evident that the offset south of the point of beginning to the initial point on the secant, and the offset north of the secant to the true parallel at the end of the sixth mile, is equal to the difference between the tangent offset in a distance of 3 miles and the tangent offset in a distance of 2 miles.

Owing to the convergency of meridians, the azimuth of the secant—a straight line in plan—varies along its length. If the secant is laid off toward the east, the direction of the secant from the point of beginning to the end of the third mile is north of true east, and beyond the end of the third mile is south of true east, the variation from true east increasing directly with the distance in either direction from the third-mile point. At the third-mile point the secant bears true east; at the point of beginning the secant bears north of east by an amount equal to the angular convergency of meridians 3 miles apart; and at the end of the sixth mile the secant bears south of east by the same amount. In Table XII are given, for various latitudes, the azimuths (measured in either direction from true north) of the secant at intervals of 1 mile.

In Table XIII are tabulated the offsets from the secant to the parallel at intervals of ½ mile.

The procedure employed in establishing a true parallel by the secant method is as follows: The initial point on the secant is located by measuring south of the beginning corner a distance equal to the secant offset for 0 mi. given in Table XIII (5 links in Fig. 23-7). The transit is set up at this point, and the direction of the secant line is established by laying off from true north the azimuth given in Table XII in the column headed 0 mi.; for the conditions illustrated by Fig. 23-7 the bearing of the secant which extends east from the point of beginning is N89°57.3′E. The secant is then projected in a straight line for 6 miles; and as each 40 chains (½ mile) is laid off along the secant the proper offset is taken to establish the corresponding section or quarter-section corner on the true parallel.

At the end of 6 miles, if it is not convenient to determine the true meridian, the succeeding secant line may be established by laying off, at the sixth-mile point, a deflection angle from the prolongation of the preceding secant to the succeeding secant, the angle being equal to the convergency of meridians 6 miles apart. Values of these deflection angles are given in the last column of Table XII. When the direction of the new secant has been thus defined, the process of measurement to establish corners on the true parallel is continued as before.

The secant method is recommended by the Bureau of Land Management for its simplicity of execution and for the proximity of the straight line (secant) to the true latitude curve. All measurements and all cutting to clear the line are substantially on the true parallel.

23-13. Establishing Township Exteriors. The exact procedure employed in establishing township boundaries depends upon factors so variable that a complete discussion of the subject is not attempted here. When practicable, the township exteriors are surveyed successively through a 24-mile quadrangle in ranges, beginning each range with the township on the south. The range lines or meridional boundaries of the townships take precedence in the order of survey and are run from south to north on true meridians, quarter-section and section corners being established alternately at intervals of 40 chains. At the end of 6 miles a temporary township corner is set, pending latitudinal measurements necessary to close the township exterior and to calculate the error of closure.

Each township line forming the north or south boundary of a township is run as a random line, as described in Art. 23-12, from the old toward the new meridional boundary, and if the error of closure is within the permissible value, the line is corrected back on a true parallel joining the two township corners. On the true parallel are established quarter-section and section corners, alternately at intervals of 40 chains, measurements being made from the boundary last run. The fractional measurement is placed in the most westerly half mile.

Where both meridional boundaries of a township are new lines or where both have been established by a previous survey, the random latitudinal boundary is run from east to west, but in other particulars the procedure is as outlined above.

A range line is terminated at its intersection with a standard parallel, the excess or deficiency in the measured distance between standard parallels being placed in the most northerly half mile. At the point of intersection between the range line and the standard parallel, a closing township corner (correction corner) is established. In order to determine the alinement of the line closed upon, the standard parallel is retraced between the two standard corners adjacent to the closing corner. The distance from the closing corner to the nearest standard corner is measured in order that the error of closure may be calculated.

Following the ideal procedure outlined above, when a full 24-mile quadrangle is to be divided into townships, the survey is usually begun at the southeast corner of the southwest township of the quadrangle (see Fig. 23-2). The range line is run 6 miles north, and the latitudinal boundary connecting the regular township corner previously established on the guide meridian or principal meridian with the 6-mile point on the range line is established as described in the preceding paragraphs. The range line is then continued another 6 miles, and a second latitudinal boundary is established in the same manner as the first, connecting the second regular township corner north of the standard parallel on the guide or principal meridian with the 12-mile point on the range line. Again the process is repeated, and then the range line is extended north of the 18-mile point to the closing township corner on the standard parallel. The most westerly range of townships is thus surveyed.

In a similar manner the boundaries of the townships forming the next range to the east are established.

Finally, the third range line, started at the southwest corner of the southeast township, is laid off as the others, but at the 6-, 12-, and 18-mile points latitudinal lines are run (1) to the west to connect with corresponding township corners, and (2) to the east to connect with the first, second, and third regular township corners north of the standard parallel on the guide meridian.

23-14. Allowable Limits of Error of Closure. The maximum allowable error of closure prescribed for the United States rectangular surveys is 1/452 provided the error of closure in either latitude or departure does not exceed 1/640. Where a survey qualifies under the latter limit, the former is bound to be satisfied. It is equivalent to a systematic error of 12½ links, in either latitude or departure, per mile of perimeter. On this basis both the latitudes and the departures for the exterior lines of a normal township should close within 3 chains; of a normal range or tier of sections within 1¾ chains; or of a normal section within ½ chain. The general requirement is applied as a test of the accuracy of the angular and linear measurements incidental to all classes of lines embraced in the division of the public lands. Whenever a closure is effected, the latitudes, departures, and error of closure of the lines composing the figure (quadrangle, township, section, meander, etc.) must be calculated, and corrective steps must be taken whenever the test discloses an error in excess of the allowable value.

In addition to the foregoing general requirement, township exteriors must be so established that the rectangular limits of township subdivisions, as discussed in the following article, are not exceeded.

Normally the boundaries of a township are considered to be established within satisfactory governing limits from which to control the subdivisional surveys when the calculated position of the section lines may be theoretically projected from the township boundaries without invading the danger zone in respect to the rectangular limits.

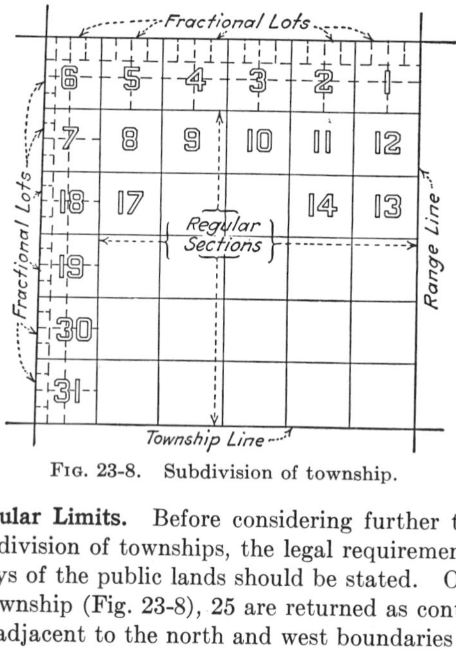

FIG. 23-8. Subdivision of township.

23-15. Rectangular Limits. Before considering further the methods employed in the subdivision of townships, the legal requirement relative to the rectangular surveys of the public lands should be stated. Of the 36 sections in each normal township (Fig. 23-8), 25 are returned as containing 640 acres each; 10 sections adjacent to the north and west boundaries (comprising sections 1–5, 7, 18, 19, 30, and 31) each contain regular subdivisions totaling 480 acres and in addition 4 *fractional lots* each containing 40 acres plus or minus definite differences to be determined in the survey; and 1 section (section 6) in the northwest corner contains regular subdivisions totaling 360 acres and in addition 7 fractional lots each containing 40 acres plus or minus certain definite differences to be determined in the survey. The aliquot parts of 640 acres, or regular subdivisions of a section, are the quarter section (½ mile square), the half-quarter or eighth section (¼ by ½ mile), and the quarter-quarter or sixteenth section (¼ mile square), the last containing 40 acres and being the legal minimum for purposes of disposal under the general land laws.

With regard to the allowable limits of precision, the "Manual of Surveying Instructions" of the Bureau of Land Management is quoted as follows:

> In the administration of the surveying laws it has been necessary to establish a definite relation between rectangularity (square miles of 640 acres, or aliquot parts thereof), as contemplated by law, and the resulting unit of subdivision consequent upon the practical application of surveying theory to the marking out of the lines on the earth's surface, wherein the ideal section is allowed to give way to one which may be termed "regular." Such relation, as applied to the boundaries of a section, has been placed at the following limits: (*a*) For alinement, not to exceed 21′ from cardinal in any part; (*b*) for measurement, the distance between regular corners to be normal according to the plan of the survey, with certain allowable adjustments not to exceed 25 links in 40 chains; and (*c*) for closure, not to exceed 50 links in either latitude or departure.
>
> Township exteriors, or portions thereof, will be considered defective when they do not qualify within the above limits. It is also necessary, in order to subdivide a township regularly, to consider a fourth limit, as follows:
>
> (*d*) For position, the corresponding section corners upon the opposite boundaries of the township to be so located that they may be connected by true lines which will not deviate more than 21′ from cardinal.

23-16. Subdivision of Townships. The procedure to be employed in the subdivision of a township into sections depends upon the regularity of the established boundaries of the township. If these boundaries are within the governing limits previously mentioned, the subdivision may proceed in a general order, the south and east boundaries of the township being the governing lines. When the township exteriors are irregular the variations in the procedure of subdivision are too numerous to allow of description here.

Following the normal plan for subdividing townships with regular boundaries, the subdivisional survey is begun on the south boundary of the township at the section corner between sections 35 and 36 (see Fig. 23-9). The line between sections 35 and 36 is run in a northerly direction parallel to the east boundary of the township, the quarter-section corner between 35 and 36 being set at 40 chains, and the section corner common to sections 25, 26, 35, and 36 being set at 80 chains. From the latter corner a random line is run eastward on a course calculated to be parallel with the south boundary of section 36, a temporary quarter-section corner being set at 40 chains. If this random line intersects the east boundary of the township exactly at the corner of sections 25 and 36, it is blazed and established as the true line, and if the linear error of closure is within the allowable limits, the temporary quarter-section corner is made permanent by shifting it to a position midway between adjacent section corners, as determined by field measurements.

If the point of intersection between the random line and east boundary falls to the north or to the south of the section corner on the township boundary, as will generally be the case, the falling is measured and from the data thus obtained the bearing of the true return course is calculated and the true line joining the section corners is blazed and established, the quarter-section corner common to sections 25 and 36 being placed midway between section corners, as described above.

This process is repeated for the successive meridional and latitudinal lines in the eastern range of sections until the north boundary of section 12 is established, the order in which the lines are surveyed being as indicated by the numbers on the section lines in Fig. 23-9.

When the northern boundary of the township is not a base line or standard parallel, the line between sections 1 and 2 is run north as a random line parallel to the east boundary, the distance to its point of intersection with the northern boundary of the township being measured. If the random line intersects the

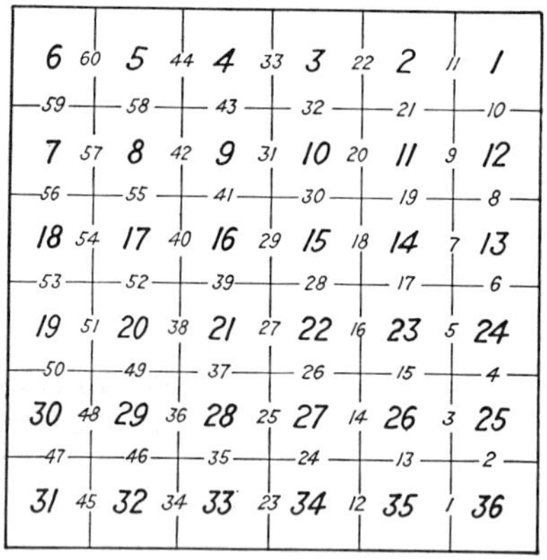

FIG. 23-9. Order of establishing section lines.

northern boundary at the corner of sections 1 and 2 and the linear error of closure of the tier of sections is within the allowable limit, the random line is blazed back and established as the true line, the fractional measurement being thrown into that portion of the line between the quarter-section corner and the north boundary of the township.

If, as is usually the case, the random line intersects the north boundary to the east or to the west of the corner of sections 1 and 2, the falling is measured, the bearing of the true return course is calculated, and the true line joining the section corners is established, the permanent quarter-section corner common to sections 1 and 2 being placed a full 40 chains from the south boundary of these sections. In this way the excess or deficiency in linear measurement is, as before, placed in that portion of the line between the permanent quarter-section corner and the north boundary of the township.

When the north boundary of the township is a base line or a standard parallel, the line between sections 1 and 2 is run as a true line parallel to the east boundary of the township, a permanent quarter-section corner being set at 40 chains, a closing section corner being established at the point of intersection of the section line and base line or standard parallel, and the distance from this closing corner to the nearest standard corner being measured.

The successive ranges of sections from east to west are surveyed in a manner identical with the procedure described in the preceding paragraphs for the most easterly range until the two most westerly ranges are reached.

The west and north boundaries of section 32 are established as for corresponding sections to the east. A random line parallel to the south boundary of the township is then run west from the corner of sections 29, 30, 31, and 32, and the point of intersection between the random line and the west boundary of the township is determined. The falling of the intersection from the true corner is then measured, the course of the true line is calculated, and the true line is blazed and established, the permanent quarter-section corner being placed on the true line at a full 40 chains from the corner of sections 29, 30, 31 and 32. Thus the deficiency due to convergency of the meridians and the excess or deficiency due to errors in linear measurements are thrown in the most westerly half mile.

The survey of the other sections comprising the two most westerly ranges is continued in similar manner, the order in which the lines are surveyed being indicated by the numbers shown in Fig. 23-9.

23-17. Subdivision of Sections. Although the public-land laws contain the fundamental provisions for the subdivision of sections into quarter sections and quarter-quarter sections, only in rare instances are the sections subdivided in the field by United States surveyors. However, certain lines of subdivision are shown upon the official plats, and the surveyor in private practice who may be employed by the entrymen or landowners to establish the lines of subdivision is compelled to correlate conditions found on the ground with those shown on the approved plat. The function of the United States surveyor is to establish the official monuments so that the officially surveyed lines may be identified and the subdivision of the section may be controlled as contemplated by law. There the duties of the United States surveyor cease, and those of the surveyor in private practice begin. In the work of subdividing sections into the parts shown on the official plat the local surveyor cannot properly serve his client unless he is familiar with the land laws regarding the subdivision of sections, nor in the event of the loss of original monuments can the surveyor expect legally to restore the same unless he understands the principles employed in the execution of the original survey.

23-18. Subdivision by Protraction. Upon the official government township plats the interior boundaries of quarter sections are shown as dashed straight lines connecting opposite quarter-section corners. The sections adjacent to the north and west boundaries of a normal township, except section 6, are further subdivided by protraction into parts containing two regular half-quarter sections and four *lots*, the latter containing the fractional areas resulting from the plan of subdivision of the normal township. Figure 23-10 illustrates the plan of the normal subdivision of sections. The regular half-quarter sections are protracted by laying off a full 20 chains from the line joining opposite quarter-section corners. The lines subdividing the fractional half-quarter sections into the fractional lots are protracted from mid-points of the opposite boundaries of the fractional quarter sections.

In section 6 the two interior quarter-quarter-section corners on the boundaries of the fractional northwest quarter are similarly fixed, one at a point 20 chains north and the other at a point 20 chains west of the center of the section,

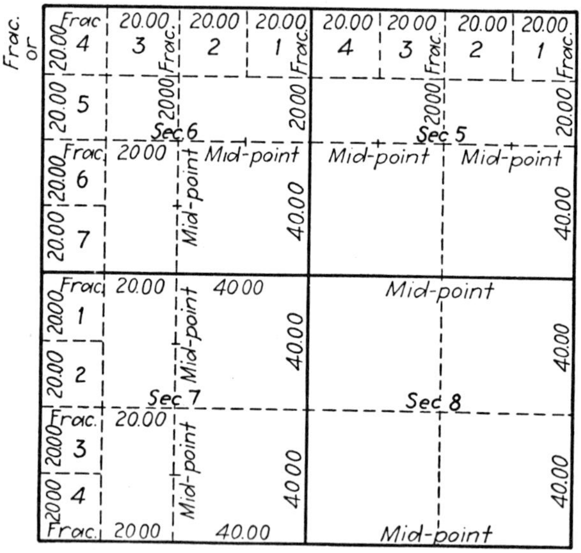

FIG. 23-10. Subdivision of sections.

from which points lines are protracted to corresponding points on the west and north boundaries of the section. Hence the subdivision of the northwest quarter of section 6 results in one regular quarter-quarter section and three lots.

In all sections bordering on the north boundary the fractional lots are numbered in succession beginning with No. 1 at the east. In all sections bordering on the west boundary the fractional lots are numbered in succession beginning with No. 1 at the north, except section 6 which, being common to both north and west boundaries, has its fractional lots numbered in progression beginning with No. 1 in the northeast corner and ending with No. 7 in the southwest corner, all as illustrated by Fig. 23-10.

Figure 23-11 illustrates a typical plat of section 6 on which the protracted areas are shown. Figure 23-12 is a similar section giving the calculated dimensions of the protracted areas.

Fractional Lots. In addition to sections made fractional by reason of their being adjacent to the north and west boundaries of a township, there are also sections made fractional on account of meanderable bodies of water (Art. 23-20), mining claims, and other segregated areas within their limits. Such sections are subdivided by protraction into such regular and fractional parts as are necessary for the entry of the undisposed public lands and to describe these lands separately from the segregated areas.

Figure 23-13 illustrates two sections made fractional by meanderable bodies of water. The practice is to number the lots in each section in sectional tiers beginning with No. 1 as the most easterly lot in the most northerly tier containing fractional sections, and to number the lots progressively toward the west in that tier, then toward the east in the tier to the south, and so on, tier by tier. This system of lot numbering is shown in both of the sections. A lot extending north and south through two or more tiers is numbered in the tier containing its greater area.

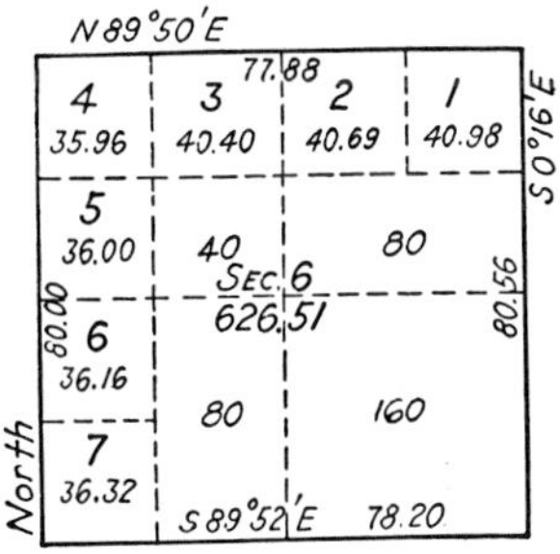

FIG. 23-11. Subdivisional areas of section 6.

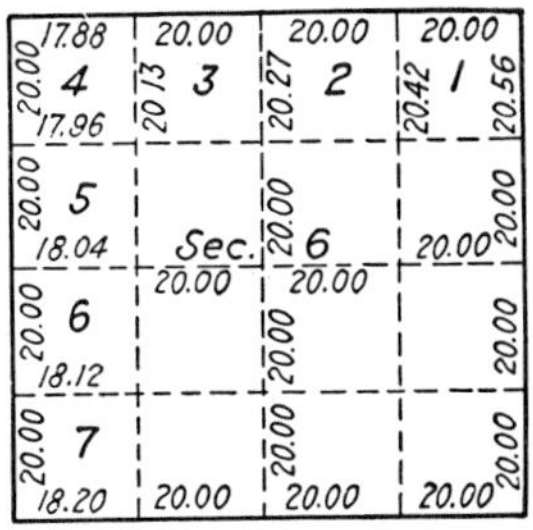

FIG. 23-12. Subdivisional dimensions of section 6.

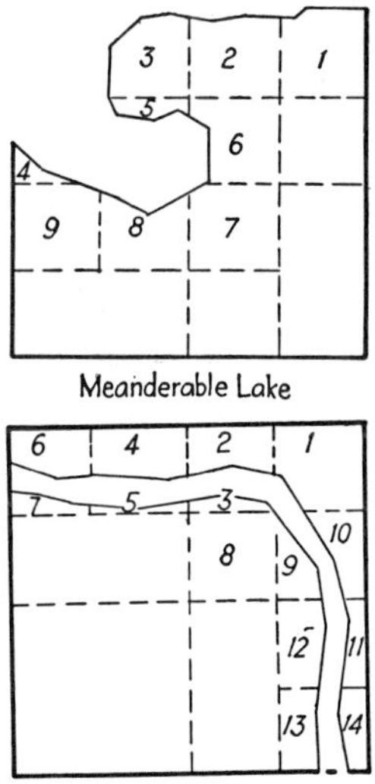

FIG. 23-13. Fractional lots.

23-19. Subdivision by Survey. The rules for the subdivision of sections given in the following paragraphs are based upon the general land laws. When an entryman has acquired title to a certain legal subdivision, he becomes the owner of the identical ground area represented by the same subdivision on the official plat. Preliminary to subdivision it is necessary to identify the actual boundaries of the section, as it cannot be legally subdivided until the section and exterior quarter-section corners have been found or have been restored to their original locations, and the resulting courses and distances have been redetermined in the field. When the opposite quarter-section corners have been located, the legal center of the section, or interior quarter-section corner, may be placed. If the boundaries of quarter-quarter sections or of fractional lots are to be established on the ground, it is necessary to measure the boundaries of the quarter section and to fix thereon the quarter-quarter-section corners at distances in proportion to those given upon the official plat; then the interior quarter-quarter-section corner may be placed.

Subdivision of Sections into Quarter Sections. According to law, the procedure to be followed in the subdivision of a section into quarter sections is to run straight lines between the established opposite quarter-section corners. The point of intersection of lines thus run is the quarter-section corner common to each of the four quarter sections into which the section is divided. It is called the *interior quarter-section corner* and is the legal center of the section.

Subdivision of Fractional Sections. Where opposite corresponding quarter-section corners of a section have not been or cannot be fixed, as is frequently the case when sections are made fractional by streams, lakes, etc., the lines of sectional subdivisions are run on courses mean between those of adjoining established section lines, or are run on courses parallel to the east, south, west, or north boundary of the section where there is no opposite section line.

Subdivision of Quarter Sections into Quarter-quarter Sections. Preliminary to the subdivision of regular quarter sections, the quarter-quarter-section (sixteenth-section) corners are established at points midway between the section and exterior quarter-section corners and between the exterior quarter-section corners and the center of the section. The quarter-quarter-section corners having thus been established, the center lines of the quarter section are run as straight lines between opposite corresponding quarter-quarter-section corners on the boundaries of the quarter section. The intersection of these lines is common to the four quarter-quarter sections into which the quarter section is divided. It is called the *interior quarter-quarter-section corner,* and it marks the legal center of the quarter section.

a. Irregular Quarter Sections. This case arises (1) when the quarter section is adjacent to the north or west boundary of a regular township and (2) when the quarter section adjoins any irregular boundary of an irregular township. The procedure is the same as that outlined in the preceding paragraph, except that the quarter-quarter-section corners on the boundaries of the quarter section which are normal to the township exterior are placed at 20 chains, proportionate measurement, counting from the regular quarter-section corner.

b. Fractional Quarter Sections. The subdivisional lines of fractional quarter sections are run from properly established quarter-quarter-section corners, with courses governed by the conditions represented upon the official plat, to the lake, water course, or reservation which renders such quarter sections fractional.

23-20. Meandering. In the process of surveying the public lands, all navigable bodies of water and other important rivers and lakes below the line of mean high water are segregated from the lands which are open to private ownership. In the process of subdivision, the regular section lines are run to an intersection with the mean high-water mark of such a body of water, at which intersection corners called *meander corners* are established. The traverse which is run between meander corners, approximately following the margin of a permanent body of water, is called a *meander line*, and the process of establishing such lines is called *meandering*. The mean high-water mark is taken as the line along which vegetation ceases (see also Art. 22-7). The fact that an irregular line must be run in tracing the boundary of a reservation does not entitle such a line to be called a meander line except where it follows closely the shore of a lake or the bank of a stream.

Meander lines are not boundaries but are lines run for the purpose of locating the water boundaries approximately, and although the official plats show fractional lots as bounded in part by meander lines, it is an established principle that ownership does not stop at such boundaries (see Arts. 22-5 and 22-9). A Supreme Court decision reads as follows:

> Meander lines are run in surveying fractional portions of the public lands bordering on navigable rivers, not as boundaries of the tract, but for the purpose of defining the sinuosities of the banks of the stream and as the means of ascertaining the quantity of land in the fraction subject to sale, which is to be paid for by the purchaser. In preparing the official plat from the field notes, the meander line is represented as the border line of the stream, and shows to a demonstration that the water-course, and not the meander line as actually run on the land, is the boundary.

In running a meander line, the surveyor begins at a meander corner and follows the bank or shore line, as closely as convenience permits, to the next meander corner, the traverse being a succession of straight lines. The true length and bearing of each of the courses of the meander line are observed with precision, but for convenience in plotting and computing areas the intermediate courses are laid off to the exact quarter degree and each intermediate transit station is placed a whole number of chains, or at least a multiple of 10 links, from the preceding station. Inasmuch as meander lines are not true boundaries, this procedure defines the sinuosities of the mean high-water line with sufficient accuracy. When a meander line is "closed" on a second meander corner, the latitudes and departures of the courses bounding the fractional lot are computed and the error of closure is determined. If this exceeds the allowable value, the line is rerun until an error in bearing or distance is discovered which will bring the closure within the specified limits (maximum error in either latitude or departure 1/640).

Rivers. Proceeding downstream, the river bank on the left hand is termed the left bank and that on the right hand the right bank. Navigable rivers and bayous as well as all rivers not embraced in the class denominated "navigable," the right-angle width of which is 3 chains and upward, are meandered on both banks, at the ordinary mean high-water mark, by taking the general courses and distances of their sinuosities.

Lakes. Regulations provide for the meandering of all lakes having an area of 25 acres or greater, the procedure being the same as for the meandering of streams. If the lake lies entirely within a section, there will be obviously no regular meander corners, and a *special meander corner* is established at the intersection of the shore of the lake with a line run from one of the quarter-section corners on a theoretical course to connect with the opposite quarter-section corner, the distance from the quarter-section corner to the special meander corner being measured. The lake is then meandered by a line beginning and ending at the special meander corner. If a meanderable lake is found to lie entirely within a quarter-section, an *auxiliary meander corner* is placed at any convenient place on its margin, and this is connected by traverse with one of the regular corners established on the boundary of the section.

Islands. In the progress of the regular surveys, every island of any meanderable body of water, except those islands which have formed in navigable streams since the admission of a state to the union, is located with respect to regular corners on section boundaries and is meandered and shown upon the official plat. Also in the survey of lands fronting on any nonnavigable body of water, any island opposite such lands is subject to survey.

23-21. Field Notes. The field notes taken in connection with the survey of public lands are required to be in narrative form and are designed to furnish not only a record of the exact surveying procedure followed in the field but also a report showing the character of the land, soil, and timber traversed by the line of subdivision and a detailed schedule of the topographic features adjacent to the lines, together with reference measurements showing the location of the lines with respect to natural objects, to improvements, and to the lines of other surveys. In this way the notes serve three purposes: (1) The field procedure is made a matter of official record, (2) the general characteristics of the territory served by the subdivision surveys are secured, and (3) the reference measurements to objects along the surveyed lines furnish evidence by which the established points and lines become practically unchangeable.

23-22. Marking Lines between Corners. As a final step in the survey of the public lands, it is the aim to fix the location of the legal lines of subdivision permanently with reference to objects on the surface of the earth. This aim is accomplished (1) by setting monuments, of a character later to be defined, at the regular corners; (2) by finding the location of the officially surveyed lines with respect to natural features of the terrain; and (3) by indicating the position of the regular lines through living timber by *blazing* and by *hack marks.*

The last method of fixing the location of the regular subdivisional lines is required by law just as definitely as is the establishment of monuments at the corners. All legal lines of the public-land surveys through timber are marked in this manner. Those trees which are on the line, called *line trees,* are marked with two horizontal notches, called *hack marks,* on each side of the tree facing the line; and an appropriate number of trees on either side of the line and within 50 links thereof are marked by flat axe marks, called *blazes,* a single blaze on each of two sides quartering toward the line.

23-23. Corners. In the subdivision of the public lands as described in the preceding articles, it is required that the United States surveyors shall permanently mark the location of the township, section, exterior quarter-section, and meander corners, as well as those quarter-quarter-section corners necessary in connection with the subdivision of fractional sections. For this purpose are employed monuments of a character specified by regulations of the Bureau of Land Management. The location of every such corner monument is, in accordance with definite rule, referred to such nearby objects as are available and suitable for this purpose; and where the corner itself cannot be marked in the ordinary manner an appropriate witness corner is established (Art. 23-24).

At the appropriate place in the field notes of the survey a record of each established monument is introduced, this record including the character and dimensions of the monument itself, the manner in which it is placed, the significance of its location, its markings, and the nature of the objects to which reference measurements are taken, together with the reference measurements.

23-24. Witness Corners. Where a true corner point falls within an unmeandered stream or lake, within a marsh, or in an inaccessible place, a witness corner is established in a convenient location nearby, preferably on one of the surveyed lines leading to the location of the regular corner. Also where the true point falls within the traveled limits of a road, a cross-marked stone is deposited below the road surface, and a witness corner is placed in a suitable location outside the roadway.

The witness corner is placed on any one of the surveyed lines leading to a corner, if a suitable place within a distance of 10 chains is available; but if there is no secure placed to be found on a surveyed line within the stated limiting distance, the witness corner may be located in any direction within a distance of 5 chains.

23-25. Corner Monuments. The Bureau of Land Management has adopted a standard iron post for monumenting the public-land surveys, which post is to be used unless exceptional circumstances warrant the use of other material. The post is made from zinc-coated wrought-iron pipe of inside diameter 2 in.; it is cut to a length of 30 in., one end is split for 4 or 5 in., and the two halves are spread outward to form a base. A brass cap is securely fastened to the top, and the pipe is filled with concrete. (Formerly 3-in. posts were specified for township corners, 2-in. posts for section corners, and 1-in. posts for quarter-section and meander corners and all other permanent points.) At the time of installation, the appropriate corner marking is stamped on the brass cap by means of steel dies. The posts are set in the ground for about three fourths of their length.

Where the procedure is duly authorized, durable native stone may be substituted for the model iron post described above, provided the stone is at least 20 in. long and at least 6 in. in its least lateral dimension. Stone may not be used as a monument for a corner whose location is among large quantities of loose rock. The required corner markings are cut with a chisel, and usually the stone is set with about three fourths of its length in the ground.

Where the ground is underlaid with rock close to the surface and it is impracticable to complete the excavations for monuments to the regular depth, the monument is placed as deep as practicable and is supported above the natural ground surface by a mound of stone. Where the solid rock is at the surface, the exact corner point is marked by a cross cut in the rock; and if it is practicable to do so, the corner monument is established in its proper location and is supported by a mound of stones.

Where the corner point falls within the trunk of a living tree which is too large to be removed readily, the tree becomes the corner monument and, as such, is scribed with the proper marks of identification.

Legal penalties are prescribed for damage to Government survey monuments or marked trees.

23-26. Marking Corners. Although to treat completely the system of marking corner monuments established in the survey of the public lands is beyond the scope of this text, a brief description of the general features of the system is here given. For further details the reader is referred to the "Manual of Surveying Instructions" of the Bureau of Land Management.

All classes of monuments are marked in accordance with a system which has been designed to provide a ready identification of the location and character of the monument on which the markings appear. Iron posts and tree corners are marked with capital letters which are themselves keys to the character of the monument and with arabic figures giving the section and township and range numbers of the adjacent subdivisions and the year in which the survey was made.

In the case of stone monuments, certain marks in the form of *notches* and *grooves* are placed on the vertical edges or faces; for an exterior corner the number of marks is made equal to the distance in miles from the adjoining township corner along the township or range line to the monument, and for an interior corner the number of marks is made equal to the distance in miles from the adjoining township boundary along section lines to the monument. These marks furnish a means of determining the number of the adjoining sections.

A witness corner and its accessories are constructed and marked similarly to a regular corner for which it stands, with the additional letters "WC" to signify *witness corner* and with an arrow pointing to the true corner.

Following is an index of the ordinary markings common to all classes of corners:

Mark	Meaning	Mark	Meaning
AMC	Auxiliary meander corner	S	Section
BO	Bearing object	S	South
BT	Bearing tree	SC	Standard corner
C	Center	SMC	Special meander corner
CC	Closing corner	T	Township
E	East	W	West
MC	Meander corner	WC	Witness corner
N	North	WP	Witness point
R	Range	¼	Quarter section
RM	Reference monument	⅟16	Quarter-quarter section

All standard township, section, and quarter-section corners on base lines and standard parallels are marked "SC." All closing township and section corners on these lines are marked "CC."

(*a*) Standard township corner. (*b*) Section corner.

FIG. 23-14. Typical markings on iron monuments.

1. *Markings on Iron Monuments.* Following are descriptions of the markings on the caps of certain of the iron-post monuments. These markings are made to read from the south side of the monument, and the year of establishing the monument is stated below the markings. Examples are shown in Fig. 23-14.

a. Standard Township Corner. The township number (as T25N) on north half, and the ranges and sections of the two adjoining subdivisions to the northeast and northwest (as R18E, S31, and R17E, S36) in the appropriate quadrants (Fig. 23-14*a*).

b. Corners Common to Four Townships. Township numbers (as T23N, T22N) on north and south halves; range numbers (as R18E, R17E) on east and west halves; section numbers (as S31, S6, S1, S36) in the four quadrants.

c. Closing Section Corners. Township and range on the half from which the closing line approaches the monument; section numbers in proper quadrants; also, if known at the time, the township, range, and section on the side of the correction line opposite the closing section line.

d. Corners Common to Four Sections on Township Exterior. Township (or range) common to the adjoining townships (as T25N); ranges (or townships) on opposite sides of the exterior (as R17E, R18E); section numbers in appropriate quadrants.

e. Interior Section Corners Common to Four Sections. Township and range in northern half; sections in appropriate quadrants.

f. Standard Quarter-section Corners. On north half marked "¼" followed by section number (as ¼S36).

g. Quarter-section Corners. On a meridional line, "¼" on north and sections on east and west halves; on a latitudinal line, "¼" on west half and sections on north and south halves.

2. *Markings on Stone Monuments.* The letters and figures on stone monuments are cut on the exposed sides of the stone and not on the top. In addition, grooves are cut in the faces of certain monuments, and notches are cut in the vertical edges of certain others. Grooves are employed where the faces are oriented to the cardinal directions, and notches are employed where the vertical edges are turned to the cardinal directions.

3. *Markings on Tree Monuments.* The system of marking tree monuments is practically the same as that employed in marking the caps of the iron monuments, already described in some detail. If *side of tree* be substituted for *quadrant of cap,* the markings given above are applicable to corresponding tree monuments. The appropriate marks are made on the trunk of the tree just above the root crown, and the series of marks on a particular side of a tree are scribed to read downward in a vertical line. The scribe marks are usually made in a vertical blaze. The marks thus made will remain long after the blaze is covered with new growth, and will in fact be destroyed only with the wood itself.

23-27. Corner Accessories. When a corner is referred by direction and distance to some other more or less permanent object near by, and the operation becomes a matter of record, it is possible to relocate the corner with respect to the object. In land surveying a recorded measurement of this kind is often called a *connection,* and the object thus located is called a *corner accessory.* It is specified that the United States surveyors in the survey of the public lands shall employ at least one accessory for every corner established, the character of the accessories to fall within the following groups: (*a*) bearing trees, or other natural objects such as notable cliffs and boulders, permanent improvements, and reference monuments, (*b*) mounds of stone, and (*c*) pits and memorials. Essentially, such an accessory is a part of the monument.

The marks on a bearing tree are made on the side nearest the corner, in the manner already described for tree-corner monuments. The mark includes the section number in which the tree stands and is terminated by the letters "BT."

Where a bearing object is of rock formation, the point to which measurements are taken is indicated by a cross, and it is marked with the letters "BO" and the section number, all marks being cut with a chisel.

Where it is impossible to make a single connection to a bearing tree or other bearing object and where a mound of stone or a pit is impracticable, a *memorial*, or durable article such as glassware, stoneware, a cross-marked stone, a charred stake, a quart of charcoal, or piece of metal is deposited alongside the base of the monument.

Where native stone is at hand, a mound of stones of sufficient size to be conspicuous is employed as an accessory.

Where accessories such as those mentioned in the preceding paragraphs are not available, pits may be used if conditions are favorable to their permanence. Where the ground is covered with sod, the soil is firm, and the slope is not steep, the pit will gradually fill with a material different in color or in texture from the original soil; and often a new species of vegetation springs up. Thus it may be possible to identify the location of a pit after the lapse of many years.

23-28. Restoration of Lost Corners. Although it has been the aim of the Bureau of Land Management in the subdivision of the public lands so to monument the established corners that there will always be physical evidence of their location, it is a matter of common experience that many corner marks become obliterated with the progress of time. It is one of the important duties of the local or county surveyor, in the relocation of property lines or in the further subdivision of lands, to examine all available evidence and to identify the official corners if they exist. Should a search of this kind result in failure, it is the duty of the surveyor to employ a process of field measurement which will result in the obliterated corner's being restored to its most probable original location (see Art. 22-15 and Ref. 7 at the end of this chapter).

As here employed, the term *corner* is used to designate a point established by a survey, while the term *monument* is used to indicate the object placed to mark the corner point upon the surface of the earth.

A corner is said to *exist* when its location within reasonable accuracy can be determined beyond reasonable doubt, either by means of the original monument, by means of the accessories to which connections were made at the time of the original survey, by the expert testimony of surveyors who may have identified the original corner and recorded connections to other accessories, or even by land owners who have indisputable knowledge of the exact location of the original monument. If the original location of a corner cannot be determined beyond reasonable doubt, the corner is said to be *lost*. If the monument of an existent corner cannot be found, the corner is said to be *obliterated*, but it is not necessarily lost.

In the absence of an original monument, either a line tree or a definite connection to natural objects or to improvements may fix a point of the original survey for both latitude and departure. The mean location of a blazed line, when identified as the original line, may sometimes help to fix a meridional line for departure, or a latitudinal line for latitude. Other calls of the original field notes in relation to various items of topography may assist materially in the recovery of the locus of the original survey. Such evidence may be developed in infinite variety.

A lost corner is restored to its original location, as nearly as possible, by processes of surveying that involve the retracement of lines leading to the corner. Restoration of a corner does not insure that it is placed exactly in its original location, and when a corner is restored the record of the survey should so state.

23-29. Proportionate Measurement. It is essential that the laying off of a given distance at the time of a resurvey to restore a lost corner should render the same absolute distance between two points on the ground as was measured during the original survey. For reasons which have been discussed in earlier chapters (Art. 7-16, etc.), the measurement of a given known line at the time of a resurvey will not in general agree with the length of the line as recorded in the original survey. Thus where linear measurements are necessary to the restoration of a lost corner, the principle of *proportionate measurement* must be employed. Proportionate measurement distributes an excess or deficiency in an over-all remeasured distance so that each of the remeasured parts will have the same ratio to the remeasured distance as the corresponding original parts had to the originally measured distance.

Single proportionate measurement consists in first comparing the resurvey measurement with the original measurement between two existing corners on opposite sides of the lost corner, and then laying off a proportionate distance from one of the existing corners to the lost corner. Double proportionate measurement consists in single proportionate measurement on each of two such lines perpendicular and intersecting at the lost corner (see also Art. 22-15 and problems 10 and 11, Art. 23-31).

23-30. Field Process of Restoration. Following are the field processes to be followed in a few of the simpler cases of the restoration of lost corners. In any event the restorative process must be in harmony with the methods employed in originally establishing the lines involved, and the preponderant lines must be given the greater weight in determining whether a corner should be relocated by single or double proportionate measurement or by some other method. Thus, standard parallels are given precedence over township exteriors, the latter are given precedence over subdivisional lines, and quarter-section corners are relocated after adjoining section corners have been restored. Detailed instructions for the relocation of lost corners are given in the manual of the Bureau of Land Management.

1. *Township Corner Common to Four Townships.* Where all the connecting lines have been established in the field, retracement is made between the nearest existing corners on the meridional line, one north and one south of the lost corner, and a temporary stake is set at the proportionate distance for the lost corner; this defines the latitude of the lost corner. Similarly measurement is made between the nearest existing corners on the latitudinal line through the point, and at the proper proportionate distance a second temporary stake is set; this marks the departure (or longitude) of the lost corner. The location of the lost corner is then found at the intersection of an east-west line through the first stake and a north-south line through the second; the corner is thus relocated by double proportionate measurement.

2. *Section Corner Common to Four Sections in Interior of Township.* Where all lines have been run, the section corner common to four sections in the interior of a township is restored by double proportionate measurement, in the manner described in (1).

3. *Regular Corner on Range Line but Not at Corner of Township.* The range line is straight between township corners. Two original corners on the 6-mile segment of the range line, one north and one south of the point sought, are identified and a line is run between them. The lost corner is relocated by a single proportionate measurement along this line. This procedure applies either to section or quarter-section corners.

4. *Regular Corner on Township Line but Not at Corner of Township.* The township line was originally run as a parallel of latitude for 6 miles. A parallel is rerun between the nearest existing corners to the east and west of the point sought, and the corner is relocated by proportionate measurements along this line.

5. *Standard Corner.* The standard corner includes any township, section, quarter-section, or meander corner, established on a base line or standard parallel at the time the line was originally run. The corner is relocated by the process explained in (4), that is, by single proportionate measurement along the parallel reestablished between the nearest existing standard corners on opposite sides of the point sought.

6. *Quarter-section Corner on Either Meridional or Latitudinal Section Line but Not on Range or Township Line.* The corner is relocated by single proportionate measurement along the straight line joining the adjacent section corners of the same section. If these section corners cannot be identified, they must be restored, as previously explained, before the quarter-section corner can be reestablished.

7. *Quarter-section Corner at Center of Section.* The corner is relocated at the intersection of meridional and latitudinal lines between opposite quarter-section corners on the boundaries of the section.

8. *Closing Corner on Standard Parallel.* The parallel is reestablished between the nearest existing corners on opposite sides of the corner sought. The lost corner is relocated by single proportionate measurement along the parallel from the nearest *standard* corners on opposite sides of the point sought.

9. *Quarter-quarter-section Corner on Section and Quarter-section Lines.* The

corner is relocated by single proportionate measurement between quarter-section and section corners on opposite sides of the point sought.

10. *Quarter-quarter-section Corner at Center of Quarter Section.* The corner is relocated at the intersection of the meridional and latitudinal lines between opposite quarter-quarter-section corners on the exterior of the quarter section.

23-31. Numerical Problems.

1. Find the angle of convergency between two meridians 6 miles apart at a mean latitude of 32°20′. Compute the linear convergency, measured along a parallel of latitude, in a distance of 6 miles.

2. Find the angle of convergency between two meridians whose distance apart is 24 miles and whose mean latitude is 45°. Compute the linear convergency in a distance of 24 miles.

3. Find the length of 1° longitude at a latitude of 40°06′20″.

4. Find the offsets between the tangent and the parallel at intervals of ½ mile over a distance of 6 miles at a latitude of 40°06′20″.

5. Find the azimuth of the secant and the offsets from the secant to the parallel at intervals of ½ mile over a distance of 6 miles at a latitude of 40°06′20″.

6. Show the dimensions and areas of the protracted subdivisions of Section 2, as required by law to be shown on the official plat, when the north, east, south, and west boundaries are respectively 80.00, 80.16, 80.00, and 80.20 chains.

7. Show the dimensions and areas of the protracted subdivisions of Section 7, as required by law to be shown on the official plat, when the north, east, south, and west boundaries are respectively 76.84, 80.00, 76.64, and 80.00 chains.

8. Show the dimensions and areas of the protracted subdivisions of Section 6, as required by law to be shown on the official plat, when the north, east, south, and west boundaries are respectively 76.36, 80.44, 76.60, and 80.00 chains.

9. A meanderable river follows a winding course from southwest to northeast across a section, the position of the regular southwest corner falling in the water. Draw a sketch assuming the described conditions; indicate thereon the positions of meander corners, witness corners, and meander lines, and indicate the numbering of the fractional lots.

10. A lost interior section corner is to be restored by a resurvey. The nearest corners which can be identified are regular section corners 1 mile north, 2 miles east, 3 miles south, and 1 mile west of the point sought. The records show the corresponding original measured distances to be 80.40, 160.56, 240.00, and 78.32 chains. The resurvey measurement between the nearest existing monuments on the meridional line through the lost corner is 320.16 chains, and that along the latitudinal line between the nearest existing corners is 238.48 chains. Calculate the proportionate measurements to be used in the relocation of the lost corner and state the procedure to be employed in its reestablishment.

11. A lost section corner on a range line is to be restored by a resurvey. One mile to the south the township corner is identified, and 2½ miles to the north the quarter-section corner is found. According to the records the corresponding distances measured at the time of the original survey were 80.00 and 200.00 chains. The resurvey distance between the existing corners is 279.64 chains. State the procedure to be followed in restoring the lost corner, and calculate the proportionate distances to be employed.

REFERENCES

1. Dodds, J. S., "Original Instructions Governing Public Land Surveys, 1815–1855," Powers Press, Ames, Iowa, 1945.

2. Horton, A. C., Jr., "Cadastral Surveys and Resurveys," *Trans. Am. Soc. Civil Engr.*, Vol. 115, pp. 52–70, 1950.
3. Stewart, Lowell O., "Public Land Surveys," Iowa State College Press, Ames, Iowa, 1935.
4. Harrington, Earl G., "Cadastral Surveys for the Public Lands of the United States," *Surveying and Mapping*, Vol. 9, No. 2, pp. 82–86, April–June, 1949.
5. U.S. Bureau of Land Management, "Ephemeris of the Sun, Polaris, and Other Selected Stars for the Year———," Government Printing Office, Washington, D.C., annual (in advance).
6. U.S. Bureau of Land Management, "Manual of Instructions for the Survey of the Public Lands of the United States," Government Printing Office, Washington, D.C., 1947.
7. U.S. Bureau of Land Management, "Restoration of Lost or Obliterated Corners and Subdivision of Sections," *Circ.* 1452, Government Printing Office, Washington, D.C., 1939.
8. U.S. Bureau of Land Management, "Standard Field Tables," 8th ed., Government Printing Office, Washington, D.C., 1956.

CHAPTER 24

TOPOGRAPHIC MAPS

24-1. General. A topographic map shows by the use of suitable symbols (1) the configuration of the earth's surface, called the *relief*, which includes such features as hills and valleys; (2) other natural features such as trees and streams; and (3) the physical changes wrought upon the earth's surface by the works of man, such as houses, roads, canals, and cultivation. The distinguishing characteristic of a topographic map, as compared with other maps, is the representation of the terrestrial relief.

Topographic maps are used in many ways. They are a necessary aid in the design of any engineering project which requires a consideration of land forms, elevations, or gradients, and they are used to supply the general information necessary to the studies of geologists, economists, and others interested in the broader aspects of the development of natural resources.

The preparation of general topographic maps is largely in the hands of governmental organizations. The principal example is the topographic map of the United States being constructed by the U.S. Geological Survey. This map is published in quadrangle sheets, which usually include territory 15′ in latitude by 15′ in longitude at a scale of 1:62,500 although they range from 7½′ by 7½′ at a scale of 1:24,000 to 4° by 12° at a scale of 1:1,000,000. A portion of a typical map the scale of which is 1:62,500 is shown in Fig. 24-7. Altogether there are more than 30 Federal agencies engaged in surveying and mapping. The central source of information regarding all Federal maps and aerial photographs is the Map Information Office, U.S. Geological Survey, Washington, D.C. Likewise, many maps are available from state, county, and city agencies. A comprehensive list of published charts and maps available from Federal and certain state agencies is given in Appendix A of Ref. 2 at the end of this chapter. (Ref. 2 also contains in Appendix B a selected bibliography on maps and mapping.)

As an aid to any survey, the surveyor should obtain available maps and/or aerial photographs of the region, even though they may not be of the particular nature or scale desired for his purpose.

24-2. Representation of Relief. Relief may be represented by *relief models, shading, hachures, form lines,* or *contour lines*. Of the symbols used on maps, only contour lines indicate elevations directly and quantitatively; they have by far the widest use. They are the principal topic of this chapter. Form lines are similar to contour lines but are not true to scale and are therefore only qualitative.

24-3. Relief Model. A relief model is a representation of ground forms done in three dimensions to suitable horizontal and vertical scales; it is a miniature of the terrain it represents. Plastic materials such as wax or clay are used; also laminated models are made by cutting cardboard sheets to the shape of successive contours and then assembling the sheets. The U.S. Army Map Service has developed a process for producing three-dimensional contour maps in the form of sheets of plastic which are flat-printed and then molded over a relief model. Cardboard relief maps of certain areas are commercially available.

The relief model is the most legible of all methods of representing relief, and it is of great value for purposes of instruction and public exhibit. It is also an aid in many of the special studies of the geologist, the geographer, and the engineer. However, its use is limited because of its cost and bulk.

24-4. Shading. Shading is a method of showing the terrestrial relief roughly in plan as it would appear from a point vertically above and with parallel rays of light flooding the landscape from a given angle, causing shadows to lie upon the less-illuminated areas. The method is pictorial and is useful in showing the general features where the relief is high and the slopes are steep. Shading is sometimes used in combination with hachures or contour lines, to render the map more legible. Improved techniques of relief shading have been developed by the U.S. Geological Survey.

24-5. Hachures. Hachures show relief more definitely but less legibly than does shading. The symbol consists of rows of short, nearly parallel lines whose spacing, weight, and direction produce an effect similar to shading but capable of more definite handling. The lines are drawn parallel to the steepest slopes, and in the best practice a standard scale of lengths and weights of lines is used to represent the various degrees of inclination of slopes. The method is illustrated in Fig. 24-1, which is a representation of a portion of the relief shown by contour lines at right center in Fig. 24-2.

24-6. Contours and Contour Lines. A *contour* is an imaginary line of constant elevation on the ground surface. It may be thought of as the trace formed by the intersection of a level surface with the ground surface, for example, the shore line of a still body of water.

If on a drawing are plotted the locations of several ground points of equal elevation, say, 720 ft. above sea level, a line on the map joining these points is called a *contour line*. Thus contours on the ground are represented by contour lines on the map. Loosely, however, the terms contour and contour line are often used interchangeably. On a given map, successive contour lines represent elevations differing by a fixed vertical distance called the *contour interval.*

The use of contour lines has the great advantage that it permits the representation of relief with much greater facility, and with far greater definiteness and accuracy, than do other symbols. It has the disadvantage that the map is not so legible to the layman.

24-7. Characteristics of Contour Lines. The principal characteristics of contour lines can be illustrated by reference to Fig. 24-2. For the purpose of this discussion the slope of the river surface is disregarded. The stage of the river at the time of the field survey was at an elevation of 510 ft., hence the shore line on the map marks the position of the 510-ft. contour line. For this map the contour interval is 5 ft. If the river were to rise through a 5-ft. stage,

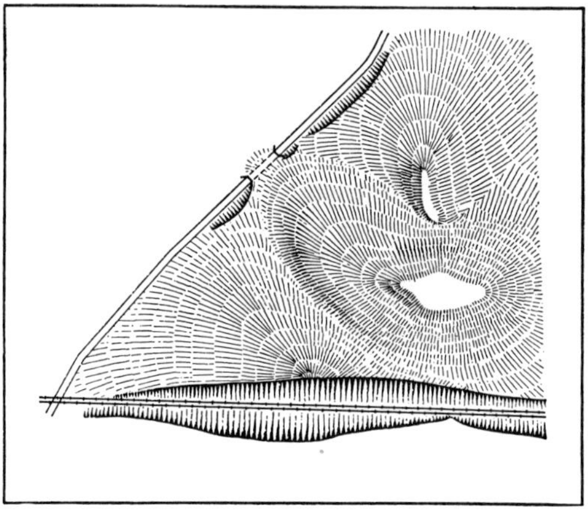

FIG. 24-1. Hachures.

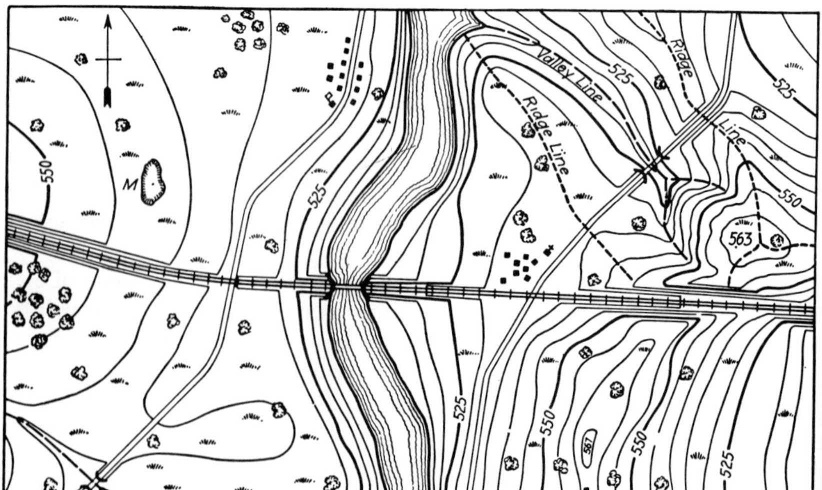

FIG. 24-2. Contour lines.

the shore line would be represented by the 515-ft. contour line; similarly, the successive contour lines at 520 ft., 525 ft., etc., represent shore lines which the river would have if it should rise farther by 5-ft. stages.

The principal characteristics of contour lines are as follows:

1. The horizontal distance between contour lines is inversely proportional to the slope. Hence on steep slopes (as at the railroad and at the river banks in Fig. 24-2) the contour lines are spaced closely.

2. On uniform slopes the contour lines are spaced uniformly.

3. Along plane surfaces (such as those of the railroad cuts and fills in Fig. 24-2) the contour lines are straight and parallel to one another.

4. As contour lines represent level lines, they are perpendicular to the lines of steepest slope. They are perpendicular to ridge and valley lines where they cross such lines.

5. As all land areas may be regarded as summits or islands above sea level, evidently all contour lines must close upon themselves either within or without the borders of the map. It follows that a closed contour line on a map always indicates either a summit or a depression. If water lines or the elevations of adjacent contour lines do not indicate which condition is represented, a depression is shown by a hachured contour line, called a *depression contour,* as shown at *M* in Fig. 24-2.

6. As contour lines represent contours of different elevation on the ground, they cannot merge or cross one another on the map, except in the rare cases of vertical surfaces (see bridge abutments of Fig. 24-2) or overhanging ground surfaces as at a cliff or a cave.

7. A single contour line cannot lie between two contour lines of higher or lower elevation.

24-8. Contour Interval. The appropriate vertical distance between contours, or *contour interval,* depends upon the purpose and scale of the map and upon the character of terrain represented. For small-scale maps of rough country, the interval may be 50 ft., 100 ft., or more; for large-scale maps of flat country, the interval may be as small as ½ ft. For maps of intermediate scale, such as are used for many engineering studies, the interval is usually 2 or 5 ft. (See also Art. 24-15.)

24-9. Contour-map Construction. Normally the construction of a topographic map consists of three operations: (1) The plotting of the horizontal control, or skeleton upon which the details of the map are hung; (2) the plotting of details, including the map location of points of known ground elevation, called *ground points,* by means of which the relief is to be indicated; and (3) the construction of contour lines at a given contour interval, the ground points being employed as guides in the proper location of the contour lines. A ground point on a contour is called a *contour point.*

The common methods of plotting both horizontal control and details are described in Chap. 18 and will not be considered further here.

Regardless of the number of ground points whose plotted locations are known, it is evident that any contour line must be drawn, to some degree, by estimation. This condition requires that the draftsman use his skill and judgment to the end that the contour lines may best represent the actual configuration of the ground surface.

Contour lines are shown for elevations which are multiples of the contour interval. They are drawn as fine smooth freehand lines of uniform width, preferably by means of a contour pen (Fig. 6-14). Usually each fifth contour line is made heavier than the rest, and sometimes these lines are drawn first in order to facilitate the location of intermediate contour lines. However, the location of intermediate lines should be considered just as important as that of the fifth lines, and as excessive degree of conformity between the contour lines indicates an inaccurate contour map.

Elevations of contours are indicated by numbers placed at appropriate intervals; usually only the fifth or heavier contour lines are numbered. The line is broken to leave a space for the number. So far as possible the numbers are faced so as to be read from one or two sides of the map; but on some maps the numbers are faced so that the top of the number is uphill. *Spot elevations* are shown by numbers at significant points such as road intersections, bridges, water surfaces, summits, and depressions.

Since contours ordinarily change direction most sharply where they cross ridge and valley lines and since the gradients of ridge and valley lines are generally fairly uniform, these lines are important aids to the correct drawing of the contour lines. Special care is taken in the field to locate the ridge and valley lines, and usually these lines are drawn first on the map. Examples of such locations are shown plotted as at point *a* in Fig. 24-3. The stream lines are drawn through those points which represent valleys, and the contour crossings are spaced along them before any attempt is made to interpolate or to draw the contour lines. This procedure aids the draftsman in his interpretation of the data. For example, in the square bounded by the points *D*–5, *D*–6, *E*–6, and *E*–5, the contour lines are made to show the head of the valley, the existence of which is indicated only by the valley line previously drawn in the square below. In the figure shown, the ridge line is rather indefinite, and but little aid would result from an attempt to sketch it on the map before drawing the contour lines.

24-10. Interpolation. The process of spacing the contour lines proportionally between plotted points is called *interpolation*. For example, in Fig. 24-3 the elevations of points *A*–2 and *B*–2 are 848.0 and 852.0 ft. respectively. The contour interval for this map has been taken as 2 ft., and the 848 and 852-ft. contour lines pass through the corresponding points. Under the assumption that the slope is uniform, the 850-ft. contour line passes through a point midway between *A*–2 and *B*–2. If a 1-ft. interval were used, then the additional 849 and 851-ft. contour lines would be drawn through the quarter points of the line from *A*–2 to *B*–2.

The procedure is usually not so simple as in the case just cited. The elevations at the corners of the other squares in the figure are mostly of such values that the contour lines do not pass through them; further, on many maps the points of known elevation are spaced irregularly. Under such conditions, the interpolations may be made by estimation on the map, by arithmetical computations, or by graphical means, as follows:

1. *Estimation.* Since each contour map is the result of more or less interpretation by the draftsman, in many cases it is not inconsistent with the

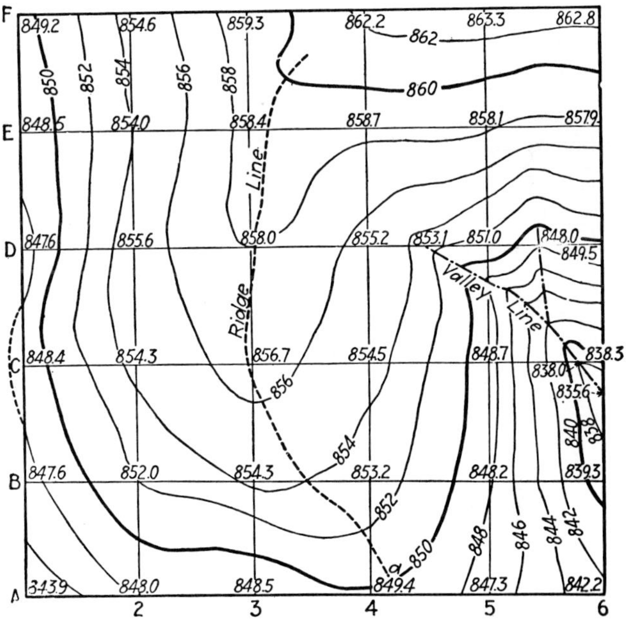

FIG. 24-3. Contour lines by checkerboard system.

other methods of map construction if the interpolation is made by careful estimation supplemented by approximate mental computations. This method is most commonly used on intermediate-scale and small-scale maps.

2. *Computation.* Where considerable precision is desired in the map, the errors of estimation may be eliminated by simple arithmetical computations made with the aid of a slide rule. For example, the elevations of points E–6 and F–6 (Fig. 24-3) are 857.9 and 862.8 ft., respectively. The contour interval is 2 ft., hence the difference in elevation between E–6 and the 858-ft. contour is 0.1 ft. Then, since the total difference in elevation is 4.9 ft., the proportional part of the distance from E–6 to F–6 to locate the 858-ft. contour line is 0.1/4.9 of the map distance between these points. Similarly, the proportional parts for the 860 and the 862-ft. contour lines

are, respectively, 2.1/4.9 and 4.1/4.9 of the distance from *E*–6 to *F*–6. The computed map distances are plotted to scale.

3. *Graphical Means.* The computations indicated in the previous paragraph become laborious if many interpolations are to be made, and accordingly various means of graphical interpolation are in use. One method is illustrated in Fig. 24-4. Parallel lines are drawn at equal intervals on tracing cloth, each fifth or tenth line being made heavier than, or of a different color from, the rest and being numbered as shown. If it is desired to interpolate the position of, say, the 52 and 54-ft. contours between *a* with elevation of 50.7 and *b* with elevation of 55.1, the line on the tracing cloth corresponding to 0.7 ft. (scale at left end) is placed over *a*, and the tracing is turned about *a* as a center until the line corresponding to 5.1 ft.

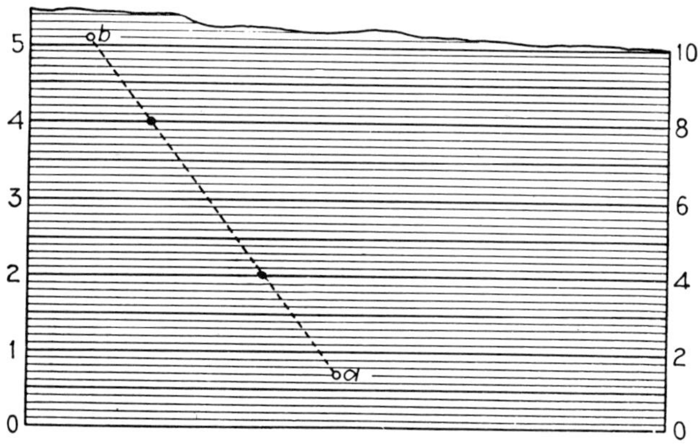

FIG. 24-4. Graphical interpolation of contour lines.

(scale at same end) covers *b*. The interpolated points are at the intersections of lines 2.0 and 4.0 (representing elevations 52 and 54) and the line *ab*, and may be pricked through the tracing cloth. Had the known points been much closer together, the scale at the right end of the tracing would have been used, and thus the value of each space would have been doubled; or if the scale were small, the contour interval large, and the topography rugged, each space might represent 1 ft. Thus by assigning different values to the spaces, a single piece of tracing cloth prepared in this way can be made to suit a variety of conditions.

Another convenient graphical means of interpolation is the use of a rubber band graduated at equal intervals with lines which form a scale similar to that just described for the tracing cloth. The band is stretched between two plotted points so that these points fall at scale divisions corresponding to their elevations. The intermediate contour points are then marked on the map.

24-11. Systems of Ground Points. Several methods of determining the location and elevation of ground points are used in topographic surveying (Chap. 25), and in topographic mapping the methods of plotting are the counterpart of the field methods. The typical systems of ground points are the *trace-contour, checkerboard, controlling-point,* and *cross-profile* systems.

1. *Tracing Contours.* A number of points on a given contour are located on the ground, and their corresponding locations are plotted on the map. The contour line is then drawn through these plotted points.

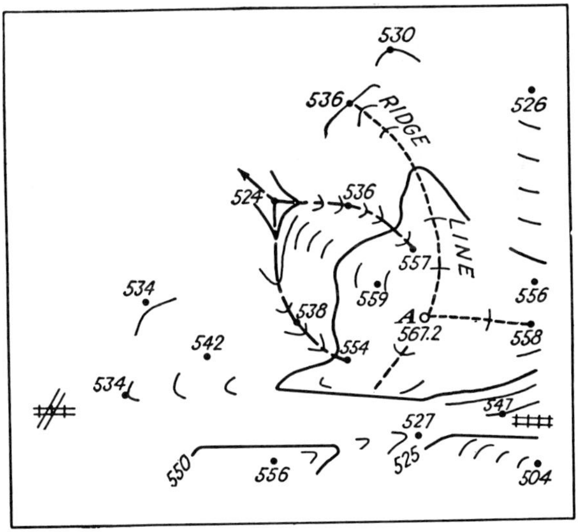

Fig. 24-5. Controlling points.

2. *Checkerboard.* A system of squares or rectangles is plotted as in Fig. 24-3, and near each corner is written its elevation. Also, the locations of valley and ridge lines are shown. The contour crossings are interpolated on the valley and ridge lines and on the sides of the squares, and the contour lines are drawn.

3. *Controlling Points.* The significance of valley and ridge lines has already been mentioned. It has also been noted that where a uniform slope exists between two ground points the intermediate contour lines on the map may be located by interpolation. Hence, if a system of ground points is chosen which locates the summits, depressions, valley and ridge lines, and important changes in slope, a contour map of the region may be drawn. Such an irregular system is illustrated in Fig. 24-5, in which are shown the points used in drawing the summit and immediate vicinity illustrated in the map of Fig. 24-2. In this sketch the ridge and valley lines have been

drawn, the contour lines have been spaced along them and between other controlling points (by interpolation), and the fifth contour lines have been sketched. It is evident that this information is of great aid in plotting and that the additional interpretation required to complete the map is simple.

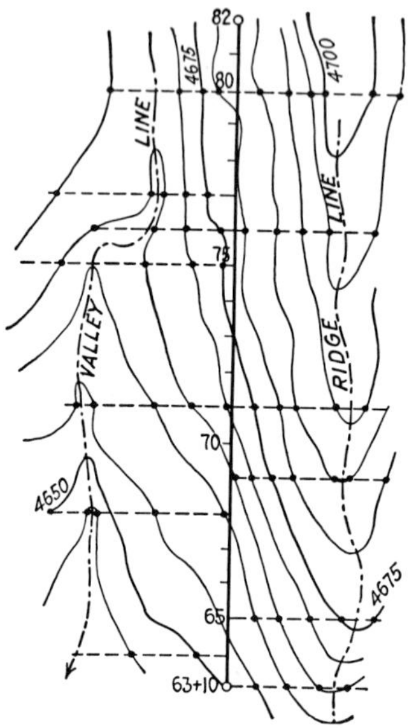

FIG. 24-6. Cross profiles.

4. *Cross Profiles.* The cross-profile method is most frequently used in connection with route surveys. The field surveys determine the location either of all contour points or of all points of change in slope, along selected lines normal to the route traverse line. The traverse, cross-profile lines, and ground points are plotted, and the contour lines are drawn. In Fig. 24-6 a transit traverse is represented as a straight line along which the 100-ft. stations are shown. The cross-profile lines are dashed, and the contour points are shown as dots. These dots obviously lie on the contour lines themselves, hence the latter may be drawn on the map as in the case of the trace-contour system. The number of points on a given contour line will usually be much less in the cross-profile system than in the trace-contour system, hence the draftsman will be called upon for a greater amount of interpretation.

24-12. Finishing the Map. In Chap. 6 the general subject of map drafting is discussed and map symbols are given. Methods of plotting are explained in Chap. 18.

Modern map drafting tends toward restraint in the use of titles and symbols. Although the use of certain colors and ornate symbols is justified by the character of some maps, these devices should be employed skillfully.

The standards most generally used for topographic map drafting and for any colors used in finishing the map are those employed by the U.S. Geological Survey (see Fig. 24-7) and the U.S. Coast and Geodetic Survey (see Fig. 30-9).

24-13. Tests for Accuracy. A topographic map can be tested for accuracy, both in plan and in elevation. In this discussion it is assumed that the errors in field measurement may be disregarded and that a graphical scale is provided on the map to render negligible any effect of shrinkage of the paper.

1. *Horizontal Dimensions.* The test for horizontal dimensions consists in comparing distances scaled from the map and distances measured on the ground between the corresponding points. The precision with which distances

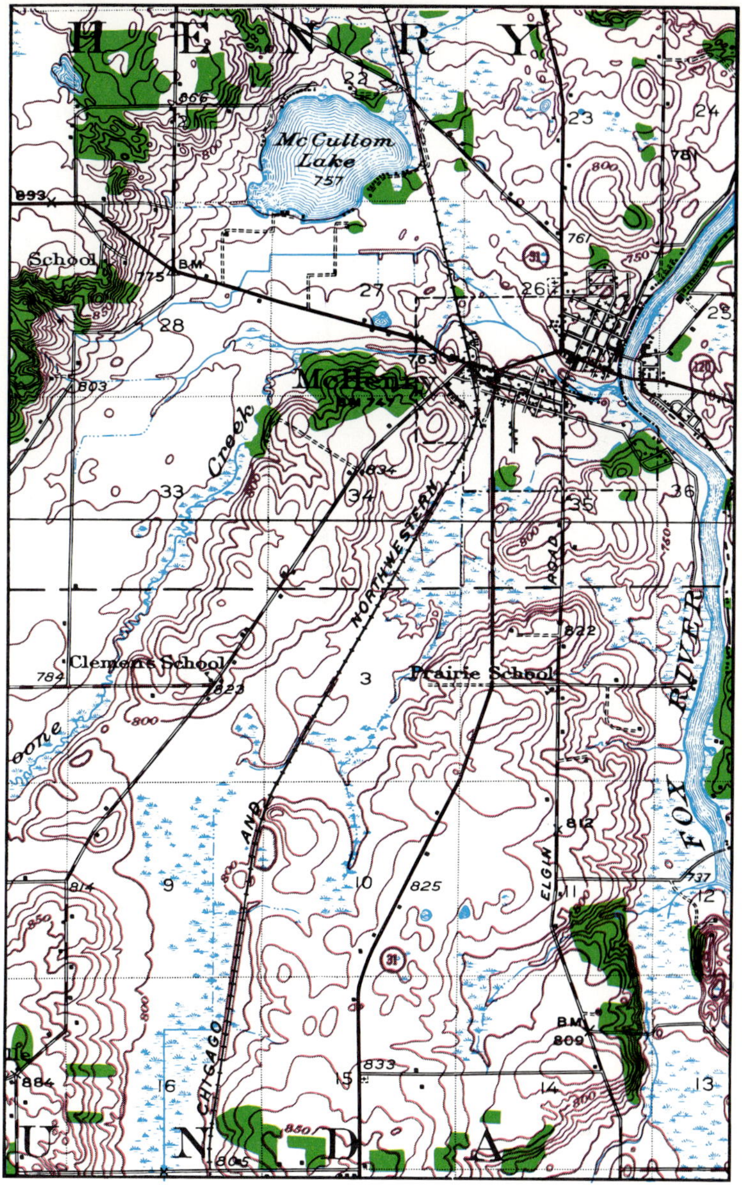

FIG. 24-7. Typical contour map of U.S. Geological Survey. Scale approximately 1 in. = 1 mile (representative fraction 1/62,500). Contour interval 10 ft.

may be scaled from a map depends upon the scale of the map and upon the size of the plotting error. Thus, if for a map scale of 1 in. = 100 ft. it is known that the error in location of any one point with respect to any other on the map is $\frac{1}{40}$ in., then the error represents 2.5 ft. on the ground.

Some surveys are made for the purpose of estimating areas as, for example, of a reservoir site. The errors in areas scaled from such maps can readily be determined from a consideration of the errors in the scaled distances, if it is remembered that the percentage of error in the area of a figure is equal to the sum of the percentages of error in the length and the breadth of the area (Art. 4-5). As an example, assume that a given area on a map is 8 by 24 in., that the scale of the map is 1 in. = 400 ft., that the average error in plotting and scaling the map distances is ± 0.03 in., and that the errors in scaled distances are independent of the lengths of the lines. The errors in the scaled dimensions of this area are then $0.03 \times 400 = 12$ ft. in each side. The percentage of error in the area is then

$$\frac{12}{8 \times 400} + \frac{12}{24 \times 400} = 0.5 \text{ per cent}$$

2. *Elevations.* One test for elevations consists in comparing, for selected points, the elevations determined by field levels and the corresponding elevations taken from the map. Usually the points are taken at 100-ft. stations along traverse lines crossing typical features of the terrain.

A more searching test is to plot selected profiles of the ground surface as determined by the field levels and the corresponding profiles taken from the map. These profiles provide a graphical record of the agreement between the map profile and the corresponding ground profile. As in the case mentioned above, the comparison between separate 100-ft. station points can be made; also, the presence of systematic errors will be evidenced if the map profile is above or below the ground profile for an undue proportion of its length. Careless work in spots will be made evident by wide divergences between the profile lines at such places.

24-14. Choice of Map Scale. From a consideration of the tests described in the preceding article it is possible to choose a map scale consistent with the purpose of the survey if the approximate size of the plotting error is known. For example, if it is known that (with reasonable care in plotting) the average error in distance between any two definite points on the map is $\frac{1}{40}$ in., and if it is known that the purpose of the survey will be met if the average error in scaled distances is 10 ft., these conditions are satisfied by a map scale of 1 in. = 400 ft.

By a similar course of reasoning, a map scale may be chosen that will represent a given *area* within desired limits of plotting error. Assume that it is desired to estimate the area within the flow line of a proposed reservoir with a permissible error not greater than 5 acres, that the area is roughly

4 mi. long and $\frac{1}{4}$ mi. wide, and that the errors in map distances will not exceed $\frac{1}{40}$ in. From these assumptions the area contains roughly 640 acres and is 1,300 ft. in width by 21,000 ft. in length, and the allowable ratio of error in the area is $\frac{5}{640}$ or $\frac{1}{128}$. It follows that the permissible ratio of error in the perimeter is $\frac{1}{256}$ (Art. 4-5). Hence the permissible errors in scaling the two sides of a rectangle which approximates the extent of the area on the map are 1,300/256 = 5 ft. and 21,000/256 = 82 ft., respectively. Therefore, the smallest permissible error in the map is 5 ft., and for a map in which the plotting error is $\frac{1}{40}$ in. the scale of the map should be 1 in. = 200 ft.

In addition to accuracy, considerations in the choice of a map scale are (1) the clarity with which features can be shown, (2) cost (the larger the scale, the greater the cost), (3) correlation of data with related maps, (4) desired size of map sheet, and (5) physical factors such as the number and character of features to be shown, the nature of the terrain, and the necessary contour interval.

24-15. Choice of Contour Interval. The contour interval may be thought of as the scale by which the vertical distances or elevations are measured on a map. The choice of a proper contour interval for a topographic survey and map is based upon four principal considerations: (1) the desired accuracy of elevations read from the map, (2) the characteristic features of the terrain, (3) the legibility of the map, and (4) cost.

1. *Accuracy.* Let it be assumed that two maps are equally accurate, so that the average error in elevations, read from the map, of points chosen at random is one half of a contour interval. Assume one map to have a contour interval of 5 ft. and the other 2 ft. It is evident that the average error in elevations of points chosen at random on one map is $2\frac{1}{2}$ ft. and on the other 1 ft. Therefore, the more refined the scale (that is, the smaller the interval), the more refined should be the measurements of the elevations of chosen points.

2. *Features.* Often field conditions exist where characteristic features require the use of a contour interval which would otherwise be inappropriate. Thus, if the shape of the terrain is such as to show much variation within a small area, or in other words, if the topography is of *fine texture,* then a smaller contour interval is required to show the greater complexity of configuration. On the other hand, if the landscape is composed of large regular forms or is of *coarse texture,* then a larger interval may be used. Where the texture varies considerably within an area to be mapped, or where it is desired to define certain parts of the area more clearly than the general area (as for construction purposes), a smaller contour interval may be used for parts of the area.

3. *Legibility.* A map otherwise excellent may be rendered useless and its appearance disfigured by a mass of contour lines which obscures other essential

features. In general, contour lines should not be spaced on the map more closely than 20 or 30 to the inch, although the legibility of the map depends largely upon the fineness and precision with which the lines are drawn. The lithographed maps of the U.S. Geological Survey and of the U.S. Coast and Geodetic Survey yield good results with much closer spacing.

4. *Cost.* The smaller the contour interval, the higher the cost, especially if the usual accuracy of one-half contour interval is maintained.

The contour interval and the scale of the map are interrelated; in general, the smaller the scale, the larger the contour interval. One rough rule is that the scale, in feet per inch, should be about 40 times the contour interval. Table 24-1 represents good practice in the choice of a contour interval for general-purpose maps under usual conditions. Maps for special purposes may depart considerably from these values (see Ref. 1 at the end of this chapter).

TABLE 24-1. RELATION BETWEEN SCALE OF MAP, SLOPE OF GROUND, AND CONTOUR INTERVAL

Scale of map	Slope of ground	Interval, ft.
Large (1 in. = 100 ft. or less)	Flat Rolling Hilly	0.5 or 1 1 or 2 2 or 5
Intermediate (1 in. = 100 ft. to 1,000 ft.)	Flat Rolling Hilly	1, 2 or 5 2 or 5 5 or 10
Small (1 in. = 1,000 ft. or more)	Flat Rolling Hilly Mountainous	2, 5, or 10 10 or 20 20 or 50 50, 100, or 200

24-16. Accuracy of Topographic Maps. The principles stated in the preceding articles provide criteria by which the accuracy of topographic maps may be specified. Thus the accuracy in horizontal dimensions may be required to be such that (1) the average errors in scaled *dimensions* between definite points chosen at random shall not exceed a stated value, or (2) the percentage of error in *areas* scaled from the map shall not exceed a stated value.

The accuracy of contour lines may be specified by assigning maximum values to (1) the average error in elevations taken from the map, (2) the maximum error indicated by random test profiles, and (3) the ratio of the length of the map test profile that lies above the ground test profile to the length that lies below the ground test profile. For example, the accuracy of a given topographic map might be specified as follows: The average error in distances between definite points as scaled from the map shall not exceed 8 ft.; the average error in elevations read from the map shall not exceed 1 ft.; the maximum error indicated by random test profiles shall not exceed 4 ft.; and the ratio of the length of the map test profile that lies above the ground test profile to the length that lies below the ground test profile shall be between ⅓ and 3.

The Federal agencies engaged in mapping have agreed on minimum requirements which entitle the following statement to be printed on a map, "This map complies with the National Standards of Map Accuracy requirements." With regard to horizontal accuracy, it is required that for maps on publication scales larger than 1:20,000, not more than 10 per cent of the well-defined points tested shall be in error by more than 1/30 in., measured on the publication scale; and for maps on publication scales of 1:20,000 or smaller, 1/50 in. Well-defined points are those that are easily visible or recoverable on the ground—in general, those which are plottable on the scale of the map within 1/100 in. With regard to vertical accuracy, it is required that not more than 10 per cent of the elevations tested shall be in error more than one-half the contour interval; the apparent vertical error may be decreased by assuming a horizontal displacement within the permissible horizontal error. The accuracy of the map may be tested by comparing points whose locations or elevations are shown on it with corresponding points as determined by surveys of a higher accuracy.

Another specification for accuracy of maps is that proposed by the committee on topographic surveys of the American Society of Civil Engineers (Ref. 3 at the end of this chapter). The requirements of this specification include the following:

	Detailed maps	General maps
Scale of map	1 in. = 200 ft. or less	1 in. = 400 to 1,000 ft.
Contour interval	1 or 2 ft.	5 or 10 ft.
Maximum error of location of definitely recognizable points	0.02 in.	0.1 in.
Maximum error of elevations determined by interpolation from contours	½ contour interval	1 contour interval
Maximum error of spot elevations	0.1 ft.	¼ contour interval
Minimum percentage of points complying with specification for accuracy	90 per cent	90 per cent

USES OF TOPOGRAPHIC MAPS

24-17. Cross-sections and Profiles from Contour Maps. Figure 24-8 shows a portion of a contour map, the purpose of which is to estimate the earthwork necessary to grade the area to a uniform slope. The full lines represent contours before earth is removed, and the dash lines represent

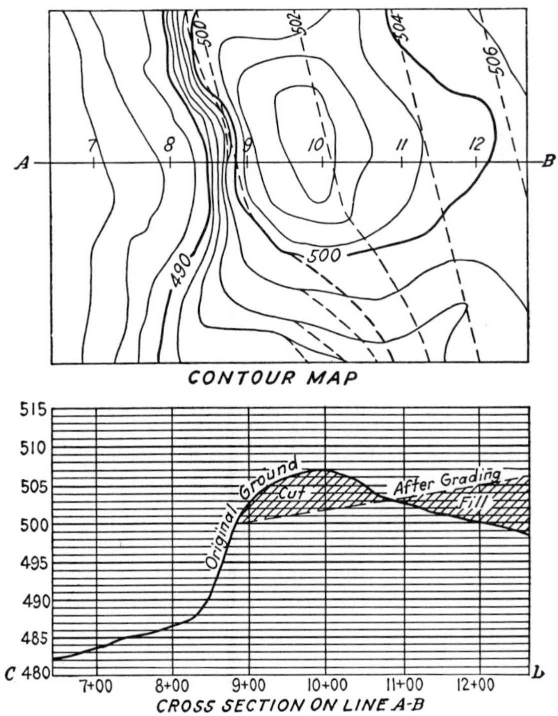

FIG. 24-8. Cross-section and profile from contour map.

contours after earthwork is completed. Below the map is a cross-section along the line AB. The horizontal location of each point marking the intersection of contours with the line AB is first projected to CD, the base line of the cross-section. The elevations are then read from the contours, and the appropriate distances are scaled up from the base line. A line (profile) drawn through the points thus plotted defines the cross-section. In the figure the full line of the cross-section shows the original surface, and the dash line shows the surface after grading is completed.

In practice, parallel lines along which the cross-sections are to be taken are drawn on the map, and the distance between them is scaled. For each cross-section, one person scales horizontal distances to contour crossings and reads contour elevations, while a second person plots the data on regular cross-section paper in the manner described in Art. 11-5. Frequently the horizontal scale to which the profile line is plotted is not the same as that used on the map, and usually the horizontal alinement is both curved and straight. For these conditions the mechanical means of plotting the profile described in the preceding paragraph cannot be used. The usual procedure is to mark the 100-ft. station points on the map, from them to scale the distances to contour crossings, and then to plot these distances on profile paper as if the elevations and stations were being taken from profile notes.

Figure 24-11 shows the profile of the ground line and of the grade line for a roadway construction. The manner of drawing the profile is similar to that for the cross-section described above.

24-18. Earthwork for Grading Areas. Quantities of earthwork to be moved in grading operations can be estimated from contour maps by (1) vertical cross-sections, (2) horizontal planes, and (3) equal-depth (or equal-height) contours. The probable errors involved are discussed in Art. 11-15.

1. *Cross-sections.* A cross-section is plotted for each of a number of parallel lines on the contour map. On each cross-section is shown the profile of (1) the original ground line and (2) the ground line after the proposed grading. To obtain the volume of a given cut or fill between adjacent cross-sections, the area between these ground lines is averaged for the two cross-sections and is multiplied by the distance between them.

2. *Horizontal Planes.* For preliminary estimates for grading areas, especially where the graded surface is itself more or less irregular, the common practice is to use the topographic map directly as a basis for calculations of volume. On the map are shown contours for the natural ground and contours for the proposed graded surface. This method consists in determining the volumes of earth to be moved between the horizontal planes at the elevations of successive contours.

The light full lines of Fig. 24-9 represent contours of the original ground, and the dash lines represent contours of the proposed graded surface. The heavy full lines are drawn through points of no cut or fill. Thus the line *abcdefa* bounds an area that is entirely in fill, and the line *dghjked* bounds an area that is entirely in cut. The "no cut or fill" lines are seen to pass through the points of intersection between full contours and the corresponding dash contours, as at *a*, *b*, *d*, *e*, *h*, and *j*. The conditions surrounding the problem make it possible to estimate the position of the lines where the cut or fill runs out between contours, as the lines *bcd, efa,* and *jke.* The cross-hatched portions are the horizontal sections of earth cut or filled at the contour elevations; thus F_1 represents the horizontal section of earth filled at elevation 96. The volume of earthwork between the two horizontal planes at the elevations of successive contours is a solid the altitude of which is the contour interval and the top and bottom bases of which are the horizontal projections of the cut or fill at the contour elevations (as, for the fill between the 94 and 96-ft. contours, the height is 2 ft. and the bases are F_2 and F_1). Where the cut or fill runs out between contours (as along line *bcd*), the height of

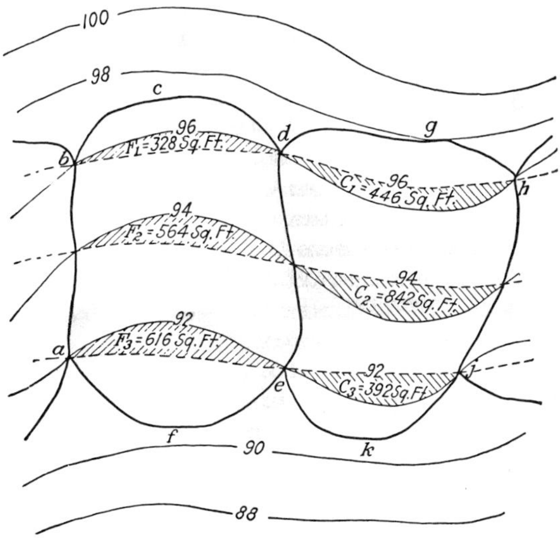

FIG. 24-9. Volume by horizontal planes.

the end volume will be less than the contour interval. This height may be estimated by assuming the slope of the ground to be uniform between contours; thus, point c is estimated to be at elevation 97.2, and the volume above the 96-ft. contour is a solid the base of which is F_1 and the altitude of which is $97.2 - 96 = 1.2$ ft. The end volumes may be considered as pyramids.

Example 1: It is desired to determine the volume of earthwork in fill bounded by the line *abcdefa* (Fig. 24-9). The intermediate volumes are to be calculated by the method of average end areas; the end volumes are to be considered as pyramids. The areas of fill at the contours are as shown in the figure. The point c is estimated to lie 1.2 ft. above the 96-ft. contour, and the point *f* is estimated to be 1.6 ft. below the 92-ft. contour. For solution, see the accompanying tabulation.

Elevation	Base area, sq. ft.	Altitude, ft.		Volume, cu. ft.
c = 97.2	0			
		1.2	⅓ × 1.2 × 328	131
96	328			
		2.0	½ × 2.0 × 892	892
94	564			
		2.0	½ × 2.0 × 1,180	1,180
92	616			
		1.6	⅓ × 1.6 × 616	329
f = 90.4	0			
			Total.................	2,532 cu. ft. or 93.8 cu. yd.

3. *Equal-depth Contours.* This method consists in determining volumes between irregularly inclined upper and lower surfaces bounding certain increments of cut or fill. In either case, horizontal projections of the inclined areas are taken from the map, usually with the planimeter, and the volume between any two successive areas is determined by multiplying the average of the two areas by the depth between them.

Figure 24-10 represents the topographic map of a portion of a tract which is to be graded by filling. The light full lines represent contours of the original ground surface, and the dash lines represent contours of the proposed graded surface. Above the dash 102-ft. contour the fill drops abruptly to the natural ground. Along the bank thus formed just above and paralleling the 102-ft. contour, actually there would be 100, 98, and 96-ft. contour lines, but to avoid confusion of lines these are not shown. At the intersection of each light full line with each of the dash lines the depth of fill (or cut) is recorded.

The heavy full lines drawn through points of equal fill are sometimes called *lines of equal fill* (or cut). The heavy outer line passes through points of zero fill and marks the limit of the fill. The next heavy line encloses the area over which the fill is a minimum of one contour interval and passes through points of intersection between a full contour and a dash contour the elevation of which is 2 ft. greater; and so on. Along the side of the bank above the dash 102-ft. contour, the heavy lines are seen to be close together and nearly parallel.

The fill between the graded surface and the surface 2 ft. below is represented by the solid the altitude of which is 2 ft. and the upper and lower surfaces of which are shown in horizontal projection by the line of zero fill and the line of 2-ft. fill, respectively. Likewise the lines of 2 and 4-ft. fill

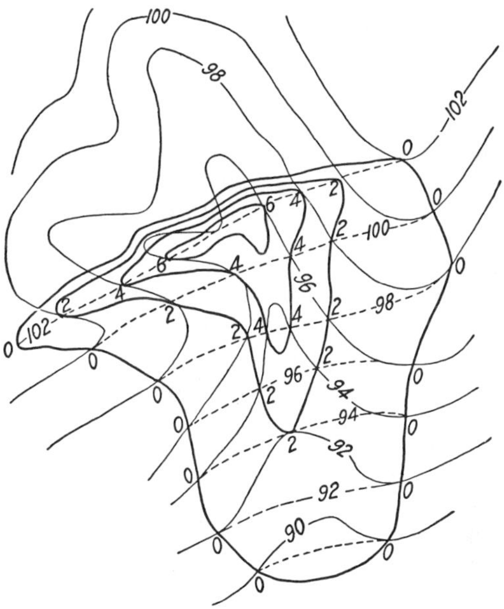

FIG. 24-10. Volume by equal-depth contours.

define the volume of fill between the depths of 2 ft. and 4 ft. from the graded surface. The volume below the innermost line of equal fill may be considered as a pyramid the base area of which is that bounded by the line and the altitude of which is estimated, being always less than the full contour interval. Though volumes are usually determined by multiplying the contour interval by the average of the areas of successive surfaces of equal cut or fill, when there is a large difference between successive areas the prismoidal formula (Art. 11-12) is sometimes used.

Example 2: An estimate of volume of earthwork in fill is to be made from a contour map similar to that of Fig. 24-10. Lines of equal fill are drawn, and the areas of the horizontal projections of surfaces of equal fill are determined by measurement with a planimeter. The altitude of the pyramid below the innermost surface of equal fill is estimated to be 1 ft. The computations are tabulated below.

Fill, ft.	Area, sq. ft.	Altitude, ft.		Volume, cu. ft.
0	101,000			
		2.0	½ × 2 × 134,000	134,000
2	33,000			
		2.0	½ × 2 × 50,000	50,000
4	17,000			
		2.0	½ × 2 × 22,000	22,000
6	5,000			
		1.0	⅓ × 1 × 5,000	2,000
7	0			

Total.............. 208,000 cu. ft.
or 7,700 cu. yd.

24-19. Earthwork for Roadway. Figure 24-11 shows (below) the contour lines for a proposed roadway (the grade line of which has been drawn on the profile) drawn dotted over the existing contour map. Above the contour map are shown a profile of the ground along the center line and the grade of the

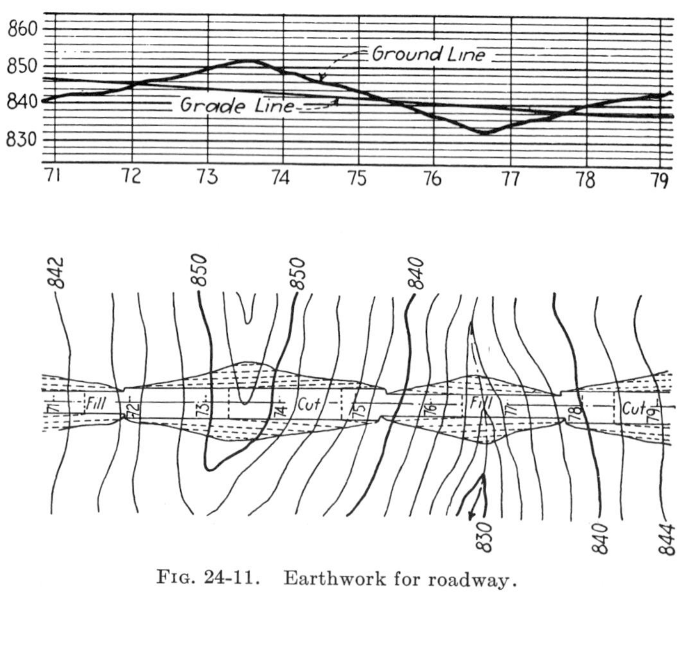

FIG. 24-11. Earthwork for roadway.

proposed roadway. The side slopes of the earthwork are 1½ to 1. The width of the roadway is 36 ft. in cut and 30 ft. in fill. From a study of these two drawings the following observations may be made:

1. The 840-ft. contour line of the proposed roadway crosses the roadway at a point on the map vertically beneath the point on the profile where the grade line crosses the 840-ft. elevation line; and similarly for the other gradient contours.

2. On the side slopes of the earthwork at any station, the distance out from the edge of the roadway to a contour line is given by the difference in elevation (between that which the contour line represents and the elevation of the grade at that station), multiplied by the side-slope ratio. Thus at station 76 + 40 the elevation of grade is 840.0 ft. and the elevation of the first contour line out from the edge of the fill is 838.0 ft.; hence the distance out is 2 × 1½ = 3 ft. (For clearness, in the illustration the lateral scale is exaggerated.)

3. As the grade line is not level, the contour lines on the earthwork slopes are not parallel to the roadway. Thus, the 844-ft. dotted contour line which crosses the roadway at station 73 + 30 is so inclined in direction that at station 74 + 80 where the elevation of grade is 842 ft. the 844-ft. contour line is out from the edge of the roadway a distance of 2 × 1½ = 3 ft.

4. The toe of a slope is drawn on the contour map by connecting the points where the dotted lines intersect the corresponding full lines.

If a line is drawn across a plotted roadway normal to the center line, a cross-section of the proposed roadway (Art. 11-5) can be plotted from the data of the contour map. Between adjacent cross-sections, the quantity of earthwork can be calculated as explained in Art. 11-11 or 11-12.

The volume of earthwork may also be estimated by means of either horizontal planes or equal-depth contours as explained in Art. 24-18.

24-20. Reservoir Areas and Volumes. A contour map can be employed to determine the capacity of a reservoir, the location of the flow line, the area of the reservoir, and the area of the drainage basin. The procedure may be illustrated by reference to the fill across the valley in Fig. 24-11, the fill being considered as a dam. If water is imagined to stand at the elevation of 834 ft., the water surface is represented by that within the full and dotted 834-ft. contour line. If the water were to rise through a 2-ft. stage to the elevation of 836 ft., the water surface would be represented by that within the full and dotted 836-ft. contour line (the full line being continued until the two parts meet). The volume of water that caused the 2-ft. rise is given by the average of the two surface areas multiplied by the vertical distance of 2 ft. Similarly, the volume of water required to cause a rise of the water surface from 836 to 838 ft. may be found, and so on. By a similar procedure the volume of any reservoir can be calculated.

The outline of the submerged area of a proposed reservoir is given by the contour line representing the maximum stage of the impounded water. The drainage area may be estimated by sketching on the map the watershed line and measuring the extent of the watershed with a planimeter.

24-21. Route Location. A contour map is useful in locating a proposed route for such projects as highways, railroads, and canals.

In roadway location it is desired to fix the center line of the proposed construction so that the subgrade will conform as nearly as practicable to the original ground surface (Chap. 26). This desired condition could be attained without difficulty if there were no limitations as to the grades, curvature, or distance, but a design of any project imposes restrictions as to these factors.

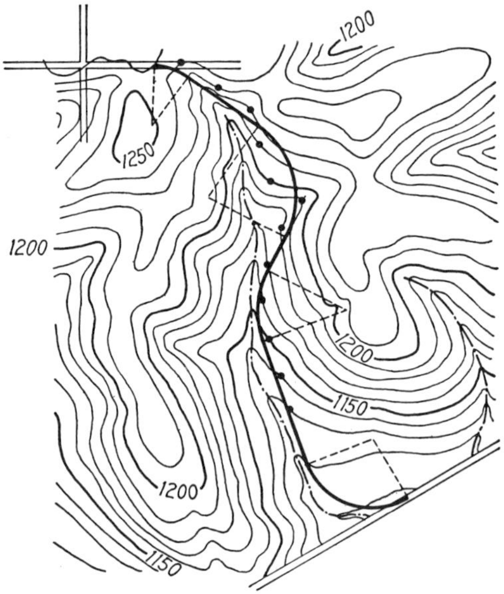

FIG. 24-12. Route location.

Suppose that a proposed highway is to be located in the valley shown in Fig. 24-12, joining the existing highways at opposite corners of the map; and let the maximum permissible gradient be 4 per cent. Beginning at the lower highway, the proposed route is projected on the map up the valley until the steep slopes require a careful study of the ground, say, to the 1,130-ft. contour line. Then a pair of dividers is set at a map distance equal to the contour interval divided by the desired gradient, in this case 10 ft. ÷ 0.04 = 250 ft. One foot of the dividers is placed on the intersection of the proposed roadway and the 1,130-ft. contour line, and the other foot of the dividers is placed on the 1,140-ft. contour line, as indicated in the figure by a heavy dot; and so on for successive contour lines. The series of dots marks on the plotted ground surface the location of a 4-per-cent grade line up the valley. Hence, the route is made to follow this line as closely as other limitations, such as the radius of curves, will permit.

The profile of this projected location may now be plotted and grade-line studies made on it, from which desirable changes will be indicated on the map location. Thus an indefinite number of projected locations might be made on the map; but the process is not carried far because, finally, the location of the line must be fitted to the ground in the field.

24-22. Numerical Problems.

1. On a map of scale 1 in. = 400 ft. with a contour interval of 5 ft., two adjacent contour lines are 0.54 in. apart. What is the slope of the ground in per cent?

2. In Fig. 24-3, assume that the corners of the squares are 100 ft. apart. Plot the ground profile along line *C*–1 to *C*–6, using a horizontal scale of 1 in. = 50 ft. and a vertical scale of 1 in. = 10 ft.

3. The accompanying tabulation gives elevations of points over the area of a 60 by 100-ft. city lot. The elevations were obtained by the checkerboard method, using 20-ft squares. Point *A*–1 is at the northwest corner of the lot, and point *F*–1 is at the southwest corner. Plot the contours, using a horizontal scale of 1 in. = 10 ft. and a contour interval of 2 ft.

Elevation, Ft.

Point	1	2	3	4
A	322.9	327.0	327.5	328.4
B	326.6	331.0	333.3	332.2
C	327.4	333.3	335.7	333.5
D	326.6	334.6	337.0	334.2
E	327.5	333.0	337.4	337.7
F	328.2	333.6	338.3	341.2

24-23. Office Problems.

Problem 1. Topographic-map Construction

Object. To construct a complete topographic map from field notes, relief being represented by contours. The data of the field problems in Chaps. 15, 25, and 26 may be used.

Procedure. (1) If the skeleton of the survey is a traverse, plot the traverse either by the method of tangents (Art. 18-5) or by the method of coordinates (Art. 18-17); if the horizontal control is in the form of rectangles or squares, plot the control by the method of coordinates. (2) Plot the details of the map by methods corresponding to those used in the field, as described in Art. 18-19. Use conventional symbols wherever applicable (Art. 6-11). (3) Mark each ground point by a dot, and mark each elevation in such location that there will be no doubt as to which point it refers, by letting the decimal point of the elevation represent the ground point. (4) Interpolate the contour crossings. (5) Place necessary notes so that they will not interfere with the map. (6) Draw a meridian arrow, and make an appropriate title. (7) Ink the map.

Problem 2. Profile from Topographic Map

Object. To plot the profile for a proposed highway or similar route from data of a contour map. It is assumed that the governing points are given and that the maximum rates of grade, the width of roadbed, and the side slopes are fixed.

Procedure. (1) Sketch in pencil a route between governing points that appears favorable. (2) Set bow dividers to measure 100 ft. or some multiple thereof at the scale of the map. From the point of beginning of the route, step off distances and read elevations as indicated by the contours. (3) Plot the corresponding profile. (4) Fix the grade line, making such readjustments of the proposed route as seem necessary to secure the most favorable location. (5) Compute the volumes of cuts and fills by the second method of Art. 11-14; check the computations by the first method of that article. (6) On each of the cuts and fills of the profile show the volume in cubic yards.

Problem 3. Volume of Earthwork from Contours

Object. To determine volumes of earthwork from a topographic map showing contours before and after grading. It is assumed (1) that a map showing contours of the original ground is assigned and (2) that other conditions attached to the problem, such as the area to be graded and the slopes of the finished surface, are given.

Procedure. (1) On the assigned map draw dash contour lines of the proposed ground surface. Draw heavy lines of no cut and fill. Ink all the foregoing lines in black. (2) With the planimeter measure the horizontal sections of earth cut and filled at each contour elevation. By method 2 of Art. 24-18, determine the volume between successive contour planes and the total volume for each cut and fill. (3) Draw all lines of equal cut and fill, and ink these lines in red. With the planimeter measure the horizontal projections of the areas enclosed by successive lines of equal cut and fill. By method 3 of Art. 24-18, determine the volume between successive surfaces of equal cut and fill. (4) Compare the total volumes given by the two methods, and show these total volumes on the drawing.

REFERENCES

1. American Congress on Surveying and Mapping, "Selection of Contour Intervals," panel discussion by Leon T. Eliel, Louis A. Woodward, Aubrey H. DaSantos, Shirley V. Griffith, and Robert H. Lyddan, *Surveying and Mapping*, Vol. 12, No. 4, pp. 344–358, October–December, 1952.
2. American Society of Civil Engineers, "Definitions of Surveying, Mapping, and Related Terms," *Manual of Engineering Practice* 34, 1954.
3. American Society of Civil Engineers, "Topographic Surveys—Progress Report of the Committee of the Surveying and Mapping Division on Topographic Surveys," *Proc. Am. Soc. Civil Engr.*, Vol. 72, No. 4, pp. 483–497, April, 1946.
4. Kane, C. V., "Contour Grading Highway Maps," *California Highways and Public Works*, pp. 1–5, September–October, 1952.
5. Raisz, Erwin, "Principles of Cartography," McGraw-Hill Book Company, New York, 1962.
6. Sloane, R. C., and J. M. Montz, "Elements of Topographic Drawing," 2d ed., McGraw-Hill Book Company, New York, 1943.
7. Thompson, Morris M., "How Accurate Is That Map?" *Surveying and Mapping*, Vol. 16, No. 2, pp. 164–173, April–June, 1956.
8. See also references at end of Chap. 25.

CHAPTER 25

TOPOGRAPHIC SURVEYING

25-1. General. The distinguishing feature of a topographic survey is the determination of the location, both in plan and in elevation, of selected ground points which are necessary to the plotting of the contour lines and to the construction of the topographic map. The topographic survey of a tract consists in (1) establishing over the area a system called the horizontal and vertical *control,* which consists of key stations connected by measurements of relatively high precision, and (2) locating the *details* (Art. 14-18), including the selected ground points, by measurements of lower precision from the control stations.

Topographic surveys fall roughly into three classes, according to the map scale to be employed, as follows:

Large scale: 1 in. = 100 ft. or less
Intermediate scale: 1 in. = 100 ft. to 1 in. = 1,000 ft.
Small scale: 1 in. = 1,000 ft. or more

Because of the range of uses of topographic maps and because of the variation in character of the areas covered, topographic surveys vary widely in character. Some are simple in plan and execution, and cover but a few acres; others are complex in plan and difficult in execution, and extend over hundreds of square miles. In this chapter will be discussed primarily ordinary ground surveys for areas of moderate size and for maps of intermediate and large scale, with special comments as necessary to cover other conditions.

Aerial photogrammetry (Chap. 31) is increasingly used for topographic surveying of all kinds. Even for ground surveys, the surveyor should secure and study aerial photographs of the area whenever they are available; examination of overlapping pairs of photographs by means of a simple stereoscope affords vision as in three dimensions and is of great aid. Other sources of information which are helpful in planning and conducting the survey are existing maps, land and survey records, and reports on geological and other features of the area.

25-2. Planning the Survey. The choice of field methods for topographic surveying is governed by (1) the intended use of the map, (2) the area of the tract, (3) the map scale, and (4) the contour interval.

1. *Intended Use of Map.* Surveys for detailed maps should be made by more refined methods than surveys for maps of a general character. For example, the earthwork estimates to be made from a topographic map by a

landscape architect must be determined from a map which represents the ground surface much more accurately in both horizontal and vertical dimensions than one to be used in estimating the storage capacity of a reservoir. Also, a survey for a bridge site should be more detailed and more accurate in the immediate vicinity of the river crossing than in areas remote therefrom.

2. *Area of Tract.* It is more difficult to maintain a desired precision in the relative location of points over a large area than over a small area. Control measurements for a large area should be more precise than those for a small area.

3. *Scale of Map.* It is sometimes considered that, if the errors in the field measurements are not greater than the errors in plotting, the former are unimportant. But since these errors may not compensate each other, the probable errors in the field measurements should be considerably less than the probable errors in plotting at the given scale. The ratio between field errors and plotting errors should be perhaps one to three.

The ease with which precision may be increased in plotting, as compared with a corresponding increase in the precision of the field measurements, points to the desirability of reducing the total cost of a survey by giving proper attention to the excellence of the work of plotting points, of interpolation, and of interpretation in drawing the map.

The choice of a suitable map scale is discussed in Art. 24-14.

4. *Contour Interval.* The smaller the contour interval, the more refined should be the field methods. The choice of a suitable contour interval is discussed in Art. 24-15.

25-3. General Field Methods. The principal instruments used are the engineer's transit, the plane table, the engineer's level, the hand level, and the clinometer. The use of the transit has advantages over the use of the plane table where there are many definite points to be located or where the ground cover limits the visibility and requires many set-ups. Conditions favorable to the use of the plane table are open country and many irregular lines to be mapped; the plane table is also advantageous for small-scale mapping. Sometimes the transit and the plane table, or the transit and the engineer's level, may be used together to advantage. Through dense woods, elevations of details are determined most advantageously by means of the hand level or the clinometer, and distances are usually determined by taping.

The horizontal control (Art. 25-5) is established by triangulation or by traversing, and the vertical control (Art. 25-8) is established by leveling, usually by direct leveling.

The details are located by methods described in Chaps. 7, 14, 17, and 24 and Arts. 25-10 to 25-15. The selected ground points used in plotting the contour lines may or may not be points on the contours, according to the field method employed. The horizontal locations of the selected ground points are determined in the same manner as for definite details, usually by the method of radiation. The elevations of ground points are determined usually by trigonometric leveling or, where the terrain is flat, by direct leveling. The stadia is used extensively except on surveys for maps of very large scale, say 1 in. = 20 ft. or less; for such surveys the errors in stadia distances are large compared with the errors of plotting. On large-scale surveys the distances to definite details are usually measured with the tape. The details may be located either at the time of establishing the control or later.

Systems of Ground Points. For the four systems of ground points commonly employed in locating details (Art. 24-11), the general field methods are as follows:

1. Where the *controlling-point* system is used (Art. 25-12), the ground points form an irregular system along ridge and valley lines and at other critical features of the terrain (Fig. 24-5). The ground points are located in plan by radiation or intersection with transit or plane table, and their elevations are determined commonly by trigonometric leveling or sometimes by direct leveling.

2. Where the *cross-profile* system is used (Art. 25-13), as on route surveys, the ground points are on relatively short lines transverse to the main traverse (Fig. 24-6). The distances from traverse to ground points are measured with the tape, and the elevations of ground points are determined by direct leveling, often with the hand level.

3. Where the *checkerboard* system is used (Art. 25-14), as where the scale is large and the tract is wooded or the topography is smooth, the tract is divided into squares or rectangles with stakes set at the corners (Figs. 25-11 and 24-3). The elevation of the ground is determined at these corners and at intermediate critical points where changes in slope occur, usually by direct leveling.

4. Where the *trace-contour* system is used (Art. 25-15), the contours are traced out on the ground. The various contour points occupied by the rod are located by radiation with transit or plane table. Frequently the engineer's level is employed as an auxiliary instrument.

Summary. The following statements summarize the use of the various systems of ground points employed in locating details:

1. *Intermediate-scale Surveys.* Usually the controlling-point system is used on hilly or rolling ground, and the cross-profile system is used on flat ground or for route surveys.

2. *Large-scale Surveys.* Usually the trace-contour system is used if the required accuracy is high and the ground is somewhat irregular in form, and the checkerboard system is used if the ground is smooth and the contour lines may be generalized to some extent.

3. *Small-scale Surveys.* The controlling-point system is used almost universally. A relatively small number of ground points are located, often by triangulation with the plane table; their elevations are determined by trig-

onometric leveling, the horizontal distances that are used in computing the differences in elevation often being scaled from the map.

CONTROL

25-4. General. Control consists of two parts: (1) *horizontal control,* for which by triangulation and/or traversing the control stations are located in plan, and (2) *vertical control,* for which by leveling the bench marks are established and the control stations are located in elevation. The control provides the skeleton of the survey which is later clothed with the *details,* or locations of such objects as roads, houses, trees, streams, ground points of known elevation, and contours.

On surveys of wide extent a relatively few stations distributed over the tract are connected by more precise measurements, forming the *primary control;* within this control system other control stations are located by less precise measurements, forming the *secondary control.* For small areas only one control system is necessary, corresponding in precision to the secondary control for larger areas. The terms "primary" and "secondary" are merely relative; for example, the degree of precision used on a secondary traverse for one survey might be sufficient for a primary traverse on another. This fact may be noted by inspection of Table 25-1, which gives approximate values of the limits of permissible error for control measurements suitable to the different map scales.

Another classification of control—either triangulation, traversing, or leveling—with regard to precision is by *orders.* The various orders are absolute, not relative. The extensive surveys executed by the Federal agencies include *first-order, second-order, third-order,* and *fourth-order* control; roughly these correspond, respectively, to primary, secondary, tertiary, and quaternary control for small-scale maps. The classification and standards of accuracy of the Federal Board of Surveys and Maps for second-order and third-order control surveys are given in Table 25-2. (See also Ref. 13 at the end of this chapter.)

25-5. Horizontal Control. The horizontal control may consist of a traverse system, a triangulation system, or a combination of the two. For an extensive survey there is first established a primary system, and this is extended by a secondary system. On surveys of less extent only the primary system is necessary, corresponding to the secondary control of large areas. The required precision of horizontal control depends upon the scale of the map and upon the size and shape of the tract. In Table 25-1 are given approximate values of permissible error on ordinary surveys; for small areas the tabulated values for secondary control may be used for primary control.

25-6. Traversing. The instruments, methods, and personnel required for traversing with the transit are described in Chaps. 14 and 15, and for traversing with the plane table in Chap. 17.

Primary Traverse. The primary traverse for an intermediate-scale map is usually run with transit and tape. Convenient routes are chosen which will

TABLE 25-1. TOPOGRAPHIC SURVEY CONTROL DATA (APPROXIMATE VALUES)

Scale of map	Kind of control, for given scale	Triangulation					Traverse			Levels	
		Length of sight, miles	Average error of closure in triangles	Distance between bases, miles	Probable error in base measure	Maximum discrepancy between bases	Length of traverse, miles	Maximum error of angles	Maximum linear error of closure	Length of circuit, miles	Maximum error of closure
Small	Primary	10 to 200	1″	100 to 500	$\frac{1}{1,000,000}$	$\frac{1}{25,000}$	50 to 500	2″	$\frac{1}{25,000}$	50 to 500	0.017 ft. $\times \sqrt{\text{miles}}$
	Secondary	5 to 20	3″	50 to 200	$\frac{1}{500,000}$	$\frac{1}{10,000}$	25 to 200	5″	$\frac{1}{10,000}$	....	0.035 ft. $\times \sqrt{\text{miles}}$
	Tertiary	1 to 10	6″	10 to 100	$\frac{1}{250,000}$	$\frac{1}{5,000}$	10 to 100	30″	$\frac{1}{5,000}$	10 to 100	0.05 ft. $\times \sqrt{\text{miles}}$
	Quaternary	½ to 2	1′ or graphical	2 to 10	....	......	1 to 2	2′ or compass	$\frac{1}{1,000}$	1 to 10	0.1 to 0.5 ft. $\times \sqrt{\text{miles}}$
Intermediate	Primary	1 to 5	10″ to 20″	5 to 50	$\frac{1}{10,000}$ to $\frac{1}{40,000}$	$\frac{1}{1,000}$ to $\frac{1}{4,000}$	1 to 20	10″ to 1′	$\frac{1}{1,000}$ to $\frac{1}{5,000}$	1 to 25	0.05 to 0.3 ft. $\times \sqrt{\text{miles}}$
	Secondary	½ to 2	Graphical	1 to 5	....	......	1 to 5	30″ to 3′	$\frac{1}{500}$ to $\frac{1}{2,500}$	1 to 5	0.1 to 0.5 ft. $\times \sqrt{\text{miles}}$
Large	Primary	1 to 5	2″ to 10″	2 to 20	$\frac{1}{20,000}$ to $\frac{1}{80,000}$	$\frac{1}{2,000}$ to $\frac{1}{5,000}$	1 to 5	30″ to 1′	$\frac{1}{5,000}$ to $\frac{1}{20,000}$	1 to 10	0.05 to 0.1 ft. $\times \sqrt{\text{miles}}$
	Secondary	¼ to 1	5″ to 20″	1 to 5	....	......	½ to 3	30″ to 2′	$\frac{1}{1,000}$ to $\frac{1}{5,000}$	1 to 3	0.05 to 0.1 ft. $\times \sqrt{\text{miles}}$

TABLE 25-2. CLASSIFICATION AND STANDARDS OF ACCURACY OF CONTROL SURVEYS*

	Second order		Third order
	Class I	Class II	
Principal uses	Area networks and supplemental cross arcs in national net	Coastal areas, inland water-ways, and engineering surveys	Topographic mapping
I. Triangulation			
Spacing of principal stations	4–10 miles	As required	As required
Strength of figure:			
Best chain: ΣR not to exceed	120	130	175
Single figure: R not to exceed	25	40	50
Base measurement: Probable error not to exceed	1/1,000,000	1/500,000	1/250,000
Triangle closure: Average not to exceed	1.5″	3″	5″
Side check: Ratio of max. diff. of logs of sides to tab. diff. for 1″ of log sine of smallest angle	2–4	4	10–12
Astronomical azimuths:			
Spacing-figures	8–10	10–12	12–15
Probable error not to exceed	0.3″	0.5″	2.0″
Closure in length after side and angle conditions have been satisfied should not exceed	1/20,000	1/10,000	1/5,000
II. Traverse			
Number of azimuth courses between azimuth checks not to exceed	25		50
Astronomical azimuth: Probable error not to exceed	2.0″		5.0″
Azimuth closure at azimuth check points not to exceed	$10'' \sqrt{\text{stations}}$ or 3.0″ per station†		$30'' \sqrt{\text{stations}}$ or 8.0″ per station†
Distance measurements accurate within	1/15,000		1/7,500
After azimuth adjustment, closing error in position not to exceed	1.67 ft. $\sqrt{\text{miles}}$ or 1/10,000†		3.34 ft. $\sqrt{\text{miles}}$ or 1/5,000†
III. Leveling			
Spacing of lines and crosslines	25–35 miles	6 miles	Not specified
Average spacing of permanently marked bench marks along lines not to exceed	1 mile	1 mile	3 miles
Length of sections	½–1 mile	½–1 mile	Not specified
Check between forward and backward running, between fixed elevations, or loop closures, not to exceed	0.035 ft. $\sqrt{\text{miles}}$	0.035 ft. $\sqrt{\text{miles}}$	0.050 ft. $\sqrt{\text{miles}}$

* Federal Board of Surveys and Maps. Selected items, Ref. 13.
† Whichever gives smaller permissible closure.

result in the advantageous location of stations; preferably the routes are taken along roads, ridge or valley lines, or public-land lines. If the area is small, the traverse is run near the perimeter of the tract.

Because of the cumulative effects of the errors in transit-tape traversing, in primary traversing it is desirable as a check to arrange closed circuits of length not to exceed perhaps 10 miles, dividing the tract into roughly equal areas. If closed circuits cannot be secured conveniently, checks for distance should be applied as the work proceeds, and checks for azimuth by astronomical observations (Chap. 21) should be applied at intervals of perhaps 10 miles. In applying the azimuth checks, it is of course necessary to take the convergency of meridians (Art. 23-11) into account.

Example 1: A specific case of primary traversing is illustrated by Figs. 25-1 and 25-2. (For portion of completed map, see Fig. 25-7.) The tract is approximately 3 by 7 miles, and the scale of the map is to be 1 in. = 500 ft. The field survey is planned with the aid of an existing small-scale map (not shown). Because of its size, the intermediate-scale map is to be drawn on a series of 15 by 25-in. sheets each representing an area 7,500 by 12,500 ft. on the ground. A system of plane rectangular coordinates is chosen such that all coordinates will be positive, by assigning x and y values of 5,000 ft. each to the center of a highway-railway crossing near the southwest corner of the tract. A coordinate map projection is then laid out to fix the coordinates of each corner of each map sheet, as shown by the straight dash lines of Fig. 25-1.

The routes for the traverses are chosen along the highway and railway lines in such a manner as to provide closed circuits not over 10 miles in length, which traverses follow roughly the perimeters of the various sheets. The transit party then runs azimuth traverses with transit and tape along the chosen routes, being governed by the condition that the permissible error of closure is 1/3,000. Permanent monuments are placed at intervals of not over 1 mile and are carefully referenced to nearby permanent objects. All streams, bridges, houses, and road crossings are located with reference to the traverse line. The trees and windmill shown in the enlarged reproduction of sheet 5 (Fig. 25-2) are examples of prominent objects to which azimuths are read from transit stations on the traverse, to serve as checks on the work.

Secondary Traverse. Wherever a secondary traverse is required to establish the instrument stations from which the details are located, the traverse may be run either simultaneously and in connection with the survey for location of details or before and separately from the location of details. Time is saved by the use of the first method, provided no serious errors or mistakes are made, and provided the accumulation of errors between primary control points is not so great as to require unduly large adjustments of the secondary-traverse measurements to effect a closure. If the details are to be located by the plane-table method, it will usually be desirable to run the secondary traverse before the location of details is begun because, in this case, there is no opportunity to adjust the secondary traverse to the primary stations if the details are mapped as the secondary traverse proceeds.

The route of the secondary traverse is selected with particular regard to the location of instrument stations that will be best situated for observing details. The route is frequently chosen along a ridge or a valley line, and in all cases the length is made such as to avoid an unduly large accumulation of errors. An area within a closed traverse of the primary control may be divided by means of secondary traverses into a series of roughly parallel strips.

Secondary traverses are usually run with the transit but are sometimes run with the plane table and occasionally with the surveyor's compass. The

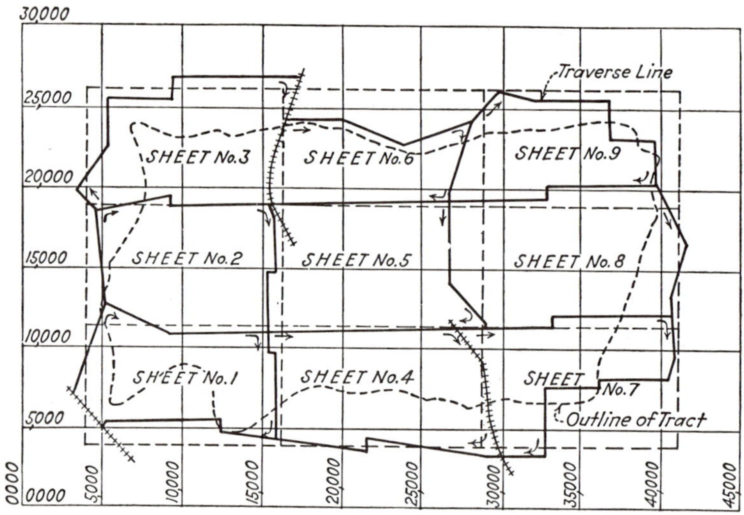

FIG. 25-1. Primary traversing for topographic survey.

lengths of the traverse lines are determined commonly by stadia or, if greater precision is required, by means of the tape.

Example 2: In the survey illustrated in Fig. 25-2, it is assumed that for a permissible error of 1/700 the method used is that of a compass-stadia traverse, and that the traverse is run before the location of details. The secondary traverse shown closes on a primary station at *O*.

A transverse having the same degree of precision as that assumed in the previous paragraph may be run with the plane table. As in the case of the compass traverse, the instrument occupies alternate stations only, and the instrument is oriented by use of the compass needle. This method may also be used for higher degrees of precision up to the limit which can be secured with stadia measurements, perhaps 1/1,000; and with taped distances a precision of perhaps 1/2,500 can be obtained. For these degrees of precision the table is oriented by backsighting, and great care is used in drawing the rays and scaling the distances.

If the precision required in the secondary traverse is 1/1,000 or greater, a transit-tape traverse is most commonly used.

25-7. Triangulation. The field conditions favorable to the use of triangulation to establish the horizontal control are (1) a fairly extended area in an open hilly region, (2) a city where traversing is difficult because of street traffic, or (3) a rugged mountainous region where traversing would be slow and laborious. Wooded regions seriously lessen the usefulness of the triangulation method; observation towers are required to establish lines of vision between stations, and the necessary expense of time and money is not justified except for surveys of considerable magnitude.

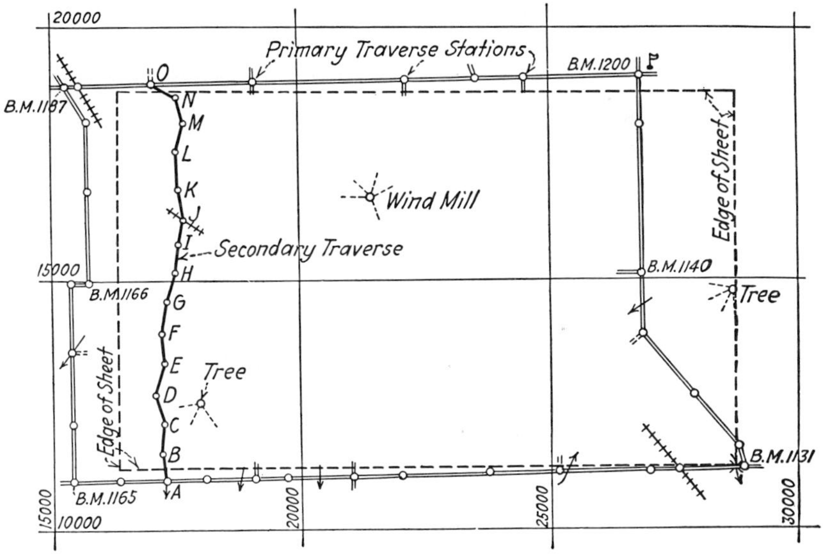

FIG. 25-2. Sheet 5 of Fig. 25-1 enlarged (see also Fig. 25-7).

A general description of the instruments, methods, and personnel for triangulation with the transit is given in Chap. 16, and graphical triangulation with the plane table is described in Art. 17-10.

Primary Triangulation. A general layout of the scheme of primary triangulation is planned on an existing small-scale map, the field stations are established on summits where visibility is good, and signals are erected. One or more base lines are established and measured, and their true azimuths are determined by astronomical observations (Chap. 21). Observations of angles are made on (1) major stations, which are marked by signals and which are to be occupied by the instrument; and (2) minor stations marked by such objects as trees, spires, and chimneys. When the field measurements have been completed, the necessary computations and adjustments are made (Chap. 16); then the coordinates of each station are computed for use in plotting.

Normally the transit is used for primary triangulation. However, for map scales smaller than 1 in. = 500 ft., usually it is possible to obtain the required map accuracy by the method of graphical triangulation with the plane table. This method has the advantage that no computations are required.

Example: The method of establishing primary triangulation with the transit is illustrated in Figs. 25-3 and 25-4. The conditions assumed for this survey are the same as those assumed in Art. 25-6.

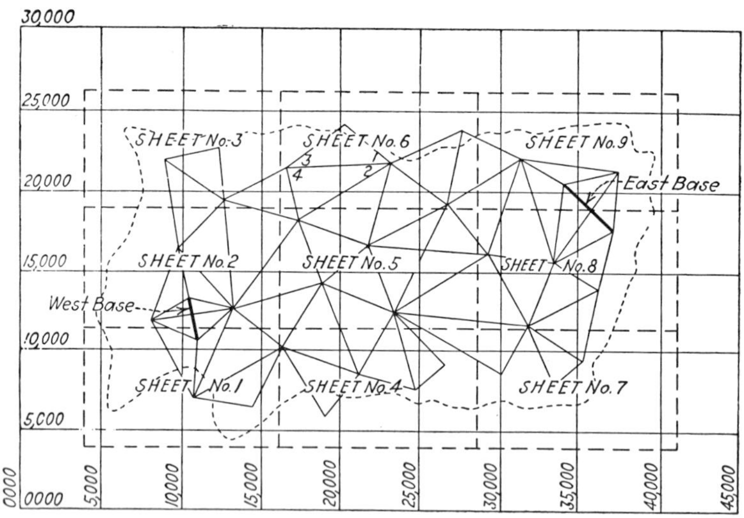

FIG. 25-3. Primary triangulation for topographic survey.

A general layout of the scheme is planned, the region is reconnoitered, and signals are erected at the selected stations.

Sites for two base lines, "West Base" and "East Base," are selected, one along a highway near one end of the tract and the other along a railroad near the other end of the tract. The base lines for this survey have lengths of about 2,500 and 4,000 ft., respectively, and are measured with a probable error not to exceed 1/10,000.

The angles are measured with such precision that each station closure, that is, the sum of the angles measured about each station, does not differ from 360° by more than 30″; and each triangle closure, that is, the sum of the angles in each triangle, does not differ from 180° by more than 1′30″.

A stellar or a solar observation is made at each base line to determine its true azimuth, and the true meridian at the south end of the West Base is taken as the reference meridian for the rectangular-coordinate, or grid, system. (The true meridian at the East Base, or at any other point in the area, will of course differ from that at the West Base by the amount of the convergency of meridians between the two locations.)

To the south end of the West Base are assigned the arbitrary coordinate values of $x = 11{,}000$ ft. and $y = 10{,}500$ ft., thus placing the tract entirely in the northeast quadrant of the coordinate system and making all coordinates positive.

The system of coordinates is projected on a series of map sheets, and the locations of all observed objects and stations are plotted by the method of coordinates (Fig. 25-3).

In Fig. 25-4 are shown to a larger scale the results of the primary triangulation as it exists on and near sheet 5 of Fig. 25-3. The instrument stations are shown by small triangles; other objects are indicated by symbols and names.

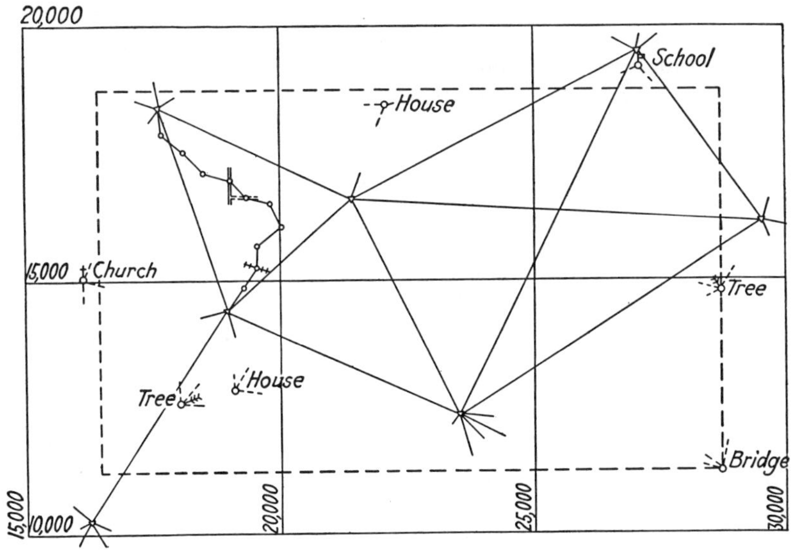

FIG. 25-4. Sheet 5 of Fig. 25-3 enlarged (see also Fig. 25-7).

The checks afforded in this work are (1) the duplicate measurements of the base lines, (2) the station and triangle closures of the angle measurements, (3) the comparison of the length of East Base as measured directly and as computed from the measured length of West Base, and (4) the comparison of the azimuth of East Base as observed directly and as computed from West Base.

Secondary Triangulation. The primary triangulation has resulted in the location of a number of transit stations, hilltops, chimneys, trees, and other prominent objects, the locations of which have been plotted on the field sheets (Fig. 25-4). The secondary control stations may now be located either by one of the methods of traversing previously explained or, where field conditions are suitable, by methods involving resection and intersection. If a secondary triangulation system is to be established, the work may be done with either the transit or the plane table. The advantage of the plane table is that it provides a ready means of solving the three-point and two-point problems. On the other hand, three-point determinations made with the transit can plotted conveniently in the office either by the use of a three-arm protractor or by the tracing-cloth method (Art. 17-14*b*).

The method of secondary triangulation has the advantage that instrument stations can be chosen at strategic points, unaffected by the cumulative errors inherent in traversing. It is sometimes employed in open rough country where chaining would be difficult.

25-8. Vertical Control. The purpose of the vertical control for a topographic survey is to establish bench marks at convenient intervals over the area, to serve (1) as points of departure and closure for the leveling operations of the topographic parties when locating details, and (2) as reference marks during subsequent construction work.

Primary and secondary level routes are required in about the same amount, and bear about the same relation to each other, as do the primary and secondary traverses or triangulation systems. Often the level routes follow the traverse lines, the traverse stations being used as bench marks. In Table 25-1 are given approximate values of permissible error on ordinary surveys.

The methods of leveling are described in Chaps. 8 and 9. Vertical control is usually accomplished by direct leveling, but for small areas or in rough country frequently the vertical control is established by trigonometric leveling (Art. 25-9).

The datum may be assumed for a given survey; but the results of governmental precise levels, referred to sea-level datum, are now available for all but the most isolated regions of the United States.

Precision. For intermediate-scale maps, four degrees of precision are commonly used in establishing the primary vertical control: (1) a maximum error expressed by the coefficient of 0.05 ft. $\sqrt{\text{distance in miles}}$, (2) 0.1 ft. $\sqrt{\text{distance in miles}}$, (3) 0.3 ft. $\sqrt{\text{distance in miles}}$, and (4) 0.5 ft. $\sqrt{\text{distance in miles}}$. The first applies to very flat regions where a contour interval of 1 ft. or less is used and on surveys which require the determination of gradients of streams, or which are to establish the grades of proposed drainage and irrigation systems. The second, third, and fourth coefficients apply to surveys in which no more exact use is made of the results than to determine the elevation of ground points for contours having 2, 5, and 10-ft. intervals, respectively. The last degree of precision listed can be reached by careful stadia leveling, which method has important advantages in hilly country.

Since the lengths of the secondary level circuits are, in general, roughly one fourth of the lengths of the primary circuits and since the errors vary approximately as the square root of the distance, the coefficients of permissible errors (for the same error of closure) are about twice the corresponding coefficients used on the primary circuits.

For large-scale maps, the contour interval is usually 1 or 2 ft. Ordinarily for these intervals the vertical control is of sufficient precision if level circuits close within 0.05 ft. $\sqrt{\text{distance in miles}}$ for extensive surveys or 0.1 ft. $\sqrt{\text{distance in miles}}$ for smaller areas.

25-9. Trigonometric Leveling. The height of the instrument, either plane table or transit, which has been oriented by the two-point or the three-point problem, is usually determined by trigonometric leveling, that is, by a vertical angle and a horizontal distance (Art. 8-5). The distance is either scaled from the map or measured directly by stadia or tape. Usually two or more stations are observed in order to increase the precision of the measurement. The method of trigonometric leveling is also applicable to field conditions where the horizontal control is established by triangulation and where a high degree of accuracy in the measured elevations is not required.

The required precision in the vertical angle bears a direct relation to that in the horizontal distance, and the permissible error in each depends upon the contour interval and the scale of the map. The following values will enable the topographer to determine the precision required in both the vertical and the horizontal distances: An error of 01′ in any vertical angle up to 10°, at a distance of 1,000 ft. produces an error in elevation of 0.3 ft. (approximately). Errors in the determination of the horizontal distances which likewise cause an error of 0.3 ft. in elevation are tabulated below:

Vertical angle	Error in horizontal distance, ft.
1°	18
3	6
6	3
9	2

Example: At a distance of 1,000 ft. and at a vertical angle of 9°, if the vertical angle is measured with a maximum error of 01′, the horizontal distance should be measured with a maximum error of 2 ft.

For distances greater than about 1,000 ft., either corrections for curvature of the earth and for atmospheric refraction should be made or preferably reciprocal measurements to eliminate natural and instrumental errors should be taken (see Art. 8-5).

LOCATION OF DETAILS

25-10. General. It is assumed in the articles which follow that the necessary horizontal and vertical control measurements have been made and that the field party is concerned with the location of details only. If the plane-table method is to be used, the horizontal control is plotted on the plane-table sheet.

The adequacy with which the resultant map meets the purpose of the survey depends largely upon the work of locating details, and the topographer should be completely informed as to the uses to be made of the map, to the end that he may give the proper emphasis to each part of the work.

The instruments used in locating details, and the four typical systems of ground points used in map construction, are discussed briefly in Art. 24-11 and more thoroughly in Arts. 25-12 to 25-15. A combination of methods may be used; for example, distances to points near the instrument may be measured by pacing or tape, and distances to more distant points by stadia, or a number of irregular controlling points may be located in a cross-profile or checkerboard survey. The aim is to locate the details with a minimum of time and effort.

Aerial photographs, or even ground views with the ordinary camera, may be of value in locating and plotting details.

25-11. Precision. The precision required in locating such definite objects as buildings, bridges, and boundary lines should be consistent with the precision of plotting, which may be assumed to be a map distance of about 1/50 in. Such less definite objects as shore lines, streams, and edges of woods are located with a precision corresponding to a map distance of perhaps 1/30 or 1/20 in. For use in maps of the same relative precision, more located points are required for a given area on large-scale surveys than on intermediate-scale surveys; hence the location of details is relatively more important on large-scale surveys.

Contours. The veracity with which contour lines represent the terrain depends upon (1) the accuracy and precision of the observations, (2) the number of observations, and (3) the distribution of the points located. Ground points are definite, but as the contour lines must necessarily be generalized to some extent it would be inappropriate to locate the ground points with refined measurements. The error of field measurement in plan should be consistent with the error in elevation, which in general should not exceed one fifth of a contour interval; thus generally the error in plan should not exceed one fifth of the horizontal distance between contours, and the error in elevation should not exceed one fifth of the vertical distance between contours. The purpose of a topographic survey will be better served by locating a greater number of points with less precision, within reasonable limits, than by locating fewer points with greater precision. Thus, if for a given survey the contour interval is 5 ft., a better map will be secured by locating with respect to each instrument station perhaps 50 points whose average error in elevation is 1 ft. than by locating 25 points whose average error is only 0.5 ft.

A general principle which should serve as a guide in the selection of ground points may be noted. As an example, let it be supposed that a given survey is to provide a map which shall be accurate to the extent that if a number of well-distributed points are chosen at random on the map, the average difference between the map elevations and ground elevations of identical points shall not exceed one half of a contour interval. Under this requirement, the attempt is made in the field to choose ground points such that a straight line between any two adjacent points will in no case pass above or below the ground by more than one contour interval. Thus, in Fig. 25-5, if the ground points were taken only at *a*, *b*, *c*, *d*, and *e* as shown, the resulting map would indicate

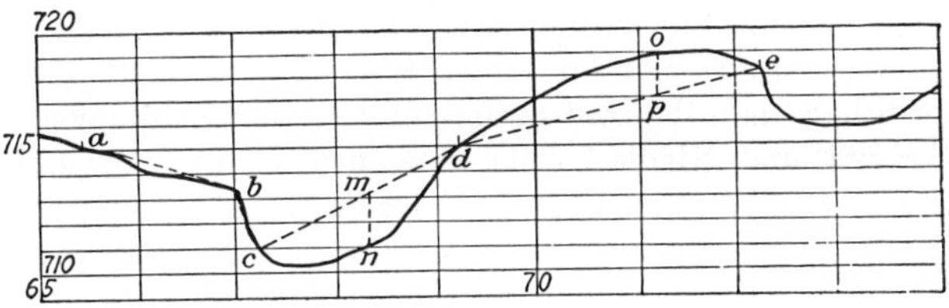

FIG. 25-5. Effect of omission of significant ground points.

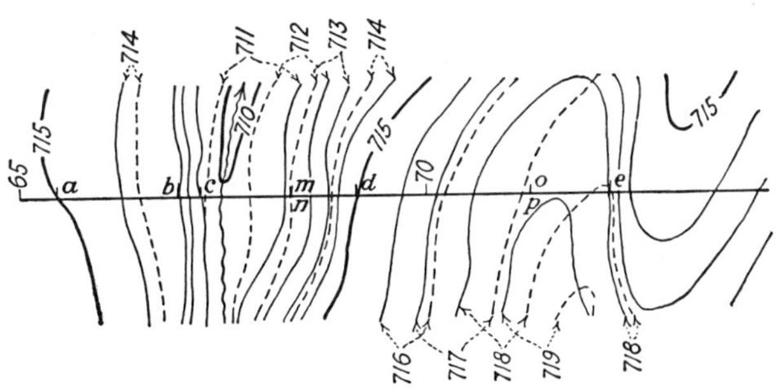

FIG. 25-6. Errors in contour lines due to insufficiency of ground points.

the straight slopes *cd* and *de*; the consequent errors in elevation of *mn* and *op* on the profile amount to two contour intervals and show that additional readings should have been taken at the points *n* and *o*. The corresponding displacement of the contours on the map is shown by dotted and full lines in Fig. 25-6.

Angles. The precision needed in the field measurements of angles to details may be readily determined by relating it to the required precision of corresponding vertical and horizontal distances. Thus for a sight at a distance of 1,000 ft., a permissible error of 0.3 ft. in elevation corresponds to a permissible error of 01′ in the vertical angle; likewise, a permissible error of 0.3 ft. in azimuth (measured along the arc from the point sighted) corresponds to a permissible error of 01′ in the horizontal angle. Values for other lengths of sight or degrees of precision are obtained by proportion; thus if it is desired to locate a point to the nearest 2 ft. in azimuth (or elevation) and if the length of the sight is 500 ft., the corresponding permissible error in the angle is $2/0.3 \times 1{,}000/500 \times 01' = 13'$.

25-12. Details by Controlling-point Method. Details may be located by the controlling-point method employing the transit and stadia (Art. 25-12*a*), the plane table (Art. 25-12*b*), or the transit and plane table together (Art. 25-12*c*). The distances are usually measured by stadia, but on large-scale surveys the distances to definite details may be measured with the tape.

If the ground is flat, such that direct rod readings over large areas are possible, the cross-profile method or the trace-contour method will usually be more expeditious than the controlling-point method employing the transit.

25-12a. Transit and Stadia. The personnel of the topography party using the transit usually consists of the transitman, recorder, and one or two rodmen. In wooded country one or more axemen are usually needed to clear the lines of vision. The organization may be modified to allow all other members of the party to work as rapidly as possible.

In locating ground points, usually the vertical angles are observed more precisely than the horizontal angles. Accordingly, the vertical circle of the instrument assumes greater importance than the horizontal circle; but because all vertical angles are measured with respect to a horizontal plane, it is important that the horizontal plate be truly horizontal and remain so without the need of constant releveling. If vertical angles are to be measured to the nearest minute of arc, a level tube of about 30″ sensitiveness should be attached to the vernier arm of the vertical arc. Many topographers prefer a stadia circle or Beaman arc for purposes of stadia leveling (Art. 15-12).

It is often advantageous to set the instrument at some isolated but strategic point which has not been located by the horizontal control surveys. The instrument may be located by means of the three-point or two-point problem if definite objects suitably located are visible and if these objects have been, or can be, observed from other stations. The elevation of the station may be determined by a stadia-leveling or trigonometric-leveling observation on one or more points within the range of vision from the occupied station.

Relatively inaccessible or distant points may at times be located by the principle of intersection (Art. 14-8). On a topographic survey of a steep canyon wall, a method of intersection employing two transits may be used. (See Ref. 6 at the end of this chapter.) The transits are set up over stations of known location and elevation, at a distance apart such that good intersections will be obtained. A man, perhaps suspended by ropes, holds a target at controlling points of the canyon wall, and the transitmen simultaneously observe the azimuth and vertical angle to each point. The location and elevation of each point are later computed for plotting on the map. A check on the elevation of each point is afforded by the two values computed from the two observed vertical angles and scaled map distances.

Field sketches are valuable aids to supplement the observed data, especially where the ground exhibits many irregular features and where many details are to be mapped. They vary from freehand sketches and notes entered in a cross-ruled book, to elaborate field drawings for which an assistant is employed and which amount to an execution of the office procedure in the field. Where details are numerous, a drawing board is employed near the

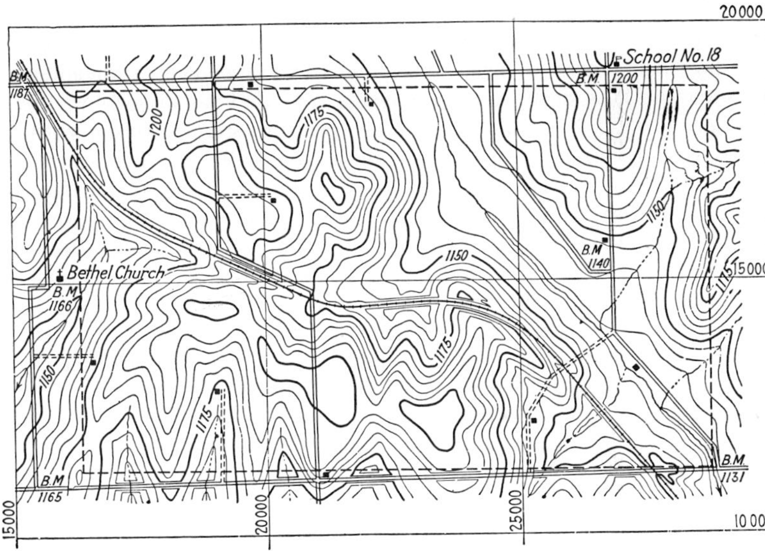

FIG. 25-7. Portion of topographic map.

transit, and as the salient points are located by the transit, they are plotted on the drawing, usually to a smaller scale than that of the map. The more complex and detailed topographic features are then sketched while the terrain is in view.

Procedure. For typical conditions on hilly ground as illustrated in the examples of Arts. 25-5 to 25-7, and for a contour interval of 5 ft. and a map scale of 1 in. = 500 ft., the procedure of locating details by the transit-stadia method will now be described. Figure 25-7 shows a portion of the finished map, being an enlarged view of sheet 5 of Figs. 25-1 and 25-3.

The transit is set up at a station on either the primary or the secondary control as at station B (Fig. 25-2). The instrumentman orients the transit by sighting on an adjacent station and locates the details in the vicinity of the station by angle and distance measurements. The elevation of station B, it is assumed, was determined by the level party which ran the secondary levels; therefore, the elevation of adjacent points can be determined by the methods explained in Art. 15-16.

The recorder keeps the record of all values given him by the instrumentman, and describes all points by remarks or sketches such that the draftsman can interpret all data correctly and draw all features properly on the map. Notes may be kept in the form shown by Fig. 15-9, supplemented by sketches if necessary.

The rodmen choose ground points along valley and ridge lines and at summits, depressions, important changes in slope, and definite details. The selection of points is important, and the rodmen should be instructed and trained for their work. They should follow a systematic arrangement of routes such that the entire area is covered and that no important objects are overlooked. They should observe the terrain carefully (often with the aid of a hand level) and report any important features which cannot be seen from the transit station. The recorder indicates by some symbol (usually the rodman's initial) whose rod is being sighted. The work may be expedited or may be conducted with one rodman less through the use of a tape recorder. In another method of identification of "side shots," as observations on detail points are called, each man in the party carries a watch, all watches being set to keep time together within a few seconds. As the instrumentman motions the rodman to a new point, both the rodman and the recorder record the time, thus making it possible to identify any reading made on either rod.

25-12b. Plane Table. The personnel of the plane-table party for mapping details by the controlling-point method usually consists of the plane-table man, computer, and one or more rodmen. The equipment usually includes a plane table, telescopic alidade (preferably with a control level on the vernier arm), scale, small triangles, 6H or 8H pencil, and stadia slide rule or stadia tables.

For the conditions indicated in Arts. 25-6 and 25-7 and Fig. 25-7, this method is as follows: Before the party goes into the field, the horizontal control, the coordinate system, and the outline of the map sheet are adjusted and plotted on the plane-table sheet as shown in Fig. 25-2. The elevations of all bench marks either are recorded on the sheet or are in the hands of the computer.

The instrumentman sets up the table at a convenient station, as at B (Fig. 25-2), and orients it usually by backsighting on an adjacent station. He then directs the rodmen to the controlling points of the terrain, as just described for the transit. When a rodman holds the rod on a ground point, the instrumentman sights on the rod, reads the stadia intercept, draws a short portion of the ray near the end of the alidade farthest from the station point, sets the cross-hair preferably on the H.I. point, and motions the rodman forward. He next centers the control bubble on the vernier arm and reads the vertical angle. He then plots the point by scaling the horizontal distance (corrected for slope, if necessary, by the computer). The computer has now calculated the elevation of the point, and the instrumentman records it on the map near the plotted point. As rapidly as sufficient data are secured, the instrumentman sketches the contour lines. Other objects of the terrain are located and are drawn either in finished form or with sufficient detail so that they may be completed in the office. A more detailed account of the procedure of taking side shots, with an alidade equipped with stadia arc, is given in Art. 17-17.

The utmost skill of the topographer is used in judging the features of the terrain and in representing them on the map with the required precision and with the least expenditure of time.

There is no need to identify the rod readings, or to use special precautions to cover the ground, as is necessary with the transit method; since all plotting is done in the field, mistakes or omissions are at once apparent, and any information brought in by the rodman can readily be incorporated in the map.

Many objects are located by the method of intersection, the elevations being determined by trigonometric leveling.

Because the plane table permits a ready solution of the three-point and two-point problems, use is made of these methods to enable the topographer to utilize advantageous instrument stations which have not been included in the control surveys, especially where the control has been established by triangulation. The elevation of such stations is determined either by stadia leveling or by trigonometric leveling.

25-12c. Transit and Plane Table. For large-scale maps and where many details are to be sighted, sometimes it is advantageous to use both the transit and the plane table. This method saves time in the field, but it may not reduce the total cost, as a larger party is required than for the plane table alone.

The transit is set up and oriented at the control station, the location of which is plotted on the plane-table sheet. The plane table is set up and oriented nearby, and its location is plotted on the map in its correct relation to the transit station. (In some cases the map distance between the transit station and the plane-table station is negligible, and the two points are regarded as identical.) When a rodman has selected a ground point, the transitman observes the stadia distance and vertical angle to it; the plane-table man sights in the direction of the point, draws a ray toward it from the plotted location of the plane-table station, plots the point at the correct distance scaled from the plotted location of the transit station, and records on the map the elevation (computed by the transitman or the computer) of the plotted point.

25-13. Details by Cross-profile Method. In the cross-profile method of locating details, the ground points are on relatively short lines transverse to the main traverse. The required lengths of sight are not great, and the hand level or the clinometer is commonly employed. For the common conditions of a 5-ft. contour interval and a map scale of 1 in. = 400 ft., the maximum lengths of hand-level or clinometer sights should be limited to about 100 ft. For smaller scales or larger intervals, longer sights up to perhaps 300 to 500 ft. may be used. If, on the other hand, the lengths of sights are limited to 50 ft. or less, the errors in elevation may be kept below a few tenths of a foot in a distance of 500 ft., and the hand level or the clinometer may be employed for surveys having a contour interval of 1 or 2 ft.

The cross-profile method is primarily suitable for route surveys. It is also sometimes used for area surveys if the ground cover is dense, because hand-level or clinometer sights can be taken through very small openings in the underbrush; the area is surveyed by means of a series of overlapping strips.

The procedure described herein applies primarily to intermediate-scale surveys of rolling or hilly country. For large-scale surveys or for flat country, the method is similar except that elevations are determined by means of the engineer's level or by the stadia method. For relatively small-scale surveys, the distances may be determined by pacing.

The party consists of a topographer and usually two men, herein called "chainmen," who act either as chainmen or as rodmen. Sometimes only one chainman is employed, and the topographer assists in chaining. The equipment consists of a leveling rod (usually a topographer's rod, Fig. 8-14), a steel or metallic tape, a hand level or a clinometer, and either a cross-ruled wide-page sketchbook or sketch sheets mounted on a clipboard or small drawing board. Sometimes a Jacob's staff or other rod about 5 ft. long is used as a support for the hand level or clinometer while sights are being taken.

The control points are the 100-ft. stations of the transit traverse. These points have been marked on the ground by stakes, and their elevations have been determined by profile leveling and have been furnished to the topography party. The ground points are either contour points or more commonly points of change in slope; in the latter case the intermediate contour points are located by interpolation.

25-13a. Contour Points with Hand Level. The party proceeds from station to station along the traverse. At each station the topographer notifies the chainmen of the elevation of the station. The head chainman carrying the rod moves out on a line estimated to be at right angles with the traverse line until the rod is on the next contour (either higher or lower) from the station, as determined by the rear chainman employing the hand level; the distance out to the contour is then measured with the tape. The rear chainman then goes out to the point occupied by the rod, and the head chainman again moves out until the next contour is reached; and so the process is repeated until all contour points are located out to the edge of the strip being surveyed. A similar procedure is followed on the other side of the traverse line. Usually the trends or directions of the contours are sketched at each crossline and along ridge and valley lines, but on the field sheets the contour lines are not sketched for their full length. Definite details are located with relation to the transit line by tape measurements. If the topography is regular, sometimes the sketches are omitted, and the distances from traverse to contour points are recorded numerically.

Example: The method is illustrated by reference to Figs. 25-8 and 25-9, which represent the field notes and the finished map, respectively, of a preliminary route survey. The contour interval is 5 ft. Suppose that the topography party has

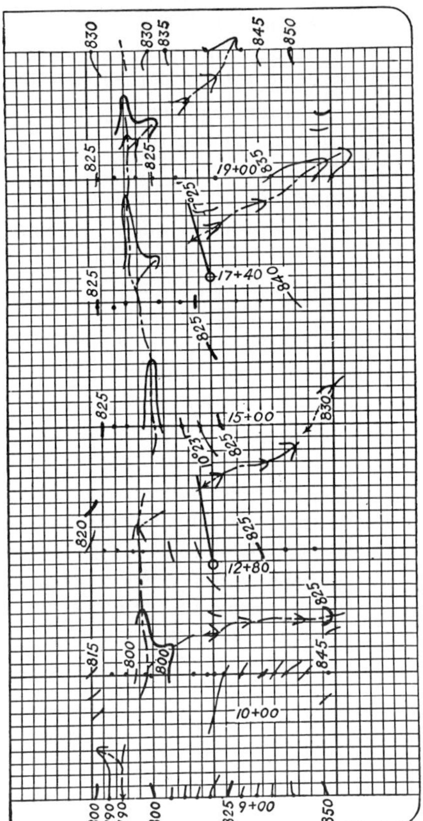

FIG. 25-8. Notes for cross-profile method.

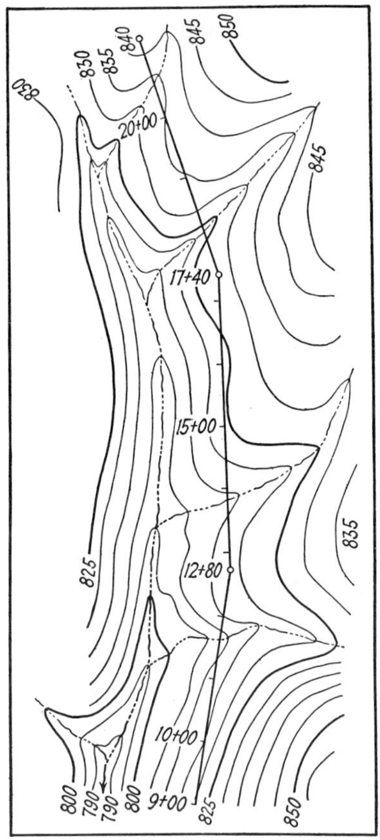

FIG. 25-9. Preliminary map for route survey.

reached station 9 + 00, where the topographer notifies the chainmen that the elevation of that station is 821.1 ft. To the left of the traverse line the ground slopes downward. To locate the 820-ft. contour, the head chainman carrying the rod (which is graduated from the bottom) and the zero end of the tape moves out to the left until the rear chainman by the use of the hand level supported, say, on a 5.0-ft. staff, reads 6.1 on the rod (821.1 + 5.0 − 6.1 = 820.0). The horizontal distance out from the station is read on the tape and called to the topographer who plots its location (9 ft. from the traverse line). To locate the 815-ft. contour, the rear chainman moves out to the 820-ft. contour, and the head chainman moves out until the rear chainman reads 10.0 on the rod (820.0 + 5.0 − 10.0 = 815.0); the horizontal distance between the two contours (26 ft.) is read from the tape, and the second point is plotted; and so on. A similar procedure is followed in going uphill, except that the head chainman carries the hand level and the rear chainman the rod.

Where the traverse follows a valley or a ridge, the width of the strip may be narrow because the range of any feasible location of the route is thereby restricted. Where the topography is comparatively flat, observations may be taken over a width of 500 ft. or more on either side of the traverse line. The width of a strip which is being surveyed with the engineer's level may be extended by supplementary measurements with the hand level.

25-13b. Points of Change in Slope by Clinometer. For relatively small-scale maps, sometimes the clinometer is employed to determine the elevations of controlling points of change in slope on the crosslines by rough indirect leveling, and the distances are determined by pacing. The method is considerably faster than that just described for the hand level and tape, but is of lower precision.

The topographer stations himself at a chaining station on the traverse line. The rodman moves out along a crossline, pacing the distance from the center line as he goes. When he reaches an important change in slope, he halts and presents his rod. The topographer sights at a point on the rod at the same height above the ground as his eye, and reads the angle of inclination. The rodman calls out the distance, and the topographer either records the angle and distance for later computation of the difference in elevation, or by means of a table of values he immediately computes the difference in elevation between the traverse station and the point indicated by the rod. By adding (or subtracting) this difference to the elevation of the center stake, the elevation of the point sighted is determined. The location of the point is then plotted, and the elevation (if computed at this time) is recorded in the sketchbook. The topographer proceeds to the point thus located, and the rodman moves forward to the next important change in slope. This process is repeated until the limit of the strip for which the topography is being taken is reached. A similar procedure is followed on the other side of the traverse line.

25-13c. Rhodes Reducing Arc. The Rhodes reducing arc is a simple instrument for locating details by the cross-profile method; by its use measured slope distances are readily reduced to horizontal distances and differences in elevation. It consists of a sighting tube (Fig. 25-10) mounted along the edge of a semicircular plate, or "arc," which in turn is mounted on a staff. The graphical scales on the arc are so arranged that when the measured slope distance (shown in the figure as, say, 50 ft.) is set off on the vertical scale the corresponding horizontal distance (as 40 ft.) is read from the scale which is perpendicular to the sighting tube, and the corresponding difference in

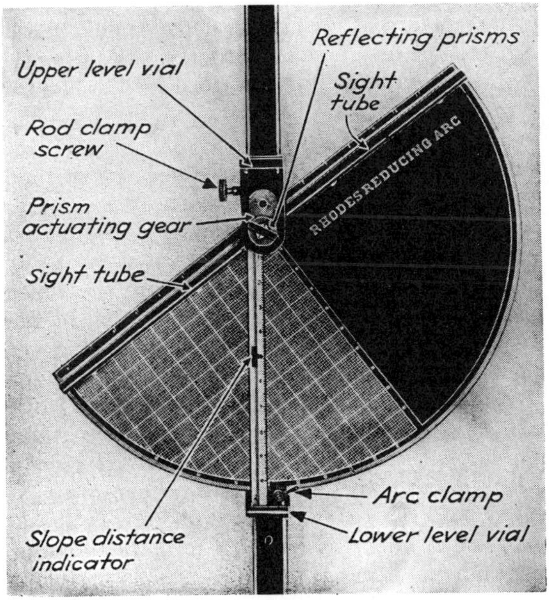

FIG. 25-10. Rhodes reducing arc.

elevation (as 30 ft.) is read from the scale which is parallel to the sighting tube.

The observer holds the instrument on a point of known location and elevation, plumbs the staff by means of an attached rod level, and sights on the target of the rod which is held on a selected ground point. Usually the target is set at a rod reading equal to the height of the center of the sighting tube above the ground. The slope distance is determined by taping, and the corresponding horizontal and vertical distances are read from the arc.

25-14. Details by Checkerboard Method. The checkerboard method of locating details is well adapted for large-scale surveys, as the points are located in plan by tape measurements. It is also useful where the tract is wooded, where the topography is smooth, and on urban surveys where blocks and lots are rectangular. The tract is staked off into squares or rectangles—usually 50-ft. or 100-ft. squares. The ground points and other details are then located with reference to the stakes and connecting lines.

The usual procedure is first to run a rectangular transit-tape or compass-tape traverse (Fig. 25-11) near the perimeter of the tract, with stakes set at each 100-ft. station. The error of closure becomes apparent in the field; if this is greater than the permissible error, the traverse hubs and the stakes are reset in such manner as to reduce the error within allowable limits. A line of profile levels is run around this traverse to close within a permissible error, thus establishing the elevation of the ground at each stake and hub, just in front of the numbered side of each stake. The elevations thus determined should be correct to the nearest 0.1 ft.

The interior stakes are set by transit-tape or compass-tape lines beginning at a stake on one side and closing on the corresponding stake on the opposite side, as from *F*–18 to *B*–1, *F*–17 to *B*–2, etc. Each stake is marked usually with a letter and a number indicating its position with respect to the coordinate axes. The elevation of the ground at each of these interior stakes is then determined with the engineer's level or with the hand level, depending upon the lengths of the lines and the accuracy required. Sketch sheets are prepared, on which are shown the elevations of the corners of the squares.

The location of ground irregularities which cannot be properly interpreted from observations on the corners of the squares is now observed by means of a hand level and tape, and the features are sketched. Other details such as fences, roads, and buildings are located by measurements either from adjacent coordinate points or from the sides of the squares. The map is constructed in the office.

Interior Corners by Tape Alone. Where lines of vision are obstructed by woods, vegetation, or buildings, the interior coordinate points may be located by tape measurements alone. Thus, beginning at station *O* (Fig. 25-11) three chainmen, *A*, *B*, and *C* with two 100-ft. tapes proceed as follows: Chainman *A* holds the zero end of his tape at the point X_0Y_1; chainman *B* holds the zero end of his tape at the point X_1Y_0; chainman *C* then stretches the two tapes, bringing the 100-ft. ends of the tapes to meet at the point X_1Y_1, and a stake is set. Next they proceed in a similar manner to set point X_1Y_2, etc. In this way all interior points can be established without the aid of transit sights.

If, however, the area is somewhat extended and if it is desired to gain the advantage of the hand level in penetrating dense undergrowth, the accuracy required does not permit the errors which would accumulate over long distances, and it is necessary to establish lines of auxiliary or secondary control at 500-ft. intervals within the limits of the area to be surveyed. Such control lines are X_0Y_5–$X'_{10}Y_5$ and X'_5Y_0–X'_5Y_{10}, for example.

This method, as applied to a rough wooded area, is described in detail in Ref. 4 at the end of this chapter.

Details by Plane Table. If many irregular features are to be mapped, the plane table may be used advantageously. Before the plane table is taken into the field, the corners of the squares are established on the ground with transit and tape, their elevations are determined by direct leveling, and the plane-table sheet is prepared showing the elevations of the corners of the squares, all as described earlier in this article. The plane table is then set up over the corner of a square and is oriented by backsighting along one of the control lines marked by stakes. Directions to details inside the squares are determined usually with a peep-sight alidade, and distances to these

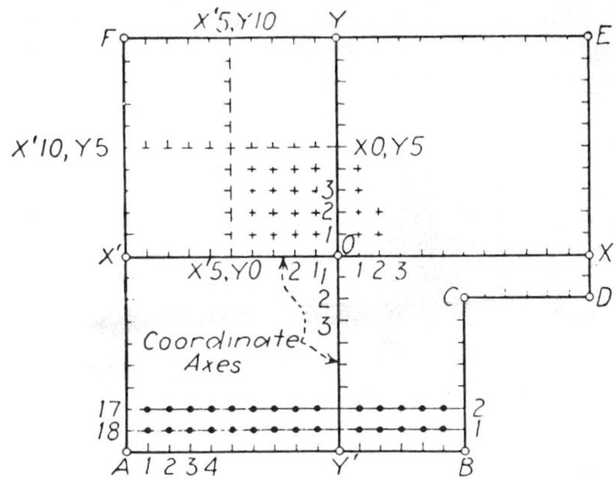

FIG. 25-11. Checkerboard system.

details are determined usually by chaining either from the instrument station or from a convenient corner or line. Only as many stations are occupied by the plane table as are necessary to cover the area.

25-15. Details by Trace-contour Method. The trace-contour method of locating contour points on the ground is commonly used on large-scale surveys, and sometimes on intermediate-scale surveys where the ground is irregular. Under these conditions, if visibility is good, the trace-contour method is more rapid and more accurate than the checkerboard method.

Although the transit may be used in this work, either alone or with the engineer's level, the plane table is commonly used because it requires fewer points to be observed and because of the saving of time in plotting.

Often the plane table and the engineer's level are used together, because the level permits greater lengths of sight and because it can be readily moved to permit direct rod readings to be taken. The party consists of a topographer at the plane table, a levelman, one or more rodmen, axemen as needed, and sometimes a computer. The levelman sets up the level at a convenient location and directs the rodman up or down the slope until a point on a given contour is located. This point is immediately sighted by the plane-table man and is plotted on the plane-table sheet. The rodman then moves to another contour point which may be either along the same contour or, on hilly ground, on the next higher or lower contour. The distances from plane-table station to contour points are measured by stadia; if the scale is large, definite objects may be located by taped distances.

25-16. Field Problems.

Elementary field problems in topographic surveying are given at the end of the chapters on stadia surveying and the plane table. In particular, field problem 3 of Chap. 15 and field problem 3 of Chap. 17 provide exercises in the controlling-point method of locating details. Field problem 1 of Chap. 26 on route location provides an exercise in the cross-profile method.

PROBLEM 1. TOPOGRAPHIC SURVEY BY CHECKERBOARD METHOD

Object. To obtain sufficient data for an accurate topographic map of large scale and small contour interval. The area to be mapped is small and possesses few details, and the topography is smooth. The data may be used in office problem 1 of Chap. 24.

Procedure. (1) With transit and tape divide the tract into 100-ft. squares, setting stakes at the corners. Letter and number each stake to conform to a coordinate system. (2) With the engineer's level run levels over the area, taking rod readings at summits, depressions, corner stakes, and points of change in slope along the sides of the squares. (3) Locate the details (either definite details or ground points) inside the squares by taking offsets, ties, etc. (Arts. 14-18 and 14-19). (4) Keep notes in a form similar to that of Fig. 10-3; identify each point by its coordinates (a letter and a number).

PROBLEM 2. TOPOGRAPHIC SURVEY BY TRACE-CONTOUR METHOD USING PLANE TABLE AND ENGINEER'S LEVEL

Object. To obtain sufficient data for an accurate topographic map of intermediate or large scale and small contour interval. The area to be mapped is small, and the topography is irregular. It is assumed that control points have been established at advantageous locations within the tract and that the control has been plotted (office problem 1 of Chap. 24) on the plane-table sheet.

Procedure. (1) The plane-table man sets up and orients the table at one or more control stations such that the entire area may be mapped. (2) The levelman sets up the engineer's level, takes a backsight on a bench mark, and computes a rod reading such that the foot of the rod will be at the elevation of a given contour. (3) The rodman moves about as directed by the levelman, locating critical contour points. When a contour point has been located, the plane-table man sights on the rod, determines the distance by stadia, and plots the contour point. (4) Definite details are located either by radiation or by intersection; distances to such details are determined either by stadia or by tape measurements, depending upon the scale of the map.

REFERENCES

1. American Society of Civil Engineers, "Technical Procedure for City Surveys," *Manual of Engineering Practice* 10, 1963.
2. American Society of Civil Engineers, "Horizontal Control Surveys to Supplement the Fundamental Net," *Manual of Engineering Practice* 20, 1940.
3. Birdseye, C. H., "Topographic Instructions of the United States Geological Survey," *U.S. Geological Survey, Bull.* 788, Government Printing Office, Washington, D.C., 1928.
4. Davis, R. E., and W. H. Rayner, "Unusual Methods Simplify Topographic Survey of Rough, Wooded Area," *Engrg. Record,* Vol. 73, No. 10, pp. 316–317, March 4, 1916.
5. Denault, M. F., "Plane Table versus Transit for Topographic Mapping," *Engrg. News-Record,* Vol. 132, No. 17, pp. 106–109, May 4, 1944, reprinted in *Surveying and Mapping,* Vol. 4, No. 3, pp. 7–12, July, 1944.
6. Fink, Howard L., "Survey Control at Glen Canyon Dam," *J. Surveying Mapping Div., Am. Soc. Civil Engr.,* Vol. 88, No. SU 1, pp. 51–61, November, 1962.
7. Gessel, Clyde D., and Dwight H. Rutledge, "Large-scale Mapping of Lake Powell," *J. Surveying Mapping Div., Am. Soc. Civil Engr.,* Vol. 88, No. SU 1, pp. 17–27, November, 1962.
8. Hodgson, C. V., "Manual of Second- and Third-order Triangulation and Traverse," *U.S. Coast and Geodetic Survey, Spec. Pub.* 145, Government Printing Office, Washington, D.C., 1935.
9. Swainson, O. W., "Topographic Manual," *U.S. Coast and Geodetic Survey, Spec. Pub.* 144, Government Printing Office, Washington, D.C., 1928.
10. Swanson, L. W., "Topographic Manual," Part II, "Photogrammetry," *U.S. Coast and Geodetic Survey, Spec. Pub.* 249, Government Printing Office, Washington, D.C., 1949.
11. U.S. Coast and Geodetic Survey, "Horizontal Control Data," *Spec. Pub.* 227, Government Printing Office, Washington, D.C., 1941.
12. U.S. Coast and Geodetic Survey, "Control Leveling," *Spec. Pub.* 226, Government Printing Office, Washington, D.C., 1961.
13. U.S. Government (Federal Board of Surveys and Maps), "Classification and Standards of Accuracy of Geodetic Control Surveys," *Surveying and Mapping,* Vol. 19, No. 2, pp. 219–224, June, 1959.

CHAPTER 26

HIGHWAY ROUTE SURVEYING

by
L. L. Funk
Formerly Advance Planning Engineer,
California Division of Highways

26-1. General. Surveys of some type are required for practically all highway construction work. For small projects involving widening or minor improvement of an existing highway, the survey may be relatively simple and may include only the obtaining of sufficient information for the design engineer to prepare plans and specifications defining the work to be done. For the more complex construction projects involving multilane highways on new location, the survey may require a myriad of details including data from specialists in related fields to determine the best location; to prepare plans, specifications, and estimates for construction; and to prepare deed descriptions and maps for appraisal and acquisition of the necessary rights of way.

In general it is the function of the survey or location engineer to gather all survey data which may be needed to make the necessary decisions concerning the particular project. This process will include obtaining the necessary information regarding terrain and land use, making surveys to determine detailed topography and the location of planimetric features, and finally the staking of the center line of the proposed highway on the ground. The survey engineer may also be given the responsibility for obtaining and utilizing data gathered by specialists in related fields such as traffic, right of way, geology, soils and foundations, and drainage.

26-2. Termini. Prior to surveys for a particular project, the beginning and end points have usually been defined in general terms in project descriptions for preliminary authorizations, planning, and so on. Termini thus defined prior to the survey should have sufficient latitude to fit varying conditions which may develop as a result of more detailed studies. This latitude is particularly important where the proposed project is on new alinement or where conversion of an existing conventional highway to a limited-access facility is planned.

During the course of the survey, the general route should be examined well beyond the tentative termini to make certain that there are no serious obstacles to future extension of the type of improvement which is proposed. After this examination has been made, definite termini can be established with assurance that the proposed project will fit any future plans for improvement of adjacent sections of the highway route.

26-3. Standards. *Standards* for highway projects relate to minimum radius of curvature, maximum grade, and width of roadway. They are closely associated with *classification,* which includes the design speed, type of traffic, and anticipated peak-hour traffic volumes for the design year, usually 20 years in the future. Design standards and classification are discussed in detail in policy statements of the American Association of State Highway Officials (Ref. 2 at the end of this chapter).

While general standards for a particular project may be predetermined by the system or class of the highway, such as interstate, primary, or secondary, the specific standards must frequently be decided during the course of the survey. They may well be dependent upon results of the accompanying traffic study and detailed examination of the terrain and upon type of land use, right-of-way costs, and special conditions which may be encountered.

Standards for the actual survey itself relate to errors of closure for traverse, triangulation, and leveling and are usually expressed in terms of the required order of accuracy. First-, second-, and third-order horizontal control accuracies have been defined by the Federal Bureau of the Budget as requiring closures of one part in 25,000, 10,000, and 5,000, respectively.

The standards to be used for a particular survey may be set by administrative policy of the agency involved. Survey standards will frequently vary widely according to the type of highway project. For example, a survey for minor improvement of an existing highway with no right-of-way acquisition involved could well have a much lower accuracy requirement than one for a project involving new location and the acquisition of valuable property.

In order to make full use of the advantages afforded by state plane-coordinate systems (Art. 26-7), some agencies require second-order accuracy for surveys for major highways on new alinement even though such accuracy may not be absolutely necessary for actual right-of-way or construction purposes. On the other hand, surveys for major bridge structures frequently require accuracies of first order or better for construction purposes alone (Ref. 1 at the end of this chapter).

26-4. Economics. Before construction of any highway project is undertaken, the necessary expenditures should be justified in terms of the improved traffic service which will be provided. For major projects, frequently justification in the form of an economic analysis is required. Such an analysis should always be prepared where studies of alternative route locations are involved in order that proper decisions as to routing can be made.

Although different types of economic analysis are used by various highway agencies, they all involve some type of monetary comparison between the traffic benefits which will be provided and the cost of the project. For example, an analysis involving comparison of alternative routes might include the value of time and distance required for the estimated traffic volumes for each alternative over a period of years as compared with the cost of rights of way, construction, and maintenance. Obviously the route having the lowest cost, taking all the above factors into account, would be best from an engineering viewpoint. In some cases other values such as social benefits to the community at large may outweigh the differences developed by the economic analysis.

As to the economics of the survey itself, the survey should be planned and carried out in such a manner that all necessary information will be obtained at the lowest possible cost. With the many types of highway projects requiring different amounts and types of information and with the many kinds of equipment and methods of making surveys which are now available, the planning and execution of highway surveys require a high degree of skill and experience. Many thousands of dollars may be saved by use of the most efficient methods.

26-5. Stages of Highway Surveys. Development of the interstate highway system and more general acceptance of the limited-access principle for major highways have resulted in more and more highway projects being constructed on new alinement, with existing facilities frequently being left to serve local traffic. Surveys for highway projects where new location is being considered start with a general study of the entire area between termini, proceed to more specific studies of possible alternative routes, and finally conclude with a detailed survey of the selected route and the staking of the final center line on the ground. These procedures are generally carried out in three stages: reconnaissance, preliminary survey, and location survey. As used in succeeding discussions in this chapter, these stages are briefly as follows;

1. *Reconnaissance* includes a general study of the entire area, the development of one or more alternative routes or corridors, and the study of each of these corridors in sufficient detail to enable the proper officials to decide which will provide the optimum location.

2. The *preliminary survey* is a survey of the selected corridor in sufficient detail to permit staking of the final center line on the ground. In some cases the preliminary survey may be complete enough that the final center line can be computed and staked in the field without variation. In other instances, minor adjustments may be required during the location survey.

3. The *location survey* consists in staking the final line and obtaining all additional information necessary to enable the design engineer to prepare completed plans, specifications, and estimates, including accurate estimates of earthwork quantities, and to prepare deed descriptions covering the rights of way to be acquired.

26-6. Aerial Photography and Topographic Mapping. Recent developments in aerial photography and in photogrammetric methods for making large-scale topographic maps have provided new techniques which can be used to advantage in almost all highway surveys. Aerial photography and mapping by photogrammetric methods are described in general in Chap. 31. Some highway departments carry out both aerial photography and mapping with their own organizations. Many others, however, contract for one or both operations with firms specializing in such work. In many areas, small-scale aerial photographs are available from government agencies.

To take full advantage of these new techniques, the survey engineer must be familiar with the products available from aerial photography and their uses in highway route surveying. For purposes of discussion it is convenient to classify these products into (1) aerial photographs, (2) small-scale and intermediate-scale maps for reconnaissance and preliminary studies, and (3) large-scale maps for final design (Ref. 3 at the end of this chapter).

1. Aerial photographs can be in the form of contact prints, enlargements, or mosaics. Contact prints are direct prints from the aerial negatives and are usually 9 by 9 in. Where the photographs are to be used for mapping purposes, a precision mapping camera having a 6-in. focal-length lens is usually employed. Aerial photography for other purposes may be taken with cameras having 8¼-, 12-, or 24-in. focal-length lenses.

A convenient method for calculating the average scale of aerial negatives or contact prints is to divide the flying height in feet by the focal length of the camera lens in inches, with the resulting scale in feet per inch. For example, a photograph taken from a height of 1,500 ft. with a 6-in. focal-length lens would have an average scale of 1 in. = 250 ft. The word "average" is used because it is important to remember that an aerial photograph is not a map and that it is subject to variations in scale due to tilt and to displacement caused by changes in relief. The relief displacement is the most important cause of scale variations on an individual aerial photograph; it is at a minimum near the center.

Satisfactory enlargements up to five diameters or more can be made from aerial negatives. Scale discrepancies in contact prints due to variations from the planned flight height can be corrected during the enlarging process by making at least one field measurement between points visible on each photograph.

Mosaics are assemblies of matched photographs usually covering an extensive area. Because of the extra copying process involved, the scale is usually limited to approximately three times as large as that of the aerial negatives. Types of mosaics and methods of assembling the photographs are described in Art. 31-19.

Scales of contact prints used in highway surveys and design range from as small as 1 in. = 2,000 ft. to as large as 1 in. = 200 ft. The smaller scales are generally used in preliminary studies, while the larger scales are more useful in developing details in the later stages of surveys and design, particularly in areas of intensive land use.

In addition to their use by the survey engineer, contact prints, enlargements, and mosaics are of great value in other phases of the location studies such as right-of-way estimates, drainage studies, geological studies, and foundation and materials site investigations as well as in all phases of design.

2. Small-scale and intermediate-scale maps for reconnaissance and preliminary studies are an invaluable aid to the survey engineer when existing maps are not available and where topographic conditions are such that route comparisons cannot be made by aerial photographs alone. The largest scale likely to be used at this stage is 1 in. = 200 ft., with a contour interval of 5 or 10 ft.

The ground control required for this type of mapping will depend upon the photogrammetric equipment available and the organization performing the work. Aerotriangulation or bridging techniques using the Balplex plotter, stereoplanigraph, or Wild autograph (Art. 31-48) may in some cases materially reduce ground-control requirements.

3. Large-scale mapping for final design will generally prove economical for any major highway project where the ground is not obscured by dense vegetation. The usual scale for such mapping is 1 in. = 50 ft. with 2-ft. contours, although scales of 1 in. = 40 ft. with 1- or 2-ft. contours and 1 in. = 100 ft. with 2- or 5-ft. contours are occasionally used.

By far the greatest volume of this type of mapping is compiled on Kelsh-type plotters (Art. 31-48). Most of these instruments have an enlargement factor of five times the photograph scale. Thus for maps at a scale of 1 in. = 50 ft. the scale of the aerial negatives should be 1 in. = 250 ft. While photogrammetric bridging techniques may be successfully employed under favorable conditions, particularly for horizontal control, conservative practice for such mapping at this time requires at least two and preferably three horizontal control points and five vertical control points set by field methods for each stereoscopic model. Field methods for establishing such control are discussed in Art. 26-14.

Most specifications for large-scale mapping by photogrammetric methods conform to National Map Accuracy Standards, which require that 90 per cent of the elevations as indicated by contours be correct within one-half contour interval. However, in view of the requirements for statistical accuracy in the computation of earthwork quantities (Art. 26-22), some organizations use specifications which limit the arithmetic mean of errors in the contours.

26-7. Basic Survey Control. Wherever economically feasible, surveys for major highway projects should be tied to national horizontal and vertical control networks of equal or higher order of accuracy (Ref. 1 at the end of this chapter). Such networks are established by the U.S. Coast and Geodetic

Survey. This organization publishes diagrams and lists of established bench marks based on the 1929 mean sea-level datum as well as diagrams and data sheets for horizontal control stations established to first-order or second-order accuracy. Data sheets show both geographic and plane-coordinate positions. The latter are based on the state coordinate systems (Art. 16-30).

Beginning and ending a survey on points whose location has been previously established have the obvious advantage of providing a built-in check which eliminates the need for a check traverse. However, the spacing of established first-order and second-order stations is so wide that, in most areas, the expense of triangulation or traverse measurement to connect to them has been a major deterrent to use of the state coordinate systems.

This obstacle has been minimized in the past few years by development of electronic distance-measuring instruments (Art. 7-6) and more widespread use of lightweight directional theodolites. Use of such equipment will, in many cases, make it economically feasible to connect highway surveys to the state coordinate system and capitalize on the many advantages thus afforded.

Details regarding use of the state coordinate systems are contained in *Special Publication* 235 of the U.S. Coast and Geodetic Survey (Ref. 8 at end of Chap. 16). In most cases the only significant differences between the use of such systems and of a local or arbitrary plane-coordinate base are that astronomical observations for azimuth must be corrected by reduction to grid azimuth and that ground-level distances must be converted to grid distances. These computations can readily be made by use of Plane Coordinate Projection Tables which are published by the U.S. Coast and Geodetic Survey for each state.

Conversion from ground-level to grid distances consists of two steps. The first, reduction from ground level to sea level, involves use of a factor which is directly proportional to the elevation. For example, the correction to be subtracted from ground-level distances at 500-ft. elevation is 0.0239 ft. per 1,000 ft. of length, while at 5,000-ft. elevation it is 0.2391 ft.; the respective factors are 0.9999761 and 0.9997609. The second step, reduction of the sea-level or geodetic distance to grid distance, is accomplished by use of a scale factor which is obtained from the projection tables previously mentioned. In practice, these steps can be combined by multiplying the two factors for direct computation of the grid distance. (The multiplication can be accomplished by adding the two factors and subtracting 1 from the result.) This combined factor is called the *grid factor*. For many surveys a mean grid factor can be used over a fairly long distance.

When engineering plans are prepared on the basis of the state coordinate system, it is customary to put a note on each sheet to the effect that distances shown are grid distances for Zone __ of the state coordinate system and to give the factor for conversion to ground distances. Thus if the factor were

1.000103, the distance to be measured on the ground would be 100.01 ft. to reproduce each 100 ft. of length shown on the plans.

RECONNAISSANCE

26-8. Purpose and Scope. On highway projects where the possibility of new location is being considered, reconnaissance consists in an extensive study of the entire area between termini in order that all feasible routes may be examined. As reconnaissance should be carried to the stage where the best location can be selected, no possible route that is overlooked at this time is likely to be found during later work. If the best route is not found during reconnaissance, nothing which can be done in the subsequent stages will compensate for the loss of a better location. The importance of studying the entire area for all possible routes which may be advantageous cannot be emphasized too strongly.

The amount of information necessary and the depth to which reconnaissance studies should be carried will depend entirely upon conditions surrounding the particular project. In some cases a rough map or small-scale aerial photographs may suffice for selection of the best route. In other instances detailed studies of many different factors will be required, and all the skill and experience of the survey engineer and the combined efforts of an entire team of specialists will be necessary to arrive at the proper solution.

26-9. Reconnaissance Information. The three most important factors which affect highway location are *traffic service, land use*, and *terrain*. The relative importance of these factors will vary according to the project to the extent that in some cases one or two of the items may be of paramount importance while in other instances the three may have equal weight. One of the first steps in the reconnaissance of any project is to evaluate rapidly the probable relative importance of these factors so that the necessary information concerning them can be developed. In any case, sufficient information must be gathered for each so that the proper decisions can be made.

1. *Traffic Service.* The benefits which will be provided by each of the various alternatives developed during reconnaissance are usually evaluated by specialists in the traffic engineering field. In some cases, where the same volume of traffic will be served by all the possible locations, a brief study to determine present and probable future volumes and the percentage of trucks may provide sufficient information.

In other more complex cases, a complete origin and destination survey may be required. After the study data have been accumulated and processed, it will be necessary for the survey engineer to provide maps showing the various alternatives to be evaluated by the traffic specialists. Tentative interchange locations or points of access must be shown so that accurate estimates of future traffic movements can be made.

2. *Land Use.* Information concerning land use may be developed from aerial photographs, maps, field inspections, and any other available sources. Aerial photographs are by far the most valuable aid in the study of land use and the development of possible locations.

If new aerial photography is to be obtained, the scale selected should depend on the width of strip to be covered and the intensity of land use. A brief field inspection and a review of existing photographs or maps will aid in selecting the scale. In areas of relatively intensive land use, scales of 1 in. = 500 ft. to 1 in. = 1000 ft. will usually be satisfactory at this stage. Scales of 1 in. = 1000 ft. to 1 in. = 2000 ft. might be used in predominantly rural areas where a wide strip is desired.

Information concerning the availability of existing maps and aerial photography from Federal agencies can be obtained from the Map Information Office of the U.S. Geological Survey, Washington, D.C.

Indications of cross-country underground pipelines in rural areas are usually visible on aerial photographs. If there is a possibility of conflict, more detailed maps should be obtained from the utility companies involved. Any existing master plans should be obtained from agencies of local government for investigation of possible conflict with proposed schools, parks, and other developments. Contacts with local, state, and Federal agencies should be made in order that all possible information can be obtained regarding plans for future reservoirs, canals, airports, and other projects which might influence the highway location.

3. *Terrain.* In many cases, aerial photographs together with field inspection will provide sufficient information concerning the terrain. Stereoscopic study of the photographs will immediately disclose controls such as low points in ridges, potential stream-crossing sites, and areas of rugged terrain which are to be avoided if possible. Approximate differences in elevation for the development of rough profiles can be determined from aerial photographs with an engineer's scale or a parallax bar used in conjunction with a stereoscope. Profiles can also be developed by using aerial photographs as a base map for locating points whose elevation can be determined with an aneroid barometer or other field methods. U.S. Geological Survey quadrangle sheets, particularly the 7½-min. series (scale 1 in. = 2,000 ft.), are also an excellent source of terrain information and in many cases will provide sufficient data for comparative profiles.

In rugged terrain where earthwork quantities may be a major factor in the comparison of alternative routes, contour maps may be necessary in order to obtain more detailed quantitative data. If the areas involved are small, field methods such as the use of the plane table or transit and stadia may be economical. However, where an extensive length along one or more routes is involved, photogrammetric methods for obtaining the necessary mapping will usually be more economical as well as permit a much greater width of mapping strip without excessive costs. While a scale of 1 in. = 400 ft. will probably be satisfactory in most cases, the possibility of obtaining mapping at a scale of 1 in. = 200 ft. for subsequent use in the preliminary-survey method described in Art. 26-15 should not be overlooked.

Additional terrain information from geological studies, foundation investigations, and drainage studies may be required in some cases. Examination of aerial photographs will frequently indicate areas where such studies should be made. The photographs will also be of assistance to the specialists making the detailed studies.

26-10. Procedure. After the necessary aerial photographs and/or available maps are assembled, reconnaissance begins with an office study of the area between termini. Brief field trips to review certain critical areas and controls may be helpful during the office study. The use of aerial photographs will reduce the number and extent of the necessary field trips to a minimum. The general objective should be to develop, where possible, a fairly direct route between termini which will fit the terrain in the best possible manner and at the same time will avoid major obstacles.

As the study progresses in office and field, certain physical controls will become apparent. These may be negative controls, to be avoided, or positive controls through which the location should pass. Examples of negative controls are expensive buildings, cemeteries, schools, parks, historical landmarks, rich agricultural land, reservoirs, utility lines, and other man-made objects associated with land use. Negative controls might also include swampy areas and other adverse terrain conditions. Examples of positive controls dictated by the terrain would include low points in ridges, suitable stream crossings, favorable exposure for snow and ice conditions, and so on. As these controls are recognized, additional trial locations should be developed to avoid the negative controls where possible and to take advantage of the positive controls.

While these trial locations are being developed, the factor of traffic service should not be overlooked. In some cases the most direct route between termini would provide the best traffic service. In other instances a location which deviated materially from the most direct feasible route in order to pass close to or through the outskirts of a town or urban area might provide sufficient additional traffic benefits more than to compensate for the added length and possible increase in right-of-way and construction costs.

After all trial locations which appear to have any particular advantages have been sketched on the aerial photographs or maps, they should be thoroughly reviewed in the field. During the field review any alternatives which are obviously more costly and possess no particular advantages should be eliminated from further consideration. Decisions should also be made regarding the need for more detailed information to determine earthwork quantities more accurately as well as the necessity for geological, foundation, or drainage investigations. If sufficient information regarding present and anticipated future volumes of traffic is available, the design standards for the highway project can be determined at this time.

Following the field review, any necessary modifications or adjustments should be made in the trial locations still under consideration. Photographs or maps are then furnished to right-of-way and traffic specialists for their use in determining right-of-way costs and traffic benefits for each alternative route. If no further detailed information or investigation is required, the preparation of estimates of construction costs can proceed. As the reconnaissance trial locations are approximate only and as the selected route or corridor will be subject to further refinement following the preliminary survey, the cost estimates should have sufficient accuracy for comparative purposes only, with unnecessary detail being avoided.

After all of the necessary estimates of costs and traffic benefits have been completed, the reconnaissance stage can be concluded with the preparation of a report and the selection of a location for further surveys leading to design and construction. In some cases the decision as to the best route or corridor is made during the course of the reconnaissance, and the report is based upon the selected location alone. More general practice, however, is to prepare the report as a general summary of information collected during the reconnaissance including a description of the alternative routes with the advantages and disadvantages of each, a discussion of the more important controls, comparative cost estimates, an economic analysis, recommendations for design standards, conclusions, and appended maps and photographs. Such a report provides a convenient means for review by top-echelon officials prior to making the final decision as to the best location.

Federal-aid regulations and highway-department procedures frequently require public hearings for new highway locations. Some highway departments follow the practice of presenting all facts to the public and obtaining the benefits of comments made at the public hearing prior to making the final decision. In any case the reconnaissance report provides the factual background for presentation of the project to the public. Aerial photographic mosaics showing the various alternative locations are particularly effective for display at public hearings as well as for discussion purposes at conferences and with interested individuals.

PRELIMINARY SURVEY

26-11. Purpose and Scope. The preliminary survey consists of a detailed survey of the route selected as a result of the reconnaissance. The width of the corridor or route strip within which information is to be obtained may be as narrow as 300 ft. or as wide as a half mile or more depending upon the terrain, the proposed right-of-way width, and the amount of detail developed during reconnaissance. Sufficient data should be gathered in the preliminary survey stage to permit positioning of the final center line on paper in accordance with design standards established as a result of the reconnaissance.

The end product of the preliminary survey is either a topographic map or a planimetric map with accompanying cross-sections. Map scales may be as large as 1 in. = 40 ft. or as small as 1 in. = 200 ft. depending upon terrain, land use, and the method used in making the survey. The corresponding range of contour intervals might be from 1 to 10 ft.

Preliminary surveys may be carried out either exclusively by field methods or by any one of a number of possible combinations of field and aerial survey methods. Regardless of the method selected, surveys of any magnitude require that four items be carried out: *traversing, bench-mark leveling, topographic details,* and *planimetric details.*

1. *Traversing* is performed by field methods in accordance with the survey standards established for the project. While often not required for preliminary survey purposes alone, the order of accuracy of the traverse should usually be at least equal to that proposed for the location survey in order that checks between the two can subsequently be made at frequent intervals. Angle measurements may be made either by transit or by theodolite. Distance measurements may be made either by taping or by the use of electronic distance-measuring instruments.

2. *Bench-mark leveling* must also be carried out by field methods. Ordinarily turning points and bench marks are read to the nearest 0.01 ft., with checks being made to existing first-order or second-order bench marks at frequent intervals. Where existing bench marks are not available at intervals of 5 miles or less, consideration should be given to use of precise leveling methods.

Where the survey corridor is relatively narrow, bench marks are usually established at intervals of approximately 1,000 ft. for use in subsequent surveys. However, in the case of a wide corridor and the survey method described in Art. 26-15, greater intervals may be used with intermediate bench marks being established during the location survey.

3. *Topographic details* may be obtained by a wide variety of either field or aerial survey methods as discussed in Arts. 26-12, 26-14, and 26-15.

4. *Planimetric details* may also be obtained by either field or aerial survey methods, which are likewise discussed in the same three articles.

In addition to these four items, *land ties* (Art. 26-24) and any pertinent *special surveys* (Art. 26-25) should be made during the preliminary survey if they are necessary in positioning the final center line. Otherwise they may be made during either the preliminary survey or the location survey according to conditions.

In the following four articles several methods of carrying out preliminary surveys are discussed. It should be emphasized that these methods are not inflexible and that all the various possible combinations of field and aerial survey methods should be considered in planning the preliminary survey. All information gathered by the reconnaissance should be reviewed during this planning, and the desirable width of the survey corridor should be established within approximate limits.

26-12. Transit-tape Method. This procedure is especially well adapted to surveys through heavily wooded country or in cases where the survey corridor is very narrow. The survey corps usually consists of a transit party, a level party, and a topography party. A field draftsman may also be included.

Customary practice is for the survey engineer to set flags along the general center of the survey corridor, using aerial photographs, maps, and other data developed during the reconnaissance as a guide. He should keep in mind the design standards to be used for both curvature and grade.

The chief of party, usually acting as transitman, runs a traverse following the flagged line. Usually the traverse is run by the method of deflection angles described in Art. 14-10, and the notes are kept up the page in the form shown in Fig. 14-5. No curves are run in at this time. Checks for azimuth are applied at intervals of several miles, either by astronomical observations (taking account of convergency of meridians) or by tying in to the state system of plane coordinates. The chainage is carried forward from the point of beginning, stakes marked with the station number being set at all full 100-ft. stations and at any plus stations that are established. Hubs are set at all angles in the traverse and at all other transit stations. To expedite the progress of the survey, a rear flagman and a stakeman are usually employed. In wooded country the line is cleared by axemen, usually under the direction of the head chainman.

The level party, consisting of a levelman and or two rodmen, runs bench-mark levels as discussed in Art. 26-11. If cross-sections are to be taken from the preliminary-survey traverse or in areas where maximum grades may be encountered, profile levels should be taken along the traverse line with ground elevations being taken at all stakes set by the transit party, at changes in slope, and at all crossings of roads and streams. If other methods are to be utilized in obtaining topographic details, all turning points should be left in place with the elevation marked on a stake for use by the topography party. Figure 10-2 illustrates a common form of level notes. However, if cross-section notes are to be recorded on the right-hand page, it may be more convenient to keep level notes *up* the page in order to orient the cross-section notes more readily as to right and left of center line and thus to avoid possible confusion in subsequent plotting.

Where maximum grades may be encountered, the preliminary profile of the traverse line should be plotted and brought up to date at the close of each day's work in order to assure that the corridor as flagged will not exceed the maximum grade. In other cases, plotting of the profile may be deferred until a trial projection of the location center line is made. The horizontal scale of the profile should be the same as that proposed for the map (Art. 26-11), with the vertical scale being selected according to the ruggedness of the terrain.

Topographic details may be obtained by the topography party by any of several different methods described as follows:

1. Cross-sectioning from the traverse line as described in Art. 10-5 is probably the most commonly used method. Distances right and left from the traverse line on each cross-section should be sufficient to cover the width of the survey corridor. The interval between cross-sections will depend upon the type of terrain; it should not be less than that required for purposes of comparing alternative trial projections within the survey corridor. In many cases 100-ft. intervals will suffice, and in easy terrain they might vary from 200 to 500 ft.

In areas of rugged terrain, particularly where the survey corridor is relatively wide, the use of the Rhodes reducing arc (Art. 25-13*c*) or the use of a transit for vertical angles and a 200-, 300-, or 500-ft. tape for slope distances may be preferable to the use of the engineer's level or hand level discussed in Art. 10-5 to obtain elevations and distances along the cross-section lines.

One form of cross-section notes is shown in Fig. 10-5. Other forms may show plus or minus differences from the center-line elevation or slope distances and vertical angles. Cross-sections are plotted as described in Art. 26-22. If a contour map is to be used for projection of trial lines, it can be plotted by scaling distances right and left from the traverse line to contours on each cross-section and by plotting the points on the preliminary-survey map.

An alternative method used by some engineers is to work directly on the cross-sections in projecting trial locations, in lieu of plotting contours. When this is done, the map is used only for plotting the traverse and planimetry and for scaling offsets to the trial lines.

2. Determining distances directly in the field from the traverse line to points where contours intersect the cross-section lines, as discussed in Art. 25-13*a*, is another method which may be employed. A form of notes used for this procedure is shown in Fig. 25-8. While this is an efficient method for obtaining information for plotting a contour map, in most cases it is not so economical as direct cross-sectioning if cross-sections are to be used in computing earthwork quantities.

3. Other field methods which may be used to obtain topographic details for plotting a contour map are the use of transit and stadia (Art. 25-12*a*) or the plane table (Art. 25-12*b*). Either of these methods may be effectively employed to run random loop traverses connected to the main traverse for obtaining topographic details in heavily wooded areas where it is desirable to minimize cutting along cross-section lines.

Planimetric details may also be obtained by any of several methods, the most efficient procedure depending upon the intensity and type of land use. Where the planimetry is rather light, the data may be obtained by the transit party and recorded as shown in Fig. 14-5. Where transit-stadia or plane-table methods are used for topography, the planimetric details are usually obtained at the same time and in the same manner as the topographic details.

In cases where the cross-section method is used for topography, the same party may obtain planimetric details by offsets from the traverse line with the center-line station, distance right or left, and description being recorded in a separate set of notes together with whatever sketches may be necessary to assist in plotting the data. In areas of intensive land use more extensive use of sketches, either diagrammetric or to scale, is desirable. In such areas, effective use may be made of large notebooks ruled with 10 by 10 lines per inch or similarly ruled 8½ by 11-in. loose-leaf sheets. Supplemental loop traverses by transit or right-angle mirror and tape may be employed to obtain data for recording either in tabular form or by sketches.

In areas of intensive land use and moderate relief, aerial photographs enlarged and ratioed or rectified to the proposed map scale provide an efficient means of obtaining planimetric details. When only the central portion of each photograph is used, displacement due to differences in relief can be minimized and accuracy within 2 or 3 per cent can usually be attained in scaling from the photographic enlargements. A sufficient number of definite features should be located in the field, and the center-line station and distance right or left from the traverse line written on the photographic enlargements to enable the draftsman to orient the photographs for tracing additional details on the preliminary-survey map. Accurate field measurements to the more important features should be obtained. Where identification may be difficult, sufficient notes should be made on the photographs to insure proper plotting.

26-13. Other Field Methods. Transit-stadia and plane table are other field methods which may occasionally be employed. They are particularly adapted to short surveys where accurate horizontal and vertical survey control is not necessary. These methods may also be useful for obtaining detailed information covering relatively small areas of difficult terrain where the direct-location method discussed in Art. 26-21 is to be used for the remainder of the survey.

For the transit-stadia method the usual procedure is to run the traverse, measuring the horizontal and vertical angles and stadia distances, and to take side shots at the same time, as described in Arts. 15-16 and 25-12*a*, with the result that the horizontal and vertical controls are established and the details are obtained in one operation. Usually the party consists of a chief who may also act as recorder, a transitman, two or more rodmen, and a recorder or draftsman. Hubs are set at transit points, but no intermediate stakes are set.

Field procedure for the plane-table method is much the same as for transit-stadia except that the map is made in the field as the work progresses. Use of the plane table for this type of work is described in Arts. 17-8, 17-9, and 25-12*b*.

26-14. Large-scale Mapping Method. The steps required for this method include an accurate control traverse, aerial photography, bench-mark levels, vertical photocontrol surveys, and stereo compilation of a large-scale topographic map. The map is usually at a scale of 1 in. = 50 ft. with a 2-ft. contour interval, although other scales and contour intervals may occasionally be employed. In any case the map should be of sufficient accuracy that the final center line can be laid out and computed with a minimum of additional field-survey work. This procedure is applicable to any project where the width of the survey corridor is generally not greater than ¼ mile and where the ground is not obscured by dense vegetation. Under these conditions the use of large-scale mapping obtained by photogrammetric methods is usually the most efficient combination of field and aerial survey methods for major highway location surveys.

Traverse methods when possible should include use of one-second theodolites for angle measurements and electronic instruments (Art. 7-6) for distance measurements. Where conditions will permit, the traverse should be on the state coordinate system and of second-order accuracy or better.

Traversing starts with an initial azimuth and distance tie to a point on the state coordinate system (Art. 16-30) and consists of angle and distance measurements between intervisible monuments positioned in the approximate center of the survey corridor at intervals of ½ mile or less. The location of these monuments should be carefully planned so that they will be conveniently available for future surveys such as ties to property corners, special site surveys, and ties to establish the final center line.

Distances measured by electronic instruments must be reduced to horizontal (Art. 7-6) and to the plane of the state coordinate system (Art. 26-7). Angle measurements with a one-second theodolite should consist of at least four direct and four reverse readings at each angle point. Distance and angle measurements may be made either concurrently or as separate operations. During the survey, intermediate checks including distance and azimuth ties should be made to any available monuments of equal or higher order on the state coordinate system. If intermediate azimuth checks cannot be obtained by this means, astronomical observations should be made at intervals of not more than 25 courses.

To provide adequate horizontal control for stereoplotting, intermediate points should be set between monuments at intervals such that at least three points will be available for each stereoscopic model (Art. 26-6). With the usual 60 per cent forward overlap the effective overlapping area of two adjacent photographs, called a stereoscopic model, covers 40 per cent of each individual photograph along the line of flight which is centered approximately on the survey corridor. Thus with 9 by 9-in. photographs each stereoscopic model is 3.6 in. in length at photoscale in a direction parallel to the survey corridor. For a map scale of 1 in. = 50 ft. and a photoscale of 1 in. = 250 ft., the resulting model length is 900 ft. Intermediate points set at intervals of 300 ft. between monuments by taping to third-order accuracy will satisfy the requirements for horizontal photocontrol.

Monuments and intermediate points should be premarked with targets of cloth or other suitable material so that they will be visible in the aerial photographs. Flight lines are then laid out on the reconnaissance photographs or maps so that, where possible, a single strip of photography will cover the survey corridor. Although each individual photograph will cover a width of 2,250 ft., accurate map compilation should be restricted to the central 7 in. of the photograph or a net width of 1,750 ft. Further allowance must be made for variations of the airplane from the exact flight line as laid out. For these reasons the flight lines should be planned to have an effective width of somewhat less than 1,750 ft. Other factors such as excessive differences in relief which may affect flight planning are discussed in Art. 31-20.

In addition to single-strip coverage of the survey corridor, aerial photography should be planned to include any special site surveys which may be necessary, such as crossings of other highways, streams, and railroads. The additional width required at such locations may be obtained either by an additional short strip of photography on each side of the central strip or by a cross-flight along the highway, stream, or railroad for whatever distance is necessary.

After completion of the aerial photography, bench-mark levels are run in the manner described in Art. 26-12. Vertical photocontrol consisting of elevations (usually read to the nearest 0.1 ft.) on the traverse points and on four points, called "wing points," of each stereomodel may be obtained either at the same time or as a separate operation. Wing-point elevations should be taken on the ground adjacent to objects which are visible in the photographs, such as a small clump of grass, a bush, or a fence post, and should be in the vicinity of the corners of the stereomodel. They should be identified in the field by stereoscopic examination of the contact prints and should be pinpricked and circled on the prints for subsequent identification by the stereoplotter operator. Elevations of the wing points, traverse monuments, and intermediate points should be written on the back of the contact prints for convenient use by the plotter operator. As the wing points are used only for vertical control, their horizontal location need not be determined in the field. In areas such as cultivated fields, where photo-identification may be difficult, it is sometimes desirable to place a row of targets on each side of the flight line for use as vertical control in lieu of natural features.

Following the completion of all or of any sizable portion of the field control and aerial photography, traverse monuments and intermediate points are carefully plotted by coordinates on the base map, sometimes called the *map manuscript*. Stereocompilation of the topographic map can then proceed. In addition to contours and planimetric features, the plotter operator should show spot elevations, usually recorded to 0.1 ft., along ridges, streams, roads, and other locations where they will be helpful during design. In comparatively level terrain where contours are widely spaced, a grid of spot elevations at intervals of 1 in. on the map should be shown.

As an alternative to plotting contours, a procedure whereby spot elevations are shown along cross-section lines on the manuscript has certain advantages on projects where much of the terrain is practically level and where the final center line can be laid out with a minor amount of field inspection (Ref. 6 at the end of this chapter). A set of the contact prints at 1 in. = 250 ft. scale, showing the targeted traverse points together with any other available data, is used in establishing the location of the final center line, which is sketched on the contact prints. Its approximate location is then transferred to the map manuscript by scaling from the traverse points. Tangents are laid out on the manuscript to fit the scaled points, the center line is computed after scaling coordinates of the P.I.'s, and curves are computed and drawn in. Full (100-ft.) stations are plotted along the center line, and cross-section lines are drawn at right angles to the center line. The stereoplotter operator reads elevations at all breaks in the slope and at specified maximum distances along each cross-section line and writes the elevation opposite a cross tick at each point read. Elevations are recorded to the nearest 0.1 ft. In addition to the 100-ft. stations, cross-sections should be taken at breaks in the terrain and at stream crossings, etc.

26-15. Intermediate-scale Mapping Method. The principal difference between this method and that of Art. 26-14 is that the map scale is smaller, generally 1 in. = 200 ft., which feature makes it possible to cover a much wider survey corridor. As a consequence, however, the information obtained is less accurate and in less detail, and the result is that more work is required during the location-survey stage.

The intermediate-scale mapping method is particularly advantageous under any or all of the following conditions:

1. Where dense vegetation precludes large-scale mapping of sufficient accuracy for earthwork quantities in the final design.

2. Where extreme differences in relief will result in excessive costs for large-scale mapping covering the necessary width of corridor.

3. In areas of difficult terrain where the reconnaissance has not made it possible to narrow the survey corridor down to limits which are practicable for large-scale mapping.

The various steps of traverse, bench-mark levels, photocontrol, aerial photography, and compilation are essentially the same as discussed in Art.

26-14 with some minor variations due to the difference in map scale. Traverse monuments should be set along the approximate center of the survey corridor at intervals of ½ to 1 mile. For a map scale of 1 in. = 200 ft. the monuments will provide sufficient targeted horizontal photocontrol. Any additional points which may be necessary can be photo-identified and their position determined during the course of the survey. It will usually be advantageous to set and target the monuments prior to aerial photography but to defer the traverse survey work until a set of contact prints of the aerial photographs is available. This procedure will permit planning the additional horizontal and vertical photocontrol so that all field survey work can be carried out at the same time. Photocontrol should be planned by an experienced photogrammetrist, and consideration should be given to the use of bridging techniques in order to minimize field-survey work.

Vertical wing-point control for this type of mapping will frequently be difficult to obtain in conjunction with bench-mark leveling. In some cases the most efficient way to obtain the necessary elevations will be by vertical angles taken with a one-second theodolite from a nearby traverse station. The slope distances can be obtained with an electronic distance-measuring instrument at the same time that traverse distances are being measured. Elevations for this type of work should be accurate to the nearest 0.5 ft. if a 5-ft. contour interval is to be used.

As in the case of large-scale mapping the stereoplotter operator should show, in addition to contours, spot elevations along ridges, streams, roads, etc., and in areas where contours will not adequately depict the terrain.

LOCATION SURVEY

26-16. Purpose and Scope. The location survey consists in staking the final center line in the field and obtaining all additional data necessary for completion of plans, specifications, and estimates and for right-of-way descriptions. Whether the center line can be staked in the field exactly as computed from the office projection or must be adjusted slightly to fit field conditions will depend upon the type of preliminary survey which was used.

As in the case of preliminary surveys, either various combinations of field and photogrammetric methods may be employed in the location survey or it may be made exclusively by field methods. The amount of detail to be obtained will also depend upon the method used in making the preliminary survey.

26-17. Office Projections. After completion of the preliminary survey, general practice is to make trial projections of various combinations of alinements and grades within the area covered by the survey corridor. The most desirable combination is one which will fit the terrain and have the lowest cost, at the same time meeting the design standards established for the project.

The usual procedure is to try various combinations of tangents and horizontal curves in order to find the one which will best fit the ground. At the same time consideration is given to other factors which will influence the final alinement, such as cost of rights of way, effect upon utilities, relation to other roads and to railroads or streams, drainage requirements, and cost of earthwork. As a general rule these same factors will influence the selection of the grade line and thus require simultaneous study of alinement and grade. In the case of major highways, intersections with other roads where traffic interchanges are proposed may be the controlling factor in the final selection of a particular combination of alinement and grade.

Where all other factors can be readily satisfied by several different alinements, the choice may well depend upon obtaining the minimum total cost for earthwork. This result will usually involve obtaining a proper balance between excavation and embankment, at the same time keeping the amounts of excavation and haul to a minimum. Earthwork quantities and haul are further discussed in Arts. 26-22 and 26-23.

Where large-scale mapping has been used for the preliminary survey (Art. 26-14) the final design and preparation of plans, specifications, and estimates can proceed as soon as the best possible combination of alinement and grade has been established. As the first step, the mathematical location of the final center line is determined by scaling coordinates of the P.I.'s or of other points on each tangent. Bearings, lengths of tangents, and curve data are then computed, and stationing is calculated for the entire line. Where other preliminary-survey methods have been used, it is usually necessary to defer computation of the center line and final design until completion of the location survey. Figure 26-1 is an example of a plan-profile sheet showing alinement and grade as developed for final design.

During the process of investigating various trial alinements and grades and during design of the selected combination, additional field-survey data may be required in certain areas. Obtaining such information is a part of the location-survey procedure and is discussed more fully in succeeding articles of this chapter.

26-18. Transit-tape Method. The work of field location starts with staking the projected center line on the ground so that it follows as closely as practicable the alinement as laid out on the preliminary-survey map. Center-line tangents are established either by scaling offsets from the preliminary-survey traverse or by scaling coordinates of points on the proposed tangents and computing ties from nearby transit points on the traverse. As the survey progresses, adjustments are made in the center line if found necessary to fit field conditions better.

Where possible, adjacent tangents are run to an intersection and the deflection angles are measured and checked for approximate agreement with the map projection. The measured deflection angle and the radius or degree

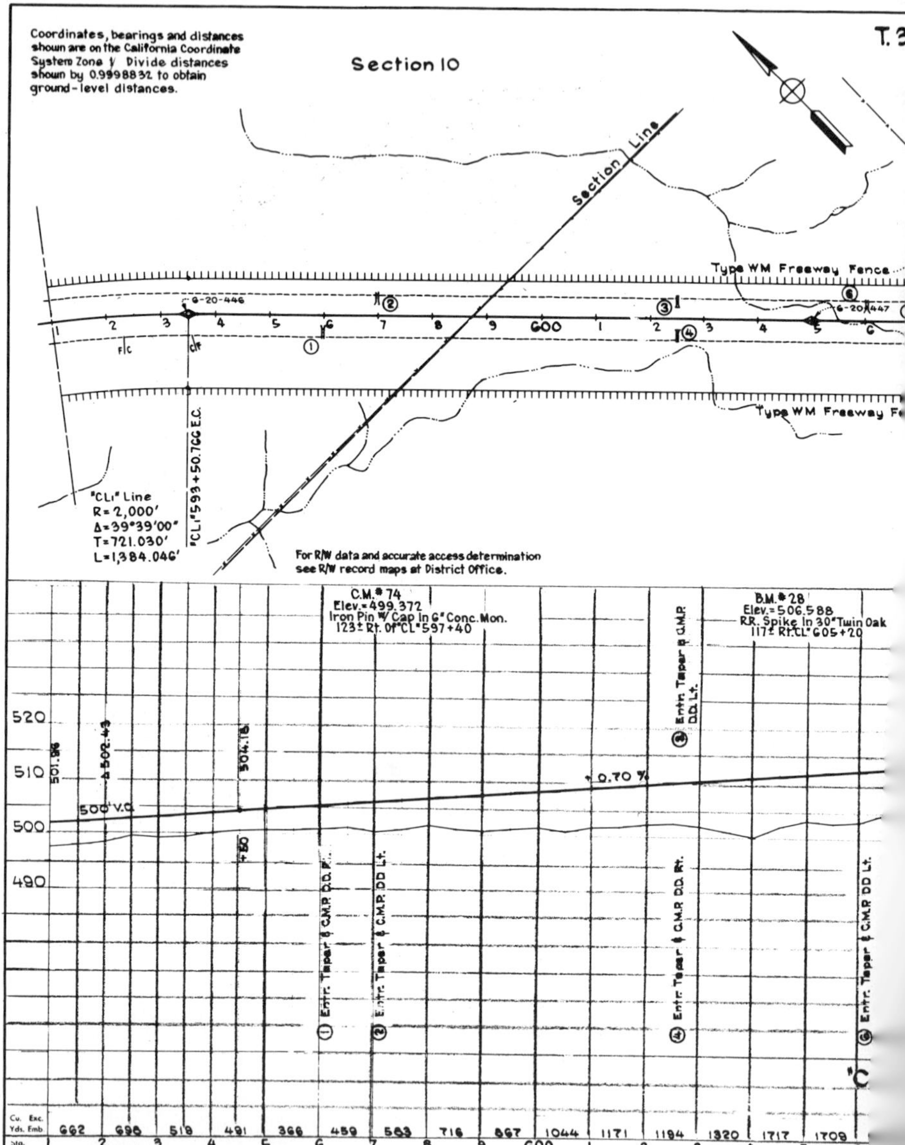

FIG. 26-1. Typical location plan and profile. *(California Division of Highways.)*

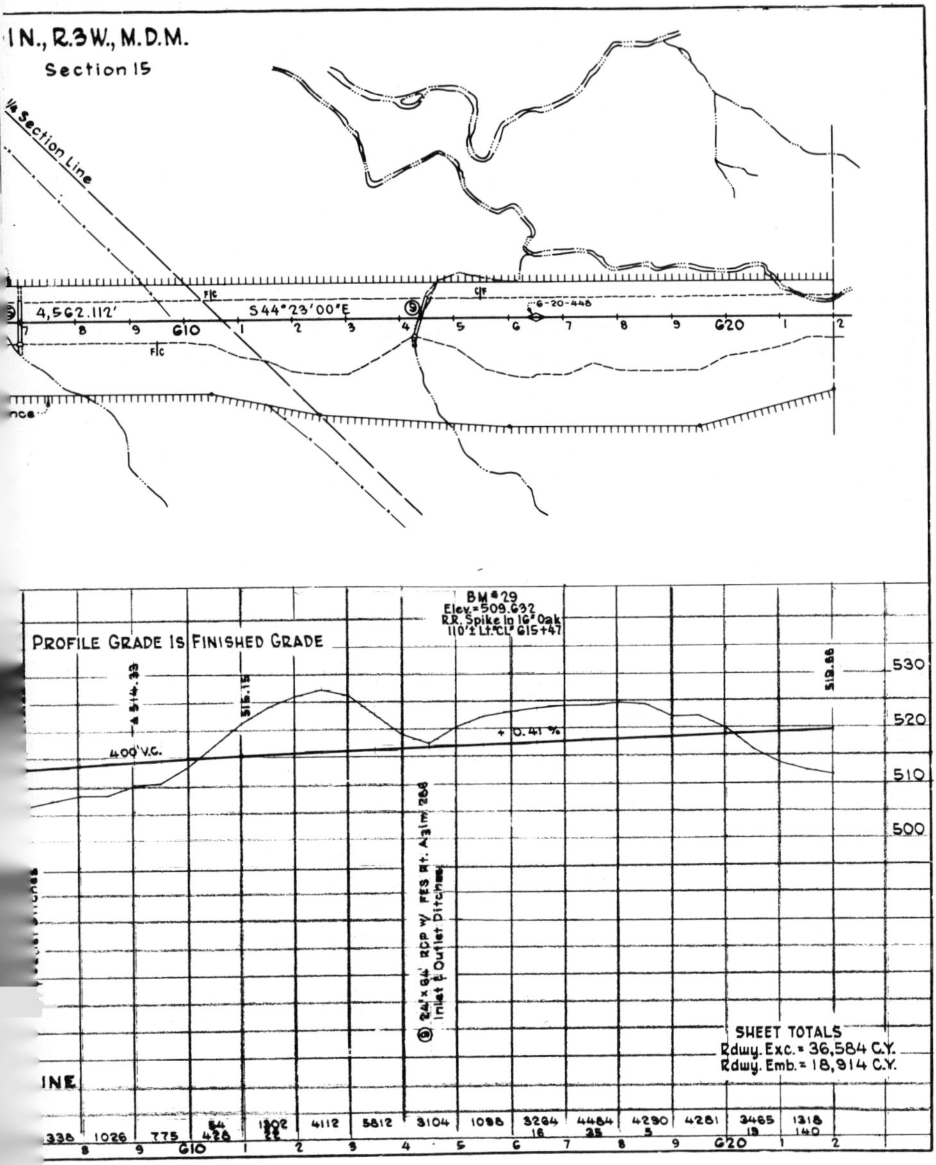

1N., R.3W., M.D.M.
Section 15
1/4 Section Line
4,562.112'
S44°23'00"E
F/C
C/F
6-20-448
8
9
610
1
2
3
4
5
6
7
8
9
620
1
2
BM#29
Elev.=509.632
R.R. Spike in 16" Oak
110'± Lt.°CL° 615+47
PROFILE GRADE IS FINISHED GRADE
400' V.C.
+0.41%
530
520
510
500
24"x64' RCP w/ FES Rt. A3lm 288
Inlet & Outlet Ditches
SHEET TOTALS
Rdwy. Exc. = 36,584 C.Y.
Rdwy. Emb. = 18,914 C.Y.
LINE
338
1026
775
54
426
1302
22
4112
5812
3104
1098
3264
16
4484
35
4290
5
4281
3465
19
1318
140

of curve from the map projection are used to compute the curve data as explained in Chap. 27.

Beginning at some point where the stationing has been established, tangents and semitangents are chained and curves are computed and run. Stakes are set at full stations (100 ft.) and in many cases at half stations or at closer intervals depending upon the terrain and upon requirements for cross-sections and future slope staking. Hubs are set at all transit points including P.I.'s, P.C.'s, and P.T.'s. Referencing of such points can be either done as a part of the location survey or deferred until construction staking, depending upon field conditions. Transit notes are kept up the page as illustrated by Fig. 14-5, and notes for curves are kept as shown by Fig. 27-7*b*. If spiral transition curves are to be used, they may be either staked at this time or left until construction staking.

As the survey is in progress, field ties should be made to the preliminary-survey traverse and closures should be computed at intervals such that any errors which may occur can be quickly isolated and corrected.

Profile levels are run, and a location profile is prepared showing the ground line and proposed grade line. Study of the profile and the preliminary map may indicate further desirable adjustments in the center line. Changes in the grade line can be made in the office during final design without the necessity for additional field work.

Usual practice is to take cross-sections from the located line at 100-ft. or closer intervals as soon as the line is staked and profile levels run. In easy terrain, cross-sections may be taken at the same time as profile levels if no further adjustments in the center line are considered likely. With the location of the center line definitely established, the width of cross-sections may generally be confined to the proposed right of way. They should be of sufficient accuracy and at sufficiently close intervals that earthwork quantities may be used for payment to the construction contractor.

At locations where the preliminary-survey cross-sections are of sufficient accuracy and where tangents of the located center line coincide with or are parallel to the preliminary traverse, cross-sections may sometimes be omitted. In the case of a narrow roadway where three-level sections would suffice, cross-sectioning may sometimes be deferred until slope stakes are set. However, on most highway projects the advantages of having plotted cross-sections or tabular data from electronic data processing on the same stations from which slope stakes are to be set are sufficient to warrant cross-sectioning during the location survey.

After the final center line has been staked, a careful field inspection with preliminary map in hand should be made for any critical planimetric feature whose erroneous location on the preliminary map would result in subsequent complications. For example, the location of a building shown as immediately adjacent to the proposed right-of-way line should be checked by a field tie from the located line. If planimetric features have been accurately located during the preliminary survey, and if the final center line has been carefully positioned with respect to the preliminary traverse, checks of the more critical features should suffice. Otherwise planimetric details should be accurately located with respect to the final center line during the location survey.

In addition to land ties (Art. 26-24) and special surveys (Art. 26-25) the location survey should also include the investigation and location of all utilities within the proposed right of way, including water lines, sanitary and storm sewers, gas and oil lines, conduits, and telephone, telegraph, and power lines. The amount of detail required by the field survey will depend upon the availability of satisfactory maps of the utility lines and the general procedure to be used in handling their relocation or rearrangement.

26-19. Large-scale Mapping Method. As this differs from other methods described in that final design can be completed for the most part from preliminary-survey information, staking of the final center line on the ground is frequently deferred until the final design is nearing completion. For this reason any additional survey data required during design are obtained by ties from the preliminary-survey base traverse. As design progresses to the point where the final center line has been computed, any short section of the line can be staked on the ground by ties computed to the base traverse.

As soon as the approximate location of the final center line has been determined in the office, the checks for critical planimetric features and the investigation and location of utility lines as discussed in Art. 26-18 should be undertaken. Field measurements should also be made to any critical topographic features which might control the exact location. For example, if a fill slope appears to intersect the ground within a few feet of the edge of a high vertical cliff or a steep bank, the exact position of the edge through the critical area should be determined by field-survey methods.

Additional field-survey work may also be required to check the contours in areas where dense vegetation completely obscures the ground. If such checking is deferred until the location of the line is established within very narrow limits, the area involved will usually be much less than if the field work were undertaken concurrently with the mapping.

After the final center line has been computed as described in Art. 26-17, staking on the ground can begin at any point and all or any portion of the center line can be staked. As the distance measurements by electronic instruments for the preliminary base line or traverse are usually of a higher order of accuracy than chainage of the located line, any minor discrepancies in stationing that occur between ties should be dropped or adjusted in the located line so that no changes are made in the computed data. Similarly, computed angles should be used.

Ties computed in the office from preliminary-traverse monuments to convenient points on tangents of the final center line are made in the field as the first step in staking the line. As computed angles and stationing are to be held, it is not necessary to extend adjacent tangents to an intersection. Instead, the P.C.'s and P.T.'s are set either by direct ties from the preliminary traverse or by chaining from points established on tangents. Where curves materially exceed ½ mile in length, one or more ties to P.O.C.'s (points on the curve) will assist in obtaining more satisfactory curve closures.

Stakes are set at full and half stations where required for slope-staking, and profile levels are run on the staked line. If organizational procedures require field cross-sections for purposes of payment to the construction contractor, the line can be cross-sectioned as discussed in Art. 26-18 either at the time it is staked or during slope-staking operations. Otherwise the profile elevations are used as a check on the accuracy of the contour map, and a decision is made as to whether further spot checks or cross-sectioning by field methods of all or any part of the line are necessary. If desired, the earthwork quantities can be adjusted by use of the field profile (Ref. 5 at the end of this chapter).

26-20. Intermediate-scale Mapping Method. Location-survey procedures where intermediate-scale (usually 1 in. = 200 ft.) mapping has been used for the preliminary survey are essentially the same as described in Art. 26-18 up to and including the running of profile levels. The relatively small scale of the map used in projecting the trial location line may, however, require greater care in adjusting the final center line in the field, particularly in areas of rugged terrain.

After the final center line has been located and staked on the ground and profile levels have been run, the next step is large-scale aerial photography for obtaining topographic and planimetric details by photogrammetric methods. Center-line stations should be targeted at 300-ft. intervals to provide horizontal photocontrol as well as vertical control along the center of the mapping. Photography should be at a scale of 1 in. = 250 ft. for mapping at a scale of 1 in. = 50 ft. The photography should be planned to include additional strips or cross-flights to cover areas where any special surveys as discussed in Art. 26-25 may be required.

After the photography is completed, wing-point photocontrol is obtained as described in Art. 26-14 and stereocompilation can proceed. Planimetric details are compiled on a map at a scale of 1 in. = 50 ft. for subsequent use in preparing final design plans. As the final center line has been established on the ground, contours are no longer needed for use in making trial projections; therefore, cross-sections of sufficient width to cover the proposed right of way are obtained for use in computing earthwork quantities.

Elevations may be written on the map at breaks along the cross-section lines for subsequent manual scaling of distances out from the center line, or special equipment for direct print-out of elevations and distances may be used (Ref. 11 at the end of this chapter). This equipment is an electronic analog-to-digital converter especially designed for use with a Kelsh-type stereoplotter (Fig. 26-2). It automatically measures and records the distance out from the center line and the elevation each time the plotter operator sets the measuring mark on a point on a cross-section line and presses a button. The measurements are recorded either in tabular form on paper or on punch cards or tape for electronic data-processing.

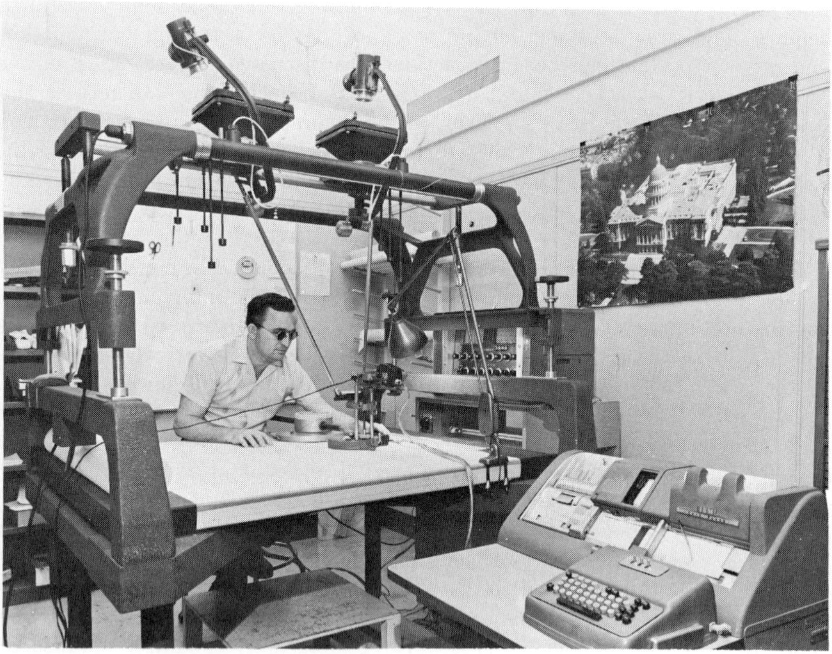

FIG. 26-2. Electronic analog-to-digital converter for use with Kelsh-type stereoplotter. (*California Division of Highways.*)

In Fig. 26-2, on the table and nearest the operator is the Auto-trol digital scaling unit. Against the wall, at his left, is the Auto-trol console which temporarily stores the horizontal and vertical measurements for transmission to the card-punch unit. At the lower right is the IBM card-punch unit.

To obtain terrain data photogrammetrically in areas where dense vegetation completely obscures the ground, two alternative procedures are available. The first is to obtain sufficient information by field methods to complete the final design and award the construction contract, with specifications requiring sufficient time between clearing and the start of grading operations to permit targeting the center line and obtaining aerial photography. More accurate cross-sections for earthwork quantities can then be obtained by the method described above without further delay in construction operations.

The second alternative is to award a clearing contract on the basis of preliminary-survey information and to defer final design and award of the major construction contract until cross-sections have been obtained, as described above, subsequent to clearing operations. Either of these methods will frequently compare favorably in cost with field methods for obtaining accurate cross-sections in areas of rugged terrain where the width between slope stakes is several hundred feet.

26-21. Direct-location Method. Direct or free location eliminates the preliminary survey and involves establishing the center line on the ground by the survey engineer as a direct location operation. It may be efficiently employed in areas of easy terrain, where no complex right-of-way problems are involved, and where the proposed roadway is not too wide to be visually fitted to the terrain in the field. Even under the most favorable conditions this method requires considerable experience and the exercise of sound judgment on the part of the survey engineer.

Tangents are flagged on the ground by the survey engineer and run out to their intersections. Tangent offsets or other methods can be used for determining the radius or degree of curve which will best fit the terrain, before the curve is actually run. Subsequent operations of bench-mark and profile leveling, cross-sectioning, and obtaining planimetric details are carried out by the procedures described in Arts. 26-12 and 26-18.

A strip of aerial photographs along the proposed route is frequently of considerable help in establishing the center line. Where detailed information is necessary for short sections of difficult terrain, the methods described in Art. 26-13 may be employed to advantage.

26-22. Earthwork Quantities. Obtaining terrain data in the form of cross-sections or a contour map and the subsequent processing of the data to obtain earthwork quantities for both the trial locations and the final line are among the most time-consuming and costly operations in the highway route survey procedure. The survey engineer should therefore be familiar with the various methods for obtaining the data, the accuracies required, and available techniques for obtaining earthwork quantities from the field data.

In general there are four uses for earthwork quantities obtained during surveys for a highway project, each having a somewhat higher accuracy requirement than the preceding. These uses are briefly discussed as follows:

1. *Reconnaissance* quantities for use in comparing alternative route corridors. Accuracy requirements are at a minimum and need be sufficient only to assure proper route selection.

2. *Preliminary* quantities from data obtained in the preliminary survey

for comparison of trial locations within the survey corridor. As the survey must cover the entire width of the corridor, the data obtained, except for the preliminary-survey method described in Art. 26-14, should be held to the minimum required for the necessary comparisons.

3. *Design* quantities for design of the grade line and appurtenant features and for estimating and bidding purposes. Terrain data are usually obtained during the location survey except when the preliminary-survey method described in Art. 26-14 is used. The accuracy should be such that adequate estimates of the amount of work to be performed are available for financing and bidding purposes and that changes in plans during construction can be avoided or held to a minimum. To avoid duplication of work, terrain data for design quantities should generally have sufficient accuracy to be used for item (4).

4. Quantities for purposes of *payment* to the construction contractor. In this case the data must be sufficiently accurate to insure equitable payment.

For reconnaissance comparison purposes a field inspection of the alternative routes may, in some cases, enable an experienced engineer to estimate earthwork quantities with sufficient accuracy. Stereoscopic study of aerial photographs and a rough profile will frequently provide further information if required. The approximate methods discussed in Art. 11-14 are also useful where more detailed comparisons are necessary. Some of these approximate methods, using data developed by the preliminary survey, may also be sufficiently accurate for preliminary quantities for comparing trial projections within the survey corridor. Further refinements at this stage include study of the contour map and use of cross-sections if necessary.

As accuracy requirements increase for design and pay quantities, further consideration must be given to sources of terrain data and methods of processing the data. In considering the accuracy of terrain data, it must be recognized that any type of data such as provided by cross-sections or a contour map can result only in approximate earthwork quantities because the data are only a sample of an infinite number of points or lines which would be necessary to represent the true surface of the earth in the area involved.

It is also important to recognize a distinction between *point* accuracy and *statistical* accuracy in terrain data. By point accuracy is meant the absolute error at a specific point. Statistical accuracy, on the other hand, relates to the mean error of a large number of points representing an area of the earth's surface. If the errors in the individual points are random in nature, the mean error will approach zero. Hence it is possible to have poor point accuracy and excellent statistical accuracy (Ref. 10 at the end of this chapter). This fact is particularly important when considering the probable accuracy of earthwork quantities where the data were obtained from a contour map compiled by photogrammetric methods (Ref. 4 at the end of this chapter).

The four sources of terrain data for calculation of earthwork quantities are cross-sections taken by field-survey methods, cross-sections taken by photogrammetric methods, contour maps made from field-survey data, and contour maps compiled by photogrammetric methods. Any of these four sources are capable of producing data for earthwork quantities of sufficient accuracy for any purpose, provided that a sufficient number of points are measured in obtaining the data and that systematic errors and blunders are held to a minimum in order to provide good statistical accuracy.

Terrain data from any of the four sources must be combined with information showing the dimensions of the proposed roadway at each station in order to form a closed area for earthwork computation. Figure 11-3 shows two such areas with the upper line in each case representing the ground surface as determined from the terrain data, the lower horizontal line being the subgrade of the finished roadbed, and the sloping lines representing 1-to-1 cut slopes. When cross-sections are plotted, the roadway may be drawn by using a plastic templet cut to the proper dimensions. In other cases, roadbed notes may be prepared showing elevations of the subgrade at center line and at each break out to the toe of cut or top of fill, with the rate of cut or fill slope beyond that point being shown.

The area at each station having been determined by any one of the methods discussed below, earthwork volumes are usually computed by the average-end-area method described in Art. 11-11. Where the roadway is narrow, the subgrade a straight line, and the terrain comparatively regular, the methods described in Arts. 11-6 to 11-8 may be used for the computation of areas. Because of large widths and more complex subgrade design, however, present-day practice usually requires either plotting cross-sections or obtaining quantities by means of electronic data-processing equipment.

Cross-sections are usually plotted on sheets or rolls of paper ruled with 10 by 10 lines per inch. A scale of 1 in. = 10 ft. is generally used, although scales as large as 1 in. = 5 ft. may be utilized for narrow roadways in easy terrain and as small as 1 in. = 20 ft. in rugged terrain where the cross-sections must cover a wide area. The ground line is plotted first, from field cross-section notes, from photogrammetric cross-section data written on the map, or from data taken from a contour map. The center-line elevation is established on a heavy vertical grid line so that the heavy horizontal lines of the 10 by 10 grid correspond to an even 10 ft. in elevation. Distances out from the center line and elevations are plotted by using the grid as a scale. If a contour map is the source of terrain data, cross-section lines are drawn on the map at right angles to the center line. One man scales distances right and left from the center line to contour lines and to breaks in the slope on each cross-section line and calls out distances and elevations to another man who plots each point. The points are then connected by an inked line which corresponds to the ground line.

After the terrain data are plotted, the subgrade elevation at the center line is plotted and the remainder of the roadway templet is drawn. End areas of excavation and embankment are then determined by measurement with a planimeter, and volumes are computed by hand, by means of a desk calculator, or from volume tables prepared for that purpose. As an alternative to the use of the planimeter for determining areas, an instrument called the *Avol rule* may be used to make cumulative measurements of equally spaced vertical ordinates on each cross-section for determining the quantities.

Recent developments in the use of electronic data-processing equipment have extended its use to many phases of highway engineering (Ref. 13 at the end of this chapter). One of the most important of these is the computation of earthwork quantities. Programs for this type of computation have been written for a wide variety of electronic computers, with the result that the use of such equipment has superseded previously used methods in many organizations. A separate computer section frequently handles all types of electronic data processing.

In most cases data similar to field cross-section notes for terrain information, together with roadbed notes representing cross-sections of the proposed work, are furnished to the computer section. In some cases programs have been written so that elevation data on the notes can be in the form of actual elevations, rod readings from a specified height of instrument, or plus or minus differences in elevation from the center line. The output from the data-processing equipment is usually a printed tabulation showing volumes of excavation and embankment between each two adjacent cross-sections. In many cases end areas are also shown, and frequently supplemental tabulations showing mass-diagram ordinates, distances out and elevations of slope stakes, and other information can be obtained. Electronic plotters have also been developed which can plot the data in form of cross-sections in the event that a visual record is considered necessary.

Methods of furnishing terrain data to the computer section depend upor the source of the data and upon the form in which it is obtained. In case where field cross-section notes are to be used, it is frequently desirable t record them on special loose-leaf forms. Roadbed notes are prepared o similar forms. Before being processed, the data are transferred to punc cards.

Where photogrammetric cross-sections are the source of terrain dat special equipment for obtaining punch cards or tape as a direct output fro the stereoplotter is discussed in Art. 26-20. In cases where the cross-sectic data have been written directly on the map manuscript by the stereoplott operator, distances out to each point may be scaled and written on a pri of the map. A narrow strip covering the area of interest may then be c from the print for use by the key-punch operator (Ref. 6 at the end of tl chapter).

Several methods are available for obtaining numerical data from conto maps for processing by electronic equipment. Probably the most commonly used is that of scaling distances out to contours and breaks in slope on each cross-section line in the same manner as that described for plotting cross-sections, except that in this case the elevations and distances out are recorded on the same forms as those used for field notes.

A more efficient method where large volumes of work are involved is the use of an analog-to-digital converter similar to that discussed in Art. 26-20. However, in this case readings are taken along each cross-section line on a print of the map rather than from the stereomodel in a Kelsh plotter. The instrument can be set so that the elevation reading will automatically increase or decrease by one contour interval each time the operator presses a button for a reading of the distance from the center line. The operator manually punches elevations at breaks other than on contours and makes the necessary changes in setting when changing from uphill to downhill or *vice versa* (Ref. 11 at the end of this chapter).

Another method for obtaining data from contour maps is the use of a transparent strip of mylar on which is a printed grid which can be used for scaling distances from the center line. The strip is positioned over the contour map longitudinally so that a crossline on the grid corresponds to the station for which a cross-section is to be taken. A tick mark is placed at each contour crossing along the cross-section line and at breaks in slope. Elevations of these points are written on the mylar strip. The strip is then shifted to the next station. Distances are read by using the grid as a scale and are written opposite each elevation. The mylar strips are furnished to the computer section for key punching (Ref. 6 at the end of this chapter).

The use of contour grading plans is a method of obtaining earthwork quantities directly from contour maps. Although largely superseded by the development of electronic data processing, it affords a rapid and accurate means of obtaining excavation and embankment quantities which is of particular value for irregularly shaped areas in rugged, broken terrain. Basically the method consists in superimposing contours of the proposed work on a contour map of the existing terrain. Volumes of excavation and embankment are obtained by planimetering the horizontal areas which will be affected at the elevation of each contour and using these areas and the contour interval to obtain volumes by the average-end-area method. This procedure is frequently used for obtaining quantities of material taken from borrow pits or for stockpile measurements. A brief explanation is given in Art. 24-18 and a more complete description in Ref. 8 at the end of this chapter.

26-23. Haul. One of the considerations in fixing the location and grade of a highway is the amount of *haul* that will be necessary to transport excavated material from cuts or borrow pits to the adjacent fills or to waste. The construction contract usually names a price per cubic yard to be paid for excavation either unclassified or for each class of material (earth, loose rock, solid rock, etc.) and for transporting this material for any distance up to a limit of free haul. Transportation of material beyond this distance is termed *overhaul* and is paid for at a rate fixed by the contract. The unit of measurement for overhaul is generally the *station yard,* one station yard being 1 cu. yd. of material transported 100 ft.

The limits of free haul are determined by fixing one point in cut and one point in the adjacent fill, at the specified free-haul distance apart, such that the included quantities of excavation and embankment will balance. The overhaul distance is computed as the distance between the center of gravity of the remaining mass of excavation and the center of gravity of the resulting fill, less the limit of free haul.

In computing the volumes of excavation required to make equal volumes of completed embankment, due allowance must be made for shrinkage of excavated earth when placed in embankment due to compaction of the fill. In rock excavation a similar allowance must be made for swell from cut to fill. Allowance must also be made where necessary for subsidence of the natural ground under fill areas.

In order to determine in advance the proper distribution of excavated material and the most economical amounts of waste and borrow and as a basis for estimating cost, a *mass diagram* is usually prepared. The abscissas in the mass diagram are the distances along the survey line, and the ordinates are the algebraic sums of earthwork quantities from the beginning to each ordinate, considering cut volumes positive and fill volumes negative and making due allowance for shrinkage, swell, and subsidence. From the mass diagram it is possible to determine by trial the earthwork distribution plan that will result in the minimum total cost for the contract pay items of excavation, overhaul, and borrow. The use of the mass diagram is discussed in detail in texts on highway and railway location and earthwork.

26-24. Land Ties. In some areas, land for highway rights of way may be acquired on the basis of easements, with a recorded metes-and-bounds description of the highway center line and the width of the required strip on each side sufficing. More common practice, however, is to acquire the necessary rights of way in fee. Land so acquired is usually described by the methods discussed in Chaps. 22 and 23. As in most instances only portions of previously described and deeded tracts of land are acquired, it is necessary that property, subdivision, or section corners be searched for during the highway route survey and that their location with respect to the center line of the highway be determined.

Such land ties may be made during either the preliminary survey or the location survey, depending upon the survey method used. They should be made to the same order of accuracy as the location survey. Where measurements involve long distances, as frequently is the case in rural areas, electronic distance-measuring instruments may be advantageously employed. In cases where the preliminary-survey traverse is to be used as the basis for all subsequent surveys, as in the method of Art. 26-14, land ties are made from the base traverse. In other cases they are frequently made from the final center line as located in the field. Ties should be made to a sufficient number of corners so that intersections of all property, subdivision, and section lines with the final highway center line can be computed.

26-25. Special Surveys. Crossings of railways and canals and of major roads and streams usually require additional survey data during the course of highway route surveys. Obtaining aerial photographic coverage for topographic mapping for such crossings is discussed in Arts. 26-14 and 26-20.

Regardless of whether topographic and planimetric details are to be obtained by field or aerial survey methods, a traverse, usually parallel to the railway, canal, road, or stream, is commonly required if any change in its alinement or grade is contemplated. Profile levels and topographic and planimetric details are obtained along the traverse for a sufficient distance on each side of the highway center line to cover any possible change. During the course of the survey, elevations of top of rail, flow line, or pavement should be be obtained with sufficient precision that the proposed work can be laid out to fit existing conditions.

Regardless of whether or not changes in alinement or grade are proposed, special site maps are usually prepared for all crossings where major structures will be involved. The amount of detail to be obtained will depend upon site conditions and the type of structure to be used. Surveys for long over-water crossings are beyond the scope of this text; they are discussed in some detail in Ref. 1 at the end of this chapter.

Terrain data for borrow pits, both prior to excavation and after the material has been removed, can be obtained by cross-sections (Arts. 10-3 and 10-4), by before-and-after elevations on a rectangular grid (Art. 11-10), or by contours for use in the contour grading method (Art. 26-22). Regardless of the method to be used for subsequent computation of quantities, the data can sometimes be obtained more economically by photogrammetric methods, particularly if several surveys may be required for measurements for progress pay estimates or where material is to be removed from the same borrow pit under several construction contracts. Where photogrammetric methods are used, the photocontrol, where possible, should be placed outside the excavation limits so that it can be used for subsequent flights.

Materials or soil surveys are frequently made during the course of the location survey as soon as a plan and profile of the proposed center line are available. They consist mainly in sampling and testing the native soils which will be encountered, for purposes of determining their suitability for use in base or subbase or as foundation material for subbase, the proper slopes to be used for cuts and fills, the proper shrinkage or swell factors to be used, the probable amount of subsidence under fills, the susceptibility to erosion, and the possibilities of slides or frost heaves. Where excavation is to be classified as earth, rock, etc., such classification is usually made as a part of the materials survey.

OTHER ROUTE SURVEYS

26-26. Railway Surveys. The procedures described in the preceding articles in this chapter apply for the most part to surveys for railway as well as highway location. In general the width of roadbed and right of way for railways is usually much less than for major highways, and the grade rates, particularly the maximum grade, are more critical. As in the case of highways, the justification for the location or relocation of a railway is based upon the added benefits to traffic as compared with the initial cost. For railways the benefits may be lower operating costs over a period of years for the estimated volumes of freight and passenger traffic, or they may be the result of new service which will be provided.

Owing primarily to the much lower rate of maximum grade usually allowable for railways, tunnels are used more frequently in railway than in highway location. Surveys for tunnels are discussed in Art. 28-11.

The economics of railway location, which depends to a great extent upon the maximum grade, is beyond the scope of this text. However, it is appropriate to state that for standard lines the grade seldom exceeds 1 or 2 per cent and generally is less than 0.5 per cent. For small branch lines in mountainous terrain the grade may sometimes be as high as 4 or 5 per cent. Maximum grades are compensated on curves to allow for curve resistance.

On a heavy-traffic high-speed railroad, every effort is made to have no curve sharper than 5 or 6° (Chap. 27). On lines of light traffic and relatively low speeds, 20° or sharper curves are used when necessary.

Railway construction surveys are discussed in Chap. 28.

26-27. Canal Surveys. The survey for location of a major canal is somewhat similar to that previously described for a highway except that the grades used are relatively flat and small differences in elevation are relatively more important.

Owing to lack of flexibility in permissible grades, the number and spread of the alternatives to be investigated during reconnaissance are usually much less than in highway location. In some cases existing maps such as U.S. Geological Survey quadrangle sheets at a scale of 1 in. = 2,000 ft., together with existing bench marks, will provide sufficient reconnaissance information when supplemented by field study and a minor amount of field leveling. If the reconnaissance is to be conducted entirely by field methods, the engineer's level is usually used with hubs being set every few hundred feet at the required grade elevation and the distances being measured by pacing or by stadia. The reconnaissance survey is run from a controlling point at one end of the line, either the selected point of diversion from the river or other canal at the upper end of the proposed canal or at the required location at the lower end, selected at sufficient elevation to place the canal above the area to be irrigated.

The grade to be used is selected so as to result in the desired velocity of flow with the chosen cross-section. For main canals a very flat grade will suffice, with a fall of 1 ft. or less per mile being all that is necessary in some cases. A velocity of 2 to 3 ft. per second is usually sufficient to prevent weeds and silt deposits. Average loamy soil will not be eroded at these velocities, while heavy soil with much gravel and rock will be safe against scouring at somewhat higher velocities. Canals in rock or lined with concrete will carry water safely at velocities up to 15 or 20 ft. per second. Any excess fall must be taken care of by drops or chutes, which are structures specially designed for that purpose.

As in the case of highways, the preliminary survey can be made either exclusively by field methods or by a combination of field and aerial survey methods. Where field methods are employed, the level party usually works ahead, setting stakes at grade as a guide for proper placing of the line. The transit or plane-table party then runs a tape or stadia traverse along the staked line and obtains sufficient topographic and planimetric detail that the final location line can be laid out in its proper location. Where large-scale mapping is to be obtained by photogrammetric methods, the procedures described in Art. 26-14 will generally apply except that in some cases it is necessary to run levels ahead of the traverse in order to insure the proper establishment of the traverse line and the mapping strip with respect to the proposed grade line.

In general the location and construction surveys for a canal are the same as those for a highway or railway. There are, however, some differences in the office projection of the center line due primarily to the shape of the cross-section. In shallow cuts, the canal cross-section is in the form of an excavated channel having an embankment constructed of the excavated material on each side (Fig. 10-10). In side-hill work the excavated material is used to form an embankment on the downhill side of the channel. Instead of a fill across low areas, as might be used in highway or railway construction, either a flume or an inverted siphon is commonly used. The water section is placed entirely in cut wherever possible.

26-28. Transmission-line Surveys. The required precision of surveys for transmission lines is generally much lower than that for a highway or a railroad. Reconnaissance can be carried out either by thorough field study or by use of aerial photographs supplemented by brief field inspections. If aerial photographs are available, tentative tower locations can be marked on the photographs and checked by field inspection if necessary.

The controlling considerations are usually economy of tower and insulator design, at the same time obtaining the shortest practicable location between termini in order to minimize power loss. Where there is no change of direction at a tower, the only loads to be considered in the design are the vertical load due to the weight of the cables, probable ice and wind loads, and possible occasional loading caused by breaking of a cable. At the end of a line of towers or where a change of directions causes similar conditions, there is a large horizontal force applied which requires special design. Therefore, the line is made as straight as possible, changes in direction being avoided wherever it is practicable to do so.

Although construction is less costly in level country, fairly heavy grades may be adopted to avoid changes in direction or unnecessarily high cost of right of way. Where possible, it is desirable to follow section lines or property lines in order to reduce the cost of right of way. If the line can be located near a highway or railway, construction cost as well as costs of patrolling and maintenance may sometimes be reduced. Rights of way are frequently obtained in the form of easements based upon a center-line description.

After the best location has been selected, a traverse is run following the proposed line as closely as practicable. Where distances between towers are relatively long, particularly in rugged terrain, electronic distance-measuring instruments (Art. 7-6) can be employed effectively. Stakes are usually necessary only at tentative tower locations, at any high points where cable clearance might be a factor, and at locations where buildings or other improvements might be affected. No curves are used in the alinement, changes in directions being made by angles at towers.

Elevations at staked points are obtained either by field leveling or by photogrammetric methods. A profile of these points is then plotted, and any necessary adjustments are made in the tentative tower locations. The selected locations are then marked on the ground, and the necessary stakes are set as a guide to placing the poles or the tower foundations.

26-29. Pipeline Surveys. Procedures for high-pressure pipeline surveys are more on the order of those required for transmission lines than for highways or railways. In addition to right-of-way costs, controlling considerations are total length and avoidance of rock excavation and expensive stream crossings. As in the case of transmission-line surveys, reconnaissance work can be minimized by the use of aerial photographs. In some cases their use will permit deferring of practically all field-survey work until the time of construction.

General practice is to run a field-survey traverse after the route is selected and to obtain elevations at summits and sags in the proposed line and at crossings of major streams where structures may be required. As an actual design of the grade line is unimportant, a complete profile is usually unnecessary.

26-30. Field Problem.

PROBLEM 1. PRELIMINARY SURVEY FOR ROAD (TOPOGRAPHY BY CROSS-PROFILE METHOD)

Object. To obtain data for a topographic map along the route of a proposed highway or railroad, locating the contours directly by means of the hand level and metallic tape. The data may be used in office problem 1 of Chap. 14. Steps 1 and 2 of the following procedure may have been accomplished in field problem 1 of Chap. 14 and field problem 1 of Chap. 10. For a preliminary survey run by an alternative method, employing transit and stadia, see field problem 3 of Chap. 15. For an exercise in setting slope stakes and taking final cross-sections, see field problem 3 of Chap. 10.

Procedure. (1) Over the assigned route, run an open deflection-angle traverse with transit and tape. (2) Establish vertical control for the route by direct profile leveling. (3) At each full station and at any necessary plus stations, locate the 5-ft. contours on a crossline extending 300 to 800 ft. on either side of the line; also locate the points where the 5-ft. contours cross the traverse line. Employ the hand level, topographer's rod, and metallic tape (see Art. 25-13). (4) Measure distances from the traverse line to other topographic features such as land lines, streams, roads, and buildings. (5) Note the quality of the land, as to whether it is clay, rock, or sand. Note the condition of the land, as to whether it is cleared land, pasture land, or woods; note the kind of trees and density of growth in woods. (6) Keep notes as shown in Fig. 14-5.

REFERENCES

1. American Society of Civil Engineers, "Report on Highway and Bridge Surveys," *Manual of Engineering Practice* 44, 1962.
2. American Association of State Highway Officials, "A Policy on Geometric Design of Rural Highways," 1954.
3. Funk, L. L., "Applications of Photogrammetry to the Location and Design of Freeways in California," *Highway Res. Board Bull.* 157, pp. 26–38, 1957.
4. Funk, L. L., "Terrain Data for Earthwork Quantities," *Highway Res. Board Bull.* 228, pp. 49–65, 1959.
5. Funk, L. L., "Adjustment of Photogrammetric Surveys," *Highway Res. Board Bull.* 228, pp. 21–27, 1959.
6. Funk, L. L., "From Map to Computer," *Highway Res. Board Bull.* 283, pp. 68–82, 1960.
7. Hickerson, Thomas F., "Route Surveys and Design," 4th ed., McGraw-Hill Book Company, New York, 1959.
8. Kane, C. V., "Contour Grading Highway Maps," *California Highways and Public Works*, pp. 1–5, September–October 1952.
9. Meyer, Carl F., "Route Surveying," 3rd ed., International Textbook Company, Scranton, Pa., 1962.
10. Miller, C. L., and T. Kaalstad, "Earthwork Data Procurement by Photogrammetric Methods," *Highway Res. Board Bull.* 199, pp. 24–39, 1958.
11. Preston, E. S., and D. E. Winsor, "Digitizing Stereoplotter Output for Preliminary Design and Construction," *Highway Res. Board Bull.* 312, pp. 43–73, 1961.
12. Pryor, William T., "Introduction to Photogrammetry and Aerial Surveys," *Highway Res. Board Bull.* 157, pp. 1–14, 1957.
13. Reynolds, F. M., "Data Processing," *California Highways and Public Works*, pp. 39–42, March–April 1960.
14. Manuals of instruction of state highway departments.
15. See also references at end of Chap. 27.

CHAPTER 27

ROUTE CURVES

27-1. General. In highway and railway location, the horizontal curves employed at points of change in direction are arcs of circles. The straight line connecting these *circular curves* are tangent to them and are therefore called *tangents*. For the completed line, the transition from tangent to circular curve and from circular curve to tangent may be accomplished gradually by means of a segment in form of a *spiral* (Arts. 27-11 to 27-13).

Vertical curves (Art. 10-15) are usually arcs of parabolas. Horizontal parabolic curves are occasionally employed in route surveying and in landscaping; they are similar to vertical curves and will not be discussed herein.

The subject of route curves is extensive, as indicated by the list of references at the end of this chapter. Herein are discussed only some of the simpler relationships. Curve calculations in terms of state plane coordinates are illustrated in Ref. 5 at the end of this chapter and in Ref. 1 at the end of Chap. 26.

CIRCULAR CURVES

27-2. General. The stationing of a route progresses around a curve in the same manner as along a tangent, as indicated in Fig. 27-1. The point where a circular curve begins is commonly called the *point of curve*, written P.C.; that where the curve ends is called the *point of tangent*, written P.T.;

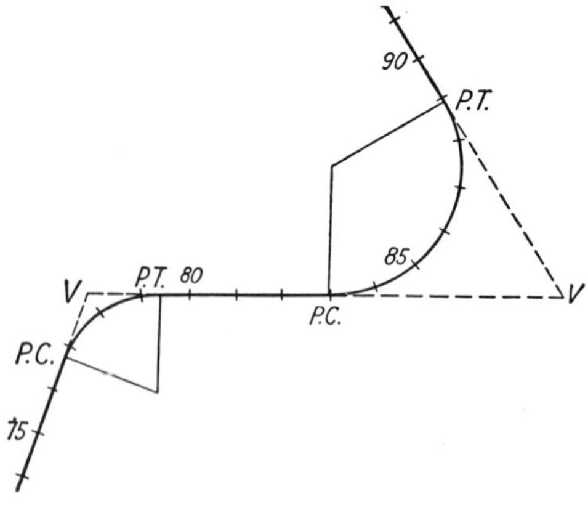

FIG. 27-1. Route stationing.

and that where two tangents produced intersect is called the *point of intersection* or the *vertex,* written P.I. or V. A point on the curve is written P.O.C. Other notations are also used; for example, the point of curve may be written T.C. signifying that the route changes from tangent to circular curve, whereas the point of tangent is written C.T. Or the beginning of curve may be written B.C. and the end of curve E.C. The point of change from tangent to spiral is written T.S., and the point of change from spiral to circular curve S.C.

In the field the distances from station to station (usually 100 ft.) on a curve are necessarily measured in straight lines, so that essentially the curve consists of a succession of 100-ft. chords. Where the curve is of long radius, as in railroad practice, the distances along the arc of the curve are considered to be the same as along the chords. In highway practice and along curved property boundaries, usually the distances are considered to be along the *arcs,* and the corresponding chord lengths are computed for measurement in the field.

The sharpness of curvature may be expressed in any of three ways:

1. *Radius.* By stating the length of the radius. This method is often employed in highway work, with the radius for a given curve taken as a multiple of 100 ft.

2. *Degree of Curve, Arc Basis.* By stating the "degree of curve," or the angle subtended at the center by an *arc* 100 ft. long. This method is usually followed in highway practice. Thus, if the degree of curve is D and the radius is R,

$$R = \left(\frac{360°}{D°}\right)\left(\frac{100}{2\pi}\right) \tag{1}$$

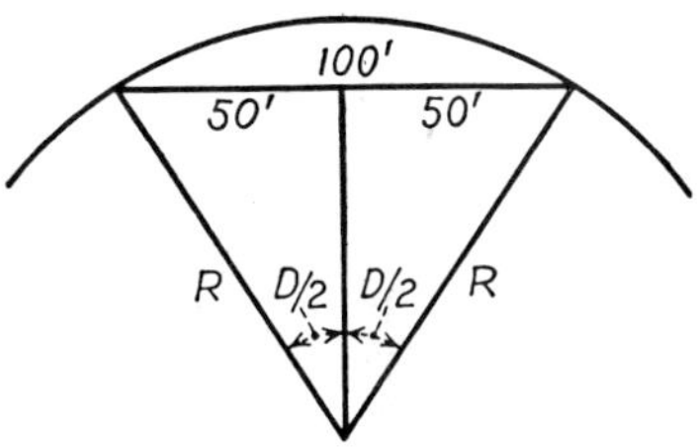

FIG. 27-2. Degree of curve.

From this relation the radius may be found if the degree of curve is known, and *vice versa.*

3. *Degree of Curve, Chord Basis.* By stating the degree of curve as the angle subtended by a *chord* of 100 ft. This method is followed in railroad practice. Thus in Fig. 27-2,

$$R = \frac{50}{\sin \frac{1}{2}D} \tag{2}$$

On the arc basis the radius of curvature varies inversely as the degree of curve; for example, the radius of a 1° curve is 5,729.58 ft. and the radius of a 10° curve is 572.96 ft. On the chord basis the radius of a 1° curve is 5,729.65 ft. and the radius of a 10° curve is 573.68 ft. The difference per 100 ft. in length between the chord and the arc is for a 1° curve less than 0.01 ft., for a 5° curve 0.03 ft., and for a 10° curve 0.13 ft.

Field measurements of the curve with the tape must, of course, be made along the chords and not along the arc. When the arc basis is used, either a correction is applied for the difference between arc length and chord length or the chords are made so short as to reduce the error to a negligible amount. In the latter case, usually 100-ft. chords are used for curves up to about 5°, 50-ft. chords from 5° to 15°, 25-ft. chords from 15° to 30°, and 10-ft. chords for curves sharper than 30°.

Except as specifically stated, hereinafter the discussions refer to the chord basis for expressing curvature.

27-3. Geometry of the Circular Curve. In discussing circular curves, the following geometrical facts are employed:

1. An inscribed angle is measured by one-half its intercepted arc, and inscribed angles having the same or equal intercepted arcs are equal. Thus in

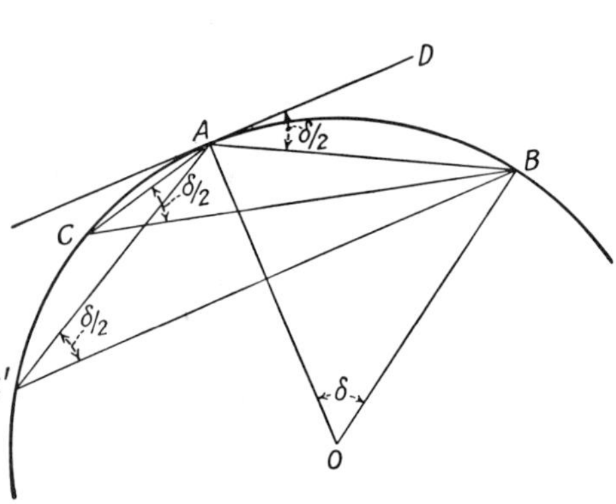

FIG. 27-3. Geometry of circular curve.

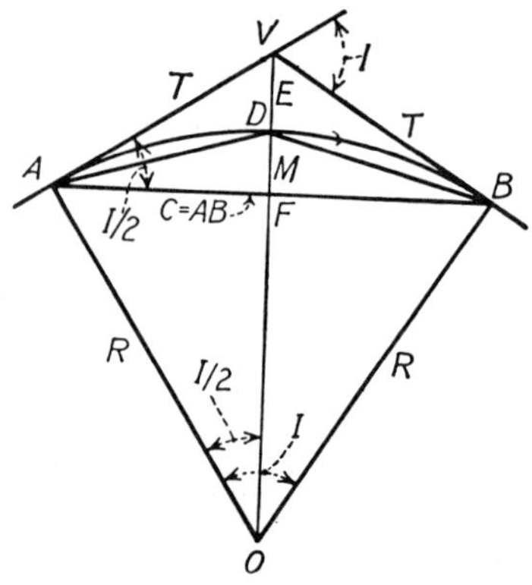

FIG. 27-4. Basis for curve formulas.

Fig. 27-3, the angle ACB (at any point C on the circumference) subtending an arc AB is one-half the central angle AOB subtending the same arc AB; and the angles at the points C and C' are equal.

2. An angle formed by a tangent and a chord is measured by one-half its intercepted arc. Thus in Fig. 27-3, the angle at the point A between AD, the tangent to the curve at that point, and the chord AB, is one-half the central angle AOB subtending the same arc AB. This is a special case of the proposition above, when the point C moves to A.

3. The two tangent distances to a circular curve, from the point of intersection of the tangents to the points of tangency, are equal.

4. Two angles are equal if their sides are perpendicular each to each, in the same order.

27-4. Curve Formulas. Figure 27-4 represents a circular curve joining two tangents. In the field the intersection angle I between the two tangents is measured. The radius of the curve is selected to fit the topography and the proposed operating conditions on the line. The line OV bisects the angles at V and at O, bisects the chord AB and the arc ADB, and is perpendicular to the chord AB at F. From the figure, $\angle AOB = I$ and

$$\angle AOV = \angle VOB = \tfrac{1}{2}I$$

The chord $AB = C$ from beginning to end of curve is called the *long chord.* The distance $AV = BV = T$ from vertex to P.C. or P.T. is called the *tangent distance.* The distance $DF = M$ from mid-point of arc to mid-point of chord is called the *middle ordinate.* The distance $DV = E$ from mid-point of arc to vertex is called the *external distance.*

Given the radius of the curve $OA = OB = R$ and the intersection angle I, then in the triangle OAV

$$\frac{T}{R} = \tan \tfrac{1}{2}I$$

$$T = R \tan \tfrac{1}{2}I = \text{tangent distance} \tag{3}$$

$$E = R \sec \tfrac{1}{2}I - R = R \operatorname{exsec} \tfrac{1}{2}I = \text{external distance} \tag{4}$$

From the triangle AOF, in which $AF = \tfrac{1}{2}C$,

$$C = 2R \sin \tfrac{1}{2}I = \text{long chord} \tag{5}$$

$$M = R - R \cos \tfrac{1}{2}I = R \operatorname{vers} \tfrac{1}{2}I = \text{middle ordinate} \tag{6}$$

From the triangle AVF, in which $\angle VAF = \tfrac{1}{2}I$ and $AF = \tfrac{1}{2}C$,

$$\frac{C}{2} = T \cos \tfrac{1}{2}I$$

$$C = 2T \cos \tfrac{1}{2}I \tag{7}$$

From the triangle ADF, in which $\angle DAF = \tfrac{1}{4}I$,

$$M = \tfrac{1}{2}C \tan \tfrac{1}{4}I \tag{8}$$

27-5. Length of Curve. The length of the circumference of a circle is $2\pi R$; this is the arc length for a full angle or 360°. As the arc length corresponding to a given radius varies in direct proportion to the central angle subtended by the arc, the length of arc for any central angle I is

$$\text{Arc} = \left(\frac{I^\circ}{360^\circ}\right) 2\pi R \tag{9}$$

in which the angle I° is expressed in degrees. This solution is simplified by the use of a table of arc lengths for various angles and for unit radius.

If the degree of curvature is expressed on the *arc* basis, from Eqs. (1) and (9) the length of curve L_a is

$$L_a = 100 \frac{I}{D} \tag{10}$$

If the degree of curvature is expressed on the *chord* basis, the length of curve is considered to be the sum of the lengths of the chords, normally each 100 ft. long. In this case also, the length of curve (on the chords) is

$$L_c = 100 \frac{I}{D} \tag{11}$$

which is somewhat less than the actual arc length. Thus if the central angle I of the curve AD (Fig. 27-5) is equal to three times the degree of curve D, as shown, then there are three 100-ft. chords between A and D, and the length of "curve" on this basis is 300 ft.

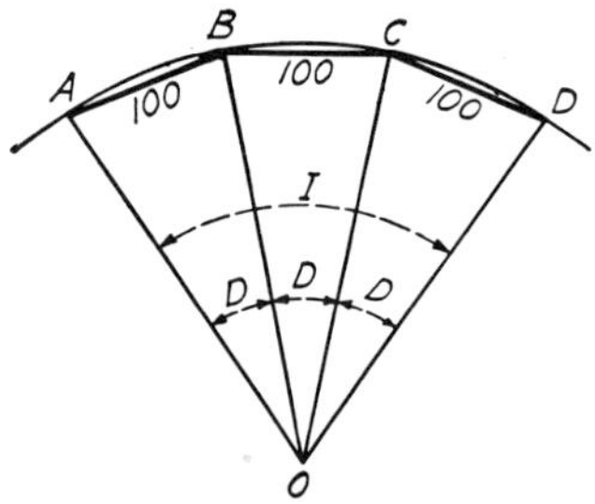

FIG. 27-5. Length of curve.

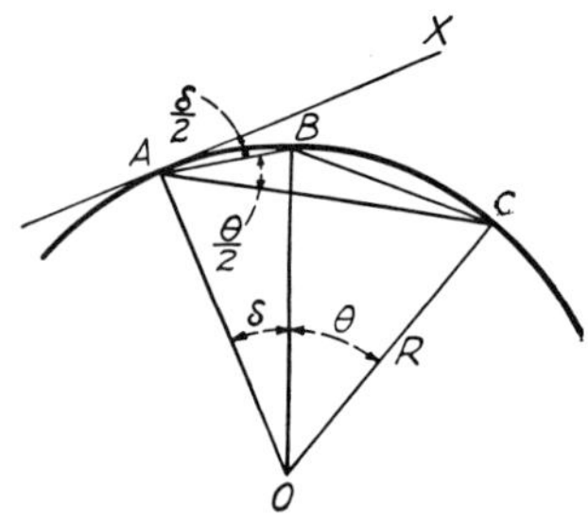

FIG. 27-6. Curve by deflection angles.

27-6. Laying Out Curve by Deflection Angles. Curves are staked out usually by the use of deflection angles turned at the P.C. from the tangent to stations along the curve together with the use of chords measured from station to station along the curve. The method is illustrated in Fig. 27-6, in which ABC represents the curve, AX the tangent to the curve at A, and angles XAB and XAC the deflection angles from the tangent to the chords AB and AC.

Assume the transit to be set up at A. Given R, δ, θ. Required to locate B and C. Considering the point B,

$$\angle XAB = \tfrac{1}{2}\delta \tag{12}$$

$$AB = 2R \sin \tfrac{1}{2}\delta \tag{13}$$

In the field, the point B is located as follows: The deflection angle $XAB = \frac{1}{2}\delta$ is set off from the tangent, the distance AB is measured from A, and the forward end of the tape at B is lined in with the transit.

Considering the point C,

$$\angle BAC = \tfrac{1}{2}\theta \tag{14}$$

$$BC = 2R \sin \tfrac{1}{2}\theta \tag{15}$$

$$\angle XAC = \tfrac{1}{2}(\delta + \theta) \tag{16}$$

In the field, the point C is located as follows: With the transit still at A, the deflection angle XAC is set off from the tangent, the distance BC is measured from B, and the forward end of the tape at C is lined in with the transit sighted along the line AC. Succeeding stations on the curve are located in similar manner.

Should the chord lengths be given instead of the central angles, then the angles are computed; for example, in Fig. 27-6 $C_1 = 2R \sin \frac{1}{2}\delta$ and $C_2 = 2R \sin \frac{1}{2}\theta$, in which the radius R and chord lengths C_1 and C_2 are known.

If B is at a full station and the distance BC is the 100-ft. distance to the next full station at C, then $\theta = D =$ degree of curve, and $\angle BAC = \frac{1}{2}D$.

A curve is located in the field normally as follows: The P.C. and P.T. are marked on the ground. The deflection angle from the P.C. is computed for each full station on the curve and for any intermediate stations that are to be located. The transit is set up at the P.C., a backsight is taken along the tangent with telescope inverted, the telescope is plunged, and each point on the curve is located by deflection angle and by distance measured from the preceding *full* station to points not more than 100 ft. ahead. The following example illustrates the procedure and gives the usual form of field notes.

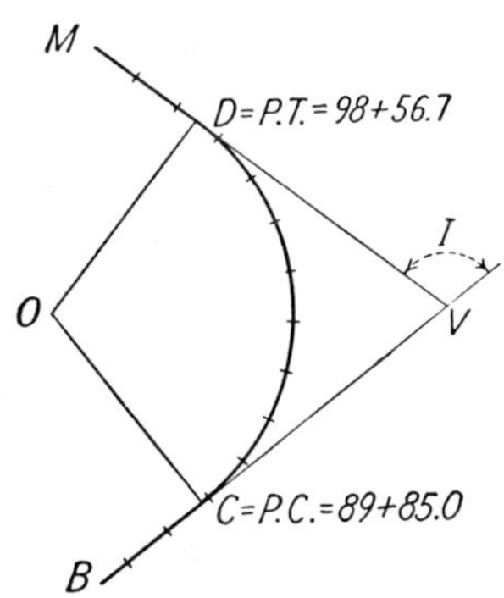

FIG. 27-7a. Laying out curve.

Example: In Fig. 27-7a, assume that stations have been set as far as B. The directions of the tangents BV and VM have been fixed by hubs, but distances along these tangents have not been measured. The degree of the curve CD is to be 12°00′. It is desired to stake out the curve CD.

The tangents BV and VM are run to an intersection at V, the transit is set at V, and the angle I is read and found to be 104°36′. With the degree of curve assumed for the curve CD the radius is determined, and the equal tangent distances CV and VD are computed, as follows:

$$R = \frac{50}{\sin \frac{1}{2}D} = 478.3 \text{ ft.}$$

$$T = R \tan \tfrac{1}{2}I = 478.3 \tan 52°18' = 618.9 \text{ ft.}$$

By measurement from V, hubs are set at C and D. Chaining is then carried forward from B, and the station and plus of C, the P.C. of the curve CD, is found to be 89 + 85.0.

Station 90, the first full station on the curve, is 0.150 station beyond C, that is, the central angle subtended by the arc from C = 89 + 85.0 to station 90 is 0.150 the degree of curve.

The central angle from C to station 90 is 12°00′ × 0.150 = 1°48′. The exact distance (along the chord) from C to station 90, computed by the formula $C = 2R \sin$

(1°48′/2), is 15.03 ft. In curves such as this, with relatively long radius, this chord would usually be assumed as proportional to the central angle, in this case 15.00 ft. long.

The length of curve (along the chords) from P.C. to P.T. is

$$L = 100\frac{I}{D} = 100\frac{104.6}{12.00} = 871.7 \text{ ft.}$$

$$\begin{aligned} \text{Station at P.C.} &= 89 + 85.0 \\ L &= 8 + 71.7 \\ \hline \text{Station at P.T.} &= 98 + 56.7 \end{aligned}$$

The deflection angle for station 90 is 1°48′/2 = 0°54′.

The angle at P.C. between station 90 and station 91 is 12°00′/2 = 6°00′. Therefore, the deflection angle from tangent to station 91 (angle V–P.C.–91) is 6°54′.

LOCATION OF CIRCULAR RAILROAD CURVE 25

Station	Point	Deflection Angle	Bearing of Tangents	Description of Curve
100				
99			N 73°10′W	
98+56.7	P.T. ⊙	52°18′		
98		48°54′		
97		42°54′		
96		36°54′		D=12°L
95		30°54′	Diff.	I=104°36′
94	⊙	24°54′	104°36′	T=618.9
93		18°54′		R=478.3
92		12°54′		L=871.7
91		6°54′		
90		0°54′		
89+85.0	P.C. ⊙	0°00′		12°L
89			N31°26′E	
88				

K&E Transit No.256 — July 31, 1964
100 Ft. Steel Tape — Clear, Warm
Line Rod — Hayle, R. ⊼
Reading Glass — Grad, W.O. Rod
Chaining Pins — Bush, S. Chain
Plumb Bobs
Hammer
Stakes, Hubs, Etc.

COMPUTATIONS

Given: $D = 12°00′$; $I = 104°36′$

Then $R = \dfrac{50}{\sin \frac{1}{2} D} = 478.3$ ft.; $T = R \tan \frac{1}{2} I = 618.9$ ft.

$L = 100 \frac{I}{D} = 871.7$ ft.

First Deflection Angle $= \frac{15}{100} \times \frac{12°00′}{2} = 0°54′$

Deflection Angle for Full Station $= D/2 = 6°00′$

Last Deflection Angle $= I/2 = 52°18′$

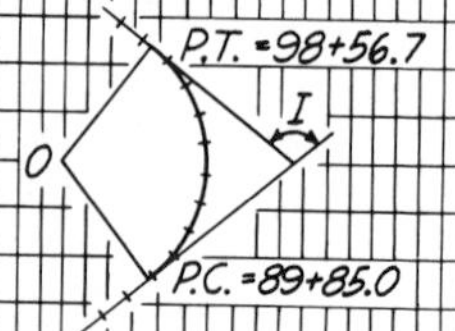

FIG. 27-7*b*. Field notes for circular curve.

Similarly the angle 91–P.C.–92 is 6°00′ and the angle V–P.C.–92 is 12°54′. By the same process the remaining deflection angles from the tangent to full stations on the curve are computed and tabulated up the page as shown in Fig. 27-7*b*.

Consider the angle 98–P.C.–P.T. The central angle 98–O–P.T. from 98 to P.T. is subtended by an arc which is 0.567 stations long; hence the angle 98–O–P.T. is 12°00′ × 0.567 = 6°48′. One half of this is 3°24′, the angle 98–P.T.–V. This added to the deflection angle for station 98 (48°54′) gives the total deflection angle 52°18′. It may be noted that the total deflection angle should equal ½I, and in this example, since the angle 52°18′ = 104°36′/2, there is given a check on the computation of all the deflection angles.

Figure 27-7b illustrates the usual form of field notes, the direction of the curve with respect to the tangent being designated by the letter R or L for right or left, following the degree of curve; thus 12°L indicates a 12° curve to the left of the tangent at the P.C. Where the field notes include measurements to details, the notes should show whether the measurements are made from the curve or from the tangent.

27-7. Transit Set-ups on the Curve. On account of obstacles, great length of curve, and so on, often it is impracticable or impossible to run all of a given curve with the transit at the P.C. In such cases, one or more set-ups are required along the curve between P.C. and P.T.

Figure 27-8 illustrates the case where the transit is set up at some intermediate point A. The curve is begun at the P.C. and is located as far as A, where a hub is set. The transit is then set up at A. A backsight (with telescope inverted) is taken on the last preceding station at which the transit was set up, in this case the P.C.; the telescope is plunged; and the angle ½α (half the central angle subtended by the chord sighted over) is turned off as shown. The line of sight is thus directed along the tangent at A. Deflections to points beyond A are turned as in the case previously explained. The angle ½α between the tangent at A and the chord A–P.C. is equal to the deflection angle at the P.C. for the point A; therefore, the vernier setting to locate point B from the transit station A is the same as that which would have been used had the transit remained at the P.C. According to this method the following procedure may be used to orient the transit at any point on the curve:

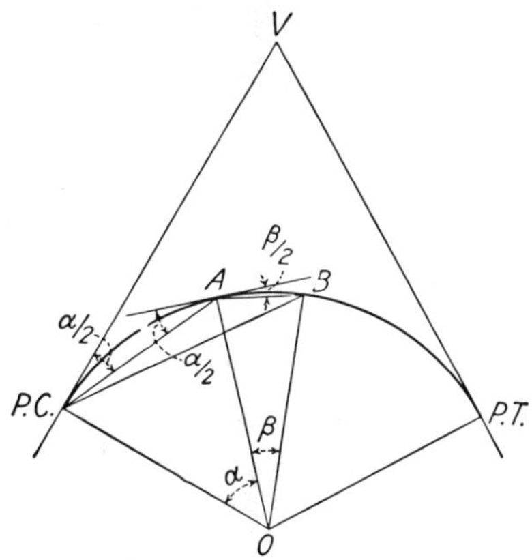

FIG. 27-8. Transit set-ups on curve.

1. Compute deflection angles as for use at the P.C.
2. When set up at any point on the curve, backsight (with telescope inverted) at any preceding transit station, with the vernier reading the computed deflection angle for the point sighted.
3. To locate other points, plunge the telescope and use the computed deflection angles.

If the backsight is to the P.C., the backsight vernier reading is zero.

27-8. Laying Out Curve by Intersection. Where the character of the topography renders chaining difficult, occasionally curves are laid out by the method of intersection, with one transit at, say, the P.C. or an intermediate point and another transit at the P.T. Each station on the curve is located by simultaneous sighting with the two transits. However, some of the angles of intersection will be so small that the precision of the method is relatively low.

27-9. Laying Out Curve by Tape Alone. Often it is convenient or necessary to lay out a circular curve by means of the tape alone. Of the various methods employed, three of the more useful are briefly described here.

Offsets from Tangent. When the angle of intersection of two tangents is small, occasionally it is convenient to establish the various points on the curve by perpendicular offsets from the tangents. For example, it is desired to establish the point A at a given station on the curve shown in Fig. 27-9,

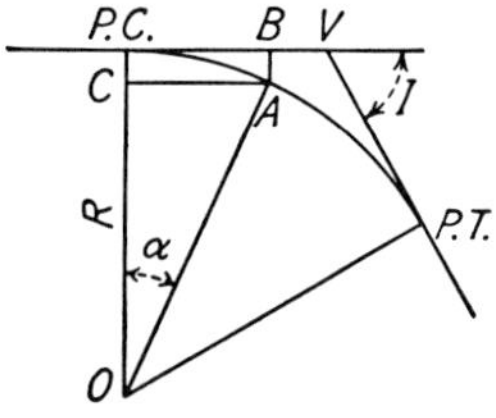

FIG. 27-9. Offsets from tangent.

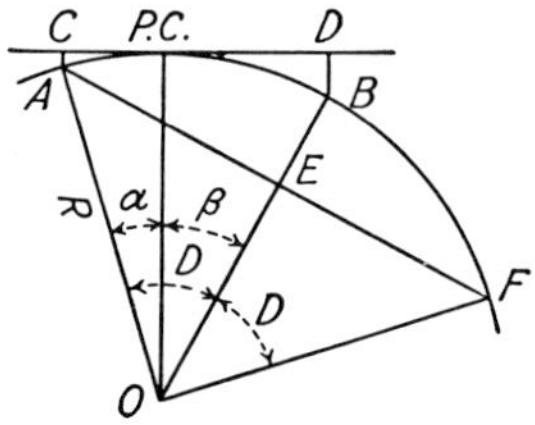

FIG. 27-10. Curve by middle ordinate.

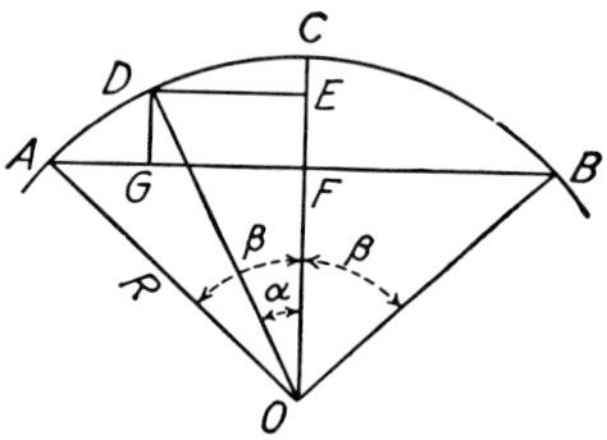

FIG. 27-11. Offsets from chord.

the intersection angle and degree of curve being known. The central angle α is equal to the distance from P.C. to A, in stations, multiplied by the degree of curve. The distance along the tangent from P.C. to B, the foot of the perpendicular offset, is equal to $R \sin \alpha$; and the length of the offset BA is equal to R vers α or $R(1 - \cos \alpha)$. In the field, the point B is established by measuring along the tangent from the P.C. If the offset distance BA is very short, the perpendicular may be established with sufficient precision by estimation; otherwise the point A is established by measuring from B and the P.C. with two tapes. Other points on the curve are established similarly, those on the second half of the curve being located by offsets from the forward tangent. A transit may be used to establish the offsets.

Horizontal parabolic curves are laid out by offsets from the tangent, in a manner similar to that described for vertical curves (Art. 10-15).

Middle Ordinate. Another method of laying out a circular curve with the tape involves the location of successive stations by use of the middle ordinate of a two-station chord. The first full station on the curve is located by offset from the tangent as just described. Thus in Fig. 27-10, B is the first full station on the curve, the distance from P.C. to B being less than one station. To start the curve, the point A is similarly established by perpendicular offset from the traverse line to the P.C., the angle α being made such that $\alpha + \beta = D$, the degree of curve, and the corresponding offset CA being equal to R vers α. Then the chord distance AB is equal to one station (on the chord basis).

The length of the mid-ordinate BE of the two-station chord AF is then computed by the relation $BE = R$ vers D. In the field, the distance BE is laid off from B along a line whose direction is estimated to be that of the radius BO of the curve. Points A and E are marked on the ground by flags. One end of the 100-ft. tape is held at B, and the forward end of the tape at F is swung until it is in line with points A and E. The full station F is marked on the ground. In a similar manner the next full station is established by means of a middle ordinate at F; and so on around the curve. The work is checked by offsets from the tangent at the P.T., similar to those used at the P.C.

Offsets from Chord. Points on a circular curve may be located by perpendicular offsets from any chord, as illustrated by Fig. 27-11. Suppose that stations A and B have been established on the ground, and that it is desired to establish a point D of known stationing by means of the chord distance AG and the offset distance GD. The line OC bisects the arc and the chord. From the known stationing, the angles α and β are computed. Then

$$AG = AF - DE = R \sin \beta - R \sin \alpha = R(\sin \beta - \sin \alpha) \qquad (17)$$

and

$$GD = FC - EC = R(\text{vers } \beta - \text{vers } \alpha) \qquad (18)$$

27-10. String-lining of Curves. Railroad track, particularly on curves, is eventually thrown out of alinement by the action of trains. *Stringlining* is a simple method of determining and applying the amounts by which the track must be moved laterally at various points to restore proper curvature. It involves the use of middle ordinates from chord to curve (see Art. 27-9); it is described in detail in various texts on route surveying. Briefly, the method is as follows: At regular intervals along the outer rail, a chord of length equal to two intervals is stretched, and the middle ordinate is measured with a scale and is recorded. For the circular portion of the curve, all middle ordinates should be equal; for the portion along which a gradual transition is made from curve to tangent, the middle ordinates should be progressively smaller by uniform increments. Irregularities in the tabulated values of middle ordinate are noted, and for each point of measurement the amount necessary to move the track is computed. Stakes are set in the ballast to serve as reference points, and the track is moved to conform with computed values.

SPIRAL CURVES

27-11. Superelevation. On a railway curve the velocity of movement of a train develops a horizontal centrifugal force. In order that the plane of the rails may be normal to the resultant of the horizontal and vertical forces acting on a car, the outer rail is *superelevated,* or elevated above the inner rail. The amount of superelevation is made equal to approximately $0.00067V^2D$ expressed in inches, in which expression V is the train speed in miles per hour and D is the degree of curve in degrees. The amount of this superelevation should not exceed 7 or 8 in. on account of the use of the track by slow trains. For a speed of 40 miles per hour, it equals in inches, a fraction more than the degree of curve in degrees. The elevation of the inner rail is maintained at grade.

Similarly, on highway curves (except very flat curves) the roadway is superelevated. Various tables giving recommended values of superelevation are published. Usually the center line of the roadway is maintained at grade, the outer edge of the pavement is raised one half of the superelevation distance, and the inner edge of the pavement is depressed by an equal amount.

Since the superelevation should be attained gradually near the end of each curve, it is desirable that the centrifugal force be built up gradually, so that there is an approximate balance between the two at all points.

27-12. Railway Spirals. On railway lines where trains are to be operated at high speed, it is common practice to insert between circular curve and tangent a curve of varying radius, called a *spiral,* in order that the degree of curvature and centrifugal force may be developed gradually. At the end of the spiral adjacent to the tangent its radius is very long; along the curve it decreases gradually until at the point where the spiral joins the circular curve the radii of the two are equal. Spiral curves are also called *easement curves* or *transition curves.*

In order to provide room for the spiral, the circular curve is offset from the main tangent, as to the position $AFGB$ of Fig. 27-12. If the two spirals EF and GH are of equal length, the offsets AC and BN are equal, and the distance $VC = VN = (R + o) \tan \frac{1}{2}I$, in which o is the length of the offset.

Many mathematical solutions of the spiral are available, and the reader is referred to these for exact values (see Ref. 2 at the end of this chapter). The following approximate and empirical solution is not greatly in error.

1. The central angle I and the degree of circular curve D are known.

2. The length of spiral L' is selected (Ref. 2); for curves likely to limit train speed L' should be not less than 240 ft.; for minor curves, L' may be 100 ft. or even less.

3. The length of the offset $o = AC = BN$ is computed. This may be assumed to be 6.50 ft. for $D = 10°$ and $L' = 300$ ft., varying directly as the degree of curve and as the square of the length of spiral. Thus for

$D = 5°$ and $L' = 200$ ft.,

$$o = \frac{5}{10} \times \frac{(2)^2}{(3)^2} \times 6.50 = 1.44 \text{ ft.}$$

4. $VC = VN = (R + o) \tan \frac{1}{2}I$.

5. $EC = NH$ = one half of the spiral length minus a correction. For curves of dimensions common in railroad practice this correction has approximately the following values: for spiral angle $\Delta = 5°$, 0.06 ft.; for $\Delta = 10°$, 0.25 ft.; for $\Delta = 15°$, 0.50 ft. For exact formula, see Ref. 2.

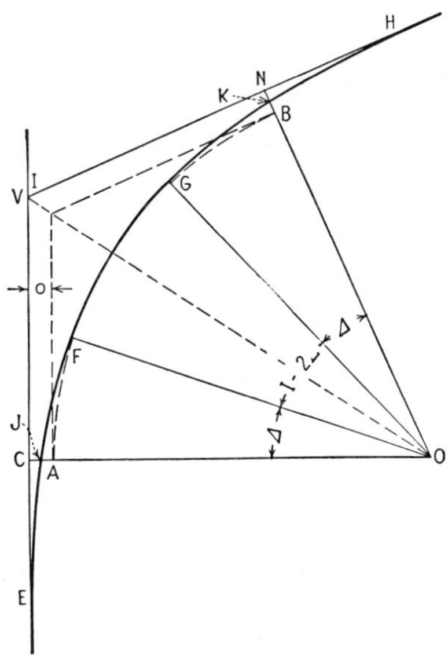

Fig. 27-12. Railway spiral.

6. Spirals bisect the offsets AC and BN so that $CJ = \frac{1}{2}AC$ and $NK = \frac{1}{2}BN$.

7. Between E and J, perpendicular offsets from the tangent to the spiral vary in proportion to the cubes of the distances from E; between J and F, radial offsets from the circular curve to the spiral vary as the cubes of the distances from F; similarly for the other spiral GH.

8. Angle AOF = angle $BOG = \Delta = DL'/200$.

9. In the field, the points N, B, C, and A are located, and the direction of each offset tangent is established by means of another and equal offset from the main tangent. The simple curve $AFGB$ is located.

The necessary offsets are made to points on the spirals. For construction surveys it is usually sufficient to offset the circular curve, leaving the staking of the spirals to be done after the line is graded.

10. The alinement with spirals is along the line *EJFGKH*.

27-13. Highway Spirals. The procedure of computing and laying out a highway spiral is similar to that for railway spirals to which reference has been made in the preceding article. Important simplifications have been made in the procedure through the publication of tables which give (1) values of the various functions involved, over a wide range; and (2) recommendations for superelevations, minimum transition lengths, safe maximum curvatures for various speeds, and widening of the pavement at the curve (see Ref. 3 at the end of this chapter).

27-14. Numerical Problems.

1. Given: $I = 34°30'$, D (chord basis) $= 3°00'$, and P.C. = station 74 + 30.0. Required: R, L, T, and E; also deflection angles arranged in notebook form for staking out this curve, using 100-ft. stations.
2. Given: $I = 92°30'$, $T = 425.00$ ft., and P.C. = station 25 + 10.00. Required: R, D (arc basis), C, E, M, and L; also deflection angles arranged in notebook form for staking out this curve, using 50-ft. stations.
3. If the curve of problem 2 represents the center line of a highway curve, suppose it is desired to set alinement stakes along two curves, one of which is to be 10 ft. outside, and the other 12 ft. inside, the center line. Required: L_1, L_2, D_1, D_2, E_1, E_2, and the deflection angles arranged in notebook form for staking out these curves.
4. Given: $I = 60°40'$, $E = 125.5$ ft. Required: R, D (arc basis), C, T, M, and L.
5. Two tangents AV and BV have an intersection angle of 45°00′. A point C is located by the coordinates $VH = 270.2$ ft. and $HC = 157.4$ ft., VH being measured along the tangent VA, and HC being measured perpendicular thereto. It is desired to connect the two tangents with a curve passing through the point C. Required: R, D (chord basis), T, L, and E.
6. Having calculated the values in problem 5, change the value of D to 2°00′. Change all other elements of the curve to agree with the new value of D and compute the deflection angles.
7. Given the data of problem 1. Make the necessary computations for the insertion of a spiral of length 250 ft. at each end of the curve.

27-15. Field Problem.

Problem 1. Laying Out a Circular Curve

Object. To lay out a circular curve, as for the curb line of a driveway, by the use of deflection angles.

Procedure. (1) From the assigned central angle and degree of curve, compute the tangent distance and the length of curve. (2) Assume that the P.C. is station 9 + 83.2, and compute deflections for each full station and +50. Prepare notes in the form shown in Fig. 27-7*b*. (3) In the field, locate two tangents making the assumed angle. Locate the P.C. and the P.T. (4) Set up the transit at the P.C., orient the instrument, and stake out the full and +50 stations by the method described in Art. 27-6. Report the error observed at the P.T. (5) Set up the transit at the P.T. and check the angle. (6) If transit set-ups on the curve are required, follow the method of Art. 27-7.

REFERENCES

1. Allen, C. F., "Railroad Curves and Earthwork; Field and Office Tables," 7th ed., McGraw-Hill Book Company, New York, 1931.
2. American Railway Engineering Association, "Manual," current edition.
3. Barnett, Joseph, "Transition Curves for Highways," U.S. Bureau of Public Roads, Government Printing Office, Washington, D.C., 1939.
4. Ives, H. C., "Highway Curves," 4th ed., by Philip Kissam, John Wiley & Sons, Inc., New York, 1952.
5. Mitchell, Hugh C., and Lansing G. Simmons, "State Plane Coordinates in Route Surveying," *Surveying and Mapping*, Vol. 12, No. 3, pp. 261–269, July–September, 1952.
6. Searles, W. H., H. C. Ives, and Philip Kissam, "Field Engineering, a Handbook of the Theory and Practice of Railway Surveying, Location, and Construction," 22d ed., John Wiley & Sons, Inc., New York, 1949.
7. Skelton, Russell R., "Route Surveys," McGraw-Hill Book Company, New York, 1949.
8. See also references at end of Chap. 26.

CHAPTER 28

CONSTRUCTION SURVEYING

28-1. General. Surveys for construction usually involve the following: (1) making a topographic survey of the site, to be used in the preparation of plans for the structure, (2) establishing on the ground a system of stakes or other markers, both in plan and in elevation, from which measurement of earthwork and structures can be taken conveniently by the construction force, (3) giving line and grade as needed either to replace stakes disturbed by construction or to reach additional points on the structure itself, and (4) making measurements necessary to verify the location of completed parts of the structure and to determine the volume of work actually performed up to a given date (usually each month), as a basis of payment to the contractor.

In connection with construction, often it is necessary to make property-line surveys as a basis for the acquisition of lands or rights of way (Chap. 22).

The detailed methods employed on construction surveys vary greatly with the type, location, and size of structure and with the preference of the engineering and construction organizations. Much depends on the ingenuity of the surveyor to the end that the correct information is given without confusion or needless effort.

The topographic survey of the structure site should include adjacent areas that are likely to be used for construction plant, roads, or auxiliary structures. Aerial photographs are useful aids for planning the construction.

28-2. Alinement. Temporary stakes or other markers are usually set at the corners of the proposed structure, as a rough guide for beginning the excavation. Outside the limits of excavation or probable disturbance but close enough to be convenient, are set permanent stations which are established with the precision required for the measurement of the structure itself. These permanent stations should be well referenced (Art. 14-17), with the reference stakes in such number and in such position that the loss of one or two will not invalidate the reference. Permanent targets or marks called *foresights* may be erected as convenient means of orienting the transit on the principal lines of the structure and for sighting along these lines by eye.

Stakes or other markers are set on all important lines in order to mark clearly the limits of the work. The number of such markers should be sufficient to avoid the necessity for many measurements by the workmen but should not be so great as to cause confusion. A simple and uniform system of designating the various points, satisfactory to the construction superintendent, should be adopted. Also the exact points, lines, and planes from which and to which measurements are to be made should be well understood.

In many cases, line and grade are given more conveniently by means of *batter boards* than by means of stakes. A batter board is a board (usually 1 by 6 in.) nailed to two substantial posts (usually 2 by 4 in.) with the board horizontal and its top edge preferably either at grade or at some whole number of feet above or below grade. The alinement is fixed by a nail driven in the top edge of the board. Between two such batter boards a stout cord or wire is stretched to define the line and grade.

Often it is impracticable to establish permanent markers on the line of the structure. Thus the face of a bridge abutment may be beyond the shore line and therefore inaccessible. Also, stakes placed at the edge of a concrete pavement would interfere with grading and with setting the forms. In such cases the survey line is established parallel to the structure line, as close as practicable and with the offset distance some whole number of feet.

28-3. Grade. A system of bench marks is established near the structure in convenient locations that will probably not be subject to disturbance, either of the bench mark itself or of the supporting ground. From time to time these bench marks should be checked against one another to detect any disturbance. Every care should be taken to preserve existing bench marks of state and Federal surveys; if construction necessitates the removal of such marks, the proper organization should be notified and the marks transferred in accordance with its instructions.

The various grades and elevations are defined on the ground by means of pegs and/or batter boards, as a guide to the workmen. The grade pegs may or may not be the same as the stakes used in giving line. When stakes are used, the vertical measurements may be taken from the top of the stake, from a keel mark or a nail on the side of the stake, or (for excavation) from the ground surface at the stake; in order to avoid mistakes, only one of these bases for measurement should be used for a given kind of work, and the basis should be made clear at the beginning of construction. When batter boards are used, the vertical measurements are taken from the top edge of the board, which is horizontal. The stake or the batter board may be set either at grade or at a fixed whole number of feet above or below grade.

When a stake is to be driven with its top at a given elevation, the rodman starts the stake and then holds the rod on the stake. The levelman reads the rod and calls out the approximate distance the stake must be driven to reach grade. The rodman drives the stake nearly the desired amount, and a second rod reading is taken; and so the process is continued until the rod reading is made equal to the difference between the height of instrument and the desired elevation. If a mark or nail on the side of the stake is to be used instead of the top of the stake, the rod is moved up or down the side of the stake until the levelman signals that the rod reading is correct. In some cases, the stake is sawed off at the desired elevation. If the grade elevation is only a short distance below the ground elevation, a hole may be dug in order that the stake may be driven to grade; this procedure avoids the necessity of measuring down from the top of the stake.

28-4. Precision. For purposes of excavation only, usually elevations are given to the nearest 0.1 ft. For points on the structure, usually elevations to 0.01 ft. are sufficiently precise. Alinement to the nearest 0.01 ft. will serve the purpose of most construction, but greater precision may be required for prefabricated steel structures or members.

It is desirable to give dimensions to the workmen in feet, inches, and fractions of an inch. Ordinarily measurements to the nearest ¼ or ⅛ in. are sufficiently precise, but certain of the measurements for the construction of buildings and bridges should be given to the nearest 1/16 in. Often it is convenient to use the relation that ⅛ in. equals approximately 0.01 ft.

28-5. Establishing Points by Intersection. Where conditions render the use of the tape difficult or impossible, often points are established at the intersection of two transit lines by simultaneous sighting with two transits in known locations. The process is the inverse of that in which the location of a ground point is determined by the method of intersection (Art. 14-8). By this method, points may be located in elevation as well as in plan. The precision of measurement is made commensurate with the requirements of construction.

28-6. Highways. Usually just prior to the beginning of construction of a section of highway, the located line is rerun as a base line for construction, missing stakes are replaced, and hubs are referenced. Borrow pits (if necessary) are staked out and cross-sectioned. Lines and grades are staked out for bridges, culverts, and other structures. If slope stakes have not already been set during the location survey, they are set except where clearing is necessary; in that case they are set when the right of way has been cleared. For purposes of clearing, only rough measurements from the center-line stakes are necessary.

The method of setting and marking slope stakes is described in Art. 10-10. Additional stakes may be offset a uniform distance away from the work, with appropriate marking to indicate the offset. If large intercepting ditches are to be placed along the cuts, these are staked out also.

Where the depth of cuts and fills does not average more than about 3 ft., the slope stakes may be omitted; in this case the line and grade for earthwork may be indicated by a line of pegs (with guard stakes) along one side of the road and offset a uniform distance such that they will not be disturbed by the grading operations. Pegs are usually placed on both sides of the road at curves, and may be so placed on tangents; when this is done, measurements for grading purposes may be taken conveniently by sighting across the two pegs or by stretching a line or tape between them.

When rough grading has been completed, a line of finishing stakes is set on both sides of the roadway. It should be understood whether these are to subgrade or to finish grade (final grade). If the slopes of cuts are terraced to provide drainage, finishing stakes are set along the terraces.

For concrete highways to give line and grade for the pavement, a line of stakes is set along each side, offset a uniform distance (usually 2 ft.) from the edge of the pavement. The grade of the top of the pavement, at the edge, is indicated either by the top of the stake or by a nail or line on the side of the stake. The alinement is indicated on one side of the roadway only, by means of a tack in the top of each stake. In another method, pegs are set so that the side forms may be placed directly upon them, and a line of stakes is set near one edge to give line for the forms. The distance between stakes in a given line is usually 100 or 50 ft. on tangents at uniform grade and half the normal distance on horizontal or vertical curves. The dimensions of the finished subgrade and of the finished pavement are checked by the construction inspector, usually by means of a templet.

For asphalt pavements, the line of finishing stakes for grading is usually sufficient.

As construction proceeds, monthly estimates are made of the work completed to date. A quantity survey is made near the close of each month, and the volumes of earthwork, etc. are classified and summarized as a basis for payment.

28-7. Streets. For street construction the procedure of surveying is similar to that just described for highways. Ordinarily the curb is built first. The line and grade for the top of each curb are indicated by pegs driven just outside the curb line, usually at 50-ft. intervals. The grade for the edge of the pavement is then marked on the face of the completed curb; or for a combined curb and gutter it is indicated by the completed gutter. Ground pegs are set on the center line of the pavement, either at the grade of the finished subgrade (in which case holes are dug when necessary to place pegs below the ground surface) or with the cut or fill indicated on the peg or on an adjacent stake. Where the street is wide, an intermediate row of pegs may be set between center line and each curb. It is usually necessary to reset the pegs after the street is graded. Where driving stakes is impractical because of hard or paved ground, nails or spikes may be driven or marks may be cut or painted on the surface.

The surveys for street location and construction should determine the location of all surface and underground utilities that may affect the project; and notification of necessary changes should be given well in advance. Information regarding the desirable location of underground utilities, together with methods of surveying and mapping, is given in *Manual* 14, "Location of Underground Utilities," of the American Society of Civil Engineers.

28-8. Railroads. Surveys for railroad earthwork are similar to those described in Art. 28-6 for highways. Prior to construction the located line is rerun, missing stakes are replaced, hubs are referenced, borrow pits are staked out, slope stakes are set, and lines and grades for structures are established on the ground. When rough grading is completed, finishing stakes are set to grade at the outer edges of the roadbed, as a guide in trimming the slopes.

When the roadbed has been graded, alinement is established precisely by setting tacked stakes along the center line at full stations on tangents and usually at fractional stations on horizontal and vertical curves. Spiral curves are staked out at this time. An additional line of pegs is set on one side of the track and perhaps 3 ft. from the proposed line of the rail, with the top of the peg usually at the elevation of the top of the rail. Track is usually laid on the subgrade and is lifted into position after the ballast has been dumped.

28-9. Sewers and Pipelines. The center line for a proposed sewer is located on the ground with stakes or other marks set usually at 50-ft. intervals where the grade is uniform and as close as 10 ft. on vertical curves. At one side of this line, just far enough from it to prevent being disturbed by the excavation, a parallel line of ground pegs is set, with the pegs at the same intervals as those on the center line. A guard stake is driven beside each peg, with the side to the line; on the side of the guard stake farthest from the line is marked the station number and offset, and on the side nearest the line is marked the cut (to the nearest ⅛ in.). In paved streets or hard roads where it is impossible to drive stakes and pegs, the line and grade are marked with spikes (driven flush), chisel marks, or paint marks.

When the trench has been excavated, batter boards are set across the trench at the intervals used for stationing. The top of the board is set at a fixed whole number of feet above the sewer invert (inside surface of bottom of sewer); and a measuring stick of the same length is prepared. A nail is driven in the top edge of each batter board to define the line. As the sewer is being laid, a cord is stretched tightly between these nails, and the free end of each section of pipe is set at the proper distance below the cord as determined by measuring with the stick.

If the trench is to be excavated by hand, the side pegs may be omitted and the batter boards set at the beginning of excavation.

For pipelines, the procedure is similar to that for sewers, but the interval between grade pegs or batter boards may be greater, and less care need be taken to lay the pipe at the exact grade.

For both sewers and pipelines, the extent of excavation in earth and in rock is measured in the trench, and the volumes of each class of excavation are computed as a basis of payment to the contractor.

The records of the survey should include the location of underground utilities crossed by, or adjacent to, the trench.

28-10. Canals. The location survey for a canal is described in Art. 26-27. Slope stakes for each bank are set as described in Arts. 10-8 and 10-10. So far as possible, the cross-section is balanced, that is, the excavated material forms the fill at the same station, and little or no material needs to be moved longitudinally.

28-11. Tunnels. Tunnel surveys are run to determine by field measurements and computations the length, direction, and slope of a line connecting given points, and to lay off this line by appropriate field measurement. The methods employed naturally vary somewhat with the purpose of the tunnel and the magnitude of the work. A coordinate system is particularly appropriate for tunnel work.

For a short tunnel such as a highway tunnel through a ridge, a traverse and a line of levels are run between the terminal points; and the length, direction, and grade of the connecting line are computed. Where practicable, the surface traverse between the terminals takes the form of a straight line. Outside the tunnel, on the center line at both ends, permanent monuments are established. Additional points are established in convenient surface locations on the center line, to fix the direction of the tunnel on each side of the ridge. As construction proceeds, the line at either end is given by setting up at the permanent monument outside the portal, taking a sight at the fixed point on line, and then setting points along the tunnel, usually in the roof. Grade is given by direct levels taken to points in either the roof or the floor, and distances are measured from the permanent monuments to stations along the tunnel (see Arts. 29-2 to 29-5). If the survey line is on the floor of the tunnel, it is usually offset from the center line to a location relatively free from traffic and disturbance; from this line a rough temporary line is given as needed by the construction force.

The dimensions of the tunnel are usually checked by some form of templet transverse to the line of the tunnel, but may be checked by direct measurement with the tape.

Railroad and aqueduct tunnels in mountainous country are often several miles in length and are not uniform in either slope or direction. Tunnels of this character are usually driven not only from the ends but also from several intermediate points where shafts are sunk or adits are driven to intersect the center line of the tunnel (see Mine Surveying, Chap. 29). The surface surveys for the control of the tunnel work usually consist of a precise triangulation system tied to monuments at the portals of the main tunnel and at the entrances of shafts and adits, and a precise system of differential levels connecting the same points. With these data as a basis the length, direction, and slope of each of the several sections of the tunnel are calculated; and construction is controlled by establishing these lines and grades as the work progresses.

28-12. Bridge Sites. Normally the location survey will provide sufficient information for use in the design of culverts and small bridges; but for long bridges and for grade-separation structures usually a special topographic survey of the site is necessary. This survey should be made as early as possible in order to allow time for design and—in the case of grade crossings or navigable streams—to permit approval of the appropriate governmental agency to be secured. The site map should show all the data of the location survey, including the line and grade of the roadway and the marking and referencing of all survey stations. The usual map scale is 1 in. = 100 ft., and the usual contour interval is 5 ft. on steep slopes and 2 ft. over flat areas.

The preliminary report submitted with the site map should give all available information necessary for economic design. For a preliminary report on a bridge site, the items listed below have been prescribed by the California Division of Highways. For a grade-separation structure, the required information is similar except that it relates to the intersecting roadway and its traffic instead of the intersecting stream and its flow. Photographs are useful adjuncts to the report.

1. *Bridge Site.* Location, stream, distance from nearest shipping point.
2. *Source of Materials* (with length of haul to site). Sand, gravel, stone, falsework timber, piling.
3. *Cost of Materials* (delivered at site). Portland cement, sand, gravel, stone, falsework timber, piling. Cost per ton mile for hauling.
4. *Waterway.* Elevation and location of nearest B.M.; drainage area (approximate); character of watershed; elevation of highest water, with date; elevation of ordinary high water; highest ice mark; elevation of low water; elevation of permanent ground water. Is stream ever dry? If so, at what months? Will all flood water pass through new structure? Can channel be cleared to afford more waterway? Is stream cutting or silting up? Is stream stable in its banks? Depth of scour? Does stream carry light, medium, or heavy drift? What clearance above high water should be allowed?
5. *Foundation.* Character of material (give data from soundings or borings if available); distance from stream bed to solid foundation; recommended depth of foundations. Should piles be used? What length?
6. *Old Bridge.* (If no bridge at present location, give data for nearest bridge over same stream.) Type; number and length of spans; area of waterway provided by old structure; adequacy of this area in flood times; excess capacity of area; foundation data (logs of borings, pile-driving data, etc.); elevation of underclearance; disposition of existing structure.
7. *Recommendations for New Structure.* Type, number, and length of spans; width of roadway; desirability of sidewalks; recommended angle of skew; necessity for approaches or fill; approximate unit cost of approach filling; necessity for maintaining traffic, and recommended method.

For long bridges over deep water, where preliminary borings may be necessary to determine foundation conditions, the location of barges from which borings are made can be determined by means of electronic measuring devices such as Raydist (Art. 30-6 and Ref. 5 at the end of this chapter).

28-13. Bridges. For a short bridge with no offshore piers, first the center line of the roadway is established, the stationing of some governing line such as the abutment face is established on the located line, and the angle of intersection of the face with the located line is turned off. This governing crossline may be established by two well-referenced transit stations at each end of the crossline beyond the limits of excavation or, if the face of the abutment is in the stream, by a similar transit line offset on the shore. Similarly, governing lines for each of the wing walls are established on shore beyond the limits of excavation, with two transit stations on the line prolonged at one, or preferably both, ends of the wing-wall line. If the faces are battered, usually one line is established for the bottom of the batter and another for the top. Stakes are set as a guide to the excavation and are replaced as necessary. When the foundation concrete has been cast, line is given on the footings for the setting of forms and then by sighting with the transit for the top of the forms. As the structure is built up, grades are carried up by leveling, with marks on the forms or on the hardened portions of the concrete. Also, the alinement is established on completed portions of the structure. The data are recorded in field books kept especially for the structure, principally by means of sketches.

For long sights or for work of high precision, as in the case of offshore piers, various transit stations are established on shore and/or on specially built towers by a system of triangulation, such that favorable intersection angles and checks will be obtained for all parts of the work. To establish the offshore piers, simultaneous sights are taken from the ends of a line of known length.

Example: It is desired to locate the central point C of a bridge pier in the middle of a stream of moderate width, on a tangent line of a roadway. From a transit station A on shore, on the center line of the roadway, a measured base line AB is laid off along the shore, of such length and azimuth that favorable intersection angles of a triangle ABC will be secured. The angle ABC is computed from the known angle CAB and the sides AB and AC of known length. Transits are set up at A and B, and at B the angle ABC is set off. The point C is then established, by simultaneous sighting, at the intersection of the line of sight BC with the line of sight along the roadway, prolonged from A. The location is checked by similar sights taken either on the other side of the roadway or on the other side of the stream. To establish the corners of the pier, a similar procedure is followed but with correspondingly different values of the angles at A and B, which must be computed in each case.

For long bridges it is necessary that distances be measured with precision even higher than that required for first-order work in order that the parts of the superstructure can be fitted together with close tolerances. Also the coordinate system adopted for horizontal control should be based upon the average elevation of the structure. The survey is accomplished in stages, the pier locations being established with somewhat lower precision than those of the fittings subsequently established on the top of the pier.

Where cofferdams are used, reference points are established on the cofferdams for measurements to the pier.

When the structure has been completed, permanent survey points are established and referenced for use in future surveys to determine the direction and extent of any movement.

28-14. Culverts. At the intersection of the center line of the culvert with the located line, the angle of intersection is turned off, and a survey line defining the direction of the culvert is projected for a short distance beyond its ends and is well referenced. At (or offset from) each end of the culvert, a line defining the face is turned off and referenced. If excavation is necessary for the channel to and from the culvert, it is staked out in a manner similar to that for a roadway cut. Bench marks are established nearby, and pegs are set for convenient leveling to the culvert. Line and grade are given as required for the particular type of structure.

28-15. Building Sites. In the preparation of his plans for a building, the architect requires a large-scale map of the site to show the information necessary to the proper location of the building both in plan and in elevation. Such maps are usually drawn to the scale of 1 in. = 10 ft. or 1 in. = 20 ft.

The party consists of an instrumentman and one or two chainmen, and the equipment includes that of a transit party. Because of the large number of elevations to be determined, frequently an engineer's level is also used. The notes may be kept in a transit or topography notebook; or a sketch board may be used to good advantage, as it provides more space for sketching and recording details than does the single page of a notebook.

First the lot corners are located, and permanent markers are set at these points. Then, with the property lines being used as reference lines, all objects are located, usually by tape measurements only, the instrumentman recording the data and drawing such sketches as are necessary. On extensive sites, or where it is not convenient for the chainmen to locate objects by coordinate measurements, transit angles and taped distances may be used.

The details should be shown and described as follows: (1) lot corners, state kind; (2) property lines, give dimensions and the distances from the walks; (3) street lines, show widths; (4) sidewalks and drives, give kind and widths; (5) pavements, give kind and widths; (6) gas and water mains, state size and show exact location; (7) manholes and storm and sanitary sewers, give size and kind of pipe; (8) trees, state kind and size; (9) poles of all kinds; (10) fire hydrants; and (11) existing structures on or near the site, state materials of construction. Also, give the elevations of (*a*) inverts of sewer outlets from manholes and the gradients of the sewers; (*b*) the reference bench mark, with description; (*c*) points along the sidewalks, curbs, and lot lines at intervals at 50 ft.; and (*d*) ground points at the corners of 50-ft. squares. The elevations of the sewer inverts, sidewalks, and curbs should be taken to hundredths, and all ground points to tenths, of a foot; also contour lines are shown if the ground is irregular.

The map should give the legal description of the tract and the other information ordinarily shown on the plot of an urban land survey. The drawing is made on tracing cloth, the size being the same as the other sheets of the architect's plans so that a print may be bound with each set of plans. A typical map of this kind is shown in Fig. 28-1.

28-16. Buildings. At the beginning of excavation, the corners of the building are marked by stakes, which will of course be lost as excavation proceeds. Sighting lines are established and referenced on each outside building line and line of columns, preferably on the center line of wall or column. A

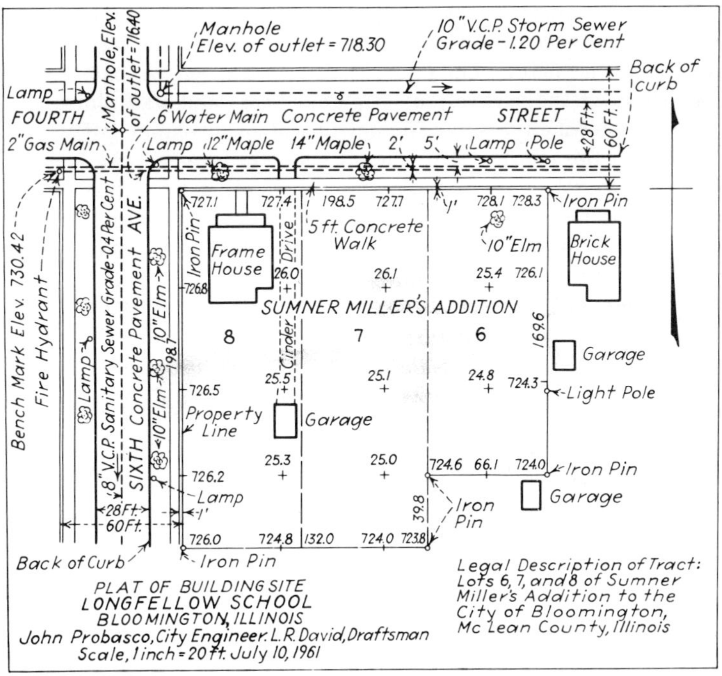

FIG. 28-1. Map of building site.

batter board is set at each end of each outside building line, about 3 ft. outside the excavation. If the ground permits, the tops of all boards are set at the same elevation; in any event the boards at opposite ends of a given line (or portion thereof) are set at the same elevation so that the cord stretched between them will be level. The elevations are chosen at some whole number of feet above the bottom of the excavation, usually that for the floor rather than that for the footings, and are established by holding the level rod on the side posts for the batter board and marking the grade on the post. When the board has been nailed on the posts, a nail is driven in the top edge of the board on the building line, which is given by the transit. Carpenter's lines stretched between opposite batter boards define both the line and the grade, and measurements can be made conveniently by the workmen for excavation, setting forms, and alining masonry and framing. If the distance between batter boards is great enough that the sag of the carpenter's line is appreciable, the grades must be taken as approximate only.

If the space around the building is obstructed so that batter boards cannot be set, other means of marking the line and grade are substituted to meet the requirements of the situation.

When excavation is completed, grades for column and wall footings are given by ground pegs driven to the elevation of either the top of the footing or the top of the floor. Lines for footings are given by batter boards set in the bottom of the excavation. Column bases and wall plates are set to grade directly by the leveler. The position of each column or wall is marked in advance on the footing; and when a concrete form, a steel member, or a first course of masonry has been placed on the footing, its alinement and grade are checked directly.

In setting the form for a concrete wall, the bottom is alined and fixed in place before the top is alined.

Similarly, at each floor level the governing lines and grade are set and checked, except that for prefabricated steel framing the structure as a whole is plumbed by means of the transit at every second or third story level. Notes are kept in a field book used especially for the purpose, principally by means of sketches.

Whenever the elevation of a floor is given, it should be clearly understood whether the value refers to the bottom of the base course, the top of the base course, or the top of the finished floor.

Throughout the construction of large buildings, selected key points are checked by means of stretched wires, plumb lines, optical plummet, transit, or level in order to detect settlement, excessive deflection of forms or members, or mistakes. Bench marks are checked to detect any disturbance.

28-17. Dams. Prior to the design of a dam, a topographic survey is made to determine the feasibility of the project, the approximate size of the reservoir, and the optimum location and height of the dam. To provide information for the design, a topographic survey of the site similar in many respects to that for a bridge (Art. 28-12), is made by field and/or photogrammetric methods. Extensive soundings and borings are made, and topography is taken in detail sufficient to define not only the dam itself but also the appurtenant structures, necessary construction plant, roads, and perhaps a branch railroad. A property-line survey is made of the area to be covered by, or directly affected by, the proposed reservoir.

Prior to construction a number of transit stations, sighting points, and bench marks are permanently established and referenced upstream and downstream from the dam, at advantageous locations and elevations for sighting on the various parts of the structure as work proceeds. These reference points are usually established by triangulation from a measured base line on one side of the valley, and all points are referred to a system of rectangular coordinates, both in plan and in elevation. To establish the horizontal location of a point on the dam, as for the purpose of setting concrete forms or of checking the alinement of the dam, simultaneous sights are taken from two transits set up at reference stations, each transit being sighted in a direction previously computed from the coordinates of the reference station and of the point to be established. The elevation of the point is usually established by direct leveling. However, it may be established by setting off on one (or, as a check, both) of the transits the computed vertical angle, the height of instrument being known. This method is the inverse of indirect leveling, described in Art. 8-5.

A traverse is run around the reservoir, above the proposed shore line, and monuments are set for use in connection with property-line surveys and for future reference. Similarly, bench marks are established at points above the shore line. The shore line may be marked out by contour leveling (Art. 10-13), with stakes set at intervals. The area to be cleared is defined with reference to these stakes. The area and volume of the reservoir may be computed as described in Art. 24-20.

INDUSTRIAL SURVEYING

28-18. General. The manufacture and assembly of large machines, aircraft, missiles, and guiding and tracking systems require extremely close tolerances of measurement for location, orientation, dimensioning, and alinement. For reasons of size as well as precision, the ordinary tools of the mechanic and the jigs and techniques of conventional tool engineering are not adequate, and it is necessary to use apparatus and methods similar to those of surveying. In recent years many optical and related devices have been developed such as special transits, telescopes, targets, linear scales, and micrometers together with equipment and systems for supporting these instruments. A precision of up to 1/200,000 can be attained, the linear measurements being made to 0.001 in. and the angular measurements and alinements to 1 second of arc.

The principles of industrial surveying are the same as those of surveying in general, but the apparatus and methods are highly specialized and are beyond the scope of this book. A thorough treatment is given in Ref. 8 at the end of this chapter. Herein are discussed briefly some of the distinctive features.

28-19. Optical Tooling. Optical tooling is the term commonly applied to precise measurement and alinement of large machines and industrial assemblies by optical and closely related means.

Apparatus. A *jig transit,* or *jig collimator,* is used to establish vertical planes. One form is shown in Fig. 28-2. It is essentially a precise transit, similar to a theodolite, but has no horizontal or vertical graduated circles and no compass. Its range of focusing is from several inches to infinity. Its level tubes are sensitive. Through its hollow spindle, vertical sights can be taken by means of a right-angle eyepiece assembly. On the telescope is an optical micrometer for measuring short vertical or horizontal distances from the line of sight without moving the telescope. At either end of the horizontal axis is mounted an optically flat mirror parallel to the line of sight for use in establishing perpendiculars to the line of sight.

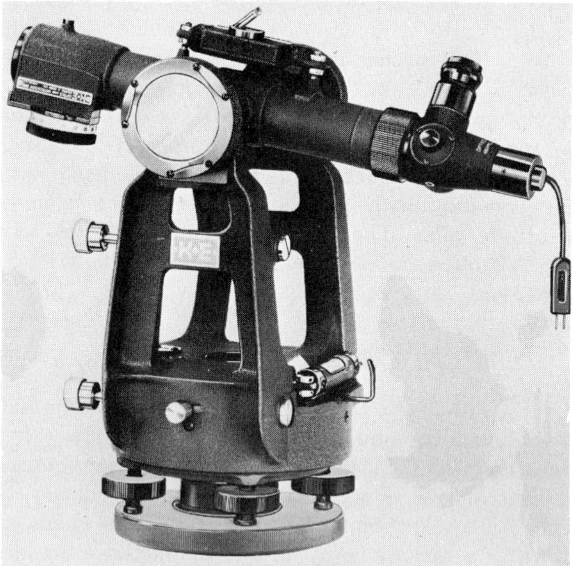

FIG. 28-2. Jig transit. (*Keuffel & Esser Co.*)

An *alinement telescope* (Fig. 28-3) has a cylindrical telescope barrel fitted with an objective lens, eyepiece, and reticule for defining the line of sight. It is provided with mutually perpendicular optical micrometers to measure short transverse distances. The barrel is precisely ground, and it may be mounted interchangeably in fixtures at various locations.

The *level* used in optical tooling is a precise engineer's level such as that used in conventional surveying and equipped with an optical micrometer.

Specially designed *targets* are used for ease and precision in sighting. *Measuring bars* and optical or screw-thread *micrometers* can be coupled to form a precise measuring rod over a range of lengths; precise *tapes* are also used. Specially graduated flat linear *scales* are provided with devices for temporary or continued attachment to points on the structure; the scales are read directly through the transit, level, or alinement telescope, the distance of the line of sight from the nearest graduation of the scale being read with an optical micrometer. Auxiliary apparatus includes special stands, frames, carriages, tracks, and brackets for supporting the instruments and targets. According to the particular needs, instruments may be attached to the structure or supported separately from it or some attached and some supported separately.

Some Methods. A base line, or *master line,* is established in a principal plane of the structure, to which line directions and transverse measurements (usually horizontal and vertical) are referred. Also a base point, or *master button,* is established near one end of the base line as a reference for linear measurements. Other reference lines and points may be established for convenience.

FIG. 28-3. Alinement telescope. (*Keuffel & Esser Co.*)

To erect a perpendicular to a base line, a jig transit is set up and lined in on the base line (or parallel to it) at the desired longitudinal distance from the master button and is sighted along the line. A suitably supported alinement telescope is directed toward the mirror on the end of the horizontal axis of the jig transit and is moved about until its line of sight is reflected back on itself; the line of sight is then perpendicular to the base line. A precise pentagonal sighting prism may be attached to a telescope to project a perpendicular to its line of sight.

To check a level surface, the engineer's level is used in connection with a graduated linear scale which is held vertical and is moved about on the surface, as in leveling with a leveling rod. To level a given surface, a similar procedure is followed and the supports of the surface are adjusted until the vertical scale reading is the same over all the surface.

To aline a wall or other vertical plane of a structure, a line of sight is established in the desired direction at a short offset distance from the desired location of the surface. A graduated linear scale is held or fastened horizontally at two or more points on the surface, and the distance from surface to line of sight is read. The wall is moved as necessary until the horizontal scale reading is the same at all points. This process is similar to direct leveling but is in vertical instead of horizontal planes.

Plumbing is accomplished by means of an optical plummet, one form of which is an alinement telescope in a vertical mounting.

28-20. Ship Surveying. Ships not built in dry docks are built on inclined ways down which they will slide at launching. Assembly of the major parts of a ship is controlled by surveying methods similar to those for the construction of a building, except that alinements and measurements are either made in or based upon planes parallel and perpendicular to the plane of the ways. Account is taken as necessary of the ratio of horizontal or vertical distances to inclined distances, which are those parallel or perpendicular to the ship's structure as shown on the plans. For measurements in longitudinal planes, this ratio is equal to the cosine of the slope of the ways.

A base line for longitudinal sighting and transverse measurement is established several feet to one side of the center line of the ways. A grade line is established several inches below the bottom of the keel (which is to be set on blocks) as a reference for measurements perpendicular to the plane of the ways, and a transverse line is established as a reference for longitudinal measurements. Sights with transit or level, in connection with rod or tape measurements, are given as necessary to locate the supporting blocks and the various portions of the lower deck. As erection proceeds, reference points and lines are projected upward from deck to deck for control of the subsequent assemblies. As large assemblies are brought in, they are centered, leveled transversely, and set at the slope of the structure. Because of differential settlement or shifting of the supports as the weight of the assembled structure is increased, frequent checks are necessary such as doubling all angles, double-checking alinements, and referring back to reference points. Measurements are usually made to 1/16 in. or to 0.005 ft.

For a longitudinal sight, in order to bring the line of sight of a transit parallel to the slope of the ways, the transit can be leveled in the usual manner and the slope (usually about 3°) set off on the vertical circle. Alternatively, the vertical axis of the transit can be made perpendicular to the plane of the ways as follows: With the telescope and one pair of opposite leveling screws in longitudinal alinement, the slope of the ways is set off on the vertical circle, and the leveling screws are manipulated until the telescope-level bubble is centered. By means of the transverse pair of leveling screws, the transverse plate-level bubble is centered. This arrangement has the advantage that, with the vertical circle set at zero, sights in a plane parallel to that of the ways can be taken in any direction.

If inclined longitudinal sights are to be taken with an engineer's level, the process is similar to that of "shooting-in" grade in grade leveling (Art. 10-12).

REFERENCES

1. AMERICAN RAILWAY ENGINEERING ASSOCIATION, "Manual," current edition.
2. AMERICAN SOCIETY OF CIVIL ENGINEERS, "Report on Highway and Bridge Surveys," *Manual of Engineering Practice* 44, 1962. (Chaps. 7 and 8 are on construction surveys.)
3. AMERICAN SOCIETY OF CIVIL ENGINEERS, "Symposium on Construction and Location Surveys" (Carquinez Bridge, Holland Tunnel, Bridges of Port of New York), *Proc. Am. Soc. Civil Engr.*, Vol. 55, No. 3, Part II, pp. 619–632, March, 1929.
4. BUSCH, ALFRED E., "Keeping Abreast of Instrumentation Development," *Surveying and Mapping*, Vol. 23, No. 2, pp. 275–282, June, 1963. (Includes instruments and methods for optical tooling.)
5. COMSTOCK, ALLEN L., and P. Z. MICHENER, "Raydist Locates Boring Sites for Chesapeake Crossing," *Civil Engrg.*, Vol. 28, No. 7, pp. 512–515, July, 1958.
6. FINK, HOWARD L., "Survey Control at Glen Canyon Dam," *J. Surveying Mapping Div., Am. Soc. Civil Engr.*, Vol. 88, No. SU 1, pp. 51–61, November, 1962.
7. HAKMAN, PETER A., "First-order Survey Pinpoints Construction across Tampa Bay," *Civil Engrg.*, Vol. 23, No. 7, pp. 461–463, 515, July, 1953.
8. KISSAM, PHILIP, "Optical Tooling for Precise Manufacture and Alignment," McGraw-Hill Book Company, New York, 1962.
9. MEARS, FREDERICK, "Surveys, Construction Methods, and a Comparison of Routes" (Cascade Tunnel), *Trans. Am. Soc. Civil Engr.*, Vol. 96, pp. 926–949, 1932.
10. Manuals of instruction of state highway departments.

CHAPTER 29

MINE SURVEYING

29-1. Definitions. Mine surveying includes *underground surveying*, as practiced in mining and tunneling operations, and *mineral-land surveying*, which involves location and patent surveys.

Several special mining and geological terms used in this chapter are defined below.

Vein or *Lode*. A relatively thin deposit of mineral between definite boundaries.

Strike. The line of intersection of the vein with a horizontal plane, also the direction of that line expressed as a bearing.

Dip. The vertical angle between the plane of the vein and a horizontal plane measured perpendicular to the strike. A vertical vein has a dip of 90°. If the vein is nearly horizontal, the dip is a small angle.

Outcrop. The portion of the vein exposed at the surface of the ground.

Drift. An inclined passage driven in a particular direction.

Heading. A passage driven into the rock or ore ahead of the main excavations.

Shaft. A vertical or nearly vertical opening into a mine.

Patent. The document issued by governmental authority granting and conveying public land.

UNDERGROUND SURVEYING

29-2. General. Underground surveying differs from surface work in that the station is usually in the roof instead of the floor of the workings; the object to be sighted and the cross-hairs of the telescope must be illuminated; distances are usually measured on the slope instead of along horizontal lines; and the transit tripod has adjustable legs to adapt its use to low workings or to very irregular or steeply inclined surfaces.

29-3. Stations. When the station is in the roof, the transit may be centered in either of two ways: (1) by first plumbing from the station mark to a point on the floor and then setting up over the point on the floor, as in surface work; or (2) by centering the transit beneath a plumb bob suspended from the roof station. When the first method is employed, usually the temporary floor point is a piece of lead into which a nail has been driven. There is always a chance that such a mark will be accidentally displaced during the process of setting up the instrument. In setting up the transit beneath a suspended plumb bob, it is necessary to have both the plate and the telescope level before the centering is done.

A station set overhead is more easily found, is less liable to disturbance, and is therefore more durable than one underfoot. It is set in the mine timbering or by driving a plug of hard wood into a hole ¾ to 1 in. in diameter, drilled several inches into the rock. The exact point is established by setting a marker called a *spad* in the timber or plug, just as a tack is driven into the transit hub in surface surveys. In the case of a roof station the object used must, of course, be something from which it is convenient to hang a plumb bob. Noncorrosive spads made especially for this purpose are sold by dealers.

29-4. Illumination. Since the field of view is dark, on long sights the cross-hairs require artificial illumination. This may be accomplished by slipping a rolled piece of paper into the sunshade, then holding a miner's lamp, electric flashlight, or other source of light in front of and a little to one side of the objective end of the telescope. By moving the source of light toward or away from the end of the telescope, the cross-hair illumination is increased or decreased until both the cross-hairs and the object sighted are visible. Some transits are equipped with special sunshades which reflect light into the telescope in much the same manner as the rolled paper. Some transits are built with a hollow horizontal axis through which light is transmitted to a reflector within the telescope.

The signal or target is usually a plumb bob hung from the roof station. To illuminate the plumb bob, either a light is held to one side of it or a piece of thin paper or tracing cloth is held behind it and illuminated by means of a lamp held beyond the paper. Also with short sights such an illuminated screen may be all that is necessary to make the cross-hairs visible. If the point sighted is not too far from the transit, good results are obtained with a piece of cardboard illuminated by a flashlight.

29-5. Distances. In underground traversing, except in nearly level workings, the distances are measured on the slope, the horizontal and vertical distances being calculated from the slope distance and the vertical angle. For this purpose a steel tape is used that will reach from one station to the other. As the stations must ordinarily be placed rather close together on account of the character of the workings, a tape of length 100 or 200 ft. will usually be sufficient. Usually the tape is graduated to hundredths of a foot throughout.

The following procedure is convenient (Fig. 29-1). The transit is set up at one station, and the vertical distance from the station to the horizontal axis of the transit is measured. Since this distance is short, possibly only a few inches, it is measured vertically from a roof station to the top of the telescope or from a floor station to the plumb-bob hook, and a constant previously determined for the instrument is added to carry the measurement to the horizontal axis. The height of instrument, or H.I., is positive if the instrument is above the station and negative if below. A plumb bob is hung at the next station, with a point on the plumb line marked by some form of clamping target at a known distance below the roof station. This distance is the *height of point,* or H.P. The H.P. is positive if the point sighted is above the station and negative if below. The vertical angle to the point so marked is measured, and the distance is taped to the same point from the end of the horizontal axis, on which the point to be used is definitely marked. While the distance is being taped, the telescope must be pointing toward the plumb line at the next station.

In underground surveying it is desirable to use the method of angular measurement called "angles to right" or "azimuths from back line" (Art. 14-12), and to double the angle. Usually the compass cannot be used for checking, and under such conditions this method is less liable to mistakes than is the deflection-angle method.

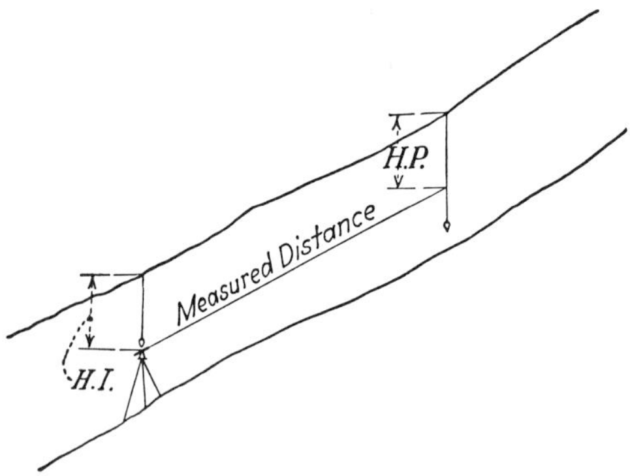

FIG. 29-1. Measurement of slope distance in mine.

The next step is to compute, from the measured inclined distance and vertical angle, the vertical distance between center of instrument and point sighted. The difference in elevation between the two stations can then be found by algebraically adding the H.I. to this vertical distance and algebraically subtracting the height of point.

Example:

H. I.	−3.45 ft.
H. P.	−4.67 ft.
Inclined distance	94.78 ft.
Vertical angle	+17°42′

$94.78 \times \sin 17°42' =$ +28.82 = vertical distance from instrument to point sighted

+(−3.45)

+25.37

−(−4.67)

+30.04 ft. = difference in elevation between the two stations

The problem is identical in principle with the one that occurs in surface surveying when a vertical angle is read to a point on a rod, either at the height of instrument or at some other height.

A special case occurs when, by leveling the telescope, the mark on the plumb line at the point sighted is set at the same elevation as the transit telescope. The vertical distance from such mark to the station plug above is then measured and used as in the general case illustrated above. The vertical angle is, of course, zero.

The horizontal distance between the two stations is the measured slope distance multiplied by the cosine of the vertical angle or angle of inclination of the line taped. Unless the vertical angle is large, this reduction may be simplified by the use of a table of versed sines, as explained in Art. 7–15.

29-6. Mining Transit. The transit commonly used underground in mine or tunnel is the ordinary engineer's transit on an extension-leg tripod. It should have a full vertical circle and a sensitive telescope bubble, and preferably should be equipped with a striding level for the horizontal axis. An instrument with a horizontal circle about 5 in. in diameter is preferable to a larger one on account of greater ease in handling. It is desirable that the vertical circle be graduated on the edge instead of the side, so that the transitman can read the circle without turning the instrument or moving around it. The center point of the transit should be definitely marked on the top of the telescope. For very steep sightings a prismatic eyepiece is convenient.

On account of the dirt and water frequently present underground, it is desirable that the vertical circle be fully enclosed, and that so far as possible the instrument be so constructed as to exclude water from the telescope, circles, compass box, and bearings.

In order that the transit may be lined in by moving its head laterally (for short distances) without moving the tripod, sometimes there is employed a "shifting center" which fits between the tripod and the foot plate; two pairs of opposing screws provide for the lateral movement.

When the slope of the underground workings requires the taking of sights along lines of large vertical angle, the transit as ordinarily constructed cannot be used on account of the fact that the horizontal plate will interfere with pointing the telescope. For such conditions an auxiliary telescope is attached either at one end of the horizontal axis or above the main telescope and at a distance therefrom somewhat more than one half of the diameter of the horizontal plate. In either type the line of sight of the auxiliary telescope is parallel to that of the main telescope. Figure 29-2 shows a transit with a side telescope attached.

Other devices sometimes employed instead of the side or top telescope include overhanging bearings for the main telescope and an arrangement by which prisms are used to offset the line of sight to an objective mounted at the end of the horizontal axis.

29-7. Use of Auxiliary Telescope. The side telescope is offset from the vertical axis of the transit, and this eccentricity affects the observed values of horizontal angles read with the side telescope. Similarly the top telescope, being offset from the horizontal axis of the instrument is eccentric in

FIG. 29-2. Mining transit with side telescope.

the vertical plane, and this eccentricity affects the observed values of vertical angles read with the top telescope. The process of computing the true angle from the observed angle is called *reduction to center,* and the value so found is called the *reduced value.* This term does not imply that the computed value is numerically smaller than the one observed; it may be greater. The difference between the observed and reduced values will here be called simply the *difference.* The difference, positive or negative as the case may be, is applied to the observed value to give the reduced value, which is the one that would have been obtained had sights been taken with the main telescope.

With the top telescope, horizontal angles will not require reduction to center, because the line of sight lies in the same vertical plane with the main telescope, and the transit is so constructed as to give the horizontal angle between vertical planes through the center of the instrument and the points sighted. For similar reasons, with the side telescope no reduction to center is necessary for vertical angles.

Top Telescope. Figure 29-3 illustrates the measurement of a vertical angle with the top telescope. The line of sight of the top telescope is offset an amount BC from the line of sight of the main telescope. If it were possible to use the main telescope, the observed vertical angle would be V, the value desired. The use of the top telescope gives a reading V'.

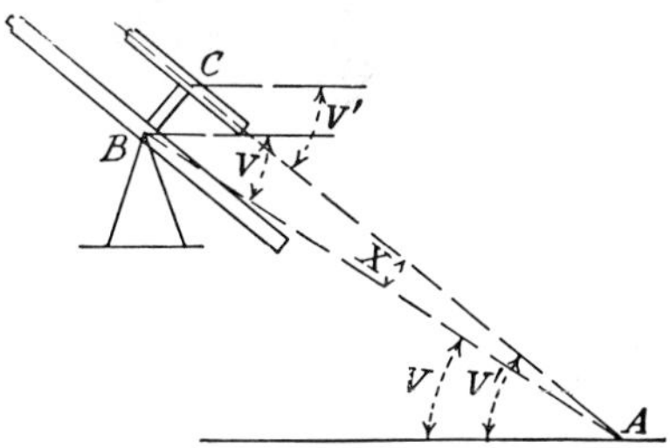

FIG. 29-3. Reduction to center, vertical angle.

$V' - V = X$, and $\sin X = BC/AB$, in which BC is the distance between the lines of sight of the two telescopes and AB is the measured distance between the horizontal axis of the transit and the point sighted. Since BC is a constant for the instrument, the angular difference X varies only with the distance AB. In practice, a table is prepared showing the values of the difference X for various distances AB. The values given in this table are then used as differences to be applied to the observed vertical angles. If the observed vertical angle is positive, the difference is added; if negative, the difference is subtracted. The field notes should indicate for what observations the top telescope was used, and whether the vertical angle read was positive or negative. The reduction to center should be computed later, in the office, the field record showing only the values read in the field. The reduced value of the vertical angle is, of course, used as if it had been read directly with the main telescope.

Example: The vertical angle observed with a top telescope is $V' = -15°23'$, and the distance between the lines of sight of main and top telescopes is $BC = 0.26$ ft. The inclined distance to the point sighted is $AB = 127.20$ ft. It is desired to find the true vertical angle.

$$\sin X = \frac{BC}{AB} = \frac{0.26}{127.20}$$

$$X = 0°07'$$

Then

$$V = -15°23' + 0°07' = -15°16'$$

Side Telescope. With the side telescope a similar reduction to center is necessary for horizontal angles, as illustrated by Fig. 29-4. If observations could be made with the main telescope, the angle $H = AOB$ would be read. When sights are taken with the eccentric side telescope, the line of sight of the auxiliary telescope, which is offset a distance OC from the center O, is always tangent to the circle with center O and radius OC, as the instrument is revolved. The line of sight of the auxiliary telescope has a direction CA when sighting upon A and a direction $C'B$ when sighting upon B. The angle H' is the difference between these two directions and is the angle through

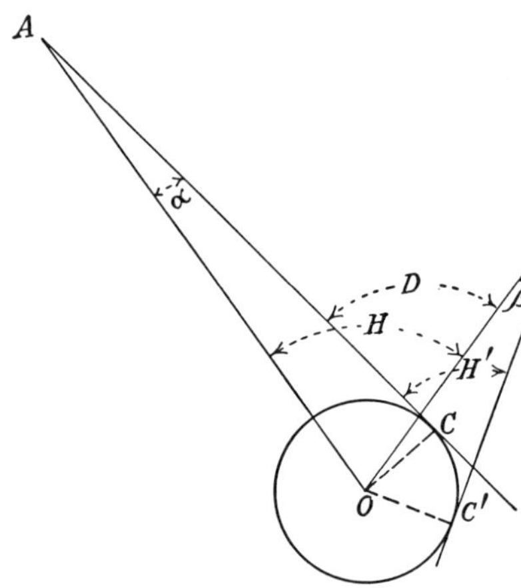

FIG. 29-4. Reduction to center, horizontal angle.

which the instrument is turned between the two sightings. This is the angle read on the horizontal circle of the transit.

To reduce the observed angle H' to the angle H at the center O, the values of the angles α and β must be used.

$$\alpha = \sin^{-1}\frac{OC}{AO} = \tan^{-1}\frac{OC}{AC}$$

Since AO and AC are practically equal, except in the case of very short lines, either distance or either function may be employed. Also

$$\beta = \sin^{-1}\frac{OC'}{OB} = \tan^{-1}\frac{OC'}{BC'}$$

Then

$$D = H' + \beta = H + \alpha$$

and

$$H = H' + (\beta - \alpha) \tag{1}$$

It will be seen that the difference to be applied to the observed angle H' to reduce it to the angle H is the difference between the angles α and β. In practice the values of α and β are taken from a table similar to that used in connection with the top telescope. Care must be taken to apply this difference with the proper algebraic sign.

Example: The horizontal angle observed with the side telescope of a mining transit is H′ = 73°19′. The distance between the lines of sight of main and side telescopes is $OC = OC' = 0.31$ ft., and the distances to points sighted from the transit are $AC = 107.31$ ft. and $BC' = 69.31$ ft. It is desired to reduce the horizontal angle to center.

$$\alpha = \tan^{-1}\frac{0.31}{107.31} = 0°10'$$

$$\beta = \tan^{-1}\frac{0.31}{69.31} = 0°15'$$

$$H = 73°19' + (0°15' - 0°10')$$
$$= 73°24'$$

A better procedure is to measure the angle a second time with a reversal of the instrument between the observations. The side telescope will be on the opposite side of the main telescope, and the angles α and β will enter into the difference with algebraic signs the reverse of those for the first observation. As a result, the mean of the two observations will be free from error of eccentricity of the telescope, and no reduction to center is necessary.

29-8. Adjustment of Auxiliary Telescope. The usual adjustments of the transit having been made, it becomes necessary so to adjust the auxiliary telescope that its line of sight lies in the same plane with and parallel to that of the main telescope. The method of mounting the telescope varies with the make of instrument, and this influences somewhat the details of adjustment. If the auxiliary telescope is rigidly mounted upon the main telescope or upon the horizontal axis, the adjustment is made by moving the cross-hairs. If the auxiliary telescope is adjustable as a whole relative to the main telescope, advantage is taken of this feature in making the adjustment.

If the work of adjustment is done on the surface, the simplest plan is to sight the main telescope on some clearly defined point several miles away, to clamp horizontal and vertical motions of the transit, and then to adjust the auxiliary telescope until its line of sight strikes the same distant point. In case a short sight must be used, either underground or on the surface, two points are marked on a vertical surface, the distance between them being made equal to the distance between the lines of sight of the two telescopes. For a top telescope, the two points are on a vertical line, and for a side telescope the two points are on a horizontal line. The line of sight of the main telescope is then directed toward one point, the horizontal and vertical motions are clamped, and the line of sight of the auxiliary telescope is adjusted until it strikes the other point.

29-9. Setting Up and Leveling the Transit. From the discussion of Art. 13-32, it is evident that the errors in horizontal angle due to errors in adjustment of horizontal axis and plate levels, and due to imperfect leveling of the instrument, increase with the magnitude of the vertical angle; hence when the sights are steeply inclined, the transit must be in excellent adjustment and carefully leveled. As an aid to precise observation, a sensitive striding level mounted on the horizontal axis is frequently employed. If the transit is not so equipped, it may be leveled by the use of the telescope level.

29-10. Connecting Surface and Underground Surveys. The methods used to accomplish a connection between surface and underground surveys depend mainly upon the character of the opening from the surface to the underground workings. In the case of a tunnel which is horizontal or nearly so, or in the case of an incline at an angle of not more than 60° or 65° with the horizontal, no special methods are necessary, except that instead of the usual tripod for supporting the transit, some special form of support may be necessary.

Where headroom is very limited or where the transit can best be supported upon a shelf built from the side or the top of the workings, it is mounted on a *trivet* instead of a tripod. A trivet consists of a modified tripod head with three short supporting pins.

For steeper inclines the transit with auxiliary telescope is used.

A point of known horizontal location at the surface can be projected down a vertical shaft by plumbing, usually with a plumb bob. Optical plummets are available capable of plumbing within a linear error of 0.01 ft. in 100 ft.

One Vertical Shaft. In the case of a vertical shaft, a vertical plane is defined by two plumb lines suspended in the shaft, in a plane of known azimuth determined by connection with the surface survey. Wire known as "electrician's banding wire" is recommended for use for the plumb lines, with bobs weighing 10 to 40 lb. suspended in oil to reduce oscillation.

Underground, a transit is set up close to the wires and in line with them, that is, in the plane of known azimuth. An angle is then turned to some other line, and two points are permanently set on this line, which is then used as a reference line of known azimuth. By this method the underground survey is referred to the same meridian as the surface survey. Great care must be taken in lining in the transit with the two plumb lines, because the distance between the plumb lines is necessarily short and a small error in orientation at the shaft will result in a considerable error in the computed locations of points some distance removed from the shaft. This linear error in the location of any point is in a direction at right angles to a line from the shaft to the point, the displacement being equal to the azimuth error (in radians) multiplied by the distance from the shaft to the point in question.

Example 1.: An azimuth is carried down a shaft by means of plumb wires illustrated at 1 and 2 in Fig. 29-5. The distance between the two wires is 5.00 ft. If one of the wires is displaced 0.005 ft. in a direction at right angles to the plane of the two wires, what angular error results in the measured direction of each of the

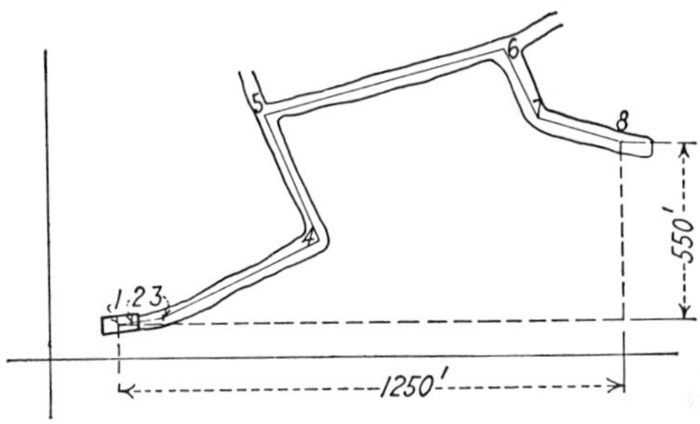

FIG. 29-5. Connecting surface and underground surveys.

lines of the underground traverse to which the known azimuth marked by the wires is connected?

If A represents the angular error, then

$$\tan A = \frac{0.005}{5.00} = 0.001$$

and

$$A = 0°03'30'' \text{ (approximate)}$$

Example 2: Calculation of the latitudes and departures of the courses of the underground traverse of example 1 shows that station 8 in the figure is 550 ft. north and 1,250 ft. east of the shaft. What is the linear error in the calculated location of that station due to the above-mentioned inaccuracy in plumbing down the shaft?

The distance from shaft to station 8 is

$$\sqrt{550^2 + 1{,}250^2} = 1{,}365 \text{ ft. (approximate)}$$

The linear error in the calculated location of the station is in direct proportion to the distance from the shaft to the station. As the error is 0.005 ft. in 5.00 ft., then by proportion:

$$\frac{\text{Error in station 8}}{1{,}365} = \frac{0.005}{5.00}$$

$$\text{Error in station 8} = 1.37 \text{ ft.}$$

Or, the linear error in location of the station is equal to the distance of the station from the shaft, multiplied by the sine or tangent of the angular error, hence

$$1{,}365 \sin 0°03'30'' = 1.37 \text{ ft.}$$

The preceding example illustrates how a relatively small error in the alinement of the plumb lines may produce a station error of a magnitude that is of real importance if a connection is to be made with other workings.

The coordinates of one of the plumb lines and the azimuth of the horizontal line joining them must be determined from the surface survey. When the transit is set up underground in the plane of the two plumb lines, the coordinates of the transit station may be found. Then the coordinates of other points in the underground traverse are determined in the ordinary way.

In the case of a shallow and wide shaft not too much filled with timbering, it is sometimes possible to set two stations at the bottom of the shaft from a transit set up at the surface. The two stations will be in a plane of known azimuth, from which the underground survey can be oriented. As in other cases of very steep sights, the transit should be in excellent adjustment and should be leveled with great care.

If possible, as usually is the case, the survey should be carried into at least two openings and should be closed within the mine. This procedure gives a check on the work. At each entrance the survey should preferably start from a line of known azimuth and location thus reducing the errors of the calculated coordinates of all underground stations.

Two Vertical Shafts. Another satisfactory method is sometimes used when two vertical shafts form the entrances to the mine. A single plumb line is suspended in each shaft. On the surface a traverse tied to or including a line of known azimuth is run between the two plumb lines, and the length and azimuth of the straight line connecting the two plumb lines are computed, as described in Art. 18-21. Underground a traverse is run from one plumb line to the other through the mine workings. Alternatively, optical plummets may be used at the entrances.

A reference meridian is arbitrarily chosen for the underground traverse, from which the length and bearing of the closing course are computed, as in the surface traverse.

The lengths of the two closing courses, surface and underground, should be equal. The bearings, however, will not agree by an amount equal to the angle between the surface reference meridian and the underground assumed meridian. The azimuths of the underground courses are now corrected to agree with the surface reference meridian, and the proper coordinates of the transit stations are computed. A disadvantage of this method is that the check afforded consists in comparing the computed lengths of the closing sides only. If in the underground traverse the error of closure should have its direction nearly perpendicular to the closing course, the length of the closing course would not be greatly affected and the error would not be detected.

29-11. Computations. From the length and vertical angle of each course the corresponding horizontal and vertical distances are computed. From the horizontal distance and azimuth of each course the latitude and departure of that course are computed. The latitude, departure, and vertical distance for each course are the coordinate differences in a three-dimensional coordinate system. Such a system is very useful in all underground work. The three coordinates of each station are computed and are recorded for future use. Later extensions to the surveys, branch lines, etc., can then be fitted easily into the general scheme; and surveys and maps of different parts of the mine can be shown in proper relation to one another.

For computing latitudes and departures from lengths and bearings, special tables called *traverse tables* are frequently used. Such tables are included in Ref. 5 at the end of this chapter. These tables give directly the latitude and departure corresponding to a given bearing angle and a given length of line. To be useful in the calculation of a transit survey, the table must give values for angles varying by single minutes. If values of latitude and departure for each minute of angle are given for distances 1, 2, 3, 4, 5, 6, 7, 8, 9 along the traverse courses, then the value for any distance is found by simple addition, moving the decimal point as may be necessary.

29-12. Field Notes and Office Records. On account of the dirt and water usually encountered underground, it is difficult to keep the notebook pages clean; and the ordinary field notebook soon becomes soiled and difficult to read. For this reason some form of loose-leaf field notebook is desirable in underground surveying. By placing a few loose leaves in a metal or heavy cardboard binder and using them underground for one day only, a more legible record is secured.

The pages of the notebook should be numbered serially to guard against loss of pages, and should be bound into an office binder for filing. From the notes the latitudes, departures, and differences in elevation are computed; these are then recorded in a special book. The three coordinates of each point are also tabulated.

The following form of record is convenient for the field notebook:

Sta.	B. S.	F. S.	H. I. + or −	Hor. Angle	H. P. + or −	Vert. Angle	Taped Dist.	Description of F. S. Point

For the office book the same headings and the following additional headings are suggested:

Calc. Brg.	Vert. Dist.	Hor. Dist.	Diff. Elev.	Lat.	Dep.	Total		
						Lat.	Dep.	Elev.

29-13. To Give Line for a Connection. In mining operations it is frequently necessary to drive an opening or connection between more or less widely separated parts of existing workings, as between A and E (Fig. 29-6). This connection may be necessary for purposes of ventilation, drainage, or haulage of excavated material or to provide for the miners a second route out of the workings in case of accident. The connection is ordinarily in a straight line between two given points. The problem is to determine the length, direction, and slope of this line in order that the work of driving the connection may be properly directed.

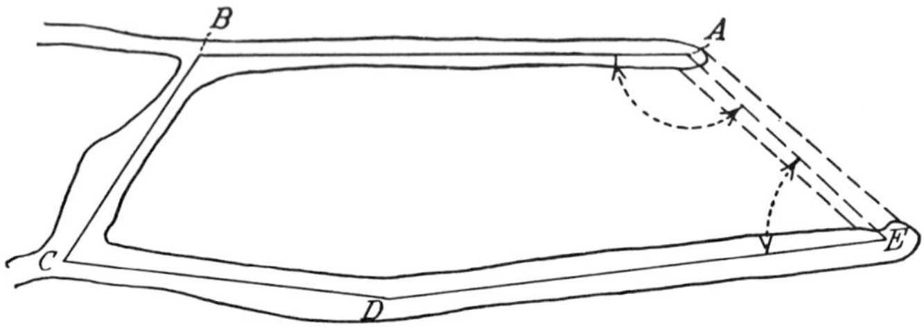

FIG. 29-6. Line for connection.

Starting at one of the two given points as A, a transit traverse as $ABCDE$ is run through the existing workings to the other point E. The length, azimuth, and slope of the connecting line AE are then computed by the usual method for a traverse with one side of unknown length and direction (see Art. 18-21). From either or both of the given points A and E, a line of the computed azimuth and slope is laid off with a transit, and thus line and grade for the connection AE are established. As the work of tunneling progresses, additional line and grade points are set at frequent intervals. Also in order that the progress of the work may be determined, measurements of distance are taken from given points to the heading. Usually the connection is driven from both ends.

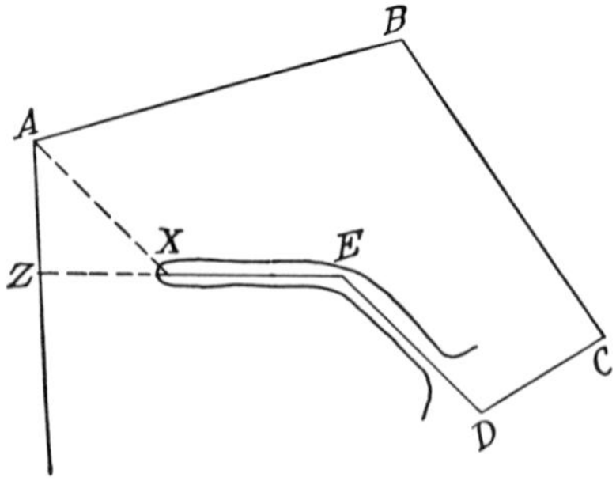

FIG. 29-7. Property boundary underground.

29-14. To Mark a Property Boundary Underground. This also is a problem in supplying the missing parts of a traverse. A survey is run from some point as A (Fig. 29-7), on the boundary line as marked on the surface, to some point as X, underground near the boundary. At A, the direction of the boundary line AZ is observed. The problem is to find the distance XZ, in a desired direction, from the underground point X to the vertical plane which defines the boundary. From the data of the survey the coordinates of A and X are available, hence the azimuth and horizontal length of the line connecting A and X may be computed, as explained in Art. 18-21. This line forms one side of a triangle, the other two sides of which are ZA on the boundary line and XZ the line of known direction from X to the boundary. The length AX and the direction of all three sides of the triangle being known, the distance XZ from X to the boundary is computed as explained in Art. 18-23.

29-15. To Measure Difference in Elevation down a Vertical Shaft. This is best done by means of the steel tape. The elevation of a point at the mouth of the shaft having been determined by ordinary differential leveling, the distance is measured vertically to some convenient point further down, and so on to the other points. If the distance between working levels is less than one tape length, the points used are conveniently placed at the levels, and elevations from these points can be carried into the various parts of the mine. The points set in the shaft for this purpose should be marked by small nails driven into the shaft timbers, but more permanent points should be set in the various working levels to serve as bench marks.

29-16. To Compute Bearing of a Drift. Relationships between vein, strike, and dip are illustrated by the following example.

Example: Find the bearing of a drift having a 3 per cent grade in a vein which has a strike of S82°00′W and a dip of 48°00′.

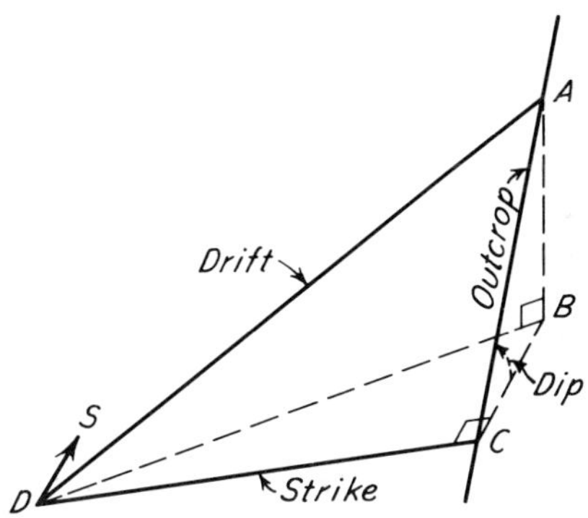

FIG. 29-8. Bearing of drift.

In Fig. 29-8 (not to scale), in which the triangle ADC is in the plane of the vein,

$$
\begin{aligned}
AC &= \text{outcrop of vein}\\
\angle BCA &= \text{dip } (48°00')\\
CD &= \text{strike (bearing S82°00'W)}\\
AD &= \text{drift (grade 3 per cent)}\\
BD &= \text{horizontal projection of the drift}\\
\angle BDC &= \text{horizontal angle between } BD \text{ and the strike}
\end{aligned}
$$

$\angle ABD$ and $\angle BCD$ are right angles

Assume $BD = 100$ ft.; then $AB = (0.03)(100) = 3.00$ ft.

$$BC = \frac{AB}{\tan \text{dip}} = \frac{3.00}{\tan 48°00'} = 2.70 \text{ ft.}$$

$$\sin BDC = \frac{BC}{BD} = \frac{2.70}{100} = 0.0270 \qquad BDC = 1°33'$$

$$\text{Bearing of drift} = 82°00' - 1°33' = \text{S}80°27'\text{W}$$

29-17. Tunnel Surveys. Tunnel surveys are run for the purpose of directing the operations of tunneling between two or more given points, either below the ground or on the surface, as described in Art. 28-11. Essentially the same process is followed in mining except that, if the tunnel is between two shafts, it is necessary to transfer elevation and direction down each shaft; also, if the tunnel is on a slope, this slope is ordinarily established with sufficient precision by laying off the vertical angle.

MINERAL–LAND SURVEYING

29-18. Ordinary Subsurface Ownership. The result of the survey of the boundaries of a piece of ordinary land is a geometrical figure in a horizontal plane. The bounding lines are usually straight, although they may be curved. The map of the survey shows this geometrical figure on a plane surface representing the horizontal plane into which all the points of the figure are projected.

The boundaries of land are marked on the surface of the earth by monuments, fences, or other objects; and ownership is often considered as applying to the surface only. But when we think of the construction of a building and realize that no part of the structure may project over the property lines, or when we remember that similar limitations apply below the surface, or that wires entirely above the ground, even if supported by poles upon other property, may not run across property without permission of the owner, we realize that ownership of land implies ownership *within vertical planes through the boundaries.* This is the rule which usually controls, unless specifically modified by laws as explained in the succeeding article.

The owner of land may deed or grant to another person or to a corporation the ownership or rights above or below some specified elevation or level, as for the purpose of driving a tunnel perhaps many feet below the surface. Any such privilege, whether it is above or below ground, if it is distinct from ownership of the soil, is known as an *easement.*

There may be a seam of coal underlying a certain piece of land. The owner of the land may sell to a mining company, operating below adjoining land, the right to mine the coal under his property. Such a sale may be for a lump sum, or the amount to be paid may be based upon the quantity of coal taken out. In the latter case it is necessary for the surveyor to establish underground the boundaries of the property in question, as previously described, in order to make possible a measurement of the quantity of mineral removed.

Where the rule of vertical planes applies, the surveys of mineral lands do not differ from other land surveys, except in so far as the shape or size of the parcel or the character of the reference points to be used for the survey may be fixed by law.

29-19. Lode Claims. To encourage the development of mining, the United States Government has passed laws modifying in certain cases the usual rule of vertical planes and specifying the manner in which the person discovering a mineral vein or lode on government land may acquire title to the vein and thereby profit by his discovery. It is provided that a mining claim of specified maximum dimensions may be located on the surface, and that after certain requirements designed to prove the serious intent of the claimant have been satisfied, the United States Governmemt will give the claimant a patent carrying a clear title to the land claimed, this title carrying ownership of the vein beneath.

The Federal laws dealing with lode claims are based upon the concept of a relatively thin vein or lode, limited between surfaces which are essentially plane. According to the law, the claim is to be located along the outcrop of the vein (its intersection with the ground surface) and is limited to a maximum length of 1,500 ft. and to a maximum width of 600 ft. Any state

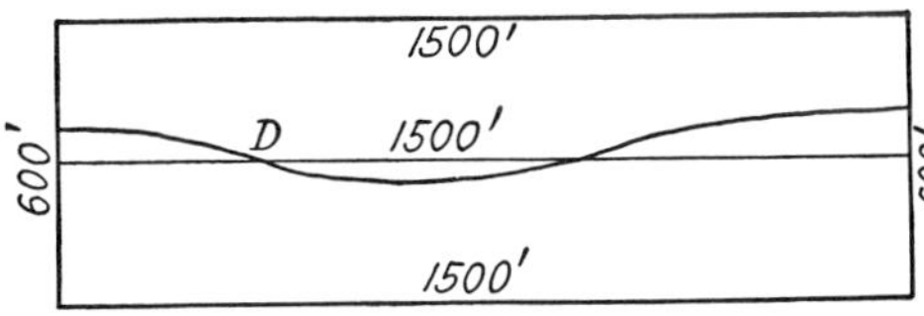

FIG. 29-9. Rectangular claim.

may by law reduce, but not increase, these dimensions. The outcrop must cross the end lines but not the side lines.

The ownership of a properly located lode claim carries with it ownership of the vein anywhere between vertical planes through the end lines. This holds even though the inclination of the vein from the vertical carries the vein underground beyond the side lines of the claim. The effect of this is to modify the usual rule of bounding vertical planes; the owner of the claim on the outcrop owns the vein even if it passes beyond a vertical plane through either of the side lines of his claim.

The end lines of the claim must be parallel straight lines, and the length of the claim must be measured along the center line of the claim. Except as limited above, the claim may be of any shape.

Special Cases. Figure 29-9 represents an ideal rectangular claim 1,500 by 600 ft. *D* is the point of discovery, and the irregular line represents the outcrop of the vein, crossing the end lines but being everywhere between the side lines. The center line and each side line are 1,500 ft. long. The end lines are parallel straight lines each 600 ft. long.

Figure 29-10 shows a four-sided trapezoidal claim, such as sometimes is necessary on account of the shape of adjoining properties. Again the center line is 1,500 ft. long, but to secure the maximum width of 600 ft. between side lines, the length of each end line must be 600/sin 60° = 692.82 ft., or one half of this distance each way from the end of the center line. The 60° angle was selected merely for illustration. The angle might, of course, have any value less than 90°, but the smaller the angle, the smaller will be the perpendicular distance between the end lines of the claim.

Occasionally the topography is such that one or more angles in the center line will be necessary if the outcrop is surely to be kept within the side lines throughout the length of the claim. Claims of this type, illustrated by Fig. 29-11, are made necessary where an irregular ground surface gives a

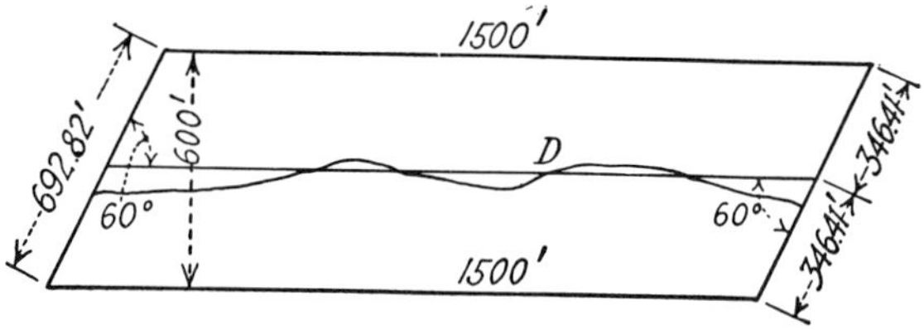

FIG. 29-10. Trapezoidal claim.

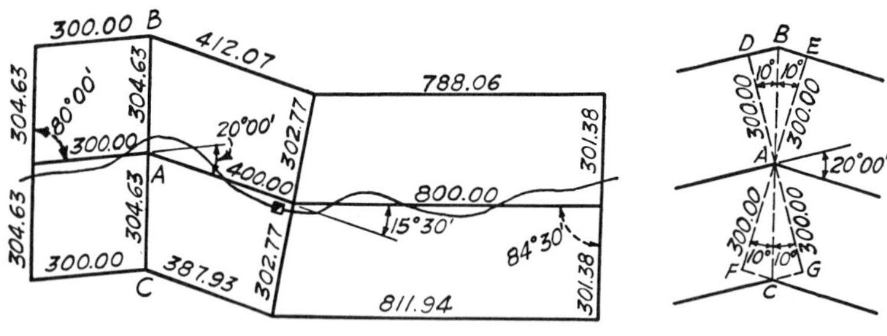

FIG. 29-11. Irregular claim, curved outcrop.

curved outcrop. This condition will be accentuated if the dip of the vein is relatively small. The lengths of the end lines are found by the method just described, and the distances from the points where the center line breaks to the corners opposite them are found as illustrated by the following example.

Example: The side lines of a claim, as shown in Fig. 29-11, lie parallel to the center line and 300 ft. from it. The center line deflects 20° at A. It is desired to locate points B and C. Corners B and C will lie on the bisector of the large angle at A, and four equal right triangles will be formed as shown. Solving one of these triangles gives the distances as follows:

$$AB = AC = \frac{300}{\cos 10°00'} = 300 + 300 \text{ exsec } 10°00' = 304.63 \text{ ft.}$$

$$BD = BE = CF = CG = 300 \tan 10°00' = 52.90 \text{ ft.}$$

The transit is now set up at A, the proper angle is turned off, and points B and C are set. By measuring the lengths of the side lines from corner to corner and comparing the measurements with the computed lengths, the location of the corners is checked.

29-20. Field Work. In locating a mining claim to secure the maximum dimensions and area allowed by law, the following procedure is suggested:

When the maximum allowable length of the center line of a claim is fixed, this line is so located as to follow the outcrop closely with as few breaks as practicable. The deflection angles are read at breaks in the center line, after establishing hubs at these points and at the two ends of the line. The sum of the various segments of the center line should be equal to the maximum allowable length of the claim (as 1,500 ft.). The transit is set up at one end of the center line, the proper angle is turned from the center line, and the corners are set at the two ends of the end line. A similar procedure is followed at the other end of the claim and at breaks on the center line. By solution of the right triangles similar to those illustrated by the preceding examples, the direction and length of each line of the traverse bounding the claim are computed. Then as a check on the location, a traverse is run through the points forming the boundary. Such care in the location survey is not a legal requirement, but it is desirable from the point of view of the locator.

The location survey may be made by the claimant or by someone employed by him. The final survey for patent must be made by a United States mineral surveyor, commissioned by the United States to do that work. If the claim is within the United States surveyed public lands, it must be referenced to a corner of the public-land surveys.

29-21. Numerical Problems.

1. A vein has a strike of N10°15′W and a dip of 43°40′. What will be the bearing of a drift in the vein having a grade of 2 per cent?
2. A vein has a strike of N27°30′E. A drift in the vein on a 3-per-cent grade has a bearing of N30°20′E. What is the dip of the vein?
3. A transit has an auxiliary side telescope, the line of sight of which is offset 0.35 ft. from that of the main telescope. In measuring a horizontal angle, the sights being taken with the side telescope to the right of the main telescope, the following measurements were taken: distance OA = 47.32 ft.; distance OB = 268.3 ft.; angle AOB = 136°42′ (point B is to the right of point A). What is the corrected angle AOB?
4. If the line of sight of a side telescope is inclined to that of the main telescope by an angle of 01′, what error in azimuth will be introduced in measuring a horizontal angle between two points, if the sight to one is horizontal and to the other is inclined 68°? 85°? Disregard reduction to center.
5. From a given station A, at the portal of a tunnel, both a tunnel traverse and a surface traverse are run, with results as follows: for the tunnel traverse, A to B, azimuth = 310°22′, distance = 320.2 ft., vertical angle = +1°20′; B to breast,

azimuth = 355°30′, distance = 286.1 ft., vertical angle = +2°01′; for the surface traverse, A to C, azimuth = 24°41′, distance = 416.8 ft., vertical angle = +2°54′; and C to D, azimuth = 343°16′, distance = 458.3 ft., vertical angle = +18°16′. A vertical shaft is to be sunk at station D, and the breast of the tunnel is to be connected with the shaft by a drift having a 2-per-cent grade.

(*a*) How deep must the shaft be?

(*b*) What will be the azimuth of the drift?

(*c*) What will be the slope distance?

6. Given the traverse of the accompanying tabulation, compute the azimuth, length (slope distance), and vertical angle of a line to connect station 28 to station 32.

Station	Object	Azimuth	Slope distance	Vertical angle	Height of point	Height of inst.
28	22	255°32′	138.07	−0°44′	−4.92	−1.09
22	28	75°32′	......	....	......	−4.92
	21	253°04′	167.48	−0°53′	−9.18	
21	22	73°04′	......	....	......	−6.22
	31	344°58′	115.78	−80°32′	−2.93	
31	21	164°58′	......	....	......	−6.78
	32	73°32′	304.02	−0°10′	0.00	

7. Three bore-holes have been sunk to a vein of ore. The depth of these holes at the points A, B, and C, and the surface measurements connecting them, are as follows: elevation of surface at A = 4,750, depth of hole = 3,500 ft.; at B, elevation = 4,920, depth = 2,860 ft.; at C, elevation = 4,790, depth = 2,080 ft.; azimuth of AC = 60°22′, distance = 1,320 ft.; azimuth of AB = 80°30′; azimuth of CB = 140°20′. Required the strike and dip of this vein.

REFERENCES

1. Durham, E. B., "Mine Surveying," McGraw-Hill Book Company, New York, 1927.
2. McNair, A. J., "Mineral Land Surveys," *Trans. Am. Soc. Civil Engr.*, Vol. 115, pp. 200–234, 1950.
3. Staley, W. W., "Introduction to Mine Surveying," Stanford University Press, Stanford University, Calif., 1939.
4. U.S. Bureau of Land Management, "Manual of Instructions for the Survey of the Public Lands of the United States," Government Printing Office, Washington, D.C., 1947.
5. U.S. Bureau of Land Management, "Standard Field Tables," Government Printing Office, Washington, D.C., 1956.
6. Wager-Smith, D. R. W., "Mineral Surveys," *Surveying and Mapping*, Vol. 19, No. 2, pp. 193–202, June, 1959.

CHAPTER 30

HYDROGRAPHIC SURVEYING AND FLOW MEASUREMENT

by
Walter Hofmann and S. E. Rantz
Hydraulic Engineers, U.S. Geological Survey

HYDROGRAPHIC SURVEYS

30-1. General. Hydrographic surveys are those which are made in relation to any considerable body of water such as a bay, harbor, lake, or river. These surveys are made for the purposes of (1) determination of channel depths for navigation, (2) determination of quantities of subaqueous excavation, (3) location of rocks, sand bars, lights, and buoys for navigation purposes, and (4) measurement of areas subject to scour or silting. In the case of rivers, surveys are made for flood control, power development, navigation, water supply, and water storage.

Since a certain amount of shore location is included in most hydrographic surveys, a single control survey is located on shore to serve both for soundings and for shore details.

30-2. Horizontal Control. As in topographic surveying, the horizontal control is a series of connected lines whose azimuths and lengths have been determined. For rough work the control may be a stadia or plane-table traverse. More precise work requires a control run with transit and tape, and for extended surveys where great precision is required, the control is based upon a system of triangulation executed within the required limits of error. For planning a system of traverses or triangulation points, no definite rules can be given. Topography, relief, wooded areas, highways, and railroads are all features which fix the character of the control. Long, narrow rivers or inlets, with shore conditions favorable to traversing, are usually surveyed from a single traverse line on one shore. If the body of water is more than ½ mile wide, it is more economical to traverse both shore lines and to connect the two traverses both in azimuth and in distance by frequent ties. Where the shore lines of rivers and lakes are obscured by woods and cannot be traversed economically, a system of triangulation is used.

In addition to a measured base line at the beginning and end of the survey, check base lines are measured every 10 or 15 miles as the work progresses. The control for large lakes and ocean shore lines consists of a network of connected triangles on shore. These are supplemented where necessary by traverse lines along the shore, connecting two or more triangulation stations.

30-3. Vertical Control. A chain of bench marks is established to serve as a vertical control. These bench marks are near the shore line and are located at frequent intervals so that water-stage gages may be set conveniently.

30-4. Shore Details. Most hydrographic surveys require the location of all irregularities in shore line, all prominent features of topography and culture, and all lighthouses, buoys, etc., in order that these points may be used for references in range-line and sounding work. These details are now generally mapped by photogrammetric methods (see Chap. 31), but they may be located by stadia and plane-table methods, which are described in Chaps. 15 and 17. If ground-survey methods are used, the shore line is located by a level party in much the same manner as any contour is traced. Points are marked only at changes in direction and are subsequently located by the stadia party.

30-5. Establishing Datum. On some tidewater surveys it is necessary to establish a datum from tidal observations. To obtain most accurate results the observations must extend over a period of several years. However, observations extending over one lunar month will give results satisfactory for all but the more precise surveys. The procedure is as follows:

1. The gage is set where it is protected from rough wave action and where the water level is not influenced by local conditions, with the gage located in sufficient depth of water to give a definite gage reading at low tide. Various types of gages are described in Arts. 30-33 to 30-37.
2. The zero of the gage is referred to a permanent bench mark on shore.
3. The elevations of high and low water are read daily for one lunar month.
4. The mean of an equal number of high and low readings gives the approximate value of mean sea level.
5. When the gage reading for mean sea level is obtained, the proper elevation for the bench mark on shore is computed.

30-6. Location of Soundings. The determination of the relief of the bottom of a body of water is made by soundings. If the survey requires relatively few soundings, the soundings are made with a rod or a line. For surveys of greater scope echo sounders (Fig. 30-8) are used. The echo sounder records a continuous profile of the bottom under the survey boat. Where the bottom has a gradual slope, it is advantageous to take soundings along parallel lines that are at right angles to the contours of the bottom. Where the bottom includes a long, narrow, steep-sided ridge or trough, a more reliable relief map of the bottom will be obtained if soundings are taken along parallel lines that cross the bottom contours at about 45°. In addition to soundings taken along these primary lines, soundings should also be taken along a secondary system of crosslines to disclose any discrepancies that may exist in the primary system of soundings. The depth of a sounding is referred to water level at the time it is made and is corrected to the datum determined by the gage. Before the corrected soundings can be plotted on the map, their location with reference to the shore traverse is determined by one of the methods listed below.

1. By taking soundings on a known range line and reading one angle either from a boat or from a fixed point on shore (Art. 30-7).
2. By propelling a boat at a uniform rate along a known range line and taking soundings at equal intervals of time (Art. 30-8).
3. By taking soundings from a boat at the intersections of known range lines (Art. 30-9).
4. By reading two angles simultaneously from two fixed points on shore (Art. 30-10).
5. By taking readings with the transit and stadia (Art. 30-11).
6. By taking soundings at known distances along a wire stretched between stations (Art. 30-12).
7. By reading two angles from a boat to three fixed points on shore (Art. 30-13) by means of the sextant (Arts. 30-14 and 30-16).

The U.S. Coast and Geodetic Survey uses electronic control equipment to a great degree in its hydrographic surveys. The three systems listed below are presently (1965) used for position finding.

1. *Raydist.* The determination of distance by the Raydist system is based on the measurement of a phase difference in two radio signals, one emanating from a transmitter aboard ship and the other from a transmitter at one of two fixed stations ashore. Raydist is reliable to distances of 100 miles or more.

2. *Shoran.* The measurement of distance by Shoran is based on the measurement of elapsed time between the transmission of a radio pulse from the sounding ship and the reception of a return signal from one of two fixed stations ashore. The use of Shoran is limited to distances of about 1½ times the line-of-sight distance between antennas.

3. *Electronic Position Indicator* (*EPI*). The measurement of distance by EPI is likewise based on the measurement of elapsed time for electromagnetic pulses to travel from the sounding ship to one of two fixed ground stations and return, but EPI is used only for measuring long distances. Its effective range is from 15 to 500 miles, but abnormal static conditions can reduce the maximum effective distance to 250 miles.

30-7. Range Line and Angle Read from Shore. The range line may be fixed either by two flags or signals on shore or by one flag on shore and a buoy set in position at some distance offshore. The term "signal" as used in hydrographic surveying designates any object, natural or artificial, used as a reference for locating the positions of the soundings that are made. If buoys are used as signals, they are located from the shore survey. The locations of all range signals must be known and plotted on the map before the locations of soundings can be plotted. Signals may be located by photogrammetry, by stadia, by transit and tape, or by triangulation.

Signals defining a range line should be far enough apart to allow easy projection of the line across the water. The intersecting ray from the transit to the boat should cross the range line at an angle as near 90° as practicable. Figure 30-1 shows one method of laying off range lines adapted to a regular shore line. Irregular shore lines leave much to the ingenuity and experience of the engineer in selecting range lines to fit the particular work at hand. Bends in rivers and curved coast lines are more conveniently laid out by range lines radiating from some fixed point on shore such as a flagstaff, church spire, or chimney. The fixed point should be sharply defined and plainly visible and should be at a sufficient distance from the shore line to make each range line approximately parallel to neighboring range lines. This method of radiating range lines reduces by one-half the number of flags needed on shore. If the range lines diverge too much before reaching the opposite shore, a few range lines crossing the radial range lines are run to fill the existing gaps.

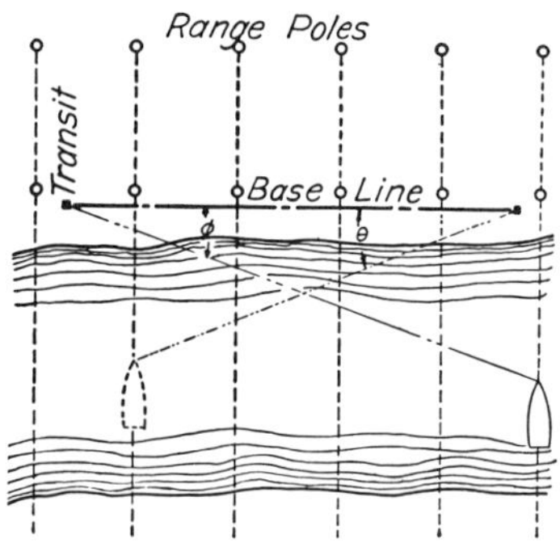

FIG. 30-1. Range lines and one angle read from shore.

A modification of this method is that in which an observer in the boat measures with a sextant the angle between the range line and a control station on shore. This modification increases the amount of office work and generally has no advantage over reading the angle from shore.

If the soundings are made with a rod or a line, each individual sounding must be referenced by angle readings. If a continuous-recording echo sounder is used, the location of the boat is referenced by angle readings taken at intervals, and each time a reading is made an identifying mark is made on the recorder chart. The intervals at which angles are read should be close enough to permit accurate mapping of soundings selected from the echo-sounder chart. The soundings to be mapped are selected at regular intervals from the echo-recorder chart and in sufficient numbers to provide a reliable representation of the contours of the bottom. Their location on the map is interpolated between referenced readings of boat location.

30-8. Range Line and Time Intervals. This method is generally used where extreme accuracy in determining the location of a sounding is not required. If the ends of the range lines are not marked with buoys whose locations have been determined, the first and last soundings are located by angle readings and the intermediate readings are interpolated according to the time intervals. In waters of low velocity where a boat can be propelled at a uniform speed, the time and space intervals will closely correspond. The boat should start at a sufficient distance back of the initial sounding to be traveling at a uniform rate when it reaches the beginning of the range line. The speed of the boat is then kept uniform, and the soundings are taken and are plotted under the assumption that a distance along the range line is proportional to the time consumed in traveling that distance. Where it is necessary to change the speed of the boat, angle readings to locate the boat should be taken at the time the change is made. This method is very convenient when a continuous-recording echo sounder is used. If soundings are made with a rod or a line, the method is applicable only where the water is relatively still, the distance short, and the required accuracy low.

30-9. Intersecting Range Lines. When the object of the survey is to determine changes in the bottom due to scour or silt or to determine the quantity of material removed by dredging, it is necessary to repeat the soundings at the same points. Fixed range lines are located on shore so that they intersect at approximately right angles and are permanently marked. The boat proceeds to the intersections and takes soundings as desired. The precision of the method depends upon the distance between intersecting range lines and upon the precision with which the points of intersection are located as soundings are taken. The system of range lines used must be adapted by the engineer to the topography of the shore and to the shape of the body of water. The time-interval method (Art. 30-8), using soundings obtained with a continuous-recording echo sounder, is also often used for work of this character.

30-10. Two Angles Read from Shore. Two transits are set up at previously determined points on the shore traverse or, more frequently, at instrument points selected to give good visibility and good intersections. In the latter situation, the instrument points are later tied to the shore traverse. The lines from the two transits to the sounding should intersect as nearly at right angles as the location will permit. Each transitman orients his transit on a known azimuth line, unclamps the upper motion, and follows the sounding rod with the vertical cross-hair. When the sounding signal is given, both transitmen simultaneously observe and record the horizontal angle and time. The time of sounding is also taken by the recorder in the boat.

Both transitmen should compare watches with the recorder twice daily, as sometimes this is the only means of identifying soundings when the transitman misses an angle or incorrectly numbers a sounding. The transitmen should check the orientation of their instruments at frequent intervals during the day.

This method is applicable where it is impossible to keep the boat on a fixed range or where the shore topography is unsuitable to the laying out of a system of intersecting range lines. The method has the disadvantage of requiring two shore observers who frequently must move to new locations to secure good intersections. Work must be suspended while new locations are being occupied, and much time is lost in this way.

30-11. Transit and Stadia. The stadia method, using a transit set up on the bank, is well adapted to smooth and shallow waters where the survey is made in connection with the topographic mapping of shore lines. Using a heavy flat-bottom boat in quiet water, it is possible to read the stadia interval with the foot of the rod resting on the bottom of the boat; however, if the water is even slightly rough, reading the stadia interval becomes both slow and uncertain. If the rod is long and sufficiently weighted to remain upright in the water, the same rod may be used both for sounding and for reading the stadia interval. The transit is set up near the water level to avoid the need for reading vertical angles.

The work proceeds in much the same manner as described under stadia surveying (Art. 15-16). At the instant the sounding is taken, the transitman reads the rod interval and turns the vertical cross-hair on the sounding pole. The azimuth is read and recorded while the boat moves to the location of the next sounding. The main advantage of the stadia method is the ease and rapidity with which the soundings can be plotted with a polar protractor or a drafting machine. This method is not suitable where soundings are taken far from shore.

30-12. Distances along Wire Stretched between Stations. On narrow channels which can be navigated by a boat and where successive soundings

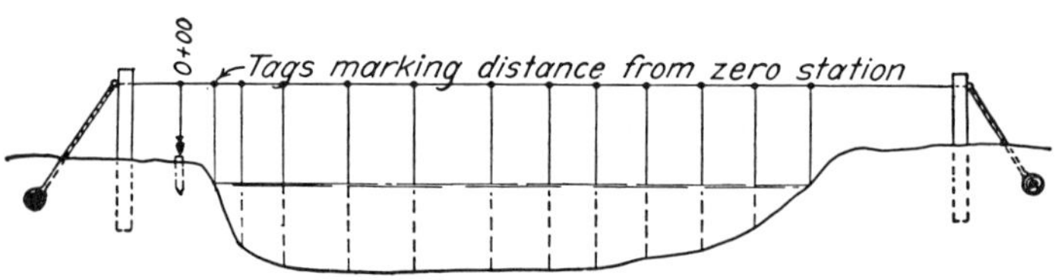

FIG. 30-2. Soundings from marked wire or cable.

on the same section are desirable, a single wire or wire cable is stretched across the channel as shown in Fig. 30-2 and is marked by metal tags at appropriate known distances along the wire from a reference point or zero station on shore, such as the plumb bob and stake shown at 0 + 00. When the wire has been disturbed and it is desired to repeat the soundings, the plumb bob is again suspended at zero on the wire and the wire is adjusted until the bob is over the zero station. This is a very precise method but is much more expensive than locating the soundings by intersecting range lines.

30-13. Two Angles Read from Boat. To reference the position of the sounding boat, two angles are simultaneously observed from the boat to three fixed points on shore whose relative positions are known, as illustrated by Fig. 30-3. This is an application of the three-point problem (see Arts. 16-29 and 17-14). In Fig. 30-3, θ and ϕ are the angles read by the observer in the boat to the known points A, B, and C on shore. Since a boat is too unstable to support a transit, the angles are read with the sextant (Arts. 30-14 and 30-16). Two angles are sufficient to locate a sounding unless the boat happens to be on the circumference of a circle passing through A, B, and C, as shown in Fig. 17-13; in such a case the location of the sounding is indeterminate.

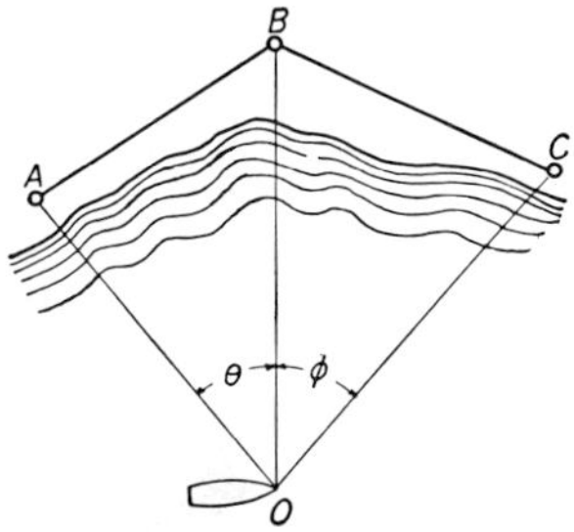

FIG. 30-3. Two angles read from boat.

The accuracy of the plotted location will vary with the relative location of the known points A, B, and C, as follows:

1. If A, B, and C are in a straight line, or if B is nearer the boat than A and C, the location is strong unless one of the angles θ or ϕ is small.
2. Extremely long sights will give small values for θ or ϕ, and the location will be weak.
3. On long or short sights small angles should be avoided, as they are difficult to plot and may give weak locations.
4. The error in the plotted location of the point, due to errors in plotting the angles, increases with the length of sight, and better results can be obtained with shorter sights.
5. The accuracy of location is poor when the point occupied approaches the circle through the three fixed points.

The method may be used in combination with the time-interval method with satisfactory results. The boat is propelled at a uniform rate along a range line, and at intervals the position of the sounding boat is located by two angles read with the sextant. Soundings taken between sextant readings are plotted in proportion to the time intervals. This method reduces the labor of plotting and speeds up the work of observing in the field.

30-14. Sextant. The transit and other instruments used in land surveys are not adapted for use in a boat, where the support is unstable. The sextant (Fig. 30-4) is well suited to hydrographic work and has the added advantage of measuring angles in any plane. It is called a "sextant" because its limb includes but one-sixth of a circle. Although the arc is limited to 60°, the instrument will measure angles to 120°. It is used principally by navigators and surveyors for measuring angles from a boat, but it is also employed on exploratory, reconnaissance, and preliminary surveys on land.

The theory of the sextant is based upon the optical principle that, if a ray of light undergoes two successive reflections in the same plane by two plane mirrors, the angle between the first and last direction of the ray is twice the angle between the mirrors.

The essential features of the sextant are illustrated in Fig. 30-5, with the instrument in position for measuring a horizontal angle *FEL*. An index mirror *I* is rigidly attached to a movable arm *ID*, which is fitted with a vernier, clamp, and tangent-screw and which moves along the graduated

FIG. 30-4. Sextant.

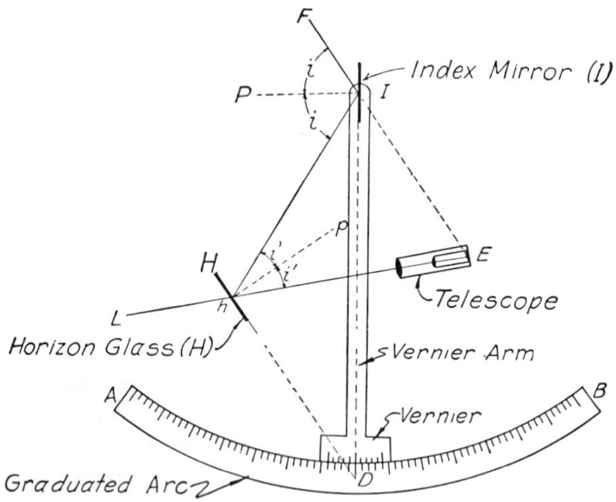

FIG. 30-5. Essential features of sextant.

arc *AB*. A second mirror *H*, called the *horizon glass*, having the lower half of the glass silvered and the upper half clear, is rigidly attached to the frame. A telescope *E*, also rigidly attached to the frame, points into the mirror *H*.

With signals at *L* and *F* and the eye at *E*, it is desired to measure the angle FEL. The ray of light from signal L passes through the clear portion of glass H on through the telescope to the eye at E. The ray of light from signal F strikes the index mirror at I and is reflected to h and then through the telescope to E. Each set of rays forms its own image on its respective half of the objective. When the arm ID is moved, these images can be made to move over each other and there will be one position in which they coincide. An observation with the sextant consists in bringing the two images into exact coincidence and reading the angle on the vernier scale on limb AB.

To prove that angle FEL equals two times angle IDh, that is, that the angle between the signals is equal to twice the angle between the mirrors: Draw IP and hp normal to the two mirrors; then the angles of incidence and reflection of the two mirrors are i and i' respectively. By trigonometry

$$\begin{aligned} FEL &= FIh - IhE \\ &= 2i - 2i' \\ &= 2(i - i') \end{aligned}$$

Also

$$\begin{aligned} IDh &= HhI - hID \\ &= (90° - i') - (90° - i) \\ &= i - i' \end{aligned}$$

Therefore, FEL (the angle between the objects) equals twice the angle IDh (the angle between the mirrors).

30-15. Adjustment of Sextant. The following are the common adjustments of the sextant:

1. *To Make the Index Mirror Perpendicular to the Plane of the Sextant.* Set the vernier arm at a reading near 30°; then observe the image of the arc in the index mirror. If the index mirror is perpendicular to the plane of the sextant, the image in the mirror will appear to form a continuous curve with the visible portion of the arc appearing outside the glass. If the image appears above the arc, the mirror leans forward; if below, it leans backward. Correct by means of adjusting screws provided for this purpose. Some instruments have no provision for this adjustment, in which case adjust by loosening the back screws and inserting thin paper shims between the mirror and the frame.

2. *To Make the Horizon Glass Perpendicular to the Plane of the Sextant.* Sight the instrument on some clearly defined horizontal line such as the roof of a building. If the reflected image of this line as seen in the silvered portion of the horizon glass does not coincide with the image as viewed through the clear portion, the horizon glass must be adjusted by tipping it backward or forward as for the index mirror. This adjustment should be made after the adjustment of the index mirror.

3. *To Make the Horizon Glass Parallel to the Index Mirror When the Vernier Reads Zero.* After the first two adjustments are made, set the vernier to read zero. Sight through the telescope and the transparent portion of the horizon glass at a well-defined distant point. If the direct and reflected images coincide, the mirrors are parallel. If not, adjust the horizon glass until these images coincide.

Modern instruments are fitted with an adjusting screw at the base of the horizon glass, but some of the older instruments have no provision for this adjustment, and an index error for the instrument must be determined. (For index error of transit, see Art. 13-21.) Sight on the distant point with the vernier clamped at zero; then bring the two images into coincidence by moving the index arm. The vernier will now read the index error which is to be applied to all observed angles.

4. *To Make the Line of Sight of the Telescope Parallel to the Plane of the Sextant.* The reticule of the telescope contains four wires which form a square near the center of the telescope. Set the instrument on a solid horizontal surface (as a table) about 20 ft. from a wall, with the plane of the graduated arc horizontal. Sight through the telescope, and on the wall set a mark that appears to be in the center of the square. Select two small blocks of wood of equal height such that when placed near the ends of the graduated limb their tops are very nearly in the same horizontal plane as the telescopic line of sight. Sight over these blocks, and on the wall set a mark in the same vertical plane with the first mark. If the two marks thus established are within ½ in. of each other, the angular error is less than 2″ and can be neglected. When this error is large enough to require adjustment, the adjustment is made by means of the screws on the telescope collar. Some sextants have no provision for this adjustment.

30-16. Measuring Angles with Sextant. The handle of the sextant is held in the right hand, and the plane of the arc is made to coincide with the plane of the two objects between which the angle is to be measured. The sextant is turned in the plane of the objects until the left-hand object can be viewed through the telescope and the clear portion of the horizon glass. With the instrument held in this position, the index arm is moved with the left hand until the images of the two objects coincide. The final setting is made with the tangent-screw, and a test for coincidence is made by twisting the sextant slightly in the hand to make the reflected image move back and forth across the position of coincidence. When the setting is thus verified, the vernier arm is clamped and the angle is read.

The possible precision of angle measurement with the sextant depends upon the size of the angle and upon the length of sight. It is evident from Fig. 30-5 that the angle FEL actually measured has its vertex E not at the eye but at the intersection of the sight rays FE and LE from the flags. The distance to this intersection will increase as the angle decreases, and for small angles the vertex may be at a considerable distance back of the observer. Hence the sextant is not an instrument of precision for small angles (say, less than 15°) and short distances (say, less than 1,000 ft.). If objects sighted are at a great distance away, the angular error is usually small.

Vertical angles can be measured with the sextant in a manner similar to that just described for horizontal angles. Most commonly the altitude of a celestial body is observed, as in navigation. The vertical angle between the body and the sea horizon is measured and is corrected for dip, or angular distance of the observer above the horizon. On land, if the sea horizon is not visible, an artificial horizon is used; this consists of a horizontal reflecting surface, such as that of a small vessel of mercury, near the sextant. The vertical angle between the celestial body and its reflection in the artificial horizon is measured; this angle is twice the altitude of the body.

30-17. Equipment for Soundings. The speed and precision of making soundings depend greatly upon the character of the equipment used. As mentioned previously, if the survey requires relatively few soundings, and if the depths to be sounded are relatively shallow, the soundings are made with a sounding rod or a sounding line. For surveys of greater scope and depth, echo sounders are used.

Sounding Rods. With depths of about 12 ft. or less and with low current velocities, rods can be used to advantage. The rod is usually made in 4-ft. sections for convenience in carrying and must be of sufficient thickness to withstand the pressure of the current. The edges are usually rounded to give minimum resistance to the flowing water. The lower end is fitted with a metal shoe of sufficient weight to hold it upright in the water and with area enough to prevent its sinking into the mud or sand. Rods are ordinarily graduated on both sides to feet and tenths, the zero being at the bottom of the shoe.

Hand Sounding Lines. Hand sounding lines are seldom used for depths greater than 150 ft. The lines may be of cotton, hemp cord, sash chain, piano wire, or small linked steel chain. To one end of the sounding line is attached a lead, and at intervals along the line are markers by means of which depths may be read.

Cotton or hemp lines must be stretched before being used. This is done by drawing the rope tightly between two posts, wetting it, and allowing it to dry. This operation is repeated several times. Finally the rope when wet is stretched taut and is then graduated, zero being at the bottom of the lead and each foot being marked with a piece of cloth drawn through the strands of the rope. The 5- and 10-ft. intervals are best marked with leather tags similar in shape to the notched brass tags used on the surveyor's chain. The rope should be kept dry when not in use. It should be soaked in water for at least an hour before being used in order to allow the rope to assume its tested length.

Although brass sash chains and small linked steel chains do not stretch, the wear on the link surfaces is appreciable, and it is necessary frequently to compare them with a steel tape and to reset the 5- and 10-ft. markings.

Another sounding line, used primarily in government work, is made of braided cotton with a phosphor-bronze stranded wire core. If a "pre-seasoning" process has been applied to the braided covering before the line is marked, the line will maintain almost constant length thereafter and will insure reliable soundings.

Sounding Leads. The weights used with sounding lines vary from 5 to 75 lb. depending upon the depth of water and the velocity of the current. For streams of moderate depth and low velocity a 10-lb. weight is usually heavy enough. The lighter sounding leads are usually made similar in shape to a window sash weight, with a slight taper toward the top or "eye" end.

They are circular in cross-section and three to four times as long as their average diameter. The heavier weights are often "torpedo" shaped, with stabilizing fins, to offer less resistance to the current. If sounding weights of more than 25 lb. are required, a wire sounding machine (Fig. 30-6) is generally used.

FIG. 30-6. **Hand sounding machine.** (*U.S. Coast and Geodetic Survey.*)

Sounding Machines. Wire sounding machines are used where heavy weights are required or where depths are great. The hand sounding machine has a bronze reel which will hold more than 900 ft. of stranded sounding wire. The brake is a wood-lined clamp operated by a small handle above the reel. The hand cranks are hinged so that they can be disengaged from the shaft while the wire runs out and can be engaged for reeling in. The machine is usually mounted at the stern of the boat, and the wire leads to a registering sheave (Fig. 30-7) overhanging the stern. The registering sheave consists of a grooved wheel mounted in a yoke so that it will revolve freely and is connected by gears to an indicator which registers the amount of wire payed out. Sounding machines that are electric-driven and that have a reel capacity of more than 5,000 ft. of stranded wire are also in use.

Echo Sounders. Continuous-recording echo sounders are now widely used. They range in size from battery-powered portable models (Fig. 30-8) that are operable to depths of 200 ft. to high-powered permanent installations aboard government survey vessels that are operable to depths of 7 miles.

Fig. 30-7. Two types of registering sheaves. (*U.S. Coast and Geodetic Survey.*)

All operate on the basic principle that a sound produced near the surface of the water will travel to the bottom and be reflected to the surface as an echo. The echo-sounding equipment is designed to produce the sound, receive and amplify the echo, measure the intervening time interval, convert the interval into units of depth measurement, and record the depths on a moving graph or strip chart.

The sound is produced by a transducer which mechanically converts an electrical impulse to a sonic wave. In permanent echo-sounder installations, the transducer is installed in the hull of the survey vessel; in portable models, the transducer is mounted on a bracket over the side of the boat with the bottom of the transducer at least 2 ft. below the water surface and preferably level with the keel. Transducers with various frequencies are used for different ranges of depth. A low frequency is used for sounding in deep water, because high-frequency signals are subject to greater absorption and require greater initial power for successful use in deep water. The transducer also picks up the echo reflected from the bottom and converts it to an electrical impulse, which is then amplified and recorded as a depth on the strip chart. Sonic waves are emitted by the transducer at very frequent time intervals; for example, a popular portable echo sounder whose maximum operating depth is 245 ft. makes soundings at the rate of 600 per minute. Because of the short time interval between soundings, the record of these soundings on the moving strip chart is a continuous line that shows the profile of the bottom under the moving survey vessel.

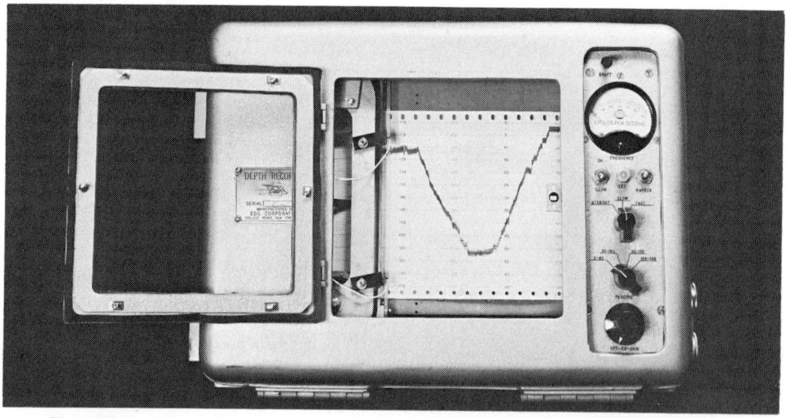

FIG. 30-8. Portable echo sounder. (*U.S. Coast and Geodetic Survey.*)

Sound travels in water at a fairly constant velocity. However, this velocity (roughly 4,800 ft. per second) varies with temperature, salinity, and depth (pressure). Echo-sounding instruments are operated for a certain assumed velocity of sound, known as the calibration velocity, and any sounding is in error by an amount directly proportional to the variation of the actual from the assumed velocity. Nomographs and tables are available giving the correction to be applied for various combinations of temperature, salinity, and depth. Many of the portable echo sounders are so equipped that the calibration velocity can be adjusted to the true velocity before the survey is started. This adjustment is done by lowering a "check bar" a known distance below the transducer and adjusting the frequency of the power supply so that the instrument records the correct depth of the check bar. Because the portable echo sounders have a limited operating depth, generally less than a few hundred feet, the variation in depth is not great enough to affect the velocity of sound seriously. Consequently, subsequent recorded soundings will not require correction. Echo sounders usually have an instrument accuracy that varies from a few hundredths of 1 per cent of the depth for the high-powered permanent installations to ½ of 1 per cent of the depth for the better portable models.

Signals and Ranges. Where practicable, existing natural or artificial objects that are conspicuous should be used as signals. Suitable natural objects include lone boulders, pinnacle rocks, and lone trees; suitable artificial objects include beacons, lighthouses, tanks, spires, and building gables. Invariably, however, it will be necessary to construct some signals and range markers.

Signal masts are usually made of pipe or timber painted white. They are firmly braced in an upright position and are fitted with flags of distinctive marking. Range poles may be either of 1 by 2-in. cross-section or round; they are fitted with an iron shoe. Marking and identification depend upon the particular work. If only a few ranges are used, colored flags may serve; otherwise, the ranges should be marked by roman numerals reading up or down the pole.

Where points marking the range are in shallow water, the range markers are usually 1 by 2-in. wooden poles, weighted at the bottom and held in a vertical position by means of guy wires.

In deep water the range points are marked by buoys. These may be constructed by framing a 55-gal. oil drum with 2 by 4-in. lumber. A mast bearing cross-banners is supported by the framing above the oil drum, and to the framing below the drum are attached a counterweight and anchor line.

30-18. Making Soundings. If an echo sounder is used, soundings are recorded continuously while the boat travels along a range line. If a leadline is used for sounding, the procedure used depends on the depth and velocity of the current. If the depth is not more than 75 ft., the sounding is made without stopping the boat. The leadsman casts the lead forward far enough to allow it to reach bottom as the line comes into a vertical position. Where the current is so swift as to make this method impracticable, a line of soundings is taken by allowing the boat to drift with the current, the leadsman lifting the lead between soundings only enough to clear obstructions. Then the boat is propelled upstream, a second line of soundings is taken paralleling the first, and so on. Soundings in deep, still water are taken by stopping the boat for each sounding. It is also necessary to stop the boat for each sounding when a wire sounding machine is used.

The field record should show the locality, the date, the names of observers, the designation of range or line, the serial number of sounding in range, the time, the two angles if sextants are used, the points sighted on shore, the depth of each sounding, the gage reading for water level, and the error or correction of the sounding line. If the angles are read from shore, each transitman records the azimuth of the sounding, the time, the range and serial designation, and any other information which might be useful in identifying his notes with those of the other observers.

Tide-gage readings are taken from the gage reader's records and entered in the field notes at the end of each day's work. The tide gage should be located as near the soundings as convenient and must be in the same tidal basin.

Soundings are sometimes taken when a lake or river is frozen over. Holes are bored in the ice with an ice auger. A marked sounding line to which is suspended a long narrow weight is lowered through the hole, and the depth is recorded. If the weather is not too severe, the soundings are best located by the transit-stadia method. In very severe weather the soundings are best located by intersecting range lines.

30-19. Reducing Soundings to Datum. Before soundings are plotted, they must be reduced to datum by subtracting (algebraically) from the sounding the corresponding gage correction. Necessary corrections for wind, current, or erroneous length of sounding line also should be made at this time.

30-20. Plotting Soundings. With any combination of range lines and angles read from the shore the plotting is relatively simple. Stations marked by range poles and buoys as well as those occupied by the transit are tied to the shore traverse, and their locations are plotted on the map. These form the control from which soundings are plotted. The lines connecting the range markers are drawn, and the intersecting transit lines are plotted with a polar protractor.

Many ingenious methods, of which items 1 and 2 below are examples, have been used in plotting soundings located by two angles read from the opposite ends of a base line on shore.

1. *Two Polar Protractors.* Two polar protractors, 6 to 10 in. in diameter, are oriented over the instrument stations in position to plot true azimuths. One end of a silk thread is glued to the center of each protractor circle. Two operators are used. Each operator draws the thread taut over the azimuth reading on his protractor representing the transit reading for that sounding. The two threads intersect at the plotted location of the sounding.

A variation of the two-protractor method is to have the traverse plotted on transparent paper or cloth. One paper protractor has its radial lines marked in black ink, and the other in colored ink, preferably red. The two protractors are placed under the tracing cloth and oriented in azimuth with their centers directly under the instrument stations. The protractor graduated in black may be white paper or cloth; it is placed under the protractor graduated in red, which must be made of transparent material, preferably a thin plastic. The intersection of the black and red lines will locate the sounding when the two protractors are set at the azimuth readings of the two instruments.

2. *Two Tangent Protractors.* If the angles are read directly by setting the transit

circle at zero when sighted along the base line, the soundings may be plotted by the use of two tangent protractors. The natural tangent of each observed angle is recorded in the field notes opposite the observed angle. Two tangent protractors are placed over the plotted location of the transit points with their zeros along the base line. When the movable arms of the protractors are set at the tangents of the angles read from the respective stations, their intersection will be the plotted location of the sounding.

3. *Tracing-cloth Method.* When the sounding is to be located by sextant angles read from the boat, the tracing-cloth method is one of the best. Lay off the angles θ and ϕ of Fig. 30-3, extending the rays OA, OB, and OC an indefinite length. Place the tracing cloth over the drawing, and shift it until rays OA, OB, and OC pass through the plotted locations of A, B, and C. The point O will then be directly over its location for plotting.

4. *Three-armed Protractor.* In this instrument the circle is graduated in both directions from 0° to 360°. The middle or fixed arm is fastened permanently in position with its ruling edge fixed at zero on the protractor circle. The two movable arms are fitted with verniers, clamps, and tangent-screws and are placed one on each side of the fixed arm. The ruling edge passes through the setting of the vernier and the center of the protractor circle. The angles θ and ϕ are set on the movable arms, and the ruling edges of the three arms are manipulated to pass through points, A, B, and C. The center of the protractor is now over the correct location of the point, which can be marked through a small hole in the protractor plate.

5. *Plotting Charts.* A graphic method of plotting soundings, developed by the U.S. Corps of Engineers, employs prepared plotting charts covering the working area (Ref. 2 at the end of this chapter). Two sextant angles are read from the boat to three signal stations ashore; the location is then plotted while the boat is progressing to the location of the next sounding.

If the sounding is on or near the circumference of a circle passing through A, B, and C, Methods 3 and 4 do not apply (Art. 17-14).

6. *Odessy Protractor.* When the hydrographic survey is controlled by electronic devices which measure distances to fixed stations, positions are plotted by an Odessy protractor. This device uses a series of closely spaced concentric circles drawn on clear plastic material to represent distances from the center in terms of miles, microseconds, or lanes to fit the system in use. The protractor must be constructed not only to fit the type of electronic equipment but also to fit the scale at which the survey is to be plotted.

30-21. Hydrographic Maps. A hydrographic map is similar to the ordinary topographic map but has its own particular symbols. These may be found in almost any book on topographic drawing or in the manual issued by the U.S. Coast and Geodetic Survey (Ref. 1 at the end of this chapter). (See also Arts. 6-11 and 6-12.) The amount and kind of information shown on a hydrographic map vary with the use of the map. A harbor map should show enough shore-line topography to locate and plan wharves, docks, warehouses, roads, and streets along the water front. A navigation chart should show only shore details which are useful aids to navigation, such as church spires, smokestacks, towers, and similar landmarks. Maps of rivers should show both low-water and high-water marks and all topography within the zone between these marks. A hydrographic map should contain the following information:

1. Datum used for elevations.
2. High-water and low-water lines.
3. Soundings, usually in feet and tenths, with the decimal point occupying the exact plotted location of the point.
4. Lines of equal depth interpolated from soundings. On navigation charts for offshore areas, the lines of equal depth are usually shown in fathoms (1 fathom equals 6 ft.); for harbors, the lines of equal depth are shown in feet.
5. Conventional signs for land features as on topographic maps.
6. Lighthouses, navigation lights, buoys, etc., either shown by conventional signs or lettered on the map.

Figure 30-9 is a portion of a typical hydrographic map of the U.S. Coast and Geodetic Survey. Soundings are shown in feet, referred to mean low water. Elevations of contours and high points on land are likewise shown in feet.

SPECIAL HYDROGRAPHIC SURVEYS

30-22. Sweep or Wire Drag. In harbors and inlets where the mean depth is only slightly greater than navigation requirements or where coral reefs and pinnacle rocks are likely to occur, there is no certainty that any system of sounding previously described will develop all the small areas dangerous to navigation. This defect has led to the introduction of the sweep or wire drag, which came into general use about the year 1900. Beginning with sweeps 200 to 1,000 ft. long, this method and its application have grown to the use of sweeps 10,000 to 15,000 ft. long with which it is possible to cover several square miles of area in a working day. The present wire drag used by the U.S. Coast and Geodetic Survey consists of a wire of suitable length which may be set at any required depth. The wire is supported at this depth by means of buoys placed at intervals and connected to the drag wire by vertical wires of adjustable length. A sinker is attached to the lower end of each of these vertical wires to maintain the drag wire at even depth. Wooden subsurface floats are attached to the drag wire at 100-ft. intervals to prevent it from sagging between buoys. The drag is pulled through the water by two power launches steering slightly divergent courses to keep the drag taut. The resistance of the water causes the drag wire and the buoys to assume a parabolic curve as long as the wire meets with no obstruction. When an obstruction is met, the buoys assume the position of two straight lines intersecting over the obstruction. This spot when found is located by sextant observations to reference points on shore, and soundings are taken for the minimum depth. The point may then be plotted, and the dragging is resumed. This method makes it possible to mark and chart coast-line navigation lanes far in advance of more detailed surveys.

30-23. Determination of Stream Slope. In all natural channels the water-surface slope for limited portions of the stream is a variable quantity. It not

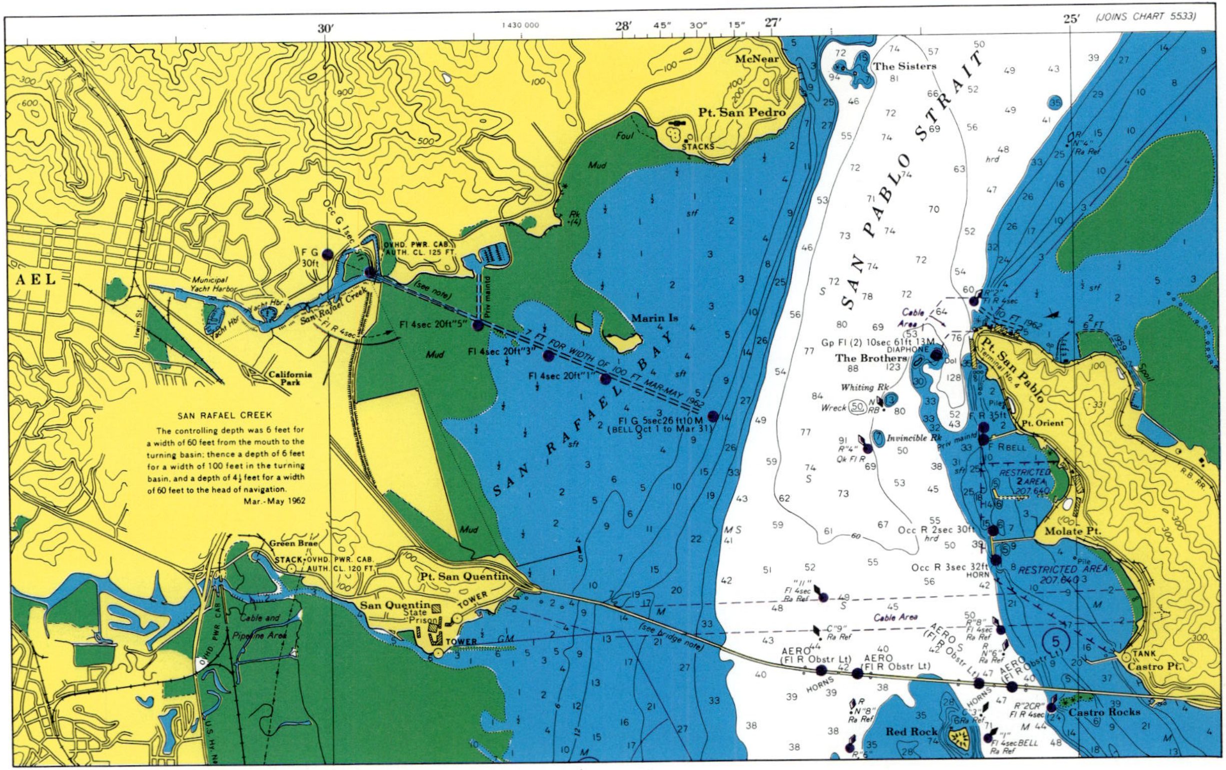

FIG. 30-9. Typical hydrographic map of U.S. Coast and Geodetic Survey. Scale 1:80,000.

only varies at different stages but also varies for the same stage at different times if local channel conditions have been changed. Because of these variations, great care must be taken in the determination of stream slope and in its use in discharge formulas. For example, a slope determination for low stages would give results grossly in error for flood stages. For reliable results the slope determination should be made at or near the stage for which the discharge is desired.

To determine the *average* water-surface slope for various stages covering a wide range of river elevations, a recording stage gage should be installed in a stilling well at each end of the reach of river. This reach should have no tributaries of appreciable size and may range in length from as little as 1,000 ft. to a few miles. The zeros of the gages are referenced to permanent bench marks on shore. The gages should be operated for a period long enough to experience the entire range of stage desired, in some cases for several months. The gage-height graphs are inspected, and concurrent values of stage at the two gages are noted for various elevations of the river. The difference in elevation between paired readings from the gage graphs represents the fall in the reach. It is important that the gage readings be selected from those portions of the gage-height graph that indicate constant or nearly constant stage. For periods of rapidly changing stage, only the peaks and troughs of the graph should be noted. The fall in the reach divided by the length of reach is the slope, usually expressed as a decimal. A fall of 2 ft. in 1,000 ft. would be expressed as an average slope of 0.002. Slopes should be computed for each of the stages selected.

An alternative method to the use of recording stage gages is to install crest-stage gages (see Art. 30-34) at each end of the reach. These gages should be visited periodically, and the peak stage that occurred between visits should be noted. From these paired readings the average slope can be computed for each stage experienced.

The average slope for a given stage, obtained from the difference in elevation at each end of a long reach, is sufficiently accurate for mapping purposes, but if the slope is to be used for computing the river discharge, a more precise determination of slope will be needed. Rarely, if ever, is the water-surface profile of a natural stream a straight line. For determining river discharge, a reasonably straight reach of stream is selected the length of which is equivalent to at least several stream widths. The length should also be such that the fall between ends of the reach is at least a few tenths of a foot. A transit survey of the reach is made, reading the elevation of the edge of water at frequent intervals along the shore to hundredths of a foot. Because the stage may change during the course of the survey, a man on each bank should quickly mark the water's edge at frequent intervals before the survey starts. These markers at the edge of the stream may be either driven stakes or large nails to which colored flagging is attached. The markers are then surveyed and are referenced to a permanent bench mark. The profile for each bank is plotted, and the portion of the reach having the most uniform profile and alinement is selected for computing the discharge.

30-24. Measurement of Surface Currents. In harbors and inlets it is often desirable to know the direction and velocity of currents at all tidal stages. These can be determined either by current meter or by locating the paths and computing the velocities of floats.

The location of current-meter observation points is referenced to fixed points on shore. Several current meters are available to measure both velocity and direction, but the Roberts radio current meter is most commonly used. The Roberts meter can be suspended from an anchored boat or buoy to record current direction and velocity for one or more tidal days (25 hr. each). Observations may be made at several depths. Momentary observations may also be made at various sites and depths with the use of a boat-mounted Roberts meter. Where surface currents only are to be observed, the simple Price current meter (see Art. 30-43) may be used for momentary observations. This meter registers velocity only, but the direction of the current can be determined at shallow depths by observing the direction in which the freely suspended meter points.

If floats are to be used, they should be designed to give minimum wind resistance and to extend under water a sufficient depth to measure the desired current. The float may be made of 2 by 4-in. wood weighted at the lower end. A length of 2 ft. is ordinarily used for surface currents, while 20 ft. is about the maximum for deep currents. For greater depths a current meter is used. The top of the float is approximately flush with the water surface, and to it is fitted a small red flag. As soon as the float attains the velocity of the current, it is located at regular time intervals with reference to fixed points on shore. It may be located by two angles read either from the shore or from a boat following the float, as previously described. From the time and distance measurements the velocity is computed.

Another method of direct measurement is sometimes used by the U.S. Coast and Geodetic Survey. A boat is anchored, and its location is determined by sextant angles read to three known points on shore. A line marked every 10.13 ft. to represent tenths of knots in current velocity is attached to a float. Time is taken at 60-sec. intervals by means of a stop watch. The float is set adrift, and the number of tag intervals of 10.13 ft. run out in 60 sec. is recorded. One-tenth of this number is the current velocity in knots, or nautical miles per hour. The nautical mile equals 6,080.2 ft.

The location of the float may be determined by sextant angles taken from the boat between known shore signals and the float. This information combined with the known distance from the anchored boat to the float will definitely fix its location and thus the direction of the current.

A rougher method sometimes used is to time the passage of a free float from one fixed range to another. The point at which the float crosses the range may be located by one angle read from the shore, and the distance may be scaled from the plotted location of the points.

At present (1965) dyes, such as Rhodamine B, are being used on a limited scale to trace the direction of surface currents. It is expected that the use of dyes for this purpose will become more widespread in the future.

30-25. Measurement of Dredged Material. Subaqueous surveys to be followed by dredging should be made by one of the methods by which soundings can be repeated at exactly the same location, that is, by intersecting ranges or by distances along a wire stretched on a fixed cross-section. Dredged material may be measured either in place or in scows. If in place, soundings on a fixed section are taken both before and after dredging, and the change in cross-sectional area is determined by calculation or by use of the planimeter. When this quantity has been determined for each section, the volume of excavation can be computed by the average-end-area method (see Art. 11-11).

If the contract calls for payment by scow measurement, each scow is numbered and the capacity of each pocket of the scow is carefully determined by mensuration. When the scow is filled to capacity, the inspector records a full measurement. If a pocket is not filled to capacity, the inspector measures the outage and deducts it from a scow measurement. When deck scows are used, the material is deposited on the deck in such shape that it can be conveniently measured and its volume computed.

Excavated material in scows is sometimes measured by the amount of water displaced in loading. It is necessary to know the dimensions of the scow, the weight per unit volume of the water in which it floats, and the weight per unit volume of the dredged material. The length of the water line and the depth of immersion must be measured with the scow both empty and loaded. When these have been determined, the yardage can be calculated for a rectangular scow by the formula

$$C = \frac{\frac{l + l'}{2}(d - d')bn}{W} \tag{1}$$

in which C = load, in cubic yards
l' = length of longitudinal water line (empty), in feet
l = length of longitudinal water line (loaded), in feet
b = width of scow, in feet
d' = depth of immersion (empty), in feet
d = depth of immersion (loaded), in feet
n = weight of 1 cu. ft. of water
W = weight of 1 cu. yd. of dredged material

For fresh water, n is taken as 62.4 lb. per cubic foot. For salt water, n is taken as 64.0 lb. per cubic foot.

30-26. Capacity of Existing Lakes or Reservoirs. The two general methods used in determining the capacity of existing lakes or reservoirs are the *contour* method and the *cross-section* method.

1. *Contour Method.* The contour method gives more reliable results. A shore traverse is run from which the water line and the desired shore topography are located by stadia. Soundings are then taken by methods suited to the particular conditions surrounding the survey. From the sounding elevations covering the immersed area, the subaqueous contours are plotted. The area enclosed by the water line and by each contour is determined by planimeter. The average of the enclosed areas at two consecutive contours multiplied by the contour interval or vertical distance between them gives the volume of water lying between the two contours. An even more reliable method of computing the volume between consecutive contours is by use of the formula for the volume of a frustrum of a general pyramid:

$$V = \frac{h}{3}(A_1 + A_2 + \sqrt{A_1 A_2}) \tag{2}$$

in which V = volume between contours
h = vertical distance between contours
A_1 = area enclosed by lower contour
A_2 = area enclosed by upper contour

The volume between the bottom contour and the deepest part is usually small and is either estimated or neglected. A summation of these partial volumes gives the capacity of the lake or reservoir. This volume is usually expressed in acre-feet, 1 acre-foot being 43,560 cu. ft.

2. *Cross-section Method.* When only a moderate degree of precision is required, the cross-section method is used. The outline of the water surface is found as in the contour method. The water outline is then plotted and divided into approximate trapezoids and triangles. The boundary lines between trapezoids or between trapezoids and triangles are on the sections which it is desired to measure. Soundings are taken on these sections by any suitable method of location. The perpendicular distances between sections and the altitudes of all triangles are determined by field measurement. The sections are plotted on cross-section paper, and the end areas are determined by planimeter. The approximate volume is computed by average end areas.

30-27. Snow Surveys. The water supply for many areas is dependent to a considerable degree upon melted snow from mountainous regions, and in these areas the determination of the amount and distribution of the snowfall is of great aid in forecasting the spring and summer runoff of streams. This information permits proper regulation and distribution of water by irrigation districts, public utilities, municipal districts, etc. The assumption is usually made that the spring and summer runoff will be approximately proportional to the water content of the snow pack on some fixed date. More refined forecasts are made from a relation based on runoff and snow-survey data for 10 or more years (Ref. 4 at the end of this chapter). Snow surveys are made annually in most of the Rocky Mountain and Pacific Coast states.

Snow courses are established at key locations which are considered representative of the entire area. They are preferably located in the early winter when some snow has fallen. A typical course forming part of the survey for a given watershed is perhaps 1,000 ft. in length, with measuring points designated at 100-ft. intervals. The course is not necessarily straight throughout its length. Its location is marked either by poles or by boards nailed to trees, and a detailed record of the location and markings is kept.

The depth and density of the snow are determined by means of a special sampling tube of metal, 1½ to 3 in. in diameter, having at the lower end a toothed cutter for drilling through hard crusts or ice layers. The cutter and contents are weighed by means of a spring scale which is calibrated to read directly in inches of water.

FLOW MEASUREMENT

30-28. General. Discharge measurements of a stream are usually made in connection with problems of water supply, power development, recreational use, fish propagation, and flood flow. Measurement of flow in canals and diversion conduits is essential for proper operation of irrigation, water supply, and hydroelectric facilities.

The discharge of a natural stream is a function of the volume and intensity of rainfall upon its drainage area and the characteristics of that area and may vary from no surface flow to violent and destructive floods. To procure an accurate knowledge of stream flow requires regular observations extending over a period of years. Much of this work has been done by the Water Resources Division of the U.S. Geological Survey cooperating with the individual states in a comprehensive study of our inland waters. The information thus collected is available to the public in the *Water-supply Papers* of the U.S. Geological Survey, Washington, D.C. Since 1961, data on stream flow are published by the individual district offices of the Geological Survey and can be obtained by writing to the district engineer of the appropriate state.

The following types of stream-flow data are collected:

1. Discharge measurements of a particular flow without regard to stage of stream.
2. Measurement of instantaneous peak discharge for all floods during a

specified period of time. From this information the magnitude and frequency of floods can be studied.

3. Continuous records of stage and discharge. This type of record requires discharge measurements for a number of different stages throughout the total range in stage in order to develop a stage-discharge relation curve (rating curve) for the station. The discharge is computed by applying this rating curve to the continuous record of water stage obtained at the gaging station.

4. Continuous record of stage only.

5. Discharge measurements to obtain formulas or coefficients for dams, weirs, or rating flumes.

All these types of data are collected by the same general methods and with the same instruments, except that observations extending over a period of time require the installation of some form of permanent gage.

30-29. Discharge Units. Discharge is the rate at which the water in a stream or conduit flows past a given section. The units of discharge commonly employed are as follows:

Cubic Feet per Second (cfs.). One cubic foot per second is equal to the discharge of a stream with a cross-section 1 ft. wide and 1 ft. deep flowing at an average velocity of 1 ft. per second.

Gallons per Minute (gpm.); Gallons per Day (gpd.) The common units for expressing flow or pumping duty for domestic water supply. For municipal supplies the unit is usually expressed in million gallons per day.

Miner's Inch. Formerly a common unit in mining and irrigation work. It is defined as the quantity of water which will flow through a orifice 1 in. square under a stated head which varies from 4 to 6½ in. in different localities. The use of this unit has led to much confusion, and its value in terms of cubic feet per second has been fixed by statute in most of the Western states. Because of the variation in values, this unit should be discarded in favor of the more accurate and clearly defined units given above.

Cubic Feet per Second per Square Mile (csm.). The ratio of the discharge in cubic feet per second at a particular section to the area, in square miles, of the drainage area above that section.

Million Gallons per Day per Square Mile (mgd. per sq. mi.). The ratio of the discharge in million gallons per day at a particular section to the area, in square miles, of the drainage area above that section.

The units of volume used in hydrographic work are:

Cubic Feet. The most convenient unit for expressing small quantities of water such as capacities of tanks or small reservoirs.

Gallons. The unit commonly used to express volumes in connection with municipal water supply. It is the standard liquid measure and is equal to 0.13368 cu. ft.

Acre-foot. The quantity of water required to cover an area of 1 acre 1 ft. deep; equal to 43,560 cu. ft. The unit is commonly used in the Western states in connection with reservoir storage and irrigation use.

Cubic Feet per Second per Day (cfs.-day). The volume of water equal to the flow of 1 cu. ft. per second for 24 hr., or 86,400 cu. ft.

Runoff in Inches. For any drainage area, the depth in inches to which the area would be covered if all the water flowing from it in a given time were uniformly distributed over the area.

30-30. Stream Gaging. Stream gaging is the process of measuring the stages, depths, areas, velocities, and rates of flow at selected sites in natural or artificial channels. Gaging stations are classified as *stage stations* or *discharge stations.* Both types provide stage records. A discharge station also provides a record of discharge through use of a relation between stage and corresponding discharge as measured at the gaging station.

The discharge measurement is the determination of the amount of water flowing past a given section in a given time. The discharge unit is usually the cubic foot per second, and the magnitude of the discharge is the product of two factors—the mean velocity of the water and the cross-sectional area at right angles to the flow.

The area of a given cross-section can be determined by methods described earlier in this chapter. However, the general practice is to sound the cross-section at the time the velocity measurements are made. Sufficient measurements of depth should be taken to define the shape of the cross-section accurately.

Mean velocity in a cross-section is difficult to determine precisely because the velocity at any point is affected by the proximity of the stream bed and banks and because even minor irregularities in the channel tend to cause cross-currents and eddies in the water. As a rule, the more point-velocity measurements taken using established procedures, the better the determination of mean velocity.

30-31. Selection of Gaging Site. A major factor affecting the value and accuracy of stream-flow data collected at a gaging station is the selection of the site. The ideal site satisfies the following criteria:

1. The general course of the stream for several hundred feet above and below the section should be straight.
2. The section should have a definite stream *control* or natural constriction to insure that the relation between stage readings and discharge remains reasonably constant (Art. 30-32).
3. There should be a pool at low stages to insure a recording of stage at extremely low flows and to eliminate high velocities at the ends of the intakes.
4. The total flow should be confined to one channel at all stages.
5. Banks should be permanent, high enough to contain floods, and clear of brush.
6. The section should be far enough upstream from the confluence with another stream to eliminate any influence the other stream may have on the stage at the gaging site.
7. The station should be so located that bridges, dams, or other works will not have a variable effect on the stage at the gaging site.
8. A satisfactory reach for measuring discharge at all stages must be available within reasonable proximity to the station.
9. The site should be readily accessible for ease in installation and operation of the gaging station.

30-32. Gaging-station Controls. A control is some physical element or characteristic of the stream channel that causes a definite relationship to exist between stage and discharge. It may be a natural feature or may be artifically constructed; it may be a simple feature occupying a single cross-section of the stream (section control), or it may be a long stretch of channel downstream from the gage (channel control). Examples of a natural section control are a riffle or rock ledge outcrop causing rapids, a natural constriction, a gravel bar, or a natural fall. The most common type of artificial control is a

FIG. 30-10. Looking upstream at a natural-section control. Gaging station in background.

small dam or a weir. The adequacy and permanence of the control are the most important factors governing the accuracy of the data obtained at a gaging station.

The location and type of control for a gaging station may change depending upon the stage. At low stage the control may be an outcropping rock ledge, a boulder or gravel riffle, or a low artificial control. At medium stages the low-water control may be submerged by a natural constriction further downstream, which constriction acts as the medium-water control. At high stages the channel characteristics for some distance downstream may control the stage-discharge relation at the gage and submerge both the low-water and medium-water controls. Seldom does a single feature act as the control at all stages.

Controls, particularly low-discharge controls, should be sensitive, that is, for any change in discharge a rather quick response to the change should be reflected at the gage, and for a small change in discharge a relatively large change in stage should occur.

FIG. 30-11. **Artificial control with notch for low discharges.**

A gaging station is located above a suitable natural control, if possible, because artificial controls are expensive to construct and maintain. Figure 30-10 shows a natural section control composed of a ledge-rock outcrop.

Where natural conditions do not provide the stability or the sensitivity required, artificial controls are used. Artificial controls are structures built in a stream channel to stabilize or constrict the channel at a section. They may be low dams, broad-crested weirs conforming to the general shape and height of the stream bed, sharp-crested weirs (on small channels), or measuring flumes such as the Parshall flume (Art. 30-70). Figure 30-11 shows an artificial control with all the water flowing through the low-water notch. This notch makes the control sensitive at low discharges.

30-33. Measurement of Water Stage. The stage of a stream or lake is the height of the water surface above a known datum. A record of stream stage may be useful in itself, but it is also needed for computing the discharge at a gaging station after the stage-discharge relation is established. Readings of stage at a gaging station are called *gage heights.*

The measurement of stage is made using a gage on which the water surface elevation or gage height can be read. The zero of the gage should be so set that the readings will never be negative. There are two principal types of gages: *nonrecording* and *recording* or automatic. Several types of each are described in the following articles.

FIG. 30-12. Vertical staff gage.

30-34. Staff Gages. A staff gage may be either *vertical* or *inclined.* The vertical gage consists of a wooden post or board fastened solidly in position and graduated to read in feet and tenths of a foot. Either the graduations are painted directly on the wood or a porcelain-enameled iron strip, graduated every 0.02 ft., is fastened to the face of the post (Fig. 30-12).

Relatively short sections of the vertical staff gage can be installed in steps up the bank of the stream or reservoir so as to cover the expected full range in stage. Care must be taken to insure that all the gages are set correctly to the proper datum.

The inclined staff gage usually consists of a heavy timber securely attached to a permanent foundation. It is constructed so as to follow the slope of the bank with a minimum projection above the ground level. The face of the timber is graduated in place with the aid of an engineer's level. The graduations can be either notched into the timber or marked by heavy staples with bronze numerals for the foot marks.

Crest-stage Gage. A useful adjunct to the staff gage is the crest-stage gage, which is a device for obtaining the highest elevation reached by a stream during the time interval between visits to the gage. The crest-stage gage is simple, economical, reliable, and easily installed. It usually consists of a length of 2-in. pipe mounted vertically at a site where its lower end will be submerged at medium and high stages. The upper end of the pipe extends a foot or two above the highest stage expected. Within the pipe is a wooden staff of the same length as the pipe and of smaller diameter. The elevation of the bottom of the staff is referenced to gage datum by levels. The pipe itself is capped at both ends. The bottom cap has a series of intake holes around its perimeter to allow water from the stream to enter the pipe when medium and high stages occur. The upper cap has a small vent hole to allow air to escape when water enters the pipe. A small handful of regranulated cork is placed in the pipe when the gage is prepared for operation.

FIG. 30-13. Type A wire-weight gage.

During a stream rise, water enters the pipe and the cork floats on the rising water within the pipe. When the water reaches its peak and starts to recede, the cork adheres to the wooden staff within the pipe, thereby leaving a record of the highest stage reached since the last gage inspection. On visits to the gage, the upper cap is removed from the gage, the wooden staff is removed from the pipe, and the elevation of the cork line on the staff is recorded. The staff is then wiped clean before being returned to the pipe, additional cork is added in the pipe if necessary, and the pipe is recapped. The crest-stage gage is then ready to record the highest stage which occurs in the interval of time that elapses before the next visit to the gage.

30-35. Wire-weight Gage. The wire-weight gage in general use is the U.S. Geological Survey type A gage, which has almost entirely replaced the chain gage and the Canfield wire-weight gage that were formerly in common use. The gage is usually mounted on a bridge, but it may be installed on the stream bank with the use of a cantilevered beam and pulley over the water.

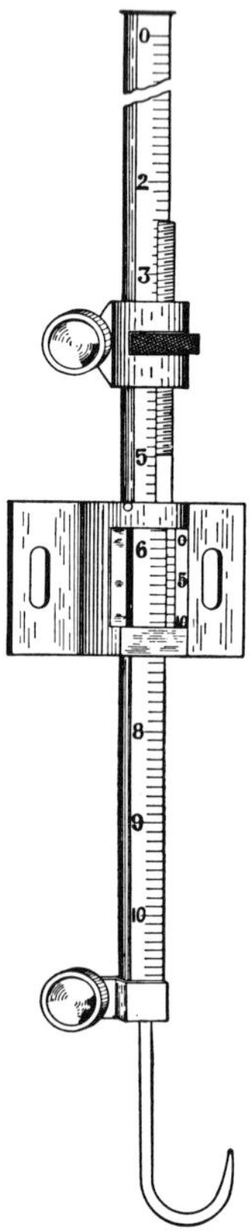

FIG. 30-14. Hook gage.

The type A gage consists of a drum wound with a single layer of cable, a bronze weight attached to the end of the cable, a graduated disk, and a Veeder counter, all enclosed in a cast-aluminum box (Fig. 30-13). The disk is graduated in tenths and hundredths of a foot and is connected to the counter and to the shaft of the drum. The diameter of the drum is such that each complete turn represents a 1-ft. movement of the weight. The graduated disk and counter of the gage are set so that the proper gage height will be indicated when the bottom of the weight is at the water surface.

30-36. Hook Gage. The hook gage shown in Fig. 30-14 is used primarily for precise measurement of the head of water flowing over a weir (Art. 30-64). The gage is generally installed in a stilling box which is either open to the main flow of water or connected to it by a pipe. The box should be located at a convenient point near the weir but at least three times the expected maximum head upstream from the weir crest. If the water surface in the weir pool is still, a box may not be necessary.

To measure the depth of water flowing over the weir, the level of the weir crest is determined with a leveling instrument, and this elevation is transferred to a mark on the inside of the stilling box. The gage scale is set to zero, and the gage is fastened to the side of the box by means of two screws through the slots shown in the illustration so that the point of the hook is at the same elevation as the mark. Thus the point of the hook is level with the crest of the weir, and the depth of water flowing over the weir is the distance from the point of the hook in this position to the exact surface of the water.

To read the gage, the hook is raised until it pierces the surface, it is then clamped in position, and by means of the slow-motion screw its height is adjusted until the water surface just shows no distortion. This position, which gives the exact elevation of the water surface, is then read on the vernier to thousandths of a foot.

30-37. Recording Gages. The conventional recording-gage installation requires the construction of a stilling well in the bank of the stream. The well can be constructed of reinforced concrete, concrete pipe, concrete block, galvanized corrugated steel pipe, or in some cases wood. It should be placed as close to the stream as possible without endangering the structure from damage by floods. The elevation of the water surface in the stream is transferred through pipe intakes to the stilling well. The bottom of the well should be at least a foot below the minimum stage anticipated, and the structure itself should be high enough not to be topped by any flood.

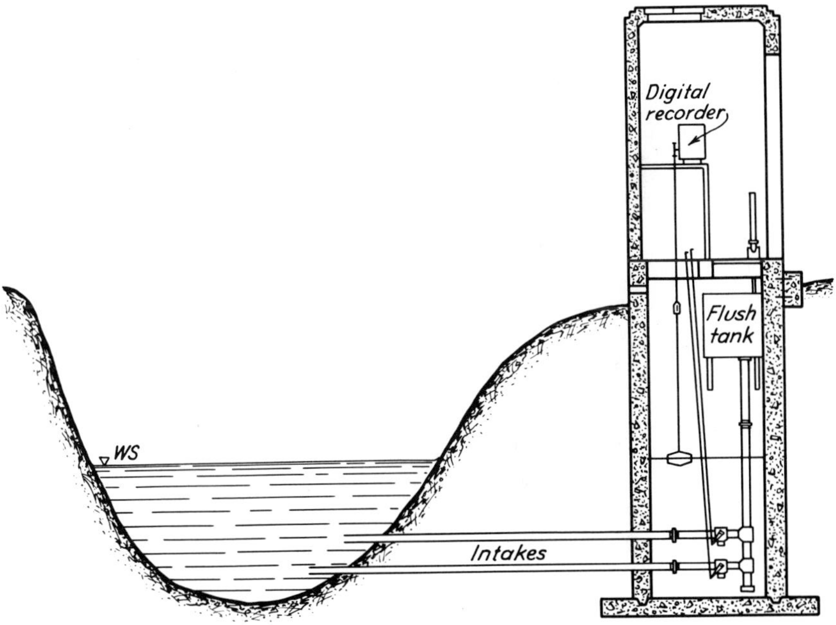

FIG. 30-15. Typical stilling-well station.

A gage shelter is installed on top of the well to house the recording instrument (Art. 30-38) and other items needed for operation of the gage. A copper float attached to the recording instrument rests on the water in the well and moves up or down as the water level rises or falls. The most convenient type of shelter is one in which a man can walk and stand. Figure 30-15 shows a section drawing of a typical reinforced-concrete stilling well and shelter, and Fig. 30-16 shows an installed galvanized corrugated steel pipe well and shelter.

At many stations the pipe intakes between the stream and the well are subject to clogging by silt or trash. At these sites it is desirable to install a flushing system consisting of a tank and riser pipes connected to the intakes with a three-way valve (Fig. 30-15). Ordinarily a hand-operated pump raises water from the well to a tank below the gage-house floor. The water is then released through the intake by proper operation of the valve.

A new type of recording-gage installation developed several years ago does not require a stilling well and intake pipes. It utilizes a bubble-type pressure system, commonly called a *bubble gage*, which consists of a specially designed servomanometer, a transistor control, and a gas-purge system (Fig. 30-17). The installation consists of an instrument shelter on the bank, a pier and

FIG. 30-16. Corrugated steel pipe stilling well and shelter.

orifice in the stream, and a small-diameter tube connecting the orifice with the bubble gage in the instrument shelter (Fig. 30-18).

Nitrogen gas is slowly discharged through the tube and orifice to prevent clogging. The orifice is located at a fixed elevation in the stream below the expected minimum stage. The pressure at the orifice, and therefore at any point in the tube, is dependent upon the head or depth of water over the orifice. This pressure is in turn transferred to the manometer and then to the recording instrument.

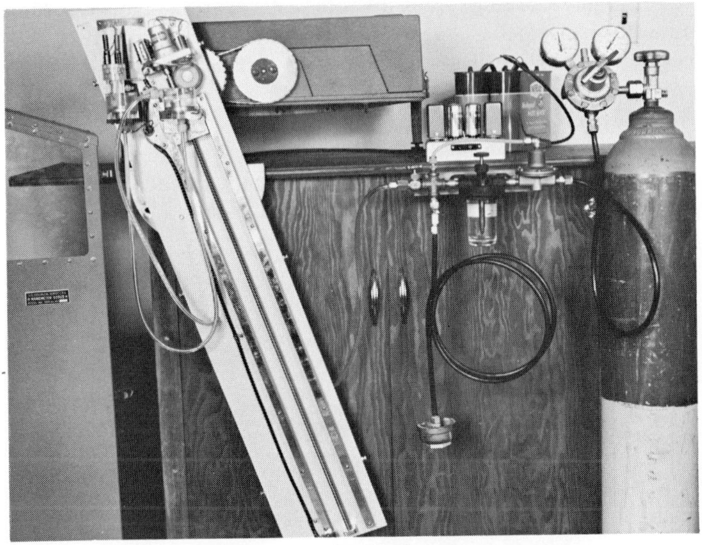

FIG. 30-17. Bubble gage.

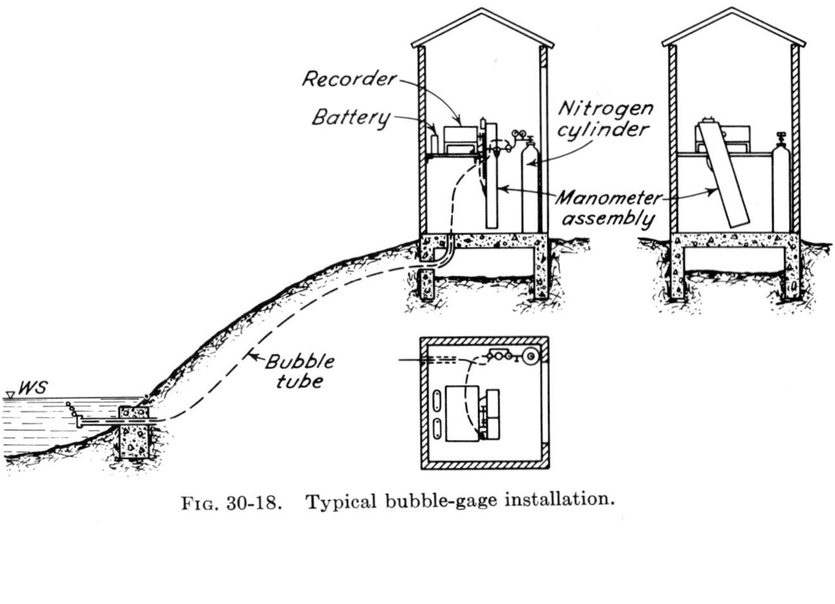

FIG. 30-18. Typical bubble-gage installation.

30-38. Stage-recording Instruments. A water-stage recorder is an instrument for producing a graphic or a punched-tape record of the rise and fall of the water level with respect to time. The recorder consists of a time element and a gage-height element. The time element is controlled by a clock which may be driven by spring, weight, or electricity; the gage-height element is actuated by a float or a bubble-type pressure system.

The graphic recorder provides a continuous trace of water stage with respect to time on a graduated chart. Usually the time element moves the

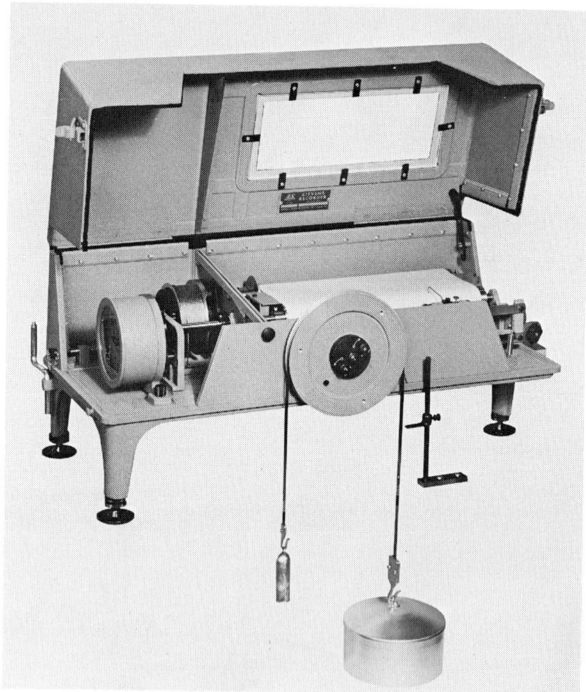

FIG. 30-19. Continuous strip-chart recorder.

chart and the gage-height element moves the pen or pencil stylus. Most graphic recorders are capable of recording an unlimited range in stage, either by use of reversing devices on the strip-chart types or by unlimited revolutions of the drum on the drum-type recorder.

Most strip-chart recorders (Fig. 30-19) will operate for several months without being serviced. The drum-type recorders should be serviced at weekly intervals. Figure 30-20 shows a commonly used drum recorder.

The modern punched-tape or digital recorder was developed to provide the stage data in a suitable format for machine computation. It is a battery-operated slow-speed paper-tape punch which records a four-digit number on a 16-channel paper tape at preselected time intervals (Fig. 30-21).

Stage or gage height is transmitted by the recorder input shaft, which is activated by a float or a bubble-type pressure system. It is recorded in binary decimal code on the paper tape in increments of hundredths of a foot, from 0 to 99.99 ft., using two coding disks.

The timer used on the digital recorder, shown in the foreground of Fig. 30-21, is an electromechanical timing device powered by the same battery

Fig. 30-20. Horizontal-drum recorder.

that operates the punch motor. It provides contact closure for activating the punching sequence at preselected time intervals of 5, 15, 30, or 60 min. by means of a different cam for each time interval.

30-39. Measuring the Cross-section. Usually the area of the cross-section of the channel is obtained as an integral part of the measurement of flow. At times, however, the cross-section must be measured independently. If possible, the cross-section should be measured during a low-water period.

Starting at a permanently referenced point above high-water level, a profile of both banks and of the underwater section as far as wading is possible is obtained by means of an engineer's level. The remaining submerged section is taken by soundings referred to water level. The distance between soundings depends upon the width of stream, the shape of stream bed, and the accuracy desired. As a general rule no less than 15 soundings should be taken, with a minimum distance of 1 ft. between soundings. Enough care must be taken to assure the observer that the profile between soundings is approximately a straight line. The horizontal distance between points on the profile is measured by means of a metallic tape, a light cable tagged at 2- or 5-ft. intervals, or stadia.

FIG. 30-21. Digital recorder.

The precision with which the measurements should be made varies with the depth, the width, the shape of the cross-section, the material of the stream bed, and the technique used in making the observations. Soundings are usually observed to the nearest tenth of a foot. Horizontal distances are usually observed to the nearest tenth of a foot in narrow streams and to the nearest half-foot or foot in wider streams.

Measurements of the cross-section made as part of the measurement of flow are limited to the section under water at the time of the measurement. Most flow measurements require 25 to 30 observations of velocity spaced at intervals across the stream. It is the practice to make a sounding of depth at each point where velocity observations are taken, using the precision criteria previously discussed. The depth and velocity observations are then recorded in a standard manner (Art. 30-55).

30-40. Measuring the Velocity. The preceding articles describe the methods and instruments used for measuring the stage and the cross-section of a stream. To compute the discharge or flow of the stream it is also necessary to measure the velocity of the water past the gaging section. Techniques and instruments for measuring velocity are described in the following articles.

30-41. Floats. Floats are used to measure velocity principally in reconnaissance work, in locating gaging stations, or in measuring flood velocities. They should be used only when facilities for making more accurate velocity measurements are lacking.

The commonly used type of float is the *surface float,* which is intended to measure surface velocities. The surface float should be made up of lightweight or hollow material in such a shape as to offer the least resistance to floating debris, ripples, eddy currents, and wind. However, the cost of having the floats specially made is not warranted because usually they cannot be recovered. The general practice is to use improvised floats of jugs, bottles, blocks of wood, flood debris, apples, oranges, etc. The surface velocity is measured by timing the travel of surface floats at various locations across the stream through a measured distance. The observed velocity must be multiplied by a coefficient to reduce it to the mean velocity. If possible, this coefficient should be determined by current meter; if not, a coefficient of 0.85 to 0.90 is commonly used.

In former years *subsurface floats* were often used to obtain a more reliable estimate of the mean velocity. The accuracy gained by the use of subsurface floats does not warrant the time and trouble involved in their preparation; consequently, subsurface floats are seldom used.

The use of floats should be restricted to straight stretches of the stream having a fairly uniform cross-section throughout the length of the reach. The distance through which the floats are timed should be at least 100 ft. and if possible as much as 500 ft. The precision of the mean velocity observation is affected by:

1. Uncertainty of the relationship between the velocity of the float and the mean velocity.
2. Interference with the movement of the float resulting from air disturbance, debris, eddy currents, and other extraneous forces.
3. Uncertainty of the exact path traveled by the float.
4. Adequacy of the reach used for the measurement.

The procedure for making a float measurement is as follows:

1. Select the reach of stream to be used, and accurately measure its length. The beginning and end of the reach should be clearly marked to aid in the observation.
2. Measure the time interval for the float to traverse the length of the reach with a stop watch at 15 to 20 locations across the stream. Usually two men are required

—one to place the floats and start the timing at the beginning of the reach, the other to indicate when the float passes the end of the reach.

3. Compute the surface velocity of the floats, and apply the appropriate coefficient to convert surface velocity to mean velocity.

4. Determine the area at a typical cross-section in the reach using the method described in Art. 30-39.

30-42. Current Meters: General. A current meter is an instrument used to measure the velocity of flowing water. The principle of operation is based on the proportionality between the velocity of the water that strikes it and the rate of rotation of the rotor of the meter. If a current meter is placed at a point in a stream and the number of revolutions of the rotor counted in a measured interval of time, the velocity of water at the point can be determined from the calibration of the meter.

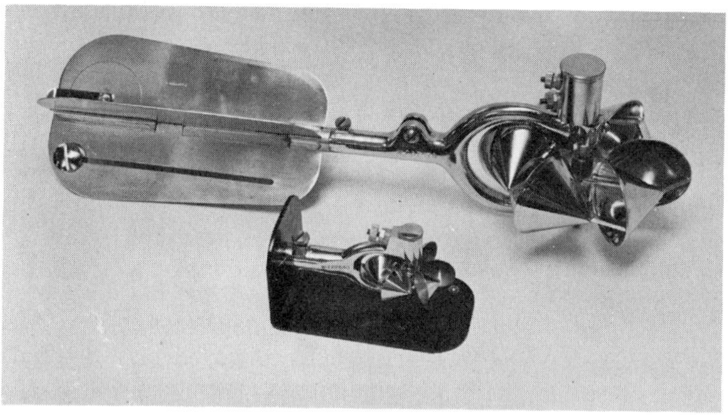

FIG. 30-22. Price current meter.

There are two main types of revolving current meters. The first type, represented by the Price meter (Figs. 30-22 and 30-29), has a vertical axis of rotation normal to the direction of stream flow, and the rotor is fitted with cup-shaped vanes, either conical or curved. The rotation is the result of the difference in dynamic forces acting on opposite sides of the cups. The second type has the axis of rotation parallel to the direction of stream flow and a rotor with spiral or helicoidal-shaped vanes. The rotation is caused by the direct impulse and pressure of water upon the vanes. Two examples of this type, the Ott meter and the Haskell meter, are shown in Figs. 30-23 and 30-24 and are described in Art. 30-45.

The advantages of vertical-axis meters are:

1. They will operate at lower velocities than the horizontal-axis meters.
2. The bearings are well protected from silty water.

3. The rotor and vanes are more rugged and less subject to damage by debris.
4. The rotor can be repaired in the field without affecting the calibration significantly.
5. A single rotor can be used through the entire range of velocities.

FIG. 30-23. Ott current meter.

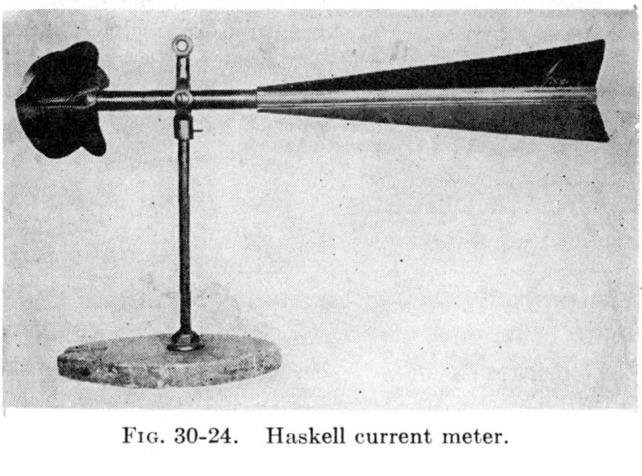

FIG. 30-24. Haskell current meter.

Horizontal-axis meters have the following advantages:

1. The forces acting on the rotor do not cause bending moments in the rotor spindle; this feature reduces the bearing friction.
2. Because of axial symmetry of flow, the vanes cause smaller disturbances than those for the vertical-axis meter.
3. The horizontal-axis rotors are less sensitive to fouling by foreign bodies carried by the stream.

A recording or indicating device is necessary for determining the number of revolutions of the rotor. Usually a combination electrical-acoustical system is used. It consists of an electrical circuit which is closed twice each revolution, once each revolution, or once each fifth revolution of the rotor by contact points in the meter head. Each closure produces an audible click in an earphone in the circuit. An electrical counter can be substituted for the earphone if desired. A stop watch is necessary for the proper timing of the observations.

30-43. Price Meter. The Price meter is the current meter commonly used in the United States. It consists of a wheel made with six conical cups fastened to a vertical shaft as shown in Fig. 30-22. The upper end of the cup shaft is fitted with either a worm gear or an eccentric that passes into the

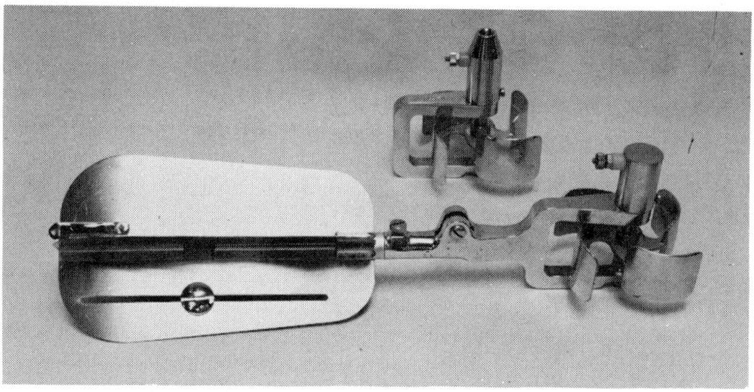

FIG. 30-25. Vertical-axis vane current meter.

cylindrical contact chamber. This chamber contains a mechanism for making mechanical or, more commonly, electrical contact which indicates by a click either each revolution or each fifth revolution. The mechanism for indicating each fifth revolution, called the *penta-count*, is used for velocities above about 6 ft. per second, because for higher velocities the ear is unable to distinguish the separate clicks for each single revolution.

A vane or tail is attached to the yoke which holds the cup wheels in place to keep the meter heading into the current. A vertical steel bar is attached through a slot in the yoke to support the weight and to supply a connection to the cable by which the meter is suspended. The meter can also be attached to a graduated wading rod for use in wading measurements.

For measurements in shallow depths a small Price meter called a *pygmy meter* has been developed. It is similar to the large meter in most respects but is only two-fifths as large, has no tailpiece, has a different type of contact chamber, and is used only for wading measurements.

30-44. Vane Meter. The U.S. Geological Survey has recently developed a four-vane vertical-axis meter (Fig. 30-25). This meter is particularly useful for measuring under ice cover (Art. 30-61) because the vanes do not fill with slush ice. Its use is also recommended for measuring from boats because it will not measure the vertical component of velocity that is often induced by the motion of an anchored boat. The yoke of the meter is especially shaped so that it can be inserted in the small-diameter holes made in the ice by the ice auger. The meter can also be used with a cable suspension.

The rotor and vanes are cast in a single piece by use of a standard mold so that units are interchangeable. The rotor shaft enters a cylindrical contact chamber similar to that of the Price meter. The main disadvantage of the

FIG. 30-26. Hoff current meter.

vane meter is that it does not register velocities below 0.5 ft. per second as accurately as other meters.

30-45. Horizontal-axis Meters. Several makes of horizontal-axis meters are used to a limited extent in the United States. Although they are similar to one another in many respects, a brief description of three of the makes follows:

1. *Ott Meter.* This meter (Fig. 30-23) is a precision-made instrument manufactured in Germany. It is not used extensively in this country because it is not so durable for general-purpose use as the Price meter. The Ott meter has a component propeller, which in oblique currents automatically registers only the velocity component in the direction parallel to the axis of the meter.

2. *Haskell Meter.* This meter (Fig. 30-24) is fitted with a conical-shaped screw propeller wheel designed to react to the direct force of the current. Propellers of different blade pitch are available for use through a wide range

of velocities—steep pitch for low velocities and flat pitch for higher velocities. The Haskell meter has been used in this country by the U.S. Lake Survey in gaging the large rivers in the Great Lakes region.

3. *Hoff Meter*. This meter (Fig. 30-26) is fitted with a hard-rubber propeller having either three or four blades. The propeller with three inclined blades is used for high-velocity measurements; the propeller with four straight blades is used for low velocities. Because the rubber propeller is lighter in weight than comparable metal propellers, it responds more readily to changes in the velocity of water and causes less bearing friction.

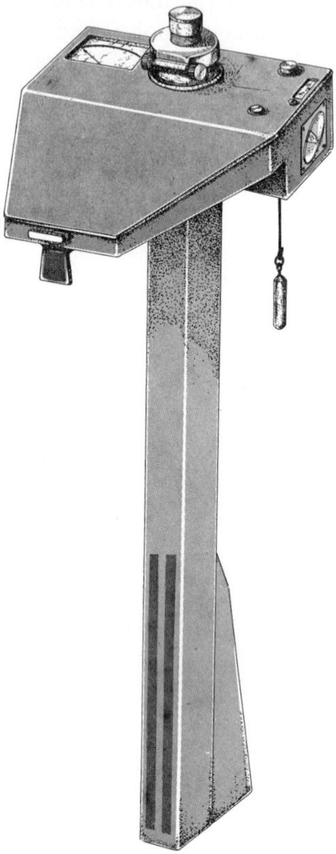

FIG. 30-27. Optical current meter.

30-46. Optical Meter. In recent years considerable effort has been expended to develop velocity meters that do not require immersion of the measuring unit, in order to eliminate the difficulties of positioning a current meter in deep high-velocity flow and preclude damage to the current meter by flood debris. The most promising meter of this type, which measures surface velocities only, is the optical current meter (Fig. 30-27). The meter is a lightweight battery-powered stroboscopic device consisting of a low-power telescope, a set of rotating mirrors, a variable-speed drive motor, and a tachometer. Velocity measurements are made from an observation point above the stream by looking down at the water surface through the meter while gradually increasing the angular speed of the rotating mirrors. As synchronization is reached, the apparent motion of the water surface, seen through the eyepiece, slows down and is finally stopped. The surface velocity of the water can then be computed using the angular speed of the mirror wheel at this null point and the vertical distance from the optical axis of the meter to the water surface.

30-47. Rating Current Meters. The current meter does not measure velocities directly but indicates the number of wheel revolutions per unit of time. The individual meter must be rated so that the revolutions per unit of time can be translated to velocity, or distance per unit of time, usually expressed in feet per second. The standard method of rating a cup, vane, or propeller meter consists in towing it through a body of still water for a

DEPARTMENT OF INTERIOR-GEOLOGICAL SURVEY

Water Resources Division

RATING TABLE FOR TYPE .*AA*... CURRENT METER NO. *W-233517*.

SUSPENSION .*Rod*.. RATED *9-21-63*

Time in seconds	Velocity in feet per second									Time in seconds
	Revolutions									
	3	5	7	10	15	20	25	30	40	
40	.190	.300	.410	.575	.850	1.12	1.40	1.68	2.22	40
41	.186	.293	.401	.562	.830	1.10	1.37	1.63	2.17	41
42	.182	.287	.392	.549	.811	1.07	1.33	1.60	2.12	42
43	.178	.281	.383	.537	.792	1.05	1.30	1.56	2.07	43
44	.175	.275	.375	.525	.775	1.02	1.28	1.52	2.02	44
45	.172	.269	.367	.514	.758	1.00	1.25	1.49	1.98	45
46	.168	.264	.360	.503	.742	.982	1.22	1.46	1.94	46
47	.165	.259	.353	.493	.727	.961	1.20	1.43	1.90	47
48	.162	.254	.346	.483	.712	.942	1.17	1.40	1.86	48
49	.160	.249	.339	.474	.698	.923	1.15	1.37	1.82	49
50	.157	.245	.333	.465	.685	.905	1.12	1.34	1.78	50
51	.154	.241	.327	.456	.672	.888	1.10	1.32	1.75	51
52	.152	.237	.321	.448	.660	.871	1.08	1.29	1.72	52
53	.150	.233	.316	.440	.648	.855	1.06	1.27	1.69	53
54	.147	.229	.310	.432	.636	.840	1.04	1.25	1.65	54
55	.145	.225	.305	.425	.625	.825	1.02	1.22	1.62	55
56	.143	.221	.300	.418	.614	.811	1.01	1.20	1.60	56
57	.141	.218	.295	.411	.604	.797	.990	1.18	1.57	57
58	.139	.215	.291	.404	.594	.784	.973	1.16	1.54	58
59	.137	.211	.286	.398	.584	.771	.957	1.14	1.52	59
60	.135	.208	.282	.392	.575	.758	.942	1.12	1.49	60
61	.133	.205	.277	.386	.566	.746	.927	1.11	1.47	61
62	.131	.202	.273	.380	.557	.735	.912	1.09	1.44	62
63	.130	.200	.269	.374	.549	.723	.898	1.07	1.42	63
64	.128	.197	.266	.369	.541	.712	.884	1.06	1.40	64
65	.127	.194	.262	.363	.533	.702	.871	1.04	1.38	65
66	.125	.192	.258	.358	.525	.692	.858	1.02	1.36	66
67	.124	.189	.255	.353	.518	.682	.846	1.01	1.34	67
68	.122	.187	.251	.349	.510	.672	.834	.996	1.32	68
69	.121	.184	.248	.344	.503	.663	.822	.982	1.30	69
70	.119	.182	.245	.339	.496	.654	.811	.968	1.28	70
	3	5	7	10	15	20	25	30	40	

FIG. 30-28. Price current meter rating table.

known distance at various known velocities and counting the number of revolutions of the wheel for each of the velocities. Usually, the meter is attached to a small car which is driven along a level track over or beside the rating flume or channel. The velocities at which the meter is moved through the flume should range from the lowest velocity at which the wheel will rotate to one approaching the maximum velocities expected in the field.

Several factors such as bearing friction, slip of the blades, inertia, and retarding effect of water influence the relation between the angular velocity of the rotor and the measured velocity of the rating car. Were it not for these factors, the relation would be a constant for all velocities and the meter rating curve would be a straight line. However, the effect of these factors diminishes as the velocities increase, so that the resulting curve is essentially a straight line except for low velocities.

When the meter is rated, the velocities in feet per second are plotted as abscissas and the revolutions of the wheel per second as ordinates, and a mean curve is drawn through the plotted points. An equation in the form $V = aR + b$, in which V is the velocity of water and R is the revolutions of the wheel per second, is then developed for the mean curve. The coefficient a is the ratio of the revolutions per second to the velocity in feet per second, and the constant b represents the velocity that will just overcome the retarding effect of the factors mentioned above. For convenience in field use the meter rating is usually transformed into a meter rating table utilizing the curve or the equation. Figure 30-28 shows a typical rating table for a Price meter.

Meter ratings should be made only at a properly equipped rating station such as that of the National Bureau of Standards, Chevy Chase, Maryland. Meters that are in constant use should have their ratings checked at intervals of about a year.

30-48. Meter Suspension. The current meter can be suspended in the water either on a rod or by a small-diameter cable. The rod suspension is used for wading measurements or measurements from footbridges where the flow is not too deep. The cable suspension is used for deeper flows.

A graduated rod with a base plate at the bottom is used for the rod suspension. The meter is mounted on a sliding support and is set at the desired position by sliding the support up or down on the rod (Fig. 30-29). The upper end of the rod is usually equipped with some type of handle and a connection for the revolution-counting system of the meter.

For the cable suspension a small-diameter high-strength cable of sufficient strength to support the meter and the sounding weight should be used. A cable of diameter about 0.10 in. is the size commonly used. The cable is fastened to a stainless-steel weight-hanger bar about 1 ft. long (Figs. 30-33 and 30-34). The meter is connected near the top of the bar and the sounding weight is fastened near the bottom so that the flow pattern around the streamlined weight has a minimal effect on the meter operation. Care should be taken to insulate the cable connection so that the revolution-indicator system of the meter will operate properly. Usually the cable is marked with one or more index tags; when used in combination with a graduated scale or tape, these tags make it possible to set the meter at the desired position.

30-49. Velocity Measurements: General. The current meter measures velocity at the point at which the meter is placed. The velocity of a stream, however, varies from point to point in a cross-section. It is therefore necessary to establish a systematic procedure for measuring sufficient point velocities to permit precise determination of the discharge.

FIG. 30-29. Current meter on wading rod.

The method of making discharge measurements at a cross-section is to consider the total section as comprising a number of subsections where area and velocity are measured. In general, a section is subdivided into 20 to 40 subsections, depending on the width of the stream, the shape of the cross-section, and the variability of velocity throughout the channel. In subdividing, the subsections need not be of equal width; the primary consideration should be that each subsection carry approximately equal discharge.

The mean velocity for each subsection is considered as the mean velocity in a vertical line at about the midpoint of the subsection. The mean velocity of the selected verticals can be obtained from many observations of point velocity in that vertical, or it can be approximated by fewer measured point velocities and use of a known relation between the velocities at those points and the mean in the vertical. The various methods used to measure the mean velocity in a vertical are described in Arts. 30-50 to 30-54.

30-50. Vertical-velocity-curve Method. Measurements of point velocity are made at each tenth of the depth from surface to bottom. However, observations should not be made closer than 0.5 ft. to either the water surface

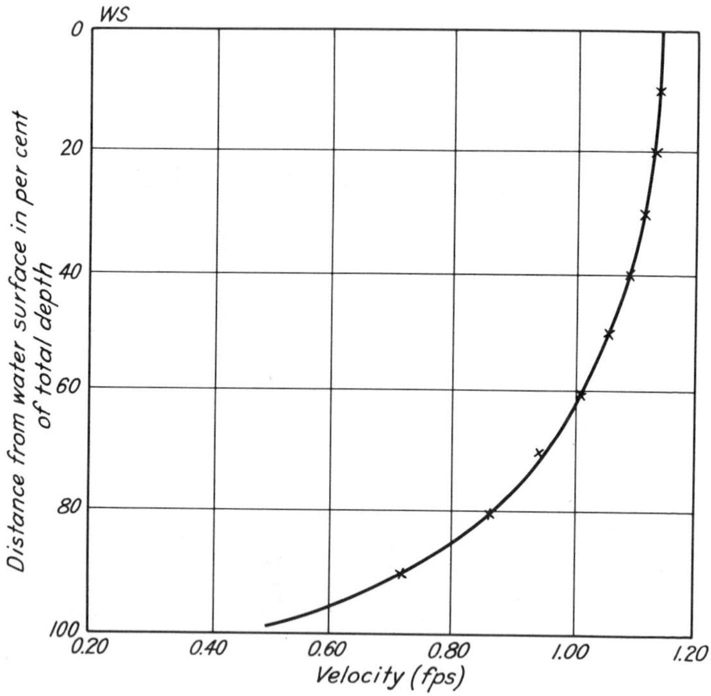

FIG. 30-30. Typical vertical-velocity curve.

or the stream bed. These measured velocities are plotted as abscissas, and the respective depths as ordinates, and a smooth curve is drawn through the points (Fig. 30-30). In order that vertical-velocity curves at different verticals may be readily compared, it is customary to plot depths as proportional parts of the total depth. The mean velocity in the vertical is obtained by measuring the area between the curve and the ordinate axis either with a planimeter or by other means, dividing the area by the length of the ordinate axis, and scaling the resulting length along the abscissa axis to obtain the mean velocity.

The vertical-velocity-curve method is the most precise method of determining mean velocities, but it requires too much time for general use. It is valuable as a basis of comparison with other methods, for determining a coefficient for the subsurface method, and for use under unusual conditions of flow.

30-51. Two-point Method. In the two-point method, velocity observations are made in the vertical at 0.2 and 0.8 depth below the water surface. The average of these two observations is taken as the mean velocity in the vertical. This method is based upon the theory that the vertical-velocity curve is a parabola and that the average of the abscissa readings at 0.2114 and 0.7886 depth below the surface gives the mean velocity in the vertical. Studies of many vertical-velocity curves indicate that this relation holds substantially true for most conditions of flow, and experience has shown that this method gives more consistent and accurate results than any of the other methods except the vertical-velocity-curve method.

The two-point method should not be used for water depths less than about 2.5 ft. because the current meter will be too close to the water surface and to the stream bed to give dependable results. When depths permit, the two-point method is the one usually employed by the U.S. Geological Survey.

30-52. Six-tenths Method. In the six-tenths method, a single measurement of the velocity is taken at 0.6 depth below the water surface and the observed value is taken as the mean velocity in the vertical. Studies of vertical-velocity curves and mathematical theory have shown that the 0.6-depth method gives reliable results. However, it should be used only either if the depths are too shallow for the two-point method (0.2 and 0.8 depth) or if other flow conditions such as drift, slush ice, etc., preclude the use of the two-point method.

30-53. Integration Method. In the vertical-integration method of measuring velocity, the current meter is lowered at a uniform rate from the water surface to the stream bed and is then raised at the same rate to the surface. The number of revolutions of the meter wheel during the complete cycle is timed. From these data the average revolutions per second can be found and the mean velocity taken from the meter rating table. This method is based on the theory that all horizontal velocities in the vertical have acted equally upon the meter wheel and that their average is the mean velocity. Current meters of the vertical-axis cup type should not be used for this method because the vertical movement affects the motion of the rotor; the horizontal-axis propeller type should be used.

30-54. Subsurface Method. This method should be used only when high velocities, large amounts of debris, or other factors make it impossible to use one of the more accurate methods. When the depth at the vertical is known, either from soundings or from a previously determined cross-section, the meter should be placed at 0.2 depth below the water surface. The velocity

9-275-F
Jan. 1956

UNITED STATES
DEPARTMENT OF THE INTERIOR
GEOLOGICAL SURVEY
WATER RESOURCES DIVISION

DISCHARGE MEASUREMENT NOTES

Meas. No. ____________
Comp. by ____________
Checked by ____________

____

Date ______________________, 19_____ Party ______________________________

Width __________ Area __________ Vel. _______ G. H. __________ Disch. ______________

Method _________ No. secs. __________ G. H. change __________ in _______ hrs. Susp. _______

Method coef. _________ Hor. angle coef. _________ Susp. coef. _________ Meter No. ____________

GAGE READINGS				
Time		Recorder	Inside	Outside
Weighted M. G. H.				
G. H. correction				
Correct M. G. H.				

Date rated ______________________ Used rating for rod __________ susp. Meter ________ ft above bottom of wt. Tags checked ________

Spin before meas. ____________ after __________

Meas. plots _____% diff. from __________ rating

Wading, cable, ice, boat, upstr., downstr., side bridge ______________ feet, mile, above, below gage, and ________________________________

Check-bar, chain found ______________________ changed to _______________ at ___________

Correct ____________________________________

Levels obtained ____________________________

Measurement rated excellent (2%), good (5%), fair (8%), poor (over 8%), based on following conditions: Cross section ____

Flow ______________________________________ Weather ______________________________

Other ____ Air _________°F@ _______

Gage ____ Water _______°F@ _______

__________________________ Record removed __________________ Intake flushed U L ____________

Observer ____

____

Control ____

____

Remarks ____

____

____

G. H. of zero flow ______________________________________ ft

FIG. 30-31. Form for discharge-measurement notes (front page).

0 .10 .20 .30 .40 .50 .60 .70 .75

Napa River at — *St. Helena, Calif.*

Angle coefficient	Dist. from initial point	Width	Depth	Observation depth	Revolutions	Time in seconds	Velocity At point	Velocity Mean in vetical	Adjusted for hor. angle or ----------	Area	Discharge
	R	WE	at		0950						
	88	2.0	0	.6	0			0		0	0
	84	4.0	0.7	.6	30	42		1.60		2.8	4.5
	80	4.0	0.6	.6	40	40		2.24		2.4	5.4
	76	4.0	0.7	.6	50	41		2.73		2.8	7.6
	72	3.5	0.8	.6	60	46		2.92		2.8	8.2
	69	3.0	1.1	.6	50	42		2.66		3.3	8.8
	66	3.0	1.2	.6	60	46		2.92		3.6	10.5
⊙											
	63	3.0	1.5	.6	50	40		2.80		4.5	12.6
	60	3.0	1.8	.2	60	43	3.12	2.68		5.4	14.5
				.8	40	40	2.24				
	57	2.5	2.2	.2	60	41	3.27	2.73		5.5	15.0
				.8	40	41	2.19				
	55	2.0	2.5	.2	60	42	3.20	2.87		5.0	14.4
				.8	50	44	2.54				
	53	2.0	2.5	.2	60	44	3.05	2.82		5.0	14.1
				.8	50	43	2.60				
	51	2.0	2.7	.2	60	45	2.98	2.70		5.4	14.6
				.8	50	46	2.43				
	49	2.0	2.8	.2	50	45	2.49	2.77		5.6	15.5
				.8	60	44	3.05				
	47	2.0	3.0	.2	60	45	2.98	2.56		6.0	15.4
				.8	40	42	2.13				
	45										

.80 .85 .90 .92 .94 .96 .97 .98 .99 1.00 .99 .98 .97 .96 .94 .92 .90 .85 .80

0 .10 .20 .30 40 .50 .60 .70 .75

FIG. 30-32. Form for discharge-measurement notes (continued).

at the 0.2 depth is then converted to the mean velocity in the vertical, using the coefficient established from previous measurements made by the two-point method or from vertical-velocity curves. The usual coefficient to adjust the 0.2-depth velocity to the mean velocity is about 0.88.

When the depth at a vertical is not known and it is impossible to sound for depth, the meter should be placed at some uniform distance below the water surface. This distance should be at least 2 ft. and preferably more for deep, swift streams to avoid the effect of surface disturbances. Coefficients are necessary to convert the subsurface velocities to the mean velocity in the vertical. These vary more than the coefficients for the 0.2-depth observations because of variations in stage, depth, and relative position as the verticals throughout the entire cross-section are measured. Usually a coefficient should be established for each vertical by obtaining vertical-velocity curves as soon as possible after the subsurface measurement is made. The commonly used coefficient for flood measurements is about 0.9.

30-55. Recording Field Observations. The data collected in the field must be recorded as they are taken, preferably on forms specially prepared for discharge-measurement notekeeping. The forms shown in Figs. 30-31 and 30-32 are those used by the U.S. Geological Survey. The front page of the notes (Fig. 30-31) should include name of stream, location of site, date, type of suspension used, meter number, time at start and finish of measurement, gage readings at beginning and end of measurement, and flow and channel conditions.

The data pertaining to the actual measurement should be recorded in the body of the notes (Fig. 30-32). Such items as the following should be included: side of stream where measurement was started, distance of each vertical from starting point, depth of each vertical, depth from surface to point where observation is made, number of revolutions of wheel during observation period, duration (in seconds) of observation, and angle between the line of flow and the cross-section if other than 90°.

30-56. Measurements with Current Meter: General. Current-meter measurements are logically divided into four classes: wading, bridge, cable, and boat measurements (Arts. 30-57 to 30-60 respectively). However, certain aspects of measuring are common to all classes. The first, and one of the most important, steps in making any type of measurement is the selection of the cross-section to be used. In selecting the cross-section the following characteristics of the reach containing the cross-section should be considered:

1. The reach should be straight, and the velocity should be uniformly distributed with elements of velocity parallel to each other.
2. The stream bed should be stable and free of large rocks, weeds, and protruding obstructions which tend to create turbulence.
3. The profile of the stream bed should be uniform to eliminate any vertical components of velocity.

The next step is to determine the width of the stream; the technique used for this step depends on the type of measurement. It is important that the cross-section be laid out and the width measured perpendicular to the direction of flow in order to avoid horizontal angles between the flow and the cross-section.

When the total width and the approximate shape of the cross-section are known, the appropriate spacing of the verticals can be determined. Usually about 25 to 30 verticals should be selected for a cross-section in a smooth natural channel. They should be so spaced that no partial section has more than 10 per cent of the total discharge and preferably that no partial section has more than 5 per cent of the total flow. Equal distances between all verticals are not recommended unless the flow is well distributed through the entire cross-section. Usually a greater distance between verticals should be taken in the shallow, slow-velocity part of the cross-section.

It is important to observe the stage of the stream while making discharge measurements and to relate the discharge to a unique value of stage. When the stage is changing rapidly, speed is essential in making the measurement, and the gage should be read several times during the measurement in order to determine the average stage for the measured flow.

30-57. Wading Method. This method, for which the equipment is illustrated in Fig. 30-29, is used whenever the depth and velocity of the stream will permit. After the cross-section is selected, a wading line is stretched across the stream at the point. The line may be either a metallic tape or a light cable marked with beads or tags at 2- or 5-ft. intervals. The meter is assembled and is mounted on the graduated wading rod, all moving parts are checked, and the screws, electrical connections, and adjustment of the meter head are examined. When the observer is ready to start the measurement, the time and the gage reading to the nearest 0.01 ft. are recorded.

To make an observation, the total depth of the vertical section is measured with the wading rod, and the depths (as 0.6 depth or 0.2 and 0.8 depth) at which the meter is to be set on the rod are computed. When more than one point in a vertical is to be measured, most observers prefer to take the bottom reading first. Care should be taken to keep the wading rod in the vertical position and the meter parallel to the direction of flow while the velocity is being observed. The observer standing in the stream will disturb the flow, and an adverse effect on the operation of the current meter may result. To minimize this effect the observer should stand facing the shore about 3 in. downstream from the wading line and 18 in. or more from the meter rod.

In making the velocity measurements, it is important that observations are made at the proper point in the vertical, that the flow is uniform, that the meter is working freely, and that the time of observation is of sufficient length to average any fluctuations in the current. Observations of velocity should be timed with a stop watch to the nearest ½ sec. over a period of 40 to 70 sec.

The number of revolutions during the time interval can be checked by noting the number at the half time and doubling this number for comparison with the final reading.

When all the observations at a station have been taken, the meter should be dismantled, dried, oiled, and repacked in its case to insure that it remains in good working order.

30-58. Bridge Method. When a stream cannot be waded, existing bridges can be used for current-meter measurements. Usually the cross-section at a bridge is not so well suited for measurements as that at a cableway constructed for this purpose (Art. 30-59). However, when a cableway is not available, a bridge with reasonably satisfactory measuring conditions can often be found. Either the upstream or downstream side of the bridge can be used for the measurement, depending on which side has the better measuring conditions. The physical features of the bridge, such as location of walkway and traffic hazards, should also be considered. Proper safety precautions should be taken at all times, particularly with respect to vehicular traffic.

The width of the stream is measured along the selected side of the bridge with a graduated tape or a tagline, and the number and location of the verticals are determined. The meter assembly is fastened to the upper end of a stainless-steel hanger bar, and a sounding weight is fastened to the lower end (Fig. 30-33). The meter and weight are suspended by a small insulated wire cable which is often marked at known intervals above the bottom of the weight by tags of different colors. The weight is selected on the basis of the depth and maximum velocities and can range from 15 to 300 lb.

Before the measurement is started, the meter and equipment are examined as described in the preceding article and the gage height is read and recorded. The hydrographer then starts measuring at one side of the stream and proceeds from vertical to vertical across the bridge. At each vertical he first measures the depth of the stream, then computes the depths (as 0.2 and 0.8 depth) at which the velocity observations will be taken, and then suspends the meter at each of these depths and observes the point velocity as indicated by the number of revolutions of the meter during the period of observation. After the entire discharge measurement is completed, the water-stage recorder or wire-weight gage is read again.

There are two types of equipment for handling the meter and weight while sounding the depth and positioning the meter. One, the hand line, consists of a rubber-covered service cord attached to the suspension wire by a small connector reel. The rubber-covered service cord has a diameter large enough to permit raising and lowering the weight and meter by hand. When the depth of a vertical is to be measured, the meter is lowered until the weight touches the stream bed. The observer then marks with his thumb the point on the service cord where it goes over the rail and slowly raises the meter until the first tag attached to the cable appears at the water surface. This tag reading plus the measured distance along the service cord from the thumb to the guardrail gives the depth of the vertical. By a similar procedure the meter is positioned to the computed depths of observation.

The other, more common, type of equipment is the sounding reel and crane (Fig. 30-33). This is used for deep or fast-moving streams where heavier sounding weights are required. The crane consists of an aluminum-alloy framework, mounted on wheels and counterweighted, which extends out over

FIG. 30-33. Equipment for measuring velocity from bridge.

the bridge rail. The sounding reel consists of a drum either 1 or 2 ft. in circumference, fitted with a friction clutch and with a depth-recording device graduated in tenths of feet. The sounding cable suspending the weight and meter is fastened directly to the drum and is raised and lowered by a hand crank, a battery-driven motor, or a gasoline engine. The procedures for sounding and for meter positioning are similar to those used for making a hand-line measurement.

When positions for the desired depth readings are computed, allowance must be made for the distance between the bottom of the weight and the center line of the meter cups.

30-59. Cableway Method. The best method of measuring discharge in streams that cannot be waded is the cableway method. Usually it is used only at continuous gaging sites because it requires the installation of a special cableway across the stream. When a site for the cableway is being located, the reach of river with the most suitable measuring conditions within reasonable distance of the gaging station should be selected.

The type and diameter of the cable to be used for the installation will depend upon the total span, maximum expected load, and maximum allow-

FIG. 30-34. Stream gaging from cable car.

able sag. The cableway should be firmly anchored at each bank to solid rock, large trees, or cast-in-place concrete anchors and should be supported by towers, A frames, or trees so as to be at about the same elevation at the points of support. It is essential that the cable supports be high enough that the bottom of the loaded cable car at the lowest point in the sag will be at least 10 ft. above the expected maximum flood stage.

The velocity and depth observations are made from a cable car attached to sheaves that roll on the cable (Fig. 30-34). The cableway is graduated at intervals of 5 or 10 ft. so that the section width can be determined and the car can be positioned for each selected vertical. A sounding reel, similar to the reel used for bridge measurements, is fastened to the car, and the same procedures as those used for bridge measurements are used for observing the depth and positioning the current meter at each vertical.

30-60. Boat Method. Measurements of discharge from boats are made at locations where no cableways or suitable bridges are available and where the stream is too deep to wade. Boats should not be used where high velocities or drift are a possible source of danger to the safe operation of the boat.

The general procedure followed in making boat measurements is to stretch a strong, lightweight cable, tagged at intervals of 5 of 10 ft., across the stream at the selected section and to anchor it firmly on each bank. After the total width is determined, the boat is moved to each selected vertical section and is clamped to the tagline at that point while the depth and velocity observations are taken. A special boat boom (Fig. 30-35) should be used to clamp the boat in position and to project the point of meter suspension upstream from flow disturbance caused by the boat.

FIG. 30-35. Equipment for measuring velocity from boat.

A conventional sounding reel is mounted on the boom, and the depth and velocity observations in each vertical are taken using the procedures described in preceding articles.

If a tagline cannot be used because of traffic on the river or excessive width of the stream, the boat can be kept in the proper cross-section by the operator who alines it with two flag markers set on each bank along the projection of the cross-section. The location of the boat in the cross-section can be determined by the use of a transit on the shore and a stadia rod held in the boat.

30-61. Discharge Measurements under Ice. When stream measurements are to be continued through the winter months, the reconnaissance for a measuring section is made prior to the onset of cold weather and sections suitable to measurement through the ice are located. Usually the most suitable measuring section will be just upstream from a riffle that creates a long pool. Before a discharge measurement is made through the ice, holes are cut in the ice at the center and at the quarter-points of the selected section. The holes should be large enough to admit the current meter, also the sounding weight if the meter is cable-supported. Observations are made in the three holes to determine if there is a measurable velocity and an absence of heavy slush ice. If conditions are unsatisfactory, another section elsewhere is tested in the same manner.

Where conditions are found satisfactory, the section is laid off on the ice and is tied to an initial point on shore. Additional holes are cut to provide at least 20 measuring points, and observations of velocity are made as described in preceding articles. The effective depth of the vertical is taken

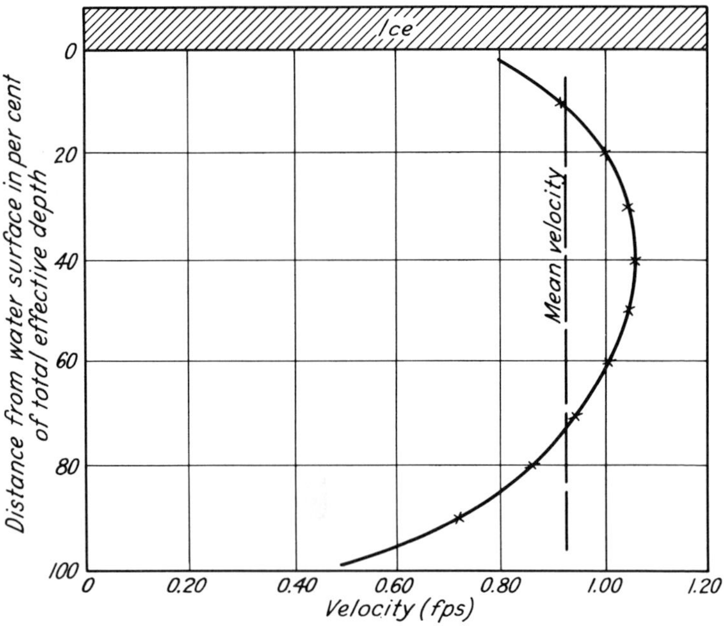

FIG. 30-36. Typical vertical-velocity curve under ice.

from the bottom of the ice to the bed of the stream. This distance is determined by measuring the depth from the surface of the water in the hole and then measuring with an L-shaped ice-measuring stick from the bottom of the ice to the water surface. The depth of the water minus the reading of the ice-measuring stick is the effective depth of the vertical.

The velocity distribution under ice cover differs from that in open water in that there is a decrease in velocity as the underside of the ice is approached (see Fig. 30-36). The most reliable method of determining the mean velocity is by use of the vertical-velocity-curve method. More commonly, however, the two-point method (at 0.2 and 0.8 depth) is used for effective depths

2.5 ft. or greater and the 0.6-depth method is used for effective depths less than 2.5 ft. Normally the average of the velocities obtained by the two-point method gives the mean velocity, but a coefficient of about 0.92 must usually be applied to the velocity obtained by the 0.6-depth method to convert it to mean velocity. Where the stream is too shallow to permit observations at 0.6 depth, it is permissible to make velocity observations at mid depth. A coefficient of 0.88 is used to convert the mid-depth velocity to mean velocity.

The rod or cable supporting the meter should be held against the upstream edge of the hole when measuring the velocity in order to keep the meter as far upstream as possible. This practice will keep the meter away from the effect of any vertical pulsation of water in the hole. Care should be taken to see that the meter is free of ice when the velocity is being observed, and for this reason the current meter should be exposed as little as possible to the cold air during the course of a measurement.

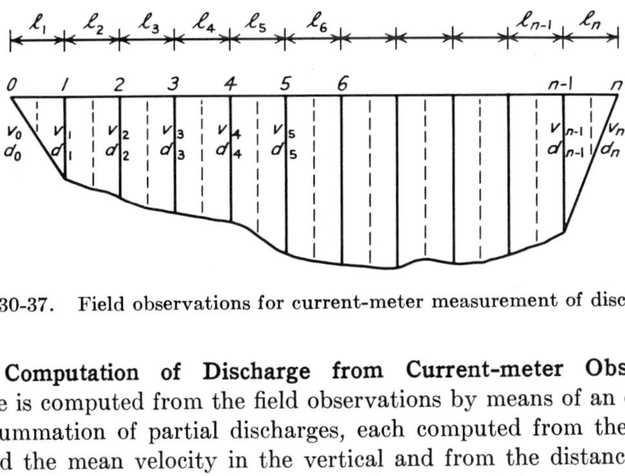

FIG. 30-37. Field observations for current-meter measurement of discharge.

30-62. Computation of Discharge from Current-meter Observations. Discharge is computed from the field observations by means of an expression for the summation of partial discharges, each computed from the observed depth and the mean velocity in the vertical and from the distance between verticals. In Fig. 30-37, d_0, d_1, d_2, . . . , d_n represent the measured depths of verticals; l_1, l_2, l_3, . . . , l_n are the respective distances between verticals; and V_0, V_1, V_2, . . . , V_n represent velocities in the verticals. Each metered vertical and its measured depth are assumed to be the mean of a zone extending from that vertical halfway to the verticals on each side. A summation of the partial discharges computed for each zone is equal to the total discharge Q, which may be expressed as follows:

$$Q = (V_0 d_0)\frac{l_1}{2} + (V_1 d_1)\frac{l_1 + l_2}{2} + (V_2 d_2)\frac{l_2 + l_3}{2} + \cdots + (V_{n-1} d_{n-1})\frac{l_{n-1} + l_n}{2} + (V_n d_n)\frac{l_n}{2} \quad (3)$$

Unless the sides of the channel are vertical, V_0, d_0, V_n, and d_n will all be equal to zero, and consequently the computed partial discharges in the two small end zones will also be equal to zero. Actually there is usually a small discharge in the two end zones, but if the verticals designated 1 and $n - 1$ in Fig. 30-37 are taken fairly close to the water's edge, no significant error will be introduced in computation of total discharge. An example of a discharge-measurement computation is given in Fig. 30-32.

30-63. Station Rating Curve. When successive discharges are plotted as abscissas and their corresponding gage heights as ordinates, the resulting graph is known as a *station rating curve.* Rating curves are commonly plotted on logarithmic graph paper. A typical curve is shown in Fig. 30-38.

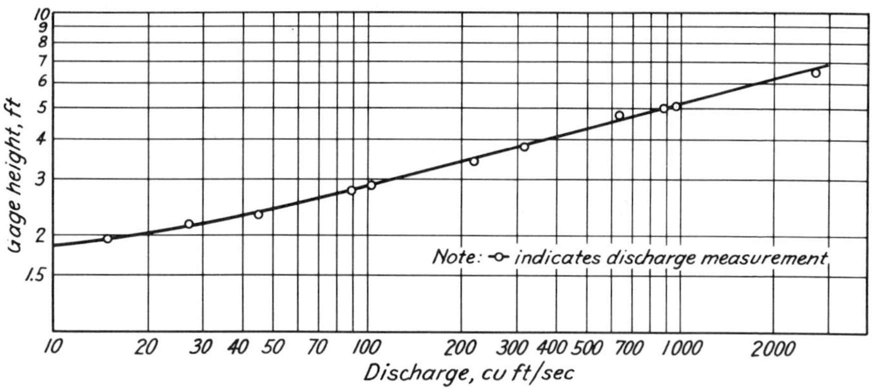

FIG. 30-38. Station rating curve for Saratoga Creek at Saratoga, California.

The reliability of such a curve will depend upon the stability of the stream channel, the availability of discharge measurements to define the entire range of the curve from low to high stages, the accuracy of the gage heights, and the precision with which the velocity and cross-sectional area are determined. Under favorable conditions, current-meter measurements of discharge should give discharges that are within 5 per cent of the true discharge. If a measured discharge, when plotted on a graph shows a large divergence from the trend of other measurements, a repeat, or "check," discharge measurement should be made to eliminate any uncertainty concerning the accuracy of the original measurement. If the original divergent measurement and check measurement agree, this agreement indicates that a change in channel conditions has occurred. A new rating curve must then be established for the stream by a new series of discharge measurements.

Having a station rating curve and a record of the stage at the gaging station, it is a simple matter to convert the stage record to one of discharge.

30-64. Weirs: General. A weir is a convenient and reliable device for measuring the flow in irrigation and power canals, large sewers, very shallow or rocky streams, and other streams not suitable to current-meter measurements. A weir is a channel obstruction of regular form over which water flows. Commonly there is a notch in the obstruction which carries all the low and medium discharge. If the weir is notched, it is classified in accordance with the shape of the notch. Thus there are *rectangular weirs, triangular* or *V-notch weirs,* and *trapezoidal weirs.*

The surface of the weir over which the water flows is the *crest* of the weir. If the crest is beveled like a chisel edge, the weir is *sharp-crested.* A sharp-crested weir is always installed with the beveled face on the downstream side. Sharp-crested weirs are useful only as a means of measuring water. Other weirs are known as *broad-crested* weirs. Broad-crested weirs are commonly incorporated in hydraulic structures, and although often they are used to measure water, this is usually a secondary function.

The information needed to compute the discharge over a weir is as follows:

1. Depth of water flowing over crest.
2. Length of crest if weir is rectangular or trapezoidal.
3. Angle of side slopes if weir is triangular or trapezoidal.
4. Whether sharp-crested or broad-crested.
5. Cross-sectional shape of crest if weir is broad-crested.
6. Height of crest above bottom of approach channel.
7. Width and depth of approach channel.
8. Velocity of approach.
9. Number and nature of end contractions.

With these data given, a formula is chosen depending upon the type of weir, and a coefficient is selected depending upon the shape of the weir crest and upon the conditions of flow. By proper substitution in the chosen formula the discharge is computed.

The basic formula for computing discharge over a weir is

$$Q = Cb(H + h_a)^n \tag{4}$$

in which Q = discharge, in cubic feet per second

C = experimentally derived coefficient that depends on the cross-sectional shape of the weir crest and, in the case of broad-crested weirs, also on the depth of flow

b = effective length of the weir crest, in feet

H = head, in feet; it is the depth of water flowing over the weir, measured from the crest elevation to the pool level of the impounded water at a distance $3H$ upstream from the weir

h_a = velocity head, in feet, in the approach channel at a distance $3H$ upstream from the weir

n = exponent which varies with the shape of the notch (whether rectangular, triangular, or trapezoidal)

Computation of Approach-velocity Head (h_a). When there is significant velocity in the approach channel upstream from the weir, the velocity head must be computed and applied in the basic weir formula. As a first step the cross-sectional area A_a at a distance $3H$ upstream from the weir must be determined. Then h_a may be computed from the formula

$$h_a = \frac{Q^2}{A_a{}^2 2g} \tag{5}$$

in which g, the acceleration due to gravity, is 32.2 ft. per second per second. Because Q will not be known at the time h_a is computed, it is necessary to

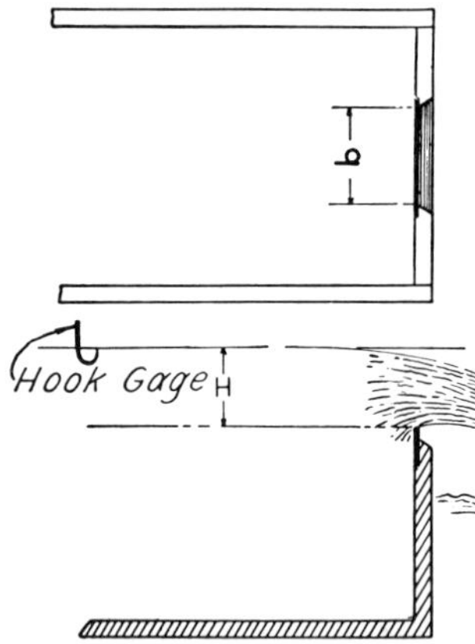

FIG. 30-39. Sharp-crested rectangular weir with end contractions.

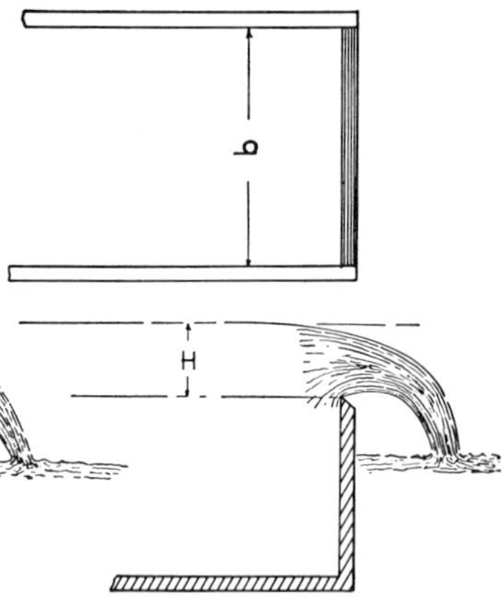

FIG. 30-40. Sharp-crested rectangular weir with end contractions suppressed.

(1) estimate a trial value of Q, (2) solve for h_a, (3) substitute this value of h_a in the basic weir formula and solve for Q, and (4) compare this computed Q with the trial value of Q. If the computed and trial values do not agree, the computations are repeated with new trial values until the computed and trial values of Q agree closely.

30-65. Rectangular Weirs. Rectangular weirs may be either sharp-crested or broad-crested; either they may have end contractions, or the end contractions may be suppressed. A rectangular weir is said to have end contractions when the sides of the notch are at a distance $3H$ or more from the sides of the approach channel (Fig. 30-39). A suppressed weir is one the sides of which coincide closely with the sides of the approach channel (Fig. 30-40).

Sharp-crested Weirs. For sharp-crested rectangular weirs the discharge formula is

$$Q = 3.33b(H + h_a)^{3/2} \tag{6}$$

If the end contractions are suppressed, the effective length b of weir crest is equal to the actual length of the crest. If there are end contractions, they reduce the effective length of the crest. The formula for effective length is

$$b = b' - 0.1NH \tag{7}$$

in which b = effective length of weir crest, in feet
b' = measured length of weir crest, in feet
N = number of end contractions
H = head, in feet

For two end contractions, $N = 2$; if contraction is suppressed at one end, $N = 1$.

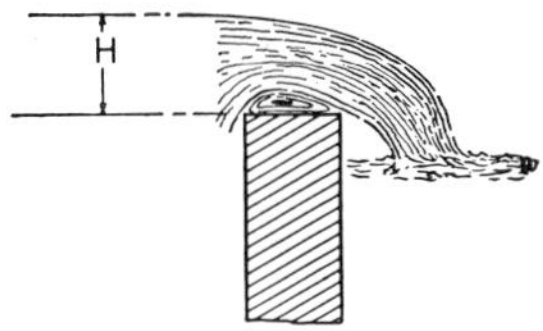

FIG. 30-41. Flat-crested weir.

The upstream face of the weir must be vertical and the crest of weir horizontal if accurate results are to be obtained from the weir formulas.

Broad-crested Weirs. As mentioned earlier, any weir not sharp-crested is classed as broad-crested. Consequently, a broad-crested weir may have one of a great variety of shapes and proportions. A simple type is shown in Fig. 30-41. The general equation for a broad-crested rectangular weir is

$$Q = Cb(H + h_a)^{3/2} \tag{8}$$

As in the case of the sharp-crested rectangular weir, the effective length b of weir crest depends on whether or not there are end contractions. The formula for effective length is that given in the preceding paragraph for sharp-crested weirs. The value of the coefficient C depends on the cross-sectional shape of the crest and on the head. If the upstream or downstream faces of the weir are sloping rather than vertical as shown in Fig. 30-41, the coefficient will be affected. Even for a flat-crested weir of the shape shown in Fig. 30-41, the coefficient will vary with the stage, being dependent on the ratio of head to width of crest (in the direction of flow). Space limitations do not permit a listing here of coefficients for the myriad shapes and proportions in which broad-crested weirs may be built. Comprehensive tabulations of these coefficients are found in Refs. 7 and 8 at the end of this chapter. To obtain the coefficients for a given broad-crested weir these references should be consulted to find a weir of similar shape and proportion the coefficient of which has been experimentally determined. It is even more desirable to obtain the desired coefficients by means of current-meter measurements of discharge made over the entire range of stage.

30-66. Triangular or V-notch Weirs. Triangular or V-notch weirs (Fig. 30-42) are usually installed solely for the purpose of measuring discharge, particularly at low flows. They are usually sharp-crested. Because the area of the notch is invariably small as compared with the cross-sectional area of the channel, the water is pooled upstream from the weir and the approach velocity is necessarily low. The approach velocity head can therefore be neglected

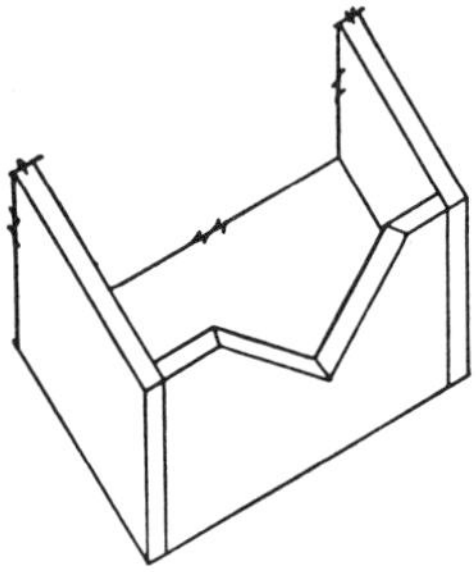

FIG. 30-42. Triangular or V-notch weir.

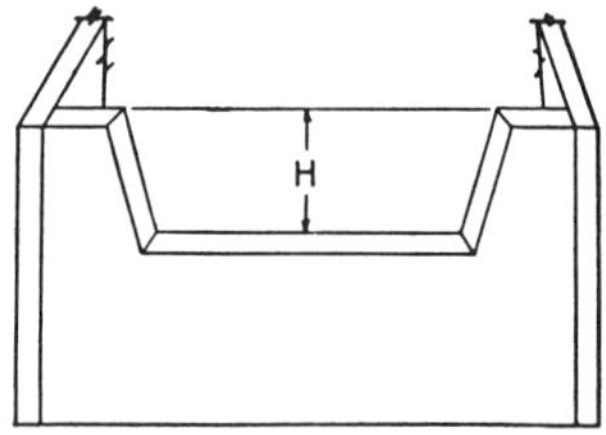

FIG. 30-43. Trapezoidal weir.

in the computation of discharge. Both the cross-sectional area of the flow passing through the notch and the velocity in the notch are functions of the head. Consequently, the general equation for a triangular weir is

$$Q = CH^n \tag{9}$$

The coefficient C varies with the central angle of the V notch.

V-notch weirs are most commonly constructed with a central angle of 90°. Much experimental work has been done with sharp-crested 90° V notches, and the recommended formula for discharge with this type of weir is

$$Q = 2.49H^{2.48} \tag{10}$$

Another central angle that is fairly common in the construction of V-notch weirs is 120°. The formula for discharge for a 120° sharp-crested weir is

$$Q = 4.40H^{2.49} \tag{11}$$

If the sharp-edged V notch has a central angle other than 90° or 120°, the coefficient should be evaluated by current-meter measurements. A rough approximation of the discharge can be obtained from the formula

$$Q = 2.5 \tan \frac{\theta}{2} H^{5/2} \tag{12}$$

in which θ is the central angle in degrees.

The upstream face of the weir must be vertical for accurate results to be obtained from the weir formulas.

30-67. Trapezoidal Weirs. Few experimental data are available for determining the discharge coefficients of trapezoidal weirs (Fig. 30-43). One exception is the *Cippoletti weir,* which is a sharp-crested trapezoidal weir whose sides have a slope of 1 horizontal to 4 vertical. This slope of the sides is approximately that required to secure a discharge through the triangular part of the weir opening that equals the decrease in discharge resulting from end contractions. Consequently, the discharge formula for a

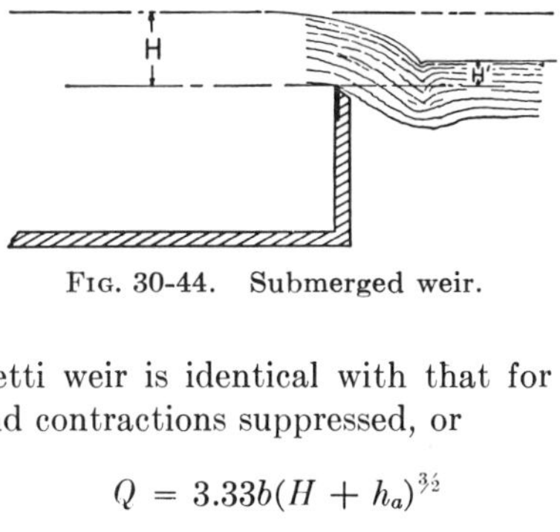

FIG. 30-44. Submerged weir.

sharp-crested Cippoletti weir is identical with that for a sharp-crested rectangular weir with end contractions suppressed, or

$$Q = 3.33b(H + h_a)^{3/2} \tag{13}$$

in which b is the length of the horizontal part of the weir crest.

For trapezoidal weirs other than Cippoletti weirs a general empirical formula for discharge is

$$Q = Cb(H + h_a)^n \tag{14}$$

The coefficient C and exponent n must be determined by current-meter measurements which cover the entire range of stage that is experienced.

30-68. Submerged Weirs. When water on the downstream side of a weir (tail water) rises above the level of the crest, the weir is said to be *submerged.* In Fig. 30-44, which illustrates the submergence of a weir, the tail-water elevation above the weir is indicated as H'. If the weir is sharp-crested, the relation between head and discharge changes soon after the tail water rises above the weir crest because the flow becomes impeded. If the weir is broad-crested, however, the relation between head and discharge is usually unaffected until the tail water reaches the stage at which the ratio of H' to H is greater than $\frac{2}{3}$.

Relatively few experimental data are available on the effect of submergence on weir coefficients, except for the case of a sharp-crested rectangular weir. Table 30-1, from Ref. 8 at the end of this chapter, lists the coefficients to be used for a sharp-crested rectangular weir for various degrees of submergence. At high submergences the values of the coefficient are less reliable than they are for lower submergences.

30-69. Portable Weir Plates. Very shallow depths and low velocities render stream-flow measurements by current meter inaccurate if not impossible to obtain. Under these conditions a useful device for discharge measurements is a portable *weir plate.*

TABLE 30-1. COEFFICIENT C TO BE USED FOR VARIOUS DEGREES OF SUBMERGENCE OF SHARP-CRESTED RECTANGULAR WEIRS IN FORMULA $Q = Cb(H + h_a)^{3/2}$

Submergence H'/H	C	Submergence H'/H	C	Submergence H'/H	C
0.00	3.33	0.35	3.05	0.70	2.32
0.05	3.33	0.40	2.98	0.75	2.16
0.10	3.33	0.45	2.90	0.80	1.96
0.15	3.31	0.50	2.81	0.85	1.72
0.20	3.26	0.55	2.71	0.90	1.45
0.25	3.19	0.60	2.59	0.95	1.08
0.30	3.13	0.65	2.46	0.98	0.70

A 90° V-notch weir is the most applicable portable weir because of its relatively high accuracy at low flows. The weir is made by cutting a 90° notch in a sheet of 10 to 16-gage metal. The notch is not beveled but is left with a flat, even edge. Plates of medium and large size should be framed on the downstream side with small angle irons for rigidity. A staff gage is attached to the upstream face of the plate with its zero at the elevation of the bottom of the notch. The staff gage should be far enough from the notch so that it is in the pool created by the weir plate and not in the region of "drawdown" of water going through the notch. To install the weir, the weir plate is pushed into the stream bed. Through eyebolts that are attached to the weir plate, rods are driven into the stream bed to stabilize the plate. Canvas is placed immediately downstream from the weir to prevent undercutting by the falling jet. A carpenter's level is used to insure that the top of the plate is horizontal and that the plate is plumb. Care must be taken to prevent leakage under and around the weir plate.

Flow should be allowed to stabilize before a measurement is made. Normally this will require about 20 min. Once the pool is stabilized, the gage height should be read several times during a 3-min. interval and a mean value used. Discharge is then computed by the triangular weir formula $Q = 2.49H^{2.48}$ [Eq. (10)].

As long as the weir is not submerged by tail water, discharge can be determined with an error no greater than 3 per cent. Discharges as small as 0.02 cu. ft. per second (cfs) can be measured with this accuracy. The upper

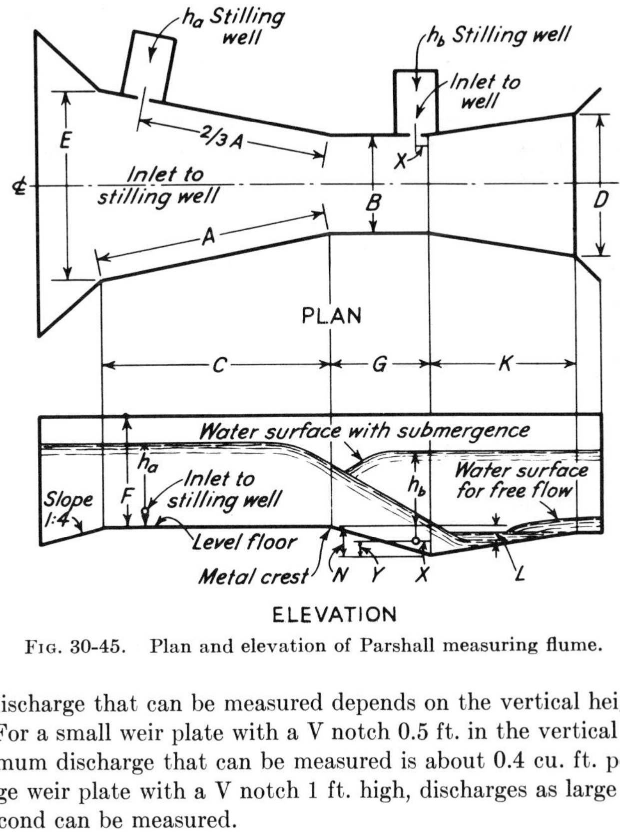

FIG. 30-45. Plan and elevation of Parshall measuring flume.

limit of discharge that can be measured depends on the vertical height of the notch. For a small weir plate with a V notch 0.5 ft. in the vertical direction, the maximum discharge that can be measured is about 0.4 cu. ft. per second. For a large weir plate with a V notch 1 ft. high, discharges as large as 2.0 cu. ft. per second can be measured.

30-70. Parshall Flume. Parshall flumes are structures installed in a stream to measure discharge (Ref. 9 at the end of this chapter). They were originally designed for use in irrigation canals in order to overcome some disadvantages that are inherent in the properties of a weir. The Parshall flume operates much more satisfactorily under conditions of high tail-water submergence than a weir does. Also, for a given discharge the flume will not pool water so high as a weir will, and therefore canal banks need not be built so high, particularly for conditions of high submergence. Debris-laden water will pass through a Parshall flume without depositing material in the upstream pool. The pools above weirs will fill with silt and debris under such circumstances, and the velocity of approach and thereby the stage-discharge relation will be changed.

Figure 30-45 shows the general design of the Parshall flume, and Table 30-2 gives the dimensions corresponding to the letters of the figure for various sizes of flumes. The large flumes consist of a converging section with level floor, a throat with floor inclined downward on a slope of 3 to 8, and a diverging section with adverse floor slope of 1 to 6. Flumes with throat width B less than 1 ft. depart from the proportions of the large flumes. Gages h_a and h_b are each set with zero of the gage at the elevation of the crest, or level floor.

Table 30-3 summarizes the relation of discharge to stage at h_a under conditions of free flow (low stage at h_b). When the stage at h_b reaches $0.7h_a$, submergence is effective in reducing discharge of the large flumes. Submergence of 50 per cent ($h_b/h_a = 0.50$) may prove critical on flumes whose throat width is less than 1 ft. Discharge under conditions of submergence is computed by subtracting a correction from the free-flow discharge. Values of the correction for the flume with throat width B equal to 1 ft. are given in Table 30.4. When multiplied by $B^{0.815}$ (Table 30-5), the resulting corrections are applicable to large flumes. The larger Parshall flumes will operate successfully under 95 per cent submergence.

Portable Parshall flumes with throat widths up to 9 in. may either be built or be purchased commercially. They are installed temporarily, in much the same manner as the portable weir plate, to measure low flows that cannot be accurately measured with a current meter. Under free-flow conditions they have an accuracy of 2 to 3 per cent.

30-71. Slope-area Determination of Discharge. The slope-area method of determining discharge is used primarily for obtaining the peak flow of a stream after the stage has receded. This method of determination is neither so simple nor so reliable as a current-meter measurement of discharge, and it should not be used if a current-meter measurement can be made. The survey work involved in the slope-area method includes a determination of (1) slope of the water surface in the selected reach at the peak stage, (2) cross-sections at each end of the reach, and (3) selection of a coefficient of roughness based on the physical characteristics of the channel.

Selection of Reach. The reach selected should be reasonably straight for a distance equivalent to at least several stream widths. The cross-section of the channel should be reasonably uniform throughout the reach and preferably should approach a regular geometric shape such as a rectangle, trapezoid, or parabola. The cross-sectional area at the downstream end of the reach should not be greater than that at the upstream end. Very rough,

TABLE 30-2. DIMENSIONS AND CAPACITIES FOR PARSHALL MEASURING FLUMES OF VARIOUS SIZES
(All dimensions in feet and inches)

Throat width B	A	C	D	E	F	G	K	L	N	X	Y	Free-flow capacity, cu. ft. per second	
												Max.	Min.
0–3	1–6⅜	1–6	0–7	0–10 3/16	1–4	0–6	1–0	0–1	0–2¼	0–1	0–1½	1.1	0.03
0–6	2–7/16	2–0	1–3⅝	1–3⅝	2–0	1–0	2–0	0–3	0–4½	0–2	0–3	3.9	0.05
0–9	2–10⅝	2–10	1–3	1–10⅝	2–6	1–0	1–6	0–3	0–4½	0–2	0–3	8.8	0.09
1–0	4–6	4–4⅞	2–0	2–9¼	3–0	2–0	3–0	0–3	0–9	0–2	0–3	16.1	0.35
1–6	4–9	4–7⅞	2–6	3–4⅜	3–0	2–0	3–0	0–3	0–9	0–2	0–3	24.6	0.51
2–0	5–0	4–10⅞	3–0	3–11½	3–0	2–0	3–0	0–3	0–9	0–2	0–3	33.1	0.66
3–0	5–6	5–4¾	4–0	5–1⅞	3–0	2–0	3–0	0–3	0–9	0–2	0–3	50.4	0.97
4–0	6–0	5–10⅝	5–0	6–4¼	3–0	2–0	3–0	0–3	0–9	0–2	0–3	67.9	1.26
6–0	7–0	6–10⅜	7–0	8–9	3–0	2–0	3–0	0–3	0–9	0–2	0–3	103.5	2.63
8–0	8–0	7–10⅛	9–0	11–1¾	3–0	2–0	3–0	0–3	0–9	0–2	0–3	139.5	4.62

TABLE 30-3. FREE-FLOW DISCHARGE IN CUBIC FEET PER SECOND FOR PARSHALL FLUMES OF VARIOUS SIZES

h_a ft.	Throat width B									
	3 in.	6 in.	9 in.	1 ft.	1.5 ft.	2 ft.	3 ft.	4 ft.	6 ft.	8 ft.
0.1	0.03	0.05	0.09							
0.2	0.08	0.16	0.26	0.35	0.51	0.66	0.97	1.26		
0.3	0.15	0.31	0.49	0.64	0.94	1.24	1.82	2.39	3.52	4.62
0.4	0.24	0.48	0.76	0.99	1.47	1.93	2.86	3.77	5.57	7.34
0.5	0.34	0.69	1.06	1.39	2.06	2.73	4.05	5.36	7.94	10.5
0.6	0.45	0.92	1.40	1.84	2.73	3.62	5.39	7.15	10.6	14.1
0.7	0.57	1.17	1.78	2.33	3.46	4.60	6.86	9.11	13.6	18.0
0.8	0.70	1.45	2.18	2.85	4.26	5.66	8.46	11.3	16.8	22.4
0.9	0.84	1.74	2.61	3.41	5.10	6.80	10.2	13.6	20.3	27.0
1.0	0.99	2.06	3.07	4.00	6.00	8.00	12.0	16.0	24.0	32.0
1.1	....	2.40	3.55	4.62	6.95	9.27	13.9	18.6	27.9	37.3
1.2	....	2.75	4.06	5.28	7.94	10.6	16.0	21.3	32.1	42.9
1.3	....	3.12	4.59	5.96	8.99	12.0	18.1	24.2	36.5	48.8
1.4	....	3.51	5.14	6.68	10.1	13.5	20.3	27.2	41.1	55.0
1.5	....	....	5.71	7.41	11.2	15.0	22.6	30.3	45.8	61.4
1.6	....	....	6.31	8.18	12.4	16.6	25.1	33.6	50.8	68.1
1.7	....	....	6.92	8.97	13.6	18.2	27.6	37.0	56.0	75.1
1.8	....	....	7.54	9.79	14.8	19.9	30.1	40.5	61.3	82.3
1.9	....	....	8.20	10.6	16.1	21.6	32.8	44.1	66.8	89.8
2.0	....	....	....	11.5	17.4	23.4	35.5	47.8	72.5	97.5
2.1	....	....	....	12.4	18.8	25.3	38.4	51.6	78.4	105.4
2.2	....	....	....	13.3	20.2	27.2	41.3	55.5	84.4	113.6
2.3	....	....	....	14.2	21.6	29.1	44.2	59.6	90.6	122.0
2.4	....	....	....	15.2	23.0	31.1	47.3	63.7	97.0	130.7
2.5	....	....	....	16.1	24.6	33.1	50.4	67.9	103.5	139.5

TABLE 30-4. SUBMERGENCE CORRECTIONS IN CUBIC FEET PER SECOND FOR 1-FT. PARSHALL FLUME

Head, h_a ft.	Submergence ratio h_b/h_a					
	0.70	0.75	0.80	0.85	0.90	0.95
0.3	0.07	0.08	0.10	0.13	0.20	0.33
0.5	0.08	0.10	0.14	0.22	0.37	0.65
1.0	0.13	0.21	0.35	0.59	1.00	1.80
1.5	0.25	0.42	0.70	1.20	2.00	3.40
2.0	0.44	0.73	1.25	2.00	3.30	5.40
2.5	0.69	1.10	1.90	2.90	4.70	7.50

steep channels, marked by riffles or large boulders and by pools, should be avoided if possible in favor of more tranquil reaches. Reaches containing heavy brush on the banks and particularly on the stream bed should also be avoided. Channels showing evidence of scour or fill should likewise be avoided because of the uncertainty as to the elevation of the stream bed at the time of peak flow. The reach should have identifiable high-water marks that define the water-surface slope at the time of peak flow. These high-water marks may be fine drift on the banks or on bushes and trees, or they may be seed lines, wash lines, or mud lines on the banks, in brush, or in nearby buildings. The high-water marks should show a fall of at least a few tenths of a foot in the reach.

Water-surface Slope. The high-water marks on each bank should be surveyed as described in Art. 30-23. From a plot of the water-surface profile for each bank the average fall Δh between the ends of the reach is determined.

TABLE 30-5. VALUES OF $B^{0.815}$

B	$B^{0.815}$
1	1.0
1.5	1.4
2	1.8
3	2.4
4	3.1
6	4.3
8	5.4

End Cross-sections. At each end of the reach the cross-section should be surveyed. From the cross-section notes, the area A and wetted perimeter at the time of the peak discharge are determined. The wetted perimeter, which is defined as that part of the cross-section which is in contact with the stream, is used to compute the hydraulic radius R for each cross-section. R is equal to the area of the cross-section (square feet) divided by its wetted perimeter (feet).

Selection of Roughness Coefficient. The roughness coefficient n is a retardation factor that depends on the roughness of the channel and on its configuration. A straight, clean channel will have a low value of n; a rough, irregular channel will have a high value. The selection of the n value for a natural stream calls for the exercise of personal judgment, which comes primarily with experience. A valuable guide, however, in the selection of proper n values is Table 30-6. This table, compiled by R. E. Horton (Ref. 8 at the end of this chapter), gives values of n for a wide range of conditions for both artificial and natural channels.

30-72. Manning Slope-area Formula. There are several formulas for computing discharge by the slope-area method, but the simplest and most

TABLE 30-6. VALUES OF n TO BE USED IN THE MANNING FORMULA

Surface	Best	Good	Fair	Bad
Uncoated cast-iron pipe	0.012	0.013	0.014	0.015
Coated cast-iron pipe	0.011	0.012*	0.013*	
Commercial wrought-iron pipe, black	0.012	0.013	0.014	0.015
Commercial wrought-iron pipe, galvanized	0.013	0.014	0.015	0.017
Smooth brass and glass pipe	0.009	0.010	0.011	0.013
Smooth lockbar and welded "OD" pipe	0.010	0.011*	0.013*	
Riveted and spiral steel pipe	0.013	0.015*	0.017*	
Vitrified sewer pipe	{0.010, 0.011}	0.013*	0.015	0.017
Common clay drainage tile	0.011	0.012*	0.014*	0.017
Glazed brickwork	0.011	0.012	0.013*	0.015
Brick in cement mortar; brick sewers	0.012	0.013	0.015*	0.017
Neat cement surfaces	0.010	0.011	0.012	0.013
Cement mortar surfaces	0.011	0.012	0.013*	0.015
Concrete pipe	0.012	0.013	0.015*	0.016
Wood stave pipe	0.010	0.011	0.012	0.013
Plank flumes:				
Planed	0.010	0.012*	0.013	0.014
Unplaned	0.011	0.013*	0.014	0.015
With battens	0.012	0.015*	0.016	
Concrete-lined channels	0.012	0.014*	0.016*	0.018
Cement-rubble surface	0.017	0.020	0.025	0.030
Dry-rubble surface	0.025	0.030	0.033	0.035
Dressed-ashlar surface	0.013	0.014	0.015	0.017
Semicircular metal flumes, smooth	0.011	0.012	0.013	0.015
Semicircular metal flumes, corrugated	0.0225	0.025	0.0275	0.030
Canals and ditches:				
Earth, straight and uniform	0.017	0.020	0.0225*	0.025
Rock cuts, smooth and uniform	0.025	0.030	0.033*	0.035
Rock cuts, jagged and irregular	0.035	0.040	0.045	
Winding sluggish canals	0.0225	0.025*	0.0275	0.030
Dredged earth channels	0.025	0.0275*	0.030	0.033
Canals with rough stony beds, weeds on earth banks	0.025	0.030	0.035*	0.040
Earth bottom, rubble sides	0.028	0.030*	0.033*	0.035
Natural stream channels:				
1. Clean, straight bank, full stage, no rifts or deep pools	0.025	0.0275	0.030	0.033
2. Same as 1, but some weeds and stones	0.030	0.033	0.035	0.040
3. Winding, some pools and shoals, clean	0.033	0.035	0.040	0.045
4. Same as 3, lower stages, more ineffective slope and sections	0.040	0.045	0.050	0.055
5. Same as 3, some weeds and stones	0.035	0.040	0.045	0.050
6. Same as 4, stony sections	0.045	0.050	0.055	0.060
7. Sluggish river reaches, rather weedy or with very deep pools	0.050	0.060	0.070	0.080
8. Very weedy reaches	0.075	0.100	0.125	0.150

* Values commonly used in designing.

widely used is the Manning formula. In accordance with this formula

$$Q = \frac{1.486}{n} AR^{2/3}S^{1/2} \tag{15}$$

in which Q = discharge, in cubic feet per second
n = roughness coefficient
A = cross-sectional area, in square feet
R = hydraulic radius, in feet
S = slope of the energy gradient; i.e., the water-surface slope in feet per foot, adjusted for change in velocity head through the reach

Because the channel dimensions at the two ends of the reach will usually differ from each other, the formula requires some manipulation. In the discussion that follows, channel properties at the upstream section will be designated by subscript A and those at the downstream section by subscript B. When the field work described in Art. 30-71 is performed the following elements in the formula will be known: n, A_A, A_B, R_A, R_B, and Δh (the difference in water-surface elevation between sections A and B).

It is customary to express the Manning formula in the form

$$Q = K_M S^{1/2} \tag{16}$$

in which $K_M = \sqrt{K_A K_B}$

$$K_A = \frac{1.486}{n} A_A R_A^{2/3}$$

$$K_B = \frac{1.486}{n} A_B R_B^{2/3}$$

It is seen that all the elements needed to compute K_M will have been obtained from the field survey.

The value of S, the slope of the energy gradient, is computed by trial from the following formula:

$$S = \frac{\Delta h + \dfrac{Q^2}{A_A^2 2g} - \dfrac{Q^2}{A_B^2 2g}}{L} \tag{17}$$

in which L is the distance, in feet, between sections A and B. Because Q will not be known at the time S is computed, it is necessary to (1) estimate a trial value of Q, (2) solve for S, (3) substitute this value of S in the formula $Q = K_M S^{1/2}$ and solve for Q, and (4) compare this computed Q with the trial value of Q. If the computed and trial values do not agree, the computations are repeated with new trial values until the computed and trial values of Q agree closely.

If the areas of the two end cross-sections are equal

$$\frac{Q^2}{A_A^2 2g} = \frac{Q^2}{A_B^2 2g}$$

and the formula for S becomes simply

$$S = \frac{\Delta h}{L} \tag{18}$$

Under these circumstances S can be computed directly without any trial solutions and the computation of Q is straightforward.

30.73. Numerical Problems.

1. The zero elevation of a staff gage is 745.41, and the gage reads 10.24 ft. when there is a depth of 3.52 ft. of water flowing over the lowest point on the control (point of zero flow). What are the elevation and gage reading of zero flow?
2. A rainfall the intensity of which is 2 in. per hour falls for a period of 4 hr. on a drainage area of 100 sq. miles. If the estimated runoff is 25 per cent of the precipitation, how many acre-feet would be impounded if the runoff could be stored?
3. The right and left water's edges of a stream are 10 and 100 ft. respectively from an initial zero point. Verticals are located at distances of 15, 25, 35, 45, 55, 65, 75, 85, and 95 ft. from the initial point. Depths of verticals are 2.0, 3.8, 4.6, 7.8, 8.4, 8.8, 8.2, 6.5, and 3.0 ft. Mean velocities in the verticals are 0.65, 1.57, 2.40, 2.88, 3.12, 3.60, 3.70, 2.10, and 0.82 ft. per second, respectively. What is the discharge of the stream in cubic feet per second? (Note: In an actual current-meter discharge measurement, about 30 verticals would be measured. This problem merely illustrates the method of *computation*.)
4. A storage dam used as a weir has a discharge equation $Q = 3.33bH^{3/2}$. If b, the length of the weir, equals 500 ft. and H, the head of water over the weir, equals 8 ft., what is the discharge over the weir in cubic feet per second?
5. A semicircular flume 15 ft. in diameter has a slope of 0.0016. The flume is built of corrugated metal in good condition and carries water that flows with a uniform depth of 4.5 ft. in the center of the flume. What is the discharge in cubic feet per second?
6. The discharge of a sewer is to be measured by a sharp-crested rectangular weir having a length of 2 ft. The weir has full contractions at both ends, and the gage shows a head of water over the crest of the weir of 5 in. Neglecting the effect of velocity of approach, what is the discharge in cubic feet per second?
7. Outline a practical method of measuring the exact discharge in gallons per minute of a spring flowing somewhere between 15 and 20 gal. per minute.
8. What method would you use in measuring the discharge to obtain a rating curve for each of the following: (*a*) a small rocky creek 8 to 10 ft. wide, (*b*) a river 150 ft. wide and 5 to 8 ft. deep, (*c*) a storage dam operating as a weir, the coefficient of which is not known, and (*d*) a river ½ mile wide and 20 ft. deep?

REFERENCES

Hydrographic Surveys

1. Jeffers, K. B., "Hydrographic Manual," *U.S. Coast and Geodetic Survey, Pub.* 20-2, Government Printing Office, Washington, D.C., 1960.
2. Pond, H. S., "Graphic Method Aids Plotting of Hydrographic Locations," *Engrg. News-Record,* Vol. 108, pp. 718–719, 1932.

3. U.S. Coast and Geodetic Survey, "Manual of Current Observations," *Spec. Pub.* 215, Government Printing Office, Washington, D.C., 1961.
4. U.S. Department of Agriculture, "Snow Surveying," *Misc. Pub.* 280, Government Printing Office, Washington, D.C., 1940.

Flow Measurement

5. Corbett, Don M., and others, "Stream-gaging Procedure," *U.S. Geological Survey, Water Supply Paper* 888, Government Printing Office, Washington, D.C., 1943.
6. Grover, N. C., and A. W. Harrington, "Stream Flow," John Wiley & Sons, Inc., New York, 1943.
7. Horton, R. E., "Weir Experiments, Coefficients, and Formulas," *U.S. Geological Survey, Water Supply Paper* 200, Government Printing Office, Washington, D.C., 1907.
8. King, H. W., and E. F. Brater, "Handbook of Hydraulics," 5th ed., McGraw-Hill Book Company, New York, 1963.
9. Parshall, R. L., "The Parshall Measuring Flume," *Colo. State Univ., Colo. Expt. Sta. Bull.* 423, 1936.

CHAPTER 31

PHOTOGRAMMETRIC SURVEYING

Revised by
Sidney W. Smith, Supervisory Civil Engineer
Topographic Division, U.S. Geological Survey

PHOTOGRAMMETRY

31-1. General. Photogrammetry is the science of measurement by means of photographs. Photogrammetric surveying is the application of photogrammetry to the operations of finding and delineating the contours, dimensions, positions, etc., of parts of the earth's surface. The principles of photogrammetry are applicable to the fields of archaeology, architecture, astronomy, ballistics, criminology, geology, hydraulics, radiology, and other sciences, but the greatest development of the science is in the field of photogrammetric surveying. The realization of photogrammetry is the mathematical or graphical analysis of single or overlapping photographs.

As the images of actual objects appear displaced and are of proportionate size according to their distance and relative position within the range of vision of the eye, so do the scale and the position of objects in photographs vary according to their distance and position relative to the camera station. Photogrammetric surveying is accomplished by the measurement of these differences in scale and displacements in position.

Photogrammetry is not a new science, but only recently has the knowledge of photogrammetric surveying become general. It is a science, gradually developed, whose basic principles and mathematical analysis have been known for about one hundred years. Its initial development was slow because it grew as a branch of a science already established. Its complete development awaited the fruition of the sciences of optics and photography and came only with the development of aviation.

31-2. Historical Development. The first record of optical projection of images is that of Aristotle who, about 350 B.C., published knowledge of the fact that the image of the sun appeared round when projected through a square hole and was amplified with increasing distance from the aperture. Leonardo da Vinci wrote of the camera obscura about A.D. 1500, and Thomas Wedgwood in 1802 printed silhouettes without fixation on leather sensitized with silver nitrate. In 1832 Wheatstone began to experiment with stereoscopy, and in 1838 he constructed the first of the present type of mirror stereoscopes as an aid to the stereoscopic observation of drawings. In 1834, Elliot observed drawings stereoscopically through a box fitted at the near end with two eyeholes and toward the far end with a central aperture through which the lines of sight from the two eyes crossed for observation of laterally transposed drawings. Stereoscopes with prisms and lenses were introduced in 1844 by Brewster and about 1852 by Helmholtz. In 1858 the principle of stereoscopic observation by means of dichromatic projection of images was demonstrated in Paris by d'Almieda. Concurrent with these developments, Arago of the French Academy of Science initiated the application of photography to topographic surveying, architecture, and archaeology, and in 1851 Aimé Laussedat of the Corps of Engineers of the French army developed the mathematical analysis of photographs as perspective projections, thereby furthering their application to topography. In 1853 Porro developed the principle of observation through lenses. About 1858, Meydenbauer began research on the application of terrestrial photogrammetry to architecture and to the design of monuments, which work was recalled by M. Deneux following the First World War for the reconstruction of monuments and of the cathedral of Reims by means of measurements from photographs.

In the field of photography, John F. W. Herschel in 1819 discovered the hyposulphites and their property of dissolving silver chloride, and Niepce (1822–1825) printed engravings on tin sensitized with bitumen. In 1835–1837, L. J. M. Daguerre evolved the method of direct photography and in 1847 Niepce made the first negative on glass. Stereoscopic photography began about 1850 with the work of Abbé Moigno, and phototheodolites were invented by Paganini in 1884. The first automatic plotting instrument for topographic surveying was developed by Deville in Canada in 1896. In 1894 Colonel von Hubl (Austria) adapted the methods of Laussedat to work in high mountains and developed the stereocomparator. Lieutenant von Orel (Austria) in 1903 transformed the stereocomparator into a plotting instrument known as the stereoautograph.

Aerial photography from balloons probably began about 1858, and a Scheimpflug eight-lens aerial camera was used in the 1911 maneuvers of the German army. The radial-line method has been undergoing constant development since the work of Adams in 1893. The development of the airplane accelerated a corresponding development in aerial-photogrammetric mapping. Wilbur Wright obtained the first photograph from an airplane in the form of a motion picture taken over Centocelli, Italy, in 1909. The first photographs taken for the specific purpose of mapping were described in 1913 by Captain Tardino in a paper presented at the International Society of Photogrammetry held in Vienna.

A landmark of modern photogrammetry was the publication in 1932 and 1942 of books by Otto Von Gruber detailing the basic principles of modern mapping methods (Refs. 20*a* and 20*b* at the end of this chapter). In the

United States the multiplex was introduced about 1935, later the Kelsh plotter (1948), and about 1955 the Balplex. These stereoscopic plotters achieved reliable contouring in the office but still required field placement of vertical control at specific points in the overlap of the stereoscopic pair. Next came first-order stereoscopic plotters of European manufacture capable of both vertical and horizontal datum extension by aerotriangulation. Compilation of planimetric features and contouring was still accomplished by instruments of lesser order and expense, such as the Kelsh plotter.

Later efforts were devoted to the adjustment of photogrammetric strips from first-order instruments of European manufacture, first by graphical methods, later through programming on electronic computers. This method enables greater latitude in the necessary basic control pattern within photogrammetric strips and the development of photogrammetric control in each stereoscopic pair sufficient for compilation by more economical instruments. The science is now (1964) on the verge of a technical breakthrough involving electronics, computers, and mathematics.

The application of photogrammetry to surveying has been rapid, and its basic principles have been so thoroughly exploited that the published works on this science should be diligently studied before an attempt is made to evolve new or startling methods or apparatus. Some of the many excellent works are included in the list of references at the end of this chapter.

31-3. Definitions. A clear understanding of the meaning of the expressions used in photogrammetric surveying is essential. Following are definitions of some of the more common terms in current use:

An *anaglyph* is a picture printed or projected in complementary colors combining the two images of a stereoscopic pair and giving a stereoscopic image when viewed through spectacles having filters of corresponding complementary colors. See also *vectograph*.

The *aperture stop* is the physical element such as a stop, diaphragm, or lens periphery of an optical system which limits the size of the pencil of rays traversing the system. The adjustment of the size of the aperture stop of a given system regulates the brightness of the image without having any necessary effect upon the size of the area covered. The *relative aperture* of a photographic or telescopic lens is defined as the ratio of the equivalent focal length to the diameter of the entrance pupil, expressed as $f/4.5$ etc., also called the *f number*.

A *camera* is a chamber or a box in which the images of exterior objects are projected upon a sensitized surface. An *aerial camera* is one specially designed for use in aircraft. A *ground camera* is one designed for use on the ground. A *phototheodolite* is a form of ground camera. A camera specially designed for the production of photographs to be used in surveying is a *surveying camera*. A *cartographic camera* is a surveying camera, as also is a *mapping camera*, although the term *surveying camera* is preferred. A *single-lens* camera is one having a single (principal) lens. Cameras having more than one (principal) lens are called *multiple-lens cameras.* A *horizon camera* is one used in conjunction with an aerial surveying camera in *vertical photography* to photograph the horizon simultaneously with the ground. The *horizon photographs* are used to indicate the *tilts* of the *vertical photographs.* Some single-lens cameras may be equipped with ancillary lenses to photograph the horizon.

A lens whose air-glass surfaces have been coated with a thin transparent film of such index of refraction as to minimize the light loss by reflections is called a *coated lens.* The reflection loss of an uncoated lens amounts to about 4 per cent per uncoated air-glass surface.

An optical instrument, usually precise, for measuring rectangular coordinates of points on any plane surface is a *comparator.* A *stereocomparator* is a stereoscopic instrument for measuring parallax and sometimes includes a means for measuring photograph coordinates of image points. The *stereocomparagraph* is a form of stereocomparator wherewith parallax is measured by means of a micrometer and *floating-mark system* and the results translated graphically onto paper.

Control is the system of relatively precise measurements by triangulation, traversing, or leveling to determine distances, directions, or differences in elevation between points on the earth. *Horizontal control* determines horizontal locations only, and *vertical control* determines elevations only. *Geodetic control* takes into account the size and shape of the earth. *Ground control* is obtained by ground surveys as distinguished from *photogrammetric control,* which is established by photogrammetric methods. Any station in a horizontal and/or vertical control system that is identified on a photograph and used for correlating the data shown on that photograph is a *control point.*

A *coordinatograph* is an instrument used to plot in plane coordinates. It may be an integral part of some stereoscopic plotting instruments whereby the planimetric motions (X and Y) of the floating mark are plotted directly. It may also be used for plotting projections and control on base sheets or manuscripts.

The *elevation* is the vertical distance above the datum, usually mean sea level, of a point or object *on* the earth's surface. Elevation is not to be confused with *altitude,* which is the vertical distance to points or objects *above* the earth's surface. The vertical distance above a given datum of an aircraft in flight or during a specified portion of a flight is the *flight altitude.* In aerial photography the datum is usually the mean ground level of the area being photographed. The *flight line* is the line drawn on a map or chart to represent the tract over which an aircraft has flown or is to be flown. The *flight map* indicates the desired lines of flight and/or the locations of exposure previous to the taking of air photographs, or it is the map on which are plotted, after photography, selected air stations and the tracks between them.

The *eye base* is the distance between the centers of rotation of the eyeballs of the observer. Eye base is synonymous with *interocular distance* and *interpupillary distance.* It is approximately 2⅝ in.

A *map* is the representation of a portion of the earth's surface on a plane surface wherein all the parts appear in their proper relationship. There are many kinds of maps. *Cadastral maps* show the extent, ownership, value, etc., of land; they usually show individual tracts of land with corners, length and bearing of boundaries, acreage, ownership, and possibly the principal cultural and drainage features. *Planimetric maps,* or *line maps,* show by conventional symbols the cultural and drainage features of land in their proper relationship in orthographic projections, or plan; relief is not depicted on planimetric maps. *Topographic maps* show by conventional symbols the culture, drainage, relief, and vegetation features of parts of the earth's surface. *Stereometric maps* are made by stereoscopic means; an example is shown in Fig. 31-1.

A *mosaic* is an assemblage of separate photographs. Mosaics are not maps but are map substitutes. The features of the part of the earth's surface shown on a mosaic are not in their proper relationship but are displaced in position and varied in image size according to the relief of the terrain and the conditions under which the photographs were made and joined together to form the mosaic. If the matching of the photographs is by image alone, it is an *uncontrolled mosaic.* If, before being laid, the individual photographs have been restituted to the horizontal plane and have been enlarged or reduced to fit predetermined locations of certain important features, it is a *controlled mosaic.* A controlled mosaic is more accurate than an uncontrolled mosaic but retains the changes in scale and displacements of image points due to differences in relief within the individual photographs. A *contoured mosaic* shows the relief by means of contours and may be either controlled or uncontrolled.

A photograph taken with the camera axis directed intentionally between the horizontal and the vertical is an *oblique photograph.* A *high-oblique* photograph is one in which the apparent horizon is shown; a *low-oblique* photograph is one in which the apparent horizon is not shown. "High" and "low" in this sense are not to be confused with relative altitudes. A *vertical photograph* is an aerial photograph made with the camera axis vertical or as nearly vertical as practicable in an aircraft.

An *oblique plotting instrument* is an optical instrument, sometimes monocular, for plotting from oblique photographs. However, instruments such as the *Zeiss stereoplanigraph* and the *Balplex,* among others, are capable of plotting from oblique photographs.

Parallax is the displacement of the images of objects with respect to other objects due to the difference in their respective distances from the observer. *Difference in parallax,* or *parallax difference,* is a measure of the distance between the objects observed. In overlapping vertical aerial photographs, the difference in parallax between two points is a measure of the difference

FIG. 31-1. Portion of topographic map of Yosemite Valley. (*U.S. Geological Survey.*)

between their elevations above sea level. The mathematical expression for the difference in parallax in terms of the difference in elevation between the points corresponding to a unit difference in parallax is the *parallax equation*. It is usual to express differences in elevation in feet and differences in parallax in millimeters, in which case the parallax equation is the number of feet difference in elevation corresponding to 1 mm. difference in parallax. For a simple and practical explanation of parallax the following example is given:

If an observer alternately looks at objects spaced at different distances from the observer first with one eye and then with the other, the objects appear to shift to the right and to the left as first one eye and then the other is opened and closed. This shift is *parallax,* and the difference in the amount of the shift is a measure of the distance from the observer to the objects and of the objects from each other. When any one of the objects is looked at, the lines of sight from the eyes converge on the object. The amount of this convergence depends on the distance between the eyes of the observer (interpupillary distance) and on the distance from the observer to the object observed. The angle at which the lines of sight intersect on the object is the *angle of convergence* or *parallactic angle,* which is different for objects at different distances from the observer. The difference in the angle of convergence is also a measure of the difference in the distance of the objects.

A *perspective projection* is the aspect of an object or objects from a common point. A photograph is a perspective projection, and the point from which the photograph is taken is the *camera station,* whether it be in the air or on the ground.

Photogrammetry is the science of measurement by means of photographs. Thus, surveying by means of photographs is an application of photogrammetry. *Aerial photogrammetry* applies to the use of aerial photographs, and *terrestrial photogrammetry* finds its application in the use of ground photographs.

In photogrammetry the projection or observation of photographs is often expressed in terms of *bundles of rays, pencils of light,* or more simply *rays of light.* The geometrical conception of a single element of light propagated in a straight line and of infinitesimal cross-section used in tracing analytically the path of light through an optical system is considered as a ray of light. A *pencil of light* is a bundle of rays originating at or directed to a single point, while a *beam of light* is a group of pencils of light. *Polarized light* is ordinary light after passage through certain polarizing media. It thereupon becomes *plane polarized* in that its vibrations are limited to a single plane. A *polaroid* is a manufactured plastic *polarizing screen.*

In photogrammetry the process of projecting a tilted or oblique photograph onto a horizontal reference plane wherein the angular relation between the photograph and the plane is determined by ground measurement is referred to as *rectification;* this should not be confused with *transformation,* which pertains to the projection of an oblique photograph onto a horizontal or nearly horizontal plane established by fixed angular relations between the photograph and the plane onto which it is projected. A *transforming printer* is one especially designed for use with a particular multiple-lens camera for the transformation of oblique or wing negatives taken by that camera.

A *spatial model* is the three-dimensional image formed in the mind of the observer as a result of the stereoscopic observation of two views of the same object. A spatial model is an *optical relief model.*

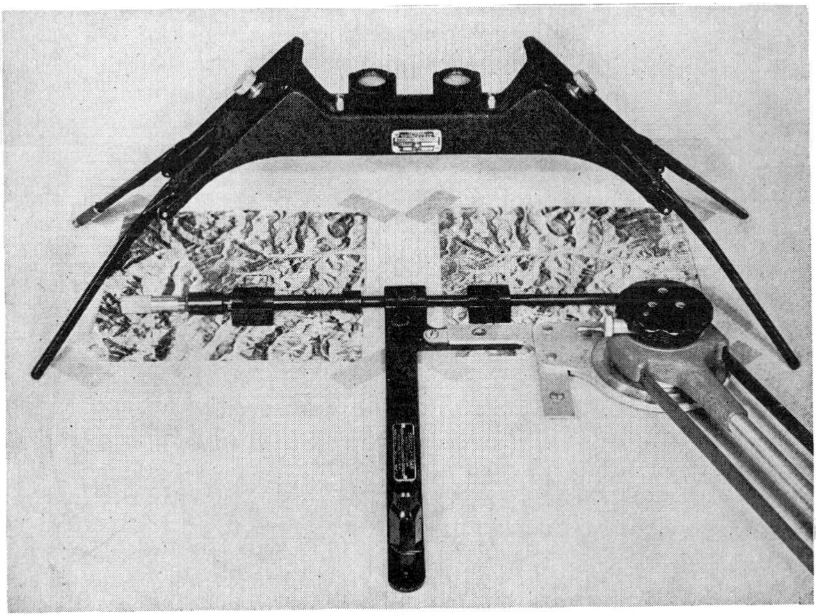

FIG. 31-2. Fairchild magnifying mirror stereoscope.

Stereoscopy is seeing as in three dimensions. *Stereoscopic measurement* is measured by means of such vision. A *stereoscope* is any mechanical device or devices used to facilitate seeing as in three dimensions. Stereoscopes may be formed of mirrors, lenses, prisms, combinations of these, pinholes, baffles, dichromatic and polaroid projection and printing, or flickering screens. The essential purpose of a stereoscope is to enable the observer to view two photographs of the same object with his two eyes in such manner as to cause the photographs to *fuse* in the mind of the observer into a single spatial model of the original object. This spatial model has the third dimension (depth) which can be measured. With the exception of observation by means of flickering screens, stereoscopic observations require binocular vision. A convenient magnifying mirror stereoscope is shown in Fig. 31-2.

Monocular vision is seeing with one eye. *Binocular vision* is seeing the same object with both eyes at the same time.

A *vectograph* is a stereoscopic photograph composed of two superimposed images which polarize light in planes 90° apart. When viewed through *polaroid spectacles* with the lens axes at right angles, a three-dimensional model is seen.

Phototriangulation is the process for the extension of horizontal and/or vertical control wherein the measurements of angles and/or distances on overlapping photographs are related into a spatial solution using the perspective principles of the photographs. Where this process involves the use of aerial photographs it is called *aerotriangulation* or *aerial triangulation*. *Analytic aerotriangulation* involves the solution of the spatial problem by computational routines. *Stereotriangulation* involves the solution of the spatial problem by means of a stereoscopic instrument through the orientation of successive stereoscopic pairs of photographs into a continuous strip. The spatial solution for the extension of horizontal and/or vertical control using these strip (or flight) coordinates may be by either graphic or computational procedures.

Radial triangulation is an aerotriangulation procedure, either graphic or analytic, in which directions from the radial center or approximate radial center of each overlapping photograph are used for horizontal-control extension by the successive intersection and resection of these direction lines.

Stereotemplet triangulation is a method of aerotriangulation utilizing stereotemplets permitting the achievement of scale solutions by area; it is not restricted to solutions along flight strips.

A *stereotemplet* is a composite of two templets derived from the leveled stereoscopic model, each covering the same model area. Radials are constructed from a different pass point for each templet to the remaining pass points and control points in the model.

Topography pertains to the features of the actual surface of the earth considered collectively as to form. A single feature as a mountain or valley is called a *topographic feature*. Topography is subdivided into *hypsography* (the relief features), *hydrography* (the water and drainage features), and *culture* (man-made features).

In recent years there has been considerable development in lenses, one of the most striking improvements being the increase in the angle of coverage of the single lens. The *angle of coverage* of a photographic lens is defined as the apex angle of the cone of rays passing through the front nodal point of the lens. The *normal-angle lens* is a lens having an angle of coverage up to 75°; the *wide-angle lens* between 75° and 100°; and the *ultrawide-angle lens* (superwide-angle) greater than 100°. The photogrammetric requirement for a photographic lens is preservation of both distortion and resolution requirements throughout the angular field; otherwise it is useless for accurate measurements.

31-4. Basic Principles. A photograph may be represented as a section of a bundle of rays cut by a plane. If the plane cuts the bundle between the perspective center and the objects photographed, the photograph is a *positive;* if on the opposite side, a *negative.* A *diapositive* is a *positive transparency,* usually on glass. A bundle of rays is *symmetrical* when it is identical on both sides of the perspective center and the composite images formed on planes cutting the bundle at equal distances on either side of the perspective center are identical. A bundle of rays passing through a distortion-free lens is symmetrical, and the resultant photograph is a true *perspective.* If the images

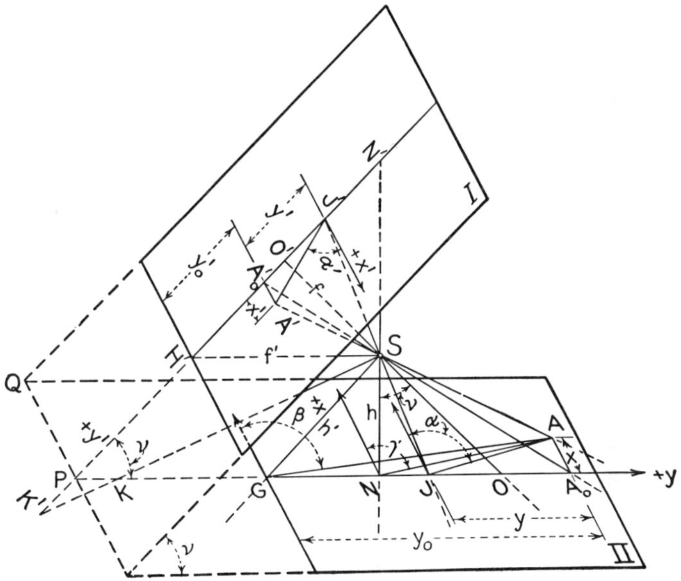

FIG. 31-3. A photograph as a perspective projection.

of a diapositive are projected back through the taking lens of a camera, the emergent bundle is a reconstruction of the original bundle and the projected image is a true representation of the objects photographed irrespective of the distortion characteristics of the lens. The method of observation of diapositives through a lens of characteristics identical with the taking lens is known as the *principle of Porro and Koppe.* To avoid the necessity for this type of observation, effort is made to produce lenses that are free from distortion. Distortions which are not perceptible to the naked eye are readily discernible in measurements with stereoscopic plotting instruments.

The consideration of a photograph as a perspective projection is shown in Fig. 31-3, wherein S is the perspective center, or camera station (Ref. 20*b* at the end of this chapter). Plane I is *the negative* (plane) and Plane II is *the positive* or the *object plane*—in aerial photography parallel to the earth's surface. Planes I and II intersect at any angle v. The perpendicular distance f from S to the plane of the photograph (in this case the negative) is the *principal distance* of the lens and is usually referred to as the *focal length* of the lens. The point of intersection O' of the *principal ray* of the bundle on the negative plane is the *principal point* of both the lens and the photograph. The plane through S normal to the line of intersection PQ of Planes I and II is the *principal plane.* Lines through points H' and G parallel to line PQ are the two vanishing lines (horizon) for corresponding images in the two planes and contain the images of the infinitely distant points in the two planes. If the angles of intersection between perpendiculars f and h to the two planes are bisected internally and externally, the points where these bisectors pierce the two planes are, respectively, J and J', K and K', which are known as the *conjugate focal points, isocenters,* or *metapoles.* Point N, the foot of the perpendicular from the perspective center to Plane II, is the *plumb point.* The isocenter J is in the principal plane and is a distance $h \tan v/2$ from the plumb point N.

In photographs taken with the axis of the camera truly vertical, points O', N', and J' are common. In this case displacements due to tip and tilt are zero, and the displacement due to relief radiates from this common point. Rarely are photographs taken with the axis of the camera truly vertical, but in aerial photography with experienced personnel, 85 to 90 per cent of the vertical photographs will have combined tip and tilt of less than 1°, and tip and tilt in excess of 3° is generally considered to be sufficient cause for rejection.

In those stereoscopic plotting instruments which have provision for observation or projection through the lens after the principle of Porro and Koppe or where the lenses are distortion-free and the photographs can be adjusted for tip and tilt, the presence of these otherwise disturbing factors is not important. In all other cases they must be either ignored or compensated graphically or analytically. Displacements due to relief radiate from the plumb point N, which except on truly vertical photographs is difficult of determination. Bearings of rays drawn from the plumb point are independent of the relative elevations of the objects photographed. Bearings on objects in any picture plane from the isocenter J of the photograph retain the same angular values only when projected onto homologous planes in a photograph. Thus bearings from the isocenter J of the photograph are affected by relief and are not true angles for photographs made in irregular terrain. However, in essentially vertical photographs (with tip and tilt less than 3°) the distances between the plumb point, the isocenter, and the principal point are so small that measured angles or rays drawn from the principal point are essentially true.

31-5. Monocular Vision. In seeing with either eye, the resultant sensation is transmitted to the brain for the experience of sight. The formation of the eye is that of a camera wherein a perspective bundle of rays is projected through a lens onto a focal plane, yet the sensation of sight is more than simple projection. The several cells of the *retina* are connected to the optic nerve, and any unusual visual happening within the space volume covered by the eye reacts upon the optic nerve and immediately draws the attention of the observer to the area where the happening occurred. The succession of such happenings or movements causes the eyes to be in constant motion. A person is conscious, however, of seeing only those objects on which the attention is fixed, although the other objects are continuously being projected onto the retina and the sense of their presence is faintly transmitted to the brain. Thus, in all space the eye sees a single area more clearly than any other. There is one spot on the retina of the eye, the *fovea centralis,* more sensitive than the remainder; the eyes move automatically in their sockets always to bring the projected image of the desired object onto focus on this sensitive spot. Objects are brought into focus by *accommodation,* that is, a physical reshaping of the eye to bring the bundle of rays from distant objects onto the retina in a sharp and distinct perspective pattern. The fovea centralis is about 0.25 mm. in diameter and at the principal distance of the eye subtends an arc of about 35′. This is the limiting angle in the eye wherein vision is most acute. The fovea centralis is composed of small bundles of nerve cells which permit discernment between sharply defined objects minutely separated. The normal eye can distinguish between sharp black lines separated by white spaces of equal width when the angle subtended by the distance between the centers of two such parallel black lines corresponds to about 1 minute of arc. At an observing distance of 12 in., the spacing between two such lines is about $\frac{1}{250}$ in. Thus, the human eye is possessed of the faculty of automatically concentrating its vision onto single objects while disregarding all others and in turn is able by concentration to differentiate between minute spacings of objects. In monocular vision the human eye is possessed of the power to project its principal ray of light where it will and to "point" it as surely as a telescope or any other physical object (such as a beam or a rod) can be directed. In the perspective projection of monocular vision, distances and the dimensions of objects are determined by their association with known objects, and the accuracy of their determination is one of judgement based upon the prior experience of the observer as well as upon the acuity of vision.

31-6. Binocular Vision. In simultaneous observation with the two eyes, the eyes not only retain their individual properties but also act together as a precise instrument of measurement whose limits may be resolved into mathematical form generally applicable to all persons of normal eyesight. With binocular vision in nature one sees a space volume whose form is transmitted to the brain by the fusion of the two perspective projections in the eyes. In scanning such a space volume, vision is successively concentrated on such objects as are drawn to the attention of the observer. In this process the principal ray of each eye is directed to a single point as in Fig. 31-4, wherein point M_1 is the object observed by the two eyes whose perspective centers are at S_L and S_R. The principal rays of the eyes are S_LM_1 and S_RM_1, which intersect at point M_1 to make an *angle of convergence* ϕ_1 with respect to each other. If the vision is shifted to point M_2, the principal rays intersect at point M_2 to form a different angle of convergence ϕ_2. The difference $\Delta\phi$ between the angles ϕ_1 and ϕ_2 is some measure of the distance between objects M_1 and M_2. Experience has shown that trained observers are able to detect differences in the angle of convergence of 8 or 10 seconds of arc and that the average observer can consistently detect angular differences of 20 seconds.

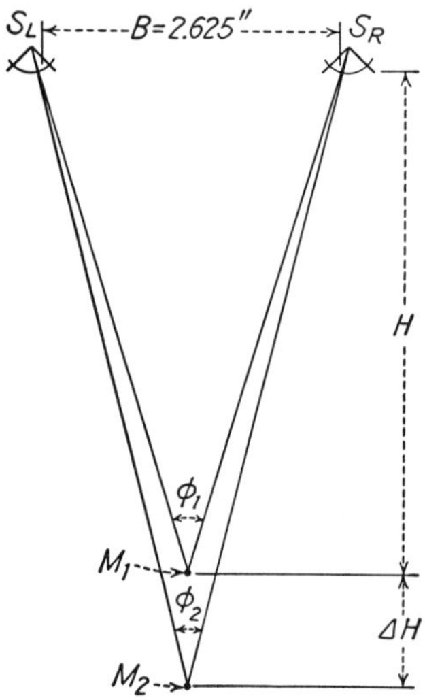

FIG. 31-4. Binocular vision.

31-7. Stereoscopic Observation. Stereoscopic observation in photogrammetric surveying is based upon binocular observation at the effective distance of normal vision, about 10 in. The theoretical limiting precision of stereoscopic measurement may be expressed as the smallest value of ΔH (Fig. 31-4) that can be discerned. If the smallest measurable angle $\Delta\phi = 20''$, $H = 10$ in., and $B = 2.625$ in., then in Fig. 31-4

$$\frac{\phi_1}{2} = \arctan\frac{B}{2H} = 7°28'38''$$

$$\frac{\phi_2}{2} = 7°28'28''$$

$$\Delta H = \frac{B}{2}\cot\frac{\phi_2}{2} - 10 \tag{1}$$

$$\Delta H = 0.004 \text{ in.} = 0.10 \text{ mm. (approx.)} \tag{2}$$

Thus at the normal observing distance the average human being can detect differences in distances of 0.004 in., or about 0.10 mm. With practice, a skilled observer can measure consistently to a limiting value of about

$$\Delta H = 0.05 \text{ mm.}$$

which is generally accepted as the standard for excellence in stereoscopic measurement. If the objects within the space volume in nature are replaced by two photographic perspective projections made from different points and these photographs are viewed with the eyes at positions corresponding to the centers of projection in such a manner that the right eye sees only the perspective projection made from the right camera station and the left eye sees only the perspective projection made from the left camera station, the physiological fusion within the brain is the same as binocular vision in nature and the two perspective projections fuse into a single spatial model identical with but at a scale usually smaller than that in nature.

FIG. 31-5. A stereogram.

Figure 31-5 is a stereogram in which the perception of depth can be obtained without the aid of viewing apparatus. If a large card is held normal to the page at the dividing line between the two photographs so that the right photograph will be seen with the right eye and the left photograph with the left eye, after a short time the view will appear in relief. The eyes should be held at the normal distance (about 10 in.) from the page. It is helpful first to direct the eyes at a distant object, then to direct them at the stereogram without changing the angle between the eyes. Both halves of the stereogram should be equally lighted.

31-8. Stereoscopes. Any device which facilitates stereoscopic observation is a stereoscope. Simple stereoscopes take the form of either the

Brewster stereoscope of Fig. 31-6*a* or the Helmholtz stereoscope of Fig. 31-6*b*; the paths of the bundle of rays are indicated by straight lines to show the manner in which the sight of each eye is directed to, and only to, the proper half of the stereogram.

Stereoscopic perception may be enhanced by increasing the stereoscopic base, by magnifying the images, or both. In the case of photographs to be used as stereograms, the stereoscopic base must be increased by changing the distance between the camera stations prior to exposure. Merely separating the pictures a greater distance once they are made will not increase the stereoscopic effect. Magnification, however, may be obtained by using a camera of increased focal length, enlarging the photographs, or observing

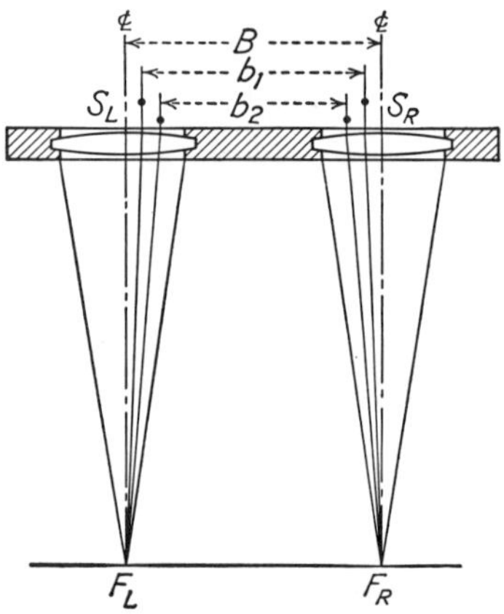

FIG. 31-6*a*. Brewster stereoscope.

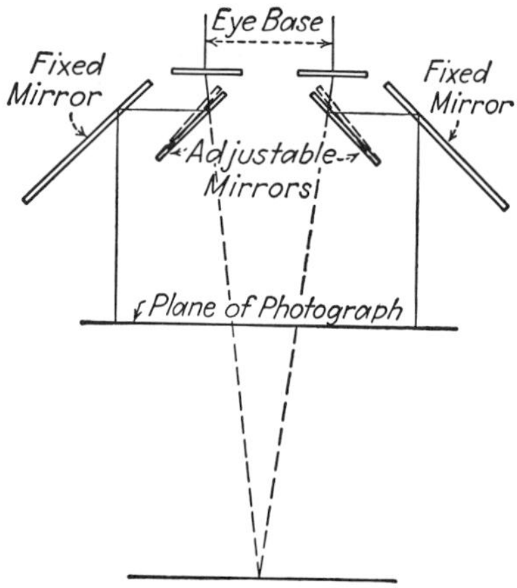

FIG. 31-6*b*. Helmholtz stereoscope.

the photographs through magnifying lenses. If the stereoscopic base is increased n times and the images are enlarged m times, the stereoscopic perception is magnified $n \times m$ times. Most photogrammetric instruments possess means for magnifying the images. Magnification is limited by optical difficulties and by the fact that gross enlargement usually results in an indistinct image.

For the proper observation of photographs under a stereoscope they must be *oriented* so that they occupy the same relative position with respect to each other as the two positions of the focal plane of the camera occupied at the time of the two exposures. In simple observation this orientation may be accomplished by rotating the photographs with respect to each other until the two lines forming the stereoscopic base (the distance between the two principal points of the photographs) are in prolongation of each other and parallel to the eye base. When the photographs are so arranged, it is then only necessary either to separate them or to bring them closer together in the same plane to permit the two eyes to observe the corresponding images without strain. When this is done, the condition for stereoscopic observation has been accomplished and anyone can see stereoscopically. If anyone with normal eyesight fails to see stereoscopically, it is usually because the photographs are not properly oriented.

31-9. Vectography. The means which facilitate stereoscopic observation are not limited strictly to mechanical machines but include such devices as *vectographs* and *anaglyphs,* both of which are simple means of providing stereoscopic observation of photographs to individuals or to large groups who are without training or experience in stereoscopy.

Ordinary light is said to radiate in all directions normal to its direction of propogation. It is susceptible to reflection in all directions from any suitable reflecting surface. Ordinary light may become polarized upon passing through a *polarizing screen;* after such passage the vibrations of the light no longer move in the dimensions of the coordinates x, y, and z, wherein z is assumed to be the axis of emission, but are limited to *planary movements* in plane zx or plane zy depending upon the physical orientation of the polarizing screen. The polarizing screen may be likened to a screen of many parallel slits through which the light can pass. Light striking such a screen in a three-dimensional mass is transformed upon passage into a series of parallel planes of light which vibrate in a single plane parallel to the slits; the light which otherwise radiated in the third dimension has been cut out. Inasmuch as the polarizing screens transform the light into planes of minute thickness, it is obvious that, if two such screens are placed one over the other with their directions of polarization at right angles to each other, they will transmit no light (or practically no light, since existing screens are not 100 per cent effective).

If the images of two overlapping photographs are projected through polarizing screens whose directions of polarization are at right angles to each other and the resulting image is viewed through complementary polarizing screens, a stereoscopic model will result. Such an image is similar to that obtained by means of the multiplex projector but has the additional faculty of being capable of projection in natural colors inasmuch as the polarizing screens are not color-absorbing.

A *vectograph* is a composite print through polarizing screens of two overlapping photographs; under ordinary light it appears at first glance to be a glossy sepia print. When it is viewed through a pair of spectacles fitted with polarizing lenses, the spatial model is brought out.

Vectographs are made by printing each of two overlapping negatives in approximate register by the imbibition process on wash-off relief film. The exposed relief films upon being properly soaked with the printing solution and with the emulsion sides registered on the opposite sides of the vectograph film are then passed through a wringer or press; after the vectograph film has absorbed the proper amount of printing solution, the relief films are stripped off and the image on the vectograph film is fixed in a photographic bath. The result is a *vectography transparency*, which may either be used in the form of a lantern slide or be made into a reflection print by painting one side with clear lacquer and the other with aluminum lacquer. Such prints may be formed into mosaics to permit the simultaneous stereoscopic observation of large areas by groups of observers or the prints may be used as single stereoscopic images as the need may require.

31-10. Dichromatic Projection and Anaglyphs. The multiplex, the Balplex, and the Kelsh stereoplotter (Art. 31-48) all use the principle of *dichromatic projection* and observation of images formed by the intersecting bundles of rays from two overlapping photographs simultaneously projected in complementary colors, usually blue-green and red.

An *anaglyph* is formed by printing the left photograph of an overlapping pair in blue-green and the right photograph in red in approximate register onto the same medium. If the composite image of the anaglyph is viewed through a pair of spectacles with a red filter over the left eye and a blue-green filter over the right eye, the resulting image will appear as a spatial model in black (or actually in grayish black due to imperfection in blending colors). This effect is accomplished by each filter passing only the light of the corresponding colors and absorbing the light of the other colors. Large quantities of anaglyphs may be obtained quickly and inexpensively by means of the ordinary halftone printing process. As in the case of vectographs, good anaglyphs are obtained only by using first-quality photographs.

TERRESTRIAL PHOTOGRAMMETRY

31-11. General. Terrestrial photogrammetry is photogrammetry by means of photographs taken with the camera supported on the ground. Photographs for terrestrial photogrammetry are taken with phototheodolites especially built for that purpose (Art. 31-12). The photographs are later inserted in an automatic plotting machine for the compilation of the map therefrom (Art. 31-15). A section of a terrestrial photograph together with the map made therefrom is shown in Fig. 31-7. The site of the Hoover Dam in the canyon of the Colorado River was surveyed by means of terrestrial photogrammetry; plotting was done on the Aerocartograph.

31-12. Phototheodolite. The Wild phototheodolite shown in Fig. 31-8 consists of a Wild model T-2 theodolite mounted on a strongly built metal frame, on the inside of which the surveying camera can be placed (Ref. 22 at the end of this chapter). The theodolite has a 3¾-in. horizontal circle and a 2-in. vertical circle graduated either in degrees and fractions thereof or in the centesimal or grad system. The reversible central telescope has a magnification of 28 times, an objective of $1\frac{9}{16}$ in., clear aperture, and internal focusing. Focusing is accomplished by turning the focusing ring around the telescope tube. The eyepiece is provided with diopter graduation. The circles are read by means of a coincidence microscope in the eyepiece; the mean reading of the two sides of the circle is given to 1 second of arc. The illumination of the horizontal circle is controlled by a revolving prism mounted on the steel camera casing. The vertical circle is illuminated by two prisms on the telescope, one for the normal and the other for the inverted position. The telescope is provided with an adjustable level which is observed through

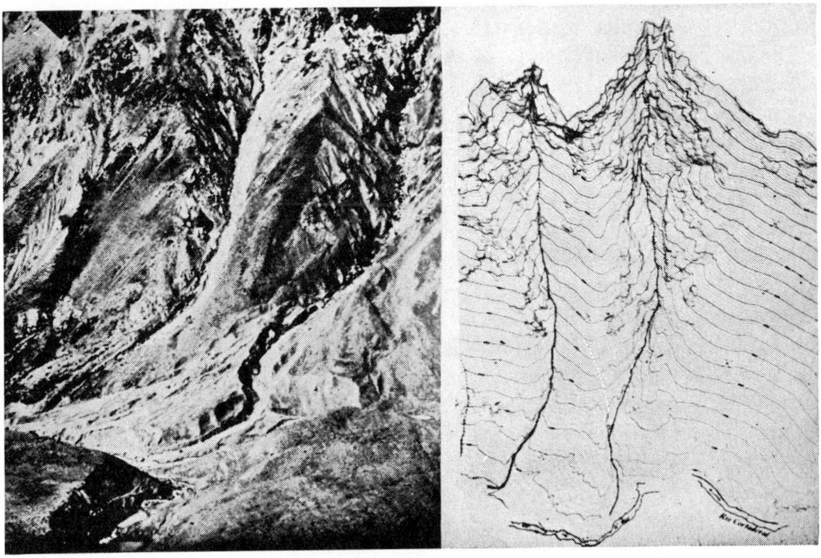

FIG. 31-7. Terrestrial photograph and map made therefrom.

a reversible prism. Two other levels serve for leveling the instrument. The theodolite can be used either as a direction instrument or as a repeating instrument.

The steel camera casing is furnished with two wyes to receive the horizontal trunnions of the camera; the trunnions are held in place with sliding covers. The focal length of the camera lens is 165 mm., and the plate size is 4 by 6 in. (100 by 150 mm.) with thickness of 2.35 to 3.35 mm. The camera lens is provided with a yellow filter, and the relative aperture is one-twelfth of the focal length. Inasmuch as phototheodolite shots are usually taken with the camera either level or directed downward, there are several adjustable tilts of the camera downward and only one upward. The following tilts are adjustable: +7, 0, −7, −14, −21, and −28 grads. Sharp focus is insured by a pressure plate which holds the negative plate firmly against the focal plane. Horizontal and vertical collimating marks are registered on the plate at the time of exposure. These marks are for the purpose of facilitating

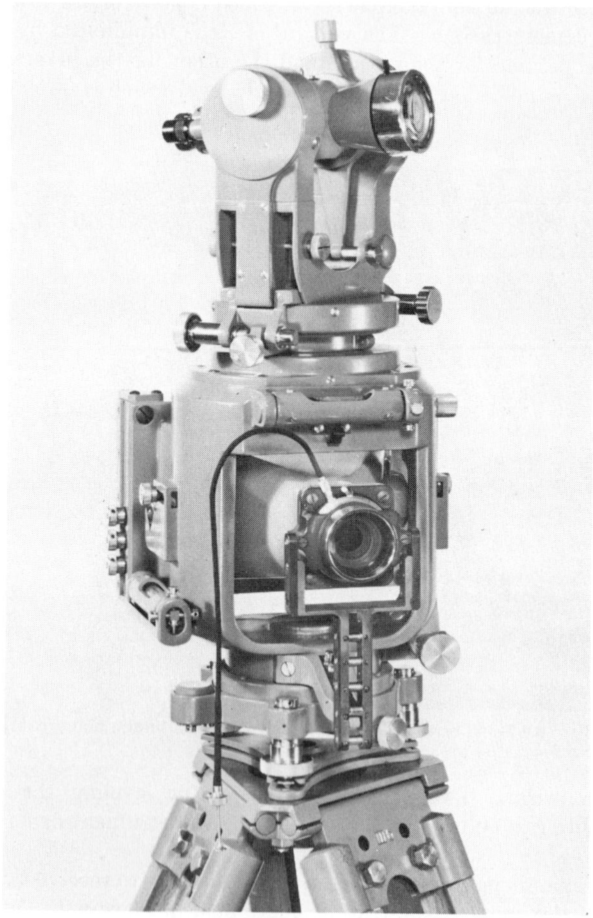

FIG. 31-8. Wild P-30 phototheodolite.

subsequent adjustment in the Wild autograph, which instrument is used for the compilation of the maps from the terrestrial photographs.

The horizontal circle of the theodolite is mounted on the camera casing in such manner that, when the circle reads zero, the optical axis of the camera and that of the telescope are in the same vertical plane. The axis of the camera can be moved to any desired angle with that of the telescope and clamped in that position.

31-13. Terrestrial Photography. For economy and speed of operation, the area to be surveyed should be covered with the minimum number of photographs from the optimum positions. This objective can be accomplished only after a thorough study has been made of the existing maps of the area, followed by a reconnaissance on the ground. In very rough terrain it is desirable to visit certain stations only once; hence the procedure of the field work should be planned in advance. Although the actual selection of the stations will depend upon the size and ruggedness of the area to be surveyed, in general the camera stations should be such that the direction of pointing is as nearly normal to the slope as possible and that the stations overlook the area.

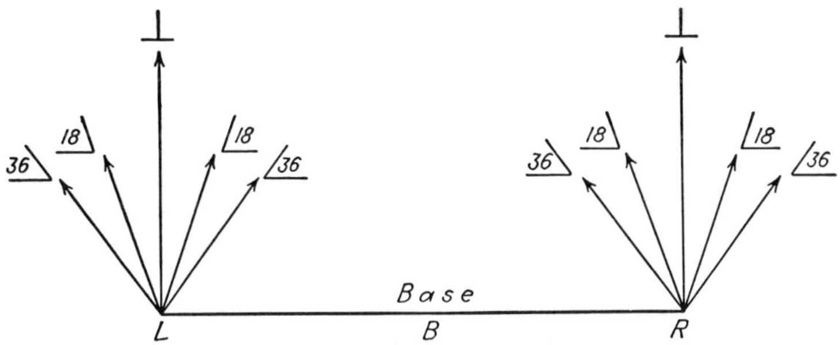

FIG. 31-9. Directions of pointings in terrestrial photography.

To meet these requirements, the camera should be directed downward rather than upward and the stations should be at the higher points in the area.

Terrestrial photographs are taken in pairs from the ends of a measured base, as shown in Fig. 31-9 wherein the arrows indicate the directions of the camera pointings. Although this figure indicates only horizontal pointings, the camera may be likewise depressed in any of the positions. Such pairs of pictures are generally made with the two positions of the camera axis parallel to each other where the camera is swung at predetermined (and usually the same) angles at each station. The minimum number of photographs is taken at each base and station to insure complete coverage of the area with the desired accuracy (see Ref. 22*a* at the end of this chapter).

31-14. Accuracy of Measurement. The accuracy of measurement depends upon the ratio of the base length to the distance of measurement, the accuracy of determination of the length of the base, and the magnitude and the accuracy of determination of the angle of parallax and of the angle of rotation of the camera.

The limit of the accuracy of measurement for automatic plotting instruments has been investigated by B. Hallert (Ref. 7) and O. Von Gruber (Refs. 20*a* and 20*b*), among others and is generally related to the flight altitude of the photography and dependent upon the nature of the terrain and the quality of the photographs. For the attainment of optimum accuracy of measurement, the system must be capable of the measurement of angles to within about 10 seconds of the true value. A similar relative accuracy may be obtained with properly taken terrestrial photographs.

It is necessary to the reconnaissance for terrestrial photography to establish base lines and camera stations which will permit photography of the area within prescribed limits of ratio of "base-length projection" to distance to objects photographed. The base-length projection is the projection of the base line on a plane normal to the direction in which the camera is pointed. Based on the attainment of an accuracy of measurement of the parallactic angle to within approximately 10 seconds and in consideration of the accuracy of field work in the determination of the base length, the angular settings of the phototheodolites, etc., it has been found that the limiting ratios of base-length projection B to photographic distance E are between $B/E = 1/20$ and $B/E = 1/4$ (page 131 of Ref. 20*a*).

Since the accuracy of the work and the facility with which the photographs may be used will depend on the quality of the photographs, it is necessary at the time of exposure to insure that the photographs are of the requisite quality.

31-15. Automatic Plotting Machines. The economy of photogrammetric surveying will rarely, if ever, permit the utilization of terrestrial photographs in methods based solely on computations or point-by-point plotting of topography; rather it is essential that terrestrial photographs be used in conjunction with an automatic plotting machine of some sort. The map shown in Fig. 31-7 was plotted by means of the Wild autograph (Art. 31-48).

AERIAL PHOTOGRAMMETRY

31-16. General. Since the First World War, aerial photogrammetry has replaced terrestrial photogrammetry for most surveying applications. The major reason for this change was the development of the airplane and its subsequent application to the photogrammetric mapping problem. Prominent in the application of aerial photogrammetry in the United States have been such organizations as the U.S. Geological Survey, the U.S. Coast and Geodetic Survey, the U.S. Army, the U.S. Navy, the U.S. Air Force, and agencies of the U.S. Department of Agriculture as well as numerous commercial companies concerned with the development of aerial cameras and related equipment and the training of photogrammetrists in the field.

Photogrammetry is now introduced into the college curricula of several civil engineering departments either as a requirement or as an elective and in some instances is offered as a separate department or option of the civil engineering curriculum. Syracuse University offered the first course in photogrammetry about 1929 under the direction of Prof. Earl Church. Technical societies within the photogrammetric field consist of one of the technical divisions of the American Society of Civil Engineers, the American Congress on Surveying and Mapping, the International Society for Photogrammetry, and the American Society of Photogrammetry. In addition to the extensive photogrammetric operations involved in the national military photogrammetric mapping, the use of photogrammetric techniques in the design and planning of civil-engineering projects is of increasing significance, particularly challenging the need for higher standards of precision.

Most of the United States has been photographed from the air, at one scale or another, by agencies of Federal, state, or local government. Prints of most of these photographs can be obtained at low cost. Detailed information concerning governmental photographic coverage of a given area can be obtained from the Map Information Office of the U.S. Geological Survey, Washington, D.C. Pertinent information available includes the date of photography, extent of area covered, type of camera used (precision or reconnaissance), flight height, focal length, approximate scale, and format. The Map Information Office can also furnish information on the additional coverage available in the possession of private organizations.

Maps are compiled in conventional symbols, and the space required to represent cartographic features on the map limits the number of features that can be shown to fewer than those on photographs of the same scale. In small-scale maps the conventional symbols are not true to scale, and it is not essential that the physical features on the photograph be at the same or an easily measurable scale. It is only necessary that each image be of sufficient size and clarity to permit correct interpretation of the photograph. It is usual for photographs with a scale of 1:12,000 to 1:50,000 to be used in the compilation of topographic maps at scales of 1:24,000, 1:31,680, and 1:62,500 with contour intervals of 5, 10, 20, 40, and 80 ft. for the National Topographic Map Series of the United States compiled by the U.S. Geological Survey.

Aerial surveys are used in the preparation of three major types of maps and photographic mosaics, namely, (1) topographic maps, (2) planimetric maps, and (3) controlled and uncontrolled mosaics.

31-17. Topographic Mapping. A topographic map (Fig. 31-1) is a map which presents the horizontal and vertical locations of the features represented; it is distinguished from a planimetric map by the addition of relief in measurable form. The topographic map usually shows the same features as the planimetric map but uses contours or comparable symbols to show mountains, valleys, and plains and, in the case of hydrographic charts, symbols and numbers to show depths in bodies of water.

For large-area topographic mapping, aerial photography is procured for use in automatic stereoscopic plotting machines in which projections of two overlapping photographs are viewed stereoscopically and are matched to plotted ground-survey or photogrammetric-control points through procedures of relative and absolute orientation. Map features are then compiled from the visual stereoscopic models thus formed.

Aerial surveying for topographic mapping consists of eight operations: advance planning, aerial photography, ground control, manuscript preparation, aerotriangulation, stereocompilation, field completion, and reproduction (See Fig. 31-10). The choice of the photogrammetric system depends upon the production of a map which will meet the map specifications for accuracy and content most economically.

1. *Advance planning* consists of a thorough examination of the map requirements and the project area with emphasis on basic source material, existing maps of the area, and the precision and density of existing horizontal and vertical control. Planning must cover the map specifications including the map scale and contour interval; the instrumentation, procedures, and methods; and the scheduling of the various phases.

2. *Aerial-photography phases* (Art. 31-21) consist of the design of the flight plan and the preparation of the flight map; the procurement of aerial photography, usually by contract; the preparation of flight indexes (Fig. 31-11) from the photoprocessing of the film; and the delivery of material from the contractor.

3. *Ground-control phases* (Art. 31-34) consist of the preparation of a control diagram on the best available map showing all available vertical and horizontal control, the transferral of photocenters to this map, and the selection and placement of additional ground-control points which will be used to supplement basic control and enable photogrammetric extension through aerotriangulation procedures. Thorough photoidentification of all ground control is essential, and this is usually done on the photographic prints. A more precise method of photoidentification is prepaneling of the ground-control points prior to aerial photography.

4. *Manuscript preparation* is usually accomplished at reproduction scale, although some organizations prefer stereomodel scale and subsequent reduction of copy for reproduction. The manuscript is prepared on dimensionally stable material, usually a transparent plastic material coated with a thin, opaque, painted surface. The grid, projection, all geodetic control, and conventional marginal information are plotted from rectangular coordinates (Fig. 31-12) at this time and then scribed by negative engraving.

5. *Aerotriangulation*, or aerial triangulation (Art. 31-34), is defined as any type of control extension accomplished by means of aerial photographs. The original conception of photogrammetric mapping necessitated ground control in every photograph, and for precise large-scale surveys on which

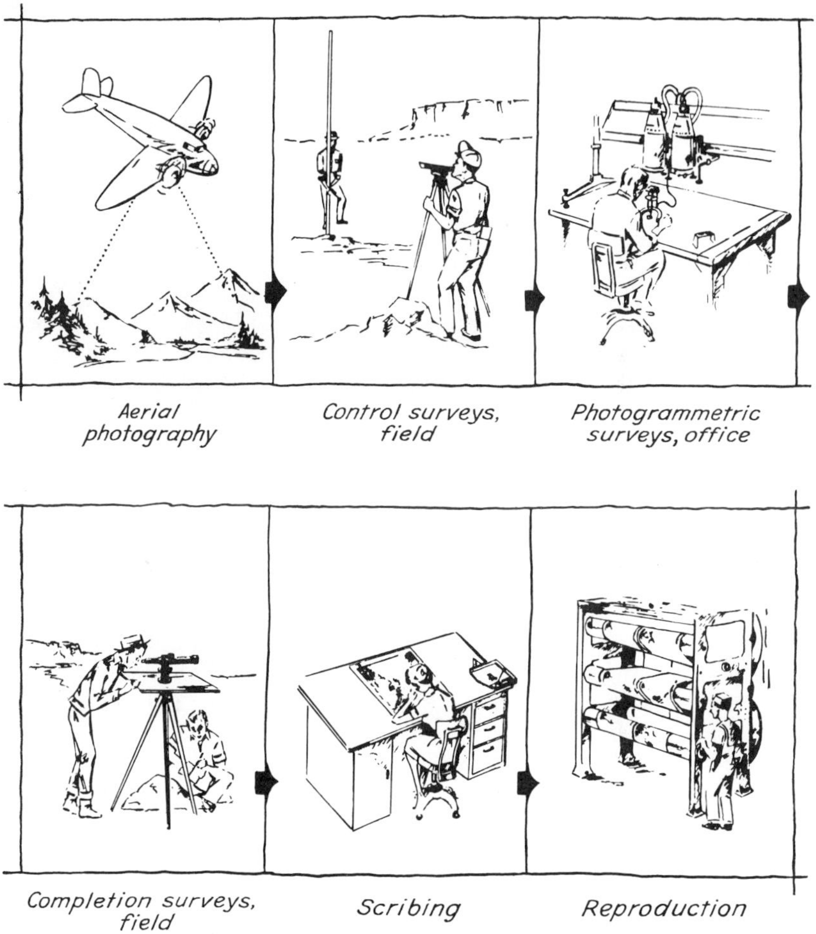

FIG. 31-10. Sequence of topographic-mapping operations.

engineering works are based this condition may still hold true. However, for surveys of medium and small scale it has been found that the ground control need not be in every photograph but may appear in every third, fifth, or in some cases as much as fifteenth photograph of a strip of photographs or at optimum positions in the several strips of an area; the intervening distances are bridged by aerial triangulation.

The principal forms of aerotriangulation are (*a*) radial triangulation (Art. 31-35), in which the radial centers of the pictures are used as angle vertices

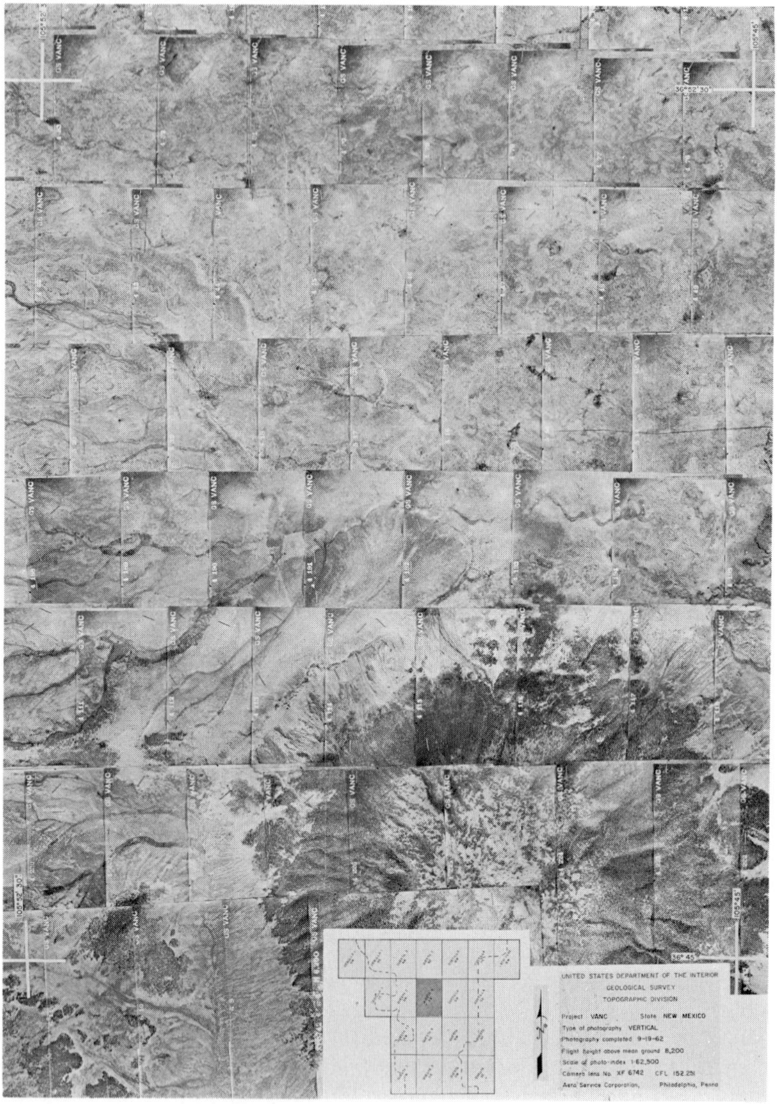

Fig. 31-11. Flight index.

and the radials are measured in the photograph plane, and (*b*) the methods of spatial aerotriangulation, in which the projection centers of the lenses (in space) serve as the vertices of pyramids. Instrumental methods of aerotriangulation include *stereotriangulation* (Art. 31-34), a procedure which uses a stereoscopic plotting instrument to obtain the successive orientation of the stereoscopic pairs of photographs into a continuous strip, and *stereotemplet triangulation* (Art. 31-34), a horizontal solution involving use of stereotemplets (Fig. 31-13) which permits an area solution not restricted to flight-strip solution (Fig. 31-14).

FIG. 31-12. Manuscript preparation.

Purely analytical methods of aerotriangulation use as basic data the rectangular coordinates, measured in a stereocomparator (Fig. 31-32), of a number of selected points in the aerial photographs; on the basis of these by comparison of distances or angles in successive photographs or in the photograph and the map, the coordinates of the nadirs of the individual photographs and hence their positions in space are determined. Analytical methods are gaining appreciably in importance owing to the progress made in the development of electronic computers. Cornell University and the National Research Council of Canada among others have developed methods for the analytical solution of both cantilever strips and bridges.

6. *Compilation of detail from photographs.* One method of stereocompilation of topographic detail is that used by the Topographic Division of the U.S.

Geological Survey in the compilation of the standard topographic quadrangles of the United States. Manuscript material consists of a dimensionally stable transparent plastic material coated with a thin, opaque, painted surface. Lines to be scribed in this surface are delineated on the manuscript components from the stereoscopic model through pantograph attachments or

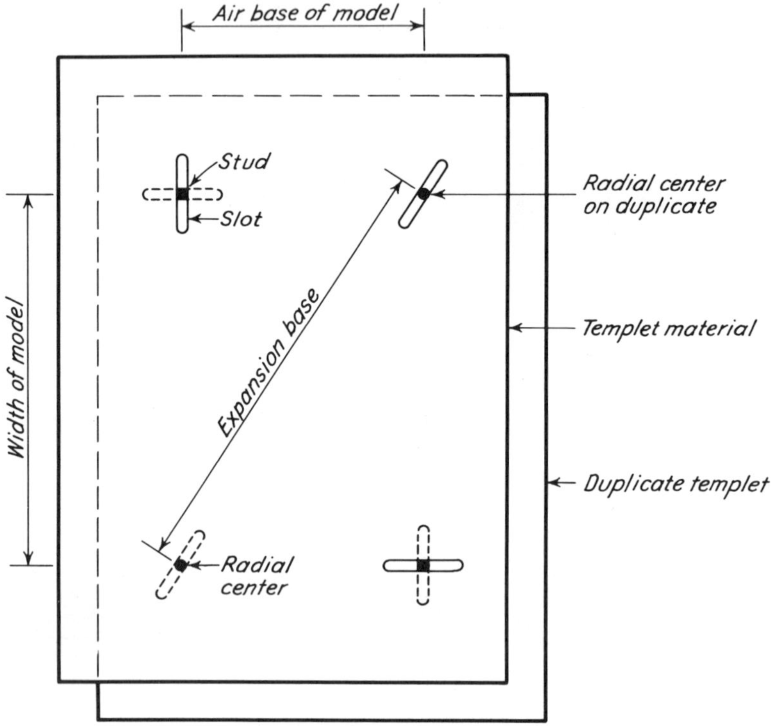

FIG. 31-13. Stereotemplet.

coordinatographs, thus enabling all stereoplotting to be accomplished at reproduction scale.

For each quadrangle or unit to be mapped, the compiler is furnished three manuscript components. The basic planimetric manuscript is plotted on a coordinate plotter at reproduction scale and has scribed upon it the projection, grid, geodetic control, and other conventional marginal information. On this manuscript will be scribed all planimetric detail to be published in red, black, or blue (Fig. 31-15, Plate 1). The second manuscript component, commonly called the "contour plate," will be scribed with all features to be published in brown (Fig. 31-15, Plate 2). The third manuscript component, called the "timber plate," will be scribed with all features to be published in green (Fig. 31-15, Plate 3).

The three manuscripts are punch-registered together by a stud punch-hole system utilizing the basic planimetric base component as a master. Flanged brass studs (Fig. 31-16) are inserted in the punch holes for the following purposes: (*a*) to hold the manuscript components together on the plotting surface (slate or coordinatograph), (*b*) in transferring scribed material photographically from one manuscript component to another, (*c*) in checking

FIG. 31-14. Stereotemplet laydown.

components over a light table for registry errors, and (*d*) to register edge material for photographic transfer to the manuscript components. It is also possible to utilize this stud-hole registry configuration for reproduction phases of quadrangle finishing.

Relative and absolute orientation of the stereomodel is achieved by orienting to a vertical datum established by either field or photogrammetric methods and to a horizontal datum established by photogrammetric extension of basic photoidentified ground control. Map features are delineated on the manuscript components from the stereomodel, usually as pencil copy which is later finish-scribed over a light source to proper line weight and symbolization.

Once absolute orientation of the stereomodel to the supplemental control is attained, the planimetric features are delineated on the master planimetric

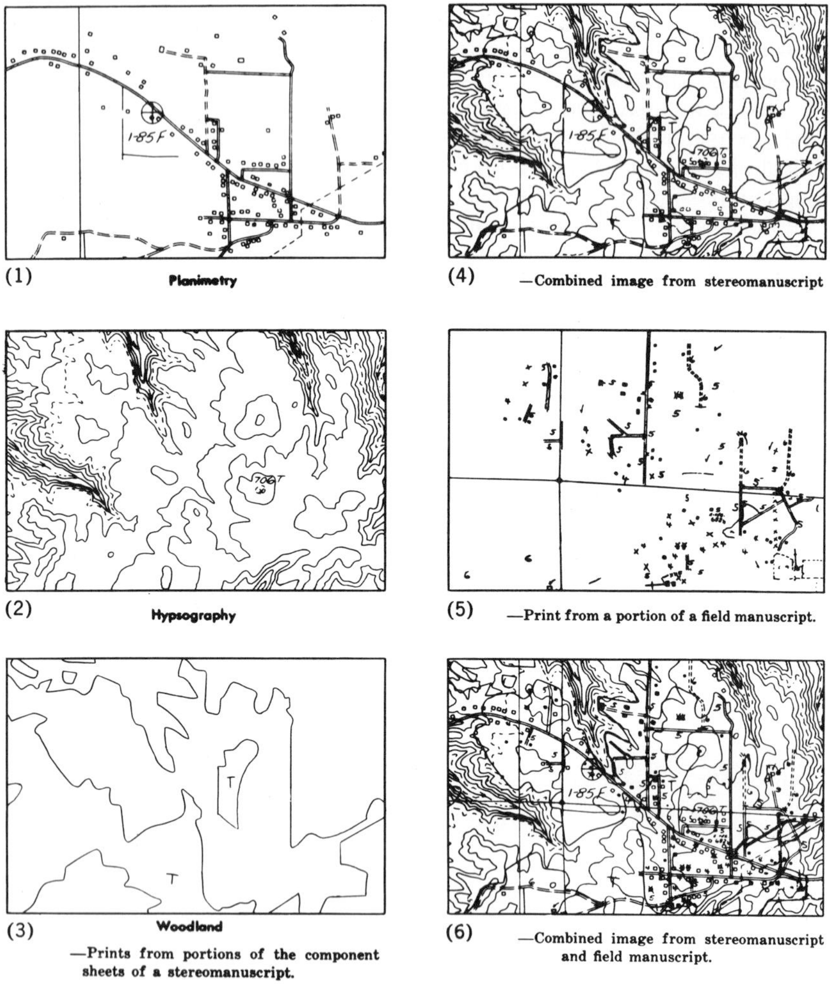

FIG. 31-15. Manuscript composites.

base component by following the model features so that the floating mark is always at ground elevation as the plotting proceeds. Planimetric features are compiled in definite order according to their prominence, as follows: roads, railroads, fence lines, buildings, drainage, shore lines, and utility lines.

The "brown" contour manuscript is then stud-registered to the master planimetric base manuscript, thus retaining the orientation of the base to the stereomodel. For the delineation of contours the floating mark is set at the predetermined contour elevation and a trace of the contour is established by a lateral movement of the tracing table while the floating mark is maintained in constant visual contact with the apparent terrain in the model. Spot photogrammetric elevations are also added at this time to supplement the information provided by the contours.

Woodland is delineated by stud-registering the "woodland" manuscript component to the master planimetric base manuscript and drawing the woodland outline in the same manner as planimetric features are depicted. Usually the planimetric base manuscript component is compiled and scribed first for the area of each stereomodel, and this scribed component is then printed photographically to the other two manuscript components, model by model, so that proper clearances between map features can be maintained when the three manuscripts are overprinted. Figure 31-15, Plates 1 to 4, depicts the three components of the stereomanuscript and the combined image

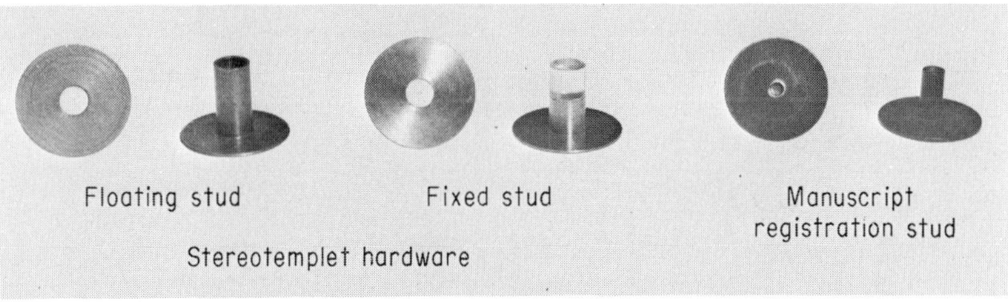

FIG. 31-16. Flanged brass studs.

from the stereomanuscript components. Figure 31-17 shows the various types of scribing tools.

7. *Field completion surveys* are usually accomplished by plane-table mapping. This phase supplies details not obtainable from aerial photographs, such as place names, public-land subdivisions, and road and stream classification. Deficiencies in the photogrammetric compilation are corrected, and a vertical and horizontal test of the map accuracy is accomplished at this time. Extensive areas of recompilation are returned for reworking by photogrammetric methods; smaller areas are completed by the field engineer on the plane-table manuscript. A composite or combined image from the stereomanuscripts is printed to the field board or field manuscript (Fig. 31-15, Plate 4), and field corrections and deletions are depicted on this manuscript (Fig. 31-15, Plate 5). Figure 31-15, Plate 6, shows the combined image from the stereomanuscript and the field manuscript, the various components of which are used for reproduction and finishing phases.

8. Where *reproduction* is to be at manuscript scale and color-separation scribing is to be used, images of the map detail are printed photographically

to sheets of dimensionally stable plastic coated with a thin film of actinically opaque material. Scribe sheets containing such images are called *guides.* On the guide, for each color, the scriber cuts through the coating with precision-ground needles or blades. Guides on which scribing is completed are called *scribed plates.* They resemble photographic negatives and are exposed to sensitized metal pressplates. The plates receive and subsequently print images of only the scribed symbols.

31-18. Planimetric Mapping. The principles of radial-line plotting are commonly applied to the extension of ground-survey control by graphical

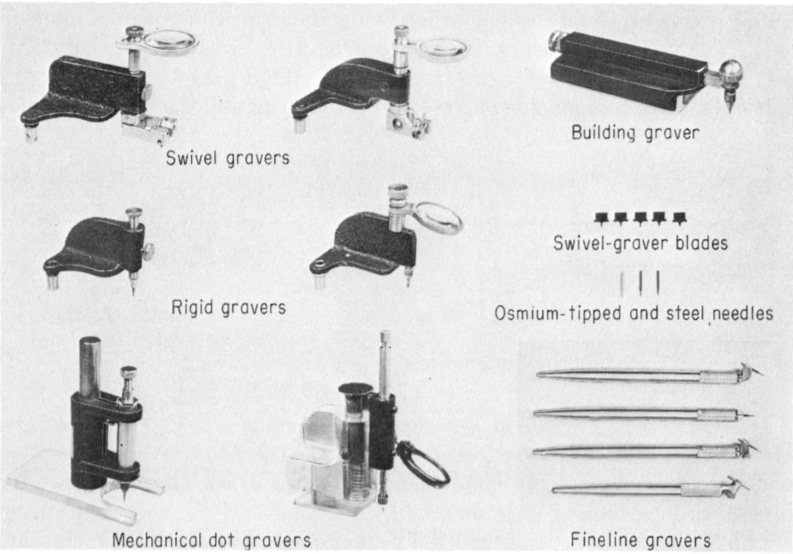

FIG. 31-17. Scribing tools.

methods and to the compilation of planimetric maps from aerial photographs. The planimetric map is a map which presents the horizontal locations only for the features represented; it is distinguished from the topographic map by the omission of relief in measurable form. An example is shown in Fig. 31-18.

Photographs used for radial-line triangulation and mapping are usually designed with 60 per cent forward overlap and 30 per cent side lap between flights. Tilts are usually limited to an average of 1 to 2°. Although not absolutely necessary, design specifications should follow closely the tolerances necessary for topographic mapping with stereoscopic plotting instruments.

Control planning for radial triangulation (radial-line plotting) involves detailed investigation of the amount, distribution, and accuracy of all existing

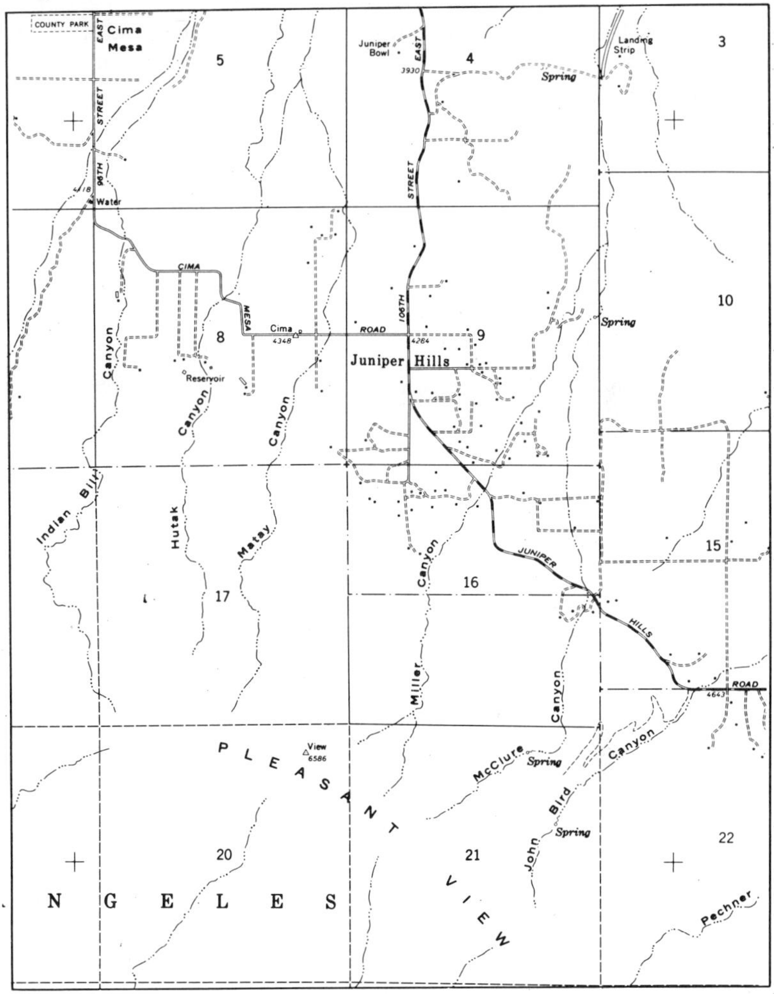

FIG. 31-18. Planimetric map.

control. The number of ground-survey control points required will depend upon the number of photographs in the assembly and the required accuracy and can be determined approximately by the empirical formula

$$C = \left(\frac{k}{e}\right)^2 t$$

in which C represents the number of ground control points required, k is a constant (0.16 from present experience), t represents the number of photographs in the assembly, and e is the allowable average error in millimeters at publication scale (see Ref. 17b at the end of this chapter).

Example: An area is photographed at an average scale of 1/24,000 for planimetric mapping at 1/24,000 publication scale. The area requires a total of 400 photographs arranged in 20 flight lines at 20 pictures per flight line. The average error of position of the radial-line plot is not to exceed $1/200$ in. at publication scale. Determine the number of photocontrol points needed for constructing the radial-line plot extension.

$$e = 1/200 \times 24{,}000 \times 25.4 \doteq 0.127 \text{ mm.}$$

The number of photocontrol points needed to control the mapping is

$$C = \left(\frac{0.16}{0.127}\right)^2 \times 400 = 636 \text{ points}$$

$$636/400 = 1.59$$

corresponding to 1.6 points per photograph or 8 points for each five photographs or 32 points per flight line.

The ideal location of ground control is obtained if the control density is uniform throughout the area. This distribution is not always practical, and a large area is often traversed along its outer boundaries and a few cross traverses added, or if placed by triangulation, a more favorable density may be obtained. Once the additional ground control has been established and all ground control photoidentified, a control base is plotted on a transparent scale-stable medium (mylar or vinyl). The control base consists of the appropriate projection or projections, plotted locations of all ground-survey control points, and additional marginal data.

In radial triangulation, the center of each vertical photograph serves as a station (called the *radial center*) from which directions to points imaged on the photographs are traced or measured and are used to extend the triangulation by intersection and by resection. A minimum of nine photogrammetric control points, including the radial center, are determined for each photograph of the assembly. The general methods of radial-line control extension are covered in Art. 31-35. From these various techniques of radial triangulation, the ground-control network is extended to establish the true map location of at least nine identified points per photograph on the base at the desired scale of the assembly. Beyond this control-extension stage the final accuracy is dependent upon the method of detail transfer, upon the nature of the topography, and upon the difference between the average photography scale and the map scale. The various techniques of compilation from the map base are covered in Art. 31-38.

31-19. Controlled and Uncontrolled Mosaics. A mosaic is an assemblage of aerial photographs the edges of which have been torn or cut and matched to form a continuous photographic reproduction of a portion of the earth's surface (Fig. 31-19). A *controlled mosaic* is defined as a mosaic oriented and scaled to horizontal ground control; usually it is assembled from rectified prints and photographs. A mosaic assembled with little or no ground-

FIG. 31-19. Aerial mosaic. (*Soil Conservation Service, U.S. Department of Agriculture.*)

survey control is termed an *uncontrolled mosaic.* Other mosaics are classified between controlled and uncontrolled and are sometimes termed *semicontrolled mosaics.*

The controlled mosaic will be dealt with here, since it is the most difficult to prepare and its construction techniques cover other less precise types. Photographs taken with a long-focal-length, narrow-angle camera are preferred for mosaic assembly in order to reduce the displacement of images due to relief and to enable procurement of photography without flying at too high an altitude. The amount of forward lap and side lap is also important; the normal requirement is for 60 and 30 per cent, respectively, but these values may be increased if increased accuracy is required.

Control planning for mosaic work involves detailed investigation of the amount, distribution and accuracy of all existing control; in the case of the controlled mosaic, the ground-control requirement necessary for extension by radial triangulation follows the procedures outlined in Arts. 31-18 and 31-35 for the planimetric map. A minimum of nine photogrammetric control points, including the radial center, is determined for each photograph.

Once the required ground control has been established and photoidentified, a control base is prepared on a scale-stable medium such as masonite, plywood, or aluminum. This base consists of the appropriate map projection, plotted location of ground-control points, and subsequently the location of additional photogrammetric control points determined usually by the radial-line method of control extension. The same base can be used for the extension by radial-line plot as that used for the mosaic assembly.

The individual photographs to be used in assembling the mosaic are enlarged or reduced and are rectified. The data furnished the photographic laboratory for the rectification include the enlargement ratio applicable to all photogrammetric control points (to make each print fit the control plot), the azimuth of the axis of tilt, the enlargement factor along the tilt axis, and the tilt angles to be used in setting the rectifier. The rectified prints must be carefully processed so that tones are matched between adjacent prints in order to give the complete assembly the appearance of a continuous photograph.

The prints are mounted on the mosaic control base, starting with the center. The first print is suitably trimmed, adhesive is applied to the back, and the print is oriented to the control, the operator using a pricker point to fit the image to the plot. The second print is then oriented to the control, a cut line is selected to give the best matching of image detail between prints, and the cut is made on the line before the print is pasted down. This procedure is continued until the entire assembly is completed.

Additional features are drafted on the mosaic as required in either black or white ink, depending upon the tone of the mosaic. These features may include grid lines, scale, authorship, north arrows, and names of features. Reproductions are made using a large copy camera; final reproductions are usually at a larger scale than the compilation.

Usually the scale of the final mosaic is larger than the scale of the aerial negatives, the maximum practical enlargement being up to four diameters, and the compilation scale is generally larger than the negative scale and smaller than the final mosaic scale. In planning mosaic photography, it is often desirable to space the photographs and flight lines closer together than the values given; thus only the center portion of the photographs need be used in the assembly, and displacement due to relief is minimized.

31-20. Flight Planning for Aerial Surveys. The latest available base map of the project area is used for flight planning. In order of preference, the planner utilizes (1) county highway maps, (2) U.S. Geological Survey quadrangle maps, (3) Army Map Service quadrangle maps, and (4) photo-indexes, strip maps, or mosaics. Specifications for stereoscopic mapping embody complete stereoscopic coverage of the project area within the usable limits of the lens system of the camera-plotting combination. Standard specifications for aerial mapping photography usually cover allowable flight line and flight height variation, spacing of photographs, allowable crab and tilt, indexing and numbering of negatives, photographic materials to be delivered by the contractor, and stipulations for inspection and payment.

Flight design depends upon the intended use of the photography, whether for mosaic, land-use study, or graphic compilation or for use with stereo-plotting equipment. Other basic considerations are the boundaries of the project; the nature, location, and extent of existing control; and the type of terrain. The discussion herein will be limited to the design of a project utilizing vertical photography taken with a 6-in. (153-mm.) focal-length mapping camera with an effective angle of coverage of 90° and the 9 by 9-in. negative format.

The amount of overlap for two adjacent photographs in the line of flight is usually between 55 and 65 per cent; that is, 60 per cent of the first picture overlaps the coverage of the second picture. For instance, if 5.4 by 9 in. of the first picture, which is 9 by 9 in. in format size, contains the same image area as the second picture with the same size of format, the forward overlap is said to be 60 per cent.

The distance between two adjacent principal points in the same flight will be $9 - (9 \times 0.60) = 3.6$ in. Likewise the distance between the principal points of photographs in adjacent flights is given by the format size perpendicular to the line of flight less the amount of the side lap. Assuming an average side lap of 30 per cent, the distance between two adjacent principal points will be $9 - (9 \times 0.30) = 6.3$ in.

Example: A flight design (Fig. 31-20) for photogrammetric mapping or mosaic work is as follows:

Focal length of camera 6 in.
Scale of desired photographs 1:12,000
Negative format size 9 by 9 in.
Area width 5.5 miles
Area length 10 miles
Direction of flights parallel to long dimension of project

Overlap (in line of flight) 60 per cent
Overlap (transverse to line of flight) 30 per cent
Flight line base map scale 1:12,000
Flight height 6,000 ft.

a. Computation of flight height:

$$\frac{1}{12{,}000} = \text{photoscale} = \frac{\text{focal length}}{\text{altitude}} = \frac{6 \text{ in.}}{6{,}000 \text{ ft.}} = \text{negative scale 1 in.} = 1{,}000 \text{ ft.}$$

Flight height = 6,000 ft.

b. Computation of distance of first flight line from boundary (side), assuming specifications state that 25 per cent of photograph must extend beyond boundary:

$$\text{Negative format size} = 9 \text{ by } 9 \text{ in.; } 25 \text{ per cent of } 9 \text{ in.} = 2.25 \text{ in.}$$

The edge of the gross coverage of the first flight line will be 2.25 in. or 2,250 ft. on the ground beyond the side boundary of the project.

c. Computation of distance between flight lines (30 per cent average side lap): 9 in. × 30 per cent = 2.7 in.; 9 − 2.7 = 6.3 in. or 6,300 ft. effective width covered by each photograph (optimum).

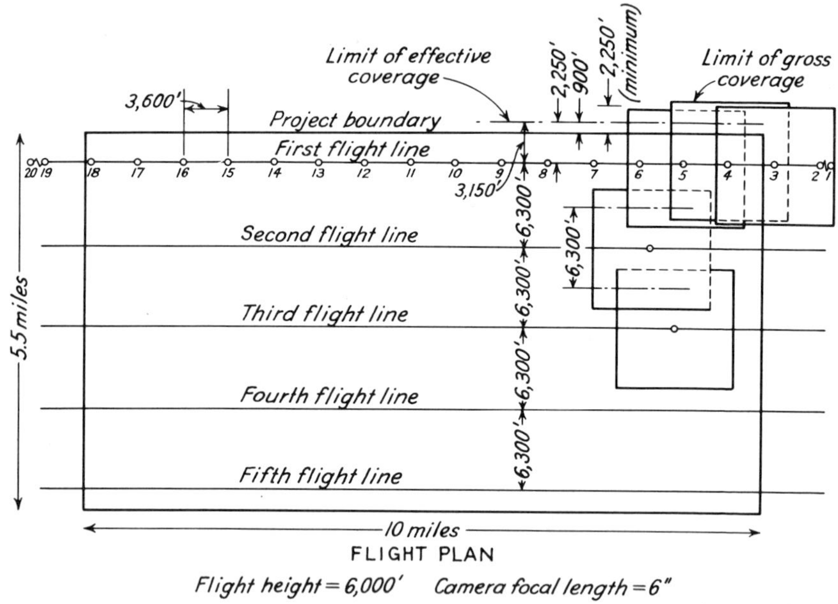

FIG. 31-20. Flight plan.

Thus the second flight line will be 6,300 + 2,250 = 8,550 ft. from the edge of the side boundary of the project, and each succeeding flight line will be 6,300 ft. from the preceding flight line.

d. Compilation of number of flight lines: 5.5 (miles) × 5,280 = 29,040 ft., total width of project. Adding 2,250 ft. at each edge of project for overlap beyond the boundary limit gives 29,040 + (2 × 2,250) = 33,540 ft., total gross width to be covered. Since the first and last flights are to be 2,250 ft. in from the project side boundary, the total effective width between outer flight centers will be 33,540 − (2 × 4,500) = 24,540 ft.

Therefore

$$\text{Number of flight lines} = \frac{\text{total effective width}}{\text{effective width of photograph}}$$

$$= \frac{24{,}540}{6{,}300} = 3.9;\ 3.9 + 1 = 5 \text{ flight lines}$$

e. Compilation of number of exposures per flight line and total number of exposures:

$$\text{Distance between exposures} = 9 - (60 \text{ per cent} \times 9) = 3.6 \text{ in. or } 3{,}600 \text{ ft.}$$
$$\text{Total length of flight line} = 10 \times 5{,}280 = 52{,}800 \text{ ft.}$$
$$\text{Number of photographs} = \frac{52{,}800}{3600} = 14.7 \text{ or } 15 \text{ exposures;}$$
$$15 + 1 = 16 \text{ exposures}$$

It is usually specified that two extra exposures be added at each end of a flight line in order to insure stereoscopic coverage; therefore, the total number of exposures per flight line = 16 + 4 = 20.

$$\text{Total number of exposures for project} = 5 \times 20 = 100$$

f. Calculation of time interval between exposures: The time interval between exposures depends upon the ground speed of the airplane and upon the distance the plane travels between exposures.

The distance between exposures in this problem is 3.6 in. or 3,600 ft. at the 1:12,000 scale of the photograph. Assuming that the aircraft speed is 150 miles per hour and converting that speed to feet per second

$$\frac{150 \times 5{,}280}{60 \times 60} = 220 \text{ ft. per second}$$
$$\text{Required exposure interval} = \frac{3{,}600}{220} = 16.4, \text{ or } 16 \text{ sec.}$$

A modern system of flight design commonly used for photography designed for use in aerotriangulation and stereocompilation utilizing stereoplotters expresses the distances between exposure stations and the distances between adjacent flights in terms of ratios of the distances between camera stations to the flight height.

H is commonly expressed as the flight height above mean ground. (The usual expression of $(H - h)$ is not used in order to simplify the expression.)

B is the distance between exposure stations measured between principal points of the pictures, also called the airbase.

W is the distance between flight strips as measured between the principal points of any two adjacent pictures of adjacent strips.

For vertical photography the B/H ratio is commonly specified as 0.63 ± 0.05 and the W/H ratio is about 1.00.

31-21. Aerial Photography: General. Aerial photography involves the utilization of photographic airplanes, aerial cameras, and accessories in the production of photographs for use in aerial surveying. Vertical photography is used almost exclusively for surveying and mapping purposes (1964), although since 1950 twin low-oblique photography has been used for certain topographic mapping applications (Art. 31-41).

Aerial photography may be classified according to orientation of camera axis (vertical or oblique), according to lens system (single or multiple), or according to special properties (black and white, continuous strip, infrared, color, radar). Vertical photography is taken with the axis of the camera lens as nearly vertical as possible at the instant of exposure, although truly vertical aerial photographs exist only in theory because conditions make absolute verticality a practical impossibility.

Oblique photography is taken with the axis of the camera lens intentionally tilted from the vertical. High obliques are those which show a portion of the horizon; low obliques do not.

31-22. Relationship of Scale, Focal Length, and Flight Altitude (Vertical Photography). In Fig. 31-21, representing truly vertical photography, let f be the focal length of the camera exposed at flight altitude H above a sea-

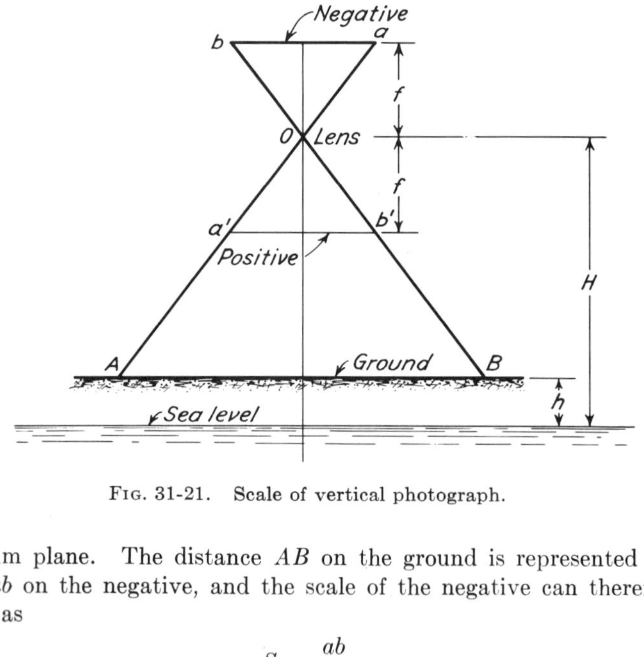

FIG. 31-21. Scale of vertical photograph.

level datum plane. The distance AB on the ground is represented by the distance ab on the negative, and the scale of the negative can therefore be expressed as

$$S = \frac{ab}{AB}$$

From similar triangles

$$\frac{ab}{AB} = \frac{f}{H - h}; \qquad S = \frac{f}{H - h}$$

In a positive print made by contact printing from the negative,

$$a'b' = ab$$

The scale of the photograph is customarily expressed as a fraction (the representative fraction) whose numerator is 1. Thus if $f = 6$ in. (0.50 ft.) and $H = 6{,}000$ ft.,

$$S = \frac{0.50}{6{,}000} = \frac{1}{12{,}000}$$

This scale can also be expressed as a ratio, 1:12,000, or as an equivalent, 1 in. = 1,000 ft.

If f and H are not known, the scale of the photograph can be calculated by first determining a distance such as AB, by scaling from an existing map or by measurement on the ground. The corresponding distance $a'b'$ is measured on the photograph, and the scale becomes

$$S = \frac{a'b'}{AB}$$

In Fig. 31-22 the effect of different focal lengths, when the flight altitude and negative format remain the same, is illustrated:

If H equals 10,000 ft., the negative scale is determined as follows: When f equals 3½ in.,

$$S = \frac{3\frac{1}{2}\text{ (in.)}}{10{,}000\text{ (ft.)}} = \frac{3\frac{1}{2}\text{ (in.)}}{120{,}000\text{ (in.)}} = \frac{1}{34{,}300} \text{ or 1 in. to 2,860 ft. or 1:34,300}$$

When f equals 6 in.,

$$S = \frac{6\text{ (in.)}}{10{,}000\text{ (ft.)}} = \frac{6\text{ (in.)}}{120{,}000\text{ (in.)}} = \frac{1}{20{,}000} \text{ or 1 in. to 1,666 ft. or 1:20,000}$$

When f equals 12 in.,

$$S = \frac{12\text{ (in.)}}{10{,}000\text{ (ft.)}} = \frac{12\text{ (in.)}}{120{,}000\text{ (in.)}} = \frac{1}{10{,}000} \text{ or 1 in. to 833 ft. or 1:10,000}$$

The standard negative format in the United States is 9 by 9 in. (23 by 23 cm.). It is usually combined with the 6-in. (153-mm.) focal-length camera and an angle of usable coverage of 90°. Other focal lengths are 3½, 8¼, and 12 in. For forestry or military applications, focal lengths of 24, 36, and 48 in. are also used.

The scale of a photograph can be computed from the relationship of the flight altitude to the focal length of the camera as well as from the relationship of a measured distance on the photograph to the corresponding distance on the ground. In Fig. 31-21, from the similarity of triangles

$$\frac{f}{H - h} = \frac{a'b'}{AB}$$

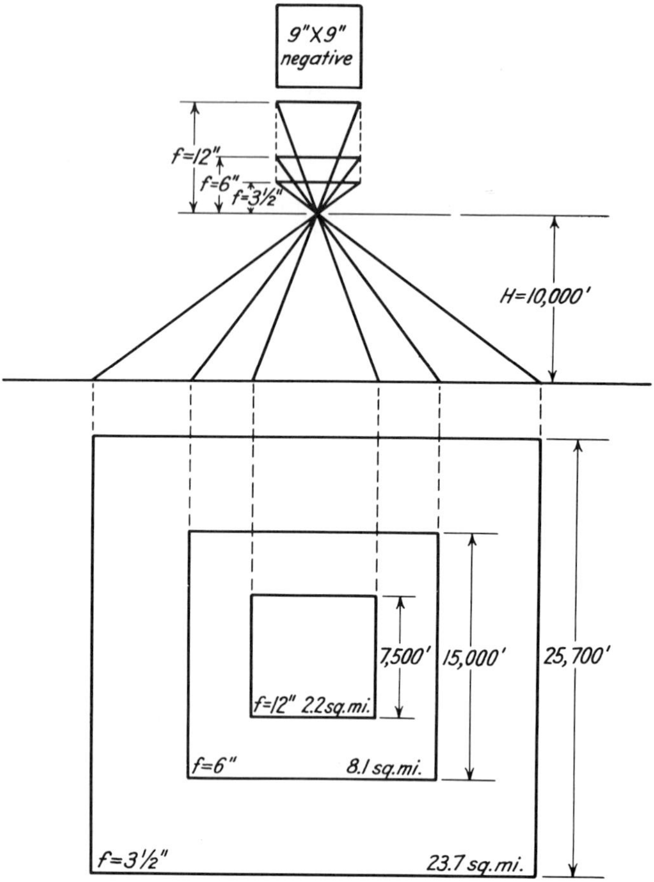

FIG. 31-22. Effect of focal length on ground coverage.

But

$$\text{Scale of photograph} = S = \frac{a'b'}{AB} = \frac{f}{H - h}$$

in which H = height of camera lens above sea level
h = average elevation of ground points above sea level
f = focal length of the aerial camera

This equation can also be expressed by the following:

$$S = \frac{1}{\dfrac{AB}{ab}} = \frac{1}{\dfrac{H - h}{f}} = \frac{f}{H - h}$$

It should be noted that the scale depends upon the elevation h of the ground; thus $(H - h)$ is not constant for usual terrain. Therefore the expression is exact only if the elevations of the ground points are exactly equal and if the photograph is not tilted. More precise methods of computation of the value of $(H - h)$ are detailed in Ref. 1 at the end of this chapter.

31-23. Determination of Flight Altitude. The flight altitude necessary to obtain photographs of the desired scale is computed by the same equation $H - h = (f \times AB)/ab$. In this case $(H - h)$ is the altitude above the mean datum of the area photographed, which datum is h ft. above mean sea level. As the ground rises and falls throughout a strip of photographs, there is usually a difference in scale between successive photographs. The altitude H above sea level is not varied to allow for the difference in the elevation of the ground, but the scales of the photographs are changed by enlargement or reduction if necessary to secure the intended result. The method of determining H is as follows, wherein it is assumed that it is desired to find the flight altitude to obtain photographs with a mean scale of 1 in. = 1,000 ft., or $S = 1{:}\,12{,}000$. If the ground elevations vary from 500 to 1,500 ft. above sea level in the area photographed, and if the focal length of the camera is 12 in. (1 ft.), then

$$H - h = \frac{f}{S} \tag{3}$$

$$H - \frac{500 + 1{,}500}{2} = \frac{1}{\dfrac{1}{12{,}000}} = 12{,}000$$

or

$$H = 13{,}000 \text{ ft. above mean sea level}$$

As a matter of convenience to the flight personnel, it is customary to express H as the true altitude above sea level instead of considering $(H - h)$ as the altitude above the ground. The flight personnel are concerned with the barometric altitude of flight as indicated by the altimeter of the airplane.

31-24. Flight Altitude Design for Topographic Mapping. For a particular photogrammetric system, the flight altitude for aerial photography is designed to consider (1) the contour interval and the limiting C factor, (2) the relationship of the flight altitude to the stereomodel scale of the plotting instrument, and (3) the limitations of the pantograph or coordinatograph attached to the stereoplotting instrument. When a plotting system is to be used for compiling contours at a given contour interval, the flight altitude is determined approximately from an empirical relationship called the C factor. For a given plotting system the C factor $= H/CI$, in which CI is the least contour interval accurately plottable. (The expression H is used for simplicity; the correct expression is $H - h$.) Commonly used values of the C factor range from 600 to 1,500 depending upon the complexity of the stereoplotting instrument. The best criterion for actual design limitations for the given photogrammetric system is actual design experience. For example, maximum design flight heights for the 10-ft. contour interval for two typical photogrammetric systems will be:

For a C factor of 600,

$$H = 600 \times 10 = 6{,}000 \text{ ft.}$$

For a C factor of 1,000,

$$H = 1{,}000 \times 10 = 10{,}000 \text{ ft.}$$

It is possible using the C-factor criterion to set a flight height beyond the limits for accurate and complete delineation of planimetric detail. In the eastern United States, this limitation is considered to be about 15,000 ft.; in the western United States it is considerably higher because of more favorable atmospheric conditions.

For direct optical-projection instruments such as Balplex or Kelsh, the flight altitude H is related to the scale of the stereomodel S by the formula

$$H \text{ (ft.)} = \frac{dS'}{304.8}$$

in which d is the optimum projection distance, in millimeters, of the plotting instrument, and S' is the denominator term of the scale of the stereoscopic model when it is expressed as a representative fraction $1/S'$. (See page 261 of Ref. 1 at the end of this chapter.)

For example, assume that an instrument of the double-projection type, having an optimum projection distance of 525 mm., is to be used; then

$$H = \frac{525S'}{304.8} = 1.72S'$$

If the flight altitude determined by the C-factor relationship is 6,000 ft. as previously derived, then $S' = 6{,}000/1.72 = 3{,}490$ or 3,500 ft. A more convenient value of $S' = 3{,}600$ ft. (1 in. to 300 ft.) is then chosen, and the flight height is modified to $H = 1.72 \times 3{,}600 = 6{,}192$ or 6,200 ft.; the working C factor becomes $6{,}200/10 = 620$.

31-25. Specifications for Aerial Photography. For mosaic construction three critical factors should be specified: camera focal length, flying height, and overlap and side-lap limitations of the photography. With this application, the requirements for a precision mapping camera are not critical. However, for use in stereoplotting instruments for topographic mapping, aerial-photography specifications should cover also the camera calibration characteristics, camera lens and format data, permissible flight line and flight-height deviations, spacing of photographs, and stereoscopic coverage requirements.

A deviation of 5 per cent of the flight altitude referred to mean ground elevation is usually permitted, and an average allowable tilt of 1° from the vertical with a maximum limit of 4° for any one photograph is usually specified. A forward overlap of 60 ± 5 per cent is commonly specified, and this relationship can also be expressed by stating that the base-height (B/H) ratio shall be 0.63 ± 0.05 providing that the wide-angle camera lens is used. The minimum side-lap overlap is usually held to 20 per cent. This spacing of flight lines can also be expressed by stating that the maximum width-height (W/H) ratio shall be 1.20 for the wide-angle camera.

Aerial photographs should not be taken when the sun is less than 3 hr. above the horizon (or, as an alternative specification, when the solar altitude is less than 30°). The best season of the year for photography depends upon the ultimate use of the film, but usually the most desirable time is when the deciduous trees are bare of leaves or in some cases where there is a light fall of snow on the ground. Certain local climatic conditions such as haze, smog, high winds, and air turbulence should also be minimized.

31-26. Aerial Cameras. Aerial cameras may be of the single-lens or the multiple-lens type. Single-lens cameras may be classified as general-purpose cameras and precise, or photogrammetric, cameras. Multiple-lens cameras are usually classified according to the number and arrangement of the lenses; they are precise cameras.

The two major factors in the selection of an aerial mapping camera, apart from the usual specifications for optical and mechanical excellence, are the angle of usable coverage and the focal length. Optical criteria for lens evaluation also include *definition*, defined as the ability to record fine detail; *distortion*, defined as geometric displacement of images from their true perspective position; and *even illumination* across the angular field of the lens. High definition, low distortion, and even illumination must be achieved with large aperture and wide angular field. Optimum achievement of these factors represents a compromise.

Camera lenses are classified according to angular field of view as normal angle, up to 75°; wide angle, from 75 to 100°; and superwide angle, over 100°. The standard negative format in the United States is 9 by 9 in. (23 by 23 cm.). Following are some widely used camera characteristics:

Classification	Format, cm.	Focal length, in.	Diagonal field, deg.
Normal angle	23 by 23	8¼ (210 mm.)	75
Wide angle	23 by 23	6 (153 mm.)	93
Superwide angle	23 by 23	3½ (88 mm.)	122

Other functional components in addition to the lens are (1) the shutter, (2) the principal-point calibration markers, (3) film storage (magazine), (4) film drive, and (5) film plane. The between-the-lens type of shutter is almost universally used in modern mapping cameras.

Multicamera systems are also used in stereoscopic mapping in order to increase the base-height ratio of the various exposures and thus enhance the accuracy of the various measurements. The Zeiss convergent camera (Fig. 31-23) consists of two normal-angle cameras combined in a single suspension mount with an angle of convergence of 27°. For low-oblique

FIG. 31-23. Zeiss convergent camera system.

photography the U.S. Geological Survey uses a system of twin wide-angle cameras combined in a special camera mount at 20° to the vertical and 40° to each other (Fig. 31-24).

31-27. Single-lens Aerial Cameras. Three of the better known aerial camera systems of the single-lens wide-angle type are the Fairchild aerial cartographic system, model FS-500 (Art. 31-28); the Zeiss RMK 15/23 (Art. 31-29); and the Wild RC-8 (Art. 31-30). The Wild RC-9 (Art. 31-30) is a newer type of camera of the single-lens superwide-angle type.

31-28. Fairchild Aerial Cartographic System. The Fairchild aerial cartographic system, model FS-500, is a precise photogrammetric camera system including the camera, suspension mount, view finder, and intervalometer (Fig. 31-25). The camera is equipped with a selected 6-in. $f/6.3$ improved

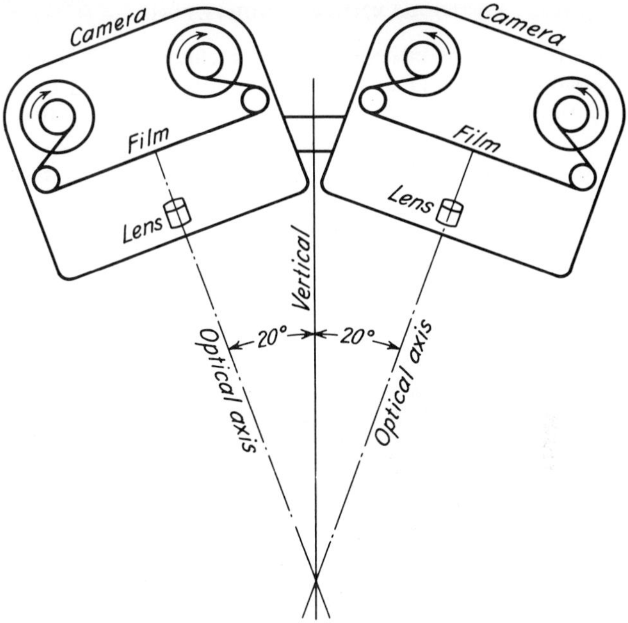

Fig. 31-24. U.S. Geological Survey camera system for twin low-oblique photography.

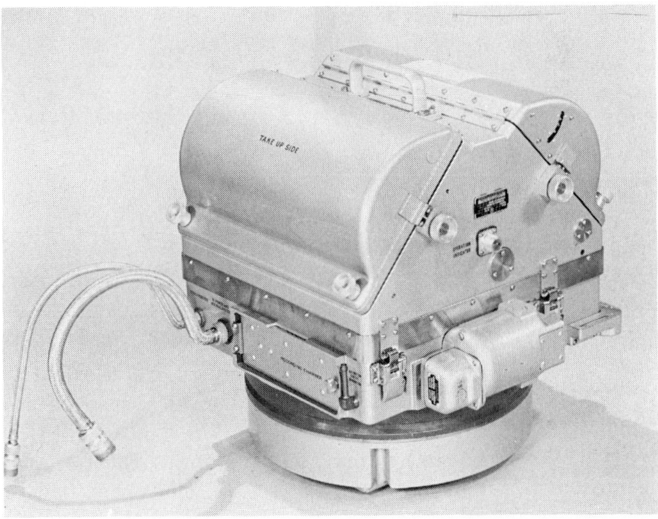

Fig. 31-25. Fairchild FS-500 camera system. (*Fairchild Space and Defense Systems.*)

planigon lens. It has a one-piece inner cone, including both the lens mount and the focal-plane frame against which the film is positioned by the magazine platen during exposure. It is equipped with the Rapidyne shutter, which can be removed from the camera without disturbing the lens mounting. Shutter speeds of 1/10 to 1/55 sec. are possible, and exposure control is provided by an adjustable diaphragm (*f*/6.3 to *f*/22) with external control; thus the operator can increase or decrease the exposure during flight. Fiducial marks, film-shrinkage reference marks, altimeter, clock, exposure number, camera number, data card, lens serial number, and calibrated focal length of the lens are recorded on each negative. The special magazine is detach-

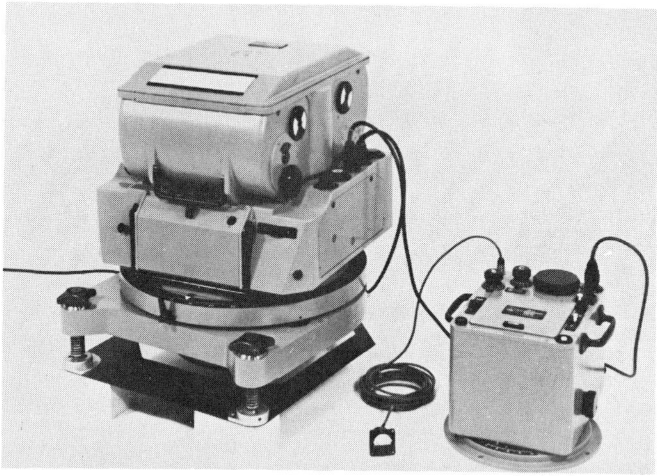

FIG. 31-26. Zeiss RMK A 15/23 camera. (*Zeiss-Aerotopograph, Munich.*)

able and holds 390 ft. of roll film 9½ in. wide, and the electrical system is operated by a 24 to 28-volt d-c system. The filter is antivignetting, yellow or clear. The camera weighs 75 lb.

31-29. Zeiss Wide-angle Survey Camera. Zeiss-Aerotopograph provides a complete group of photogrammetric cameras using either the Zeiss Topar (normal-angle) or the Zeiss Pleogon (wide-angle) lens. In the United States the wide-angle Zeiss mapping camera most widely used is the RMK 15/23 (Fig. 31-26). This camera is furnished with the Universal suspension mount and intervalometer. It uses the Zeiss Pleogon 6-in. (153-mm.) *f*/5.6 lens and the 9 by 9-in. (23 by 23-cm.) format, and the optics, shutter, and focal-plane frame are combined in one unit. The exterior camera body houses the drive motors for the shutter, the film advance, and the vacuum pump of the film-flattening system. Recording instruments for the auxiliary flight data are mounted to one side of the camera housing, and the photograph number and the fiducial markers are also recorded on the negative. Color filters as well as infrared filters are available, and shutter speeds of $\frac{1}{100}$ to $\frac{1}{1,000}$ sec. are possible. The detachable magazine holds 400 ft. of film.

31-30. Wild Wide-angle and Superwide-angle Cameras. Wild Heerbrugg, Ltd., provides a group of cameras using the Wild Aviotar (normal-angle), the Wild Aviogon and Infragon (wide-angle), or the Wild Super Aviogon and

FIG. 31-27. Wild RC-8 camera. (*Aero Service Corporation.*)

Infragon (superwide-angle) lens. The Wild RC-8 camera (Fig. 31-27) is widely used in the United States; four interchangeable lens cones are available with focal lengths of $4\frac{1}{2}$, 6, and $8\frac{1}{2}$ in. The wide-angle lens cone with the Aviogon $f/5.6$ lens, $f = 6$ in. (153 mm.), and 9 by 9-in. (23 by 23-cm.) format is widely used for aerial photography to be used in stereoplotting machines. Fiducial marks in the corners of the picture are recorded within the format, and other auxiliary flight data are recorded on the negative outside the format. Shutter speeds of $\frac{1}{700}$ to $\frac{1}{100}$ sec. are possible, and interchangeable filters and an adjustable iris diaphragm are provided. Two film magazines with

200 ft. of film each are provided. A special wide-angle lens cone with the Infragon lens, $f/5.6$, $f = 6$ in. (153 mm.), format 9 by 9 in. (23 by 23 cm.), is available for infrared photography.

The Wild RC-9 Super Wide Angle camera (Fig. 31-28) is a newer type of camera designed for small-scale photogrammetric mapping. It is furnished with the Super Aviogon lens, $f = 3\frac{1}{2}$ in. (88 mm.), and the 9 by 9-in. (23 by

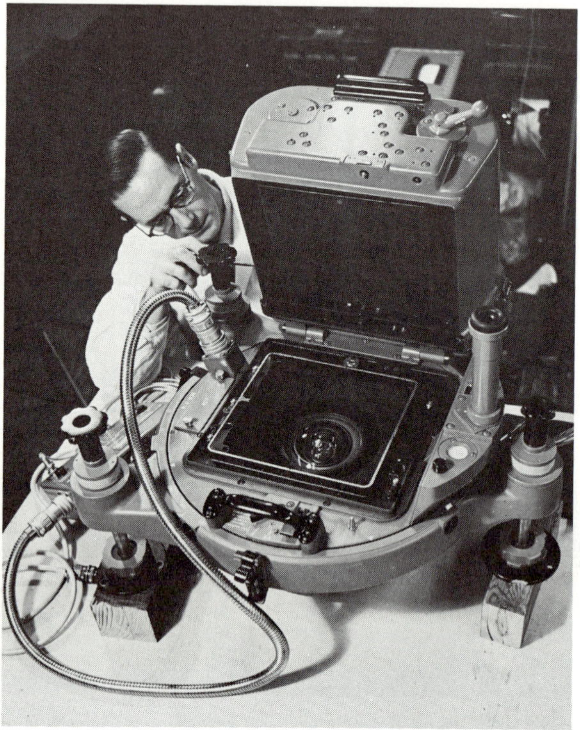

FIG. 31-28. Wild RC-9 camera. (*Aero Service Corporation.*)

23-cm.) format with provision for infrared photography if desired. This type of photography can be used in the Wild A-9 autograph for both aerotriangulation and plotting and in the B-8 and B-9 Aviograph for plotting. The main advantage of this system is the greater area coverage (three times) of the superwide-angle photograph as compared with the wide-angle photograph. A saving in density of ground control as well as in photogrammetric costs for a given area and flight height is possible with this photogrammetric system as compared with wide-angle camera systems.

31-31. Multiple-lens Camera Systems. Multiple-lens camera systems consist of two to nine lenses placed in a single camera, all of which form images simultaneously either on separate negatives or on one master negative. The success of such lens systems is largely dependent upon the accuracy with which the pictures can be rectified and transformed into a single composite photograph. Two camera systems fall into this category: the trimetrogon camera and the U.S. Coast and Geodetic Survey nine-lens camera.

FIG. 31-29. Trimetrogon camera. (*U.S. Air Force.*)

31-32. Trimetrogon Camera. Trimetrogon aerial photography is used for the preparation of small-scale maps and charts of reconnaissance accuracy. Originally the photographs were obtained by the simultaneous exposure of the film of three separate single-lens aerial cameras fitted with metrogon lenses and suitably mounted in the airplane, from which the system gets its name, but more recently the trimetrogon aerial camera shown in Fig. 31-29 has been developed. Contact prints of both vertical and oblique photographs are ordinarily used for charting, although oblique photographs are sometimes rectified in an oblique printer. Compilation of data is accomplished by using the vertical and oblique sketch-master or the Rectoblique plotter (Fig. 31-39). Horizontal control is ordinarily extended by the radial-line method or by means of one of the mechanical adaptations of this method.

31-33. U.S. Coast and Geodetic Survey Nine-lens Camera. To avoid the necessity for using nine separate rolls of film as in the tandem T-3A aerial camera and for other reasons, the U.S. Coast and Geodetic Survey developed the nine-lens aerial camera shown in Fig. 31-30. This camera uses nine lenses, all of which point vertically downward. Eight of them are mounted in a circle around one central lens which views the ground directly. Rays from

FIG. 31-30. U.S. Coast and Geodetic Survey nine-lens aerial camera.

the ground are reflected through the other eight from highly polished steel mirrors. The lenses are of 8¼-in. focal length, and all project their images onto a single roll of film 24 in. wide. The photographs are printed onto a single paper by means of a rectifying printer (Fig. 31-31) to form a nine-lens composite photograph approximately 36 in. square.

31-34. Map Control. Maps prepared by methods of photogrammetric surveying are based upon two kinds of control: field or ground control, which may be obtained in a variety of ways, and secondary or photocontrol, which is generally obtained by phototriangulation or aerotriangulation but sometimes is obtained in the field. Ground control is classified in four orders of precision, as follows:

ORDERS OF GROUND CONTROL

Order	Maximum error in horizontal location	Maximum error in elevation, ft.
First.......	1/25,000	$0.017 \sqrt{\text{length of line in miles}}$
Second.....	1/10,000	$0.035 \sqrt{\text{length of line in miles}}$
Third......	1/5,000	$0.050 \sqrt{\text{length of line in miles}}$
Fourth.....	1/2,000 (approx.)	1

Fourth-order ground control is usually referred to as *secondary* control but is sometimes referred to as *picture-point* control or *supplemental* control.

FIG. 31-31. U.S. Coast and Geodetic Survey rectifying printer.

A control point obtained for photogrammetric surveys is also called a photo-control point, a picture-control point, or a ground-control point.

First-order control is established by the U.S. Coast and Geodetic Survey, and second-order and third-order control by the U.S. Geological Survey, the Corps of Engineers, and other mapping agencies of the Federal government, states, municipalities, or other surveying agencies. The locations and elevations for first, second, and third-order control are permanently monumented and published. The published locations and elevations of the control can usually be obtained by inquiry addressed to the head of the department or bureau which established the control. Such control should be used whenever possible, and it is excellent practice to tie all surveys to as much of this control as may be in the vicinity. Fourth-order ground-control locations are not usually published, nor are the stations permanently monumented.

Secondary or supplemental control is the control upon which the actual survey is based. It is separate from and additional to the field control of third or higher order, which is referred to as *primary* control. In photogrammetric surveying, secondary control may be said to tie the pictures to the ground. Secondary control (except for fourth-order ground control) may be established by radial triangulation, stereotemplet triangulation, or stereotriangulation; these processes are defined under aerotriangulation or aerial triangulation.

If a controlled mosaic is to be assembled, a group of photoimages the horizontal locations of which are known is required for orientation of each photograph, but these necessary points can usually be obtained by photogrammetric methods and usually little vertical control is required. Topographic photogrammetric mapping with stereoplotting instruments requires that two supplemental points for horizontal control and four supplemental points for vertical control be furnished in the overlap area of each successive pair of photographs. These supplemental points can be obtained either by ground methods or by one or a combination of the following methods of aerotriangulation:

1. Stereotriangulation methods
   a. Horizontal bridging
   b. Vertical bridging
2. Stereotemplet triangulation
3. Analytical methods
4. Graphical radial triangulation
   a. Radial line
   b. Slotted templet
   c. Spider templets

Secondary-control (photocontrol) density to be placed by ground-survey methods depends upon the scale of the map, the contour interval, the photogrammetric instrumentation, and the method of stereocompilation to be used. The objective of the supplemental control planning is to enable the map compiler to delineate map details within the limits of accuracy required by the map specifications. The National Map Accuracy Standards define limits for horizontal and vertical map accuracy as follows:

1. Horizontal accuracy: For maps on publication scales larger than 1:20,000, not more than 10 per cent of the points tested shall be in error by more than $\frac{1}{30}$ in. measured on the publication scale; for maps on publication scales of 1:20,000 or smaller, $\frac{1}{50}$ in. These limitations apply to well-defined points only.

2. Vertical accuracy, as applied to contour maps on all publication scales, should be such that not more than 10 per cent of the elevations tested shall be in error more than one-half the contour interval. The apparent vertical error in checking elevations taken from the map may be decreased by assuming a horizontal displacement within the permissible error for a map of that scale.

The foregoing criteria are generally applied to specifications for all reconnaissance mapping, but with the addition of a "brush clause." This clause recognizes that map accuracy generally suffers when the ground is obscured and permits a reduced accuracy in such areas.

For attainment of these standards of accuracy in the compilation it is common to set allowable tolerances for the placement of supplemental ground control. For topographic mapping the allowable error in elevation for supplemental vertical control is usually one-tenth the contour interval; for supplemental horizontal control the horizontal points are to appear on the final map without error greater than $\frac{1}{200}$ in.

Stereotriangulation Methods. Stereotriangulation methods utilize a stereoscopic instrument to establish both horizontal and vertical control data by means of the successive orientation of the stereoscopic pairs of photographs in a continuous strip. Orientation of the initial model is by reference to ground control established by survey methods.

1. Horizontal bridging. The photogrammetric extension of horizontal control to obtain supplemental horizontal control points (secondary control) or "pass points" may be accomplished with the multiplex, the Balplex, or any of the universal types of heavy plotters of European manufacture such as the Wild autograph A-7 or the Zeiss stereoplanigraph C-8 (Art. 31-48). Several methods are possible with the Balplex projector, and a widely used technique is to support a series of projectors utilizing vertical photography on a reinforced long bar or a supported frame. The projectors are brought to a common relative orientation and adjusted to a common scale within the limits imposed by the allowable variation in projection distance afforded by the projectors.

Low-oblique photography, both convergent and transverse, can also be bridged in the same manner with the Twinplex plotter developed by the U.S. Geological Survey (Art. 31-41).

2. Vertical bridging. Supplemental elevations to control the vertical datum of individual stereomodels may be established by a photogrammetric bridge of vertical photographs in multiplex, Balplex, or European instruments of the universal plotter type such as the Wild autograph A-7 or the Zeiss stereoplanigraph C-8. Low-oblique photography can also be vertically bridged by the Twinplex plotter as well as the Wild autograph A-7 and the Zeiss stereoplanigraph C-8 (Art. 31-48).

FIG. 31-32. Stereocomparator (O.M.I.-Nistri TA3). (*O.M.I. Corporation of America.*)

Stereotemplet Triangulation. Aerotriangulation with stereotemplets is a more recent method of achieving precise area horizontal scale solutions through photogrammetric mapping procedures. A stereotemplet (Fig. 31-13) is basically a composite slotted templet mechanically representing the horizontal plot of a stereoscopic optical model formed in a stereoscopic plotting instrument. To develop control for an area covered by a number of models, the stereotemplets representing the various models are adjusted to the desired common scale when the templets are assembled to satisfy horizontal control (Fig. 31-14).

Analytical Aerotriangulation. Purely analytical methods of aerotriangulation use as basic data the rectangular coordinates, measured in a stereocomparator (Fig. 31-32), of a number of selected points in the aerial photographs; on the basis of these, by comparison of distances or angles in successive photographs or in the photograph and the map, the coordinates of the nadirs of the individual photographs and hence their position in space is determined. Analytical methods are gaining appreciably in importance owing to the progress made in the development and utilization of electronic computers.

31-35. Radial-line Method. The radial-line method of providing secondary map control from vertical aerial photographs is based upon the following perspective properties of such photographs: (1) that points near the center of a photograph are nearly free from errors of tilt; (2) that all errors due to small amounts of tilt and to differences of ground elevation are, within the limits of graphical measurement, radial from the principal point of each photograph; and (3) that objects included in properly overlapping photographs may be located by rays drawn to them from the principal points of photographs, the location of the objects being at the intersection of such lines. As in plane-table operations, the locations of points on aerial photographs are found at the intersections of rays drawn from two or more stations.

Marking of Photographs. In applying the radial-line method, a group of several consecutive photographs which include at least two ground-control points—one near each end of the group—is selected. Upon each photograph certain points (objects) are selected and marked and lines are drawn, as follows:

1. The principal point of each photograph is plotted.
2. Beginning with the first photograph of the group, as, for example, 51, Fig. 31-33, a definite object called the "substitute center" (51*M*), which also appears on the adjoining photograph (52), is chosen near the principal point 51*C* and is marked on both photographs.
3. Points 51*R* and 51*L* are chosen at objects near the right and left edges, respectively, of the photograph, which objects also appear on photograph 52.
4. Similarly, points 52*R* and 52*L* are chosen at objects near the lower corners of photograph 51, and these two points must appear on the two succeeding photographs (52 and 53) opposite the center of 52 and near the upper corners of 53.
5. 52*M* is chosen as a point which appears in photograph 52, near its center and also in photographs 51 and 53, near the lower edge of 51 and near the upper edge of 53.

This procedure of selecting and marking points having been carried through the group of photographs, it will be seen that each photograph of the group, except the first and last photographs, will have nine marked points in addition to the principal point. The first and last photographs will each have six marked points and the principal point as shown on 51 of Fig. 31-33, and the two photographs at each end of the group must also include at least one control point.

Compilation of Control. The method of combining the data of the separate photographs into a map showing the correct relative locations of the selected points and the two control points is as follows:

A sheet of transparent film base (cellulose acetate) large enough to span the

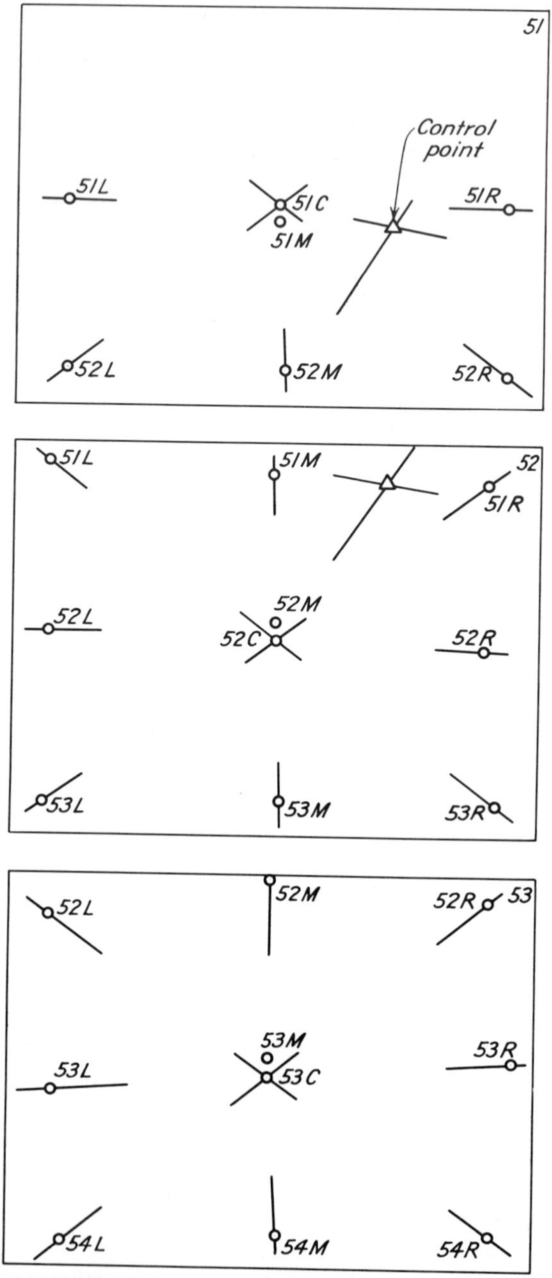

FIG. 31-33. Photographs marked for radial-line control.

entire group of photographs when matched together is laid over the first photograph of the series, and it is so placed that each photograph in turn can be laid under the film base in correct relation. Three points, namely, the principal point 51*C* and the points 51*M* and 52*M* are traced. The radial line to each of the other points, including the control point, is also traced. Photograph 51 is removed, the film base is placed over photograph 52 so that the traced position of 52*M* falls on its position as it appears on photograph 52, and the film base is swung about this point until the traced position of 51*M* falls on its radial line on the photograph. The film base is now held with weights in this correctly oriented position. Radial lines to the other points are

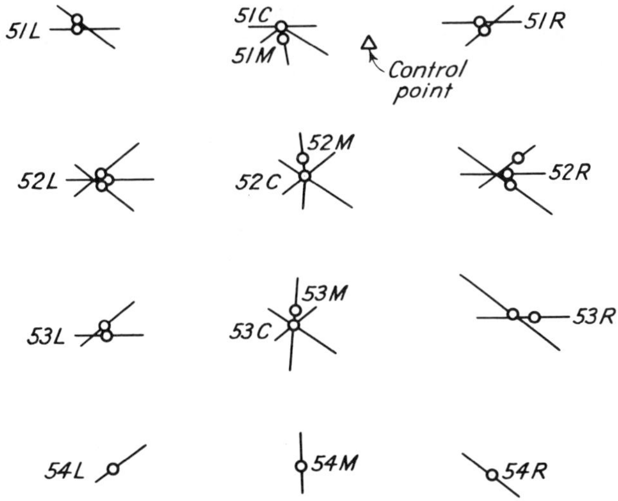

FIG. 31-34. Plotted map-control points.

traced. These lines will give the location of all marked points that appear on photograph 51 and radial lines to three other points (53*L*, 53*M*, and 53*R*) which appear on photograph 52.

The location of a point thus determined may or may not coincide with its pictured location on either photograph, depending upon the tilt and relief displacements in the photograph. If the pictures are without errors due to tilt, etc., the pictured location of each point should lie at the intersection of the two lines drawn toward it, but if displacement errors are present the pictured location of any point will not coincide with the intersection of the radial lines. This condition is shown in Fig. 31-34, in which the pictured locations of the various points are shown by small circles and the correct locations are shown by the intersections of the corresponding radial lines.

When the location of points on the film base is begun, no attempt is made to use any particular scale, but a definite scale, as yet unknown, is established by the distance between the two central points, 51*M* and 52*M*, which have been traced as they appear on photograph 51. It is essential to accuracy that each central point, such as 51*M* or 52*M*, be selected as a point which lies close to the line connecting adjacent principal points and which lies close to its corresponding principal point and that the control point fall within the zone of overlap so that it can be intersected. The scale of the plotting in this case is the scale of the first photograph. As the procedure is continued it is evident that, as each succeeding photograph is treated, there have been located on the film base three points which appear along the upper border of the photograph and a direction line to the next central point. The film base is shifted until (1) the plotted locations of the three points appearing along the upper border of the photograph fall on their radial lines of the photograph and (2) the radial line on the film base to the principal point of the photograph now being treated falls on the point as marked on that photograph. When these conditions are satisfied, the film base is correctly placed with respect to the photograph; the principal point is then traced, and radial lines are drawn. This procedure is similar to the graphical solution of the three-point problem (Art. 17-14).

This process is carried through the group of photographs until the second control point is located. The scale of the data assembled on the film base can then be determined by measuring the distance between the two control points on the film base. Proper enlargement or reduction can be accomplished with the pantograph or graphically.

Plotting at Fixed Scale. The more general application of the radial-line method presumes at least three control points in the overlap of the first pair of photographs. With a minimum of three control points in the first photograph, the radial-line plotting may be run at a fixed scale which may be selected at will; however, it should be very near the mean scale of the strip of photographs. In plotting by the radial-line method to such a fixed scale, the following procedure should be followed:

1. Mark the control points on the photographs and draw rays from the central point (either principal point or substitute center) through the control points as for the points in the previous explanation.
2. Continue marking points and drawing rays throughout the strip until the next control point is reached.
3. Plot the control points on the film base at the desired arbitrary scale.
4. Place the film base, so marked, over the first photograph containing the control points, and orient the film base until the rays on the first photograph pass through the plotted locations of the control points.
5. Trace the rays onto the film base and continue as before.

If upon reaching the next control point there is a difference between its resected location and its plotted location, a graphical adjustment may be applied. In a long extension a slight adjustment may be expected, as the short base (the distance between the control points in the first photographs) has been expanded along the strip and any error in the adjustment of the first photograph has been magnified.

Topographical Features. The radial-line method is used even more extensively in the determination of the true locations of topographic features to be depicted on the final map. This application usually follows the adjustment of the radial-line strip in order to avoid shifting the locations of more than the minimum number of points. If control is in every photograph, the radial-line method of intersection may be used to "cut in" additional points, which operation can be done more expeditiously in the drafting room than in the field.

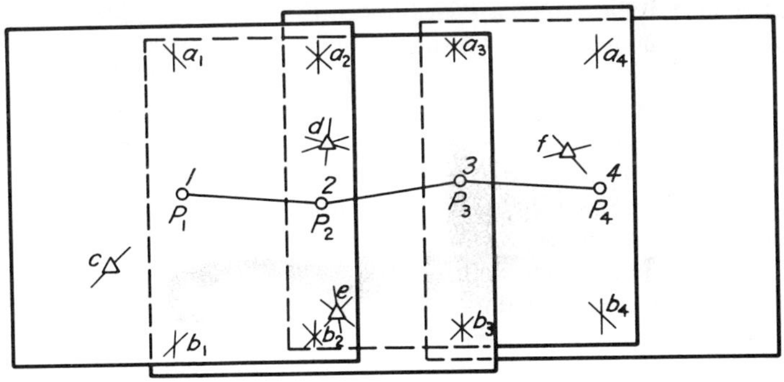

FIG. 31-35. Extension of plot by radial intersection (hand templets).

Hand Templets. In this method of control extension by radial-line methods, supplemental horizontal control points are established by the use of transparent templets, called "hand templets," one representing each photograph. On each templet, all selected control points are represented by lines through them radiating from the principal point. The templets are then assembled over a control base, and the principal points and the supplemental horizontal-control points are located on the base by resection and intersection, respectively. If a sufficient number of known plotted control positions are included so that the assembly is true to scale, the intersections of the radials will be the true locations of the supplemental control points. Figure 31-35 represents the method of extending the plot for an assembly of four photographs.

Precision. With good photographs, the combined average tip and tilt is less than 2° for extensive areas and not more than 3° anywhere in the area. This standard is so generally accepted in radial-line work that it is customary to ignore the presence of tip and tilt altogether and either to consider the physical center of the photograph (located with reference to the collimating marks on the camera) as both the plumb point and the isocenter or, as a convenience in expediting the work, to use the substitute center as previously explained.

31-36. Slotted-templet Method. The slotted templet method of control is a variation of the radial-line method. The utility and efficiency of the radial-line method have been greatly enhanced by the development of mechanical methods of adjusting the radial-line intersections to known primary control points in order to obtain a system of secondary control to which the map or chart compilation can be referenced. Instead of the rays being drawn on the photographs as heretofore explained, the points to be resected are selected, marked on the photographs, and transferred to cardboard templets. Slots representing rays radiating from the nadir point to the selected photopoints are cut into the templets by means of a mechanical slot cutter; the slotted templets are approximately oriented on the manuscript by placing the nadir point of each in its approximate location in the flight line at the scale of the manuscript, and movable metal studs are inserted through these slots representing the rays to each selected photopoint.

Those photopoint studs which correspond to known primary control points are then fixed in position on the manuscript by pins driven through the studs, and the system of slotted templets is shifted slightly about these fixed points until the arrangement having the least apparent residual strain is found. The positions thus found for the movable studs are then the most probable positions for the corresponding photopoints, and these positions are plotted on the manuscript for use as secondary (photocontrol) control points. A slotted-templet assembly is shown in Fig. 31-36.

31-37. Spider-templet Method. An alternative slotted-templet method of control involves the use of slotted strips of spring steel radiating about a center bolt which corresponds to the nadir point of the photograph. These strips are fixed in positions corresponding to the radial lines from the nadir point to the selected photopoints. The assembled metal templets, called "spiders," are then oriented on the manuscript in positions corresponding approximately to the exposure stations of the photographs; metal studs are inserted in the rays for each photopoint; points corresponding to known primary control positions are fixed on the manuscript; and the metal-templet system is shifted and adjusted about these fixed positions in much the same manner as the cardboard-templet system heretofore described. The probable positions of the selected photopoints thus obtained are then plotted for use as secondary control in compiling the map or chart. An assembled metal-templet system called a "laydown" is illustrated in Fig. 31-37.

A large proportion of slotted-templet control operations is performed for the purpose of controlling map and chart compilations from oblique aerial photographs, especially those compilations made from trimetrogon (three-camera) photography. Determination of true horizontal directions from the nadir point of the oblique aerial photograph is necessary in these operations in

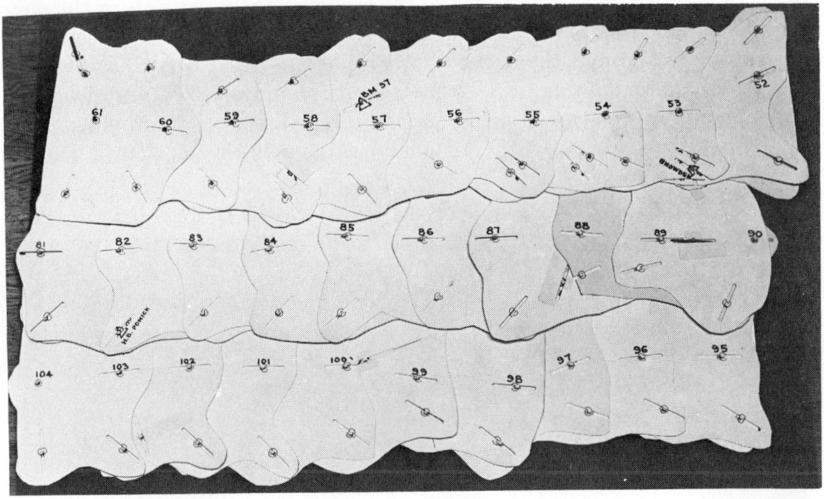

FIG. 31-36. Slotted-templet assembly.

FIG. 31-37. Spider-templet laydown. (*Abrams Aerial Survey Corporation.*)

order to permit proper orientation of the cardboard-templet slots or the metal-templet radial arms, since the map or chart is an orthographic projection on a horizontal datum and the aerial photographs are tilted from the vertical by amounts approximating (in trimetrogon obliques) 60°. A mechanical trigonometrical determination of the required directions is provided by the photoangulator (Fig. 31-38). An analytical geometric solution of the same

FIG. 31-38. Photoangulator. (*U.S. Army.*)

problem is offered by an instrument known as the rectoblique plotter (Fig. 31-39).

31-38. Compilation of Detail from Radial-line Method of Control. The first step in radial-line compilation is the transfer of secondary control to the compilation manuscript. A transparency must be used for compilation by this method. Perhaps the simplest approach assumes that the map-compilation scale is equal to the average scale of the photographs. Then the compilation manuscript is oriented over the radial-line plot so that the primary control points coincide on both manuscripts, and the supplemental horizontal-control points are traced onto the compilation manuscript. The transparent map base is then oriented over each photograph, and the manuscript is oriented and positioned to fit each control point in turn. The features in the vicinity of each control point are then traced. This method works fairly well when the relief is mild and the flying height of the photography is fairly constant over the entire area of the project.

Other methods of compilation involve the use of instruments such as the Fairchild Rectoplanigraph (Fig. 31-45), a device by which the base plot and the photograph are viewed simultaneously, or the Reed reflecting projector (Fig. 31-46), a device in which the photograph is placed in a holder at the top of the instrument and reflected to the base plot for tracing. Another device of more complexity is the radial planimetric plotter (Fig. 31-40) wherein two overlapping photographs are viewed stereoscopically, the details in their common area of overlap being transferred to the map through a mechanical

FIG. 31-39. Rectoblique plotter.

linkage. The photographs are fastened on small platforms and are oriented for exact stereoscopic viewing by the operator.

The radial planimetric plotter is the only one that provides location of details theoretically free from errors due to relief displacement. Methods of transfer of detail from the single photograph all suffer from the inevitable relief displacement of images between control points. One method of reducing this effect is to locate additional points by radial-line intersections on the radial-control network, thus reducing the distances between adjacent points and providing for the selection of points at locations which give more uniform slopes.

31-39. Contouring. There are two general methods of completing the contours on a map made from photographs: by field methods (usually plane-table mapping) and by application of the stereoscopic principle to the overlapping photographs.

The field method usually involves plane-table mapping wherein the plane-table board or worksheet is a copy of the radial plot reproduced on metal-mounted paper or coated, dimensionally stable, plastic. Aerial contact prints, enlargements, and precise mosaics are also used as plane-table sheets in reasonably flat country where the picture or mosaic, properly rectified, becomes a true map.

The other general method involves setting up the stereo pairs, oriented to ground or photogrammetric control, in stereoscopic plotting machines at proper scale and orientation (Art. 31-17). With this method there must be sufficient control to fix the datum of each model, either through photogrammetric extension or by ground-survey methods.

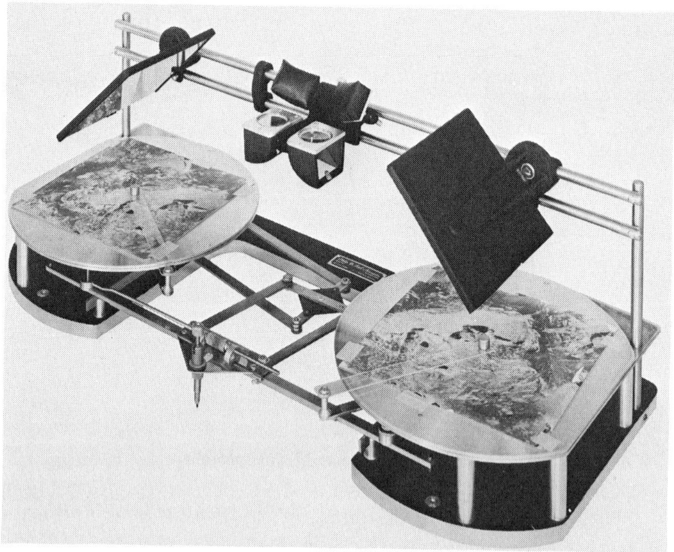

FIG. 31-40. Radial planimetric plotter. (*Philip B. Kail Associates, Inc.*)

Contouring may also be accomplished on the stereocomparagraph (see Fig. 31-59) on a separate templet made for each photograph. Later the contours are compiled on the planimetric base in the same manner as the detail of the photograph. Automatic plotting machines such as the Zeiss stereoplanigraph (Art. 31-48) compile all the detail of the map in one operation; the radial-line technique is not used.

31-40. Mapping from Oblique Aerial Photography. Oblique photographs are usually classified as *high oblique,* in which the apparent horizon is shown, or *low oblique,* in which the apparent horizon is not shown.

Low-oblique photography has been used for precise topographic mapping by the U.S. Geological Survey since about 1950. The system utilizes a special twin camera mount and twin wide-angle mapping cameras, exposed simultaneously, with their optical axes in a common plane forming an angle of 20° with a plumb line and 40° with each other (Fig. 31-41). Aerotriangulation and compilation can be accomplished by the Twinplex plotter (Art. 31-41), and compilation can be accomplished by instruments such as the Balplex (Art. 31-48). Zeiss-Aerotopograph provides a twin-camera system of like nature (Fig. 31-23), the camera using the Zeiss Topar (normal-angle) lens, $f = 8\frac{1}{4}$ in., $f/4$ and special gears for shutter synchronization. The cameras are mounted at 27° to the vertical and 54° to each other, with their optical axes in a common plane; this system is commonly applied to cadastral surveys. The major advantages of this type of photography are

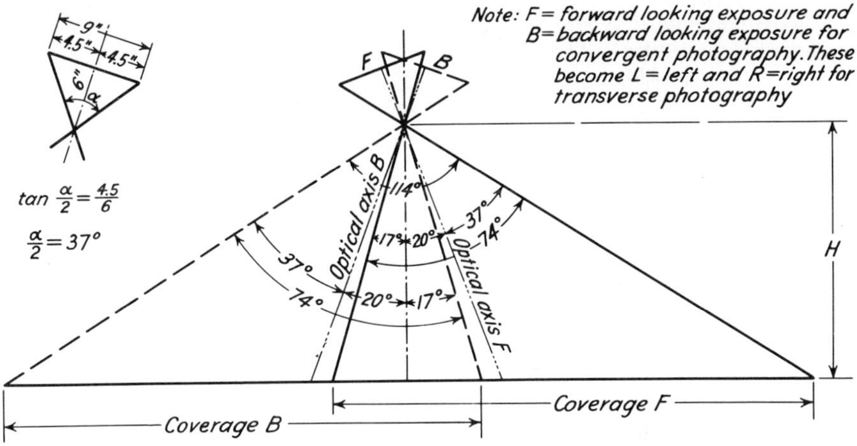

FIG. 31-41. Angular coverage of U.S. Geological Survey twin low-oblique photography.

higher stereoscopic measuring accuracy, a much greater picture area, and a lessening in the density of the ground-control requirement as compared with vertical photography taken at the same flight height.

Oblique photographs also find application where either aeronautical charts or exploratory maps of low accuracy are required for vast, relatively inaccessible areas. Certain parts of northern Canada and Alaska have been mapped by extensive use of oblique aerial photography.

Several methods for using oblique photography in the preparation of maps have been developed, each method to meet a variation in conditions. Three principal systems are in use: the perspective-grid method, the single-photograph oblique plotter, and the oblique stereoplotter. The perspective-grid method is used most advantageously in mapping terrain of relatively low relief. In essence, a precomputed perspective grid is placed upon the oblique photograph, and planimetric detail is copied, grid by grid, from the aerial photograph to the corresponding rectangular grids of the map manuscript. This method is not applicable to topographic mapping, and in areas of moderate relief the accuracy of the planimetric map suffers accordingly. In areas where relief affects the accuracy of the perspective-grid method, various oblique-plotting instruments have been developed to overcome the difficulties introduced. These instruments in principle are phototheodolites which occupy the perspective center of the aerial photograph. Map locations and elevations of points are determined by intersection in a manner analogous to plane-table surveying. Instruments of this type are the Wilson photoalidade, the Miller oblique plotting instrument, and the Canadian oblique plotter.

Stereoplotting instruments used for mapping with low-oblique photographs either are standard types, with the exception of the Twinplex plotter, or are universal in nature. The Zeiss stereoplanigraph and the Santoni stereocartograph are of the universal type; the multiplex, the Balplex, and certain models of the Kelsh plotter are modified either by use of special brackets or with special diapositive holders. These instruments, when used with low-oblique photographs obtained by precision cameras, can accomplish mapping of standard accuracy when oriented to adequate ground control. Stereoplotting instruments are not ordinarily used with high-oblique photography for reconnaissance mapping, but their application will give more precise results than the perspective-grid method or the use of the single-photograph oblique plotter as previously described.

31-41. Twinplex Plotter. A method of photogrammetric mapping using low-oblique photography taken by twin wide-angle cameras exposed simultaneously at each camera station and oriented either along the flight line (convergent low oblique) or transverse to the flight line (transverse low oblique) was developed about 1950 by the U.S. Geological Survey. The cameras have the optical axis in a common vertical plane forming an angle of 20° with a plumb line and 40° with each other. Stereoplotting instruments such as the Balplex with modified slide and yoke assembly can plot with this type of photography, and the Twinplex plotter (Figs. 31-42*a* and 31-42*b*) was developed to facilitate bridging procedures.

The advantages claimed for this type of convergent low-oblique photography, if the exposure stations are so spaced that the convergent photographs overlap each other by 100 per cent, are an effective base-height ratio of 1.23 and an effective width-height ratio of 1.27. Comparable values for vertical photography are base-height ratio of 0.63 and width-height ratio of 1.15. The convergent stereomodel covers approximately 2.2 times the area of a corresponding stereomodel from vertical photography, assuming equal flight heights. Fewer supplemental vertical control points are required for the convergent application as compared with vertical photography covering the same area at the same flight height. In addition to this advantage, a

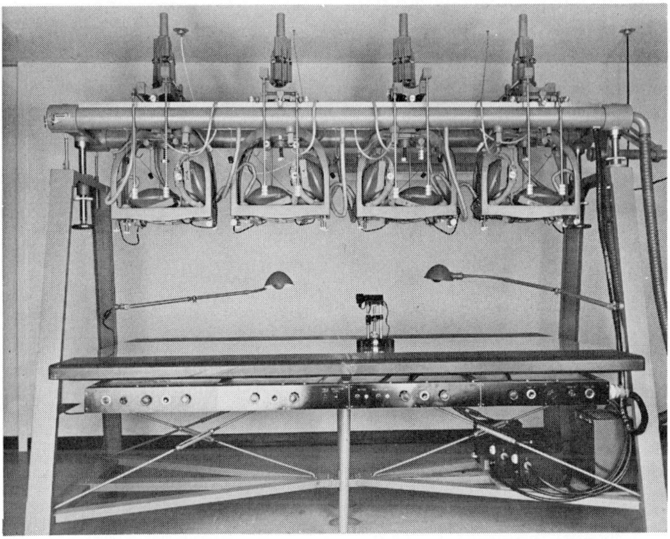

FIG. 31-42*a*. U.S. Geological Survey Twinplex plotter (convergent low-oblique application).

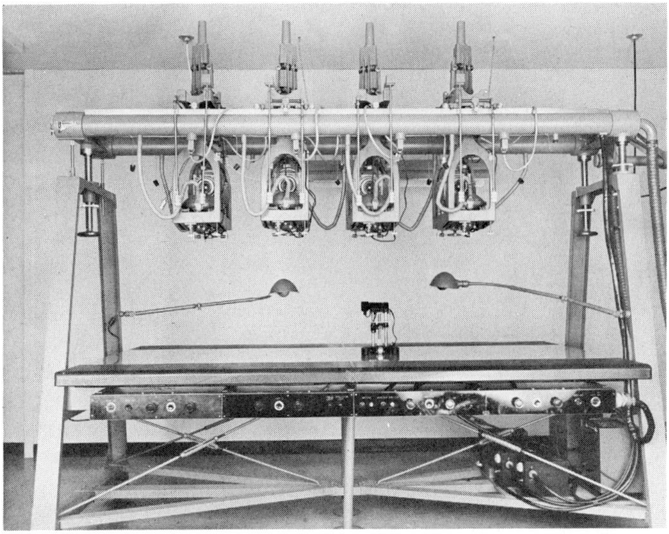

FIG. 31-42*b*. U.S. Geological Survey Twinplex plotter (transverse low-oblique application).

somewhat higher C factor is claimed due in part to a stronger intersection of conjugate image rays derived from the increase in base and the more exaggerated stereoscopic impression of relief.

The transverse low-oblique application of this camera system is of value particularly in military or reconnaissance applications. The width-height ratio of 2.31 decreases the required number of flight lines for a given area of coverage as compared with vertical photography with the same base-height ratio (0.63).

31-42. Displacement of Image Points Due to Relief. Figure 31-43 represents vertical photographs of irregular terrain taken from two camera

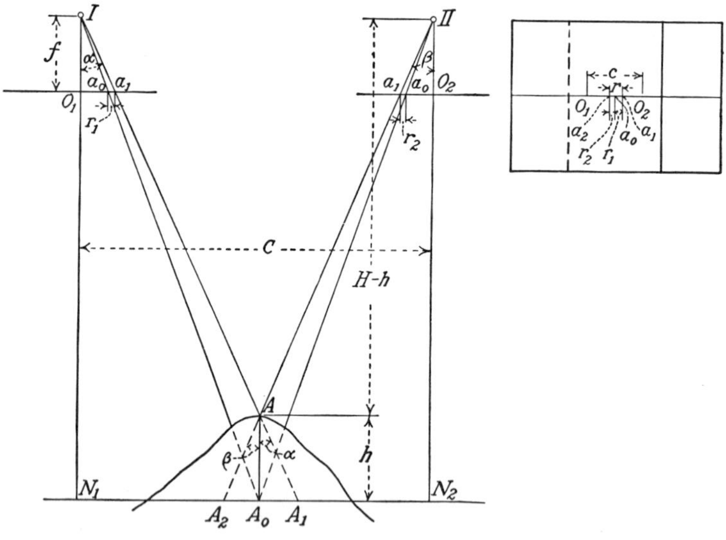

FIG. 31-43. Displacement due to relief.

stations I and II with a camera of focal length f at an altitude of $(H - h)$ above some point A. Rays from A toward the perspective center I pierce the focal plane at point a_1. Had point A been at an elevation $h = 0$, that is, either at sea level or in some other established reference plane, the rays would have been reflected from point A_0 and would have pierced the focal plane of camera I at point a_0. The distance a_1a_0 on photograph I is the displacement r_1 of the image point of A due to its relief h above the datum. On photograph II the displacement r_2 due to relief is the distance a_2a_0. From the figure, $O_1a_1/f = \tan \alpha$, in which O_1a_1 is the distance on the photograph from the principal point to the pictured location of A and α is the angle of ray IA from the vertical.

Also

$$\frac{r_1}{A_0A_1} = \frac{f}{H} \qquad \text{and} \qquad A_0A_1 = h \tan \alpha$$

Hence

$$r_1 = \frac{hf \tan \alpha}{H}$$

and since $H/f = S$ is the scale of the photographs,

$$r_1 = \frac{h}{S} \tan \alpha \tag{4}$$

Example: Let $h = 200$ ft., $f = 12$ in., $H = 10{,}000$ ft., and $O_1a_1 = 4.30$ in. Find the displacement r_1 of the pictured location of point A due to elevation h.

$$S = \frac{10{,}000}{12} = 833 \text{ ft. per in.} \qquad \text{and} \qquad \tan \alpha = \frac{4.30}{12} = 0.358$$

Then

$$r_1 = \frac{200 \times 0.358}{833} = 0.086 \text{ in.}$$

The displacement r_2 of the image of point A on photograph II is equal to the displacement r_1 on photograph I only when $\tan \alpha = \tan \beta$. The total displacement on the two photographs $(r_1 + r_2)$ is the *parallax displacement due to relief* and may be expressed as

$$r = \frac{h(\tan \alpha + \tan \beta)}{S} \tag{5}$$

Computations of the relief displacement on single photographs is a necessary step (1) in the rectification and enlargement in ratio printers of essentially vertical photographs, wherein it is necessary first to establish the scale of the photographs at a very exact value and at the same time to eliminate the effect of tip and tilt of the camera, and (2) in arriving at a decision concerning the usability of particular photographs for a particular purpose. To compute and use the values of displacement of all the image points of important objects in the photograph for purposes of surveying would result in an enormous amount of mathematical computation, and this practice is not usually followed. *It should be emphasized that it is impossible to remove the displacement due to relief from an aerial photograph. It is a natural phenomenon and should not be considered an error.* Were it not for the parallax displacement due to relief, stereoscopic measurement would be impossible.

31-43. Effect of Tip and Tilt. Rotation of the camera about a horizontal axis normal to the line of flight is usually referred to as *tip*, and rotation of the camera about the line of flight—the stereoscopic base—is called *tilt*. By this system of reference, the pitch of a ship at sea would be called the tip and the roll of the ship would be called the tilt. However, in the consideration of single photographs the combination of these two rotational effects is often called tilt. The displacements due to tilt are referred to as "errors," and the relief displacements are referred to either as "relief displacement" or as "parallax."

Figure 31-43*a* is a generalized perspective showing ground stations *A*, *B*, and *C*; the plumb point *N*; the photographic picture planes *MM* (assumed to be horizontal) and *TT* (assumed to be tilted through an angle α, the direction of tilt being in the vertical plane passing through the principal line *vv*); the perspective center *I*; and the rays from the ground stations to the perspective

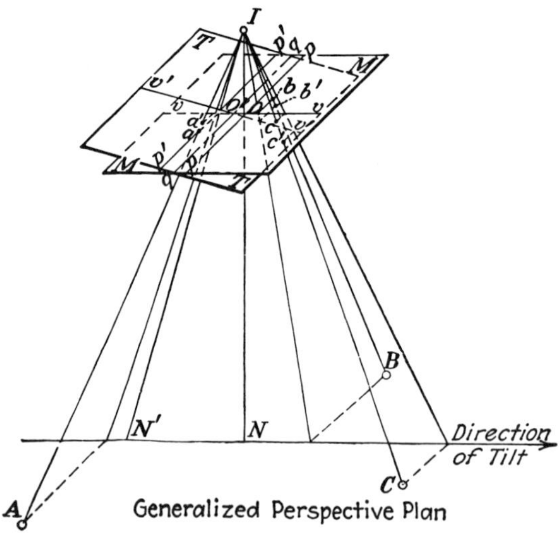

FIG. 31-43*a*. Effect of tip and tilt in an aerial photograph.

center. In Fig. 31-43*b* is a side view of the photographic picture planes, this view being an orthographic projection on a vertical plane (the principal plane) which is parallel to the direction of flight (stereoscopic base). An end view and a plan view of the picture planes are shown in Fig. 31-43*b*; in the plan view it is assumed that the plane *TT* is rotated into the plane *MM* about an intersecting trace *qq*. The locations of all points in planes *MM* and *TT* along line *qq* are without error, and line *qq* may be considered as the line of equal scale, or the axis of tilt, of the photographs.

In the case of photographs taken with a single-lens camera of focal length 8¼ in. at an altitude of 10,000 ft. with a combined tip and tilt of 3°, directions measured from the principal point as an origin are distorted less than 3 minutes of arc. Displacements of objects are essentially zero at the center of the photograph and are approximately 75 ft. at the corners. Sides of squares on the photograph, as, for example, sections of land, are diverged approximately 2½°. A displacement of 75 ft. on the ground is represented by a distance of about 0.06 in. in the photograph. Displacements due to relief vary from zero at the center of the photograph to a distance equal to the height of the object at 45° from the principal ray of the photograph. Thus, the image of an

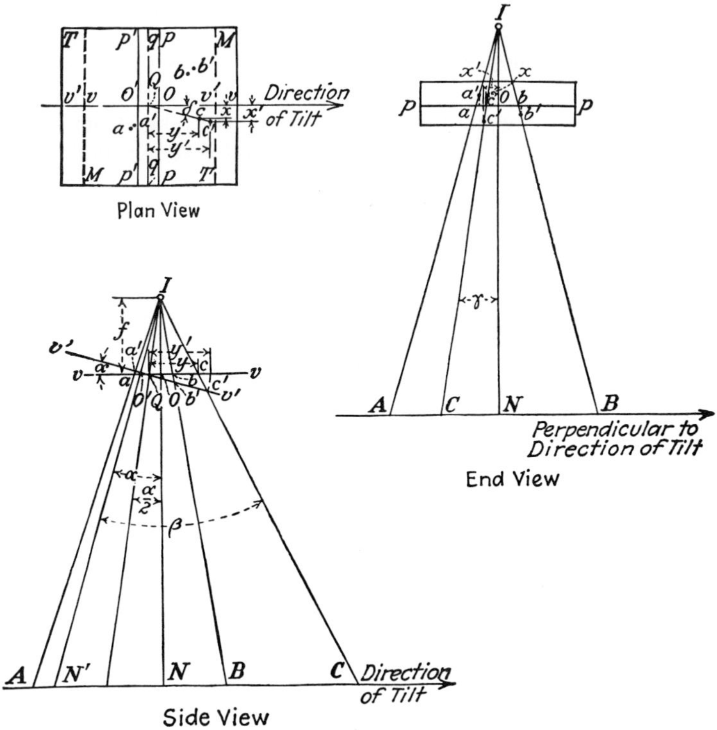

FIG. 31-43*b*. Effect of tip and tilt in an aerial photograph.

object 75 ft. high appearing at an angle of 45° from the center of the photograph would be displaced 0.06 in. on the photograph. Variations in elevation of points on aerial photographs are generally much greater than this amount, and the displacements due to relief are usually considered to be greater than the displacements due to tip and tilt. In any event, both types of displacement are indeterminate until the relative elevations of the points are known, and on most aerial photographs the distances and areas should be scaled with caution except as an approximation.

The effects of tilt of the photograph may be reduced to mathematical forms (see pages 319 to 322 of Ref. 1 at the end of this chapter). However, in most applications the amount and direction of tilt are readily determined by means of plotting instruments; it is not necessary to solve the problem mathematically.

31-44. Determination of Elevation by Measurement of Parallax. When an object is observed from two different locations, there is an apparent displacement of the position of the object with respect to a reference system. Such an apparent displacement is known as "parallax." In photogrammetry, the term "x parallax" generally is used to denote a quantity more precisely designated as "absolute stereoscopic parallax." Considering a pair of truly vertical photographs of equal principal distances and taken from equal flight heights, the absolute stereoscopic parallax of a point is the algebraic difference,

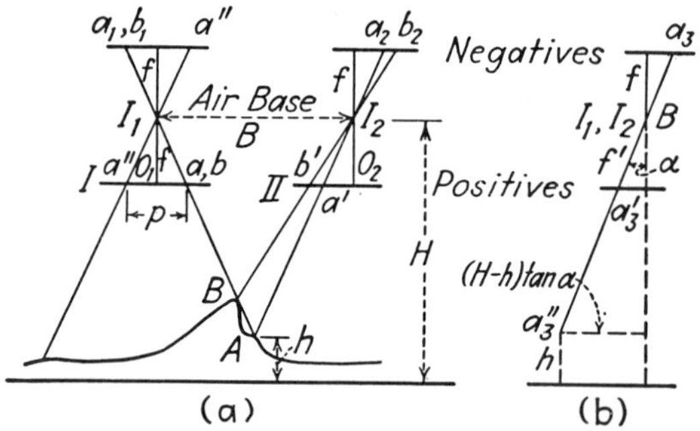

FIG. 31-44. Determination of elevation by measurement of parallax.

parallel to the air base, of the distances of the two images from their respective principal points. The term x parallax is also used to denote similar measurements when the theoretical conditions are not strictly attained, as, for example, when measuring parallax on unrectified aerial photographs.

The "y parallax" of a point is the difference of the perpendicular distance of its two images from the vertical plane containing the air base. The existence of y parallax is an indication of tilt in either or both photographs, a difference in flying height, and/or improper orientation of the pictures in the plotting machine. This y parallax is the key to the orientation motions to be applied to the stereoscopic plotting instrument in order to recover the exact relationship of the two pictures that they had at the instant of exposure.

The object of measuring the parallax difference between two points is to determine the difference in their elevations and thus to determine their respective heights above sea level. In Fig. 31-44a, which is a variation of Fig. 31-43, it is desired to measure the ground elevations above sea level of points A and B. The figure shows the profile of a landscape wherein any two ground points A and B appear on *negatives* I and II in positions a_1, b_1 and a_2, b_2, respectively and on the corresponding *positives* at positions a, b and a', b', respectively. Line I_1a'' is constructed parallel to line I_2a', making a triangle $a''I_1a$ similar to triangle I_1AI_2. Then

$$\frac{p}{f} = \frac{B}{H - h}$$

or

$$p = \frac{fB}{H - h} \tag{6}$$

in which H is the altitude of flight, h is the elevation of point A above sea level, B is the "air base," that is, the distance the plane flew between exposures, and the distance p is the *absolute stereoscopic parallax* of point A. If a line is drawn through I_1 parallel to I_2b', the absolute stereoscopic parallax of point B may be similarly determined. The difference between the absolute stereoscopic parallax of A and that of B depends upon and is a measure of the difference between their elevations.

The rate of change of parallax with respect to changes in h may be expressed as

$$\frac{dp}{dh} = fB(H - h)^{-2} = \frac{fB}{(H - h)^2} \tag{7}$$

To determine the parallax in terms of the measured stereoscopic base of the photographs, let B_m (in millimeters) be the measured stereoscopic base of the photographs which is equal to the air base B (in feet) divided by the mean scale of the photograph, that is, by $(H - h)/f$. Or

$$B = \frac{B_m\,(H - h)}{f} \tag{8}$$

in which both H and f are in feet.

By substituting this value of B in Eq. (7),

$$\frac{dp}{dh} = \frac{fB_m(H - h)}{f(H - h)^2} = \frac{B_m}{H - h}$$

or

$$dp = \frac{B_m\,dh}{H - h} \tag{9}$$

For small values of change in h compared with H, as, for example, the vertical distance between two contours as compared with the flight altitude,

$$\Delta p = \frac{B_m\,\Delta h}{H - h} \tag{10}$$

This equation expresses the value of the change in parallax (in millimeters) for a corresponding change in elevation between two points or between two contours on the photograph. It is to be noted in this equation that for changes in elevation the corresponding difference in parallax is independent of the focal length of the camera, the over-all size of the photographs, and the percentage of overlap.

Although the equation for Δp was developed from a profile taken vertically through the camera stations and the plumb points, it is equally applicable to the entire area of overlap of the photographs as long as the measurement of parallax is limited to that component parallel to the stereoscopic base. For example, in Fig. 31-44*b*, let it be assumed that point A is at a distance $(H - h)\tan \alpha$ to the left (or to the right) of the principal plane of the photographs containing the stereoscopic base. Then

$$\frac{p}{B} = \frac{f'}{Ia_3''} = \frac{f/\cos \alpha}{(H - h)/\cos \alpha} \tag{11}$$

or

$$\frac{p}{B} = \frac{f}{H - h} \tag{12}$$

as before.

31-45. Parallax Tables. Parallax tables may be computed for values of $\Delta h = 20$ ft. (or any other suitable contour interval) and $B_m = 100$ mm. for changes in flight altitudes throughout the range of flight altitudes encountered in aerial photography, as, for example, between 5,000 and 25,000 ft. Such tables are useful in the compilation of elevation differences and in contouring directly from the photographs with instruments such as the stereocomparagraph (Fig. 31-59). The value of $B_m = 100$ mm. is taken arbitrarily, and in the use of the tables it is necessary only to consider the mean value of the actual measured stereoscopic bases of the photographs as a percentage and to multiply the corresponding parallax by this percentage to obtain the parallax due to the change in contour interval for the particular photographs under consideration.

Such a parallax table would be constructed by solving Eq. (10) for an assumed stereoscopic base of 100 mm., for 20-ft. increments of difference in elevation, and for variations of $(H - h)$ for the chosen contour interval throughout the probable range of flight altitudes likely to be encountered.

For the computation of the parallax table, Eq. (10) has the form

$$\Delta p \text{ (mm.)} = \frac{100 \text{ (mm.)} \times 20 \text{ (ft.)}}{(H - h) \text{ (ft.)}} \tag{13}$$

The values of parallax difference are in millimeters and are convenient in the use of the stereocomparagraph whereon the differences in parallax of the photographs are indicated in millimeters.

The more conventional form of this interpretation of the parallax equation would be

$$\Delta p = \frac{100 \times 20}{H - h} \tag{14}$$

valid for 5,000 ft. $< (H - h) <$ 25,000 ft., in which it is assumed that 5,000 and 25,000 ft. are the limits of range of likely flight altitudes.

A master parallax table and a more complete explanation of its computation and use are given in pages 616 to 622 of Ref. 1 at the end of this chapter.

PHOTOGRAMMETRIC PLOTTING INSTRUMENTS

31-46. General. Photogrammetric instruments are used to solve the equations of the loci of points on photographs automatically; they permit rapid and accurate compilation of the photographic detail in orthographic projection on a selected datum. Both monocular or single-photo, instruments and stereoscopic instruments are used for this purpose.

31-47. Monocular and Single-photo Instruments. Monocular and single-photo instruments are used mainly in the original compilation of small-scale planimetric charts and in the revision of existing map coverage where the planimetry is obsolete. They fall into two general classifications; those based upon the camera-lucida principle and those employing optical projection. These are exemplified by the Fairchild rectoplanograph and the Reed Focalmatic projector respectively.

The Fairchild rectoplanograph (Fig. 31-45) is based upon the principle of the camera lucida, with which the eye receives two superimposed images, one from the photograph and the other from a map manuscript. The instrument also provides a superior optical system in the form of a high-quality double prism, as well as a group of adjustments allowing the map maker to plot from any aerial photograph up to a 9 by 9-in. format. Photographs taken with any focal length of camera from 4 to 12 in. can be accommodated, the scale of the image can be varied to match that of the map, and angles of tip and tilt in the photographs can be compensated.

The Reed Focalmatic, an automatic-focusing opaque projector, is a vertical reflecting projector which remains in sharp focus with minimum distortion between a ⅓ × reduction and a 2.0 × magnification (Fig. 31-46). It will project a 7 by 7-in. area of opaque copy in sharp focus on the horizontal plotting board so that detail in maps or photographs can be copied in exact scale. The lens is a Wollensak 162 mm., *f*/4.5, and the light source is 325 watts, 115 volts.

31-48. Stereoscopic Plotting Instruments. Stereoscopic plotting instruments are used when it is desired to measure elevations and trace contours and when the requirements are such as to demand greater precision in the compilation than can be obtained with single-photo instruments.

Basically there are three problems to be solved in the compilation of topographic maps from stereoscopic photography. First, some means of compensation for tilt in the photography must be provided. Second, means must be provided for measuring elevations and delineating contours. Third, means must be provided for compiling the information in the photographs, which are perspective projections, onto the map manuscript in orthographic projection at one common scale. When all three of these problems are solved in the stereoscopic plotter and the compilation is done in one continuous operation, the instruments are termed *automatic* plotting instruments.

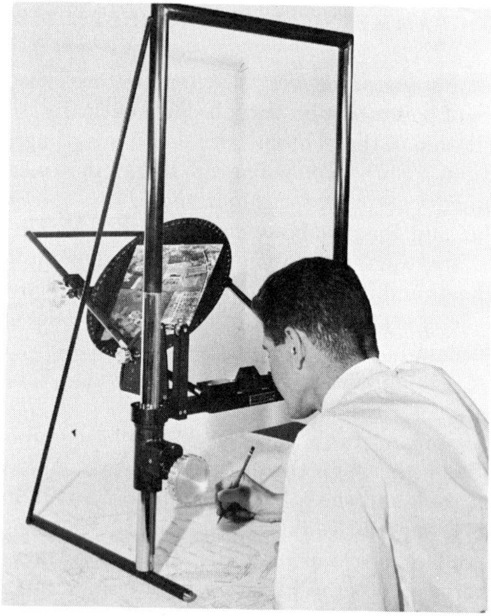

FIG. 31-45. Fairchild rectoplanigraph. (*Fairchild Space and Defense Systems.*)

Automatic Stereoscopic Plotting Instruments. The three main elements in all automatic stereoscopic plotting machines are (1) the projection system, (2) the viewing system, and (3) the measuring system. Another element often classified is the tracing system which, in projection-type instruments such as the multiplex and the Balplex, consists of a pantograph attached to the tracing table, and which, in the more complex plotters of the Wild autograph or Zeiss stereoplanigraph family, consists of a coordinatograph. These devices enable the stereomodel at optimum model scale to be translated into a common manuscript scale, usually at the reproduction scale of the map.

Projection systems can be further divided into (1) optical projector systems such as the multiplex, Balplex, and Kelsh plotters; (2) mechanical projection systems such as the Wild autograph A-7; and (3) combinations of mechanical and optical projections such as the aerocartograph (pages 642 to 643 of Ref. 1 at the end of this chapter). Viewing systems are generally of two types. In one, the image is viewed through a binocular microscope so arranged that

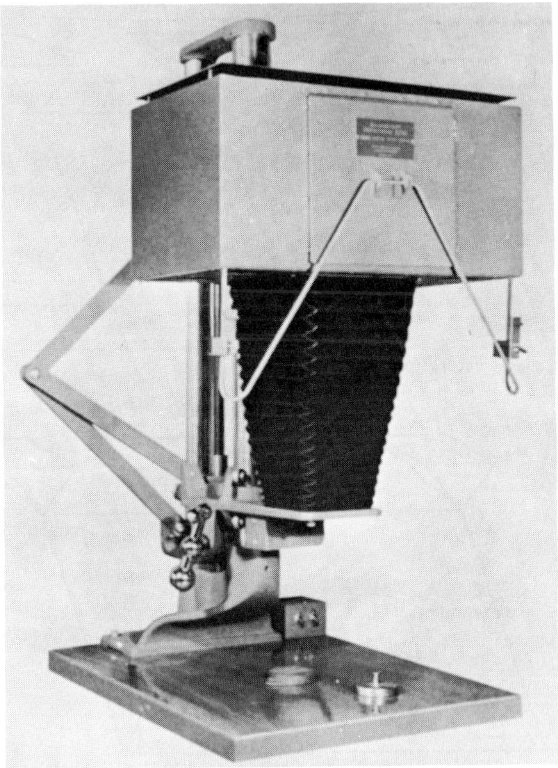

FIG. 31-46. Reed Focalmatic projector.

the vision of one eye is directed to one photograph and the vision of the other eye is directed to the second photograph of a stereoscopic pair. In the second system, the projections of the photograph are viewed on a clear white reflecting surface which also contains the index mark. In this system, image separation is obtained either by the use of either colored or polarizing filters or by the use of a flicker system which alternately exposes the left and right photographs to view.

The measuring system in many instruments incorporates a simple device known as the Zeiss parallelogram (Fig. 31-47) which enables either projector of the instrument to be used for the left or right photograph of a pair. This system makes possible the use of a two-projector instrument for the extension or bridging of control.

1. Anaglyphic Projection-type Stereoplotters. The multiplex (Fig. 31-48), the Balplex (Fig. 31-49), and the Kelsh plotter (Fig. 31-50), are three examples

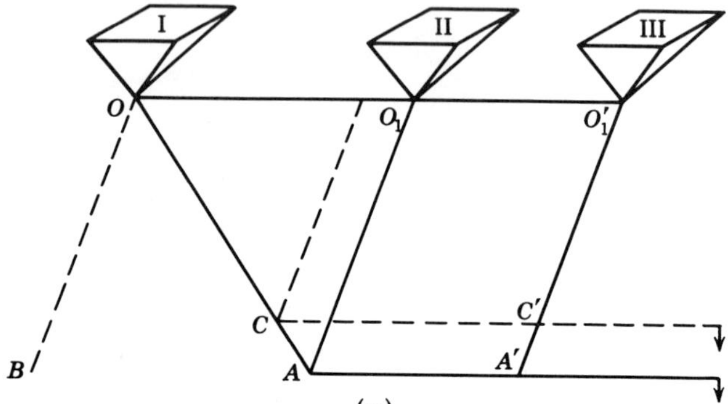

(a)

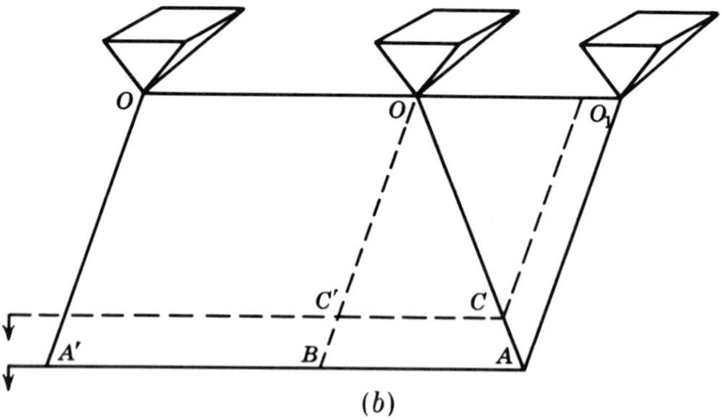

(b)

FIG. 31-47. Zeiss parallelogram.

of optical projection systems wherein the projections of the photograph are viewed on a clear white reflecting surface which also contains the index mark. While the appearance and construction of the instruments differ, they all employ the basic principle of projecting two overlapping photographs upon a plotting table; image separation is achieved by the use of colored filters, and the operator views the image through special spectacles. This method of viewing the projected images is called the *anaglyph* system of viewing reflected light.

The main components of this type of plotter are (1) the supporting frame, including the supporting projector bar; (2) the projectors (which project the photographs either at original size or at a reduction); (3) the tracing table (stand) with platen and floating mark for the measurement of elevations and the tracing of planimetric details and contours; and (4) the spectacles for viewing the projection stereoscopically. Usually with this type of instrument a variable-ratio pantograph is attached to the tracing table to make

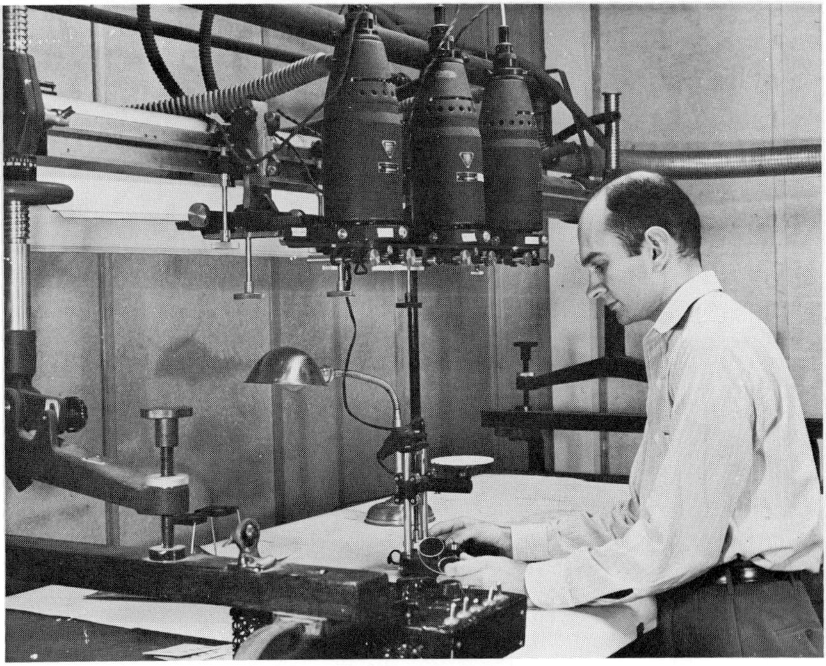

FIG. 31-48. Multiplex.

possible a reduction from model scale to manuscript scale. This device produces the line drawing at a location removed from the position of direct orthographic projection of the model.

The principal distance of the Balplex (both 525 and 760) projector is 55 mm. compared with the 30-mm. principal distance of the multiplex projector (U.S. Geological Survey type). A special reduction printer 153/55, shown in Fig. 31-51, is used for Balplex diapositives; it may also be used for multiplex diapositives if provided with an interchangeable printer head. Kelsh diapositives are negative format size, and for precision mapping, a 153/153 printer with optical distortion compensation is used for diapositive preparation.

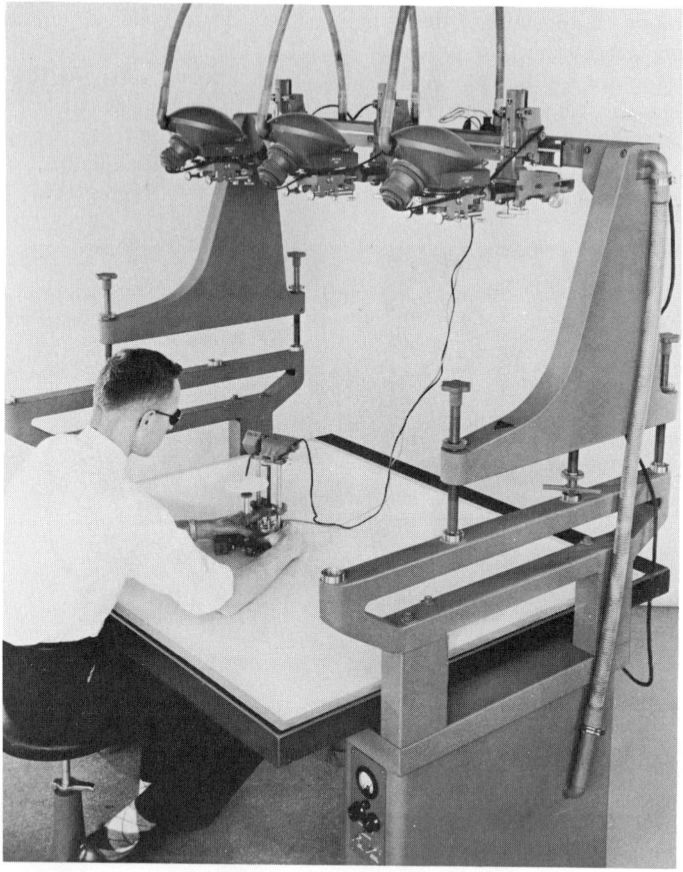

FIG. 31-49. Balplex.

The accompanying table compares certain features of all four types of projectors.

Projector	Balplex 525	Balplex 760	Kelsh	Multiplex
Optimum projection distance, mm	525	760	760	360
Range of vertical focus, mm	405–645	650–870	685–915	270–450
Diapositive size, mm	110 × 110	110 × 110	241 × 241	64 × 64

The Kelsh plotter can be used for stereocompilation from vertical and certain convergent low-oblique photography; stereotemplet triangulation can be used for horizontal extension of ground control. The multiplex can be used for stereotriangulation and compilation from vertical photography. The Balplex can be used for stereotriangulation and compilation from vertical photography and compilation from convergent and lateral oblique photography. Convergent-projector tilts of 20° and lateral-projector tilts of 10° to 60° are possible through a special slide and yoke assembly.

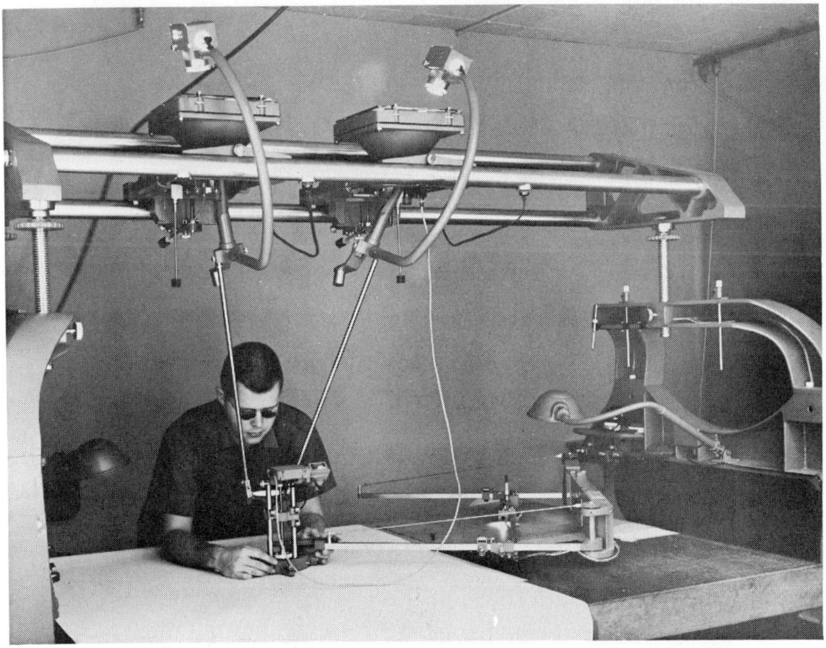

FIG. 31-50. Kelsh plotter (model K-5). (*U.S. Geological Survey.*)

The operation of all three instruments is roughly equivalent in complexity. The diapositives are first inserted in the projectors, and the projectors are then oriented through procedures called relative and absolute orientation so that their projected images will intersect in space above the plotting table and form a spatial model which is a true small-scale reproduction or image of the area common to two successive photographs of a strip (Fig. 31-52). Once absolute orientation is achieved, the procedure for compilation of the topographic-map manuscript is the same for all three types of plotters. For the delineation of contours, the floating mark of the tracing-table platen is set at a predetermined contour elevation, and a trace of the contour is established by a lateral movement of the tracing table while the floating mark is maintained in constant visual contact with the apparent terrain in the model. Planimetric features are delineated by following the feature

FIG. 31-51. Diapositive reduction printer (153/55).

with the floating mark and adjusting the elevation of the platen so that the floating mark is always at ground elevation as the plotting proceeds.

2. Zeiss Stereoplanigraph C-8. The Zeiss stereoplanigraph C-8 is a precision stereoscopic plotting instrument of German manufacture capable of first-order precision. Plotting can be accomplished with this instrument from stereoscopic pairs of terrestrial photographs or aerial photographs with either

low or high tilt, and the instrument may also be used for aerial triangulation. Figure 31-53 is a photograph of the model C-8 stereoplanigraph, and Fig. 31-54 is a schematic diagram which shows the principle of the instrument. Large-scale precision plotting for cadastral surveys and engineering applications is possible, and a complete range of vertical, oblique, and convergent photography can be accommodated as well as terrestrial stereo pairs up to a negative format

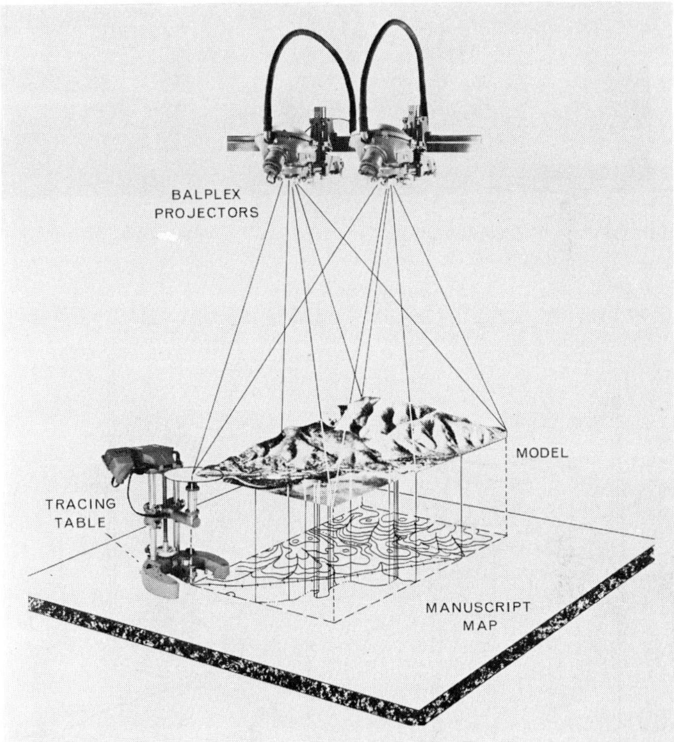

FIG. 31-52. Model representation.

of 23 by 23 cm. (9 by 9 in.). Standard equipment includes a printing counter which records coordinate values on a record sheet. Other attachments allow coordinate values to be recorded on punched tape. A profile-measuring device for photogrammetric route surveys is also available.

The stereoplanigraph C-8 utilizes the principle of optical projection, and the projection system consists of two universal photocarriers with spotlight illumination, the auxiliary lens system, and measuring mark mirrors. The lens holders and the plate holders are interchangeable in the photocarriers. Plotting lenses are selected according to the focal length of the aerial camera,

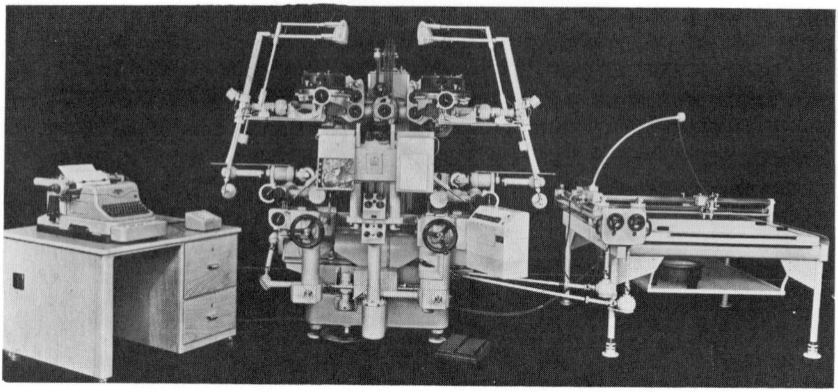

FIG. 31-53. Stereoplanigraph C-8. (*Zeiss-Aerotopograph, Munich.*)

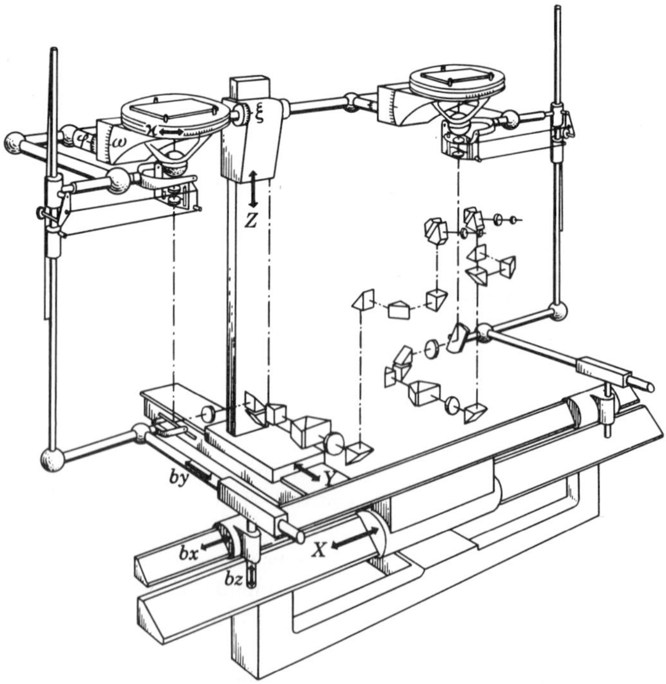

FIG. 31-54. Stereoplanigraph C-8 (exploded view).

and differences of distortion can be individually eliminated with the aid of correction plates mounted in the plate holders. It is possible to plot photographs taken with any usual normal or wide-angle mapping camera with focal lengths of 90 to 310 mm. (3¾ to 12 in.).

The observation system consists of the eyepiece, or ocular head, and the vertical guide rods; it features inverting prisms for unreverted observation of diapositives and negatives and thus makes possible the interchange of projections for bridging. The measuring system consists of the optical measuring marks, the three-dimensional compound slide with guide spindles, the drive mechanism, and the printing counter for automatic recording. The measuring system is operated by two handwheels and a pedal disk; gross

FIG. 31-55. Wild autograph A-7. (*Wild-Heerbrugg Instruments Inc.*)

adjustment can be made by means of an electric moter. A rectangular coordinatograph can be attached to the instrument to facilitate model-scale to manuscript-scale relationships at various gear ratios. The Porro-Koppe principle is utilized in the design of the instrument, and the operator views the stereoscopic pair perspectively.

3. Wild Autograph, Models A-7 and A-9. The Wild autograph model A-7 (Fig. 31-55) is a precision stereoscopic plotting instrument of Swiss manufacture capable of first-order precision. It will accommodate vertical, terrestrial, and lateral oblique photographs tilted from 0° to 30° and (by inversion of the Y and Z axes) from 60° to 90°. Aerial triangulation can also be accomplished with this instrument. Negative formats of up to 23 by 23 cm. (9 by 9 in.) can be accommodated.

Model A-7 utilizes the principle of mechanical projection by means of space rods which move in any direction through a central point corresponding to the perspective center of the photograph. The Zeiss parallelogram principle is used in the design so that the projectors, or cameras, can be set at a fixed separation. The viewing is by binocular microscope through a lens-prism system providing orthogonal observation throughout the range of the format. Photographs at focal lengths of 98 to 215 mm. (4 to 8½ in.) taken with any precision narrow- or wide-angle camera can be accommodated, and optical distortion is compensated by means of correction plate carriers. In the observation microscope the floating marks are fitted directly below the picture carriers and exactly above the suspension points of the two space rods. During the plotting, the floating marks always move parallel to the picture planes and illumination of the photographs is provided by guided 6-volt lamps with color filters.

The observation system is provided with a reversing device for correct observation of negatives or diapositives. For aerotriangulation through consecutive stereomodels, optical interchange of left and right pictures in the eyepieces and optical turning of the image is possible through dove prisms. The eyepiece separation is adjustable for width, and errors of the eye axes can be compensated. The y parallax can also be converted to x parallax through a tumbling device.

Instrument coordinates from aerotriangulation procedures or numerical cadastral work are recorded by an electric coordinate printer if desired. A rectangular coordinatograph can be furnished to facilitate stereomodel to manuscript scale changes by means of interchangeable drive-gear ratios. Many other attachments are possible for special applications.

The Wild autograph model A-9 (Fig. 31-56) is a first-order stereoplotting instrument for plotting from vertical superwide-angle photographs, although it will also accommodate vertical wide-angle photographs. It is suitable for aerial triangulation covering large areas and for small-scale mapping. Negative formats of 9 by 9 in. size are reduced to 4½ by 4½ in. in the Wild U-3 diapositive printer, which also compensates for optical distortion. Diapositives can also be printed without compensation for optical distortion, and correction can be accomplished by correction plate carriers.

As in the Wild autograph A-7, the photographs are viewed orthogonally; the floating marks are located directly below the picture carriers and exactly above the suspension points of the guide rods. During the plotting, the floating marks move exactly parallel to the picture planes, and frontal observation of the pictures through binocular microscopes is provided at 10× magnification. In order to facilitate continuous aerotriangulation by means of consecutive stereopairs, the left and right images can be interchanged in the eyepieces and the images can be rotated optically by means of dove prisms. The x and y coordinates are read on counters to 0.01 mm., z or height coordinates are read on interchangeable glass scales directly in meters or feet, and the angular elements of relative orientation can be read or set on scales graduated to 0.01 grad. Plotting is done on a separate 28 by 28-in. drawing table with an electromagnetic drawing-point lifter directly connected to the instrument through a transmission gear. A larger drawing table (40 by 40 in.) is also furnished if desired, and for aerotriangulation the Wild EK-5 coordinate printer is used.

4. Analytical Plotters (Orthographic Direct Viewing). The analytical plotter is a stereograph (a parallax bar with a pencil attachment used for

FIG. 31-56. Wild autograph stereoplotter. (*Aero Service Corporation.*)

plotting contours from a stereogram) with auxiliary means provided for removing the effect of tilt and/or relief from the stereogram. Analytical plotters are all distinguished by the facts that (1) the photographs are mounted in a common plane and (2) a parallax bar provides the basic means of measuring elevations. The Zeiss stereotope and the Nistri-O.M.I. analytical stereoplotter are examples.

The *Zeiss stereotope* (Fig. 31-57) can be used to plot topography free from the effects of tilt and relief when used with vertical photographs of any principal distance and angular field. Tilt or tip can be accommodated up to

4 grads and negative formats of up to 9 by 9 in. A pantograph provides a means of transition from a given average picture scale to a definite mapping scale ranging from 1:5 to about 2.5:1. The plotter uses contact prints which are moved on their carriages underneath floating-mark plates which remain stationary in the field of view. Instead of the photographs being oriented in their relative position as to tilt, they remain level on the carriage, and a correction is made to the parallax reading in the common overlap area by means of mechanical rectifying computers. The computers correct vertical errors due to deformation and warping of the model caused by deviation of the optical axis from the vertical, radial point displacements (horizontal position errors) due to the perspective projection, and horizontal position displacement

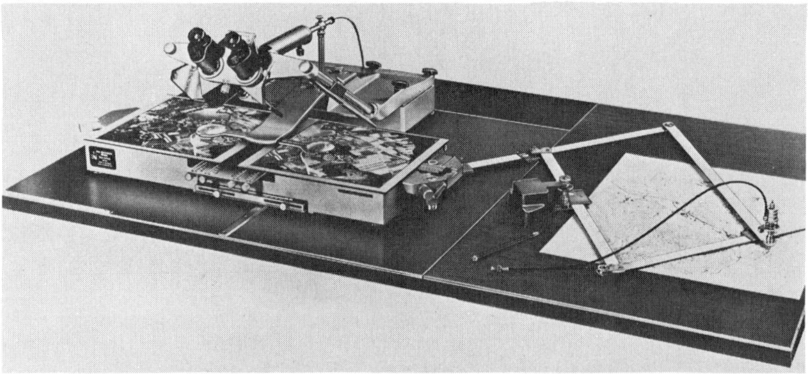

FIG. 31-57. Zeiss stereotope. (*Zeiss-Aerotopograph, Munich.*)

due to picture tilt. The mechanical rectifying computers are set with the aid of horizontal and vertical photocontrol points whose horizontal and vertical positions are known.

The *Nistri-O.M.I. analytical plotter* is a more complex analytical plotter of first-order accuracy based upon the original concept of U. Helava of the National Research Council of Canada; it is the commercial model AP/C of the Nistri-O.M.I. analytical stereoplotter. The plotter (Fig. 31-58) consists of three basic components: the stereoplotter, the electronic computer (Bendix), and the coordinatograph. The stereoplotter optics consist of a direct orthographic viewing binocular train with optical magnification of 6 ×, 10 ×, and 14 × the photoscale. Glass plate diapositives are used; normal, wide-angle, and superwide-angle photography with focal lengths of 3 to 12 in. can be accommodated, and angular deviations of ±5° in photographic tip, tilt, and swing are possible.

Conventional handwheels are used for the x and y motions, and a footwheel for the z or elevation motion. Each photograph is independently mounted on the photocarrier and is free to move in increments of dx, dy, and dK (swing). Differential motions of the photocarriers are executed by appropriate servomechanisms governed by the electronic computer. They are programmed in the computers so as to compensate automatically for y parallax and for image displacement due to scale, tilt, earth curvature, lens distortion, film shrinkage, and so on. Interior, relative, and absolute orientation can be performed either by a mathematically precise automatic solution or by conventional methods. Three types of output are available from the stereomodel: a

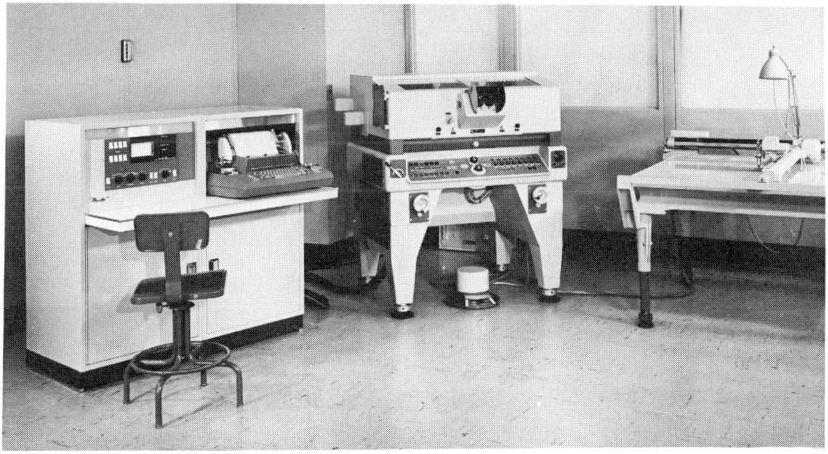

FIG. 31-58. O.M.I. analytical stereoplotter model AP/C. (*O.M.I. Corporation of America.*)

coordinatograph for drawing visual analog tracings, a bank of Nixie-type tubes for visual digital read-out of any constant or model value, and a punch paper tape for recording digital data. In connection with the punch paper tape, a hard-copy typewriter is available.

Stereometer-type Plotting Instruments. For certain mapping applications where accuracy can be sacrificed in favor of low first cost and simplicity and mobility of the equipment, the simple stereometer-type instrument can be used satisfactorily. Differences in elevation are determined with these devices by measuring the differences in absolute stereoscopic parallax in the stereopair and converting these to differences in elevation. For most accurate results, the photography should be rectified to remove tilt distortions. Two examples of this type of instrument are the Fairchild stereocomparagraph and the Zeiss Stereopret.

1. Stereocomparagraph (Fairchild). The stereocomparagraph (Fig. 31-59) was invented by Benjamin B. Talley, the preceding author of this chapter, in 1935. It consists of a drawing attachment, a reflecting stereoscope, a parallax bar, a light fixture, and a parallel-motion device to keep the instrument in proper alinement. The stereoscope is equipped with matching lenses for the magnification of photographic detail. The measuring system consists of two index marks, one etched at the center of each of the two lenses attached to the base of the instrument; these marks may be fused when observed in the stereoscope. The right-hand mark is attached to

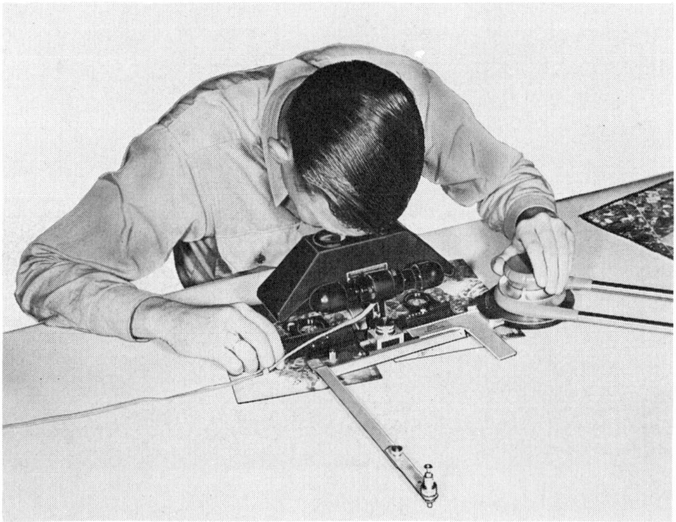

FIG. 31-59. Stereocomparagraph (Fairchild). (*Fairchild Space and Defense Systems.*)

a micrometer screw to change and measure the spacing between the marks. The micrometer is graduated from 0 to 25 mm., and reads directly to 1/100 mm. when clamped for reading. The right-hand mark may also be adjusted through a small range in a direction perpendicular to the micrometer screw. The left-hand mark is held in place by a locating pin and may be adjusted in increments of 5 mm. through a range of 25 mm. Nominal spacing between the marks is 150 mm.

2. Zeiss Stereopret. The Zeiss Stereopret (Fig. 31-60) is an intermediate-stage instrument between a mirror stereoscope with tracing stereometer and third-order plotting instruments. No provision is made for removing image displacements caused by tilt or relief or for correcting errors due to relative orientation. This type of instrument consists essentially of a reflecting

stereoscope, a parallax bar, a drawing attachment, a lighting fixture, and a parallel-motion device to keep the instrument in proper alinement. The drawing attachment for the Stereopret consists of a pantograph attached to the parallax bar; it can be set at any convenient plotting scale from 0.2 to 2.5 times the photoscale.

Orthographic Plotting Instruments. Orthographic plotting instruments provide means for delineating contours and converting to an orthographic projection in one continuous operation, but means external to the instrument must be provided to compensate for tilt in the photography. These instruments give a correct solution only when truly vertical or rectified photography

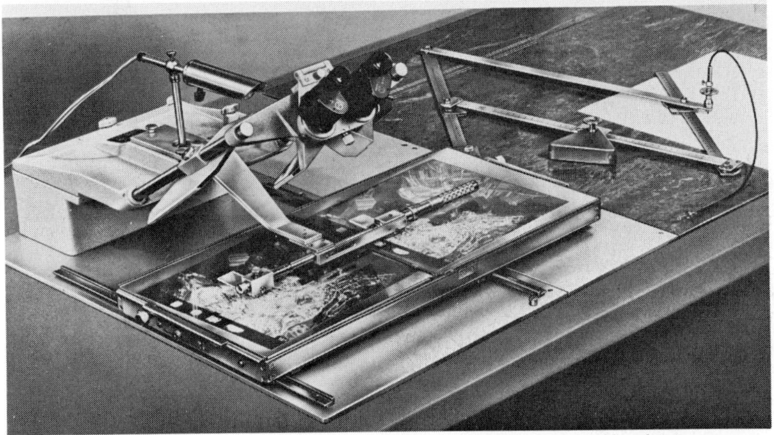

FIG. 31-60. Zeiss Stereopret. (*Zeiss Aerotopograph, Munich.*)

is used. An example of this type of instrument is the KEK stereoscopic plotter.

The KEK plotter (Fig. 31-61) consists of a mirror stereoscope mounted at a fixed height above two photoholders, which may be simultaneously raised or lowered, and a parallel-motion device supporting two measuring marks above the photographs. The stereoscope is designed so that the operator can view the entire overlapping of a pair of 9 by 9-in. photographs. The photoholders may be tilted in any direction, pivoting about a point $8\frac{1}{4}$ in. above the photographs, or they may be rotated in their planes about the holder center. The scale may be adjusted either by varying the distance between the floating marks or by means of the pantograph attached to the measuring-mark carriage. Elevations are measured by raising or lowering the photoholders and thus causing the stereoscopic model to appear to rise and fall with respect to the floating marks. The vertical scale is variable, since the perspective distance from eye to photograph is changed for each change in elevation.

Orthographic Pictorial Representation (*Uniform-scale Photography*). The instrument for this type of representation produces an orthophotograph which covers the area of one stereoscopic model or one photograph. A mosaic of such orthophotographs can be prepared to cover any size of area desired. Horizontal distances, true to scale, can be measured directly from the orthophotograph or the orthophotomosaic. An instrument of this type is the Orthophotoscope (Fig. 31-62), which was developed about 1955 by the U.S. Geological Survey and which converts conventional perspective photographs to the equivalent of orthographic photographs. With this instrument, photographs are produced having no displacement of images due to tilt or

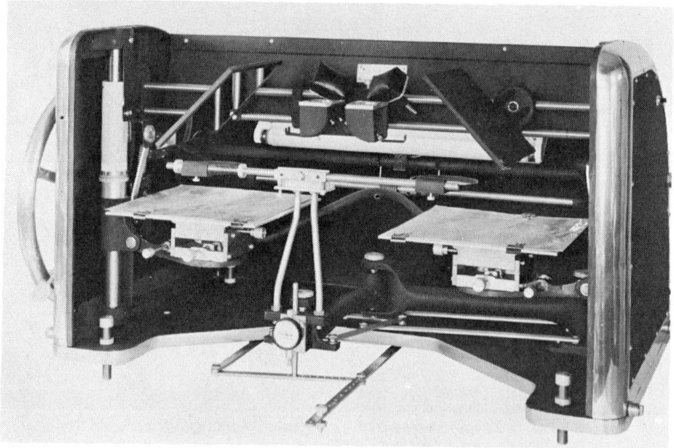

FIG. 31-61. KEK plotter. (*Philip B. Kail Associates, Inc.*)

relief, so that horizontal distances can be measured accurately on the photograph. Such uniform-scale photographs have many important uses in engineering, geology, forestry, and other scientific fields. The instrument is operated in conjunction with an oriented model formed by any type of double-projection stereoscopic plotter using the anaglyphic principle. The present device utilizes the Balplex projector.

In operation, a sensitized film is "scanned" by a slit in a screen on which the model is projected. The elevation of the film is varied according to the terrain as the scanning proceeds. The sensitized surface is an orthographic film which is sensitive only to the blue light and not to the red light of the usual anaglyphic projection. When the film scanning is complete, the film is developed as a negative and the "orthophotographs" are printed from this negative in any quantity or scale. Figure 31-63 compares corresponding portions of a perspective photograph (left) and an orthophotoscope (right). In the perspective photograph the power line appears to be crooked, while in the orthophotoscope it is shown in its true straight alinement.

Some form of horizontal and vertical control is required for absolute orientation of each model. Control placed for stereoplotting instruments can be used, and since the accuracy requirement for vertical control is not so critical as that for horizontal control, vertical orientation based upon inspection of the terrain may be adequate for some applications.

FIG. 31-62. Orthophotoscope. (*U.S. Geological Survey.*)

31-49. Automation of Mapping Instruments. Major efforts toward automation of photogrammetric instrumentation proceed along two general lines of attack. One effort aims at automating the map-drawing operation itself; the other at utilizing modified mapping equipment in combination with electronic computing equipment to produce engineering data directly from the stereoscopic model without requiring the complete map as an intermediary.

The first approach has not as yet become economically feasible, although the technical literature is rich in potential. A summary of advances to date is given in Ref. 6*a* at the end of this chapter.

An example of the second approach is the terrain data translator (TDT), shown in Fig. 31-64, which makes possible the automatic digital recording of the elevation and horizontal location in a three-dimensional stereomodel produced by projection-type plotters such as the multiplex, Kelsh, and Balplex.

The TDT system is in two parts: a first-order precision reading unit to which a modified tracing table is connected and an electronic unit for keeping count of the horizontal position of the plotting table and for reading the data into punched cards, punched tape, or hard copy (remote-controlled typewriter).

This type of device is commonly applied to earthwork computations for highway route surveys where alternative routes are under consideration.

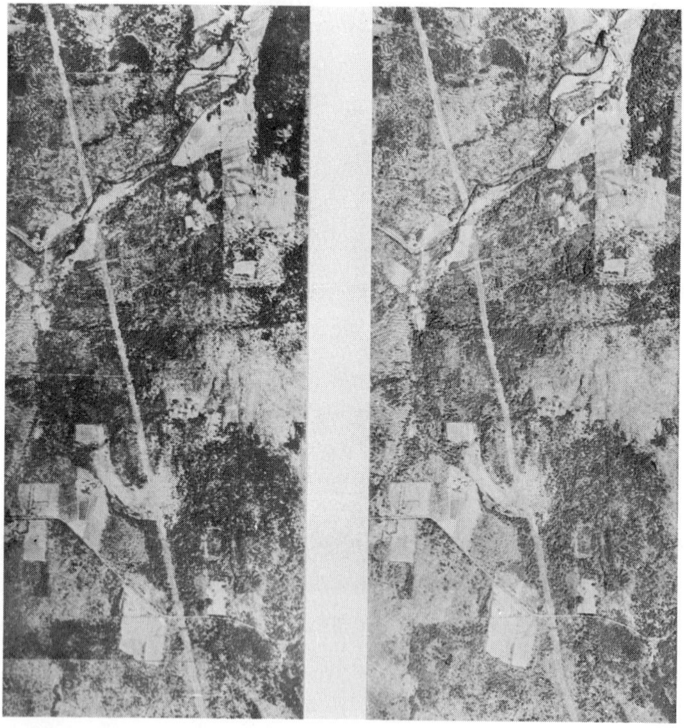

FIG. 31-63. Comparison of perspective photograph (left) and orthophotograph. (*U.S. Geological Survey.*)

In this application the modified tracing table of a conventional stereoscopic plotting instrument is hooked up to the readout device, which in turn feeds data into an electronic computer. Position and height readings of points in the highway cross-section (or profile) in the stereoscopic model are fed to the computer, which is programmed to convert these readings to volume computations.

Another approach to automation in mapping aside from that applicable to the stereoplotting process is the Electroplotter II (Fig. 31-65), a digital input graph plotter designed for plotting points or alphanumeric symbols and for drawing single or multicolor lines. Input data may be supplied from computer-generated magnetic tapes, punch cards, or punched tape, or directly on-line from the computer output coupler. The usable plotting area is 30 by 30 in.,

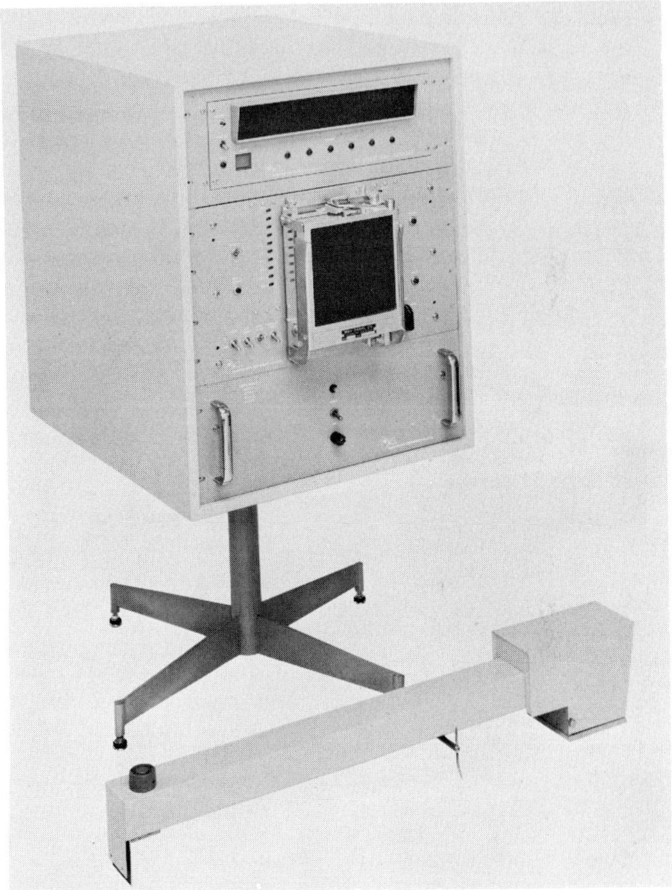

FIG. 31-64. Terrain data translator. (*Benson-Lehner Corporation.*)

and paper, mylar, or acetate sheets can be accommodated. Sheets are held down by means of a vacuum system using a multichamber honeycomb table with pinhole outlets to the plotting surface.

31-50. Other Applications of Photogrammetry. Although the most common application of photogrammetry is in the preparation of topographic maps, photogrammetry is by no means limited to this field.

During recent years aerial photographs and photogrammetry have become widely used in the preparation of property-ownership maps for tax-equalization studies. Through their use the petroleum geologist, who formerly spent about 90 per cent of his time and effort in keeping himself located and oriented on the ground and 10 per cent of his time on geology, is now able to reverse these percentages. Aerial photographs are used to study and

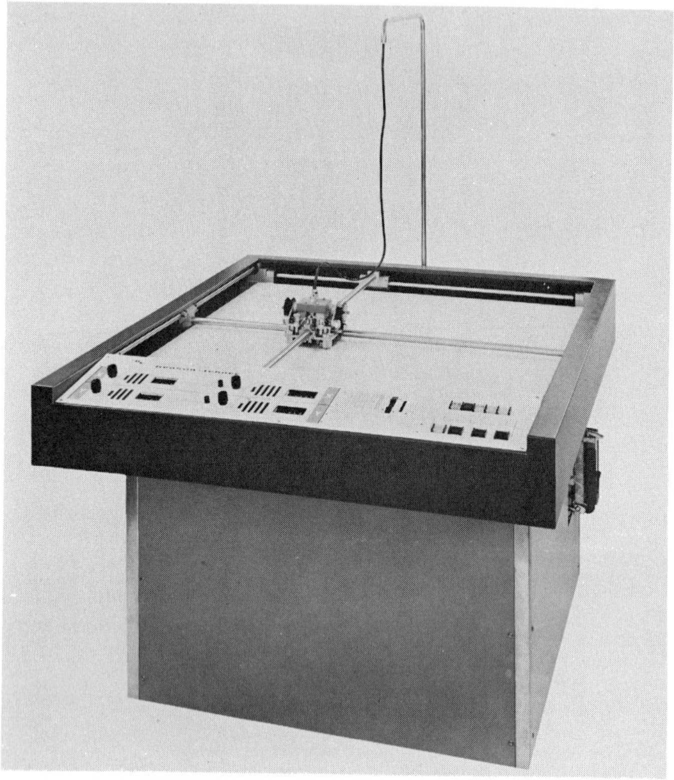

FIG. 31-65. Electroplotter II. (*Benson-Lehner Corporation.*)

catalogue geological formations and land classifications in mining and in aerial exploration. In forestry they are used for the classification of growing timber, for the determination of tree heights for the estimation of merchantable timber, and for other studies. The Federal government has used tremendous numbers of photographs for rural-rehabilitation studies and in connection with soil-conservation and erosion projects. Aerial photographs are increasingly used in the determination of correct land usages for agricultural purposes. Aerial maps find a wide application in city, county, and regional planning and development and in general engineering studies such as highway, pipeline, and transmission-line locations.

One of the more recent developments and one which is becoming world-wide in its scope is the use of photogrammetry in connection with geophysical prospecting by means of the magnetometer. In unmapped areas, the movement of the airplane is determined by shoran, and over land or shallow water the ground is photographed with either a shutter-type camera or a continuous-strip camera (page 195 of Ref. 1 at the end of this chapter). This work is done at altitudes of 500 to 2,500 ft. above the ground by special photographic techniques. Color photography is often used in this work, since the varying tints of the geological formations are more readily discernible in natural color than in black-and-white photographs.

Strictly military applications of aerial photography and photogrammetry have been omitted, since they are myriad in extent and usefulness.

31-51. Numerical Problems.

1. A photograph which contains a plateau the elevation of which is 5,000 ft. and a valley which is 3,000 ft. lower was taken with a camera having an 8.25-in. focal length at an altitude of 20,000 ft. above sea level. (*a*) What would be the scale, on the photograph, of the plateau? (*b*) What would be the scale, on the photograph, of the valley?

2. How many feet on the ground would 1 in. represent on a photograph taken with a 6-in.-focal-length camera at an altitude of 9,000 ft.?

3. A picture is taken from a plane traveling 300 miles per hour, with the shutter speed of the camera set at $\frac{1}{300}$ sec. (*a*) How far would the camera move during the exposure time? (*b*) What effect does this have on the photograph?

4. Assuming truly vertical photographs and level terrain, what would be the scale of the following photographs: (*a*) flight height above sea level 12,000 ft., elevation of level terrain 2,000 ft. above sea level, and focal length of camera 6 in.; (*b*) same, except camera focal length $8\frac{1}{4}$ in.; (*c*) same, except camera focal length 4 in.

5. The ground distance between two crossroads which have the same elevation is 1 mile. The distance between the images of these two crossroads on a truly vertical photograph is 3 in. What is the scale of the photograph? If the camera which took the photograph had a focal length of 153 mm. and the road elevation was 1,500 ft. above sea level; what would be the flight height above sea level?

6. What would be the displacement due to relief on a truly vertical photograph for the following cases: (*a*) The Washington Monument is imaged on a vertical photograph. The ground position of the nadir point is 3,000 ft. from the center of the monument. The flight height is 12,000 ft. above sea level. The elevation of the base of the monument is 55 ft. and the elevation of the top of the monument is 610 ft. The focal length of the camera is 153 mm. (*b*) Under the same conditions, except that a camera with focal length of $8\frac{1}{4}$ in. is used.

7. In both cases in problem 6, what would the displacement amount to on the ground? Have you ever heard it said, "Long-focal-length cameras give less displacement due to relief?" What do *you* think?

8. What would be the image scale at the top and at the bottom of the monument under the condition of problem 6*a*?

9. In a stereopair of photographs, what produces parallax?

10. What effect does rectification have upon displacement due to tilt and relief?

11. Given the following conditions applicable to a photogrammetric survey: size of photographs 9 by 9 in., overlap 60 per cent, side lap 30 per cent, flight altitude 10,000 ft, focal length of camera lens 12 in. (*a*) Determine the minimum number of photographs required to cover an area of 50 sq. miles. (*b*) Determine the actual number of photographs that would be obtained if the area was 5 by 10 miles and was flown in strips parallel to the side 5 miles long. (*c*) Determine the number of photographs that would be obtained if the area was flown in strips parallel to the 10-mile length. (*d*) Prepare schematic maps corresponding to requirements *b* and *c* above.

12. Given the conditions of problem 11. If the speed of the airplane is 120 miles per hour, what is the time interval between exposures?

13. Assuming the cost of a photogrammetric survey to vary directly as the number of photographs involved, determine the relative costs of conducting the survey of problem 12, with photographs taken with the same camera at altitudes of 5,000, 10,000, and 15,000 ft.

14. For a certain photograph the following conditions are given: scale of photograph 1/10,000, focal length 12 in., distance on the photograph from principal point to pictured location of an object 3.5 in., object 500 ft. above the datum plane. (*a*) What is the linear displacement of this point on the photograph? (*b*) What would be the linear displacement had the photograph been taken under the foregoing conditions except that the focal length of the camera was 8¼ in.? (*c*) Which of the photographs would be more suitable for a mosaic? For contouring?

15. Another point in the photograph of problem 14 is located at a distance from the principal point of 4.2 in., and its linear displacement is found to be 0.063 in. away from the principal point. What is the elevation of the point with respect to the datum?

16. Determine the maximum true flight altitude of photography over terrain with elevations ranging from sea level to 800 ft., with an 8¼-in. lens covering a 9 by 9-in. photograph and with a 60 per cent overlap that will permit accurate contouring at a contour interval of 20 ft. using Balplex (ER-55) projectors.

17. Determine the number of Balplex (ER-55) projectors necessary to bridge a distance of 10 miles between control points if the photographs are taken at a flight altitude of 12,000 ft. with an aerial camera of 6-in. focal length and a 9 by 9-in. format, with 60 per cent overlap in the line of flight. What is the total area that could be compiled under the foregoing conditions?

18. Compute a parallax table for a contour interval of 40 ft. applicable to a 9 by 9-in. photograph taken with a 6-in.-focal-length camera at 10,000 ft. flight altitude, with 60 per cent overlap, over terrain with elevations ranging from sea level to 600 ft., for use in plotting contours with the stereocomparagraph.

19. Compile a planimetric line map by the radial-line method of an area in the vicinity of the campus (or any level area for which suitable ground control and aerial photographs are available) using a strip of five vertical aerial photographs having horizontal control in the overlap of the first and last overlapping pairs of photographs.

REFERENCES

1. AMERICAN SOCIETY OF PHOTOGRAMMETRY, "Manual of Photogrammetry," 2d ed. (bibliography, pp. 845–860), Washington, D.C., 1952.

2. American Society of Photogrammetry, "Manual of Photographic Interpretation," Washington, D.C., 1960.
3. Anderson, R. O., "Applied Photogrammetry," 4th ed., Edwards Brothers, Inc., Ann Arbor, Mich., 1946.
4. Bagley, J. W., "Aerophotography and Aerosurveying," McGraw-Hill Book Company, New York, 1941.
5. Baker, Wilfred H., "Elements of Photogrammetry," The Ronald Press Company, New York, 1960.
6. Church, E., "Elements of Aerial Photogrammetry," Syracuse University Press, Syracuse, N.Y., 1944.
6a. De Meter, "Latest Advances in Automatic Mapping," *Photogrammetric Engrg.*, Vol. XXIX, No. 6, pp. 1027–1036, November 1963.
7. Hallert, B., "Photogrammetry," McGraw-Hill Book Company, New York, 1960.
8. International Training Center for Aerial Surveys, "International Bibliography of Photogrammetry," Delft, Holland, current edition.
9. Judge, A. W., "Stereoscopic Photography, Its Application to Science, Industry, and Education," Chapman and Hall, Ltd., London, 1950.
10. Lyon, Duane, "Basic Metrical Photogrammetry," published by the author, St. Louis, 1959.
11. McNeil, Gomer T., "Photographic Measurements; Problems and Solutions," Pitman Publishing Corporation, New York, 1954.
12. Merritt, Everett L., "Analytical Photogrammetry," Pitman Publishing Corporation, New York, 1958.
13. Moffitt, Francis H., "Photogrammetry," International Textbook Company, Scranton, Pa., 1959.
14. Schwiderfsky, K., "An Outline of Photogrammetry" (translated by John Fosberry), Pitman Publishing Corporation, New York, 1959.
15a. Sharp, H. O., "Photogrammetry," 3d ed., John Wiley & Sons, Inc., New York, 1943.
15b. Sharp, H. O., "Practical Photogrammetry," The Macmillan Company, New York, 1951.
16. Talley, B. B., "Engineering Applications of Aerial and Terrestrial Photogrammetry," Pitman Publishing Corporation, New York, 1938.
17a. Trory, Lyle G., "Handbook of Aerial Mapping and Photogrammetry," 2d ed., Cambridge University Press, New York, 1952.
17b. Trory, Lyle G., "Slotted Templet Error," *Photogrammetric Engrg.*, Vol. XIII, No. 2, pp. 227–229, June, 1947.
18a. U.S. Coast and Geodetic Survey, "Topographic Manual, Part II, Photogrammetry," by L. W. Swanson, *Spec. Pub.* 249, Government Printing Office, Washington, D.C., 1951.
18b. U.S. Coast and Geodetic Survey, "Notes on the Compilation of Planimetric Line Maps from Five-lens Aerial Photographs," Washington, D.C., 1933.
19. U.S. Geological Survey, "Topographic Instructions," Government Printing Office, Washington, D.C. (published in separate volumes according to subject; list available from USGS).
20a. Von Gruber, Otto, "Photogrammetry," American Photographic Publishing Co., Boston, 1932.
20b. Von Gruber, Otto, "Photogrammetry—Collected Papers and Essays," American Photographic Publishing Co., Boston, 1942.
21a. Whitmore, G. D., "Elements of Photogrammetry," International Textbook Company, Scranton, Pa., 1941.

21*b*. WHITMORE, G. D., "Advanced Surveying and Mapping," International Textbook Company, Scranton, Pa., 1949.
22. ZELLER, M., "Textbook of Photogrammetry" (translated by E. A. Miskin and R. Powell), H. K. Lewis & Co. Ltd., London, 1952.

PERIODICALS

23. *Photogrammetric Engineering,* published five times a year by the American Society of Photogrammetry, Washington, D.C.
24. *Surveying and Mapping,* published quarterly by the American Congress on Surveying and Mapping, Washington, D.C.
25. *The Canadian Surveyor,* published quarterly by the Canadian Institute of Surveying and Photogrammetry, Ottawa, Canada.
26. *Photogrammetria,* published semiannually by the International Society of Photogrammetry, Delft, Holland.

CHAPTER 32

MAP PROJECTIONS

32-1. Maps of Small Areas. In plotting a map of a small area the curvature of the earth need not be considered. A level surface is assumed to be a plane, and points are usually plotted on the map by linear rectangular coordinates, or given distances from selected north-south and east-west coordinate axes.

32-2. Maps of Large Areas. For maps of larger areas this simple method is not satisfactory because of the sphericity of the earth. (Actually, spheroidicity; however, the earth is very nearly a sphere, since the polar diameter is only one third of 1 per cent shorter than the equatorial diameter.) It is impossible to represent the surface of a sphere on a plane surface without distortion, just as it is impossible to flatten a section of orange peel without tearing it. In consequence, any plane map of a relatively large area of spherical surface must be distorted to some degree. For example, a great circle on the surface of the earth should appear as a straight line on the map, but on most plane maps a great circle is represented by a curved line in order to avoid undue distortions in shape and area of the territory represented. Again, it has been shown in Chapter 23 that meridians are not parallel but that they converge toward the poles, the angular convergency varying with the latitude; nevertheless the horizontal projections of all meridians are straight lines and the curved parallels of latitude are always perpendicular to them.

32-3. Map Projection Defined. In maps of large areas where curvature becomes important, it is necessary to locate points by coordinates which are the geographical latitudes and longitudes expressed in angular units; for example, New York is at a latitude 40°45′ north of the equator and at a longitude 74°00′ west of Greenwich. Points are plotted with respect to a series of lines representing the earth's parallels and meridians. Any system of representing these parallels and meridians on a plane surface is called a *map projection.*

32-4. Ideal vs. Practicable Projection. On a theoretically perfect map, without distortion, the following conditions would be satisfied: (1) All distances and areas would have correct relative magnitudes, (2) all azimuths and angles would be correctly shown, (3) all great circles would appear as straight lines, and (4) geographic latitudes and longitudes of all points would be correctly shown. Although in a plane map not all of these requirements can be satisfied at the same time, one or more conditions may be satisfied, as follows:

1. An *equal-area* projection results in a map showing all areas in proper relative *size,* although these areas may be much out of shape and the map may have other defects.

2. A *conformal* or *orthomorphic* projection results in a map showing the correct angle between any pair of short intersecting lines, thus making small areas appear in correct *shape.* As the scale varies from point to point, the shapes of larger areas are incorrect.

3. An *azimuthal* projection results in a map showing the correct *direction* or azimuth of any point from one central point.

32-5. Types of Projections. In the following paragraphs a few of the more important types of projections are briefly described. These descrip-

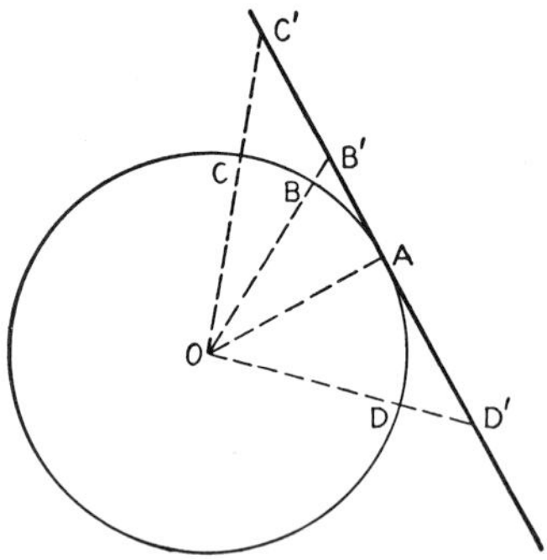

FIG. 32-1. Gnomonic projection.

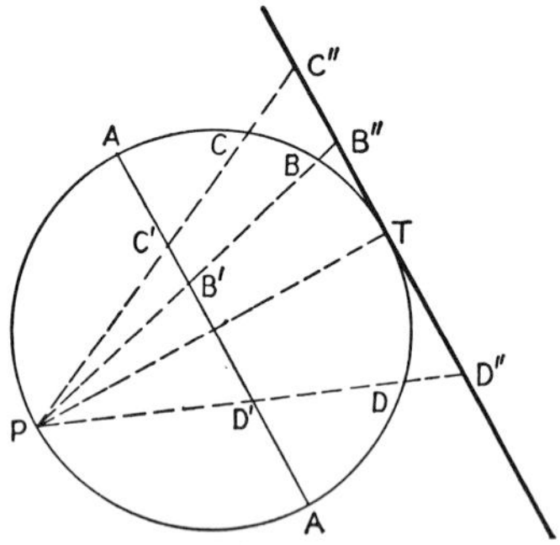

FIG. 32-2. Stereographic projection.

tions are of a general nature, involving little mathematical treatment, and no attempt has been made to make the list complete. For more detailed treatment the student is referred to the references listed at the end of this chapter, particularly to *Special Publication* 68, "Elements of Map Projection," of the U.S. Coast and Geodetic Survey. Some of the projections are true projections in the geometrical sense; others are map projections in the sense that they represent the parallels and meridians on a plane surface, although they cannot be obtained by any perspective or geometric projecting process. Some of the more useful map projections are of this second kind.

32-6. Gnomonic Projection. This is a geometric projection to a plane which is tangent to the sphere at any point A (Fig. 32-1). Radiating lines from the earth's center O through such points (B, C, etc.) on the surface of the earth as are to be shown on the map are produced to an intersection with the tangent plane, where the point in question is plotted. In the figure, B is plotted at B', C at C', and so on. These plottings in the tangent plane give in that plane a map constructed on the gnomonic projection. The important property of maps made on this projection is that they show great circles as straight lines, which renders them useful for navigational purposes; otherwise they are not particularly useful except for the part of the map near the point of tangency. The shapes and sizes of areas are much distorted except near the tangent point.

32-7. Stereographic Projection. This is a geometric projection to a plane. Let A–A (Fig. 32-2) represent any circle of the earth (in practice, generally a great circle), and let P represent the pole of that circle. Then by lines radiating from P to points B, C, and so on, these points are projected to form a map in the plane A–A, and are represented on that map by the points B', C', etc. If the plane of projection is tangent to the sphere, as at T (Fig. 32-2), the points are plotted at B'', C'', etc. This is a conformal projection and is an excellent one for general maps showing a hemisphere; its main defect is that areas are not correctly shown.

32-8. Orthographic Projection. This is a geometric projection to a plane tangent to the sphere at any point; the projecting lines are parallel and are perpendicular to the tangent plane in which the map is constructed. If the central tangent point of the map is at one of the poles of the earth, each parallel of latitude is shown correctly to scale, but the distance between parallels becomes rapidly smaller as we depart farther from the center of the map. The map is true to scale along the parallels but not along the meridians. Different but comparable results are obtained if the tangent plane touches the earth's surface at some point other than the pole. Maps of the surface of the moon are usually constructed on this projection.

32-9. Geometric Projections to a Cylinder. The surface of a cylinder is curved in one direction only and can be developed into a plane. Advantage is taken of this fact in the so-called cylindrical projections. The cylinder used may cut the sphere but is usually tangent along a great circle, usually the equator. The projecting lines used may radiate from the center of the sphere or may all be parallel to the equatorial plane. These particular projections are little used, but a modification of the cylindrical projection, called the *Mercator projection*, possesses some valuable properties (see Arts. 32-13 and 32-14).

32-10. Geometric Projections to a Cone. Like the surface of a cylinder, the surface of a cone is capable of development without distortion into a plane. On this account a number of conical projections have been devised, using sometimes a cone tangent along a parallel of latitude and sometimes a cone cutting the sphere along two parallels. The lines of projection may either emanate from a central point or be parallel to each other and to the plane of the chosen parallel. But here again, the more important and valuable projections are not of the geometric-projection type.

32-11. Polyconic Projection. Instead of a single cone, a series of conical surfaces may be used, points on the surface of the earth being considered as projected to a series of frustums of cones which are fitted together. These conical surfaces are then developed each way from a central meridian. Owing to differences in radii, the resulting strips would not exactly fit together when laid flat, but spaces would appear between them, such spaces increasing in width as the distance from the central meridian increases (Fig. 32-3*b*). To avoid such spaces, the north-south scale must be modified along the various meridians. Upon such a system of lines points are plotted by latitude and longitude.

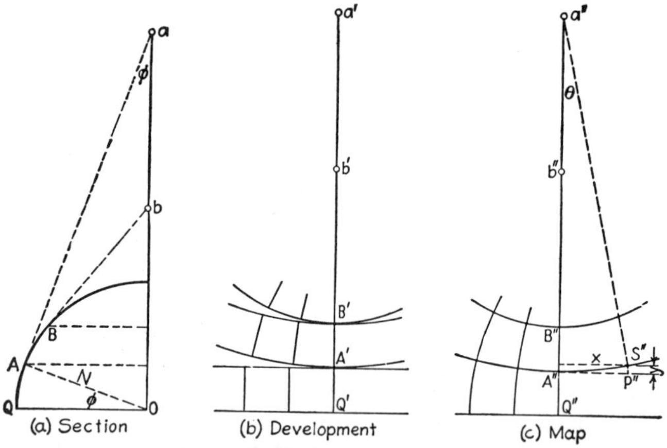

FIG. 32-3. Polyconic projection.

In Fig. 32-3, it is seen that each parallel of latitude appears on the map as the arc of a circle having as radius the corresponding tangent distance; the parallel through A has a radius Aa, that through B has a radius Bb, and so on. The centers of these circles all lie on the central meridian of the map. The length of each tangent distance Aa, etc., is $N \cot \phi$ in which N is the length of the normal or vertical at latitude ϕ extended to its intersection with the earth's axis. For the assumption that the earth is a sphere, N is equal to the radius of the sphere. More precisely,

$$N = \frac{a}{\sqrt{1 - e^2 \sin^2 \phi}} \tag{1}$$

in which a is the earth's equatorial radius and e is the eccentricity of the ellipse in the meridian section ($e^2 = 0.00676866$). The length of the tangent distance $N \cot \phi$ varies with the latitude.

The distances $Q'A'$, $A'B'$, etc., along the central meridian on the map (Fig. 32-3*b*) are true scale representations of the corresponding arc distances QA, AB, etc. in the meridian section (Fig. 32-3*a*). The parallels drawn on the map may be selected with as small a difference in latitude as may be desired, and each one is drawn with its own particular radius as shown, with the center of the arc on the central meridian at the proper tangent distance (to scale) above the point where the parallel cuts the central meridian.

It should be observed that the method of drawing these arcs of the parallels of latitude on the map is such that each parallel is separately developed as the circumference of the base of its own distinct cone, and that the spacing between them increases with increasing differences of longitude from the central meridian, thereby changing the north-south scale of the map from place to place as the longitude difference increases.

The arc distance $A''S''$ in the map (Fig. 32-3*c*) represents to true scale the difference in longitude between the points A'' and S''. The angle $A''a''S''$ is the angle θ of Art. 23-11, and from Eq. (1) of that article

$$\theta = \lambda \sin \phi$$

in which λ is the arc $A''S''$. The rectangular coordinates of the point S'' referred to A'' as origin are

$$x = A''P'' = a''S'' \sin \theta = N \cot \phi \sin \theta \qquad (2)$$

$$y = P''S'' = a''S'' \operatorname{vers} \theta = N \cot \phi \operatorname{vers} \theta \qquad (3)$$

and if the chord $A''S''$ is drawn, in the triangle $A''S''P''$, $S''P'' = A''P''$ $\tan P''A''S''$, or $y = x \tan (\theta/2)$.

Values of x and y have been computed and are tabulated in *Special Publication* 5 of the U.S. Coast and Geodetic Survey.

As many points as desired along the parallels on the map, like point S'', are plotted by the use of the table. Then the meridians are drawn through such points. These meridians are curved, concave toward the central meridian, but if the parallels are drawn close enough together each meridian may be drawn as a series of straight lines from parallel to parallel. On the network of parallels and meridians so prepared, points are plotted by latitude and longitude. Near the central meridian there is little error in such a map, but the error increases in proportion to the square of the difference in longitude along any one parallel. The variation with difference in latitude is not in direct proportion.

It is to be noted that along the central meridian and along every parallel the map is true to scale; that along the other meridians the scale is somewhat changed; that near the central meridian the parallels and meridians intersect nearly at right angles; and that areas of great extent north and south may be mapped with a very small distortion.

Although better adapted to mapping an area of great extent in latitude than for an area of great extent east and west, the polyconic projection is sufficiently accurate for maps of considerable areas, and it is widely used by the U.S. Geological Survey and the U.S. Coast and Geodetic Survey.

32-12. Lambert Conformal Conic Projection. Attention was called to this excellent projection by its use for the French battle maps during the First World War. Tables for its construction have been published by the U.S. Coast and Geodetic Survey. It is used for the state plane coordinate systems of states (or zones thereof) of greater east-west than north-south extent (Art. 16-30).

This is a simple conic projection, the cone used being imagined to cut the surface of the earth along two parallels of latitude, called *standard parallels* (Fig. 32-4). When points on the earth's surface are projected to such a cone, there is a slight compression or decrease of scale between the standard

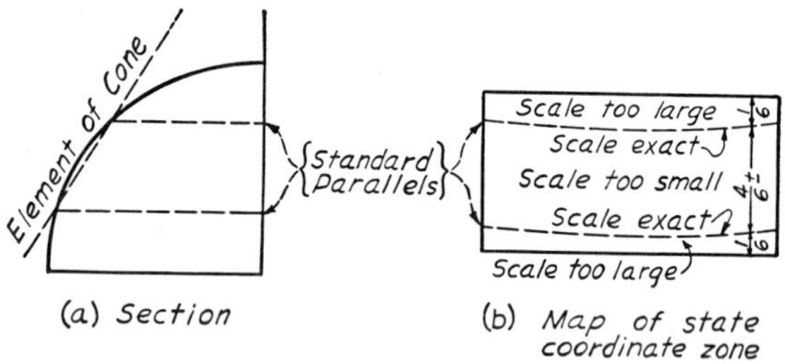

FIG. 32-4. Lambert conformal conic projection.

parallels, and a stretching or increase of scale outside the standard parallels. Only a slight adjustment of scales is necessary to make the map conformal. It has been shown that, for a map of the United States, scale errors need not exceed 2 per cent at any point. For details the reader is referred to publications of the U.S. Coast and Geodetic Survey dealing with this projection.

32-13. Mercator Projection. As previously stated, this projection is cylindrical, but it cannot be constructed as a geometrical projection.

In a cylindrical projection, formed by means of a cylinder touching the earth along the equator, all meridians appear as straight parallel lines. But on the sphere any two such meridians are a maximum distance apart at the equator and converge toward the poles. Showing them parallel, therefore, results in a systematically increasing scale along the parallels of latitude as we pass from the equator toward the pole, with resulting distortion of all areas shown on the map. The change of scale along the parallel, varying with the latitude, is readily computed (still assuming the earth to be spherical) by means of the formula

$$S' = S \cos \phi \tag{4}$$

in which S is the scale at the equator and S' is the scale at any latitude ϕ. For example, if the equatorial scale of the map is 1,000 miles per inch, then at latitude 60° (since the cosine of 60° is ½) the scale is 500 miles to the inch.

The particular feature of the Mercator projection is that the scale along the meridian is varied to agree with the scale along the parallel, so that although the scale varies from point to point on the map at any given point the scale is the same in all directions. The map is therefore conformal. It has also the important property that a line of constant true bearing or *rhumb line*, appears straight, which property renders it invaluable for purposes of navigation. The shortest course between two points is determined by drawing on a gnomonic chart a great circle, which there appears as a straight line. Selected points, at convenient distances apart, of this great circle are then plotted on the Mercator chart, after making any necessary corrections on account of shoals, wind, currents, etc. The rhumb line connecting any two adjacent points indicates the true bearing of the course, which is read by means of a protractor. This true bearing, corrected for magnetic declination, gives the compass bearing to be used in steering.

Owing to the rapid variation of scale, maps constructed on the Mercator projection give very inaccurate information as to relative sizes of areas in widely different latitudes. For example, on the map Greenland appears larger than South America, whereas in fact South America is nine times as large as Greenland. Consequently, such a map is not suited to general use, although because of its many other advantages it is widely published.

32-14. Transverse Mercator Projection. A transverse Mercator projection is the ordinary Mercator projection turned through an angle of 90° so that it is related to a central meridian in the same way that the ordinary Mercator projection is related to the equator. This projection is used for the state plane coordinate systems of states (or zones thereof) of greater north-south than east-west extent (Art. 16-30). For the state systems the Mercator projection cylinder is made to cut the surface of the sphere along two standard lines parallel to the central meridian instead of being tangent to the sphere as in the ordinary Mercator projection.

32-15. Spheroidicity of the Earth. In the foregoing discussion it has been assumed that the earth is spherical, as it is very nearly. Actually, however, the meridian section of the earth is an ellipse, the polar diameter being some twenty-seven miles shorter than the equatorial diameter. This fact is recognized in the mathematical solutions of the problems involved and in the preparation of tables for the various map projections. The radius of curvature in the meridian is different for different latitudes, as is also the length of the normal from the surface terminating in the polar axis. Other dimensions depart correspondingly from those of a true sphere.

It is evident that this variation from the truly spherical shape does not change the nature of the various map projections that have been discussed, although it does make necessary certain corrections to and changes in the numerical values of the quantities used for plotting the different projections. A discussion of these refinements is beyond the scope of this book, and the reader is referred to the publications dealing with map projections and with geodesy in general. A few of these publications are listed as references at the end of this chapter.

REFERENCES

1. NATIONAL GEOGRAPHIC SOCIETY, "The Round Earth on Flat Paper," *Nat. Geographic Soc., Monograph*, 1947.
2. RAISZ, ERWIN, "Principles of Cartography," McGraw-Hill Book Company, New York, 1962.
3. STEERS, J. S., "An Introduction to the Study of Map Projections," 3d ed., University of London Press, London, England, 1933.
4. U.S. COAST AND GEODETIC SURVEY, *Spec. Pubs.* 5, 57, 68, 193, 194, 195, 205, 235, Government Printing Office, Washington, D.C.

GENERAL TABLES

TABLE I. CORRECTION FOR REFRACTION AND PARALLAX, TO BE SUBTRACTED FROM THE OBSERVED ALTITUDE OF THE SUN

Barometric pressure, 29.5 in.

App't alt.	Temperature										App't alt.
	−10° C. +14° F.	−5° C. +23° F.	0° C. +32° F.	+5° C. +41° F.	+10° C. +50° F.	+15° C. +59° F.	+20° C. +68° F.	+25° C. +77° F.	+30° C. +86° F.	+35° C. +95° F.	
°	′	′	′	′	′	′	′	′	′	′	°
10	5.52	5.42	5.30	5.20	5.10	5.00	4.92	4.83	4.75	4.67	10
11	5.02	4.92	4.82	4.73	4.63	4.55	4.47	4.38	4.32	4.23	11
12	4.60	4.50	4.42	4.33	4.25	4.17	4.10	4.03	3.97	3.88	12
13	4.23	4.15	4.07	4.00	3.92	3.85	3.78	3.72	3.65	3.58	13
14	3.92	3.83	3.77	3.70	3.62	3.55	3.50	3.45	3.37	3.32	14
15	3.65	3.58	3.50	3.43	3.37	3.32	3.25	3.20	3.13	3.08	15
16	3.43	3.35	3.30	3.23	3.17	3.12	3.07	3.00	2.95	2.90	16
17	3.22	3.15	3.10	3.03	2.98	2.92	2.88	2.82	2.77	2.72	17
18	3.02	2.95	2.90	2.85	2.80	2.75	2.70	2.65	2.60	2.55	18
19	2.83	2.78	2.73	2.68	2.63	2.58	2.53	2.48	2.43	2.40	19
20	2.68	2.63	2.58	2.53	2.48	2.43	2.38	2.33	2.30	2.27	20
21	2.53	2.48	2.43	2.38	2.35	2.30	2.27	2.22	2.17	2.13	21
22	2.38	2.35	2.30	2.25	2.22	2.18	2.13	2.08	2.05	2.02	22
23	2.28	2.25	2.20	2.15	2.12	2.08	2.03	1.98	1.95	1.93	23
24	2.17	2.13	2.08	2.05	2.02	1.98	1.93	1.88	1.87	1.83	24
25	2.07	2.03	1.98	1.95	1.92	1.88	1.83	1.80	1.77	1.75	25
26	1.99	1.95	1.90	1.87	1.83	1.80	1.75	1.72	1.70	1.67	26
27	1.88	1.85	1.82	1.78	1.75	1.72	1.68	1.63	1.62	1.60	27
28	1.80	1.77	1.72	1.70	1.67	1.63	1.60	1.57	1.53	1.52	28
29	1.72	1.68	1.65	1.63	1.60	1.57	1.53	1.50	1.47	1.46	29
30	1.65	1.62	1.58	1.57	1.53	1.50	1.47	1.45	1.42	1.40	30
32	1.53	1.50	1.47	1.45	1.42	1.38	1.35	1.33	1.30	1.28	32
34	1.41	1.37	1.35	1.32	1.30	1.27	1.25	1.23	1.20	1.18	34
36	1.30	1.27	1.25	1.22	1.20	1.18	1.15	1.13	1.10	1.08	36
38	1.20	1.18	1.15	1.13	1.12	1.10	1.07	1.05	1.02	1.02	38
40	1.11	1.10	1.07	1.05	1.03	1.02	0.98	0.97	0.95	0.93	40
42	1.03	1.00	0.98	0.97	0.95	0.93	0.90	0.88	0.87	0.87	42
44	0.96	0.93	0.92	0.90	0.88	0.87	0.85	0.83	0.82	0.80	44
46	0.89	0.88	0.87	0.85	0.83	0.82	0.80	0.78	0.77	0.75	46
48	0.83	0.82	0.80	0.78	0.77	0.75	0.73	0.72	0.70	0.68	48
50	0.77	0.75	0.73	0.72	0.70	0.68	0.67	0.67	0.65	0.63	50
55	0.63	0.62	0.60	0.60	0.58	0.57	0.57	0.55	0.53	0.52	55
60	0.52	0.52	0.50	0.50	0.48	0.47	0.47	0.45	0.45	0.43	60
65	0.42	0.40	0.40	0.40	0.38	0.38	0.37	0.37	0.35	0.33	65
70	0.32	0.32	0.32	0.30	0.30	0.30	0.28	0.28	0.28	0.27	70
75	0.23	0.23	0.23	0.22	0.22	0.22	0.20	0.20	0.20	0.18	75
80	0.15	0.15	0.13	0.13	0.13	0.13	0.13	0.12	0.12	0.12	80
85	0.07	0.07	0.07	0.07	0.07	0.07	0.07	0.05	0.05	0.05	85
90	0.00	0.00	0.00	0.00	0.00	0.00	0.00	0.00	0.00	0.00	90

Table II. Correction for Refraction, to Be Subtracted from the Observed Altitude of a Star

Barometric pressure, 29.5 in.

App't alt.	Temperature										App't alt.
	−10° C. +14° F.	−5° C. +23° F.	0° C. +32° F.	+5° C. +41° F.	+10° C. +50° F.	+15° C. +59° F.	+20° C. +68° F.	+25° C. +77° F.	+30° C. +86° F.	+35° C. +95° F.	
°	′	′	′	′	′	′	′	′	′	′	°
10	5.67	5.57	5.45	5.35	5.25	5.15	5.07	4.98	4.90	4.82	10
11	5.17	5.07	4.97	4.88	4.78	4.70	4.62	4.53	4.47	4.38	11
12	4.75	4.65	4.57	4.48	4.40	4.32	4.25	4.18	4.12	4.03	12
13	4.38	4.30	4.22	4.15	4.07	4.00	3.93	3.87	3.80	3.73	13
14	4.06	3.97	3.91	3.84	3.76	3.69	3.64	3.59	3.51	3.46	14
15	3.79	3.72	3.64	3.57	3.51	3.46	3.39	3.34	3.27	3.22	15
16	3.57	3.49	3.44	3.37	3.31	3.26	3.21	3.14	3.09	3.04	16
17	3.36	3.29	3.24	3.17	3.12	3.06	3.02	2.96	2.91	2.86	17
18	3.16	3.09	3.04	2.99	2.94	2.89	2.84	2.79	2.74	2.69	18
19	2.97	2.92	2.87	2.82	2.77	2.72	2.67	2.62	2.57	2.54	19
20	2.82	2.77	2.72	2.67	2.62	2.57	2.52	2.47	2.44	2.41	20
21	2.67	2.62	2.57	2.52	2.49	2.44	2.41	2.36	2.31	2.27	21
22	2.52	2.49	2.44	2.39	2.36	2.32	2.27	2.22	2.19	2.16	22
23	2.42	2.39	2.34	2.29	2.26	2.22	2.17	2.12	2.09	2.07	23
24	2.31	2.27	2.22	2.19	2.16	2.12	2.07	2.02	2.01	1.97	24
25	2.21	2.17	2.12	2.09	2.06	2.02	1.97	1.94	1.91	1.89	25
26	2.12	2.08	2.03	2.00	1.96	1.93	1.88	1.85	1.83	1.80	26
27	2.01	1.98	1.95	1.91	1.88	1.85	1.81	1.76	1.75	1.73	27
28	1.93	1.90	1.85	1.83	1.80	1.76	1.73	1.70	1.66	1.65	28
29	1.85	1.81	1.78	1.76	1.73	1.70	1.66	1.63	1.60	1.59	29
30	1.78	1.75	1.71	1.70	1.66	1.63	1.60	1.58	1.55	1.53	30
32	1.65	1.62	1.59	1.57	1.54	1.50	1.47	1.45	1.42	1.40	32
34	1.53	1.49	1.47	1.44	1.42	1.39	1.35	1.35	1.32	1.30	34
36	1.42	1.39	1.37	1.34	1.32	1.30	1.27	1.25	1.22	1.20	36
38	1.32	1.30	1.27	1.25	1.24	1.22	1.19	1.17	1.14	1.14	38
40	1.22	1.21	1.18	1.16	1.14	1.13	1.09	1.08	1.06	1.04	40
42	1.14	1.11	1.09	1.08	1.06	1.04	1.01	0.99	0.98	0.98	42
44	1.07	1.04	1.03	1.01	0.99	0.98	0.96	0.94	0.93	0.91	44
46	0.99	0.98	0.97	0.95	0.93	0.92	0.90	0.88	0.87	0.85	46
48	0.93	0.92	0.90	0.88	0.87	0.85	0.83	0.82	0.80	0.78	48
50	0.86	0.84	0.82	0.81	0.79	0.77	0.76	0.76	0.74	0.72	50
55	0.72	0.71	0.69	0.69	0.67	0.66	0.66	0.64	0.62	0.61	55
60	0.59	0.59	0.57	0.57	0.55	0.54	0.54	0.52	0.52	0.50	60
65	0.48	0.46	0.46	0.46	0.44	0.44	0.43	0.43	0.41	0.39	65
70	0.37	0.37	0.37	0.35	0.35	0.35	0.33	0.33	0.33	0.32	70
75	0.27	0.27	0.27	0.26	0.26	0.26	0.24	0.24	0.24	0.22	75
80	0.18	0.18	0.16	0.16	0.16	0.16	0.16	0.15	0.15	0.15	80
85	0.08	0.08	0.08	0.08	0.08	0.08	0.08	0.06	0.06	0.06	85
90	0.00	0.00	0.00	0.00	0.00	0.00	0.00	0.00	0.00	0.00	90

TABLE III. REFRACTION CORRECTIONS TO BE APPLIED TO APPARENT DECLINATIONS

To be used with solar attachment

January	Hour angle	Refraction correction lat. 40°	February	Hour angle	Refraction correction lat. 40°	March	Hour angle	Refraction correction lat. 40°	April	Hour angle	Refraction correction lat. 40°	May	Hour angle	Refraction correction lat. 40°	June	Hour angle	Refraction correction lat. 40°
		′ ″		See Jan. 30				′ ″			′ ″			′ ″			′ ″
1	1	1 58	1		′ ″	1	1	1 03	1	3	0 57	1	3	0 39	1	4	0 44
2	2	2 16	2	3	2 13	2	2	1 10	2	4	1 19	2	4	0 55	2	5	1 11
3	3	3 04	3	4	3 41	3	3	1 27	3	5	2 18	3	5	1 30	3		
4	4	6 23	4			4	4	2 06	4			4			4	1	0 19
5			5	1	1 26	5			5	1	0 39	5	1	0 26	5	2	0 23
	1	1 54		2	1 37		1	0 59		2	0 44		2	0 30		3	0 30
6	2	2 11	6	3	2 04	6	2	1 06	6	3	0 54	6	3	0 37	6	4	0 43
7	3	2 59	7	4	3 21	7	3	1 21	7	4	1 14	7	4	0 53	7	5	1 10
8	4	6 01	8			8	4	1 56	8	5	2 08	8	5	1 26	8		
9			9	1	1 21	9	5	4 04	9			9			9	1	0 18
10			10	2	1 31	10			10	1	0 36	10	1	0 25	10	2	0 22
	1	1 50		3	1 56		1	0 55		2	0 41		2	0 29		3	0 29
11	2	2 06	11	4	3 04	11	2	1 02	11	3	0 51	11	3	0 36	11	4	0 43
12	3	2 49	12			12	3	1 15	12	4	1 10	12	4	0 51	12	5	1 09
13	4	5 33	13			13	4	1 47	13	5	1 58	13	5	1 22	13		
14			14	1	1 16	14	5	3 34	14			14			14	1	0 18
15			15	2	1 25	15			15	1	0 34	15	1	0 23	15	2	0 22
	1	1 46		3	1 48		1	0 52		2	0 38		2	0 27		3	0 29
16	2	2 01	16	4	2 47	16	2	0 58	16	3	0 48	16	3	0 34	16	4	0 42
17	3	2 40	17	5	8 39	17	3	1 10	17	4	1 06	17	4	0 49	17	5	1 08
18	4	5 00	18			18	4	1 39	18	5	1 49	18	5	1 18	18		
19			19	1	1 12	19	5	3 08	19			19			19	1	0 18
20			20	2	1 20	20			20	1	0 32	20	1	0 22	20	2	0 22
	1	1 42		3	1 40		1	0 48		2	0 36		2	0 26		3	0 29
21	2	1 56	21	4	2 31	21	2	0 54	21	3	0 45	21	3	0 33	21	4	0 42
22	3	2 31	22	5	6 49	22	3	1 05	22	4	1 02	22	4	0 47	22	5	1 08
23	4	4 35	23			23	4	1 32	23	5	1 42	23	5	1 15	23		
24			24	1	1 07	24	5	2 51	24			24			24	1	0 18
25	1	1 37	25	2	1 15	25			25	1	0 30	25	1	0 21	25	2	0 22
	2	1 50		3	1 33		1	0 45		2	0 34		2	0 25		3	0 29
26	3	2 22	26	4	2 18	26	2	0 50	26	3	0 42	26	3	0 32	26	4	0 42
27	4	4 07	27	5	5 29	27	3	1 01	27	4	0 58	27	4	0 46	27	5	1 08
28			28			28	4	1 25	28	5	1 36	28	5	1 13	28		
29			29	See Mar. 1		29	5	2 34	29			29			29	1	0 18
30	1	1 32				30			30	1	0 28	30	1	0 20	30	2	0 22
31	2	1 44				31	1	0 42		2	0 32	31	2	0 24		3	0 29
							2	0 47					3	0 31			

For latitudes other than 40° multiply by latitude coefficient, Table IV.

TABLE III. REFRACTION CORRECTIONS TO BE APPLIED TO APPARENT DECLINATIONS.—*Concluded*

To be used with solar attachment

July	Hour angle	Refraction correction lat. 40°	August	Hour angle	Refraction correction lat. 40°	September	Hour angle	Refraction correction lat. 40°	October	Hour angle	Refraction correction lat. 40°	November	Hour angle	Refraction correction lat. 40°	December	Hour angle	Refraction correction lat. 40°
		′ ″			′ ″			′ ″			′ ″			′ ″			′ ″
1	4	0 43	1	5	1 22	1	1	0 39	1	1	0 56	1			1	1	1 54
2	5	1 09	2			2	2	0 44	2	2	1 06	2	1	1 26	2	2	2 11
3			3	1	0 26	3	3	0 54	3	3	1 21	3	2	1 37	3	3	2 59
4	1	0 19	4	2	0 30	4	4	1 14	4	4	1 56	4	3	2 04	4	4	6 01
5	2	0 23	5	3	0 37	5	5	2 08	5	5	4 04	5	4	3 21	5		
	3	0 30		4	0 53												
6	4	0 43	6	5	1 26	6	1	0 42	6	1	1 03	6	1	1 32	6	1	1 58
7	5	1 10	7			7	2	0 47	7	2	1 10	7	2	1 44	7	2	2 16
8			8	1	0 28	8	3	0 57	8	3	1 27	8	3	2 13	8	3	3 04
9	1	0 20	9	2	0 32	9	4	1 19	9	4	2 06	9	4	3 41	9	4	6 23
10	2	0 24	10	3	0 39	10	5	2 18	10	5	4 39	10			10		
	3	0 31		4	0 55												
11	4	0 44	11	5	1 30	11	1	0 45	11	1	1 07	11	1	1 37	11	1	2 00
12	5	1 11	12			12	2	0 50	12	2	1 15	12	2	1 50	12	2	2 19
13			13	1	0 30	13	3	1 01	13	3	1 33	13	3	2 22	13	3	3 09
14	1	0 21	14	2	0 34	14	4	1 25	14	4	2 18	14	4	4 07	14	4	6 43
15	2	0 25	15	3	0 42	15	5	2 34	15	5	5 29	15			15		
	3	0 32		4	0 58												
16	4	0 46	16	5	1 36	16	1	0 48	16	1	1 12	16	1	1 42	16	1	2 01
17	5	1 13	17			17	2	0 54	17	2	1 20	17	2	1 56	17	2	2 20
18			18	1	0 32	18	3	1 05	18	3	1 40	18	3	2 31	18	3	3 11
19	1	0 22	19	2	0 36	19	4	1 32	19	4	2 31	19	4	4 35	19	4	6 47
20	2	0 26	20	3	0 45	20	5	2 51	20	5	6 49	20			20		
	3	0 33		4	1 02												
21	4	0 47	21	5	1 42	21	1	0 52	21	1	1 16	21	1	1 46	21	1	2 01
22	5	1 15	22			22	2	0 58	22	2	1 25	22	2	2 01	22	2	2 20
23			23	1	0 34	23	3	1 10	23	3	1 48	23	3	2 40	23	3	3 11
24	1	0 23	24	2	0 38	24	4	1 39	24	4	2 47	24	4	4 59	24	4	6 49
25	2	0 27	25	3	0 48	25	5	3 08	25	5	8 39	25			25		
	3	0 34		4	1 06												
26	4	0 49	26	5	1 49	26	1	0 55	26	1	1 21	26	1	1 50	26	1	2 00
27	5	1 18	27			27	2	1 02	27	2	1 31	27	2	2 06	27	2	2 19
28			28	1	0 36	28	3	1 15	28	3	1 56	28	3	2 49	28	3	3 09
29	1	0 25	29	2	0 41	29	4	1 47	29	4	3 04	29	4	5 33	29	4	6 43
30	2	0 29	30	3	0 51	30	5	3 34	30	5	11 01	30			30		
31	3	0 36	31	4	1 10				31	See Nov. 1					31	See Jan. 1	
	4	0 51		5	1 58												

For latitudes other than 40° multiply by latitude coefficient, Table IV.

TABLE IV. LATITUDE COEFFICIENTS

Latitude	Coefficient	Latitude	Coefficient	Latitude	Coefficient
15°	0.30	30°	0.65	45°	1.20
16	0.32	31	0.68	46	1.24
17	0.34	32	0.71	47	1.29
18	0.36	33	0.75	48	1.33
19	0.38	34	0.78	49	1.38
20	0.40	35	0.82	50	1.42
21	0.42	36	0.85	51	1.47
22	0.44	37	0.89	52	1.53
23	0.46	38	0.92	53	1.58
24	0.48	39	0.96	54	1.64
25	0.50	**40**	**1.00**	55	1.70
26	0.53	41	1.04	56	1.76
27	0.56	42	1.08	57	1.82
28	0.59	43	1.12	58	1.88
29	0.62	44	1.16	59	1.94

To obtain the refraction correction (to be applied to declination) for any other latitude than 40°, multiply the refraction correction for latitude 40° (Table III) by the coefficient corresponding to the latitude of observation.

TABLE V. FOR REDUCING TO ELONGATION OF POLARIS OBSERVATIONS MADE NEAR ELONGATION

Azimuth at elongation / * Time	1° 0′	1° 10′	1° 20′	1° 30′	1° 40′	1° 50′	2° 0′	2° 10′	Azimuth at elongation / * Time
m	″	″	″	″	″	″	″	″	m
0	0.0	0.0	0.0	0.0	0.0	0.0	0.0	0.0	0
1	0.0	0.0	0.0	+ 0.1	+ 0.1	+ 0.1	+ 0.1	+ 0.1	1
2	+ 0.1	+ 0.2	+ 0.2	0.2	0.2	0.3	0.3	0.3	2
3	0.3	0.4	0.4	0.5	0.5	0.6	0.6	0.7	3
4	0.5	0.6	0.7	0.8	0.9	1.0	1.1	1.2	4
5	+ 0.9	+ 1.0	+ 1.1	+ 1.3	+ 1.4	+ 1.6	+ 1.7	+ 1.9	5
6	1.2	1.4	1.6	1.8	2.1	2.3	2.5	2.7	6
7	1.7	2.0	2.2	2.5	2.8	3.1	3.4	3.7	7
8	2.2	2.6	2.9	3.3	3.7	4.0	4.4	4.8	8
9	2.8	3.2	3.7	4.2	4.6	5.1	5.6	6.0	9
10	+ 3.4	+ 4.0	+ 4.6	+ 5.1	+ 5.7	+ 6.3	+ 6.9	+ 7.4	10
11	4.1	4.8	5.5	6.2	6.9	7.6	8.3	9.0	11
12	4.9	5.8	6.6	7.4	8.2	9.0	9.9	10.7	12
13	5.8	6.8	7.7	8.7	9.7	10.6	11.6	12.6	13
14	6.7	7.8	9.0	10.1	11.2	12.3	13.4	14.6	14
15	+ 7.7	+ 9.0	+10.3	+11.6	+12.8	+14.1	+15.4	+16.7	15
16	8.8	10.2	11.7	13.2	14.6	16.1	17.5	19.0	16
17	9.9	11.5	13.2	14.9	16.5	18.2	19.8	21.5	17
18	11.1	12.9	14.8	16.7	18.5	20.4	22.2	24.1	18
19	12.4	14.4	16.5	18.6	20.6	22.7	24.7	26.8	19
20	+13.7	+16.0	+18.3	+20.6	+22.8	+25.1	+27.4	+29.7	20
21	15.1	17.6	20.1	22.7	25.2	27.7	30.2	32.7	21
22	16.6	19.3	22.1	24.9	27.6	30.4	33.2	35.9	22
23	18.1	21.1	24.2	27.2	30.2	33.2	36.2	39.3	23
24	19.7	23.0	26.3	29.6	32.9	36.2	39.5	42.8	24
25	+21.4	+25.0	+28.5	+32.1	+35.7	+39.2	+42.8	+46.4	25

* Sidereal time from elongation.

TABLE V. FOR REDUCING TO ELONGATION OF POLARIS OBSERVATIONS MADE NEAR ELONGATION.—*Concluded*

Azimuth at elongation / *Time	2° 10′	2° 20′	2° 30′	2° 40′	2° 50′	3° 0′	3° 10′	3° 20′	Azimuth at elongation / *Time
m	″	″	″	″	″	″	″	″	m
0	0.0	0.0	0.0	0.0	0.0	0.0	0.0	0.0	0
1	+ 0.1	+ 0.1	+ 0.1	+ 0.1	+ 0.1	+ 0.1	+ 0.1	+ 0.1	1
2	0.3	0.3	0.4	0.4	0.4	0.4	0.4	0.5	2
3	0.7	0.7	0.8	0.8	0.9	0.9	1.0	1.0	3
4	1.2	1.3	1.4	1.5	1.6	1.6	1.7	1.8	4
5	+ 1.9	+ 2.0	+ 2.1	+ 2.3	+ 2.4	+ 2.6	+ 2.7	+ 2.9	5
6	2.7	2.9	3.1	3.3	3.5	3.7	3.9	4.1	6
7	3.7	3.9	4.2	4.5	4.8	5.0	5.3	5.6	7
8	4.8	5.1	5.5	5.9	6.2	6.6	7.0	7.3	8
9	6.0	6.5	7.0	7.4	7.9	8.3	8.8	9.3	9
10	+ 7.4	+ 8.0	+ 8.6	+ 9.2	+ 9.7	+10.3	+10.9	+11.4	10
11	9.0	9.7	10.4	11.1	11.8	12.4	13.1	13.8	11
12	10.7	11.5	12.3	13.2	14.0	14.8	15.6	16.5	12
13	12.6	13.5	14.5	15.4	16.4	17.4	18.4	19.3	13
14	14.6	15.7	16.8	17.9	19.0	20.2	21.3	22.4	14
15	+16.7	+18.0	+19.3	+20.6	+21.9	+23.1	+24.4	+25.7	15
16	19.0	20.5	21.9	23.4	24.9	26.3	27.8	29.3	16
17	21.5	23.1	24.8	26.4	28.1	29.7	31.4	33.0	17
18	24.1	25.9	27.8	29.6	31.5	33.3	35.2	37.0	18
19	26.8	28.9	30.9	33.0	35.1	37.1	39.2	41.3	19
20	+29.7	+32.0	+34.3	+36.6	+38.8	+41.1	+43.4	+45.7	20
21	32.7	35.3	37.8	40.3	42.8	45.3	47.9	50.4	21
22	35.9	38.7	41.5	44.2	47.0	49.8	52.5	55.3	22
23	39.3	42.3	45.3	48.3	51.4	54.4	57.4	60.4	23
24	42.8	46.0	49.3	52.6	55.9	59.2	62.5	65.8	24
25	+46.4	+49.9	+53.5	+57.1	+60.7	+64.2	+67.8	+71.4	25

*Sidereal time from elongation.

Table VI. Trigonometric Formulas

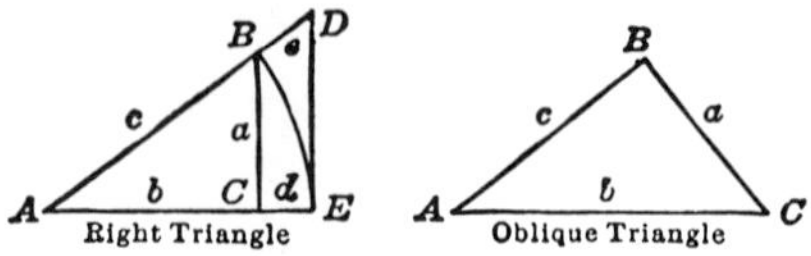

Right Triangles

$$\sin A = \frac{a}{c} = \cos B \qquad \sec A = \frac{c}{b} = \operatorname{cosec} B$$

$$\cos A = \frac{b}{c} = \sin B \qquad \operatorname{cosec} A = \frac{c}{a} = \sec B$$

$$\tan A = \frac{a}{b} = \cot B \qquad \operatorname{vers} A = \frac{c - b}{c} = \frac{d}{c}$$

$$\cot A = \frac{b}{a} = \tan B \qquad \operatorname{exsec} A = \frac{e}{c}$$

$$a = c \sin A = c \cos B = b \tan A = b \cot B = \sqrt{c^2 - b^2}$$

$$b = c \cos A = c \sin B = a \cot A = a \tan B = \sqrt{c^2 - a^2}$$

$$c = \frac{a}{\sin A} = \frac{a}{\cos B} = \frac{b}{\sin B} = \frac{b}{\cos A} = \frac{d}{\operatorname{vers} A} = \frac{e}{\operatorname{exsec} A} = \sqrt{a^2 + b^2}$$

$$d = c \operatorname{vers} A \qquad e = c \operatorname{exsec} A$$

TABLE VI. TRIGONOMETRIC FORMULAS.—*Concluded*

OBLIQUE TRIANGLES

Given	Sought	Formulas
A, B, a	b, c	$b = \frac{a}{\sin A}\cdot\sin B$ $\quad c = \frac{a}{\sin A}\cdot\sin (A + B)$
A, a, b	B, c	$\sin B = \frac{\sin A}{a}\cdot b$ $\quad c = \frac{a}{\sin A}\cdot\sin C$
C, a, b	$\frac{1}{2}(A + B)$	$\frac{1}{2}(A + B) = 90^\circ - \frac{1}{2}C$
	$\frac{1}{2}(A - B)$	$\tan\frac{1}{2}(A - B) = \frac{a - b}{a + b}\cdot\tan\frac{1}{2}(A + B)$
a, b, c	A	If $s = \frac{1}{2}(a + b + c)$, $\sin\frac{1}{2}A = \sqrt{\frac{(s - b)(s - c)}{bc}}$
		$\cos\frac{1}{2}A = \sqrt{\frac{s(s - a)}{bc}}$, $\tan\frac{1}{2}A = \sqrt{\frac{(s - b)(s - c)}{s(s - a)}}$
		$\sin A = 2\frac{\sqrt{s(s - a)(s - b)(s - c)}}{bc}$
		$\text{vers } A = \frac{2(s - b)(s - c)}{bc}$
	area	$area = \sqrt{s(s - a)(s - b)(s - c)}$
C, a, b	*area*	$area = \frac{1}{2}ab\sin C$
A, B, C, a	*area*	$area = \frac{a^2\sin B\cdot\sin C}{2\sin A}$

TABLE VII. FACTORS FOR DETERMINING STRENGTH OF FIGURE
(*U.S. Coast and Geodetic Survey*)

Values of $(\delta_A^2 + \delta_A\delta_B + \delta_B^2)$ for various combinations of distance angles A and B of a triangle

	10°	12°	14°	16°	18°	20°	22°	24°	26°	28°	30°	35°	40°	45°	50°	55°	60°	65°	70°	75°	80°	85°	90°
10°	428	359																					
12	359	295	253																				
14	315	253	214	187																			
16	284	225	187	162	143																		
18	262	204	168	143	126	113																	
20°	245	189	153	130	113	100	91																
22	232	177	142	119	103	91	81	74															
24	221	167	134	111	95	83	74	67	61														
26	213	160	126	104	89	77	68	61	56	51													
28	206	153	120	99	83	72	63	57	51	47	43												
30°	199	148	115	94	79	68	59	53	48	43	40	33											
35	188	137	106	85	71	60	52	46	41	37	33	27	23										
40	179	129	99	79	65	54	47	41	36	32	29	23	19	16									
45	172	124	93	74	60	50	43	37	32	28	25	20	16	13	11								
50°	167	119	89	70	57	47	39	34	29	26	23	18	14	11	9	8							
55	162	115	86	67	54	44	37	32	27	24	21	16	12	10	8	7	5						
60	159	112	83	64	51	42	35	30	25	22	19	14	11	9	7	5	4	4					
65	155	109	80	62	49	40	33	28	24	21	18	13	10	7	6	5	4	3	2				
70°	152	106	78	60	48	38	32	27	23	19	17	12	9	7	5	4	3	2	2	1			
75	150	104	76	58	46	37	30	25	21	18	16	11	8	6	4	3	2	2	1	1	1		
80	147	102	74	57	45	36	29	24	20	17	15	10	7	5	4	3	2	1	1	1	0	0	
85	145	100	73	55	43	34	28	23	19	16	14	10	7	5	3	2	2	1	1	0	0	0	0

90°	143	98	71	54	42	33	27	22	19	16	13	9	6	4	3	2	1	1	1	0	0	0	0
95°	140	96	70	53	41	32	26	22	18	15	13	9	6	4	3	2	1	1	0	0	0	0	
100	138	95	68	51	40	31	25	21	17	14	12	8	6	4	3	2	1	1	0	0	0		
105	136	93	67	50	39	30	25	20	17	14	12	8	5	4	2	2	1	1	0	0			
110	134	91	65	49	38	30	24	19	16	13	11	7	5	3	2	2	1	1	1				
115°	132	89	64	48	37	29	23	19	15	13	11	7	5	3	2	2	1	1					
120	129	88	62	46	36	28	22	18	15	12	10	7	5	3	2	2	1						
125	127	86	61	45	35	27	22	18	14	12	10	7	5	4	3	2							
130	125	84	59	44	34	26	21	17	14	12	10	7	5	4	3								
135°	122	82	58	43	33	26	21	17	14	12	10	7	5	4									
140	119	80	56	42	32	25	20	17	14	12	10	8	6										
145	116	77	55	41	32	25	21	17	15	13	11	9											
150	112	75	54	40	32	26	21	18	16	15	13												
152°	111	75	53	40	32	26	22	19	17	16													
154	110	74	53	41	33	27	23	21	19														
156	108	74	54	42	34	28	25	22															
158	107	74	54	43	35	30	27																
160	107	74	56	45	38	33																	
162°	107	76	59	48	42																		
164	109	79	63	54																			
166	113	86	71																				
168	122	98																					
170	143																						

TABLE VIII. DAILY VARIATION OF MAGNETIC DECLINATION
(*U.S. Coast and Geodetic Survey*)

Mean values for 10 quiet days for each month, 1918–1928, at observatories at Sitka, Alaska; Cheltenham, Maryland; and Tucson, Arizona.

A plus sign indicates that the west declination is greater or the east declination is less than the mean for the day.

Hour, local mean time	January, February, November, December			March, April, September, October			May, June, July, August		
	Sitka	Chel-ten-ham	Tucson	Sitka	Chel-ten-ham	Tucson	Sitka	Chel-ten-ham	Tucson
	′	′	′	′	′	′	′	′	′
1 a.m.	+0.2	+0.1	+0.2	+0.1	−0.5	0.0	+0.9	−0.3	0.0
2 a.m.	0.0	+0.2	+0.3	0.0	−0.4	−0.1	+0.9	−0.3	−0.1
3 a.m.	−0.1	+0.1	+0.2	−0.2	−0.7	−0.2	+0.4	−0.5	−0.3
4 a.m.	−0.2	−0.2	+0.2	−0.5	−1.0	−0.3	−0.6	−1.0	−0.6
5 a.m.	−0.4	−0.5	+0.1	−1.3	−1.4	−0.6	−2.9	−2.1	−1.3
6 a.m.	−0.7	−0.8	−0.2	−2.4	−2.3	−1.5	−4.9	−4.1	−2.6
7 a.m.	−1.2	−1.2	−0.8	−3.7	−3.7	−3.2	−6.8	−5.7	−4.5
8 a.m.	−1.8	−2.5	−1.9	−4.8	**−4.7**	**−4.0**	−7.7	**−5.9**	**−4.7**
9 a.m.	**−2.3**	**−3.2**	**−2.6**	**−5.0**	−3.9	−3.3	**−7.8**	−4.7	−3.4
10 a.m.	−1.7	−2.5	−2.3	−3.6	−2.2	−1.4	−4.8	−1.3	−0.4
11 a.m.	−0.9	−0.4	−0.6	−1.6	+1.0	+0.8	−1.3	+2.2	+1.8
Noon	+0.2	+2.0	+1.1	+0.6	+3.7	+2.3	+1.8	+4.7	+3.2
1 p.m.	+0.9	+3.1	+2.1	+2.1	**+4.7**	**+3.0**	+3.8	**+5.6**	**+3.6**
2 p.m.	+1.5	**+3.2**	**+2.2**	+2.9	**+4.7**	+2.8	+5.2	+5.4	+3.2
3 p.m.	**+1.8**	+2.5	+1.7	**+3.3**	+3.9	+2.2	**+5.6**	+4.1	+2.4
4 p.m.	+1.6	+1.5	+1.0	+3.2	+2.2	+1.4	+5.1	+2.5	+1.4
5 p.m.	+1.3	+0.7	+0.4	+2.9	+1.2	+0.9	+3.8	+1.0	+0.6
6 p.m.	+0.9	+0.2	0.0	+2.3	+0.6	+0.6	+2.5	+0.2	+0.3
7 p.m.	+0.6	−0.2	−0.2	+1.8	+0.3	+0.3	+1.6	+0.1	+0.3
8 p.m.	+0.3	−0.5	−0.3	+1.3	0.0	+0.2	+1.0	+0.4	+0.4
9 p.m.	+0.1	−0.7	−0.2	+1.0	−0.2	+0.2	+0.8	+0.1	+0.3
10 p.m.	0.0	−0.6	−0.2	+0.8	−0.3	+0.1	+1.0	0.0	+0.2
11 p.m.	0.0	−0.4	0.0	+0.6	−0.3	0.0	+1.3	−0.2	+0.1
Midnight	+0.1	−0.2	+0.1	+0.3	−0.4	0.0	+1.1	−0.2	+0.1
Range	4.1	6.4	4.8	8.3	9.4	7.0	13.4	11.5	8.3

TABLE IX.* HORIZONTAL DISTANCES AND ELEVATIONS FROM STADIA READINGS

	0°		1°		2°		3°	
Minutes	Hor. Dist.	Diff. Elev.	Hor. Dist.	Diff. Elev.	Hor. Dist.	Diff. Elev.	Hor. Dist.	Diff. Elev.
0	100.00	0.00	99.97	1.74	99.88	3.49	99.73	5.23
2	100.00	0.06	99.97	1.80	99.87	3.55	99.72	5.28
4	100.00	0.12	99.97	1.86	99.87	3.60	99.71	5.34
6	100.00	0.17	99.96	1.92	99.87	3.66	99.71	5.40
8	100.00	0.23	99.96	1.98	99.86	3.72	99.70	5.46
10	100.00	0.29	99.96	2.04	99.86	3.78	99.69	5.52
12	100.00	0.35	99.96	2.09	99.85	3.84	99.69	5.57
14	100.00	0.41	99.95	2.15	99.85	3.90	99.68	5.63
16	100.00	0.47	99.95	2.21	99.84	3.95	99.68	5.69
18	100.00	0.52	99.95	2.27	99.84	4.01	99.67	5.75
20	100.00	0.58	99.95	2.33	99.83	4.07	99.66	5.80
22	100.00	0.64	99.94	2.38	99.83	4.13	99.66	5.86
24	100.00	0.70	99.94	2.44	99.82	4.18	99.65	5.92
26	99.99	0.76	99.94	2.50	99.82	4.24	99.64	5.98
28	99.99	0.81	99.93	2.56	99.81	4.30	99.63	6.04
30	99.99	0.87	99.93	2.62	99.81	4.36	99.63	6.09
32	99.99	0.93	99.93	2.67	99.80	4.42	99.62	6.15
34	99.99	0.99	99.93	2.73	99.80	4.48	99.62	6.21
36	99.99	1.05	99.92	2.79	99.79	4.53	99.61	6.27
38	99.99	1.11	99.92	2.85	99.79	4.59	99.60	6.33
40	99.99	1.16	99.92	2.91	99.78	4.65	99.59	6.38
42	99.99	1.22	99.91	2.97	99.78	4.71	99.59	6.44
44	99.98	1.28	99.91	3.02	99.77	4.76	99.58	6.50
46	99.98	1.34	99.90	3.08	99.77	4.82	99.57	6.56
48	99.98	1.40	99.90	3.14	99.76	4.88	99.56	6.61
50	99.98	1.45	99.90	3.20	99.76	4.94	99.56	6.67
52	99.98	1.51	99.89	3.26	99.75	4.99	99.55	6.73
54	99.98	1.57	99.89	3.31	99.74	5.05	99.54	6.78
56	99.97	1.63	99.89	3.37	99.74	5.11	99.53	6.84
58	99.97	1.69	99.88	3.43	99.73	5.17	99.52	6.90
60	99.97	1.74	99.88	3.49	99.73	5.23	99.51	6.96
C = 0.75	0.75	0.01	0.75	0.02	0.75	0.03	0.75	0.05
C = 1.00	1.00	0.01	1.00	0.03	1.00	0.04	1.00	0.06
C = 1.25	1.25	0.02	1.25	0.03	1.25	0.05	1.25	0.08

* From "Theory and Practice of Surveying," by J. B. Johnson. By permission of the publishers, John Wiley & Sons, Inc., New York.

TABLE IX. HORIZONTAL DISTANCES AND ELEVATIONS FROM STADIA READINGS.—*Continued*

	4°		5°		6°		7°	
Minutes	Hor. Dist.	Diff. Elev.	Hor. Dist.	Diff. Elev.	Hor. Dist.	Diff. Elev.	Hor. Dist.	Diff. Elev.
0	99.51	6.96	99.24	8.68	98.91	10.40	98.51	12.10
2	99.51	7.02	99.23	8.74	98.90	10.45	98.50	12.15
4	99.50	7.07	99.22	8.80	98.88	10.51	98.48	12.21
6	99.49	7.13	99.21	8.85	98.87	10.57	98.47	12.26
8	99.48	7.19	99.20	8.91	98.86	10.62	98.46	12.32
10	99.47	7.25	99.19	8.97	98.85	10.68	98.44	12.38
12	99.46	7.30	99.18	9.03	98.83	10.74	98.43	12.43
14	99.46	7.36	99.17	9.08	98.82	10.79	98.41	12.49
16	99.45	7.42	99.16	9.14	98.81	10.85	98.40	12.55
18	99.44	7.48	99.15	9.20	98.80	10.91	98.39	12.60
20	99.43	7.53	99.14	9.25	98.78	10.96	98.37	12.66
22	99.42	7.59	99.13	9.31	98.77	11.02	98.36	12.72
24	99.41	7.65	99.11	9.37	98.76	11.08	98.34	12.77
26	99.40	7.71	99.10	9.43	98.74	11.13	98.33	12.83
28	99.39	7.76	99.09	9.48	98.73	11.19	98.31	12.88
30	99.38	7.82	99.08	9.54	98.72	11.25	98.29	12.94
32	99.38	7.88	99.07	9.60	98.71	11.30	98.28	13.00
34	99.37	7.94	99.06	9.65	98.69	11.36	98.27	13.05
36	99.36	7.99	99.05	9.71	98.68	11.42	98.25	13.11
38	99.35	8.05	99.04	9.77	98.67	11.47	98.24	13.17
40	99.34	8.11	99.03	9.83	98.65	11.53	98.22	13.22
42	99.33	8.17	99.01	9.88	98.64	11.59	98.20	13.28
44	99.32	8.22	99.00	9.94	98.63	11.64	98.19	13.33
46	99.31	8.28	98.99	10.00	98.61	11.70	98.17	13.39
48	99.30	8.34	98.98	10.05	98.60	11.76	98.16	13.45
50	99.29	8.40	98.97	10.11	98.58	11.81	98.14	13.50
52	99.28	8.45	98.96	10.17	98.57	11.87	98.13	13.56
54	99.27	8.51	98.94	10.22	98.56	11.93	98.11	13.61
56	99.26	8.57	98.93	10.28	98.54	11.98	98.10	13.67
58	99.25	8.63	98.92	10.34	98.53	12.04	98.08	13.73
60	99.24	8.68	98.91	10.40	98.51	12.10	98.06	13.78
C = 0.75	0.75	0.06	0.75	0.07	0.75	0.08	0.74	0.10
C = 1.00	1.00	0.08	0.99	0.09	0.99	0.11	0.99	0.13
C = 1.25	1.25	0.10	1.24	0.11	1.24	0.14	1.24	0.16

TABLE IX. HORIZONTAL DISTANCES AND ELEVATIONS FROM STADIA READINGS. — *Continued*

	8°		9°		10°		11°	
Minutes	Hor. Dist.	Diff. Elev.	Hor. Dist.	Diff. Elev.	Hor. Dist.	Diff. Elev.	Hor. Dist.	Diff. Elev.
0	98.06	13.78	97.55	15.45	96.98	17.10	96.36	18.73
2	98.05	13.84	97.53	15.51	96.96	17.16	96.34	18.78
4	98.03	13.89	97.52	15.56	96.94	17.21	96.32	18.84
6	98.01	13.95	97.50	15.62	96.92	17.26	96.29	18.89
8	98.00	14.01	97.48	15.67	96.90	17.32	96.27	18.95
10	97.98	14.06	97.46	15.73	96.88	17.37	96.25	19.00
12	97.97	14.12	97.44	15.78	96.86	17.43	96.23	19.05
14	97.95	14.17	97.43	15.84	96.84	17.48	96.21	19.11
16	97.93	14.23	97.41	15.89	96.82	17.54	96.18	19.16
18	97.92	14.28	97.39	15.95	96.80	17.59	96.16	19.21
20	97.90	14.34	97.37	16.00	96.78	17.65	96.14	19.27
22	97.88	14.40	97.35	16.06	96.76	17.70	96.12	19.32
24	97.87	14.45	97.33	16.11	96.74	17.76	96.09	19.38
26	97.85	14.51	97.31	16.17	96.72	17.81	96.07	19.43
28	97.83	14.56	97.29	16.22	96.70	17.86	96.05	19.48
30	97.82	14.62	97.28	16.28	96.68	17.92	96.03	19.54
32	97.80	14.67	97.26	16.33	96.66	17.97	96.00	19.59
34	97.78	14.73	97.24	16.39	96.64	18.03	95.98	19.64
36	97.76	14.79	97.22	16.44	96.62	18.08	95.96	19.70
38	97.75	14.84	97.20	16.50	96.60	18.14	95.93	19.75
40	97.73	14.90	97.18	16.55	96.57	18.19	95.91	19.80
42	97.71	14.95	97.16	16.61	96.55	18.24	95.89	19.86
44	97.69	15.01	97.14	16.66	96.53	18.30	95.86	19.91
46	97.68	15.06	97.12	16.72	96.51	18.35	95.84	19.96
48	97.66	15.12	97.10	16.77	96.49	18.41	95.82	20.02
50	97.64	15.17	97.08	16.83	96.47	18.46	95.79	20.07
52	97.62	15.23	97.06	16.88	96.45	18.51	95.77	20.12
54	97.61	15.28	97.04	16.94	96.42	18.57	95.75	20.18
56	97.59	15.34	97.02	16.99	96.40	18.62	95.72	20.23
58	97.57	15.40	97.00	17.05	96.38	18.68	95.70	20.28
60	97.55	15.45	96.98	17.10	96.36	18.73	95.68	20.34
C = 0.75	0.74	0.11	0.74	0.12	0.74	0.14	0.73	0.15
C = 1.00	0.99	0.15	0.99	0.16	0.98	0.18	0.98	0.20
C = 1.25	1.23	0.18	1.23	0.21	1.23	0.23	1.22	0.25

TABLE IX. HORIZONTAL DISTANCES AND ELEVATIONS FROM STADIA READINGS. — *Continued*

	12°		13°		14°		15°	
Minutes	Hor. Dist.	Diff. Elev.	Hor. Dist.	Diff. Elev.	Hor. Dist.	Diff. Elev.	Hor. Dist.	Diff. Elev.
0	95.68	20.34	94.94	21.92	94.15	23.47	93.30	25.00
2	95.65	20.39	94.91	21.97	94.12	23.52	93.27	25.05
4	95.63	20.44	94.89	22.02	94.09	23.58	93.24	25.10
6	95.61	20.50	94.86	22.08	94.07	23.63	93.21	25.15
8	95.58	20.55	94.84	22.13	94.04	23.68	93.18	25.20
10	95.56	20.60	94.81	22.18	94.01	23.73	93.16	25.25
12	95.53	20.66	94.79	22.23	93.98	23.78	93.13	25.30
14	95.51	20.71	94.76	22.28	93.95	23.83	93.10	25.35
16	95.49	20.76	94.73	22.34	93.93	23.88	93.07	25.40
18	95.46	20.81	94.71	22.39	93.90	23.93	93.04	25.45
20	95.44	20.87	94.68	22.44	93.87	23.99	93.01	25.50
22	95.41	20.92	94.66	22.49	93.84	24.04	92.98	25.55
24	95.39	20.97	94.63	22.54	93.81	24.09	92.95	25.60
26	95.36	21.03	94.60	22.60	93.79	24.14	92.92	25.65
28	95.34	21.08	94.58	22.65	93.76	24.19	92.89	25.70
30	95.32	21.13	94.55	22.70	93.73	24.24	92.86	25.75
32	95.29	21.18	94.52	22.75	93.70	24.29	92.83	25.80
34	95.27	21.24	94.50	22.80	93.67	24.34	92.80	25.85
36	95.24	21.29	94.47	22.85	93.65	24.39	92.77	25.90
38	95.22	21.34	94.44	22.91	93.62	24.44	92.74	25.95
40	95.19	21.39	94.42	22.96	93.59	24.49	92.71	26.00
42	95.17	21.45	94.39	23.01	93.56	24.55	92.68	26.05
44	95.14	21.50	94.36	23.06	93.53	24.60	92.65	26.10
46	95.12	21.55	94.34	23.11	93.50	24.65	92.62	26.15
48	95.09	21.60	94.31	23.16	93.47	24.70	92.59	26.20
50	95.07	21.66	94.28	23.22	93.45	24.75	92.56	26.25
52	95.04	21.71	94.26	23.27	93.42	24.80	92.53	26.30
54	95.02	21.76	94.23	23.32	93.39	24.85	92.49	26.35
56	94.99	21.81	94.20	23.37	93.36	24.90	92.46	26.40
58	94.97	21.87	94.17	23.42	93.33	24.95	92.43	26.45
60	94.94	21.92	94.15	23.47	93.30	25.00	92.40	26.50
C = 0.75	0.73	0.16	0.73	0.17	0.73	0.19	0.72	0.20
C = 1.00	0.98	0.22	0.97	0.23	0.97	0.25	0.96	0.27
C = 1.25	1.22	0.27	1.21	0.29	1.21	0.31	1.20	0.34

TABLE IX. HORIZONTAL DISTANCES AND ELEVATIONS FROM STADIA READINGS. — *Continued*

	16°		17°		18°		19°	
Minutes	Hor. Dist.	Diff. Elev.	Hor. Dist.	Diff. Elev.	Hor. Dist.	Diff. Elev.	Hor. Dist.	Diff. Elev.
0	92.40	26.50	91.45	27.96	90.45	29.39	89.40	30.78
2	92.37	26.55	91.42	28.01	90.42	29.44	89.36	30.83
4	92.34	26.59	91.39	28.06	90.38	29.48	89.33	30.87
6	92.31	26.64	91.35	28.10	90.35	29.53	89.29	30.92
8	92.28	26.69	91.32	28.15	90.31	29.58	89.26	30.97
10	92.25	26.74	91.29	28.20	90.28	29.62	89.22	31.01
12	92.22	26.79	91.26	28.25	90.24	29.67	89.18	31.06
14	92.19	26.84	91.22	28.30	90.21	29.72	89.15	31.10
16	92.15	26.89	91.19	28.34	90.18	29.76	89.11	31.15
18	92.12	26.94	91.16	28.39	90.14	29.81	89.08	31.19
20	92.09	26.99	91.12	28.44	90.11	29.86	89.04	31.24
22	92.06	27.04	91.09	28.49	90.07	29.90	89.00	31.28
24	92.03	27.09	91.06	28.54	90.04	29.95	88.96	31.33
26	92.00	27.13	91.02	28.58	90.00	30.00	88.93	31.38
28	91.97	27.18	90.99	28.63	89.97	30.04	88.89	31.42
30	91.93	27.23	90.96	28.68	89.93	30.09	88.86	31.47
32	91.90	27.28	90.92	28.73	89.90	30.14	88.82	31.51
34	91.87	27.33	90.89	28.77	89.86	30.19	88.78	31.56
36	91.84	27.38	90.86	28.82	89.83	30.23	88.75	31.60
38	91.81	27.43	90.82	28.87	89.79	30.28	88.71	31.65
40	91.77	27.48	90.79	28.92	89.76	30.32	88.67	31.69
42	91.74	27.52	90.76	28.96	89.72	30.37	88.64	31.74
44	91.71	27.57	90.72	29.01	89.69	30.41	88.60	31.78
46	91.68	27.62	90.69	29.06	89.65	30.46	88.56	31.83
48	91.65	27.67	90.66	29.11	89.61	30.51	88.53	31.87
50	91.61	27.72	90.62	29.15	89.58	30.55	88.49	31.92
52	91.58	27.77	90.59	29.20	89.54	30.60	88.45	31.96
54	91.55	27.81	90.55	29.25	89.51	30.65	88.41	32.01
56	91.52	27.86	90.52	29.30	89.47	30.69	88.38	32.05
58	91.48	27.91	90.48	29.34	89.44	30.74	88.34	32.09
60	91.45	27.96	90.45	29.39	89.40	30.78	88.30	32.14
C = 0.75	0.72	0.21	0.72	0.23	0.71	0.24	0.71	0.25
C = 1.00	0.96	0.28	0.95	0.30	0.95	0.32	0.94	0.33
C = 1.25	1.20	0.35	1.19	0.38	1.19	0.40	1.18	0.42

TABLE IX. HORIZONTAL DISTANCES AND ELEVATIONS FROM STADIA READINGS. — *Continued*

	20°		21°		22°		23°	
Minutes.	Hor. Dist.	Diff. Elev.	Hor. Dist.	Diff. Elev.	Hor. Dist.	Diff. Elev.	Hor. Dist.	Diff. Elev.
0	88.30	32.14	87.16	33.46	85.97	34.73	84.73	35.97
2	88.26	32.18	87.12	33.50	85.93	34.77	84.69	36.01
4	88.23	32.23	87.08	33.54	85.89	34.82	84.65	36.05
6	88.19	32.27	87.04	33.59	85.85	34.86	84.61	36.09
8	88.15	32.32	87.00	33.63	85.80	34.90	84.57	36.13
10	88.11	32.36	86.96	33.67	85.76	34.94	84.52	36.17
12	88.08	32.41	86.92	33.72	85.72	34.98	84.48	36.21
14	88.04	32.45	86.88	33.76	85.68	35.02	84.44	36.25
16	88.00	32.49	86.84	33.80	85.64	35.07	84.40	36.29
18	87.96	32.54	86.80	33.84	85.60	35.11	84.35	36.33
20	87.93	32.58	86.77	33.89	85.56	35.15	84.31	36.37
22	87.89	32.63	86.73	33.93	85.52	35.19	84.27	36.41
24	87.85	32.67	86.69	33.97	85.48	35.23	84.23	36.45
26	87.81	32.72	86.65	34.01	85.44	35.27	84.18	36.49
28	87.77	32.76	86.61	34.06	85.40	35.31	84.14	36.53
30	87.74	32.80	86.57	34.10	85.36	35.36	84.10	36.57
32	87.70	32.85	86.53	34.14	85.31	35.40	84.06	36.61
34	87.66	32.89	86.49	34.18	85.27	35.44	84.01	36.65
36	87.62	32.93	86.45	34.23	85.23	35.48	83.97	36.69
38	87.58	32.98	86.41	34.27	85.19	35.52	83.93	36.73
40	87.54	33.02	86.37	34.31	85.15	35.56	83.89	36.77
42	87.51	33.07	86.33	34.35	85.11	35.60	83.84	36.80
44	87.47	33.11	86.29	34.40	85.07	35.64	83.80	36.84
46	87.43	33.15	86.25	34.44	85.02	35.68	83.76	36.88
48	87.39	33.20	86.21	34.48	84.98	35.72	83.72	36.92
50	87.35	33.24	86.17	34.52	84.94	35.76	83.67	36.96
52	87.31	33.28	86.13	34.57	84.90	35.80	83.63	37.00
54	87.27	33.33	86.09	34.61	84.86	35.85	83.59	37.04
56	87.24	33.37	86.05	34.65	84.82	35.89	83.54	37.08
58	87.20	33.41	86.01	34.69	84.77	35.93	83.50	37.12
60	87.16	33.46	85.97	34.73	84.73	35.97	83.46	37.16
C = 0.75	0.70	0.26	0.70	0.27	0.69	0.29	0.69	0.30
C = 1.00	0.94	0.35	0.93	0.37	0.92	0.38	0.92	0.40
C = 1.25	1.17	0.44	1.16	0.46	1.15	0.48	1.15	0.50

TABLE IX. HORIZONTAL DISTANCES AND ELEVATIONS FROM STADIA READINGS. — *Continued*

Minutes	24°		25°		26°		27°	
	Hor. Dist.	Diff. Elev.	Hor. Dist.	Diff. Elev.	Hor. Dist.	Diff Elev.	Hor. Dist.	Diff. Elev.
0	83.46	37.16	82.14	38.30	80.78	39.40	79.39	40.45
2	83.41	37.20	82.09	38.34	80.74	39.44	79.34	40.49
4	83.37	37.23	82.05	38.38	80.69	39.47	79.30	40.52
6	83.33	37.27	82.01	38.41	80.65	39.51	79.25	40.55
8	83.28	37.31	81.96	38.45	80.60	39.54	79.20	40.59
10	83.24	37.35	81.92	38.49	80.55	39.58	79.15	40.62
12	83.20	37.39	81.87	38.53	80.51	39.61	79.11	40.66
14	83.15	37.43	81.83	38.56	80.46	39.65	79.06	40.69
16	83.11	37.47	81.78	38.60	80.41	39.69	79.01	40.72
18	83.07	37.51	81.74	38.64	80.37	39.72	78.96	40.76
20	83.02	37.54	81.69	38.67	80.32	39.76	78.92	40.79
22	82.98	37.58	81.65	38.71	80.28	39.79	78.87	40.82
24	82.93	37.62	81.60	38.75	80.23	39.83	78.82	40.86
26	82.89	37.66	81.56	38.78	80.18	39.86	78.77	40.89
28	82.85	37.70	81.51	38.82	80.14	39.90	78.73	40.92
30	82.80	37.74	81.47	38.86	80.09	39.93	78.68	40.96
32	82.76	37.77	81.42	38.89	80.04	39.97	78.63	40.99
34	82.72	37.81	81.38	38.93	80.00	40.00	78.58	41.02
36	82.67	37.85	81.33	38.97	79.95	40.04	78.54	41.06
38	82.63	37.89	81.28	39.00	79.90	40.07	78.49	41.09
40	82.58	37.93	81.24	39.04	79.86	40.11	78.44	41.12
42	82.54	37.96	81.19	39.08	79.81	40.14	78.39	41.16
44	82.49	38.00	81.15	39.11	79.76	40.18	78.34	41.19
46	82.45	38.04	81.10	39.15	79.72	40.21	78.30	41.22
48	82.41	38.08	81.06	39.18	79.67	40.24	78.25	41.26
50	82.36	38.11	81.01	39.22	79.62	40.28	78.20	41.29
52	82.32	38.15	80.97	39.26	79.58	40.31	78.15	41.32
54	82.27	38.19	80.92	39.29	79.53	40.35	78.10	41.35
56	82.23	38.23	80.87	39.33	79.48	40.38	78.06	41.39
58	82.18	38.26	80.83	39.36	79.44	40.42	78.01	41.42
60	82.14	38.30	80.78	39.40	79.39	40.45	77.96	41.45
C = 0.75	0.68	0.31	0.68	0.32	0.67	0.33	0.66	0.35
C = 1.00	0.91	0.41	0.90	0.43	0.89	0.45	0.89	0.46
C = 1.25	1.14	0.52	1.13	0.54	1.12	0.56	1.11	0.58

TABLE IX. HORIZONTAL DISTANCES AND ELEVATIONS FROM STADIA READINGS. — *Concluded*

	28°		29°		30°	
Minutes	Hor. Dist.	Diff. Elev.	Hor. Dist.	Diff. Elev.	Hor. Dist.	Diff. Elev.
0	77.96	41.45	76.50	42.40	75.00	43.30
2	77.91	41.48	76.45	42.43	74.95	43.33
4	77.86	41.52	76.40	42.46	74.90	43.36
6	77.81	41.55	76.35	42.49	74.85	43.39
8	77.77	41.58	76.30	42.53	74.80	43.42
10	77.72	41.61	76.25	42.56	74.75	43.45
12	77.67	41.65	76.20	42.59	74.70	43.47
14	77.62	41.68	76.15	42.62	74.65	43.50
16	77.57	41.71	76.10	42.65	74.60	43.53
18	77.52	41.74	76.05	42.68	74.55	43.56
20	77.48	41.77	76.00	42.71	74.49	43.59
22	77.42	41.81	75.95	42.74	74.44	43.62
24	77.38	41.84	75.90	42.77	74.39	43.65
26	77.33	41.87	75.85	42.80	74.34	43.67
28	77.28	41.90	75.80	42.83	74.29	43.70
30	77.23	41.93	75.75	42.86	74.24	43.73
32	77.18	41.97	75.70	42.89	74.19	43.76
34	77.13	42.00	75.65	42.92	74.14	43.79
36	77.09	42.03	75.60	42.95	74.09	43.82
38	77.04	42.06	75.55	42.98	74.04	43.84
40	76.99	42.09	75.50	43.01	73.99	43.87
42	76.94	42.12	75.45	43.04	73.93	43.90
44	76.89	42.15	75.40	43.07	73.88	43.93
46	76.84	42.19	75.35	43.10	73.83	43.95
48	76.79	42.22	75.30	43.13	73.78	43.98
50	76.74	42.25	75.25	43.16	73.73	44.01
52	76.69	42.28	75.20	43.18	73.68	44.04
54	76.64	42.31	75.15	43.21	73.63	44.07
56	76.59	42.34	75.10	43.24	73.58	44.09
58	76.55	42.37	75.05	43.27	73.52	44.12
60	76.50	42.40	75.00	43.30	73.47	44.15
C = 0.75	0.66	0.36	0.65	0.37	0.65	0.38
C = 1.00	0.88	0.48	0.87	0.49	0.86	0.51
C = 1.25	1.10	0.60	1.09	0.62	1.08	0.64

TABLE X.* MINUTES IN DECIMALS OF A DEGREE

′	0″	10″	15″	20″	30″	40″	45″	50″	′
0	.00000	.00278	.00417	.00556	.00833	.01111	.01250	.01389	0
1	.01667	.01944	.02083	.02222	.02500	.02778	.02917	.03055	1
2	.03333	.03611	.03750	.03889	.04167	.04444	.04583	.04722	2
3	.05000	.05278	.05417	.05556	.05833	.06111	.06250	.06389	3
4	.06667	.06944	.07083	.07222	.07500	.07778	.07917	.08056	4
5	.08333	.08611	.08750	.08889	.09167	.09444	.09583	.09722	5
6	.10000	.10278	.10417	.10556	.10833	.11111	.11250	.11389	6
7	.11667	.11944	.12083	.12222	.12500	.12778	.12917	.13056	7
8	.13333	.13611	.13750	.13889	.14167	.14444	.14583	.14722	8
9	.15000	.15278	.15417	.15556	.15833	.16111	.16250	.16389	9
10	.16667	.16944	.17083	.17222	.17500	.17778	.17917	.18056	10
11	.18333	.18611	.18750	.18889	.19167	.19444	.19583	.19722	11
12	.20000	.20278	.20417	.20556	.20833	.21111	.21250	.21389	12
13	.21667	.21944	.22083	.22222	.22500	.22778	.22917	.23056	13
14	.23333	.23611	.23750	.23889	.24167	.24444	.24583	.24722	14
15	.25000	.25278	.25417	.25556	.25833	.26111	.26250	.26389	15
16	.26667	.26944	.27083	.27222	.27500	.27778	.27917	.28056	16
17	.28333	.28611	.28750	.28889	.29167	.29444	.29583	.29722	17
18	.30000	.30278	.30417	.30556	.30833	.31111	.31250	.31389	18
19	.31667	.31944	.32083	.32222	.32500	.32778	.32917	.33056	19
20	.33333	.33611	.33750	.33889	.34167	.34444	.34583	.34722	20
21	.35000	.35278	.35417	.35556	.35833	.36111	.36250	.36389	21
22	.36667	.36944	.37083	.37222	.37500	.37778	.37917	.38056	22
23	.38333	.38611	.38750	.38889	.39167	.39444	.39583	.39722	23
24	.40000	.40278	.40417	.40556	.40833	.41111	.41250	.41389	24
25	.41667	.41944	.42083	.42222	.42500	.42778	.42917	.43056	25
26	.43333	.43611	.43750	.43889	.44167	.44444	.44583	.44722	26
27	.45000	.45278	.45417	.45556	.45833	.46111	.46250	.46389	27
28	.46667	.46944	.47083	.47222	.47500	.47778	.47917	.48056	28
29	.48333	.48611	.48750	.48889	.49167	.49444	.49583	.49722	29
30	.50000	.50278	.50417	.50556	.50833	.51111	.51250	.51389	30
31	.51667	.51944	.52083	.52222	.52500	.52778	.52917	.53056	31
32	.53333	.53611	.53750	.53889	.54167	.54444	.54583	.54722	32
33	.55000	.55278	.55417	.55556	.55833	.56111	.56250	.56389	33
34	.56667	.56944	.57083	.57222	.57500	.57778	.57917	.58056	34
35	.58333	.58611	.58750	.58889	.59167	.59444	.59583	.59722	35
36	.60000	.60278	.60417	.60556	.60833	.61111	.61250	.61389	36
37	.61667	.61944	.62083	.62222	.62500	.62778	.62917	.63056	37
38	.63333	.63611	.63750	.63889	.64167	.64444	.64583	.64722	38
39	.65000	.65278	.65417	.65556	.65833	.66111	.66250	.66389	39
40	.66667	.66944	.67083	.67222	.67500	.67778	.67917	.68056	40
41	.68333	.68611	.68750	.68889	.69167	.69444	.69583	.69722	41
42	.70000	.70278	.70417	.70556	.70833	.71111	.71250	.71389	42
43	.71667	.71944	.72083	.72222	.72500	.72778	.72917	.73056	43
44	.73333	.73611	.73750	.73889	.74167	.74444	.74583	.74722	44
45	.75000	.75278	.75417	.75556	.75833	.76111	.76250	.76389	45
46	.76667	.76944	.77083	.77222	.77500	.77778	.77917	.78056	46
47	.78333	.78611	.78750	.78889	.79167	.79444	.79583	.79722	47
48	.80000	.80278	.80417	.80556	.80833	.81111	.81250	.81389	48
49	.81667	.81944	.82083	.82222	.82500	.82778	.82917	.83056	49
50	.83333	.83611	.83750	.83889	.84167	.84444	.84583	.84722	50
51	.85000	.85278	.85417	.85556	.85833	.86111	.86250	.86389	51
52	.86667	.86944	.87083	.87222	.87500	.87778	.87917	.88056	52
53	.88333	.88611	.88750	.88889	.89167	.89444	.89583	.89722	53
54	.90000	.90278	.90417	.90556	.90833	.91111	.91250	.91389	54
55	.91667	.91944	.92083	.92222	.92500	.92778	.92917	.93056	55
56	.93333	.93611	.93750	.93889	.94167	.94444	.94583	.94722	56
57	.95000	.95278	.95417	.95556	.95833	.96111	.96250	.96389	57
58	.96667	.96944	.97083	.97222	.97500	.97778	.97917	.98056	58
59	.98333	.98611	.98750	.98889	.99167	.99444	.99583	.99722	59
′	0″	10″	15″	20″	30″	40″	45″	50″	′

* From "Field Engineering" by Searles and Ives. By permission of the publishers, John Wiley & Sons, Inc., New York.

TABLE XI. CONVERGENCY OF MERIDIANS, SIX MILES LONG AND SIX MILES APART, AND DIFFERENCES OF LATITUDE AND LONGITUDE
(*U.S. Bureau of Land Management*)

Lat.	Convergency		Difference of longitude per range		Difference of latitude for—	
	On the parallel	Angle	In arc	In time	1 mi.	1 Tp.
°	*Lks.*	′ ″	′ ″	*Seconds*	′	′
25	33.9	2 25	5 44.34	22.96		
26	35.4	2 32	5 47.20	23.15		
27	37.0	2 39	5 50.22	23.35	0.871	5.229
28	38.6	2 46	5 53.40	23.56		
29	40.2	2 53	5 56.74	23.78		
30	41.9	3 0	6 0.26	24.02		
31	43.6	3 7	6 3.97	24.26		
32	45.4	3 15	6 7.87	24.52	0.871	5.225
33	47.2	3 23	6 11.96	24.80		
34	49.1	3 30	6 16.26	25.08		
35	50.9	3 38	6 20.78	25.39		
36	52.7	3 46	6 25.53	25.70		
37	54.7	3 55	6 30.52	26.03	0.870	5.221
38	56.8	4 4	6 35.76	26.38		
39	58.8	4 13	6 41.27	26.75		
40	60.9	4 22	6 47.06	27.14		
41	63.1	4 31	6 53.15	27.54		
42	65.4	4 41	6 59.56	27.97	0.869	5.216
43	67.7	4 51	7 6.29	28.42		
44	70.1	5 1	7 13.39	28.89		
45	72.6	5 12	7 20.86	29.39		
46	75.2	5 23	7 28.74	29.92		
47	77.8	5 34	7 37.04	30.47	0.869	5.211
48	80.6	5 46	7 45.80	31.05		
49	83.5	5 59	7 55.05	31.67		
50	86.4	6 12	8 4.83	32.32		
51	89.6	6 25	8 15.17	33.03		
52	92.8	6 39	8 26.13	33.74	0.868	5.207
53	96.2	6 54	8 37.75	34.52		
54	99.8	7 9	8 50.07	35.34		
55	103.5	7 25	9 3.18	36.22		
56	107.5	7 42	9 17.12	37.14		
57	111.6	8 0	9 31.97	38.13	0.867	5.202
58	116.0	8 19	9 47.83	39.19		
59	120.6	8 38	10 4.78	40.32		
60	125.5	8 59	10 22.94	41.52		
61	130.8	9 22	10 42.42	42.83		
62	136.3	9 46	11 3.38	44.22	0.866	5.198
63	142.2	10 11	11 25.97	45.73		
64	148.6	10 38	11 50.37	47.36		
65	155.0	11 8	12 16.82	49.12		
66	162.8	11 39	12 45.55	51.04		
67	170.7	12 13	13 16.88	53.12	0.866	5.195
68	179.3	12 51	13 51.15	55.41		
69	188.7	13 31	14 28.77	57.92		
70	199.1	14 15	15 10.26	60.68	0.866	5.193

TABLE XII. AZIMUTHS OF THE SECANT
(U.S. Bureau of Land Management)

Lat.	0 mi.	1 mi.	2 mi.	3 mi.	Deflection angle 6 mi.
°	° ′	° ′	° ′		′ ″
25	89 58.8	89 59.2	89 59.6	90°	2 25
26	58.7	59.2	59.6	E or W.	2 32
27	58.7	59.1	59.6	" " "	2 39
28	58.6	59.1	59.5	" " "	2 46
29	58.6	59.0	59.5	" " "	2 53
30	58.5	59.0	59.5	" " "	3 0
31	58.4	59.0	59.5	" " "	3 7
32	58.4	58.9	59.5	" " "	3 15
33	58.3	58.9	59.4	" " "	3 23
34	58.2	58.8	59.4	" " "	3 30
35	58.2	58.8	59.4	" " "	3 38
36	58.1	58.7	59.4	" " "	3 46
37	58.0	58.7	59.3	" " "	3 55
38	58.0	58.6	59.3	" " "	4 4
39	57.9	58.6	59.3	" " "	4 13
40	57.8	58.5	59.3	" " "	4 22
41	57.7	58.5	59.2	" " "	4 31
42	57.7	58.4	59.2	" " "	4 41
43	57.6	58.4	59.2	" " "	4 51
44	57.5	58.3	59.2	" " "	5 1
45	57.4	58.3	59.1	" " "	5 12
46	57.3	58.2	59.1	" " "	5 23
47	57.2	58.1	59.1	" " "	5 34
48	57.1	58.1	59.0	" " "	5 46
49	57.0	58.0	59.0	" " "	5 59
50	56.9	57.9	59.0	" " "	6 12
51	56.8	57.9	58.9	" " "	6 25
52	56.7	57.8	58.9	" " "	6 39
53	56.6	57.7	58.8	" " "	6 54
54	56.4	57.6	58.8	" " "	7 9
55	56.3	57.5	58.8	" " "	7 25
56	56.2	57.4	58.7	" " "	7 42
57	56.0	57.3	58.7	" " "	8 0
58	55.8	57.2	58.6	" " "	8 19
59	55.7	57.1	58.6	" " "	8 38
60	55.5	57.0	58.5	" " "	8 59
61	55.3	56.9	58.4	" " "	9 22
62	55.1	56.7	58.4	" " "	9 46
63	54.9	56.6	58.3	" " "	10 11
64	54.7	56.5	58.2	" " "	10 38
65	54.4	56.3	58.1	" " "	11 8
66	54.2	56.1	58.1	" " "	11 39
67	53.9	55.9	58.0	" " "	12 13
68	53.6	55.7	57.9	" " "	12 51
69	53.2	55.5	57.8	" " "	13 31
70	89° 52′.9	89° 55′.3	89° 57′.6	" " "	14′ 15″
	6 mi.	5 mi.	4 mi.	3 mi.	

TABLE XIII. OFFSETS, IN LINKS, FROM THE SECANT TO THE PARALLEL
(*U.S. Bureau of Land Management*)

Lat.	0 mi.	½ mi.	1 mi.	1½ mi.	2 mi.	2½ mi.	3 mi.
°							
25	2 N.	1 N.	0	1 S.	1 S.	2 S.	2 S.
26	2	1	0	1	1	2	2
27	3	1	0	1	2	2	2
28	3	1	0	1	2	2	2
29	3	1	0	1	2	2	2
30	3	1	0	1	2	2	2
31	3	1	0	1	2	2	2
32	3	1	0	1	2	2	3
33	3	1	0	1	2	2	3
34	3	2	0	1	2	3	3
35	4	2	0	1	2	3	3
36	4	2	0	1	2	3	3
37	4	2	0	1	2	3	3
38	4	2	0	1	2	3	3
39	4	2	0	1	2	3	3
40	4	2	0	1	3	3	3
41	4	2	0	2	3	3	4
42	5	2	0	2	3	3	4
43	5	2	0	2	3	4	4
44	5	2	0	2	3	4	4
45	5	2	0	2	3	4	4
46	5	2	0	2	3	4	4
47	5	2	0	2	3	4	4
48	6	3	0	2	3	4	4
49	6	3	0	2	3	4	5
50	6	3	0	2	4	4	5
51	6	3	0	2	4	5	5
52	6	3	0	2	4	5	5
53	7	3	0	2	4	5	5
54	7	3	0	2	4	5	6
55	7	3	0	3	4	5	6
56	7	3	0	3	4	6	6
57	8	3	0	3	5	6	6
58	8	4	0	3	5	6	6
59	8	4	0	3	5	6	7
60	9	4	0	3	5	7	7
61	9	4	0	3	5	7	7
62	9	4	0	3	6	7	8
63	10	4	0	3	6	7	8
64	10	5	0	4	6	8	8
65	11	5	0	4	6	8	9
66	11	5	0	4	7	8	9
67	12	5	0	4	7	9	9
68	12	6	0	4	7	9	10
69	13	6	0	5	8	10	10
70	14 N.	6 N.	0	5 S.	8 S.	10 S.	11 S.
	6 mi.	5½ mi.	5 mi.	4½ mi.	4 mi.	3½ mi.	3 mi.

No. 100   No. 109
Log. 000   TABLE XIV. LOGARITHMS OF NUMBERS   Log. 040

N.	0	1	2	3	4	5	6	7	8	9	Diff.
100	00 0000	0434	0868	1301	1734	2166	2598	3029	3461	3891	432
1	4321	4751	5181	5609	6038	6466	6894	7321	7748	8174	428
2	8600	9026	9451	9876	0300	0724	1147	1570	1993	2415	424
3	01 2837	3259	3680	4100	4521	4940	5360	5779	6197	6616	420
4	7033	7451	7868	8284	8700	9116	9532	9947	0361	0775	416
105	02 1189	1603	2016	2428	2841	3252	3664	4075	4486	4896	412
6	5306	5715	6125	6533	6942	7350	7757	8164	8571	8978	408
7	9384	9789	0195	0600	1004	1408	1812	2216	2619	3021	404
8	03 3424	3826	4227	4628	5029	5430	5830	6230	6629	7028	400
9	7426	7825	8223	8620	9017	9414	9811	0207	0602	0998	397
	04										

PROPORTIONAL PARTS

Diff.	1	2	3	4	5	6	7	8	9
434	43.4	86.8	130.2	173.6	217.0	260.4	303.8	347.2	390.6
433	43.3	86.6	129.9	173.2	216.5	259.8	303.1	346.4	389.7
432	43.2	86.4	129.6	172.8	216.0	259.2	302.4	345.6	388.8
431	43.1	86.2	129.3	172.4	215.5	258.6	301.7	344.8	387.9
430	43.0	86.0	129.0	172.0	215.0	258.0	301.0	344.0	387.0
429	42.9	85.8	128.7	171.6	214.5	257.4	300.3	343.2	386.1
428	42.8	85.6	128.4	171.2	214.0	256.8	299.6	342.4	385.2
427	42.7	85.4	128.1	170.8	213.5	256.2	298.9	341.6	384.3
426	42.6	85.2	127.8	170.4	213.0	255.6	298.2	340.8	383.4
425	42.5	85.0	127.5	170.0	212.5	255.0	297.5	340.0	382.5
424	42.4	84.8	127.2	169.6	212.0	254.4	296.8	339.2	381.6
423	42.3	84.6	126.9	169.2	211.5	253.8	296.1	338.4	380.7
422	42.2	84.4	126.6	168.8	211.0	253.2	295.4	337.6	379.8
421	42.1	84.2	126.3	168.4	210.5	252.6	294.7	336.8	378.9
420	42.0	84.0	126.0	168.0	210.0	252.0	294.0	336.0	378.0
419	41.9	83.8	125.7	167.6	209.5	251.4	293.3	335.2	377.1
418	41.8	83.6	125.4	167.2	209.0	250.8	292.6	334.4	376.2
417	41.7	83.4	125.1	166.8	208.5	250.2	291.9	333.6	375.3
416	41.6	83.2	124.8	166.4	208.0	249.6	291.2	332.8	374.4
415	41.5	83.0	124.5	166.0	207.5	249.0	290.5	332.0	373.5
414	41.4	82.8	124.2	165.6	207.0	248.4	289.8	331.2	372.6
413	41.3	82.6	123.9	165.2	206.5	247.8	289.1	330.4	371.7
412	41.2	82.4	123.6	164.8	206.0	247.2	288.4	329.6	370.8
411	41.1	82.2	123.3	164.4	205.5	246.6	287.7	328.8	369.9
410	41.0	82.0	123.0	164.0	205.0	246.0	287.0	328.0	369.0
409	40.9	81.8	122.7	163.6	204.5	245.4	286.3	327.2	368.1
408	40.8	81.6	122.4	163.2	204.0	244.8	285.6	326.4	367.2
407	40.7	81.4	122.1	162.8	203.5	244.2	284.9	325.6	366.3
406	40.6	81.2	121.8	162.4	203.0	243.6	284.2	324.8	365.4
405	40.5	81.0	121.5	162.0	202.5	243.0	283.5	324.0	364.5
404	40.4	80.8	121.2	161.6	202.0	242.4	282.8	323.2	363.6
403	40.3	80.6	120.9	161.2	201.5	241.8	282.1	322.4	362.7
402	40.2	80.4	120.6	160.8	201.0	241.2	281.4	321.6	361.8
401	40.1	80.2	120.3	160.4	200.5	240.6	280.7	320.8	360.9
400	40.0	80.0	120.0	160.0	200.0	240.0	280.0	320.0	360.0
399	39.9	79.8	119.7	159.6	199.5	239.4	279.3	319.2	359.1
398	39.8	79.6	119.4	159.2	199.0	238.8	278.6	318.4	358.2
397	39.7	79.4	119.1	158.8	198.5	238.2	277.9	317.6	357.3
396	39.6	79.2	118.8	158.4	198.0	237.6	277.2	316.8	356.4
395	39.5	79.0	118.5	158.0	197.5	237.0	276.5	316.0	351.5

No. 110
Log. 041

TABLE XIV.—*Continued*

No. 119
Log. 078

N.	0	1	2	3	4	5	6	7	8	9	Diff.
110	04 1393	1787	2182	2576	2969	3362	3755	4148	4540	4932	393
1	5323	5714	6105	6495	6885	7275	7664	8053	8442	8830	390
2	9218	9606	9993	0380	0766	1153	1538	1924	2309	2694	386
3	05 3078	3463	3846	4230	4613	4996	5378	5760	6142	6524	383
4	6905	7286	7666	8046	8426	8805	9185	9563	9942	0320	379
115	06 0698	1075	1452	1829	2206	2582	2958	3333	3709	4083	376
6	4458	4832	5206	5580	5953	6326	6699	7071	7443	7815	373
7	8186	8557	8928	9298	9668	0038	0407	0776	1145	1514	370
8	07 1882	2250	2617	2985	3352	3718	4085	4451	4816	5182	366
9	5547	5912	6276	6640	7004	7368	7731	8094	8457	8819	363

PROPORTIONAL PARTS

Diff.	1	2	3	4	5	6	7	8	9
395	39.5	79.0	118.5	158.0	197.5	237.0	276.5	316.0	355.5
394	39.4	78.8	118.2	157.6	197.0	236.4	275.8	315.2	354.6
393	39.3	78.6	117.9	157.2	196.5	235.8	275.1	314.4	353.7
392	39.2	78.4	117.6	156.8	196.0	235.2	274.4	313.6	352.8
391	39.1	78.2	117.3	156.4	195.5	234.6	273.7	312.8	351.9
390	39.0	78.0	117.0	156.0	195.0	234.0	273.0	312.0	351.0
389	38.9	77.8	116.7	155.6	194.5	233.4	272.3	311.2	350.1
388	38.8	77.6	116.4	155.2	194.0	232.8	271.6	310.4	349.2
387	38.7	77.4	116.1	154.8	193.5	232.2	270.9	309.6	348.3
386	38.6	77.2	115.8	154.4	193.0	231.6	270.2	308.8	347.4
385	38.5	77.0	115.5	154.0	192.5	231.0	269.5	308.0	346.5
384	38.4	76.8	115.2	153.6	192.0	230.4	268.8	307.2	345.6
383	38.3	76.6	114.9	153.2	191.5	229.8	268.1	306.4	344.7
382	38.2	76.4	114.6	152.8	191.0	229.2	267.4	305.6	343.8
381	38.1	76.2	114.3	152.4	190.5	228.6	266.7	304.8	342.9
380	38.0	76.0	114.0	152.0	190.0	228.0	266.0	304.0	342.0
379	37.9	75.8	113.7	151.6	189.5	227.4	265.3	303.2	341.1
378	37.8	75.6	113.4	151.2	189.0	226.8	264.6	302.4	340.2
377	37.7	75.4	113.1	150.8	188.5	226.2	263.9	301.6	339.3
376	37.6	75.2	112.8	150.4	188.0	225.6	263.2	300.8	338.4
375	37.5	75.0	112.5	150.0	187.5	225.0	262.5	300.0	337.5
374	37.4	74.8	112.2	149.6	187.0	224.4	261.8	299.2	336.6
373	37.3	74.6	111.9	149.2	186.5	223.8	261.1	298.4	335.7
372	37.2	74.4	111.6	148.8	186.0	223.2	260.4	297.6	334.8
371	37.1	74.2	111.3	148.4	185.5	222.6	259.7	296.8	333.9
370	37.0	74.0	111.0	148.0	185.0	222.0	259.0	296.0	333.0
369	36.9	73.8	110.7	147.6	184.5	221.4	258.3	295.2	332.1
368	36.8	73.6	110.4	147.2	184.0	220.8	257.6	294.4	331.2
367	36.7	73.4	110.1	146.8	183.5	220.2	256.9	293.6	330.3
366	36.6	73.2	109.8	146.4	183.0	219.6	256.2	292.8	329.4
365	36.5	73.0	109.5	146.0	182.5	219.0	255.7	292.0	328.5
364	36.4	72.8	109.2	145.6	182.0	218.4	254.8	291.2	327.6
363	36.3	72.6	108.9	145.2	181.5	217.8	254.1	290.4	326.7
362	36.2	72.4	108.6	144.8	181.0	217.2	253.4	289.6	325.8
361	36.1	72.2	108.3	144.4	180.5	216.6	252.7	288.8	324.9
360	36.0	72.0	108.0	144.0	180.0	216.0	252.0	288.0	324.0
359	35.9	71.8	107.7	143.6	179.5	215.4	251.3	287.2	323.1
358	35.8	71.6	107.4	143.2	179.0	214.8	250.6	286.4	322.2
357	35.7	71.4	107.1	142.8	178.5	214.2	249.9	285.6	321.3
356	35.6	71.2	106.8	142.4	178.0	213.6	249.2	284.8	320.4

No. 120 Log. 079 TABLE XIV.—*Continued* No. 134 Log. 130

N.	0	1	2	3	4	5	6	7	8	9	Diff.
120	07 9181	9543	9904	0266	0626	0987	1347	1707	2067	2426	360
1	08 2785	3144	3503	3861	4219	4576	4934	5291	5647	6004	357
2	6360	6716	7071	7426	7781	8136	8490	8845	9198	9552	355
3	9905	0258	0611	0963	1315	1667	2018	2370	2721	3071	352
4	09 3422	3772	4122	4471	4820	5169	5518	5866	6215	6562	349
125	6910	7257	7604	7951	8298	8644	8990	9335	9681	0026	346
6	10 0371	0715	1059	1403	1747	2091	2434	2777	3119	3462	343
7	3804	4146	4487	4828	5169	5510	5851	6191	6531	6871	341
8	7210	7549	7888	8227	8565	8903	9241	9579	9916	0253	338
9	11 0590	0926	1263	1599	1934	2270	2605	2940	3275	3609	335
130	3943	4277	4611	4944	5278	5611	5943	6276	6608	6940	333
1	7271	7603	7934	8265	8595	8926	9256	9586	9915	0245	330
2	12 0574	0903	1231	1560	1888	2216	2544	2871	3198	3525	328
3	3852	4178	4504	4830	5156	5481	5806	6131	6456	6781	325
4	7105	7429	7753	8076	8399	8722	9045	9368	9690	0012	323
	13										

PROPORTIONAL PARTS

Diff.	1	2	3	4	5	6	7	8	9
355	35.5	71.0	106.5	142.0	177.5	213.0	248.5	284.0	319.5
354	35.4	70.8	106.2	141.6	177.0	212.4	247.8	283.2	318.6
353	35.3	70.6	105.9	141.2	176.5	211.8	247.1	282.4	317.7
352	35.2	70.4	105.6	140.8	176.0	211.2	246.4	281.6	316.8
351	35.1	70.2	105.3	140.4	175.5	210.6	245.7	280.8	315.9
350	35.0	70.0	105.0	140.0	175.0	210.0	245.0	280.0	315.0
349	34.9	69.8	104.7	139.6	174.5	209.4	244.3	279.2	314.1
348	34.8	69.6	104.4	139.2	174.0	208.8	243.6	278.4	313.2
347	34.7	69.4	104.1	138.8	173.5	208.2	242.9	277.6	312.3
346	34.6	69.2	103.8	138.4	173.0	207.6	242.2	276.8	311.4
345	34.5	69.0	103.5	138.0	172.5	207.0	241.5	276.0	310.5
344	34.4	68.8	103.2	137.6	172.0	206.4	240.8	275.2	309.6
343	34.3	68.6	102.9	137.2	171.5	205.8	240.1	274.4	308.7
342	34.2	68.4	102.6	136.8	171.0	205.2	239.4	273.6	307.8
341	34.1	68.2	102.3	136.4	170.5	204.6	238.7	272.8	306.9
340	34.0	68.0	102.0	136.0	170.0	204.0	238.0	272.0	306.0
339	33.9	67.8	101.7	135.6	169.5	203.4	237.3	271.2	305.1
338	33.8	67.6	101.4	135.2	169.0	202.8	236.6	270.4	304.2
337	33.7	67.4	101.1	134.8	168.5	202.2	235.9	269.6	303.3
336	33.6	67.2	100.8	134.4	168.0	201.6	235.2	268.8	302.4
335	33.5	67.0	100.5	134.0	167.5	201.0	234.5	268.0	301.5
334	33.4	66.8	100.2	133.6	167.0	200.4	233.8	267.2	300.6
333	33.3	66.6	99.9	133.2	166.5	199.8	233.1	266.4	299.7
332	33.2	66.4	99.6	132.8	166.0	199.2	232.4	265.6	298.8
331	33.1	66.2	99.3	132.4	165.5	198.6	231.7	264.8	297.9
330	33.0	66.0	99.0	132.0	165.0	198.0	231.0	264.0	297.0
329	32.9	65.8	98.7	131.6	164.5	197.4	230.3	263.2	296.1
328	32.8	65.6	98.4	131.2	164.0	196.8	229.6	262.4	295.2
327	32.7	65.4	98.1	130.8	163.5	196.2	228.9	261.6	294.3
326	32.6	65.2	97.8	130.4	163.0	195.6	228.2	260.8	293.4
325	32.5	65.0	97.5	130.0	162.5	195.0	227.5	260.0	292.5
324	32.4	64.8	97.2	129.6	162.0	194.4	226.8	259.2	291.6
323	32.3	64.6	96.9	129.2	161.5	193.8	226.1	258.4	290.7
322	32.2	64.4	96.6	128.8	161.0	193.2	225.4	257.6	289.8

No. 135
Log. 130

TABLE XIV.—*Continued*

No. 149
Log. 175

N.	0	1	2	3	4	5	6	7	8	9	Diff.
135	13 0334	0655	0977	1298	1619	1939	2260	2580	2900	3219	321
6	3539	3858	4177	4496	4814	5133	5451	5769	6086	6403	318
7	6721	7037	7354	7671	7987	8303	8618	8934	9249	9564	316
8	9879	0194	0508	0822	1136	1450	1763	2076	2389	2702	314
9	14 3015	3327	3639	3951	4263	4574	4885	5196	5507	5818	311
140	6128	6438	6748	7058	7367	7676	7985	8294	8603	8911	309
1	9219	9527	9835	0142	0449	0756	1063	1370	1676	1982	307
2	15 2288	2594	2900	3205	3510	3815	4120	4424	4728	5032	305
3	5336	5640	5943	6246	6549	6852	7154	7457	7759	8061	303
4	8362	8664	8965	9266	9567	9868	0168	0469	0769	1068	301
145	16 1368	1667	1967	2266	2564	2863	3161	3460	3758	4055	299
6	4353	4650	4947	5244	5541	5838	6134	6430	6726	7022	297
7	7317	7613	7908	8203	8497	8792	9086	9380	9674	9968	295
8	17 0262	0555	0848	1141	1434	1726	2019	2311	2603	2895	293
9	3186	3478	3769	4060	4351	4641	4932	5222	5512	5802	291

PROPORTIONAL PARTS

Diff.	1	2	3	4	5	6	7	8	9
321	32.1	64.2	96.3	128.4	160.5	192.6	224.7	256.8	288.9
320	32.0	64.0	96.0	128.0	160.0	192.0	224.0	256.0	288.0
319	31.9	63.8	95.7	127.6	159.5	191.4	223.3	255.2	287.1
318	31.8	63.6	95.4	127.2	159.0	190.8	222.6	254.4	286.2
317	31.7	63.4	95.1	126.8	158.5	190.2	221.9	253.6	285.3
316	31.6	63.2	94.8	126.4	158.0	189.6	221.2	252.8	284.4
315	31.5	63.0	94.5	126.0	157.5	189.0	220.5	252.0	283.5
314	31.4	62.8	94.2	125.6	157.0	188.4	219.8	251.2	282.6
313	31.3	62.6	93.9	125.2	156.5	187.8	219.1	250.4	281.7
312	31.2	62.4	93.6	124.8	156.0	187.2	218.4	249.6	280.8
311	31.1	62.2	93.3	124.4	155.5	186.6	217.7	248.8	279.9
310	31.0	62.0	93.0	124.0	155.0	186.0	217.0	248.0	279.0
309	30.9	61.8	92.7	123.6	154.5	185.4	216.3	247.2	278.1
308	30.8	61.6	92.4	123.2	154.0	184.8	215.6	246.4	277.2
307	30.7	61.4	92.1	122.8	153.5	184.2	214.9	245.6	276.3
306	30.6	61.2	91.8	122.4	153.0	183.6	214.2	244.8	275.4
305	30.5	61.0	91.5	122.0	152.5	183.0	213.5	244.0	274.5
304	30.4	60.8	91.2	121.6	152.0	182.4	212.8	243.2	273.6
303	30.3	60.6	90.9	121.2	151.5	181.8	212.1	242.4	272.7
302	30.2	60.4	90.6	120.8	151.0	181.2	211.4	241.6	271.8
301	30.1	60.2	90.3	120.4	150.5	180.6	210.7	240.8	270.9
300	30.0	60.0	90.0	120.0	150.0	180.0	210.0	240.0	270.0
299	29.9	59.8	89.7	119.6	149.5	179.4	209.3	239.2	269.1
298	29.8	59.6	89.4	119.2	149.0	178.8	208.6	238.4	268.2
297	29.7	59.4	89.1	118.8	148.5	178.2	207.9	237.6	267.3
296	29.6	59.2	88.8	118.4	148.0	177.6	207.2	236.8	266.4
295	29.5	59.0	88.5	118.0	147.5	177.0	206.5	236.0	265.5
294	29.4	58.8	88.2	117.6	147.0	176.4	205.8	235.2	264.6
293	29.3	58.6	87.9	117.2	146.5	175.8	205.1	234.4	263.7
292	29.2	58.4	87.6	116.8	146.0	175.2	204.4	233.6	262.8
291	29.1	58.2	87.3	116.4	145.5	174.6	203.7	232.8	261.9
290	29.0	58.0	87.0	116.0	145.0	174.0	203.0	232.0	261.0
289	28.9	57.8	86.7	115.6	144.5	173.4	202.3	231.2	260.1
288	28.8	57.6	86.4	115.2	144.0	172.8	201.6	230.4	259.2
287	28.7	57.4	86.1	114.8	143.5	172.2	200.9	229.6	258.3
286	28.6	57.2	85.8	114.4	143.0	171.6	200.2	228.8	257.4

TABLE XIV.—*Continued*

N.	0	1	2	3	4	5	6	7	8	9	Diff.
150	17 6091	6381	6670	6959	7248	7536	7825	8113	8401	8689	289
1	8977	9264	9552	9839	0126	0413	0699	0986	1272	1558	287
2	18 1844	2129	2415	2700	2985	3270	3555	3839	4123	4407	285
3	4691	4975	5259	5542	5825	6108	6391	6674	6956	7239	283
4	7521	7803	8084	8366	8647	8928	9209	9490	9771	0051	281
155	19 0332	0612	0892	1171	1451	1730	2010	2289	2567	2846	279
6	3125	3403	3681	3959	4237	4514	4792	5069	5346	5623	278
7	5900	6176	6453	6729	7005	7281	7556	7832	8107	8382	276
8	8657	8932	9206	9481	9755	0029	0303	0577	0850	1124	274
9	20 1397	1670	1943	2216	2488	2761	3033	3305	3577	3848	272
160	4120	4391	4663	4934	5204	5475	5746	6016	6286	6556	271
1	6826	7096	7365	7634	7904	8173	8441	8710	8979	9247	269
2	9515	9783	0051	0319	0586	0853	1121	1388	1654	1921	267
3	21 2188	2454	2720	2986	3252	3518	3783	4049	4314	4579	266
4	4844	5109	5373	5638	5902	6166	6430	6694	6957	7221	264
165	7484	7747	8010	8273	8536	8798	9060	9323	9585	9846	262
6	22 0108	0370	0631	0892	1153	1414	1675	1936	2196	2456	261
7	2716	2976	3236	3496	3755	4015	4274	4533	4792	5051	259
8	5309	5568	5826	6084	6342	6600	6858	7115	7372	7630	258
9	7887	8144	8400	8657	8913	9170	9426	9682	9938	0193	256
	23										

PROPORTIONAL PARTS

Diff.	1	2	3	4	5	6	7	8	9
285	28.5	57.0	85.5	114.0	142.5	171.0	199.5	228.0	256.5
284	28.4	56.8	85.2	113.6	142.0	170.4	198.8	227.2	255.6
283	28.3	56.6	84.9	113.2	141.5	169.8	198.1	226.4	254.7
282	28.2	56.4	84.6	112.8	141.0	169.2	197.4	225.6	253.8
281	28.1	56.2	84.3	112.4	140.5	168.6	196.7	224.8	252.9
280	28.0	56.0	84.0	112.0	140.0	168.0	196.0	224.0	252.0
279	27.9	55.8	83.7	111.6	139.5	167.4	195.3	223.2	251.1
278	27.8	55.6	83.4	111.2	139.0	166.8	194.6	222.4	250.2
277	27.7	55.4	83.1	110.8	138.5	166.2	193.9	221.6	249.3
276	27.6	55.2	82.8	110.4	138.0	165.6	193.2	220.8	248.4
275	27.5	55.0	82.5	110.0	137.5	165.0	192.5	220.0	247.5
274	27.4	54.8	82.2	109.6	137.0	164.4	191.8	219.2	246.6
273	27.3	54.6	81.9	109.2	136.5	163.8	191.1	218.4	245.7
272	27.2	54.4	81.6	108.8	136.0	163.2	190.4	217.6	244.8
271	27.1	54.2	81.3	108.4	135.5	162.6	189.7	216.8	243.9
270	27.0	54.0	81.0	108.0	135.0	162.0	189.0	216.0	243.0
269	26.9	53.8	80.7	107.6	134.5	161.4	188.3	215.2	242.1
268	26.8	53.6	80.4	107.2	134.0	160.8	187.6	214.4	241.2
267	26.7	53.4	80.1	106.8	133.5	160.2	186.9	213.6	240.3
266	26.6	53.2	79.8	106.4	133.0	159.6	186.2	212.8	239.4
265	26.5	53.0	79.5	106.0	132.5	159.0	185.5	212.0	238.5
264	26.4	52.8	79.2	105.6	132.0	158.4	184.8	211.2	237.6
263	26.3	52.6	78.9	105.2	131.5	157.8	184.1	210.4	236.7
262	26.2	52.4	78.6	104.8	131.0	157.2	183.4	209.6	235.8
261	26.1	52.2	78.3	104.4	130.5	156.6	182.7	208.8	234.9
260	26.0	52.0	78.0	104.0	130.0	156.0	182.0	208.0	234.0
259	25.9	51.8	77.7	103.6	129.5	155.4	181.3	207.2	233.1
258	25.8	51.6	77.4	103.2	129.0	154.8	180.6	206.4	232.2
257	25.7	51.4	77.1	102.8	128.5	154.2	179.9	205.6	231.3
256	25.6	51.2	76.8	102.4	128.0	153.6	179.2	204.8	230.4
255	25.5	51.0	76.5	102.0	127.5	153.0	178.5	204.0	229.5

No. 170   No. 189
Log. 230   Log. 278

TABLE XIV.—*Continued*

N.	0	1	2	3	4	5	6	7	8	9	Diff.
170	23 0449	0704	0960	1215	1470	1724	1979	2234	2488	2742	255
1	2996	3250	3504	3757	4011	4264	4517	4770	5023	5276	253
2	5528	5781	6033	6285	6537	6789	7041	7292	7544	7795	252
3	8046	8297	8548	8799	9049	9299	9550	9800	0050	0300	250
4	24 0549	0799	1048	1297	1546	1795	2044	2293	2541	2790	249
175	3038	3286	3534	3782	4030	4277	4525	4772	5019	5266	248
6	5513	5759	6006	6252	6499	6745	6991	7237	7482	7728	246
7	7973	8219	8464	8709	8954	9198	9443	9687	9932	0176	245
8	25 0420	0664	0908	1151	1395	1638	1881	2125	2368	2610	243
9	2853	3096	3338	3580	3822	4064	4306	4548	4790	5031	242
180	5273	5514	5755	5996	6237	6477	6718	6958	7198	7439	241
1	7679	7918	8158	8398	8637	8877	9116	9355	9594	9833	239
2	26 0071	0310	0548	0787	1025	1263	1501	1739	1976	2214	238
3	2451	2688	2925	3162	3399	3636	3873	4109	4346	4582	237
4	4818	5054	5290	5525	5761	5996	6232	6467	6702	6937	235
185	7172	7406	7641	7875	8110	8344	8578	8812	9046	9279	234
6	9513	9746	9980	0213	0446	0679	0912	1144	1377	1609	233
7	27 1842	2074	2306	2538	2770	3001	3233	3464	3696	3927	232
8	4158	4389	4620	4850	5081	5311	5542	5772	6002	6232	230
9	6462	6692	6921	7151	7380	7609	7838	8067	8296	8525	229

PROPORTIONAL PARTS

Diff.	1	2	3	4	5	6	7	8	9
255	25.5	51.0	76.5	102.0	127.5	153.0	178.5	204.0	229.5
254	25.4	50.8	76.2	101.6	127.0	152.4	177.8	203.2	228.6
253	25.3	50.6	75.9	101.2	126.5	151.8	177.1	202.4	227.7
252	25.2	50.4	75.6	100.8	126.0	151.2	176.4	201.6	226.8
251	25.1	50.2	75.3	100.4	125.5	150.6	175.7	200.8	225.9
250	25.0	50.0	75.0	100.0	125.0	150.0	175.0	200.0	225.0
249	24.9	49.8	74.7	99.6	124.5	149.4	174.3	199.2	224.1
248	24.8	49.6	74.4	99.2	124.0	148.8	173.6	198.4	223.2
247	24.7	49.4	74.1	98.8	123.5	148.2	172.9	197.6	222.3
246	24.6	49.2	73.8	98.4	123.0	147.6	172.2	196.8	221.4
245	24.5	49.0	73.5	98.0	122.5	147.0	171.5	196.0	220.5
244	24.4	48.8	73.2	97.6	122.0	146.4	170.8	195.2	219.6
243	24.3	48.6	72.9	97.2	121.5	145.8	170.1	194.4	218.7
242	24.2	48.4	72.6	96.8	121.0	145.2	169.4	193.6	217.8
241	24.1	48.2	72.3	96.4	120.5	144.6	168.7	192.8	216.9
240	24.0	48.0	72.0	96.0	120.0	144.0	168.0	192.0	216.0
239	23.9	47.8	71.7	95.6	119.5	143.4	167.3	191.2	215.1
238	23.8	47.6	71.4	95.2	119.0	142.8	166.6	190.4	214.2
237	23.7	47.4	71.1	94.8	118.5	142.2	165.9	189.6	213.3
236	23.6	47.2	70.8	94.4	118.0	141.6	165.2	188.8	212.4
235	23.5	47.0	70.5	94.0	117.5	141.0	164.5	188.0	211.5
234	23.4	46.8	70.2	93.6	117.0	140.4	163.8	187.2	210.6
233	23.3	46.6	69.9	93.2	116.5	139.8	163.1	186.4	209.7
232	23.2	46.4	69.6	92.8	116.0	139.2	162.4	185.6	208.8
231	23.1	46.2	69.3	92.4	115.5	138.6	161.7	184.8	207.9
230	23.0	46.0	69.0	92.0	115.0	138.0	161.0	184.0	207.0
229	22.9	45.8	68.7	91.6	114.5	137.4	160.3	183.2	206.1
228	22.8	45.6	68.4	91.2	114.0	136.8	159.6	182.4	205.2
227	22.7	45.4	68.1	90.8	113.5	136.2	158.9	181.6	204.3
226	22.6	45.2	67.8	90.4	113.0	135.6	158.2	180.8	203.4

No. 190
Log. 278

TABLE XIV.—*Continued*

No. 214
Log. 332

N.	0	1	2	3	4	5	6	7	8	9	Diff.
190	27 8754	8982	9211	9439	9667	9895	0123	0351	0578	0806	228
1	28 1033	1261	1488	1715	1942	2169	2396	2622	2849	3075	227
2	3301	3527	3753	3979	4205	4431	4656	4882	5107	5332	226
3	5557	5782	6007	6232	6456	6681	6905	7130	7354	7578	225
4	7802	8026	8249	8473	8696	8920	9143	9366	9589	9812	223
195	29 0035	0257	0480	0702	0925	1147	1369	1591	1813	2034	222
6	2256	2478	2699	2920	3141	3363	3584	3804	4025	4246	221
7	4466	4687	4907	5127	5347	5567	5787	6007	6226	6446	220
8	6665	6884	7104	7323	7542	7761	7979	8198	8416	8635	219
9	8853	9071	9289	9507	9725	9943	0161	0378	0595	0813	218
200	30 1030	1247	1464	1681	1898	2114	2331	2547	2764	2980	217
1	3196	3412	3628	3844	4059	4275	4491	4706	4921	5136	216
2	5351	5566	5781	5996	6211	6425	6639	6854	7068	7282	215
3	7496	7710	7924	8137	8351	8564	8778	8991	9204	9417	213
4	9630	9843	0056	0268	0481	0693	0906	1118	1330	1542	212
205	31 1754	1966	2177	2389	2600	2812	3023	3234	3445	3656	211
6	3867	4078	4289	4499	4710	4920	5130	5340	5551	5760	210
7	5970	6180	6390	6599	6809	7018	7227	7436	7646	7854	209
8	8063	8272	8481	8689	8898	9106	9314	9522	9730	9938	208
9	32 0146	0354	0562	0769	0977	1184	1391	1598	1805	2012	207
210	2219	2426	2633	2839	3046	3252	3458	3665	3871	4077	206
1	4282	4488	4694	4899	5105	5310	5516	5721	5926	6131	205
2	6336	6541	6745	6950	7155	7359	7563	7767	7972	8176	204
3	8380	8583	8787	8991	9194	9398	9601	9805	0008	0211	203
4	33 0414	0617	0819	1022	1225	1427	1630	1832	2034	2236	202

PROPORTIONAL PARTS

Diff.	1	2	3	4	5	6	7	8	9
225	22.5	45.0	67.5	90.0	112.5	135.0	157.5	180.0	202.5
224	22.4	44.8	67.2	89.6	112.0	134.4	156.8	179.2	201.6
223	22.3	44.6	66.9	89.2	111.5	133.8	156.1	178.4	200.7
222	22.2	44.4	66.6	88.8	111.0	133.2	155.4	177.6	199.8
221	22.1	44.2	66.3	88.4	110.5	132.6	154.7	176.8	198.9
220	22.0	44.0	66.0	88.0	110.0	132.0	154.0	176.0	198.0
219	21.9	43.8	65.7	87.6	109.5	131.4	153.3	175.2	197.1
218	21.8	43.6	65.4	87.2	109.0	130.8	152.6	174.4	196.2
217	21.7	43.4	65.1	86.8	108.5	130.2	151.9	173.6	195.3
216	21.6	43.2	64.8	86.4	108.0	129.6	151.2	172.8	194.4
215	21.5	43.0	64.5	86.0	107.5	129.0	150.5	172.0	193.5
214	21.4	42.8	64.2	85.6	107.0	128.4	149.8	171.2	192.6
213	21.3	42.6	63.9	85.2	106.5	127.8	149.1	170.4	191.7
212	21.2	42.4	63.6	84.8	106.0	127.2	148.4	169.6	190.8
211	21.1	42.2	63.3	84.4	105.5	126.6	147.7	168.8	189.9
210	21.0	42.0	63.0	84.0	105.0	126.0	147.0	168.0	189.0
209	20.9	41.8	62.7	83.6	104.5	125.4	146.3	167.2	188.1
208	20.8	41.6	62.4	83.2	104.0	124.8	145.6	166.4	187.2
207	20.7	41.4	62.1	82.8	103.5	124.2	144.9	165.6	186.3
206	20.6	41.2	61.8	82.4	103.0	123.6	144.2	164.8	185.4
205	20.5	41.0	61.5	82.0	102.5	123.0	143.5	164.0	184.5
204	20.4	40.8	61.2	81.6	102.0	122.4	142.8	163.2	183.6
203	20.3	40.6	60.9	81.2	101.5	121.8	142.1	162.4	182.7
202	20.2	40.4	60.6	80.8	101.0	121.2	141.4	161.6	181.8

No. 215 No. 239
Log. 332 Table XIV.—*Continued* Log. 380

N.	0	1	2	3	4	5	6	7	8	9	Diff.
215	33 2438	2640	2842	3044	3246	3447	3649	3850	4051	4253	202
6	4454	4655	4856	5057	5257	5458	5658	5859	6059	6260	201
7	6460	6660	6860	7060	7260	7459	7659	7858	8058	8257	200
8	8456	8656	8855	9054	9253	9451	9650	9849	0047	0246	199
9	34 0444	0642	0841	1039	1237	1435	1632	1830	2028	2225	198
220	2423	2620	2817	3014	3212	3409	3606	3802	3999	4196	197
1	4392	4589	4785	4981	5178	5374	5570	5766	5962	6157	196
2	6353	6549	6744	6939	7135	7330	7525	7720	7915	8110	195
3	8305	8500	8694	8889	9083	9278	9472	9666	9860	0054	194
4	35 0248	0442	0636	0829	1023	1216	1410	1603	1796	1989	193
225	2183	2375	2568	2761	2954	3147	3339	3532	3724	3916	193
6	4108	4301	4493	4685	4876	5068	5260	5452	5643	5834	192
7	6026	6217	6408	6599	6790	6981	7172	7363	7554	7744	191
8	7935	8125	8316	8506	8696	8886	9076	9266	9456	9646	190
9	9835	0025	0215	0404	0593	0783	0972	1161	1350	1539	189
230	36 1728	1917	2105	2294	2482	2671	2859	3048	3236	3424	188
1	3612	3800	3988	4176	4363	4551	4739	4926	5113	5301	188
2	5488	5675	5862	6049	6236	6423	6610	6796	6983	7169	187
3	7356	7542	7729	7915	8101	8287	8473	8659	8845	9030	186
4	9216	9401	9587	9772	9958	0143	0328	0513	0698	0883	185
235	37 1068	1253	1437	1622	1806	1991	2175	2360	2544	2728	184
6	2912	3096	3280	3464	3647	3831	4015	4198	4382	4565	184
7	4748	4932	5115	5298	5481	5664	5846	6029	6212	6394	183
8	6577	6759	6942	7124	7306	7488	7670	7852	8034	8216	182
9	8398	8580	8761	8943	9124	9306	9487	9668	9849	0030	181
	38										

PROPORTIONAL PARTS

Diff.	1	2	3	4	5	6	7	8	9
202	20.2	40.4	60.6	80.8	101.0	121.2	141.4	161.6	181.8
201	20.1	40.2	60.3	80.4	100.5	120.6	140.7	160.8	180.9
200	20.0	40.0	60.0	80.0	100.0	120.0	140.0	160.0	180.0
199	19.9	39.8	59.7	79.6	99.5	119.4	139.3	159.2	179.1
198	19.8	39.6	59.4	79.2	99.0	118.8	138.6	158.4	178.2
197	19.7	39.4	59.1	78.8	98.5	118.2	137.9	157.6	177.3
196	19.6	39.2	58.8	78.4	98.0	117.6	137.2	156.8	176.4
195	19.5	39.0	58.5	78.0	97.5	117.0	136.5	156.0	175.5
194	19.4	38.8	58.2	77.6	97.0	116.4	135.8	155.2	174.6
193	19.3	38.6	57.9	77.2	96.5	115.8	135.1	154.4	173.7
192	19.2	38.4	57.6	76.8	96.0	115.2	134.4	153.6	172.8
191	19.1	38.2	57.3	76.4	95.5	114.6	133.7	152.8	171.9
190	19.0	38.0	57.0	76.0	95.0	114.0	133.0	152.0	171.0
189	18.9	37.8	56.7	75.6	94.5	113.4	132.3	151.2	170.1
188	18.8	37.6	56.4	75.2	94.0	112.8	131.6	150.4	169.2
187	18.7	37.4	56.1	74.8	93.5	112.2	130.9	149.6	168.3
186	18.6	37.2	55.8	74.4	93.0	111.6	130.2	148.8	167.4
185	18.5	37.0	55.5	74.0	92.5	111.0	129.5	148.0	166.5
184	18.4	36.8	55.2	73.6	92.0	110.4	128.8	147.2	165.6
183	18.3	36.6	54.9	73.2	91.5	109.8	128.1	146.4	164.7
182	18.2	36.4	54.6	72.8	91.0	109.2	127.4	145.6	163.8
181	18.1	36.2	54.3	72.4	90.5	108.6	126.7	144.8	162.9
180	18.0	36.0	54.0	72.0	90.0	108.0	126.0	144.0	162.0
179	17.9	35.8	53.7	71.6	89.5	107.4	125.3	143.2	161.1

No. 240
Log. 380

TABLE XIV.—*Continued*

No. 269
Log. 431

N.	0	1	2	3	4	5	6	7	8	9	Diff.
240	38 0211	0392	0573	0754	0934	1115	1296	1476	1656	1837	181
1	2017	2197	2377	2557	2737	2917	3097	3277	3456	3636	180
2	3815	3995	4174	4353	4533	4712	4891	5070	5249	5428	179
3	5606	5785	5964	6142	6321	6499	6677	6856	7034	7212	178
4	7390	7568	7746	7924	8101	8279	8456	8634	8811	8989	178
245	9166	9343	9520	9698	9875	0051	0228	0405	0582	0759	177
6	39 0935	1112	1288	1464	1641	1817	1993	2169	2345	2521	176
7	2697	2873	3048	3224	3400	3575	3751	3926	4101	4277	176
8	4452	4627	4802	4977	5152	5326	5501	5676	5850	6025	175
9	6199	6374	6548	6722	6896	7071	7245	7419	7592	7766	174
250	7940	8114	8287	8461	8634	8808	8981	9154	9328	9501	173
1	9674	9847	0020	0192	0365	0538	0711	0883	1056	1228	173
2	40 1401	1573	1745	1917	2089	2261	2433	2605	2777	2949	172
3	3121	3292	3464	3635	3807	3978	4149	4320	4492	4663	171
4	4834	5005	5176	5346	5517	5688	5858	6029	6199	6370	171
255	6540	6710	6881	7051	7221	7391	7561	7731	7901	8070	170
6	8240	8410	8579	8749	8918	9087	9257	9426	9595	9764	169
7	9933	0102	0271	0440	0609	0777	0946	1114	1283	1451	169
8	41 1620	1788	1956	2124	2293	2461	2629	2796	2964	3132	168
9	3300	3467	3635	3803	3970	4137	4305	4472	4639	4806	167
260	4973	5140	5307	5474	5641	5808	5974	6141	6308	6474	167
1	6641	6807	6973	7139	7306	7472	7638	7804	7970	8135	166
2	8301	8467	8633	8798	8964	9129	9295	9460	9625	9791	165
3	9956	0121	0286	0451	0616	0781	0945	1110	1275	1439	165
4	42 1604	1768	1933	2097	2261	2426	2590	2754	2918	3082	164
265	3246	3410	3574	3737	3901	4065	4228	4392	4555	4718	164
6	4882	5045	5208	5371	5534	5697	5860	6023	6186	6349	163
7	6511	6674	6836	6999	7161	7324	7486	7648	7811	7973	162
8	8135	8297	8459	8621	8783	8944	9106	9268	9429	9591	162
9	9752	9914	0075	0236	0398	0559	0720	0881	1042	1203	161
	43										

PROPORTIONAL PARTS

Diff.	1	2	3	4	5	6	7	8	9
178	17.8	35.6	53.4	71.2	89.0	106.8	124.6	142.4	160.2
177	17.7	35.4	53.1	70.8	88.5	106.2	123.9	141.6	159.3
176	17.6	35.2	52.8	70.4	88.0	105.6	123.2	140.8	158.4
175	17.5	35.0	52.5	70.0	87.5	105.0	122.5	140.0	157.5
174	17.4	34.8	52.2	69.6	87.0	104.4	121.8	139.2	156.6
173	17.3	34.6	51.9	69.2	86.5	103.8	121.1	138.4	155.7
172	17.2	34.4	51.6	68.8	86.0	103.2	120.4	137.6	154.8
171	17.1	34.2	51.3	68.4	85.5	102.6	119.7	136.8	153.9
170	17.0	34.0	51.0	68.0	85.0	102.0	119.0	136.0	153.0
169	16.9	33.8	50.7	67.6	84.5	101.4	118.3	135.2	152.1
168	16.8	33.6	50.4	67.2	84.0	100.8	117.6	134.4	151.2
167	16.7	33.4	50.1	66.8	83.5	100.2	116.9	133.6	150.3
166	16.6	33.2	49.8	66.4	83.0	99.6	116.2	132.8	149.4
165	16.5	33.0	49.5	66.0	82.5	99.0	115.5	132.0	148.5
164	16.4	32.8	49.2	65.6	82.0	98.4	114.8	131.2	147.6
163	16.3	32.6	48.9	65.2	81.5	97.8	114.1	130.4	146.7
162	16.2	32.4	48.5	64.8	81.0	97.2	113.4	129.6	145.8
161	16.1	32.2	48.3	64.4	80.5	96.6	112.7	128.8	144.9

No. 270
Log. 431

Table XIV.—*Continued*

No. 299
Log. 476

N.	0	1	2	3	4	5	6	7	8	9	Diff.
270	43 1364	1525	1685	1846	2007	2167	2328	2488	2649	2809	161
1	2969	3130	3290	3450	3610	3770	3930	4090	4249	4409	160
2	4569	4729	4888	5048	5207	5367	5526	5685	5844	6004	159
3	6163	6322	6481	6640	6799	6957	7116	7275	7433	7592	159
4	7751	7909	8067	8226	8384	8542	8701	8859	9017	9175	158
275	9333	9491	9648	9806	9964	0122	0279	0437	0594	0752	158
6	44 0909	1066	1224	1381	1538	1695	1852	2009	2166	2323	157
7	2480	2637	2793	2950	3106	3263	3419	3576	3732	3889	157
8	4045	4201	4357	4513	4669	4825	4981	5137	5293	5449	156
9	5604	5760	5915	6071	6226	6382	6537	6692	6848	7003	155
280	7158	7313	7468	7623	7778	7933	8088	8242	8397	8552	155
1	8706	8861	9015	9170	9324	9478	9633	9787	9941	0095	154
2	45 0249	0403	0557	0711	0865	1018	1172	1326	1479	1633	154
3	1786	1940	2093	2247	2400	2553	2706	2859	3012	3165	153
4	3318	3471	3624	3777	3930	4082	4235	4387	4540	4692	153
285	4845	4997	5150	5302	5454	5606	5758	5910	6062	6214	152
6	6366	6518	6670	6821	6973	7125	7276	7428	7579	7731	152
7	7882	8033	8184	8336	8487	8638	8789	8940	9091	9242	151
8	9392	9543	9694	9845	9995	0146	0296	0447	0597	0748	151
9	46 0898	1048	1198	1348	1499	1649	1799	1948	2098	2248	150
290	2398	2548	2697	2847	2997	3146	3296	3445	3594	3744	150
1	3893	4042	4191	4340	4490	4639	4788	4936	5085	5234	149
2	5383	5532	5680	5829	5977	6126	6274	6423	6571	6719	149
3	6868	7016	7164	7312	7460	7608	7756	7904	8052	8200	148
4	8347	8495	8643	8790	8938	9085	9233	9380	9527	9675	148
295	9822	9969	0116	0263	0410	0557	0704	0851	0998	1145	147
6	47 1292	1438	1585	1732	1878	2025	2171	2318	2464	2610	146
7	2756	2903	3049	3195	3341	3487	3633	3779	3925	4071	146
8	4216	4362	4508	4653	4799	4944	5090	5235	5381	5526	146
9	5671	5816	5962	6107	6252	6397	6542	6687	6832	6976	145

PROPORTIONAL PARTS

Diff.	1	2	3	4	5	6	7	8	9
161	16.1	32.2	48.3	64.4	80.5	96.6	112.7	128.8	144.9
160	16.0	32.0	48.0	64.0	80.0	96.0	112.0	128.0	144.0
159	15.9	31.8	47.7	63.6	79.5	95.4	111.3	127.2	143.1
158	15.8	31.6	47.4	63.2	79.0	94.8	110.6	126.4	142.2
157	15.7	31.4	47.1	62.8	78.5	94.2	109.9	125.6	141.3
156	15.6	31.2	46.8	62.4	78.0	93.6	109.2	124.8	140.4
155	15.5	31.0	46.5	62.0	77.5	93.0	108.5	124.0	139.5
154	15.4	30.8	46.2	61.6	77.0	92.4	107.8	123.2	138.6
153	15.3	30.6	45.9	61.2	76.5	91.8	107.1	122.4	137.7
152	15.2	30.4	45.6	60.8	76.0	91.2	106.4	121.6	136.8
151	15.1	30.2	45.3	60.4	75.5	90.6	105.7	120.8	135.9
150	15.0	30.0	45.0	60.0	75.0	90.0	105.0	120.0	135.0
149	14.9	29.8	44.7	59.6	74.5	89.4	104.3	119.2	134.1
148	14.8	29.6	44.4	59.2	74.0	88.8	103.6	118.4	133.2
147	14.7	29.4	44.1	58.8	73.5	88.2	102.9	117.6	132.3
146	14.6	29.2	43.8	58.4	73.0	87.6	102.2	116.8	131.4
145	14.5	29.0	43.5	58.0	72.5	87.0	101.5	116.0	130.5
144	14.4	28.8	43.2	57.6	72.0	86.4	100.8	115.2	129.6
143	14.3	28.6	42.9	57.2	71.5	85.8	100.1	114.4	128.7
142	14.2	28.4	42.6	56.8	71.0	85.2	99.4	113.6	127.8
141	14.1	28.2	42.3	56.4	70.5	84.6	98.7	112.8	126.9
140	14.0	28.0	42.0	56.0	70.0	84.0	98.0	112.0	126.0

No. 300 | No. 339
Log. 477 | Log. 531

TABLE XIV.—*Continued*

N.	0	1	2	3	4	5	6	7	8	9	Diff.
300	47 7121	7266	7411	7555	7700	7844	7989	8133	8278	8422	145
1	8566	8711	8855	8999	9143	9287	9431	9575	9719	9863	144
2	48 0007	0151	0294	0438	0582	0725	0869	1012	1156	1299	144
3	1443	1586	1729	1872	2016	2159	2302	2445	2588	2731	143
4	2874	3016	3159	3302	3445	3587	3730	3872	4015	4157	143
305	4300	4442	4585	4727	4869	5011	5153	5295	5437	5579	142
6	5721	5863	6005	6147	6289	6430	6572	6714	6855	6997	142
7	7138	7280	7421	7563	7704	7845	7986	8127	8269	8410	141
8	8551	8692	8833	8974	9114	9255	9396	9537	9677	9818	141
9	9958	0099	0239	0380	0520	0661	0801	0941	1081	1222	140
310	49 1362	1502	1642	1782	1922	2062	2201	2341	2481	2621	140
1	2760	2900	3040	3179	3319	3458	3597	3737	3876	4015	139
2	4155	4294	4433	4572	4711	4850	4989	5128	5267	5406	139
3	5544	5683	5822	5960	6099	6238	6376	6515	6653	6791	139
4	6930	7068	7206	7344	7483	7621	7759	7897	8035	8173	138
315	8311	8448	8586	8724	8862	8999	9137	9275	9412	9550	138
6	9687	9824	9962	0099	0236	0374	0511	0648	0785	0922	137
7	50 1059	1196	1333	1470	1607	1744	1880	2017	2154	2291	137
8	2427	2564	2700	2837	2973	3109	3246	3382	3518	3655	136
9	3791	3927	4063	4199	4335	4471	4607	4743	4878	5014	136
320	5150	5286	5421	5557	5693	5828	5964	6099	6234	6370	136
1	6505	6640	6776	6911	7046	7181	7316	7451	7586	7721	135
2	7856	7991	8126	8260	8395	8530	8664	8799	8934	9068	135
3	9203	9337	9471	9606	9740	9874	0009	0143	0277	0411	134
4	51 0545	0679	0813	0947	1081	1215	1349	1482	1616	1750	134
325	1883	2017	2151	2284	2418	2551	2684	2818	2951	3084	133
6	3218	3351	3484	3617	3750	3883	4016	4149	4282	4415	133
7	4548	4681	4813	4946	5079	5211	5344	5476	5609	5741	133
8	5874	6006	6139	6271	6403	6535	6668	6800	6932	7064	132
9	7196	7328	7460	7592	7724	7855	7987	8119	8251	8382	132
330	8514	8646	8777	8909	9040	9171	9303	9434	9566	9697	131
1	9828	9959	0090	0221	0353	0484	0615	0745	0876	1007	131
2	52 1138	1269	1400	1530	1661	1792	1922	2053	2183	2314	131
3	2444	2575	2705	2835	2966	3096	3226	3356	3486	3616	130
4	3746	3876	4006	4136	4266	4396	4526	4656	4785	4915	130
335	5045	5174	5304	5434	5563	5693	5822	5951	6081	6210	129
6	6339	6469	6598	6727	6856	6985	7114	7243	7372	7501	129
7	7630	7759	7888	8016	8145	8274	8402	8531	8660	8788	129
8	8917	9045	9174	9302	9430	9559	9687	9815	9943	0072	128
9	53 0200	0328	0456	0584	0712	0840	0968	1096	1223	1351	128

PROPORTIONAL PARTS

Diff.	1	2	3	4	5	6	7	8	9
139	13.9	27.8	41.7	55.6	69.5	83.4	97.3	111.2	125.1
138	13.8	27.6	41.4	55.2	69.0	82.8	96.6	110.4	124.2
137	13.7	27.4	41.1	54.8	68.5	82.2	95.9	109.6	123.3
136	13.6	27.2	40.8	54.4	68.0	81.6	95.2	108.8	122.4
135	13.5	27.0	40.5	54.0	67.5	81.0	94.5	108.0	121.5
134	13.4	26.8	40.2	53.6	67.0	80.4	93.8	107.2	120.6
133	13.3	26.6	39.9	53.2	66.5	79.8	93.1	106.4	119.7
132	13.2	26.4	39.6	52.8	66.0	79.2	92.4	105.6	118.8
131	13.1	26.2	39.3	52.4	65.5	78.6	91.7	104.8	117.9
130	13.0	26.0	39.0	52.0	65.0	78.0	91.0	104.0	117.0
129	12.9	25.8	38.7	51.6	64.5	77.4	90.3	103.2	116.1
128	12.8	25.6	38.4	51.2	64.0	76.8	89.6	102.4	115.2
127	12.7	25.4	38.1	50.8	63.5	76.2	88.9	101.6	114.3

No. 340
Log. 531

TABLE XIV.—*Continued*

No. 379
Log. 579

N.	0	1	2	3	4	5	6	7	8	9	Diff.
340	53 1479	1607	1734	1862	1990	2117	2245	2372	2500	2627	128
1	2754	2882	3009	3136	3264	3391	3518	3645	3772	3899	127
2	4026	4153	4280	4407	4534	4661	4787	4914	5041	5167	127
3	5294	5421	5547	5674	5800	5927	6053	6180	6306	6432	126
4	6558	6685	6811	6937	7063	7189	7315	7441	7567	7693	126
345	7819	7945	8071	8197	8322	8448	8574	8699	8825	8951	126
6	9076	9202	9327	9452	9578	9703	9829	9954	0079	0204	125
7	54 0329	0455	0580	0705	0830	0955	1080	1205	1330	1454	125
8	1579	1704	1829	1953	2078	2203	2327	2452	2576	2701	125
9	2825	2950	3074	3199	3323	3447	3571	3696	3820	3944	124
350	4068	4192	4316	4440	4564	4688	4812	4936	5060	5183	124
1	5307	5431	5555	5678	5802	5925	6049	6172	6296	6419	124
2	6543	6666	6789	6913	7036	7159	7282	7405	7529	7652	123
3	7775	7898	8021	8144	8267	8389	8512	8635	8758	8881	123
4	9003	9126	9249	9371	9494	9616	9739	9861	9984	0106	123
355	55 0228	0351	0473	0595	0717	0840	0962	1084	1206	1328	122
6	1450	1572	1694	1816	1938	2060	2181	2303	2425	2547	122
7	2668	2790	2911	3033	3155	3276	3398	3519	3640	3762	121
8	3883	4004	4126	4247	4368	4489	4610	4731	4852	4973	121
9	5094	5215	5336	5457	5578	5699	5820	5940	6061	6182	121
360	6303	6423	6544	6664	6785	6905	7026	7146	7267	7387	120
1	7507	7627	7748	7868	7988	8108	8228	8349	8469	8589	120
2	8709	8829	8948	9068	9188	9308	9428	9548	9667	9787	120
3	9907	0026	0146	0265	0385	0504	0624	0743	0863	0982	119
4	56 1101	1221	1340	1459	1578	1698	1817	1936	2055	2174	119
365	2293	2412	2531	2650	2769	2887	3006	3125	3244	3362	119
6	3481	3600	3718	3837	3955	4074	4192	4311	4429	4548	119
7	4666	4784	4903	5021	5139	5257	5376	5494	5612	5730	118
8	5848	5966	6084	6202	6320	6437	6555	6673	6791	6909	118
9	7026	7144	7262	7379	7497	7614	7732	7849	7967	8084	118
370	8202	8319	8436	8554	8671	8788	8905	9023	9140	9257	117
1	9374	9491	9608	9725	9842	9959	0076	0193	0309	0426	117
2	57 0543	0660	0776	0893	1010	1126	1243	1359	1476	1592	117
3	1709	1825	1942	2058	2174	2291	2407	2523	2639	2755	116
4	2872	2988	3104	3220	3336	3452	3568	3684	3800	3915	116
375	4031	4147	4263	4379	4494	4610	4726	4841	4957	5072	116
6	5188	5303	5419	5534	5650	5765	5880	5996	6111	6226	115
7	6341	6457	6572	6687	6802	6917	7032	7147	7262	7377	115
8	7492	7607	7722	7836	7951	8066	8181	8295	8410	8525	115
9	8639	8754	8868	8983	9097	9212	9326	9441	9555	9669	114

PROPORTIONAL PARTS

Diff.	1	2	3	4	5	6	7	8	9
128	12.8	25.6	38.4	51.2	64.0	76.8	89.6	102.4	115.2
127	12.7	25.4	38.1	50.8	63.5	76.2	88.9	101.6	114.3
126	12.6	25.2	37.8	50.4	63.0	75.6	88.2	100.8	113.4
125	12.5	25.0	37.5	50.0	62.5	75.0	87.5	100.0	112.5
124	12.4	24.8	37.2	49.6	62.0	74.4	86.8	99.2	111.6
123	12.3	24.6	36.9	49.2	61.5	73.8	86.1	98.4	110.7
122	12.2	24.4	36.6	48.8	61.0	73.2	85.4	97.6	109.8
121	12.1	24.2	36.3	48.4	60.5	72.6	84.7	96.8	108.9
120	12.0	24.0	36.0	48.0	60.0	72.0	84.0	96.0	108.0
119	11.9	23.8	35.7	47.6	59.5	71.4	83.3	95.2	107.1

No. 380   No. 414
Log. 579   Log. 617

TABLE XIV.—*Continued*

N.	0	1	2	3	4	5	6	7	8	9	Diff.
380	57 9784	9898	0012	0126	0241	0355	0469	0583	0697	0811	114
1	58 0925	1039	1153	1267	1381	1495	1608	1722	1836	1950	
2	2063	2177	2291	2404	2518	2631	2745	2858	2972	3085	
3	3199	3312	3426	3539	3652	3765	3879	3992	4105	4218	
4	4331	4444	4557	4670	4783	4896	5009	5122	5235	5348	113
385	5461	5574	5686	5799	5912	6024	6137	6250	6362	6475	
6	6587	6700	6812	6925	7037	7149	7262	7374	7486	7599	
7	7711	7823	7935	8047	8160	8272	8384	8496	8608	8720	112
8	8832	8944	9056	9167	9279	9391	9503	9615	9726	9838	
9	9950	0061	0173	0284	0396	0507	0619	0730	0842	0953	
390	59 1065	1176	1287	1399	1510	1621	1732	1843	1955	2066	
1	2177	2288	2399	2510	2621	2732	2843	2954	3064	3175	111
2	3286	3397	3508	3618	3729	3840	3950	4061	4171	4282	
3	4393	4503	4614	4724	4834	4945	5055	5165	5276	5386	
4	5496	5606	5717	5827	5937	6047	6157	6267	6377	6487	
395	6597	6707	6817	6927	7037	7146	7256	7366	7476	7586	110
6	7695	7805	7914	8024	8134	8243	8353	8462	8572	8681	
7	8791	8900	9009	9119	9228	9337	9446	9556	9665	9774	
8	9883	9992	0101	0210	0319	0428	0537	0646	0755	0864	109
9	60 0973	1082	1191	1299	1408	1517	1625	1734	1843	1951	
400	2060	2169	2277	2386	2494	2603	2711	2819	2928	3036	
1	3144	3253	3361	3469	3577	3686	3794	3902	4010	4118	108
2	4226	4334	4442	4550	4658	4766	4874	4982	5089	5197	
3	5305	5413	5521	5628	5736	5844	5951	6059	6166	6274	
4	6381	6489	6596	6704	6811	6919	7026	7133	7241	7348	
405	7455	7562	7669	7777	7884	7991	8098	8205	8312	8419	107
6	8526	8633	8740	8847	8954	9061	9167	9274	9381	9488	
7	9594	9701	9808	9914	0021	0128	0234	0341	0447	0554	
8	61 0660	0767	0873	0979	1086	1192	1298	1405	1511	1617	
9	1723	1829	1936	2042	2148	2254	2360	2466	2572	2678	106
410	2784	2890	2996	3102	3207	3313	3419	3525	3630	3736	
1	3842	3947	4053	4159	4264	4370	4475	4581	4686	4792	
2	4897	5003	5108	5213	5319	5424	5529	5634	5740	5845	
3	5950	6055	6160	6265	6370	6476	6581	6686	6790	6895	105
4	7000	7105	7210	7315	7420	7525	7629	7734	7839	7943	

PROPORTIONAL PARTS

Diff.	1	2	3	4	5	6	7	8	9
118	11.8	23.6	35.4	47.2	59.0	70.8	82.6	94.4	106.2
117	11.7	23.4	35.1	46.8	58.5	70.2	81.9	93.6	105.3
116	11.6	23.2	34.8	46.4	58.0	69.6	81.2	92.8	104.4
115	11.5	23.0	34.5	46.0	57.5	69.0	80.5	92.0	103.5
114	11.4	22.8	34.2	45.6	57.0	68.4	79.8	91.2	102.6
113	11.3	22.6	33.9	45.2	56.5	67.8	79.1	90.4	101.7
112	11.2	22.4	33.6	44.8	56.0	67.2	78.4	89.6	100.8
111	11.1	22.2	33.3	44.4	55.5	66.6	77.7	88.8	99.9
110	11.0	22.0	33.0	44.0	55.0	66.0	77.0	88.0	99.0
109	10.9	21.8	32.7	43.6	54.5	65.4	76.3	87.2	98.1
108	10.8	21.6	32.4	43.2	54.0	64.8	75.6	86.4	97.2
107	10.7	21.4	32.1	42.8	53.5	64.2	74.9	85.6	96.3
106	10.6	21.2	31.8	42.4	53.0	63.6	74.2	84.8	95.4
105	10.5	21.0	31.5	42.0	52.5	63.0	73.5	84.0	94.5
104	10.4	20.8	31.2	41.6	52.0	62.4	72.8	83.2	93.6

No. 415
Log. 618

TABLE XIV.—*Continued*

No. 459
Log. 662

N.	0	1	2	3	4	5	6	7	8	9	Diff.
415	61 8048	8153	8257	8362	8466	8571	8676	8780	8884	8989	105
6	9093	9198	9302	9406	9511	9615	9719	9824	9928	0032	
7	62 0136	0240	0344	0448	0552	0656	0760	0864	0968	1072	104
8	1176	1280	1384	1488	1592	1695	1799	1903	2007	2110	
9	2214	2318	2421	2525	2628	2732	2835	2939	3042	3146	
420	3249	3353	3456	3559	3663	3766	3869	3973	4076	4179	
1	4282	4385	4488	4591	4695	4798	4901	5004	5107	5210	
2	5312	5415	5518	5621	5724	5827	5929	6032	6135	6238	103
3	6340	6443	6546	6648	6751	6853	6956	7058	7161	7263	
4	7366	7468	7571	7673	7775	7878	7980	8082	8185	8287	
425	8389	8491	8593	8695	8797	8900	9002	9104	9206	9308	102
6	9410	9512	9613	9715	9817	9919	0021	0123	0224	0326	
7	63 0428	0530	0631	0733	0835	0936	1038	1139	1241	1342	
8	1444	1545	1647	1748	1849	1951	2052	2153	2255	2356	
9	2457	2559	2660	2761	2862	2963	3064	3165	3266	3367	
430	3468	3569	3670	3771	3872	3973	4074	4175	4276	4376	101
1	4477	4578	4679	4779	4880	4981	5081	5182	5283	5383	
2	5484	5584	5685	5785	5886	5986	6087	6187	6287	6388	
3	6488	6588	6688	6789	6889	6989	7089	7189	7290	7390	
4	7490	7590	7690	7790	7890	7990	8090	8190	8290	8389	100
435	8489	8589	8689	8789	8888	8988	9088	9188	9287	9387	
6	9486	9586	9686	9785	9885	9984	0084	0183	0283	0382	
7	64 0481	0581	0680	0779	0879	0978	1077	1177	1276	1375	
8	1474	1573	1672	1771	1871	1970	2069	2168	2267	2366	
9	2465	2563	2662	2761	2860	2959	3058	3156	3255	3354	99
440	3453	3551	3650	3749	3847	3946	4044	4143	4242	4340	
1	4439	4537	4636	4734	4832	4931	5029	5127	5226	5324	
2	5422	5521	5619	5717	5815	5913	6011	6110	6208	6306	
3	6404	6502	6600	6698	6796	6894	6992	7089	7187	7285	98
4	7383	7481	7579	7676	7774	7872	7969	8067	8165	8262	
445	8360	8458	8555	8653	8750	8848	8945	9043	9140	9237	
6	9335	9432	9530	9627	9724	9821	9919	0016	0113	0210	
7	65 0308	0405	0502	0599	0696	0793	0890	0987	1084	1181	
8	1278	1375	1472	1569	1666	1762	1859	1956	2053	2150	97
9	2246	2343	2440	2536	2633	2730	2826	2923	3019	3116	
450	3213	3309	3405	3502	3598	3695	3791	3888	3984	4080	
1	4177	4273	4369	4465	4562	4658	4754	4850	4946	5042	
2	5138	5235	5331	5427	5523	5619	5715	5810	5906	6002	96
3	6098	6194	6290	6386	6482	6577	6673	6769	6864	6960	
4	7056	7152	7247	7343	7438	7534	7629	7725	7820	7916	
455	8011	8107	8202	8298	8393	8488	8584	8679	8774	8870	
6	8965	9060	9155	9250	9346	9441	9536	9631	9726	9821	
7	9916	0011	0106	0201	0296	0391	0486	0581	0676	0771	95
8	66 0865	0960	1055	1150	1245	1339	1434	1529	1623	1718	
9	1813	1907	2002	2096	2191	2286	2380	2475	2569	2663	

PROPORTIONAL PARTS

Diff.	1	2	3	4	5	6	7	8	9
105	10.5	21.0	31.5	42.0	52.5	63.0	73.5	84.0	94.5
104	10.4	20.8	31.2	41.6	52.0	62.4	72.8	83.2	93.6
103	10.3	20.6	30.9	41.2	51.5	61.8	72.1	82.4	92.7
102	10.2	20.4	30.6	40.8	51.0	61.2	71.4	81.6	91.8
101	10.1	20.2	30.3	40.4	50.5	60.6	70.7	80.8	90.9
100	10.0	20.0	30.0	40.0	50.0	60.0	70.0	80.0	90.0
99	9.9	19.8	29.7	39.6	49.5	59.4	69.3	79.2	89.1

No. 460 Log. 662 TABLE XIV.—*Continued* No. 499 Log. 698

N.	0	1	2	3	4	5	6	7	8	9	Diff.
460	66 2758	2852	2947	3041	3135	3230	3324	3418	3512	3607	
1	3701	3795	3889	3983	4078	4172	4266	4360	4454	4548	
2	4642	4736	4830	4924	5018	5112	5206	5299	5393	5487	94
3	5581	5675	5769	5862	5956	6050	6143	6237	6331	6424	
4	6518	6612	6705	6799	6892	6986	7079	7173	7266	7360	
465	7453	7546	7640	7733	7826	7920	8013	8106	8199	8293	
6	8386	8479	8572	8665	8759	8852	8945	9038	9131	9224	
7	9317	9410	9503	9596	9689	9782	9875	9967	0060	0153	93
8	67 0246	0339	0431	0524	0617	0710	0802	0895	0988	1080	
9	1173	1265	1358	1451	1543	1636	1728	1821	1913	2005	
470	2098	2190	2283	2375	2467	2560	2652	2744	2836	2929	
1	3021	3113	3205	3297	3390	3482	3574	3666	3758	3850	
2	3942	4034	4126	4218	4310	4402	4494	4586	4677	4769	92
3	4861	4953	5045	5137	5228	5320	5412	5503	5595	5687	
4	5778	5870	5962	6053	6145	6236	6328	6419	6511	6602	
475	6694	6785	6876	6968	7059	7151	7242	7333	7424	7516	
6	7607	7698	7789	7881	7972	8063	8154	8245	8336	8427	
7	8518	8609	8700	8791	8882	8973	9064	9155	9246	9337	91
8	9428	9519	9610	9700	9791	9882	9973	0063	0154	0245	
9	68 0336	0426	0517	0607	0698	0789	0879	0970	1060	1151	
480	1241	1332	1422	1513	1603	1693	1784	1874	1964	2055	
1	2145	2235	2326	2416	2506	2596	2686	2777	2867	2957	
2	3047	3137	3227	3317	3407	3497	3587	3677	3767	3857	90
3	3947	4037	4127	4217	4307	4396	4486	4576	4666	4756	
4	4845	4935	5025	5114	5204	5294	5383	5473	5563	5652	
485	5742	5831	5921	6010	6100	6189	6279	6368	6458	6547	
6	6636	6726	6815	6904	6994	7083	7172	7261	7351	7440	
7	7529	7618	7707	7796	7886	7975	8064	8153	8242	8331	89
8	8420	8509	8598	8687	8776	8865	8953	9042	9131	9220	
9	9309	9398	9486	9575	9664	9753	9841	9930	0019	0107	
490	69 0196	0285	0373	0462	0550	0639	0728	0816	0905	0993	
1	1081	1170	1258	1347	1435	1524	1612	1700	1789	1877	
2	1965	2053	2142	2230	2318	2406	2494	2583	2671	2759	
3	2847	2935	3023	3111	3199	3287	3375	3463	3551	3639	88
4	3727	3815	3903	3991	4078	4166	4254	4342	4430	4517	
495	4605	4693	4781	4868	4956	5044	5131	5219	5307	5394	
6	5482	5569	5657	5744	5832	5919	6007	6094	6182	6269	
7	6356	6444	6531	6618	6706	6793	6880	6968	7055	7142	
8	7229	7317	7404	7491	7578	7665	7752	7839	7926	8014	
9	8100	8188	8275	8362	8449	8535	8622	8709	8796	8883	87

PROPORTIONAL PARTS

Diff.	1	2	3	4	5	6	7	8	9
98	9.8	19.6	29.4	39.2	49.0	58.8	68.6	78.4	88.2
97	9.7	19.4	29.1	38.8	48.5	58.2	67.9	77.6	87.3
96	9.6	19.2	28.8	38.4	48.0	57.6	67.2	76.8	86.4
95	9.5	19.0	28.5	38.0	47.5	57.0	66.5	76.0	85.5
94	9.4	18.8	28.2	37.6	47.0	56.4	65.8	75.2	84.6
93	9.3	18.6	27.9	37.2	46.5	55.8	65.1	74.4	83.7
92	9.2	18.4	27.6	36.8	46.0	55.2	64.4	73.6	82.8
91	9.1	18.2	27.3	36.4	45.5	54.6	63.7	72.8	81.9
90	9.0	18.0	27.0	36.0	45.0	54.0	63.0	72.0	81.0
89	8.9	17.8	26.7	35.6	44.5	53.4	62.3	71.2	80.1
88	8.8	17.6	26.4	35.2	44.0	52.8	61.6	70.4	79.2
87	8.7	17.4	26.1	34.8	43.5	52.2	60.9	69.6	78.3
86	8.6	17.2	25.8	34.4	43.0	51.6	60.2	68.8	77.4

No. 500
Log. 698

TABLE XIV.—*Continued*

No. 544
Log. 736

N.	0	1	2	3	4	5	6	7	8	9	Diff.
500	69 8970	9057	9144	9231	9317	9404	9491	9578	9664	9751	
1	9838	9924	0011	0098	0184	0271	0358	0444	0531	0617	
2	70 0704	0790	0877	0963	1050	1136	1222	1309	1395	1482	
3	1568	1654	1741	1827	1913	1999	2086	2172	2258	2344	
4	2431	2517	2603	2689	2775	2861	2947	3033	3119	3205	
505	3291	3377	3463	3549	3635	3721	3807	3893	3979	4065	86
6	4151	4236	4322	4408	4494	4579	4665	4751	4837	4922	
7	5008	5094	5179	5265	5350	5436	5522	5607	5693	5778	
8	5864	5949	6035	6120	6206	6291	6376	6462	6547	6632	
9	6718	6803	6888	6974	7059	7144	7229	7315	7400	7485	
510	7570	7655	7740	7826	7911	7996	8081	8166	8251	8336	85
1	8421	8506	8591	8676	8761	8846	8931	9015	9100	9185	
2	9270	9355	9440	9524	9609	9694	9779	9863	9948	0033	
3	71 0117	0202	0287	0371	0456	0540	0625	0710	0794	0879	
4	0963	1048	1132	1217	1301	1385	1470	1554	1639	1723	
515	1807	1892	1976	2060	2144	2229	2313	2397	2481	2566	
6	2650	2734	2818	2902	2986	3070	3154	3238	3323	3407	84
7	3491	3575	3659	3742	3826	3910	3994	4078	4162	4246	
8	4330	4414	4497	4581	4665	4749	4833	4916	5000	5084	
9	5167	5251	5335	5418	5502	5586	5669	5753	5836	5920	
520	6003	6087	6170	6254	6337	6421	6504	6588	6671	6754	
1	6838	6921	7004	7088	7171	7254	7338	7421	7504	7587	
2	7671	7754	7837	7920	8003	8086	8169	8253	8336	8419	
3	8502	8585	8668	8751	8834	8917	9000	9083	9165	9248	
4	9331	9414	9497	9580	9663	9745	9828	9911	9994	0077	83
525	72 0159	0242	0325	0407	0490	0573	0655	0738	0821	0903	
6	0986	1068	1151	1233	1316	1398	1481	1563	1646	1728	
7	1811	1893	1975	2058	2140	2222	2305	2387	2469	2552	
8	2634	2716	2798	2881	2963	3045	3127	3209	3291	3374	82
9	3456	3538	3620	3702	3784	3866	3948	4030	4112	4194	
530	4276	4358	4440	4522	4604	4685	4767	4849	4931	5013	
1	5095	5176	5258	5340	5422	5503	5585	5667	5748	5830	
2	5912	5993	6075	6156	6238	6320	6401	6483	6564	6646	
3	6727	6809	6890	6972	7053	7134	7216	7297	7379	7460	
4	7541	7623	7704	7785	7866	7948	8029	8110	8191	8273	
535	8354	8435	8516	8597	8678	8759	8841	8922	9003	9084	81
6	9165	9246	9327	9408	9489	9570	9651	9732	9813	9893	
7	9974	0055	0136	0217	0298	0378	0459	0540	0621	0702	
8	73 0782	0863	0944	1024	1105	1186	1266	1347	1428	1508	
9	1589	1669	1750	1830	1911	1991	2072	2152	2233	2313	
540	2394	2474	2555	2635	2715	2796	2876	2956	3037	3117	
1	3197	3278	3358	3438	3518	3598	3679	3759	3839	3919	
2	3999	4079	4160	4240	4320	4400	4480	4560	4640	4720	80
3	4800	4880	4960	5040	5120	5200	5279	5359	5439	5519	
4	5599	5679	5759	5838	5918	5998	6078	6157	6237	6317	

PROPORTIONAL PARTS

Diff.	1	2	3	4	5	6	7	8	9
87	8.7	17.4	26.1	34.8	43.5	52.2	60.9	69.6	78.3
86	8.6	17.2	25.8	34.4	43.0	51.6	60.2	68.8	77.4
85	8.5	17.0	25.5	34.0	42.5	51.0	59.5	68.0	76.5
84	8.4	16.8	25.2	33.6	42.0	50.4	58.8	67.2	75.6

No. 545 — Log. 736 TABLE XIV.—*Continued* No. 584 — Log. 767

N.	0	1	2	3	4	5	6	7	8	9	Diff.
545	73 6397	6476	6556	6635	6715	6795	6874	6954	7034	7113	
6	7193	7272	7352	7431	7511	7590	7670	7749	7829	7908	
7	7987	8067	8146	8225	8305	8384	8463	8543	8622	8701	
8	8781	8860	8939	9018	9097	9177	9256	9335	9414	9493	
9	9572	9651	9731	9810	9889	9968	0047	0126	0205	0284	79
550	74 0363	0442	0521	0600	0678	0757	0836	0915	0994	1073	
1	1152	1230	1309	1388	1467	1546	1624	1703	1782	1860	
2	1939	2018	2096	2175	2254	2332	2411	2489	2568	2647	
3	2725	2804	2882	2961	3039	3118	3196	3275	3353	3431	
4	3510	3588	3667	3745	3823	3902	3980	4058	4136	4215	
555	4293	4371	4449	4528	4606	4684	4762	4840	4919	4997	
6	5075	5153	5231	5309	5387	5465	5543	5621	5699	5777	78
7	5855	5933	6011	6089	6167	6245	6323	6401	6479	6556	
8	6634	6712	6790	6868	6945	7023	7101	7179	7256	7334	
9	7412	7489	7567	7645	7722	7800	7878	7955	8033	8110	
560	8188	8266	8343	8421	8498	8576	8653	8731	8808	8885	
1	8963	9040	9118	9195	9272	9350	9427	9504	9582	9659	
2	9736	9814	9891	9968	0045	0123	0200	0277	0354	0431	
3	75 0508	0586	0663	0740	0817	0894	0971	1048	1125	1202	
4	1279	1356	1433	1510	1587	1664	1741	1818	1895	1972	
565	2048	2125	2202	2279	2356	2433	2509	2586	2663	2740	77
6	2816	2893	2970	3047	3123	3200	3277	3353	3430	3506	
7	3583	3660	3736	3813	3889	3966	4042	4119	4195	4272	
8	4348	4425	4501	4578	4654	4730	4807	4883	4960	5036	
9	5112	5189	5265	5341	5417	5494	5570	5646	5722	5799	
570	5875	5951	6027	6103	6180	6256	6332	6408	6484	6560	
1	6636	6712	6788	6864	6940	7016	7092	7168	7244	7320	76
2	7396	7472	7548	7624	7700	7775	7851	7927	8003	8079	
3	8155	8230	8306	8382	8458	8533	8609	8685	8761	8836	
4	8912	8988	9063	9139	9214	9290	9366	9441	9517	9592	
575	9668	9743	9819	9894	9970	0045	0121	0196	0272	0347	
6	76 0422	0498	0573	0649	0724	0799	0875	0950	1025	1101	
7	1176	1251	1326	1402	1477	1552	1627	1702	1778	1853	
8	1928	2003	2078	2153	2228	2303	2378	2453	2529	2604	
9	2679	2754	2829	2904	2978	3053	3128	3203	3278	3353	75
580	3428	3503	3578	3653	3727	3802	3877	3952	4027	4101	
1	4176	4251	4326	4400	4475	4550	4624	4699	4774	4848	
2	4923	4998	5072	5147	5221	5296	5370	5445	5520	5594	
3	5669	5743	5818	5892	5966	6041	6115	6190	6264	6338	
4	6413	6487	6562	6636	6710	6785	6859	6933	7007	7082	

PROPORTIONAL PARTS

Diff.	1	2	3	4	5	6	7	8	9
83	8.3	16.6	24.9	33.2	41.5	49.8	58.1	66.4	74.7
82	8.2	16.4	24.6	32.8	41.0	49.2	57.4	65.6	73.8
81	8.1	16.2	24.3	32.4	40.5	48.6	56.7	64.8	72.9
80	8.0	16.0	24.0	32.0	40.0	48.0	56.0	64.0	72.0
79	7.9	15.8	23.7	31.6	39.5	47.4	55.3	63.2	71.1
78	7.8	15.6	23.4	31.2	39.0	46.8	54.6	62.4	70.2
77	7.7	15.4	23.1	30.8	38.5	46.2	53.9	61.6	69.3
76	7.6	15.2	22.8	30.4	38.0	45.6	53.2	60.8	68.4
75	7.5	15.0	22.5	30.0	37.5	45.0	52.5	60.0	67.5
74	7.4	14.8	22.2	29.6	37.0	44.4	51.8	59.2	66.6

No. 585
Log. 767

Table XIV.—*Continued*

No. 629
Log. 799

N.	0	1	2	3	4	5	6	7	8	9	Diff.
585	76 7156	7230	7304	7379	7453	7527	7601	7675	7749	7823	
6	7898	7972	8046	8120	8194	8268	8342	8416	8490	8564	74
7	8638	8712	8786	8860	8934	9008	9082	9156	9230	9303	
8	9377	9451	9525	9599	9673	9746	9820	9894	9968	0042	
9	77 0115	0189	0263	0336	0410	0484	0557	0631	0705	0778	
590	0852	0926	0999	1073	1146	1220	1293	1367	1440	1514	
1	1587	1661	1734	1808	1881	1955	2028	2102	2175	2248	
2	2322	2395	2468	2542	2615	2688	2762	2835	2908	2981	
3	3055	3128	3201	3274	3348	3421	3494	3567	3640	3713	
4	3786	3860	3933	4006	4079	4152	4225	4298	4371	4444	73
595	4517	4590	4663	4736	4809	4882	4955	5028	5100	5173	
6	5246	5319	5392	5465	5538	5610	5683	5756	5829	5902	
7	5974	6047	6120	6193	6265	6338	6411	6483	6556	6629	
8	6701	6774	6846	6919	6992	7064	7137	7209	7282	7354	
9	7427	7499	7572	7644	7717	7789	7862	7934	8006	8079	
600	8151	8224	8296	8368	8441	8513	8585	8658	8730	8802	
1	8874	8947	9019	9091	9163	9236	9308	9380	9452	9524	
2	9596	9669	9741	9813	9885	9957	0029	0101	0173	0245	72
3	78 0317	0389	0461	0533	0605	0677	0749	0821	0893	0965	
4	1037	1109	1181	1253	1324	1396	1468	1540	1612	1684	
605	1755	1827	1899	1971	2042	2114	2186	2258	2329	2401	
6	2473	2544	2616	2688	2759	2831	2902	2974	3046	3117	
7	3189	3260	3332	3403	3475	3546	3618	3689	3761	3832	
8	3904	3975	4046	4118	4189	4261	4332	4403	4475	4546	
9	4617	4689	4760	4831	4902	4974	5045	5116	5187	5259	
610	5330	5401	5472	5543	5615	5686	5757	5828	5899	5970	
1	6041	6112	6183	6254	6325	6396	6467	6538	6609	6680	71
2	6751	6822	6893	6964	7035	7106	7177	7248	7319	7390	
3	7460	7531	7602	7673	7744	7815	7885	7956	8027	8098	
4	8168	8239	8310	8381	8451	8522	8593	8663	8734	8804	
615	8875	8946	9016	9087	9157	9228	9299	9369	9440	9510	
6	9581	9651	9722	9792	9863	9933	0004	0074	0144	0215	
7	79 0285	0356	0426	0496	0567	0637	0707	0778	0848	0918	
8	0988	1059	1129	1199	1269	1340	1410	1480	1550	1620	
9	1691	1761	1831	1901	1971	2041	2111	2181	2252	2322	
620	2392	2462	2532	2602	2672	2742	2812	2882	2952	3022	70
1	3092	3162	3231	3301	3371	3441	3511	3581	3651	3721	
2	3790	3860	3930	4000	4070	4139	4209	4279	4349	4418	
3	4488	4558	4627	4697	4767	4836	4906	4976	5045	5115	
4	5185	5254	5324	5393	5463	5532	5602	5672	5741	5811	
625	5880	5949	6019	6088	6158	6227	6297	6366	6436	6505	
6	6574	6644	6713	6782	6852	6921	6990	7060	7129	7198	
7	7268	7337	7406	7475	7545	7614	7683	7752	7821	7890	
8	7960	8029	8098	8167	8236	8305	8374	8443	8513	8582	
9	8651	8720	8789	8858	8927	8996	9065	9134	9203	9272	69

PROPORTIONAL PARTS

Diff.	1	2	3	4	5	6	7	8	9
75	7.5	15.0	22.5	30.0	37.5	45.0	52.5	60.0	67.5
74	7.4	14.8	22.2	29.6	37.0	44.4	51.8	59.2	66.6
73	7.3	14.6	21.9	29.2	36.5	43.8	51.1	58.4	65.7
72	7.2	14.4	21.6	28.8	36.0	43.2	50.4	57.6	64.8
71	7.1	14.2	21.3	28.4	35.5	42.6	49.7	56.8	63.9
70	7.0	14.0	21.0	28.0	35.0	42.0	49.0	56.0	63.0
69	6.9	13.8	20.7	27.6	34.5	41.4	48.3	55.2	62.1

TABLE XIV.—*Continued*

N.	0	1	2	3	4	5	6	7	8	9	Diff.
630	79 9341	9409	9478	9547	9616	9685	9754	9823	9892	9961	
1	80 0029	0098	0167	0236	0305	0373	0442	0511	0580	0648	
2	0717	0786	0854	0923	0992	1061	1129	1198	1266	1335	
3	1404	1472	1541	1609	1678	1747	1815	1884	1952	2021	
4	2089	2158	2226	2295	2363	2432	2500	2568	2637	2705	
635	2774	2842	2910	2979	3047	3116	3184	3252	3321	3389	
6	3457	3525	3594	3662	3730	3798	3867	3935	4003	4071	
7	4139	4208	4276	4344	4412	4480	4548	4616	4685	4753	
8	4821	4889	4957	5025	5093	5161	5229	5297	5365	5433	**68**
9	5501	5569	5637	5705	5773	5841	5908	5976	6044	6112	
640	80 6180	6248	6316	6384	6451	6519	6587	6655	6723	6790	
1	6858	6926	6994	7061	7129	7197	7264	7332	7400	7467	
2	7535	7603	7670	7738	7806	7873	7941	8008	8076	8143	
3	8211	8279	8346	8414	8481	8549	8616	8684	8751	8818	
4	8886	8953	9021	9088	9156	9223	9290	9358	9425	9492	
645	9560	9627	9694	9762	9829	9896	9964	0031	0098	0165	
6	81 0233	0300	0367	0434	0501	0569	0636	0703	0770	0837	
7	0904	0971	1039	1106	1173	1240	1307	1374	1441	1508	**67**
8	1575	1642	1709	1776	1843	1910	1977	2044	2111	2178	
9	2245	2312	2379	2445	2512	2579	2646	2713	2780	2847	
650	2913	2980	3047	3114	3181	3247	3314	3381	3448	3514	
1	3581	3648	3714	3781	3848	3914	3981	4048	4114	4181	
2	4248	4314	4381	4447	4514	4581	4647	4714	4780	4847	
3	4913	4980	5046	5113	5179	5246	5312	5378	5445	5511	
4	5578	5644	5711	5777	5843	5910	5976	6042	6109	6175	
655	6241	6308	6374	6440	6506	6573	6639	6705	6771	6838	
6	6904	6970	7036	7102	7169	7235	7301	7367	7433	7499	
7	7565	7631	7698	7764	7830	7896	7962	8028	8094	8160	
8	8226	8292	8358	8424	8490	8556	8622	8688	8754	8820	
9	8885	8951	9017	9083	9149	9215	9281	9346	9412	9478	**66**
660	9544	9610	9676	9741	9807	9873	9939	0004	0070	0136	
1	82 0201	0267	0333	0399	0464	0530	0595	0661	0727	0792	
2	0858	0924	0989	1055	1120	1186	1251	1317	1382	1448	
3	1514	1579	1645	1710	1775	1841	1906	1972	2037	2103	
4	2168	2233	2299	2364	2430	2495	2560	2626	2691	2756	
665	2822	2887	2952	3018	3083	3148	3213	3279	3344	3409	
6	3474	3539	3605	3670	3735	3800	3865	3930	3996	4061	
7	4126	4191	4256	4321	4386	4451	4516	4581	4646	4711	
8	4776	4841	4906	4971	5036	5101	5166	5231	5296	5361	**65**
9	5426	5491	5556	5621	5686	5751	5815	5880	5945	6010	
670	6075	6140	6204	6269	6334	6399	6464	6528	6593	6658	
1	6723	6787	6852	6917	6981	7046	7111	7175	7240	7305	
2	7369	7434	7499	7563	7628	7692	7757	7821	7886	7951	
3	8015	8080	8144	8209	8273	8338	8402	8467	8531	8595	
4	8660	8724	8789	8853	8918	8982	9046	9111	9175	9239	

PROPORTIONAL PARTS

Diff.	1	2	3	4	5	6	7	8	9
68	6.8	13.6	20.4	27.2	34.0	40.8	47.6	54.4	61.2
67	6.7	13.4	20.1	26.8	33.5	40.2	46.9	53.6	60.3
66	6.6	13.2	19.8	26.4	33.0	39.6	46.2	52.8	59.4
65	6.5	13.0	19.5	26.0	32.5	39.0	45.5	52.0	58.5
64	6.4	12.8	19.2	25.6	32.0	38.4	44.8	51.2	57.6

No. 675 | No. 719
Log. 829 | Log. 857

TABLE XIV.—*Continued*

N.	0	1	2	3	4	5	6	7	8	9	Diff.
675	82 9304	9368	9432	9497	9561	9625	9690	9754	9818	9882	
6	9947	0011	0075	0139	0204	0268	0332	0396	0460	0525	
7	83 0589	0653	0717	0781	0845	0909	0973	1037	1102	1166	
8	1230	1294	1358	1422	1486	1550	1614	1678	1742	1806	64
9	1870	1934	1998	2062	2126	2189	2253	2317	2381	2445	
680	2509	2573	2637	2700	2764	2828	2892	2956	3020	3083	
1	3147	3211	3275	3338	3402	3466	3530	3593	3657	3721	
2	3784	3848	3912	3975	4039	4103	4166	4230	4294	4357	
3	4421	4484	4548	4611	4675	4739	4802	4866	4929	4993	
4	5056	5120	5183	5247	5310	5373	5437	5500	5564	5627	
685	5691	5754	5817	5881	5944	6007	6071	6134	6197	6261	
6	6324	6387	6451	6514	6577	6641	6704	6767	6830	6894	
7	6957	7020	7083	7146	7210	7273	7336	7399	7462	7525	
8	7588	7652	7715	7778	7841	7904	7967	8030	8093	8156	
9	8219	8282	8345	8408	8471	8534	8597	8660	8723	8786	63
690	8849	8912	8975	9038	9101	9164	9227	9289	9352	9415	
1	9478	9541	9604	9667	9729	9792	9855	9918	9981	0043	
2	84 0106	0169	0232	0294	0357	0420	0482	0545	0608	0671	
3	0733	0796	0859	0921	0984	1046	1109	1172	1234	1297	
4	1359	1422	1485	1547	1610	1672	1735	1797	1860	1922	
695	1985	2047	2110	2172	2235	2297	2360	2422	2484	2547	
6	2609	2672	2734	2796	2859	2921	2983	3046	3108	3170	
7	3233	3295	3357	3420	3482	3544	3606	3669	3731	3793	
8	3855	3918	3980	4042	4104	4166	4229	4291	4353	4415	
9	4477	4539	4601	4664	4726	4788	4850	4912	4974	5036	
700	5098	5160	5222	5284	5346	5408	5470	5532	5594	5656	62
1	5718	5780	5842	5904	5966	6028	6090	6151	6213	6275	
2	6337	6399	6461	6523	6585	6646	6708	6770	6832	6894	
3	6955	7017	7079	7141	7202	7264	7326	7388	7449	7511	
4	7573	7634	7696	7758	7819	7881	7943	8004	8066	8128	
705	8189	8251	8312	8374	8435	8497	8559	8620	8682	8743	
6	8805	8866	8928	8989	9051	9112	9174	9235	9297	9358	
7	9419	9481	9542	9604	9665	9726	9788	9849	9911	9972	
8	85 0033	0095	0156	0217	0279	0340	0401	0462	0524	0585	
9	0646	0707	0769	0830	0891	0952	1014	1075	1136	1197	
710	1258	1320	1381	1442	1503	1564	1625	1686	1747	1809	
1	1870	1931	1992	2053	2114	2175	2236	2297	2358	2419	
2	2480	2541	2602	2663	2724	2785	2846	2907	2968	3029	61
3	3090	3150	3211	3272	3333	3394	3455	3516	3577	3637	
4	3698	3759	3820	3881	3941	4002	4063	4124	4185	4245	
715	4306	4367	4428	4488	4549	4610	4670	4731	4792	4852	
6	4913	4974	5034	5095	5156	5216	5277	5337	5398	5459	
7	5519	5580	5640	5701	5761	5822	5882	5943	6003	6064	
8	6124	6185	6245	6306	6366	6427	6487	6548	6608	6668	
9	6729	6789	6850	6910	6970	7031	7091	7152	7212	7272	

PROPORTIONAL PARTS

Diff.	1	2	3	4	5	6	7	8	9
65	6.5	13.0	19.5	26.0	32.5	39.0	45.5	52.0	58.5
64	6.4	12.8	19.2	25.6	32.0	38.4	44.8	51.2	57.6
63	6.3	12.6	18.9	25.2	31.5	37.8	44.1	50.4	56.7
62	6.2	12.4	18.6	24.8	31.0	37.2	43.4	49.6	55.8
61	6.1	12.2	18.3	24.4	30.5	36.6	42.7	48.8	54.9
60	**6.0**	**12.0**	**18.0**	**24.0**	**30.0**	**36.0**	**42.0**	**48.0**	**54.0**

No. 720 Log. 857 | No. 764 Log. 883

TABLE XIV.—*Continued*

N.	0	1	2	3	4	5	6	7	8	9	Diff.
720	85 7332	7393	7453	7513	7574	7634	7694	7755	7815	7875	
1	7935	7995	8056	8116	8176	8236	8297	8357	8417	8477	
2	8537	8597	8657	8718	8778	8838	8898	8958	9018	9078	60
3	9138	9198	9258	9318	9379	9439	9499	9559	9619	9679	
4	9739	9799	9859	9918	9978	0038	0098	0158	0218	0278	
725	86 0338	0398	0458	0518	0578	0637	0697	0757	0817	0877	
6	0937	0996	1056	1116	1176	1236	1295	1355	1415	1475	
7	1534	1594	1654	1714	1773	1833	1893	1952	2012	2072	
8	2131	2191	2251	2310	2370	2430	2489	2549	2608	2668	
9	2728	2787	2847	2906	2966	3025	3085	3144	3204	3263	
730	3323	3382	3442	3501	3561	3620	3680	3739	3799	3858	
1	3917	3977	4036	4096	4155	4214	4274	4333	4392	4452	
2	4511	4570	4630	4689	4748	4808	4867	4926	4985	5045	
3	5104	5163	5222	5282	5341	5400	5459	5519	5578	5637	
4	5696	5755	5814	5874	5933	5992	6051	6110	6169	6228	
735	6287	6346	6405	6465	6524	6583	6642	6701	6760	6819	59
6	6878	6937	6996	7055	7114	7173	7232	7291	7350	7409	
7	7467	7526	7585	7644	7703	7762	7821	7880	7939	7998	
8	8056	8115	8174	8233	8292	8350	8409	8468	8527	8586	
9	8644	8703	8762	8821	8879	8938	8997	9056	9114	9173	
740	9232	9290	9349	9408	9466	9525	9584	9642	9701	9760	
1	9818	9877	9935	9994	0053	0111	0170	0228	0287	0345	
2	87 0404	0462	0521	0579	0638	0696	0755	0813	0872	0930	
3	0989	1047	1106	1164	1223	1281	1339	1398	1456	1515	
4	1573	1631	1690	1748	1806	1865	1923	1981	2040	2098	
745	2156	2215	2273	2331	2389	2448	2506	2564	2622	2681	
6	2739	2797	2855	2913	2972	3030	3088	3146	3204	3262	
7	3321	3379	3437	3495	3553	3611	3669	3727	3785	3844	
8	3902	3960	4018	4076	4134	4192	4250	4308	4366	4424	58
9	4482	4540	4598	4656	4714	4772	4830	4888	4945	5003	
750	5061	5119	5177	5235	5293	5351	5409	5466	5524	5582	
1	5640	5698	5756	5813	5871	5929	5987	6045	6102	6160	
2	6218	6276	6333	6391	6449	6507	6564	6622	6680	6737	
3	6795	6853	6910	6968	7026	7083	7141	7199	7256	7314	
4	7371	7429	7487	7544	7602	7659	7717	7774	7832	7889	
755	7947	8004	8062	8119	8177	8234	8292	8349	8407	8464	
6	8522	8579	8637	8694	8752	8809	8866	8924	8981	9039	
7	9096	9153	9211	9268	9325	9383	9440	9497	9555	9612	
8	9669	9726	9784	9841	9898	9956	0013	0070	0127	0185	
9	88 0242	0299	0356	0413	0471	0528	0585	0642	0699	0756	
760	0814	0871	0928	0985	1042	1099	1156	1213	1271	1328	
1	1385	1442	1499	1556	1613	1670	1727	1784	1841	1898	
2	1955	2012	2069	2126	2183	2240	2297	2354	2411	2468	57
3	2525	2581	2638	2695	2752	2809	2866	2923	2980	3037	
4	3093	3150	3207	3264	3321	3377	3434	3491	3548	3605	

PROPORTIONAL PARTS

Diff.	1	2	3	4	5	6	7	8	9
59	5.9	11.8	17.7	23.6	29.5	35.4	41.3	47.2	53.1
58	5.8	11.6	17.4	23.2	29.0	34.8	40.6	46.4	52.2
57	5.7	11.4	17.1	22.8	28.5	34.2	39.9	45.6	51.3
56	5.6	11.2	16.8	22.4	28.0	33.6	39.2	44.8	50.4

TABLE XIV.—*Continued*

N.	0	1	2	3	4	5	6	7	8	9	Diff.
765	88 3661	3718	3775	3832	3888	3945	4002	4059	4115	4172	
6	4229	4285	4342	4399	4455	4512	4569	4625	4682	4739	
7	4795	4852	4909	4965	5022	5078	5135	5192	5248	5305	
8	5361	5418	5474	5531	5587	5644	5700	5757	5813	5870	
9	5926	5983	6039	6096	6152	6209	6265	6321	6378	6434	
770	6491	6547	6604	6660	6716	6773	6829	6885	6942	6998	
1	7054	7111	7167	7223	7280	7336	7392	7449	7505	7561	
2	7617	7674	7730	7786	7842	7898	7955	8011	8067	8123	
3	8179	8236	8292	8348	8404	8460	8516	8573	8629	8685	
4	8741	8797	8853	8909	8965	9021	9077	9134	9190	9246	
775	9302	9358	9414	9470	9526	9582	9638	9694	9750	9806	56
6	9862	9918	9974	0030	0086	0141	0197	0253	0309	0365	
7	89 0421	0477	0533	0589	0645	0700	0756	0812	0868	0924	
8	0980	1035	1091	1147	1203	1259	1314	1370	1426	1482	
9	1537	1593	1649	1705	1760	1816	1872	1928	1983	2039	
780	2095	2150	2206	2262	2317	2373	2429	2484	2540	2595	
1	2651	2707	2762	2818	2873	2929	2985	3040	3096	3151	
2	3207	3262	3318	3373	3429	3484	3540	3595	3651	3706	
3	3762	3817	3873	3928	3984	4039	4094	4150	4205	4261	
4	4316	4371	4427	4482	4538	4593	4648	4704	4759	4814	
785	4870	4925	4980	5036	5091	5146	5201	5257	5312	5367	
6	5423	5478	5533	5588	5644	5699	5754	5809	5864	5920	
7	5975	6030	6085	6140	6195	6251	6306	6361	6416	6471	
8	6526	6581	6636	6692	6747	6802	6857	6912	6967	7022	
9	7077	7132	7187	7242	7297	7352	7407	7462	7517	7572	
790	7627	7682	7737	7792	7847	7902	7957	8012	8067	8122	55
1	8176	8231	8286	8341	8396	8451	8506	8561	8615	8670	
2	8725	8780	8835	8890	8944	8999	9054	9109	9164	9218	
3	9273	9328	9383	9437	9492	9547	9602	9656	9711	9766	
4	9821	9875	9930	9985	0039	0094	0149	0203	0258	0312	
795	90 0367	0422	0476	0531	0586	0640	0695	0749	0804	0859	
6	0913	0968	1022	1077	1131	1186	1240	1295	1349	1404	
7	1458	1513	1567	1622	1676	1731	1785	1840	1894	1948	
8	2003	2057	2112	2166	2221	2275	2329	2384	2438	2492	
9	2547	2601	2655	2710	2764	2818	2873	2927	2981	3036	
800	3090	3144	3199	3253	3307	3361	3416	3470	3524	3578	
1	3633	3687	3741	3795	3849	3904	3958	4012	4066	4120	
2	4174	4229	4283	4337	4391	4445	4499	4553	4607	4661	
3	4716	4770	4824	4878	4932	4986	5040	5094	5148	5202	
4	5256	5310	5364	5418	5472	5526	5580	5634	5688	5742	54
805	5796	5850	5904	5958	6012	6066	6119	6173	6227	6281	
6	6335	6389	6443	6497	6551	6604	6658	6712	6766	6820	
7	6874	6927	6981	7035	7089	7143	7196	7250	7304	7358	
8	7411	7465	7519	7573	7626	7680	7734	7787	7841	7895	
9	7949	8002	8056	8110	8163	8217	8270	8324	8378	8431	

PROPORTIONAL PARTS

Diff.	1	2	3	4	5	6	7	8	9
57	5.7	11.4	17.1	22.8	28.5	34.2	39.9	45.6	51.3
56	5.6	11.2	16.8	22.4	28.0	33.6	39.2	44.8	50.4
55	5.5	11.0	16.5	22.0	27.5	33.0	38.5	44.0	49.5
54	5.4	10.8	16.2	21.6	27.0	32.4	37.8	43.2	48.6

No. 810 Log. 908 | No. 854 Log. 931

TABLE XIV.—*Continued*

N.	0	1	2	3	4	5	6	7	8	9	Diff.
810	90 8485	8539	8592	8646	8699	8753	8807	8860	8914	8967	
1	9021	9074	9128	9181	9235	9289	9342	9396	9449	9503	
2	9556	9610	9663	9716	9770	9823	9877	9930	9984	0037	
3	91 0091	0144	0197	0251	0304	0358	0411	0464	0518	0571	
4	0624	0678	0731	0784	0838	0891	0944	0998	1051	1104	
815	1158	1211	1264	1317	1371	1424	1477	1530	1584	1637	
6	1690	1743	1797	1850	1903	1956	2009	2063	2116	2169	
7	2222	2275	2328	2381	2435	2488	2541	2594	2647	2700	
8	2753	2806	2859	2913	2966	3019	3072	3125	3178	3231	
9	3284	3337	3390	3443	3496	3549	3602	3655	3708	3761	53
820	3814	3867	3920	3973	4026	4079	4132	4184	4237	4290	
1	4343	4396	4449	4502	4555	4608	4660	4713	4766	4819	
2	4872	4925	4977	5030	5083	5136	5189	5241	5294	5347	
3	5400	5453	5505	5558	5611	5664	5716	5769	5822	5875	
4	5927	5980	6033	6085	6138	6191	6243	6296	6349	6401	
825	6454	6507	6559	6612	6664	6717	6770	6822	6875	6927	
6	6980	7033	7085	7138	7190	7243	7295	7348	7400	7453	
7	7506	7558	7611	7663	7716	7768	7820	7873	7925	7978	
8	8030	8083	8135	8188	8240	8293	8345	8397	8450	8502	
9	8555	8607	8659	8712	8764	8816	8869	8921	8973	9026	
830	9078	9130	9183	9235	9287	9340	9392	9444	9496	9549	
1	9601	9653	9706	9758	9810	9862	9914	9967	0019	0071	
2	92 0123	0176	0228	0280	0332	0384	0436	0489	0541	0593	
3	0645	0697	0749	0801	0853	0906	0958	1010	1062	1114	
4	1166	1218	1270	1322	1374	1426	1478	1530	1582	1634	52
835	1686	1738	1790	1842	1894	1946	1998	2050	2102	2154	
6	2206	2258	2310	2362	2414	2466	2518	2570	2622	2674	
7	2725	2777	2829	2881	2933	2985	3037	3089	3140	3192	
8	3244	3296	3348	3399	3451	3503	3555	3607	3658	3710	
9	3762	3814	3865	3917	3969	4021	4072	4124	4176	4228	
840	4279	4331	4383	4434	4486	4538	4589	4641	4693	4744	
1	4796	4848	4899	4951	5003	5054	5106	5157	5209	5261	
2	5312	5364	5415	5467	5518	5570	5621	5673	5725	5776	
3	5828	5879	5931	5982	6034	6085	6137	6188	6240	6291	
4	6342	6394	6445	6497	6548	6600	6651	6702	6754	6805	
845	6857	6908	6959	7011	7062	7114	7165	7216	7268	7319	
6	7370	7422	7473	7524	7576	7627	7678	7730	7781	7832	
7	7883	7935	7986	8037	8088	8140	8191	8242	8293	8345	
8	8396	8447	8498	8549	8601	8652	8703	8754	8805	8857	
9	8908	8959	9010	9061	9112	9163	9215	9266	9317	9368	
850	9419	9470	9521	9572	9623	9674	9725	9776	9827	9879	51
1	9930	9981	0032	0083	0134	0185	0236	0287	0338	0389	
2	93 0440	0491	0542	0592	0643	0694	0745	0796	0847	0898	
3	0949	1000	1051	1102	1153	1204	1254	1305	1356	1407	
4	1458	1509	1560	1610	1661	1712	1763	1814	1865	1915	

PROPORTIONAL PARTS

Diff.	1	2	3	4	5	6	7	8	9
53	5.3	10.6	15.9	21.2	26.5	31.8	37.1	42.4	47.7
52	5.2	10.4	15.6	20.8	26.0	31.2	36.4	41.6	46.8
51	5.1	10.2	15.3	20.4	25.5	30.6	35.7	40.8	45.9
50	5.0	10.0	15.0	20.0	25.0	30.0	35.0	40.0	45.0

No. 855
Log. 931

TABLE XIV.—*Continued*

No. 899
Log. 954

N.	0	1	2	3	4	5	6	7	8	9	Diff.
855	93 1966	2017	2068	2118	2169	2220	2271	2322	2372	2423	
6	2474	2524	2575	2626	2677	2727	2778	2829	2879	2930	
7	2981	3031	3082	3133	3183	3234	3285	3335	3386	3437	
8	3487	3538	3589	3639	3690	3740	3791	3841	3892	3943	
9	3993	4044	4094	4145	4195	4246	4296	4347	4397	4448	
860	4498	4549	4599	4650	4700	4751	4801	4852	4902	4953	
1	5003	5054	5104	5154	5205	5255	5306	5356	5406	5457	
2	5507	5558	5608	5658	5709	5759	5809	5860	5910	5960	
3	6011	6061	6111	6162	6212	6262	6313	6363	6413	6463	
4	6514	6564	6614	6665	6715	6765	6815	6865	6916	6966	
865	7016	7066	7116	7167	7217	7267	7317	7367	7418	7468	
6	7518	7568	7618	7668	7718	7769	7819	7869	7919	7969	
7	8019	8069	8119	8169	8219	8269	8320	8370	8420	8470	
8	8520	8570	8620	8670	8720	8770	8820	8870	8920	8970	50
9	9020	9070	9120	9170	9220	9270	9320	9369	9419	9469	
870	9519	9569	9619	9669	9719	9769	9819	9869	9918	9968	
1	94 0018	0068	0118	0168	0218	0267	0317	0367	0417	0467	
2	0516	0566	0616	0666	0716	0765	0815	0865	0915	0964	
3	1014	1064	1114	1163	1213	1263	1313	1362	1412	1462	
4	1511	1561	1611	1660	1710	1760	1809	1859	1909	1958	
875	2008	2058	2107	2157	2207	2256	2306	2355	2405	2455	
6	2504	2554	2603	2653	2702	2752	2801	2851	2901	2950	
7	3000	3049	3099	3148	3198	3247	3297	3346	3396	3445	
8	3495	3544	3593	3643	3692	3742	3791	3841	3890	3939	
9	3989	4038	4088	4137	4186	4236	4285	4335	4384	4433	
880	4483	4532	4581	4631	4680	4729	4779	4828	4877	4927	
1	4976	5025	5074	5124	5173	5222	5272	5321	5370	5419	
2	5469	5518	5567	5616	5665	5715	5764	5813	5862	5912	
3	5961	6010	6059	6108	6157	6207	6256	6305	6354	6403	
4	6452	6501	6551	6600	6649	6698	6747	6796	6845	6894	
885	6943	6992	7041	7090	7139	7189	7238	7287	7336	7385	
6	7434	7483	7532	7581	7630	7679	7728	7777	7826	7875	
7	7924	7973	8022	8070	8119	8168	8217	8266	8315	8364	49
8	8413	8462	8511	8560	8608	8657	8706	8755	8804	8853	
9	8902	8951	8999	9048	9097	9146	9195	9244	9292	9341	
890	9390	9439	9488	9536	9585	9634	9683	9731	9780	9829	
1	9878	9926	9975	0024	0073	0121	0170	0219	0267	0316	
2	95 0365	0414	0462	0511	0560	0608	0657	0706	0754	0803	
3	0851	0900	0949	0997	1046	1095	1143	1192	1240	1289	
4	1338	1386	1435	1483	1532	1580	1629	1677	1726	1775	
895	1823	1872	1920	1969	2017	2066	2114	2163	2211	2260	
6	2308	2356	2405	2453	2502	2550	2599	2647	2696	2744	
7	2792	2841	2889	2938	2986	3034	3083	3131	3180	3228	
8	3276	3325	3373	3421	3470	3518	3566	3615	3663	3711	
9	3760	3808	3856	3905	3953	4001	4049	4098	4146	4194	

PROPORTIONAL PARTS

Diff.	1	2	3	4	5	6	7	8	9
51	5.1	10.2	15.3	20.4	25.5	30.6	35.7	40.8	45.9
50	5.0	10.0	15.0	20.0	25.0	30.0	35.0	40.0	45.0
49	4.9	9.8	14.7	19.6	24.5	29.4	34.3	39.2	44.1
48	4.8	9.6	14.4	19.2	24.0	28.8	33.6	38.4	43.2

No. 900 No. 944
Log. 954 Log. 975

TABLE XIV.—*Continued*

N.	0	1	2	3	4	5	6	7	8	9	Diff.
900	95 4243	4291	4339	4387	4435	4484	4532	4580	4628	4677	
1	4725	4773	4821	4869	4918	4966	5014	5062	5110	5158	
2	5207	5255	5303	5351	5399	5447	5495	5543	5592	5640	
3	5688	5736	5784	5832	5880	5928	5976	6024	6072	6120	
4	6168	6216	6265	6313	6361	6409	6457	6505	6553	6601	48
905	6649	6697	6745	6793	6840	6888	6936	6984	7032	7080	
6	7128	7176	7224	7272	7320	7368	7416	7464	7512	7559	
7	7607	7655	7703	7751	7799	7847	7894	7942	7990	8038	
8	8086	8134	8181	8229	8277	8325	8373	8421	8468	8516	
9	8564	8612	8659	8707	8755	8803	8850	8898	8946	8994	
910	9041	9089	9137	9185	9232	9280	9328	9375	9423	9471	
1	9518	9566	9614	9661	9709	9757	9804	9852	9900	9947	
2	9995	0042	0090	0138	0185	0233	0280	0328	0376	0423	
3	96 0471	0518	0566	0613	0661	0709	0756	0804	0851	0899	
4	0946	0994	1041	1089	1136	1184	1231	1279	1326	1374	
915	1421	1469	1516	1563	1611	1658	1706	1753	1801	1848	
6	1895	1943	1990	2038	2085	2132	2180	2227	2275	2322	
7	2369	2417	2464	2511	2559	2606	2653	2701	2748	2795	
8	2843	2890	2937	2985	3032	3079	3126	3174	3221	3268	
9	3316	3363	3410	3457	3504	3552	3599	3646	3693	3741	
920	3788	3835	3882	3929	3977	4024	4071	4118	4165	4212	
1	4260	4307	4354	4401	4448	4495	4542	4590	4637	4684	
2	4731	4778	4825	4872	4919	4966	5013	5061	5108	5155	
3	5202	5249	5296	5343	5390	5437	5484	5531	5578	5625	
4	5672	5719	5766	5813	5860	5907	5954	6001	6048	6095	47
925	6142	6189	6236	6283	6329	6376	6423	6470	6517	6564	
6	6611	6658	6705	6752	6799	6845	6892	6939	6986	7033	
7	7080	7127	7173	7220	7267	7314	7361	7408	7454	7501	
8	7548	7595	7642	7688	7735	7782	7829	7875	7922	7969	
9	8016	8062	8109	8156	8203	8249	8296	8343	8390	8436	
930	8483	8530	8576	8623	8670	8716	8763	8810	8856	8903	
1	8950	8996	9043	9090	9136	9183	9229	9276	9323	9369	
2	9416	9463	9509	9556	9602	9649	9695	9742	9789	9835	
3	9882	9928	9975	0021	0068	0114	0161	0207	0254	0300	
4	97 0347	0393	0440	0486	0533	0579	0626	0672	0719	0765	
935	0812	0858	0904	0951	0997	1044	1090	1137	1183	1229	
6	1276	1322	1369	1415	1461	1508	1554	1601	1647	1693	
7	1740	1786	1832	1879	1925	1971	2018	2064	2110	2157	
8	2203	2249	2295	2342	2388	2434	2481	2527	2573	2619	
9	2666	2712	2758	2804	2851	2897	2943	2989	3035	3082	
940	3128	3174	3220	3266	3313	3359	3405	3451	3497	3543	
1	3590	3636	3682	3728	3774	3820	3866	3913	3959	4005	
2	4051	4097	4143	4189	4235	4281	4327	4374	4420	4466	
3	4512	4558	4604	4650	4696	4742	4788	4834	4880	4926	
4	4972	5018	5064	5110	5156	5202	5248	5294	5340	5386	46

PROPORTIONAL PARTS

Diff.	1	2	3	4	5	6	7	8	9
47	4.7	9.4	14.1	18.8	23.5	28.2	32.9	37.6	42.3
46	4.6	9.2	13.8	18.4	23.0	27.6	32.2	36.8	41.4

No. 945
Log. 975

TABLE XIV.—*Continued*

No. 989
Log. 995

N.	0	1	2	3	4	5	6	7	8	9	Diff.
945	97 5432	5478	5524	5570	5616	5662	5707	5753	5799	5845	
6	5891	5937	5983	6029	6075	6121	6167	6212	6258	6304	
7	6350	6396	6442	6488	6533	6579	6625	6671	6717	6763	
8	6808	6854	6900	6946	6992	7037	7083	7129	7175	7220	
9	7266	7312	7358	7403	7449	7495	7541	7586	7632	7678	
950	7724	7769	7815	7861	7906	7952	7998	8043	8089	8135	
1	8181	8226	8272	8317	8363	8409	8454	8500	8546	8591	
2	8637	8683	8728	8774	8819	8865	8911	8956	9002	9047	
3	9093	9138	9184	9230	9275	9321	9366	9412	9457	9503	
4	9548	9594	9639	9685	9730	9776	9821	9867	9912	9958	
955	98 0003	0049	0094	0140	0185	0231	0276	0322	0367	0412	
6	0458	0503	0549	0594	0640	0685	0730	0776	0821	0867	
7	0912	0957	1003	1048	1093	1139	1184	1229	1275	1320	
8	1366	1411	1456	1501	1547	1592	1637	1683	1728	1773	
9	1819	1864	1909	1954	2000	2045	2090	2135	2181	2226	
960	2271	2316	2362	2407	2452	2497	2543	2588	2633	2678	
1	2723	2769	2814	2859	2904	2949	2994	3040	3085	3130	
2	3175	3220	3265	3310	3356	3401	3446	3491	3536	3581	
3	3626	3671	3716	3762	3807	3852	3897	3942	3987	4032	
4	4077	4122	4167	4212	4257	4302	4347	4392	4437	4482	
965	4527	4572	4617	4662	4707	4752	4797	4842	4887	4932	45
6	4977	5022	5067	5112	5157	5202	5247	5292	5337	5382	
7	5426	5471	5516	5561	5606	5651	5696	5741	5786	5830	
8	5875	5920	5965	6010	6055	6100	6144	6189	6234	6279	
9	6324	6369	6413	6458	6503	6548	6593	6637	6682	6727	
970	6772	6817	6861	6906	6951	6996	7040	7085	7130	7175	
1	7219	7264	7309	7353	7398	7443	7488	7532	7577	7622	
2	7666	7711	7756	7800	7845	7890	7934	7979	8024	8068	
3	8113	8157	8202	8247	8291	8336	8381	8425	8470	8514	
4	8559	8604	8648	8693	8737	8782	8826	8871	8916	8960	
975	9005	9049	9094	9138	9183	9227	9272	9316	9361	9405	
6	9450	9494	9539	9583	9628	9672	9717	9761	9806	9850	
7	9895	9939	9983	0028	0072	0117	0161	0206	0250	0294	
8	99 0339	0383	0428	0472	0516	0561	0605	0650	0694	0738	
9	0783	0827	0871	0916	0960	1004	1049	1093	1137	1182	
980	1226	1270	1315	1359	1403	1448	1492	1536	1580	1625	
1	1669	1713	1758	1802	1846	1890	1935	1979	2023	2067	
2	2111	2156	2200	2244	2288	2333	2377	2421	2465	2509	
3	2554	2598	2642	2686	2730	2774	2819	2863	2907	2951	
4	2995	3039	3083	3127	3172	3216	3260	3304	3348	3392	
985	3436	3480	3524	3568	3613	3657	3701	3745	3789	3833	
6	3877	3921	3965	4009	4053	4097	4141	4185	4229	4273	
7	4317	4361	4405	4449	4493	4537	4581	4625	4669	4713	
8	4757	4801	4845	4889	4933	4977	5021	5065	5108	5152	44
9	5196	5240	5284	5328	5372	5416	5460	5504	5547	5591	

PROPORTIONAL PARTS

Diff.	1	2	3	4	5	6	7	8	9
46	4.6	9.2	13.8	18.4	23.0	27.6	32.2	36.8	41.4
45	4.5	9.0	13.5	18.0	22.5	27.0	31.5	36.0	40.5
44	4.4	8.8	13.2	17.6	22.0	26.4	30.8	35.2	39.6
43	4.3	8.6	12.9	17.2	21.5	25.8	30.1	34.4	38.7

No. 990 Log. 995

TABLE XIV.—*Concluded*

No. 999 Log. 999

N.	0	1	2	3	4	5	6	7	8	9	Diff.
990	99 5635	5679	5723	5767	5811	5854	5898	5942	5986	6030	
1	6074	6117	6161	6205	6249	6293	6337	6380	6424	6468	44
2	6512	6555	6599	6643	6687	6731	6774	6818	6862	6906	
3	6949	6993	7037	7080	7124	7168	7212	7255	7299	7343	
4	7386	7430	7474	7517	7561	7605	7648	7692	7736	7779	
995	7823	7867	7910	7954	7998	8041	8085	8129	8172	8216	
6	8259	8303	8347	8390	8434	8477	8521	8564	8608	8652	
7	8695	8739	8782	8826	8869	8913	8956	9000	9043	9087	
8	9131	9174	9218	9261	9305	9348	9392	9435	9479	9522	
9	9565	9609	9652	9696	9739	9783	9826	9870	9913	9957	43

TABLE XV(*a*). VALUES OF *S*, *T*, AND *C* IN TABLE XV, FOR ANGLES BETWEEN 0° AND 2° AND BETWEEN 88° AND 90°

If we were to plot the values of the logarithmic functions given in Table XV as ordinates and corresponding minutes as abscissas, it would be found that the points for each function were on a curve with variable radius. It would be noted further that the curves for sines, tangents, and cotangents were of comparatively small radii when the angles were small; that the curves for cosines, cotangents, and tangents of angles near 90°, respectively, had the same shape as the curves for sines, tangents, and cotangents of the complements of the angles; and that other than the portions of the curves just mentioned were nearly straight lines for short distances.

When seconds are involved, it will be sufficiently accurate to interpolate in the ordinary manner between adjacent values in the tables —in other words, to assume that the curve joining two adjacent points is a straight line—for all functions between 2° and 88°, and also for sines of angles between 88° and 90° and for cosines of angles between 0° and 2°. The values in the columns headed *S*, *T*, and *C* provide a means (1) of accurately determining for any given angle between 0° and 2° the logarithmic sine, tangent, or cotangent, and for any given angle between 88° and 90°, the logarithmic cosine, cotangent, or tangent; or (2) for a given value of the logarithmic sine, tangent, or cotangent of accurately determining the angle when it lies between 0° and 2°, and for any given value of the cosine, cotangent, or tangent, the angle when it lies between 88° and 90°.

TABLE XV(*a*).—*Concluded*

Given: angle. Required: logarithmic function.

$$\left.\begin{aligned}\log\sin\alpha &= \log\alpha\text{ (in seconds)} + S\\ \log\tan\alpha &= \log\alpha\text{ (in seconds)} + T\\ \log\cot\alpha &= C - \log\alpha\text{ (in seconds)}\end{aligned}\right\}\text{In which }\alpha\text{ is less than }2^\circ.$$

$$\left.\begin{aligned}\log\cos\beta &= \log(90^\circ-\beta)\text{ (in seconds)} + S\\ \log\tan\beta &= C - \log(90^\circ-\beta)\text{ (in seconds)}\\ \log\cot\beta &= \log(90^\circ-\beta)\text{ (in seconds)} + T\end{aligned}\right\}\text{In which }\beta\text{ lies between }88^\circ\text{ and }90^\circ.$$

Given: logarithmic function. Required: angle.

$$\left.\begin{aligned}\log\alpha\text{ (in seconds)} &= \log\sin\alpha - S\\ &= \log\tan\alpha - T\\ &= C - \log\cot\alpha\end{aligned}\right\}\text{In which }\alpha\text{ is less than }2^\circ.$$

$$\left.\begin{aligned}\log(90^\circ-\beta)\text{ (in seconds)} &= \log\cos\beta - S\\ &= C - \log\tan\beta\\ &= \log\cot\beta - T\end{aligned}\right\}\text{In which }\beta\text{ lies between }88^\circ\text{ and }90^\circ.$$

EXAMPLES

Given: angle.

Required: logarithmic function.

$$\text{When}\left\{\begin{matrix}\alpha\\ 90^\circ-\beta\end{matrix}\right\} = 19'\,22'' = 1162''$$

$$\log 1162'' = 3.065206$$
$$S\text{ (for }19') = 4.685573$$
$$\left.\begin{matrix}\log\sin 19'\,22''\\ \log\cos 89^\circ\,40'\,38''\end{matrix}\right\} = 7.750779$$

$$\text{When}\left\{\begin{matrix}\alpha\\ 90^\circ-\beta\end{matrix}\right\} = 23'\,21'' = 1401''$$

$$\log 1401'' = 3.146438$$
$$C\text{ (for }23') = 15.314419$$
$$\left.\begin{matrix}\log\cot 23'\,21''\\ \log\tan 89^\circ\,36'\,39''\end{matrix}\right\} = 12.167981$$

Given: logarithmic function.

Required: angle.

$$\text{When}\left\{\begin{matrix}\log\sin\alpha\\ \log\cos\beta\end{matrix}\right\} = 7.750779$$
$$\text{S (for }19') = 4.685573$$
$$\log\left\{\begin{matrix}\alpha\\ 90^\circ-\beta\end{matrix}\right\}\text{ (in seconds)} = 3.065206$$

$$\text{or}\quad\begin{aligned}\alpha &= 1162'' = 19'\,22''\\ \beta &= 89^\circ\,40'\,38''\end{aligned}$$

$$\text{When}\left\{\begin{matrix}\log\cot\alpha\\ \log\tan\beta\end{matrix}\right\} = 12.167981$$
$$C\text{ (for }23') = 15.314419$$

$$\log\left\{\begin{matrix}\alpha\\ 90^\circ-\beta\end{matrix}\right\}\text{ (in seconds)} = 3.146438$$

$$\text{or}\quad\begin{aligned}\alpha &= 1401'' = 23'\,21''\\ \beta &= 89^\circ\,36'\,39''\end{aligned}$$

TABLE XV. LOGARITHMIC SINES, COSINES, TANGENTS, AND COTANGENTS

0° 179°

"	'	Sine.	S.*	T.*	Tang.	Cotang.	C.*	D. 1".	Cosine.	'
			4.685				15.314			
0	0	Inf. neg.	575	575	Inf. neg.	Inf. pos.	425		10.00 0000	60
60	1	6.46 3726	575	575	6.46 3726	13.53 6274	425		0000	59
120	2	.76 4756	575	575	.76 4756	.23 5244	425		0000	58
180	3	6.94 0847	575	575	6.94 0847	13.05 9153	425		0000	57
240	4	7.06 5786	575	575	7.06 5786	12.93 4214	425		0000	56
300	5	7.16 2696	575	575	7.16 2696	12.83 7304	425	.02	10.00 0000	55
360	6	.24 1877	575	575	.24 1878	.75 8122	425	.00	9.99 9999	54
420	7	.30 8824	575	575	.30 8825	.69 1175	425	.00	9999	53
480	8	.36 6816	574	576	.36 6817	.63 3183	424	.00	9999	52
540	9	.41 7968	574	576	.41 7970	.58 2030	424	.02	9999	51
600	10	7.46 3726	574	576	7.46 3727	12.53 6273	424	.00	9.99 9998	50
660	11	.50 5118	574	576	.50 5120	.49 4880	424	.02	9998	49
720	12	.54 2906	574	577	.54 2909	.45 7091	423	.00	9997	48
780	13	.57 7668	574	577	.57 7672	.42 2328	423	.02	9997	47
840	14	.60 9853	574	577	.60 9857	.39 0143	423	.00	9996	46
900	15	7.63 9816	573	578	7.63 9820	12.36 0180	422	.02	9.99 9996	45
960	16	.66 7845	573	578	.66 7849	.33 2151	422	.00	9995	44
1020	17	.69 4173	573	578	.69 4179	.30 5821	422	.02	9995	43
1080	18	.71 8997	573	579	.71 9003	.28 0997	421	.02	9994	42
1140	19	.74 2478	573	579	.74 2484	.25 7516	421	.00	9993	41
1200	20	7.76 4754	572	580	7.76 4761	12.23 5239	420	.02	9.99 9993	40
1260	21	.78 5943	572	580	.78 5951	.21 4049	420	.02	9992	39
1320	22	.80 6146	572	581	.80 6155	.19 3845	419	.02	9991	38
1380	23	.82 5451	572	581	.82 5460	.17 4540	419	.02	9990	37
1440	24	.84 3934	571	582	.84 3944	.15 6056	418	.00	9989	36
1500	25	7.86 1662	571	583	7.86 1674	12.13 8326	417	.02	9.99 9989	35
1560	26	.87 8695	571	583	.87 8708	.12 1292	417	.02	9988	34
1620	27	.89 5085	570	584	.89 5099	.10 4901	416	.02	9987	33
1680	28	.91 0879	570	584	.91 0894	.08 9106	416	.02	9986	32
1740	29	.92 6119	570	585	.92 6134	.07 3866	415	.03	9985	31
1800	30	7.94 0842	569	586	7.94 0858	12.05 9142	414	.02	9.99 9983	30
1860	31	.95 5082	569	587	.95 5100	.04 4900	413	.02	9982	29
1920	32	.96 8870	569	587	.96 8889	.03 1111	413	.02	9981	28
1980	33	.98 2233	568	588	.98 2253	.01 7747	412	.02	9980	27
2040	34	7.99 5198	568	589	7.99 5219	12.00 4781	411	.03	9979	26
2100	35	8.00 7787	567	590	8.00 7809	11.99 2191	410	.02	9.99 9977	25
2160	36	.02 0021	567	591	.02 0044	.97 9956	409	.02	9976	24
2220	37	.03 1919	566	592	.03 1945	.96 8055	408	.03	9975	23
2280	38	.04 3501	566	593	.04 3527	.95 6473	407	.02	9973	22
2340	39	.05 4781	566	593	.05 4809	.94 5191	407	.02	9972	21
2400	40	8.06 5776	565	594	8.06 5806	11.93 4194	406	.03	9.99 9971	20
2460	41	.07 6500	565	595	.07 6531	.92 3469	405	.02	9969	19
2520	42	.08 6965	564	596	.08 6997	.91 3003	404	.03	9968	18
2580	43	.09 7183	564	598	.09 7217	.90 2783	402	.03	9966	17
2640	44	.10 7167	563	599	.10 7203	.89 2797	401	.02	9964	16
2700	45	8.11 6926	562	600	8.11 6963	11.88 3037	400	.03	9.99 9963	15
2760	46	.12 6471	562	601	.12 6510	.87 3490	399	.03	9961	14
2820	47	.13 5810	561	602	.13 5851	.86 4149	398	.02	9959	13
2880	48	.14 4953	561	603	.14 4996	.85 5004	397	.03	9958	12
2940	49	.15 3907	560	604	.15 3952	.84 6048	396	.03	9956	11
3000	50	8.16 2681	560	605	8.16 2727	11.83 7273	395	.03	9.99 9954	10
3060	51	.17 1280	559	607	.17 1328	.82 8672	393	.03	9952	9
3120	52	.17 9713	558	608	.17 9763	.82 0237	392	.03	9950	8
3180	53	.18 7985	558	609	.18 8036	.81 1964	391	.03	9948	7
3240	54	.19 6102	557	611	.19 6156	.80 3844	389	.03	9946	6
3300	55	8.20 4070	556	612	8.20 4126	11.79 5874	388	.03	9.99 9944	5
3360	56	.21 1895	556	613	.21 1953	.78 8047	387	.03	9942	4
3420	57	.21 9581	555	615	.21 9641	.78 0359	385	.03	9940	3
3480	58	.22 7134	554	616	.22 7195	.77 2805	384	.03	9938	2
3540	59	.23 4557	554	618	.23 4621	.76 5379	382	.03	9936	1
3600	60	8.24 1855	553	619	8.24 1921	11.75 8079	381		9.99 9934	0
			4.685				15.314			
"	'	Cosine.	S.*	T.*	Cotang.	Tang.	C.*	D. 1".	Sine.	'

90° 89°

* For use of *S*, *T*, and *C* see Table XV(*a*), page 998.

1° **TABLE XV.**—*Continued* 178°

″	′	Sine.	S.*	T.*	Tang.	Cotang.	C.*	D. 1″.	Cosine.	′
			4.685				15.314			
3600	**0**	**8.24 1855**	553	619	**8.24 1921**	**11.75 8079**	381	.03	**9.99 9934**	**60**
3660	1	.24 9033	552	620	.24 9102	.75 0898	380	.05	9932	59
3720	2	.25 6094	551	622	.25 6165	.74 3835	378	.03	9929	58
3780	3	.26 3042	551	623	.26 3115	.73 6885	377	.03	9927	57
3840	4	.26 9881	550	625	.26 9956	.73 0044	375	.05	9925	56
3900	**5**	**8.27 6614**	549	627	**8.27 6691**	**11.72 3309**	373	.03	**9.99 9922**	**55**
3960	6	.28 3243	548	628	.28 3323	.71 6677	372	.03	9920	54
4020	7	.28 9773	547	630	.28 9856	.71 0144	370	.05	9918	53
4080	8	.29 6207	546	632	.29 6292	.70 3708	368	.03	9915	52
4140	9	.30 2546	546	633	.30 2634	.69 7366	367	.05	9913	51
4200	**10**	**8.30 8794**	545	635	**8.30 8884**	**11.69 1116**	365	.05	**9.99 9910**	**50**
4260	11	.31 4954	544	637	.31 5046	.68 4954	363	.03	9907	49
4320	12	.32 1027	543	638	.32 1122	.67 8878	362	.05	9905	48
4380	13	.32 7016	542	640	.32 7114	.67 2886	360	.05	9902	47
4440	14	.33 2924	541	642	.33 3025	.66 6975	358	.03	9899	46
4500	**15**	**8.33 8753**	540	644	**8.33 8856**	**11.66 1144**	356	.05	**9.99 9897**	**45**
4560	16	.34 4504	539	646	.34 4610	.65 5390	354	.05	9894	44
4620	17	.35 0181	539	648	.35 0289	.64 9711	352	.05	9891	43
4680	18	.35 5783	538	649	.35 5895	.64 4105	351	.05	9888	42
4740	19	.36 1315	537	651	.36 1430	.63 8570	349	.05	9885	41
4800	**20**	**8.36 6777**	536	653	**8.36 6895**	**11.63 3105**	347	.05	**9.99 9882**	**40**
4860	21	.37 2171	535	655	.37 2292	.62 7708	345	.05	9879	39
4920	22	.37 7499	534	657	.37 7622	.62 2378	343	.05	9876	38
4980	23	.38 2762	533	659	.38 2889	.61 7111	341	.05	9873	37
5040	24	.38 7962	532	661	.38 8092	.61 1908	339	.05	9870	36
5100	**25**	**8.39 3101**	531	663	**8.39 3234**	**11.60 6766**	337	.05	**9.99 9867**	**35**
5160	26	.39 8179	530	666	.39 8315	.60 1685	334	.05	9864	34
5220	27	.40 3199	529	668	.40 3338	.59 6662	332	.05	9861	33
5280	28	.40 8161	527	670	.40 8304	.59 1696	330	.07	9858	32
5340	29	.41 3068	526	672	.41 3213	.58 6787	328	.05	9854	31
5400	**30**	**8.41 7919**	525	674	**8.41 8068**	**11.58 1932**	326	.05	**9.99 9851**	**30**
5460	31	.42 2717	524	676	.42 2869	.57 7131	324	.07	9848	29
5520	32	.42 7462	523	679	.42 7618	.57 2382	321	.05	9844	28
5580	33	.43 2156	522	681	.43 2315	.56 7685	319	.05	9841	27
5640	34	.43 6800	521	683	.43 6962	.56 3038	317	.07	9838	26
5700	**35**	**8.44 1394**	520	685	**8.44 1560**	**11.55 8440**	315	.05	**9.99 9834**	**25**
5760	36	.44 5941	518	688	.44 6110	.55 3890	312	.07	9831	24
5820	37	.45 0440	517	690	.45 0613	.54 9387	310	.05	9827	23
5880	38	4893	516	693	5070	4930	307	.07	9824	22
5940	39	.45 9301	515	695	.45 9481	.54 0519	305	.07	9820	21
6000	**40**	**8.46 3665**	514	697	**8.46 3849**	**11.53 6151**	303	.05	**9.99 9816**	**20**
6060	41	.46 7985	512	700	.46 8172	.53 1828	300	.07	9813	19
6120	42	.47 2263	511	702	.47 2454	.52 7546	298	.07	9809	18
6180	43	.47 6498	510	705	.47 6693	.52 3307	295	.07	9805	17
6240	44	.48 0693	509	707	.48 0892	.51 9108	293	.07	9801	16
6300	**45**	**8.48 4848**	507	710	**8.48 5050**	**11.51 4950**	290	.05	**9.99 9797**	**15**
6360	46	.48 8963	506	713	.48 9170	.51 0830	287	.07	9794	14
6420	47	.49 3040	505	715	.49 3250	.50 6750	285	.07	9790	13
6480	48	.49 7078	503	718	.49 7293	.50 2707	282	.07	9786	12
6540	49	.50 1080	502	720	.50 1298	.49 8702	280	.07	9782	11
6600	**50**	**8.50 5045**	501	723	**8.50 5267**	**11.49 4733**	277	.07	**9.99 9778**	**10**
6660	51	.50 8974	499	726	.50 9200	.49 0800	274	.08	9774	9
6720	52	.51 2867	498	729	.51 3098	.48 6902	271	.07	9769	8
6780	53	.51 6726	497	731	.51 6961	.48 3039	269	.07	9765	7
6840	54	.52 0551	495	734	.52 0790	.47 9210	266	.07	9761	6
6900	**55**	**8.52 4343**	494	737	**8.52 4586**	**11.47 5414**	263	.07	**9.99 9757**	**5**
6960	56	.52 8102	492	740	.52 8349	.47 1651	260	.08	9753	4
7020	57	.53 1828	491	743	.53 2080	.46 7920	257	.07	9748	3
7080	58	5523	490	745	5779	4221	255	.07	9744	2
7140	59	.53 9186	488	748	.53 9447	.46 0553	252	.08	9740	1
7200	**60**	**8.54 2819**	487	751	**8.54 3084**	**11.45 6916**	249		**9.99 9735**	**0**
			4.685				15.314			
″	′	Cosine.	S.*	T.*	Cotang.	Tang.	C.*	D. 1″.	Sine.	′

91° **88°**

* For use of *S*, *T*, and *C* see Table XV(*a*), page 998.

2° TABLE XV.—*Continued* 177°

′	Sine.	D. 1″.	Cosine.	D.1″.	Tang.	D.1″.	Cotang.	′
0	**8.54 2819**	60.05	**9.99 9735**	.07	**8.54 3084**	60.12	**11.45 6916**	**60**
1	6422	59.55	9731	.08	.54 6691	59.62	.45 3309	59
2	.54 9995	59.07	9726	.07	.55 0268	59.15	.44 9732	58
3	.55 3539	58.58	9722	.08	3817	58.65	6183	57
4	.55 7054	58.10	9717	.07	.55 7336	58.20	.44 2664	56
5	**8.56 0540**	57.65	**9.99 9713**	.08	**8.56 0828**	57.72	**11.43 9172**	**55**
6	3999	57.20	9708	.07	4291	57.27	5709	54
7	.56 7431	56.75	9704	.08	.56 7727	56.83	.43 2273	53
8	.57 0836	56.30	9699	.08	.57 1137	56.38	.42 8863	52
9	4214	55.87	9694	.08	4520	55.95	5480	51
10	**8.57 7566**	55.43	**9.99 9689**	.07	**8.57 7877**	55.52	**11.42 2123**	**50**
11	.58 0892	55.02	9685	.08	.58 1208	55.10	.41 8792	49
12	4193	54.60	9680	.08	4514	54.68	5486	48
13	.58 7469	54.20	9675	.08	.58 7795	54.27	.41 2205	47
14	.59 0721	53.78	9670	.08	.59 1051	53.87	.40 8949	46
15	**8.59 3948**	53.40	**9.99 9665**	.08	**8.59 4283**	53.48	**11.40 5717**	**45**
16	.59 7152	53.00	9660	.08	.59 7492	53.08	.40 2508	44
17	.60 0332	52.62	9655	.08	.60 0677	52.70	.39 9323	43
18	3489	52.23	9650	.08	3839	52.32	6161	42
19	6623	51.85	9645	.08	.60 6978	51.93	.39 3022	41
20	**8.60 9734**	51.48	**9.99 9640**	.08	**8.61 0094**	51.58	**11.38 9906**	**40**
21	.61 2823	51.13	9635	.10	3189	51.22	6811	39
22	5891	50.77	9629	.08	6262	50.85	3738	38
23	.61 8937	50.42	9624	.08	.61 9313	50.50	.38 0687	37
24	.62 1962	50.05	9619	.08	.62 2343	50.15	.37 7657	36
25	**8.62 4965**	49.72	**9.99 9614**	.10	**8.62 5352**	49.80	**11.37 4648**	**35**
26	.62 7948	49.38	9608	.08	.62 8340	49.47	.37 1660	34
27	.63 0911	49.05	9603	.10	.63 1308	49.13	.36 8692	33
28	3854	48.70	9597	.08	4256	48.80	5744	32
29	6776	48.40	9592	.10	.63 7184	48.48	.36 2816	31
30	**8.63 9680**	48.05	**9.99 9586**	.08	**8.64 0093**	48.15	**11.35 9907**	**30**
31	.64 2563	47.75	9581	.10	2982	47.85	7018	29
32	5428	47.43	9575	.08	5853	47.52	4147	28
33	.64 8274	47.13	9570	.10	.64 8704	47.22	.35 1296	27
34	.65 1102	46.82	9564	.10	.65 1537	46.92	.34 8463	26
35	**8.65 3911**	46.52	**9.99 9558**	.08	**8.65 4352**	46.62	**11.34 5648**	**25**
36	6702	46.22	9553	.10	7149	46.32	2851	24
37	.65 9475	45.92	9547	.10	.65 9928	46.02	.34 0072	23
38	.66 2230	45.63	9541	.10	.66 2689	45.73	.33 7311	22
39	4968	45.35	9535	.10	5433	45.45	4567	21
40	**8.66 7689**	45.07	**9.99 9529**	.08	**8.66 8160**	45.17	**11.33 1840**	**20**
41	.67 0393	44.78	9524	.10	.67 0870	44.88	.32 9130	19
42	3080	44.52	9518	.10	3563	44.60	6437	18
43	5751	44.23	9512	.10	6239	44.35	3761	17
44	.67 8405	43.97	9506	.10	.67 8900	44.07	.32 1100	16
45	**8.68 1043**	43.70	**9.99 9500**	.12	**8.68 1544**	43.80	**11.31 8456**	**15**
46	3665	43.45	9493	.10	4172	43.53	5828	14
47	6272	43.18	9487	.10	6784	43.28	3216	13
48	.68 8863	42.92	9481	.10	.68 9381	43.03	.31 0619	12
49	.69 1438	42.67	9475	.10	.69 1963	42.77	.30 8037	11
50	**8.69 3998**	42.42	**9.99 9469**	.10	**8.69 4529**	42.53	**11.30 5471**	**10**
51	6543	42.17	9463	.12	7081	42.27	2919	9
52	.69 9073	41.93	9456	.10	.69 9617	42.03	.30 0383	8
53	.70 1589	41.68	9450	.12	.70 2139	41.78	.29 7861	7
54	4090	41.45	9443	.10	4646	41.57	5354	6
55	**8.70 6577**	41.20	**9.99 9437**	.10	**8.70 7140**	41.30	**11.29 2860**	**5**
56	.70 9049	40.97	9431	.12	.70 9618	41.08	.29 0382	4
57	.71 1507	40.75	9424	.10	.71 2083	40.85	.28 7917	3
58	3952	40.52	9418	.12	4534	40.63	5466	2
59	6383	40.28	9411	.12	6972	40.40	3028	1
60	**8.71 8800**		**9.99 9404**		**8.71 9396**		**11.28 0604**	**0**
′	Cosine.	D.1″.	Sine.	D.1″.	Cotang.	D.1″.	Tang.	′

92° 87°

3° TABLE XV.—*Continued* 176°

′	Sine.	D.1″.	Cosine.	D.1″.	Tang.	D.1″.	Cotang.	′
0	8.71 8800	40.07	9.99 9404	.10	8.71 9396	40.17	11.28 0604	60
1	.72 1204	39.85	9398	.12	.72 1806	39.97	.27 8194	59
2	3595	39.62	9391	.12	4204	39.73	5796	58
3	5972	39.42	9384	.10	6588	39.52	3412	57
4	.72 8337	39.18	9378	.12	.72 8959	39.30	.27 1041	56
5	8.73 0688	38.98	9.99 9371	.12	8.73 1317	39.10	11.26 8683	55
6	3027	38.78	9364	.12	3663	38.88	6337	54
7	5354	38.55	9357	.12	5996	38.68	4004	53
8	7667	38.37	9350	.12	.73 8317	38.48	.26 1683	52
9	.73 9969	38.17	9343	.12	.74 0626	38.27	.25 9374	51
10	8.74 2259	37.95	9.99 9336	.12	8.74 2922	38.08	11.25 7078	50
11	4536	37.77	9329	.12	5207	37.87	4793	49
12	6802	37.55	9322	.12	7479	37.68	2521	48
13	.74 9055	37.37	9315	.12	.74 9740	37.48	.25 0260	47
14	.75 1297	37.18	9308	.12	.75 1989	37.30	.24 8011	46
15	8.75 3528	36.98	9.99 9301	.12	8.75 4227	37.10	11.24 5773	45
16	5747	36.80	9294	.12	6453	36.92	3547	44
17	.75 7955	36.60	9287	.13	.75 8668	36.73	.24 1332	43
18	.76 0151	36.43	9279	.12	.76 0872	36.55	.23 9128	42
19	2337	36.23	9272	.12	3065	36.35	6935	41
20	8.76 4511	36.07	9.99 9265	.13	8.76 5246	36.18	11.23 4754	40
21	6675	35.88	9257	.12	7417	36.02	2583	39
22	.76 8828	35.70	9250	.13	.76 9578	35.82	.23 0422	38
23	.77 0970	35.52	9242	.12	.77 1727	35.65	.22 8273	37
24	3101	35.37	9235	.13	3866	35.48	6134	36
25	8.77 5223	35.17	9.99 9227	.12	8.77 5995	35.32	11.22 4005	35
26	7333	35.02	9220	.13	.77 8114	35.13	.22 1886	34
27	.77 9434	34.83	9212	.12	.78 0222	34.97	.21 9778	33
28	.78 1524	34.68	9205	.13	2320	34.80	7680	32
29	3605	34.50	9197	.13	4408	34.63	5592	31
30	8.78 5675	34.35	9.99 9189	.13	8.78 6486	34.47	11.21 3514	30
31	7736	34.18	9181	.12	.78 8554	34.32	.21 1446	29
32	.78 9787	34.02	9174	.13	.79 0613	34.15	.20 9387	28
33	.79 1828	33.85	9166	.13	2662	33.98	7338	27
34	3859	33.70	9158	.13	4701	33.83	5299	26
35	8.79 5881	33.55	9.99 9150	.13	8.79 6731	33.68	11.20 3269	25
36	7894	33.38	9142	.13	.79 8752	33.52	.20 1248	24
37	.79 9897	33.25	9134	.13	.80 0763	33.37	.19 9237	23
38	.80 1892	33.07	9126	.13	2765	33.22	7235	22
39	3876	32.93	9118	.13	4758	33.07	5242	21
40	8.80 5852	32.78	9.99 9110	.13	8.80 6742	32.92	11.19 3258	20
41	7819	32.63	9102	.13	.80 8717	32.77	.19 1283	19
42	.80 9777	32.48	9094	.13	.81 0683	32.63	.18 9317	18
43	.81 1726	32.35	9086	.15	2641	32.47	7359	17
44	3667	32.20	9077	.13	4589	32.33	5411	16
45	8.81 5599	32.05	9.99 9069	.13	8.81 6529	32.20	11.18 3471	15
46	7522	31.90	9061	.13	.81 8461	32.05	.18 1539	14
47	.81 9436	31.78	9053	.15	.82 0384	31.90	.17 9616	13
48	.82 1343	31.62	9044	.13	2298	31.78	7702	12
49	3240	31.50	9036	.15	4205	31.63	5795	11
50	8.82 5130	31.35	9.99 9027	.13	8.82 6103	31.48	11.17 3897	10
51	7011	31.22	9019	.15	7992	31.37	2008	9
52	.82 8884	31.08	9010	.13	.82 9874	31.23	.17 0126	8
53	.83 0749	30.97	9002	.15	.83 1748	31.08	.16 8252	7
54	2607	30.82	8993	.15	3613	30.97	6387	6
55	8.83 4456	30.68	9.99 8984	.13	8.83 5471	30.83	11.16 4529	5
56	6297	30.55	8976	.15	7321	30.70	2679	4
57	8130	30.43	8967	.15	.83 9163	30.58	.16 0837	3
58	.83 9956	30.30	8958	.13	.84 0998	30.45	.15 9002	2
59	.84 1774	30.18	8950	.15	2825	30.32	7175	1
60	8.84 3585		9.99 8941		8.84 4644		11.15 5356	0
′	Cosine.	D.1″.	Sine.	D.1″.	Cotang.	D.1″.	Tang.	′

93° 86°

4° **Table XV.**—*Continued* 175°

′	Sine.	D.1″.	Cosine.	D.1″.	Tang.	D.1″.	Cotang.	′
0	**8.84 3585**	30.03	**9.99 8941**	.15	**8.84 4644**	30.18	**11.15 5356**	**60**
1	5387	29.93	8932	.15	6455	30.08	3545	59
2	7183	29.80	8923	.15	.84 8260	29.95	.15 1740	58
3	.84 8971	29.67	8914	.15	.85 0057	29.82	.14 9943	57
4	.85 0751	29.57	8905	.15	1846	29.70	8154	56
5	**8.85 2525**	29.43	**9.99 8896**	.15	**8.85 3628**	29.58	**11.14 6372**	**55**
6	4291	29.30	8887	.15	5403	29.47	4597	54
7	6049	29.20	8878	.15	7171	29.35	2829	53
8	7801	29.08	8869	.15	.85 8932	29.23	.14 1068	52
9	.85 9546	28.95	8860	.15	.86 0686	29.12	.13 9314	51
10	**8.86 1283**	28.85	**9.99 8851**	.17	**8.86 2433**	29.00	**11.13 7567**	**50**
11	3014	28.73	8841	.15	4173	28.88	5827	49
12	4738	28.62	8832	.15	5906	28.77	4094	48
13	6455	28.50	8823	.17	7632	28.65	2368	47
14	8165	28.38	8813	.15	.86 9351	28.55	.13 0649	46
15	**8.86 9868**	28.28	**9.99 8804**	.15	**8.87 1064**	28.43	**11.12 8936**	**45**
16	.87 1565	28.17	8795	.17	2770	28.32	7230	44
17	3255	28.05	8785	.15	4469	28.22	5531	43
18	4938	27.95	8776	.17	6162	28.12	3838	42
19	6615	27.83	8766	.15	7849	28.00	2151	41
20	**8.87 8285**	27.73	**9.99 8757**	.17	**8.87 9529**	27.88	**11.12 0471**	**40**
21	.87 9949	27.63	8747	.15	.88 1202	27.78	.11 8798	39
22	.88 1607	27.52	8738	.17	2869	27.68	7131	38
23	3258	27.42	8728	.17	4530	27.58	5470	37
24	4903	27.32	8718	.17	6185	27.47	3815	36
25	**8.88 6542**	27.20	**9.99 8708**	.15	**8.88 7833**	27.38	**11.11 2167**	**35**
26	8174	27.12	8699	.17	.88 9476	27.27	.11 0524	34
27	.88 9801	27.00	8689	.17	.89 1112	27.17	.10 8888	33
28	.89 1421	26.90	8679	.17	2742	27.07	7258	32
29	3035	26.80	8669	.17	4366	26.97	5634	31
30	**8.89 4643**	26.72	**9.99 8659**	.17	**8.89 5984**	26.87	**11.10 4016**	**30**
31	6246	26.60	8649	.17	7596	26.78	2404	29
32	7842	26.50	8639	.17	.89 9203	26.67	.10 0797	28
33	.89 9432	26.42	8629	.17	.90 0803	26.58	.09 9197	27
34	.90 1017	26.32	8619	.17	2398	26.48	7602	26
35	**8.90 2596**	26.22	**9.99 8609**	.17	**8.90 3987**	26.38	**11.09 6013**	**25**
36	4169	26.12	8599	.17	5570	26.28	4430	24
37	5736	26.02	8589	.18	7147	26.20	2853	23
38	7297	25.93	8578	.17	.90 8719	26.10	.09 1281	22
39	.90 8853	25.85	8568	.17	.91 0285	26.02	.08 9715	21
40	**8.91 0404**	25.75	**9.99 8558**	.17	**8.91 1846**	25.92	**11.08 8154**	**20**
41	1949	25.65	8548	.18	3401	25.83	6599	19
42	3488	25.57	8537	.17	4951	25.73	5049	18
43	5022	25.47	8527	.18	6495	25.63	3505	17
44	6550	25.38	8516	.17	8034	25.57	1966	16
45	**8.91 8073**	25.30	**9.99 8506**	.18	**8.91 9568**	25.47	**11.08 0432**	**15**
46	.91 9591	25.20	8495	.17	.92 1096	25.38	.07 8904	14
47	.92 1103	25.12	8485	.18	2619	25.28	7381	13
48	2610	25.03	8474	.17	4136	25.22	5864	12
49	4112	24.95	8464	.18	5649	25.12	4351	11
50	**8.92 5609**	24.85	**9.99 8453**	.18	**8.92 7156**	25.03	**11.07 2844**	**10**
51	7100	24.78	8442	.18	.92 8658	24.95	.07 1342	9
52	.92 8587	24.68	8431	.17	.93 0155	24.87	.06 9845	8
53	.93 0068	24.60	8421	.18	1647	24.78	8353	7
54	1544	24.52	8410	.18	3134	24.70	6866	6
55	**8.93 3015**	24.43	**9.99 8399**	.18	**8.93 4616**	24.62	**11.06 5384**	**5**
56	4481	24.35	8388	.18	6093	24.53	3907	4
57	5942	24.27	8377	.18	7565	24.45	2435	3
58	7398	24.20	8366	.18	.93 9032	24.37	.06 0968	2
59	.93 8850	24.10	8355	.18	.94 0494	24.30	.05 9506	1
60	**8.94 0296**		**9.99 8344**		**8.94 1952**		**11.05 8048**	**0**
′	Cosine.	D.1″.	Sine.	D.1″.	Cotang.	D.1″.	Tang.	′

94° **85°**

5° TABLE XV.—*Continued* 174°

′	Sine.	D.1″.	Cosine.	D.1″.	Tang.	D.1″.	Cotang.	′
0	**8.94 0296**	24.03	**9.99 8344**	.18	**8.94 1952**	24.20	**11.05 8048**	**60**
1	1738	23.93	8333	.18	3404	24.13	6596	59
2	3174	23.87	8322	.18	4852	24.05	5148	58
3	4606	23.80	8311	.18	6295	23.98	3705	57
4	6034	23.70	8300	.18	7734	23.90	2266	56
5	**8.94 7456**	23.63	**9.99 8289**	.20	**8.94 9168**	23.82	**11.05 0832**	**55**
6	.94 8874	23.55	8277	.18	.95 0597	23.73	.04 9403	54
7	.95 0287	23.48	8266	.18	2021	23.67	7979	53
8	1696	23.40	8255	.20	3441	23.58	6559	52
9	3100	23.32	8243	.18	4856	23.52	5144	51
10	**8.95 4499**	23.25	**9.99 8232**	.20	**8.95 6267**	23.45	**11.04 3733**	**50**
11	5894	23.17	8220	.18	7674	23.35	2326	49
12	7284	23.10	8209	.20	.95 9075	23.30	.04 0925	48
13	.95 8670	23.03	8197	.18	.96 0473	23.22	.03 9527	47
14	.96 0052	22.95	8186	.20	1866	23.15	8134	46
15	**8.96 1429**	22.87	**9.99 8174**	.18	**8.96 3255**	23.07	**11.03 6745**	**45**
16	2801	22.82	8163	.20	4639	23.00	5361	44
17	4170	22.73	8151	.20	6019	22.92	3981	43
18	5534	22.65	8139	.18	7394	22.87	2606	42
19	6893	22.60	8128	.20	.96 8766	22.78	.03 1234	41
20	**8.96 8249**	22.52	**9.99 8116**	.20	**8.97 0133**	22.72	**11.02 9867**	**40**
21	.96 9600	22.45	8104	.20	1496	22.65	8504	39
22	.97 0947	22.37	8092	.20	2855	22.57	7145	38
23	2289	22.32	8080	.20	4209	22.52	5791	37
24	3628	22.23	8068	.20	5560	22.43	4440	36
25	**8.97 4962**	22.18	**9.99 8056**	.20	**8.97 6906**	22.37	**11.02 3094**	**35**
26	6293	22.10	8044	.20	8248	22.30	1752	34
27	7619	22.03	8032	.20	.97 9586	22.25	.02 0414	33
28	.97 8941	21.97	8020	.20	.98 0921	22.17	.01 9079	32
29	.98 0259	21.90	8008	.20	2251	22.10	7749	31
30	**8.98 1573**	21.83	**9.99 7996**	.20	**8.98 3577**	22.03	**11.01 6423**	**30**
31	2883	21.77	7984	.20	4899	21.97	5101	29
32	4189	21.72	7972	.22	6217	21.92	3783	28
33	5491	21.63	7959	.20	7532	21.83	2468	27
34	6789	21.57	7947	.20	.98 8842	21.78	.01 1158	26
35	**8.98 8083**	21.52	**9.99 7935**	.22	**8.99 0149**	21.70	**11.00 9851**	**25**
36	.98 9374	21.43	7922	.20	1451	21.65	8549	24
37	.99 0660	21.38	7910	.22	2750	21.58	7250	23
38	1943	21.32	7897	.20	4045	21.53	5955	22
39	3222	21.25	7885	.22	5337	21.45	4663	21
40	**8.99 4497**	21.18	**9.99 7872**	.20	**8.99 6624**	21.40	**11.00 3376**	**20**
41	5768	21.13	7860	.22	7908	21.33	2092	19
42	7036	21.05	7847	.20	8.99 9188	21.28	11.00 0812	18
43	8299	21.02	7835	.22	9.00 0465	21.22	10.99 9535	17
44	8.99 9560	20.93	7822	.22	1738	21.15	8262	16
45	**9.00 0816**	20.88	**9.99 7809**	.20	**9.00 3007**	21.08	**10.99 6993**	**15**
46	2069	20.82	7797	.22	4272	21.03	5728	14
47	3318	20.75	7784	.22	5534	20.97	4466	13
48	4563	20.70	7771	.22	6792	20.92	3208	12
49	5805	20.65	7758	.22	8047	20.85	1953	11
50	**9.00 7044**	20.57	**9.99 7745**	.22	**9.00 9298**	20.80	**10.99 0702**	**10**
51	8278	20.53	7732	.22	.01 0546	20.73	.98 9454	9
52	.00 9510	20.45	7719	.22	1790	20.68	8210	8
53	.01 0737	20.42	7706	.22	3031	20.62	6969	7
54	1962	20.33	7693	.22	4268	20.57	5732	6
55	**9.01 3182**	20.30	**9.99 7680**	.22	**9.01 5502**	20.50	**10.98 4498**	**5**
56	4400	20.22	7667	.22	6732	20.45	3268	4
57	5613	20.18	7654	.22	7959	20.40	2041	3
58	6824	20.12	7641	.22	.01 9183	20.33	.98 0817	2
59	8031	20.07	7628	.23	.02 0403	20.28	.97 9597	1
60	**9.01 9235**		**9.99 7614**		**9.02 1620**		**10.97 8380**	**0**
′	Cosine.	D.1″.	Sine.	D.1″.	Cotang.	D.1″.	Tang.	′

95° **84°**

6° **TABLE XV.**—*Continued* 173°

′	Sine.	D.1″.	Cosine.	D.1″.	Tang.	D.1″.	Cotang.	′
0	**9.01 9235**	20.00	**9.99 7614**	.22	**9.02 1620**	20.23	**10.97 8380**	**60**
1	.02 0435	19.95	7601	.22	2834	20.17	7166	59
2	1632	19.88	7588	.23	4044	20.12	5956	58
3	2825	19.85	7574	.22	5251	20.07	4749	57
4	4016	19.78	7561	.23	6455	20.00	3545	56
5	**9.02 5203**	19.72	**9.99 7547**	.22	**9.02 7655**	19.95	**10.97 2345**	**55**
6	6386	19.68	7534	.23	.02 8852	19.90	.97 1148	54
7	7567	19.62	7520	.22	.03 0046	19.85	.96 9954	53
8	8744	19.57	7507	.23	1237	19.80	8763	52
9	.02 9918	19.52	7493	.22	2425	19.73	7575	51
10	**9.03 1089**	19.47	**9.99 7480**	.23	**9.03 3609**	19.70	**10.96 6391**	**50**
11	2257	19.40	7466	.23	4791	19.63	5209	49
12	3421	19.35	7452	.22	5969	19.58	4031	48
13	4582	19.32	7439	.23	7144	19.53	2856	47
14	5741	19.25	7425	.23	8316	19.48	1684	46
15	**9.03 6896**	19.20	**9.99 7411**	.23	**9.03 9485**	19.43	**10.96 0515**	**45**
16	8048	19.15	7397	.23	.04 0651	19.37	.95 9349	44
17	.03 9197	19.08	7383	.23	1813	19.33	8187	43
18	.04 0342	19.05	7369	.23	2973	19.28	7027	42
19	1485	19.00	7355	.23	4130	19.23	5870	41
20	**9.04 2625**	18.95	**9.99 7341**	.23	**9.04 5284**	19.17	**10.95 4716**	**40**
21	3762	18.88	7327	.23	6434	19.13	3566	39
22	4895	18.85	7313	.23	7582	19.08	2418	38
23	6026	18.80	7299	.23	8727	19.03	1273	37
24	7154	18.75	7285	.23	.04 9869	18.98	.95 0131	36
25	**9.04 8279**	18.68	**9.99 7271**	.23	**9.05 1008**	18.93	**10.94 8992**	**35**
26	.04 9400	18.65	7257	.25	2144	18.88	7856	34
27	.05 0519	18.60	7242	.23	3277	18.83	6723	33
28	1635	18.57	7228	.23	4407	18.80	5593	32
29	2749	18.50	7214	.25	5535	18.73	4465	31
30	**9.05 3859**	18.45	**9.99 7199**	.23	**9.05 6659**	18.70	**10.94 3341**	**30**
31	4966	18.42	7185	.25	7781	18.65	2219	29
32	6071	18.35	7170	.23	.05 8900	18.60	.94 1100	28
33	7172	18.32	7156	.25	.06 0016	18.57	.93 9984	27
34	8271	18.27	7141	.23	1130	18.50	8870	26
35	**9.05 9367**	18.22	**9.99 7127**	.25	**9.06 2240**	18.47	**10.93 7760**	**25**
36	.06 0460	18.18	7112	.23	3348	18.42	6652	24
37	1551	18.13	7098	.25	4453	18.38	5547	23
38	2639	18.08	7083	.25	5556	18.32	4444	22
39	3724	18.03	7068	.25	6655	18.28	3345	21
40	**9.06 4806**	17.98	**9.99 7053**	.23	**9.06 7752**	18.25	**10.93 2248**	**20**
41	5885	17.95	7039	.25	8846	18.20	1154	19
42	6962	17.90	7024	.25	.06 9938	18.15	.93 0062	18
43	8036	17.85	7009	.25	.07 1027	18.10	.92 8973	17
44	.06 9107	17.82	6994	.25	2113	18.07	7887	16
45	**9.07 0176**	17.77	**9.99 6979**	.25	**9.07 3197**	18.02	**10.92 6803**	**15**
46	1242	17.73	6964	.25	4278	17.97	5722	14
47	2306	17.67	6949	.25	5356	17.93	4644	13
48	3366	17.63	6934	.25	6432	17.88	3568	12
49	4424	17.60	6919	.25	7505	17.85	2495	11
50	**9.07 5480**	17.55	**9.99 6904**	.25	**9.07 8576**	17.80	**10.92 1424**	**10**
51	6533	17.50	6889	.25	.07 9644	17.77	.92 0356	9
52	7583	17.47	6874	.27	.08 0710	17.72	.91 9290	8
53	8631	17.42	6858	.25	1773	17.67	8227	7
54	.07 9676	17.38	6843	.27	2833	17.63	7167	6
55	**9.08 0719**	17.33	**9.99 6828**	.27	**9.08 3891**	17.60	**10.91 6109**	**5**
56	1759	17.30	6812	.25	4947	17.55	5053	4
57	2797	17.25	6797	.25	6000	17.50	4000	3
58	3832	17.20	6782	.27	7050	17.47	2950	2
59	4864	17.17	6766	.25	8098	17.43	1902	1
60	**9.08 5894**		**9.99 6751**		**9.08 9144**		**10.91 0856**	**0**
′	Cosine.	D.1″.	Sine.	D.1″.	Cotang.	D.1″.	Tang.	′

96° **83°**

7° TABLE XV.—*Continued* 172°

′	Sine.	D.1″.	Cosine.	D.1″.	Tang.	D.1″.	Cotang.	′
0	**9.08 5894**	17.13	**9.99 6751**	.27	**9.08 9144**	17.38	**10.91 0856**	**60**
1	6922	17.08	6735	.25	.09 0187	17.35	.90 9813	59
2	7947	17.05	6720	.27	1228	17.30	8772	58
3	8970	17.00	6704	.27	2266	17.27	7734	57
4	.08 9990	16.97	6688	.25	3302	17.23	6698	56
5	**9.09 1008**	16.93	**9.99 6673**	.27	**9.09 4336**	17.18	**10.90 5664**	**55**
6	2024	16.88	6657	.27	5367	17.13	4633	54
7	3037	16.83	6641	.27	6395	17.12	3605	53
8	4047	16.82	6625	.25	7422	17.07	2578	52
9	5056	16.77	6610	.27	8446	17.03	1554	51
10	**9.09 6062**	16.72	**9.99 6594**	.27	**9.09 9468**	16.98	**10.90 0532**	**50**
11	7065	16.68	6578	.27	.10 0487	16.95	.89 9513	49
12	8066	16.65	6562	.27	1504	16.92	8496	48
13	.09 9065	16.62	6546	.27	2519	16.88	7481	47
14	.10 0062	16.57	6530	.27	3532	16.83	6468	46
15	**9.10 1056**	16.53	**9.99 6514**	.27	**9.10 4542**	16.80	**10.89 5458**	**45**
16	2048	16.48	6498	.27	5550	16.77	4450	44
17	3037	16.47	6482	.28	6556	16.72	3444	43
18	4025	16.42	6465	.27	7559	16.68	2441	42
19	5010	16.37	6449	.27	8560	16.65	1440	41
20	**9.10 5992**	16.35	**9.99 6433**	.27	**9.10 9559**	16.62	**10.89 0441**	**40**
21	6973	16.30	6417	.28	.11 0556	16.58	.88 9444	39
22	7951	16.27	6400	.27	1551	16.53	8449	38
23	8927	16.23	6384	.27	2543	16.50	7457	37
24	.10 9901	16.20	6368	.28	3533	16.47	6467	36
25	**9.11 0873**	16.15	**9.99 6351**	.27	**9.11 4521**	16.43	**10.88 5479**	**35**
26	1842	16.12	6335	.28	5507	16.40	4493	34
27	2809	16.08	6318	.27	6491	16.35	3509	33
28	3774	16.05	6302	.28	7472	16.33	2528	32
29	4737	16.02	6285	.27	8452	16.28	1548	31
30	**9.11 5698**	15.97	**9.99 6269**	.28	**9.11 9429**	16.25	**10.88 0571**	**30**
31	6656	15.95	6252	.28	.12 0404	16.22	.87 9596	29
32	7613	15.90	6235	.27	1377	16.18	8623	28
33	8567	15.87	6219	.28	2348	16.15	7652	27
34	.11 9519	15.83	6202	.28	3317	16.12	6683	26
35	**9.12 0469**	15.80	**9.99 6185**	.28	**9.12 4284**	16.08	**10.87 5716**	**25**
36	1417	15.75	6168	.28	5249	16.03	4751	24
37	2362	15.73	6151	.28	6211	16.02	3789	23
38	3306	15.70	6134	.28	7172	15.97	2828	22
39	4248	15.65	6117	.28	8130	15.95	1870	21
40	**9.12 5187**	15.63	**9.99 6100**	.28	**9.12 9087**	15.90	**10.87 0913**	**20**
41	6125	15.58	6083	.28	.13 0041	15.88	.86 9959	19
42	7060	15.55	6066	.28	0994	15.83	9006	18
43	7993	15.53	6049	.28	1944	15.82	8056	17
44	8925	15.48	6032	.28	2893	15.77	7107	16
45	**9.12 9854**	15.45	**9.99 6015**	.28	**9.13 3839**	15.75	**10.86 6161**	**15**
46	.13 0781	15.42	5998	.30	4784	15.70	5216	14
47	1706	15.40	5980	.28	5726	15.68	4274	13
48	2630	15.35	5963	.28	6667	15.63	3333	12
49	3551	15.32	5946	.30	7605	15.62	2395	11
50	**9.13 4470**	15.28	**9.99 5928**	.28	**9.13 8542**	15.57	**10.86 1458**	**10**
51	5387	15.27	5911	.28	.13 9476	15.55	.86 0524	9
52	6303	15.22	5894	.30	.14 0409	15.52	.85 9591	8
53	7216	15.20	5876	.28	1340	15.48	8660	7
54	8128	15.15	5859	.30	2269	15.45	7731	6
55	**9.13 9037**	15.12	**9.99 5841**	.30	**9.14 3196**	15.42	**10.85 6804**	**5**
56	.13 9944	15.10	5823	.28	4121	15.38	5879	4
57	.14 0850	15.07	5806	.30	5044	15.37	4956	3
58	1754	15.02	5788	.28	5966	15.32	4034	2
59	2655	15.00	5771	.30	6885	15.30	3115	1
60	**9.14 3555**		**9.99 5753**		**9.14 7803**		**10.85 2197**	**0**
′	Cosine.	D.1″.	Sine.	D.1″.	Cotang.	D.1″.	Tang.	′

97° **82°**

8° **TABLE XV.—*Continued*** 171°

′	Sine.	D.1″.	Cosine.	D.1″.	Tang.	D.1″.	Cotang.	′
0	**9.14 3555**	14.97	**9.99 5753**	.30	**9.14 7803**	15.25	**10.85 2197**	**60**
1	4453	14.93	5735	.30	8718	15.23	1282	**59**
2	5349	14.90	5717	.30	.14 9632	15.20	.85 0368	**58**
3	6243	14.88	5699	.30	.15 0544	15.17	.84 9456	**57**
4	7136	14.83	5681	.28	1454	15.15	8546	**56**
5	**9.14 8026**	14.82	**9.99 5664**	.30	**9.15 2363**	15.10	**10.84 7637**	**55**
6	8915	14.78	5646	.30	3269	15.08	6731	54
7	.14 9802	14.73	5628	.30	4174	15.05	5826	53
8	.15 0686	14.72	5610	.32	5077	15.02	4923	52
9	1569	14.70	5591	.30	5978	14.98	4022	51
10	**9.15 2451**	14.65	**9.99 5573**	.30	**9.15 6877**	14.97	**10.84 3123**	**50**
11	3330	14.63	5555	.30	7775	14.93	2225	49
12	4208	14.58	5537	.30	8671	14.90	1329	48
13	5083	14.57	5519	.30	.15 9565	14.87	.84 0435	47
14	5957	14.55	5501	.32	.16 0457	14.83	.83 9543	46
15	**9.15 6830**	14.50	**9.99 5482**	.30	**9.16 1347**	14.82	**10.83 8653**	**45**
16	7700	14.48	5464	.30	2236	14.78	7764	44
17	8569	14.43	5446	.32	3123	14.75	6877	43
18	.15 9435	14.43	5427	.30	4008	14.73	5992	42
19	.16 0301	14.38	5409	.32	4892	14.70	5108	41
20	**9.16 1164**	14.35	**9.99 5390**	.30	**9.16 5774**	14.67	**10.83 4226**	**40**
21	2025	14.33	5372	.32	6654	14.63	3346	39
22	2885	14.30	5353	.32	7532	14.62	2468	38
23	3743	14.28	5334	.30	8409	14.58	1591	37
24	4600	14.23	5316	.32	.16 9284	14.55	.83 0716	36
25	**9.16 5454**	14.22	**9.99 5297**	.32	**9.17 0157**	14.53	**10.82 9843**	**35**
26	6307	14.20	5278	.30	1029	14.50	8971	34
27	7159	14.15	5260	.32	1899	14.47	8101	33
28	8008	14.13	5241	.32	2767	14.45	7233	32
29	8856	14.10	5222	.32	3634	14.42	6366	31
30	**9.16 9702**	14.08	**9.99 5203**	.32	**9.17 4499**	14.38	**10.82 5501**	**30**
31	.17 0547	14.03	5184	.32	5362	14.37	4638	29
32	1389	14.02	5165	.32	6224	14.33	3776	28
33	2230	14.00	5146	.32	7084	14.30	2916	27
34	3070	13.97	5127	.32	7942	14.28	2058	26
35	**9.17 3908**	13.93	**9.99 5108**	.32	**9.17 8799**	14.27	**10.82 1201**	**25**
36	4744	13.90	5089	.32	.17 9655	14.22	.82 0345	24
37	5578	13.88	5070	.32	.18 0508	14.20	.81 9492	23
38	6411	13.85	5051	.32	1360	14.18	8640	22
39	7242	13.83	5032	.32	2211	14.13	7789	21
40	**9.17 8072**	13.80	**9.99 5013**	.33	**9.18 3059**	14.13	**10.81 6941**	**20**
41	8900	13.77	4993	.32	3907	14.08	6093	19
42	.17 9726	13.75	4974	.32	4752	14.08	5248	18
43	.18 0551	13.72	4955	.33	5597	14.03	4403	17
44	1374	13.70	4935	.32	6439	14.02	3561	16
45	**9.18 2196**	13.67	**9.99 4916**	.33	**9.18 7280**	14.00	**10.81 2720**	**15**
46	3016	13.63	4896	.32	8120	13.97	1880	14
47	3834	13.62	4877	.33	8958	13.93	1042	13
48	4651	13.58	4857	.32	.18 9794	13.92	.81 0206	12
49	5466	13.57	4838	.33	.19 0629	13.88	.80 9371	11
50	**9.18 6280**	13.53	**9.99 4818**	.33	**9.19 1462**	13.87	**10.80 8538**	**10**
51	7092	13.52	4798	.32	2294	13.83	7706	9
52	7903	13.48	4779	.33	3124	13.82	6876	8
53	8712	13.45	4759	.33	3953	13.78	6047	7
54	.18 9519	13.43	4739	.32	4780	13.77	5220	6
55	**9.19 0325**	13.42	**9.99 4720**	.33	**9.19 5606**	13.73	**10.80 4394**	**5**
56	1130	13.38	4700	.33	6430	13.72	3570	4
57	1933	13.35	4680	.33	7253	13.68	2747	3
58	2734	13.33	4660	.33	8074	13.67	1926	2
59	3534	13.30	4640	.33	8894	13.65	1106	1
60	**9.19 4332**		**9.99 4620**		**9.19 9713**		**10.80 0287**	**0**
′	Cosine.	D.1″.	Sine.	D.1″.	Cotang.	D.1″.	Tang.	′

98° **81°**

9° TABLE XV.—*Continued* 170°

′	Sine.	D.1″.	Cosine.	D.1″.	Tang.	D.1″.	Cotang.	′
0	9.19 4332	13.28	9.99 4620	.33	9.19 9713	13.60	10.80 0287	60
1	5129	13.27	4600	.33	.20 0529	13.60	.79 9471	59
2	5925	13.23	4580	.33	1345	13.57	8655	58
3	6719	13.20	4560	.33	2159	13.53	7841	57
4	7511	13.18	4540	.35	2971	13.52	7029	56
5	9.19 8302	13.15	9.99 4519	.33	9.20 3782	13.50	10.79 6218	55
6	9091	13.13	4499	.33	4592	13.47	5408	54
7	.19 9879	13.12	4479	.33	5400	13.45	4600	53
8	.20 0666	13.08	4459	.35	6207	13.43	3793	52
9	1451	13.05	4438	.33	7013	13.40	2987	51
10	9.20 2234	13.05	9.99 4418	.33	9.20 7817	13.37	10.79 2183	50
11	3017	13.00	4398	.35	8619	13.35	1381	49
12	3797	13.00	4377	.33	.20 9420	13.33	.79 0580	48
13	4577	12.95	4357	.35	.21 0220	13.30	.78 9780	47
14	5354	12.95	4336	.33	1018	13.28	8982	46
15	9.20 6131	12.92	9.99 4316	.35	9.21 1815	13.27	10.78 8185	45
16	6906	12.88	4295	.35	2611	13.23	7389	44
17	7679	12.88	4274	.33	3405	13.22	6595	43
18	8452	12.83	4254	.35	4198	13.18	5802	42
19	9222	12.83	4233	.35	4989	13.18	5011	41
20	9.20 9992	12.80	9.99 4212	.35	9.21 5780	13.13	10.78 4220	40
21	.21 0760	12.77	4191	.33	6568	13.13	3432	39
22	1526	12.75	4171	.35	7356	13.10	2644	38
23	2291	12.73	4150	.35	8142	13.07	1858	37
24	3055	12.72	4129	.35	8926	13.07	1074	36
25	9.21 3818	12.68	9.99 4108	.35	9.21 9710	13.03	10.78 0290	35
26	4579	12.65	4087	.35	.22 0492	13.00	.77 9508	34
27	5338	12.65	4066	.35	1272	13.00	8728	33
28	6097	12.62	4045	.35	2052	12.97	7948	32
29	6854	12.58	4024	.35	2830	12.95	7170	31
30	9.21 7609	12.57	9.99 4003	.35	9.22 3607	12.92	10.77 6393	30
31	8363	12.55	3982	.37	4382	12.90	5618	29
32	9116	12.53	3960	.35	5156	12.88	4844	28
33	.21 9868	12.50	3939	.35	5929	12.85	4071	27
34	.22 0618	12.48	3918	.35	6700	12.85	3300	26
35	9.22 1367	12.47	9.99 3897	.37	9.22 7471	12.80	10.77 2529	25
36	2115	12.43	3875	.35	8239	12.80	1761	24
37	2861	12.42	3854	.37	9007	12.77	0993	23
38	3606	12.38	3832	.35	.22 9773	12.77	.77 0227	22
39	4349	12.38	3811	.37	.23 0539	12.72	.76 9461	21
40	9.22 5092	12.35	9.99 3789	.35	9.23 1302	12.72	10.76 8698	20
41	5833	12.33	3768	.37	2065	12.68	7935	19
42	6573	12.30	3746	.35	2826	12.67	7174	18
43	7311	12.28	3725	.37	3586	12.65	6414	17
44	8048	12.27	3703	.37	4345	12.63	5655	16
45	9.22 8784	12.23	9.99 3681	.35	9.23 5103	12.60	10.76 4897	15
46	.22 9518	12.23	3660	.37	5859	12.58	4141	14
47	.23 0252	12.20	3638	.37	6614	12.57	3386	13
48	0984	12.18	3616	.37	7368	12.53	2632	12
49	1715	12.15	3594	.37	8120	12.53	1880	11
50	9.23 2444	12.13	9.99 3572	.37	9.23 8872	12.50	10.76 1128	10
51	3172	12.12	3550	.37	.23 9622	12.48	.76 0378	9
52	3899	12.10	3528	.37	.24 0371	12.45	.75 9629	8
53	4625	12.07	3506	.37	1118	12.45	8882	7
54	5349	12.07	3484	.37	1865	12.42	8135	6
55	9.23 6073	12.03	9.99 3462	.37	9.24 2610	12.40	10.75 7390	5
56	6795	12.00	3440	.37	3354	12.38	6646	4
57	7515	12.00	3418	.37	4097	12.37	5903	3
58	8235	11.97	3396	.37	4839	12.33	5161	2
59	8953	11.95	3374	.38	5579	12.33	4421	1
60	9.23 9670		9.99 3351		9.24 6319		10.75 3681	0
′	Cosine.	D.1″.	Sine.	D.1″.	Cotang.	D.1″.	Tang.	′

99° 80°

10° TABLE XV.—*Continued* 169°

′	Sine.	D.1″.	Cosine.	D.1″.	Tang.	D.1″.	Cotang.	′
0	**9.23 9670**	11.93	**9.99 3351**	.37	**9.24 6319**	12.30	**10.75 3681**	**60**
1	.24 0386	11.92	3329	.37	7057	12.28	2943	59
2	1101	11.88	3307	.38	7794	12.27	2206	58
3	1814	11.87	3284	.37	8530	12.23	1470	57
4	2526	11.85	3262	.37	9264	12.23	0736	56
5	**9.24 3237**	11.83	**9.99 3240**	.38	**9.24 9998**	12.20	**10.75 0002**	**55**
6	3947	11.82	3217	.37	.25 0730	12.18	.74 9270	54
7	4656	11.78	3195	.38	1461	12.17	8539	53
8	5363	11.77	3172	.38	2191	12.15	7809	52
9	6069	11.77	3149	.37	2920	12.13	7080	51
10	**9.24 6775**	11.72	**9.99 3127**	.38	**9.25 3648**	12.10	**10.74 6352**	**50**
11	7478	11.72	3104	.38	4374	12.10	5626	49
12	8181	11.70	3081	.37	5100	12.07	4900	48
13	8883	11.67	3059	.38	5824	12.05	4176	47
14	.24 9583	11.65	3036	.38	6547	12.03	3453	46
15	**9.25 0282**	11.63	**9.99 3013**	.38	**9.25 7269**	12.02	**10.74 2731**	**45**
16	0980	11.62	2990	.38	7990	12.00	2010	44
17	1677	11.60	2967	.38	8710	11.98	1290	43
18	2373	11.57	2944	.38	.25 9429	11.95	.74 0571	42
19	3067	11.57	2921	.38	.26 0146	11.95	.73 9854	41
20	**9.25 3761**	11.53	**9.99 2898**	.38	**9.26 0863**	11.92	**10.73 9137**	**40**
21	4453	11.52	2875	.38	1578	11.90	8422	39
22	5144	11.50	2852	.38	2292	11.88	7708	38
23	5834	11.48	2829	.38	3005	11.87	6995	37
24	6523	11.47	2806	.38	3717	11.85	6283	36
25	**9.25 7211**	11.45	**9.99 2783**	.40	**9.26 4428**	11.83	**10.73 5572**	**35**
26	7898	11.42	2759	.38	5138	11.82	4862	34
27	8583	11.42	2736	.38	5847	11.80	4153	33
28	9268	11.38	2713	.38	6555	11.77	3445	32
29	.25 9951	11.37	2690	.40	7261	11.77	2739	31
30	**9.26 0633**	11.35	**9.99 2666**	.38	**9.26 7967**	11.73	**10.73 2033**	**30**
31	1314	11.33	2643	.40	8671	11.73	1329	29
32	1994	11.32	2619	.38	.26 9375	11.70	.73 0625	28
33	2673	11.30	2596	.40	.27 0077	11.70	.72 9923	27
34	3351	11.27	2572	.38	0779	11.67	9221	26
35	**9.26 4027**	11.27	**9.99 2549**	.40	**9.27 1479**	11.65	**10.72 8521**	**25**
36	4703	11.23	2525	.40	2178	11.63	7822	24
37	5377	11.23	2501	.38	2876	11.62	7124	23
38	6051	11.20	2478	.40	3573	11.60	6427	22
39	6723	11.20	2454	.40	4269	11.58	5731	21
40	**9.26 7395**	11.17	**9.99 2430**	.40	**9.27 4964**	11.57	**10.72 5036**	**20**
41	8065	11.15	2406	.40	5658	11.55	4342	19
42	8734	11.13	2382	.38	6351	11.53	3649	18
43	.26 9402	11.12	2359	.40	7043	11.52	2957	17
44	.27 0069	11.10	2335	.40	7734	11.50	2266	16
45	**9.27 0735**	11.08	**9.99 2311**	.40	**9.27 8424**	11.48	**10.72 1576**	**15**
46	1400	11.07	2287	.40	9113	11.47	0887	14
47	2064	11.03	2263	.40	.27 9801	11.45	.72 0199	13
48	2726	11.03	2239	.42	.28 0488	11.43	.71 9512	12
49	3388	11.02	2214	.40	1174	11.40	8826	11
50	**9.27 4049**	10.98	**9.99 2190**	.40	**9.28 1858**	11.40	**10.71 8142**	**10**
51	4708	10.98	2166	.40	2542	11.38	7458	9
52	5367	10.97	2142	.40	3225	11.37	6775	8
53	6025	10.93	2118	.42	3907	11.35	6093	7
54	6681	10.93	2093	.40	4588	11.33	5412	6
55	**9.27 7337**	10.90	**9.99 2069**	.42	**9.28 5268**	11.32	**10.71 4732**	**5**
56	7991	10.90	2044	.40	5947	11.28	4053	4
57	8645	10.87	2020	.40	6624	11.28	3376	3
58	9297	10.85	1996	.42	7301	11.27	2699	2
59	.27 9948	10.85	1971	.40	7977	11.25	2023	1
60	**9.28 0599**		**9.99 1947**		**9.28 8652**		**10.71 1348**	**0**
′	Cosine.	D.1″.	Sine.	D.1″.	Cotang.	D.1″.	Tang.	′

100° **79°**

11° TABLE XV.—*Continued* 168°

′	Sine.	D.1″.	Cosine.	D.1″.	Tang.	D.1″.	Cotang.	′
0	**9.28 0599**	10.82	**9.99 1947**	.42	**9.28 8652**	11.23	**10.71 1348**	**60**
1	1248	10.82	1922	.42	9326	11.22	0674	59
2	1897	10.78	1897	.40	.28 9999	11.20	.71 0001	58
3	2544	10.77	1873	.42	.29 0671	11.18	.70 9329	57
4	3190	10.77	1848	.42	1342	11.18	8658	56
5	**9.28 3836**	10.73	**9.99 1823**	.40	**9.29 2013**	11.15	**10.70 7987**	**55**
6	4480	10.73	1799	.42	2682	11.13	7318	54
7	5124	10.70	1774	.42	3350	11.12	6650	53
8	5766	10.70	1749	.42	4017	11.12	5983	52
9	6408	10.67	1724	.42	4684	11.08	5316	51
10	**9.28 7048**	10.67	**9.99 1699**	.42	**9.29 5349**	11.07	**10.70 4651**	**50**
11	7688	10.63	1674	.42	6013	11.07	3987	49
12	8326	10.63	1649	.42	6677	11.03	3323	48
13	8964	10.60	1624	.42	7339	11.03	2661	47
14	.28 9600	10.60	1599	.42	8001	11.02	1999	46
15	**9.29 0236**	10.57	**9.99 1574**	.42	**9.29 8662**	11.00	**10.70 1338**	**45**
16	0870	10.57	1549	.42	9322	10.97	0678	44
17	1504	10.55	1524	.43	.29 9980	10.97	.70 0020	43
18	2137	10.52	1498	.42	.30 0638	10.95	.69 9362	42
19	2768	10.52	1473	.42	1295	10.93	8705	41
20	**9.29 3399**	10.50	**9.99 1448**	.43	**9.30 1951**	10.93	**10.69 8049**	**40**
21	4029	10.48	1422	.42	2607	10.90	7393	39
22	4658	10.47	1397	.42	3261	10.88	6739	38
23	5286	10.45	1372	.43	3914	10.88	6086	37
24	5913	10.43	1346	.42	4567	10.85	5433	36
25	**9.29 6539**	10.42	**9.99 1321**	.43	**9.30 5218**	10.85	**10.69 4782**	**35**
26	7164	10.40	1295	.42	5869	10.83	4131	34
27	7788	10.40	1270	.43	6519	10.82	3481	33
28	8412	10.37	1244	.43	7168	10.80	2832	32
29	9034	10.35	1218	.42	7816	10.78	2184	31
30	**9.29 9655**	10.35	**9.99 1193**	.43	**9.30 8463**	10.77	**10.69 1537**	**30**
31	.30 0276	10.32	1167	.43	9109	10.75	0891	29
32	0895	10.32	1141	.43	.30 9754	10.75	.69 0246	28
33	1514	10.30	1115	.42	.31 0399	10.72	.68 9601	27
34	2132	10.27	1090	.43	1042	10.72	8958	26
35	**9.30 2748**	10.27	**9.99 1064**	.43	**9.31 1685**	10.70	**10.68 8315**	**25**
36	3364	10.25	1038	.43	2327	10.68	7673	24
37	3979	10.23	1012	.43	2968	10.67	7032	23
38	4593	10.23	0986	.43	3608	10.65	6392	22
39	5207	10.20	0960	.43	4247	10.63	5753	21
40	**9.30 5819**	10.18	**9.99 0934**	.43	**9.31 4885**	10.63	**10.68 5115**	**20**
41	6430	10.18	0908	.43	5523	10.60	4477	19
42	7041	10.15	0882	.45	6159	10.60	3841	18
43	7650	10.15	0855	.43	6795	10.58	3205	17
44	8259	10.13	0829	.43	7430	10.57	2570	16
45	**9.30 8867**	10.12	**9.99 0803**	.43	**9.31 8064**	10.55	**10.68 1936**	**15**
46	.30 9474	10.10	0777	.45	8697	10.55	1303	14
47	.31 0080	10.08	0750	.43	9330	10.52	0670	13
48	0685	10.07	0724	.45	.31 9961	10.52	.68 0039	12
49	1289	10.07	0697	.43	.32 0592	10.50	.67 9408	11
50	**9.31 1893**	10.03	**9.99 0671**	.43	**9.32 1222**	10.48	**10.67 8778**	**10**
51	2495	10.03	0645	.45	1851	10.47	8149	9
52	3097	10.02	0618	.45	2479	10.45	7521	8
53	3698	9.98	0591	.43	3106	10.45	6894	7
54	4297	10.00	0565	.45	3733	10.42	6267	6
55	**9.31 4897**	9.97	**9.99 0538**	.45	**9.32 4358**	10.42	**10.67 5642**	**5**
56	5495	9.95	0511	.43	4983	10.40	5017	4
57	6092	9.95	0485	.45	5607	10.40	4393	3
58	6689	9.92	0458	.45	6231	10.37	3769	2
59	7284	9.92	0431	.45	6853	10.37	3147	1
60	**9.31 7879**		**9.99 0404**		**9.32 7475**		**10.67 2525**	**0**
′	Cosine.	D.1″.	Sine.	D.1″.	Cotang.	D.1″.	Tang.	′

101° 78°

12° **Table XV.—*Continued*** 167°

′	Sine.	D.1″.	Cosine.	D.1″.	Tang.	D.1″.	Cotang.	′
0	**9.31 7879**	9.90	**9.99 0404**	.43	**9.32 7475**	10.33	**10.67 2525**	**60**
1	8473	9.88	0378	.45	8095	10.33	1905	59
2	9066	9.87	0351	.45	8715	10.32	1285	58
3	.31 9658	9.85	0324	.45	9334	10.32	0666	57
4	.32 0249	9.85	0297	.45	.32 9953	10.28	.67 0047	56
5	**9.32 0840**	9.83	**9.99 0270**	.45	**9.33 0570**	10.28	**10.66 9430**	**55**
6	1430	9.82	0243	.47	1187	10.27	8813	54
7	2019	9.80	0215	.45	1803	10.25	8197	53
8	2607	9.78	0188	.45	2418	10.25	7582	52
9	3194	9.77	0161	.45	3033	10.22	6967	51
10	**9.32 3780**	9.77	**9.99 0134**	.45	**9.33 3646**	10.22	**10.66 6354**	**50**
11	4366	9.73	0107	.47	4259	10.20	5741	49
12	4950	9.73	0079	.45	4871	10.18	5129	48
13	5534	9.72	0052	.45	5482	10.18	4518	47
14	6117	9.72	.99 0025	.47	6093	10.15	3907	46
15	**9.32 6700**	9.68	**9.98 9997**	.45	**9.33 6702**	10.15	**10.66 3298**	**45**
16	7281	9.68	9970	.47	7311	10.13	2689	44
17	7862	9.67	9942	.45	7919	10.13	2081	43
18	8442	9.65	9915	.47	8527	10.10	1473	42
19	9021	9.63	9887	.45	9133	10.10	0867	41
20	**9.32 9599**	9.62	**9.98 9860**	.47	**9.33 9739**	10.08	**10.66 0261**	**40**
21	.33 0176	9.62	9832	.47	.34 0344	10.07	.65 9656	39
22	0753	9.60	9804	.45	0948	10.07	9052	38
23	1329	9.57	9777	.47	1552	10.05	8448	37
24	1903	9.58	9749	.47	2155	10.03	7845	36
25	**9.33 2478**	9.55	**9.98 9721**	.47	**9.34 2757**	10.02	**10.65 7243**	**35**
26	3051	9.55	9693	.47	3358	10.00	6642	34
27	3624	9.52	9665	.47	3958	10.00	6042	33
28	4195	9.53	9637	.45	4558	9.98	5442	32
29	4767	9.50	9610	.47	5157	9.97	4843	31
30	**9.33 5337**	9.48	**9.98 9582**	.48	**9.34 5755**	9.97	**10.65 4245**	**30**
31	5906	9.48	9553	.47	6353	9.93	3647	29
32	6475	9.47	9525	.47	6949	9.93	3051	28
33	7043	9.45	9497	.47	7545	9.93	2455	27
34	7610	9.43	9469	.47	8141	9.90	1859	26
35	**9.33 8176**	9.43	**9.98 9441**	.47	**9.34 8735**	9.90	**10.65 1265**	**25**
36	8742	9.42	9413	.47	9329	9.88	0671	24
37	9307	9.40	9385	.48	.34 9922	9.87	.65 0078	23
38	.33 9871	9.38	9356	.47	.35 0514	9.87	.64 9486	22
39	.34 0434	9.37	9328	.47	1106	9.85	8894	21
40	**9.34 0996**	9.37	**9.98 9300**	.48	**9.35 1697**	9.83	**10.64 8303**	**20**
41	1558	9.35	9271	.47	2287	9.82	7713	19
42	2119	9.33	9243	.48	2876	9.82	7124	18
43	2679	9.33	9214	.47	3465	9.80	6535	17
44	3239	9.30	9186	.48	4053	9.78	5947	16
45	**9.34 3797**	9.30	**9.98 9157**	.48	**9.35 4640**	9.78	**10.64 5360**	**15**
46	4355	9.28	9128	.47	5227	9.77	4773	14
47	4912	9.28	9100	.48	5813	9.75	4187	13
48	5469	9.25	9071	.48	6398	9.73	3602	12
49	6024	9.25	9042	.47	6982	9.73	3018	11
50	**9.34 6579**	9.25	**9.98 9014**	.48	**9.35 7566**	9.72	**10.64 2434**	**10**
51	7134	9.22	8985	.48	8149	9.70	1851	9
52	7687	9.22	8956	.48	8731	9.70	1269	8
53	8240	9.20	8927	.48	9313	9.67	0687	7
54	8792	9.18	8898	.48	.35 9893	9.68	.64 0107	6
55	**9.34 9343**	9.17	**9.98 8869**	.48	**9.36 0474**	9.65	**10.63 9526**	**5**
56	.34 9893	9.17	8840	.48	1053	9.65	8947	4
57	.35 0443	9.15	8811	.48	1632	9.63	8368	3
58	0992	9.13	8782	.48	2210	9.62	7790	2
59	1540	9.13	8753	.48	2787	9.62	7213	1
60	**9.35 2088**		**9.98 8724**		**9.36 3364**		**10.63 6636**	**0**
′	Cosine.	D.1″.	Sine.	D.1″.	Cotang.	D.1″.	Tang.	′

102° **77°**

13° TABLE XV.—*Continued* 166°

′	Sine.	D.1″.	Cosine.	D.1″.	Tang.	D.1″.	Cotang.	′
0	9.35 2088	9.12	9.98 8724	.48	9.36 3364	9.60	10.63 6636	60
1	2635	9.10	8695	.48	3940	9.58	6060	59
2	3181	9.08	8666	.50	4515	9.58	5485	58
3	3726	9.08	8636	.48	5090	9.57	4910	57
4	4271	9.07	8607	.48	5664	9.55	4336	56
5	9.35 4815	9.05	9.98 8578	.50	9.36 6237	9.55	10.63 3763	55
6	5358	9.05	8548	.48	6810	9.53	3190	54
7	5901	9.03	8519	.50	7382	9.52	2618	53
8	6443	9.02	8489	.48	7953	9.52	2047	52
9	6984	9.00	8460	.50	8524	9.50	1476	51
10	9.35 7524	9.00	9.98 8430	.48	9.36 9094	9.48	10.63 0906	50
11	8064	8.98	8401	.50	.36 9663	9.48	.63 0337	49
12	8603	8.97	8371	.48	.37 0232	9.45	.62 9768	48
13	9141	8.95	8342	.50	0799	9.47	9201	47
14	.35 9678	8.95	8312	.50	1367	9.43	8633	46
15	9.36 0215	8.95	9.98 8282	.50	9.37 1933	9.43	10.62 8067	45
16	0752	8.92	8252	.48	2499	9.42	7501	44
17	1287	8.92	8223	.50	3064	9.42	6936	43
18	1822	8.90	8193	.50	3629	9.40	6371	42
19	2356	8.88	8163	.50	4193	9.38	5807	41
20	9.36 2889	8.88	9.98 8133	.50	9.37 4756	9.38	10.62 5244	40
21	3422	8.87	8103	.50	5319	9.37	4681	39
22	3954	8.85	8073	.50	5881	9.35	4119	38
23	4485	8.85	8043	.50	6442	9.35	3558	37
24	5016	8.83	8013	.50	7003	9.33	2997	36
25	9.36 5546	8.82	9.98 7983	.50	9.37 7563	9.32	10.62 2437	35
26	6075	8.82	7953	.52	8122	9.32	1878	34
27	6604	8.78	7922	.50	8681	9.30	1319	33
28	7131	8.80	7892	.50	9239	9.30	0761	32
29	7659	8.77	7862	.50	.37 9797	9.28	.62 0203	31
30	9.36 8185	8.77	9.98 7832	.52	9.38 0354	9.27	10.61 9646	30
31	8711	8.75	7801	.50	0910	9.27	9090	29
32	9236	8.75	7771	.52	1466	9.23	8534	28
33	.36 9761	8.72	7740	.50	2020	9.25	7980	27
34	.37 0285	8.72	7710	.52	2575	9.23	7425	26
35	9.37 0808	8.70	9.98 7679	.50	9.38 3129	9.22	10.61 6871	25
36	1330	8.70	7649	.52	3682	9.20	6318	24
37	1852	8.68	7618	.50	4234	9.20	5766	23
38	2373	8.68	7588	.52	4786	9.18	5214	22
39	2894	8.67	7557	.52	5337	9.18	4663	21
40	9.37 3414	8.65	9.98 7526	.50	9.38 5888	9.17	10.61 4112	20
41	3933	8.65	7496	.52	6438	9.15	3562	19
42	4452	8.63	7465	.52	6987	9.15	3013	18
43	4970	8.62	7434	.52	7536	9.13	2464	17
44	5487	8.60	7403	.52	8084	9.12	1916	16
45	9.37 6003	8.60	9.98 7372	.52	9.38 8631	9.12	10.61 1369	15
46	6519	8.60	7341	.52	9178	9.10	0822	14
47	7035	8.57	7310	.52	.38 9724	9.10	.61 0276	13
48	7549	8.57	7279	.52	.39 0270	9.08	.60 9730	12
49	8063	8.57	7248	.52	0815	9.08	9185	11
50	9.37 8577	8.53	9.98 7217	.52	9.39 1360	9.05	10.60 8640	10
51	9089	8.53	7186	.52	1903	9.07	8097	9
52	.37 9601	8.53	7155	.52	2447	9.03	7553	8
53	.38 0113	8.52	7124	.53	2989	9.03	7011	7
54	0624	8.50	7092	.52	3531	9.03	6469	6
55	9.38 1134	8.48	9.98 7061	.52	9.39 4073	9.02	10.60 5927	5
56	1643	8.48	7030	.53	4614	9.00	5386	4
57	2152	8.48	6998	.52	5154	9.00	4846	3
58	2661	8.45	6967	.52	5694	8.98	4306	2
59	3168	8.45	6936	.53	6233	8.97	3767	1
60	9.38 3675		9.98 6904		9.39 6771		10.60 3229	0
′	Cosine.	D.1″.	Sine.	D.1″.	Cotang.	D.1″.	Tang.	′

103° 76°

14° TABLE XV.—*Continued* 165°

′	Sine.	D.1″.	Cosine.	D.1″.	Tang.	D.1″.	Cotang.	′
0	**9.38 3675**	8.45	**9.98 6904**	.52	**9.39 6771**	8.97	**10.60 3229**	**60**
1	4182	8.42	6873	.53	7309	8.95	2691	59
2	4687	8.42	6841	.53	7846	8.95	2154	58
3	5192	8.42	6809	.52	8383	8.93	1617	57
4	5697	8.40	6778	.53	8919	8.93	1081	56
5	**9.38 6201**	8.38	**9.98 6746**	.53	**9.39 9455**	8.92	**10.60 0545**	**55**
6	6704	8.38	6714	.52	.39 9990	8.90	.60 0010	54
7	7207	8.37	6683	.53	.40 0524	8.90	.59 9476	53
8	7709	8.35	6651	.53	1058	8.88	8942	52
9	8210	8.35	6619	.53	1591	8.88	8409	51
10	**9.38 8711**	8.33	**9.98 6587**	.53	**9.40 2124**	8.87	**10.59 7876**	**50**
11	9211	8.33	6555	.53	2656	8.85	7344	49
12	.38 9711	8.32	6523	.53	3187	8.85	6813	48
13	.39 0210	8.30	6491	.53	3718	8.85	6282	47
14	0708	8.30	6459	.53	4249	8.82	5751	46
15	**9.39 1206**	8.28	**9.98 6427**	.53	**9.40 4778**	8.83	**10.59 5222**	**45**
16	1703	8.27	6395	.53	5308	8.80	4692	44
17	2199	8.27	6363	.53	5836	8.80	4164	43
18	2695	8.27	6331	.53	6364	8.80	3636	42
19	3191	8.23	6299	.55	6892	8.78	3108	41
20	**9.39 3685**	8.23	**9.98 6266**	.53	**9.40 7419**	8.77	**10.59 2581**	**40**
21	4179	8.23	6234	.53	7945	8.77	2055	39
22	4673	8.22	6202	.55	8471	8.75	1529	38
23	5166	8.20	6169	.53	8996	8.75	1004	37
24	5658	8.20	6137	.55	.40 9521	8.73	.59 0479	36
25	**9.39 6150**	8.18	**9.98 6104**	.53	**9.41 0045**	8.73	**10.58 9955**	**35**
26	6641	8.18	6072	.55	0569	8.72	9431	34
27	7132	8.15	6039	.53	1092	8.72	8908	33
28	7621	8.17	6007	.55	1615	8.70	8385	32
29	8111	8.15	5974	.53	2137	8.68	7863	31
30	**9.39 8600**	8.13	**9.98 5942**	.55	**9.41 2658**	8.68	**10.58 7342**	**30**
31	9088	8.12	5909	.55	3179	8.67	6821	29
32	.39 9575	8.12	5876	.55	3699	8.67	6301	28
33	.40 0062	8.12	5843	.53	4219	8.65	5781	27
34	0549	8.10	5811	.55	4738	8.65	5262	26
35	**9.40 1035**	8.08	**9.98 5778**	.55	**9.41 5257**	8.63	**10.58 4743**	**25**
36	1520	8.08	5745	.55	5775	8.63	4225	24
37	2005	8.07	5712	.55	6293	8.62	3707	23
38	2489	8.05	5679	.55	6810	8.60	3190	22
39	2972	8.05	5646	.55	7326	8.60	2674	21
40	**9.40 3455**	8.05	**9.98 5613**	.55	**9.41 7842**	8.60	**10.58 2158**	**20**
41	3938	8.03	5580	.55	8358	8.58	1642	19
42	4420	8.02	5547	.55	8873	8.57	1127	18
43	4901	8.02	5514	.57	9387	8.57	0613	17
44	5382	8.00	5480	.55	.41 9901	8.57	.58 0099	16
45	**9.40 5862**	7.98	**9.98 5447**	.55	**9.42 0415**	8.55	**10.57 9585**	**15**
46	6341	7.98	5414	.55	0927	8.55	9073	14
47	6820	7.98	5381	.57	1440	8.53	8560	13
48	7299	7.97	5347	.55	1952	8.52	8048	12
49	7777	7.95	5314	.57	2463	8.52	7537	11
50	**9.40 8254**	7.95	**9.98 5280**	.55	**9.42 2974**	8.50	**10.57 7026**	**10**
51	8731	7.93	5247	.57	3484	8.48	6516	9
52	9207	7.93	5213	.55	3993	8.50	6007	8
53	.40 9682	7.92	5180	.57	4503	8.47	5497	7
54	.41 0157	7.92	5146	.55	5011	8.47	4989	6
55	**9.41 0632**	7.90	**9.98 5113**	.57	**9.42 5519**	8.47	**10.57 4481**	**5**
56	1106	7.88	5079	.57	6027	8.45	3973	4
57	1579	7.88	5045	.57	6534	8.45	3466	3
58	2052	7.87	5011	.55	7041	8.43	2959	2
59	2524	7.87	4978	.57	7547	8.42	2453	1
60	**9.41 2996**		**9.98 4944**		**9.42 8052**		**10.57 1948**	**0**
′	Cosine.	D.1″.	Sine.	D.1″.	Cotang.	D.1″.	Tang.	′

104° 75°

15° **Table XV.**—*Continued* 164°

′	Sine.	D.1″.	Cosine.	D.1″.	Tang.	D.1″.	Cotang.	′
0	**9.41 2996**	7.85	**9.98 4944**	.57	**9.42 8052**	8.43	**10.57 1948**	**60**
1	3467	7.85	4910	.57	8558	8.40	1442	59
2	3938	7.83	4876	.57	9062	8.40	0938	58
3	4408	7.83	4842	.57	.42 9566	8.40	.57 0434	57
4	4878	7.82	4808	.57	.43 0070	8.38	.56 9930	56
5	**9.41 5347**	7.80	**9.98 4774**	.57	**9.43 0573**	8.37	**10.56 9427**	**55**
6	5815	7.80	4740	.57	1075	8.37	8925	54
7	6283	7.80	4706	.57	1577	8.37	8423	53
8	6751	7.77	4672	.57	2079	8.35	7921	52
9	7217	7.78	4638	.58	2580	8.33	7420	51
10	**9.41 7684**	7.77	**9.98 4603**	.57	**9.43 3080**	8.33	**10.56 6920**	**50**
11	8150	7.75	4569	.57	3580	8.33	6420	49
12	8615	7.73	4535	.58	4080	8.32	5920	48
13	9079	7.75	4500	.57	4579	8.32	5421	47
14	.41 9544	7.72	4466	.57	5078	8.30	4922	46
15	**9.42 0007**	7.72	**9.98 4432**	.58	**9.43 5576**	8.28	**10.56 4424**	**45**
16	0470	7.72	4397	.57	6073	8.28	3927	44
17	0933	7.70	4363	.58	6570	8.28	3430	43
18	1395	7.70	4328	.57	7067	8.27	2933	42
19	1857	7.68	4294	.58	7563	8.27	2437	41
20	**9.42 2318**	7.67	**9.98 4259**	.58	**9.43 8059**	8.25	**10.56 1941**	**40**
21	2778	7.67	4224	.57	8554	8.23	1446	39
22	3238	7.65	4190	.58	9048	8.25	0952	38
23	3697	7.65	4155	.58	.43 9543	8.22	.56 0457	37
24	4156	7.65	4120	.58	.44 0036	8.22	.55 9964	36
25	**9.42 4615**	7.63	**9.98 4085**	.58	**9.44 0529**	8.22	**10.55 9471**	**35**
26	5073	7.62	4050	.58	1022	8.20	8978	34
27	5530	7.62	4015	.57	1514	8.20	8486	33
28	5987	7.60	3981	.58	2006	8.18	7994	32
29	6443	7.60	3946	.58	2497	8.18	7503	31
30	**9.42 6899**	7.58	**9.98 3911**	.60	**9.44 2988**	8.18	**10.55 7012**	**30**
31	7354	7.58	3875	.58	3479	8.15	6521	29
32	7809	7.57	3840	.58	3968	8.17	6032	28
33	8263	7.57	3805	.58	4458	8.15	5542	27
34	8717	7.55	3770	.58	4947	8.13	5053	26
35	**9.42 9170**	7.55	**9.98 3735**	.58	**9.44 5435**	8.13	**10.55 4565**	**25**
36	.42 9623	7.53	3700	.60	5923	8.13	4077	24
37	.43 0075	7.53	3664	.58	6411	8.12	3589	23
38	0527	7.52	3629	.58	6898	8.10	3102	22
39	0978	7.52	3594	.60	7384	8.10	2616	21
40	**9.43 1429**	7.50	**9.98 3558**	.58	**9.44 7870**	8.10	**10.55 2130**	**20**
41	1879	7.50	3523	.60	8356	8.08	1644	19
42	2329	7.48	3487	.58	8841	8.08	1159	18
43	2778	7.47	3452	.60	9326	8.07	0674	17
44	3226	7.48	3416	.58	.44 9810	8.07	.55 0190	16
45	**9.43 3675**	7.45	**9.98 3381**	.60	**9.45 0294**	8.05	**10.54 9706**	**15**
46	4122	7.45	3345	.60	0777	8.05	9223	14
47	4569	7.45	3309	.60	1260	8.05	8740	13
48	5016	7.43	3273	.58	1743	8.03	8257	12
49	5462	7.43	3238	.60	2225	8.02	7775	11
50	**9.43 5908**	7.42	**9.98 3202**	.60	**9.45 2706**	8.02	**10.54 7294**	**10**
51	6353	7.42	3166	.60	3187	8.02	6813	9
52	6798	7.40	3130	.60	3668	8.00	6332	8
53	7242	7.40	3094	.60	4148	8.00	5852	7
54	7686	7.38	3058	.60	4628	7.98	5372	6
55	**9.43 8129**	7.38	**9.98 3022**	.60	**9.45 5107**	7.98	**10.54 4893**	**5**
56	8572	7.37	2986	.60	5586	7.97	4414	4
57	9014	7.37	2950	.60	6064	7.97	3936	3
58	9456	7.35	2914	.60	6542	7.95	3458	2
59	.43 9897	7.35	2878	.60	7019	7.95	2981	1
60	**9.44 0338**		**9.98 2842**		**9.45 7496**		**10.54 2404**	**0**
′	Cosine.	D.1″.	Sine.	D.1″.	Cotang.	D.1″.	Tang.	′

105° **74°**

16° TABLE XV.—*Continued* 163°

′	Sine.	D.1″.	Cosine.	D.1″.	Tang.	D.1″.	Cotang.	′
0	**9.44 0338**	7.33	**9.98 2842**	.62	**9.45 7496**	7.95	**10.54 2504**	**60**
1	0778	7.33	2805	.60	7973	7.93	2027	59
2	1218	7.33	2769	.60	8449	7.93	1551	58
3	1658	7.30	2733	.62	8925	7.92	1075	57
4	2096	7.32	2696	.60	9400	7.92	0600	56
5	**9.44 2535**	7.30	**9.98 2660**	.60	**9.45 9875**	7.90	**10.54 0125**	**55**
6	2973	7.28	2624	.62	.46 0349	7.90	.53 9651	54
7	3410	7.28	2587	.60	0823	7.90	9177	53
8	3847	7.28	2551	.62	1297	7.88	8703	52
9	4284	7.27	2514	.62	1770	7.87	8230	51
10	**9.44 4720**	7.25	**9.98 2477**	.60	**9.46 2242**	7.88	**10.53 7758**	**50**
11	5155	7.25	2441	.62	2715	7.85	7285	49
12	5590	7.25	2404	.62	3186	7.87	6814	48
13	6025	7.23	2367	.60	3658	7.83	6342	47
14	6459	7.23	2331	.62	4128	7.85	5872	46
15	**9.44 6893**	7.22	**9.98 2294**	.62	**9.46 4599**	7.83	**10.53 5401**	**45**
16	7326	7.22	2257	.62	5069	7.83	4931	44
17	7759	7.20	2220	.62	5539	7.82	4461	43
18	8191	7.20	2183	.62	6008	7.82	3992	42
19	8623	7.18	2146	.62	6477	7.80	3523	41
20	**9.44 9054**	7.18	**9.98 2109**	.62	**9.46 6945**	7.80	**10.53 3055**	**40**
21	9485	7.17	2072	.62	7413	7.78	2587	39
22	.44 9915	7.17	2035	.62	7880	7.78	2120	38
23	.45 0345	7.17	1998	.62	8347	7.78	1653	37
24	0775	7.15	1961	.62	8814	7.77	1186	36
25	**9.45 1204**	7.13	**9.98 1924**	.63	**9.46 9280**	7.77	**10.53 0720**	**35**
26	1632	7.13	1886	.62	.46 9746	7.75	.53 0254	34
27	2060	7.13	1849	.62	.47 0211	7.75	.52 9789	33
28	2488	7.12	1812	.63	0676	7.75	9324	32
29	2915	7.12	1774	.62	1141	7.73	8859	31
30	**9.45 3342**	7.10	**9.98 1737**	.62	**9.47 1605**	7.73	**10.52 8395**	**30**
31	3768	7.10	1700	.63	2069	7.72	7931	29
32	4194	7.08	1662	.62	2532	7.72	7468	28
33	4619	7.08	1625	.63	2995	7.70	7005	27
34	5044	7.08	1587	.63	3457	7.70	6543	26
35	**9.45 5469**	7.07	**9.98 1549**	.62	**9.47 3919**	7.70	**10.52 6081**	**25**
36	5893	7.05	1512	.63	4381	7.68	5619	24
37	6316	7.05	1474	.63	4842	7.68	5158	23
38	6739	7.05	1436	.62	5303	7.67	4697	22
39	7162	7.03	1399	.63	5763	7.67	4237	21
40	**9.45 7584**	7.03	**9.98 1361**	.63	**9.47 6223**	7.67	**10.52 3777**	**20**
41	8006	7.02	1323	.63	6683	7.65	3317	19
42	8427	7.02	1285	.63	7142	7.65	2858	18
43	8848	7.00	1247	.63	7601	7.63	2399	17
44	9268	7.00	1209	.63	8059	7.63	1941	16
45	**9.45 9688**	7.00	**9.98 1171**	.63	**9.47 8517**	7.63	**10.52 1483**	**15**
46	.46 0108	6.98	1133	.63	8975	7.62	1025	14
47	0527	6.98	1095	.63	9432	7.62	0568	13
48	0946	6.97	1057	.63	.47 9889	7.60	.52 0111	12
49	1364	6.97	1019	.63	.48 0345	7.60	.51 9655	11
50	**9.46 1782**	6.95	**9.98 0981**	.65	**9.48 0801**	7.60	**10.51 9199**	**10**
51	2199	6.95	0942	.63	1257	7.58	8743	9
52	2616	6.93	0904	.63	1712	7.58	8288	8
53	3032	6.93	0866	.65	2167	7.57	7833	7
54	3448	6.93	0827	.63	2621	7.57	7379	6
55	**9.46 3864**	6.92	**9.98 0789**	.65	**9.48 3075**	7.57	**10.51 6925**	**5**
56	4279	6.92	0750	.63	3529	7.55	6471	4
57	4694	6.90	0712	.65	3982	7.55	6018	3
58	5108	6.90	0673	.63	4435	7.53	5565	2
59	5522	6.88	0635	.65	4887	7.53	5113	1
60	**9.46 5935**		**9.98 0596**		**9.48 5339**		**10.51 4661**	**0**
′	Cosine.	D.1″.	Sine.	D.1″.	Cotang.	D.1″.	Tang.	′

106° **73°**

17° TABLE XV.—*Continued* 162°

′	Sine.	D.1″.	Cosine.	D.1″.	Tang.	D.1″.	Cotang.	′
0	9.46 5935	6.88	9.98 0596	.63	9.48 5339	7.53	10.51 4661	60
1	6348	6.88	0558	.65	5791	7.52	4209	59
2	6761	6.87	0519	.65	6242	7.52	3758	58
3	7173	6.87	0480	.63	6693	7.50	3307	57
4	7585	6.85	0442	.65	7143	7.50	2857	56
5	9.46 7996	6.85	9.98 0403	.65	9.48 7593	7.50	10.51 2407	55
6	8407	6.83	0364	.65	8043	7.48	1957	54
7	8817	6.83	0325	.65	8492	7.48	1508	53
8	9227	6.83	0286	.65	8941	7.48	1059	52
9	.46 9637	6.82	0247	.65	9390	7.47	0610	51
10	9.47 0046	6.82	9.98 0208	.65	9.48 9838	7.47	10.51 0162	50
11	0455	6.80	0169	.65	.49 0286	7.45	.50 9714	49
12	0863	6.80	0130	.65	0733	7.45	9267	48
13	1271	6.80	0091	.65	1180	7.45	8820	47
14	1679	6.78	0052	.67	1627	7.43	8373	46
15	9.47 2086	6.77	9.98 0012	.65	9.49 2073	7.43	10.50 7927	45
16	2492	6.77	.97 9973	.65	2519	7.43	7481	44
17	2898	6.77	9934	.65	2965	7.42	7035	43
18	3304	6.77	9895	.67	3410	7.40	6590	42
19	3710	6.75	9855	.65	3854	7.42	6146	41
20	9.47 4115	6.73	9.97 9816	.67	9.49 4299	7.40	10.50 5701	40
21	4519	6.73	9776	.65	4743	7.38	5257	39
22	4923	6.73	9737	.67	5186	7.40	4814	38
23	5327	6.72	9697	.65	5630	7.38	4370	37
24	5730	6.72	9658	.67	6073	7.37	3927	36
25	9.47 6133	6.72	9.97 9618	.65	9.49 6515	7.37	10.50 3485	35
26	6536	6.70	9579	.67	6957	7.37	3043	34
27	6938	6.70	9539	.67	7399	7.37	2601	33
28	7340	6.68	9499	.67	7841	7.35	2159	32
29	7741	6.68	9459	.65	8282	7.33	1718	31
30	9.47 8142	6.67	9.97 9420	.67	9.49 8722	7.35	10.50 1278	30
31	8542	6.67	9380	.67	9163	7.33	0837	29
32	8942	6.67	9340	.67	.49 9603	7.32	.50 0397	28
33	9342	6.65	9300	.67	.50 0042	7.32	.49 9958	27
34	.47 9741	6.65	9260	.67	0481	7.32	9519	26
35	9.48 0140	6.65	9.97 9220	.67	9.50 0920	7.32	10.49 9080	25
36	0539	6.63	9180	.67	1359	7.30	8641	24
37	0937	6.62	9140	.67	1797	7.30	8203	23
38	1334	6.62	9100	.68	2235	7.28	7765	22
39	1731	6.62	9059	.67	2672	7.28	7328	21
40	9.48 2128	6.62	9.97 9019	.67	9.50 3109	7.28	10.49 6891	20
41	2525	6.60	8979	.67	3546	7.27	6454	19
42	2921	6.58	8939	.68	3982	7.27	6018	18
43	3316	6.60	8898	.67	4418	7.27	5582	17
44	3712	6.58	8858	.68	4854	7.25	5146	16
45	9.48 4107	6.57	9.97 8817	.67	9.50 5289	7.25	10.49 4711	15
46	4501	6.57	8777	.67	5724	7.25	4276	14
47	4895	6.57	8737	.68	6159	7.23	3841	13
48	5289	6.55	8696	.68	6593	7.23	3407	12
49	5682	6.55	8655	.67	7027	7.23	2973	11
50	9.48 6075	6.53	9.97 8615	.68	9.50 7460	7.22	10.49 2540	10
51	6467	6.55	8574	.68	7893	7.22	2107	9
52	6860	6.52	8533	.67	8326	7.22	1674	8
53	7251	6.53	8493	.68	8759	7.20	1241	7
54	7643	6.52	8452	.68	9191	7.18	0809	6
55	9.48 8034	6.50	9.97 8411	.68	9.50 9622	7.20	10.49 0378	5
56	8424	6.50	8370	.68	.51 0054	7.18	.48 9946	4
57	8814	6.50	8329	.68	0485	7.18	9515	3
58	9204	6.48	8288	.68	0916	7.17	9084	2
59	9593	6.48	8247	.68	1346	7.17	8654	1
60	9.48 9982		9.97 8206		9.51 1776		10.48 8224	0
′	Cosine.	D.1″.	Sine.	D.1″.	Cotang.	D.1″.	Tang.	′

107° 72°

18° TABLE XV.—*Continued* 161°

′	Sine.	D.1″.	Cosine.	D.1″.	Tang.	D.1″.	Cotang.	′
0	**9.48 9982**	6.48	**9.97 8206**	.68	**9.51 1776**	7.17	**10.48 8224**	**60**
1	.49 0371	6.47	8165	.68	2206	7.15	7794	59
2	0759	6.47	8124	.68	2635	7.15	7365	58
3	1147	6.47	8083	.68	3064	7.15	6936	57
4	1535	6.45	8042	.68	3493	7.13	6507	56
5	**9.49 1922**	6.43	**9.97 8001**	.70	**9.51 3921**	7.13	**10.48 6079**	**55**
6	2308	6.45	7959	.68	4349	7.13	5651	54
7	2695	6.43	7918	.68	4777	7.12	5223	53
8	3081	6.42	7377	.70	5204	7.12	4796	52
9	3466	6.42	7835	.68	5631	7.10	4369	51
10	**9.49 3851**	6.42	**9.97 7794**	.70	**9.51 6057**	7.12	**10.48 3934**	**50**
11	4236	6.42	7752	.68	6484	7.10	3516	49
12	4621	6.40	7711	.70	6910	7.08	3090	48
13	5005	6.38	7669	.68	7335	7.10	2665	47
14	5388	6.40	7628	.70	7761	7.08	2239	46
15	**9.49 5772**	6.37	**9.97 7586**	.70	**9.51 8186**	7.07	**10.48 1814**	**45**
16	6154	6.38	7544	.68	8610	7.07	1390	44
17	6537	6.37	7503	.70	9034	7.07	0966	43
18	6919	6.37	7461	.70	9458	7.07	0542	42
19	7301	6.35	7419	.70	.51 9882	7.05	.48 0118	41
20	**9.49 7682**	6.35	**9.97 7377**	.70	**9.52 0305**	7.05	**10.47 9695**	**40**
21	8064	6.33	7335	.70	0728	7.05	9272	39
22	8444	6.35	7293	.70	1151	7.03	8849	38
23	8825	6.32	7251	.70	1573	7.03	8427	37
24	9204	6.33	7209	.70	1995	7.03	8005	36
25	**9.49 9584**	6.32	**9.97 7167**	.70	**9.52 2417**	7.02	**10.47 7583**	**35**
26	.49 9963	6.32	7125	.70	2838	7.02	7162	34
27	.50 0342	6.32	7083	.70	3259	7.02	6741	33
28	0721	6.30	7041	.70	3680	7.00	6320	32
29	1099	6.28	6999	.70	4100	7.00	5900	31
30	**9.50 1476**	6.30	**9.97 6957**	.72	**9.52 4520**	7.00	**10.47 5480**	**30**
31	1854	6.28	6914	.70	4940	6.98	5060	29
32	2231	6.27	6872	.70	5359	6.98	4641	28
33	2607	6.28	6830	.72	5778	6.98	4222	27
34	2984	6.27	6787	.70	6197	6.97	3803	26
35	**9.50 3360**	6.25	**9.97 6745**	.72	**9.52 6615**	6.97	**10.47 3385**	**25**
36	3735	6.25	6702	.70	7033	6.97	2967	24
37	4110	6.25	6660	.72	7451	6.95	2549	23
38	4485	6.25	6617	.72	7868	6.95	2132	22
39	4860	6.23	6574	.70	8285	6.95	1715	21
40	**9.50 5234**	6.23	**9.97 6532**	.72	**9.52 8702**	6.95	**10.47 1298**	**20**
41	5608	6.22	6489	.72	9119	6.93	0881	19
42	5981	6.22	6446	.70	9535	6.93	0465	18
43	6354	6.22	6404	.72	.52 9951	6.92	.47 0049	17
44	6727	6.20	6361	.72	.53 0366	6.92	.46 9634	16
45	**9.50 7099**	6.20	**9.97 6318**	.72	**9.53 0781**	6.92	**10.46 9219**	**15**
46	7471	6.20	6275	.72	1196	6.92	8804	14
47	7843	6.18	6232	.72	1611	6.90	8389	13
48	8214	6.18	6189	.72	2025	6.90	7975	12
49	8585	6.18	6146	.72	2439	6.90	7561	11
50	**9.50 8956**	6.17	**9.97 6103**	.72	**9.53 2853**	6.88	**10.46 7147**	**10**
51	9326	6.17	6060	.72	3266	6.88	6734	9
52	.50 9696	6.15	6017	.72	3679	6.88	6321	8
53	.51 0065	6.15	5974	.73	4092	6.87	5908	7
54	0434	6.15	5930	.72	4504	6.87	5496	6
55	**9.51 0803**	6.15	**9.97 5887**	.72	**9.53 4916**	6.87	**10.46 5084**	**5**
56	1172	6.13	5844	.73	5328	6.85	4672	4
57	1540	6.12	5800	.72	5739	6.85	4261	3
58	1907	6.13	5757	.72	6150	6.85	3850	2
59	2275	6.12	5714	.73	6561	6.85	3439	1
60	**9.51 2642**		**9.97 5670**		**9.53 6972**		**10.46 3028**	**0**
′	Cosine.	D.1″.	Sine.	D.1″.	Cotang.	D.1″.	Tang.	′

108° 71°

19° **TABLE XV.**—*Continued* 160°

′	Sine.	D.1″.	Cosine.	D.1″.	Tang.	D.1″.	Cotang.	′
0	**9.51 2642**	6.12	**9.97 5670**	.72	**9.53 6972**	6.83	**10.46 3028**	**60**
1	3009	6.10	5627	.73	7382	6.83	2618	59
2	3375	6.10	5583	.73	7792	6.83	2208	58
3	3741	6.10	5539	.72	8202	6.82	1798	57
4	4107	6.08	5496	.73	8611	6.82	1389	56
5	**9.51 4472**	6.08	**9.97 5452**	.73	**9.53 9020**	6.82	**10.46 0980**	**55**
6	4837	6.08	5408	.72	9429	6.80	0571	54
7	5202	6.07	5365	.73	.53 9837	6.80	.46 0163	53
8	5566	6.07	5321	.73	.54 0245	6.80	.45 9755	52
9	5930	6.07	5277	.73	0653	6.80	9347	51
10	**9.51 6294**	6.05	**9.97 5233**	.73	**9.54 1061**	6.78	**10 45 8939**	**50**
11	6657	6.05	5189	.73	1468	6.78	8532	49
12	7020	6.03	5145	.73	1875	6.77	8125	48
13	7382	6.05	5101	.73	2281	6.78	7719	47
14	7745	6.03	5057	.73	2688	6.77	7312	46
15	**9.51 8107**	6.02	**9.97 5013**	.73	**9.54 3094**	6.75	**10.45 6906**	**45**
16	8468	6.02	4969	.73	3499	6.77	6501	44
17	8829	6.02	4925	.75	3905	6.75	6095	43
18	9190	6.02	4880	.73	4310	6.75	5690	42
19	9551	6.00	4836	.73	4715	6.73	5285	41
20	**9.51 9911**	6.00	**9.97 4792**	.73	**9.54 5119**	6.75	**10.45 4881**	**40**
21	.52 0271	6.00	4748	.75	5524	6.73	4476	39
22	0631	5.98	4703	.73	5928	6.72	4072	38
23	0990	5.98	4659	.75	6331	6.73	3669	37
24	1349	5.97	4614	.73	6735	6.72	3265	36
25	**9.52 1707**	5.98	**9.97 4570**	.75	**9.54 7138**	6.70	**10.45 2862**	**35**
26	2066	5.97	4525	.73	7540	6.72	2460	34
27	2424	5.95	4481	.75	7943	6.70	2057	33
28	2781	5.95	4436	.75	8345	6.70	1655	32
29	3138	5.95	4391	.73	8747	6.70	1253	31
30	**9.52 3495**	5.95	**9.97 4347**	.75	**9.54 9149**	6.68	**10.45 0851**	**30**
31	3852	5.93	4302	.75	9550	6.68	0450	29
32	4208	5.93	4257	.75	.54 9951	6.68	.45 0049	28
33	4564	5.93	4212	.75	.55 0352	6.67	.44 9648	27
34	4920	5.92	4167	.75	0752	6.68	9248	26
35	**9.52 5275**	5.92	**9.97 4122**	.75	**9.55 1153**	6.65	**10.44 8847**	**25**
36	5630	5.90	4077	.75	1552	6.67	8448	24
37	5984	5.92	4032	.75	1952	6.65	8048	23
38	6339	5.90	3987	.75	2351	6.65	7649	22
39	6693	5.88	3942	.75	2750	6.65	7250	21
40	**9.52 7046**	5.90	**9.97 3897**	.75	**9.55 3149**	6.65	**10.44 6851**	**20**
41	7400	5.88	3852	.75	3548	6.63	6452	19
42	7753	5.87	3807	.77	3946	6.63	6054	18
43	8105	5.88	3761	.75	4344	6.62	5656	17
44	8458	5.87	3716	.75	4741	6.63	5259	16
45	**9.52 8810**	5.85	**9.97 3671**	.77	**9.55 5139**	6.62	**10.44 4861**	**15**
46	9161	5.87	3625	.75	5536	6.62	4464	14
47	9513	5.85	3580	.75	5933	6.60	4067	13
48	.52 9864	5.85	3535	.77	6329	6.60	3671	12
49	.53 0215	5.83	3489	.75	6725	6.60	3275	11
50	**9.53 0565**	5.83	**9.97 3444**	.77	**9.55 7121**	6.60	**10.44 2879**	**10**
51	0915	5.83	3398	.77	7517	6.60	2483	9
52	1265	5.82	3352	.75	7913	6.58	2087	8
53	1614	5.82	3307	.77	8308	6.58	1692	7
54	1963	5.82	3261	.77	8703	6.57	1297	6
55	**9.53 2312**	5.82	**9.97 3215**	.77	**9.55 9097**	6.57	**10.44 0903**	**5**
56	2661	5.80	3169	.75	9491	6.57	0509	4
57	3009	5.80	3124	.77	.55 9885	6.57	.44 0115	3
58	3357	5.78	3078	.77	.56 0279	6.57	.43 9721	2
59	3704	5.80	3032	.77	0673	6.55	9327	1
60	**9.53 4052**		**9.97 2986**		**9.56 1066**		**10.43 8934**	**0**
′	Cosine.	D.1″.	Sine.	D.1″.	Cotang.	D.1″.	Tang.	′

109° **70°**

20° TABLE XV.—*Continued* 159°

′	Sine.	D.1″.	Cosine.	D.1″.	Tang.	D.1″.	Cotang.	′
0	**9.53 4052**	5.78	**9.97 2986**	.77	**9.56 1066**	6.55	**10.43 8934**	**60**
1	4399	5.77	2940	.77	1459	6.53	8541	59
2	4745	5.78	2894	.77	1851	6.55	8149	58
3	5092	5.77	2848	.77	2244	6.53	7756	57
4	5438	5.75	2802	.78	2636	6.53	7364	56
5	**9.53 5783**	5.77	**9.97 2755**	.77	**9.56 3028**	6.52	**10.43 6972**	**55**
6	6129	5.75	2709	.77	3419	6.53	6581	54
7	6474	5.73	2663	.77	3811	6.52	6189	53
8	6818	5.75	2617	.78	4202	6.52	5798	52
9	7163	5.73	2570	.77	4593	6.50	5407	51
10	**9.53 7507**	5.73	**9.97 2524**	.77	**9.56 4983**	6.50	**10.43 5017**	**50**
11	7851	5.72	2478	.78	5373	6.50	4627	49
12	8194	5.73	2431	.77	5763	6.50	4237	48
13	8538	5.70	2385	.78	6153	6.48	3847	47
14	8880	5.72	2338	.78	6542	6.50	3458	46
15	**9.53 9223**	5.70	**9.97 2291**	.77	**9.56 6932**	6.47	**10.43 3068**	**45**
16	9565	5.70	2245	.78	7320	6.48	2680	44
17	.53 9907	5.68	2198	.78	7709	6.48	2291	43
18	.54 0249	5.68	2151	.77	8098	6.47	1902	42
19	0590	5.68	2105	.78	8486	6.45	1514	41
20	**9.54 0931**	5.68	**9.97 2058**	.78	**9.56 8873**	6.47	**10.43 1127**	**40**
21	1272	5.68	2011	.78	9261	6.45	0739	39
22	1613	5.67	1964	.78	.56 9648	6.45	.43 0352	38
23	1953	5.67	1917	.78	.57 0035	6.45	.42 9965	37
24	2293	5.65	1870	.78	0422	6.45	9578	36
25	**9.54 2632**	5.65	**9.97 1823**	.78	**9.57 0809**	6.43	**10.42 9191**	**35**
26	2971	5.65	1776	.78	1195	6.43	8805	34
27	3310	5.65	1729	.78	1581	6.43	8419	33
28	3649	5.63	1682	.78	1967	6.42	8033	32
29	3987	5.63	1635	.78	2352	6.43	7648	31
30	**9.54 4325**	5.63	**9.97 1588**	.80	**9.57 2738**	6.42	**10.42 7262**	**30**
31	4663	5.62	1540	.78	3123	6.40	6877	29
32	5000	5.63	1493	.78	3507	6.42	6493	28
33	5338	5.60	1446	.80	3892	6.40	6108	27
34	5674	5.62	1398	.78	4276	6.40	5724	26
35	**9.54 6011**	5.60	**9.97 1351**	.80	**9.57 4660**	6.40	**10.42 5340**	**25**
36	6347	5.60	1303	.78	5044	6.38	4956	24
37	6683	5.60	1256	.80	5427	6.38	4573	23
38	7019	5.58	1208	.78	5810	6.38	4190	22
39	7354	5.58	1161	.80	6193	6.38	3807	21
40	**9.54 7689**	5.58	**9.97 1113**	.78	**9.57 6576**	6.38	**10.42 3424**	**20**
41	8024	5.58	1066	.80	6959	6.37	3041	19
42	8359	5.57	1018	.80	7341	6.37	2659	18
43	8693	5.57	0970	.80	7723	6.35	2277	17
44	9027	5.55	0922	.80	8104	6.37	1896	16
45	**9.54 9360**	5.55	**9.97 0874**	.78	**9.57 8486**	6.35	**10.42 1514**	**15**
46	.54 9693	5.55	0827	.80	8867	6.35	1133	14
47	.55 0026	5.55	0779	.80	9248	6.35	0752	13
48	0359	5.55	0731	.80	.57 9629	6.33	.42 0371	12
49	0692	5.53	0683	.80	.58 0009	6.33	.41 9991	11
50	**9.55 1024**	5.53	**9.97 0635**	.82	**9.58 0389**	6.33	**10.41 9611**	**10**
51	1356	5.52	0586	.80	0769	6.33	9231	9
52	1687	5.52	0538	.80	1149	6.32	8851	8
53	2018	5.52	0490	.80	1528	6.32	8472	7
54	2349	5.52	0442	.80	1907	6.32	8093	6
55	**9.55 2680**	5.50	**9.97 0394**	.82	**9.58 2286**	6.32	**10.41 7714**	**5**
56	3010	5.52	0345	.80	2665	6.32	7335	4
57	3341	5.48	0297	.80	3044	6.30	6956	3
58	3670	5.50	0249	.82	3422	6.30	6578	2
59	4000	5.48	0200	.80	3800	6.28	6200	1
60	**9.55 4329**		**9.97 0152**		**9.58 4177**		**10.41 5823**	**0**
′	Cosine.	D.1″.	Sine.	D.1″.	Cotang.	D.1″.	Tang.	′

110° 69°

21° TABLE XV.—*Continued* 158°

′	Sine.	D.1″.	Cosine.	D.1″.	Tang.	D.1″.	Cotang.	′
0	**9.55 4329**	5.48	**9.97 0152**	.82	**9.58 4177**	6.30	**10.41 5823**	**60**
1	4658	5.48	0103	.80	4555	6.28	5445	59
2	4987	5.47	0055	.82	4932	6.28	5068	58
3	5315	5.47	.97 0006	.82	5309	6.28	4691	57
4	5643	5.47	.96 9957	.80	5686	6.27	4314	56
5	**9.55 5971**	5.47	**9.96 9909**	.82	**9.58 6062**	6.28	**10.41 3938**	**55**
6	6299	5.45	9860	.82	6439	6.27	3561	54
7	6626	5.45	9811	.82	6815	6.25	3185	53
8	6953	5.45	9762	.80	7190	6.27	2810	52
9	7280	5.43	9714	.82	7566	6.25	2434	51
10	**9.55 7606**	5.43	**9.96 9665**	.82	**9.58 7941**	6.25	**10.41 2059**	**50**
11	7932	5.43	9616	.82	8316	6.25	1684	49
12	8258	5.42	9567	.82	8691	6.25	1309	48
13	8583	5.43	9518	.82	9066	6.23	0934	47
14	8909	5.42	9469	.82	9440	6.23	0560	46
15	**9.55 9234**	5.40	**9.96 9420**	.83	**9.58 9814**	6.23	**10.41 0186**	**45**
16	9558	5.42	9370	.82	.59 0188	6.23	.40 9812	44
17	.55 9883	5.40	9321	.82	0562	6.22	9438	43
18	.56 0207	5.40	9272	.82	0935	6.22	9065	42
19	0531	5.40	9223	.83	1308	6.22	8692	41
20	**9.56 0855**	5.38	**9.96 9173**	.82	**9.59 1681**	6.22	**10.40 8319**	**40**
21	1178	5.38	9124	.82	2054	6.20	7946	39
22	1501	5.38	9075	.83	2426	6.22	7574	38
23	1824	5.37	9025	.82	2799	6.20	7201	37
24	2146	5.37	8976	.83	3171	6.18	6829	36
25	**9.56 2468**	5.37	**9.96 8926**	.82	**9.59 3542**	6.20	**10.40 6458**	**35**
26	2790	5.37	8877	.83	3914	6.18	6086	34
27	3112	5.35	8827	.83	4285	6.18	5715	33
28	3433	5.37	8777	.82	4656	6.18	5344	32
29	3755	5.33	8728	.83	5027	6.18	4973	31
30	**9.56 4075**	5.35	**9.96 8678**	.83	**9.59 5398**	6.17	**10.40 4602**	**30**
31	4396	5.33	8628	.83	5768	6.17	4232	29
32	4716	5.33	8578	.83	6138	6.17	3862	28
33	5036	5.33	8528	.82	6508	6.17	3492	27
34	5356	5.33	8479	.83	6878	6.15	3122	26
35	**9.56 5676**	5.32	**9.96 8429**	.83	**9.59 7247**	6.15	**10.40 2753**	**25**
36	5995	5.32	8379	.83	7616	6.15	2384	24
37	6314	5.30	8329	.85	7985	6.15	2015	23
38	6632	5.32	8278	.83	8354	6.13	1646	22
39	6951	5.30	8228	.83	8722	6.15	1278	21
40	**9.56 7269**	5.30	**9.96 8178**	.83	**9.59 9091**	6.13	**10.40 0909**	**20**
41	7587	5.28	8128	.83	9459	6.13	0541	19
42	7904	5.30	8078	.85	.59 9827	6.12	.40 0173	18
43	8222	5.28	8027	.83	.60 0194	6.13	.39 9806	17
44	8539	5.28	7977	.83	0562	6.12	9438	16
45	**9.56 8856**	5.27	**9.96 7927**	.85	**9.60 0929**	6.12	**10.39 9071**	**15**
46	9172	5.27	7876	.83	1296	6.12	8704	14
47	9488	5.27	7826	.85	1663	6.10	8337	13
48	.56 9804	5.27	7775	.83	2029	6.10	7971	12
49	.57 0120	5.25	7725	.85	2395	6.10	7605	11
50	**9.57 0435**	5.27	**9.96 7674**	.83	**9.60 2761**	6.10	**10.39 7239**	**10**
51	0751	5.25	7624	.85	3127	6.10	6873	9
52	1066	5.23	7573	.85	3493	6.08	6507	8
53	1380	5.25	7522	.85	3858	6.08	6142	7
54	1695	5.23	7471	.83	4223	6.08	5777	6
55	**9.57 2009**	5.23	**9.96 7421**	.85	**9.60 4588**	6.08	**10.39 5412**	**5**
56	2323	5.22	7370	.85	4953	6.07	5047	4
57	2636	5.23	7319	.85	5317	6.08	4683	3
58	2950	5.22	7268	.85	5682	6.07	4318	2
59	3263	5.20	7217	.85	6046	6.07	3954	1
60	**9.57 3575**		**9.96 7166**		**9.60 6410**		**10.39 3590**	**0**
′	Cosine.	D.1″.	Sine.	D.1″.	Cotang.	D.1″.	Tang.	′

111° **68°**

22° TABLE XV.—*Continued* 157°

′	Sine.	D.1″.	Cosine.	D.1″.	Tang.	D.1″.	Cotang.	′
0	**9.57 3575**	5.22	**9.96 7166**	.85	**9.60 6410**	6.05	**10.39 3590**	**60**
1	3888	5.20	7115	.85	6773	6.07	3227	59
2	4200	5.20	7064	.85	7137	6.05	2863	58
3	4512	5.20	7013	.87	7500	6.05	2500	57
4	4824	5.20	6961	.85	7863	6.03	2137	56
5	**9.57 5136**	5.18	**9.96 6910**	.85	**9.60 8225**	6.05	**10.39 1775**	**55**
6	5447	5.18	6859	.85	8588	6.03	1412	54
7	5758	5.18	6808	.87	8950	6.03	1050	53
8	6069	5.17	6756	.85	9312	6.03	0688	52
9	6379	5.17	6705	.87	.60 9674	6.03	.39 0326	51
10	**9.57 6689**	5.17	**9.96 6653**	.85	**9.61 0036**	6.02	**10.38 9964**	**50**
11	6999	5.17	6602	.87	0397	6.03	9603	49
12	7309	5.15	6550	.85	0759	6.02	9241	48
13	7618	5.15	6499	.87	1120	6.00	8880	47
14	7927	5.15	6447	.87	1480	6.02	8520	46
15	**9.57 8236**	5.15	**9.96 6395**	.85	**9.61 1841**	6.00	**10.38 8159**	**45**
16	8545	5.13	6344	.87	2201	6.00	7799	44
17	8853	5.15	6292	.87	2561	6.00	7439	43
18	9162	5.13	6240	.87	2921	6.00	7079	42
19	9470	5.12	6188	.87	3281	6.00	6719	41
20	**9.57 9777**	5.13	**9.96 6136**	.85	**9.61 3641**	5.98	**10.38 6359**	**40**
21	.58 0085	5.12	6085	.87	4000	5.98	6000	39
22	0392	5.12	6033	.87	4359	5.98	5641	38
23	0699	5.10	5981	.87	4718	5.98	5282	37
24	1005	5.12	5929	.88	5077	5.97	4923	36
25	**9.58 1312**	5.10	**9.96 5876**	.87	**9.61 5435**	5.97	**10.38 4565**	**35**
26	1618	5.10	5824	.87	5793	5.97	4207	34
27	1924	5.08	5772	.87	6151	5.97	3849	33
28	2229	5.10	5720	.87	6509	5.97	3491	32
29	2535	5.08	5668	.88	6867	5.95	3133	31
30	**9.58 2840**	5.08	**9.96 5615**	.87	**9.61 7224**	5.97	**10.38 2776**	**30**
31	3145	5.07	5563	.87	7582	5.95	2418	29
32	3449	5.08	5511	.88	7939	5.93	2061	28
33	3754	5.07	5458	.87	8295	5.95	1705	27
34	4058	5.05	5406	.88	8652	5.93	1348	26
35	**9.58 4361**	5.07	**9.96 5353**	.87	**9.61 9008**	5.93	**10.38 0992**	**25**
36	4665	5.05	5301	.88	9364	5.93	0636	24
37	4968	5.07	5248	.88	.61 9720	5.93	.38 0280	23
38	5272	5.03	5195	.87	.62 0076	5.93	.37 9924	22
39	5574	5.05	5143	.88	0432	5.92	9568	21
40	**9.58 5877**	5.03	**9.96 5090**	.88	**9.62 0787**	5.92	**10.37 9213**	**20**
41	6179	5.05	5037	.88	1142	5.92	8858	19
42	6482	5.02	4984	.88	1497	5.92	8503	18
43	6783	5.03	4931	.87	1852	5.92	8148	17
44	7085	5.02	4879	.88	2207	5.90	7793	16
45	**9.58 7386**	5.03	**9.96 4826**	.88	**9.62 2561**	5.90	**10.37 7439**	**15**
46	7688	5.02	4773	.88	2915	5.90	7085	14
47	7989	5.00	4720	.90	3269	5.90	6731	13
48	8289	5.02	4666	.88	3623	5.88	6377	12
49	8590	5.00	4613	.88	3976	5.90	6024	11
50	**9.58 8890**	5.00	**9.96 4560**	.88	**9.62 4330**	5.88	**10.37 5670**	**10**
51	9190	4.98	4507	.88	4683	5.88	5317	9
52	9489	5.00	4454	.90	5036	5.87	4964	8
53	.58 9789	4.98	4400	.88	5388	5.88	4612	7
54	.59 0088	4.98	4347	.88	5741	5.87	4259	6
55	**9.59 0387**	4.98	**9.96 4294**	.90	**9.62 6093**	5.87	**10.37 3907**	**5**
56	0686	4.97	4240	.88	6445	5.87	3555	4
57	0984	4.97	4187	.90	6797	5.87	3203	3
58	1282	4.97	4133	.88	7149	5.87	2851	2
59	1580	4.97	4080	.90	7501	5.85	2499	1
60	**9.59 1878**		**9.96 4026**		**9.62 7852**		**10.37 2148**	**0**
′	Cosine.	D.1″.	Sine.	D.1″.	Cotang.	D.1″.	Tang.	′

112° 67°

23° TABLE XV.—*Continued* 156°

′	Sine.	D.1″.	Cosine.	D.1″.	Tang.	D.1″.	Cotang.	′
0	**9.59 1878**	4.97	**9.96 4026**	.90	**9.62 7825**	5.85	**10.37 2148**	**60**
1	2176	4.95	3972	.88	8203	5.85	1797	59
2	2473	4.95	3919	.90	8554	5.85	1446	58
3	2770	4.95	3865	.90	8905	5.83	1095	57
4	3067	4.93	3811	.90	9255	5.85	0745	56
5	**9.59 3363**	4.93	**9.96 3757**	.88	**9.62 9606**	5.83	**10.37 0394**	**55**
6	3659	4.93	3704	.90	.62 9956	5.83	.37 0044	54
7	3955	4.93	3650	.90	.63 0306	5.83	.36 9694	53
8	4251	4.93	3596	.90	0656	5.82	9344	52
9	4547	4.92	3542	.90	1005	5.83	8995	51
10	**9.59 4842**	4.92	**9.96 3488**	.90	**9.63 1355**	5.82	**10.36 8645**	**50**
11	5137	4.92	3434	.92	1704	5.82	8296	49
12	5432	4.92	3379	.90	2053	5.82	7947	48
13	5727	4.90	3325	.90	2402	5.80	7598	47
14	6021	4.90	3271	.90	2750	5.82	7250	46
15	**9.59 6315**	4.90	**9.96 3217**	.90	**9.63 3099**	5.80	**10.36 6901**	**45**
16	6609	4.90	3163	.92	3447	5.80	6553	44
17	6903	4.88	3108	.90	3795	5.80	6205	43
18	7196	4.90	3054	.92	4143	5.78	5857	42
19	7490	4.88	2999	.90	4490	5.80	5510	41
20	**9.59 7783**	4.87	**9.96 2945**	.92	**9.63 4838**	5.78	**10.36 5162**	**40**
21	8075	4.88	2890	.90	5185	5.78	4815	39
22	8368	4.87	2836	.92	5532	5.78	4468	38
23	8660	4.87	2781	.90	5879	5.78	4121	37
24	8952	4.87	2727	.92	6226	5.77	3774	36
25	**9.59 9244**	4.87	**9.96 2672**	.92	**9.63 6572**	5.78	**10.36 3428**	**35**
26	9536	4.85	2617	.92	6919	5.77	3081	34
27	.59 9827	4.85	2562	.90	7265	5.77	2735	33
28	.60 0118	4.85	2508	.92	7611	5.75	2389	32
29	0409	4.85	2453	.92	7956	5.77	2044	31
30	**9.60 0700**	4.83	**9.96 2398**	.92	**9.63 8302**	5.75	**10.36 1698**	**30**
31	0990	4.83	2343	.92	8647	5.75	1353	29
32	1280	4.83	2288	.92	8992	5.75	1008	28
33	1570	4.83	2233	.92	9337	5.75	0663	27
34	1860	4.83	2178	.92	.63 9682	5.75	.36 0318	26
35	**9.60 2150**	4.82	**9.96 2123**	.93	**9.64 0027**	5.73	**10.35 9973**	**25**
36	2439	4.82	2067	.92	0371	5.75	9629	24
37	2728	4.82	2012	.92	0716	5.73	9284	23
38	3017	4.80	1957	.92	1060	5.73	8940	22
39	3305	4.82	1902	.93	1404	5.72	8596	21
40	**9.60 3594**	4.80	**9.96 1846**	.92	**9.64 1747**	5.73	**10.35 8253**	**20**
41	3882	4.80	1791	.93	2091	5.72	7909	19
42	4170	4.78	1735	.92	2434	5.72	7566	18
43	4457	4.80	1680	.93	2777	5.72	7223	17
44	4745	4.78	1624	.92	3120	5.72	6880	16
45	**9.60 5032**	4.78	**9.96 1569**	.93	**9.64 3463**	5.72	**10.35 6537**	**15**
46	5319	4.78	1513	.92	3806	5.70	6194	14
47	5606	4.77	1458	.93	4148	5.70	5852	13
48	5892	4.78	1402	.93	4490	5.70	5510	12
49	6179	4.77	1346	.93	4832	5.70	5168	11
50	**9.60 6465**	4.77	**9.96 1290**	.92	**9.64 5174**	5.70	**10.35 4826**	**10**
51	6751	4.75	1235	.93	5516	5.68	4484	9
52	7036	4.77	1179	.93	5857	5.70	4143	8
53	7322	4.75	1123	.93	6199	5.68	3801	7
54	7607	4.75	1067	.93	6540	5.68	3460	6
55	**9.60 7892**	4.75	**9.96 1011**	.93	**9.64 6881**	5.68	**10.35 3119**	**5**
56	8177	4.73	0955	.93	7222	5.67	2778	4
57	8461	4.73	0899	.93	7562	5.68	2438	3
58	8745	4.73	0843	.95	7903	5.67	2097	2
59	9029	4.73	0786	.93	8243	5.67	1757	1
60	**9.60 9313**		**9.96 0730**		**9.64 8583**		**10.35 1417**	**0**
′	Cosine.	D.1″.	Sine.	D.1″.	Cotang.	D.1″.	Tang.	′

113° **66°**

24° TABLE XV.—*Continued* 155°

′	Sine.	D.1″.	Cosine.	D.1″.	Tang.	D.1″.	Cotang.	′
0	**9.60 9313**	4.73	**9.96 0730**	.93	**9.64 8583**	5.67	**10.35 1417**	**60**
1	9597	4.72	0674	.93	8923	5.67	1077	59
2	.60 9880	4.73	0618	.95	9263	5.65	0737	58
3	.61 0164	4.72	0561	.93	9602	5.67	0398	57
4	0447	4.70	0505	.95	.64 9942	5.65	.35 0058	56
5	**9.61 0729**	4.72	**9.96 0448**	.93	**9.65 0281**	5.65	**10.34 9719**	**55**
6	1012	4.70	0392	.95	0620	5.65	9380	54
7	1294	4.70	0335	.93	0959	5.63	9041	53
8	1576	4.70	0279	.95	1297	5.65	8703	52
9	1858	4.70	0222	.95	1636	5.63	8364	51
10	**9.61 2140**	4.68	**9.96 0165**	.93	**9.65 1974**	5.63	**10.34 8026**	**50**
11	2421	4.68	0109	.95	2312	5.63	7688	49
12	2702	4.68	.96 0052	.95	2650	5.63	7350	48
13	2983	4.68	.95 9995	.95	2988	5.63	7012	47
14	3264	4.68	9938	.93	3326	5.62	6674	46
15	**9.61 3545**	4.67	**9.95 9882**	.95	**9.65 3663**	5.62	**10.34 6337**	**45**
16	3825	4.67	9825	.95	4000	5.62	6000	44
17	4105	4.67	9768	.95	4337	5.62	5663	43
18	4385	4.67	9711	.95	4674	5.62	5326	42
19	4665	4.65	9654	.97	5011	5.62	4989	41
20	**9.61 4944**	4.65	**9.95 9596**	.95	**9.65 5348**	5.60	**10.34 4652**	**40**
21	5223	4.65	9539	.95	5684	5.60	4316	39
22	5502	4.65	9482	.95	6020	5.60	3980	38
23	5781	4.65	9425	.95	6356	5.60	3644	37
24	6060	4.63	9368	.97	6692	5.60	3308	36
25	**9.61 6338**	4.63	**9.95 9310**	.95	**9.65 7028**	5.60	**10.34 2972**	**35**
26	6616	4.63	9253	.97	7364	5.58	2636	34
27	6894	4.63	9195	.95	7699	5.58	2301	33
28	7172	4.63	9138	.97	8034	5.58	1966	32
29	7450	4.62	9080	.95	8369	5.58	1631	31
30	**9.61 7727**	4.62	**9.95 9023**	.97	**9.65 8704**	5.58	**10.34 1296**	**30**
31	8004	4.62	8965	.95	9039	5.57	0961	29
32	8281	4.62	8908	.97	9373	5.58	0627	28
33	8558	4.60	8850	.97	.65 9708	5.57	.34 0292	27
34	8834	4.60	8792	.97	.66 0042	5.57	.33 9958	26
35	**9.61 9110**	4.60	**9.95 8734**	.95	**9.66 0376**	5.57	**10.33 9624**	**25**
36	9386	4.60	8677	.97	0710	5.55	9290	24
37	9662	4.60	8619	.97	1043	5.57	8957	23
38	.61 9938	4.58	8561	.97	1377	5.55	8623	22
39	.62 0213	4.58	8503	.97	1710	5.55	8290	21
40	**9.62 0488**	4.58	**9.95 8445**	.97	**9.66 2043**	5.55	**10.33 7957**	**20**
41	0763	4.58	8387	.97	2376	5.55	7624	19
42	1038	4.58	8329	.97	2709	5.55	7291	18
43	1313	4.57	8271	.97	3042	5.55	6958	17
44	1587	4.57	8213	.98	3375	5.53	6625	16
45	**9.62 1861**	4.57	**9.95 8154**	.97	**9.66 3707**	5.53	**10.33 6293**	**15**
46	2135	4.57	8096	.97	4039	5.53	5961	14
47	2409	4.55	8038	.98	4371	5.53	5629	13
48	2682	4.57	7979	.97	4703	5.53	5297	12
49	2956	4.55	7921	.97	5035	5.52	4965	11
50	**9.62 3229**	4.55	**9.95 7863**	.98	**9.66 5366**	5.53	**10.33 4634**	**10**
51	3502	4.53	7804	.97	5698	5.52	4302	9
52	3774	4.55	7746	.98	6029	5.52	3971	8
53	4047	4.53	7687	.98	6360	5.52	3640	7
54	4319	4.53	7628	.97	6691	5.50	3309	6
55	**9.62 4591**	4.53	**9.95 7570**	.98	**9.66 7021**	5.52	**10.33 2979**	**5**
56	4863	4.53	7511	.98	7352	5.50	2648	4
57	5135	4.52	7452	.98	7682	5.52	2318	3
58	5406	4.52	7393	.97	8013	5.50	1987	2
59	5677	4.52	7335	.98	8343	5.50	1657	1
60	**9.62 5948**		**9.95 7276**		**9.66 8673**		**10.33 1327**	**0**
′	Cosine.	D.1″.	Sine.	D.1″.	Cotang.	D.1″.	Tang.	′

114° **65°**

25° TABLE XV.—*Continued* 154°

′	Sine.	D.1″.	Cosine.	D.1″.	Tang.	D.1″.	Cotang.	′
0	9.62 5948	4.52	9.95 7276	.98	9.66 8673	5.48	10.33 1327	60
1	6219	4.52	7217	.98	9002	5.50	0998	59
2	6490	4.50	7158	.98	9332	5.48	0668	58
3	6760	4.50	7099	.98	9661	5.50	0339	57
4	7030	4.50	7040	.98	.66 9991	5.48	.33 0009	56
5	9.62 7300	4.50	9.95 6981	1.00	9.67 0320	5.48	10.32 9680	55
6	7570	4.50	6921	.98	0649	5.47	9351	54
7	7840	4.48	6862	.98	0977	5.48	9023	53
8	8109	4.48	6803	.98	1306	5.48	8694	52
9	8378	4.48	6744	1.00	1635	5.47	8365	51
10	9.62 8647	4.48	9.95 6684	.98	9.67 1963	5.47	10.32 8037	50
11	8916	4.48	6625	.98	2291	5.47	7709	49
12	9185	4.47	6566	1.00	2619	5.47	7381	48
13	9453	4.47	6506	.98	2947	5.45	7053	47
14	9721	4.47	6447	1.00	3274	5.47	6726	46
15	9.62 9989	4.47	9.95 6387	1.00	9.67 3602	5.45	10.32 6398	45
16	.63 0257	4.45	6327	.98	3929	5.47	6071	44
17	0524	4.47	6268	1.00	4257	5.45	5743	43
18	0792	4.45	6208	1.00	4584	5.45	5416	42
19	1059	4.45	6148	.98	4911	5.43	5089	41
20	9.63 1326	4.45	9.95 6089	1.00	9.67 5237	5.45	10.32 4763	40
21	1593	4.43	6029	1.00	5564	5.43	4436	39
22	1859	4.43	5969	1.00	5890	5.45	4110	38
23	2125	4.45	5909	1.00	6217	5.43	3783	37
24	2392	4.43	5849	1.00	6543	5.43	3457	36
25	9.63 2658	4.42	9.95 5789	1.00	9.67 6869	5.42	10.32 3131	35
26	2923	4.43	5729	1.00	7194	5.43	2806	34
27	3189	4.42	5669	1.00	7520	5.43	2480	33
28	3454	4.42	5609	.98	7846	5.42	2154	32
29	3719	4.42	5548	1.00	8171	5.42	1829	31
30	9.63 3984	4.42	9.95 5488	1.00	9.67 8496	5.42	10.32 1504	30
31	4249	4.42	5428	1.00	8821	5.42	1179	29
32	4514	4.40	5368	1.02	9146	5.42	0854	28
33	4778	4.40	5307	1.00	9471	5.40	0529	27
34	5042	4.40	5247	1.02	.67 9795	5.42	.32 0205	26
35	9.63 5306	4.40	9.95 5186	1.00	9.68 0120	5.40	10.31 9880	25
36	5570	4.40	5126	1.02	0444	5.40	9556	24
37	5834	4.38	5065	1.00	0768	5.40	9232	23
38	6097	4.38	5005	1.02	1092	5.40	8908	22
39	6360	4.38	4944	1.02	1416	5.40	8584	21
40	9.63 6623	4.38	9.95 4883	1.00	9.68 1740	5.38	10.31 8260	20
41	6886	4.37	4823	1.02	2063	5.40	7937	19
42	7148	4.38	4762	1.02	2387	5.38	7613	18
43	7411	4.37	4701	1.02	2710	5.38	7290	17
44	7673	4.37	4640	1.02	3033	5.38	6967	16
45	9.63 7935	4.37	9.95 4579	1.02	9.68 3356	5.38	10.31 6644	15
46	8197	4.35	4518	1.02	3679	5.37	6321	14
47	8458	4.37	4457	1.02	4001	5.38	5999	13
48	8720	4.35	4396	1.02	4324	5.37	5676	12
49	8981	4.35	4335	1.02	4646	5.37	5354	11
50	9.63 9242	4.35	9.95 4274	1.02	9.68 4968	5.37	10.31 5032	10
51	9503	4.35	4213	1.02	5290	5.37	4710	9
52	.63 9764	4.33	4152	1.03	5612	5.37	4388	8
53	.64 0024	4.33	4090	1.02	5934	5.35	4066	7
54	0284	4.33	4029	1.02	6255	5.37	3745	6
55	9.64 0544	4.33	9.95 3968	1.03	9.68 6577	5.35	10.31 3423	5
56	0804	4.33	3906	1.02	6898	5.35	3102	4
57	1064	4.33	3845	1.03	7219	5.35	2781	3
58	1324	4.32	3783	1.02	7540	5.35	2460	2
59	1583	4.32	3722	1.03	7861	5.35	2139	1
60	9.64 1842		9.95 3660		9.68 8182		10.31 1818	0
′	Cosine.	D.1″.	Sine.	D.1″.	Cotang.	D.1″.	Tang.	′

115° 64°

26° TABLE XV.—*Continued* 153°

′	Sine.	D.1″.	Cosine.	D.1″.	Tang.	D.1″.	Cotang.	′
0	**9.64 1842**	4.32	**9.95 3660**	1.02	**9.68 8182**	5.33	**10.31 1818**	**60**
1	2101	4.32	3599	1.03	8502	5.32	1498	59
2	2360	4.30	3537	1.03	8823	5.33	1177	58
3	2618	4.32	3475	1.03	9143	5.33	0857	57
4	2877	4.30	3413	1.02	9463	5.33	0537	56
5	**9.64 3135**	4.30	**9.95 3352**	1.03	**9.68 9783**	5.33	**10.31 0217**	**55**
6	3393	4.28	3290	1.03	.69 0103	5.33	.30 9897	54
7	3650	4.30	3228	1.03	0423	5.32	9577	53
8	3908	4.28	3166	1.03	0742	5.33	9258	52
9	4165	4.30	3104	1.03	1062	5.32	8938	51
10	**9.64 4423**	4.28	**9.95 3042**	1.03	**9.69 1381**	5.32	**10.30 8619**	**50**
11	4680	4.27	2980	1.03	1700	5.32	8300	49
12	4936	4.28	2918	1.05	2019	5.32	7981	48
13	5193	4.28	2855	1.03	2338	5.30	7662	47
14	5450	4.27	2793	1.03	2656	5.32	7344	46
15	**9.64 5706**	4.27	**9.95 2731**	1.03	**9.69 2975**	5.30	**10.30 7025**	**45**
16	5962	4.27	2669	1.05	3293	5.32	6707	44
17	6218	4.27	2606	1.03	3612	5.30	6388	43
18	6474	4.25	2544	1.05	3930	5.30	6070	42
19	6729	4.25	2481	1.03	4228	5.30	5752	41
20	**9.64 6984**	4.27	**9.95 2419**	1.05	**9.69 4566**	5.28	**10.30 5434**	**40**
21	7240	4.23	2356	1.03	4883	5.30	5117	39
22	7494	4.25	2294	1.05	5201	5.28	4799	38
23	7749	4.25	2231	1.05	5518	5.30	4482	37
24	8004	4.23	2168	1.03	5836	5.28	4164	36
25	**9.64 8258**	4.23	**9.95 2106**	1.05	**9.69 6153**	5.28	**10.30 3847**	**35**
26	8512	4.23	2043	1.05	6470	5.28	3530	34
27	8766	4.23	1980	1.05	6787	5.27	3213	33
28	9020	4.23	1917	1.05	7103	5.28	2897	32
29	9274	4.22	1854	1.05	7420	5.27	2580	31
30	**9.64 9527**	4.23	**9.95 1791**	1.05	**9.69 7736**	5.28	**10.30 2264**	**30**
31	.64 9781	4.22	1728	1.05	8053	5.27	1947	29
32	.65 0034	4.22	1665	1.05	8369	5.27	1631	28
33	0287	4.20	1602	1.05	8685	5.27	1315	27
34	0539	4.22	1539	1.05	9001	5.25	0999	26
35	**9.65 0792**	4.20	**9.95 1476**	1.07	**9.69 9316**	5.27	**10.30 0684**	**25**
36	1044	4.22	1412	1.05	9632	5.25	0368	24
37	1297	4.20	1349	1.05	.69 9947	5.27	.30 0053	23
38	1549	4.18	1286	1.07	.70 0263	5.25	.29 9737	22
39	1800	4.20	1222	1.05	0578	5.25	9422	21
40	**9.65 2052**	4.20	**9.95 1159**	1.05	**9.70 0893**	5.25	**10.29 9107**	**20**
41	2304	4.18	1096	1.07	1208	5.25	8792	19
42	2555	4.18	1032	1.07	1523	5.23	8477	18
43	2806	4.18	0968	1.05	1837	5.25	8163	17
44	3057	4.18	0905	1.07	2152	5.23	7848	16
45	**9.65 3308**	4.17	**9.95 0841**	1.05	**9.70 2466**	5.25	**10.29 7534**	**15**
46	3558	4.17	0778	1.07	2781	5.23	7219	14
47	3808	4.18	0714	1.07	3095	5.23	6905	13
48	4059	4.17	0650	1.07	3409	5.22	6591	12
49	4309	4.15	0586	1.07	3722	5.23	6278	11
50	**9.65 4558**	4.17	**9.95 0522**	1.07	**9.70 4036**	5.23	**10.29 5964**	**10**
51	4808	4.17	0458	1.07	4350	5.22	5650	9
52	5058	4.15	0394	1.07	4663	5.22	5337	8
53	5307	4.15	0330	1.07	4976	5.23	5024	7
54	5556	4.15	0266	1.07	5290	5.22	4710	6
55	**9.65 5805**	4.15	**9.95 0202**	1.07	**9.70 5603**	5.22	**10.29 4397**	**5**
56	6054	4.13	0138	1.07	5916	5.20	4084	4
57	6302	4.15	0074	1.07	6228	5.22	3772	3
58	6551	4.13	.95 0010	1.08	6541	5.22	3459	2
59	6799	4.13	.94 9945	1.07	6854	5.20	3146	1
60	**9.65 7047**		**9.94 9881**		**9.70 7166**		**10.29 2834**	**0**
′	Cosine.	D.1″.	Sine.	D.1″.	Cotang.	D.1″.	Tang.	′

116° **63°**

27° TABLE XV.—*Continued* 152°

′	Sine.	D.1″.	Cosine.	D.1″.	Tang.	D.1″.	Cotang.	′
0	**9.65 7047**	4.13	**9.94 9881**	1.08	**9.70 7166**	5.20	**10.29 2834**	**60**
1	7295	4.12	9816	1.07	7478	5.20	2522	59
2	7542	4.13	9752	1.07	7790	5.20	2210	58
3	7790	4.12	9688	1.08	8102	5.20	1898	57
4	8037	4.12	9623	1.08	8414	5.20	1586	56
5	**9.65 8284**	4.12	**9.94 9558**	1.07	**9.70 8726**	5.18	**10.29 1274**	**55**
6	8531	4.12	9494	1.08	9037	5.20	0963	54
7	8778	4.12	9429	1.08	9349	5.18	0651	53
8	9025	4.10	9364	1.07	9660	5.18	0340	52
9	9271	4.10	9300	1.08	9971	5.18	.29 0029	51
10	**9.65 9517**	4.10	**9.94 9235**	1.08	**9.71 0282**	5.18	**10.28 9718**	**50**
11	.65 9763	4.10	9170	1.08	0593	5.18	9407	49
12	.66 0009	4.10	9105	1.08	0904	5.18	9096	48
13	0255	4.10	9040	1.08	1215	5.17	8785	47
14	0501	4.08	8975	1.08	1525	5.18	8475	46
15	**9.66 0746**	4.08	**9.94 8910**	1.08	**9.71 1836**	5.17	**10.28 8164**	**45**
16	0991	4.08	8845	1.08	2146	5.17	7854	44
17	1236	4.08	8780	1.08	2456	5.17	7544	43
18	1481	4.08	8715	1.08	2766	5.17	7234	42
19	1726	4.07	8650	1.10	3076	5.17	6924	41
20	**9.66 1970**	4.07	**9.94 8584**	1.08	**9.71 3386**	5.17	**10.28 6614**	**40**
21	2214	4.08	8519	1.08	3696	5.15	6304	39
22	2459	4.07	8454	1.10	4005	5.15	5995	38
23	2703	4.05	8388	1.08	4314	5.17	5686	37
24	2946	4.07	8323	1.10	4624	5.15	5376	36
25	**9.66 3190**	4.05	**9.94 8257**	1.08	**9.71 4933**	5.15	**10.28 5067**	**35**
26	3433	4.07	8192	1.10	5242	5.15	4758	34
27	3677	4.05	8126	1.10	5551	5.15	4449	33
28	3920	4.05	8060	1.08	5860	5.13	4140	32
29	4163	4.05	7995	1.10	6168	5.15	3832	31
30	**9.66 4406**	4.03	**9.94 7929**	1.10	**9.71 6477**	5.13	**10.28 3523**	**30**
31	4648	4.05	7863	1.10	6785	5.13	3215	29
32	4891	4.03	7797	1.10	7093	5.13	2907	28
33	5133	4.03	7731	1.10	7401	5.13	2599	27
34	5375	4.03	7665	1.08	7709	5.13	2291	26
35	**9.66 5617**	4.03	**9.94 7600**	1.12	**9.71 8017**	5.13	**10.28 1983**	**25**
36	5859	4.02	7533	1.10	8325	5.13	1675	24
37	6100	4.03	7467	1.10	8633	5.12	1367	23
38	6342	4.02	7401	1.10	8940	5.13	1060	22
39	6583	4.02	7335	1.10	9248	5.12	0752	21
40	**9.66 6824**	4.02	**9.94 7269**	1.10	**9.71 9555**	5.12	**10.28 0445**	**20**
41	7065	4.00	7203	1.12	.71 9862	5.12	.28 0138	19
42	7305	4.02	7136	1.10	.72 0169	5.12	.27 9831	18
43	7546	4.00	7070	1.10	0476	5.12	9524	17
44	7786	4.02	7004	1.12	0783	5.10	9217	16
45	**9.66 8027**	4.00	**9.94 6937**	1.10	**9.72 1089**	5.12	**10.27 8911**	**15**
46	8267	3.98	6871	1.12	1396	5.10	8604	14
47	8506	4.00	6804	1.10	1702	5.12	8298	13
48	8746	4.00	6738	1.12	2009	5.10	7991	12
49	8986	3.98	6671	1.12	2315	5.10	7685	11
50	**9.66 9225**	3.98	**9.94 6604**	1.10	**9.72 2621**	5.10	**10.27 7379**	**10**
51	9464	3.98	6538	1.12	2927	5.08	7073	9
52	9703	3.98	6471	1.12	3232	5.10	6768	8
53	.66 9942	3.98	6404	1.12	3538	5.10	6462	7
54	.67 0181	3.97	6337	1.12	3844	5.08	6156	6
55	**9.67 0419**	3.98	**9.94 6270**	1.12	**9.72 4149**	5.08	**10.27 5851**	**5**
56	0658	3.97	6203	1.12	4454	5.10	5546	4
57	0896	3.97	6136	1.12	4760	5.08	5240	3
58	1134	3.97	6069	1.12	5065	5.08	4935	2
59	1372	3.95	6002	1.12	5370	5.07	4630	1
60	**9.67 1609**		**9.94 5935**		**9.72 5674**		**10.27 4326**	**0**
′	Cosine.	D.1″.	Sine.	D.1″.	Cotang.	D.1″.	Tang.	′

117° **62°**

28° TABLE XV.—*Continued* 151°

′	Sine.	D.1″.	Cosine.	D.1″.	Tang.	D.1″.	Cotang.	′
0	**9.67 1609**	3.97	**9.94 5935**	1.12	**9.72 5674**	5.08	**10.27 4326**	**60**
1	1847	3.95	5868	1.13	5979	5.08	4021	59
2	2084	3.95	5800	1.12	6284	5.07	3716	58
3	2321	3.95	5733	1.12	6588	5.07	3412	57
4	2558	3.95	5666	1.13	6892	5.05	3108	56
5	**9.67 2795**	3.95	**9.94 5598**	1.12	**9.72 7197**	5.07	**10.27 2803**	**55**
6	3032	3.93	5531	1.12	7501	5.07	2499	54
7	3268	3.95	5464	1.13	7805	5.07	2195	53
8	3505	3.93	5396	1.13	8109	5.05	1891	52
9	3741	3.93	5328	1.12	8412	5.07	1588	51
10	**9.67 3977**	3.93	**9.94 5261**	1.13	**9.72 8716**	5.07	**10.27 1284**	**50**
11	4213	3.92	5193	1.13	9020	5.05	0980	49
12	4448	3.93	5125	1.12	9323	5.05	0677	48
13	4684	3.92	5058	1.13	9626	5.05	0374	47
14	4919	3.93	4990	1.13	.72 9929	5.07	.27 0071	46
15	**9.67 5155**	3.92	**9.94 4922**	1.13	**9.73 0233**	5.03	**10.26 9767**	**45**
16	5390	3.90	4854	1.13	0535	5.05	9465	44
17	5624	3.92	4786	1.13	0838	5.05	9162	43
18	5859	3.92	4718	1.13	1141	5.05	8859	42
19	6094	3.90	4650	1.13	1444	5.03	8556	41
20	**9.67 6328**	3.90	**9.94 4582**	1.13	**9.73 1746**	5.03	**10.26 8254**	**40**
21	6562	3.90	4514	1.13	2048	5.05	7952	39
22	6796	3.90	4446	1.15	2351	5.03	7649	38
23	7030	3.90	4377	1.13	2653	5.03	7347	37
24	7264	3.90	4309	1.13	2955	5.03	7045	36
25	**9.67 7498**	3.88	**9.94 4241**	1.15	**9.73 3257**	5.02	**10.26 6743**	**35**
26	7731	3.88	4172	1.13	3558	5.03	6442	34
27	7964	3.88	4104	1.13	3860	5.03	6140	33
28	8197	3.88	4036	1.15	4162	5.02	5838	32
29	8430	3.88	3967	1.13	4463	5.02	5537	31
30	**9.67 8663**	3.87	**9.94 3899**	1.15	**9.73 4764**	5.03	**10.26 5236**	**30**
31	8895	3.88	3830	1.15	5066	5.02	4934	29
32	9128	3.87	3761	1.13	5367	5.02	4633	28
33	9360	3.87	3693	1.15	5668	5.02	4332	27
34	9592	3.87	3624	1.15	5969	5.00	4031	26
35	**9.67 9824**	3.87	**9.94 3555**	1.15	**9.73 6269**	5.02	**10.26 3731**	**25**
36	.68 0056	3.87	3486	1.15	6570	5.00	3430	24
37	0288	3.85	3417	1.15	6870	5.02	3130	23
38	0519	3.85	3348	1.15	7171	5.00	2829	22
39	0750	3.87	3279	1.15	7471	5.00	2529	21
40	**9.68 0982**	3.85	**9.94 3210**	1.15	**9.73 7771**	5.00	**10.26 2229**	**20**
41	1213	3.83	3141	1.15	8071	5.00	1929	19
42	1443	3.85	3072	1.15	8371	5.00	1629	18
43	1674	3.85	3003	1.15	8671	5.00	1329	17
44	1905	3.83	2934	1.17	8971	5.00	1029	16
45	**9.68 2135**	3.83	**9.94 2864**	1.15	**9.73 9271**	4.98	**10.26 0729**	**15**
46	2365	3.83	2795	1.15	9570	5.00	0430	14
47	2595	3.83	2726	1.17	.73 9870	4.98	.26 0130	13
48	2825	3.83	2656	1.15	.74 0169	4.98	.25 9831	12
49	3055	3.82	2587	1.17	0468	4.98	9532	11
50	**9.68 3284**	3.83	**9.94 2517**	1.15	**9.74 0767**	4.98	**10.25 9233**	**10**
51	3514	3.82	2448	1.17	1066	4.98	8934	9
52	3743	3.82	2378	1.17	1365	4.98	8635	8
53	3972	3.82	2308	1.15	1664	4.97	8336	7
54	4201	3.82	2239	1.17	1962	4.98	8038	6
55	**9.68 4430**	3.80	**9.94 2169**	1.17	**9.74 2261**	4.97	**10.25 7739**	**5**
56	4658	3.82	2099	1.17	2559	4.98	7441	4
57	4887	3.80	2029	1.17	2858	4.97	7142	3
58	5115	3.80	1959	1.17	3156	4.97	6844	2
59	5343	3.80	1889	1.17	3454	4.97	6546	1
60	**9.68 5571**		**9.94 1819**		**9.74 3752**		**10.25 6248**	**0**
′	Cosine.	D.1″.	Sine.	D.1″.	Cotang.	D.1″.	Tang.	′

118° **61°**

29° TABLE XV.—*Continued* 150°

′	Sine.	D.1″.	Cosine.	D.1″.	Tang.	D.1″.	Cotang.	′
0	**9.68 5571**	3.80	**9.94 1819**	1.17	**9.74 3752**	4.97	**10.25 6248**	**60**
1	5799	3.80	1749	1.17	4050	4.97	5950	59
2	6027	3.78	1679	1.17	4348	4.95	5652	58
3	6254	3.80	1609	1.17	4645	4.97	5355	57
4	6482	3.78	1539	1.17	4943	4.95	5057	56
5	**9.68 6709**	3.78	**9.94 1469**	1.18	**9.74 5240**	4.97	**10.25 4760**	**55**
6	6936	3.78	1398	1.17	5538	4.95	4462	54
7	7163	3.77	1328	1.17	5835	4.95	4165	53
8	7389	3.78	1258	1.18	6132	4.95	3868	52
9	7616	3.78	1187	1.17	6429	4.95	3571	51
10	**9.68 7843**	3.77	**9.94 1117**	1.18	**9.74 6726**	4.95	**10.25 3274**	**50**
11	8069	3.77	1046	1.18	7023	4.93	2977	49
12	8295	3.77	0975	1.17	7319	4.95	2681	48
13	8521	3.77	0905	1.18	7616	4.95	2384	47
14	8747	3.75	0834	1.18	7913	4.93	2087	46
15	**9.68 8972**	3.77	**9.94 0763**	1.17	**9.74 8209**	4.93	**10.25 1791**	**45**
16	9198	3.75	0693	1.18	8505	4.93	1495	44
17	9423	3.75	0622	1.18	8801	4.93	1199	43
18	9648	3.75	0551	1.18	9097	4.93	0903	42
19	.68 9873	3.75	0480	1.18	9393	4.93	0607	41
20	**9.69 0098**	3.75	**9.94 0409**	1.18	**9.74 9689**	4.93	**10.25 0311**	**40**
21	0323	3.75	0338	1.18	.74 9985	4.93	.25 0015	39
22	0548	3.73	0267	1.18	.75 0281	4.92	.24 9719	38
23	0772	3.73	0196	1.18	0576	4.93	9424	37
24	0996	3.73	0125	1.18	0872	4.92	9128	36
25	**9.69 1220**	3.73	**9.94 0054**	1.20	**9.75 1167**	4.92	**10.24 8833**	**35**
26	1444	3.73	.93 9982	1.18	1462	4.92	8538	34
27	1668	3.73	9911	1.18	1757	4.92	8243	33
28	1892	3.72	9840	1.20	2052	4.92	7948	32
29	2115	3.73	9768	1.18	2347	4.92	7653	31
30	**9.69 2339**	3.72	**9.93 9697**	1.20	**9.75 2642**	4.92	**10.24 7358**	**30**
31	2562	3.72	9625	1.18	2937	4.90	7063	29
32	2785	3.72	9554	1.20	3231	4.92	6769	28
33	3008	3.72	9482	1.20	3526	4.90	6474	27
34	3231	3.70	9410	1.18	3820	4.92	6180	26
35	**9.69 3453**	3.72	**9.93 9339**	1.20	**9.75 4115**	4.90	**10.24 5885**	**25**
36	3676	3.70	9267	1.20	4409	4.90	5591	24
37	3898	3.70	9195	1.20	4703	4.90	5297	23
38	4120	3.70	9123	1.18	4997	4.90	5003	22
39	4342	3.70	9052	1.20	5291	4.90	4709	21
40	**9.69 4564**	3.70	**9.93 8980**	1.20	**9.75 5585**	4.88	**10.24 4415**	**20**
41	4786	3.68	8908	1.20	5878	4.90	4122	19
42	5007	3.70	8836	1.22	6172	4.88	3828	18
43	5229	3.68	8763	1.20	6465	4.90	3535	17
44	5450	3.68	8691	1.20	6759	4.88	3241	16
45	**9.69 5671**	3.68	**9.93 8619**	1.20	**9.75 7052**	4.88	**10.24 2948**	**15**
46	5892	3.68	8547	1.20	7345	4.88	2655	14
47	6113	3.68	8475	1.22	7638	4.88	2362	13
48	6334	3.67	8402	1.20	7931	4.88	2069	12
49	6554	3.68	8330	1.20	8224	4.88	1776	11
50	**9.69 6775**	3.67	**9.93 8258**	1.22	**9.75 8517**	4.88	**10.24 1483**	**10**
51	6995	3.67	8185	1.20	8810	4.87	1190	9
52	7215	3.67	8113	1.22	9102	4.88	0898	8
53	7435	3.65	8040	1.22	9395	4.87	0605	7
54	7654	3.67	7967	1.20	9687	4.87	0313	6
55	**9.69 7874**	3.67	**9.93 7895**	1.22	**9.75 9979**	4.88	**10.24 0021**	**5**
56	8094	3.65	7822	1.22	.76 0272	4.87	.23 9728	4
57	8313	3.65	7749	1.22	0564	4.87	9436	3
58	8532	3.65	7676	1.20	0856	4.87	9144	2
59	8751	3.65	7604	1.22	1148	4.85	8852	1
60	**9.69 8970**		**9.93 7531**		**9.76 1439**		**10.23 8561**	**0**
′	Cosine.	D.1″.	Sine.	D.1″.	Cotang.	D.1″.	Tang.	′

119° **60°**

30° TABLE XV.—*Continued* 149°

′	Sine.	D.1″.	Cosine.	D.1″.	Tang.	D.1″.	Cotang.	′
0	**9.69 8970**	3.65	**9.93 7531**	1.22	**9.76 1439**	4.87	**10.23 8561**	**60**
1	9189	3.63	7458	1.22	1731	4.87	8269	59
2	9407	3.65	7385	1.22	2023	4.85	7977	58
3	9626	3.63	7312	1.23	2314	4.87	7686	57
4	.69 9844	3.63	7238	1.22	2606	4.85	7394	56
5	**9.70 0062**	3.63	**9.93 7165**	1.22	**9.76 2897**	4.85	**10.23 7103**	**55**
6	0280	3.63	7092	1.22	3188	4.85	6812	54
7	0498	3.63	7019	1.22	3479	4.85	6521	53
8	0716	3.62	6946	1.23	3770	4.85	6230	52
9	0933	3.63	6872	1.22	4061	4.85	5939	51
10	**9.70 1151**	3.62	**9.93 6799**	1.23	**9.76 4352**	4.85	**10.23 5648**	**50**
11	1368	3.62	6725	1.22	4643	4.83	5357	49
12	1585	3.62	6652	1.23	4933	4.85	5067	48
13	1802	3.62	6578	1.22	5224	4.83	4776	47
14	2019	3.62	6505	1.23	5514	4.85	4486	46
15	**9.70 2236**	3.60	**9.93 6431**	1.23	**9.76 5805**	4.83	**10.23 4195**	**45**
16	2452	3.62	6357	1.22	6095	4.83	3905	44
17	2669	3.60	6284	1.23	6385	4.83	3615	43
18	2885	3.60	6210	1.23	6675	4.83	3325	42
19	3101	3.60	6136	1.23	6965	4.83	3035	41
20	**9.70 3317**	3.60	**9.93 6062**	1.23	**9.76 7255**	4.83	**10.23 2745**	**40**
21	3533	3.60	5988	1.23	7545	4.82	2455	39
22	3749	3.58	5914	1.23	7834	4.83	2166	38
23	3964	3.58	5840	1.23	8124	4.83	1876	37
24	4179	3.60	5766	1.23	8414	4.82	1586	36
25	**9.70 4395**	3.58	**9.93 5692**	1.23	**9.76 8703**	4.82	**10.23 1297**	**35**
26	4610	3.58	5618	1.25	8992	4.82	1008	34
27	4825	3.58	5543	1.23	9281	4.83	0719	33
28	5040	3.57	5469	1.23	9571	4.82	0429	32
29	5254	3.58	5395	1.25	.76 9860	4.80	.23 0140	31
30	**9.70 5469**	3.57	**9.93 5320**	1.23	**9.77 0148**	4.82	**10.22 9852**	**30**
31	5683	3.58	5246	1.25	0437	4.82	9563	29
32	5898	3.57	5171	1.23	0726	4.82	9274	28
33	6112	3.57	5097	1.25	1015	4.80	8985	27
34	6326	3.55	5022	1.23	1303	4.82	8697	26
35	**9.70 6539**	3.57	**9.93 4948**	1.25	**9.77 1592**	4.80	**10.22 8408**	**25**
36	6753	3.57	4873	1.25	1880	4.80	8120	24
37	6967	3.55	4798	1.25	2168	4.82	7832	23
38	7180	3.55	4723	1.23	2457	4.80	7543	22
39	7393	3.55	4649	1.25	2745	4.80	7255	21
40	**9.70 7606**	3.55	**9.93 4574**	1.25	**9.77 3033**	4.80	**10.22 6967**	**20**
41	7819	3.55	4499	1.25	3321	4.78	6679	19
42	8032	3.55	4424	1.25	3608	4.80	6392	18
43	8245	3.55	4349	1.25	3896	4.80	6104	17
44	8458	3.53	4274	1.25	4184	4.78	5816	16
45	**9.70 8670**	3.53	**9.93 4199**	1.27	**9.77 4471**	4.80	**10.22 5529**	**15**
46	8882	3.53	4123	1.25	4759	4.78	5241	14
47	9094	3.53	4048	1.25	5046	4.78	4954	13
48	9306	3.53	3973	1.25	5333	4.80	4667	12
49	9518	3.53	3898	1.27	5621	4.78	4379	11
50	**9.70 9730**	3.52	**9.93 3822**	1.25	**9.77 5908**	4.78	**10.22 4092**	**10**
51	.70 9941	3.53	3747	1.27	6195	4.78	3805	9
52	.71 0153	3.52	3671	1.25	6482	4.77	3518	8
53	0364	3.52	3596	1.27	6768	4.78	3232	7
54	0575	3.52	3520	1.25	7055	4.78	2945	6
55	**9.71 0786**	3.52	**9.93 3445**	1.27	**9.77 7342**	4.77	**10.22 2658**	**5**
56	0997	3.52	3369	1.27	7628	4.78	2372	4
57	1208	3.52	3293	1.27	7915	4.77	2085	3
58	1419	3.50	3217	1.27	8201	4.78	1799	2
59	1629	3.50	3141	1.25	8488	4.77	1512	1
60	**9.71 1839**		**9.93 3066**		**9.77 8774**		**10.22 1226**	**0**
′	Cosine.	D.1″.	Sine.	D.1″.	Cotang.	D.1″.	Tang.	′

120° 59°

31° TABLE XV.—*Continued* 148°

′	Sine.	D.1″.	Cosine.	D.1″.	Tang.	D.1″.	Cotang.	′
0	**9.71 1839**	3.52	**9.93 3066**	1.27	**9.77 8774**	4.77	**10.22 1226**	**60**
1	2050	3.50	2990	1.27	9060	4.77	0940	59
2	2260	3.48	2914	1.27	9346	4.77	0654	58
3	2469	3.50	2838	1.27	9632	4.77	0368	57
4	2679	3.50	2762	1.28	.77 9918	4.75	.22 0082	56
5	**9.71 2889**	3.48	**9.93 2685**	1.27	**9.78 0203**	4.77	**10.21 9797**	**55**
6	3098	3.50	2609	1.27	0489	4.77	9511	54
7	3308	3.48	2533	1.27	0775	4.75	9225	53
8	3517	3.48	2457	1.28	1060	4.77	8940	52
9	3726	3.48	2380	1.27	1346	4.75	8654	51
10	**9.71 3935**	3.48	**9.93 2304**	1.27	**9.78 1631**	4.75	**10.21 8369**	**50**
11	4144	3.47	2228	1.28	1916	4.75	8084	49
12	4352	3.48	2151	1.27	2201	4.75	7799	48
13	4561	3.47	2075	1.28	2486	4.75	7514	47
14	4769	3.48	1998	1.28	2771	4.75	7229	46
15	**9.71 4978**	3.47	**9.93 1921**	1.27	**9.78 3056**	4.75	**10.21 6944**	**45**
16	5186	3.47	1845	1.28	3341	4.75	6659	44
17	5394	3.47	1768	1.28	3626	4.73	6374	43
18	5602	3.45	1691	1.28	3910	4.75	6090	42
19	5809	3.47	1614	1.28	4195	4.73	5805	41
20	**9.71 6017**	3.45	**9.93 1537**	1.28	**9.78 4479**	4.75	**10.21 5521**	**40**
21	6224	3.47	1460	1.28	4764	4.73	5236	39
22	6432	3.45	1383	1.28	5048	4.73	4952	38
23	6639	3.45	1306	1.28	5332	4.73	4668	37
24	6846	3.45	1229	1.28	5616	4.73	4384	36
25	**9.71 7053**	3.43	**9.93 1152**	1.28	**9.78 5900**	4.73	**10.21 4100**	**35**
26	7259	3.45	1075	1.28	6184	4.73	3816	34
27	7466	3.45	0998	1.28	6468	4.73	3532	33
28	7673	3.43	0921	1.30	6752	4.73	3248	32
29	7879	3.43	0843	1.28	7036	4.72	2964	31
30	**9.71 8085**	3.43	**9.93 0766**	1.30	**9.78 7319**	4.73	**10.21 2681**	**30**
31	8291	3.43	0688	1.28	7603	4.72	2397	29
32	8497	3.43	0611	1.30	7886	4.73	2114	28
33	8703	3.43	0533	1.28	8170	4.72	1830	27
34	8909	3.42	0456	1.30	8453	4.72	1547	26
35	**9.71 9114**	3.43	**9.93 0378**	1.30	**9.78 8736**	4.72	**10.21 1264**	**25**
36	9320	3.42	0300	1.28	9019	4.72	0981	24
37	9525	3.42	0223	1.30	9302	4.72	0698	23
38	9730	3.42	0145	1.30	9585	4.72	0415	22
39	.71 9935	3.42	.93 0067	1.30	.78 9868	4.72	.21 0132	21
40	**9.72 0140**	3.42	**9.92 9989**	1.30	**9.79 0151**	4.72	**10.20 9849**	**20**
41	0345	3.40	9911	1.30	0434	4.70	9566	19
42	0549	3.42	9833	1.30	0716	4.72	9284	18
43	0754	3.40	9755	1.30	0999	4.70	9001	17
44	0958	3.40	9677	1.30	1281	4.70	8719	16
45	**9.72 1162**	3.40	**9.92 9599**	1.30	**9.79 1563**	4.72	**10.20 8437**	**15**
46	1366	3.40	9521	1.32	1846	4.70	8154	14
47	1570	3.40	9442	1.30	2128	4.70	7872	13
48	1774	3.40	9364	1.30	2410	4.70	7590	12
49	1978	3.38	9286	1.32	2692	4.70	7308	11
50	**9.72 2181**	3.40	**9.92 9207**	1.30	**9.79 2974**	4.70	**10.20 7026**	**10**
51	2385	3.38	9129	1.32	3256	4.70	6744	9
52	2588	3.38	9050	1.30	3538	4.68	6462	8
53	2791	3.38	8972	1.32	3819	4.70	6181	7
54	2994	3.38	8893	1.30	4101	4.70	5899	6
55	**9.72 3197**	3.38	**9.92 8815**	1.32	**9.79 4383**	4.68	**10.20 5617**	**5**
56	3400	3.38	8736	1.32	4664	4.70	5336	4
57	3603	3.37	8657	1.32	4946	4.68	5054	3
58	3805	3.37	8578	1.32	5227	4.68	4773	2
59	4007	3.38	8499	1.32	5508	4.68	4492	1
60	**9.72 4210**		**9.92 8420**		**9.79 5789**		**10.20 4211**	**0**
′	Cosine.	D.1″.	Sine.	D.1″.	Cotang.	D.1″.	Tang.	′

121° **58°**

32° TABLE XV.—*Continued* 147°

'	Sine.	D.1″.	Cosine.	D.1″.	Tang.	D.1″.	Cotang.	'
0	9.72 4210	3.37	9.92 8420	1.30	9.79 5789	4.68	10.20 4211	60
1	4412	3.37	8342	1.32	6070	4.68	3930	59
2	4614	3.37	8263	1.33	6351	4.68	3649	58
3	4816	3.35	8183	1.32	6632	4.68	3368	57
4	5017	3.37	8104	1.32	6913	4.68	3087	56
5	9.72 5219	3.35	9.92 8025	1.32	9.79 7194	4.67	10.20 2806	55
6	5420	3.37	7946	1.32	7474	4.68	2526	54
7	5622	3.35	7867	1.33	7755	4.68	2245	53
8	5823	3.35	7787	1.32	8036	4.67	1964	52
9	6024	3.35	7708	1.32	8316	4.67	1684	51
10	9.72 6225	3.35	9.92 7629	1.33	9.79 8596	4.68	10.20 1404	50
11	6426	3.33	7549	1.32	8877	4.67	1123	49
12	6626	3.35	7470	1.33	9157	4.67	0843	48
13	6827	3.33	7390	1.33	9437	4.67	0563	47
14	7027	3.35	7310	1.32	9717	4.67	0283	46
15	9.72 7228	3.33	9.92 7231	1.33	9.79 9997	4.67	10.20 0003	45
16	7428	3.33	7151	1.33	.80 0277	4.67	.19 9723	44
17	7628	3.33	7071	1.33	0557	4.65	9443	43
18	7828	3.32	6991	1.33	0836	4.67	9164	42
19	8027	3.33	6911	1.33	1116	4.67	8884	41
20	9.72 8227	3.33	9.92 6831	1.33	9.80 1396	4.65	10.19 8604	40
21	8427	3.32	6751	1.33	1675	4.67	8325	39
22	8626	3.32	6671	1.33	1955	4.65	8045	38
23	8825	3.32	6591	1.33	2234	4.65	7766	37
24	9024	3.32	6511	1.33	2513	4.65	7487	36
25	9.72 9223	3.32	9.92 6431	1.33	9.80 2792	4.67	10.19 7208	35
26	9422	3.32	6351	1.35	3072	4.65	6928	34
27	9621	3.32	6270	1.33	3351	4.65	6649	33
28	.72 9820	3.30	6190	1.33	3630	4.65	6370	32
29	.73 0018	3.32	6110	1.35	3909	4.63	6091	31
30	9.73 0217	3.30	9.92 6029	1.33	9.80 4187	4.65	10.19 5813	30
31	0415	3.30	5949	1.35	4466	4.65	5534	29
32	0613	3.30	5868	1.33	4745	4.63	5255	28
33	0811	3.30	5788	1.35	5023	4.65	4977	27
34	1009	3.28	5707	1.35	5302	4.63	4698	26
35	9.73 1206	3.30	9.92 5626	1.35	9.80 5580	4.65	10.19 4420	25
36	1404	3.30	5545	1.33	5859	4.63	4141	24
37	1602	3.28	5465	1.35	6137	4.63	3863	23
38	1799	3.28	5384	1.35	6415	4.63	3585	22
39	1996	3.28	5303	1.35	6693	4.63	3307	21
40	9.73 2193	3.28	9.92 5222	1.35	9.80 6971	4.63	10.19 3029	20
41	2390	3.28	5141	1.35	7249	4.63	2751	19
42	2587	3.28	5060	1.35	7527	4.63	2473	18
43	2784	3.27	4979	1.37	7805	4.63	2195	17
44	2980	3.28	4897	1.35	8083	4.63	1917	16
45	9.73 3177	3.27	9.92 4816	1.35	9.80 8361	4.62	10.19 1639	15
46	3373	3.27	4735	1.35	8638	4.63	1362	14
47	3569	3.27	4654	1.37	8916	4.62	1084	13
48	3765	3.27	4572	1.35	9193	4.63	0807	12
49	3961	3.27	4491	1.37	9471	4.62	0529	11
50	9.73 4157	3.27	9.92 4409	1.35	9.80 9748	4.62	10.19 0252	10
51	4353	3.27	4328	1.37	.81 0025	4.62	.18 9975	9
52	4549	3.25	4246	1.37	0302	4.63	9698	8
53	4744	3.25	4164	1.35	0580	4.62	9420	7
54	4939	3.27	4083	1.37	0857	4.62	9143	6
55	9.73 5135	3.25	9.92 4001	1.37	9.81 1134	4.60	10.18 8866	5
56	5330	3.25	3919	1.37	1410	4.62	8590	4
57	5525	3.23	3837	1.37	1687	4.62	8313	3
58	5719	3.25	3755	1.37	1964	4.62	8036	2
59	5914	3.25	3673	1.37	2241	4.60	7759	1
60	9.73 6109		9.92 3591		9.81 2517		10.18 7483	0
'	Cosine.	D.1″.	Sine.	D.1″.	Cotang.	D.1″.	Tang.	'

122° 57°

33° TABLE XV.—*Continued* 146°

′	Sine.	D.1″.	Cosine.	D.1″.	Tang.	D.1″.	Cotang.	′
0	9.73 6109	3.23	9.92 3591	1.37	9.81 2517	4.62	10.18 7483	60
1	6303	3.25	3509	1.37	2794	4.60	7206	59
2	6498	3.23	3427	1.37	3070	4.62	6930	58
3	6692	3.23	3345	1.37	3347	4.60	6653	57
4	6886	3.23	3263	1.37	3623	4.60	6377	56
5	9.73 7080	3.23	9.92 3181	1.38	9.81 3899	4.62	10.18 6101	55
6	7274	3.22	3098	1.37	4176	4.60	5824	54
7	7467	3.23	3016	1.38	4452	4.60	5548	53
8	7661	3.23	2933	1.37	4728	4.60	5272	52
9	7855	3.22	2851	1.38	5004	4.60	4996	51
10	9.73 8048	3.22	9.92 2768	1.37	9.81 5280	4.58	10.18 4720	50
11	8241	3.22	2686	1.38	5555	4.60	4445	49
12	8434	3.22	2603	1.38	5831	4.60	4169	48
13	8627	3.22	2520	1.37	6107	4.58	3893	47
14	8820	3.22	2438	1.38	6382	4.60	3618	46
15	9.73 9013	3.22	9.92 2355	1.38	9.81 6658	4.58	10.18 3342	45
16	9206	3.20	2272	1.38	6933	4.60	3067	44
17	9398	3.20	2189	1.38	7209	4.58	2791	43
18	9590	3.22	2106	1.38	7484	4.58	2516	42
19	9783	3.20	2023	1.38	7759	4.60	2241	41
20	9.73 9975	3.20	9.92 1940	1.38	9.81 8035	4.58	10.18 1965	40
21	.74 0167	3.20	1857	1.38	8310	4.58	1690	39
22	0359	3.18	1774	1.38	8585	4.58	1415	38
23	0550	3.20	1691	1.40	8860	4.58	1140	37
24	0742	3.20	1607	1.38	9135	4.58	0865	36
25	9.74 0934	3.18	9.92 1524	1.38	9.81 9410	4.57	10.18 0590	35
26	1125	3.18	1441	1.40	9684	4.58	0310	34
27	1316	3.20	1357	1.38	.81 9959	4.58	.18 0041	33
28	1508	3.18	1274	1.40	.82 0234	4.57	.17 9766	32
29	1699	3.17	1190	1.38	0508	4.58	9492	31
30	9.74 1889	3.18	9.92 1107	1.40	9.82 0783	4.57	10.17 9217	30
31	2080	3.18	1023	1.40	1057	4.58	8943	29
32	2271	3.18	0939	1.38	1332	4.57	8668	28
33	2462	3.17	0856	1.40	1606	4.57	8394	27
34	2652	3.17	0772	1.40	1880	4.57	8120	26
35	9.74 2842	3.18	9.92 0688	1.40	9.82 2154	4.58	10.17 7846	25
36	3033	3.17	0604	1.40	2429	4.57	7571	24
37	3223	3.17	0520	1.40	2703	4.57	7297	23
38	3413	3.15	0436	1.40	2977	4.57	7023	22
39	3602	3.17	0352	1.40	3251	4.55	6749	21
40	9.74 3792	3.17	9.92 0268	1.40	9.82 3524	4.57	10.17 6476	20
41	3982	3.15	0184	1.42	3798	4.57	6202	19
42	4171	3.17	0099	1.40	4072	4.55	5928	18
43	4361	3.15	.92 0015	1.40	4345	4.57	5655	17
44	4550	3.15	.91 9931	1.42	4619	4.57	5381	16
45	9.74 4739	3.15	9.91 9846	1.40	9.82 4893	4.55	10.17 5107	15
46	4928	3.15	9762	1.42	5166	4.55	4834	14
47	5117	3.15	9677	1.40	5439	4.57	4561	13
48	5306	3.13	9593	1.42	5713	4.55	4287	12
49	5494	3.15	9508	1.40	5986	4.55	4014	11
50	9.74 5683	3.13	9.91 9424	1.42	9.82 6254	4.55	10.17 3741	10
51	5871	3.15	9339	1.42	6532	4.55	3468	9
52	6060	3.13	9254	1.42	6805	4.55	3195	8
53	6248	3.13	9169	1.40	7078	4.55	2922	7
54	6436	3.13	9085	1.42	7351	4.55	2649	6
55	9.74 6624	3.13	9.91 9000	1.42	9.82 7624	4.55	10.17 2376	5
56	6812	3.12	8915	1.42	7897	4.55	2103	4
57	6999	3.13	8830	1.42	8170	4.53	1830	3
58	7187	3.12	8745	1.43	8442	4.55	1558	2
59	7374	3.13	8659	1.42	8715	4.53	1285	1
60	9.74 7562		9.91 8574		9.82 8987		10.17 1013	0
′	Cosine.	D.1″.	Sine.	D.1″.	Cotang.	D.1″.	Tang.	′

123° 56°

34° TABLE XV.—*Continued* 145°

′	Sine.	D.1″	Cosine.	D.1″	Tang.	D.1″	Cotang.	′
0	**9.74 7562**	3.12	**9.91 8574**	1.42	**9.82 8987**	4.55	**10.17 1013**	**60**
1	7749	3.12	8489	1.42	9260	4.53	0740	59
2	7936	3.12	8404	1.43	9532	4.55	0468	58
3	8123	3.12	8318	1.42	.82 9805	4.53	.17 0195	57
4	8310	3.12	8233	1.43	.83 0077	4.53	.16 9923	56
5	**9.74 8497**	3.10	**9.91 8147**	1.42	**9.83 0349**	4.53	**10.16 9651**	**55**
6	8683	3.12	8062	1.43	0621	4.53	9379	54
7	8870	3.10	7976	1.42	0893	4.53	9107	53
8	9056	3.12	7891	1.43	1165	4.53	8835	52
9	9243	3.10	7805	1.43	1437	4.53	8563	51
10	**9.74 9429**	3.10	**9.91 7719**	1.42	**9.83 1709**	4.53	**10.16 8291**	**50**
11	9615	3.10	7634	1.43	1981	4.53	8019	49
12	9801	3.10	7548	1.43	2253	4.53	7747	48
13	.74 9987	3.08	7462	1.43	2525	4.52	7475	47
14	.75 0172	3.10	7376	1.43	2796	4.53	7204	46
15	**9.75 0358**	3.08	**9.91 7290**	1.43	**9.83 3068**	4.52	**10.16 6932**	**45**
16	0543	3.10	7204	1.43	3339	4.53	6661	44
17	0729	3.08	7118	1.43	3611	4.52	6389	43
18	0914	3.08	7032	1.43	3882	4.53	6118	42
19	1099	3.08	6946	1.45	4154	4.52	5846	41
20	**9.75 1284**	3.08	**9.91 6859**	1.43	**9.83 4425**	4.52	**10.16 5575**	**40**
21	1469	3.08	6773	1.43	4696	4.52	5304	39
22	1654	3.08	6687	1.45	4967	4.52	5033	38
23	1839	3.07	6600	1.43	5238	4.52	4762	37
24	2023	3.08	6514	1.45	5509	4.52	4491	36
25	**9.75 2208**	3.07	**9.91 6427**	1.43	**9.83 5780**	4.52	**10.16 4220**	**35**
26	2392	3.07	6341	1.45	6051	4.52	3949	34
27	2576	3.07	6254	1.45	6322	4.52	3678	33
28	2760	3.07	6167	1.43	6593	4.52	3407	32
29	2944	3.07	6081	1.45	6864	4.50	3136	31
30	**9.75 3128**	3.07	**9.91 5994**	1.45	**9.83 7134**	4.52	**10.16 2866**	**30**
31	3312	3.05	5907	1.45	7405	4.50	2595	29
32	3495	3.07	5820	1.45	7675	4.52	2325	28
33	3679	3.07	5733	1.45	7946	4.50	2054	27
34	3862	3.07	5646	1.45	8216	4.52	1784	26
35	**9.75 4046**	3.05	**9.91 5559**	1.45	**9.83 8487**	4.50	**10.16 1513**	**25**
36	4229	3.05	5472	1.45	8757	4.50	1243	24
37	4412	3.05	5385	1.47	9027	4.50	0973	23
38	4595	3.05	5297	1.45	9297	4.52	0703	22
39	4778	3.03	5210	1.45	9568	4.50	0432	21
40	**9.75 4960**	3.05	**9.91 5123**	1.47	**9.83 9838**	4.50	**10.16 0162**	**20**
41	5143	3.05	5035	1.45	.84 0108	4.50	.15 9892	19
42	5326	3.03	4948	1.47	0378	4.50	9622	18
43	5508	3.03	4860	1.45	0648	4.48	9352	17
44	5690	3.03	4773	1.47	0917	4.50	9083	16
45	**9.75 5872**	3.03	**9.91 4685**	1.45	**9.84 1187**	4.50	**10.15 8813**	**15**
46	6054	3.03	4598	1.47	1457	4.50	8543	14
47	6236	3.03	4510	1.47	1727	4.48	8273	13
48	6418	3.03	4422	1.47	1996	4.50	8004	12
49	6600	3.03	4334	1.47	2266	4.48	7734	11
50	**9.75 6782**	3.02	**9.91 4246**	1.47	**9.84 2535**	4.50	**10.15 7465**	**10**
51	6963	3.02	4158	1.47	2805	4.48	7195	9
52	7144	3.03	4070	1.47	3074	4.48	6926	8
53	7326	3.02	3982	1.47	3343	4.48	6657	7
54	7507	3.02	3894	1.47	3612	4.50	6388	6
55	**9.75 7688**	3.02	**9.91 3806**	1.47	**9.84 3882**	4.48	**10.15 6118**	**5**
56	7869	3.02	3718	1.47	4151	4.48	5849	4
57	8050	3.00	3630	1.48	4420	4.48	5580	3
58	8230	3.02	3541	1.47	4689	4.48	5311	2
59	8411	3.00	3453	1.47	4958	4.48	5042	1
60	**9.75 8591**		**9.91 3365**		**9.84 5227**		**10.15 4773**	**0**
′	Cosine.	D.1″.	Sine.	D.1″.	Cotang.	D.1″.	Tang.	′

124° **55°**

35° **Table XV.**—*Continued* 144°

′	Sine.	D.1″.	Cosine.	D.1″.	Tang.	D.1″.	Cotang.	′
0	**9.75 8591**	3.02	**9.91 3365**	1.48	**9.84 5227**	4.48	**10.15 4773**	**60**
1	8772	3.00	3276	1.48	5496	4.47	4504	59
2	8952	3.00	3187	1.47	5764	4.48	4236	58
3	9132	3.00	3099	1.48	6033	4.48	3967	57
4	9312	3.00	3010	1.47	6302	4.47	3698	56
5	**9.75 9492**	3.00	**9.91 2922**	1.48	**9.84 6570**	4.48	**10.15 3430**	**55**
6	9672	3.00	2833	1.48	6839	4.48	3161	54
7	.75 9852	2.98	2744	1.48	7108	4.47	2892	53
8	.76 0031	3.00	2655	1.48	7376	4.47	2624	52
9	0211	2.98	2566	1.48	7644	4.48	2356	51
10	**9.76 0390**	2.98	**9.91 2477**	1.48	**9.84 7913**	4.47	**10.15 2087**	**50**
11	0569	2.98	2388	1.48	8181	4.47	1819	49
12	0748	2.98	2299	1.48	8449	4.47	1551	48
13	0927	2.98	2210	1.48	8717	4.48	1283	47
14	1106	2.98	2121	1.50	8986	4.47	1014	46
15	**9.76 1285**	2.98	**9.91 2031**	1.48	**9.84 9254**	4.47	**10.15 0746**	**45**
16	1464	2.97	1942	1.48	9522	4.47	0478	44
17	1642	2.98	1853	1.50	.84 9790	4.45	.15 0210	43
18	1821	2.97	1763	1.48	.85 0057	4.47	.14 9943	42
19	1999	2.97	1674	1.50	0325	4.47	9675	41
20	**9.76 2177**	2.98	**9.91 1584**	1.48	**9.85 0593**	4.47	**10.14 9407**	**40**
21	2356	2.97	1495	1.50	0861	4.47	9139	39
22	2534	2.97	1405	1.50	1129	4.45	8871	38
23	2712	2.95	1315	1.48	1396	4.47	8604	37
24	2889	2.97	1226	1.50	1664	4.45	8336	36
25	**9.76 3067**	2.97	**9.91 1136**	1.50	**9.85 1931**	4.47	**10.14 8069**	**35**
26	3245	2.95	1046	1.50	2199	4.45	7801	34
27	3422	2.97	0956	1.50	2466	4.45	7534	33
28	3600	2.95	0866	1.50	2733	4.47	7267	32
29	3777	2.95	0776	1.50	3001	4.45	6999	31
30	**9.76 3954**	2.95	**9.91 0686**	1.50	**9.85 3268**	4.45	**10.14 6732**	**30**
31	4131	2.95	0596	1.50	3535	4.45	6465	29
32	4308	2.95	0506	1.52	3802	4.45	6198	28
33	4485	2.95	0415	1.50	4069	4.45	5931	27
34	4662	2.93	0325	1.50	4336	4.45	5664	26
35	**9.76 4838**	2.95	**9.91 0235**	1.52	**9.85 4603**	4.45	**10.14 5397**	**25**
36	5015	2.93	0144	1.50	4870	4.45	5130	24
37	5191	2.93	.91 0054	1.52	5137	4.45	4863	23
38	5367	2.95	.90 9963	1.50	5404	4.45	4596	22
39	5544	2.93	9873	1.52	5671	4.45	4329	21
40	**9.76 5720**	2.93	**9.90 9782**	1.52	**9.85 5938**	4.43	**10.14 4062**	**20**
41	5896	2.93	9691	1.50	6204	4.45	3796	19
42	6072	2.92	9601	1.52	6471	4.43	3529	18
43	6247	2.93	9510	1.52	6737	4.45	3263	17
44	6423	2.92	9419	1.52	7004	4.43	2996	16
45	**9.76 6598**	2.93	**9.90 9328**	1.52	**9.85 7270**	4.45	**10.14 2730**	**15**
46	6774	2.92	9237	1.52	7537	4.43	2463	14
47	6949	2.92	9146	1.52	7803	4.43	2197	13
48	7124	2.93	9055	1.52	8069	4.45	1931	12
49	7300	2.92	8964	1.52	8336	4.43	1664	11
50	**9.76 7475**	2.90	**9.90 8873**	1.53	**9.85 8602**	4.43	**10.14 1398**	**10**
51	7649	2.92	8781	1.52	8868	4.43	1132	9
52	7824	2.92	8690	1.52	9134	4.43	0866	8
53	7999	2.90	8599	1.53	9400	4.43	0600	7
54	8173	2.92	8507	1.52	9666	4.43	0334	6
55	**9.76 8348**	2.90	**9.90 8416**	1.53	**9.85 9932**	4.43	**10.14 0068**	**5**
56	8522	2.92	8324	1.52	.86 0198	4.43	.13 9802	4
57	8697	2.90	8233	1.53	0464	4.43	9536	3
58	8871	2.90	8141	1.53	0730	4.42	9270	2
59	9045	2.90	8049	1.52	0995	4.43	9005	1
60	**9.76 9219**		**9.90 7958**		**9.86 1261**		**10.13 8739**	**0**
′	Cosine.	D.1″.	Sine.	D.1″.	Cotang.	D.1″.	Tang.	′

125° **54°**

36° **Table XV.**—*Continued* 143°

′	Sine.	D.1″.	Cosine.	D.1″.	Tang.	D.1″.	Cotang.	′
0	**9.76 9219**	2.90	**9.90 7958**	1.53	**9.86 1261**	4.43	**10.13 8739**	**60**
1	9393	2.88	7866	1.53	1527	4.42	8473	59
2	9566	2.90	7774	1.53	1792	4.43	8208	58
3	9740	2.88	7682	1.53	2058	4.42	7942	57
4	.76 9913	2.90	7590	1.53	2323	4.43	7677	56
5	**9.77 0087**	2.88	**9.90 7498**	1.53	**9.86 2589**	4.42	**10.13 7411**	**55**
6	0260	2.88	7406	1.53	2854	4.42	7146	54
7	0433	2.88	7314	1.53	3119	4.43	6881	53
8	0606	2.88	7222	1.55	3385	4.42	6615	52
9	0779	2.88	7129	1.53	3650	4.42	6350	51
10	**9.77 0952**	2.88	**9.90 7037**	1.53	**9.86 3915**	4.42	**10.13 6085**	**50**
11	1125	2.88	6945	1.55	4180	4.42	5820	49
12	1298	2.87	6852	1.53	4445	4.42	5555	48
13	1470	2.88	6760	1.55	4710	4.42	5290	47
14	1643	2.87	6667	1.53	4975	4.42	5025	46
15	**9.77 1815**	2.87	**9.90 6575**	1.55	**9.86 5240**	4.42	**10.13 4760**	**45**
16	1987	2.87	6482	1.55	5505	4.42	4495	44
17	2159	2.87	6389	1.55	5770	4.42	4230	43
18	2331	2.87	6296	1.53	6035	4.42	3965	42
19	2503	2.87	6204	1.55	6300	4.40	3700	41
20	**9.77 2675**	2.87	**9.90 6111**	1.55	**9.86 6564**	4.42	**10.13 3436**	**40**
21	2847	2.85	6018	1.55	6829	4.42	3171	39
22	3018	2.87	5925	1.55	7094	4.40	2906	38
23	3190	2.85	5832	1.55	7358	4.42	2642	37
24	3361	2.87	5739	1.57	7623	4.40	2377	36
25	**9.77 3533**	2.85	**9.90 5645**	1.55	**9.86 7887**	4.42	**10.13 2113**	**35**
26	3704	2.85	5552	1.55	8152	4.40	1848	34
27	3875	2.85	5459	1.55	8416	4.40	1584	33
28	4046	2.85	5366	1.57	8680	4.42	1320	32
29	4217	2.85	5272	1.55	8945	4.40	1055	31
30	**9.77 4388**	2.83	**9.90 5179**	1.57	**9.86 9209**	4.40	**10.13 0791**	**30**
31	4558	2.85	5085	1.55	9473	4.40	0527	29
32	4729	2.83	4992	1.57	.86 9737	4.40	.13 0263	28
33	4899	2.85	4898	1.57	.87 0001	4.40	.12 9999	27
34	5070	2.83	4804	1.55	0265	4.40	9735	26
35	**9.77 5240**	2.83	**9.90 4711**	1.57	**9.87 0529**	4.40	**10.12 9471**	**25**
36	5410	2.83	4617	1.57	0793	4.40	9207	24
37	5580	2.83	4523	1.57	1057	4.40	8943	23
38	5750	2.83	4429	1.57	1321	4.40	8679	22
39	5920	2.83	4335	1.57	1585	4.40	8415	21
40	**9.77 6090**	2.82	**9.90 4241**	1.57	**9.87 1849**	4.38	**10.12 8151**	**20**
41	6259	2.83	4147	1.57	2112	4.40	7888	19
42	6429	2.82	4053	1.57	2376	4.40	7624	18
43	6598	2.83	3959	1.58	2640	4.38	7360	17
44	6768	2.82	3864	1.57	2903	4.40	7097	16
45	**9.77 6937**	2.82	**9.90 3770**	1.57	**9.87 3167**	4.38	**10.12 6833**	**15**
46	7106	2.82	3676	1.58	3430	4.40	6570	14
47	7275	2.82	3581	1.57	3694	4.38	6306	13
48	7444	2.82	3487	1.58	3957	4.38	6043	12
49	7613	2.80	3392	1.57	4220	4.40	5780	11
50	**9.77 7781**	2.82	**9.90 3298**	1.58	**9.87 4484**	4.38	**10.12 5516**	**10**
51	7950	2.82	3203	1.58	4747	4.38	5253	9
52	8119	2.80	3108	1.57	5010	4.38	4990	8
53	8287	2.80	3014	1.58	5273	4.40	4727	7
54	8455	2.82	2919	1.58	5537	4.38	4463	6
55	**9.77 8624**	2.80	**9.90 2824**	1.58	**9.87 5800**	4.38	**10.12 4200**	**5**
56	8792	2.80	2729	1.58	6063	4.38	3937	4
57	8960	2.80	2634	1.58	6326	4.38	3674	3
58	9128	2.78	2539	1.58	6589	4.38	3411	2
59	9295	2.80	2444	1.58	6852	4.37	3148	1
60	**9.77 9463**		**9.90 2349**		**9.87 7114**		**10.12 2886**	**0**
′	Cosine.	D.1″.	Sine.	D.1″.	Cotang.	D.1″.	Tang.	′

126° **53°**

37° TABLE XV.—*Continued* 142°

′	Sine.	D.1″.	Cosine.	D.1″.	Tang.	D.1″.	Cotang.	′
0	9.77 9463	2.80	9.90 2349	1.60	9.87 7114	4.38	10.12 2886	60
1	9631	2.78	2253	1.58	7377	4.38	2623	59
2	9798	2.80	2158	1.58	7640	4.38	2360	58
3	.77 9966	2.78	2063	1.60	7903	4.37	2097	57
4	.78 0133	2.78	1967	1.58	8165	4.38	1835	56
5	9.78 0300	2.78	9.90 1872	1.60	9.87 8428	4.38	10.12 1572	55
6	0467	2.78	1776	1.58	8691	4.37	1309	54
7	0634	2.78	1681	1.60	8953	4.38	1047	53
8	0801	2.78	1585	1.58	9216	4.37	0784	52
9	0968	2.77	1490	1.60	9478	4.38	0522	51
10	9.78 1134	2.78	9.90 1394	1.60	9.87 9741	4.37	10.12 0259	50
11	1301	2.78	1298	1.60	.88 0003	4.37	.11 9997	49
12	1468	2.77	1202	1.60	0265	4.38	9735	48
13	1634	2.77	1106	1.60	0528	4.37	9472	47
14	1800	2.77	1010	1.60	0790	4.37	9210	46
15	9.78 1966	2.77	9.90 0914	1.60	9.88 1052	4.37	10.11 8948	45
16	2132	2.77	0818	1.60	1314	4.38	8686	44
17	2298	2.77	0722	1 60	1577	4.37	8423	43
18	2464	2.77	0626	1.62	1839	4.37	8161	42
19	2630	2.77	0529	1.60	2101	4.37	7899	41
20	9.78 2796	2.75	9.90 0433	1.60	9.88 2363	4.37	10.11 7637	40
21	2961	2.77	0337	1.62	2625	4.37	7375	39
22	3127	2.75	0240	1.60	2887	4.35	7113	38
23	3292	2.77	0144	1.62	3148	4.37	6852	37
24	3458	2.75	.90 0047	1.60	3410	4.37	6590	36
25	9.78 3623	2.75	9.89 9951	1.62	9.88 3672	4.37	10.11 6328	35
26	3788	2.75	9854	1.62	3934	4.37	6066	34
27	3953	2.75	9757	1.62	4196	4.35	5804	33
28	4118	2.73	9660	1.60	4457	4.37	5543	32
29	4282	2.75	9564	1.62	4719	4.35	5281	31
30	9.78 4447	2.75	9.89 9467	1.62	9.88 4980	4.37	10.11 5020	30
31	4612	2.73	9370	1.62	5242	4.37	4758	29
32	4776	2.75	9273	1.62	5504	4.35	4496	28
33	4941	2.73	9176	1.63	5765	4.35	4235	27
34	5105	2.73	9078	1.62	6026	4.37	3974	26
35	9.78 5269	2.73	9.89 8981	1.62	9.88 6288	4.35	10.11 3712	25
36	5433	2.73	8884	1.62	6549	4.37	3451	24
37	5597	2.73	8787	1.63	6811	4.35	3189	23
38	5761	2.73	8689	1.62	7072	4.35	2928	22
39	5925	2.73	8592	1.63	7333	4.35	2667	21
40	9.78 6089	2.72	9.89 8494	1.62	9.88 7594	4.35	10.11 2406	20
41	6252	2.73	8397	1.63	7855	4.35	2145	19
42	6416	2.72	8299	1.62	8116	4.37	1884	18
43	6579	2.72	8202	1.63	8378	4.35	1622	17
44	6742	2.73	8104	1.63	8639	4.35	1361	16
45	9.78 6906	2.72	9.89 8006	1.63	9.88 8900	4.35	10.11 1100	15
46	7069	2.72	7908	1.63	9161	4.33	0839	14
47	7232	2.72	7810	1.63	9421	4.35	0579	13
48	7395	2.70	7712	1.63	9682	4.35	0318	12
49	7557	2.72	7614	1.63	.88 9943	4.35	.11 0057	11
50	9.78 7720	2.72	9.89 7516	1.63	9.89 0204	4.35	10.10 9796	10
51	7883	2.70	7418	1.63	0465	4.33	9535	9
52	8045	2.72	7320	1.63	0725	4.35	9275	8
53	8208	2.70	7222	1.65	0986	4.35	9014	7
54	8370	2.70	7123	1.63	1247	4.33	8753	6
55	9.78 8532	2.70	9.89 7025	1.65	9.89 1507	4.35	10.10 8493	5
56	8694	2.70	6926	1.63	1768	4.33	8232	4
57	8856	2.70	6828	1.65	2028	4.35	7972	3
58	9018	2.70	6729	1.63	2289	4.33	7711	2
59	9180	2.70	6631	1.65	2549	4.35	7451	1
60	9.78 9342		9.89 6532		9.89 2810		10.10 7190	0
′	Cosine.	D.1″.	Sine.	D.1″.	Cotang.	D.1″.	Tang.	′

127° 52°

38° TABLE XV.—*Continued* 141°

′	Sine.	D.1″.	Cosine.	D.1″.	Tang.	D.1″.	Cotang.	′
0	9.78 9342	2.70	9.89 6532	1.65	9.89 2810	4.33	10.10 7190	60
1	9504	2.68	6433	1.63	3070	4.35	6930	59
2	9665	2.70	6335	1.65	3331	4.33	6669	58
3	9827	2.68	6236	1.65	3591	4.33	6409	57
4	.78 9988	2.68	6137	1.65	3851	4.33	6149	56
5	9.79 0149	2.68	9.89 6038	1.65	9.89 4111	4.35	10.10 5889	55
6	0310	2.68	5939	1.65	4372	4.33	5628	54
7	0471	2.68	5840	1.65	4632	4.33	5368	53
8	0632	2.68	5741	1.67	4892	4.33	5108	52
9	0793	2.68	5641	1.65	5152	4.33	4848	51
10	9.79 0954	2.68	9.89 5542	1.65	9.89 5412	4.33	10.10 4588	50
11	1115	2.67	5443	1.67	5672	4.33	4328	49
12	1275	2.68	5343	1.65	5932	4.33	4068	48
13	1436	2.67	5244	1.65	6192	4.33	3808	47
14	1596	2.68	5145	1.67	6452	4.33	3548	46
15	9.79 1757	2.67	9.89 5045	1.67	9.89 6712	4.32	10.10 3288	45
16	1917	2.67	4945	1.65	6971	4.33	3029	44
17	2077	2.67	4846	1.67	7231	4.33	2769	43
18	2237	2.67	4746	1.67	7491	4.33	2509	42
19	2397	2.67	4646	1.67	7751	4.32	2249	41
20	9.79 2557	2.65	9.89 4546	1.67	9.89 8010	4.33	10.10 1990	40
21	2716	2.67	4446	1.67	8270	4.33	1730	39
22	2876	2.65	4346	1.67	8530	4.32	1470	38
23	3035	2.67	4246	1.67	8789	4.33	1211	37
24	3195	2.65	4146	1.67	9049	4.32	0951	36
25	9.79 3354	2.67	9.89 4046	1.67	9.89 9308	4.33	10.10 0692	35
26	3514	2.65	3946	1.67	9568	4.32	0432	34
27	3673	2.65	3846	1.68	.89 9827	4.33	.10 0173	33
28	3832	2.65	3745	1.67	.90 0087	4.32	.09 9913	32
29	3991	2.65	3645	1.68	0346	4.32	9654	31
30	9.79 4150	2.63	9,89 3544	1.67	9.90 0605	4.32	10.09 9395	30
31	4308	2.65	3444	1.68	0864	4.33	9136	29
32	4467	2.65	3343	1.67	1124	4.32	8876	28
33	4626	2.63	3243	1.68	1383	4.32	8617	27
34	4784	2.63	3142	1.68	1642	4.32	8358	26
35	9.79 4942	2.65	9.89 3041	1.68	9.90 1901	4.32	10.09 8099	25
36	5101	2.63	2940	1.68	2160	4.33	7840	24
37	5259	2.63	2839	1.67	2420	4.32	7580	23
38	5417	2.63	2739	1.68	2679	4.32	7321	22
39	5575	2.63	2638	1.70	2938	4.32	7062	21
40	9.79 5733	2.63	9.89 2536	1.68	9.90 3197	4.32	10.09 6803	20
41	5891	2.63	2435	1.68	3456	4.30	6544	19
42	6049	2.62	2334	1.68	3714	4.32	6286	18
43	6206	2.63	2233	1.68	3973	4.32	6027	17
44	6364	2.62	2132	1.70	4232	4.32	5768	16
45	9.79 6521	2.63	9.89 2030	1.68	9.90 4491	4.32	10.09 5509	15
46	6679	2.62	1929	1.70	4750	4.30	5250	14
47	6836	2.62	1827	1.68	5008	4.32	4992	13
48	6993	2.62	1726	1.70	5267	4.32	4733	12
49	7150	2.62	1624	1.68	5526	4.32	4474	11
50	9.79 7307	2.62	9.89 1523	1.70	9.90 5785	4.30	10.09 4215	10
51	7464	2.62	1421	1.70	6043	4.32	3957	9
52	7621	2.60	1319	1.70	6302	4.30	3698	8
53	7777	2.62	1217	1.70	6560	4.32	3440	7
54	7934	2.62	1115	1.70	6819	4.30	3181	6
55	9.79 8091	2.60	9.89 1013	1.70	9.90 7077	4.32	10.09 2923	5
56	8247	2.60	0911	1.70	7336	4.30	2664	4
57	8403	2.62	0809	1.70	7594	4.32	2406	3
58	8560	2.60	0707	1.70	7853	4.30	2147	2
59	8716	2.60	0605	1.70	8111	4.30	1889	1
60	9.79 8872		9.89 0503		9.90 8369		10.09 1631	0
′	Cosine.	D.1″.	Sine.	D.1″.	Cotang.	D.1″.	Tang.	′

128° 51°

39° TABLE XV.—*Continued* 140°

′	Sine.	D.1″.	Cosine.	D.1″.	Tang.	D.1″.	Cotang.	′
0	**9.79 8872**	2.60	**9.89 0503**	1.72	**9.90 8369**	4.32	**10.09 1631**	**60**
1	9028	2.60	0400	1.70	8628	4.30	1372	59
2	9184	2.58	0298	1.72	8886	4.30	1114	58
3	9339	2.60	0195	1.70	9144	4.30	0856	57
4	9495	2.60	.89 0093	1.72	9402	4.30	0598	56
5	**9.79 9651**	2.58	**9.88 9990**	1.70	**9.90 9660**	4.30	**10.09 0340**	**55**
6	9806	2.60	9888	1.72	.90 9918	4.32	.09 0082	54
7	.79 9962	2.58	9785	1.72	.91 0177	4.30	.08 9823	53
8	.80 0117	2.58	9682	1.72	0435	4.30	9565	52
9	0272	2.58	9579	1.70	0693	4.30	9307	51
10	**9.80 0427**	2.58	**9.88 9477**	1.72	**9.91 0951**	4.30	**10.08 9049**	**50**
11	0582	2.58	9374	1.72	1209	4.30	8791	49
12	0737	2.58	9271	1.72	1467	4.30	8533	48
13	0892	2.58	9168	1.73	1725	4.28	8275	47
14	1047	2.57	9064	1.72	1982	4.30	8018	46
15	**9.80 1201**	2.58	**9.88 8961**	1.72	**9.91 2240**	4.30	**10.08 7760**	**45**
16	1356	2.58	8858	1.72	2498	4.30	7502	44
17	1511	2.57	8755	1.73	2756	4.30	7244	43
18	1665	2.57	8651	1.72	3014	4.28	6986	42
19	1819	2.57	8548	1.73	3271	4.30	6729	41
20	**9.80 1973**	2.58	**9.88 8444**	1.72	**9.91 3529**	4.30	**10.08 6471**	**40**
21	2128	2.57	8341	1.73	3787	4.28	6213	39
22	2282	2.57	8237	1.72	4044	4.30	5956	38
23	2436	2.55	8134	1.73	4302	4.30	5698	37
24	2589	2.57	8030	1.73	4560	4.28	5440	36
25	**9.80 2743**	2.57	**9.88 7926**	1.73	**9.91 4817**	4.30	**10.08 5183**	**35**
26	2897	2.55	7822	1.73	5075	4.28	4925	34
27	3050	2.57	7718	1.73	5332	4.30	4668	33
28	3204	2.55	7614	1.73	5590	4.28	4410	32
29	3357	2.57	7510	1.73	5847	4.28	4153	31
30	**9.80 3511**	2.55	**9.88 7406**	1.73	**9.91 6104**	4.30	**10.08 3896**	**30**
31	3664	2.55	7302	1.73	6362	4.28	3638	29
32	3817	2.55	7198	1.75	6619	4.30	3381	28
33	3970	2.55	7093	1.73	6877	4.28	3123	27
34	4123	2.55	6989	1.73	7134	4.28	2866	26
35	**9.80 4276**	2.55	**9.88 6885**	1.75	**9.91 7391**	4.28	**10.08 2609**	**25**
36	4428	2.55	6780	1.73	7648	4.30	2352	24
37	4581	2.55	6676	1.75	7906	4.28	2094	23
38	4734	2.53	6571	1.75	8163	4.28	1837	22
39	4886	2.55	6466	1.73	8420	4.28	1580	21
40	**9.80 5039**	2.53	**9.88 6362**	1.75	**9.91 8677**	4.28	**10.08 1323**	**20**
41	5191	2.53	6257	1.75	8934	4.28	1066	19
42	5343	2.53	6152	1.75	9191	4.28	0809	18
43	5495	2.53	6047	1.75	9448	4.28	0552	17
44	5647	2.53	5942	1.75	9705	4.28	0295	16
45	**9.80 5799**	2.53	**9.88 5837**	1.75	**9.91 9962**	4.28	**10.08 0038**	**15**
46	5951	2.53	5732	1.75	.92 0219	4.28	.07 9781	14
47	6103	2.52	5627	1.75	0476	4.28	9524	13
48	6254	2.53	5522	1.77	0733	4.28	9267	12
49	6406	2.52	5416	1.75	0990	4.28	9010	11
50	**9.80 6557**	2.58	**9.88 5311**	1.77	**9.92 1247**	4.27	**10.07 8753**	**10**
51	6709	2.52	5205	1.75	1503	4.28	8497	9
52	6860	2.52	5100	1.77	1760	4.28	8240	8
53	7011	2.53	4994	1.75	2017	4.28	7983	7
54	7163	2.52	4889	1.77	2274	4.27	7726	6
55	**9.80 7314**	2.52	**9.88 4783**	1.77	**9.92 2530**	4.28	**10.07 7470**	**5**
56	7465	2.50	4677	1.75	2787	4.28	7213	4
57	7615	2.52	4572	1.77	3044	4.27	6956	3
58	7766	2.52	4466	1.77	3300	4.28	6700	2
59	7917	2.50	4360	1.77	3557	4.28	6443	1
60	**9.80 8067**		**9.88 4254**		**9.92 3814**		**10.07 6186**	**0**
′	Cosine.	D.1″.	Sine.	D.1″.	Cotang.	D.1″.	Tang.	′

129° 50°

40° TABLE XV.—*Continued* 139°

′	Sine.	D.1″.	Cosine.	D.1″.	Tang.	D.1″.	Cotang.	′
0	**9.80 8067**	2.52	**9.88 4254**	1.77	**9.92 3814**	4.27	**10.07 6186**	**60**
1	8218	2.50	4148	1.77	4070	4.28	5930	59
2	8368	2.52	4042	1.77	4327	4.27	5673	58
3	8519	2.50	3936	1.78	4583	4.28	5417	57
4	8669	2.50	3829	1.77	4840	4.27	5160	56
5	**9.80 8819**	2.50	**9.88 3723**	1.77	**9.92 5096**	4.27	**10.07 4904**	**55**
6	8969	2.50	3617	1.78	5352	4.28	4648	54
7	9119	2.50	3510	1.77	5609	4.27	4391	53
8	9269	2.50	3404	1.78	5865	4.28	4135	52
9	9419	2.50	3297	1.77	6122	4.27	3878	51
10	**9.80 9569**	2.48	**9.88 3191**	1.78	**9.92 6378**	4.27	**10.07 3622**	**50**
11	9718	2.50	3084	1.78	6634	4.27	3366	49
12	.80 9868	2.48	2977	1.77	6890	4.28	3110	48
13	.81 0017	2.50	2871	1.78	7147	4.27	2853	47
14	0167	2.48	2764	1.78	7403	4.27	2597	46
15	**9.81 0316**	2.48	**9.88 2657**	1.78	**9.92 7659**	4.27	**10.07 2341**	**45**
16	0465	2.48	2550	1.78	7915	4.27	2085	44
17	0614	2.48	2443	1.78	8171	4.27	1829	43
18	0763	2.48	2336	1.78	8427	4.28	1573	42
19	0912	2.48	2229	1.80	8684	4.27	1316	41
20	**9.81 1061**	2.48	**9.88 2121**	1.78	**9.92 8940**	4.27	**10.07 1060**	**40**
21	1210	2.47	2014	1.78	9196	4.27	0804	39
22	1358	2.48	1907	1.80	9452	4.27	0548	38
23	1507	2.47	1799	1.78	9708	4.27	0292	37
24	1655	2.48	1692	1.80	.92 9964	4.27	.07 0036	36
25	**9.81 1804**	2.47	**9.88 1584**	1.78	**9.93 0220**	4.25	**10.06 9780**	**35**
26	1952	2.47	1477	1.80	0475	4.27	9525	34
27	2100	2.47	1369	1.80	0731	4.27	9269	33
28	2248	2.47	1261	1.80	0987	4.27	9013	32
29	2396	2.47	1153	1.78	1243	4.27	8757	31
30	**9.81 2544**	2.47	**9.88 1046**	1.80	**9.93 1499**	4.27	**10.06 8501**	**30**
31	2692	2.47	0938	1.80	1755	4.25	8245	29
32	2840	2.47	0830	1.80	2010	4.27	7990	28
33	2988	2.45	0722	1.82	2266	4.27	7734	27
34	3135	2.47	0613	1.80	2522	4.27	7478	26
35	**9.81 3283**	2.45	**9.88 0505**	1.80	**9.93 2778**	4.25	**10.06 7222**	**25**
36	3430	2.47	0397	1.80	3033	4.27	6967	24
37	3578	2.45	0289	1.82	3289	4.27	6711	23
38	3725	2.45	0180	1.80	3545	4.25	6455	22
39	3872	2.45	.88 0072	1.82	3800	4.27	6200	21
40	**9.81 4019**	2.45	**9.87 9963**	1.80	**9.93 4056**	4.25	**10.06 5944**	**20**
41	4166	2.45	9855	1.82	4311	4.27	5689	19
42	4313	2.45	9746	1.82	4567	4.25	5433	18
43	4460	2.45	9637	1.80	4822	4.27	5178	17
44	4607	2.43	9529	1.82	5078	4.25	4922	16
45	**9.81 4753**	2.45	**9.87 9420**	1.82	**9.93 5333**	4.27	**10.06 4667**	**15**
46	4900	2.43	9311	1.82	5589	4.25	4411	14
47	5046	2.45	9202	1.82	5844	4.27	4156	13
48	5193	2.43	9093	1.82	6100	4.25	3900	12
49	5339	2.43	8984	1.82	6355	4.27	3645	11
50	**9.81 5485**	2.45	**9.87 8875**	1.82	**9.93 6611**	4.25	**10.06 3389**	**10**
51	5632	2.43	8766	1.83	6866	4.25	3134	9
52	5778	2.43	8656	1.82	7121	4.27	2879	8
53	5924	2.42	8547	1.82	7377	4.25	2623	7
54	6069	2.43	8438	1.83	7632	4.25	2368	6
55	**9.81 6215**	2.43	**9.87 8328**	1.82	**9.93 7887**	4.25	**10.06 2113**	**5**
56	6361	2.43	8219	1.83	8142	4.27	1858	4
57	6507	2.42	8109	1.83	8398	4.25	1602	3
58	6652	2.43	7999	1.82	8653	4.25	1347	2
59	6798	2.42	7890	1.83	8908	4.25	1092	1
60	**9.81 6943**		**9.87 7780**		**9.93 9163**		**10.06 0837**	**0**
′	Cosine.	D.1″.	Sine.	D.1″.	Cotang.	D.1″.	Tang.	′

130° 49°

41° TABLE XV.—*Continued* 138°

′	Sine.	D.1″.	Cosine.	D.1″.	Tang.	D.1″.	Cotang.	′
0	9.81 **6943**	2.42	9.87 **7780**	1.83	9.93 **9163**	4.25	10.06 **0837**	**60**
1	7088	2.42	7670	1.83	9418	4.25	0582	59
2	7233	2.43	7560	1.83	9673	4.25	0327	58
3	7379	2.42	7450	1.83	.93 9928	4.25	.06 0072	57
4	7524	2.40	7340	1.83	.94 0183	4.27	.05 9817	56
5	9.81 **7668**	2.42	9.87 **7230**	1.83	9.94 **0439**	4.25	10.05 **9561**	**55**
6	7813	2.42	7120	1.83	0694	4.25	9306	54
7	7958	2.42	7010	1.85	0949	4.25	9051	53
8	8103	2.40	6899	1.83	1204	4.25	8796	52
9	8247	2.42	6789	1.85	1459	4.23	8541	51
10	9.81 **8392**	2.40	9.87 **6678**	1.83	9.94 **1713**	4.25	10.05 **8287**	**50**
11	8536	2.42	6568	1.85	1968	4.25	8032	49
12	8681	2.40	6457	1.83	2223	4.25	7777	48
13	8825	2.40	6347	1.85	2478	4.25	7522	47
14	8969	2.40	6236	1.85	2733	4.25	7267	46
15	9.81 **9113**	2.40	9.87 **6125**	1.85	9.94 **2988**	4.25	10.05 **7012**	**45**
16	9257	2.40	6014	1.83	3243	4.25	6757	44
17	9401	2.40	5904	1.85	3498	4.23	6502	43
18	9545	2.40	5793	1.85	3752	4.25	6248	42
19	9689	2.38	5682	1.85	4007	4.25	5993	41
20	9.81 **9832**	2.40	9.87 **5571**	1.87	9.94 **4262**	4.25	10.05 **5738**	**40**
21	.81 9976	2.40	5459	1.85	4517	4.23	5483	39
22	.82 0120	2.38	5348	1.85	4771	4.25	5229	38
23	0263	2.38	5237	1.85	5026	4.25	4974	37
24	0406	2.40	5126	1.87	5281	4.23	4719	36
25	9.82 **0550**	2.38	9.87 **5014**	1.85	9.94 **5535**	4.25	10.05 **4465**	**35**
26	0693	2.38	4903	1.87	5790	4.25	4210	34
27	0836	2.38	4791	1.85	6045	4.23	3955	33
28	0979	2.38	4680	1.87	6299	4.25	3701	32
29	1122	2.38	4568	1.87	6554	4.23	3446	31
30	9.82 **1265**	2.37	9.87 **4456**	1.87	9.94 **6808**	4.25	10.05 **3192**	**30**
31	1407	2.38	4344	1.87	7063	4.25	2937	29
32	1550	2.38	4232	1.85	7318	4.23	2682	28
33	1693	2.37	4121	1.87	7572	4.25	2428	27
34	1835	2.37	4009	1.88	7827	4.23	2173	26
35	9.82 **1977**	2.38	9.87 **3896**	1.87	9.94 **8081**	4.23	10.05 **1919**	**25**
36	2120	2.37	3784	1.87	8335	4.25	1665	24
37	2262	2.37	3672	1.87	8590	4.23	1410	23
38	2404	2.37	3560	1.87	8844	4.25	1156	22
39	2546	2.37	3448	1.88	9099	4.23	0901	21
40	9.82 **2688**	2.37	9.87 **3335**	1.87	9.94 **9353**	4.25	10.05 **0647**	**20**
41	2830	2.37	3223	1.88	9608	4.23	0392	19
42	2972	2.37	3110	1.87	.94 9862	4.23	.05 0138	18
43	3114	2.35	2998	1.88	.95 0116	4.25	.04 9884	17
44	3255	2.37	2885	1.88	0371	4.23	9629	16
45	9.82 **3397**	2.37	9.87 **2772**	1.88	9.95 **0625**	4.23	10.04 **9375**	**15**
46	3539	2.35	2659	1.87	0879	4.23	9121	14
47	3680	2.35	2547	1.88	1133	4.25	8867	13
48	3821	2.37	2434	1.88	1388	4.23	8612	12
49	3963	2.35	2321	1.88	1642	4.23	8358	11
50	9.82 **4104**	2.35	9.87 **2208**	1.88	9.95 **1896**	4.23	10.04 **8104**	**10**
51	4245	2.35	2095	1.90	2150	4.25	7850	9
52	4386	2.35	1981	1.88	2405	4.23	7595	8
53	4527	2.35	1868	1.88	2659	4.23	7341	7
54	4668	2.33	1755	1.90	2913	4.23	7087	6
55	9.82 **4808**	2.35	9.87 **1641**	1.88	9.95 **3167**	4.23	10.04 **6833**	**5**
56	4949	2.35	1528	1.90	3421	4.23	6579	4
57	5090	2.33	1414	1.88	3675	4.23	6325	3
58	5230	2.35	1301	1.90	3929	4.23	6071	2
59	5371	2.33	1187	1.90	4183	4.23	5817	1
60	9.82 **5511**		9.87 **1073**		9.95 **4437**		10.04 **5563**	**0**
′	Cosine.	D.1″.	Sine.	D.1″.	Cotang.	D.1″.	Tang.	′

131° **48°**

42° TABLE XV.—*Continued* 137°

′	Sine.	D.1″.	Cosine.	D.1″.	Tang.	D.1″.	Cotang.	′
0	9.82 5511	2.33	9.87 1073	1.88	9.95 4437	4.23	10.04 5563	60
1	5651	2.33	0960	1.90	4691	4.25	5309	59
2	5791	2.33	0846	1.90	4946	4.23	5054	58
3	5931	2.33	0732	1.90	5200	4.23	4800	57
4	6071	2.33	0618	1.90	5454	4.23	4546	56
5	9.82 6211	2.33	9.87 0504	1.90	9.95 5708	4.22	10.04 4292	55
6	6351	2.33	0390	1.90	5961	4.23	4039	54
7	6491	2.33	0276	1.92	6215	4.23	3785	53
8	6631	2.32	0161	1.90	6469	4.23	3531	52
9	6770	2.33	.87 0047	1.90	6723	4.23	3277	51
10	9.82 6910	2.32	9.86 9933	1.92	9.95 6977	4.23	10.04 3023	50
11	7049	2.33	9818	1.90	7231	4.23	2769	49
12	7189	2.32	9704	1.92	7485	4.23	2515	48
13	7328	2.32	9589	1.92	7739	4.23	2261	47
14	7467	2.32	9474	1.90	7993	4.23	2007	46
15	9.82 7606	2.32	9.86 9360	1.92	9.95 8247	4.22	10.04 1753	45
16	7745	2.32	9245	1.92	8500	4.23	1500	44
17	7884	2.32	9130	1.92	8754	4.23	1246	43
18	8023	2.32	9015	1.92	9008	4.23	0992	42
19	8162	2.32	8900	1.92	9262	4.23	0738	41
20	9.82 8301	2.30	9.86 8785	1.92	9.95 9516	4.22	10.04 0484	40
21	8439	2.32	8670	1.92	.95 9769	4.23	.04 0231	39
22	8578	2.30	8555	1.92	.96 0023	4.23	.03 9977	38
23	8716	2.32	8440	1.93	0277	4.22	9723	37
24	8855	2.30	8324	1.92	0530	4.23	9470	36
25	9.82 8993	2.30	9.86 8209	1.93	9.96 0784	4.23	10.03 9216	35
26	9131	2.30	8093	1.92	1038	4.23	8962	34
27	9269	2.30	7978	1.93	1292	4.22	8708	33
28	9407	2.30	7862	1.92	1545	4.23	8455	32
29	9545	2.30	7747	1.93	1799	4.22	8201	31
30	9.82 9683	2.30	9.86 7631	1.93	9.96 2052	4.23	10.03 7948	30
31	9821	2.30	7515	1.93	2306	4.23	7694	29
32	.82 9959	2.30	7399	1.93	2560	4.22	7440	28
33	.83 0097	2.28	7283	1.93	2813	4.23	7187	27
34	0234	2.30	7167	1.93	3067	4.22	6933	26
35	9.83 0372	2.28	9.86 7051	1.93	9.96 3320	4.23	10.03 6680	25
36	0509	2.28	6935	1.93	3574	4.23	6426	24
37	0646	2.30	6819	1.93	3828	4.22	6172	23
38	0784	2.28	6703	1.95	4081	4.23	5919	22
39	0921	2.28	6586	1.93	4335	4.22	5665	21
40	9.83 1058	2.28	9.86 6470	1.95	9.96 4588	4.23	10.03 5412	20
41	1195	2.28	6353	1.93	4842	4.22	5158	19
42	1332	2.28	6237	1.95	5095	4.23	4905	18
43	1469	2.28	6120	1.93	5349	4.22	4651	17
44	1606	2.27	6004	1.95	5602	4.22	4398	16
45	9.83 1742	2.28	9.86 5887	1.95	9.96 5855	4.23	10.03 4145	15
46	1879	2.27	5770	1.95	6109	4.22	3891	14
47	2015	2.28	5653	1.95	6362	4.23	3638	13
48	2152	2.27	5536	1.95	6616	4.22	3384	12
49	2288	2.28	5419	1.95	6869	4.23	3131	11
50	9.83 2425	2.27	9.86 5302	1.95	9.96 7123	4.22	10.03 2877	10
51	2561	2.27	5185	1.95	7376	4.22	2624	9
52	2697	2.27	5068	1.97	7629	4.23	2371	8
53	2833	2.27	4950	1.95	7883	4.22	2117	7
54	2969	2.27	4833	1.95	8136	4.22	1864	6
55	9.83 3105	2.27	9.86 4716	1.97	9.96 8389	4.23	10.03 1611	5
56	3241	2.27	4598	1.95	8643	4.22	1357	4
57	3377	2.25	4481	1.97	8896	4.22	1104	3
58	3512	2.27	4363	1.97	9149	4.23	0851	2
59	3648	2.25	4245	1.97	9403	4.22	0597	1
60	9.83 3783		9.86 4127		9.96 9656		10.03 0344	0
′	Cosine.	D.1″.	Sine.	D.1″.	Cotang.	D.1″.	Tang.	′

132° 47°

43° **TABLE XV.—*Continued*** 136°

′	Sine.	D.1″.	Cosine.	D.1″.	Tang.	D.1″.	Cotang.	′
0	**9.83 3783**	2.27	**9.86 4127**	1.95	**9.96 9656**	4.22	**10.03 0344**	**60**
1	3919	2.25	4010	1.97	.96 9909	4.22	.03 0091	59
2	4054	2.25	3892	1.97	.97 0162	4.23	.02 9838	58
3	4189	2.27	3774	1.97	0416	4.22	9584	57
4	4325	2.25	3656	1.97	0669	4.22	9331	56
5	**9.83 4460**	2.25	**9.86 3538**	1.98	**9.97 0922**	4.22	**10.02 9078**	**55**
6	4595	2.25	3419	1.97	1175	4.23	8825	54
7	4730	2.25	3301	1.97	1429	4.22	8571	53
8	4865	2.23	3183	1.98	1682	4.22	8318	52
9	4999	2.25	3064	1.97	1935	4.22	8065	51
10	**9.83 5134**	2.25	**9.86 2946**	1.98	**9.97 2188**	4.22	**10.02 7812**	**50**
11	5269	2.23	2827	1.97	2441	4.23	7559	49
12	5403	2.25	2709	1.98	2695	4.22	7305	48
13	5538	2.23	2590	1.98	2948	4.22	7052	47
14	5672	2.25	2471	1.97	3201	4.22	6799	46
15	**9.83 5807**	2.23	**9.86 2353**	1.98	**9.97 3454**	4.22	**10.02 6546**	**45**
16	5941	2.23	2234	1.98	3707	4.22	6293	44
17	6075	2.23	2115	1.98	3960	4.22	6040	43
18	6209	2.23	1996	1.98	4213	4.22	5787	42
19	6343	2.23	1877	1.98	4466	4.23	5534	41
20	**9.83 6477**	2.23	**9.86 1758**	2.00	**9.97 4720**	4.22	**10.02 5280**	**40**
21	6611	2.23	1638	1.98	4973	4.22	5027	39
22	6745	2.22	1519	1.98	5226	4.22	4774	38
23	6878	2.23	1400	2.00	5479	4.22	4521	37
24	7012	2.23	1280	1.98	5732	4.22	4268	36
25	**9.83 7146**	2.22	**9.86 1161**	2.00	**9.97 5985**	4.22	**10.02 4015**	**35**
26	7279	2.22	1041	1.98	6238	4.22	3762	34
27	7412	2.23	0922	2.00	6491	4.22	3509	33
28	7546	2.22	0802	2.00	6744	4.22	3256	32
29	7679	2.22	0682	2.00	6997	4.22	3003	31
30	**9.83 7812**	2.22	**9.86 0562**	2.00	**9.97 7250**	4.22	**10.02 2750**	**30**
31	7945	2.22	0442	2.00	7503	4.22	2497	29
32	8078	2.22	0322	2.00	7756	4.22	2244	28
33	8211	2.22	0202	2.00	8009	4.22	1991	27
34	8344	2.22	.86 0082	2.00	8262	4.22	1738	26
35	**9.83 8477**	2.22	**9.85 9962**	2.00	**9.97 8515**	4.22	**10.02 1485**	**25**
36	8610	2.20	9842	2.02	8768	4.22	1232	24
37	8742	2.22	9721	2.00	9021	4.22	0979	23
38	8875	2.20	9601	2.02	9274	4.22	0726	22
39	9007	2.22	9480	2.00	9527	4.22	0473	21
40	**9.83 9140**	2.20	**9.85 9360**	2.02	**9.97 9780**	4.22	**10.02 0220**	**20**
41	9272	2.20	9239	2.00	.98 0033	4.22	.01 9967	19
42	9404	2.20	9119	2.02	0286	4.20	9714	18
43	9536	2.20	8998	2.02	0538	4.22	9462	17
44	9668	2.20	8877	2.02	0791	4.22	9209	16
45	**9.83 9800**	2.20	**9.85 8756**	2.02	**9.98 1044**	4.22	**10.01 8956**	**15**
46	.83 9932	2.20	8635	2.02	1297	4.22	8703	14
47	.84 0064	2.20	8514	2.02	1550	4.22	8450	13
48	0196	2.20	8393	2.02	1803	4.22	8197	12
49	0328	2.18	8272	2.02	2056	4.22	7944	11
50	**9.84 0459**	2.20	**9.85 8151**	2.03	**9.98 2309**	4.22	**10.01 7691**	**10**
51	0591	2.18	8029	2.02	2562	4.20	7438	9
52	0722	2.20	7908	2.03	2814	4.22	7186	8
53	0854	2.18	7786	2.02	3067	4.22	6933	7
54	0985	2.18	7665	2.03	3320	4.22	6680	6
55	**9.84 1116**	2.18	**9.85 7543**	2.02	**9.98 3573**	4.22	**10.01 6427**	**5**
56	1247	2.18	7422	2.03	3826	4.22	6174	4
57	1378	2.18	7300	2.03	4079	4.22	5921	3
58	1509	2.18	7178	2.03	4332	4.20	5668	2
59	1640	2.18	7056	2.03	4584	4.22	5416	1
60	**9.84 1771**		**9.85 6934**		**9.98 4837**		**10.01 5163**	**0**
′	Cosine.	D.1″.	Sine.	D.1″.	Cotang.	D.1″.	Tang.	′

133° **46°**

44° TABLE XV.—*Concluded* 135°

′	Sine.	D.1″.	Cosine.	D.1″.	Tang.	D.1″.	Cotang.	′
0	**9.84 1771**	2.18	**9.85 6934**	2.03	**9.98 4837**	4.22	**10.01 5163**	60
1	1902	2.18	6812	2.03	5090	4.22	4910	59
2	2033	2.17	6690	2.03	5343	4.22	4657	58
3	2163	2.18	6568	2.03	5596	4.20	4404	57
4	2294	2.17	6446	2.05	5848	4.22	4152	56
5	**9.84 2424**	2.18	**9.85 6323**	2.03	**9.98 6101**	4.22	**10.01 3899**	55
6	2555	2.17	6201	2.05	6354	4.22	3646	54
7	2685	2.17	6078	2.03	6607	4.22	3393	53
8	2815	2.18	5956	2.05	6860	4.20	3140	52
9	2946	2.17	5833	2.03	7112	4.22	2888	51
10	**9.84 3076**	2.17	**9.85 5711**	2.05	**9.98 7365**	4.22	**10.01 2635**	50
11	3206	2.17	5588	2.05	7618	4.22	2382	49
12	3336	2.17	5465	2.05	7871	4.20	2129	48
13	3466	2.15	5342	2.05	8123	4.22	1877	47
14	3595	2.17	5219	2.05	8376	4.22	1624	46
15	**9.84 3725**	2.17	**9.85 5096**	2.05	**9.98 8629**	4.22	**10.01 1371**	45
16	3855	2.15	4973	2.05	8882	4.20	1118	44
17	3984	2.17	4850	2.05	9134	4.22	0866	43
18	4114	2.15	4727	2.07	9387	4.22	0613	42
19	4243	2.15	4603	2.05	9640	4.22	0360	41
20	**9.84 4372**	2.17	**9.85 4480**	2.07	**9.98 9893**	4.20	**10.01 0107**	40
21	4502	2.15	4356	2.05	.99 0145	4.22	.00 9855	39
22	4631	2.15	4233	2.07	0398	4.22	9602	38
23	4760	2.15	4109	2.05	0651	4.20	9349	37
24	4889	2.15	3986	2.07	0903	4.22	9097	36
25	**9.84 5018**	2.15	**9.85 3862**	2.07	**9.99 1156**	4.22	**10.00 8844**	35
26	5147	2.15	3738	2.07	1409	4.22	8591	34
27	5276	2.15	3614	2.07	1662	4.20	8338	33
28	5405	2.13	3490	2.07	1914	4.22	8086	32
29	5533	2.15	3366	2.07	2167	4.22	7833	31
30	**9.84 5662**	2.13	**9.85 3242**	2.07	**9.99 2420**	4.20	**10.00 7580**	30
31	5790	2.15	3118	2.07	2672	4.22	7328	29
32	5919	2.13	2994	2.08	2925	4.22	7075	28
33	6047	2.13	2869	2.07	3178	4.22	6822	27
34	6175	2.15	2745	2.08	3431	4.20	6569	26
35	**9.84 6304**	2.13	**9.85 2620**	2.07	**9.99 3683**	4.22	**10.00 6317**	25
36	6432	2.13	2496	2.08	3936	4.22	6064	24
37	6560	2.13	2371	2.07	4189	4.20	5811	23
38	6688	2.13	2247	2.08	4441	4.22	5559	22
39	6816	2.13	2122	2.08	4694	4.22	5306	21
40	**9.84 6944**	2.12	**9.85 1997**	2.08	**9.99 4947**	4.20	**10.00 5053**	20
41	7071	2.13	1872	2.08	5199	4.22	4801	19
42	7199	2.13	1747	2.08	5452	4.22	4548	18
43	7327	2.12	1622	2.08	5705	4.20	4295	17
44	7454	2.13	1497	2.08	5957	4.22	4043	16
45	**9.84 7582**	2.12	**9.85 1372**	2.10	**9.99 6210**	4.22	**10.00 3790**	15
46	7709	2.12	1246	2.08	6463	4.20	3537	14
47	7836	2.13	1121	2.08	6715	4.22	3285	13
48	7964	2.12	0996	2.10	6968	4.22	3032	12
49	8091	2.12	0870	2.08	7221	4.20	2779	11
50	**9.84 8218**	2.12	**9.85 0745**	2.10	**9.99 7473**	4.22	**10.00 2527**	10
51	8345	2.12	0619	2.10	7726	4.22	2274	9
52	8472	2.12	0493	2.08	7979	4.20	2021	8
53	8599	2.12	0368	2.10	8231	4.22	1769	7
54	8726	2.10	0242	2.10	8484	4.22	1516	6
55	**9.84 8852**	2.12	**9.85 0116**	2.10	**9.99 8737**	4.20	**10.00 1263**	5
56	8979	2.12	84 9990	2.10	8989	4.22	1011	4
57	9106	2.10	9864	2.10	9242	4.22	0758	3
58	9232	2.12	9738	2.12	9495	4.20	0505	2
59	9359	2.10	9611	2.10	9.99 9747	4.22	0253	1
60	**9.84 9485**		**9.84 9485**		10.00 0000		10.00 0000	0
′	Cosine.	D.1″.	Sine.	D.1″.	Cotang.	D.1″.	Tang.	′

134° 45°

TABLE XVI. NATURAL SINES AND COSINES

′	0°		′	′	0°		′	′	0°		′
′	SINE	COSINE	′	′	SINE	COSINE	′	′	SINE	COSINE	′
0	.00000	1	60	21	.00611	.99998	39	41	.01193	.99993	19
1	.00029	1	59	22	.00640	.99998	38	42	.01222	.99993	18
2	.00058	1	58	23	.00669	.99998	37	43	.01251	.99992	17
3	.00087	1	57	24	.00698	.99998	36	44	.01280	.99992	16
4	.00116	1	56	25	.00727	.99997	35	45	.01309	.99991	15
5	.00145	1	55	26	.00756	.99997	34	46	.01338	.99991	14
6	.00175	1	54	27	.00785	.99997	33	47	.01367	.99991	13
7	.00204	1	53	28	.00814	.99997	32	48	.01396	.99990	12
8	.00233	1	52	29	.00844	.99996	31	49	.01425	.99990	11
9	.00262	1	51	30	.00873	.99996	30	50	.01454	.99989	10
10	.00291	1	50	31	.00902	.99996	29	51	.01483	.99989	9
11	.00320	.99999	49	32	.00931	.99996	28	52	.01513	.99989	8
12	.00349	.99999	48	33	.00960	.99995	27	53	.01542	.99988	7
13	.00378	.99999	47	34	.00989	.99995	26	54	.01571	.99988	6
14	.00407	.99999	46	35	.01018	.99995	25	55	.01600	.99987	5
15	.00436	.99999	45	36	.01047	.99995	24	56	.01629	.99987	4
16	.00465	.99999	44	37	.01076	.99994	23	57	.01658	.99986	3
17	.00495	.99999	43	38	.01105	.99994	22	58	.01687	.99986	2
18	.00524	.99999	42	39	.01134	.99994	21	59	.01716	.99985	1
19	.00553	.99998	41	40	.01164	.99993	20	60	.01745	.99985	0
20	.00582	.99998	40								
′	COSINE	SINE	′	′	COSINE	SINE	′	′	COSINE	SINE	′
	89°				89°				89°		

TABLE XVI.—*Continued*

′	1° Sine	1° Cosine	2° Sine	2° Cosine	3° Sine	3° Cosine	4° Sine	4° Cosine	′
0	.01745	.99985	.03490	.99939	.05234	.99863	.06976	.99756	60
1	.01774	.99984	.03519	.99938	.05263	.99861	.07005	.99754	59
2	.01803	.99984	.03548	.99937	.05292	.99860	.07034	.99752	58
3	.01832	.99983	.03577	.99936	.05321	.99858	.07063	.99750	57
4	.01862	.99983	.03606	.99935	.05350	.99857	.07092	.99748	56
5	.01891	.99982	.03635	.99934	.05379	.99855	.07121	.99746	55
6	.01920	.99982	.03664	.99933	.05408	.99854	.07150	.99744	54
7	.01949	.99981	.03693	.99932	.05437	.99852	.07179	.99742	53
8	.01978	.99980	.03723	.99931	.05466	.99851	.07208	.99740	52
9	.02007	.99980	.03752	.99930	.05495	.99849	.07237	.99738	51
10	.02036	.99979	.03781	.99929	.05524	.99847	.07266	.99736	50
11	.02065	.99979	.03810	.99927	.05553	.99846	.07295	.99734	49
12	.02094	.99978	.03839	.99926	.05582	.99844	.07324	.99731	48
13	.02123	.99977	.03868	.99925	.05611	.99842	.07353	.99729	47
14	.02152	.99977	.03897	.99924	.05640	.99841	.07382	.99727	46
15	.02181	.99976	.03926	.99923	.05669	.99839	.07411	.99725	45
16	.02211	.99976	.03955	.99922	.05698	.99838	.07440	.99723	44
17	.02240	.99975	.03984	.99921	.05727	.99836	.07469	.99721	43
18	.02269	.99974	.04013	.99919	.05756	.99834	.07498	.99719	42
19	.02298	.99974	.04042	.99918	.05785	.99833	.07527	.99716	41
20	.02327	.99973	.04071	.99917	.05814	.99831	.07556	.99714	40
21	.02356	.99972	.04100	.99916	.05844	.99829	.07585	.99712	39
22	.02385	.99972	.04129	.99915	.05873	.99827	.07614	.99710	38
23	.02414	.99971	.04159	.99913	.05902	.99826	.07643	.99708	37
24	.02443	.99970	.04188	.99912	.05931	.99824	.07672	.99705	36
25	.02472	.99969	.04217	.99911	.05960	.99822	.07701	.99703	35
26	.02501	.99969	.04246	.99910	.05989	.99821	.07730	.99701	34
27	.02530	.99968	.04275	.99909	.06018	.99819	.07759	.99699	33
28	.02560	.99967	.04304	.99907	.06047	.99817	.07788	.99696	32
29	.02589	.99966	.04333	.99906	.06076	.99815	.07817	.99694	31
30	.02618	.99966	.04362	.99905	.06105	.99813	.07846	.99692	30
31	.02647	.99965	.04391	.99904	.06134	.99812	.07875	.99689	29
32	.02676	.99964	.04420	.99902	.06163	.99810	.07904	.99687	28
33	.02705	.99963	.04449	.99901	.06192	.99808	.07933	.99685	27
34	.02734	.99963	.04478	.99900	.06221	.99806	.07962	.99683	26
35	.02763	.99962	.04507	.99898	.06250	.99804	.07991	.99680	25
36	.02792	.99961	.04536	.99897	.06279	.99803	.08020	.99678	24
37	.02821	.99960	.04565	.99896	.06308	.99801	.08049	.99676	23
38	.02850	.99959	.04594	.99894	.06337	.99799	.08078	.99673	22
39	.02879	.99959	.04623	.99893	.06366	.99797	.08107	.99671	21
40	.02908	.99958	.04653	.99892	.06395	.99795	.08136	.99668	20
41	.02938	.99957	.04682	.99890	.06424	.99793	.08165	.99666	19
42	.02967	.99956	.04711	.99889	.06453	.99792	.08194	.99664	18
43	.02996	.99955	.04740	.99888	.06482	.99790	.08223	.99661	17
44	.03025	.99954	.04769	.99886	.06511	.99788	.08252	.99659	16
45	.03054	.99953	.04798	.99885	.06540	.99786	.08281	.99657	15
46	.03083	.99952	.04827	.99883	.06569	.99784	.08310	.99654	14
47	.03112	.99952	.04856	.99882	.06598	.99782	.08339	.99652	13
48	.03141	.99951	.04885	.99881	.06627	.99780	.08368	.99649	12
49	.03170	.99950	.04914	.99879	.06656	.99778	.08397	.99647	11
50	.03199	.99949	.04943	.99878	.06685	.99776	.08426	.99644	10
51	.03228	.99948	.04972	.99876	.06714	.99774	.08455	.99642	9
52	.03257	.99947	.05001	.99875	.06743	.99772	.08484	.99639	8
53	.03286	.99946	.05030	.99873	.06773	.99770	.08513	.99637	7
54	.03316	.99945	.05059	.99872	.06802	.99768	.08542	.99635	6
55	.03345	.99944	.05088	.99870	.06831	.99766	.08571	.99632	5
56	.03374	.99943	.05117	.99869	.06860	.99764	.08600	.99630	4
57	.03403	.99942	.05146	.99867	.06889	.99762	.08629	.99627	3
58	.03432	.99941	.05175	.99866	.06918	.99760	.08658	.99625	2
59	.03461	.99940	.05205	.99864	.06947	.99758	.08687	.99622	1
60	.03490	.99939	.05234	.99863	.06976	.99756	.08716	.99619	0
′	Cosine 88°	Sine 88°	Cosine 87°	Sine 87°	Cosine 86°	Sine 86°	Cosine 85°	Sine 85°	′

TABLE XVI.—*Continued*

′	5° Sine	5° Cosine	6° Sine	6° Cosine	7° Sine	7° Cosine	8° Sine	8° Cosine	′
0	.08716	.99619	.10453	.99452	.12187	.99255	.13917	.99027	60
1	.08745	.99617	.10482	.99449	.12216	.99251	.13946	.99023	59
2	.08774	.99614	.10511	.99446	.12245	.99248	.13975	.99019	58
3	.08803	.99612	.10540	.99443	.12274	.99244	.14004	.99015	57
4	.08831	.99609	.10569	.99440	.12302	.99240	.14033	.99011	56
5	.08860	.99607	.10597	.99437	.12331	.99237	.14061	.99006	55
6	.08889	.99604	.10626	.99434	.12360	.99233	.14090	.99002	54
7	.08918	.99602	.10655	.99431	.12389	.99230	.14119	.98998	53
8	.08947	.99599	.10684	.99428	.12418	.99226	.14148	.98994	52
9	.08976	.99596	.10713	.99424	.12447	.99222	.14177	.98990	51
10	.09005	.99594	.10742	.99421	.12476	.99219	.14205	.98986	50
11	.09034	.99591	.10771	.99418	.12504	.99215	.14234	.98982	49
12	.09063	.99588	.10800	.99415	.12533	.99211	.14263	.98978	48
13	.09092	.99586	.10829	.99412	.12562	.99208	.14292	.98973	47
14	.09121	.99583	.10858	.99409	.12591	.99204	.14320	.98969	46
15	.09150	.99580	.10887	.99406	.12620	.99200	.14349	.98965	45
16	.09179	.99578	.10916	.99402	.12649	.99197	.14378	.98961	44
17	.09208	.99575	.10945	.99399	.12678	.99193	.14407	.98957	43
18	.09237	.99572	.10973	.99396	.12706	.99189	.14436	.98953	42
19	.09266	.99570	.11002	.99393	.12735	.99186	.14464	.98948	41
20	.09295	.99567	.11031	.99390	.12764	.99182	.14493	.98944	40
21	.09324	.99564	.11060	.99386	.12793	.99178	.14522	.98940	39
22	.09353	.99562	.11089	.99383	.12822	.99175	.14551	.98936	38
23	.09382	.99559	.11118	.99380	.12851	.99171	.14580	.98931	37
24	.09411	.99556	.11147	.99377	.12880	.99167	.14608	.98927	36
25	.09440	.99553	.11176	.99374	.12908	.99163	.14637	.98923	35
26	.09469	.99551	.11205	.99370	.12937	.99160	.14666	.98919	34
27	.09498	.99548	.11234	.99367	.12966	.99156	.14695	.98914	33
28	.09527	.99545	.11263	.99364	.12995	.99152	.14723	.98910	32
29	.09556	.99542	.11291	.99360	.13024	.99148	.14752	.98906	31
30	.09585	.99540	.11320	.99357	.13053	.99144	.14781	.98902	30
31	.09614	.99537	.11349	.99354	.13081	.99141	.14810	.98897	29
32	.09642	.99534	.11378	.99351	.13110	.99137	.14838	.98893	28
33	.09671	.99531	.11407	.99347	.13139	.99133	.14867	.98889	27
34	.09700	.99528	.11436	.99344	.13168	.99129	.14896	.98884	26
35	.09729	.99526	.11465	.99341	.13197	.99125	.14925	.98880	25
36	.09758	.99523	.11494	.99337	.13226	.99122	.14954	.98876	24
37	.09787	.99520	.11523	.99334	.13254	.99118	.14982	.98871	23
38	.09816	.99517	.11552	.99331	.13283	.99114	.15011	.98867	22
39	.09845	.99514	.11580	.99327	.13312	.99110	.15040	.98863	21
40	.09874	.99511	.11609	.99324	.13341	.99106	.15069	.98858	20
41	.09903	.99508	.11638	.99320	.13370	.99102	.15097	.98854	19
42	.09932	.99506	.11667	.99317	.13399	.99098	.15126	.98849	18
43	.09961	.99503	.11696	.99314	.13427	.99094	.15155	.98845	17
44	.09990	.99500	.11725	.99310	.13456	.99091	.15184	.98841	16
45	.10019	.99497	.11754	.99307	.13485	.99087	.15212	.98836	15
46	.10048	.99494	.11783	.99303	.13514	.99083	.15241	.98832	14
47	.10077	.99491	.11812	.99300	.13543	.99079	.15270	.98827	13
48	.10106	.99488	.11840	.99297	.13572	.99075	.15299	.98823	12
49	.10135	.99485	.11869	.99293	.13600	.99071	.15327	.98818	11
50	.10164	.99482	.11898	.99290	.13629	.99067	.15356	.98814	10
51	.10192	.99479	.11927	.99286	.13658	.99063	.15385	.98809	9
52	.10221	.99476	.11956	.99283	.13687	.99059	.15414	.98805	8
53	.10250	.99473	.11985	.99279	.13716	.99055	.15442	.98800	7
54	.10279	.99470	.12014	.99276	.13744	.99051	.15471	.98796	6
55	.10308	.99467	.12043	.99272	.13773	.99047	.15500	.98791	5
56	.10337	.99464	.12071	.99269	.13802	.99043	.15529	.98787	4
57	.10366	.99461	.12100	.99265	.13831	.99039	.15557	.98782	3
58	.10395	.99458	.12129	.99262	.13860	.99035	.15586	.98778	2
59	.10424	.99455	.12158	.99258	.13889	.99031	.15615	.98773	1
60	.10453	.99452	.12187	.99255	.13917	.99027	.15643	.98769	0
′	84° Cosine	84° Sine	83° Cosine	83° Sine	82° Cosine	82° Sine	81° Cosine	81° Sine	′

TABLE XVI.—*Continued*

′	9° Sine	9° Cosine	10° Sine	10° Cosine	11° Sine	11° Cosine	12° Sine	12° Cosine	′
0	.15643	.98769	.17365	.98481	.19081	.98163	.20791	.97815	60
1	.15672	.98764	.17393	.98476	.19109	.98157	.20820	.97809	59
2	.15701	.98760	.17422	.98471	.19138	.98152	.20848	.97803	58
3	.15730	.98755	.17451	.98466	.19167	.98146	.20877	.97797	57
4	.15758	.98751	.17479	.98461	.19195	.98140	.20905	.97791	56
5	.15787	.98746	.17508	.98455	.19224	.98135	.20933	.97784	55
6	.15816	.98741	.17537	.98450	.19252	.98129	.20962	.97778	54
7	.15845	.98737	.17565	.98445	.19281	.98124	.20990	.97772	53
8	.15873	.98732	.17594	.98440	.19309	.98118	.21019	.97766	52
9	.15902	.98728	.17623	.98435	.19338	.98112	.21047	.97760	51
10	.15931	.98723	.17651	.98430	.19366	.98107	.21076	.97754	50
11	.15959	.98718	.17680	.98425	.19395	.98101	.21104	.97748	49
12	.15988	.98714	.17708	.98420	.19423	.98096	.21132	.97742	48
13	.16017	.98709	.17737	.98414	.19452	.98090	.21161	.97735	47
14	.16046	.98704	.17766	.98409	.19481	.98084	.21189	.97729	46
15	.16074	.98700	.17794	.98404	.19509	.98079	.21218	.97723	45
16	.16103	.98695	.17823	.98399	.19538	.98073	.21246	.97717	44
17	.16132	.98690	.17852	.98394	.19566	.98067	.21275	.97711	43
18	.16160	.98686	.17880	.98389	.19595	.98061	.21303	.97705	42
19	.16189	.98681	.17909	.98383	.19623	.98056	.21331	.97698	41
20	.16218	.98676	.17937	.98378	.19652	.98050	.21360	.97692	40
21	.16246	.98671	.17966	.98373	.19680	.98044	.21388	.97686	39
22	.16275	.98667	.17995	.98368	.19709	.98039	.21417	.97680	38
23	.16304	.98662	.18023	.98362	.19737	.98033	.21445	.97673	37
24	.16333	.98657	.18052	.98357	.19766	.98027	.21474	.97667	36
25	.16361	.98652	.18081	.98352	.19794	.98021	.21502	.97661	35
26	.16390	.98648	.18109	.98347	.19823	.98016	.21530	.97655	34
27	.16419	.98643	.18138	.98341	.19851	.98010	.21559	.97648	33
28	.16447	.98638	.18166	.98336	.19880	.98004	.21587	.97642	32
29	.16476	.98633	.18195	.98331	.19908	.97998	.21616	.97636	31
30	.16505	.98629	.18224	.98325	.19937	.97992	.21644	.97630	30
31	.16533	.98624	.18252	.98320	.19965	.97987	.21672	.97623	29
32	.16562	.98619	.18281	.98315	.19994	.97981	.21701	.97617	28
33	.16591	.98614	.18309	.98310	.20022	.97975	.21729	.97611	27
34	.16620	.98609	.18338	.98304	.20051	.97969	.21758	.97604	26
35	.16648	.98604	.18367	.98299	.20079	.97963	.21786	.97598	25
36	.16677	.98600	.18395	.98294	.20108	.97958	.21814	.97592	24
37	.16706	.98595	.18424	.98288	.20136	.97952	.21843	.97585	23
38	.16734	.98590	.18452	.98283	.20165	.97946	.21871	.97579	22
39	.16763	.98585	.18481	.98277	.20193	.97940	.21899	.97573	21
40	.16792	.98580	.18509	.98272	.20222	.97934	.21928	.97566	20
41	.16820	.98575	.18538	.98267	.20250	.97928	.21956	.97560	19
42	.16849	.98570	.18567	.98261	.20279	.97922	.21985	.97553	18
43	.16878	.98565	.18595	.98256	.20307	.97916	.22013	.97547	17
44	.16906	.98561	.18624	.98250	.20336	.97910	.22041	.97541	16
45	.16935	.98556	.18652	.98245	.20364	.97905	.22070	.97534	15
46	.16964	.98551	.18681	.98240	.20393	.97899	.22098	.97528	14
47	.16992	.98546	.18710	.98234	.20421	.97893	.22126	.97521	13
48	.17021	.98541	.18738	.98229	.20450	.97887	.22155	.97515	12
49	.17050	.98536	.18767	.98223	.20478	.97881	.22183	.97508	11
50	.17078	.98531	.18795	.98218	.20507	.97875	.22212	.97502	10
51	.17107	.98526	.18824	.98212	.20535	.97869	.22240	.97496	9
52	.17136	.98521	.18852	.98207	.20563	.97863	.22268	.97489	8
53	.17164	.98516	.18881	.98201	.20592	.97857	.22297	.97483	7
54	.17193	.98511	.18910	.98196	.20620	.97851	.22325	.97476	6
55	.17222	.98506	.18938	.98190	.20649	.97845	.22353	.97470	5
56	.17250	.98501	.18967	.98185	.20677	.97839	.22382	.97463	4
57	.17279	.98496	.18995	.98179	.20706	.97833	.22410	.97457	3
58	.17308	.98491	.19024	.98174	.20734	.97827	.22438	.97450	2
59	.17336	.98486	.19052	.98168	.20763	.97821	.22467	.97444	1
60	.17365	.98481	.19081	.98163	.20791	.97815	.22495	.97437	0
′	80° Cosine	80° Sine	79° Cosine	79° Sine	78° Cosine	78° Sine	77° Cosine	77° Sine	′

TABLE XVI.—*Continued*

′	13° Sine	13° Cosine	14° Sine	14° Cosine	15° Sine	15° Cosine	16° Sine	16° Cosine	′
0	.22495	.97437	.24192	.97030	.25882	.96593	.27564	.96126	60
1	.22523	.97430	.24220	.97023	.25910	.96585	.27592	.96118	59
2	.22552	.97424	.24249	.97015	.25938	.96578	.27620	.96110	58
3	.22580	.97417	.24277	.97008	.25966	.96570	.27648	.96102	57
4	.22608	.97411	.24305	.97001	.25994	.96562	.27676	.96094	56
5	.22637	.97404	.24333	.96994	.26022	.96555	.27704	.96086	55
6	.22665	.97398	.24362	.96987	.26050	.96547	.27731	.96078	54
7	.22693	.97391	.24390	.96980	.26079	.96540	.27759	.96070	53
8	.22722	.97384	.24418	.96973	.26107	.96532	.27787	.96062	52
9	.22750	.97378	.24446	.96966	.26135	.96524	.27815	.96054	51
10	.22778	.97371	.24474	.96959	.26163	.96517	.27843	.96046	50
11	.22807	.97365	.24503	.96952	.26191	.96509	.27871	.96037	49
12	.22835	.97358	.24531	.96945	.26219	.96502	.27899	.96029	48
13	.22863	.97351	.24559	.96937	.26247	.96494	.27927	.96021	47
14	.22892	.97345	.24587	.96930	.26275	.96486	.27955	.96013	46
15	.22920	.97338	.24615	.96923	.26303	.96479	.27983	.96005	45
16	.22948	.97331	.24644	.96916	.26331	.96471	.28011	.95997	44
17	.22977	.97325	.24672	.96909	.26359	.96463	.28039	.95989	43
18	.23005	.97318	.24700	.96902	.26387	.96456	.28067	.95981	42
19	.23033	.97311	.24728	.96894	.26415	.96448	.28095	.95972	41
20	.23062	.97304	.24756	.96887	.26443	.96440	.28123	.95964	40
21	.23090	.97298	.24784	.96880	.26471	.96433	.28150	.95956	39
22	.23118	.97291	.24813	.96873	.26500	.96425	.28178	.95948	38
23	.23146	.97284	.24841	.96866	.26528	.96417	.28206	.95940	37
24	.23175	.97278	.24869	.96858	.26556	.96410	.28234	.95931	36
25	.23203	.97271	.24897	.96851	.26584	.96402	.28262	.95923	35
26	.23231	.97264	.24925	.96844	.26612	.96394	.28290	.95915	34
27	.23260	.97257	.24954	.96837	.26640	.96386	.28318	.95907	33
28	.23288	.97251	.24982	.96829	.26668	.96379	.28346	.95898	32
29	.23316	.97244	.25010	.96822	.26696	.96371	.28374	.95890	31
30	.23345	.97237	.25038	.96815	.26724	.96363	.28402	.95882	30
31	.23373	.97230	.25066	.96807	.26752	.96355	.28429	.95874	29
32	.23401	.97223	.25094	.96800	.26780	.96347	.28457	.95865	28
33	.23429	.97217	.25122	.96793	.26808	.96340	.28485	.95857	27
34	.23458	.97210	.25151	.96786	.26836	.96332	.28513	.95849	26
35	.23486	.97203	.25179	.96778	.26864	.96324	.28541	.95841	25
36	.23514	.97196	.25207	.96771	.26892	.96316	.28569	.95832	24
37	.23542	.97189	.25235	.96764	.26920	.96308	.28597	.95824	23
38	.23571	.97182	.25263	.96756	.26948	.96301	.28625	.95816	22
39	.23599	.97176	.25291	.96749	.26976	.96293	.28652	.95807	21
40	.23627	.97169	.25320	.96742	.27004	.96285	.28680	.95799	20
41	.23656	.97162	.25348	.96734	.27032	.96277	.28708	.95791	19
42	.23684	.97155	.25376	.96727	.27060	.96269	.28736	.95782	18
43	.23712	.97148	.25404	.96719	.27088	.96261	.28764	.95774	17
44	.23740	.97141	.25432	.96712	.27116	.96253	.28792	.95766	16
45	.23769	.97134	.25460	.96705	.27144	.96246	.28820	.95757	15
46	.23797	.97127	.25488	.96697	.27172	.96238	.28847	.95749	14
47	.23825	.97120	.25516	.96690	.27200	.96230	.28875	.95740	13
48	.23853	.97113	.25545	.96682	.27228	.96222	.28903	.95732	12
49	.23882	.97106	.25573	.96675	.27256	.96214	.28931	.95724	11
50	.23910	.97100	.25601	.96667	.27284	.96206	.28959	.95715	10
51	.23938	.97093	.25629	.96660	.27312	.96198	.28987	.95707	9
52	.23966	.97086	.25657	.96653	.27340	.96190	.29015	.95698	8
53	.23995	.97079	.25685	.96645	.27368	.96182	.29042	.95690	7
54	.24023	.97072	.25713	.96638	.27396	.96174	.29070	.95681	6
55	.24051	.97065	.25741	.96630	.27424	.96166	.29098	.95673	5
56	.24079	.97058	.25769	.96623	.27452	.96158	.29126	.95664	4
57	.24108	.97051	.25798	.96615	.27480	.96150	.29154	.95656	3
58	.24136	.97044	.25826	.96608	.27508	.96142	.29182	.95647	2
59	.24164	.97037	.25854	.96600	.27536	.96134	.29209	.95639	1
60	.24192	.97030	.25882	.96593	.27564	.96126	.29237	.95630	0
′	Cosine 76°	Sine 76°	Cosine 75°	Sine 75°	Cosine 74°	Sine 74°	Cosine 73°	Sine 73°	′

TABLE XVI.—*Continued*

′	17° Sine	17° Cosine	18° Sine	18° Cosine	19° Sine	19° Cosine	20° Sine	20° Cosine	′
0	.29237	.95630	.30902	.95106	.32557	.94552	.34202	.93969	60
1	.29265	.95622	.30929	.95097	.32584	.94542	.34229	.93959	59
2	.29293	.95613	.30957	.95088	.32612	.94533	.34257	.93949	58
3	.29321	.95605	.30985	.95079	.32639	.94523	.34284	.93939	57
4	.29348	.95596	.31012	.95070	.32667	.94514	.34311	.93929	56
5	.29376	.95588	.31040	.95061	.32694	.94504	.34339	.93919	55
6	.29404	.95579	.31068	.95052	.32722	.94495	.34366	.93909	54
7	.29432	.95571	.31095	.95043	.32749	.94485	.34393	.93899	53
8	.29460	.95562	.31123	.95033	.32777	.94476	.34421	.93889	52
9	.29487	.95554	.31151	.95024	.32804	.94466	.34448	.93879	51
10	.29515	.95545	.31178	.95015	.32832	.94457	.34475	.93869	50
11	.29543	.95536	.31206	.95006	.32859	.94447	.34503	.93859	49
12	.29571	.95528	.31233	.94997	.32887	.94438	.34530	.93849	48
13	.29599	.95519	.31261	.94988	.32914	.94428	.34557	.93839	47
14	.29626	.95511	.31289	.94979	.32942	.94418	.34584	.93829	46
15	.29654	.95502	.31316	.94970	.32969	.94409	.34612	.93819	45
16	.29682	.95493	.31344	.94961	.32997	.94399	.34639	.93809	44
17	.29710	.95485	.31372	.94952	.33024	.94390	.34666	.93799	43
18	.29737	.95476	.31399	.94943	.33051	.94380	.34694	.93789	42
19	.29765	.95467	.31427	.94933	.33079	.94370	.34721	.93779	41
20	.29793	.95459	.31454	.94924	.33106	.94361	.34748	.93769	40
21	.29821	.95450	.31482	.94915	.33134	.94351	.34775	.93759	39
22	.29849	.95441	.31510	.94906	.33161	.94342	.34803	.93748	38
23	.29876	.95433	.31537	.94897	.33189	.94332	.34830	.93738	37
24	.29904	.95424	.31565	.94888	.33216	.94322	.34857	.93728	36
25	.29932	.95415	.31593	.94878	.33244	.94313	.34884	.93718	35
26	.29960	.95407	.31620	.94869	.33271	.94303	.34912	.93708	34
27	.29987	.95398	.31648	.94860	.33298	.94293	.34939	.93698	33
28	.30015	.95389	.31675	.94851	.33326	.94284	.34966	.93688	32
29	.30043	.95380	.31703	.94842	.33353	.94274	.34993	.93677	31
30	.30071	.95372	.31730	.94832	.33381	.94264	.35021	.93667	30
31	.30098	.95363	.31758	.94823	.33408	.94254	.35048	.93657	29
32	.30126	.95354	.31786	.94814	.33436	.94245	.35075	.93647	28
33	.30154	.95345	.31813	.94805	.33463	.94235	.35102	.93637	27
34	.30182	.95337	.31841	.94795	.33490	.94225	.35130	.93626	26
35	.30209	.95328	.31868	.94786	.33518	.94215	.35157	.93616	25
36	.30237	.95319	.31896	.94777	.33545	.94206	.35184	.93606	24
37	.30265	.95310	.31923	.94768	.33573	.94196	.35211	.93596	23
38	.30292	.95301	.31951	.94758	.33600	.94186	.35239	.93585	22
39	.30320	.95293	.31979	.94749	.33627	.94176	.35266	.93575	21
40	.30348	.95284	.32006	.94740	.33655	.94167	.35293	.93565	20
41	.30376	.95275	.32034	.94730	.33682	.94157	.35320	.93555	19
42	.30403	.95266	.32061	.94721	.33710	.94147	.35347	.93544	18
43	.30431	.95257	.32089	.94712	.33737	.94137	.35375	.93534	17
44	.30459	.95248	.32116	.94702	.33764	.94127	.35402	.93524	16
45	.30486	.95240	.32144	.94693	.33792	.94118	.35429	.93514	15
46	.30514	.95231	.32171	.94684	.33819	.94108	.35456	.93503	14
47	.30542	.95222	.32199	.94674	.33846	.94098	.35484	.93493	13
48	.30570	.95213	.32227	.94665	.33874	.94088	.35511	.93483	12
49	.30597	.95204	.32254	.94656	.33901	.94078	.35538	.93472	11
50	.30625	.95195	.32282	.94646	.33929	.94068	.35565	.93462	10
51	.30653	.95186	.32309	.94637	.33956	.94058	.35592	.93452	9
52	.30680	.95177	.32337	.94627	.33983	.94049	.35619	.93441	8
53	.30708	.95168	.32364	.94618	.34011	.94039	.35647	.93431	7
54	.30736	.95159	.32392	.94609	.34038	.94029	.35674	.93420	6
55	.30763	.95150	.32419	.94599	.34065	.94019	.35701	.93410	5
56	.30791	.95142	.32447	.94590	.34093	.94009	.35728	.93400	4
57	.30819	.95133	.32474	.94580	.34120	.93999	.35755	.93389	3
58	.30846	.95124	.32502	.94571	.34147	.93989	.35782	.93379	2
59	.30874	.95115	.32529	.94561	.34175	.93979	.35810	.93368	1
60	.30902	.95106	.32557	.94552	.34202	.93969	.35837	.93358	0
′	72° Cosine	72° Sine	71° Cosine	71° Sine	70° Cosine	70° Sine	69° Cosine	69° Sine	′

TABLE XVI.—*Continued*

′	21° Sine	21° Cosine	22° Sine	22° Cosine	23° Sine	23° Cosine	24° Sine	24° Cosine	′
0	.35837	.93358	.37461	.92718	.39073	.92050	.40674	.91355	60
1	.35864	.93348	.37488	.92707	.39100	.92039	.40700	.91343	59
2	.35891	.93337	.37515	.92697	.39127	.92028	.40727	.91331	58
3	.35918	.93327	.37542	.92686	.39153	.92016	.40753	.91319	57
4	.35945	.93316	.37569	.92675	.39180	.92005	.40780	.91307	56
5	.35973	.93306	.37595	.92664	.39207	.91994	.40806	.91295	55
6	.36000	.93295	.37622	.92653	.39234	.91982	.40833	.91283	54
7	.36027	.93285	.37649	.92642	.39260	.91971	.40860	.91272	53
8	.36054	.93274	.37676	.92631	.39287	.91959	.40886	.91260	52
9	.36081	.93264	.37703	.92620	.39314	.91948	.40913	.91248	51
10	.36108	.93253	.37730	.92609	.39341	.91936	.40939	.91236	50
11	.36135	.93243	.37757	.92598	.39367	.91925	.40966	.91224	49
12	.36162	.93232	.37784	.92587	.39394	.91914	.40992	.91212	48
13	.36190	.93222	.37811	.92576	.39421	.91902	.41019	.91200	47
14	.36217	.93211	.37838	.92565	.39448	.91891	.41045	.91188	46
15	.36244	.93201	.37865	.92554	.39474	.91879	.41072	.91176	45
16	.36271	.93190	.37892	.92543	.39501	.91868	.41098	.91164	44
17	.36298	.93180	.37919	.92532	.39528	.91856	.41125	.91152	43
18	.36325	.93169	.37946	.92521	.39555	.91845	.41151	.91140	42
19	.36352	.93159	.37973	.92510	.39581	.91833	.41178	.91128	41
20	.36379	.93148	.37999	.92499	.39608	.91822	.41204	.91116	40
21	.36406	.93137	.38026	.92488	.39635	.91810	.41231	.91104	39
22	.36434	.93127	.38053	.92477	.39661	.91799	.41257	.91092	38
23	.36461	.93116	.38080	.92466	.39688	.91787	.41284	.91080	37
24	.36488	.93106	.38107	.92455	.39715	.91775	.41310	.91068	36
25	.36515	.93095	.38134	.92444	.39741	.91764	.41337	.91056	35
26	.36542	.93084	.38161	.92432	.39768	.91752	.41363	.91044	34
27	.36569	.93074	.38188	.92421	.39795	.91741	.41390	.91032	33
28	.36596	.93063	.38215	.92410	.39822	.91729	.41416	.91020	32
29	.36623	.93052	.38241	.92399	.39848	.91718	.41443	.91008	31
30	.36650	.93042	.38268	.92388	.39875	.91706	.41469	.90996	30
31	.36677	.93031	.38295	.92377	.39902	.91694	.41496	.90984	29
32	.36704	.93020	.38322	.92366	.39928	.91683	.41522	.90972	28
33	.36731	.93010	.38349	.92355	.39955	.91671	.41549	.90960	27
34	.36758	.92999	.38376	.92343	.39982	.91660	.41575	.90948	26
35	.36785	.92988	.38403	.92332	.40008	.91648	.41602	.90936	25
36	.36812	.92978	.38430	.92321	.40035	.91636	.41628	.90924	24
37	.36839	.92967	.38456	.92310	.40062	.91625	.41655	.90911	23
38	.36867	.92956	.38483	.92299	.40088	.91613	.41681	.90899	22
39	.36894	.92945	.38510	.92287	.40115	.91601	.41707	.90887	21
40	.36921	.92935	.38537	.92276	.40141	.91590	.41734	.90875	20
41	.36948	.92924	.38564	.92265	.40168	.91578	.41760	.90863	19
42	.36975	.92913	.38591	.92254	.40195	.91566	.41787	.90851	18
43	.37002	.92902	.38617	.92243	.40221	.91555	.41813	.90839	17
44	.37029	.92892	.38644	.92231	.40248	.91543	.41840	.90826	16
45	.37056	.92881	.38671	.92220	.40275	.91531	.41866	.90814	15
46	.37083	.92870	.38698	.92209	.40301	.91519	.41892	.90802	14
47	.37110	.92859	.38725	.92198	.40328	.91508	.41919	.90790	13
48	.37137	.92849	.38752	.92186	.40355	.91496	.41945	.90778	12
49	.37164	.92838	.38778	.92175	.40381	.91484	.41972	.90766	11
50	.37191	.92827	.38805	.92164	.40408	.91472	.41998	.90753	10
51	.37218	.92816	.38832	.92152	.40434	.91461	.42024	.90741	9
52	.37245	.92805	.38859	.92141	.40461	.91449	.42051	.90729	8
53	.37272	.92794	.38886	.92130	.40488	.91437	.42077	.90717	7
54	.37299	.92784	.38912	.92119	.40514	.91425	.42104	.90704	6
55	.37326	.92773	.38939	.92107	.40541	.91414	.42130	.90692	5
56	.37353	.92762	.38966	.92096	.40567	.91402	.42156	.90680	4
57	.37380	.92751	.38993	.92085	.40594	.91390	.42183	.90668	3
58	.37407	.92740	.39020	.92073	.40621	.91378	.42209	.90655	2
59	.37434	.92729	.39046	.92062	.40647	.91366	.42235	.90643	1
60	.37461	.92718	.39073	.92050	.40674	.91355	.42262	.90631	0
′	Cosine	Sine	Cosine	Sine	Cosine	Sine	Cosine	Sine	′
	68°		67°		66°		65°		

TABLE XVI.—*Continued*

′	25° Sine	25° Cosine	26° Sine	26° Cosine	27° Sine	27° Cosine	28° Sine	28° Cosine	′
0	.42262	.90631	.43837	.89879	.45399	.89101	.46947	.88295	60
1	.42288	.90618	.43863	.89867	.45425	.89087	.46973	.88281	59
2	.42315	.90606	.43889	.89854	.45451	.89074	.46999	.88267	58
3	.42341	.90594	.43916	.89841	.45477	.89061	.47024	.88254	57
4	.42367	.90582	.43942	.89828	.45503	.89048	.47050	.88240	56
5	.42394	.90569	.43968	.89816	.45529	.89035	.47076	.88226	55
6	.42420	.90557	.43994	.89803	.45554	.89021	.47101	.88213	54
7	.42446	.90545	.44020	.89790	.45580	.89008	.47127	.88199	53
8	.42473	.90532	.44046	.89777	.45606	.88995	.47153	.88185	52
9	.42499	.90520	.44072	.89764	.45632	.88981	.47178	.88172	51
10	.42525	.90507	.44098	.89752	.45658	.88968	.47204	.88158	50
11	.42552	.90495	.44124	.89739	.45684	.88955	.47229	.88144	49
12	.42578	.90483	.44151	.89726	.45710	.88942	.47255	.88130	48
13	.42604	.90470	.44177	.89713	.45736	.88928	.47281	.88117	47
14	.42631	.90458	.44203	.89700	.45762	.88915	.47306	.88103	46
15	.42657	.90446	.44229	.89687	.45787	.88902	.47332	.88089	45
16	.42683	.90433	.44255	.89674	.45813	.88888	.47358	.88075	44
17	.42709	.90421	.44281	.89662	.45839	.88875	.47383	.88062	43
18	.42736	.90408	.44307	.89649	.45865	.88862	.47409	.88048	42
19	.42762	.90396	.44333	.89636	.45891	.88848	.47434	.88034	41
20	.42788	.90383	.44359	.89623	.45917	.88835	.47460	.88020	40
21	.42815	.90371	.44385	.89610	.45942	.88822	.47486	.88006	39
22	.42841	.90358	.44411	.89597	.45968	.88808	.47511	.87993	38
23	.42867	.90346	.44437	.89584	.45994	.88795	.47537	.87979	37
24	.42894	.90334	.44464	.89571	.46020	.88782	.47562	.87965	36
25	.42920	.90321	.44490	.89558	.46046	.88768	.47588	.87951	35
26	.42946	.90309	.44516	.89545	.46072	.88755	.47614	.87937	34
27	.42972	.90296	.44542	.89532	.46097	.88741	.47639	.87923	33
28	.42999	.90284	.44568	.89519	.46123	.88728	.47665	.87909	32
29	.43025	.90271	.44594	.89506	.46149	.88715	.47690	.87896	31
30	.43051	.90259	.44620	.89493	.46175	.88701	.47716	.87882	30
31	.43077	.90246	.44646	.89480	.46201	.88688	.47741	.87868	29
32	.43104	.90233	.44672	.89467	.46226	.88674	.47767	.87854	28
33	.43130	.90221	.44698	.89454	.46252	.88661	.47793	.87840	27
34	.43156	.90208	.44724	.89441	.46278	.88647	.47818	.87826	26
35	.43182	.90196	.44750	.89428	.46304	.88634	.47844	.87812	25
36	.43209	.90183	.44776	.89415	.46330	.88620	.47869	.87798	24
37	.43235	.90171	.44802	.89402	.46355	.88607	.47895	.87784	23
38	.43261	.90158	.44828	.89389	.46381	.88593	.47920	.87770	22
39	.43287	.90146	.44854	.89376	.46407	.88580	.47946	.87756	21
40	.43313	.90133	.44880	.89363	.46433	.88566	.47971	.87743	20
41	.43340	.90120	.44906	.89350	.46458	.88553	.47997	.87729	19
42	.43366	.90108	.44932	.89337	.46484	.88539	.48022	.87715	18
43	.43392	.90095	.44958	.89324	.46510	.88526	.48048	.87701	17
44	.43418	.90082	.44984	.89311	.46536	.88512	.48073	.87687	16
45	.43445	.90070	.45010	.89298	.46561	.88499	.48099	.87673	15
46	.43471	.90057	.45036	.89285	.46587	.88485	.48124	.87659	14
47	.43497	.90045	.45062	.89272	.46613	.88472	.48150	.87645	13
48	.43523	.90032	.45088	.89259	.46639	.88458	.48175	.87631	12
49	.43549	.90019	.45114	.89245	.46664	.88445	.48201	.87617	11
50	.43575	.90007	.45140	.89232	.46690	.88431	.48226	.87603	10
51	.43602	.89994	.45166	.89219	.46716	.88417	.48252	.87589	9
52	.43628	.89981	.45192	.89206	.46742	.88404	.48277	.87575	8
53	.43654	.89968	.45218	.89193	.46767	.88390	.48303	.87561	7
54	.43680	.89956	.45243	.89180	.46793	.88377	.48328	.87546	6
55	.43706	.89943	.45269	.89167	.46819	.88363	.48354	.87532	5
56	.43733	.89930	.45295	.89153	.46844	.88349	.48379	.87518	4
57	.43759	.89918	.45321	.89140	.46870	.88336	.48405	.87504	3
58	.43785	.89905	.45347	.89127	.46896	.88322	.48430	.87490	2
59	.43811	.89892	.45373	.89114	.46921	.88308	.48456	.87476	1
60	.43837	.89879	.45399	.89101	.46947	.88295	.48481	.87462	0
′	64° Cosine	64° Sine	63° Cosine	63° Sine	62° Cosine	62° Sine	61° Cosine	61° Sine	′

TABLE XVI.—*Continued*

′	29° Sine	29° Cosine	30° Sine	30° Cosine	31° Sine	31° Cosine	32° Sine	32° Cosine	′
0	.48481	.87462	.50000	.86603	.51504	.85717	.52992	.84805	60
1	.48506	.87448	.50025	.86588	.51529	.85702	.53017	.84789	59
2	.48532	.87434	.50050	.86573	.51554	.85687	.53041	.84774	58
3	.48557	.87420	.50076	.86559	.51579	.85672	.53066	.84759	57
4	.48583	.87406	.50101	.86544	.51604	.85657	.53091	.84743	56
5	.48608	.87391	.50126	.86530	.51628	.85642	.53115	.84728	55
6	.48634	.87377	.50151	.86515	.51653	.85627	.53140	.84712	54
7	.48659	.87363	.50176	.86501	.51678	.85612	.53164	.84697	53
8	.48684	.87349	.50201	.86486	.51703	.85597	.53189	.84681	52
9	.48710	.87335	.50227	.86471	.51728	.85582	.53214	.84666	51
10	.48735	.87321	.50252	.86457	.51753	.85567	.53238	.84650	50
11	.48761	.87306	.50277	.86442	.51778	.85551	.53263	.84635	49
12	.48786	.87292	.50302	.86427	.51803	.85536	.53288	.84619	48
13	.48811	.87278	.50327	.86413	.51828	.85521	.53312	.84604	47
14	.48837	.87264	.50352	.86398	.51852	.85506	.53337	.84588	46
15	.48862	.87250	.50377	.86384	.51877	.85491	.53361	.84573	45
16	.48888	.87235	.50403	.86369	.51902	.85476	.53386	.84557	44
17	.48913	.87221	.50428	.86354	.51927	.85461	.53411	.84542	43
18	.48938	.87207	.50453	.86340	.51952	.85446	.53435	.84526	42
19	.48964	.87193	.50478	.86325	.51977	.85431	.53460	.84511	41
20	.48989	.87178	.50503	.86310	.52002	.85416	.53484	.84495	40
21	.49014	.87164	.50528	.86295	.52026	.85401	.53509	.84480	39
22	.49040	.87150	.50553	.86281	.52051	.85385	.53534	.84464	38
23	.49065	.87136	.50578	.86266	.52076	.85370	.53558	.84448	37
24	.49090	.87121	.50603	.86251	.52101	.85355	.53583	.84433	36
25	.49116	.87107	.50628	.86237	.52126	.85340	.53607	.84417	35
26	.49141	.87093	.50654	.86222	.52151	.85325	.53632	.84402	34
27	.49166	.87079	.50679	.86207	.52175	.85310	.53656	.84386	33
28	.49192	.87064	.50704	.86192	.52200	.85294	.53681	.84370	32
29	.49217	.87050	.50729	.86178	.52225	.85279	.53705	.84355	31
30	.49242	.87036	.50754	.86163	.52250	.85264	.53730	.84339	30
31	.49268	.87021	.50779	.86148	.52275	.85249	.53754	.84324	29
32	.49293	.87007	.50804	.86133	.52299	.85234	.53779	.84308	28
33	.49318	.86993	.50829	.86119	.52324	.85218	.53804	.84292	27
34	.49344	.86978	.50854	.86104	.52349	.85203	.53828	.84277	26
35	.49369	.86964	.50879	.86089	.52374	.85188	.53853	.84261	25
36	.49394	.86949	.50904	.86074	.52399	.85173	.53877	.84245	24
37	.49419	.86935	.50929	.86059	.52423	.85157	.53902	.84230	23
38	.49445	.86921	.50954	.86045	.52448	.85142	.53926	.84214	22
39	.49470	.86906	.50979	.86030	.52473	.85127	.53951	.84198	21
40	.49495	.86892	.51004	.86015	.52498	.85112	.53975	.84182	20
41	.49521	.86878	.51029	.86000	.52522	.85096	.54000	.84167	19
42	.49546	.86863	.51054	.85985	.52547	.85081	.54024	.84151	18
43	.49571	.86849	.51079	.85970	.52572	.85066	.54049	.84135	17
44	.49596	.86834	.51104	.85956	.52597	.85051	.54073	.84120	16
45	.49622	.86820	.51129	.85941	.52621	.85035	.54097	.84104	15
46	.49647	.86805	.51154	.85926	.52646	.85020	.54122	.84088	14
47	.49672	.86791	.51179	.85911	.52671	.85005	.54146	.84072	13
48	.49697	.86777	.51204	.85896	.52696	.84989	.54171	.84057	12
49	.49723	.86762	.51229	.85881	.52720	.84974	.54195	.84041	11
50	.49748	.86748	.51254	.85866	.52745	.84959	.54220	.84025	10
51	.49773	.86733	.51279	.85851	.52770	.84943	.54244	.84009	9
52	.49798	.86719	.51304	.85836	.52794	.84928	.54269	.83994	8
53	.49824	.86704	.51329	.85821	.52819	.84913	.54293	.83978	7
54	.49849	.86690	.51354	.85806	.52844	.84897	.54317	.83962	6
55	.49874	.86675	.51379	.85792	.52869	.84882	.54342	.83946	5
56	.49899	.86661	.51404	.85777	.52893	.84866	.54366	.83930	4
57	.49924	.86646	.51429	.85762	.52918	.84851	.54391	.83915	3
58	.49950	.86632	.51454	.85747	.52943	.84836	.54415	.83899	2
59	.49975	.86617	.51479	.85732	.52967	.84820	.54440	.83883	1
60	.50000	.86603	.51504	.85717	.52992	.84805	.54464	.83867	0
′	60° Cosine	60° Sine	59° Cosine	59° Sine	58° Cosine	58° Sine	57° Cosine	57° Sine	′

TABLE XVI.—*Continued*

′	33° Sine	33° Cosine	34° Sine	34° Cosine	35° Sine	35° Cosine	36° Sine	36° Cosine	′
0	.54464	.83867	.55919	.82904	.57358	.81915	.58779	.80902	60
1	.54488	.83851	.55943	.82887	.57381	.81899	.58802	.80885	59
2	.54513	.83835	.55968	.82871	.57405	.81882	.58826	.80867	58
3	.54537	.83819	.55992	.82855	.57429	.81865	.58849	.80850	57
4	.54561	.83804	.56016	.82839	.57453	.81848	.58873	.80833	56
5	.54586	.83788	.56040	.82822	.57477	.81832	.58896	.80816	55
6	.54610	.83772	.56064	.82806	.57501	.81815	.58920	.80799	54
7	.54635	.83756	.56088	.82790	.57524	.81798	.58943	.80782	53
8	.54659	.83740	.56112	.82773	.57548	.81782	.58967	.80765	52
9	.54683	.83724	.56136	.82757	.57572	.81765	.58990	.80748	51
10	.54708	.83708	.56160	.82741	.57596	.81748	.59014	.80730	50
11	.54732	83692	.56184	.82724	.57619	.81731	.59037	.80713	49
12	.54756	.83676	.56208	.82708	.57643	.81714	.59061	.80696	48
13	.54781	.83660	.56232	.82692	.57667	.81698	.59084	.80679	47
14	.54805	.83645	.56256	.82675	.57691	.81681	.59108	.80662	46
15	.54829	.83629	.56280	.82659	.57715	.81664	.59131	.80644	45
16	.54854	.83613	.56305	.82643	.57738	.81647	.59154	.80627	44
17	.54878	.83597	.56329	.82626	.57762	.81631	.59178	.80610	43
18	.54902	.83581	.56353	.82610	.57786	.81614	.59201	.80593	42
19	.54927	.83565	.56377	.82593	.57810	.81597	.59225	.80576	41
20	.54951	.83549	.56401	.82577	.57833	.81580	.59248	.80558	40
21	.54975	.83533	.56425	.82561	.57857	.81563	.59272	.80541	39
22	.54999	.83517	.56449	.82544	.57881	.81546	.59295	.80524	38
23	.55024	.83501	.56473	.82528	.57904	.81530	.59318	.80507	37
24	.55048	.83485	.56497	.82511	.57928	.81513	.59342	.80489	36
25	.55072	.83469	.56521	.82495	.57952	.81496	.59365	.80472	35
26	.55097	.83453	.56545	.82478	.57976	.81479	.59389	.80455	34
27	.55121	.83437	.56569	.82462	.57999	.81462	.59412	.80438	33
28	.55145	.83421	.56593	.82446	.58023	.81445	.59436	.80420	32
29	.55169	.83405	.56617	.82429	.58047	.81428	.59459	.80403	31
30	.55194	.83389	.56641	.82413	.58070	.81412	.59482	.80386	30
31	.55218	.83373	.56665	.82396	.58094	.81395	.59506	.80368	29
32	.55242	.83356	.56689	.82380	.58118	.81378	.59529	.80351	28
33	.55266	.83340	.56713	.82363	.58141	.81361	.59552	.80334	27
34	.55291	.83324	.56736	.82347	.58165	.81344	.59576	.80316	26
35	.55315	.83308	.56760	.82330	.58189	.81327	.59599	.80299	25
36	.55339	.83292	.56784	.82314	.58212	.81310	.59622	.80282	24
37	.55363	.83276	.56808	.82297	.58236	.81293	.59646	.80264	23
38	.55388	.83260	.56832	.82281	.58260	.81276	.59669	.80247	22
39	.55412	.83244	.56856	.82264	.58283	.81259	.59693	.80230	21
40	.55436	.83228	.56880	.82248	.58307	.81242	.59716	.80212	20
41	.55460	.83212	.56904	.82231	.58330	.81225	.59739	.80195	19
42	.55484	.83195	.56928	.82214	.58354	.81208	.59763	.80178	18
43	.55509	.83179	.56952	.82198	.58378	.81191	.59786	.80160	17
44	.55533	.83163	.56976	.82181	.58401	.81174	.59809	.80143	16
45	.55557	.83147	.57000	.82165	.58425	.81157	.59832	.80125	15
46	.55581	.83131	.57024	.82148	.58449	.81140	.59856	.80108	14
47	.55605	.83115	.57047	.82132	.58472	.81123	.59879	.80091	13
48	.55630	.83098	.57071	.82115	.58496	.81106	.59902	.80073	12
49	.55654	.83082	.57095	.82098	.58519	.81089	.59926	.80056	11
50	.55678	.83066	.57119	.82082	.58543	.81072	.59949	.80038	10
51	.55702	.83050	.57143	.82065	.58567	.81055	.59972	.80021	9
52	.55726	.83034	.57167	.82048	.58590	.81038	.59995	.80003	8
53	.55750	.83017	.57191	.82032	.58614	.81021	.60019	.79986	7
54	.55775	.83001	.57215	.82015	.58637	.81004	.60042	.79968	6
55	.55799	.82985	.57238	.81999	.58661	.80987	.60065	.79951	5
56	.55823	.82969	.57262	.81982	.58684	.80970	.60089	.79934	4
57	.55847	.82953	.57286	.81965	.58708	.80953	.60112	.79916	3
58	.55871	.82936	.57310	.81949	.58731	.80936	.60135	.79899	2
59	.55895	.82920	.57334	.81932	.58755	.80919	.60158	.79881	1
60	.55919	.82904	.57358	.81915	.58779	.80902	.60182	.79864	0
′	56° Cosine	56° Sine	55° Cosine	55° Sine	54° Cosine	54° Sine	53° Cosine	53° Sine	′

TABLE XVI.—*Continued*

′	37° Sine	37° Cosine	38° Sine	38° Cosine	39° Sine	39° Cosine	40° Sine	40° Cosine	′
0	.60182	.79864	.61566	.78801	.62932	.77715	.64279	.76604	60
1	.60205	.79846	.61589	.78783	.62955	.77696	.64301	.76586	59
2	.60228	.79829	.61612	.78765	.62977	.77678	.64323	.76567	58
3	.60251	.79811	.61635	.78747	.63000	.77660	.64346	.76548	57
4	.60274	.79793	.61658	.78729	.63022	.77641	.64368	.76530	56
5	.60298	.79776	.61681	.78711	.63045	.77623	.64390	.76511	55
6	.60321	.79758	.61704	.78694	.63068	.77605	.64412	.76492	54
7	.60344	.79741	.61726	.78676	.63090	.77586	.64435	.76473	53
8	.60367	.79723	.61749	.78658	.63113	.77568	.64457	.76455	52
9	.60390	.79706	.61772	.78640	.63135	.77550	.64479	.76436	51
10	.60414	.79688	.61795	.78622	.63158	.77531	.64501	.76417	50
11	.60437	.79671	.61818	.78604	.63180	.77513	.64524	.76398	49
12	.60460	.79653	.61841	.78586	.63203	.77494	.64546	.76380	48
13	.60483	.79635	.61864	.78568	.63225	.77476	.64568	.76361	47
14	.60506	.79618	.61887	.78550	.63248	.77458	.64590	.76342	46
15	.60529	.79600	.61909	.78532	.63271	.77439	.64612	.76323	45
16	.60553	.79583	.61932	.78514	.63293	.77421	.64635	.76304	44
17	.60576	.79565	.61955	.78496	.63316	.77402	.64657	.76286	43
18	.60599	.79547	.61978	.78478	.63338	.77384	.64679	.76267	42
19	.60622	.79530	.62001	.78460	.63361	.77366	.64701	.76248	41
20	.60645	.79512	.62024	.78442	.63383	.77347	.64723	.76229	40
21	.60668	.79494	.62046	.78424	.63406	.77329	.64746	.76210	39
22	.60691	.79477	.62069	.78405	.63428	.77310	.64768	.76192	38
23	.60714	.79459	.62092	.78387	.63451	.77292	.64790	.76173	37
24	.60738	.79441	.62115	.78369	.63473	.77273	.64812	.76154	36
25	.60761	.79424	.62138	.78351	.63496	.77255	.64834	.76135	35
26	.60784	.79406	.62160	.78333	.63518	.77236	.64856	.76116	34
27	.60807	.79388	.62183	.78315	.63540	.77218	.64878	.76097	33
28	.60830	.79371	.62206	.78297	.63563	.77199	.64901	.76078	32
29	.60853	.79353	.62229	.78279	.63585	.77181	.64923	.76059	31
30	.60876	.79335	.62251	.78261	.63608	.77162	.64945	.76041	30
31	.60899	.79318	.62274	.78243	.63630	.77144	.64967	.76022	29
32	.60922	.79300	.62297	.78225	.63653	.77125	.64989	.76003	28
33	.60945	.79282	.62320	.78206	.63675	.77107	.65011	.75984	27
34	.60968	.79264	.62342	.78188	.63698	.77088	.65033	.75965	26
35	.60991	.79247	.62365	.78170	.63720	.77070	.65055	.75946	25
36	.61015	.79229	.62388	.78152	.63742	.77051	.65077	.75927	24
37	.61038	.79211	.62411	.78134	.63765	.77033	.65100	.75908	23
38	.61061	.79193	.62433	.78116	.63787	.77014	.65122	.75889	22
39	.61084	.79176	.62456	.78098	.63810	.76996	.65144	.75870	21
40	.61107	.79158	.62479	.78079	.63832	.76977	.65166	.75851	20
41	.61130	.79140	.62502	.78061	.63854	.76959	.65188	.75832	19
42	.61153	.79122	.62524	.78043	.63877	.76940	.65210	.75813	18
43	.61176	.79105	.62547	.78025	.63899	.76921	.65232	.75794	17
44	.61199	.79087	.62570	.78007	.63922	.76903	.65254	.75775	16
45	.61222	.79069	.62592	.77988	.63944	.76884	.65276	.75756	15
46	.61245	.79051	.62615	.77970	.63966	.76866	.65298	.75738	14
47	.61268	.79033	.62638	.77952	.63989	.76847	.65320	.75719	13
48	.61291	.79016	.62660	.77934	.64011	.76828	.65342	.75700	12
49	.61314	.78998	.62683	.77916	.64033	.76810	.65364	.75680	11
50	.61337	.78980	.62706	.77897	.64056	.76791	.65386	.75661	10
51	.61360	.78962	.62728	.77879	.64078	.76772	.65408	.75642	9
52	.61383	.78944	.62751	.77861	.64100	.76754	.65430	.75623	8
53	.61406	.78926	.62774	.77843	.64123	.76735	.65452	.75604	7
54	.61429	.78908	.62796	.77824	.64145	.76717	.65474	.75585	6
55	.61451	.78891	.62819	.77806	.64167	.76698	.65496	.75566	5
56	.61474	.78873	.62842	.77788	.64190	.76679	.65518	.75547	4
57	.61497	.78855	.62864	.77769	.64212	.76661	.65540	.75528	3
58	.61520	.78837	.62887	.77751	.64234	.76642	.65562	.75509	2
59	.61543	.78819	.62909	.77733	.64256	.76623	.65584	.75490	1
60	.61566	.78801	.62932	.77715	.64279	.76604	.65606	.75471	0
′	Cosine 52°	Sine 52°	Cosine 51°	Sine 51°	Cosine 50°	Sine 50°	Cosine 49°	Sine 49°	′

TABLE XVI.—*Concluded*

′	41° Sine	41° Cosine	42° Sine	42° Cosine	43° Sine	43° Cosine	44° Sine	44° Cosine	′
0	.65606	.75471	.66913	.74314	.68200	.73135	.69466	.71934	60
1	.65628	.75452	.66935	.74295	.68221	.73116	.69487	.71914	59
2	.65650	.75433	.66956	.74276	.68242	.73096	.69508	.71894	58
3	.65672	.75414	.66978	.74256	.68264	.73076	.69529	.71873	57
4	.65694	.75395	.66999	.74237	.68285	.73056	.69549	.71853	56
5	.65716	.75375	.67021	.74217	.68306	.73036	.69570	.71833	55
6	.65738	.75356	.67043	.74198	.68327	.73016	.69591	.71813	54
7	.65759	.75337	.67064	.74178	.68349	.72996	.69612	.71792	53
8	.65781	.75318	.67086	.74159	.68370	.72976	.69633	.71772	52
9	.65803	.75299	.67107	.74139	.68391	.72957	.69654	.71752	51
10	.65825	.75280	.67129	.74120	.68412	.72937	.69675	.71732	50
11	.65847	.75261	.67151	.74100	.68434	.72917	.69696	.71711	49
12	.65869	.75241	.67172	.74080	.68455	.72897	.69717	.71691	48
13	.65891	.75222	.67194	.74061	.68476	.72877	.69737	.71671	47
14	.65913	.75203	.67215	.74041	.68497	.72857	.69758	.71650	46
15	.65935	.75184	.67237	.74022	.68518	.72837	.69779	.71630	45
16	.65956	.75165	.67258	.74002	.68539	.72817	.69800	.71610	44
17	.65978	.75146	.67280	.73983	.68561	.72797	.69821	.71590	43
18	.66000	.75126	.67301	.73963	.68582	.72777	.69842	.71569	42
19	.66022	.75107	.67323	.73944	.68603	.72757	.69862	.71549	41
20	.66044	.75088	.67344	.73924	.68624	.72737	.69883	.71529	40
21	.66066	.75069	.67366	.73904	.68645	.72717	.69904	.71508	39
22	.66088	.75050	.67387	.73885	.68666	.72697	.69925	.71488	38
23	.66109	.75030	.67409	.73865	.68688	.72677	.69946	.71468	37
24	.66131	.75011	.67430	.73846	.68709	.72657	.69966	.71447	36
25	.66153	.74992	.67452	.73826	.68730	.72637	.69987	.71427	35
26	.66175	.74973	.67473	.73806	.68751	.72617	.70008	.71407	34
27	.66197	.74953	.67495	.73787	.68772	.72597	.70029	.71386	33
28	.66218	.74934	.67516	.73767	.68793	.72577	.70049	.71366	32
29	.66240	.74915	.67538	.73747	.68814	.72557	.70070	.71345	31
30	.66262	.74896	.67559	.73728	.68835	.72537	.70091	.71325	30
31	.66284	.74876	.67580	.73708	.68857	.72517	.70112	.71305	29
32	.66306	.74857	.67602	.73688	.68878	.72497	.70132	.71284	28
33	.66327	.74838	.67623	.73669	.68899	.72477	.70153	.71264	27
34	.66349	.74818	.67645	.73649	.68920	.72457	.70174	.71243	26
35	.66371	.74799	.67666	.73629	.68941	.72437	.70195	.71223	25
36	.66393	.74780	.67688	.73610	.68962	.72417	.70215	.71203	24
37	.66414	.74760	.67709	.73590	.68983	.72397	.70236	.71182	23
38	.66436	.74741	.67730	.73570	.69004	.72377	.70257	.71162	22
39	.66458	.74722	.67752	.73551	.69025	.72357	.70277	.71141	21
40	.66480	.74703	.67773	.73531	.69046	.72337	.70298	.71121	20
41	.66501	.74683	.67795	.73511	.69067	.72317	.70319	.71100	19
42	.66523	.74664	.67816	.73491	.69088	.72297	.70339	.71080	18
43	.66545	.74644	.67837	.73472	.69109	.72277	.70360	.71059	17
44	.66566	.74625	.67859	.73452	.69130	.72257	.70381	.71039	16
45	.66588	.74606	.67880	.73432	.69151	.72236	.70401	.71019	15
46	.66610	.74586	.67901	.73413	.69172	.72216	.70422	.70998	14
47	.66632	.74567	.67923	.73393	.69193	.72196	.70443	.70978	13
48	.66653	.74548	.67944	.73373	.69214	.72176	.70463	.70957	12
49	.66675	.74528	.67965	.73353	.69235	.72156	.70484	.70937	11
50	.66697	.74509	.67987	.73333	.69256	.72136	.70505	.70916	10
51	.66718	.74489	.68008	.73314	.69277	.72116	.70525	.70896	9
52	.66740	.74470	.68029	.73294	.69298	.72095	.70546	.70875	8
53	.66762	.74451	.68051	.73274	.69319	.72075	.70567	.70855	7
54	.66783	.74431	.68072	.73254	.69340	.72055	.70587	.70834	6
55	.66805	.74412	.68093	.73234	.69361	.72035	.70608	.70813	5
56	.66827	.74392	.68115	.73215	.69382	.72015	.70628	.70793	4
57	.66848	.74373	.68136	.73195	.69403	.71995	.70649	.70772	3
58	.66870	.74353	.68157	.73175	.69424	.71974	.70670	.70752	2
59	.66891	.74334	.68179	.73155	.69445	.71954	.70690	.70731	1
60	.66913	.74314	.68200	.73135	.69466	.71934	.70711	.70711	0
′	48° Cosine	48° Sine	47° Cosine	47° Sine	46° Cosine	46° Sine	45° Cosine	45° Sine	′

TABLE XVII. NATURAL TANGENTS AND COTANGENTS

	0°		1°		2°		3°		
′	Tan.	Co-tan.	Tan.	Co-tan.	Tan.	Co-tan.	Tan.	Co-tan.	′
0	.00000	Infinite.	.01746	57.2900	.03492	28.6363	.05241	19.0811	60
1	.00029	3437.750	.01775	56.3506	.03521	28.3994	.05270	18.9755	59
2	.00058	1718.870	.01804	55.4415	.03550	28.1664	.05299	18.8711	58
3	.00087	1145.920	.01833	54.5613	.03579	27.9372	.05328	18.7678	57
4	.00116	859.436	.01862	53.7086	.03609	27.7117	.05357	18.6656	56
5	.00145	687.549	.01891	52.8821	.03638	27.4899	.05387	18.5645	55
6	.00175	572.957	.01920	52.0807	.03667	27.2715	.05416	18.4645	54
7	.00204	491.106	.01949	51.3032	.03696	27.0566	.05445	18.3655	53
8	.00233	429.718	.01978	50.5485	.03725	26.8450	.05474	18.2677	52
9	.00262	381.971	.02007	49.8157	.03754	26.6367	.05503	18.1708	51
10	.00291	343.774	.02036	49.1039	.03783	26.4316	.05533	18.0750	50
11	.00320	312.521	.02066	48.4121	.03812	26.2296	.05562	17.9802	49
12	.00349	286.478	.02095	47.7395	.03842	26.0307	.05591	17.8863	48
13	.00378	264.441	.02124	47.0853	.03871	25.8348	.05620	17.7934	47
14	.00407	245.552	.02153	46.4489	.03900	25.6418	.05649	17.7015	46
15	.00436	229.182	.02182	45.8294	.03929	25.4517	.05678	17.6106	45
16	.00465	214.858	.02211	45.2261	.03958	25.2644	.05708	17.5205	44
17	.00495	202.219	.02240	44.6386	.03987	25.0798	.05737	17.4314	43
18	.00524	190.984	.02269	44.0661	.04016	24.8978	.05766	17.3432	42
19	.00553	180.932	.02298	43.5081	.04046	24.7185	.05795	17.2558	41
20	.00582	171.885	.02328	42.9641	.04075	24.5418	.05824	17.1693	40
21	.00611	163.700	.02357	42.4335	.04104	24.3675	.05854	17.0837	39
22	.00640	156.259	.02386	41.9158	.04133	24.1957	.05883	16.9990	38
23	.00669	149.465	.02415	41.4106	.04162	24.0263	.05912	16.9150	37
24	.00698	143.237	.02444	40.9174	.04191	23.8593	.05941	16.8319	36
25	.00727	137.507	.02473	40.4358	.04220	23.6945	.05970	16.7496	35
26	.00756	132.219	.02502	39.9655	.04250	23.5321	.05999	16.6681	34
27	.00785	127.321	.02531	39.5059	.04279	23.3718	.06029	16.5874	33
28	.00814	122.774	.02560	39.0568	.04308	23.2137	.06058	16.5075	32
29	.00844	118.540	.02589	38.6177	.04337	23.0577	.06087	16.4283	31
30	.00873	114.589	.02619	38.1885	.04366	22.9038	.06116	16.3499	30
31	.00902	110.892	.02648	37.7686	.04395	22.7519	.06145	16.2722	29
32	.00931	107.426	.02677	37.3579	.04424	22.6020	.06175	16.1952	28
33	.00960	104.171	.02706	36.9560	.04454	22.4541	.06204	16.1190	27
34	.00989	101.107	.02735	36.5627	.04483	22.3081	.06233	16.0435	26
35	.01018	98.2179	.02764	36.1776	.04512	22.1640	.06262	15.9687	25
36	.01047	95.4895	.02793	35.8006	.04541	22.0217	.06291	15.8945	24
37	.01076	92.9085	.02822	35.4313	.04570	21.8813	.06321	15.8211	23
38	.01105	90.4633	.02851	35.0695	.04599	21.7426	.06350	15.7483	22
39	.01135	88.1436	.02881	34.7151	.04628	21.6056	.06379	15.6762	21
40	.01164	85.9398	.02910	34.3678	.04658	21.4704	.06408	15.6048	20
41	.01193	83.8435	.02939	34.0273	.04687	21.3369	.06437	15.5340	19
42	.01222	81.8470	.02968	33.6935	.04716	21.2049	.06467	15.4638	18
43	.01251	79.9434	.02997	33.3662	.04745	21.0747	.06496	15.3943	17
44	.01280	78.1263	.03026	33.0452	.04774	20.9460	.06525	15.3254	16
45	.01309	76.3900	.03055	32.7303	.04803	20.8188	.06554	15.2571	15
46	.01338	74.7292	.03084	32.4213	.04832	20.6932	.06584	15.1893	14
47	.01367	73.1390	.03114	32.1181	.04862	20.5691	.06613	15.1222	13
48	.01396	71.6151	.03143	31.8205	.04891	20.4465	.06642	15.0557	12
49	.01425	70.1533	.03172	31.5284	.04920	20.3253	.06671	14.9898	11
50	.01455	68.7501	.03201	31.2416	.04949	20.2056	.06700	14.9244	10
51	.01484	67.4019	.03230	30.9599	.04978	20.0872	.06730	14.8596	9
52	.01513	66.1055	.03259	30.6833	.05007	19.9702	.06759	14.7954	8
53	.01542	64.8580	.03288	30.4116	.05037	19.8546	.06788	14.7317	7
54	.01571	63.6567	.03317	30.1446	.05066	19.7403	.06817	14.6685	6
55	.01600	62.4992	.03346	29.8823	.05095	19.6273	.06847	14.6059	5
56	.01629	61.3829	.03376	29.6245	.05124	19.5156	.06876	14.5438	4
57	.01658	60.3058	.03405	29.3711	.05153	19.4051	.06905	14.4823	3
58	.01687	59.2659	.03434	29.1220	.05182	19.2959	.06934	14.4212	2
59	.01716	58.2612	.03463	28.8771	.05212	19.1879	.06963	14.3607	1
60	.01746	57.2900	.03492	28.6363	.05241	19.0811	.06993	14.3007	0
′	Co-tan.	Tan.	Co-tan.	Tan.	Co-tan.	Tan.	Co-tan.	Tan.	′
	89°		88°		87°		86°		

TABLE XVII.—*Continued*

′	4° Tan.	4° Co-tan.	5° Tan.	5° Co-tan.	6° Tan.	6° Co-tan.	7° Tan.	7° Co-tan.	′
0	.06993	14.3007	.08749	11.4301	.10510	9.51436	.12278	8.14435	60
1	.07022	14.2411	.08778	11.3919	.10540	9.48781	.12308	8.12481	59
2	.07051	14.1821	.08807	11.3540	.10569	9.46141	.12338	8.10536	58
3	.07080	14.1235	.08837	11.3163	.10599	9.43515	.12367	8.08600	57
4	.07110	14.0655	.08866	11.2789	.10628	9.40904	.12397	8.06674	56
5	.07139	14.0079	.08895	11.2417	.10657	9.38307	.12426	8.04756	55
6	.07168	13.9507	.08925	11.2048	.10687	9.35724	.12456	8.02848	54
7	.07197	13.8940	.08954	11.1681	.10716	9.33154	.12485	8.00948	53
8	.07227	13.8378	.08983	11.1316	.10746	9.30599	.12515	7.99058	52
9	.07256	13.7821	.09013	11.0954	.10775	9.28058	.12544	7.97176	51
10	.07285	13.7267	.09042	11.0594	.10805	9.25530	.12574	7.95302	50
11	.07314	13.6719	.09071	11.0237	.10834	9.23016	.12603	7.93438	49
12	.07344	13.6174	.09101	10.9882	.10863	9.20516	.12633	7.91582	48
13	.07373	13.5634	.09130	10.9529	.10893	9.18028	.12662	7.89734	47
14	.07402	13.5098	.09159	10.9178	.10922	9.15554	.12692	7.87895	46
15	.07431	13.4566	.09189	10.8829	.10952	9.13093	.12722	7.86064	45
16	.07461	13.4039	.09218	10.8483	.10981	9.10646	.12751	7.84242	44
17	.07490	13.3515	.09247	10.8139	.11011	9.08211	.12781	7.82428	43
18	.07519	13.2996	.09277	10.7797	.11040	9.05789	.12810	7.80622	42
19	.07548	13.2480	.09306	10.7457	.11070	9.03379	.12840	7.78825	41
20	.07578	13.1969	.09335	10.7119	.11099	9.00983	.12869	7.77035	40
21	.07607	13.1461	.09365	10.6783	.11128	8.98598	.12899	7.75254	39
22	.07636	13.0958	.09394	10.6450	.11158	8.96227	.12929	7.73480	38
23	.07665	13.0458	.09423	10.6118	.11187	8.93867	.12958	7.71715	37
24	.07695	12.9962	.09453	10.5789	.11217	8.91520	.12988	7.69957	36
25	.07724	12.9469	.09482	10.5462	.11246	8.89185	.13017	7.68208	35
26	.07753	12.8981	.09511	10.5136	.11276	8.86862	.13047	7.66466	34
27	.07782	12.8496	.09541	10.4813	.11305	8.84551	.13076	7.64732	33
28	.07812	12.8014	.09570	10.4491	.11335	8.82252	.13106	7.63005	32
29	.07841	12.7536	.09600	10.4172	.11364	8.79964	.13136	7.61287	31
30	.07870	12.7062	.09629	10.3854	.11394	8.77689	.13165	7.59575	30
31	.07899	12.6591	.09658	10.3538	.11423	8.75425	.13195	7.57872	29
32	.07929	12.6124	.09688	10.3224	.11452	8.73172	.13224	7.56176	28
33	.07958	12.5660	.09717	10.2913	.11482	8.70931	.13254	7.54487	27
34	.07987	12.5199	.09746	10.2602	.11511	8.68701	.13284	7.52806	26
35	.08017	12.4742	.09776	10.2294	.11541	8.66482	.13313	7.51132	25
36	.08046	12.4288	.09805	10.1988	.11570	8.64275	.13343	7.49465	24
37	.08075	12.3838	.09834	10.1683	.11600	8.62078	.13372	7.47806	23
38	.08104	12.3390	.09864	10.1381	.11629	8.59893	.13402	7.46154	22
39	.08134	12.2946	.09893	10.1080	.11659	8.57718	.13432	7.44509	21
40	.08163	12.2505	.09923	10.0780	.11688	8.55555	.13461	7.42871	20
41	.08192	12.2067	.09952	10.0483	.11718	8.53402	.13491	7.41240	19
42	.08221	12.1632	.09981	10.0187	.11747	8.51259	.13521	7.39616	18
43	.08251	12.1201	.10011	9.98931	.11777	8.49128	.13550	7.37999	17
44	.08280	12.0772	.10040	9.96007	.11806	8.47007	.13580	7.36389	16
45	.08309	12.0346	.10069	9.93101	.11836	8.44896	.13609	7.34786	15
46	.08339	11.9923	.10099	9.90211	.11865	8.42795	.13639	7.33190	14
47	.08368	11.9504	.10128	9.87338	.11895	8.40705	.13669	7.31600	13
48	.08397	11.9087	.10158	9.84482	.11924	8.38625	.13698	7.30018	12
49	.08427	11.8673	.10187	9.81641	.11954	8.36555	.13728	7.28442	11
50	.08456	11.8262	.10216	9.78817	.11983	8.34496	.13758	7.26873	10
51	.08485	11.7853	.10246	9.76009	.12013	8.32446	.13787	7.25310	9
52	.08514	11.7448	.10275	9.73217	.12042	8.30406	.13817	7.23754	8
53	.08544	11.7045	.10305	9.70441	.12072	8.28376	.13846	7.22204	7
54	.08573	11.6645	.10334	9.67680	.12101	8.26355	.13876	7.20661	6
55	.08602	11.6248	.10363	9.64935	.12131	8.24345	.13906	7.19125	5
56	.08632	11.5853	.10393	9.62205	.12160	8.22344	.13935	7.17594	4
57	.08661	11.5461	.10422	9.59490	.12190	8.20352	.13965	7.16071	3
58	.08690	11.5072	.10452	9.56791	.12219	8.18370	.13995	7.14553	2
59	.08720	11.4685	.10481	9.54106	.12249	8.16398	.14024	7.13042	1
60	.08749	11.4301	.10510	9.51436	.12278	8.14435	.14054	7.11537	0
′	85° Co-tan.	85° Tan.	84° Co-tan.	84° Tan.	83° Co-tan.	83° Tan.	82° Co-tan.	82° Tan.	′

TABLE XVII.—*Continued*

′	8° Tan.	8° Co-tan.	9° Tan.	9° Co-tan.	10° Tan.	10° Co-tan.	11° Tan.	11° Co-tan.	′
0	.14054	7.11537	.15838	6.31375	.17633	5.67128	.19438	5.14455	60
1	.14084	7.10038	.15868	6.30189	.17663	5.66165	.19468	5.13658	59
2	.14113	7.08546	.15898	6.29007	.17693	5.65205	.19498	5.12862	58
3	.14143	7.07059	.15928	6.27829	.17723	5.64248	.19529	5.12069	57
4	.14173	7.05579	.15958	6.26655	.17753	5.63295	.19559	5.11279	56
5	.14202	7.04105	.15988	6.25486	.17783	5.62344	.19589	5.10490	55
6	.14232	7.02637	.16017	6.24321	.17813	5.61397	.19619	5.09704	54
7	.14262	7.01174	.16047	6.23160	.17843	5.60452	.19649	5.08921	53
8	.14291	6.99718	.16077	6.22003	.17873	5.59511	.19680	5.08139	52
9	.14321	6.98268	.16107	6.20851	.17903	5.58573	.19710	5.07360	51
10	.14351	6.96823	.16137	6.19703	.17933	5.57638	.19740	5.06584	50
11	.14381	6.95385	.16167	6.18559	.17963	5.56706	.19770	5.05809	49
12	.14410	6.93952	.16196	6.17419	.17993	5.55777	.19801	5.05037	48
13	.14440	6.92525	.16226	6.16283	.18023	5.54851	.19831	5.04267	47
14	.14470	6.91104	.16256	6.15151	.18053	5.53927	.19861	5.03499	46
15	.14499	6.89688	.16286	6.14023	.18083	5.53007	.19891	5.02734	45
16	.14529	6.88278	.16316	6.12899	.18113	5.52090	.19921	5.01971	44
17	.14559	6.86874	.16346	6.11779	.18143	5.51176	.19952	5.01210	43
18	.14588	6.85475	.16376	6.10664	.18173	5.50264	.19982	5.00451	42
19	.14618	6.84082	.16405	6.09552	.18203	5.49356	.20012	4.99695	41
20	.14648	6.82694	.16435	6.08444	.18233	5.48451	.20042	4.98940	40
21	.14678	6.81312	.16465	6.07340	.18263	5.47548	.20073	4.98188	39
22	.14707	6.79936	.16495	6.06240	.18293	5.46648	.20103	4.97438	38
23	.14737	6.78564	.16525	6.05143	.18323	5.45751	.20133	4.96690	37
24	.14767	6.77199	.16555	6.04051	.18353	5.44857	.20164	4.95945	36
25	.14796	6.75838	.16585	6.02962	.18383	5.43966	.20194	4.95201	35
26	.14826	6.74483	.16615	6.01878	.18414	5.43077	.20224	4.94460	34
27	.14856	6.73133	.16645	6.00797	.18444	5.42192	.20254	4.93721	33
28	.14886	6.71789	.16674	5.99720	.18474	5.41309	.20285	4.92984	32
29	.14915	6.70450	.16704	5.98646	.18504	5.40429	.20315	4.92249	31
30	.14945	6.69116	.16734	5.97576	.18534	5.39552	.20345	4.91516	30
31	.14975	6.67787	.16764	5.96510	.18564	5.38677	.20376	4.90785	29
32	.15005	6.66463	.16794	5.95448	.18594	5.37805	.20406	4.90056	28
33	.15034	6.65144	.16824	5.94390	.18624	5.36936	.20436	4.89330	27
34	.15064	6.63831	.16854	5.93335	.18654	5.36070	.20466	4.83605	26
35	.15094	6.62523	.16884	5.92283	.18684	5.35206	.20497	4.87882	25
36	.15124	6.61219	.16914	5.91235	.18714	5.34345	.20527	4.87162	24
37	.15153	6.59921	.16944	5.90191	.18745	5.33487	.20557	4.86444	23
38	.15183	6.58627	.16974	5.89151	.18775	5.32631	.20588	4.85727	22
39	.15213	6.57339	.17004	5.88114	.18805	5.31778	.20618	4.85013	21
40	.15243	6.56055	.17033	5.87080	.18835	5.30928	.20648	4.84300	20
41	.15272	6.54777	.17063	5.86051	.18865	5.30080	.20679	4.83590	19
42	.15302	6.53503	.17093	5.85024	.18895	5.29235	.20709	4.82882	18
43	.15332	6.52234	.17123	5.84001	.18925	5.28393	.20739	4.82175	17
44	.15362	6.50970	.17153	5.82982	.18955	5.27553	.20770	4.81471	16
45	.15391	6.49710	.17183	5.81966	.18986	5.26715	.20800	4.80769	15
46	.15421	6.48456	.17213	5.80953	.19016	5.25880	.20830	4.80068	14
47	.15451	6.47206	.17243	5.79944	.19046	5.25048	.20861	4.79370	13
48	.15481	6.45961	.17273	5.78938	.19076	5.24218	.20891	4.78673	12
49	.15511	6.44720	.17303	5.77936	.19106	5.23391	.20921	4.77978	11
50	.15540	6.43484	.17333	5.76937	.19136	5.22566	.20952	4.77286	10
51	.15570	6.42253	.17363	5.75941	.19166	5.21744	.20982	4.76595	9
52	.15600	6.41026	.17393	5.74949	.19197	5.20925	.21013	4.75906	8
53	.15630	6.39804	.17423	5.73960	.19227	5.20107	.21043	4.75219	7
54	.15660	6.38587	.17453	5.72974	.19257	5.19293	.21073	4.74534	6
55	.15689	6.37374	.17483	5.71992	.19287	5.18480	.21104	4.73851	5
56	.15719	6.36165	.17513	5.71013	.19317	5.17671	.21134	4.73170	4
57	.15749	6.34961	.17543	5.70037	.19347	5.16863	.21164	4.72490	3
58	.15779	6.33761	.17573	5.69064	.19378	5.16058	.21195	4.71813	2
59	.15809	6.32566	.17603	5.68094	.19408	5.15256	.21225	4.71137	1
60	.15838	6.31375	.17633	5.67128	.19438	5.14455	.21256	4.70463	0
′	Co-tan. 81°	Tan. 81°	Co-tan. 80°	Tan. 80°	Co-tan. 79°	Tan. 79°	Co-tan. 78°	Tan. 78°	′

TABLE XVII.—*Continued*

′	12° Tan.	12° Co-tan.	13° Tan.	13° Co-tan.	14° Tan.	14° Co-tan.	15° Tan.	15° Co-tan.	′
0	.21256	4.70463	.23087	4.33148	.24933	4.01078	.26795	3.73205	60
1	.21286	4.69791	.23117	4.32573	.24964	4.00582	.26826	3.72771	59
2	.21316	4.69121	.23148	4.32001	.24995	4.00086	.26857	3.72338	58
3	.21347	4.68452	.23179	4.31430	.25026	3.99592	.26888	3.71907	57
4	.21377	4.67786	.23209	4.30860	.25056	3.99099	.26920	3.71476	56
5	.21408	4.67121	.23240	4.30291	.25087	3.98607	.26951	3.71046	55
6	.21438	4.66458	.23271	4.29724	.25118	3.98117	.26982	3.70616	54
7	.21469	4.65797	.23301	4.29159	.25149	3.97627	.27013	3.70188	53
8	.21499	4.65138	.23332	4.28595	.25180	3.97139	.27044	3.69761	52
9	.21529	4.64480	.23363	4.28032	.25211	3.96651	.27076	3.69335	51
10	.21560	4.63825	.23393	4.27471	.25242	3.96165	.27107	3.68909	50
11	.21590	4.63171	.23424	4.26911	.25273	3.95680	.27138	3.68485	49
12	.21621	4.62518	.23455	4.26352	.25304	3.95196	.27169	3.68061	48
13	.21651	4.61868	.23485	4.25795	.25335	3.94713	.27201	3.67638	47
14	.21682	4.61219	.23516	4.25239	.25366	3.94232	.27232	3.67217	46
15	.21712	4.60572	.23547	4.24685	.25397	3.93751	.27263	3.66796	45
16	.21743	4.59927	.23578	4.24132	.25428	3.93271	.27294	3.66376	44
17	.21773	4.59283	.23608	4.23580	.25459	3.92793	.27326	3.65957	43
18	.21804	4.58641	.23639	4.23030	.25490	3.92316	.27357	3.65538	42
19	.21834	4.58001	.23670	4.22481	.25521	3.91839	.27388	3.65121	41
20	.21864	4.57363	.23700	4.21933	.25552	3.91364	.27419	3.64705	40
21	.21895	4.56726	.23731	4.21387	.25583	3.90890	.27451	3.64289	39
22	.21925	4.56091	.23762	4.20842	.25614	3.90417	.27482	3.63874	38
23	.21956	4.55458	.23793	4.20298	.25645	3.89945	.27513	3.63461	37
24	.21986	4.54826	.23823	4.19756	.25676	3.89474	.27545	3.63048	36
25	.22017	4.54196	.23854	4.19215	.25707	3.89004	.27576	3.62636	35
26	.22047	4.53568	.23885	4.18675	.25738	3.88536	.27607	3.62224	34
27	.22078	4.52941	.23916	4.18137	.25769	3.88068	.27638	3.61814	33
28	.22108	4.52316	.23946	4.17600	.25800	3.87601	.27670	3.61405	32
29	.22139	4.51693	.23977	4.17064	.25831	3.87136	.27701	3.60996	31
30	.22169	4.51071	.24008	4.16530	.25862	3.86671	.27732	3.60588	30
31	.22200	4.50451	.24039	4.15997	.25893	3.86208	.27764	3.60181	29
32	.22231	4.49832	.24069	4.15465	.25924	3.85745	.27795	3.59775	28
33	.22261	4.49215	.24100	4.14934	.25955	3.85284	.27826	3.59370	27
34	.22292	4.48600	.24131	4.14405	.25986	3.84824	.27858	3.58966	26
35	.22322	4.47986	.24162	4.13877	.26017	3.84364	.27889	3.58562	25
36	.22353	4.47374	.24193	4.13350	.26048	3.83906	.27920	3.58160	24
37	.22383	4.46764	.24223	4.12825	.26079	3.83449	.27952	3.57758	23
38	.22414	4.46155	.24254	4.12301	.26110	3.82992	.27983	3.57357	22
39	.22444	4.45548	.24285	4.11778	.26141	3.82537	.28015	3.56957	21
40	.22475	4.44942	.24316	4.11256	.26172	3.82083	.28046	3.56557	20
41	.22505	4.44338	.24347	4.10736	.26203	3.81630	.28077	3.56159	19
42	.22536	4.43735	.24377	4.10216	.26235	3.81177	.28109	3.55761	18
43	.22567	4.43134	.24408	4.09699	.26266	3.80726	.28140	3.55364	17
44	.22597	4.42534	.24439	4.09182	.26297	3.80276	.28172	3.54968	16
45	.22628	4.41936	.24470	4.08666	.26328	3.79827	.28203	3.54573	15
46	.22658	4.41340	.24501	4.08152	.26359	3.79378	.28234	3.54179	14
47	.22689	4.40745	.24532	4.07639	.26390	3.78931	.28266	3.53785	13
48	.22719	4.40152	.24562	4.07127	.26421	3.78485	.28297	3.53393	12
49	.22750	4.39560	.24593	4.06616	.26452	3.78040	.28329	3.53001	11
50	.22781	4.38969	.24624	4.06107	.26483	3.77595	.28360	3.52609	10
51	.22811	4.38381	.24655	4.05599	.26515	3.77152	.28391	3.52219	9
52	.22842	4.37793	.24686	4.05092	.26546	3.76709	.28423	3.51829	8
53	.22872	4.37207	.24717	4.04586	.26577	3.76268	.28454	3.51441	7
54	.22903	4.36623	.24747	4.04081	.26608	3.75828	.28486	3.51053	6
55	.22934	4.36040	.24778	4.03578	.26639	3.75388	.28517	3.50666	5
56	.22964	4.35459	.24809	4.03075	.26670	3.74950	.28549	3.50279	4
57	.22995	4.34879	.24840	4.02574	.26701	3.74512	.28580	3.49894	3
58	.23026	4.34300	.24871	4.02074	.26733	3.74075	.28612	3.49509	2
59	.23056	4.33723	.24902	4.01576	.26764	3.73640	.28643	3.49125	1
60	.23087	4.33148	.24933	4.01078	.26795	3.73205	.28675	3.48741	0
′	Co-tan.	Tan.	Co-tan.	Tan.	Co-tan.	Tan.	Co-tan.	Tan.	′
	77°		76°		75°		74°		

TABLE XVII.—*Continued*

′	16° Tan.	16° Co-tan.	17° Tan.	17° Co-tan.	18° Tan.	18° Co-tan.	19° Tan.	19° Co-tan.	′
0	.28675	3.48741	.30573	3.27085	.32492	3.07768	.34433	2.90421	60
1	.28706	3.48359	.30605	3.26745	.32524	3.07464	.34465	2.90147	59
2	.28738	3.47977	.30637	3.26406	.32556	3.07160	.34498	2.89873	58
3	.28769	3.47596	.30669	3.26067	.32588	3.06857	.34530	2.89600	57
4	.28800	3.47216	.30700	3.25729	.32621	3.06554	.34563	2.89327	56
5	.28832	3.46837	.30732	3.25392	.32653	3.06252	.34596	2.89055	55
6	.28864	3.46458	.30764	3.25055	.32685	3.05950	.34628	2.88783	54
7	.28895	3.46080	.30796	3.24719	.32717	3.05649	.34661	2.88511	53
8	.28927	3.45703	.30828	3.24383	.32749	3.05349	.34693	2.88240	52
9	.28958	3.45327	.30860	3.24049	.32782	3.05049	.34726	2.87970	51
10	.28990	3.44951	.30891	3.23714	.32814	3.04749	.34758	2.87700	50
11	.29021	3.44576	.30923	3.23381	.32846	3.04450	.34791	2.87430	49
12	.29053	3.44202	.30955	3.23048	.32878	3.04152	.34824	2.87161	48
13	.29084	3.43829	.30987	3.22715	.32911	3.03854	.34856	2.86892	47
14	.29116	3.43456	.31019	3.22384	.32943	3.03556	.34889	2.86624	46
15	.29147	3.43084	.31051	3.22053	.32975	3.03260	.34922	2.86356	45
16	.29179	3.42713	.31083	3.21722	.33007	3.02963	.34954	2.86089	44
17	.29210	3.42343	.31115	3.21392	.33040	3.02667	.34987	2.85822	43
18	.29242	3.41973	.31147	3.21063	.33072	3.02372	.35019	2.85555	42
19	.29274	3.41604	.31178	3.20734	.33104	3.02077	.35052	2.85289	41
20	.29305	3.41236	.31210	3.20406	.33136	3.01783	.35085	2.85023	40
21	.29337	3.40869	.31242	3.20079	.33169	3.01489	.35117	2.84758	39
22	.29368	3.40502	.31274	3.19752	.33201	3.01196	.35150	2.84494	38
23	.29400	3.40136	.31306	3.19426	.33233	3.00903	.35183	2.84229	37
24	.29432	3.39771	.31338	3.19100	.33266	3.00611	.35216	2.83965	36
25	.29463	3.39406	.31370	3.18775	.33298	3.00319	.35248	2.83702	35
26	.29495	3.39042	.31402	3.18451	.33330	3.00028	.35281	2.83439	34
27	.29526	3.38679	.31434	3.18127	.33363	2.99738	.35314	2.83176	33
28	.29558	3.38317	.31466	3.17804	.33395	2.99447	.35346	2.82914	32
29	.29590	3.37955	.31498	3.17481	.33427	2.99158	.35379	2.82653	31
30	.29621	3.37594	.31530	3.17159	.33460	2.98868	.35412	2.82391	30
31	.29653	3.37234	.31562	3.16838	.33492	2.98580	.35445	2.82130	29
32	.29685	3.36875	.31594	3.16517	.33524	2.98292	.35477	2.81870	28
33	.29716	3.36516	.31626	3.16197	.33557	2.98004	.35510	2.81610	27
34	.29748	3.36158	.31658	3.15877	.33589	2.97717	.35543	2.81350	26
35	.29780	3.35800	.31690	3.15558	.33621	2.97430	.35576	2.81091	25
36	.29811	3.35443	.31722	3.15240	.33654	2.97144	.35608	2.80833	24
37	.29843	3.35087	.31754	3.14922	.33686	2.96858	.35641	2.80574	23
38	.29875	3.34732	.31786	3.14605	.33718	2.96573	.35674	2.80316	22
39	.29906	3.34377	.31818	3.14288	.33751	2.96288	.35707	2.80059	21
40	.29938	3.34023	.31850	3.13972	.33783	2.96004	.35740	2.79802	20
41	.29970	3.33670	.31882	3.13656	.33816	2.95721	.35772	2.79545	19
42	.30001	3.33317	.31914	3.13341	.33848	2.95437	.35805	2.79289	18
43	.30033	3.32965	.31946	3.13027	.33881	2.95155	.35838	2.79033	17
44	.30065	3.32614	.31978	3.12713	.33913	2.94872	.35871	2.78778	16
45	.30097	3.32264	.32010	3.12400	.33945	2.94590	.35904	2.78523	15
46	.30128	3.31914	.32042	3.12087	.33978	2.94309	.35937	2.78269	14
47	.30160	3.31565	.32074	3.11775	.34010	2.94028	.35969	2.78014	13
48	.30192	3.31216	.32106	3.11464	.34043	2.93748	.36002	2.77761	12
49	.30224	3.30868	.32139	3.11153	.34075	2.93468	.36035	2.77507	11
50	.30255	3.30521	.32171	3.10842	.34108	2.93189	.36068	2.77254	10
51	.30287	3.30174	.32203	3.10532	.34140	2.92910	.36101	2.77002	9
52	.30319	3.29829	.32235	3.10223	.34173	2.92632	.36134	2.76750	8
53	.30351	3.29483	.32267	3.09914	.34205	2.92354	.36167	2.76498	7
54	.30382	3.29139	.32299	3.09606	.34238	2.92076	.36199	2.76247	6
55	.30414	3.28795	.32331	3.09298	.34270	2.91799	.36232	2.75996	5
56	.30446	3.28452	.32363	3.08991	.34303	2.91523	.36265	2.75746	4
57	.30478	3.28109	.32396	3.08685	.34335	2.91246	.36298	2.75496	3
58	.30509	3.27767	.32428	3.08379	.34368	2.90971	.36331	2.75246	2
59	.30541	3.27426	.32460	3.08073	.34400	2.90696	.36364	2.74997	1
60	.30573	3.27085	.32492	3.07768	.34433	2.90421	.36397	2.74748	0
′	73° Co-tan.	73° Tan.	72° Co-tan.	72° Tan.	71° Co-tan.	71° Tan.	70° Co-tan.	70° Tan.	′

TABLE XVII.—*Continued*

′	20°		21°		22°		23°		′
	TAN.	CO-TAN.	TAN.	CO-TAN.	TAN.	CO-TAN.	TAN.	CO-TAN.	
0	.36397	2.74748	.38386	2.60509	.40403	2.47509	.42447	2.35585	60
1	.36430	2.74499	.38420	2.60283	.40436	2.47302	.42482	2.35395	59
2	.36463	2.74251	.38453	2.60057	.40470	2.47095	.42516	2.35205	58
3	.36496	2.74004	.38487	2.59831	.40504	2.46888	.42551	2.35015	57
4	.36529	2.73756	.38520	2.59606	.40538	2.46682	.42585	2.34825	56
5	.36562	2.73509	.38553	2.59381	.40572	2.46476	.42619	2.34636	55
6	.36595	2.73263	.38587	2.59156	.40606	2.46270	.42654	2.34447	54
7	.36628	2.73017	.38620	2.58932	.40640	2.46065	.42688	2.34258	53
8	.36661	2.72771	.38654	2.58708	.40674	2.45860	.42722	2.34069	52
9	.36694	2.72526	.38687	2.58484	.40707	2.45655	.42757	2.33881	51
10	.36727	2.72281	.38721	2.58261	.40741	2.45451	.42791	2.33693	50
11	.36760	2.72036	.38754	2.58038	.40775	2.45246	.42826	2.33505	49
12	.36793	2.71792	.38787	2.57815	.40809	2.45043	.42860	2.33317	48
13	.36826	2.71548	.38821	2.57593	.40843	2.44839	.42894	2.33130	47
14	.36859	2.71305	.38854	2.57371	.40877	2.44636	.42929	2.32943	46
15	.36892	2.71062	.38888	2.57150	.40911	2.44433	.42963	2.32756	45
16	.36925	2.70819	.38921	2.56928	.40945	2.44230	.42998	2.32570	44
17	.36958	2.70577	.38955	2.56707	.40979	2.44027	.43032	2.32383	43
18	.36991	2.70335	.38988	2.56487	.41013	2.43825	.43067	2.32197	42
19	.37024	2.70094	.39022	2.56266	.41047	2.43623	.43101	2.32012	41
20	.37057	2.69853	.39055	2.56046	.41081	2.43422	.43136	2.31826	40
21	.37090	2.69612	.39089	2.55827	.41115	2.43220	.43170	2.31641	39
22	.37124	2.69371	.39122	2.55608	.41149	2.43019	.43205	2.31456	38
23	.37157	2.69131	.39156	2.55389	.41183	2.42819	.43239	2.31271	37
24	.37190	2.68892	.39190	2.55170	.41217	2.42618	.43274	2.31086	36
25	.37223	2.68653	.39223	2.54952	.41251	2.42418	.43308	2.30902	35
26	.37256	2.68414	.39257	2.54734	.41285	2.42218	.43343	2.30718	34
27	.37289	2.68175	.39290	2.54516	.41319	2.42019	.43378	2.30534	33
28	.37322	2.67937	.39324	2.54299	.41353	2.41819	.43412	2.30351	32
29	.37355	2.67700	.39357	2.54082	.41387	2.41620	.43447	2.30167	31
30	.37388	2.67462	.39391	2.53865	.41421	2.41421	.43481	2.29984	30
31	.37422	2.67225	.39425	2.53648	.41455	2.41223	.43516	2.29801	29
32	.37455	2.66989	.39458	2.53432	.41490	2.41025	.43550	2.29619	28
33	.37488	2.66752	.39492	2.53217	.41524	2.40827	.43585	2.29437	27
34	.37521	2.66516	.39526	2.53001	.41558	2.40629	.43620	2.29254	26
35	.37554	2.66281	.39559	2.52786	.41592	2.40432	.43654	2.29073	25
36	.37588	2.66046	.39593	2.52571	.41626	2.40235	.43689	2.28891	24
37	.37621	2.65811	.39626	2.52357	.41660	2.40038	.43724	2.28710	23
38	.37654	2.65576	.39660	2.52142	.41694	2.39841	.43758	2.28528	22
39	.37687	2.65342	.39694	2.51929	.41728	2.39645	.43793	2.28348	21
40	.37720	2.65109	.39727	2.51715	.41763	2.39449	.43828	2.28167	20
41	.37754	2.64875	.39761	2.51502	.41797	2.39253	.43862	2.27987	19
42	.37787	2.64642	.39795	2.51289	.41831	2.39058	.43897	2.27806	18
43	.37820	2.64410	.39829	2.51076	.41865	2.38862	.43932	2.27626	17
44	.37853	2.64177	.39862	2.50864	.41899	2.38668	.43966	2.27447	16
45	.37887	2.63945	.39896	2.50652	.41933	2.38473	.44001	2.27267	15
46	.37920	2.63714	.39930	2.50440	.41968	2.38279	.44036	2.27088	14
47	.37953	2.63483	.39963	2.50229	.42002	2.38084	.44071	2.26909	13
48	.37986	2.63252	.39997	2.50018	.42036	2.37891	.44105	2.26730	12
49	.38020	2.63021	.40031	2.49807	.42070	2.37697	.44140	2.26552	11
50	.38053	2.62791	.40065	2.49597	.42105	2.37504	.44175	2.26374	10
51	.38086	2.62561	.40098	2.49386	.42139	2.37311	.44210	2.26196	9
52	.38120	2.62332	.40132	2.49177	.42173	2.37118	.44244	2.26018	8
53	.38153	2.62103	.40166	2.48967	.42207	2.36925	.44279	2.25840	7
54	.38186	2.61874	.40200	2.48758	.42242	2.36733	.44314	2.25663	6
55	.38220	2.61646	.40234	2.48549	.42276	2.36541	.44349	2.25486	5
56	.38253	2.61418	.40267	2.48340	.42310	2.36349	.44384	2.25309	4
57	.38286	2.61190	.40301	2.48132	.42345	2.36158	.44418	2.25132	3
58	.38320	2.60963	.40335	2.47924	.42379	2.35967	.44453	2.24956	2
59	.38353	2.60736	.40369	2.47716	.42413	2.35776	.44488	2.24780	1
60	.38386	2.60509	.40403	2.47509	.42447	2.35585	.44523	2.24604	0
′	CO-TAN.	TAN.	CO-TAN.	TAN.	CO-TAN.	TAN.	CO-TAN.	TAN.	′
	69°		68°		67°		66°		

TABLE XVII.—*Continued*

′	24° Tan.	24° Co-tan.	25° Tan.	25° Co-tan.	26° Tan.	26° Co-tan.	27° Tan.	27° Co-tan.	
0	.44523	2.24604	.46631	2.14451	.48773	2.05030	.50953	1.96261	60
1	.44558	2.24428	.46666	2.14288	.48809	2.04879	.50989	1.96120	59
2	.44593	2.24252	.46702	2.14125	.48845	2.04728	.51026	1.95979	58
3	.44627	2.24077	.46737	2.13963	.48881	2.04577	.51063	1.95838	57
4	.44662	2.23902	.46772	2.13801	.48917	2.04426	.51099	1.95698	56
5	.44697	2.23727	.46808	2.13639	.48953	2.04276	.51136	1.95557	55
6	.44732	2.23553	.46843	2.13477	.48989	2.04125	.51173	1.95417	54
7	.44767	2.23378	.46879	2.13316	.49026	2.03975	.51209	1.95277	53
8	.44802	2.23204	.46914	2.13154	.49062	2.03825	.51246	1.95137	52
9	.44837	2.23030	.46950	2.12993	.49098	2.03675	.51283	1.94997	51
10	.44872	2.22857	.46985	2.12832	.49134	2.03526	.51319	1.94858	50
11	.44907	2.22683	.47021	2.12671	.49170	2.03376	.51356	1.94718	49
12	.44942	2.22510	.47056	2.12511	.49206	2.03227	.51393	1.94579	48
13	.44977	2.22337	.47092	2.12350	.49242	2.03078	.51430	1.94440	47
14	.45012	2.22164	.47128	2.12190	.49278	2.02929	.51467	1.94301	46
15	.45047	2.21992	.47163	2.12030	.49315	2.02780	.51503	1.94162	45
16	.45082	2.21819	.47199	2.11871	.49351	2.02631	.51540	1.94023	44
17	.45117	2.21647	.47234	2.11711	.49387	2.02483	.51577	1.93885	43
18	.45152	2.21475	.47270	2.11552	.49423	2.02335	.51614	1.93746	42
19	.45187	2.21304	.47305	2.11392	.49459	2.02187	.51651	1.93608	41
20	.45222	2.21132	.47341	2.11233	.49495	2.02039	.51688	1.93470	40
21	.45257	2.20961	.47377	2.11075	.49532	2.01891	.51724	1.93332	39
22	.45292	2.20790	.47412	2.10916	.49568	2.01743	.51761	1.93195	38
23	.45327	2.20619	.47448	2.10758	.49604	2.01596	.51798	1.93057	37
24	.45362	2.20449	.47483	2.10600	.49640	2.01449	.51835	1.92920	36
25	.45397	2.20278	.47519	2.10442	.49677	2.01302	.51872	1.92782	35
26	.45432	2.20108	.47555	2.10284	.49713	2.01155	.51909	1.92645	34
27	.45467	2.19938	.47590	2.10126	.49749	2.01008	.51946	1.92508	33
28	.45502	2.19769	.47626	2.09969	.49786	2.00862	.51983	1.92371	32
29	.45537	2.19599	.47662	2.09811	.49822	2.00715	.52020	1.92235	31
30	.45573	2.19430	.47698	2.09654	.49858	2.00569	.52057	1.92098	30
31	.45608	2.19261	.47733	2.09498	.49894	2.00423	.52094	1.91962	29
32	.45643	2.19092	.47769	2.09341	.49931	2.00277	.52131	1.91826	28
33	.45678	2.18923	.47805	2.09184	.49967	2.00131	.52168	1.91690	27
34	.45713	2.18755	.47840	2.09028	.50004	1.99986	.52205	1.91554	26
35	.45748	2.18587	.47876	2.08872	.50040	1.99841	.52242	1.91418	25
36	.45784	2.18419	.47912	2.08716	.50076	1.99695	.52279	1.91282	24
37	.45819	2.18251	.47948	2.08560	.50113	1.99550	.52316	1.91147	23
38	.45854	2.18084	.47984	2.08405	.50149	1.99406	.52353	1.91012	22
39	.45889	2.17916	.48019	2.08250	.50185	1.99261	.52390	1.90876	21
40	.45924	2.17749	.48055	2.08094	.50222	1.99116	.52427	1.90741	20
41	.45960	2.17582	.48091	2.07939	.50258	1.98972	.52464	1.90607	19
42	.45995	2.17416	.48127	2.07785	.50295	1.98828	.52501	1.90472	18
43	.46030	2.17249	.48163	2.07630	.50331	1.98684	.52538	1.90337	17
44	.46065	2.17083	.48198	2.07476	.50368	1.98540	.52575	1.90203	16
45	.46101	2.16917	.48234	2.07321	.50404	1.98396	.52613	1.90069	15
46	.46136	2.16751	.48270	2.07167	.50441	1.98253	.52650	1.89935	14
47	.46171	2.16585	.48306	2.07014	.50477	1.98110	.52687	1.89801	13
48	.46206	2.16420	.48342	2.06860	.50514	1.97966	.52724	1.89667	12
49	.46242	2.16255	.48378	2.06706	.50550	1.97823	.52761	1.89533	11
50	.46277	2.16090	.48414	2.06553	.50587	1.97680	.52798	1.89400	10
51	.46312	2.15925	.48450	2.06400	.50623	1.97538	.52836	1.89266	9
52	.46348	2.15760	.48486	2.06247	.50660	1.97395	.52873	1.89133	8
53	.46383	2.15596	.48521	2.06094	.50696	1.97253	.52910	1.89000	7
54	.46418	2.15432	.48557	2.05942	.50733	1.97111	.52947	1.88867	6
55	.46454	2.15268	.48593	2.05790	.50769	1.96969	.52984	1.88734	5
56	.46489	2.15104	.48629	2.05637	.50806	1.96827	.53022	1.88602	4
57	.46525	2.14940	.48665	2.05485	.50843	1.96685	.53059	1.88469	3
58	.46560	2.14777	.48701	2.05333	.50879	1.96544	.53096	1.88337	2
59	.46595	2.14614	.48737	2.05182	.50916	1.96402	.53134	1.88205	1
60	.46631	2.14451	.48773	2.05030	.50953	1.96261	.53171	1.88073	0
′	65° Co-tan.	65° Tan.	64° Co-tan.	64° Tan.	63° Co-tan.	63° Tan.	62° Co-tan.	62° Tan.	′

TABLE XVII.—*Continued*

	28°		29°		30°		31°		
	Tan.	Co-tan.	Tan.	Co-tan.	Tan.	Co-tan.	Tan.	Co-tan.	
0	.53171	1.88073	.55431	1.80405	.57735	1.73205	.60086	1.66428	60
1	.53208	1.87941	.55469	1.80281	.57774	1.73089	.60126	1.66318	59
2	.53246	1.87809	.55507	1.80158	.57813	1.72973	.60165	1. 6209	58
3	.53283	1.87677	.55545	1.80034	.57851	1.72857	.60205	1.66099	57
4	.53320	1.87546	.55583	1.79911	.57890	1.72741	.60245	1.65990	56
5	.53358	1.87415	.55621	1.79788	.57929	1.72625	.60284	1.65881	55
6	.53395	1.87283	.55659	1.79665	.57968	1.72509	.60324	1.65772	54
7	.53432	1.87152	.55697	1.79542	.58007	1.72393	.60364	1.65668	53
8	.53470	1.87021	.55736	1.79419	.58046	1.72278	.60403	1.65534	52
9	.53507	1.86891	.55774	1.79296	.58085	1.72163	.60443	1.65445	51
10	.53545	1.86760	.55812	1.79174	.58124	1.72047	.60483	1.65337	50
11	.53582	1.86630	.55850	1.79051	.58162	1.71932	.60522	1.65228	49
12	.53620	1.86499	.55888	1.78929	.58201	1.71817	.60562	1.65120	48
13	.53657	1.86369	.55926	1.78807	.58240	1.71702	.60602	1.65011	47
14	.53694	1.86239	.55964	1.78685	.58279	1.71588	.60642	1.64903	46
15	.53732	1.86109	.56003	1.78563	.58318	1.71473	.60681	1.64795	45
16	.53769	1.85979	.56041	1.78441	.58357	1.71358	.60721	1.64687	44
17	.53807	1.85850	.56079	1.78319	.58396	1.71244	.60761	1.64579	43
18	.53844	1.85720	.56117	1.78198	.58435	1.71129	.60801	1.64471	42
19	.53882	1.85591	.56156	1.78077	.58474	1.71015	.60841	1.64363	41
20	.53920	1.85462	.56194	1.77955	.58513	1.70901	.60881	1.64256	40
21	.53957	1.85333	.56232	1.77834	.58552	1.70787	.60921	1.64148	39
22	.53995	1.85204	.56270	1.77713	.58591	1.70673	.60960	1.64041	38
23	.54032	1.85075	.56309	1.77592	.58631	1.70560	.61000	1.63934	37
24	.54070	1.84946	.56347	1.77471	.58670	1.70446	.61040	1.63826	36
25	.54107	1.84818	.56385	1.77351	.58709	1.70332	.61080	1.63719	35
26	.54145	1.84689	.56424	1.77230	.58748	1.70219	.61120	1.63612	34
27	.54183	1.84561	.56462	1.77110	.58787	1.70106	.61160	1.63505	33
28	.54220	1.84433	.56500	1.76990	.58826	1.69992	.61200	1.63398	32
29	.54258	1.84305	.56539	1.76869	.58865	1.69879	.61240	1.63292	31
30	.54296	1.84177	.56577	1.76749	.58904	1.69766	.61280	1.63185	30
31	.54333	1.84049	.56616	1.76630	.58944	1.69653	.61320	1.63079	29
32	.54371	1.83922	.56654	1.76510	.58983	1.69541	.61360	1.62972	28
33	.54409	1.83794	.56693	1.76390	.59022	1.69428	.61400	1.62866	27
34	.54446	1.83667	.56731	1.76271	.59061	1.69316	.61440	1.62760	26
35	.54484	1.83540	.56769	1.76151	.59101	1.69203	.61480	1.62654	25
36	.54522	1.83413	.56808	1·76032	.59140	1.69091	.61520	1.62548	24
37	.54560	1.83286	.56846	1.75913	.59179	1.68979	.61561	1.62442	23
38	.54597	1.83159	.56885	1.75794	.59218	1.68866	.61601	1.62336	22
39	.54635	1.83033	.56923	1.75675	.59258	1.68754	.61641	1.62230	21
40	.54673	1.82906	.56962	1.75556	.59297	1.68643	.61681	1.62125	20
41	.54711	1.82780	.57000	1.75437	.59336	1.68531	.61721	1.62019	19
42	.54748	1.82654	.57039	1.75319	.59376	1.68419	.61761	1.61914	18
43	.54786	1.82528	.57078	1.75200	.59415	1.68308	.61801	1.61808	17
44	.54824	1.82402	.57116	1.75082	.59454	1.68196	.61842	1.61703	16
45	.54862	1.82276	.57155	1.74964	.59494	1.68085	.61882	1.61598	15
46	.54900	1.82150	.57193	1.74846	.59533	1.67974	.61922	1.61493	14
47	.54938	1.82025	.57232	1.74728	.59573	1.67863	.61962	1.61388	13
48	.54975	1.81899	.57271	1.74610	.59612	1.67752	.62003	1.61283	12
49	.55013	1.81774	.57309	1.74492	.59651	1.67641	.62043	1.61179	11
50	.55051	1.81649	.57348	1.74375	.59691	1.67530	.62083	1.61074	10
51	.55089	1.81524	.57386	1.74257	.59730	1.67419	.62124	1.60970	9
52	.55127	1.81399	.57425	1.74140	.59770	1.67309	.62164	1.60865	8
53	.55165	1.81274	.57464	1.74022	.59809	1.67198	.62204	1.60761	7
54	.55203	1.81150	.57503	1.73905	.59849	1.67088	.62245	1.60657	6
55	.55241	1.81025	.57541	1.73788	.59888	1.66978	.62285	1.60553	5
56	.55279	1.80901	.57580	1.73671	.59928	1.66867	.62325	1.60449	4
57	.55317	1.80777	.57619	1.73555	.59967	1.66757	.62366	1.60345	3
58	.55355	1.80653	.57657	1.73438	.60007	1.66647	.62406	1.60241	2
59	.55393	1.80529	.57696	1.73321	.60046	1.66538	.62446	1.60137	1
60	.55431	1.80405	.57735	1.73205	.60086	1.66428	.62487	1.60033	0
	Co-tan.	Tan.	Co-tan.	Tan.	Co-tan.	Tan.	Co-tan.	Tan.	
	61°		60°		59°		58°		

TABLE XVII.—*Continued*

′	32° Tan.	32° Co-tan.	33° Tan.	33° Co-tan.	34° Tan.	34° Co-tan.	35° Tan.	35° Co-tan.	′
0	.62487	1.60033	.64941	1.53986	.67451	1.48256	.70021	1.42815	60
1	.62527	1.59930	.64982	1.53888	.67493	1.48163	.70064	1.42726	59
2	.62568	1.59826	.65023	1.53791	.67536	1.48070	.70107	1.42638	58
3	.62608	1.59723	.65065	1.53693	.67578	1.47977	.70151	1.42550	57
4	.62649	1.59620	.65106	1.53595	.67620	1.47885	.70194	1.42462	56
5	.62689	1.59517	.65148	1.53497	.67663	1.47792	.70238	1.42374	55
6	.62730	1.59414	.65189	1.53400	.67705	1.47699	.70281	1.42286	54
7	.62770	1.59311	.65231	1.53302	.67748	1.47607	.70325	1.42198	53
8	.62811	1.59208	.65272	1.53205	.67790	1.47514	.70368	1.42110	52
9	.62852	1.59105	.65314	1.53107	.67832	1.47422	.70412	1.42022	51
10	.62892	1.59002	.65355	1.53010	.67875	1.47330	.70455	1.41934	50
11	.62933	1.58900	.65397	1.52913	.67917	1.47238	.70499	1.41847	49
12	.62973	1.58797	.65438	1.52816	.67960	1.47146	.70542	1.41759	48
13	.63014	1.58695	.65480	1.52719	.68002	1.47053	.70586	1.41672	47
14	.63055	1.58593	.65521	1.52622	.68045	1.46962	.70629	1.41584	46
15	.63095	1.58490	.65563	1.52525	.68088	1.46870	.70673	1.41497	45
16	.63136	1.58388	.65604	1.52429	.68130	1.46778	.70717	1.41409	44
17	.63177	1.58286	.65646	1.52332	.68173	1.46686	.70760	1.41322	43
18	.63217	1.58184	.65688	1.52235	.68215	1.46595	.70804	1.41235	42
19	.63258	1.58083	.65729	1.52139	.68258	1.46503	.70848	1.41148	41
20	.63299	1.57981	.65771	1.52043	.68301	1.46411	.70891	1.41061	40
21	.63340	1.57879	.65813	1.51946	.68343	1.46320	.70935	1.40974	39
22	.63380	1.57778	.65854	1.51850	.68386	1.46229	.70979	1.40887	38
23	.63421	1.57676	.65896	1.51754	.68429	1.46137	.71023	1.40800	37
24	.63462	1.57575	.65938	1.51658	.68471	1.46046	.71066	1.40714	36
25	.63503	1.57474	.65980	1.51562	.68514	1.45955	.71110	1.40627	35
26	.63544	1.57372	.66021	1.51466	.68557	1.45864	.71154	1.40540	34
27	.63584	1.57271	.66063	1.51370	.68600	1.45773	.71198	1.40454	33
28	.63625	1.57170	.66105	1.51275	.68642	1.45682	.71242	1.40367	32
29	.63666	1.57069	.66147	1.51179	.68685	1.45592	.71285	1.40281	31
30	.63707	1.56969	.66189	1.51084	.68728	1.45501	.71329	1.40195	30
31	.63748	1.56868	.66230	1.50988	.68771	1.45410	.71373	1.40109	29
32	.63789	1.56767	.66272	1.50893	.68814	1.45320	.71417	1.40022	28
33	.63830	1.56667	.66314	1.50797	.68857	1.45229	.71461	1.39936	27
34	.63871	1.56566	.66356	1.50702	.68900	1.45139	.71505	1.39850	26
35	.63912	1.56466	.66398	1.50607	.68942	1.45049	.71549	1.39764	25
36	.63953	1.56366	.66440	1.50512	.68985	1.44958	.71593	1.39679	24
37	.63994	1.56265	.66482	1.50417	.69028	1.44868	.71637	1.39593	23
38	.64035	1.56165	.66524	1.50322	.69071	1.44778	.71681	1.39507	22
39	.64076	1.56065	.66566	1.50228	.69114	1.44688	.71725	1.39421	21
40	.64117	1.55966	.66608	1.50133	.69157	1.44598	.71769	1.39336	20
41	.64158	1.55866	.66650	1.50038	.69200	1.44508	.71813	1.39250	19
42	.64199	1.55766	.66692	1.49944	.69243	1.44418	.71857	1.39165	18
43	.64240	1.55666	.66734	1.49849	.69286	1.44329	.71901	1.39079	17
44	.64281	1.55567	.66776	1.49755	.69329	1.44239	.71946	1.38994	16
45	.64322	1.55467	.66818	1.49661	.69372	1.44149	.71990	1.38909	15
46	.64363	1.55368	.66860	1.49566	.69416	1.44060	.72034	1.38824	14
47	.64404	1.55269	.66902	1.49472	.69459	1.43970	.72078	1.38738	13
48	.64446	1.55170	.66944	1.49378	.69502	1.43881	.72122	1.38653	12
49	.64487	1.55071	.66986	1.49284	.69545	1.43792	.72166	1.38568	11
50	.64528	1.54972	.67028	1.49190	.69588	1.43703	.72211	1.38484	10
51	.64569	1.54873	.67071	1.49097	.69631	1.43614	.72255	1.38399	9
52	.64610	1.54774	.67113	1.49003	.69675	1.43525	.72299	1.38314	8
53	.64652	1.54675	.67155	1.48909	.69718	1.43436	.72344	1.38229	7
54	.64693	1.54576	.67197	1.48816	.69761	1.43347	.72388	1.38145	6
55	.64734	1.54478	.67239	1.48722	.69804	1.43258	.72432	1.38060	5
56	.64775	1.54379	.67282	1.48629	.69847	1.43169	.72477	1.37976	4
57	.64817	1.54281	.67324	1.48536	.69891	1.43080	.72521	1.37891	3
58	.64858	1.54183	.67366	1.48442	.69934	1.42992	.72565	1.37807	2
59	.64899	1.54085	.67409	1.48349	.69977	1.42903	.72610	1.37722	1
60	.64941	1.53986	.67451	1.48256	.70021	1.42815	.72654	1.37638	0
′	57° Co-tan.	57° Tan.	56° Co-tan	56° Tan.	55° Co-tan.	55° Tan.	54° Co-tan.	54° Tan.	′

TABLE XVII.—*Continued*

′	36° Tan.	36° Co-tan.	37° Tan.	37° Co-tan.	38° Tan.	38° Co-tan.	39° Tan.	39° Co-tan.	′
0	.72654	1.37638	.75355	1.32704	.78129	1.27994	.80978	1.23490	60
1	.72699	1.37554	.75401	1.32624	.78175	1.27917	.81027	1.23416	59
2	.72743	1.37470	.75447	1.32544	.78222	1.27841	.81075	1.23343	58
3	.72788	1.37386	.75492	1.32464	.78269	1.27764	.81123	1.23270	57
4	.72832	1.37302	.75538	1.32384	.78316	1.27688	.81171	1.23196	56
5	.72877	1.37218	.75584	1.32304	.78363	1.27611	.81220	1.23123	55
6	.72921	1.37134	.75629	1.32224	.78410	1.27535	.81268	1.23050	54
7	.72966	1.37050	.75675	1.32144	.78457	1.27458	.81316	1.22977	53
8	.73010	1.36967	.75721	1.32064	.78504	1.27382	.81364	1.22904	52
9	.73055	1.36883	.75767	1.31984	.78551	1.27306	.81413	1.22831	51
10	.73100	1.36800	.75812	1.31904	.78598	1.27230	.81461	1.22758	50
11	.73144	1.36716	.75858	1.31825	.78645	1.27153	.81510	1.22685	49
12	.73189	1.36633	.75904	1.31745	.78692	1.27077	.81558	1.22612	48
13	.73234	1.36549	.75950	1.31666	.78739	1.27001	.81606	1.22539	47
14	.73278	1.36466	.75996	1.31586	.78786	1.26925	.81655	1.22467	46
15	.73323	1.36383	.76042	1.31507	.78834	1.26849	.81703	1.22394	45
16	.73368	1.36300	.76088	1.31427	.78881	1.26774	.81752	1.22321	44
17	.73413	1.36217	.76134	1.31348	.78928	1.26698	.81800	1.22249	43
18	.73457	1.36133	.76180	1.31269	.78975	1.26622	.81849	1.22176	42
19	.73502	1.36051	.76226	1.31190	.79022	1.26546	.81898	1.22104	41
20	.73547	1.35968	.76272	1.31110	.79070	1.26471	.81946	1.22031	40
21	.73592	1.35885	.76318	1.31031	.79117	1.26395	.81995	1.21959	39
22	.73637	1.35802	.76364	1.30952	.79164	1.26319	.82044	1.21886	38
23	.73681	1.35719	.76410	1.30873	.79212	1.26244	.82092	1.21814	37
24	.73726	1.35637	.76456	1.30795	.79259	1.26169	.82141	1.21742	36
25	.73771	1.35554	.76502	1.30716	.79306	1.26093	.82190	1.21670	35
26	.73816	1.35472	.76548	1.30637	.79354	1.26018	.82238	1.21598	34
27	.73861	1.35389	.76594	1.30558	.79401	1.25943	.82287	1.21526	33
28	.73906	1.35307	.76640	1.30480	.79449	1.25867	.82336	1.21454	32
29	.73951	1.35224	.76686	1.30401	.79496	1.25792	.82385	1.21382	31
30	.73996	1.35142	.76733	1.30323	.79544	1.25717	.82434	1.21310	30
31	.74041	1.35060	.76779	1.30244	.79591	1.25642	.82483	1.21238	29
32	.74086	1.34978	.76825	1.30166	.79639	1.25567	.82531	1.21166	28
33	.74131	1.34896	.76871	1.30087	.79686	1.25492	.82580	1.21094	27
34	.74176	1.34814	.76918	1.30009	.79734	1.25417	.82629	1.21023	26
35	.74221	1.34732	.76964	1.29931	.79781	1.25343	.82678	1.20951	25
36	.74267	1.34650	.77010	1.29853	.79829	1.25268	.82727	1.20879	24
37	.74312	1.34568	.77057	1.29775	.79877	1.25193	.82776	1.20808	23
38	.74357	1.34487	.77103	1.29696	.79924	1.25118	.82825	1.20736	22
39	.74402	1.34405	.77149	1.29618	.79972	1.25044	.82874	1.20665	21
40	.74447	1.34323	.77196	1.29541	.80020	1.24969	.82923	1.20593	20
41	.74492	1.34242	.77242	1.29463	.80067	1.24895	.82972	1.20522	19
42	.74538	1.34160	.77289	1.29385	.80115	1.24820	.83022	1.20451	18
43	.74583	1.34079	.77335	1.29307	.80163	1.24746	.83071	1.20379	17
44	.74628	1.33998	.77382	1.29229	.80211	1.24672	.83120	1.20308	16
45	.74674	1.33916	.77428	1.29152	.80258	1.24597	.83169	1.20237	15
46	.74719	1.33835	.77475	1.29074	.80306	1.24523	.83218	1.20166	14
47	.74764	1.33754	.77521	1.28997	.80354	1.24449	.83268	1.20095	13
48	.74810	1.33673	.77568	1.28919	.80402	1.24375	.83317	1.20024	12
49	.74855	1.33592	.77615	1.28842	.80450	1.24301	.83366	1.19953	11
50	.74900	1.33511	.77661	1.28764	.80498	1.24227	.83415	1.19882	10
51	.74946	1.33430	.77708	1.28687	.80546	1.24153	.83465	1.19811	9
52	.74991	1.33349	.77754	1.28610	.80594	1.24079	.83514	1.19740	8
53	.75037	1.33268	.77801	1.28533	.80642	1.24005	.83564	1.19669	7
54	.75082	1.33187	.77848	1.28456	.80690	1.23931	.83613	1.19599	6
55	.75128	1.33107	.77895	1.28379	.80738	1.23858	.83662	1.19528	5
56	.75173	1.33026	.77941	1.28302	.80786	1.23784	.83712	1.19457	4
57	.75219	1.32946	.77988	1.28225	.80834	1.23710	.83761	1.19387	3
58	.75264	1.32865	.78035	1.28148	.80882	1.23637	.83811	1.19316	2
59	.75310	1.32785	.78082	1.28071	.80930	1.23563	.83860	1.19246	1
60	.75355	1.32704	.78129	1.27994	.80978	1.23490	.83910	1.19175	0
′	Co-tan. 53°	Tan. 53°	Co-tan. 52°	Tan. 52°	Co-tan. 51°	Tan. 51°	Co-tan. 50°	Tan 50°	′

TABLE XVII.—*Continued*

′	40° Tan.	40° Co-tan.	41° Tan.	41° Co-tan.	42° Tan.	42° Co-tan.	43° Tan.	43° Co-tan.	′
0	.83910	1.19175	.86929	1.15037	.90040	1.11061	.93252	1.07237	60
1	.83960	1.19105	.86980	1.14969	.90093	1.10996	.93306	1.07174	59
2	.84009	1.19035	.87031	1.14902	.90146	1.10931	.93360	1.07112	58
3	.84059	1.18964	.87082	1.14834	.90199	1.10867	.93415	1.07049	57
4	.84108	1.18894	.87133	1.14767	.90251	1.10802	.93469	1.06987	56
5	.84158	1.18824	.87184	1.14699	.90304	1.10737	.93524	1.06925	55
6	.84208	1.18754	.87236	1.14632	.90357	1.10672	.93578	1.06862	54
7	.84258	1.18684	.87287	1.14565	.90410	1.10607	.93633	1.06800	53
8	.84307	1.18614	.87338	1.14498	.90463	1.10543	.93688	1.06738	52
9	.84357	1.18544	.87389	1.14430	.90516	1.10478	.93742	1.06676	51
10	.84407	1.18474	.87441	1.14363	.90569	1.10414	.93797	1.06613	50
11	.84457	1.18404	.87492	1.14296	.90621	1.10349	.93852	1.06551	49
12	.84507	1.18334	.87543	1.14229	.90674	1.10285	.93906	1.06489	48
13	.84556	1.18264	.87595	1.14162	.90727	1.10220	.93961	1.06427	47
14	.84606	1.18194	.87646	1.14095	.90781	1.10156	.94016	1.06365	46
15	.84656	1.18125	.87698	1.14028	.90834	1.10091	.94071	1.06303	45
16	.84706	1.18055	.87749	1.13961	.90887	1.10027	.94125	1.06241	44
17	.84756	1.17986	.87801	1.13894	.90940	1.09963	.94180	1.06179	43
18	.84806	1.17916	.87852	1.13828	.90993	1.09899	.94235	1.06117	42
19	.84856	1.17846	.87904	1.13761	.91046	1.09834	.94290	1.06056	41
20	.84906	1.17777	.87955	1.13694	.91099	1.09770	.94345	1.05994	40
21	.84956	1.17708	.88007	1.13627	.91153	1.09706	.94400	1.05932	39
22	.85006	1.17638	.88059	1.13561	.91206	1.09642	.94455	1.05870	38
23	.85057	1.17569	.88110	1.13494	.91259	1.09578	.94510	1.05809	37
24	.85107	1.17500	.88162	1.13428	.91313	1.09514	.94565	1.05747	36
25	.85157	1.17430	.88214	1.13361	.91366	1.09450	.94620	1.05685	35
26	.85207	1.17361	.88265	1.13295	.91419	1.09386	.94676	1.05624	34
27	.85257	1.17292	.88317	1.13228	.91473	1.09322	.94731	1.05562	33
28	.85307	1.17223	.88369	1.13162	.91526	1.09258	.94786	1.05501	32
29	.85358	1.17154	.88421	1.13096	.91580	1.09195	.94841	1.05439	31
30	.85408	1.17085	.88473	1.13029	.91633	1.09131	.94896	1.05378	30
31	.85458	1.17016	.88524	1.12963	.91687	1.09067	.94952	1.05317	29
32	.85509	1.16947	.88576	1.12897	.91740	1.09003	.95007	1.05255	28
33	.85559	1.16878	.88628	1.12831	.91794	1.08940	.95062	1.05194	27
34	.85609	1.16809	.88680	1.12765	.91847	1.08876	.95118	1.05133	26
35	.85660	1.16741	.88732	1.12699	.91901	1.08813	.95173	1.05072	25
36	.85710	1.16672	.88784	1.12633	.91955	1.08749	.95229	1.05010	24
37	.85761	1.16603	.88836	1.12567	.92008	1.08686	.95284	1.04949	23
38	.85811	1.16535	.88888	1.12501	.92062	1.08622	.95340	1.04888	22
39	.85862	1.16466	.88940	1.12435	.92116	1.08559	.95395	1.04827	21
40	.85912	1.16398	.88992	1.12369	.92170	1.08496	.95451	1.04766	20
41	.85963	1.16329	.89045	1.12303	.92224	1.08432	.95506	1.04705	19
42	.86014	1.16261	.89097	1.12238	.92277	1.08369	.95562	1.04644	18
43	.86064	1.16192	.89149	1.12172	.92331	1.08306	.95618	1.04583	17
44	.86115	1.16124	.89201	1.12106	.92385	1.08243	.95673	1.04522	16
45	.86166	1.16056	.89253	1.12041	.92439	1.08179	.95729	1.04461	15
46	.86216	1.15987	.89306	1.11975	.92493	1.08116	.95785	1.04401	14
47	.86267	1.15919	.89358	1.11909	.92547	1.08053	.95841	1.04340	13
48	.86318	1.15851	.89410	1.11844	.92601	1.07990	.95897	1.04279	12
49	.86368	1.15783	.89463	1.11778	.92655	1.07927	.95952	1.04218	11
50	.86419	1.15715	.89515	1.11713	.92709	1.07864	.96008	1.04158	10
51	.86470	1.15647	.89567	1.11648	.92763	1.07801	.96064	1.04097	9
52	.86521	1.15579	.89620	1.11582	.92817	1.07738	.96120	1.04036	8
53	.86572	1.15511	.89672	1.11517	.92872	1.07676	.96176	1.03976	7
54	.86623	1.15443	.89725	1.11452	.92926	1.07613	.96232	1.03915	6
55	.86674	1.15375	.89777	1.11387	.92980	1.07550	.96288	1.03855	5
56	.86725	1.15308	.89830	1.11321	.93034	1.07487	.96344	1.03794	4
57	.86776	1.15240	.89883	1.11256	.93088	1.07425	.96400	1.03734	3
58	.86827	1.15172	.89935	1.11191	.93143	1.07362	.96457	1.03674	2
59	.86878	1.15104	.89988	1.11126	.93197	1.07299	.96513	1.03613	1
60	.86929	1.15037	.90040	1.11061	.93252	1.07237	.96569	1.03553	0
′	49° Co-tan.	49° Tan.	48° Co-tan.	48° Tan.	47° Co-tan.	47° Tan.	46° Co-tan.	46° Tan.	′

Table XVII.—*Concluded*

	44°				44°				44°		
′	Tan.	Co-tan.	′	′	Tan.	Co-tan.	′	′	Tan.	Co-tan.	′
0	.96569	1.03553	60	21	.97756	1.02295	39	41	.98901	1.01112	19
1	.96625	1.03493	59	22	.97813	1.02236	38	42	.98958	1.01053	18
2	.96681	1.03433	58	23	.97870	1.02176	37	43	.99016	1.00994	17
3	.96738	1.03372	57	24	.97927	1.02117	36	44	.99073	1.00935	16
4	.96794	1.03312	56	25	.97984	1.02057	35	45	.99131	1.00876	15
5	.96850	1.03252	55	26	.98041	1.01998	34	46	.99189	1.00818	14
6	.96907	1.03192	54	27	.98098	1.01939	33	47	.99247	1.00759	13
7	.96963	1.03132	53	28	.98155	1.01879	32	48	.99304	1.00701	12
8	.97020	1.03072	52	29	.98213	1.01820	31	49	.99362	1.00642	11
9	.97076	1.03012	51	30	.98270	1.01761	30	50	.99420	1.00583	10
10	.97133	1.02952	50	31	.98327	1.01702	29	51	.99478	1.00525	9
11	.97189	1.02892	49	32	.98384	1.01642	28	52	.99536	1.00467	8
12	.97246	1.02832	48	33	.98441	1.01583	27	53	.99594	1.00408	7
13	.97302	1.02772	47	34	.98499	1.01524	26	54	.99652	1.00350	6
14	.97359	1.02713	46	35	.98556	1.01465	25	55	.99710	1.00291	5
15	.97416	1.02653	45	36	.98613	1.01406	24	56	.99768	1.00233	4
16	.97472	1.02593	44	37	.98671	1.01347	23	57	.99826	1.00175	3
17	.97529	1.02533	43	38	.98728	1.01288	22	58	.99884	1.00116	2
18	.97586	1.02474	42	39	.98786	1.01229	21	59	.99942	1.00058	1
19	.97643	1.02414	41	40	.98843	1.01170	20	60	1	1	0
20	.97700	1.02355	40								
′	Co-tan.	Tan.	′	′	Co-tan.	Tan.	′	′	Co-tan.	Tan.	′
	45°				45°				45°		

INDEX